Aerodynamics of Road Vehicles

Other SAE books of interest:

Theory and Applications of Aerodynamics for Ground Vehicles
By T. Yomi Obidi
(Product Code: R-392)

Fundamentals of Automobile Body Structure Design
By Donald E. Malen
(Product Code: R-394)

For more information or to order a book, contact
SAE International at
400 Commonwealth Drive,
Warrendale, PA 15096-0001, USA;

Phone: 877-606-7323 (U.S. and Canada only)
or 724-776-4970 (outside U.S. and Canada);
Fax: 724-776-0790;
Email: CustomerService@sae.org
Website: http://books.sae.org

Aerodynamics of Road Vehicles, Fifth Edition

By Thomas Schuetz

Warrendale, Pennsylvania
USA

400 Commonwealth Drive
Warrendale, PA 15096-0001 USA
E-mail: CustomerService@sae.org
Phone: +1 877-606-7323 (inside USA and Canada)
        +1 724-776-4970 (outside USA)
Fax: +1 724-776-0790

Copyright © 2016 SAE International. All rights reserved.

No part of this publication may be reproduced, stored in a retrieval system, distributed, or transmitted, in any form or by any means without the prior written permission of SAE International. For permission and licensing requests, contact SAE Permissions, 400 Commonwealth Drive, Warrendale, PA 15096-0001 USA; email: copyright@sae.org; phone: 724-772-4028; fax: 724-772-9765.

SAE Order Number R-430
http://dx.doi.org/10.4271/r-430

Library of Congress Cataloging-in-Publication Data

Aerodynamik des Automobils. English
  Aerodynamics of road vehicles.—Fifth Edition/by Thomas Schuetz.
  pages   cm
  ISBN 978-0-7680-7977-7
1. Motor vehicles—Aerodynamics.   I. Schuetz, Thomas, 1982- editor.   II. Title.
  TL245.A4813 2015
  629.2'31--dc23

                                                                      2015030298

Information contained in this work has been obtained by SAE International from sources believed to be reliable. However, neither SAE International nor its authors guarantee the accuracy or completeness of any information published herein and neither SAE International nor its authors shall be responsible for any errors, omissions, or damages arising out of use of this information. This work is published with the understanding that SAE International and its authors are supplying information, but are not attempting to render engineering or other professional services. If such services are required, the assistance of an appropriate professional should be sought.

ISBN-Print 978-0-7680-7977-7
ISBN-PDF 978-0-7680-8253-1
ISBN-epub 978-0-7680-8255-5
ISBN-prc 978-0-7680-8254-8

To purchase bulk quantities, please contact:

SAE Customer Service
Email: CustomerService@sae.org
Phone: +1 877-606-7323 *(inside USA and Canada)*
         +1 724-776-4970 *(outside USA)*
Fax: +1 724-776-0790

Visit the SAE International Bookstore at books.sae.org

Contents

Preface xv

Acknowledgments xix

Chapter 1: Introduction to Automobile Aerodynamics 1
- 1.1 Scope 1
  - 1.1.1 The Role of Aerodynamics in Vehicle Design 1
  - 1.1.2 The Character of Vehicle Aerodynamics 2
  - 1.1.3 Related Fields 5
- 1.2 Historical Development of Vehicle Aerodynamics 7
  - 1.2.1 Literature 7
  - 1.2.2 Ahead of Its Time 9
  - 1.2.3 Dominant Reference Number c_D 10
  - 1.2.4 Creating Shapes for Cars 11
  - 1.2.5 "Borrowed" Shapes 12
  - 1.2.6 The "Streamline" Era 14
  - 1.2.7 Early Investigations with Parameters 25
- 1.3 From Horseless Carriage to Automobile 26
  - 1.3.1 Stamping 26
  - 1.3.2 "Bathtub" Body 27
  - 1.3.3 One-Volume Bodies 30
- 1.4 Rear-End Shapes 34
  - 1.4.1 Kamm-Back 34
  - 1.4.2 Fastback 37
  - 1.4.3 Notchback 38
  - 1.4.4 Hatchback 39
- 1.5 Directional Stability 39
- 1.6 Commercial Vehicles 42
- 1.7 Motorcycles and Helmets 46
- 1.8 Internal Flows 47
  - 1.8.1 Engine Compartment 47
  - 1.8.2 Passenger Compartment 48
- 1.9 Development Strategies 49
  - 1.9.1 Detail Optimization 49
  - 1.9.2 Shape Optimization 53
  - 1.9.3 Limit Strategy 57
- 1.10 Design and Aerodynamics 60
- 1.11 Development Tools 63
  - 1.11.1 Wind Tunnels 63
  - 1.11.2 Rating 65

	1.11.3	Classical Approach	68
	1.11.4	CFD—Integrated Methods	70

Chapter 2: The Physical Principles of Aerodynamics 75

- 2.1 Basic Equations in Fluid Dynamics 75
  - 2.1.1 Conservation Laws 75
  - 2.1.2 Kinematics and Dynamics of Flow Fields 76
  - 2.1.3 The Continuity Equation 81
  - 2.1.4 The Euler Equation 82
  - 2.1.5 The Bernoulli Equation 83
  - 2.1.6 Potential Theory 84
  - 2.1.7 The Navier-Stokes Equation 85
  - 2.1.8 Integral Forms of the Conservation Laws 88
- 2.2 Dynamics of Inviscid Flow 93
  - 2.2.1 Interpreting Streamline Patterns 93
  - 2.2.2 Planar Model Flows 95
  - 2.2.3 Vortex Flows 105
- 2.3 The Dynamics of the Frictional Flow 111
  - 2.3.1 The Reynolds Number 111
  - 2.3.2 The Prandtl Boundary Layer Concept 112
  - 2.3.3 Boundary Layer Separation 115
  - 2.3.4 Boundary Layer Turbulence 118
  - 2.3.5 Drag of Simple Bodies 123
  - 2.3.6 Multi-Body Systems 131
  - 2.3.7 Pipe Systems with Internal Flow 134
- 2.4 Appendix 146
  - 2.4.1 Density and Viscosity of Air 146
  - 2.4.2 Compressibility Effects 148

Chapter 3: Consumption and Performance 151

- 3.1 The Significance of Aerodynamic Drag 151
- 3.2 Theory of Driving Resistance 154
  - 3.2.1 Rolling Resistance 154
  - 3.2.2 Aerodynamic Drag 156
  - 3.2.3 Grade Resistance 158
  - 3.2.4 Acceleration 158
  - 3.2.5 Overall Driving Resistance 159
  - 3.2.6 Example 159
- 3.3 Performance 160
  - 3.3.1 Acceleration and Elasticity 160
  - 3.3.2 Ascending Ability 162
  - 3.3.3 Top Speed 163
- 3.4 Consumption 165
  - 3.4.1 Calculating Fuel Consumption 166

	3.4.2	Consumption Measurement and CO_2 and Energy Equivalents. . 168

- 3.5 Driving Cycles. 170
  - 3.5.1 History 171
  - 3.5.2 New European Driving Cycle (NEDC) 171
  - 3.5.3 NEDC Cycle for Hybrid Drives 173
  - 3.5.4 United States Cycles 173
  - 3.5.5 Asian Cycles. 175
  - 3.5.6 WLTP—Worldwide Harmonized Light Vehicles Test Procedure 176
  - 3.5.7 Realistic Driving Cycles 177
- 3.6 Possibilities for Reducing Fuel Consumption 177
  - 3.6.1 Energy Flow Diagram. 177
  - 3.6.2 Engine Efficiency and Engine Maps. 179
  - 3.6.3 Ancillary Components 181
  - 3.6.4 Transmission 183
  - 3.6.5 Vehicle Mass. 185
  - 3.6.6 Rolling Resistance 186
  - 3.6.7 Aerodynamic Drag 186
- 3.7 Reducing Aerodynamic Drag 187
  - 3.7.1 Possibilities for Reducing Resistance 188
  - 3.7.2 Weight Equivalency 189
  - 3.7.3 Amortization Analysis 189
- 3.8 CO_2 Legislation and Labels 191
  - 3.8.1 EU Legislation 191
  - 3.8.2 CO_2 Labels in the EU 192
  - 3.8.3 Legislation in the United States. 193
  - 3.8.4 CO_2 Labels in the United States 194
  - 3.8.5 Legislation in Asia. 196

Chapter 4: Aerodynamic Forces and Their Influence on Passenger Vehicles 197

- 4.1 Aerodynamic Forces and Force Coefficients. 201
  - 4.1.1 Buckingham Π Theorem 205
- 4.2 Flow Field Around Cars. 206
  - 4.2.1 Dead Wake 208
  - 4.2.2 Longitudinal Vortices 215
  - 4.2.3 Internal Flow 216
  - 4.2.4 Environmental Influences 217
  - 4.2.5 Influence of the Reynolds Number. 219
- 4.3 Analysis of Aerodynamic Drag Components. 220
  - 4.3.1 Pressure and Friction Drag 220
  - 4.3.2 Microdrag. 223
  - 4.3.3 Analysis by Individual Components 226

4.4 Other Components of Aerodynamic Force and Aerodynamic Moment 246
4.4.1 Lift and Pitching Moment 246
4.4.2 Side Force and Yaw Moment 248
4.4.3 Roll Moment 250
4.5 Influence on Aerodynamic Forces 250
4.5.1 The Influence of Basic Shape 252
4.5.2 Cooling Air Effect 320
4.5.3 Add-On Parts 334
4.5.4 Interference 367
4.6 The Aerodynamic Development Process 380
4.6.1 Goal Definition 382
4.6.2 Project Milestones and Tools 383
4.6.3 Examples 388
4.7 Drag and Lift of Passenger Cars in Production 390
4.7.1 Overview of Competitors by Vehicle Class 390
4.7.2 Drag Surface Area $c_D \cdot A_x$ 392
4.7.3 Intercomparison in Accordance with EADE 394
4.7.4 Influence of Vehicle Concepts 396
4.7.5 Influence of Equipment and Engine 400
4.7.6 Driving on the Ceiling? 402
4.8 Future Development 403
4.9 Reference Bodies 406
4.9.1 SAE Reference Body 406
4.9.2 Ahmed Body 409
4.9.3 DrivAer Body 410

Chapter 5: Aerodynamics and Driving Stability 413
5.1 Unsteady Aerodynamic Forces and Moments 414
5.1.1 Overtaking Maneuvers 414
5.1.2 Side Wind 416
5.2 Dynamic Driving Effects 445
5.2.1 Single-Track Model 445
5.2.2 Reaction to Lift Forces 449
5.2.3 Aerodynamic Axle Load Relief Settings 459
5.2.4 Reaction to Crosswinds 461

Chapter 6: Functionality, Safety, and Comfort 473
6.1 Component Loads 474
6.1.1 Component Loads and Pinpointing Them 474
6.1.2 Doors, Flaps, and Outside Mirrors 476
6.1.3 Windshield Wipers 480
6.2 Comfort When Driving with the Top Down 487
6.2.1 Objective 487

		6.2.2	Airflow with Convertible Top Open 488
		6.2.3	Wind Noise.. 489
		6.2.4	Thermal Comfort... 489
		6.2.5	Design Solutions for Convertibles 496
		6.2.6	Design Solutions for Sunroofs........................... 501
	6.3	Prevention of Vehicle Soiling 503
		6.3.1	Basics of Vehicle Soiling 503
		6.3.2	External Contamination..................................... 508
		6.3.3	Vehicle-Induced Contamination 523

Chapter 7: Cooling and Internal Flow 527

	7.1	Cooling Requirements 527
		7.1.1	Representative Operating Conditions 528
		7.1.2	Components and Systems 529
		7.1.3	Other Requirements 531
	7.2	Cooling System 532
		7.2.1	Engine Cooling System Circuit 532
		7.2.2	Fundamentals of Heat Transfer.......................... 533
		7.2.3	Design of Heat Exchanger 535
		7.2.4	Heat Exchangers in the Vehicle......................... 537
	7.3	Internal Flow 541
		7.3.1	Operating Conditions.. 541
		7.3.2	Cooling Module.. 544
		7.3.3	Fan 545
	7.4	Optimizing the Overall System 547
		7.4.1	Calculation of the Cooling Air Mass Flow 548
		7.4.2	Influence Parameters of the Internal Flow 551
		7.4.3	Air Intakes and Cooling Air Ducts.................... 553
		7.4.4	Cooling Matrix.. 554
		7.4.5	Fan 559
		7.4.6	Engine Compartment 561
		7.4.7	Cooling Air Outlets... 562
	7.5	Measurement Technology for Cooling Airflow 563
		7.5.1	Vane Anemometers.. 564
		7.5.2	Pressure Measurements 564
		7.5.3	Optical Measuring Methods 565
		7.5.4	Hot-Wire Anemometry...................................... 566

Chapter 8: Aeroacoustics 569

	8.1	The Influence of Airflow on the Interior and Exterior Noise of Motor Vehicles.. 569
	8.2	Aerodynamic Noise Generation.................................... 573
	8.3	Aeroacoustic Measuring Systems.................................. 574
		8.3.1	Aeroacoustic Wind Tunnels............................... 574

		8.3.2	Measuring Interior Noise. 576

        8.3.2   Measuring Interior Noise.................................. 576
        8.3.3   Measuring Exterior Noise 576
        8.3.4   Measuring Structure-Borne Sound 581
        8.3.5   Sound Source Location with Special Instruments 582
  8.4   Main Noise Sources and Options for Their Reduction 583
        8.4.1   Leaks 583
        8.4.2   Rear View Mirrors 584
        8.4.3   Windshield Wipers 586
        8.4.4   Antennas... 586
        8.4.5   A-Pillar 587
        8.4.6   Cavity Resonances 589
        8.4.7   Sun Roof Opening Noise 591
        8.4.8   Wheel Housings 591
        8.4.9   Underbody... 592
        8.4.10 Reduction of Interior Noise by Using Special Acoustic
                  Glass Windows 593
        8.4.11 Convertibles... 593
  8.5   Psycho-Acoustic Aspects 594
        8.5.1   Assessing Different Behavior Under Yaw Conditions....... 596
        8.5.2   Simulation with Static Vortex Generators 597
        8.5.3   Simulation with Dynamic Vortex Generators.............. 597
        8.5.4   Noise Synthesis 597

Chapter 9: High-Performance Vehicles 601

  9.1   Introduction.. 601
        9.1.1   Definition 601
        9.1.2   Preview 602
  9.2   Outline of the History 602
        9.2.1   Racing Cars 602
        9.2.2   Record-Breaking Vehicles................................ 609
        9.2.3   Sports Cars 618
  9.3   Vehicle Classes... 623
  9.4   Race Tracks.. 629
  9.5   Regulations 631
  9.6   Aerodynamics, Performance, and
        Handling Characteristics.. 635
        9.6.1   Drag.. 635
        9.6.2   Downforce 639
        9.6.3   Balance 643
        9.6.4   Road Performance...................................... 649
        9.6.5   Efficiency 651
        9.6.6   Cooling and Ventilation 654
        9.6.7   Oblique Incident Flow 656
        9.6.8   Slipstream.. 659

    9.7   Aerodynamics of Components 661
            9.7.1   Basic Body 662
            9.7.2   Wings... 666
            9.7.3   Spoiler and Gurneys 677
            9.7.4   Ground Effect.. 682
            9.7.5   Diffusers 686
            9.7.6   Inlets and Outlets 696
            9.7.7   Air Guiding Elements (Vanes)......................... 703
            9.7.8   Wheels.. 707

Chapter 10: Commercial Vehicles 711

10.1 Target Group 711
10.2 Driving Resistances and Fuel Consumption...................... 713
10.3 History of Commercial Vehicle Aerodynamics 718
10.4 Principles of Commercial Vehicle Aerodynamics 721
        10.4.1   Straight/Oblique Flow................................ 723
        10.4.2   Legislative Framework............................... 727
10.5 Tools for Optimizing Commercial Vehicle Aerodynamics 728
        10.5.1   Challenges Posed by Commercial Vehicles............. 728
        10.5.2   Model-Scale Wind Tunnel 728
        10.5.3   Full-Scale Wind Tunnel.............................. 732
        10.5.4   CFD Simulation...................................... 735
        10.5.5   Test Drives with Wheel Hub Measurement Device 736
10.6 Optimizing Aerodynamic Drag on Trucks 739
        10.6.1   Characteristic Airflow and Pressure Conditions 739
        10.6.2   Cab.. 741
        10.6.3   Mirrors and Attachments on the Cab.................. 748
        10.6.4   Airflow Through the Engine Compartment 751
        10.6.5   Chassis 754
        10.6.6   Semitrailers and Bodies 758
        10.6.7   Concept Vehicles 769
10.7 Optimizing Aerodynamic Drag on Buses and Coaches 772
        10.7.1   Characteristic Airflow and Pressure Conditions 772
        10.7.2   Front 773
        10.7.3   Rear View Mirrors................................... 775
        10.7.4   Windscreen Wipers 777
        10.7.5   Underbody 778
        10.7.6   Wheels and Wheel Covers............................ 778
        10.7.7   Airflow Through the Engine Compartment 780
        10.7.8   Rear 781
10.8 Aerodynamic Interaction....................................... 782
        10.8.1   Nose-to-Tail Driving................................. 782
        10.8.2   Tipping and Susceptibility to Side Winds 784
        10.8.3   Aerodynamic Loads on Components................... 786

Contents

	10.8.4	Dust Turbulence	786
	10.8.5	Intake of Warm Air	787
	10.8.6	Management of Exhaust Gas	788
10.9	Vehicle Soiling		789
	10.9.1	Task Description and Testing Methods	789
	10.9.2	Foreign Soiling	791
	10.9.3	Self-Soiling	792

Chapter 11: Motorcycle Aerodynamics ... 795

- 11.1 Introduction ... 795
- 11.2 Historical Review and Current Types ... 796
  - 11.2.1 History of Motorcycle Aerodynamics ... 796
  - 11.2.2 Current Motorcycle Categories ... 801
  - 11.2.3 Special Bikes ... 807
- 11.3 Aerodynamic Tasks ... 811
  - 11.3.1 Aerodynamic Forces and Moments ... 811
  - 11.3.2 Aerodynamics and Longitudinal Dynamics ... 813
  - 11.3.3 Aerodynamics and Lateral Dynamics ... 821
  - 11.3.4 Cooling and Internal Flow ... 831
  - 11.3.5 Wind and Weather Protection ... 835
  - 11.3.6 Aeroacoustics ... 837
- 11.4 Development Methods ... 838
  - 11.4.1 Development Process ... 838
  - 11.4.2 Simulation (CFD) ... 839
  - 11.4.3 Wind Tunnel ... 853
  - 11.4.4 Road Test ... 864
  - 11.4.5 Outlook—The Future of Development Methods ... 868
- 11.5 Aerodynamic Design—Practical Examples ... 869
  - 11.5.1 Measures to Optimize Drag and Lift ... 869
  - 11.5.2 Design of Internal Flows, Cooling, and Heat Protection ... 872
  - 11.5.3 Measures for Wind and Weather Protection ... 875
- 11.6 Outlook ... 876

Chapter 12: Helmets ... 877

- 12.1 Head Protection Technology ... 877
- 12.2 Motorcycle Helmets ... 879
  - 12.2.1 Aerodynamics ... 879
  - 12.2.2 Aeroacoustics ... 887
  - 12.2.3 Ventilation and Rain Tests ... 896
- 12.3 Helmets for Open Race Cars ... 901
  - 12.3.1 History ... 901
  - 12.3.2 Aerodynamics and Ventilation ... 901
  - 12.3.3 Acoustics ... 903

	12.4	Measurement and Simulation Technology	906
		12.4.1 Introduction	906
		12.4.2 Wind Tunnel	906
		12.4.3 Aerodynamic Forces	907
		12.4.4 Aeroacoustic and Artificial-Head-Measurement Technology	909
		12.4.5 Computational Fluid Dynamics (CFD)	909

Chapter 13: Wind Tunnels and Measurement Technique 913

13.1	Scope of Wind Tunnels	913
13.2	Wind Tunnel Physics	916
	13.2.1 Design and Function of Wind Tunnels	916
	13.2.2 Wind Tunnel Nozzle	920
	13.2.3 The Test Section	926
	13.2.4 The Collector	934
	13.2.5 Plenum	937
	13.2.6 Diffusers	940
	13.2.7 Turning Vanes	941
	13.2.8 Flow Conditioning Screens	941
	13.2.9 Honeycombs	942
	13.2.10 Acoustic and Anti-Buffeting Measures	942
	13.2.11 Ground Simulation	955
	13.2.12 Unsteady Flow and Gust Simulation	963
	13.2.13 Wind Tunnel Correction Methods	967
13.3	Wind Tunnel Measurements	981
	13.3.1 Test Sequence	981
	13.3.2 Measurement of Flow Velocity	984
	13.3.3 Pressure Measurements	990
	13.3.4 Measurements of Aerodynamic Loads	997
	13.3.5 Flow Visualization	1006
	13.3.6 Investigation of Vehicle Soiling	1011
	13.3.7 Engine Cooling Tests	1014
	13.3.8 Heating and Climatization Tests	1017
	13.3.9 On-Road Measurements	1021
	13.3.10 Additional Equipment in Climatic and Thermal Wind Tunnels	1032
13.4	Model Testing—Dimensionless Numbers	1034
13.5	Existing Wind Tunnels for Motor Vehicles	1039
	13.5.1 Full-Scale Wind Tunnels	1041
	13.5.2 Model-Scale Wind Tunnels	1048
	13.5.3 Climatic and Thermal Wind Tunnels	1050
	13.5.4 Overview and Correlation Measurements	1054
13.6	Outlook	1062

Chapter 14: Numerical Methods 1065
    14.1 Simulation of Three-Dimensional Viscous Flows 1066
            14.1.1   Requirements and Properties of CFD—Methods.......... 1068
            14.1.2   Basics of Kinetic Theory 1070
            14.1.3   Lattice Methods...................................... 1072
            14.1.4   Navier-Stokes Methods 1084
            14.1.5   Potential Flow Methods (BEM)..........................1103
            14.1.6   One-Dimensional Methods for Cooling Module Design 1113
            14.1.7   Rotating Geometries (Wheels, Fans) 1119
            14.1.8   Porous Media (Heat Exchanger).........................1121
            14.1.9   The Solution Process...................................1123
            14.1.10 Hardware and Benchmarking...........................1138
            14.1.11 Integration of CFD in the Development Process 1144
            14.1.12 Outlook..1146
    14.2 Computational Aeroacoustics for Motor Vehicles 1148
            14.2.1   Introduction 1148
            14.2.2   Calculation of Aerodynamic and Aeroacoustic Sources 1151
            14.2.3   Sources and Exterior Sound Field........................1153
            14.2.4   Transfer into the Vehicle Cabin 1157
            14.2.5   Examples of Applications 1161
            14.2.6   Conclusion and Outlook 1172

Abbreviations...1175

Symbols..1179

Literature..1193

The Authors..1245

Index..1253

Preface

The performance, handling, and comfort of an automobile are significantly affected by its aerodynamic properties. Low aerodynamic drag is a decisive prerequisite for improving fuel economy. This long-established correlation has become more widely recognized as fuel prices increase and more stringent legal regulations are imposed.

However, there are other aspects of vehicle aerodynamics that are no less important for the quality of an automobile. These include side wind stability, wind noise, and soiling of the body, lights, and windows. Finally, the cooling of the engine, the gear box, and the brakes all depend on the flow through and around the vehicle.

Vehicle aerodynamics is still an empirical science, if not an art. Whereas other technical disciplines such as aeronautics, naval architecture, and turbomachinery are governed by well-established theoretical and experimental methods of fluid mechanics, no consistent design procedures are yet available for bluff bodies like road vehicles. This is partly due to the complexity of the flow field around a vehicle, which is characterized by large separated regions. This means that the vehicle aerodynamicist must refer to a large amount of data from earlier development work. Their success depends on the ability to transfer these results to their own problem and to combine results originating from many different earlier developments to find a consistent solution.

It is the intention of this book to introduce the vehicle engineer to this approach. The topics are focused on three aspects:

- The fundamentals of fluid mechanics as related to vehicle aerodynamics;
- The essential experimental results presented as ground rules of fluid mechanics and brought to general validity wherever possible;
- Design strategies, which show how many existing individual results can be combined together to provide general solutions.

The aerodynamics of passenger cars, commercial vehicles, motorcycles, sports cars, and race cars is dealt with in detail. Not only the external flow field is covered; the problems associated the internal flow systems are examined as well. Because the external and the internal flow fields are interrelated, both have to be considered at the same time.

The related testing techniques are described in detail, emphasizing the correlation between the wind tunnel (which is one of two main tools used) and on the road, which is the real-world condition when it is in a customer's hands. Despite major advances in computational fluid dynamics (CFD), the development progress using experimental testing remains an irreplaceable part of the aerodynamicist's work. At the heart is the wind tunnel testing, which is discussed in detail. The limitations of re-creating the on-road conditions must be considered in this simulated environment. Only if these can be quantified can the test results from vehicle development in the tunnel be evaluated properly. Measurement techniques that have been developed continuously in conjunction with wind tunnel developments over the years are also examined.

Numerical aerodynamics methods are also covered in detail. Even if some deficits in computation fluid dynamics still exist, the progress that has been achieved has led to numerical simulations having a significant contribution to vehicle development. This has mainly been due to the exponential increase in computing resources in recent times. The basic process for vehicle development has evolved into optimizing the design using numerical methods (in conjunction with empirical information), and then verifying to refine the results with correlating experiments.

This book is intended for vehicle engineers in industry and research, at universities and in engineering departments. It is also aimed at stylists and designers, students, and professional writers in the car world. The individual chapters are designed so that each can be understood on their own. In-depth knowledge of aerodynamics is not required since the relevant principles are treated in a separate section at the beginning of the book. When it comes to an overlap of topics at some point, this shouldn't be considered a disadvantage. The redundancy rather supports the combination of the individual subjects, repetition, and consolidation. The selection of references was not intended to be a complete record but rather to quote the essential works with which the reader is able to go further in depth. This edition is equipped with an overall and numbered bibliography that helps with a very fast and efficient search. Particular attention was paid on the uniform nomenclature; symbols and other names have been standardized across all chapters with a few exceptions.

The first German edition of this book was originally based on a course given by the authors at the 'Haus der Technik,' Essen, Germany, under the aegis of Dr. H. Hahn. The first edition was published in 1981 was translated into Russian and Polish. Then, the second edition (1986) was published in English; after that, the book has been continuously renewed and extended. Alternating between English and German, it has been followed by four more editions.

The seventh English edition as well as the sixth German edition were directed by a new editor, and a comprehensive review was made of the entire fifth edition. Particular attention has been paid to the latest advances in numerical aerodynamics and wind tunnel technology. So more than ever it is now possible to model the fluid mechanical processes around the automobile realistically to analyze and understand the flow phenomena. Necessary chapter updates have been included due to the introduction of more stringent legislation on emissions, new rules in racing, and the increasing applied use of aerodynamics on commercial vehicles and motorcycles.

The editor would like to thank most sincerely the helpers and supporters for translating the individual chapters, especially (in alphabetic order) Dr. Brad Duncan, Dr. Maximilian Grosse, Dr. Shaun Johnson, Dr. Riccardo Pagliarella, Dr. Robert Spence, Dr. Joel Walter, and Gudrun and Mark Waskett.

The editor also thanks Dr. Wolf-Heinrich Hucho for his trust to hand over his work, which has matured for decades and is well known and respected among all experts. The chance to continue the work done so far is a privilege. This new edition benefits from the vast experience of Dr. Hucho, which is reflected in the historical overview in the first chapter.

Thomas Schuetz
Munich, Winter 2014

Acknowledgments

The editor would like to thank most sincerely all the persons who made this work possible. First of all, many thanks to Dr. Wolf-Heinrich Hucho for his trust to hand over his work, which has matured for decades and is well known and respected among all experts. The opportunity to continue his work is a privilege.

All contributions made by the helpers and supporters for translating the individual chapters are greatly appreciated. Especially the work done by (in alphabetic order) EXA Corp. and Dr. Brad Duncan, Dr. Maximilian Grosse, Dr. Shaun Johnson, Dr. Riccardo Pagliarella, Dr. Robert Spence, Dr. Joel Walter, Gudrun Waskett, and Mark Waskett.

Employers BMW Group and AUDI are gratefully acknowledged for kindly endorsing the work; especially managers Sven Klussmann, Holger Winkelmann, Dr. Moni Islam, and Norbert Lindener for their understanding and support during the development period of this book.

Chapter 1
Introduction to Automobile Aerodynamics

Wolf-Heinrich Hucho

1.1 Scope

1.1.1 The Role of Aerodynamics in Vehicle Design

Essential properties of an automobile are affected by aerodynamics, first in its performance in terms of fuel economy (which is closely related to emissions) and top speed. The latter seems to be of minor importance today, at least in official advertisements; however, many customers remain interested in top speed. Anyhow, fuel economy and top speed are determined by aerodynamic drag.

However, vehicle aerodynamics comprises much more than drag. This is elucidated by Figure 1.1. The flow around a vehicle is responsible for its directional stability: straight-line stability, dynamic passive steering, and response to crosswind all depend on the external flow-field. Furthermore, the outer flow should be tailored to prevent droplets of rainwater from accumulating on windows and outside mirrors, to keep the headlights free of dirt, to minimize wind noise, to prevent the windshield wipers from lifting off, and to cool the engine's oil pan, muffler, and brakes and so on. The internal flow has to accommodate the heat losses of the engine. It must ensure with the aid of the radiator and fan that its wasted heat is carried away under all driving conditions. Auxiliaries must be prevented from overheating. Finally, another internal flow system has to provide a comfortable climate inside the passenger compartment. Stratification of temperature- and flow-field are optimal for keeping heads cool and feet warm.

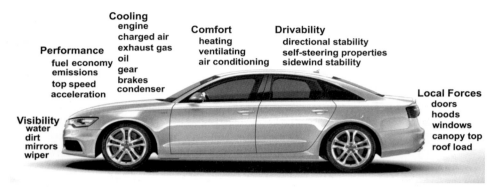

Figure 1.1 Spectrum of tasks for vehicle aerodynamics.

Apart from race cars and high-performance vehicles, the shape of a car is not dictated primarily by the properties shown in Figure 1.1 as, for example, they would be for an aircraft. Aircrafts are designed to achieve a given lift while keeping drag as low as possible. Automobile design is primarily dictated by aesthetic principles; aerodynamic properties have to be considered but need not be as apparent.

Automobile design has to provide technical solutions within an attractive overall shape that fits in with current style sensibilities. Car designs have to appear fresh and new. However, as will be discussed in section 1.10, there are many constraints and demands on automobile design. It is important to emphasize the styling of the brand and to make the various types of a model range of one and the same brand distinguishable at the same time. Finally, it is necessary to make the continuity of style and its progress visible. Two outcomes can be derived for automotive aerodynamics:

1. The aerodynamic properties of a vehicle are primarily determined by its shape and not by aerodynamic arguments.

2. The target is not a single ultimate shape, as is the case for modern transport aircraft. Vehicle aerodynamics is confronted with ever new shapes and has to make the best of it. However, no doubt, it has had, and will continue to have, retrospective influence on design.

1.1.2 The Character of Vehicle Aerodynamics

The flow processes to which a moving vehicle is subjected fall into three categories:

1. Flow of air around the vehicle;
2. Flow of air through the vehicle's body;
3. Flow processes within the vehicle's machinery.

The first two flow fields are closely related. For example, the flow of air through the engine compartment depends on the flow around the vehicle. Both flow fields must be considered simultaneously. On the other hand, the flow processes within engine and transmission are not directly connected with the first two. They are subject to fluid dynamics, not called aerodynamics, and are not treated here.

In terms of aerodynamics, vehicles are bluff bodies. Their shapes are many and diverse, and they are, as symbolized with Figure 1.2, very complex with regard to their geometry. Their aspect ratio—width or height as related to their length—is small, and they operate close to the ground. The flow around their body is fully three-dimensional, and boundary layers are turbulent. Separations are typical.

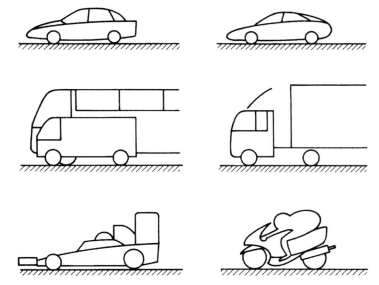

Figure 1.2 Diversity of shapes that vehicle aerodynamics has to deal with.

With regard to their kinematics, two kinds of separation must be distinguished (see Figure 1.3): Separation may start from a (more or less) straight line or a sharp edge perpendicular to the local flow (Figure 1.3a). Generally, this type of flow results in a wake, either closed like a bubble or wide open downstream. Inside this wake or bubble, the flow is unsteady, if not chaotic, and pressure is moderately low. This kind of separation is to be observed at a blunt rear end of a square back. Reattachment will occur on a notchback, or in front of the windshield, when it is fairly steep. In both cases a closed bubble will be formed.

The other type of separation is displayed in Figure 1.3b. The flow separates from a line oblique to the oncoming flow. A shear layer sheds and then rolls up, forming a longitudinal vortex, which is stable for some distance downstream, before finally dissipating. This vortex induces low pressure on panels close to it. Typically, a pair of such longitudinal vortices sheds from the C-pillars of a "fastback," and on the A pillars as well. For a long time, the character and effect of this kind of vortex was poorly understood if not totally overlooked in vehicle aerodynamics.

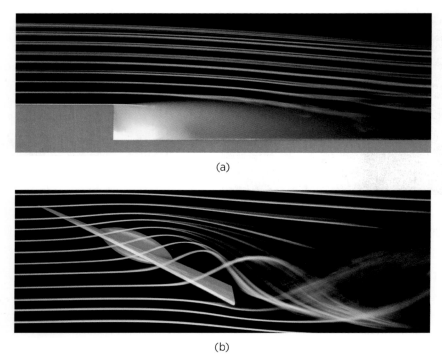

Figure 1.3 Two types of flow separation: a) at an edge (or line) perpendicular to the oncoming flow; b) at an edge (or line) which is oblique to the local flow. Photos Sönke Hucho.

Both kinds of separation, wake and longitudinal vortices, interact with each other. They are responsible for a major part of aerodynamic drag of a vehicle. Tailoring them makes up an essential part of aerodynamic design work, be it in a wind tunnel or on a computer. Additional flows in partly open cavities such as wheelhouses and closed ducts such as the cooling air duct must also be matched to the flow field around the vehicle.

The complexity of the flow past a vehicle can be made visible either in a wind tunnel or on the monitor of a computer. A striking example is displayed in Figure 1.4. The smoke trails,[1] Figure 1.4a, are fed to the oncoming flow in the central longitudinal cross section of the vehicle by a rake of tubes with small diameter. Flow remains attached over a long distance on its way over the body, even where the local curvature is comparatively sharp. Where the distance between the streamlines is small, local flow speed is high and pressure is low; lift is generated. Remarkable is the separation of the flow at the trailing edge of the roof, after which a large closed region of separation is generated behind the vehicle. Although not done here, this region can be made visible by inserting smoke into the flow field at the rear of the vehicle. However, only a "global" impression of the wake flow can be gained in this way. Deeper insight in the flow of the near wake is supplied by CFD, as shown with Figure 1.4b.

---

1. In reality what is called "smoke" actually is fog as used in stage productions.

(a)   (b)

Figure 1.4 Flow visualization: a) with smoke (fog) in a wind tunnel, b) by numerical fluid dynamics (CFD). Photo and figure Daimler AG.

1.1.3 Related Fields

There are other fields of applied fluid mechanics that have similarities to vehicle aerodynamics with regard to physics as well as to working methods. For vehicle aerodynamicists, it is worth having a look at them, because they may gain firm insights from these neighboring disciplines.[2]

The relationship is closest to railway aerodynamics. The flow fields around a road vehicle and a train are similar. The most essential difference results from the fact that a train is made up from many cars, thus generating a body which is extremely long in relation to its height or width. As a consequence, the boundary layer at the end of a train is extremely thick.

The essential targets for railway aerodynamics are the following:

1. Moderating the compressive "head wave" in front of a train is important when passing a station, entering a tunnel, and meeting a train in a tunnel;
2. Low aerodynamic drag is needed for high-speed traffic;
3. The oscillating vortex shedding at the end car is annoying for the passengers seated there and must be avoided;
4. Low wind noise to meet environmental regulations;
5. Low sensitivity to crosswinds;
6. Matching external and internal flow for engine cooling and air conditioning.

Modern high-speed trains are much faster than most road vehicles. How this difference in operating speed shifts the priorities through aerodynamic shape development may be elucidated by considering the development of the front and rear end of buses and trains, where low drag is important for each type of long-distance coach because of fuel

---

2. See Hucho (2011 [372]).

economy. For a bus, the requirement can easily be met with comparatively small radii at the A-pillars, the leading edges of the roof, and to the underbody flow. The result is a blunt front end. Quite contrary, the main criterion for the layout of the front end of a high-speed train is to moderate the compression pressure wave mentioned above. The faster the train, the more slender its head has to be. The rear end of a bus does not give much room for shaping it for low drag. The train is quite different: the shape of its rear is identical to its front end. However, it should be designed for low drag and a steady, non-oscillating flow. How far the resulting shape can be made to comply with the shape of the front end is an open question.

The parallels to the aerodynamics of buildings are numerous as well:

1. Buildings are bluff bodies;
2. Their flow fields are characterized by large separations;
3. Buildings are on the ground or closely above;
4. They are subjected to interference effects of neighboring buildings;
5. They are exposed to natural wind, which is gusty and has the properties of a turbulent boundary layer.

Their aerodynamic development also has similar objectives:

1. Determination of forces on a building as a whole;
2. Determination of static and dynamic forces on components like roof, tiles, facade, windows, and antennas;
3. Shaping of the flow field near the ground to prevent pedestrians from being annoyed or even blown over;
4. Matching the internal flow to the external flow in order to provide for comfortable air conditioning.

Of specific interest for buildings are aero-elastic effects.[3] These may also come into focus for vehicles as more super-light constructions are applied. Large and almost plain surfaces, like hoods, roof, and doors, could be stimulated to flutter by the surrounding flow. The same is true for the roof of a convertible.

With progress in numerical fluid mechanics, the links between vehicle aerodynamics and hydrodynamics of ships have become more pronounced. Cooperation had already been established in developing the numerical description of body surfaces.

Like vehicles, ships have the following:

1. Bluff bodies;
2. Their flow field—at least near the stern—is dominated by separation;
3. The main objective of their shape development is low drag (more precisely, maximum propulsive efficiency);

---

3. See Ruscheweyh (1982 [692]) and Försching (1974 [254]).

4. Large super structures (typical for cruising ships) are susceptible to crosswinds.

Even for the design of interior flow fields, the vehicle aerodynamicist can make use of experience gained in other disciplines. As an example, the layout of cooling-air ducts has benefited from results achieved with the oil coolers of aircraft. As another example, the huge amount of physiological data gathered over the years for the purposes of air-conditioning in buildings can be used in designing an air-conditioning system of a vehicle. However, there are differences that should not be overlooked. For example, the physiological data mentioned have been gathered in large rooms, and no direct radiation from the sun was taken into account. The passenger compartment of a car or even a bus is anything but large, and the influence of the sun cannot be neglected.

Further common interest among the "related fields" exists in experimental and numerical techniques applied: wind tunnels, water tunnels, towing tanks, blockage, ground simulation, gust simulation, artificially thickened boundary layers, and so forth.

1.2 Historical Development of Vehicle Aerodynamics

1.2.1 Literature

According to the author's knowledge, a comprehensive account of the development of the history of vehicle aerodynamics does not exist. An overall survey requires one to go back to individual publications either dedicated to the work of outstanding inventors, respected researchers, or to papers that try to give an overview—mainly under national aspects.[4] For a closer look, one has to dig through widely spread publications on specific details.

In the following sections, we attempt to segregate the threads of many aerodynamic ideas into the specific ideas related to the evolution of low drag. This will focus on two aspects: what was really new at the time in the field of aerodynamics and what ultimately proved to be suitable for mass-produced vehicles.

In the endeavor for low drag, two different approaches can be made out. In the first, vehicles were designed and built based on one or the other aerodynamic idea. As compiled in Figure 1.5 (aside from various exotic vehicles such as Buckminster Fuller's Dymaxion),[5] unique sedans were developed over the years—the Citroen DS 19 and NSU Ro 80 are examples—and although produced for years, they remained peculiar specimens. In the second approach, referring to right side of the schematic in Figure 1.6, the weak points of a given design were analyzed, and aerodynamicists were able to develop unobtrusive measures by numerous (but minor) modifications of shape details. Both routes to low drag are described in the following sections.

---

4. Some of the latter are Koenig-Fachsenfeld (1951 [448]), Bröhl (1978 [101]), Kieselbach (1982/83 [429], [430]), Graf Metternich (1985 [569]), Mutoh (1985 [600]), Barreau & Boutin (2008 [42]), Barnard (2008 [39]), Vivarelli (2009 [828]).
5. See Krause & Lichtenstein ([465] 1999).

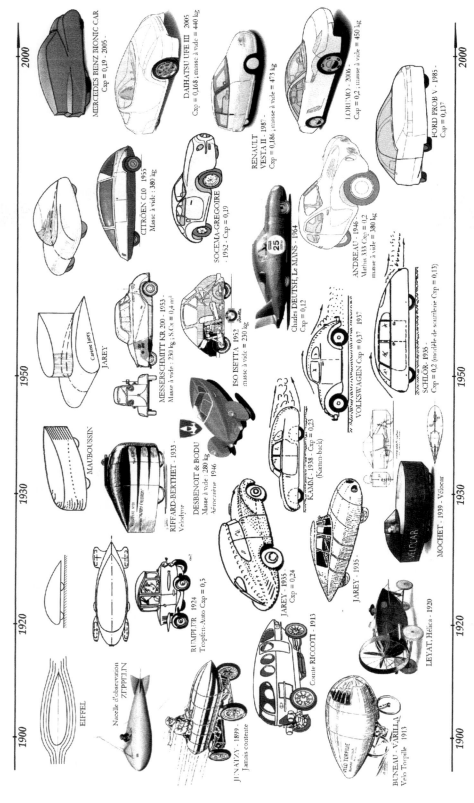

Figure 1.5 "Panel Historique" of vehicle aerodynamics in view of Barreau & Laurent (2008).

Introduction to Automobile Aerodynamics

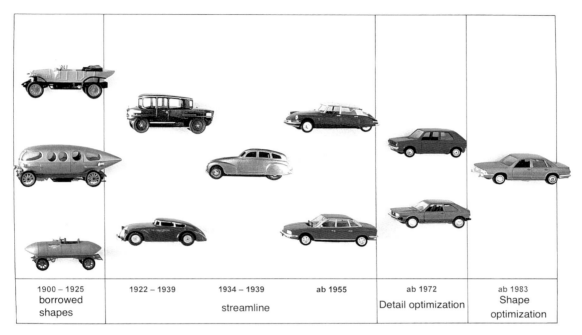

Figure 1.6 Convergence of car aerodynamics by systematic development; in view of W.-H. Hucho (2005). Showcase Sönke Hucho; scale of models 1:43.

1.2.2 Ahead of Its Time

Initially, aerodynamics was accepted only with hesitation in the automotive world; the merger of design and aerodynamics developed slowly. Synthesis of the two has been successful only after several tries. This is surprising insofar as in adjacent disciplines of traffic technology, namely naval architecture and aeronautics, making use of fluid mechanics turned out to be very fruitful. Evidently, the designers of ships and airplanes were in a better position, having found their originals in nature: fish and birds. From those natural shapes, they took on many essential features. In contrast, the automobile had no such ideal. Hence, the car designers tried to borrow shapes from ships, airships, and airplanes, which must have appeared attractive to them. This approach turned out to be wrong, and breakthroughs in automobile aerodynamics came only when they broke away from these improper examples.

Another reason for the early and repeated failures of aerodynamics with vehicles was that it advanced far too early. The first automobiles were quite slow. On the poor roads of those days, streamlined bodies must have looked ridiculous. Protecting the driver and passengers from wind, mud, and rain was accomplished well with the traditional design of horse-drawn carriages. Later, when roads allowed for higher speeds, the prejudice that streamlined bodies were something for aficionados overrode the economic benefits of aerodynamics.

1.2.3 Dominant Reference Number c_D

As already mentioned, the history of vehicle aerodynamics will be told following the development of the drag coefficient c_D. This figure c_D was, and still is, the focus of the entire subject, and for a time it was almost taken as a synonym for it. However, as emphasized with Figure 1.1, c_D is only a part of it.

As worked out in detail in section 3.2, the aerodynamic drag F_D—and in the same way all other components of the resulting aerodynamic forces and moments—increases with the square of the cruising speed v_F:

$$F_D \sim v_F^2 \tag{1.1}$$

With a medium-size car, aerodynamic drag typically accounts for about 75%–80% of the total resistance to motion at v_F = 100 kph (62 mph) on a level road. Hence, reducing aerodynamic drag contributes significantly to improving the fuel economy and reducing the CO_2 emissions of a car. For this reason, drag remains the focal point of vehicle aerodynamics. For a long time, top speed was the motivation for reducing drag in many countries, but this is of minor priority today—at least officially. The complete expression for equation (1.1) is

$$F_D = c_D A_x \frac{\rho}{2} v_F^2 \tag{1.2}$$

where c_D is the nondimensional drag coefficient; A_x is the projected frontal area of the vehicle, as defined in Figure 1.7, and ρ is the density of the ambient air, dependent on temperature; some numbers are plotted in chapter 2.

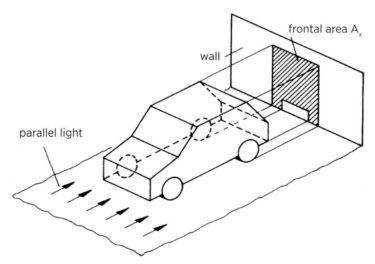

Figure 1.7 Definition of the frontal area A_x of a vehicle.

The aerodynamic drag F_D of a vehicle therefore is determined by its size, which is pretty well defined by the frontal area A_x, and by its shape, the aerodynamic quality of which is characterized by the drag coefficient c_D. Generally, the size of a vehicle, and hence its frontal area A_x, is determined by specifications; consequently, effort to reduce aerodynamic drag is concentrated on reducing the drag coefficient by properly shaping the body.

1.2.4 Creating Shapes for Cars

A survey over the development of aerodynamics in vehicle technology is summarized in Figure 1.6, for cars only. In contrast to the vigorous scene displayed in Figure 1.5, a line can be made out along which four periods may be distinguished.[6] Following this path, subsequent discussion will focus on fluid dynamic effects: which was discovered when, and how it was transformed to application. Erroneous methods will also be described; these, in particular, make clear what is required to convert fluid dynamics into vehicle-aerodynamics.

During the first two periods, which are called "borrowed shapes" and "streamline," refer to the left side of Figure 1.6. During these periods, aerodynamic development was done by individuals, most of them coming from outside the car industry. These individuals tried to transfer basic principles from aircraft aerodynamics to cars. How their findings were applied by the many automobile manufacturers of those days has been documented in detail by Kieselbach (1982/83 [432], [429], [430]) and will not be repeated here. During the latter two periods shown on the right side of Figure 1.6, the discipline of vehicle aerodynamics was taken over by the car companies and was integrated into product development. Since then, teams, not individual inventors, have been responsible for aerodynamic development of vehicles.

Whenever data from the literature are compared—of course drag is the focus—they must be viewed with caution, and not only those from older publications. Drag has been and still is measured by various methods, including model-scale or full-scale wind tunnels, coast down, and maximum speed. These methods (and others) do not necessarily give identical results. Even today's drag data from various full-scale wind tunnels sometimes show discrepancies that are difficult to fully explain.

Wherever possible, "historic" data should ideally be supplemented by new measurements. The nostalgic interest of old-timers is a help; measurements with originals of historical vehicles in modern full-scale wind tunnels provide valuable information. Consider one example: the Tatra 87, designed by Ledwinka in 1937 (see Figure 1.8). A scale model (1:5) was investigated in a DVL wind tunnel in Berlin, Adlershof; the result was c_D = 0.244.[7] At the same time, from measured top speed and engine power c_D = 0.31 was deduced. Years later, c_D = 0.36 was measured in the full-scale climatic wind tunnel of Volkswagen AG.

---

6. Neither time-wise nor substantially, they are as strictly to be distinguished from each other as shown in the schematic.
7. Published by Koenig-Fachsenfeld (1941 [449]).

Figure 1.8 Tatra 87 MY 1937, designed by Ledwinka, in the Volkswagen climatic-wind tunnel. Photo VW, exhibit Traffic Center, German Museum, Munich.

1.2.5 "Borrowed" Shapes

For the very first attempts to design automobiles following the rules of aerodynamics of aerodynamics—as far as they were known those days—it was typical to borrow shapes that had proven to be efficient in naval applications and airships. They were adopted almost unchanged and led to torpedoes and airships on wheels. No doubt, these vehicles had far lower drag than their contemporary competitors, which still looked like horse-drawn carriages. However, with regard to fluid mechanics they were far from perfect, in spite of their streamlined contours. They ignored the fact that close to the ground the flow around a body of revolution loses its rotational symmetry, so that drag increases. Furthermore, fully exposed wheels and an exposed undercarriage disturbed the flow.

Certainly the first vehicle developed with a specific objective of low air drag was built by Camille Jenatzy (see Figure 1.9). With this electrically driven record vehicle, he was the first to exceed 100 kph (62 mph), the "magic" speed limit of that time. On April 29,1899, he reached 105.9 kph (65.8 mph).[8] With a length-to-diameter ratio $l/d \approx 4$, the torpedo-shaped body itself was fairly well streamlined, but the exposed wheels, axles, and driver must certainly have disturbed its otherwise good flow properties. Jenatzy's record-breaking car may be seen as the predecessor of single-seat (monoposto) race cars, even though its body was positioned above the wheels rather than between them.

---

8. See Frankenberg & Matteucci (1973 [258]).

Figure 1.9 Camille Jenatzy's record vehicle. A replica is on exhibit in the Technical Museum, Berlin. Photo Chambre Syndical des Constructeurs Automobiles Français.

A vehicle with a body like an airship is shown in Figure 1.10. It was built in 1913 on the chassis of an Alfa Romeo on behalf of Count Ricotti. With a length-to-diameter ratio of $l/d \approx 3$, its shape was much fuller than that of Jenatzy's vehicle, and the passengers no longer had to sit in the open air. Earlier, already in 1911, similar vehicles but with entirely integrated wheels had been drafted by Bergmann.[9] An Italian patent from 1912 described a unique design for how to adapt the chassis to the demands of aerodynamics. The four wheels were arranged in a rhombus: one single wheel in front and the other in the rear were positioned on the longitudinal centerline with the remaining two on both sides in the middle of the wheel base. With this arrangement, the body could be shaped like a symmetrical air foil, standing upright.[10] If low drag had been the sole objective of vehicle aerodynamics, the technology would indeed have arrived at its target very early.

Figure 1.10 The airship-like car of Count Ricotti, 1913. A replica is on exhibit in the Museum of Alfa Romeo, Arese (near Milan), Italy.

---

9. See Koenig-Fachsenfeld (1951 [448]).
10. See Vivarelli (2009 [828]).

Chapter 1

A contrast to the integrated shapes used by Jenatzy, Bergmann, and Count Ricotti is the so-called "boat-tail." A typical example, the Audi-Alpensieger from 1913, is shown in Figure 1.11b. Hence, the car looked more elegant, even faster, than its predecessor with a blunt tail (see Figure 1.11a). Therefore, the boat-tail was frequently applied to sport cars and roadsters to give them a dynamic appearance. In terms of aerodynamics, however, it is absolutely ineffective: the flow, separating at the front and from the fenders, will not reattach on the "boat-tailed" rear end. This is a typical example of how aerodynamic arguments were and sometimes still are used—and abused—to justify stylistic curiosities. A trend in tail fins peaked of the middle of the last century, and thereafter several spoiler designs also were not far from being aerodynamically useless.

Audi Alpensieger 1913 and predecessor 1912

Figure 1.11 The "boat-tailed" rear end: a) Audi 1912, on exhibit in the Audi Museum, Zwickau, Saxony; b) Audi Alpensieger, 1913, on exhibit in the German Museum, Munich. Photos by the author.

1.2.6 The "Streamline" Era

While intuitive approaches were dominant during the initial phase of vehicle aerodynamics, a more orderly development started after World War I, supported from three different sides:

1. The analysis of tractive resistances by Riedler (1911 [671]) clearly identified the importance of aerodynamic drag.

2. The more Prandtl and Eiffel worked out the nature of aerodynamic drag, the more this knowledge was applied to vehicles.[11] However, getting away from Newton's impact theory was a very slow process.

---

11. A typical example is the work done by Aston (1911 [18]).

Introduction to Automobile Aerodynamics

3. Looking for other fields of activity, aeronautical engineers in Germany turned to vehicle design. They tried to transfer their specific know-how to the automobile. At different places, "streamline cars" were developed.

This took place on four different paths, all starting from an airfoil, as sketched in Figure 1.12. Beginning in 1919, Rumpler, the well-known manufacturer of the famous "Rumpler Taube" (dove), developed several cars which he designated "tear drop cars."[12] A typical Rumpler sedan is shown in Figure. 1.13. Indeed, viewed from above, this car had a shape like an airfoil. To make use of the otherwise useless narrow space in the slender tail, Rumpler decided for a rear-mounted engine.[13]

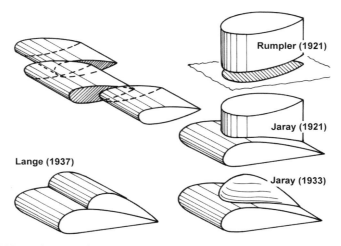

Figure 1.12 Derivation of car shapes from a (symmetrical) airfoil; in diagram form.

Figure 1.13 The Rumpler "tear drop car" in the climatic wind tunnel of Volkswagen; on exhibit in German Museum, Munich. Another version is on exhibit in the Technical Museum, Berlin.

---

12. At his time a falling drop was thought to have a perfect shape in sense of fluid dynamics, and even today many people believe in this saying. However, the falling drop's shape is anything else but "streamlined."
13. Technical details of his design have been reported by Heller (1921 [337]), by Eppinger (1921 [232]), and by Rumpler (1924 [690]) himself.

With a model of scale 1:7.5, Rumpler in 1922 performed wind tunnel measurements in the Aerodynamic Laboratory (AVA) in Göttingen. The only information remaining from these experiments is that the drag of Rumpler's car was approximately one-third the value of its contemporary competitors. But Rumpler was not out for low drag alone, as is well documented in his advertisements of the day. With sketches like the one in Figure 1.14, he pointed out that his car stirred up far less dust and mud as others.

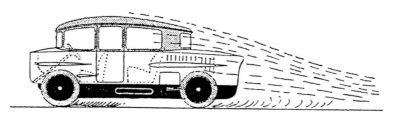

Figure 1.14 Early advertising: Less dust whirled up by a "tear drop car" as compared to a conventional car from that time. Drawing Rumpler (1921).

Measurements with an original Rumpler car, performed in 1979 in the climatic wind tunnel of Volkswagen AG, gave the following results: drag coefficient c_D = 0.28; frontal area 2.57m². Only decades later was this extremely low drag figure again attained with a production vehicle: the Opel Omega MY 1986. The value of Rumpler's c_D is particularly noteworthy because the wheels of Rumpler's vehicle are totally uncovered. As can be deduced from measurements published by Klemperer [443], this must have increased drag by 50%.

Rumpler invested most of the money he had earned with his airplanes in an entirely new plant and started to manufacture his "teardrop car."[14] However, he foundered, as his cars could hardly be sold. Repeatedly it was said that the unconventional shape was not accepted by the buying public. Perhaps this is true. But it is also true that Rumpler's car was full of unproven innovations, as for instance the pendulum rear axle that he

---

14. Details of this story have been told in Graf Metternich (1985 [569]) and Kubisch (1985 [470]).

Introduction to Automobile Aerodynamics

invented. The quality problems with these new components alone would have been enough to prevent his car's success.

About the same time as Rumpler, his competitor Jaray began a development that was aimed at a "streamlined car," a term he coined in his famous paper "The streamlined car, a new shape for an automobile body" (1922 [410]). The basic analysis of flow around the bluff body is summarized in Figure 1.15[15] and discussed in more detail below. Much later, this approach turned out to be a signpost for modern vehicle aerodynamics, designated as "body optimization," which will be considered further in section 1.9.2.

On behalf of Jaray, Klemperer (1922 [443]) carried out the measurements in the wind tunnel of Count Zeppelin at Friedrichshafen, Germany. The results are summarized in Figure 1.15. The starting point was a body of revolution with a slenderness (ratio of length l to diameter d) $\lambda = l/d = 5$. Off the ground, its drag was very low: $c_D = 0.045$. A step-by-step movement of this body toward the ground was accompanied by an increase in drag, first moderately but then significantly at a ground clearance typical for cars. Jaray and Klemperer observed that the flow past this symmetrical body lost its rotational symmetry as it approached the ground, and eventually a strong separation occurred along the upper aft surface. They concluded that this separation was the major cause for the drag increase.

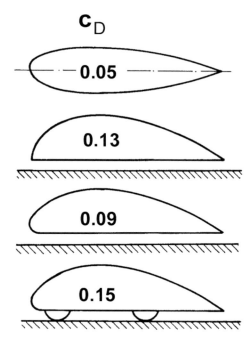

Figure 1.15 Derivation of a low drag "half body" close to the ground from a body of revolution flying in free air; measurement Klemperer (1922).

---

15. Jaray's work has been reported by Bröhl (1978 [101]).

At the limit of zero ground clearance, a rotationally symmetric flow could be recovered when the body of revolution was replaced by a halfbody. Together with its mirror image—produced by imaging the halfbody below the road surface—an effective body of revolution was generated. When the body was raised to a ground clearance necessary for an automobile, drag increased again. The reason for this was flow separation at the sharp leading edge of the underside. By rounding this edge, the drag increase was partially compensated. The result was $c_D = 0.09$. When wheels were added—without forming hollow wheel houses—drag went up to $c_D = 0.15$, a value three times that for the body of revolution far off the ground. However, in comparison to the cars of those days, for which $c_D \approx 0.7$ was typical, this figure was very encouraging: so much so that Jaray used this modified halfbody as the starting point for the development of "real" streamlined automobiles. This real halfbody was matched to the requirements of an automobile by replacing the half circle forming its main rib by a rectangle having rounded upper edges. The idea of using the shape of a halfbody directly to a car was so attractive that it was tried by several designers and finally led to the van (more about this will follow in section 1.3.3).

Jaray recognized that the drag increase from an isolated body of revolution to his halfbody in ground proximity was mainly due to the flow separation at the rear. He therefore sought rear-body shapes that allowed for as much pressure recovery as possible. However, because a boundary layer is able to sustain only moderate adverse pressure gradients, streamlined bodies like airfoils or airship hulls must end with long slender "tails." For automobiles, such long tails are unsuitable because the length of a vehicle is limited for a variety of practical reasons. Jaray tried to achieve sufficient pressure recovery with a shorter length by distributing the pressure recovery on two planes. He invented the "combination form," as seen on the right side of Figure 1.12. He conceived two wing sections, one positioned horizontally and the other, shorter one turned by 90° and put on top of the first. Whether or not the flow follows his hypothesis has never been investigated. There is doubt, however, because a three-dimensional boundary layer in a corner is prone to separation, and two such corners are formed by the perpendicularly mounted profiles.

Nevertheless, with his combination shape, Jaray succeeded in cutting the aerodynamic drag by half, as compared to box-type contemporary cars. Subsequently, from 1922 to 1923, the German car manufacturers Audi (Figure 1.16), Dixi, and Ley exhibited prototypes with shapes based on the combination form. In the United States, Chrysler followed in 1928. However, none of these cars were accepted by the public, and consequently they were not put into mass production. However, most surprisingly from today's point of view, it seems no one complained that all of these cars looked very much the same.

Introduction to Automobile Aerodynamics

Figure 1.16 One of the first cars with Jaray lines, the Audi Type K14, 50 hp, 1923, designed by Gläser at Dresden, Saxony. Photo archive R.J.F. Kieselbach.

Early in the 1930s, when the construction of the German Autobahn (freeway system) was started, Jaray undertook a second approach. He modified his concept and now placed half a body of revolution on top of the lower horizontal profile, as is sketched at the bottom right of Figure 1.12. This gave his cars a rather sporty look. However, in terms of aerodynamics, they were not as good as expected. Nevertheless, this approach was realized with the BMW 328 equipped with a body made by Wendler (see Figure. 1.17). Despite a good-looking flow pattern, the drag was no better than $c_D = 0.44$. This may be due to flow separation at the rear and to high airflow through the large openings in the front (cooling flow drag).

Figure 1.17 BMW 328, 1938; $c_D = 0.44$, designed and built by Wendler; exhibit in German Museum, Munich. Photo taken in the climatic wind tunnel of Volkswagen AG.

How far Jaray was ahead of his time becomes evident when prototypes designed according to his principles are compared with contemporary cars; Figure 1.18 gives a typical example: the Stuttgart model from Mercedes-Benz. Jaray's shapes were revolutionary and anticipated what only became standard practice after World War II: a smooth body surface, integrated fenders and headlamps, and a cambered windshield. He clearly distanced the shape of an automobile from that of a horse-drawn carriage. However, he did not succeed in overcoming the most serious drawback of all low-drag car concepts developed to that point: the tail that was still too long (as can be seen by comparing the rear overhang of the two cars).

Figure 1.18 Daimler Benz type Stuttgart, 1928; c_D = 0.66; A_x = 2.53. Photo taken in the Untertürkheim wind tunnel, Daimler AG.

The stylists, now designers, who in the 1930s began to separate themselves from the body engineers, solved the problem of the long tail in a manner typical for them: they cut it off and made the slope of the rear steeper. The so called "pseudo-Jaray" shape was born. Numerous cars were built according to this form; four typical examples are compiled in Figure 1.19. All these cars had a comparatively high drag coefficient. Nevertheless, they were called "fastback" in the United States.[16]

---

16. A typical example for lacking know-how in physics is overrun dialectically.

Introduction to Automobile Aerodynamics

Lincoln Zephyr V12 1926

Volvo PV 544 1955

Opel Kapitän 1938

VW 1600 TL

Figure 1.19 Cars with pseudo-Jaray shape. Photos from manufacturer.

The high aerodynamic drag of the fastback cars remained a mystery for a long time. Didn't the parallel wool tufts on the rear slope to be seen in Figure 1.20 indicate an orderly, attached flow down to the license plate? Shouldn't this be an indication of low drag? Only during the development of the first Volkswagen Golf (Rabbit) in 1974 was an answer to this enigma found, as will be described later.

Figure 1.20 Wool tufts from an Adler test car with a typical pseudo-Jaray rear end. Photo taken during a test run on road. Source Koenig-Fachsenfeld (1951).

However, there was one car that closely followed the ideas of Jaray and was successful in the market: the Tatra 87, shown previously in Figure 1.8. Production was started in 1936 and continued until 1950. With a length-to-height ratio $l/h = 2.9$, it was less slender than typical Jaray shapes. However, by shifting the greenhouse farther forward compared to average cars, a rather moderate slope of the rear end was created, close to the shape of Jaray. The engine was placed in the rear, so no long forward hood was needed. The resulting profile contradicted the trend up to that time, which was to demonstrate performance with a long engine hood. Nevertheless, the consequent design of the Tatra 87 was harmonic. It is a centerpiece of the German Museum in Munich. Whether or not the idea of Jaray, the "combination shape," actually worked in the Tatra 87 has never been investigated. Perhaps, for example, the air intake for the rear engine, acting as a boundary layer suction, helped to keep the flow attached on the sloping rear. A different approach to derive an automobile shape from an airfoil was made by Mauboussin (1933 [535]), who had his roots in aeronautics. He followed a route similar to Jaray's: he designed a streamlined body for Chenard and Walcker and called it the "Mistral." The left side of Figure 1.21 is a reproduction of his drawing; the right exhibits the prototype of Mistral. In contrast to Jaray, the trailing edge of this car was vertical, providing a large fin to improve crosswind stability.

Introduction to Automobile Aerodynamics

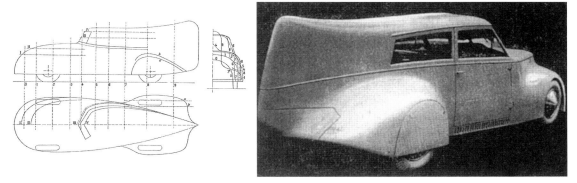

Figure 1.21 Cars with pseudo-Jaray shape. Photos by manufacturer.

A further path to make use of know-how in aircraft design was undertaken by Andreau (1946 [17]) with his car, the Mathis-Andreau 333: three wheels, three passengers, three liters per 100 km. His unique design can be seen in Figure 1.22. A more advanced version with a large vertical fin, Code SIA 1935, was planned. Details may be found in Barreau and Boutin (2008 [42]) who also reported on low-drag cars developed by Panhard (Dynavia, 1948), Renault (Vesta, 1987), Citroen (Eco, 2000), Volkswagen (Concept I, 2002), and Daihatsu (UFE II, III, 2005).

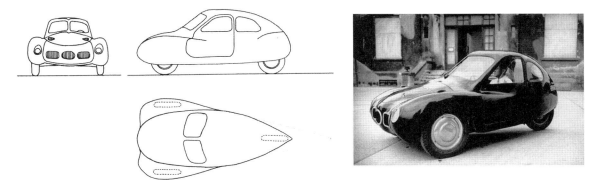

Figure 1.22 Three wheeler Mathis-Andreau 333. Archive Barreau.

A different attempt to develop a low-drag car with the aid of aircraft aerodynamics was started at the Aerodynamische Versuchsanstalt (AVA) at Göttingen under the direction of Prandtl; the work was performed by Lange (1937 [485]). Again the body was made up of two airfoils (see Figure 1.12, bottom left). This time an upper profile was placed horizontally on a lower one, which was horizontal as well. The upper profile started at the windshield and had a common trailing edge with the lower profile. After rounding off all edges, the model shown in Figure 1.23a was finished. Despite the fact that these lines were never used to build a real car, this model is known as the Lange Car. The drag coefficient published by Lange, $c_D = 0.14$, was nearly confirmed much later: a one-fifth scale model built by Volkswagen came up with $c_D = 0.16$. This very good result has to be considered in light of the fact that the model was completely smooth, with no recesses for windows, no real wheelhouses, and had no undercarriage at all. With a length-to-height ratio of $l/h = 3.52$, its tail was still very long, suited only for sports cars with rear-mounted engines. After more than 70 years, however, the Lange shape still "lives" through the Porsche 911 displayed in Figure 1.23b.

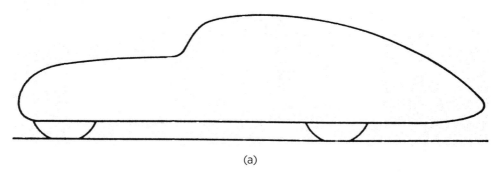

(a)

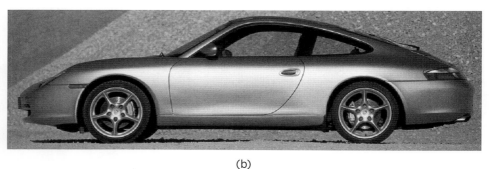

(b)

Figure 1.23 a) Lange car, silhouette, $c_D = 0.14$; b) Porsche 911 (2004), $c_D = 0.28$, $A_x = 2.0$ m². Photo Porsche AG.

Along the various paths to transfer solutions from aircraft design, Rumpler, Jaray, Mauboussin, Andreau, and Lange all made the same mistake: they took shapes developed for aeronautical applications and applied them to an automobile with as few changes as possible. They subordinated the requirements of packaging and consumer appeal to the demands of aerodynamics. They looked rigidly at the "ideal" drag figure of $c_D = 0.15$, measured by Klemperer with a half body close to ground, and they tried to come as near to it as possible. A drag coefficient of $c_D = 0.30$ was set as a target—which was met only in the early 1980s with production cars. No one came up with the idea of reducing the drag of contemporary cars, of, say $c_D = 0.65$, using a step-by-step with less dramatic changes. This resulted in a polarization: in terms of aerodynamics, the "standard" car remained poor but it was accepted by the market. Aerodynamic cars came with exotic shapes and technical constraints, and the public refused to buy them.

1.2.7 Early Investigations with Parameters

The canyon separating aerodynamics from vehicle technology was only bridged when vehicle engineers themselves took an interest in aerodynamics. This happened at almost the same time and independently in two places: in the United States by Lay[17] and in Germany by Kamm.[18]

Lay [490] was the first to carry out a systematic variation of parameters, and in 1933 he published the results in his famous paper, "Is 50 Miles per Gallon Possible with Correct Streamlining?" Some key results from his paper are compiled in Figure 1.24. By systematically modifying the shape of a generic car model at the front and rear, Lay isolated several individual aerodynamic effects. His investigations revealed a strong interaction between the flow fields of a car's forebody and its afterbody. For example, a car with a good aerodynamic design at the rear turned out to be effective only if the flow around the front of the car remained attached, a fact mentioned previously in the discussion of the boat tail (see Figure 1.11). A steep windshield can be very unfortunate for an otherwise well-designed body and can result in high drag. However, if the drag of a model is already high, for instance because of separation at the rear, the influence of windshield slope on drag is minor.

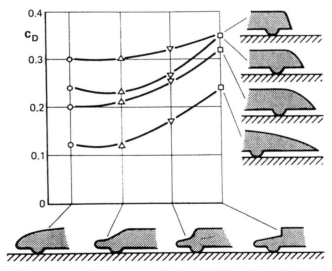

Figure. 1.24 Brief summary of the test results Lay (1933) achieved with set a of models. Effect of main body parameters on drag, and their interaction.

Unfortunately, Lay's model was too far away from the geometry of a real car in significant details—a mistake repeated by many others to follow. His model was built up from segments and had parallel side walls and sharp edges, which resulted in a fairly high

---

17. Professor at the University of Michigan.
18. Professor at the Technical University (TH) Stuttgart and director of the Research Institute for Road Vehicles and Engines (FKFS); see his biography written by Potthoff & Schmid (2012 [653]).

drag. Flow separations and the roll-up of vortices may have masked the influence of individual shape details. His findings were therefore of limited applicability, and consequently his work hardly received recognition in car design. Kamm solved the length problem of the Jaray shape. This will be dealt with together with other rear-end shapes in section 1.4.

1.3 From Horseless Carriage to Automobile

1.3.1 Stamping

The aerodynamicists' endeavor for low drag was supported by a development unrelated to aerodynamics: the metamorphosis from a carriage to an automobile (Figure 1.25). As long as the body made use of the same technique as a carriage, it was built up on a steel frame. The skeleton was made from wood, with a skin of fabric, plywood, or sheet metal. The surfaces were only moderately bent cylindrically, the transitions relatively sharp. Running boards and curved fenders were attached, and headlamps were free standing.

(a)

(b)

Figure 1.25 Metamorphosis from carriage to automobile: a) FIAT Barilla MY 1932; b) FIAT 500 since 1938. Photos Barker & Harding (1992).

Introduction to Automobile Aerodynamics

This traditional skeleton-and-skin technique was replaced by a self supporting body, usually from steel, occasionally from aluminum. Stamping allowed for three-dimensional curvature; initially large radii were preferred. The upright standing radiator was hidden behind a grill. Later, fenders and headlamps were integrated. The angled box thus transitioned to a smooth body with well-rounded transitions, a much better shape for smooth flow.

1.3.2 "Bathtub" Body

When automobile production was resumed in 1945, it started with pre-war models. U.S. models were one generation younger than those in Europe because the development of new cars in the United States was "frozen" two years later than in Europe. With the new Pontoon-shape—its nick name was "bathtub"—they set a high standard. The mutation of the pseudo-Jaray shape with streamlined fenders to the bathtub took place in two steps:

1. A luggage compartment was attached to a sloping rear. The result was the notchback. The growth in overall length was well accepted by consumers.
2. Fenders and headlamps were integrated into the body, and running boards were omitted.

"Form followed function," leading to a three-volume configuration consisting of one box each for the engine, passengers, and luggage (see Figure 1.26a). Early in the 1940s, this design concept was developed in the United States by Darrin, and Kaiser was the first to apply it in model year 1946. A popular example is seen in Figure 1.26b: the 1949 Ford Lincoln Continental. Some typical post-war pontoon cars from the United States and Europe are shown in Figure 1.27.

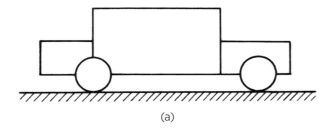

(a)

(b)

Figure 1.26 Three-volume configuration: a) schematic; b) Ford Lincoln Continental, 1949. Photo Ford Motor Co.

Chapter 1

Hanomag „Kommisbrot" 1924

Kaiser-Frazer 1947
Design Howard Darrin

Borgward Hansa 1949

Mercedes Benz 1953

Figure. 1.27 Typical pontoon cars from the United States and in Europe. Photos from manufacturers.

Originally, the pontoon body left little room for aerodynamic shaping. However, its flush panels, integrated headlamps and fenders, well-rounded "edges," and the cambered and sloped windshield all helped to improve the flow around a car. Consequently, the average drag coefficient came down from 0.55 to 0.45 (i.e., by almost 20%). In addition, the frontal area shrank and came closer to its ergonomic limit. With only a little assistance from aerodynamicists, the aerodynamics of cars was improved considerably.

For a long time, aerodynamics persisted at this level: $c_D \approx 0.45$–0.50. Within the car industry it was commonly assumed that lower drag could only be achieved with shapes resembling the streamliners of the 1930s, which, as everyone remembered, were rejected by the buying public. One manufacturer dared to break away from this prejudice: the French brand Citroen. Some of its model lineup through the years is sketched in Figure 1.28. With the Citroen DS 19, a car was launched making its aerodynamic design obvious (Figure 1.29a). It was followed by the CX, a luxury car, in an aero-looking style as well. The latter was even less accepted than the former. However, when streamlining finally came into fashion, Citroen changed to a more angled design (Figure 1.29b). Curiously enough, the angled BX with $c_D = 0.33$ had a lower drag coefficient than the streamlined DS, which was $c_D = 0.38$! Eventually, Citroen lost its independence as an innovator,[19] and Panhard, its competitor with its self-willed Dyna, disappeared from the scene as well.

	Model Year	A_x in m^2	c_D
 DS 19	1956	2.14	0.38
 GS	1970	1.77	0.37
 CX 2000	1974	1.96	0.40
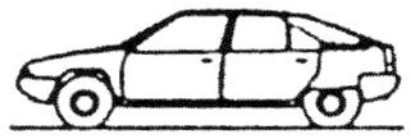 BX	1982	1.89	0.33–0.34

Figure 1.28 Lineup of aerodynamic cars from Citroen (1956–1982).

Figure 1.29 Citroen cars: top DS 19, 1956; bottom CX,1982. Archive Kieselbach.

NSU, the maker of small cars, dared a test to escape from the uniform pontoon shape with its unique Ro 80 (see Figure 1.30). With its high (as compared to the notchbacks

---

19. For what reasons, aerodynamics and/or quality?

of the time) rear end, this car was a forerunner of the wedge shape, which dominated model lines of the next generations.[20]

Figure 1.30 NSU Ro 80, MY 1976; c_D = 0.38; A_x = 1.99 m². Photo Volkswagen AG.

1.3.3 One-Volume Bodies

Despite the difficulties marketing streamlined cars, some aero-enthusiasts went even further. Because the excellent drag coefficient of c_D = 0.30 for the streamlined cars remained a good distance away from the "ideal" figure of c_D = 0.15 that Klemperer had demonstrated with a halfbody, they tried to further exploit this potential. Not only scientists but designers as well rated the one-volume concept as the ultimate configuration for a car. In the United States, Norman Bel Geddes (1931) and Richard Buckminster Fuller (1933) and in France, Emile Claveau (1925) and Andre Dubonnet (1935) followed the example of Count Ricotti and had special bodies built. Their shapes were created rather intuitively and probably never "saw" a wind tunnel. A unique example is the rear-wheel-steered three-wheeler of Buckminster Fuller (see Figure 1.31); prototypes were built in three versions.

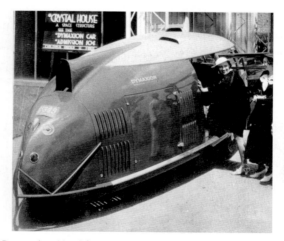

Figure 1.31 Dymaxion No. 1 from Krausse & Lichtenstein Eds: Your private sky, R. Buckminster Fuller. Lars Müller Publishers, Baden Swiss, 1999. Photo Krausse & Lichtenstein (1999).

---

20. Its success in the market was limited, however, not because of the aero look but due to problems with its Wankel engine.

In contrast, in 1922 Persu ([634]) designed a car whose shape he matched as closely as possible to the "ideal" halfbody he knew from Jaray's work. Because of limited space in the rear, he decided for a mid-engine layout. To have a drivable prototype he built a test car from parts he had on hand. The result is seen in Figure 1.32b. This rather hideous rolling test bed, unsurprisingly, failed to promote Persu's idea of a one-volume car. The vehicle is said to be on exhibit in Bucharest, Romania. A car according to his design, in Figure 1.32a, was never built.

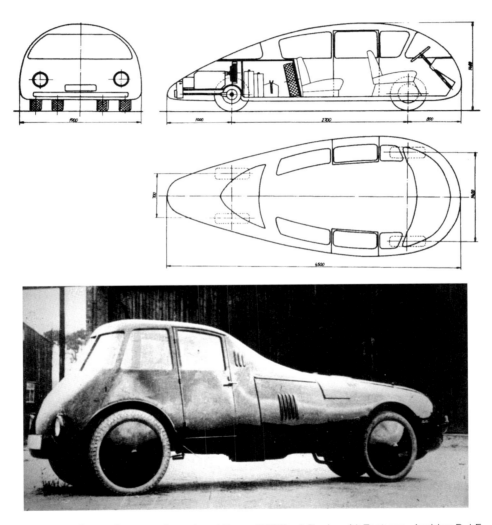

Figure 1.32 One-volume car from Aurel Persu (1922): a) Design; b) Test car. Archive R.J.F. Kieselbach.

Later, in the 1930s, several American engineers turned to the one-volume concept. A summary from the work of Fishleigh (1931 [249]), Heald (1933 [319]), Lay (1933 [490]), and

Reid (1935 [662]) is compiled in Figure 1.33. To assess their results achieved models of varying fidelity and different scales, each halfbody shape is compared with a contemporary sedan model tested by the same author. With the exception of the extremely long tail included in the tests by Lay, all halfbody models had a drag coefficient approximately one-third that of contemporary sedans.

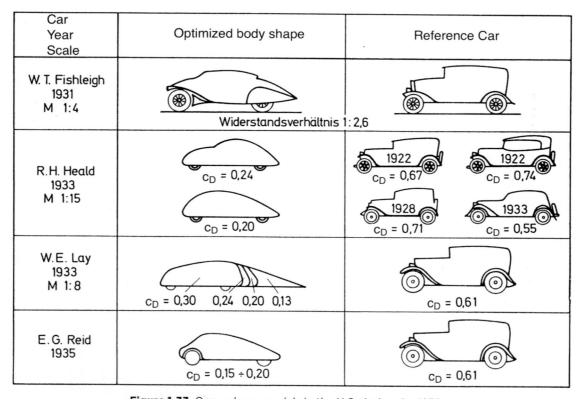

Figure 1.33 One-volume models in the U.S. during the 1930s.

While the shapes of all these models appear to have been created by intuition, the contours of the one-volume car developed at the AVA at Göttingen, in contrast, were derived from low-drag airfoils. Prior to this, an analysis of the flow past the Lange model by Schlör (1938 [717]) showed that this model was not nearly as ideal as had been presumed. With a carefully designed one-volume shape, it should be possible to overcome the drawbacks of the Lange shape: separation ahead of the windshield and in the rear. Schlör built up the contour of the longitudinal center plane of his car (see Figure 1.34a) by combining two wing sections from the Göttingen family, both of which had a low drag coefficient of $c_D = 0.125$.

Over the years, the Schlör car has been investigated in three different scales: a 1:5 model and the full-scale car at the AVA, and a 1:4 model reproduced by Volkswagen using the lines in Figure 1.34a. The drag measurements are compiled in Figure 1.35; the abscissa is the ground clearance in reference to the overall height of the car. Far off ground, the model had a drag coefficient even lower than that of the wing sections from which it was derived, a typical 3D effect. The actual car, measured in the large AVA wind tunnel, had $c_D = 0.186$, comparing fairly well with $c_D = 0.189$ from coast-down tests performed at the Technical University, Hannover. This low-drag coefficient should be considered together with the very large frontal area of the Schlör car, $A_x = 2.54$ m², which resulted mainly from the large width of 2.10 m necessary to allow for a sufficiently large steering angle with the fully covered front wheels. This large frontal area must be seen as part of the concept of Schlör's car.

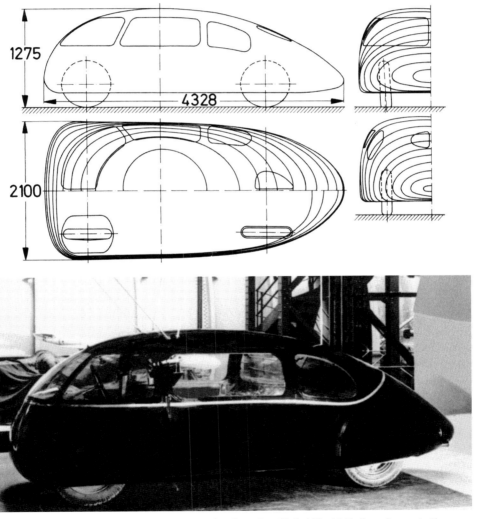

Figure 1.34 Schlör-car (1938): a) Body plan after K. Schlör; b) Full-scale car in the large AVA wind tunnel (nozzle with elliptical cross section, 7.0 m × 4.5 m) at Göttingen. Archive DLR.

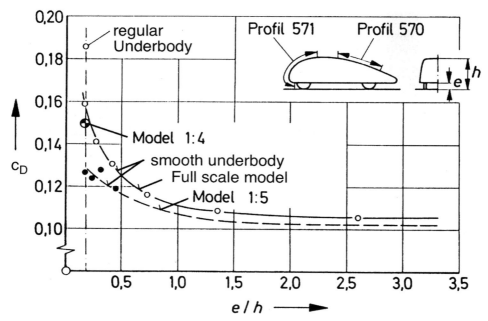

Figure 1.35 Drag measurements carried out with models of the Schlör car in different scales, and full scale.

The one volume shape could only assert itself in narrow market niches. A famous mini car for the German market was the BMW Isetta, and is now the Daimler Smart Car. Even spacious vehicles have been built with one volume body, specifically, multipurpose vehicles (minivans), as for instance the Dodge Caravan and the Honda Odyssey.

1.4 Rear-End Shapes

1.4.1 Kamm-Back

Jaray's "combination shape" was unable to solve the problem with the long tail. It remained unknown whether it even did what was expected, namely to recover pressure along the flow path to the rear. Only in the middle of the 1930s did Kamm and his team succeeded in developing a blunt—and thus shorter—rear end shape with lower drag. The shape and function of this rear end can be explained with the help of Figure 1.36. Starting at the maximum cross section (main span), the downstream body contours are carefully tapered to keep the flow attached. This produces a steady increase of static pressure. Just ahead of the location where the flow would separate, the body is truncated vertically. This leads to a plane base whose cross section is small compared to the frontal area. The near wake behind the vehicle is narrow, and the negative static pressure on the flat base (at the rear) is moderate—due to the upstream pressure recovery just described. Both contribute to a low drag. This shape was named "Kamm-back," or briefly "K-back."

Introduction to Automobile Aerodynamics

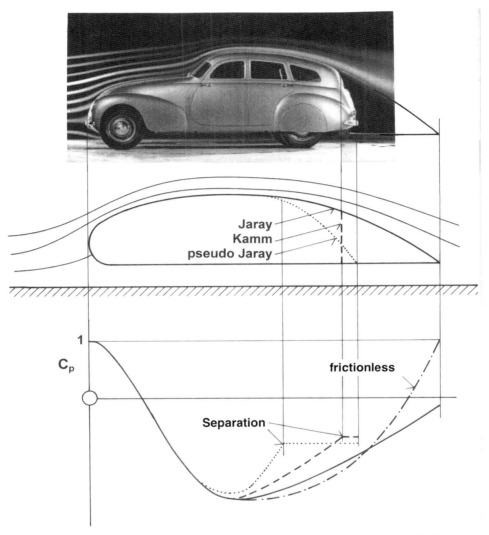

Figure 1.36 Flow mechanism of the Kamm-back in comparison to a Jaray-back and a pseudo-version therefrom, schematic: a) Kamm-car K2 in the Volkswagen climatic wind tunnel; b) contours of centerline, and the related pressure distributions, the latter schematic. Photo Volkswagen AG.

Under the direction of Kamm, several prototypes were built according to his ideas. The first, K1, seems to have been the most progressive. The body was absolutely smooth, and front wheels as well as rear wheels were covered. In Figure 1.37 the flow over a one-fifth scale model of the K1 is compared with that of the Daimler-Benz W158, a prototype similar to the Daimler-Benz 170V, which was produced by Mercedes-Benz until the 1950s. The wool-tuft pictures clearly show the difference in the flow pattern of the two models. Accordingly, the drag figures differed significantly. While the drag coefficient for the W158 was $c_D = 0.51$, for the K1 it was 0.21, both models being measured without cooling airflow.

Figure 1.37 Comparison of the flow past Kamm-car K1 and a contemporary Mercedes W 158, model scale 1:5. Photos FKFS.

The rear end of the K1 was still fairly long. A more typical Kamm-back is seen in Figure 1.36a, which displays the K2. The smoke trails clearly make visible how far the flow is able to follow the contour of the roof. The comparison of the longitudinal cross sections shown in Figure 1.38 emphasizes the advantage of Kamm's lines compared to Tatra 87 (Jaray) and Adler Trumpf (pseudo-Jaray). The measurements on K2 made in the large climatic wind tunnel of Volkswagen AG resulted in $c_D = 0.37$. While not a bad result at all, it was nevertheless much worse than the one Kamm published: from coast down tests he concluded $c_D = 0.24$. A second look at Figure 1.38 makes evident that the Tatra 87 is only slightly longer than the K2. This was achieved by moving the greenhouse farther to the front, thus giving room for a moderate slope at the back. This was possible because the Tatra 87 had a rear engine. However, a short front hood was—and still is—against the styling trend, as performance is emphasized by a long engine hood.

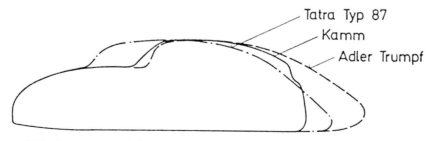

Figure 1.38 Comparison of the center lines of Kamm car, the Tatra 87 (rear fin omitted), and an Adler test-car with pseudo-Jaray back.

A unique shape was created at FIAT: they merged Jaray's combination shape with a Kamm-back. The result can be seen in Figure 1.39: a FIAT 508 Mille Miglia with a modified rear end. The sharp end of the vertical profile after Jaray was replaced by a blunt end according to Kamm. While the production car was reported by Vivarelli (2009 [828]) to have $c_D = 0.35$, the prototype with the Jaray-Kamm shape had, according to the same source, $c_D = 0.23$. A synthesis of the results from Stuttgart and Göttingen (i.e., the adaption of the Kamm-back to the cars from Lange and Schlör) was never tried.

Figure 1.39 FIAT 508 C Mille Miglia (MM), a unique combination of Jaray-shape and Kamm-back. Courtesy Curzio Vivarelli (2009).

1.4.2 Fastback

The effort to create rear end shapes for low drag was intensified in the early 1970s, initiated—by chance—from an observation in the course of development of Volkswagen's Golf I. It lead to systematic investigations first with prismatic bodies ending with a slanted back, and subsequently on cars that chapter 4 will report on in detail. Figure 1.40 anticipates the key result.

The well-ordered wool tufts to be seen in Figure 1.20 seem to indicate attached flow on the slanted back of the Adler test car. However, as already mentioned, this alone is not an indication of low drag. Taking a closer look, one will realize that the direction of the flows on the side wall roof are converging along the C-pillar. This means that the flow from the side is separating from the body along this line and is rolling up as sketched in Figure 1.40b (see also Figure 1.3b). Comparing this flow pattern with a wing of low aspect ratio[21] uncovers a striking similarity. On both sides the vortices roll up, pull down the flow coming over the roof, and keep it attached until the lower end of the slant.

---

21. The aspect ratio of a wing b^2/A; b is the wing span, and A its projected ground area.

This pair of vortices induces a drag whose magnitude depends on the slant angle φ, as shown in Figure 1.40c: Starting from a squareback, $\varphi = 0$, the drag decreases with increasing φ. Passing a minimum at $\varphi = 10°$ drag starts to increase again, at first slightly, than steeply. At $\varphi = 30°$ drag falls off abruptly. The reason: the C-pillar vortices burst and no longer induce drag. Surprisingly, the slope angle of the fastbacks of the 1950s was between $\varphi = 25°$ and $30°$, which is exactly the range where drag is highest. The mystery of the high drag of fastbacks was resolved. Today's fastbacks come with $\varphi \approx 15°$. They are aerodynamically well tuned, and nobody complains about their length—on the contrary, it is recognized as a sign of elegance.

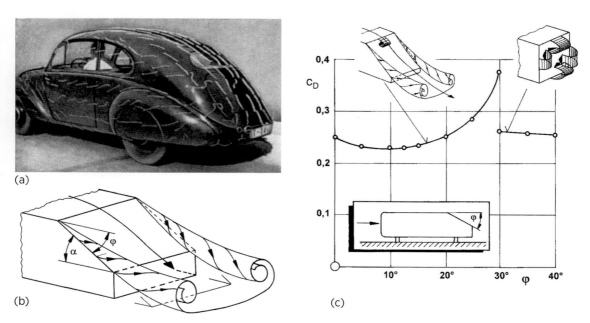

Figure 1.40 Flow past a slanted back: a) and b) comparison with a slender wing, schematic, and the back of a car, see also Figure 1.20, after Hucho; c) effect of slope angle φ on drag coefficient c_D, after Ahmed (1984).

1.4.3 Notchback

The flow pattern past a notchback is by far more complex than that of a fastback. The backlight, which generally is inclined, can be compared to a wing of large aspect ratio. The pair of trailing vortices shed from the C-pillars have lower circulation, and their ability to pull the flow coming over the roof down is less pronounced. The flow separating at the trailing edge of the roof may or may not reattach on the trunk lid. In any case, a large near wake is formed behind the car.

Carr's (1983 [131]) attempt to find a relation of drag and the two parameters φ and φ_e identified in Figure 1.41a, right sketch, had no success. Rather, the three single parameters inclination φ of backlight, length Δx, and height Δz have to be matched to each other for every individual car, as shown in Figure 1.41b, after Buchheim et al. (1983 [107]).

Introduction to Automobile Aerodynamics

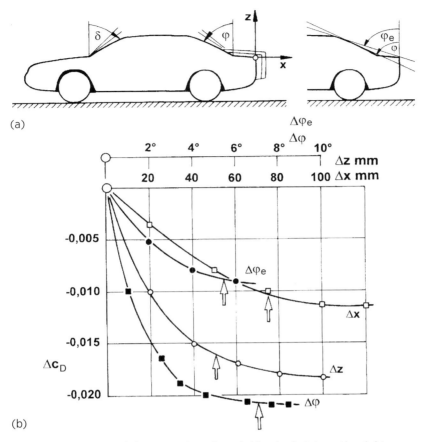

Figure 1.41 a) Description of the geometry of a notchback, sketch on the right as proposed by Carr (1974) with the angles φ and φ_e; b) Matching of three geometric parameters φ, z, and x of the trunk for the Audi 100 III, related to previous model. The arrows indicate how far design could follow aerodynamics; after Buchheim et al. (1983).

1.4.4 Hatchback

The hatchback is an appropriate measure to increase the transport capacity of a car. Depending on the priority given to low drag, the roof and sides are (slightly) drawn in. Ultimately, one would arrive at a Kamm-back. Sloping the hatch door in the range of 30° $< \varphi < 90°$ has no effect on drag but makes the car's appearance more attractive. Some call this shape a "sportsback."

1.5 Directional Stability

The first aerodynamic investigations on vehicles started with symmetrical flow condition, as is the case when a car moves through quiescent air. However, early in the 1920s they were extended to oblique flow, which is present under crosswind. Already Klemperer (1922 [443]) observed that drag increased when he placed the car at a yaw angle β in the wind tunnel: the higher the yawing angle, the higher the drag. He also noticed that the drag increase was less with better flow over the car. At very high yawing angles, the drag declined until it became negative: "The body of the car is acting like a sail when a sail boat is hard on wind" (Klemperer). This effect is used by strand sailors, not cars, because it is only present at very low cruising speed where air drag is low, and it is negligible in any case.

The faster the cars became, the more directional stability came into focus. Heald (1933 [319]) reported that the sideforce due to oblique flow increases almost linearly until $\beta \approx 20°$. Surprisingly, he said nothing about the off-turning yawing moment. How important this is for stability was first taken seriously by Mauboussin (1933 [535]), who consequently stabilized his "voiture aérodynamique Mistral" with a large vertical fin well integrated into the greenhouse, as seen in Figure 1.21.

From his own measurements with several cars Kamm (1933) concluded: "The stability properties of a car in an oblique airflow are the worse, the better its shape is for low drag." Consequently, together with his team, he developed rear fins. Already the Kamm car K1 was equipped with a pair of slotted fins, as described by Sawatzki (1941 [707]); see Figure 1.42. With these fins the off-turning yawing moment could be converted into a neutral or even moderately stabilizing influence. Thanks to the rear end contour of the K1, the fins remained within the main measures length l and height h of the car and did not reduce the rearward view—thus proving its suitability for use in public traffic.

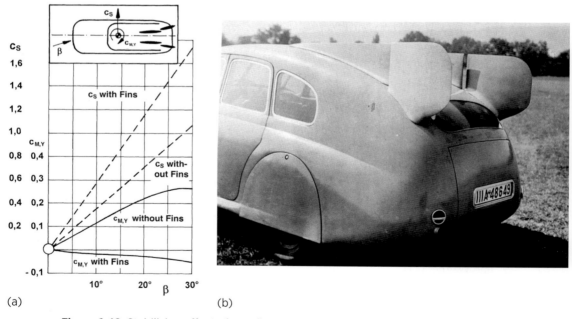

Figure 1.42 Stabilizing effect of rear fins: a) Kamm car K1 with a pair of slotted fins; b) effect of these fins on the coefficients of yawing moment c_{MY} and sideforce c_S versus yawing angel β, after Sawatzki (1941).

Introduction to Automobile Aerodynamics

The first mass-produced car equipped with a rear fin was the Tatra 77, the forerunner of the famous Tatra 87 shown in Figure 1.8. This small fin may have reduced the off-turning yawing moment only slightly, however, for neutralizing it was far too small. Fins suited to do so are difficult to design (see Figure 1.21). Instead, stylists in the United States added fins as a design feature to large cars in the 1950s. Efficient fins have only been used for record vehicles—cars and motorcycles (see section 1.7).

The effect of lift on driving stability was recognized fairly early. The rocket-driven car RAK from Opel, which was developed by Fritz v. Opel and Max Valier in 1928, was equipped with wings on both sides to produce a downforce (see Figure 1.43) from [729]. Because these wings had positive[22] camber they had to be at a negative angle of attack. Today, wings and spoilers, sometimes moveable, are used on sports cars and race cars to generate high downforce—which comes with increased (induced) drag. How the two forces, downforce and drag, are balanced against each other is explained in sections 9.6 and 9.7.2.

Figure 1.43 The rocked-driven car RAK (1928) from Opel, equipped with wings on both sides, producing a downforce. Au volant Fritz v. Opel. Photo manufacturer.

Whether or not a car is sensitive to crosswinds is not only a matter of its aerodynamic properties. Rather, the entirety of the vehicle's dynamics has to be taken into account: location of center of gravity, moment of inertia, kinematics of axles, tires, and so on, and, last but not least, the driver. For a long time, this system of vehicle and driver was looked at in the time domain. Under fixed control, the prevailing approach was to try to exclude the influence of the driver, thus isolating and quantifying the phenomenon crosswind stability of a vehicle alone. However, investigations by Wallentowitz (1978 [839]) and Wagner (2003 [833]) have demonstrated that the driver has a strong influence on the reaction of a car in crosswinds. For details, see chapter 5. Accordingly, today the complete system, including the vehicle driver, is considered, preferably in the frequency domain, be it on the road, the test track, or with numerical simulation.

---

22. Upward.

It also has to be taken in consideration that the natural wind is unsteady in both speed and direction. Artificially generated gusts, generated by the structure of the landscape alongside the road (buildings, trees, hills), are superimposed on the natural gusts. As future cars become lighter, their crosswind stability in such settings has to be observed more carefully. On the other hand, the crosswind may also be controlled by measures along the road sides. Noise protection walls may be helpful. However, gaps in these fences must be avoided, and their beginning and end should not be abrupt.

1.6 Commercial Vehicles

The need for high-speed trucks and buses arose with the construction of high-speed road systems in the 1930s. Prior to the construction of modern road systems such as the Autobahn (Germany), motorway (U.K.), and the interstate highway system (U.S.), the mass transport of goods and people was accomplished primarily by rail. The first trucks and buses were nothing but elongated passenger cars. Consequently, the same aerodynamic design principles were applied: at first Jaray profiles, later the Kamm-back. Kieselbach (1983 [430]) has documented this period with many photographs and design drawings.

With the introduction of the "tram bus" by Gaubschat in 1936, the shape of buses broke away from cars. With the engine underneath the floor—or later in the rear—more seats could be placed within an overall specified length. The long engine hood could be omitted. The front of the "tram bus" was extremely well rounded (see Figure 1.44), much rounder than necessary for low drag. Presumably Gaubschat was not aware of the results Pawlowski (1930 [629]) had already published in 1930. As can be seen from Figure 1.45, which is a reproduction from the original, comparatively small front radii are sufficient to yield minimum drag for a box-shaped vehicle, and this "optimum" radius depends on Reynolds number. Although his results were confirmed by Lay in 1933 [490], even including road tests, and this finding has been repeated several times, it was not applied for a long time.

Figure 1.44 "Tram-Bus" designed by Gaubschat (1936), on a chassis from Büssing. Archive Kieselbach.

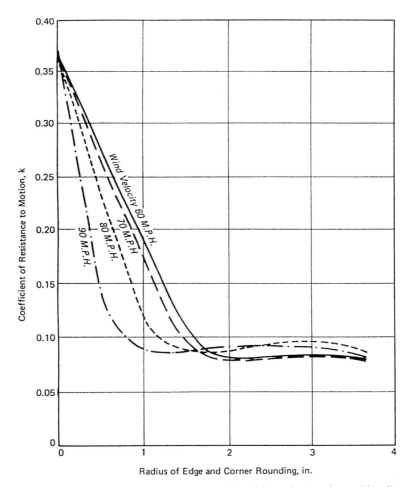

Figure 1.45 Effect of vehicle speed (i.e., Reynoldszahl, on the "optimum" leading edge radius). Copy of the original from Pawlowski (1930); $k = c_D/2$.

Originally, the Kamm-back was developed for buses. It was introduced to bus design in 1936, based on the experiments of Koenig-Fachsenfeld et al. (1936 [450]) mentioned previously. Because it allowed for one more row of seats compared to the Jaray back (see Figure 1.46), it was well accepted in practice. Today, a flat rear is typical for all kinds of buses. However, this should not be called a Kamm-back. Rather, taper at the rear—boat-tailing—is not applied because it would lead to the loss of one or two seats in the last row.

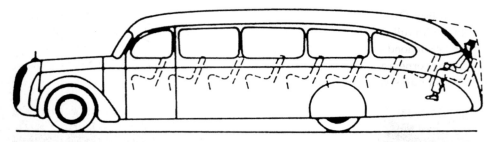

Figure 1.46 Comparison of rear end design for a bus: full-line pseudo-Jaray, dotted-line Kamm-back; after Koenig-Fachsenfeld (1939).

A milestone in the aerodynamics of commercial vehicles was achieved by the front-end design of the first Volkswagen van by Möller (1951 [578]). Apart from the drastic drag reduction, there were two other reasons for the wide recognition of this work: first, the unique market position held by this van all over the world for a long time, and second, because the reference to it made by Schlichting in his famous book *Boundary Layer Theory* [716], from which Figure 1.47 is reproduced.

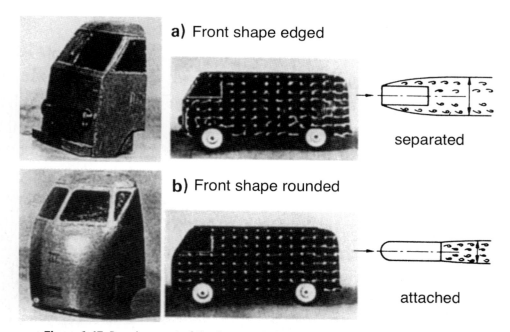

Figure 1.47 Development of the front end of the first VW van; after Möller (1951).

However, even the Volkswagen (VW) van did not make use of the earlier work of Pawlowski. The front end of the first VW van was much rounder than necessary to keep the flow attached and achieve the corresponding low drag. The first light-duty vehicle—to the knowledge of this author—designed according to the ideas of Pawlowski was the Volkswagen LT (Light Truck, now Crafter). The basic work was done on a quarter-scale model in 1969. Owing to the long lead time of this vehicle, its data were not published—together with full-scale measurements—until 1976. As can be clearly concluded from Figure 1.48, a nondimensional radius of $r/b = 0.045$ is sufficient to keep the flow around its leading corners attached. The "optimization" of front radii now has become standard practice for buses, cabs of trucks, and sometimes even for trailers.

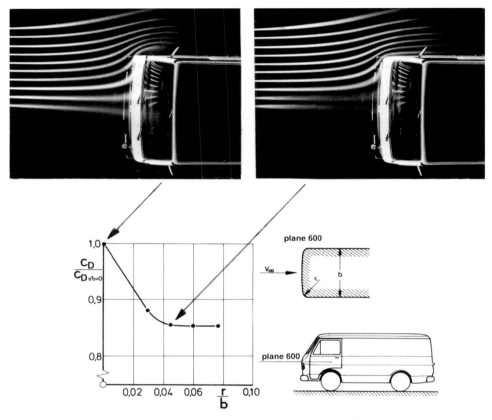

Figure 1.48 Effect of leading-edge radius on drag of VW-LT. Photos and Data Volkswagen AG.

A further step in improving truck aerodynamics was the invention of the cab spoiler. Road tests carried out by Sherwood (1953 [760]) at the University of Maryland indicated that a drag reduction was possible by using a roof fairing to guide the air from the roof of the cab to the roof of the trailer without flow separation. Because of relative motion between cab and trailer, the particular fairing was impractical. In 1961 the U.S. company Rudkin Wiley was the first to come out with a practical "wing" on top of the cab, which they called the Airshield. Actually, comparable fairings were known for a long time. The results of Frey (1933 [259]) revealed how flow could be influenced by guide vanes, and guide vanes for steam locomotives, invented by Betz as early as 1922, were in widespread use. How well such fairings can guide the flow over the open space between cab and trailer is demonstrated in Figure 1.49. The ease with which such fairings can be applied and matched to an individual vehicle has promoted their wide use in daily trucking.

Figure 1.49 Air vane on the cab of a MAN tractor-trailer truck. Courtesy MAN.

1.7 Motorcycles and Helmets

The designers of motorcycles discovered the potential of aerodynamics only fairly late. For speed record cycles, the advantages of a complete fairing became obvious—but so did its limitations with respect to handling. In addition, the faired motorcycles turned out to be extremely sensitive to crosswinds. On behalf of NSU AG, wind tunnel experiments were carried out with 1:5 scale models in the course of preparing for a speed record attempt. Figure 1.50 documents the gain in performance achieved with a "fish-like" fairing.[23] The long tail fin was necessary to reduce crosswind sensitivity to a level that could be controlled by the driver. As is outlined in chapter 11, touring cycles followed the pattern of record cycles later, but took on only partial fairings. Fairing or not—this is almost a question of philosophy for bikers.

---

23. See Schlichting (1953 [714]) and Scholz (1951 [735], 1953 [736]).

Introduction to Automobile Aerodynamics

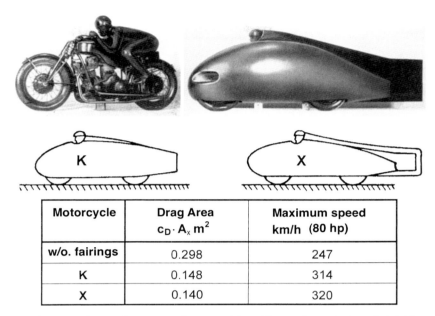

Motorcycle	Drag Area $c_D \cdot A_x \, m^2$	Maximum speed km/h (80 hp)
w/o. fairings	0.298	247
K	0.148	314
X	0.140	320

Figure 1.50 Improving performance of a record-breaking motorcycle by a full-fairing, after Schlichting (1950).

The helmet has also become object of aerodynamic development. Four items are important, to be discussed in chapter 12:

1. forces and moments on the head via the helmet;
2. the climate inside the helmet;
3. the aeroacoustic environment,
4. the interference with the vehicle, be it motorcycle or race car.

For the development of helmets it is helpful that only comparatively small wind tunnels are needed. However, they should allow for high wind speed and low noise.

1.8 Internal Flows

1.8.1 Engine Compartment

There were two trends in the design of cars that drew the attention of the aerodynamicists to the engine compartment: the radiator and the engine itself. Initially, both were exposed to free air. However, they soon disappeared under a hood, and the engine's power was growing steadily. Already in 1922, in his first tests with 1:10 scale models, Klemperer provided for an airflow through the engine compartment. It was he who observed that this flow interfered with the flow around the model, and was the origin for an additional aerodynamic drag. Fiedler & Kamm (1940 [246]) pointed out how this additional drag could be kept small. Later, findings from work on oil coolers for airplanes were transferred to automobiles.

Initially, it was sufficient to quantify the additional aerodynamic drag $c_{D,C}$ with the help of so-called sink drag, according to Betz. The face velocity u_f could be estimated following an approach of Taylor (1948 [791]), which agreed well with experiments as shown in Figure 1.51. Over the years, more heat exchangers were added: condenser of air conditioning, coolers for oil and charged air, and so on. This, and the fact that the flow in front of all these coolers no longer was uniform, called for a more detailed treatment, which was contributed by Emmenthal & Hucho (1974 [230])[24]. Much later, the three-dimensional flow within the engine compartment could be simulated numerically.

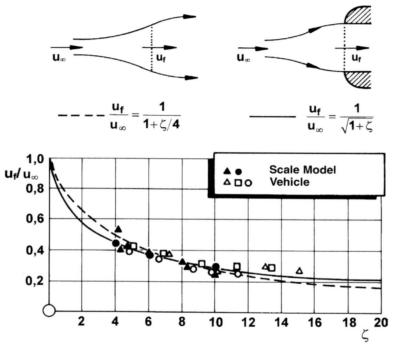

Figure 1.51 Speed of cooling airflow (face velocity u_f) versus pressure-loss coefficient ζ in the cooling-air duct; theory Taylor (1974).

1.8.2 Passenger Compartment

The airflow inside a passenger compartment was also investigated by Kamm and his team. Focus was on the interference of the volume flow through the compartment and the flow around the body. At first, the flux into the car was generated by stagnation—ram air—either through a duct from the zone near the stagnation point at the front of the car or by a connection from the cowl (scuttle) to the compartment. The outflow of the air was left to leaks in the body. Later, fans were added, first with axial flow, which were quite noisy, and subsequently with radial flow and forward bent blades. The air-release was, at first, performed by ducts leading to an area of low pressure on the outside of the

---

24. Later, this numerical approach was elaborated to a commercially available software; see sections 7.2. and 14.1.5

body, for instance behind the C-Pillar. Later, the openings for air inlets and outlets were placed in areas where static pressure is approximately the same as in the undisturbed environment. Together with a continuously running fan, the airflow through the interior could be made independent of cruising speed.

From research on thermal comfort it is known that the temperature felt by a person can be reduced by blowing very low speed air at the person—within limits, of course. The results of several authors are compiled in Figure 1.52. They show a common trend. Traditionally, the airflow into the passenger compartment is introduced by "mixing ventilation" (MV). Initially, with jets as used in airplanes, and today with small cascades, the driver and passengers can be blown on directly. With relatively high speed—which cannot be tolerated for long—it soon will be reduced or turned away. Recently, an alternative ventilation system has been tested in airplanes, the "cabin displacement ventilation" (CDV).[25] Fresh air is fed at low speed (< 0.1 m/s) to the cabin over a large region from top or bottom and drawn off from the opposite side. Higher comfort is expected, and, important for aircraft, less effort, weight, and fuel are required. CDV can be expected to be useful for buses and railway cars.

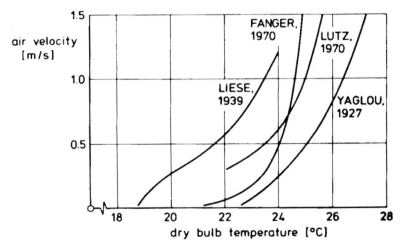

Figure 1.52 Lines of equal comfort—speed of blown air versus dry bulb temperature—measured by different authors; see Temming & Hucho (1979).

1.9 Development Strategies

1.9.1 Detail Optimization

Late in the 1960s, aerodynamicists freed themselves from the idea of creating aerodynamic cars by forming a unified whole—which in any case had not been accepted by the stylists. Cars like the Citroen DS (Figure 1.29) or NSU RO 80 (Figure 1.30) remained exceptions to the rule. However, the aerodynamicists started to develop a completely different approach: They analyzed the flow past cars in detail, as sketched in Figure

---

25. See Bosbach et al. (2012 [88]).

1.53. In doing so, they observed that the flow could be influenced by "optimizing" many shape elements. Very often, they could do this with very small modifications—so small that they did not mesh with design. The VW Golf I is a striking example of how far one could come with this strategy. The design model crossed the Alps[26] with $c_D = 0.51$—at that time the drag coefficient of the VW Beetle was $c_D = 0.49$—and as the car was ready to drive, it left the Volkswagen wind tunnel with $c_D = 0.41$, and with no objection from its (famous) designer.[27]

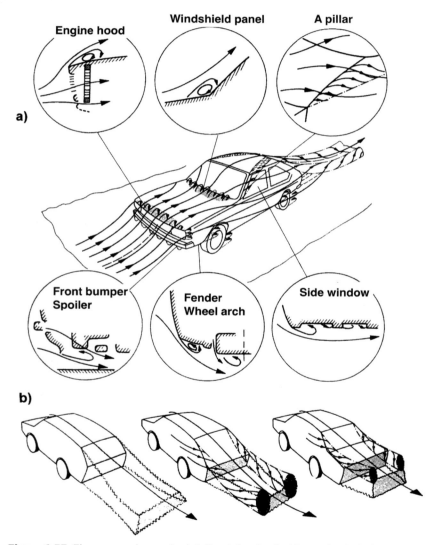

Figure 1.53 Flow around a car in details: a) forebody; b) rear body in its variants hatchback, fastback, and notchback; schematic.

---

26. On its way from Torino to Wolfsburg.
27. For details see Janssen & Hucho (1975 [408]).

Figure 1.54 offers an insight into this strategy called detail optimization. Without impairing the styling of the squareback model, the drag coefficient could be reduced by 21%. The five details that had to be optimized are shown by the cross sections on the left side of the figure. However, some of the drag improvement had to be given away. If the rule of "no change to the design" had been stretched a bit more, the drag could have been reduced even further, in this case by 33% overall. Of course, how much the drag can be reduced by applying this strategy depends on the style of the original model. If it starts at a comparatively low drag, the gain of a detail optimization will only be moderate.

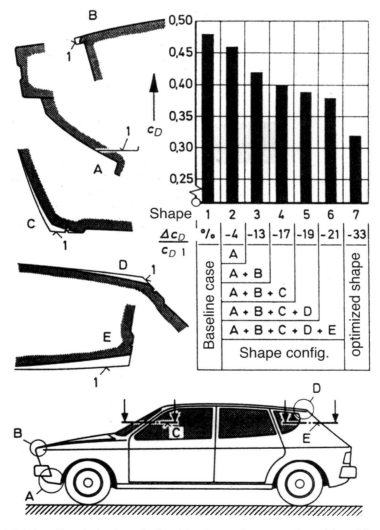

Figure 1.54 Detail optimization of a hatchback car; shown are the achieved "optimum" changes of key body details; the many steps necessary to arrive at them are not shown.

From test results of numerous detail optimizations, four types of characteristic functions were identified by Hucho (1979 [378]): "saturation," "asymptote," "minimum," and "jump." They relate drag coefficient c_D to vectors ρ_i which describe individual shape elements i; see Figure 1.55. These vectors ρ_i can be radii, heights, lengths, and angles; the first three are referenced to a characteristic dimension, in this case the vehicle length l (i.e., $\rho_i = r_i/l$). c_{D0} is the drag coefficient before implementing a change in shape. Later, these empirical functions were approximated analytically and incorporated into a rating method[28].

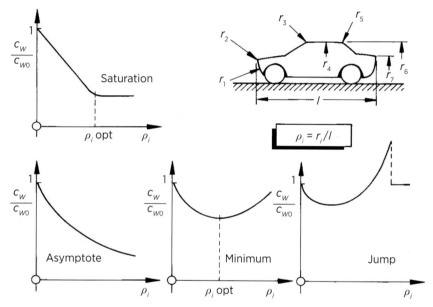

Figure 1.55 Typical drag and geometry relationships, schematic, according to Hucho (1979).

The power of the detail optimization is demonstrated by Figure 1.56. Two sports cars are compared: the 1968 Opel GT and the 1974 VW Sirocco I. While keeping its angled style, drag coefficient of Sirocco I was reduced to the same low value as the well-streamlined Opel GT.

However, the strategy of detail optimization soon came to its limits; $c_D < 0.40$ could hardly be realized. The reason for this was the boundary condition "no change in styling." Even so, the strategy of step-wise modifications to the body details was accepted as standard procedure in the course of developing new vehicles. It provided a firm basis for discussions between design and aerodynamics, and it prepared the way for the next strategy: shape optimization. For vehicles where drag is a lower priority, as is for instance true for cross-country vehicles or pickup trucks, the detail optimization is still a valid strategy.

---

28. See section 1.11.2.

Figure 1.56 Comparison of two sportive cars with identical drag coefficient: a) Opel GT, MY. 1968, c_D = 0.41; A_x = 1.51 m²; b) VW Sirocco I, MY 1974, c_D = 0.41; A_x = 1.73 m². Photos by manufacturers.

1.9.2 Shape Optimization

As a consequence of the first energy crisis in 1973/1974, the fuel economy of automobiles came into focus. With it, the readiness of designers and engineers grew to make better use of the potential offered by aerodynamics. Some aerodynamicists were well prepared for this situation. When it became evident that detail optimization under the boundary condition "no change in design" had reached its limits, they had begun research which directly tied on the work of Jaray & Klemperer in the 1920s. This led to the fourth phase, shown in the right column in Figure 1.6, which was named "shape optimization."

At first, low drag "basic bodies" were developed. However, in contrast with earlier investigations, only those bodies having main dimensions (length, width, and height) identical to the later vehicle were accepted. In a number of small steps, similar to detail optimization, basic models were derived from these basic bodies—which served as a starting point for design.

The development of the shape of the basic body started with a free-flying body of revolution. When this is brought close to ground with a clearance typical for cars, its drag goes up, as already observed by Jaray & Klemperer (1922) (see Figure 1.15). Two geometrical parameters can be considered responsible for this drag increase: effective camber and effective thickness. Close to ground, the flow past a body of revolution loses its rotational symmetry. The flow on the upper side separates, as if the body were cambered (see Figure 1.57a), and the drag increases. At the same time, the effective thickness increases (Figure 1.57b), and this causes a drag increase as well. Actually, the same is true for real cars and car-like bodies (see Figure 1.35), if their shape is not matched to the flow pattern in close ground proximity.

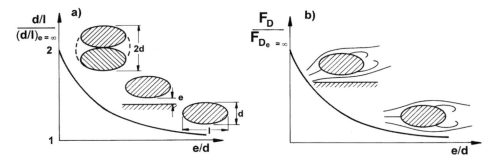

Figure 1.57 Bluff body approaching the ground, schematic: a) the effective bluntness is increasing; b) the flow is losing its rotational symmetry, and the drag is increasing.

The influence of thickness on drag is documented with Figure 1.58, where the drag coefficient c_D is plotted versus the slenderness $\lambda = l/d$ (i.e., the reciprocal of the thickness). The diagram goes back to Hoerner (1965 [350]). Starting with the sphere, $\lambda = 1$, the drag goes down with increasing slenderness as long as the decrease of pressure drag overrides the increase of friction drag. The minimum drag is reached at $\lambda \approx 3$. When the slenderness rises further, the drag is increased again because of the growth of friction drag. With an effective slenderness of $\lambda \approx 1.5$, the majority of cars are on the left side of the minimum. If they were more slender, their drag would be lower.

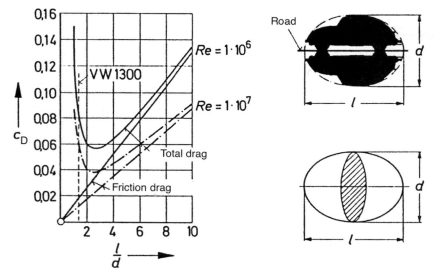

Figure 1.58 Drag of bodies of revolution versus degree of slenderness $\lambda = l/d$; effect of Reynolds number.

However, for "average" cars, slenderness and ground clearance are given. The development of the shape of the basic body must be directed on compensating the effects of both thickness and camber. This comes with the advantage that lift will be reduced, mainly on the rear part of the body. Morelli et al. (1976 [590]) were the first to follow this idea. Applying slender-body theory, Morelli arrived at a unique body (see the last line in the table in Figure 1.59, where results from basic-body research are compiled). For an average ratio of length l to height h, $l/h = 3$ (effective[29] $\lambda = 1.5$) a drag coefficient $c_D = 0.07$ to 0.09 can be achieved; this is in accordance with data from Jaray & Klemperer (see Figure 1.15). Adding wheels, this range doubles to $0.14 < c_D < 0.18$, accentuating that the flow around the wheels has to be included even in the shape development of basic bodies.

Shape	l/h	c_D	Annotation	Author
	4.0	0.05	flying	Hoerner (1965)
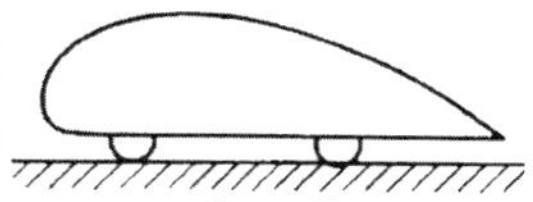	4.0	0.15	wheels shaped	Klemperer (1923)
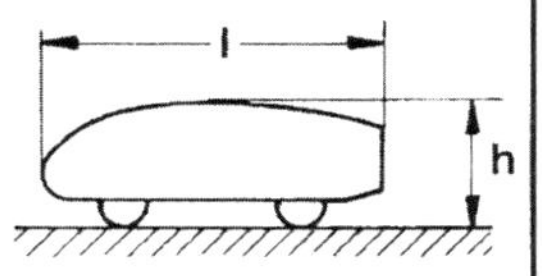	3.0	0.16	wheels shaped	Hucho & Jansen (1972)
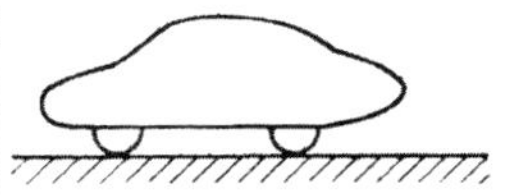	3.1	0.07	w/o. wheels	Fioravanti et al. (1976)
		0.18	with wheels	
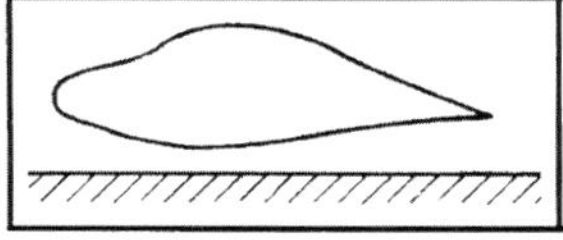	3.9	0.05	w/o. wheels	Morelli et al. (1976)
		0.09		Buchheim et al. (1981)

Figure 1.59 Drag coefficient of basic bodies, with and without wheels.

The first car whose shape was derived from a basic body was the "Auto 2000," a research car from Volkswagen. The milestones of this development are lined up in Figure 1.60. The many steps between these are ignored here. The work was started with a basic body, scale 1:4: $c_D = 0.160$. In the next step, the basic model, measurements were already performed in full scale. The model was equipped with three specific shape elements: sloped engine hood and windscreen and a low drag fastback. Drag went up to $c_D = 0.180$. By working in details required by the design, adding a real under-body panel, and opening the cooling air duct, the drag increased further to $c_D = 0.240$. Finally, after working out all details, the ready-to-drive prototype was measured at $c_D = 0.290$.

---

29. Due to mirror image simulating the ground.

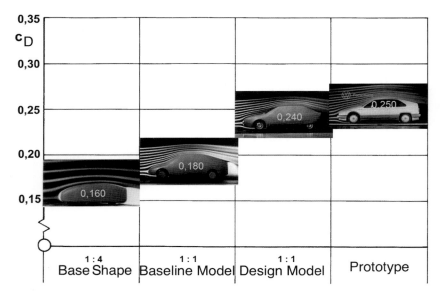

Figure 1.60 Shape optimization of the VW Research Car 2000. Data and smoke-trail photos Volkswagen AG.

Following the same procedure, the 2001 Audi 100 III, a notchback car, was developed. The production car came out with $c_D = 0.300$; it is shown in Figure 1.61 a.[30] In comparison to its successor, the 2011 Audi A6, it appears rather angled, hardly distinguishable from a detail-optimized shape.

(a)   (b)

Figure 1.61 a) Result of shape optimization of Audi 100 III, "World Champion," after R. Buchheim et al. (1983); b) in comparison to its follower Audi A6, MY 2011. Photos courtesy Audi AG.

With both kinds of optimization—detail and shape—aerodynamicists have two strategies at their disposal to achieve very low drag, as symbolized with Figure 1.62. They either can approach their target by optimizing details of a given design or they can start from a low-drag basic body. The first route is daily practice in both the wind tunnel and

---

30. It was declared as "World Champion."

on a computer. How far in any specific case the modification of details can go is a matter for discussion and mutual compromise between design and aerodynamics.

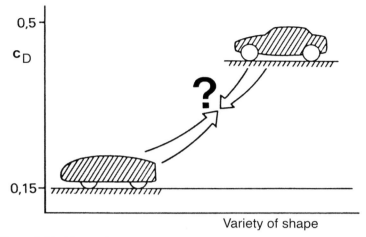

Figure 1.62 Alternative routes to develop a low-drag shape; schematic.

The second route—shape optimization—does not need to be repeated for every new type. Instead, it lends itself to further exploit the potential of aerodynamics whenever a novel concept is investigated, allowing for unaccustomed shapes. It is a strategy for research and advanced engineering.

1.9.3 Limit Strategy

What has been achieved so far to reduce air-drag with the strategies described in the previous sections is summarized with Figure 1.63. The drag coefficient of European passenger cars is plotted versus time (MY). The cars standing out make evident how aerodynamics has made progress in becoming an integral part of the design. As already stated in section 1.3, the reduction of drag until the 1960s in particular must be attributed to modern body technologies and novel design trends—stamping large, well-rounded parts, self-supporting body, and pontoon style—and only in a minor part to aerodynamics. From then on, systematic drag reduction was rather moderate. Only Citroen DS 19 and NSU Ro 80 demonstrated what was possible already at that time. The first energy crisis, in the winter of 1973/1974, increased pressure on the automobile industry to improve fuel economy. And indeed, making better use of aerodynamics was discovered as an economic measure to meet this urgent demand. Starting in the mid-1970s, air drag began to decrease, initially fast, but soon only slowly again. Over two decades progress was only marginal. With $c_D \approx 0.30$, a lower limit seemed to be reached; $c_D = 0.25$ appeared to be an asymptote. Only recently was it undercut—by Mercedes CLA, a coupé-like four-door car: $c_D = 0.22$; $A_x = 2.22$ m^2.

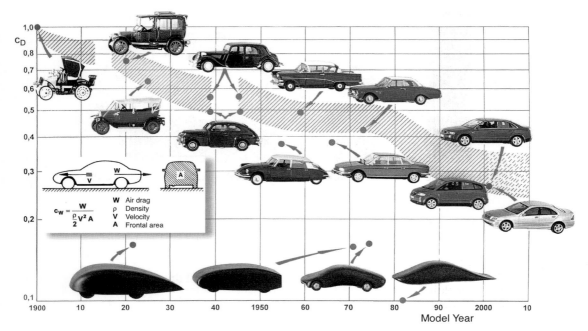

Figure 1.63 History of drag coefficient c_D of European cars. The car-like bodies in black lined up in the lower part of the diagram reveal the latent potential of vehicle aerodynamics. Show-case Sönke Hucho; scale of models 1:43.

The black models in the lower part of the diagram point out what the scope of physics still has to offer. This has, of course, been well known for a long time, and from time to time this potential has been unlocked for concept cars. A recent one is shown in Figure 1.64: the 1-liter car VW XL 1. How its c_D = 0.186 was achieved has been reported by Repmann (2012 [667]). Since then, it was decided to produce the XL1 in a low-volume series. A striking detail is its forward position of the greenhouse and the correspondingly short engine hood, similar to the Tatra 87. The former gives room for a smooth slope off the back, which allows for the pressure recovery needed for low drag, as demonstrated by Figure 1.40c.

The potential mentioned above can be used by the so-called "limit strategy." This is founded on the concept that at first for each detail and for every region the contribution to drag is minimized with no regard of technical feasibility. Thereafter, technical solutions are developed to come as close to the limit as possible. This, of course is nothing new. What is new is the consequence with which this idea is followed.

Introduction to Automobile Aerodynamics

Figure 1.64 The 1-liter car VW-XL1 (2011); c_D = 0.186; A_x = 1.50 m². Data and photo Volkswagen AG.

With the aid of an almost classical example it will be demonstrated how this limit-strategy is applied. In Figure 1.65 the main steps of the development of the front end of VW Golf I are compiled. In a first test, the maximum drag reduction possible with the front end alone was determined. For this purpose the front end was covered with a Styrofoam molding, well rounded to ensure attached flow.[31] In this case, the gain in drag was Δc_D = –0.050. After removing this molding, the leading edge of the engine hood was slightly rounded in three steps M1 to M3, and in the same way the leading edge of the fenders, K1 to K3. By combining M3 with K3, 90% of the drag reduction with the molding could be accomplished. The appearance of the front end was still angled and was accepted by the designer. Other regions where the limit strategy should be applied include the underbody, (front) wheels, and rear end.[32]

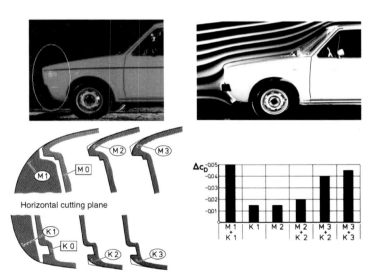

Figure 1.65 Typical—if not classic—example for the "limit strategy" development of the front end of the VW Golf I, after Janssen & Hucho (1975).

---

31. With reference to a proposal of Carr (1963 [126]).
32. Details have been described by Hucho (2009 [379]).

Modern front ends are designed much rounder than those "optimized." They are close to the shape of the molding applied above. The well-rounded front end is not overdone. It helps to improve the flow farther downstream; the boundary layers are more highly energized and thus are able to overcome a larger pressure increase without separating. Along with this, the base pressure will be increased and drag reduced.

1.10 Design and Aerodynamics

It is clear that when a new vehicle is developed, the proposals made by aerodynamics are helpful only if they are accepted by design. This has to be noted by the aerodynamicists. The better they are at getting into the designer's place, the sooner they will find a partner for discussions.[33] However, aerodynamicists must resist the temptation to do the design. Design is a category of applied arts; familiarity with art is not sufficient to be an artist.

Similarly, designers should not underestimate the technical world. The two examples in Figure 1.66 demonstrate what happens when they can't resist doing so. The model in part a, drafted by the architect Le Corbusier, is equipped with a pseudo-Jaray rear end with sharp edges. These will cause a pair of strong longitudinal vortices, which induce high drag. Later, this rear end was adopted by the Citroen 2CV ("duck"), where aerodynamics must not have been taken seriously at all. The streamlined prototype in part b, designed by Bertone, must have enhanced the vortex motion in the rear, thus increasing drag.

(a) Le Corbusier "Voiture Minimum" 1936

(b) Alpha Romeo 1900 BAT 5 concept 1 1953

Figure 1.66 Designers trying to make aerodynamics: a) Le Corbusier, Voiture Minimum (1936), Photo MIT Press; b) Nucio Bertone, Alfa Romeo 1900 BAT concept 1 (1953), Photo Bertone.

---

33. Books to help the engineer to get an insight into the world of design are Kieselbach (1998 [431]), Silk et al. (1984 [762]), Lamm & Holls (1996 [483]), Seeger (2012 [754]).

Design must communicate with the social and technological environment. How this is practiced has recently been discussed by G. Schmidt ([721]) : and if this was all, one would arrive at target pretty quickly, as documented on the left photo in Figure 1.67. However, this definition does not go far enough. The result of a design process must be beautiful, fascinating. What is perceived as beautiful, so much so as to tempt an interested person to buy, is a matter of taste, and, as is well known, taste covers a wide field of individual differences: "One man's meat is another man's poison." The designer has to live with this situation. He must take heart to dare not to satisfy everyone. There is no such thing as "everybody's darling" (Kapitza, 1992) [416a]. On the other hand, there is the risk of overdoing and finally ending up with styling, as to be seen on the photo to the right of Figure 1.67. Quite contrary: "A harmonious product does not require decoration. It should enhance itself by its pure shape" (F. A. Porsche). This leaves the designer with the question of whether the variety of "pure shapes" is wide enough for the many types to significantly differentiate among them.

Figure 1.67 Two members of the Daimler SL family. Left: the oldest, W 194, 1951/52, c_D = 0.376, A_x = 1.78 m^2; right: the youngest, R 231, c_D = 0.27, A_x = 2.12 m^2. Data and photos Daimler AG.

The aerodynamicist has to recognize and accept that the design process is by far the most demanding part in the course of developing a new car. This is not only true during the development but also in the assessment of the result. Criteria of extremely different categories, objectively measurable data, and emotional impressions, must be brought into balance. In spite of numerous restrictions with the inherent effect of leveling out, the designer is expected to create novelties while at the same time facing several subjective challenges:

1. He has to differentiate several models within a brand—"class" A, B, and C. The classes must be easily distinguishable, but each must be seen as a member of a single brand.
2. The design should be innovative but nevertheless should not allow the predecessor to look (too) old.
3. The brand must be recognizable across different kind of vehicles—cars, sports cars, sport utility vehicles (SUVs), and others.

Design has to bond and it has to differentiate—simultaneously. This in anticipation of the next 8 to 10 years: life-span for a model 6 years; development 3 to 4 years. And, shortly after the start of the development, the design is "frozen."

While performing his work, the designer not only has it out with aerodynamicists. Others who come up against his demands are symbolized in Figure 1.68: manufacturing, safety, packaging, regulations, and so on. These disciplines set requirements that design must take as given. In contrast, the nature of the "conflict" between design and aerodynamics is different; no such definite borders are present. The solution must be found in discussion. While design articulates itself with much more than shape—as there are various matters of "graphics"—aerodynamics is almost left to shape alone.

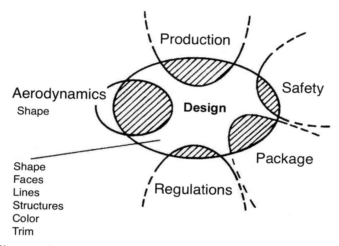

Figure 1.68 Shape as intersection of design and aerodynamics and all the other disciplines needed to develop and produce a vehicle.

For a long time aerodynamicists and designers interacted with great difficulty; their relationship was characterized by misunderstandings. As a discipline of its own, design has slowly developed. As previously, in the age of carriage, design remained a part of body construction. Later, when the self-supporting body asserted itself, design and development of the body were performed by the car manufacturer "in house." Some of the draftsmen transformed to stylists, later designers. They became members of auto companies; some even founded their own design studios.

The first aerodynamicists who turned their attention to the automobile had their roots in aeronautics. It did not come to their mind that in the auto world the role of aerodynamics was different—and not nearly as dominant. They created shapes according to what they thought were the demands of fluid mechanics. All other requirements, as for instance packaging, were subordinated. To the designer, they left no room at all for aesthetic expression. Typical results from this attitude were Rumpler's "tear-drop" car, Figure 1.13,

Mauboussin's "Mistral," Figure 1.21, and the one-volume car of Schlör, Figure 1.34. Only Jaray's streamline cars, Figure 1.8, came closer to the requirements of car design.

From the designer's point of view, for many years aerodynamics was not much more than a marginal matter, and an inconvenience as well. At best, they used the shapes derived by aerodynamics as articulations of fashion. For them, streamlines symbolized the dynamics of the spirit of the times, and functionality (with respect to aerodynamics) was beyond their comprehension. Consequently they used—and very often misused—aerodynamics as an expression of style. The already mentioned boat's stern, Figure 1.11, the fastback, Figure 1.19, the large tailfins of the 1950s, and some spoilers of today are witnesses to the misunderstanding of aerodynamics by designers.

This discord between design and aerodynamics—art and physics—was resolved only when the automobile manufacturers started to perform the aerodynamics by themselves. As members of engineering, aerodynamicists were converted into vehicle engineers. The daily contact with design gave them an understanding of this kind of art. The aerodynamicists learned to respect the designer's creativity, and they realized how much the designer's freedom is already constrained by requirements, technical or legal. The designers, on the other hand, realized that aerodynamics is not a black art, but rather a rational discipline, a science. The strategy of detail optimization, the step-by-step procedure, provided a sound basis for cooperation. Both the need to improve fuel economy and the desire to make the economics of a car visible by its style made designers open to aerodynamics. The results are measurable: formerly, when design models were rolled from the studio into the wind tunnel, they came with $c_D \approx 0.45$. Today, as an average, they come in with $c_D \approx 0.30$. However, the designers must get ready for more stringent guidelines for drag, which will demand extreme shapes. The VW XL1, Figure 1.64, is an example for what this means for them.

1.11 Development Tools

1.11.1 Wind Tunnels

For a long time, the aerodynamic development of automobiles was executed almost exclusively in wind tunnels, initially in those built for aeronautical purposes. It began with small scale models. Klemperer (1923) used scale 1:10, and with their mirror image to simulate the road. Test runs on roads were—and, of course still are—performed to confirm the results.

When wind tunnels for aircraft were big enough, they were employed for cars in larger scale: typically 3/8-scale in the United States and 1:5 or 1:4 in Europe. Whenever possible, full-scale models were used for two reasons: First the results of small-scale measurement turned out to be not always as reliable as needed because of Reynolds number effects, which are difficult to predict, see Figure 1.69.[34] Second, design prefers full scale for modeling. Only full-scale mockups can be properly assessed.

---

34. For more results see section 13.4.

Chapter 1

Over the years wind tunnels were built specifically laid out for automotive needs. A milestone was the full-scale wind tunnel built in 1938 by Kamm in Stuttgart-Untertürkheim (see Figure 13.90).[35]

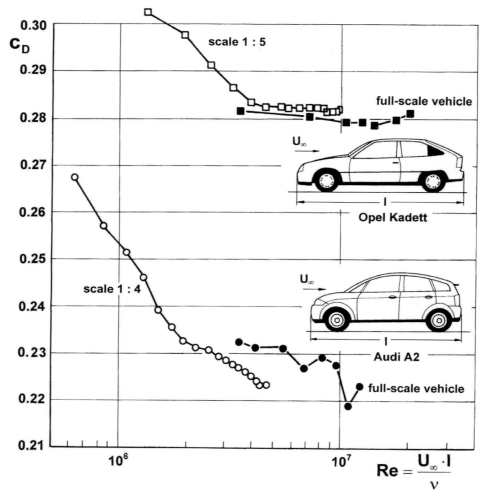

Figure 1.69 Effect of Reynolds number on measured drag coefficient with models in scale 1:5, 1:4, and full-scale cars; from Dietz (2000).

Before long, the field of thermal management—engine cooling and climate control—had testing requirements in addition to aerodynamics. Taking this into account, the large climatic wind tunnel of Volkswagen AG was built and went into service in 1968 (see Mörchen 1968 [579]). Instead of an "all in one" test-bed, separate tunnels were built,

---

35. Which is still being operated by Daimler AG.

each specialized in either aerodynamics or thermal tests. FIAT was the first to do so (see Antonucci et al. 1977 [20]). Further specialization led to aeroacoustic wind tunnels, the first being built in 1989 by BMW (see Figure 13.91), after a design by Janssen.[36]

As the air drag became lower, the importance of reproducing the relative motion between vehicle and road and the rotation of the wheels grew. The (broad) rolling road was introduced by Kamm for scale models in the 1930s.[37] In order to circumvent the necessity of lifting the model (slightly, to provide play for weighing), Potthoff (1960s) replaced it by a narrow belt, running between the (stationary) wheels. From this, the five-belt system for full-scale testing was derived.[38] Presently, a single-rolling road is tested, which is nearly as broad as the test section.[39]

1.11.2 Rating

Initially, aerodynamic development of a new vehicle was performed by "cut and try," guided by the intuition of the test engineer. However, as more vehicles had to be investigated and improved, a more systematic course of action was needed. A first attempt was derived from a rating system, which was elaborated originally for a quantitative prediction of drag coefficients: the MIRA rating system. It was formulated in 1967 by White [861], correlating shape details with their contribution to drag. It was based on measurements made on 118 passenger cars (including station wagons) in the MIRA full-scale wind tunnel. As outlined in Figure 1.70a, the method divided a vehicle into nine zones. According to its contribution to drag, each zone was assessed ("rated") by points P_i. A streamlined design detail would have a low score. Highly disruptive elements, such as a double-flanged seam on the A-pillar, were taken into account by adding extra points. The cooling airflow was not considered. The drag coefficient c_D is computed from the total points P_i as follows:

$$c_D = a \sum_{i=1}^{9} P_i \tag{1.3}$$

where a is an empirical constant. This linear relationship is represented in Figure 1.70b, together with the specified tolerance of ±7% for this method. The range of drag coefficients for common vehicles of the day was $0.3 < c_D < 0.6$; the rating method was adequate for differentiating cars into broad categories. The point system assumed comprehensive experience on the part of the user and was, in any event, subjective. Eventually, the approach proved not very practical. However, by identifying and weighting critical zones, the method did give designers indications of how to design an aerodynamic car. The merits of the method lay in this inspiration.

---

36. See Lindener & Kaltenhauser (2004 [504]).
37. See Potthoff & Schmid (2012 [653]).
38. See Potthoff et al. (2004 [652]).
39. First results have been published by Petz and Charwat (2012 [636]).

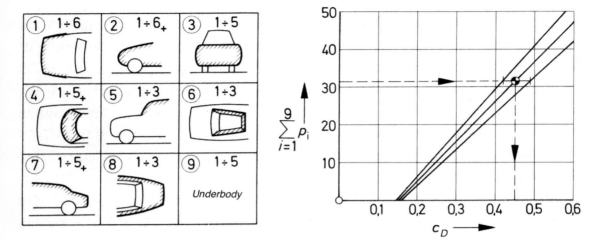

Figure 1.70 The first MIRA rating method: a) Assessment of individual body details with regard to their contribution to aerodynamic drag; b) linear correlation of rated points and drag coefficient; after White (1967).

A much more refined rating method was presented in 1987 by Carr [130]. From a great number of measurements he derived empirical functions like those schematized in Figure 1.55, which are sketched in more detail in Figure 1.71. The vehicle was structured in seven "drag zones": front end, rear end, friction, underbody, wheels and wheel housings, attachments and cooling air. From a total of 52 dimensions, such as angles, radii and areas, as identified in Figure 1.72, drag components are calculated and combined to produce an overall drag coefficient c_D:

$$c_D = \sum_i c_{Dti} \qquad (1.4)$$

$$c_{D,ti} = \sum K_j p_i \qquad (1.5)$$

Here, c_{Dti} is the calculated drag contribution of "part" i, which is described by vector p_i, p_j, and K_j, are empirical constants. Verification on 20 common cars in the MIRA wind tunnel gave a maximum deviation of ±5%; this is close to the accuracy of quarter-scale measurements.

Introduction to Automobile Aerodynamics

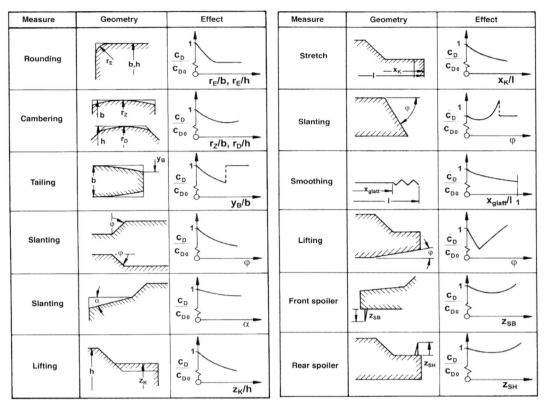

Figure 1.71 The essential relations between shape modifications of individual geometric details and aerodynamic drag, schematic, after Hucho (2011).

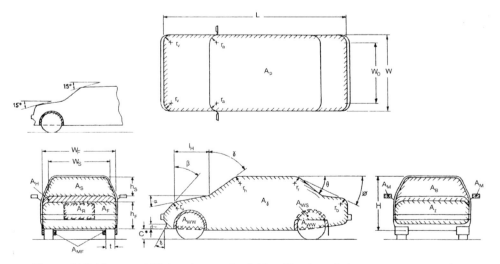

Figure 1.72 The new MIRA rating method; identification of shape parameters, after Carr (1987).

Compared to the first method of White, the new method of Carr has the advantage of being entirely objective. The required dimensions can be read from a drawing. The method can be (and now is) integrated into a CAD system,[40] and the drag coefficient c_D can be called up for every draft of a vehicle design. However, two reservations should be kept in mind:

First, interference between the individual zones and interaction between individual parameters in the same zone are ignored. The method could be extended in this respect but it would then be nonlinear. Secondly, by definition the method is based on past experience. It is only able to evaluate what is already known and what has been included in the empirical functions. Novel ideas cannot be assessed by this method. In order to prevent unintentional misuse, the applicability limits of the method should be clearly defined when it is integrated into a CAD system.

1.11.3 Classical Approach

When, beginning in the late 1960s, numerical methods were pushed forward in most disciplines of vehicle technology, aerodynamicists also came under pressure to join this trend. The motivation was similar in all cases: to arrive at better results sooner, for lower cost. Because at that time solving the Navier-Stokes equations for a geometrical configuration as complex as a car seemed out of reach, the practical approach was to fall back on methods that had proven themselves in other applications of fluid dynamics such as aeronautics, turbomachinery, and naval architecture. From those the "classical" scheme for numerical treatment sketched in Figure 1.73 could be applied.

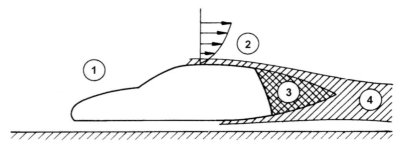

Figure 1.73 "Classical" dividing up of a flow field in (1) potential flow, (2) boundary layer, (3) near wake, and (4) far wake.

The body to be investigated is divided into four zones. First, (1), the potential flow is computed; fully three-dimensional bodies can be handled with the panel method.[41] It had been applied on vehicles previously with success.[42] A prominent example is the development of the head of the German high-speed train ICE I. The head shape was to be designed to keep the pressure wave[43] small when two trains meet each other, when

---
40. Elaborated by Morén (2007 [591]).
41. Elaborated by Hess & Smith (1967 [342]).
42. See Ahmed & Hucho (1977 [10]).
43. Actually this phenomenon is not a *wave*, but a pressure increase followed by a sudden decrease.

a tunnel mouth is entered, or when a train passes through a railway station without stopping. In any case, the head was to be shaped to keep the flow attached. (The same is true for automobiles when they are laid out for low drag.) Therefore, in domain (1) the small effect of viscosity (skin friction) could be neglected.[44]

Even for race cars the panel method did good work as for instance in the layout of wings in the fore and aft positions (see Katz & Dykstra 1989 [423]). Lift and induced drag could be computed in ideal flow; the effect of friction was assumed to be approximately the same for all positions. Furthermore, the panel method allows for wind tunnel corrections, globally and locally, as established by Steinbach (1993 [778]). In comparison to the integrated methods, which are discussed in the next section, the great advantage of the panel method is this: only a surface grid is needed, and the calculation speed is very fast.

The viscous flow near the walls, domain (2), can be described with integrated quantities of displacement and momentum thicknesses, which can be computed by proven semi-empirical methods. However, the computation of three-dimensional boundary layers is charged with uncertainty: separation and reattachment, in particular, are not predictable with classical three-dimensional boundary layer methods.

In the rear, (3), the Kutta-condition valid for wings is replaced by a near wake, which has to be modeled. This problem was solved by Dilgen (1995 [199]). Based on a universal pressure distribution, which he derived from a model by Roshko (1955 [683]), he developed an inverse panel method. An example was published by Papenfuß & Dilgen (1993 [625]) and is reproduced with Figure 1.74. On a squareback car, the pressure distribution was in good agreement with the wind tunnel measurements, even at the base. Force measurements were not performed; the model was moved by a catapult.

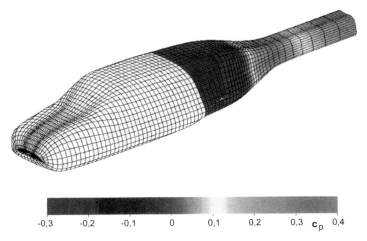

Figure 1.74 Shape of near wake computed with the numerical model derived by Dilgen (1995), from Papenfuß (1993).

---

44. For details see Mackrodt et al. (1980 [522]).

Domain (4) contains the longitudinal vortices (see Figure 1.53b), if they exist for the vehicle under consideration, and the far wake downstream. Generally, the effect of the latter on the upstream vehicle is small and can be neglected. The formation of longitudinal vortices at the rear of a slanted back was modeled by Ramm & Hummel (1992 [657]) (see Figure 1.75). However, their interaction with the near wake has not been investigated yet and remains to be the missing link of the "classical" approach so far. Despite of its relative simplicity and its advantage being very fast, the classical approach was never developed systematically and was eventually abandoned in the middle of the 1990s.

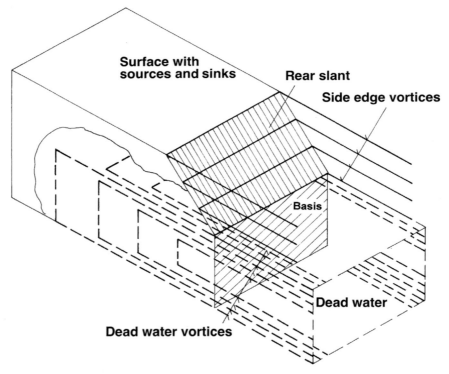

Figure 1.75 Vortex model for a slanted back (simplified), vortex position a priori unknown (!), after Ramm & Hummel (1992).

1.11.4 CFD—Integrated Methods

In an integrated (holistic) approach a flow field is described in its entirety. For more than a century, the Navier-Stokes equations—without any restrictions—have been ready for this. However, for the time being, numerical solutions of technically relevant problems are not feasible. Only far-reaching simplifications made them solvable.[45]

---

45. See section 14.1.4.

Initially, integrated methods were developed—and commercialized—for problems of industrial fluid dynamics, mostly for internal flows where the possibility of dividing the flow field up around a rigid body is rare or does not exist at all. In vehicle aerodynamics, these methods got a chance rather tentatively, because it was not easy to meet the high demands in accuracy, especially for drag, and turnaround time. However, finally they were accepted. In the following, some milestones on their route to success will be briefly described.[46]

To the knowledge of the author, Hirt & Ramshaw (1978 [347]) were the first who published a numerical simulation past a vehicle. They solved the full Navier-Stokes equations with a finite difference method for the laminar flow around a generic trailer-truck with sharp edges; the symmetric half-model had 12,540 cells. To ensure numerical stability, they increased viscosity; the Reynolds number was extremely low with $Re_l = 10^3$. Nevertheless, the results showed the essential structures of the flow past a complex body: separation at sharp edges, reattachment, near and far wake.

The first step on the road to applying computational fluid dynamics (CFD) in engineering projects was performed by Spalding (1969). Based on research work performed in Imperial College, he and his school developed problem-specific CFD codes to solve the Reynolds averaged Navier-Stokes equations (RANS) for turbulent flows. To close the system of equations Spalding developed a two-equation model taking account for eddy viscosity, the k-ε turbulence model. The flow adjacent to the wall was described by the logarithmic law-of-the-wall. The computational domain was structured by a finite volume grid, body fitted, and distorted. Ultimately, the universal general-purpose code PHOENICS was developed. It was commercialized in 1981 and was applied immediately to automobiles[47]:

In early cases, in order to limit the effort, the flow field was divided into two sections with an overlap (see Figure 1.76) from Rawnsley & Glynn (1985 [659]). The forward part was treated as potential flow with a boundary layer. Only in the rearward section, the back of the car, and the near wake, the RANS equations were solved. A representative result is reproduced in Figure 1.77, from Greaves (1987 [295]), computed with $9 \cdot 10^4$ cells for the half-model. The pressure distribution in the centerline cross section is close to experiment. However, differences must be recognized at the foot of the wind shield (cowl, scuttle), and over the backlight, both regions where separation and reattachment occur. In spite of these differences drag agreement was fairly good, lift less so. Soon, with progress in computer performance, the dividing of the flow field was abandoned (see Rawnsley & Tatchell 1986 [660]). In summary, in the middle of the 1980s proof was furnished that the essential phenomena of the flow past a vehicle could be numerically simulated with RANS codes, although at first only qualitatively.

---

46. Detailed treatment is given in chapter 14.
47. Under particular appreciation of the work of Spalding this development was described in depth by Runchal: Runchal, D., Akshai, K., Brian Spalding: CFD & Reality. CHAM. [Online] 2008.

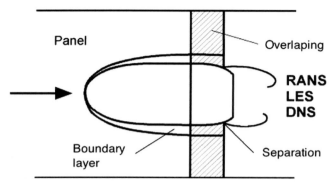

Figure 1.76 Dividing up of the numerical domain according to the first version of Code PHONICS. Forward part Panel and boundary layer, rearward part RANS, after Rawnsley & Glynn (1985).

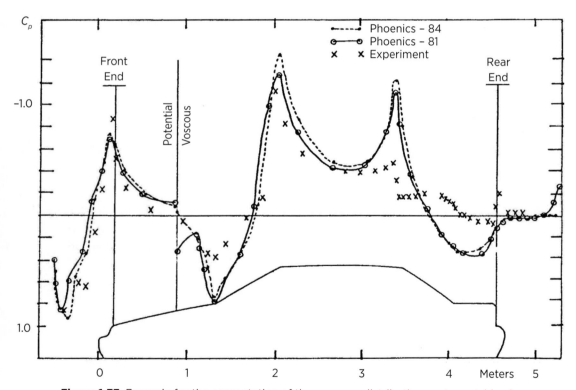

Figure 1.77 Example for the computation of the pressure distribution past a notchback car with Code PHOENICS, versions 1981 and 1984, comparison with experiment, after Greaves (1987).

Only since nonstructured grids were developed (see Figure 1.78d) and further performance improvements in computers have RANS codes become tools helpful for the aerodynamic development of vehicles: in 1983, FLUENT® was launched by Creare (now

ANSYS); in 1987, Gosman followed with STAR CD® (adapco). A completely different approach was taken by EXA with their PowerFLOW® code. It is based on the Lattice-Boltzmann method, in which the fluid is no longer treated as continuum, but rather as a collection of discrete particles. Similar to the kinetic gas theory, their dynamics are followed with statistical methods.

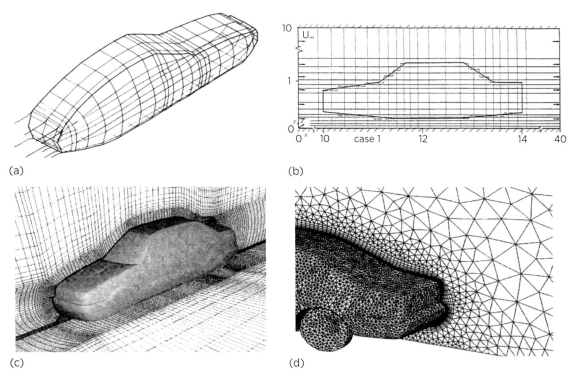

Figure 1.78 Grids for computation: a) Panel (Porsche 924), after Summa & Maskew (1983); b) Cartesian, after Demuren & Rodi (1982); c) structured, body-fitted (VW-Passat), after Hupertz (1998); d) nonstructured, after Khandia et al. (2000).

How CFD gained its place in vehicle aerodynamics has been reported by Sebben (2006 [753]), based on her experience at Volvo. Here are the cornerstones: at the end of the 1980s Volvo started investigating RANS codes; in 1991 a CFD team was built up. As a first 3D example, a very simple generic car model was selected; the grid consisted of 10^5 cells. It took four months to generate the grid and solve and analyze the results. By 1993, the turnaround time was reduced to six weeks. The breakthrough was reached in 1996. Since then, CFD has been applied on every design. In 2000, $\Delta c_D = \pm 0.015$ as compared to full-scale measurement was achieved. The best figure currently claimed by another car manufacturer is $\Delta c_D = \pm 0.005$, excluding experiments on small-scale models. Today, fully 3D models are computed with a grid of $50 \cdot 10^6$ cells.

Chapter 2
The Physical Principles of Aerodynamics

Andreas Dillmann

2.1 Basic Equations in Fluid Dynamics

2.1.1 Conservation Laws

The physical basis of fluid dynamics is the application of Newton's axioms of mechanics [117] to a volumetric element assumed to be located within the flow field. It is postulated that the fluid mass contained by this element is constant for all points in time. Further, it is assumed that the element's spatial dimension can be chosen to be arbitrarily small without consideration of the fluid's actual molecular structure. This is the continuum hypothesis. The requirement of temporal invariance of the mass Δm of the fluid element then establishes the simple relationship

$$\frac{d(\Delta m)}{dt} = 0 \tag{2.1}$$

Newton's second axiom, however, states that the product of mass and acceleration dv/dt is equal to the sum of forces $\boldsymbol{F}$ acting on the fluid element:

$$\boldsymbol{F} = \Delta m \frac{d\boldsymbol{v}}{dt} \tag{2.2}$$

In classical aerodynamics, relations (2.1) and (2.2), which are commonly denoted as the conservation laws for mass and momentum in the field of fluid dynamics, are entirely sufficient for a complete description of the fluid-flow state. If thermodynamic effects are also of importance, mass and momentum conservation must be supplemented by the property of conservation of total energy (from the first law of thermodynamics), which will not be considered here.

2.1.2 Kinematics and Dynamics of Flow Fields

2.1.2.1 Basic Terms in Continuum Mechanics

From the point of view of continuum mechanics, the state of motion of a fluidic medium is fully described when the velocity vector $v = (u,v,w)$, the hydrodynamic pressure p, and the density ρ are known in each point in space, represented by the Cartesian coordinates (x,y,z), and at every point in time t. The former implies that physical quantities are understood as fields instead of properties of the individual fluid element [117]. We now formulate comprehensive governing equations from the relations (2.1) and (2.2), which requires relating the concepts of force, mass, and time change to the corresponding field quantities.

The mass Δm of a fluid element for a given volume ΔV can be computed readily from the density field as

$$\Delta m = \rho \Delta V \qquad (2.3)$$

where, in classic aerodynamics, the density ρ is usually regarded as a constant, which is a valid simplification when the flow velocities are significantly smaller than the speed of sound.[1] The density therefore plays the role of a material constant.[2]

Two fundamentally distinct categories of forces acting on the fluid element have to be considered. The volumetric forces affect the fluid externally, are proportional to its mass Δm and act on the element through a prescribed acceleration vector g. The most fundamental example is the force of gravity, which creates hydrostatic pressure gradients. Note, however, that the resulting action (hydrostatic lift) is typically of no interest in aerodynamics and therefore is usually ignored. In contrast to this, the surface forces, exerted on the fluid element by neighbor particles, play a fundamental role. Surface forces are proportional to the surface area on which they impinge and are represented by mechanical stress fields, which are themselves divided into normal $\tau_{i,i}$ and shear stresses $\tau_{i,j}$, acting perpendicular and tangentially to the surface, respectively. Note that the index i denotes the unit vector normal to the surface and the index j the unit vector in the direction of the resulting force. A stress value therefore has a positive sign if, for a positively (negatively) oriented surface normal vector, the resultant force vector equally points in the positive (negative) coordinate direction (cf. Figure 2.1). Note that in an inviscid fluid, only the pressure p is available to act as a normal stress, independent of direction ($\tau_{x,x} = \tau_{y,y} = \tau_{z,z} = -p$), but in a viscous fluid both normal and shear stresses can occur. In this case, the stresses are related to the velocity field by material laws.

---

1. For more information and a simple relation for estimating compressibility effects, see section 2.4.2.
2. Numerical data on the dependency of air density on pressure and temperature can be found in section 2.4.1.

The Physical Principles of Aerodynamics

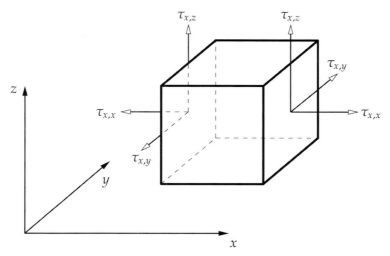

Figure 2.1 Sign convention of the stresses, depicted for the normal and shear stresses acting on surfaces with $x =$ const. The arrows indicate the direction of the force created by the respective stress.

It can be generally shown that the 3×3 matrix of stresses, in a force and momentum equilibrium for a fluid element, corresponds to a linear mapping, which uniquely associates a force vector to any given surface normal. In this mapping, shear stresses are found to be pairwise identical when indices are interchanged, that is,

$$\tau_{i,j} = \tau_{j,i} \tag{2.4}$$

Mathematically, the stress matrix is a symmetric tensor of second order and is hence also referred to as the stress tensor [117].

Of particular importance is finally the representation of the temporal variability of a quantity f for a fluid element moving through the field $f(x, y, z, t)$. The total differential firstly yields the change df of f for infinitesimally small changes of x, y, z, and t as

$$df = \frac{\partial f}{\partial t}dt + \frac{\partial f}{\partial x}dx + \frac{\partial f}{\partial y}dy + \frac{\partial f}{\partial z}dz \tag{2.5}$$

The latter directly yields the temporal variation of f seen by an observer moving along an arbitrary trajectory $(x(t), y(t), z(t))$:

$$\frac{df}{dt} = \frac{\partial f}{\partial t} + \dot{x}\frac{\partial f}{\partial x} + \dot{y}\frac{\partial f}{\partial y} + \dot{z}\frac{\partial f}{\partial z}, \tag{2.6}$$

The first term on the right-hand side (2.5) describes the local rate of change of f with time, as perceived by a fixed observer. The remaining terms arise when the observer is moving with velocity $v_B = (\dot{x}, \dot{y}, \dot{z})$ through the inhomogeneous spatial field, which is why they are also referred to as convective derivatives. For the special case of an observer moving exactly with the fluid element, the local flow velocity $v = (u,$

v, w) is used as an arbitrarily selectable observer velocity v_B, yielding the material derivative [886]

$$\frac{Df}{Dt} := \frac{\partial f}{\partial t} + u\frac{\partial f}{\partial x} + v\frac{\partial f}{\partial y} + w\frac{\partial f}{\partial z} \tag{2.7}$$

or by using the gradient $\nabla f = (\partial f/\partial x, \partial f/\partial y, \partial f/\partial z)$

$$\frac{Df}{Dt} = \frac{\partial f}{\partial t} + \mathbf{v} \cdot \nabla f . \tag{2.8}$$

In contrast to (2.6), the symbol D/Dt in (2.7) or (2.8) indicates a particle-fixed time change. As can be deduced from (2.8), the convective component in spatially inhomogeneous fields vanishes only when the movement is tangential to an isoline of f, to which the local gradient vector of f is always perpendicular.

2.1.2.2 Kinematics of the Velocity Field

A flowing fluid element generally does not move like a rigid body, but can change its geometrical shape during the movement, where the deformation is determined by the velocity field in which the motion occurs. Formulating the conservation equations requires knowledge of the individual deformation components, which can be separated using the following procedure, based on Helmholtz [117].

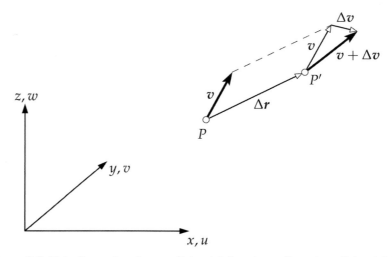

Figure 2.2 Velocity vectors in a spatial point P and an adjacent spatial point P'.

For this purpose, we consider the relative velocity $\Delta v = (\Delta u, \Delta v, \Delta w)$ between two adjacent points P and P' of a flow field, their spatial distance $\Delta r = (\Delta x, \Delta y, \Delta z)$ being assumed to be small (Figure 2.2). If we label the partial derivatives of the velocity components u, v, w in accordance with spatial coordinates x, y, z by means of subscript indices, the difference in velocity $\Delta \mathbf{v}$ can be formulated as the product of a 3×3 matrix with distance vector Δr [117]:

$$\Delta \mathbf{v} = \begin{pmatrix} \Delta u \\ \Delta v \\ \Delta w \end{pmatrix} = \begin{pmatrix} u_x \Delta x + u_y \Delta y + u_z \Delta z \\ v_x \Delta x + v_y \Delta y + v_z \Delta z \\ w_x \Delta x + w_y \Delta y + w_z \Delta z \end{pmatrix} = \begin{pmatrix} u_x & u_y & u_z \\ v_x & v_y & v_z \\ w_x & w_y & w_z \end{pmatrix} \cdot \begin{pmatrix} \Delta x \\ \Delta y \\ \Delta z \end{pmatrix} \quad (2.9)$$

where the matrix of the velocity derivatives assigns the respective velocity vector to each distance vector $\Delta \mathbf{r}$ and thereby represents a tensor. Since every tensor A, by addition or subtraction of its transposed[3] tensor A^T, that is

$$A = \frac{1}{2}(A + A^T) + \frac{1}{2}(A - A^T) \quad (2.10)$$

can be broken down into a symmetric and an anti-symmetric component, and the scalar product of an anti-symmetric tensor can always be written as a vectorial cross product, we obtain from (2.9), using rotation $\nabla \mathbf{v}$ of the velocity field, the following decomposition:

$$\Delta \mathbf{v} = \frac{1}{2} \begin{pmatrix} 2u_x & u_y + v_x & u_z + w_x \\ v_x + u_y & 2v_y & v_z + w_y \\ w_x + u_z & w_y + v_z & 2w_z \end{pmatrix} \cdot \Delta \mathbf{r} + \frac{1}{2}(\nabla \mathbf{v})\Delta \mathbf{r}, \quad (2.11)$$

where the second term corresponds to a mere solid-body rotation around the point P at the angular velocity

$$\omega = \frac{1}{2}\nabla \mathbf{v} \quad (2.12)$$

while the first expression describes the deformation of the fluid element [117].

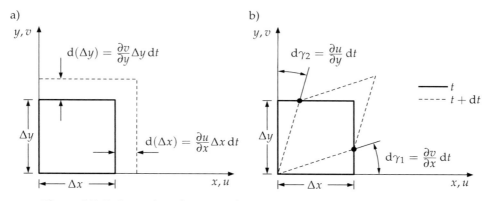

Figure 2.3 Deformation of a rectangle in the x,y-plane under the influence of velocity gradients.

The elements of the deformation tensor can easily be interpreted geometrically. Figure 2.3 shows, as an example, the deformation of a rectangle with initial side lengths $\Delta x, \Delta y$ in the x,y-plane during an infinitesimal interval dt. If the gradient $\partial u / \partial x$ of

---

3. The transposed tensor A^T is obtained from the tensor A by interchanging lines and columns.

the u-component in the x-direction and the gradient $\partial v/\partial y$ of the component v in the y-direction are positive, the opposite sides of the fluid element remain parallel, while the lengths of the sides increase according to the differences in velocity $\partial u/\partial x \cdot \Delta x$ and $\partial v/\partial y \cdot \Delta y$ (Figure 2.3a). The relative increase in length per unit of time is therefore:

$$\frac{1}{\Delta x}\frac{D(\Delta x)}{Dt} = \frac{\partial u}{\partial x}, \qquad \frac{1}{\Delta y}\frac{D(\Delta y)}{Dt} = \frac{\partial v}{\partial y}, \qquad (2.13)$$

where the symbol for the material derivative is used in order to indicate that (2.13) refers to a fluid element under motion. The main diagonal elements of the deformation tensor therefore describe the relative temporal rate of change of the side lengths of a fluid element.

Accordingly, if the gradient $\partial u/\partial y$ of the u-component in the y-direction and the gradient $\partial v/\partial x$ of the v-component in the x-direction are positive, the corner points of the rectangle are displaced due to the velocity differences $\partial u/\partial y \cdot \Delta y$ and $\partial v/\partial x \cdot \Delta x$ such that the original rectangle is deformed into a parallelogram (Figure 2.3b). The temporal rate of change of the edge angle $D\gamma/Dt$ from the original 90° is obtained as the sum of the two angle changes $D\gamma_1/Dt$ and $D\gamma_2/Dt$ and reads

$$\frac{D\gamma}{Dt} = \frac{\partial u}{\partial u} + \frac{\partial v}{\partial x} \qquad (2.14)$$

Thus, the minor diagonal elements of the deformation tensor describe the temporal shear rate of a fluid element.

2.1.2.3 Streamline, Stream Surface, and Stream Tube

The path curve whose tangent at each of its points coincides with the instantaneous velocity vector **v** is the essential concept for describing the mechanics of motion of a point mass. In fluid mechanics, this role is performed by the streamline, which is also tangential to the local flow velocity at every spatial point. However, in contrast to the path curve, time is considered to be frozen, that is, we observe a snapshot of the velocity field (Figure 2.4a). The streamlines therefore only coincide with the actual trajectories of the flow particles in a steady flow and can then be made visible by the local addition of dye or smoke. The streamline image provides a good overview of the flow field kinematics and allows qualitative conclusions to be drawn with regard to the pressure and velocity fields (section 2.2.1).

The stream surface represents a three-dimensional generalization and is formed by all streamlines that pass through a space curve C and with which they have no more than one common point (Figure 2.4b). If this curve is closed, the result is a stream tube (Figure 2.4c) that exhibits flow only through the end faces and, by definition, no flow through the lateral surface consisting of streamlines. Stream tubes play an important part in stream filament theory (sections 2.1.8.3 and 2.3.7.1).

The Physical Principles of Aerodynamics

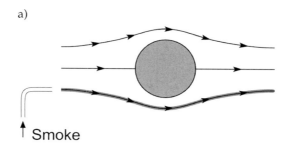

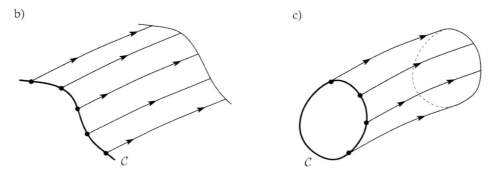

Figure 2.4 Definition of a) streamline, b) stream surface and c) stream tube.

2.1.3 The Continuity Equation

The mass conservation principle formulated in (2.1) requires that the mass $\Delta m = \rho \Delta V$ of a fluid element remains constant under motion in the flow field. In case of an incompressible fluid of constant density, this requirement is equivalent to the constraint of constant volume

$$\frac{D(\Delta V)}{Dt} = 0 \,. \tag{2.15}$$

For a cuboid volumetric element with $\Delta V = \Delta x \cdot \Delta y \cdot \Delta z$, straightforward time derivation and utilization of the product rule yields [296]

$$\frac{1}{\Delta V}\frac{D(\Delta V)}{Dt} = \frac{1}{\Delta x}\frac{D(\Delta x)}{Dt} + \frac{1}{\Delta y}\frac{D(\Delta y)}{Dt} + \frac{1}{\Delta z}\frac{D(\Delta z)}{Dt} \tag{2.16}$$

that is, the relative change of the volume of the cuboid element is precisely the sum of the relative changes in the side lengths, which are given by the diagonal elements of the deformation tensor as described above. The relative change of ΔV is therefore calculated as the divergence of the velocity vector as

$$\frac{1}{\Delta V}\frac{D(\Delta V)}{Dt} = \frac{\partial u}{\partial x} + \frac{\partial v}{\partial y} + \frac{\partial w}{\partial z} = \nabla \cdot \mathbf{v} \tag{2.17}$$

and the mass conservation requirement of equation (2.15) is reduced to the incompressible continuity equation

$$\nabla \cdot v = 0, \qquad (2.18)$$

according to which the velocity field must be a divergence-free vector field.

2.1.4 The Euler Equation

In an inviscid fluid, the only force of action impinging on a fluid element is the pressure force, which is always perpendicular to the impacted (fluid) surface and independent of its spatial orientation. According to Newton's second axiom (cf. eq. (2.2)), the product of mass and acceleration therefore equals the resultant forces acting on the fluid element. If we initially examine only the x-component of the motion equation, we obtain in accordance with Figure 2.5:

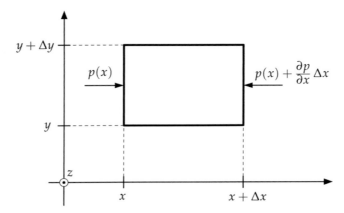

Figure 2.5 Forces acting on a fluid element in the x-direction.

$$\rho \Delta x \Delta y \Delta z \cdot \frac{Du}{Dt} = p(x) \cdot \Delta y \Delta z - \left[p(x) + \frac{\partial p}{\partial x} \Delta x \right] \cdot \Delta y \Delta z . \qquad (2.19)$$

The first two terms on the right-hand side cancel, so that an accelerative force only results if there is a gradient in pressure p in the x-direction. Using the definition (2.8) for the material derivative, we obtain from (2.19), after division by the mass $\Delta m = \rho \Delta x \Delta y \Delta z$ of the fluid element, the simple result:

$$\frac{\partial u}{\partial t} + \mathbf{v} \cdot \nabla u = -\frac{1}{\rho} \frac{\partial p}{\partial x} . \qquad (2.20)$$

Analogous relationships can be obtained for the y- and z-directions, with u replaced by v or w and $\partial p/\partial x$ by $\partial p/\partial y$ or $\partial p/\partial z$. By combining these results in vector form, we obtain Euler's equation of motion for the inviscid fluid:

$$\frac{\partial \mathbf{v}}{\partial t} + (\mathbf{v} \cdot \nabla)\mathbf{v} = -\frac{1}{\rho} \nabla p , \qquad (2.21)$$

in which the operator

$$(\mathbf{v} \cdot \nabla) := u\frac{\partial}{\partial x} + v\frac{\partial}{\partial y} + w\frac{\partial}{\partial z} \qquad (2.22)$$

is to be applied component-wise to **v**. Finally, a representation useful for theoretical investigations is obtained by converting (2.21) with the aid of the known vector-analytical identity [296]

$$\nabla(a \cdot b) = (b \cdot \nabla)a + (a \cdot \nabla)b + a(\nabla b) + b(\nabla a), \qquad (2.23)$$

using $a = b = \mathbf{v}$, into the "second vector form" of the Euler equation [886]:

$$\frac{\partial \mathbf{v}}{dt} + \nabla\left(\frac{\mathbf{v}^2}{2}\right) - \mathbf{v}(\nabla \mathbf{v}) = -\frac{1}{\rho}\nabla p. \qquad (2.24)$$

The continuity equation (2.18) and the Euler equation (2.21) or (2.24) together form a system of four partial differential equations for the four unknowns $\mathbf{v} = (u,v,w)$ and p and are therefore entirely sufficient for a mathematical description of the flow of an incompressible inviscid fluid. Since the system is of first order in the spatial derivatives, only a single boundary condition can be imposed at the borders of the fluid domain. For wall surfaces without in- or outflow, for example, one requires the normal component of **v** to vanish.

2.1.5 The Bernoulli Equation

From rigid body mechanics, we know that the integral of Newton's equation of motion (2.2) along the path curve in a conservative force field (e.g., a gravity field) leads to the energy theorem according to which the sum of kinetic and potential energies is always constant during the entire motion [117]. If we restrict ourselves to the case of steady incompressible flows, an analogous relationship can also be determined by integrating the Euler equation along a streamline whose vectorial line element $d\mathbf{r}$ coincides in every point with the local velocity vector (Figure 2.6). If we apply a scalar multiplication with $d\mathbf{r}$ to the second vector form of the steady Euler equation, that is

$$\nabla\left(\frac{\mathbf{v}^2}{2}\right) \cdot d\mathbf{r} - [\mathbf{v} \times (\nabla \times \mathbf{v})] \cdot d\mathbf{r} = -\frac{1}{\rho}\nabla p \cdot d\mathbf{r}, \qquad (2.25)$$

the second scalar product on the left side vanishes, since the cross-product of the two vectors **v** and $\nabla \mathbf{v}$ is always perpendicular to the plane they span and is therefore always orthogonal to the velocity **v**. The two remaining terms, on account of the identity [296]

$$\nabla f \cdot d\mathbf{r} = \frac{\partial f}{\partial x}dx + \frac{\partial f}{\partial y}dy + \frac{\partial f}{\partial z}dz = df \qquad (2.26)$$

can be integrated by elementary operations to obtain the Bernoulli equation

$$p + \frac{\rho}{2}\mathbf{v}^2 = \text{const.}, \qquad (2.27)$$

where p and $\rho/2\mathbf{v}^2$ are often referred to as static and dynamic pressure, respectively, the sum of which is constant along a streamline in an inviscid steady flow. Note that the integration constant, also referred to as the total pressure, can differ from one streamline to another. An acceleration of the flow is therefore always equivalent to a reduction in pressure, while, conversely, deceleration of the flow always leads to a pressure increase. As can be seen directly from the nature of the formula, this corresponds to the energy theorem of mechanics, with the pressure p taking on the role of potential energy.

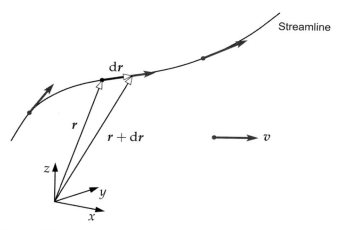

Figure 2.6 Integration of the Euler equation along a streamline.

2.1.6 Potential Theory

In the case of an inviscid incompressible flow, the continuity equation (2.18) and the Euler equation (2.24) are sufficient to determine the flow field completely. However their partial non-linearity prevents the application of simple solution methods. The associated difficulties can be drastically reduced by an additional simple kinematic assumption.

As can be deduced from having a spherical fluid element in mind, the pressure, which is the only active force in an inviscid flow, is not capable of changing the angular momentum of a fluid particle, since it always acts normal to the surface of the sphere and its resultant force therefore always acts at the center of gravity of the fluid element. Accordingly for any motion of an ideal inviscid fluid, commencing from a static state, the angular velocity (2.12) of the fluid elements at each spatial point must vanish, that is, the velocity field must fulfill the condition of rotational freedom:

$$\nabla \times \mathbf{v} = 0, \qquad (2.28)$$

which leads directly to the form

$$\mathbf{v} = \nabla \phi \qquad (2.29)$$

since every irrotational vector field **v** always has a scalar potential ϕ. If we now substitute (2.29) into the continuity equation $\nabla \cdot \mathbf{v} = 0$ of the incompressible flow, the Laplace equation for the velocity potential is obtained directly on account of $\nabla \cdot \nabla \phi = \Delta \phi$ as:

$$\Delta \phi = 0, \qquad (2.30)$$

whereas substituting the rotational freedom in the second vector form of the steady Euler equation

$$\nabla\left(\frac{\mathbf{v}^2}{2}\right) - \mathbf{v} \times (\nabla \times \mathbf{v}) = -\frac{1}{\rho} \nabla p \qquad (2.31)$$

gives us the Bernoulli equation (2.27) again on account of the absence of the second term. However, unlike in the general case, the integration constant is identical for all streamlines. For typical aerodynamic applications in which a constant flow condition u_∞, p_∞ is typically present at a large distance from the body to which the flow is applied, the Bernoulli equation (2.27) thus becomes

$$p + \frac{\rho}{2}\mathbf{v}^2 = p_\infty + \frac{\rho}{2}u_\infty^2, \qquad (2.32)$$

which, by introducing the dimensionless pressure coefficient c_p, can also be written in the more familiar form

$$c_p := \frac{p - p_\infty}{\frac{\rho}{2}u_\infty^2} = 1 - \frac{\mathbf{v}^2}{u_\infty^2} \qquad (2.33)$$

In this way, the solution to a given flow problem is basically reduced to solving the Laplace equation (2.30), the characteristics of which are well known in mathematics and for which a large number of solution methods exist [296]. If the velocity potential ϕ associated with the preceding problem has been determined, we can obtain the velocity and pressure directly from (2.29) and (2.32) or (2.33), and hence all flow variables are known.

2.1.7 The Navier-Stokes Equation

Although the idealized model of an inviscid fluid often provides statements that are suitable for practical use in many cases, real flows always involve friction. In order to take into account the friction effects, which play an important part in aerodynamics, an additional term must be added to Euler's equation of motion.

Frictional fluids are characterized by the fact that the fluid elements resist shear deformation. In gases, this resistance is due to the continuous exchange of molecules between adjacent fluid elements, which is a result of thermal molecular movement. The latter causes slower fluid elements to be accelerated by faster neighboring ones and, conversely, faster elements to be decelerated by slower ones. At fixed walls, the molecules take on the wall velocity in the statistical average, which is why the macroscopic fluid elements seemingly adhere to the wall.

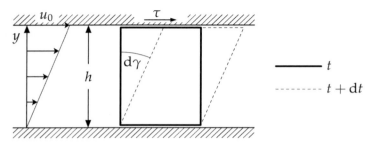

Figure 2.7 Deformation of a fluid element in the gap between two flat, parallel plates.

If a frictional fluid is introduced into a gap of height h between two flat, parallel plates, with the lower one at rest and the upper plate moving at a constant velocity u_0 (Figure 2.7), the result is a linear velocity profile in which the fluid elements are deformed at the constant shear rate

$$\frac{d\gamma}{dt} = \frac{u_0}{h} = \frac{du}{dy} \tag{2.34}$$

For many industrially relevant fluids such as water, air, and oil, the result is a linear relationship, also referred to as the Newtonian shear stress model, between the shear stress τ needed for deformation and the shear rate $d\gamma/dt$:[4]

$$\tau = \mu \frac{d\gamma}{dt} = \mu \frac{du}{dy}. \tag{2.35}$$

Here, the dynamic viscosity μ is purely a material parameter dependent primarily on the temperature. Since the intensity of molecular momentum exchange in gases increases with rising temperature, the viscosity also increases, whereby for most gases $\mu \sim \sqrt{T}$ applies in the first approximation.[5] Information on the dynamic viscosity of air can be found in section 2.4.1.

In order to obtain a generalized three-dimensional version of Newton's shear stress model, it is obvious to assume the stress tensor $\tau_{i,k}$ proportional to the deformation tensor (2.11), since the latter's elements correspond exactly to the shear rate of a spatial fluid element:

$$\begin{pmatrix} \tau_{x,x} & \tau_{x,y} & \tau_{x,z} \\ \tau_{y,x} & \tau_{y,y} & \tau_{xy,z} \\ \tau_{z,x} & \tau_{z,y} & \tau_{z,z} \end{pmatrix} = \mu \begin{pmatrix} 2u_x & u_y + v_x & u_z + w_x \\ v_x + u_y & 2v_y & v_z + w_y \\ w_x + u_z & w_y + v_z & 2w_z \end{pmatrix}. \tag{2.36}$$

---

4. Fluids that conform to the linear relationship (2.35) between shear stress and shear rate are also referred to as Newtonian fluids.
5. In liquids, the properties of which are determined to a large extent by intermolecular forces, there is typically a reduction in μ as the temperature increases caused by a gradual relaxation of these bonds.

Here, the subscript indices of the velocity components correspond to the partial spatial derivatives (e.g., $u_x = \partial u/\partial x$). As can be easily verified, for the special case this general approach yields $u = u(y)$, $v = w = 0$, which is exactly the Newtonian model (2.35).

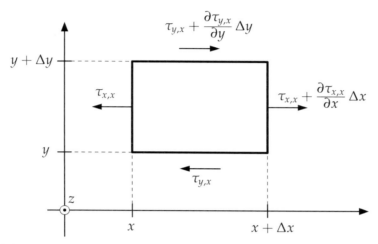

Figure 2.8 Frictional forces acting on a fluid element in the x-direction. For the sake of clarity the contributory effects of $\tau_{z,x}$ are omitted.

It is now straightforward to determine the frictional force R needed to supplement the Euler equation, by balancing the forces at the fluid element. For the x component R_x, according to Figure 2.8, we obtain:

$$R_x = -\tau_{x,x}\Delta y \Delta z + \left(\tau_{x,x} + \frac{\partial \tau_{x,x}}{\partial x}\Delta x\right)\Delta y \Delta z$$
$$-\tau_{y,x}\Delta x \Delta z + \left(\tau_{y,x} + \frac{\partial \tau_{y,x}}{\partial y}\Delta y\right)\Delta x \Delta z \qquad (2.37)$$
$$-\tau_{z,x}\Delta x \Delta y + \left(\tau_{z,x} + \frac{\partial \tau_{z,x}}{\partial z}\Delta z\right)\Delta x \Delta y$$

or, after removing terms that mutually eliminate each other

$$R_x = \left(\frac{\partial \tau_{x,x}}{\partial x} + \frac{\partial \tau_{y,x}}{\partial y} + \frac{\partial \tau_{z,x}}{\partial z}\right)\Delta x \Delta y \Delta z . \qquad (2.38)$$

Substituting the corresponding elements of the stress tensor (2.36) in the latter, we initially obtain:

$$R_x = \mu\left[\left(\frac{\partial^2 u}{\partial x^2} + \frac{\partial^2 u}{\partial y^2} + \frac{\partial^2 u}{\partial z^2}\right) + \frac{\partial}{\partial x}\left(\frac{\partial u}{\partial x} + \frac{\partial v}{\partial y} + \frac{\partial w}{\partial z}\right)\right]\Delta x \Delta y \Delta z , \qquad (2.39)$$

where the second term vanishes for incompressible fluids on account of the continuity equation (2.18). Dividing by $\Delta m = \rho \Delta V$ yields the required frictional force per unit of mass to supplement the x component (2.20) of the Euler equation:

$$\frac{\partial u}{\partial t} + \mathbf{v} \cdot \nabla u = -\frac{1}{\rho}\frac{\partial p}{\partial x} + \frac{\mu}{\rho}\left(\frac{\partial^2 u}{\partial x^2} + \frac{\partial^2 u}{\partial y^2} + \frac{\partial^2 u}{\partial z^2}\right). \tag{2.40}$$

Similar expressions are obtained for the *y*- and *z*-components of the frictional force when *u* is replaced by *v* and *w*. Using the Laplace operator

$$\Delta := \frac{\partial^2}{\partial x^2} + \frac{\partial^2}{\partial y^2} + \frac{\partial^2}{\partial z^2} \tag{2.41}$$

and after introducing the kinematic viscosity

$$\nu = \frac{\mu}{\rho}, \tag{2.42}$$

we finally obtain the Navier-Stokes equation in vector form:

$$\frac{\partial v}{\partial t} + (v \cdot \nabla)\mathbf{v} = -\frac{1}{\rho}\nabla p + \nu \Delta \mathbf{v}, \tag{2.43}$$

which is the sought after equation of motion of an incompressible frictional medium.

Since the Navier-Stokes equation is of second order in the spatial derivatives, two boundary conditions now have to be fulfilled at the flow domain boundaries. This is entirely compatible with the no-slip condition along the walls, so that in addition to the requirement of a vanishing wall-normal velocity component, the tangential component now also has to vanish.

2.1.8 Integral Forms of the Conservation Laws

The mass and momentum conservation laws derived in the preceding sections were formulated for an infinitesimal fluid element and can therefore be understood as a local balance of these physical parameters at each point of the fluid-filled space. For many applications, however, a global formulation is more favorable: one that balances the conservation of mass and momentum for a spatially fixed control volume with internal flow and the boundaries of which can be more or less arbitrarily defined by the user. Such balance equations can be obtained with the aid of Gauss's divergence theorem by volumetric integration of the continuity equation and the Euler or Navier-Stokes equations.

2.1.8.1 Gauss's Divergence Theorem

Gauss's divergence theorem enables a spatial integral over a volume *V* to be converted into a surface integral through the closed boundary surface *D* of *V*. The vector surface element *dS* of surface *S*, which faces in the direction of the outer surface normal, has the magnitude of the surface area $|dS| = dS$ (Figure 2.9). Use of the ∇ symbol allows the simple operator notation

$$\int_V dV \nabla = \int_S dS, \tag{2.44}$$

which means that all the symbols on the right side of the ∇ symbol in the volume integral must appear to the right of *dS* in the surface integral. The integration volume and its surface can be either moving or spatially fixed. The latter case is always assumed below.

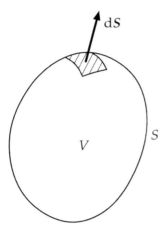

Figure 2.9 Vector surface element *dS* of a volume *S* enclosed by the surface *V*. The vector *dS* points in the direction of the local surface normal; its magnitude |*dS*| is identical to the area of the shaded surface.

2.1.8.2 The Integral Form of the Continuity Equation

If Gauss's divergence theorem (2.19) is applied to the volumetric integral of divergence $\nabla \cdot \mathbf{v}$, we obtain the following directly:

$$\int_V \nabla \cdot \mathbf{v}\, dV = \int_S \mathbf{v} \cdot d\mathbf{S}, \tag{2.45}$$

where the scalar product $\mathbf{v} \cdot d\mathbf{S}$ is the product of the surface area dS and the surface-normal velocity component, and therefore corresponds to the differential partial volumetric flow through the surface element. If the flow is in the direction of the outer normal, the partial volumetric flow has a positive sign; otherwise the sign is negative. The integral form of the incompressible continuity equation $\nabla \cdot \mathbf{v} = 0$ then acquires the simple form

$$\int_S \mathbf{v} \cdot d\mathbf{S} = 0, \tag{2.46}$$

which indicates that the sum of all partial volumetric flows through the surface of the control volume must vanish at all times.

2.1.8.3 Inviscid Stream Filament Theory

In the special case of a simple stream tube with perpendicular flow through the two end faces A_1 and A_2 at constant velocities v_1 and v_2, equation (2.46) reduces to the one-dimensional form of the continuity equation $v_1 A_1 = v_2 A_2$ or

$$v \cdot A = \text{const.} \tag{2.47}$$

according to which the volumetric flow must be constant at any point of the stream tube. Together with the Bernoulli equation

$$p + \frac{\rho}{2} v^2 = \text{const.} \tag{2.48}$$

we thus have a simple means of handling one-dimensional inviscid flow processes. This method is also referred to in fluid mechanics as "stream filament theory" and plays an important part in the qualitative interpretation of streamline patterns (section 2.2.1). By incorporating suitable supplementary terms, it can also be extended to viscous flows (section 2.3.7.1).

2.1.8.4 The Momentum Theorem for Inviscid Flows

The center of mass theorem plays an important part in the mechanics of rigid body systems [117]. It is based on the summation of equations of motion for an ensemble of point masses that are subjected both to the influence of external forces and to mutually interacting forces (e.g., gravitation) that comply with the principle "actio est reaction," that is, that consist of identical pairs. For this reason, the interaction forces vanish when the sum is formed, and we obtain the motion equation for the total mass, which is solely under the influence of external forces and moves as if these forces were acting on the center of mass of the system.

The equivalent of the center of mass theorem in fluid mechanics is the momentum theorem, which is obtained by integrating the fluid's motion equation over a control volume fixed in space. For steady flows, this yields a momentum balance for which it is only necessary to know the flow data on the surface of the control volume, while the internal processes are no longer important.

If we restrict ourselves initially to inviscid fluids, the task is then to convert the volumetric integral of the slightly modified steady Euler equation (2.21)

$$\int_V r(\mathbf{v} \cdot \nabla)\mathbf{v} \, dV + \int_V \nabla p \, dV = 0 \tag{2.49}$$

into a surface integral using Gauss's divergence theorem. While (2.44) can be directly applied to the second term in (2.49), the first term has to be modified by applying (2.44) by components, using the known identity [296]

$$\nabla(\lambda \mathbf{v}) = \mathbf{v} \cdot \nabla\lambda + \lambda \nabla \cdot \mathbf{v}, \tag{2.50}$$

where λ indicates an arbitrary scalar function and the second term in this case vanishes on account of the continuity equation $\nabla \cdot \mathbf{v} = 0$. We thereby obtain the momentum theorem for the steady flow of an inviscid incompressible fluid:

$$\int_S \mathbf{v}(\rho \mathbf{v} \cdot d\mathbf{S}) + \int_S p \, d\mathbf{S} = 0, \tag{2.51}$$

the statement of which is easily interpreted in physical terms. As is apparent from Figure 2.10, the second term of the sum is equivalent to the pressure forces $-pd\mathbf{S}$ on the surface elements $d\mathbf{S}$, whereas the first term is the product of the (scalar) mass flow $d\dot{m} = \rho\mathbf{v} \cdot d\mathbf{S}$ passing through $d\mathbf{S}$ with local (vector) velocity $\mathbf{v}$, and therefore describes the momentum flow through $d\mathbf{S}$. If the last components are interpreted formally as momentum forces, then the momentum theorem consists of the simple statement that the sum of the momentum and pressure forces on the surface must be in equilibrium, with both forces always being directed toward the interior of the control volume.

The Physical Principles of Aerodynamics

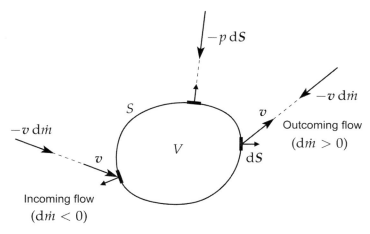

Figure 2.10 Direction of pressure and momentum forces on the surface S of the control volume V.

As a simple application, we may consider the 90° redirection of a flow through a pipe bend, whose two end faces A_1 and A_2 are crossed perpendicularly by the fluid flow at the constant velocities v_1 and v_2 and at constant pressures p_1 and p_2 (Figure 2.11). The pressure and momentum forces $(p_1 + \rho v_1^2)A_1$ and $(p_2 + \rho v_2^2)A_2$, occurring at the two end faces, are in equilibrium with the pressure force $\boldsymbol{F}_M = \int_M p\, d\boldsymbol{S}$ exerted by the wall of the pipe bend on the lateral surface M of the control volume, through which no flow passes, and precisely equals the negative reaction force $\boldsymbol{F}_R$ exerted by the fluid on the wall of the pipe bend.

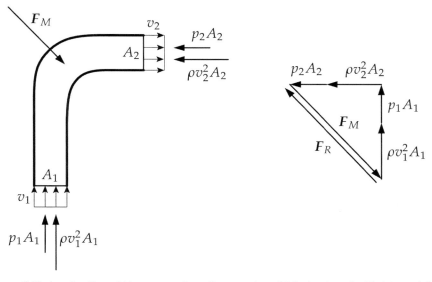

Figure 2.11 Application of the momentum theorem to a 90° pipe bend with internal flow.

2.1.8.5 The Momentum Theorem with Friction

Extending the momentum theorem to inviscid flows formally requires volumetric integration of the friction term proportional to $\Delta \mathbf{v}$ from the Navier-Stokes equation. If we apply Gauss's divergence theorem, for example, to the x-component Δu, we obtain

$$\int_V \Delta u \, dV = \int_V \nabla \cdot \nabla u \, dV = \int_S \nabla u \cdot d\mathbf{S}, \tag{2.52}$$

that is, the velocity gradients on the surface of the control volume must be known in order to determine the integral (2.52). In contrast to the inviscid form (2.51) of the momentum theorem, for which only the pressure and velocity themselves are needed, this often generates some additional complication, which can be avoided by placing the control surface as far as possible in flow regions in which no significant velocity gradients occur. Hence, for the partial surfaces located there, the inviscid form (2.51) of the momentum theorem can be used. On fixed walls, however, the product $\mu \nabla u \cdot d\mathbf{S}$ corresponds to the frictional force exerted by the local wall shear stress on the surface element so that the integral precisely represents the total frictional resistance of the wall.

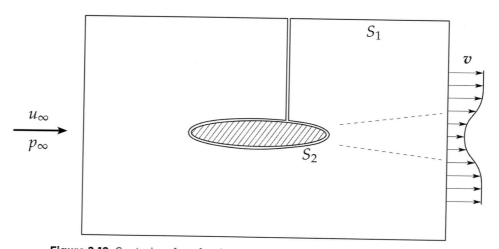

Figure 2.12 Control surface for determining forces acting on a body in flow.

As an elementary example, we examine the force acting on a body in a frictional flow using the momentum theorem. The control surface S is chosen in accordance with Figure 2.12 such that a partial surface S_2 coincides with the surface of the body under consideration and therefore has no internal flow, while the other partial surface S_1 surrounds the body at a sufficiently large distance such that friction effects can be neglected. From the momentum theorem we then obtain

$$\int_{S_1} \mathbf{v}(\rho \mathbf{v} \cdot d\mathbf{S}) + \int_{S_1} p \, d\mathbf{S} - F_{D,S_2} - F_{R,S_2} = 0. \tag{2.53}$$

Here, the first two terms are the (negative) momentum and pressure forces on S_1, while the last two terms are the pressure and frictional forces, respectively, which are exerted in total by the body on the fluid at the surface S_2. Since the sum of these two last terms, in accordance with the "actio est reactio" principle, equals the overall force $\mathbf{F}$ exerted by the fluid on the body exactly, the following simple result is obtained:

$$F = -\int_{S_1} \mathbf{v}(\rho \mathbf{v} \cdot d\mathbf{S}) - \int_{S_1} p\, d\mathbf{S}, \qquad (2.54)$$

according to which the force exerted on the body can be calculated from the pressure and velocity values alone that act on S_1, as may be determined, for example, by wind tunnel measurements. Since incident flow values are usually known, measurements in the wake flow of the body are sufficient in most cases. This principle, based on (2.54), is used in practice, for example, to analyze the vehicle drag in a wind tunnel test (cf. section 4.3.2).

2.2 Dynamics of Inviscid Flow

Even though, in reality, the flow is always frictional and three-dimensional, the planar potential flows discussed in detail below are an important basis for aerodynamics, since they permit a particularly simple mathematical description of many fundamental phenomena, upon which the concrete "reading" of streamline patterns can be exercised. Further on, this section will also put an emphasis on understanding the behavior of boundary layers, which can be described using methods from vortex dynamics and which play a particularly important part in blunt body aerodynamics [372].

2.2.1 Interpreting Streamline Patterns

The streamlines introduced in Section 2.1.2.3 are not only an important theoretical concept, which is used, for example, in the derivation of the Bernoulli equation, but are also a classic tool for the aerodynamics engineer when rendering flows visible and attempting to understand them in physical terms. A typical streamline pattern usually contains a large number of streamlines, from whose position and geometry, with the aid of the continuity equation and the Bernoulli equation, important conclusions regarding the flow's pressure and velocity fields can be drawn. Interpretation of a streamline pattern becomes particularly easy if the streamlines are in one plane, so that in a first approximation the flow can be construed as two-dimensional.[6]

Two adjacent streamlines, for example, can be construed as a planar stream tube, the narrowing of which in the flow direction always leads to the acceleration of the flow and therefore to a drop in pressure, whereas an expansion in the flow direction causes deceleration and therefore an increase in pressure. In this way, direct conclusions can be drawn about the longitudinal pressure and velocity gradients from the streamline spacing (Figure 2.13a).

---

6. Examples of this are the flow in the longitudinal or center section of a vehicle, as shown, for example, in Figure 4.91.

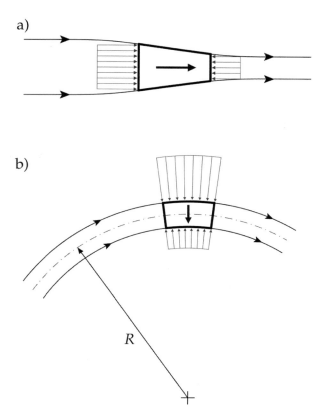

Figure 2.13 Force action on a fluid element due to pressure gradients a) in the longitudinal direction in a constricted stream tube and b) in the transverse direction in a curved stream tube. Red arrows symbolize the pressure distribution on the side faces of the fluid element, while the blue arrow denotes the direction of the respective resultant pressure force.

Information about the transverse pressure and velocity gradients across a streamline is obtained from the curvature of the line, which is described by the local radius of curvature R along the curve (Figure 2.13b). This is because a fluid element can only move on a curved trajectory if the resulting outward-oriented centrifugal force is balanced by a centripetal force of equal size acting in the opposite direction which, in an inviscid flow, can only be generated by pressure forces. A pressure gradient therefore forms transverse to a curved streamline with the pressure gradient toward the "inner side" of the curve, thus creating a resultant pressure force acting centripetally. The smaller the radius of curvature, the steeper the pressure gradient and larger the outward-oriented centrifugal force.

In conclusion, several simple rules can be formulated and easily verified with the aid of the streamline patterns in section 2.2.2:

1. Converging streamlines depict acceleration and a decrease in pressure.
2. Diverging streamlines depict deceleration and an increase in pressure.

3. If the streamlines are curved, the pressure rises in the centrifugal direction.
4. There is no pressure change transverse to straight, parallel streamlines ($R \to \infty$).
5. If a streamline has a kink point ($R \to 0$), which is not simultaneously a stagnation point ($v = 0$), pressure and velocity become infinite at that point ($p = -\infty$, $|v| = +\infty$).

2.2.2 Planar Model Flows

2.2.2.1 Use of Complex Functions

A very elegant and extensive method of solving the planar Laplace equation

$$\frac{\partial^2 F}{\partial x^2} + \frac{\partial^2 F}{\partial y^2} = 0 \quad (2.55)$$

is based on the use of complex functions, which only have to be differentiable (so-called analytical functions). If we differentiate any complex function

$$F(z) = F(x + iy) = \phi(x, y) + i\psi(x, y) \quad (2.56)$$

with respect to x and y by utilizing the chain rule:

$$\frac{\partial F}{\partial x} = \frac{\partial F}{\partial z}\frac{\partial z}{\partial x} = F'(z), \quad \frac{\partial F}{\partial y} = \frac{\partial F}{\partial z}\frac{\partial z}{\partial y} = iF'(z) \quad (2.57)$$

and by comparing the coefficients of the two derivatives, we find that their real part $\phi(x,y)$ and imaginary part $\psi(x,y)$ are not independent of one another, but must satisfy the Cauchy-Riemann differential equations:

$$\frac{\partial \phi}{\partial x} = \frac{\partial \psi}{\partial y}, \quad \frac{\partial \phi}{\partial y} = -\frac{\partial \psi}{\partial x} \quad (2.58)$$

By differentiating again with respect to x or y, it follows directly that the real as well as the imaginary part of each analytical function are always solutions of the Laplace equation. Further, with the aid of equation (2.58), it can easily be shown that the scalar product of the two gradients $\nabla \phi$ and $\nabla \psi$ always vanishes, so that contour lines ϕ = const. and ψ = const. form an orthogonal mesh. If the real part ϕ is therefore regarded as the velocity potential of a flow, the lines ψ = const. are the associated streamlines, since $\mathbf{v} = \nabla \phi$ is always tangential to them (cf. Figure 2.14). The first derivative of $F(z)$ with respect to the complex argument then provides the complex conjugate flow velocity $\bar{\mathbf{v}}$ by way of (2.57) and (2.58):

$$F'(z) = \frac{\partial F}{\partial x} = \frac{\partial \phi}{\partial x} + i\frac{\partial \psi}{\partial x} = \frac{\partial \phi}{\partial x} - i\frac{\partial \phi}{\partial y} = u - iv = \bar{\mathbf{v}}. \quad (2.59)$$

Thus, determining the streamline pattern and velocity in a given complex potential function $F(z)$ is reduced to elementary computing operations in complex variables. If the velocity is known, the pressure is then determined by using the Bernoulli equations (2.32) or (2.33).

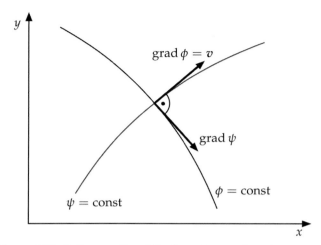

Figure 2.14 Orthogonality of the lines ϕ = const. and ψ = const.

An important tool in obtaining complex flow functions is the conformal map, which uses a complex mapping function $\zeta(z)$ to map each point of a flow in the z-plane uniquely to a point in the ζ-plane, with the result that from a given flow in the z-plane a new flow with altered geometry arises in the ζ-plane. Since conformal mapping formally creates a flow function $F(\zeta)$, we obtain the complex conjugate flow velocity $\bar{v}$ in the ζ-plane using the chain rule [296] as:

$$\bar{v} = F'(\zeta) = \frac{dF}{dz}\frac{dz}{d\zeta} = \frac{F'(z)}{\zeta'(z)}, \qquad (2.60)$$

from which the pressure can be determined again in the familiar way via the Bernoulli equation. Since, according to the Riemann mapping theorem [296], a conformal map exists for every simply connected region, which maps this region to a circular area, the opposite also applies, namely that the flow around any given contour can be obtained in principle from the flow around a circular cylinder, provided that the corresponding mapping function is known. For many industrially interesting configurations, the mapping functions can be obtained from the literature [73].

2.2.2.2 Elementary Flows

The complex functions immediately provide an abundance of possible solutions to the planar Laplace equation, and certain fundamental characteristics of inviscid flows can in fact already be discussed by means of elementary solutions. Some of the following solutions will be used later as modules to create models of more complex flow cases by linear superposition and conformal mapping.

2.2.2.2.1 Parallel Flow

The simplest nontrivial example is the linear function

$$F(z) = c \cdot z, \qquad (2.61)$$

which describes a homogeneous parallel flow at the constant velocity $\bar{v} = F'(z) = c$. For the special case of a real-valued constant $c = u_\infty$, this flow is parallel to the x-axis in the positive coordinate direction and has horizontal streamlines $\psi = $ const. (Figure 2.15). This simple flow function is already of major practical significance in describing the flow around a body.

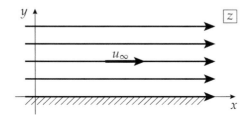

Figure 2.15 Planar parallel flow.

2.2.2.2.2 Stagnation Point Flow

If the x axis is interpreted as a wall streamline, the quadratic function

$$F(z) = \frac{c}{2} \cdot z^2 \tag{2.62}$$

with a positive real constant c, describes the flow against a flat plate, which, on account of $\bar{v} = F'(z) = c \cdot z$, has a stagnation point ($v = 0$) at its origin (Figure 2.16). Owing to the redirection caused by the plate, a pressure field forms with circular isobars around this point, where the maximum pressure p_s is reached. The pressure field is calculated from the velocity field using the Bernoulli equation (2.32) as

$$p = p_s - \frac{\rho}{2}c^2\left(x^2 + y^2\right) \tag{2.63}$$

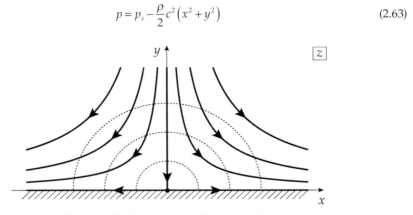

Figure 2.16 Planar stagnation-point flow.

The streamlines $\psi = $ const. are hyperbolae of equal length, where the streamline along the y-axis and dividing to left and right at the stagnation point represents a limiting case. This simple example illustrates very clearly the rules construed in section 2.2.1 for the relationship between pressure field and streamline pattern.

2.2.2.2.3 Flow Around a Sharp Edge

Another case that is essential for a fundamental understanding of the physics is the flow function

$$F(z) = \frac{3}{2} c \cdot z^{\frac{2}{3}} \qquad (2.64)$$

which describes the flow around a protruding sharp edge for the real-valued c (Figure 2.17). The result is once again hyperbolic streamlines and circular isobars. However, due to the vanishing radius of curvature of the wall streamline at $z = 0$, $p = -\infty$ applies for pressure and $|v| = +\infty$ for velocity. Such a solution is of course physically unrealistic. In the real flow, such singularities are avoided in that the wall streamline does not follow the pattern predetermined by the contour, but continues straight, resulting in the formation of a separation plane between the fluid moving straight ahead and the stationary fluid behind the edge.[7] The separation plane is not stable but disintegrates very quickly with the formation of vortices. This process, which is referred to as flow separation, plays a fundamental role in aerodynamics and is discussed in more detail in section 2.2.3.3 and 2.3.3.

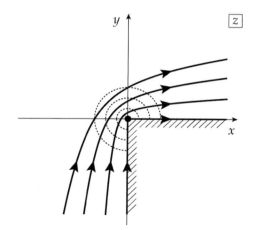

Figure 2.17 Planar flow around a sharp edge.

2.2.2.2.4 Source and Vortex

The complex logarithmic potential function

$$F(z) = c \cdot \ln z, \qquad (2.65)$$

which has a singularity at the origin $z = 0$, is of particular importance for the generation of flows around a body when utilizing the superposition method described in the following section. If we select the constant c to be real, then

---

[7]. In liquids, the flow around a sharp edge may also cause the pressure to fall below the vapor pressure, resulting locally in the formation of vapor bubbles. This process is referred to in engineering as cavitation.

$$F(z) = \frac{Q}{2\pi} \ln z \qquad (2.66)$$

describes a mass source ($Q > 0$) or a mass sink ($Q < 0$) in the origin, which ejects or swallows a volumetric flow Q, whereby the flow moves in a purely radial direction and the velocity is inversely proportional to the distance from the source (Figure 2.18a). As a result, the isobars are circles around the origin $z = 0$ at which, owing to continuous mass production, the divergence freedom $\nabla \cdot \mathbf{v} = 0$ is violated at the singularity and $|\mathbf{v}| \to +\infty, p \to -\infty$ applies. However, if we select the constant c to be purely imaginary, then

$$F(z) = \frac{\Gamma}{2\pi i} \ln z \qquad (2.67)$$

describes a counterclockwise ($\Gamma > 0$) or clockwise ($\Gamma < 0$) rotating potential vortex, in which the fluid particles move on circular streamlines around the center $z = 0$ at which the rotational freedom $\nabla \mathbf{v} = 0$ is now violated at the singularity (Figure 2.18b). Since the purely azimuthally[8] directed velocity is inversely proportional to the distance from the vortex center, the isobars form circles around the origin in which the same singular pressure and velocity values occur as in the source flow.

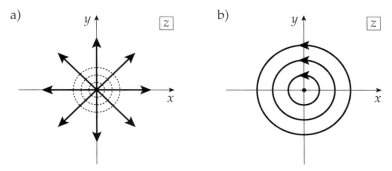

Figure 2.18 Planar source and vortex flow.

2.2.2.3 Flows Around Bodies

Owing to the linearity of the Laplace equation, new solutions can always be obtained by additive superposition of known solutions. This "superposition method" is characterized by particular clarity and, in conjunction with the conformal mapping method, proves very useful when obtaining a basic physical understanding of the flow around simple bodies based on simple solutions. In addition to the parallel flow case, the singular solutions for the source/sink flow and the potential vortex are elementary solutions from which flows around bodies can be synthesized in an easy manner.

---

8. The velocity does not have a radial component, but only a component in the circumferential direction.

2.2.2.3.1 Semi-Infinite Bodies

The simplest example of such a flow is the superposition of a parallel flow u_∞ in the direction of the x-axis with a source of intensity Q in the origin $z = 0$. The corresponding complex flow function

$$F(z) = u_\infty z + \frac{Q}{2\pi} \ln z \qquad (2.68)$$

directly provides the flow pattern in Figure 2.19, in which the internal flow exiting from the source pushes apart the streamlines of the incoming parallel flow, where the branching stagnation streamline $\psi = 0$ separates the inner and outer flows from one another. If we imagine this streamline as a solid wall, the inner flow can be replaced easily by a solid body, without any changes to the outer flow pattern, and we obtain the flow around a semi-infinite body. The end diameter D is related to the source strength Q and the incident flow velocity u_∞ via a simple volumetric flow balance for $x \to \infty$:

$$Q = u_\infty D, \qquad (2.69)$$

whereas the location x_s of the stagnation point on the x axis is derived from the condition $u(x, y = 0) = 0$ as

$$x_s = -\frac{Q}{2\pi u_\infty} \qquad (2.70)$$

Since the streamlines are forced apart in front of the body, an overpressure region (red) forms upstream, which is subsequently followed by an underpressure region (blue) at the sides of the body, where the streamlines are bent toward the body. The simple example illustrates very clearly the typical physical effects caused by the displacement effect of a body in flow.

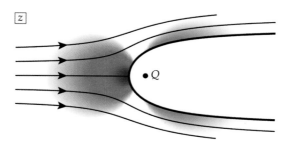

Figure 2.19 Flow field around a planar semi-infinite body.

2.2.2.3.2 Semi-Infinite Body in Ground Proximity

The "method of images" can be used to model the influence of the ground on the flow around a semi-infinite body, which is important in vehicle aerodynamics. For this purpose, we assign to a source located at $(x = 0, y = +h)$ a second source of equal strength at $(x = 0, y = -h)$ such that the vertical velocities of the two sources cancel each other on the x-axis as the line of symmetry. The complex potential function

$$F(z) = u_\infty z + \frac{Q}{2\pi}\ln(z-ih) + \frac{Q}{2\pi}\ln(z+ih) \qquad (2.71)$$

then directly yields the flow pattern depicted in Figure 2.20. The streamline pattern and the pressure distribution are significantly changed compared to the free-flying semi-infinite body by the influence of the mirror source. In particular, the additional velocities induced produce significantly lower pressures on the body, and this effect is particularly pronounced in close proximity to the ground, where the pressure disturbances double (to a first approximation). By integrating the ground pressure distribution, the force exerted on the vehicle can be easily determined using the momentum equation (2.54), and we obtain the total negative lift that is directed to the ground

$$F_L = \frac{\rho}{4}\frac{Q^2}{\pi}\frac{1}{h} \qquad (2.72)$$

that is, the body is increasingly sucked down with decreasing distance to the ground. The underside of the vehicle and the ground form a channel that act like a constricting nozzle. This Venturi effect, named in reference to the constricted pipe of the same name, is opposed in general by lift forces depending on the vehicle shape, so that the result may equally be an overall positive or negative lift [372].

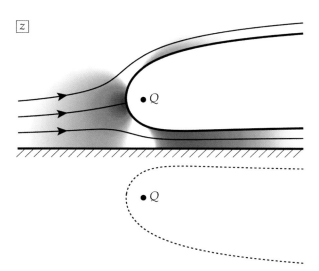

Figure 2.20 Flow field around a planar semi-infinite body in ground proximity (using the mirror method).

2.2.2.3.3 Flow Around a Circular Cylinder and d'Alembert's Paradox

A fully closed body can be created quite trivially using the superposition method by placing a sink of strength Q at $z = 0$ and a source of equal but opposite strength $-Q$ at $z = l$, which swallows the complete mass ejected from the source:

$$F(z) = u_\infty z + \frac{Q}{2\pi}\ln z - \frac{Q}{2\pi}\ln(z-l). \qquad (2.73)$$

If we let the distance l of the source/sink pair in (2.73) go to zero such that the "dipole moment" $M = Q \cdot l$ remains finite in the limiting process $l \to 0$, we find the complex potential function of a circular cylinder with the radius $R = \sqrt{M/2\pi u_\infty}$:

$$F(z) = u_\infty \left[z + \frac{R^2}{z} \right], \tag{2.74}$$

where the second term arises from the limiting process from the two logarithmic components in (2.54). The source/sink pair is also referred to as dipole flow as it effectively combines the source and sink in one point. The flow pattern resulting from (2.74) is shown in Figure 2.21 and the velocity and pressure distributions on the cylinder contour have the simple functional forms

$$|v| = 2u_\infty \sin \varphi \tag{2.75}$$

and

$$c_p = 1 - 4 \sin^2 \varphi. \tag{2.76}$$

As we can see, the local flow velocity at the equator $\varphi = 90°$ of the cylinder is twice as high as the incident flow velocity, which causes considerable underpressures at this point. The great practical significance of the cylinder flow is that it is the starting point for modeling many important flows using conformal maps.

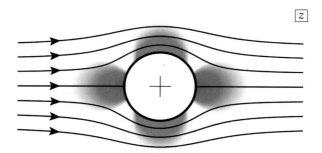

Figure 2.21 Flow field around a circular cylinder.

As we can immediately see from Figure 2.21, the result for the circular cylinder is a completely symmetrical streamline pattern and therefore a completely symmetrical pressure distribution with regard to the x- and y-axes, and its integral over the body's surface yields zero. The body is thus not subjected to any forces. This fact, though surprising at first glance (known also as d'Alembert's paradox after its discoverer), is entirely plausible on closer examination, because in an inviscid flow there is no energy-consuming mechanism that could be an equivalent for the work performed to overcome this resistance. It can therefore generally be shown that d'Alembert's paradox applies to any body in a planar or spatial flow, as long as the flow is completely inviscid and therefore irrotational [886]. In practice, it follows that the contribution of pressure forces to the total drag of a body can be minimized by realizing the pattern of the stream lines calculated from potential flow theory as well as possible (section 2.3.5.4).

2.2.2.3.4 Elliptic Cylinders at Zero Incidence

An elementary example of a solution generated by a conformal map is the flow around an elliptic cylinder, which can be generated from the circular cylinder flow (2.74) by mapping each point of the z-plane, using Joukowski mapping function

$$\zeta = z + \frac{a^2}{z} \qquad (2.77)$$

with $a \leq R$ into the ζ plane (Figure 2.22). By substituting into (2.77), the mapped contour of the circular cylinder $z = Re^{i\varphi}$ is derived directly as

$$\zeta = R\left(1 + \frac{a^2}{R^2}\right)\cos\varphi + iR\left(1 - \frac{a^2}{R^2}\right)\sin\varphi \qquad (2.78)$$

and is therefore correspond to an ellipse with the axis ratio

$$k = \frac{1 - \dfrac{a^2}{R^2}}{1 + \dfrac{a^2}{R^2}}. \qquad (2.79)$$

Figure 2.23 shows the calculated velocity distributions k on the elliptic contour for different axis ratios v_k. It is apparent that the maximum over-velocity

$$v_{k\max} = (1+k)u_\infty \qquad (2.80)$$

will decrease constantly with increasing slenderness ratio, while at the same time the velocity distribution, and therefore the pressure distribution, is almost constant over wide areas of the body. Owing to the symmetry of the pressure distribution, the elliptic cylinder also experiences no resultant force in inviscid flow. In frictional flows, however, it can be shown that slender ellipses have distinct advantages over the circular contour due to their shallower pressure distribution, since they have a lower tendency to flow separation (section 2.3.3).

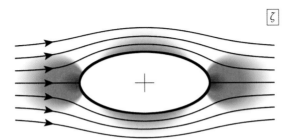

Figure 2.22 Flow field around an elliptic cylinder.

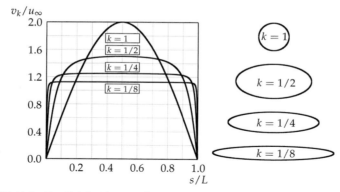

Figure 2.23 Velocity distribution on the contour of ellipses with different slenderness ratios (s: arc length, L: half circumference).

2.2.2.3.5 Elliptic Cylinders with Oblique Incident Flow

It is also possible to realize the case of an elliptic cylinder in a free stream at an angle of attack α by replacing the complex argument z with $ze^{-i\alpha}$ in the complex flow function (2.74) of the circular cylinder flow, which corresponds to a counterclockwise coordinate rotation by the angle α. Figure 2.24 shows the case of an ellipse with incident flow at $\alpha = 30°$ that can be construed as a simple model for a vehicle under the influence of side-wind. As can be seen directly from the streamline pattern and the pressure distribution, no resultant lateral force occurs, but a clockwise torque tends to move the body transverse to the incident flow. Therefore, vehicle shapes that are designed to achieve optimized flow behavior by approximating the potential-theory streamline pattern as much as possible may have increased sensitivity to side-winds [372].

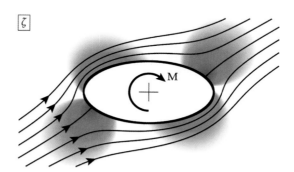

Figure 2.24 Flow field for an elliptic cylinder with oblique incident flow.

2.2.2.3.6 Wing Profile

In addition to torques, oblique flows may also cause transverse forces if, for example, by attaching a separation edge, the flow conditions are changed such that circulation around the body is achieved. The simplest example of this is the trailing sharp edge of a symmetric wing airfoil, which avoids a flow around this point and instead ensures a singularity-free smooth run-off (Figure 2.25). Using potential theory, this flow can be modeled by conformal mapping of a circular cylinder whose center is no longer at the origin but on the x-axis at $x_0 = -(R - a)$. This requires superimposing a clockwise potential vortex over the circular cylinder's flow function to set the correct position of the rear stagnation point [886]:

$$F(z) = u_\infty \left[z + \frac{R^2}{z} \right] + \frac{\Gamma}{2\pi i} \ln z \qquad (2.81)$$

where

$$\Gamma = -4\pi R u_\infty \sin \alpha, \qquad (2.82)$$

which, both on the circular cylinder and the wing profile, generates a lift force $F_L = -\rho u_\infty \Gamma = + \rho u_\infty^2 4\pi R \sin \alpha$ perpendicular to the incident flow, which increases linearly with α for small angles of attack.

This example shows that comparatively small modifications to the trailing edge of a body in flow can have drastic effects on the aerodynamic forces, demonstrating that problems can potentially occur when shapes tried and tested in aviation are applied without reservation to vehicles, as was the case, for example, with the Rumpler teardrop car (cf. Figure 1.13).

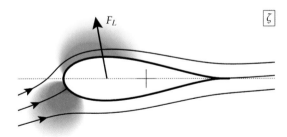

Figure 2.25 Flow field around a symmetrical wing profile.

Lift forces can also occur at zero angle of attack if circulation-generating separation edges are connected to an asymmetric body shape. A simple example of this is the curved wing airfoil, which has a relatively flat bottom side with a highly arched top side. It is immediately obvious that it is exactly the vehicle shapes according to Jaray, discussed in section 1.4, that are likely to raise significant problems in this respect.

2.2.3 Vortex Flows

The flows around bodies discussed in the previous section are all characterized in that they are completely free from separation (the wall streamlines follow the pattern predetermined by the body contour over their entire length). Note, however, that in reality this is basically true only for very slender body shapes. On the contrary, the flow around a "blunt" body is largely determined by separation processes and the formation of vortices, which is why the dynamics of vortex systems are an important key to understanding the aerodynamics of the blunt body [372].

2.2.3.1 The Vortex Filament Model

Vortex flows are characterized by a flow field that is not completely free of rotation. Since pressure forces cannot change the angular momentum of a fluid particle, the rotational areas of the flow must have developed as the result of frictional forces. If these areas are spatially limited, so that the conditions for the existence of a velocity potential are still fulfilled in most of the flow region, such flows can be described using potential theory.

The simplest example of a vortex flow is the potential vortex introduced in section 2.2.2.2.4, where freedom from rotation is always ensured away from the center, while it is violated in the center itself due to the occurrence of a singularity. Its three-dimensional generalization is the vortex filament in which the rotational area is limited to a spatial curve (Figure 2.26). In full analogy to the magnetic field of a current-carrying conductor, it induces a velocity field in its vicinity, which can be determined using the Biot-Savart law known from electrodynamics [886]:

$$\mathbf{v} = \frac{\Gamma}{4\pi} \oint \frac{d\mathbf{s} \mathbf{r}}{|\mathbf{r}|^3}. \tag{2.83}$$

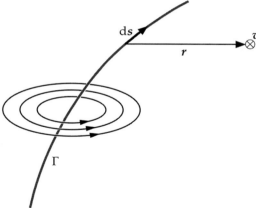

Figure 2.26 Three-dimensional vortex filament.

In the latter, Γ describes the circulation of the vortex filament, $d\mathbf{s}$ its vectorial line element, and $\mathbf{r}$ the distance from $d\mathbf{s}$ to the measuring point at which the contribution of $d\mathbf{s}$ to the velocity v is to be determined, whereby the line integral is to be extended over the entire vortex filament. For the special case of an infinitely long straight vortex filament, (2.83) yields exactly the velocity field of the planar potential vortex, the magnitude of which is given by

$$|v| = \frac{\Gamma}{2\pi a} \tag{2.84}$$

where a is the perpendicular distance from the vortex filament. According to the Helmholtz vortex theorems, a vortex filament inside the fluid cannot end anywhere, has a temporally and spatially constant circulation, and always consists of the same fluid particles [886].

2.2.3.2 Vortex Induction

If there are multiple vortex filaments in a fluid, the motion of a single vortex filament is fully determined, owing to its masslessness, by the induced velocities of the other vortices which superimpose each other without interaction [886]. If we consider, as the simplest example, the motion of two infinitely long parallel vortex filaments Γ_1 and Γ_2 that behave in any plane perpendicular to their axis like two planar potential vortices, then each vortex induces at the site of the other vortex a velocity (2.84) perpendicular to the connecting line of the two vortices, which is why their mutual distance remains unaltered (Figure 2.27). Both vortices therefore perform a circular motion at constant angular speed around the point S at which their induced velocities eliminate each other. If their rotational direction is identical, S lies between the two vortices, whereas if the vortices rotate in opposite directions, it is outside the connecting line; for the special case $\Gamma_1 = -\Gamma_2$ the rotation point is at infinity so that both vortices perform a pure translational motion.

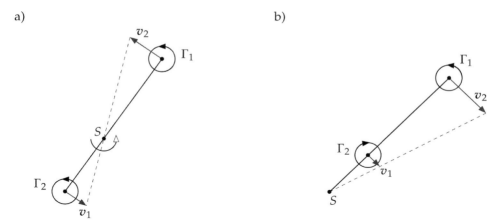

Figure 2.27 Motion of two planar point vortices under the influence of their mutual induction: a) same direction of rotation; b) opposite direction of rotation.

For the aerodynamics of bodies with flow separation, the dynamic behavior of separation planes (at which the flow velocity changes erratically) plays a major role. Such separation planes can be modeled by "vortex sheets" that obey the laws of vortex induction and thus provide a simple model for understanding the separation plane dynamics and the development of large-scale vortex structures. If we arrange many identical parallel vortex filaments to form an equidistant chain, as shown in Figure 2.28, then with increasing vortex density, the flow that is directed above the chain to the right and below the chain to the left becomes more and more parallel. If we now let the number of vortices per unit of length approach infinity, while the total vorticity per unit length remains constant, the discrete sheet becomes a steady vortex layer separating a rightward-moving homogeneous parallel flow from a leftward-moving flow so that an unsteady jump in velocity Δu takes place over the vortex layer. Since it is always possible to superimpose a homogeneous parallel flow of any velocity over this unsteady flow, the vortex layer and vortex sheets are in fact models of a separation plane.

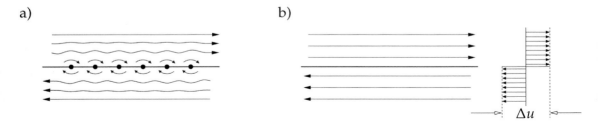

Figure 2.28 Models for separation planes with velocity jump: (a) chain of discrete individual vortices: (b) continuous vortex layer.

2.2.3.3 Dynamic Behavior of Separation Planes

If we initially consider an infinite linear sheet of identical, equidistantly distributed vortices as a model of an infinitely extended separation plane, each set of adjacent vortices always induces a pair of identical, opposite velocities on any given individual vortex so that the chain as a whole remains at rest (Figure 2.29a). However, this balance is unstable against small wave-shaped perturbations of the chain. The induced velocities now lead to an accumulation of vortices on the sloping parts of the wave line (Figure 2.29b). Owing to their decreasing distance, the vortices begin to rotate faster and faster around each other under the influence of their mutual induction so that the chain eventually rolls into a periodic spiral structure, with the accumulated vortices merging increasingly into a single vortex with increasing proximity (Figure 2.29c). Exactly the same dynamic behavior is displayed by a longitudinal separation plane, as produced, for example, by flow separation when flowing over a sharp edge.

An example of separation planes caused by transverse velocity jumps is the vortex system of a wing of finite span, which is formed by the pressure equalization between the top and bottom sides via the wing ends (Figure 2.30). On the bottom, the streamlines tend to move apart to the left and right under the influence of the overpressure, while they are bent together on the top side by the underpressure. At the trailing edge of the wing, a transverse velocity component toward the wing ends forms symmetrically to the wing center on the high-pressure side, while a corresponding component toward the center is present on the suction side. The resulting separation plane may be construed in any plane perpendicular to the wing axis as a finite vortex sheet. Since only upward-directed velocity components are induced to the vortices at the ends of the chain, the chain rolls up at the edges to form two strong wing tip vortices, whose shape remains unchanged over time under steady incident flow. The downward velocities induced by these steady vortex systems change the pressure distribution on the wing surface such that, in addition to a reduction of the lift, "induced" drag [713] occurs, which contributes substantially to the total drag of the aircraft.

The Physical Principles of Aerodynamics

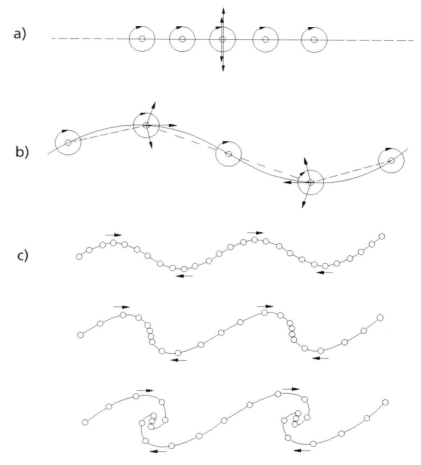

Figure 2.29 Dynamic behavior of a chain of equidistantly distributed single vortices of identical circulation: (a) linear chain, unstable balance; (b) wave-shaped chain, unstable; (c) rolling of the wave-shaped chain to a regular spiral structure.

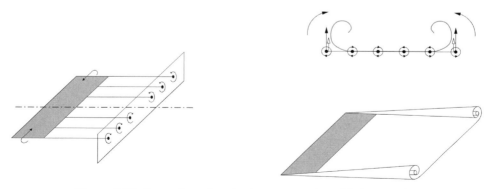

Figure 2.30 Formation of vortex bags on a wing of finite span by rolling up the transversal separation layer.

As Küchemann and Weber [472] have shown, such steady vortex structures can arise whenever the unsteady rolling process of a separation layer in the plane is made into a steady process by superimposing a perpendicular velocity component, whereby the third spatial dimension assumes, to a certain extent, the role of time. Steady wing tip vortices are therefore always observed when a fluid flows around a separation edge at an angle such that one velocity component occurs parallel to the edge. A very well studied example in aircraft aerodynamics is the delta wing (Figure 2.31a). At large angles of attack, flow separation at its leading edge causes the development of two strong steady wing tip vortices, which remain stable at angles of attack of up to approximately 30°. At larger angles of attack, the vortex structures become unsteady, which is also referred to as vortex breakdown [386].

Owing to strong underpressures in the vortex center, steady wing tip vortices can develop a considerable force that contributes significantly to the total force exerted on the body. Figure 2.31a shows the pressure distribution on a section of a delta wing in which the two suction peaks caused by the wing tip vortices are clearly visible; they contribute significantly to the lift of the wing over a wide range of angles of attack.

The aerodynamics of vehicles is equally strongly influenced by steady wing tip vortices, as they can develop, for example, in the flow around the fastback (Figure 2.31b) of a vehicle (sections 4.5.1.2.2 and 4.5.1.3). As can be seen from the pressure distribution of the rear flow, considerable suction pressure peaks also occur in these cases, which have a negative impact on the drag balance of the vehicle [372].

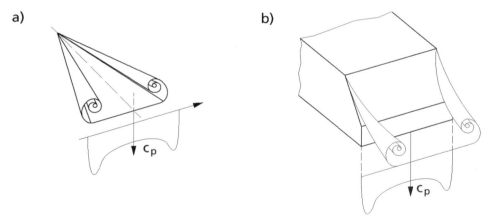

Figure 2.31 Generation of negative pressure peaks by steady wing tip vortices: a) on the delta wing; b) on the fastback of a vehicle.

The Physical Principles of Aerodynamics

2.3 The Dynamics of the Frictional Flow

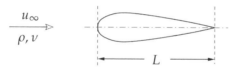

Figure 2.32 Incident flow of a body.

2.3.1 The Reynolds Number

Aerodynamic problems are usually characterized by the fact that a fluid flows against a body of a typical dimension L (e.g., length) at a free stream velocity u_∞, where the fluid is sufficiently characterized in incompressible flow by the density ρ and the kinematic viscosity ν (Figure 2.32). It is therefore useful to nondimensionalize the Navier-Stokes equation using these parameters by normalizing the spatial coordinates with L, the time with L/u_∞, the velocity with u_∞, and the pressure with the stagnation pressure $\rho u_\infty^2 / 2$. If we denote the dimensionless quantities with an apostrophe, we initially obtain the following by substitution:

$$\frac{u_\infty}{L/u_\infty} \frac{D\mathbf{v}'}{Dt'} = -\frac{1}{\rho} \frac{\frac{1}{2}\rho u_\infty^2}{L} \nabla' p' + \nu \frac{u_\infty}{L^2} \Delta' \mathbf{v}', \tag{2.85}$$

which yields, after multiplication with L/u_∞^2, the dimensionless form of the Navier-Stokes equation

$$\frac{D\mathbf{v}'}{Dt'} = -\frac{1}{2}\nabla' p' + \frac{1}{\mathrm{Re}}\Delta' \mathbf{v}' \tag{2.86}$$

with the dimensionless parameter

$$\mathrm{Re} = \frac{u_\infty^2/L}{\nu u_\infty/L^2} = \frac{u_\infty L}{\nu} \tag{2.87}$$

which is referred to as the Reynolds number and describes the physical relation between inertia (u_∞^2/L) to frictional terms ($\nu u_\infty/L^2$). In aerodynamics, owing to the low kinematic viscosity ν of air ($\nu_L = 1.5 \cdot 10^{-5}$ m²/s at 20 °C), we generally deal with very large Reynolds numbers (Re = $10^5 - 10^8$), where the inertia terms play a dominant role.[9]

As the Reynolds number is the only parameter in the dimensionless Navier-Stokes equation (2.86), derived integral parameters, such as the drag F_D of the body in flow, can only be a function of Re alone. Hence, they must themselves have a dimensionless representation that is based on the same normalization factors. As can be clearly seen, we can nondimensionalize the drag F_D using the dynamic pressure and a surface, where we,

---

9. For a car (L = 5 m) traveling at a speed, for example, of u_∞ = 200 km/h, the Reynolds number is Re = $2 \cdot 10^7$.

instead of the obvious factor L^2, conventionally select the front face[10] A_s of the body in flow, which is in a fixed numerical proportion to L^2 for a given body. The drag coefficient c_D defined in this manner then depends only on the Reynolds number:

$$c_D := \frac{F_D}{\frac{\rho}{2} u_\infty^2 A_s} = f(\text{Re}), \qquad (2.88)$$

that is, for all incident flow velocities u_∞, body lengths L and material properties ρ, ν, the values of c_D lie on a single, universal curve that is typical for each body.

Such dimensionless relationships, which can be obtained directly even without knowledge of equations but with the aid of dimensional analysis (cf. section 4.1), are key to conducting experiments on scaled models, which are often employed when measurements on the original body are not possible for dimensional or cost reasons. It is then sufficient, as per (2.88), to ensure that the Reynolds number of the model coincides with that of the original body (or is at least sufficiently close) to be able, for example, to convert the drag measured on the model to the drag of the original by means of the c_D value.

There are several ways to achieve the correct value of the Reynolds number. For example, one could increase the incident flow velocity u_∞ for a scaled-down model (reduced L), although compressibility effects that occur at high speeds present an upper limit. The other alternative is to reduce ν, for which one can either use a different flow medium (e.g., water with $\nu_{H_2O}/\nu_L \approx 1/15$) or utilize the dependency of the kinematic viscosity of air on pressure and temperature. The latter option is used in cryogenic and high-pressure wind tunnels.

2.3.2 The Prandtl Boundary Layer Concept

In contrast to the Euler equations, the Navier-Stokes equations are difficult to access using analytical solution methods, and only a few closed-form solutions are known [886]. However, given the typically large Reynolds numbers in aerodynamics and the therefore small factor $1/\text{Re}$ in the dimensionless Navier-Stokes equations (2.86), the natural question arises whether simplification possibilities exist for the limiting case $\text{Re} \to \infty$ for which friction effects may not be completely ignored.

One of the oldest known exact solutions of the Navier-Stokes equation is the problem of an infinite plate suddenly set in motion at time $t = 0$ and instantaneously accelerated to the constant velocity $-u_\infty$ (Figure 2.33). The resulting velocity profile [886]

$$u(y,t) = u_\infty \left[\text{erf}\left(\frac{y}{2\sqrt{\nu t}}\right) - 1 \right], \qquad (2.89)$$

---

10. The front face A_s is the area of the surface which is generated by parallel projection of the body outline in the flow direction onto a perpendicular wall behind the body (cf. Figure 1.7). For road vehicles the projection in x-direction is used: $A_s = A_x$.

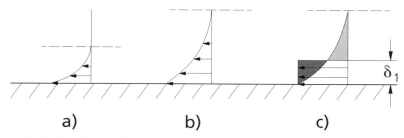

Figure 2.33 Velocity profile on an infinite plate suddenly set in motion: (a) shortly after being set in motion; (b) at a slightly later time; (c) substitute profile to define the displacement thickness.

where erf(x) is the Gaussian integral [296], assumes the value of the plate velocity at $y = 0$ and asymptotically approaches zero for $y \to \infty$, while the thickness of the friction layer increases proportional to $\sqrt{vt}$. An exact definition of the thickness of this friction layer, also known according to Prandtl as the "boundary layer," does not result from the solution itself. However, it is useful to define this "displacement thickness" such that the total volumetric flow $\int_0^\infty u\, dy$ induced by the plate corresponds to the fictitious volumetric flow $-u_\infty \cdot \delta_1$ of a plate thickened by δ_1:[11]

$$\delta_1 = -\int_0^\infty \frac{u}{u_\infty} dy, \tag{2.90}$$

from which, by substituting the velocity profile (2.89), we obtain the boundary layer thickness of the infinite plate suddenly set in motion as

$$\delta_1 = \frac{2}{\sqrt{\pi}}\sqrt{vt} = 1.128\sqrt{vt}. \tag{2.91}$$

If we now theoretically replace the infinitely extended plate with a very thin finite plate of length L, there will be a fixed location x where the boundary layer thickness can be related to the local "effective time" $T = x/u_\infty$ of the friction, which corresponds to the time before the leading edge of the plate has passed the location x (Figure 2.34). Ignoring the gradients occurring in the x-direction, we can gain an initial estimate of the boundary layer thickness by substituting the time T with x/u_∞ in (2.0), while ignoring the insignificant numerical factor:

$$\delta_1(x) \approx \sqrt{\frac{vx}{u_\infty}}, \tag{2.92}$$

resulting in a boundary layer thickness that grows with the square root of the run length x.

---

11. In boundary layer theory, other definitions of a boundary layer thickness are in use that, instead of being based on the equivalent mass flow, are based on the momentum flow (momentum loss thickness) or the flow of kinetic energy (energy loss thickness). More information can be found in [715] and [716].

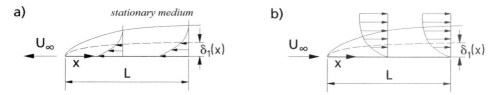

Figure 2.34 Boundary layer on a thin plate of finite length that is towed through a quiescent medium: a) in the laboratory-fixed coordinate system, b) in the fixed-plate coordinate system.

With the help of (2.92), it is very easy to obtain, in terms of magnitude, the relative boundary layer thickness at the end of the plate $x = L$ as a function of the Reynolds number. If we define

$$\varepsilon := \frac{\delta_1(L)}{L} \approx \sqrt{\frac{\nu}{u_\infty L}} = \frac{1}{\sqrt{\text{Re}}}, \tag{2.93}$$

even for Re = 10^4, we will obtain a boundary layer thickness of only $\varepsilon = 1\%$ of the plate length. For typical applications in aerodynamics in the Reynolds number range between 10^5 and 10^8, we can therefore assume, in fact, extremely thin boundary layers in which the essential friction processes take place.

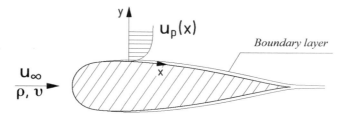

Figure 2.35 The Prandtl boundary layer concept.

This finding leads directly to the Prandtl boundary layer concept (Figure 2.35) in which the boundary layer encloses a body in flow like a thin skin. Outside the boundary layer, the flow behaves as inviscid and can therefore be calculated using potential theory by ignoring the boundary layer thickness. This is not a contradiction to the Navier-Stokes equation since the friction term $\Delta \mathbf{v}$ vanishes identically for all potential flows ($\mathbf{v} = \nabla \phi$ with $\Delta \phi = 0$) because of the linearity of the differential operators involved.

The thin boundary layer itself can be treated as a flat plate flow, on which the pressure and velocity distribution determined by the body contour is exerted from the outside. Since the two variables are coupled by the Bernoulli equation, the calculation of frictional flow is therefore split into determining the potential-theory velocity $u_p(x)$ at the body surface and the subsequent calculation of a plate flow using this velocity distribution for $y \to 0$, which is a significant simplification of the original problem.

The Physical Principles of Aerodynamics

In addition, the assumption $\varepsilon \ll 1$ provides further simplifications of the Navier-Stokes equation (2.43) within the boundary layer. For the most important case of the two-dimensional steady flow, we obtain, as the only remaining component of the boundary layer equation ([715], [716]):

$$u\frac{\partial u}{\partial x} + v\frac{\partial u}{\partial y} = u_P \frac{du_P}{dx} + v\frac{\partial^2 u}{\partial y^2}, \qquad (2.94)$$

which, together with the continuity equation $\partial u/\partial x$, builds a system of two coupled partial differential equations for determining u and v. If we want to calculate the displacement thickness δ_1 corresponding to the velocity profile $u(x,y)$ calculated using (2.94), we have to note that this equation system is formulated for the plate-fixed coordinate system. This is why, in the relationship (2.90) which applies only to a moving plate, we have to substitute the velocity $-u$ with $u_P - u$:

$$\delta_1 = \int_0^\infty \left(1 - \frac{u}{u_P}\right) dy. \qquad (2.95)$$

For the flat plate with $u_P = u_\infty = $ const., the solution of (2.94) was first calculated by Blasius [715]. It is one of the few cases in which the drag of a body in frictional flow can be calculated in a closed form (section 2.3.5.1). Substituting the calculated velocity profile into (2.95), we obtain for the displacement thickness of the plate boundary layer

$$\delta_1 = 1.721 \sqrt{\frac{vx}{u_\infty}}, \qquad (2.96)$$

which, except for the numerical factor, matches the estimation (2.92) and thus justifies a posteriori the estimated magnitude (2.94) underlying the derivation of (2.93).

2.3.3 Boundary Layer Separation

The calculation of frictional flow using the boundary layer concept assumes that the boundary layer always adheres to the body like a thin skin and does not interact with the surrounding potential flow. However, this applies only to slender bodies such as thin airfoils or bodies of revolution, where the potential flow methods are used with great success. However, for "bluff bodies," the thickness of which can no longer be considered to be small in comparison to their length, there is a strong interaction between the boundary layer and the potential flow, which is known as flow separation. Although the boundary layer concept cannot be used for flow calculation in this case, it is still of great practical value and thereby forms the basis for physical understanding of the phenomena involved.

If we consider, as another thought experiment, instead of a plate, the sudden motion of a cylinder as a model for a bluff body, the sudden movement from rest initially produces the potential-theory pressure distribution (2.76), which is characterized by a strong pressure rise in the rear region (Figure 2.36). Owing to the initial pressure drop that occurs, a fluid particle near the wall is first accelerated from the front stagnation point A, reaching twice the incident flow velocity at the minimum pressure at B, to be decelerated again to rest by the subsequent pressure rise in the rear stagnation point C. Since the sum of kinetic energy and potential energy in the motion remains constant, this process corresponds to the inviscid motion of a sphere rolling from rest into a trough, which reaches its maximum speed at the lowest point and reaches the opposite edge of the trough again at the speed zero.

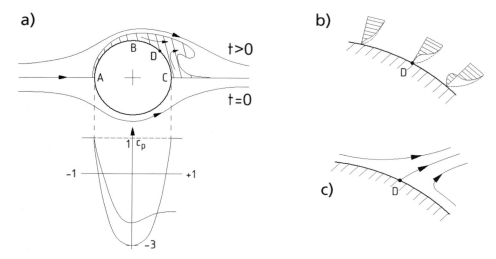

Figure 2.36 Flow separation at a cylinder suddenly set in motion: a) temporal development of flow pattern and pressure distribution; b) near-wall velocity profiles; c) near-wall streamline pattern in the vicinity of the separation point D.

The boundary layer that develops after the start of motion progressively removes energy from the fluid particles by friction, so that the kinetic energy gained between A and B is no longer sufficient to reach the rear stagnation point against the pressure rise between B and C. Rather, they come to rest at an intermediate point D, and reverse flow occurs behind D. This forms a stagnation point at D, from which the decelerated boundary layer material pushes out into the outer flow, forcing the flow away from the body. The change in the streamline pattern causes the pressure distribution to lose its initial symmetry. On the rear of the body, the flow can no longer realize the potential-theory pressure pattern, and in the steady final state the space behind the body is filled by decelerated material (dead wake) in which underpressure typically prevails. This lack of pressure recovery overall leads to the occurrence of considerable pressure drag. A separated layer of fluid has formed between the outer flow and dead wake which, according to the principles described in section 2.2.3.3, rolls up into vortices that float away with the flow behind the body. The extent of dead wake is finite due to the resulting suction; it merges downstream behind the body, mixes with the ambient medium, and then transitions to the wake flow of the body.

Through the separation mechanism, the boundary layer can have a significant impact on the global flow structure despite its initial restriction to a small flow region. Depending on the size of the dead wake region, the pressure drag it causes often exceeds the frictional resistance considerably, such that the drag coefficient of a bluff body may well be higher by one or two orders of magnitude than that of a slender body of the same flow-facing area.

As a rule of thumb, we can assume that flow separation always occurs where, following potential theory, a steep pressure rise is expected in the vicinity of the wall. Since in potential theory the contour of a body is always also the streamline, it is often possible to qualitatively determine areas prone to flow separation from the body shape alone based on the rules in section 2.2.1.

Thus, for example, it is immediately clear why flow separation occurs, even at very low speeds, in the flow around a sharp body edge behind which the flow would have to run against a theoretically infinitely steep pressure gradient (Figure 2.37a). As the examples in section 4.5.1.1 demonstrate, however, this effect can be prevented by slightly rounding the edge in flow, which often has positive effects on the drag of the body. However, the effect can also be taken advantage of in the form of separation edges, specifically to force the separation of flow if adherence of flow would be detrimental. An elementary example of this is the aircraft wing, for which lift-generating circulation would not be possible without flow separation at the sharp trailing edge (section 2.2.2.3.6).

For bodies without sharp edges, the location of separation is essentially determined by the body shape, where slender bodies, for which the pressure rise calculated from potential theory at the rear stagnation point occurs very late, are clearly at an advantage over more blunt body shapes. Figure 2.37b shows an example of two ellipses with different slenderness ratios. While the flow separates shortly after the maximum thickness on the thick ellipse, it adheres longer to the slender ellipse. Owing to the smaller dead wake extension of the slender body, the lack of pressure recovery caused by flow separation therefore has a significantly lower impact on the drag than is the case with the thick body. The advantage is even more pronounced for a teardrop-shaped wing profile as in Figure 2.25, which further minimizes the rear dead-wake area thanks to a sharp trailing edge, which is why such body shapes provide by far the lowest drag coefficients (section 2.3.5.4).

Separation may also occur in the stagnation area of a body if the pressure rise in the incident flow is along a wall surface (Figure 2.37c). On a cylindrical obstacle placed on the ground, the resulting stagnation point vortex propagates to the sides as a horseshoe vortex, closing around the cylinder like a collar. This complies with the Helmholtz vortex theorems by which a vortex filament cannot simply end inside the fluid [886]. Horseshoe vortices occur on vehicles, for example, at add-on parts (e.g., exterior mirrors) and on the cowl where the flat engine hood transitions to the steeper windshield (section 4.5.1.1).

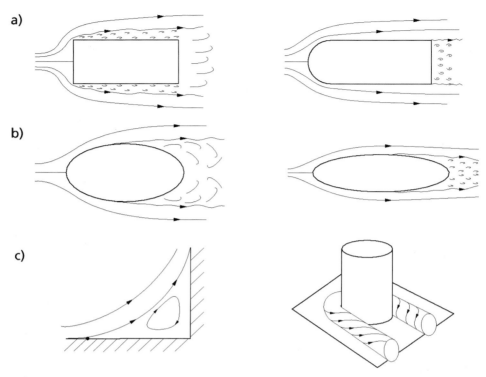

Figure 2.37 Flow separation: a) at sharp edges; b) at bluff bodies; c) in the stagnation area of a body mounted on a wall.

2.3.4 Boundary Layer Turbulence

Turbulence is defined as the transition of a flow from flowing in an orderly steady state referred to as laminar flow to an unstable state in which high-frequency chaotic fluctuations, due to the dynamics of small vortices, are superimposed on all flow parameters. Although the Navier-Stokes equations, in principle, also describe this type of flow, its direct numerical solution, at least for industrially relevant turbulent flows, is beyond the capacity of even the most powerful computers today, so we have to rely on empirical models of turbulence.

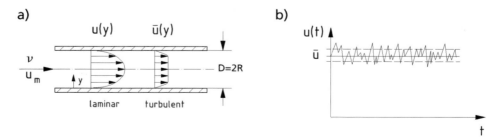

Figure 2.38 Turbulent flow in a pipe: a) geometry and velocity profiles; b) typical time signal with fixed-location velocity measurement.

The historically oldest example of a systematic study providing an understanding of many important laws of turbulent flows is by Reynolds, who examined the turbulence in a circular cylindrical pipe of cross-sectional area $A = \pi D^2/4$ with a constant internal volumetric flow $\dot{V}$ (Figure 2.38a). Calculating the Reynolds number using the pipe diameter D and the volumetrically averaged velocity $u_m = \dot{V}/A$, the "critical Reynolds number" [715] was found by experiment to be

$$\mathrm{Re}_{\mathrm{crit}} \approx 2300, \tag{2.97}$$

at which point the pipe flow transitions from the laminar to the turbulent state. Figure 2.38b shows a typical time signal of a fixed-location measurement of the velocity u, which is composed of an average velocity $\bar{u}$, constant over time, and a stochastic component u' whose RMS value can be used to define a dimensionless degree of turbulence

$$Tu = \frac{\sqrt{(u')^2}}{\bar{u}} \tag{2.98}$$

which gives the ratio of the root mean square fluctuation amplitude to the average velocity. Since the turbulent velocity fluctuations occur in all three spatial directions, they lead to violent mixing motions normal to the pipe axis, which tend to smooth the average velocity profile over the cross section. Instead of the parabolic velocity profile of the laminar pipe flow, this yields a much fuller profile, which in the medium range of Reynolds numbers up to approximately $\mathrm{Re} = 10^5$ can well be described by the power law [715]

$$\bar{u}(y) = \bar{u}_{\max} \left(\frac{y}{D/2} \right)^{\frac{1}{7}} \tag{2.99}$$

where y is the distance from the wall.[12] The steeper velocity gradients at the wall compared to laminar flow result in a significantly higher wall shear stress, which manifests itself in an increased pressure drop in a pipe with turbulent internal flow. In addition, at high Reynolds numbers, the turbulent flow becomes increasingly sensitive to wall roughness, which significantly increases the pressure drop compared to a smooth pipe. The laminar flow, by contrast, remains completely unaffected by wall roughness.

---

12. For higher Reynolds numbers, the semi-empirical logarithmic wall law, which is based on theoretical considerations and therefore has a universal character, is a better approximation (cf. Schlichting and Gersten [716]).

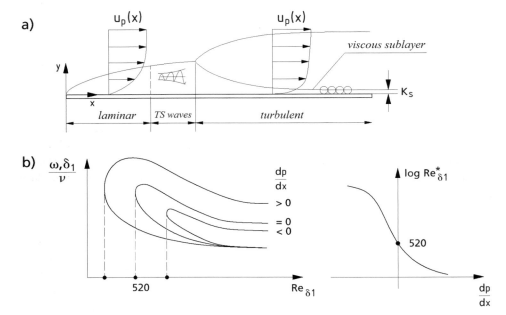

Figure 2.39 Boundary layer on a plate with pressure gradient: a) schematic structure; b) stability diagram of laminar flow; c) schematic curve of the indifference Reynolds number.

Since we can interpret the boundary layer as a flow channel, it is not surprising that we observe the same phenomena as in pipe flow. In contrast to the pipe flow, however, the Reynolds number is not constant in a boundary layer but increases with the axial distance x due to increasing boundary layer thickness. We can use this analogy to compare pipe flow and boundary layer flow. The comparison is restricted, however, to boundary layers without a pressure gradient because of the constant velocity in the middle of the pipe. Calculating the local Reynolds number with the displacement thickness δ_1, using (2.96) we obtain:

$$\mathrm{Re}_{\delta_1} = \frac{u_\infty \delta_1}{\nu} = 1.721\sqrt{\frac{u_\infty x}{\nu}} = 1.721\sqrt{\mathrm{Re}_x} , \qquad (2.100)$$

which can be used to establish a direct relationship to the pipe flow by equating Re_{δ_1} with the Reynolds number obtained with the pipe radius $\mathrm{Re}_R = \mathrm{Re}_d/2$. According to this approach, the sudden change from laminar to turbulent flow of the boundary layer flow would always occur at the constant value

$$\mathrm{Re}_{x,\mathrm{crit}} = \left(\frac{\mathrm{Re}_{\delta_1,\mathrm{crit}}}{1.721}\right)^2 \triangleq \left(\frac{\mathrm{Re}_{R,\mathrm{crit}}}{1.721}\right)^2 = \left(\frac{1150}{1.721}\right)^2 = 4.5 \cdot 10^5 \qquad (2.101)$$

Experiments show that the transition occurs at

$$\mathrm{Re}_{x,\mathrm{crit}} = 5 \cdot 10^5 , \qquad (2.102)$$

which is a surprisingly good level of agreement in view of the quite different geometries [715]. In fact, the two turbulent flow patterns show such a large degree of similarity that it is even possible to convert measured pressure drop coefficients of the pipe flow directly into drag coefficients of a plate at zero incidence (cf., Schlichting [715] and Schlichting and Gersten [716]).

The principal structure of a plate boundary layer with transition from laminar to turbulent flow is shown schematically in Figure 2.39. Starting from the plate leading edge, the boundary layer flow is initially laminar; however, with increasing boundary layer thickness it increasingly becomes unstable against perturbations, whereby the start of the instability depends on the pressure gradient $\partial p/\partial x$ which is exerted on the boundary layer by the inviscid outer flow. After the transition to turbulence, the boundary layer thickness rapidly increases and the velocity profile becomes more uniform as a result of the turbulent momentum exchange. In the immediate vicinity of the wall, however, the turbulent transverse fluctuations cannot develop, leaving a viscous sublayer in which the viscosity forces dominate. This also explains the dependency of the turbulent flow on the wall roughness observed in tubes: if roughness elements, which can be thought of, for example, as grains of sand, remain within the viscous sublayer, the turbulent boundary layer remains unaffected, but if they extend beyond the sublayer, they can interact with the turbulent flow and affect its structure significantly.

The diagram in Figure 2.39b schematically shows the stability of the laminar boundary layer flow for pressure rise, constant pressure, and pressure drop as a function of the Reynolds number Re_{δ_1}. The stability curves enclose the areas in which the boundary layer is unstable against small perturbations, which, for example, can be introduced into the boundary layer by an object oscillating harmonically at a frequency ω. As can be seen, there is a Reynolds number $Re_{\delta_1}^*$ marking the "indifference point" (point of neutral stability) for each pressure gradient above which the boundary layer is potentially unstable, so that random perturbations are magnified to Tollmien-Schlichting waves, which shortly afterwards cause transition to turbulence [715], [716].

In the case of a plate at zero incidence without pressure gradient ($\partial p/\partial x = 0$), the "indifference" Reynolds number is $Re_{\delta_1}^* = 520$, which, according to (2.100), corresponds to $Re_x^* = 9 \cdot 10^4$ The difference to $Re_{x,crit} = 5 \cdot 10^5$ corresponds to the axial distance required to transition the resulting Tollmien-Schlichting waves into turbulence. With negative pressure gradient ($\partial p/\partial x < 0$), the indifference point moves downstream and the region of instability becomes narrower, such that an accelerated boundary layer is always stable over a longer distance than a constant pressure boundary layer (Figure 2.39c). If the boundary layer, on the other hand, runs up against a positive pressure gradient ($\partial p/\partial x > 0$), the indifference point moves upstream and the region of instability increases significantly. A boundary layer with positive pressure gradient is thus very much less stable than a constant-pressure boundary layer, such that transition to turbulence can always be expected in an area of increasing pressure.

Whether the boundary layer does in fact become unstable over the entire run length L of the plate depends on whether the indifference Reynolds number $\text{Re}_{\delta_1}^*$ or Re_x^* is reached for $x < L$. Using (2.100) we can estimate the relationship between Re_{δ_1} and the Reynolds number Re_L, obtained using the plate length:

$$\text{Re}_{\delta_1} = 1.721\sqrt{\text{Re}_x} = 1.721\sqrt{\text{Re}_L} \cdot \sqrt{\frac{x}{L}}, \qquad (2.103)$$

that is, at sufficiently low Reynolds numbers Re_L, Re_{δ_1} remains below the indifference threshold and there will be no transition, while with increasing Re_L the indifference point is reached increasingly earlier, thus moving the transition point closer toward the plate leading edge.

The pressure gradient imposed on the boundary layer by the outer flow therefore not only determines the process of flow separation but also the transition of the boundary layer from the laminar to the turbulent state. Three possible scenarios arise here:

- At low Reynolds numbers, the laminar separation point is located usually far upstream of the indifference point (i.e., the flow separates in a laminar fashion). Turbulence can at best occur in the separated boundary layer but not at the body surface.

- At sufficiently high Reynolds numbers, the laminar separation point is far downstream of the indifference point and the flow transitions from laminar into turbulence before separation. Owing to the fuller velocity profile of the turbulent flow, the boundary layer has more kinetic energy than in the laminar state and can therefore withstand the positive pressure gradient for a significantly longer distance. Turbulent separation occurs therefore later, such that the surface pressure distribution approximates the ideal state defined by potential-theory more closely (i.e., the pressure drag decreases). Depending on the body shape, this effect may more than compensate for the increase in the frictional resistance, such that the total drag is significantly reduced by the effect of turbulence. This positive effect of turbulence, which can be found, for example, in the sphere (section 2.3.5.3), is utilized for the industrial design of the shape of bodies with low drag (section 2.3.5.4).

- If the laminar separation point and the indifference point are close together, this may lead to laminar separation with subsequent turbulent reattachment of the flow, which is referred to as a separation bubble (Figure 2.40). This occurs, for example, in model airplane or glider profiles, where an attempt is made to avoid it by the selective use of turbulence generators. A similar situation occurs in the flow around sharp edges where, due to the very steep pressure gradient, separation and turbulence start simultaneously, which can also cause turbulent reattachment.

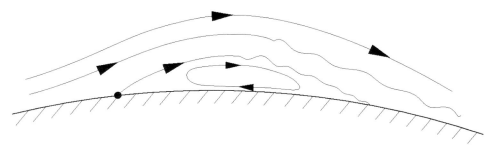

Figure 2.40 Formation of a separation bubble by laminar separation and turbulent reattachment.

2.3.5 Drag of Simple Bodies
2.3.5.1 Plate at Zero Incidence

A plate at zero incidence is the simplest case of a body with pure frictional resistance and is a characteristic model of the drag of slender bodies (Figure 2.41). It is historically the oldest case of a body for which drag can be calculated theoretically in the case of laminar flow by using the Prandtl boundary layer concept [715], [716]. The calculation, first carried out by Blasius, supplies the following simple expression for the drag coefficient c_f:

$$c_f = \frac{1.328}{\sqrt{Re}}, \tag{2.104}$$

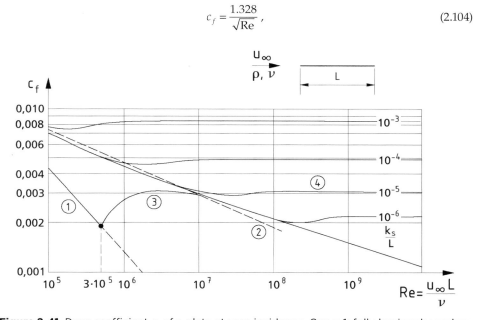

Figure 2.41 Drag coefficient c_f of a plate at zero incidence. Curve 1: fully laminar boundary layer, equation (2.104); curve 2: fully turbulent boundary layer, equation (2.106); curve 3: turbulent boundary layer with laminar entrance length, equation (2.108). Asymptotic values of the roughness curves according to equation (2.110). Source: Schlichting [715].

where Re is the Reynolds number calculated using the plate length L and, by convention, the wetted surface area is used as a normalization area, so that the same drag coefficient is found for plates wetted on one and two sides.[13] The displacement thickness δ_1, after simple transformation of (2.96), can be written in the following form:

$$\frac{\delta_1}{L} = 1.721 \, \mathrm{Re}^{-\frac{1}{2}} \left(\frac{x}{L}\right)^{\frac{1}{2}}. \tag{2.105}$$

For the case of fully turbulent flow, which currently cannot be handled theoretically, there is at least the possibility of conversion between pipe flow and plate flow. Based on an empirical relationship of the pressure drop coefficient of circular cylindrical tubes identified by Blasius (cf. Section 2.3.7.5.1), Prandtl provided the following simple equation for the drag coefficient c_f: [715]

$$c_f = \frac{0.074}{\mathrm{Re}^{1/5}} \tag{2.106}$$

From the same relationship on which the velocity law (2.99) is based, we can determine a form for the displacement thickness [715] that is analogous to (2.105):

$$\frac{\delta_1}{L} = 0.046 \, \mathrm{Re}^{-\frac{1}{5}} \left(\frac{x}{L}\right)^{\frac{4}{5}}, \tag{2.107}$$

which shows that the turbulent boundary layer thickness increases much faster than in the case of laminar flow because of the increased mixing of momentum. Equations (2.106) and (2.107) agree with measured values very well up to $\mathrm{Re} = 5 \cdot 10^7$, which is sufficient for most applications.[14] As can be inferred from Figure 2.41, the turbulent c_f values according to (2.106) are considerably larger than those according to (2.104) for the laminar flow, which is due to the increased wall shear stress caused by turbulent mixing.

Prandtl also provides a formula for the transition region of c_f that includes the laminar entrance length of a plate in a partially turbulent flow. A fixed transition Reynolds number of $\mathrm{Re}_{x,\mathrm{crit}} = 5 \cdot 15^7$ therefore yields the simple relationship

$$c_f = \frac{0.074}{\mathrm{Re}^{1/5}} - \frac{1700}{\mathrm{Re}}, \tag{2.108}$$

which covers the Reynolds number range from $\mathrm{Re} = 5 \cdot 10^5$ to $\mathrm{Re} = 5 \cdot 10^7$ in good agreement with experiments. If the incident flow is already turbulent, as is the case in many technical applications, the transition Reynolds number shifts to lower values.

Figure 2.41 also displays curves for different sand roughnesses of the relative grain size k_s/L, which deviate from the curve for the smooth plate. As the grains extend outside of the sublayer, which becomes thinner as Re increases, the curves quickly lead into a horizontal pattern. The independence of the drag coefficient of the Reynolds number in

---

13. To clarify this difference compared to the usual convention, the symbol c_f (friction) is used instead of the usual symbol c_D.
14. Detailed relationships for the range of higher Reynolds numbers can be found in Schlichting and Gersten [716].

the limiting case of fully developed roughness flow is an indication that the transition is driven by pressure drag occurring at the roughness elements and that viscous effects are no longer relevant. It is striking that the transition from the smooth to the rough curve always begins when the product of Reynolds number and relative roughness height approximately approaches the value 10^2. As a criterion for a hydraulically smooth plate, which as yet exhibits no roughness influence, we are therefore able to apply the empirical criterion

$$\mathrm{Re} \cdot \frac{k_s}{L} = \frac{u_\infty k_s}{\nu} \leq 100 \qquad (2.109)$$

according to which the Reynolds number formed using the roughness height may not exceed 100 [715]. For Re = 10^7 and a plate length of 5 m, this results in a maximum permissible roughness height of k_s = 0.05 mm, which is certainly technically feasible. For the part of the roughness curves that are independent of the Reynolds number, Kármán has provided a simple relationship for the drag coefficient [715]:

$$c_f = \left[1.89 - 1.82 \log \frac{k_s}{L} \right]^{-2.5}, \qquad (2.110)$$

where, in the case of industrial roughnesses that, in contrast to sand roughness, tend to have a statistical particle size distribution, a suitable mean value must be entered for k_s.

The drag characteristic of a plate at zero incidence demonstrates how complex the physics of frictional resistance is, even for such a basic example. The additional influence of a pressure gradient is discussed using the example of a sphere (section 2.3.5.3) and a streamlined body (section 2.3.5.4).

2.3.5.2 Plate in Cross-Flow

By turning a plate at zero incidence by 90°, we obtain a plate in cross-flow and hence the simplest case of a body with pure pressure resistance, which serves as a model for a blunt body. This simple example clearly shows that the answer to the question whether a body is to be regarded as slender or blunt not only depends on its shape but also on its orientation to the incident flow.

Figure 2.42 shows the drag curve for a circular disk, for which, in contrast to a plate at zero incidence, turbulence or roughness effects play no significant role [350]. For very small Reynolds numbers, the curve initially decreases. Since the separation point is geometrically fixed, the curve runs rapidly to the limit value $c_{D\infty}$ = 1.11 and maintains the limit value even for large Reynolds numbers such that the frictional drag, except for the industrially irrelevant case of creeping flow (Re → 0), is characterized by a single number. This extremely simple behavior is typical of many blunt bodies with defined separation edges and pronounced dead wake. Corresponding limit values $c_{D\infty}$ for numerous bodies are tabulated in manuals (e.g., Hoerner [350]). A similar drag characteristic is displayed by many vehicle designs, which can also be construed as blunt bodies (cf., section 4.7).

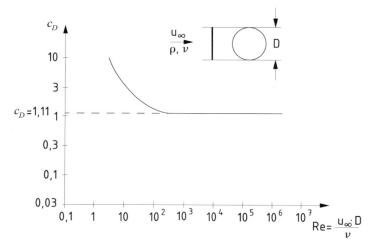

Figure 2.42 Drag coefficient of a circular disk in cross-flow. Source: Hoerner [350].

The large drag coefficient indicates that considerable underpressure must prevail in the dead wake behind the plate. However, according to Hucho [372], this is not constant in the flow direction but reaches its highest value just behind the plate and then decreases asymptotically to zero further downstream as transition from the dead wake to the wake flow occurs.

According to Table 2.1, for rectangular plates (side lengths a, b), the values for $c_{D\infty}$ increase with increasing aspect ratio a/b and, for the limiting case $a/b \to \infty$ (flat plate), attains the value $c_{D\infty} = 2.01$, which is one of the highest realizable drag coefficients [350]:

Table 2.1 Drag Coefficients $c_{D\infty}$ of Rectangular Plates (Aspect Ratio a/b) According to Hoerner [350]					
a/b	1	2	4	10	∞
$c_{D\infty}$	1.10	1.15	1.19	1.29	2.01

The reason for this is the increasingly worse ventilation of the dead wake via the side edges of the plate as the aspect ratio increases, which causes an ever higher suction in the dead wake. This effect is exploited industrially, for example, in the front spoiler of a car, which reduces the effective inlet flow to the rough underbody of the vehicle by its blockage effect, thereby decreasing its drag while creating a downward force onto the front axle via its strong backside underpressure (section 4.5.3.1).

2.3.5.3 Flow Around a Sphere

A sphere, defined by its diameter D as the only length dimension, is an elementary example of a blunt body, demonstrating the pressure-controlled interaction between separation and turbulence and its effect on the body's drag. At very low Reynolds numbers, the drag curve (Figure 2.43a) initially follows the analytical solution by Stokes for creeping flow, which has the simple dimensionless form

$$c_D = \frac{24}{\text{Re}} \qquad (2.111)$$

Above Re = 1, laminar separation occurs, in which the separation point is initially at the rear area of the sphere, moving to the equator for increasing Reynolds numbers, whereby the end position is slightly upstream of the equator. Once this position is reached, the c_D value temporarily exhibits almost no change. The drag curve of the sphere in this range is therefore qualitatively similar to that of the circular disk; however, the transition to constant drag takes place much more slowly due to the moving separation point [350].

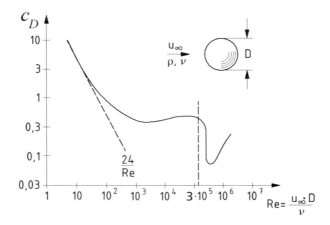

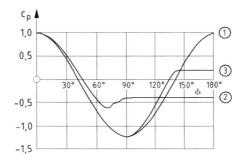

Figure 2.43 a) drag coefficient c_D of a sphere for $\text{Re}_{\text{Crit}} = 3.0 \cdot 10^5$; b) surface pressure distribution. Curve 1: potential theory; curve 2: subcritical, Re = $1.6 \cdot 10^5$; curve 3: supercritical, Re = $1.1 \cdot 10^6$. Source: Schlichting [715].

At a critical Reynolds number, which for a smooth sphere and largely nonturbulent outer flow is around $\text{Re}_{\text{Crit}} \approx 3 \cdot 10^5$, a sudden change occurs. The boundary layer flow abruptly transitions from the laminar to the turbulent state before a laminar separation can occur. Since the more energy-rich turbulent boundary layer can tolerate much greater pressure rises, the turbulent separation point moves backward by almost 40°, which also significantly changes the surface pressure distribution shown in Figure 2.43b. The pressure distribution is now much closer to the pattern given by potential flow theory, and the pressure recovery in the rear region is much larger than in the laminar case, such that the pressure drag decreases significantly. The c_D value falls from 0.47 to 0.09, which corresponds to a drag reduction of more than 80% and is an indication that the increase in frictional resistance is far overcompensated by the decrease in pressure drag. With a further increase in the Reynolds number, the drag increases again slowly, as the turbulent separation point now moves toward the equator.

As Prandtl demonstrated experimentally by placing a thin wire ring ("tripwire") in front of the equator, the drop in drag can be advanced to lower Reynolds numbers by perturbation of the boundary layer [372]. The same effect can also be achieved by extensively distributed sand roughnesses which, however, as in the case of a plate at zero incidence, leads to a significantly increased drag in the supercritical area (Figure 2.44a). Extensively distributed "dimples" are also extremely effective, as used, for example, for a golf ball. At a typical Reynolds number of $1.5 \cdot 10^5$, there is a drag reduction of about 50% according to Figure 2.44.

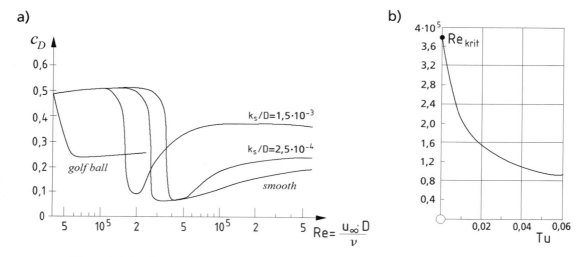

Figure 2.44 Influence of roughness and turbulence on the critical Reynolds number: a) c_w pattern for sand-grain rough surfaces and golf ball; b) critical Reynolds number depending on the degree of turbulence Tu of the outer flow. Source: Hucho [372].

The presence of turbulence in the incident flow also results in a lowering of the critical Reynolds number. As shown in Figure 2.44b, for the nonturbulent case (e.g., free flight through laminar air) a value of $Re_{Crit} = 3.8 \cdot 10^5$ is obtained, which decreases with increasing turbulence intensity and falls to $Re_{Crit} = 1 \cdot 10^5$ at a value of $Tu = 5\%$ [372]. The high sensitivity of Re_{Crit} to Tu is used in wind tunnels to measure the degree of turbulence of the incident flow.

2.3.5.4 Streamlined Body

The shape of a streamlined body is designed exclusively under aerodynamic considerations such that the streamlines follow the pattern predicted by potential flow theory as closely as possible. Hence, the body ideally displays only frictional resistance. The boundary layer effects in this case are controlled by the body shape such that separation is avoided and turbulence is generated only where it has a positive effect (i.e., in the region of pressure rise towards the rear of the body).

Figure 2.45 shows, as an example, a body of this kind with its experimentally and from potential theory obtained pressure distribution according to [715]. It has a rounded nose in the front area, which provides homogeneous acceleration of the boundary layer such that the boundary layer initially remains laminar. Its maximum thickness is formed in a manner that a pronounced pressure minimum occurs with a relatively steep initial pressure rise, causing a reliable boundary layer transition to turbulence at this point (at the latest). Hence, laminar separation or formation of a separation bubble is safely avoided. This is followed by a gentle, largely constant pressure rise toward the rear of the body, which can be easily overcome by the now turbulent boundary layer. Separation and formation of the dead wake therefore occur only at the very end of the body where the slight underpressure of the dead wake, caused by the lack of pressure recovery, is now acting on a very small surface, such that the body displays almost no pressure resistance. In this way, very low c_D values can be achieved, which can be estimated, according to Schlichting and Truckenbrodt [713], for airfoils with the simple empirical formula

$$c_D = 2c_f \left(1 + 2.5 \cdot \frac{d}{l}\right) \qquad (2.112)$$

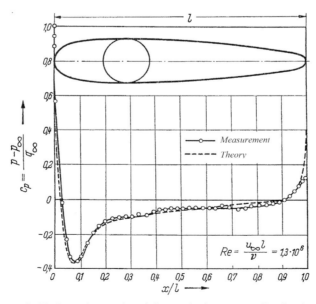

Figure 2.45 Experimental and theoretical pressure distribution on the surface of a streamlined body.

while Hoerner [350] provides the following relationship for slender bodies of revolution:

$$c_D = c_f \left(3\frac{l}{d} + 4.5\left(\frac{d}{l}\right)^{\frac{1}{2}} + 21\left(\frac{d}{l}\right)^2 \right) \qquad (2.113)$$

Here d/l is the ratio of the thickness or diameter to the length of the body and c_f is the turbulent friction coefficient (2.106) of the plate at zero incidence, which shows that the achievable drag coefficients are indeed very low compared to blunt bodies. For the body of rotation in Figure 2.45 with $d/l = 0.2$, for example, (2.113) at Re = 10^7 yields a c_D value of 0.05, which corresponds to a circular disk whose end face is less than one-twentieth of that of the body of revolution.

In addition, by shifting the maximum thickness and thus the minimum pressure back as far as possible, we are able to maximize the laminar entrance length (laminar profiles). There is a limit to this, however, since the pressure gradient in the rear part increases by the process and can become so large that it may no longer be overcome even by the turbulent boundary layer, resulting in separation. Hence, there is an optimum location for the maximum thickness that has to be found via suitable design strategies. The example of a laminar profile NACA 634-021, which is shown in Figure 2.46 with a cylinder of the same drag [716], demonstrates how drag coefficients can be minimized using this method. A glider airfoil with $d = 20$ cm thus has the same drag as a wire of only 1.2 mm diameter!

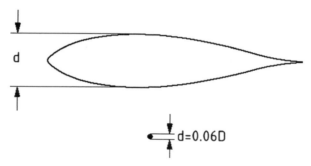

Figure 2.46 To-scale comparison of the laminar profile NACA 634-021 with a cylinder of the same drag. Source: [716].

While aerodynamics is the "supreme discipline" in aircraft engineering and takes precedence over most other requirements, this does not apply in other applications such as automotive engineering. In automotive engineering, there are numerous other constraints of a technical and nontechnical nature, which are often more important than the demands of aerodynamics and thus make it impossible to realize the ideal streamlined shape. The vast majority of vehicle shapes are therefore blunt bodies with significantly higher drag coefficients. Nevertheless, many of the physical findings from the streamlined body can be applied to achieve targeted shape and detail optimization, as demonstrated in detail in section 4.5.

The Physical Principles of Aerodynamics

2.3.6 Multi-Body Systems

If multiple bodies are in the same flow, their pressure and velocity fields may interact with each other, influencing the drag of each body positively or negatively. Examples in vehicle aerodynamics include the ground effect, effects occurring when add-on parts are attached or while driving in a convoy or during overtaking. The interaction is determined via the interference drag,

$$\Delta W = W_{1+2} - (W_1 + W_2), \qquad (2.114)$$

which describes the difference between the drag W_{1+2} of the combined arrangement and the drags W_1 and W_2 of each individual body in flow. Depending on the nature of the interaction, the interference drag can be positive, negative, or zero. Slender and blunt bodies behave very differently in this respect, which will be illustrated via some basic examples. Extensive discussion of this topic can be found in Hoerner [350] and Hucho [372].

2.3.6.1 Streamlined Bodies Side by Side

If two streamlined bodies are located side by side (Figure 2.47), the resulting flow can be construed as a simple model of the ground effect, since this arrangement represents an experimental realization of the principle behind the "method of images" (section 2.2.2.3.2). Therefore, it also approximates the mutual interaction of vehicles during overtaking when both vehicles are next to each other.

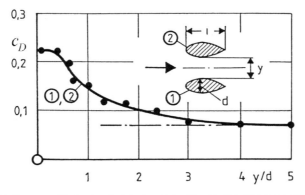

Figure 2.47 Drag coefficients of side-by-side slender bodies according to [350]. (Re = 4.0 · 10^5, thickness ratio d/l = 1/3, drag of individual profile c_w = 0.07)

If the two bodies are at a large lateral distance, they exert no influence on each other and their respective drag equals that of the single body. As the bodies get closer to each other, the drag increases, reaching its maximum when the two bodies touch. The interference drag (2.114) of this arrangement is thus always positive. A major reason for the increase in drag at lower distances is the enhanced acceleration of the flow between the two bodies and the resulting increase in suction on the sides facing each other. Such suction pressures not only generate mutual attraction due to the Venturi effect (section 2.2.2.3.2) but also cause steeper pressure rises toward the rear and therefore separation of the flow. In addition, the closer the two bodies are to one another, the more they behave like a single body having twice the diameter with a significantly steeper pressure rise in the rear region of the outer sides, such that separation may also occur in this case. This example clearly demonstrates that finding the best vehicle shapes is not just a matter of taking successful shapes from aviation and "putting them on wheels" (section 1.2). However, by flattening the bottom of the body, the conditions can be significantly improved.

2.3.6.2 Streamlined Bodies in a Row

When two streamlined bodies are arranged in a row (Figure 2.48), they similarly do not affect each other at large distances (i.e., the drag coefficients of the two bodies are equal and correspond to those of individual bodies in free flow). As the distance decreases, the drag of the front body diminishes because it is influenced by the stagnation area of the rear body. At small distances, the front body even experiences negative drag (i.e., it is pushed by the rear body). However, the influence of the rear body also steepens the pressure rise in the rear region of the front body, such that separation may occur, which impairs the favorable effect on the front body.

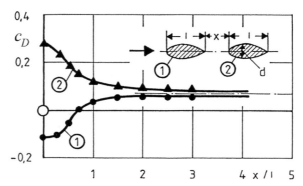

Figure 2.48 Drag coefficients of slender bodies in a row according to [350] (Re = $4.0 \cdot 10^5$, thickness ratio $d/l = 1/3$, drag of individual profile $c_w = 0.07$).

Simultaneously, the drag increases for the rear body, because its boundary layer is influenced by the wake flow of the front body, in that it takes up lower-energy, already decelerated fluid, which is more susceptible to separation. As can be seen from Figure 2.48, the drag increase of the rear body approximately matches the drag decrease of the front body, such that the interference drag (2.114) of the arrangement is virtually zero.

2.3.6.3 Bluff Bodies in a Row

The interaction of blunt bodies is much more complex than that of slender bodies, because the influences of the pressure field and wake flow are augmented by the influence of the dead wake. Figure 2.49 a shows the drag curve for two circular disks of the same diameter arranged in a row. While the front disk remains completely unaffected by convergence, the rear disk is already located in the front disk's dead wake at relatively large distances, so that its effective incident flow velocity is reduced and thus its drag is decreased ("slipstream effect"). If the distance decreases further, the rear disk eventually enters the low-pressure field of the dead wake (cf., section 2.3.5.2), whose underpressure acting on its front side eventually overcomes the underpressure at the rear, causing its drag to become negative. Hence in contrast to the behavior of streamlined bodies, the rear body is dragged by the front body, while the front body experiences virtually no influence. If the distance increases further, this effect is lost again until the two disks finally touch each other and the drag of the rear disk falls to zero. For all distances, the overall arrangement has a negative interference drag (2.114).

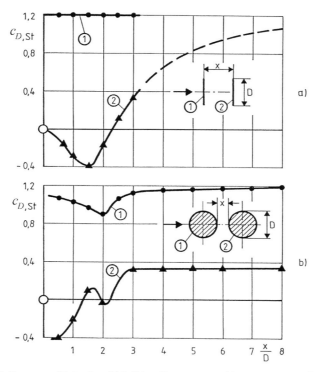

Figure 2.49 Drag coefficients of bluff bodies arranged in a row according to [350]: a) circular disks; b) circular cylinders.

The flow around two circular cylinders arranged in a row (Figure 2.49b) exhibits similar behavior as the two circular disks; however, in the case of subcritical flow considered here, the turbulence-generating effect of the dead wake and the associated impact on the laminar-turbulent transition further adds to the complexity. Even at greater distances, the rear cylinder is in the turbulent wake flow of the front cylinder, such that its boundary layer also becomes turbulent, while that of the front cylinder remains laminar. Furthermore, in contrast to the two circular disks, some interaction occurs between the front and rear bodies at low distances if the rear body enters the dead wake of the front body. Similar to the two streamlined bodies, the stagnation effect of the rear body temporarily reduces the drag on the front body.

2.3.7 Pipe Systems with Internal Flow

2.3.7.1 Stream Filament Theory Including Friction and Energy Input

Unlike problems involving flows around bodies, which generally require a multidimensional approach, in the case of pipe systems a quasi–one-dimensional approach is often sufficient. Here, the complicated flow channel shapes are modeled as stream tubes and dealt with using stream filament theory. This requires that all changes in the flow parameters transverse to the main flow direction can be ignored, such that we can employ the same cross-section averaged velocity within the stream tube for all streamlines.

In an internal flow, separation between the outer inviscid flow and the boundary layer is generally no longer possible; the friction processes in fact dominate the entire interior of the stream tube. The inviscid theory of stream filaments presented in section 2.1.8.3 is therefore largely unsuitable for the treatment of industrial problems. It needs to be complemented by appropriate terms for flow losses caused by friction and for the input of mechanical energy by continuous-flow machines (usually work machines such as pumps, fans, and compressors).

Since the same volumetric flow $\dot{V} = v \cdot A$ entering a component at a given cross section needs to exit again at another cross section, the associated exit and entry velocities are constrained by the continuity equation, that is, energy input and energy output cannot manifest themselves by a change in velocity (kinetic energy), but only in the form of changes in pressure (potential energy).

It is therefore useful to extend the Bernoulli equation (2.48), which in fact is an energy theorem for inviscid flows, by the addition of pressure terms. If the pressure change due to the input of mechanical energy is Δp_M and the pressure change due to friction losses is Δp_V, the extended Bernoulli equation for a stream tube with the inlet cross section 1 and the outlet cross section 2 (cf., Figure 2.50 a) modifies to

$$\frac{\rho}{2}v_1^2 + p_1 + \Delta p_M = \frac{\rho}{2}v_2^2 + p_2 + \Delta p_V, \tag{2.115}$$

while the continuity equation remains unchanged as

$$\dot{V}_1 = v_1 A_1 = v_2 A_2 = \dot{V}_2. \tag{2.116}$$

A simple example of a linear flow line is the ventilation system of a vehicle driven by a fan. The total pressure loss Δp_V then occurs during internal flow through the air conditioning and the vehicle and is composed of numerous individual loss elements in series or in parallel. The relevant principles are explained in section 2.3.7.4.

Such lines can also be connected to form complex networks and therefore, to an extent, are the basic elements of the extended theory of stream filaments. A special case arises if the start point 1 and end point 2 fall together. We then have a closed flow circuit, for which (2.115) assumes the simple form $\Delta p_M = \Delta p_V$, that is, energy input by the continuous-flow machine just matches the energy loss due to friction (Figure 2.50b). A simple example of this is the closed cooling circuit of a water-cooled internal combustion engine.

The Physical Principles of Aerodynamics

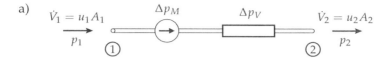

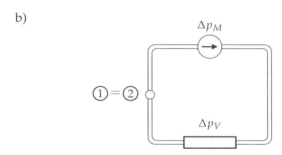

Figure 2.50 Stream filament with pressure loss Δp_V as the result of friction and pressure gain Δp_M by energy input: a) open flow line; b) closed flow circuit.

2.3.7.2 Pressure Loss of Components with Internal Flow

According to Reynolds's similarity law discussed in section 2.3.1, the pressure loss caused by friction in a component with internal flow, or even a complete system, depends only on the Reynolds number and therefore needs to be normalized using the stagnation pressure $\rho v^2/2$ such that the dimensionless representation

$$\frac{\Delta p_v}{\frac{\rho}{2} v^2} = \zeta(\text{Re}) \tag{2.117}$$

applies at an appropriate reference velocity v and the pressure loss coefficient $\zeta(\text{Re})$ that is dependent only on the Reynolds number. Values of ζ are tabulated for many standard components in relevant handbooks ([350], [187]) or are provided by the manufacturers of those components. It proves to be particularly convenient that in many cases ζ is independent of the Reynolds number, so that (as in the case of the c_w value for blunt bodies) the specification of a single number is all that is needed to characterize the component. Reference values for the pressure loss coefficients of basic system components are listed in section 2.3.7.5.

For some investigations, it is useful to formulate the pressure loss in terms of the volumetric flow and to combine the component-related variables in a flow resistance R. Using $\dot{V} = v \cdot A$, (2.54) yields

$$\Delta p_V = \frac{\rho}{2} R \dot{V}^2 \tag{2.118}$$

where

$$R = \frac{\zeta(\text{Re})}{A^2}, \tag{2.119}$$

which has a similar construction to Ohm's law if we identify the pressure drop at the component with the drop in electric voltage across a resistor. The volume flow then corresponds to the electric current; however, the pressure drop Δp_V depends on the square of $\dot{V}$.

2.3.7.3 System Characteristic and Operating Point

In practice, the typical task is to determine the required delivery pressure $\Delta p_M(\dot{V})$ if the flow resistance R_{ges} of the entire system and the states in the cross sections 1 and 2 of the stream tube are known. Solving (2.115) into Δp_M, taking into account (2.93), we obtain:

$$\Delta p_M = (p_2 - p_1) + \frac{\rho}{2}(v_2^2 - v_1^2) + \frac{\rho}{2} R_{ges} \dot{V}^2 \qquad (2.120)$$

and thereby a characteristic curve $\Delta p_M(\dot{V})$ of parabolic shape, provided R_{ges} is independent of the Reynolds number. As can be seen from Figure 2.51 as the slope gets steeper with increasing flow losses (e.g., due to throttling by means of control elements) and, in the special case of a closed flow circuit, it starts in the origin of the diagram.

The manufacturer of the continuous-flow machine provides the machine characteristic $\Delta p_M(n, \dot{V})$, which specifies the pressure Δp_M built up by the machine as a function of the volumetric flow and the rotational speed n. A distinction is made between two basic classes of characteristics depending on the design of the machine [187]. Displacement machines (e.g., piston pumps, gear pumps, and diaphragm pumps) deliver a volumetric flow for incompressible fluids that is uniquely determined by the working space volume and the rotational speed (Figure 2.51b). With an increase in flow losses, the volumetric flow is forced to remain constant by a strong rise in pressure, so that excessive throttling may even destroy the system under certain circumstances. Fluid machines (fans, compressors, and centrifugal pumps), in contrast, transfer the supplied power to the fluid via a rotating impeller. For constant speed, their characteristics are parabolas open at the bottom (Figure 2.51c). With an increase in flow losses, the pressure rises and the volumetric flow decreases. At sufficiently strong throttling, $\dot{V}$ may drop to zero, although the pressure remains limited, in contrast to the displacement machine.

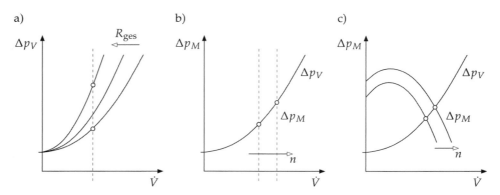

Figure 2.51 System and machine characteristics: a) variation of the system characteristic depending on the flow resistance. Its intersection with the characteristic b) of a displacement machine or c) of a fluid machine yields the operating point of the system.

The intersection of the system characteristic and the machine characteristic is the operating point of the system. It must be selected such that the required volumetric flow is guaranteed at all times and the pressures in the system are below the maximum design safety limit everywhere. The power P_M required of the machine, taking into account efficiency $\eta < 1$ dependent on the design and operating state, is determined by [187]

$$P_M = \frac{1}{\eta} \Delta p_M \dot{V} \tag{2.121}$$

and is also provided by the manufacturer in the form of a specific characteristic.

2.3.7.4 Series and Parallel Connection of Loss Elements

Typically, a system with internal flow contains not just a single flow resistance but can be construed, similar to electrical networks, as a series or parallel connection of loss elements. The corresponding relationships can be easily deduced using the flow resistance defined in (2.119).

In series connection of flow resistances (Figure 2.52a), the pressure losses accumulate while the same volumetric internal flow passes through all the components. Accordingly, using the statement

$$\Delta p_V = \Delta p_{V1} + \Delta p_{V2} \tag{2.122}$$
$$= R_1 \frac{\rho}{2} \dot{V}^2 + R_2 \frac{\rho}{2} \dot{V}^2$$
$$= R_{ges} \frac{\rho}{2} \dot{V}^2$$

we obtain the total resistance of the series connection as

$$R_{ges} = R_1 + R_2, \tag{2.123}$$

which corresponds to the series connection of electrical resistors. In parallel connection (Figure 2.52b), the same pressure loss occurs in both flow resistances, and the volumetric flow $\dot{V}$ is split into the two partial flows $\dot{V}_1$ and $\dot{V}_2$. The statement

$$\dot{V} = \dot{V}_1 + \dot{V}_2 \tag{2.124}$$
$$= \sqrt{\frac{2}{\rho} \frac{\Delta p_V}{R_1}} + \sqrt{\frac{2}{\rho} \frac{\Delta p_V}{R_1}}$$
$$= \sqrt{\frac{2}{\rho} \frac{\Delta p_V}{R_{ges}}}$$

yields the total resistance of the parallel connection as

$$\frac{1}{\sqrt{R_{ges}}} = \frac{1}{\sqrt{R_1}} + \frac{1}{\sqrt{R_2}}, \tag{2.125}$$

which differs from the corresponding formula for electrical resistors by the root signs that appear here due to the squared relationship between voltage and current.

In contrast to electric currents, losses also occur when fluid flows diverge and converge, each partial flow incurring a loss of its own. In the equivalent circuit diagram (Figure 2.52c), the resistances R_{ij}, which arise during transition of the flow $\dot{V}_i$ into the flow $\dot{V}_j$, therefore need to be applied to the respective partial flows, whereby, for example, the adapted equivalent circuit diagram of the parallel flow takes the form shown in Figure 2.52d. The total resistance of the parallel connection is then calculated by combined application of the laws for series and parallel connections, taking into account, however, the dependence of the branching resistances on the volumetric flow (section 2.3.7.5.5). The calculation must therefore be iterative.

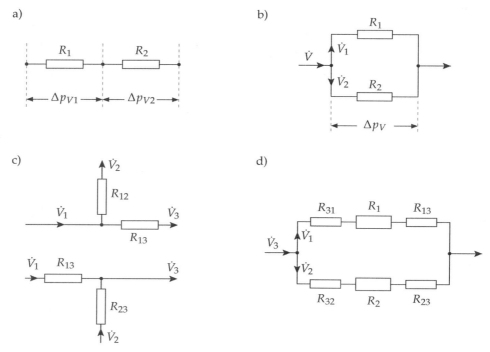

Figure 2.52 Equivalent circuit diagrams for loss elements: a) series connection; b) parallel connection without branching losses; c) branching losses; d) parallel connection with branching losses.

2.3.7.5 Pressure Loss Coefficients of System Components

The causes of the development of pressure losses in components with internal flow can be largely explained by the same physical mechanisms that lead to resistance (drag) of bodies in flow: wall friction, separation, and turbulence. The focus of the following discussion is therefore rather on the presentation of these loss mechanisms than on specifying exact pressure loss coefficients, which depend, apart from the Reynolds number, on many details such as the component geometry, the quality of walls, and so forth and generally have to be determined for each individual component in complex experiments. Generally applicable loss coefficients are available only for simple geometries such as the circular cylindrical pipe or the Carnot diffuser. The following pressure loss coefficients are thus in many cases only references that can be used for comparative assessments or rough calculations but do not replace more detailed specifications by the manufacturer or from relevant tables.

2.3.7.5.1 Losses in Linear Flow

Section 2.3.4 discussed the analogy between a plate flow and the flow in a circular cylindrical pipe, which can be construed geometrically as a flat plate rolled up to a cylinder. At the pipe inlet, a boundary layer forms that grows toward the center of the pipe after traveling a certain inlet region, such that the pipe cross section is completely filled by boundary layer after this point. This is referred to as a fully developed pipe flow in which, except for the linear pressure drop, no other changes occur in the flow direction. The dimensionless pressure loss coefficient is hence proportional to the ratio of pipe length L and pipe diameter D:

$$\zeta = \lambda\left(\text{Re}, \frac{k}{D}\right) \cdot \frac{L}{D}. \tag{2.126}$$

In this case, $\lambda(\text{Re})$ is the pipe resistance coefficient, which is proportional to the wall shear stress τ_W and therefore has a qualitatively similar dependence on the Reynolds number Re and the wall roughness k/D as the friction coefficient c_f of the plate boundary layer. The Reynolds number

$$\text{Re} = \frac{u_m D}{\nu} \tag{2.127}$$

of the pipe flow is conventionally formed using the average velocity u_m obtained from the volumetric flow $\dot{V}$ and the cross-sectional area $A = \rho D^2/4$:

$$u_m = \frac{\dot{V}}{A}. \tag{2.128}$$

Figure 2.53 shows, according to Schlichting [715], the relationship $\lambda\left(\text{Re}, \frac{k_s}{D}\right)$ for the case of sand-grain rough tubes (grain size k_s), which was determined experimentally by Nikuradse. In the region of laminar flow (Re < 2300), the Hagen-Poiseuille law applies, which is an exact solution of the Navier-Stokes equation [715] and has the simple dimensionless form

$$\lambda = \frac{64}{\text{Re}} \tag{2.129}$$

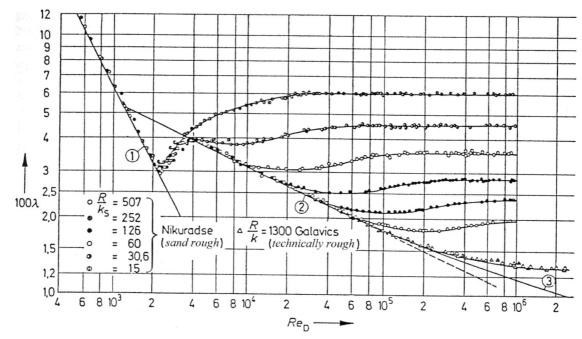

Figure 2.53 Resistance factor λ for sand-grain rough tubes: Curve 1: laminar flow, equation (2.129); curve 2: empirical relationship according to Blasius [715]; curve 3: resistance law for smooth tubes according to equation (2.130). The values for technically rough tubes also follow equation (2.130).

There is no dependence on roughness in this case; the pipe behaves as hydraulically smooth. In the turbulent region (Re > 2300), it is striking that the curves for sand-grain rough tubes have an inflection point in the transition region and only lead into the horizontal pattern after passing through a minimum, while the measured values by Galavics, which were determined on technically rough tubes, follow a monotonically descending pattern without an inflection point. According to Eck [221], this is due to the fact that in sand roughness all the roughness elements have the same size and therefore they all emerge from the thinning viscous sublayer simultaneously with increasing Reynolds number. However, in the case of industrial roughness with associated statistical size distribution, initially only a few peaks are in the turbulent flow and smaller roughnesses appear only gradually. Since many industrial applications are precisely in this transition region, it is recommended, according to [221], to obtain the pipe resistance coefficient for turbulent flow not from Figure 2.53 but to calculate it using a formula specified by Colebrook, which was developed specially for industrial roughness and covers the entire turbulent Reynolds number range for smooth and rough tubes[15]

---

15. This implicit equation for λ must be evaluated iteratively by inserting an initial value on the right side and then repeating this procedure with the result a number of times. The simple method usually converges very quickly.

$$\frac{1}{\sqrt{\lambda}} = -2\log\left[\frac{2.51}{\text{Re}\sqrt{\lambda}} + 0.27\frac{k}{D}\right]. \tag{2.130}$$

For the roughness parameter k/D, a mean value must be used, which is tabulated for many important surfaces in engineering. Typical values are between 0 and 0.05 for new tubes but may increase for heavy contamination to 0.2 and above [221].

The remarkable insensitivity of turbulent flow to the geometry, already mentioned in section 2.3.4, is again reflected in the fact that the Colebrook formula (2.130) can also readily be applied to noncircular cross-section tubes, if—instead of the pipe diameter D—the so-called hydraulic diameter

$$D_h = \frac{4A}{l} \tag{2.131}$$

is used, with A as the cross-sectional area facing the internal flow and l as the wetted circumference of this area [221]. For a circular cross section, this yields $D_h = D$, while for a square cross section (side length a) this is $D_h = a$.

2.3.7.5.2 Redirection Losses

The change in flow direction in a pipe bend is always associated with losses. The streamline curvature due to the redirection produces a pressure gradient transverse to the flow, such that the pressure is higher on the outside of the bend and lower on the inside than in the incoming flow (Figure 2.54a). The pressure rises close to the wall, causing flow separation, which occurs at the entry into the bend on the outside and at the exit from the bend on the inside. Moreover, the redirection of the centrifugal forces causes deflection of the rapid fluid particles close to the axis toward the outside, while the slow particles close to the wall under the influence of the pressure gradient flow toward the inside. This results in a circulation-like secondary flow perpendicular to the main flow, depriving the main flow of even more energy in addition to the separation effects. The sum of the energy losses is reflected in the pressure loss coefficient ζ.

Figure 2.54b shows typical values of the pressure loss coefficient ζ for circular pipe bends and elbow pipes [63]. As can be seen, the greater the change in direction of flow and the smaller the bending radius of the pipe bend, the higher the redirection losses. By far the highest losses occur in elbow pipes with a rectangular cross section in which both separation and secondary flow effects are most pronounced.

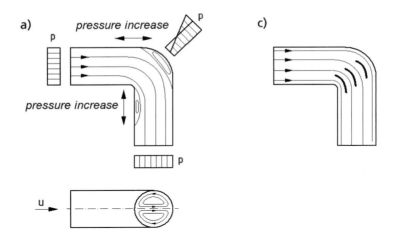

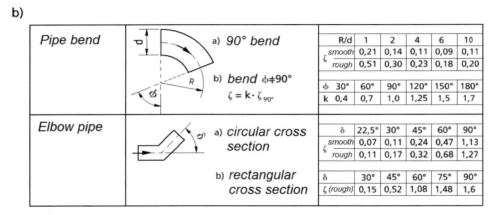

Figure 2.54 Flow losses from redirection: a) separation in a pipe bend; b) pressure loss coefficients for pipe bends and elbow pipes (source: [63]); c) reduction of redirection losses by baffles (source: [904]).

The pressure loss of elbow pipes can be significantly reduced by installing baffles, which effectively prevent both flow separation and secondary flow when arranged appropriately [904]. This method is used, for example, in the redirection corners of wind tunnels to achieve a cost-saving design while avoiding redirection losses associated with high energy costs. Simply by installing a single nonprofiled guide blade, redirection losses can be almost halved, while installing a profiled cascade (Figure 2.54c) can reduce them by more than 90%. Suppressing the secondary flow also delivers a significantly better quality of channel flow.

2.3.7.5.3 Losses Due to Change in Cross Section

2.3.7.5.3.1 Increase in Cross Section

Since an increase in cross section in the flow direction results in a decrease in flow velocity and thus in an increase in pressure, such components, also referred to as diffusers, are always highly prone to separation, and considerable pressure losses occur even at low divergence angles. Figure 2.55a shows typical values of the pressure loss coefficient ζ for conical diffusers depending on the diameter ratio D_2/D_1 and the opening angle φ, which is determined by the overall length [63]. It can be seen that, with an increase in divergence angle from 4° to 10°, the loss almost doubles, while in case of further increase to 12° it even almost quadruples.

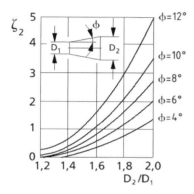

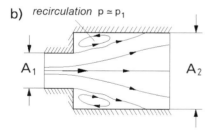

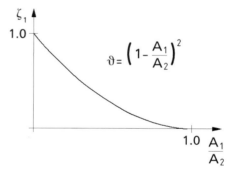

Figure 2.55 Pressure loss coefficients on increase of cross section: a) steady cross-sectional expansion (conical diffuser); b) unsteady cross-sectional expansion (Carnot diffuser). Source: [63].

The limiting case $\varphi = 90°$ of an unsteady cross-sectional jump is called a Carnot diffuser (Figure 2.55b). It also yields a recovery in pressure due to the continuity-related deceleration of the flow, but causes high losses as a result of flow separation. Using the momentum theorem (cf., section 2.1.8.5), it is possible to theoretically determine the pressure loss coefficient, ignoring the wall friction [62]:

$$\zeta = \left[1 - \frac{A_1}{A_2}\right]^2. \tag{2.132}$$

While the losses for small cross-sectional jumps are still acceptable, for a decreasing cross-section ratio A_1/A_2 they rise considerably. The special case $A_2 \to \infty$ with the highest pressure loss coefficient $\zeta = 1$ here corresponds to a jet freely blowing out of a wall, in which the entire kinetic energy is not converted into pressure recovery but is dissipated into the surrounding medium.

2.3.7.5.3.2 Decrease in Cross Section

In the event of continuous cross-sectional constriction in a nozzle, the flow is accelerated and the pressure decreases in the flow direction. There is therefore no risk of separation and no pressure loss (except for friction losses). However, this is not true in the case of a sudden cross-sectional constriction where the accumulation in front of the cross-sectional jump and the overflow of the sharp edge result in pressure losses that depend on the cross section ratio A_2/A_1 (Figure 2.56). The special case $A_1 \to \infty$ here corresponds to a flow into a sharp-edged inlet with a pressure loss coefficient of $\zeta = 0.5$. By simple beveling of the inlet edge, this can be lowered to $\zeta = 0.2$ while careful rounding can reduce it to $\zeta \approx 0$ [63].

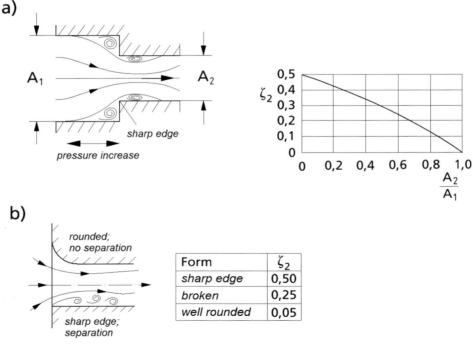

Figure 2.56 Pressure loss with unsteady cross section decrease: a) for finite cross section ratio; b) in the limiting case $A_1/A_2 \to 0$ (source: [63]).

2.3.7.5.4 Losses Caused by Fixtures and Fluid-Control Devices

Fixtures such as flaps, sliders, or apertures are used to control, adjust, or measure the flow in channels. They are obstacles that always cause separation phenomena in the flow around them (Figure 2.57). For example, the accumulation in front of a slider or aperture causes separation in front of the flow obstacle. The pressure rise after passing through the cross-sectional constriction causes loss-producing separation phenomena not only at the previously mentioned components but also in a damper flap that does not cause a stagnation effect near the wall. Flow separation also occurs wherever the fixtures have sharp edges.

The pressure losses that occur are significant, particularly in the nearly closed state of fluid-control devices, whose characteristics are always nonlinear without special structural measures (Figure 2.57).

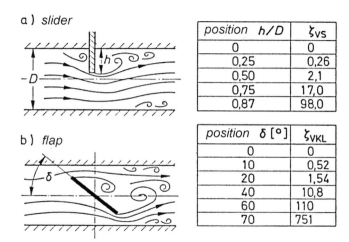

Figure 2.57 Pressure losses caused by regulating and fluid-control devices.

a) slider

position h/D	ζ_{VS}
0	0
0,25	0,26
0,50	2,1
0,75	17,0
0,87	98,0

b) flap

position δ [°]	ζ_{VKL}
0	0
10	0,52
20	1,54
40	10,8
60	110
70	751

2.3.7.5.5 Losses Caused by Divergence and Convergence

Divergence and convergence of fluid flows produce complex separation phenomena and turbulent mixing processes in which the partial flows exchange energy with each other. Since the pattern of these processes depends strongly on the size of each partial flow, this also applies to the associated loss coefficients ζ_{ij}. As Figure 2.58 demonstrates, this can also result in negative ζ values, that is, the branching point acts as a continuous-flow machine (jet pump) for the particular partial flow. This effect is especially pronounced when two fluid flows converge at an angle of 90° for the input partial flow if that partial flow is much smaller than the continuous main flow [221].

a) divergence

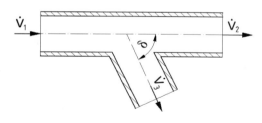

δ=45°

$\dot{V}_2/\dot{V}_3$	0	0,2	0,4	0,6	0,8	1,0
ζ_{12}	0,9	0,66	0,47	0,33	0,29	0,35
ζ_{13}	0,04	-0,06	-0,04	0,07	0,20	0,33

δ=90°

$\dot{V}_2/\dot{V}_3$	0	0,2	0,4	0,6	0,8	1,0
ζ_{12}	0,96	0,88	0,89	0,96	1,10	1,29
ζ_{13}	0,05	-0,08	-0,04	0,07	0,21	0,35

b) convergence

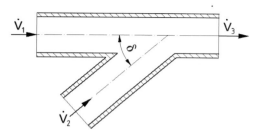

δ=45°

$\dot{V}_2/\dot{V}_3$	0	0,2	0,4	0,6	0,8	1,0
ζ_{13}	0,05	0,17	0,18	0,05	-0,20	-0,57
ζ_{23}	-0,9	-0,37	0	0,22	0,37	0,38

δ=90°

$\dot{V}_2/\dot{V}_3$	0	0,2	0,4	0,6	0,8	1,0
ζ_{13}	0,06	0,18	0,3	0,4	0,5	0,6
ζ_{23}	-1,20	-0,4	0,1	0,47	0,73	0,92

Figure 2.58 Flow losses from branching in tubes of the same cross section: a) divergence; b) convergence. Source: [221].

2.4 Appendix

2.4.1 Density and Viscosity of Air

Ignoring air humidity, the density of air can be calculated using the ideal gas equation

$$\rho_L = \frac{p}{R_L T} \qquad (2.133)$$

as a function of pressure and temperature, where

$$R_L = 287.0 \; \frac{\text{J}}{\text{kg K}} \qquad (2.134)$$

is the specific gas constant of air [640]. Figure 2.59a shows the calculated density of air as a function of temperature for different barometric pressures; the curve for the standard pressure $p = 1013.25$ hPa is highlighted.

a)

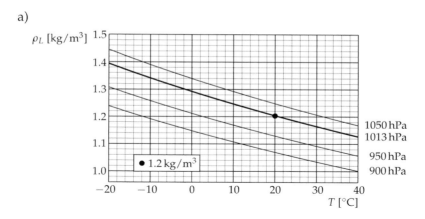

b)

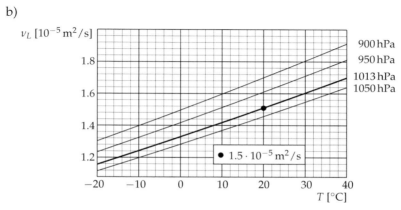

Figure 2.59 (a Density ρ_L and (b kinematic viscosity ν_L of air as a function of temperature at different barometric pressures.

According to the Sutherland model, the dynamic viscosity μ_L of air as an ideal gas only depends on the temperature and is calculated as per the relationship

$$\mu_L(T) = \mu_{0L} \left(\frac{T}{T_0}\right)^{\frac{3}{2}} \frac{T_0 + T_s}{T + T_s}, \qquad (2.135)$$

where $T_0 = 273.15$ K and $\mu_{0L} = 1.717 \cdot 10^{-5}$ Ns/m² denote the reference state. The material parameter, known as the Sutherland temperature, has a value of $T_s = 123.6$ K for air [640].

The kinematic viscosity ν is required much more often in practice and is defined by

$$\nu = \frac{\mu}{\rho}. \qquad (2.136)$$

This is a function of temperature and pressure that can be determined from (2.133) and (2.135). Figure 2.59b shows the calculated values of the kinematic viscosity ν_L of air as a function of temperature for different barometric pressures; the curve for the standard pressure $p = 1013.25$ hPa is highlighted.

2.4.2 Compressibility Effects

In applying incompressible aerodynamics, we are often faced with the problem of estimating the change in air density caused by compressibility, when the velocity changes from a given value u_∞ of the incident flow to a specific value u. At stagnation points in particular, but also at those points at which the highest over-velocities occur due to the body's displacement effect, the velocity changes may well be of the same order of magnitude as u_∞ itself.

Elementary gas dynamics [904] provides the well-known relation for the change in density of a compressible medium at isentropic change along a streamline:

$$\frac{\rho}{\rho_0} = \left[1 - \frac{\kappa-1}{2} Ma^2\right]^{\frac{1}{\kappa-1}} = 1 + \frac{1}{2} Ma^2 + O(Ma^4), \tag{2.137}$$

where the Mach number $Ma = u/c$ denotes the ratio of the local flow velocity to the local speed of sound

$$c = \sqrt{\kappa R_L T} \tag{2.138}$$

and the adiabatic exponent κ, which only depends on the number of the molecular degrees of freedom, assumes the value $\kappa = 1.4$ for diatomic gases such as air. In this case, the static air density ρ_0 is not the density ρ_∞ of the incident flow but the density at the stagnation point ($Ma = 0$). For the case of $Ma \ll 1$ that is predominantly relevant in vehicle aerodynamics, we can in good approximation use the quadratic approximation of (2.137) instead of the exact relationship.

However, (2.137) is not particularly well suited for a quick estimate of compressibility effects, because neither the value of the static air density ρ_0 nor the value of the local speed of sound c, which also changes with the density based on the isentropic relationship [904]

$$\frac{T}{T_0} = \left(\frac{\rho}{\rho_0}\right)^{\kappa-1} \tag{2.139}$$

are generally known without additional calculation. However, by setting the velocity u in relation to the speed of sound c_∞ of the incident flow, we can easily show, using (2.137), (2.138), and (2.139), that the square of this variable, except for higher order terms, is equal to the square of the Mach number Ma:

$$\left(\frac{u}{c_\infty}\right)^2 = \left(\frac{u}{c}\right)^2 \left(\frac{c}{c_0}\right)^2 \left(\frac{c_0}{c_\infty}\right)^2 = Ma^2 + O(Ma^4) \tag{2.140}$$

with which, using the quadratic approximation of (2.137), we obtain a very simple relationship for the density change $\Delta \rho$ dependent on u:

$$\frac{\Delta \rho(u)}{\rho_\infty} = \frac{\rho}{\rho_0} \frac{\rho_0}{\rho_\infty} - 1 = -\frac{1}{2} \frac{u^2 - u_\infty^2}{c_\infty^2} \tag{2.141}$$

which allows us to estimate the change in density without great computational effort. To facilitate the practical application of this formula, Figure 2.60 shows the speed of sound in air as a function of temperature, calculated using (2.138).

Example:

A cylinder is exposed to an incident flow at speed $u_\infty = 50$ m/s at $T_\infty = 300$K, where the flow velocity at the maximum thickness increases to $u = 2u_\infty = 100$ m/s (cf. (2.75). At a speed of sound of $c_\infty = 347$ m/s, (2.141) yields a density increase of +1.0%, at the stagnation point ($u = 0$), while at the points of maximum velocity we see a density decrease of −3.1%. The maximum expected density differences in the flow field are therefore 4.1%.

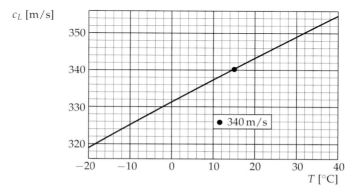

Figure 2.60 Speed of sound c_L of air as a function of temperature.

Chapter 3
Consumption and Performance

Teddy Woll

3.1 The Significance of Aerodynamic Drag

Aerodynamic drag is the resistance factor that most affects a vehicle as speed increases. At low speeds, this factor is negligible, but when vehicle speed exceeds approximately 60 km/h, it surpasses rolling resistance and becomes the most dominating factor at higher speeds. Aerodynamic drag thus not only greatly impacts fuel consumption but also the performance of a vehicle, particularly in the high-speed range (refer to sections 3.2 and 3.3). Today's increased engine outputs now enable sufficiently high driving performance. Measures that further minimize drag are therefore implemented almost entirely to reduce fuel consumption. The average weighted fuel consumption (fleet consumption) of German passenger cars[1] has thus continued to drop, although average engine output and vehicle weight have increased (refer to Figure 3.1).

---

1. This increase, which began in the late 1980s, can primarily be attributed to the introduction of the catalytic converter and the end of the "oil crisis" (refer to the progressional trend toward higher engine outputs), while the drop experienced in 2009 is the result of the "Cash for Clunkers" program sponsored by the federal government.

Chapter 3

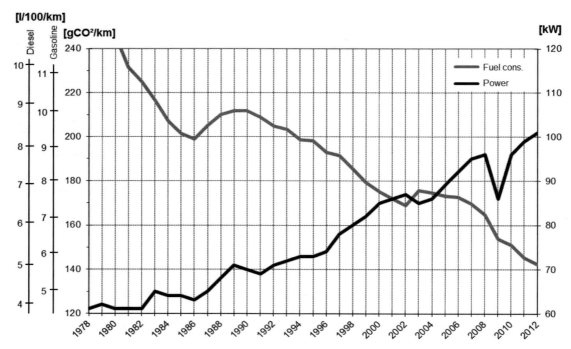

Figure 3.1 Development of fleet consumption and average engine output of all German passenger cars (VDA 2003 and 2012).

The gradual increase in vehicle weight is a direct result of efforts made to improve safety and comfort, leverage cost-reducing strategies that favor the use of standardized parts and platforms[2] more powerful engines, and, in recent times, hybridize powertrains. While many vehicles experienced only moderate weight gains, some vehicles almost doubled their curb weight in six generations. In the majority of cases, vehicle weight has also been driven by the pronounced trend toward larger cars, which can be observed in mild form in every segment and is exacerbated by the unwavering demand for taller vehicles such as small and large vans, all sorts of sport utility vehicles (SUVs), and crossover concepts. Nevertheless, several brand-new vehicles are now becoming lighter than their predecessors, a development that has not been seen for decades.

The increase in frontal area reveals a very similar progression (refer to Figure 3.2, right) as the driving forces behind it are very related: comfort dimensions (e.g., head and elbow room, especially in the back seats), safety (due to side-impact protection and other factors), and lightweight construction, since lightweight materials such as aluminum can only tap their full potential when the cross-section widths of load-bearing structures (e.g., longitudinal side rail, rocker panels, pillars) are made larger.

---

2. Critical parts and components of the platform, such as the bodyshell and axles, must be designed to handle the loads of the heaviest vehicle for which they will be used. The lighter and less powerful vehicles are then automatically at a disadvantage because they do not require parts of this strength.

Consumption and Performance

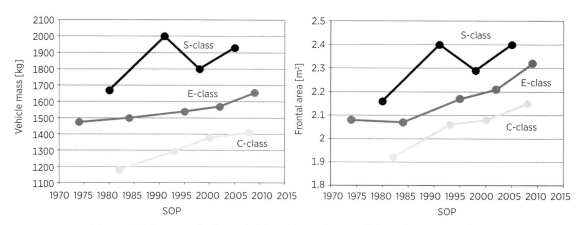

Figure 3.2 Increase in the weight and frontal area of Mercedes-Benz sedans.

Several manufacturers have still been able to find a way to compensate for frontal area growth by reducing the coefficient of drag (c_D) by a disproportionate amount. Figure 3.2, right, for example, shows that the C-Class that experienced the largest growth has profited from drag-reducing measures (c_D) (from 0.34 to 0.30 and 0.26) that brought its $c_D \cdot A_x$ performance figure down to 0.54 m² from 0.65 m² in the same timeframe.

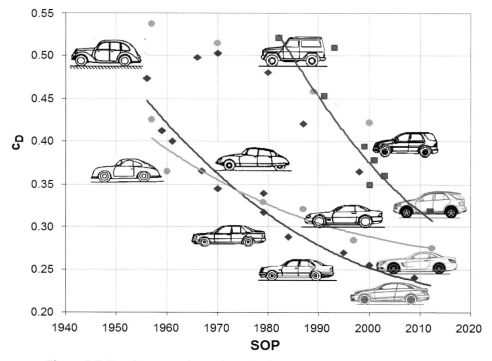

Figure 3.3 Development of aerodynamic drag across key vehicle categories.

153

The asymptotic progression of the c_D figures shows that finding potential areas in which additional improvements can be made are becoming increasingly difficult. There are several reasons for this:

- Dimensional concepts are mature such that further aerodynamic measures would only conflict with usability (e.g., trunk width, roof height, rear seat width) as well as with current design trends (e.g., wide rear stance, short overhangs, larger wheels).
- Heavier curb weights lead to stronger engines and greater cooling requirements; larger tires that frequently have wider cross sections and lower profiles must also be fitted.
- Hardly any reserve potential to be tapped in the underfloor area and cooling-air passageways (refer to section 3.7).

The reduction in aerodynamic drag across all classes and categories of vehicle thus explains only one part of the remarkable average drop in fuel consumption of 2%% each year, as Figure 3.1 illustrates. The greatest contribution toward achieving this objective has been made by improving the operating efficiency of engines and transmissions while fitting more advanced tires that reduce rolling resistance.

The target objective for European Union (EU) fleet consumption of 95 g/km CO_2 by 2020 (refer to Figure 3.1 and section 3.8) will therefore necessitate further intense development work to reduce fuel consumption. The role that can be played by lowering aerodynamic drag is analyzed in detail in the following sections of this chapter.

3.2 Theory of Driving Resistance

The longitudinal movement of a vehicle is influenced by four external forces:

- Rolling resistance F_R
- Aerodynamic drag F_D
- Grade resistance F_H
- Acceleration F_B

While aerodynamic drag and rolling resistance create irreversible losses, grade resistance and acceleration are reversible in principle. The potential or kinetic energy generated by a vehicle under increased load conditions can thus be partially recovered through the use of a hybrid or electric drive. Further electrifying the powertrain then also means that rolling resistance and aerodynamic drag in particular become more important.

3.2.1 Rolling Resistance

Rolling resistance primarily results from the hysteresis loss effect as the tire structure deforms when it rotates, which is largely due to the compression and bending of the tread and the sidewall (80 to 95%). This intensity of this effect is determined by the flexing amplitude (wheel load, internal tire pressure, etc.) and the flexing frequency (driving speed). Adding to this are frictional losses between the tire and road (micro slip, up to 5%) and vibrations along the sidewall and tread surface, which increase considerably as speed rises. Resistance created by the axle geometry (e.g., toe-in resistance, which also depends on the tire sideslip rigidity and a low camber influence) and frictional losses in the wheel bearings and brakes likewise play a role. Air turbulence at the tires (up to 15%) is yet another factor, and although it is most closely aligned with air resistance, it is not easy to separate it from or view independently of rolling resistance (e.g., during rolling measurements) (refer to Figure 3.4).

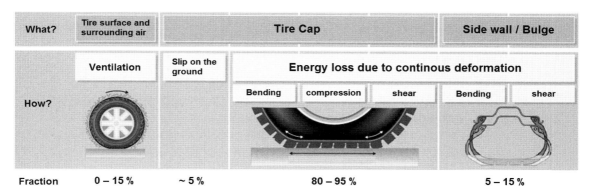

Figure 3.4 The factors that determine rolling resistance [494], [325].

Rolling resistance F_R (in N) can be calculated from mass and uplift force:

$$F_R = \mu_R \cdot \left(m_F \cdot g - c_L \cdot A_x \cdot \frac{\rho_L}{2} \cdot v^2 \right) \tag{3.1}$$

μ_R refers to the rolling resistance coefficient, m_F the vehicle weight or mass in kg, g the acceleration of gravity (9.81 m/s²), and c_L the coefficient of lift—refer to identical equation (3.65). for the remaining variables. Since vehicles typically have differing axle loads, it can prove useful to calculate the rolling resistance separately for each axle. Axle loads also increase or decrease in relation to the different coefficients of lift at the front and rear axles as speed increases (refer to section 3.3).

The rolling resistance coefficient is not a constant but varies according to the wheel load, internal tire pressure, and temperature, and increases along with speed due to the speed-dependent factor of hysteresis loss and deformation. This increase depends on the type of construction: Y tires (rated to 300 km/h) and W tires[3] (rated to 270 km/h) feature additional circumferential filaments for reinforcement that limit or restrict tire deformation and prevent rolling resistance from compounding at high speeds but have higher

---

3. The letter designation describes the speed index that specifies the maximum travel speed permitted. It is included as the last character in the tire designation or label and can be read on the tire sidewall. The speed index is determined based on the rated top speed for the vehicle as listed in the vehicle's registration documentation ($v_{max,perm.}$). Added to this is a safety buffer, which is calculated using the following formula: $v_{max,perm.} \cdot 0.01 + 6.5$ km/h.

rolling resistance than H tires (rated to 210 km/h) or V tires (rated to 240 km/h) at lower speeds (refer to Figure 3.5).

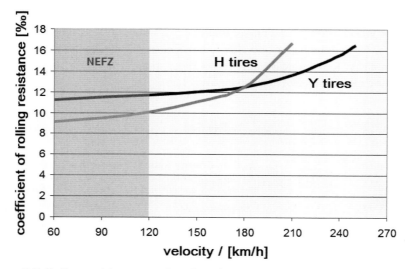

Figure 3.5 Rolling resistance as a function of speed and type of construction [572].

As a result of new rubber compounds (e.g., silica technology) and filaments (e.g., Kevlar), rolling resistance can be reduced to under 7% in the future without having a serious impact on other characteristics such as mileage, braking performance, grip under wet conditions, and sideslip rigidity, for example.

3.2.2 Aerodynamic Drag

The air resistance of a vehicle is primarily caused by differences in pressure around the vehicle in the direction of flow (e.g., pressure resistance, > 80%), friction on the surface of the vehicle (< 10%), and pulse and frictional loss as air passes through the radiator, engine compartment, and interior compartment (approximately 10%). These three resistance factors together account for the aerodynamic drag force, in N, which can be calculated based on the dimensionless c_D figure using the following formula:

$$F_D = c_D \cdot A_x \cdot \frac{\rho_L}{2} \cdot v^2 \tag{3.2}$$

A_x designates the frontal area, in m², and v designates the speed, in m/s. The air density, ρ_L, does not represent a constant but changes depending on the temperature and air pressure, which in turn is influenced by the weather and altitude (refer to Figure 3.6):

$$\rho_L = 1{,}293 \cdot \frac{273}{T+273} \cdot \frac{p}{1013} \cdot \frac{\text{kg}}{\text{m}^3} \tag{3.3}$$

T designates the temperature, in °C, and p designates the air pressure, in mbar.

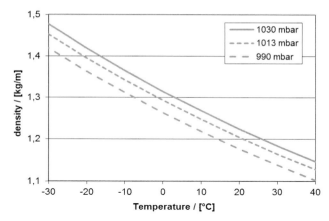

Figure 3.6 Air density as a function of temperature and air pressure.

Airflow velocity cannot be regarded as a constant in practice, even if the vehicle is driven at a constant speed, since it is also influenced and affected by natural and artificial gusts of wind (e.g., created by other vehicles). The resulting airflow velocity is determined by adding the vectors of driving speed and wind speed. In most cases, the resistance created by side winds is minimal, since the c_D figure increases only gradually near small oblique airflow angles and the wind speeds are negligible in comparison to the vehicle's travel speed. Taking the new B-Class as an example (refer to Figure 3.7), no increase in the c_D figure can be measured up to an oblique airflow angle of 2.5 degrees. At 5 degrees, a 5 to 8% increase can be observed, depending on the side of the vehicle. Only when this threshold or marker point is exceeded does resistance become significant; such oblique airflow values are rarely encountered in practice, however (refer to section 4.2.4).

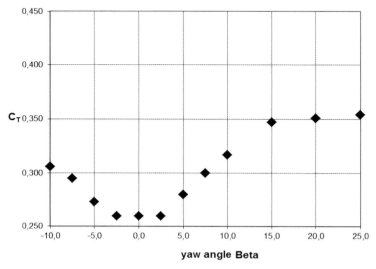

Figure 3.7 c_D figure as a function of the flow angle for the new B-Class from Mercedes-Benz.

3.2.3 Grade Resistance

Grade resistance F_H, which comes into play on hills and inclines, is calculated as follows in conjunction with slope angle α (in degrees) and incline N (in percent, 100% = 45 degrees):

$$F_H = m_F \cdot g \cdot \sin(\alpha) = m_F \cdot g \cdot \sin(\arctan N) \tag{3.4}$$

Grades are currently not taken into account when manufacturers determine fuel consumption on the test rig or in a simulated environment. Exceptions to this are select real-world driving cycles such as the AMS cycle (street circuit in the northern part of the Black Forest—refer to "Auto-Motor und Sport" and section 3.5.7). The energy stored in the vehicle as it travels through hilly terrain in particular (Alpine countries as well as hilly areas such as the Swabian Alb or city traffic in Stuttgart—refer to Figure 3.8) has a considerable influence on fuel consumption and must not be overlooked or underestimated. This especially applies in the context of hybrid and electric drives, which are capable of recovering and utilizing potential energy to a certain extent.

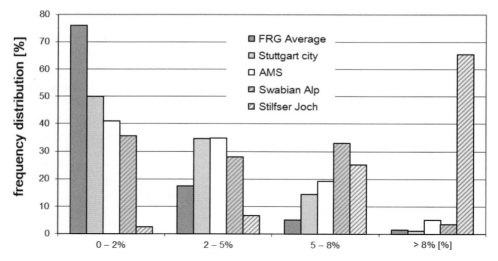

Figure 3.8 Frequency distribution of inclines on various roads.

3.2.4 Acceleration

Kinetic energy is also stored in the vehicle as a result of acceleration. More specifically, this energy relates to the overall translatory mass of the vehicle and all rotating objects throughout the powertrain such as wheels, shafts, flywheels, and gear wheels, and rotating parts in the engine. Acceleration F_B (in N) can be approximated using the following formula:

$$F_B = \left(m_F + \frac{J_R + i^2 \cdot J_M}{r_{stat} \cdot r_{dyn}} \right) \cdot \frac{dv}{dt} = m_{res} \cdot \frac{dv}{dt} \tag{3.5}$$

J refers to the moment of inertia (of tires, engine) in kg · m²/s², i refers to the gear ratio, and r refers to the tire radius (static, dynamic) in m. The mass moment of inertia of the rotating parts is very noticeable: Depending on the gear engaged (the lower the gear, the more significant) and the size of the rotating masses (e.g., engine and tire size), overall mass m_{res} exceeds the vehicle mass by up to 30%.

3.2.5 Overall Driving Resistance

Overall driving resistance F_{tot}, given in N, is the product of adding each individual resistance factor:

$$F_{tot} = F_R + F_D + F_H + F_B \qquad (3.6)$$

$$= \mu_R \cdot m_F \cdot g + c_D \cdot A_x \cdot \frac{\rho_L}{2} \cdot v^2 + m_F \cdot g \cdot \sin(\arctan N) + m_{res} \cdot \frac{dv}{dt}$$

The overall power requirement, given in W, is calculated by multiplying the overall driving resistance by the travel speed of the vehicle:

$$P_{tot} = F_{tot} \cdot v = (F_R + F_D + F_H + F_B) \cdot v \qquad (3.7)$$

3.2.6 Example

To illustrate the forces described in the previous sections, the example below is of an actual vehicle together with its corresponding data:

Table 3.1 Technical Data of the Mercedes-Benz B-Class as a Sample Vehicle	
m_F = 1500 kg	c_D = 0.26
μ_R = 0.0085 (with v increasing as per Figure 3.5)	A_x = 2.42 m²

When dv/dt equals 0, the relative speed yields the rolling and wind resistance curve as shown in Figure 3.9. The forces exhibited during constant acceleration at 1 m/s² (0 to 100 km/h in 27.8 seconds) on a 5% incline can be used for comparison purposes. The following findings can be made based on the curves plotted:

- With this vehicle, aerodynamic drag exceeds rolling resistance at approximately 60 km/h.
- When the top speed of 190 km/h is reached, aerodynamic drag generates about 10 times the resistance than is experienced at 60 km/h.
- A 5% incline generates as much resistance as air at approximately 150 km/h.
- An acceleration of 1 m/s² is comparable to the aerodynamic drag experienced at 230 km/h.

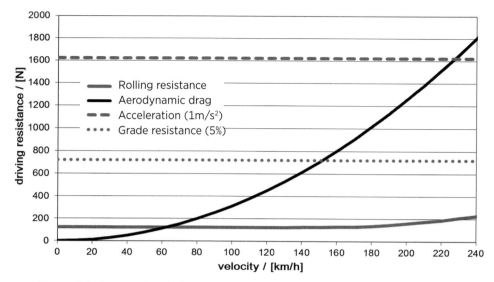

Figure 3.9 Progressional plotted curvature of air and rolling resistance as well as acceleration and grade resistance in relation to speed using the B-Class from Mercedes-Benz as an example.

3.3 Performance

3.3.1 Acceleration and Elasticity

Acceleration as it applies to vehicle testing is understood to mean the time that must elapse to compensate for or balance out a difference in speed, including the respective shift points. The same holds true for elasticity but without factoring in the shift points. The default reference for acceleration is typically the 0 to 100 km/h test, whereas elasticity tests are usually conducted between 60 and 100 km/h and 80 to 120 km/h in the highest or second highest gear (refer to Table 3.1).

When values have been specified for the speed and gradient or incline, the available reserve tractive force or pulling power for acceleration or elasticity can be accurately characterized in a tractive force diagram (cf. Figure 3.10) or calculated from equation 3.8. This figure is derived from the tractive force applied at wheel FZ minus the driving resistance:

$$F_{Res} = F_Z - (F_R + F_D + F_H) \tag{3.8}$$

Correlation of F_{Res} with equation (3.5) can determine momentary acceleration

$$a = \frac{F_Z - (F_R + F_D + F_H)}{m_{res}} \tag{3.9}$$

and integration can determine the increase in speed after a specified time t has elapsed:

$$v\left(\frac{\text{km}}{\text{h}}\right) = 3{,}6 \cdot \int_0^t a\, dt \qquad (3.10)$$

Tractive force can be approximated based on output PA (multiplied by drive efficiency η) and speed v_F:

$$F_Z = \frac{\eta \cdot P_A}{v_F} \qquad (3.11)$$

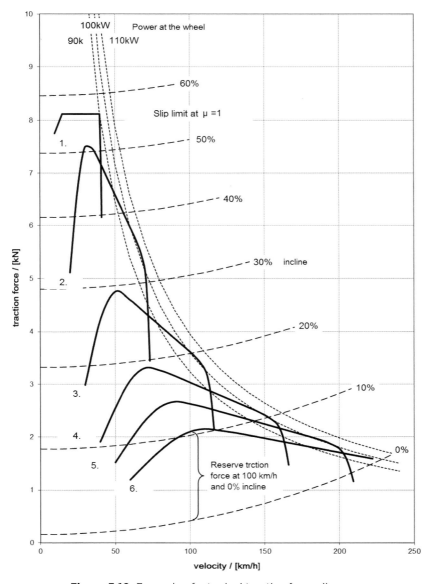

Figure 3.10 Example of a typical tractive force diagram.

Figure 3.11 shows the progression of speed over time for vehicle elasticity in fifth gear and for full-load acceleration, including the interruptions or breaks in acceleration required to shift gears. These interruptions vary depending on the driver and, in the case of an automatic transmission, on the application and drive program selected (e.g., Comfort or Sport).

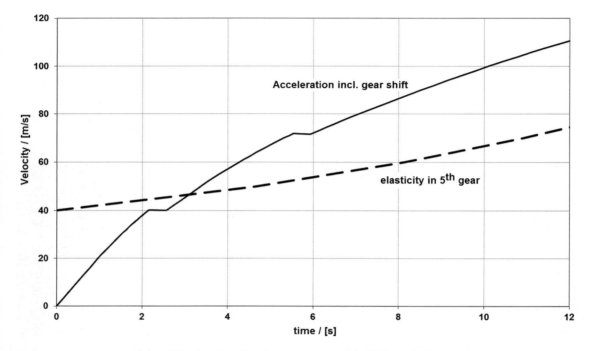

Figure 3.11 Acceleration from a stop and elasticity in fifth gear.

3.3.2 Ascending Ability

The ascending ability of a vehicle is essentially determined by the gear engaged and the drive torque available directly at the wheels, as Figure 3.10 illustrates. The maximum ascending ability in first gear, for example, exceeds the gradient climbing ability from a standstill due to the torque curve of the engine, whereby the maximum ascending ability shown in the example is limited by the traction offered by the tire. When higher gears are engaged, the torque curve of the engine and, thus, the ascending ability of the vehicle, play an even more important role and dictate whether the driver needs to downshift on a steep highway or freeway incline such as experienced in hilly or mountainous terrain that ascends by approximately 7%. The drive axle fitted to the vehicle is therefore typically geared much lower than required for driving on a flat and level surface, including in high-torque diesel engine applications. Acceleration capabilities improve as a result, although fuel consumption increases considerably.

3.3.3 Top Speed

At top speed, tractive force equals the sum total of all natural forces that resist driving the vehicle. Assuming a constant rolling resistance coefficient and no gradient, equations (3.1), (3.2), and (3.6) can be used to derive the following equation [298]:

$$\frac{\eta \cdot P_{max}}{v_{max}} = c_D \cdot A_x \cdot \frac{\rho_L}{2} \cdot v_{max}^2 + \mu_{R,f} \cdot \left(m_{F,r} \cdot g - c_{l,f} \cdot A_x \cdot \frac{\rho_L}{2} \cdot v_{max}^2 \right) \quad (3.12)$$
$$+ \mu_{R,h} \cdot \left(m_{F,r} \cdot g - c_{l,r} \cdot A_x \cdot \frac{\rho_L}{2} \cdot v_{max}^2 \right)$$

At the right of the equation, the first summand represents aerodynamic drag, the second rolling resistance at the front axle, and the third rolling resistance at the rear axle. The following equation can be derived as a polynomial with speed written as a variable:

$$\frac{\eta \cdot P_{max}}{v_{max}} = A_x \cdot \frac{\rho_L}{2} \cdot \left(c_D - c_{l,f} \cdot \mu_{R,f} - c_{l,r} \cdot \mu_{R,r} \right) v_{max}^2 + g \cdot m_{F,f} \cdot \mu_{R,f} + g \cdot m_{F,r} \cdot \mu_{R,r} \quad (3.13)$$

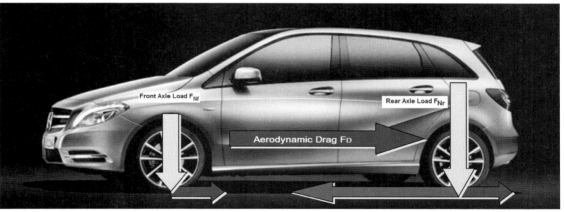

Figure 3.12 Forces acting on a vehicle at high speeds.

Equation 3.0 leads to a cubic equation for vmax:

$$0 = v_{max}^3 + a_1 \cdot v_{max} + a_0 \quad (3.14)$$

with

$$a_1 = \frac{g \cdot m_{F,f} \cdot \mu_{R,f} + g \cdot m_{F,r} \cdot \mu_{R,r}}{A_x \cdot \frac{\rho_L}{2} \left(c_D - c_{l,f} \cdot \mu_{R,f} - c_{l,r} \cdot \mu_{R,r} \right)} \quad \text{and} \quad a_0 = \frac{-\eta \cdot P_{max}}{A_x \cdot \frac{\rho_L}{2} \left(c_D - c_{l,f} \cdot \mu_{R,f} - c_{l,r} \cdot \mu_{R,r} \right)}$$

The practical solution for v_{max} is (working under the mathematical assumption that $(a_0/2)^2 + (a_1/3)^3 > 0$):

$$v_{max} = \sqrt[3]{-\frac{a_0}{2} + \sqrt{\left(\frac{a_0}{2}\right)^2 + \left(\frac{a_1}{3}\right)^3}} + \sqrt[3]{-\frac{a_0}{2} - \sqrt{\left(\frac{a_0}{2}\right)^2 + \left(\frac{a_1}{3}\right)^3}} \quad (3.15)$$

Without factoring in rolling resistance ($\mu_R = 0$) and when traveling on a flat and level surface ($\alpha = 0$), the following simplified equation applies:

$$v_{max} = \sqrt[3]{\frac{2\eta \cdot P_{max}}{\rho_L c_D \cdot A_x}} \qquad (3.16)$$

If a tractive force diagram (refer to Figure 3.10) is available, the top speed can also simply be determined based on the intersection of the tractive force curve plotted in the highest gear together with the respective power hyperbola under 0% incline conditions. The following Table 3.2 summarizes all performance and consumption values for the vehicle example used in section 3.2.6:

Table 3.2 Driving Performance and Consumption Values for the B180 CDI

Basic Data		Consumption	
Curb weight	1,630 kg	City	8.5 l/100 km
$c_D \cdot A_x$	0.27 · 2.21 m²	EUDC	4.8 l/100 km
Output in kW/rpm	110/4,200	NEDC	6.1 l/100 km
Torque in Nm/rpm	340/2,000	Const. 90 km/h (approx.)	4.3 l/100 km
Performance		Const. 120 km/h (approx.)	5.6 l/100 km
Acceleration from 0 to 100 km/h	10.1 s	Const. 150 km/h (approx.)	7.4 l/100 km
Acceleration over distance of 1 km	31.6 s	Const. 180 km/h (approx.)	9.7 l/100 km
Elasticity from 60 to 120 km/h in fifth gear	14.7 s	Const. 210 km/h (approx.)	13.5 l/100 km
Elasticity from 80 to 120 km/h in fifth/sixth gear	9.5/13.0 s	Mot cycle (test value)	5.6 l/100 km
Top speed	216 km/h (approx.)	AMS cycle (test value)	6.0 l/100 km

For the example used in Table 3.2, working under the assumption of a drive efficiency rating of 90% and an air density of 1.3 kg/m³ (cf. Figure 3.6), a top speed of 201 km/h can be derived from equation (3.16). This means that even in the absence of rolling resistance, the vmax can be approximated with relative accuracy. If the top speed and output of the vehicle are known factors and uplift can be neglected, the following equation (derived from eq. (3.13)) can be used to determine the rolling resistance at vmax:

$$\mu_R = \frac{\dfrac{\eta \cdot P_{max}}{v_{max}} - c_D \cdot A_x \cdot \dfrac{\rho_L}{2}}{m_F \cdot g} \qquad (3.17)$$

Correlating the data from Table 3.2, a rolling resistance coefficient of 0.0155 is calculated, which is very consistent with the values from Figure 3.5.

As a result of the parallel increase in engine output ratings (Figure 3.1) and the decrease in effective aerodynamic drag, the majority of vehicles registered in Germany can reach speeds in excess of 200 km/h. At the same time, however, traffic density is rising and speed limits are broadened in their reach (cf. Table 3.3). The importance or significance of top speed will therefore gradually decline, although it continues to serve as a benchmark performance figure for sports cars and driving aficionados.

Table 3.3 Speed Limits in Different Countries			
Europe (km/h)	City	Country road	Highway
Norway	50	80/90	90/100
Sweden, Finland, Denmark	50	80/90	110, 120, 130
Great Britain	48	96	112
Russia	60	90	110
Switzerland, Hungary, Ireland, Serbia	50	80/100	120
Belgium, Greece, Spain, Portugal	50	90	120
France, Croatia, Holland, Romania	50	90/100	130
Italy, Turkey, Luxembourg, Slovakia, Czech Republic	50	90	130
Bulgaria, Poland	50	90	140
Germany, Austria	50	100	∞, 130
Asia/Oceania (km/h)	City	Country road	Highway
Japan	40	50	80
China	—	80	120
India	—	80	100
Australia/New Zealand	50	100	120
USA (in mph)	Other roads	Urban interstates	Interstates
Kentucky/Maryland/Massachusetts/NY	55	65	65
Kansas/Louisiana/Tennessee	65	70	70
Colorado/Montana/Nebraska/Utah	55	65	75
New Mexico/North & South Dakota	55	75	80

3.4 Consumption

As mentioned in the introduction, lowering fuel consumption is the primary objective of all efforts expended to reduce aerodynamic drag. The following sections therefore go into detail on calculating and measuring consumption by leveraging driving cycles, influencing factors, and the possibilities available for reducing consumption and complying with legislative requirements.

3.4.1 Calculating Fuel Consumption

A model of the vehicle and its longitudinally dynamic characteristics are required for every consumption simulation exercise. To this end, the mathematical point mass model and the driving resistance equation it yields, equation (3.6) from section 3.2.5, are used. A detailed model of the powertrain as depicted in Figure 3.13 serves to generate the tractive force required to counteract the influencing driving resistance factors.

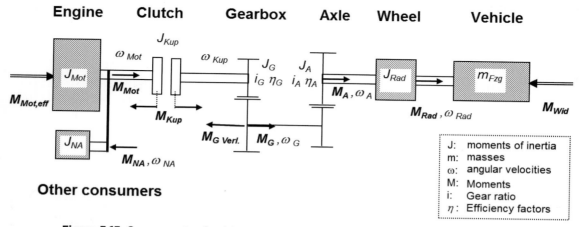

Figure 3.13 Components of a drive system with a manual transmission and its mapped performance in the simulation exercise [611].

A driving profile is also required for the calculation that not only includes a vector speed profile but also information regarding the shifting strategy and number and required power input of ancillary consumers as well as a potential gradient. In Europe, the NEDC is typically used and provides all of this information along with the shifting point requirements. Details are introduced in section 3.5.2.

Using this information as a basis, various possibilities arise for calculating or computing fuel consumption. One method is rooted in energy flow analysis and average efficiency (e.g., transmission efficiency of 85%) or power loss (e.g., ancillary consumers take away 6% of drive energy) for the individual components as monitored across the respective cycles. The downside of this method is that these efficiency scores and performance losses can vary quite extensively, depending on the driving cycle, and must therefore be determined with exacting precision; Sovran and Dwight [766] have opted for this approach. The benefit, however, is that one can see at which sections or areas of the driving curve efficiency is particularly poor and why. Less graphic but more useful, more accurate, and implemented in many software products is a calculation routine for characterizing the so-called load spectrum[4] (cf. Figure 3.14):

- The driving profile is subdivided into small, quasi-stationary time increments
- For every time increment, the output required at the wheel is determined and then—via the model in Figure 3.13 on moments of inertia (if acceleration is present), efficiency, and loss in the powertrain (e.g., transmission friction torque)—a paired value for the required engine torque and speed;
- Focusing on the engine map (cf. section 3.6.2), the fuel consumption for this time increment is then calculated; and
- adding all time increments allows the total consumption for the journey (e.g., for the NEDC cycle) and, thus, the consumption in l/100 km, to be computed.

The quasi-stationary calculation is based on the assumption that the speed profile can be check-driven with exacting precision and that the actual speed corresponds to the predefined target speed at all times. This approximation is permissible and compliant as long as highly dynamic driving profiles that involve considerable slip (in the clutch coupling or torque converter and between the tires and roadway surface) do not have to be simulated. In this context, the calculation method then represents a very good compromise between computational outlay and accurate results.

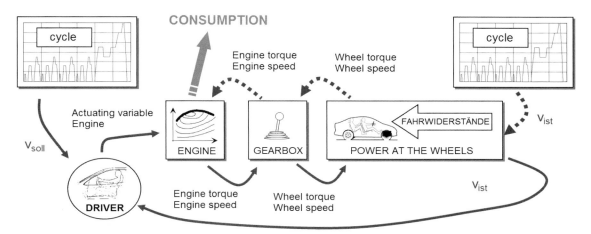

Figure 3.14 Quasi-stationary (····) and dynamic (—) consumption calculation [611].

In dynamic simulation, the model is augmented by a driver model as shown in Figure 3.14. The driver model assumes the function of a regulator that translates the respective speed signal into the engine torque input variable. The calculation is now made in exactly the reverse order as for the quasi-stationary calculation. This calculation method

---

4. A load spectrum is a convergence of paired values relating to engine torque and engine speed for a given cycle.

requires time and computationally elaborate iterations but can also map highly dynamic processes very accurately.

3.4.2 Consumption Measurement and CO₂ and Energy Equivalents

Measuring fuel consumption not only assists in computing fuel consumption figures, but much more importantly, it helps during the certification stage, which must be successfully passed before a vehicle model can be sold and operated and is a critical component in specifying fuel consumption figures on data sheets and in calculating fleet consumption (cf. section 3.8). To this end, the driving cycle in question is realized, or driven through, on a test rig (cf. section 3.5). The exhaust gas produced by the vehicle is trapped in a balloon and analyzed. The pollutant emissions then measured and quantified must undershoot the legal limits specified for the respective emissions class (Table 3.4):

Table 3.4 Exhaust Gas Limit Values in Europe (Exhaust Emissions in the NEDC Following a Cold Start)

Vehicles < 2,500 kg		Validity for vehicles	HC [g/km]	CO [g/km]	NO_x [g/km]	$HC+NO_x$ [g/km]	PM [g/km]	p [1/km]	Ve [g/Test]
Gasoline incl. hybrid	Euro 4	New, 1/1/2005 All, 1/1/2006	0.1	1.0	0.08	—	—	—	2.0
	Euro 5	New, 9/1/2009 All, 1/1/2011	0.1	0.68	0.06	—	0.005	—	2.0
	Euro 6	New, 9/1/2014 All, 9/1/2015	0.1	0.68	0.06	—	0.0045	—	2.0
Diesel incl. hybrid	Euro 4	New, 1/1/2005 All, 1/1/2006	—	0.5	0.25	0.30	0.025	—	—
	Euro 5	New, 9/1/2009 All, 1/1/2011	—	0.5	0.18	0.23	0.005	6*10¹¹	—
	Euro 6	New, 9/1/2014 All, 9/1/2015	—	0.5	0.08	0.17	0.0045	6*10¹¹	—

Fuel consumption bS in l/100 km is calculated from the exhaust test. The following applies for gasoline:

$$b_S = 0.118 \cdot \frac{0.848 \cdot HC + 0.429 \cdot CO + 0.273 \cdot CO_2}{\rho_K} \tag{3.18}$$

The following applies for diesel:

$$b_S = 0.116 \cdot \frac{0.861 \cdot HC + 0.429 \cdot CO + 0.273 \cdot CO_2}{\rho_K} \tag{3.19}$$

Consumption and Performance

Fuel density ρ_K is given in g/ml, and HC, CO, and CO_2 are given in g/km. The CO_2 emissions per km can be converted directly into l/100 km using the following multiplicators:

- 1 liter of gasoline per 100 km corresponds to 23.4 g/km CO_2
- 1 liter of diesel per 100 km corresponds to 26.5 g/km CO_2.

This discrepancy results from the difference in carbon content and, thus, the energy density of the fuels:

- 1 liter of gasoline contains 8.967 kWh of energy.
- 1 liter of diesel contains 9.943 kWh of energy.

This also explains why diesel-powered vehicles are always more fuel efficient (with a view to consumption in l/100 km) than their gasoline-powered equivalents. The lower consumption of diesel-powered vehicles typically goes beyond the 11% higher energy density, however, as the diesel engine still operates more efficiently than a gasoline engine.

Since the vehicle is strapped down to the test rig and only the rotating masses (cf. section 3.2.4) are moved, the effective translatory mass and aerodynamic drag of the vehicle must be simulated. These parameters are usually determined by downhill coasting tests and then simulated on the test rig. In one downhill coasting test, the vehicle is repeatedly accelerated up to a certain speed (e.g., 140 km/h), at which point the drivetrain is decoupled. The downward sloping speed-time progressional curves can then be correlated with statistical methods to determine the individual driving resistance forces (cf. section 3.2). Mass as a whole is broken down into large sectional areas, or categories, for reasons of simplicity, and vehicles are assigned to so-called inertia mass classes (Table 3.5) (cf. section 3.6.5):

Table 3.5 Six Inertia Mass Classes in Europe, Japan, and the United States

IM = equivalent inertia mass (to be aligned and configured on the test rig); RM = reference mass of the vehicle. In the EU: Curb mass of the vehicle with the fuel tank filled to 90% plus 100 kg to account for the driver and payload; in the United States: Mass of the operational vehicle plus 136 kg

EU (NEDC)		Japan (10–15 mode)			USA (FTP 75)		
Reference mass RM [kg]	SM [kg]	Reference mass RM [kg]	Reference mass RM [kg]	SM [kg]	Reference mass RM [kg]	Reference mass RM [kg]	SM [kg]
RM ≤ 480	455	RM ≤ 562	RM ≤ 480	455	RM ≤ 562	RM ≤ 480	455
480 < RM ≤ 540	510	562 < RM ≤ 687	480 < RM ≤ 540	510	562 < RM ≤ 687	480 < RM ≤ 540	510
540 < RM ≤ 595	570	687 < RM ≤ 812	540 < RM ≤ 595	570	687 < RM ≤ 812	540 < RM ≤ 595	570
595 < RM ≤ 650	625	812 < RM ≤ 937	595 < RM ≤ 650	625	812 < RM ≤ 937	595 < RM ≤ 650	625

Table 3.5 Six Inertia Mass Classes in Europe, Japan, and the United States *(Cont.)*

EU (NEDC)		Japan (10–15 mode)			USA (FTP 75)		
Reference mass RM [kg]	SM [kg]	Reference mass RM [kg]	Reference mass RM [kg]	SM [kg]	Reference mass RM [kg]	Reference mass RM [kg]	SM [kg]
710 < RM ≤ 765	740	1,125 < RM ≤ 1,375	710 < RM ≤ 765	740	1,125 < RM ≤ 1,375	710 < RM ≤ 765	740
765 < RM ≤ 850	800	1,375 < RM ≤ 1,625	765 < RM ≤ 850	800	1,375 < RM ≤ 1,625	765 < RM ≤ 850	800
850 < RM ≤ 965	910	1,625 < RM ≤ 1,875	850 < RM ≤ 965	910	1,625 < RM ≤ 1,875	850 < RM ≤ 965	910
965 < RM ≤ 1,080	1,020	1,875 < RM ≤ 2,125	965 < RM ≤ 1,080	1,020	1,875 < RM ≤ 2,125	965 < RM ≤ 1,080	1,020
1,080 < RM ≤ 1,190	1,130	2,125 < RM ≤ 2,375	1,080 < RM ≤ 1,190	1,130	2,125 < RM ≤ 2,375	1,080 < RM ≤ 1,190	1,130
1,190 < RM ≤ 1,305	1,250	2,375 < RM ≤ 2,625	1,190 < RM ≤ 1,305	1,250	2,375 < RM ≤ 2,625	1,190 < RM ≤ 1,305	1,250
1,305 < RM ≤ 1,420	1,360	2,625 < RM ≤ 2,875	1,305 < RM ≤ 1,420	1,360	2,625 < RM ≤ 2,875	1,305 < RM ≤ 1,420	1,360
1,420 < RM ≤ 1,530	1,470	2,875 < RM ≤ 3,250	1,420 < RM ≤ 1,530	1,470	2,875 < RM ≤ 3,250	1,420 < RM ≤ 1,530	1,470
1,530 < RM ≤ 1,640	1,590	Continued in increments of 500	1,530 < RM ≤ 1,640	1,590	Continued in increments of 500	1,530 < RM ≤ 1,640	
1,640 < RM ≤ 1,760	1,700		1,640 < RM ≤ 1,760	1,700		1,640 < RM ≤ 1,760	1,700
1,760 < RM ≤ 1,870	1,810		1,760 < RM ≤ 1,870	1,810		1,760 < RM ≤ 1,870	1,810
1,870 < RM ≤ 1,980	1,930		1,870 < RM ≤ 1,980	1,930		1,870 < RM ≤ 1,980	1,930
1,980 < RM ≤ 2,100	2,040		1,980 < RM ≤ 2,100	2,040		1,980 < RM ≤ 2,100	2,040
2,100 < RM ≤ 2,210	2,150		2,100 < RM ≤ 2,210	2,150		2,100 < RM ≤ 2,210	2,150
2,210 < RM ≤ 2,380	2,270		2,210 < RM ≤ 2,380	2,270		2,210 < RM ≤ 2,380	2,270
2,380 < RM ≤ 2,610	2,270		2,380 < RM ≤ 2,610	2,270		2,380 < RM ≤ 2,610	2,270
2,610 < RM	2,270		2,610 < RM	2,270		2,610 < RM	2,270

3.5 Driving Cycles

The previous section revealed that the way in which the vehicle is operated has a large affect on consumption in addition to the individual vehicle parameters. Driving cycles therefore pursue the objective of representing the behavioral driving characteristics of vehicle operators. In so doing, they not only establish a basis for comparing the consumption of one vehicle model to another, but also provide important reference points for developers and engineers so that they make the right decisions throughout the development process. None of this is possible, however, if a customer's behavioral driving characteristics are not simulated or replicated at least somewhat accurately. The following shows why it is exactly this approximation that presents such an imposing problem.

A basic distinction can be made between synthetic and realistic driving cycles, which are extracted from real dynamic load collectives. Genuine real-world driving cycles, for example, that involve highly dynamic movements and gradients (cf. section 3.5.7) cannot be replicated with exacting results on test rigs, which in turn compromises comparability.

3.5.1 History

Up until 1978 in Germany, the relatively simple DIN 70020 standard was applied for determining consumption, which was measured and calculated at a constant speed that represented 75% of the vehicle's rated top speed but did not exceed 110 km/h. Added to this value was a factor of 10% for "transient" driving. The fuel consumption figures determined in this manner were far too low when compared with what was achieved in practice. This marked the start of the so-called Euro Mix,[5] which took effect from 1978 to 1996 in Europe and accounted for the combined fuel consumption observed during one-third Economic Commission for Europe (ECE) city cycle driving (cf. Figure 3.15), one-third constant-speed driving at 90 km/h, and one-third constant-speed driving at 120 km/h. This standard was also criticized because real-world fuel consumption was still much higher than the test results. Reasons for this included the following:

- Constant travel at 90 (country road) and 120 km/h (highway/freeway) is not representative of actual, more dynamic real-world driving, which dramatically increases fuel consumption
- The one-third split represents too rough of an approximation

The Euro Mix was then replaced in 1996 by the NEDC (New European Driving Cycle).

3.5.2 New European Driving Cycle (NEDC)

The NEDC, which takes about 20 minutes to complete and covers 11 km, incorporated the ECE city cycle and supplemented it by adding the Extra Urban Driving Cycle (EUDC) (cf. Figure 3.15).

---

5. Only the three individual consumption points were defined by law, which were used to unofficially determine average fuel consumption in line with Euro Mix criteria.

NEDC

Temperature: 20 °C
No AC and other consumers
Duration: 1180 s
Lenght: 11,010 km
Stop duration: 280 s
Stops: 24 %
Av. velocity: 33.6 km/h
vmax: 120 km/h

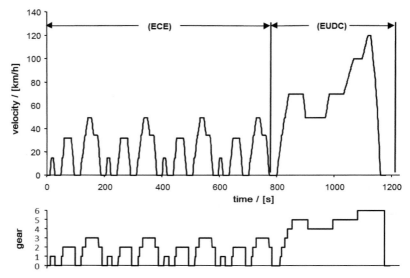

Figure 3.15 New European Driving Cycle; BP/EP: Beginning/end of sample testing.

Before vehicles are positioned on the test rig to run through the NEDC, they are preconditioned at 20 °C for a minimum of 8 hours. During measurement, the air-conditioning system and all electric ancillary consumers[6] are switched off and the alternator is operated under base load conditions only.

Although the inherent discrepancy with real-world consumption performance has gotten smaller thanks to the NEDC cycle and its lack of a driving segment carried out under constant-speed conditions (as was the case with the Euro Mix), the main reason for this closer approximation is that the period during which the engine idles has increased dramatically by accounting for 24% of the cycle time and the inefficient slow acceleration phases are overvalued; highway or freeway driving is all but ignored in the NEDC. The average speed observed in an NEDC simulation is only 33.6 km/h, while in Germany, for example, official average speeds of 15 km/h in the city (roughly equivalent with the ECE city cycle), 60 km/h on country roads (roughly equivalent with the EUDC), and 100 km/h on highways and freeways[7] are cited.

The consequence of this low travel speed in the NEDC is an undervaluation of a vehicle's driving characteristics that dominate its fuel consumption at high speeds. Among these characteristics, one stands out in particular: Aerodynamic drag. Tall gear ratios also play a role, however. Both enter the equation for a mere 10 seconds, at which time the NEDC

---

6. Only those consumers that are deemed essential for operating the engine and vehicle, such as the engine control unit, the instrument cluster, and all other mission-critical control units and sensors are supplied with electric power.
7. In countries that experience less traffic as a result of toll roads, for example, the average speed traveled on highways and freeways is most likely higher despite posted speed limit restrictions.

vehicle is accelerated up to 120 km/h. Because of their limited exposure, the effects they cause are almost imperceptible. The outcome of this is misleading information for customers who look to published documentation for accurate figures and can purchase the wrong vehicle for their needs as a result of this trust and the fact that vehicle developers and engineers make incorrect decisions for those customers that drive at speeds higher than those tested under NEDC conditions. The Worldwide Harmonized Light Vehicles Test Procedures (WLTP) (cf. section 3.5.6) currently in discussion provide a better, more accurate basis for characterization.

3.5.3 NEDC Cycle for Hybrid Drives

The measurement of fuel consumption for vehicles that feature a plug-in hybrid drive[8] is defined and specified in ECE standard R 101, whereby the vehicles are required to complete the NEDC cycle twice: first with a fully recharged battery then with a fully drained battery. If the drive (for achieving NEDC speeds and acceleration numbers) and battery (see below for the 11 km journey) are dimensioned sufficiently, a plug-in hybrid can be authenticated for the first cycle with 0 g/km CO_2 and a consumption of 0 l/100 km as the energy required to create electrical power has previously not been taken into account as a relevant factor.

The total fuel consumption to be specified (in l/100 km) is calculated using the following formula:

$$b_S = \frac{b_{S,1} \cdot s_e + b_{S,2} \cdot s_{av}}{s_e + s_{av}} \qquad (3.20)$$

$b_{S,1}$ represents the fuel consumption when the battery is fully recharged, $b_{S,2}$ the fuel consumption when the battery is depleted, s_e the vehicle's range on electricity only and s_{av} the 25 km journey as the reference distance between two recharging points. As soon as a plug-in hybrid has an electric range of 25 km and offers sufficient reserve capacity for the NEDC, it can be rated as consuming half the fuel of a vehicle powered by a conventional engine only. Should the electric range exceed this base figure, consumption drops still further. This explains the extremely low fuel consumption data given for some plug-in hybrids such as the BMW Vision Efficient Dynamics, Toyota Prius Plug-In, Audi A1 eTron, and VW XL 1.

3.5.4 United States Cycles

In the United States, up to 2010, fuel consumption was measured in accordance with the US combined driving cycle as shown in Figure 3.16. This cycle comprises 55% city and 45% highway or freeway driving at average speeds of 32 and 77 km/h and maximum

---

8. The hybrid battery in these vehicles can also be recharged by connecting to the main power grid via a built-in charger so that the powertrain can be operated on electricity only during small, inner-city trips.

speeds of 91 and 96 km/h. The cycle is less "schematic" than the NEDC because it is based on realistic speed profiles and it appears to accurately reflect American driving habits.

FTP75/City

Temperature: 24 °C
No other consumers
Duration: 2x 765 s
Length: 2x 10.22 miles
Av. velocity: 48.1 mph
vmax: 59.9 mph

Highway

Temperature: 20 - 30 °C
No other consumers
Duration: 1877 s
Length: 11.1 miles
Av. velocity: 19.7 mph
vmax: 56.7 mph

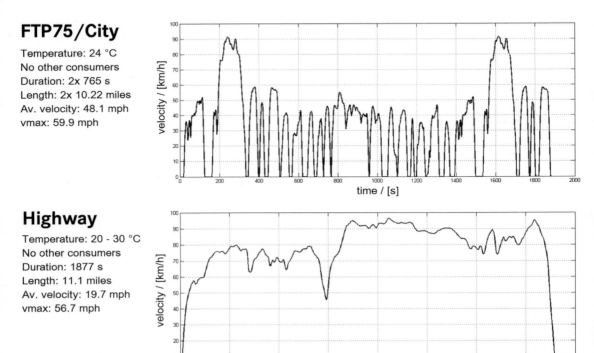

Figure 3.16 US combined driving cycle: City cycle above, highway cycle below.

Since 2011, the US combined driving cycle has been supplemented with three additional cycles for a total of five cycles (refer to Figure 3.17): One fast highway cycle at up to 80 mph (US06), one hot cycle conducted under a prevailing ambient temperature of 35 °C with the air-conditioning system on (SC03), one cold cycle conducted under a prevailing ambient temperature of −7 °C, and one heat on cycle (cold CO), all of which are transitioned through in sequential order. For all cycles and the two extreme cycles in particular, the vehicles tested must be conditioned to the correct temperature in each case for a minimum of 12 hours.

US-Combined
+ US06
Temperature: 20 - 30 °C
No other consumers

+ SC03
Temperature: 35 °C
AC on 22 °C Auto
Solar radiation 840 W/m²
Humidity max. 40 %

+ ColdCO
Temperature: 6.7 °C
Heat on 22 °C Auto

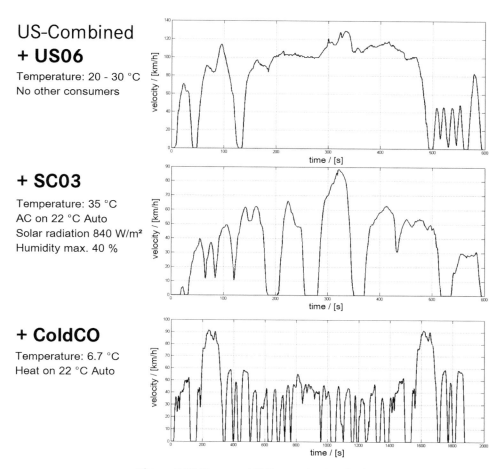

Figure 3.17 The new US five-cycle standard.

3.5.5 Asian Cycles

Japan currently uses the 10–15 mode cycle (cf. Figure 3.18). Characterized by an average speed of 22.7 km/h and a maximum speed of 70 km/h, it is the slowest out of all the cycles. On March 1, 2013, it will be replaced by the JC 08 emissions test cycle, which is based on realistic driving profiles and incorporates the maximum speed limit in Japan of 80 km/h (cf. Figure 3.18).

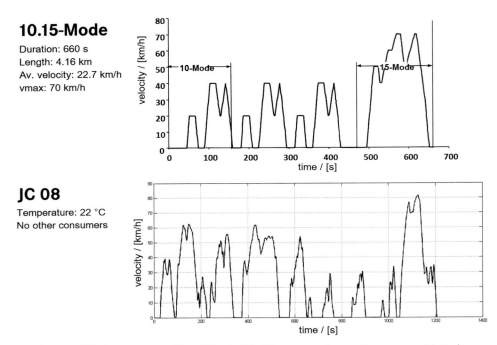

Figure 3.18 Japan cycles: The still valid 10–15 mode cycle and the new JC 08 cycle.

China currently uses the NEDC and is not expected to adapt to the WLTP program but instead develop its own cycle. In India, on the other hand, official fuel consumption figures are based on a modified version of the NEDC otherwise known as the Modified Indian Driving Cycle (MIDC). Consumption is measured in km/l.

3.5.6 WLTP—Worldwide Harmonized Light Vehicles Test Procedure

For several years now, there has been debate about the importance of implementing an internationally valid and more realistic test cycle, with discussion surrounding the WLTP program. This program has since achieved a respectable level of stability in its fifth version but has not yet been finalized. It is scheduled to be introduced from 2017 and comprises four sub-cycles: city, country road, highway, and freeway driving (cf. Figure 3.19) and—like the American cycles and the new JC 08 cycle in Japan—integrates cycle-based driving derived from realistic driving profiles that involve higher travel speeds as well as a large portion of driving at speeds higher than defined in the NEDC. This approach thereby also elevates the importance or significance of aerodynamic drag by one-third as compared to NEDC testing conditions (cf. section 3.5.2).

Consumption and Performance

WLTP V5

Duration: 1800 s
Length: 23.141 km
Stop duration: 234 s
Stop: 13 %
Av. velocity: 46.3 km/h
vmax: 131.3 km/h

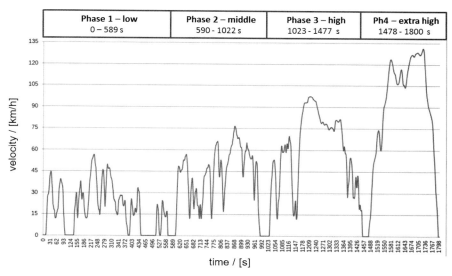

Figure 3.19 WLTP—Worldwide Harmonized Light Vehicles Test Procedure [368].

3.5.7 Realistic Driving Cycles

The explanations given for the various cycles defined have shown that there is currently no cycle that is truly optimal. Furthermore, when the different driving behaviors observed around the world are taken into account, it becomes clear that it is not even possible to design a broad-reaching cycle that accurately applies in all cases and scenarios. Each country and vehicle operator has its own characteristic usage profile. As such, it comes as no surprise that not only every auto manufacturer in Germany has developed and safeguards or protects its own internal fuel consumption testing cycles; trade journals such as *AutoBild*, *Auto Motor Sport*, and *Mot* have also orchestrated testing schemes that reflect what they believe to be an accurate depiction of real-world fuel consumption (cf. Table 3.2).

One possible approach for avoiding the dilemma of providing inaccurate consumption figures but that is also not unproblematic when it comes to observing data protection laws is to have every auto manufacturer evaluate the driving profiles of a representative number of vehicles in each class and leverage the results to develop and engineer vehicles that more specifically target customer requirements.

3.6 Possibilities for Reducing Fuel Consumption

This section is designed to indicate which properties or characteristics of a vehicle affect fuel consumption and in what way as well as point out the areas in which efforts expended by vehicle developers and engineers to reduce consumption can really pay off.

3.6.1 Energy Flow Diagram

A good entry point in the context of the example vehicle referenced earlier in section 3.2.6 is to examine the flow of energy in the NEDC cycle (cf. Figure 3.20). Due to the comparably low load involved, the direct-injected gasoline engine as shown here achieves a median efficiency level of 28% under NEDC testing conditions. At its best operating point (cf. section 3.2.6), a modern gasoline engine can achieve an efficiency level of approximately 36%, whereas its diesel engine equivalent is capable of surpassing the 40% mark. Some 5% of energy is wasted at idle speed (the time at which a start/stop system reaches its maximum energy-saving potential) and approximately 67% of the energy stored in the fuel itself is lost in the form of frictional heat and wall-heat transfer into the coolant (a fact that at least proves beneficial during the cold winter months) as well as exhaust-gas heat. The remaining 28% of the fuel energy can thus be utilized to generate a mechanical force at the crankshaft, of which approximately 10% (or 3% of the total energy) is required by ancillary consumers and is lost in the powertrain (transmission, drive shaft). Eighty percent of the mechanical force generated at the crankshaft or 22% of the fuel energy is then ultimately transferred directly to the wheels as mechanically usable output energy to propel the vehicle. Of this figure, almost half (45% or 10 % of the fuel energy) is expended during acceleration, and 30% (6.7 %) is used to overcome aerodynamic drag—despite the low average speed of 33.6 km/h. Exactly one quarter (5.6 %) of the energy available at the wheel is converted to heat in the tire (cf. section 3.2.1 and Figure 3.20).

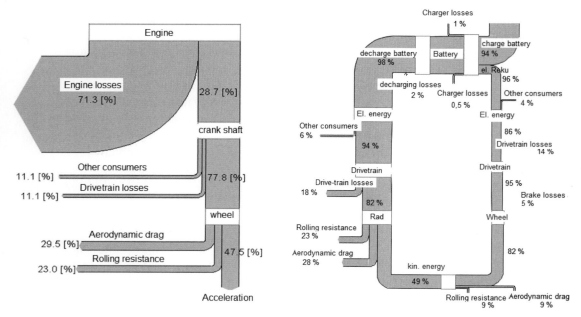

Figure 3.20 Energy flows in the NEDC: To the left, for a vehicle with a combustion engine; to the right, for a vehicle with a battery-electric drive.

In the case of an electric drive (Figure 3.20, right), this energy flow diagram appears totally different; the electric motor has a fantastic mean efficiency of 94%. The much higher gear reduction leads to powertrain losses that exceed those of combustion drivetrains. And the losses created by optimized electric ancillary consumers are less severe.

Due to the higher amount of mass involved (the weight of the battery pack is the main contributing factor here), almost exactly half of the mechanical energy is required for acceleration. However, recuperation or regenerative braking is capable of capturing two-thirds of this acceleration energy and feeding it back to the battery, while the losses attributed to aerodynamic drag and rolling resistance are irreversible (cf. section 3.2).

In considering both drive systems, it becomes clear that the significance of aerodynamic drag as it pertains to the NEDC is comparably low. When faster cycles are run through, however, aerodynamic drag plays a considerable role and reaches a percentage rating of almost 90% at high freeway speeds despite the fact that rolling resistance has also increased (cf. section 3.2.6).

3.6.2 Engine Efficiency and Engine Maps

As is evident in Figure 3.20 (left), the combustion engine[9]—having being optimized for over 100 years now—still bears the greatest potential for reducing fuel consumption. A so-called conchoid diagram shows the correlations between engine speed (abscissa), torque or mean pressure in the engine (ordinate), and specific fuel consumption, which is displayed using contour lines in the shape of an efficiency curve that resembles something of a hiking trail map. Such a graphical depiction also plots the maximum torque (in red) and the power hyperbola in 10 kW increments (in blue). The comparably small plateau reveals that gasoline engines in particular reach their lowest specific fuel consumption or highest efficiency in a small operating range. Modern gasoline engines consume approximately 240 g/kWh, whereas their diesel engine equivalents already undershoot the 200 g/kWh mark for specific fuel consumption. This can be converted to efficiency using the following formula:

$$\eta(\%) = \frac{100}{\text{specific fuel consumption}\left(\frac{\text{kg}}{\text{kWh}}\right) \times \text{Thermal value of fuel}\left(\frac{\text{kWh}}{\text{kg}}\right)} \quad (3.21)$$

The thermal values for diesel (11.8 kWh/kg) and gasoline (11.5 kWh/kg) lead to a maximum efficiency rating of approximately 36% for gasoline engines and 42% for diesel engines.

This optimum efficiency is achieved under relatively high load conditions for diesel engines (80 to 85% of maximum torque or mean pressure) and at relatively low engine speeds (2,000 to 2,500 rpm); for gasoline engines, optimum efficiency is achieved under lower load conditions (60 to 65%) and slightly higher engine speeds (3,000 to 3,500 rpm). Modern, consumption-optimized engines, however, are characterized by the fact that they achieve approximately 90% of their maximum efficiency at low operating speeds and loads which, during NEDC testing (thickly bordered area) and normal driving, helps to lower fuel consumption (cf. Figure 3.21).

---

9. The combustion engine is a heat engine, which is why its efficiency is, according to Carnot $\eta = 1 - T_V/T_A$ ($T_{V/A}$ = combustion, exhaust temperature), limited to a maximum of approximately 60%. Large-scale marine diesel engines achieve $\eta = 0.50$.

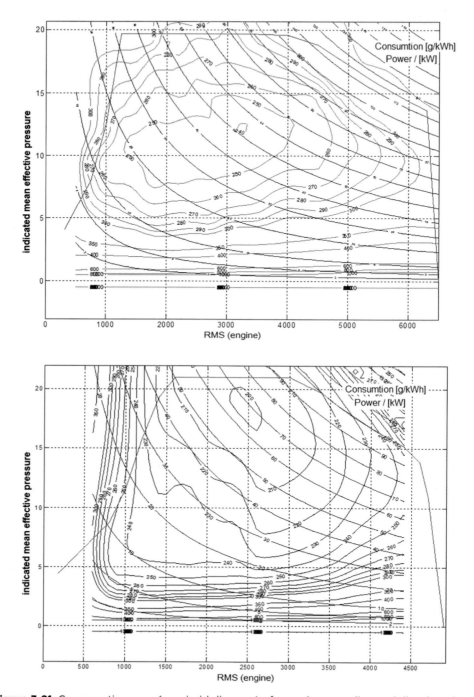

Figure 3.21 Consumption map (conchoid diagram) of a modern gasoline and diesel engine

Engine developers therefore continue to pursue the objective of increasing torque output at low speeds and improving efficiency at low mean pressures in particular, which can be realized through the use of direct injection and/or downsizing and turbocharging. Engines that have higher torque at low speeds can be operated with longer gear ratios (cf. section 3.6.4) to increase load and reduce fuel consumption.

Figure 3.21 makes it clear that driving under full-load conditions at high engine operating speeds must, by default, lead to very high fuel consumption due to the driving resistance forces involved and the fact that the specific consumption of the engines starkly rises as output and operating speed climb. In terms of diesel engines, this rise does not become that apparent under full-load conditions (approximately 240 g/kWh or 20% worse than optimal performance) when compared to gasoline engines (over 300 g/kWh and, therefore, almost 40% worse than optimal performance). Diesel-powered vehicles thus have much lower fuel consumption at very high freeway speeds than their gasoline-powered equivalents.

3.6.3 Ancillary Components

Ancillary components primarily refer to those systems that secure or safeguard operation of the engine and include the oil and water pump, fuel pump, injection system, ignition system, engine electronics, and at high engine temperatures, the electric fan. Adding to this are comfort or convenience-oriented systems such as the power steering and air-conditioning systems.[10]

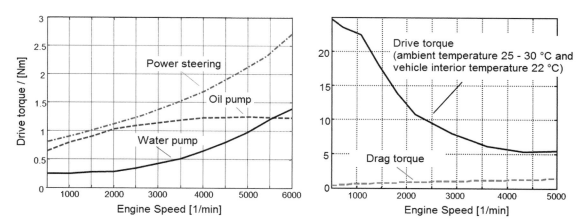

Figure 3.22 Drive torques for oil, water, and hydraulic power steering pump (left) and air-conditioning system (right) as a function of engine speed.

Many systems that just a few years ago were driven by the belt drive of the engine have since been electrified for consumption reasons. The electric water pump and electric power steering system, for example, can be actively regulated in line with current

---

10. The power steering pump of the power steering system is always coupled directly to the engine; the A/C compressor can also sometimes be disconnected via a magnetic coupling.

operating requirements and therefore be run much more efficiently. The drag torque of the hydraulic power steering pump alone increased fuel consumption by approximately 0.2 l/100 km and has been reduced to under 0.1 l/100 km under optimal conditions, thanks to the integration of an electric power steering system.[11]

The largest consumer of energy among ancillary components is by far the air-conditioning system, which has yet to be taken into account by NEDC testing procedures, although it is also operated before and after the summer months on most vehicles to dehumidify air in the passenger compartment, for example. Depending on the difference between the outside temperature and the desired inside temperature as well as the driving cycle itself, the air-conditioning system can increase consumption at a rate that is barely perceptible or cause the vehicle to consume several additional liters of fuel per 100 km. Having said this, however, adjusting the air-conditioning system to an efficient, moderate setting on the highway or freeway can prove to be more efficient than driving down the road with the windows open, which in turn increases aerodynamic drag. In addition, heat-reflecting glass, better insulated passenger compartments, and even more efficient air-conditioning systems (e.g., CO_2-based units) can tap further areas in which fuel consumption can be reduced in the future. This is of special interest when viewed in the context of electric vehicles, since high temperatures (the air-conditioning system must work extra hard) and low temperatures (the electric motor generates only minimal heat, which is why the vehicle must be additionally heated using electricity drawn from the battery) greatly reduce range.

All electrical, safety-relevant (e.g., lights, wipers, heated door mirrors, washer nozzles, and rear windows) and comfort and convenience-oriented systems (e.g., heated seats, ventilation, electric heater booster, power seat adjustment, and additional seat-based systems) are supplied by the alternator, whose comparably low efficiency[12] leads to an additional fuel consumption of up to 0.1 l/100 km (2.5 g/km CO_2), depending on the driving profile, for each 100 watts of electric power.

Modern alternator management systems can reduce the load placed on the engine and thus reduce fuel consumption by having the alternator work under maximum load during braking (similar to the regenerative braking that occurs in electric and hybrid vehicles) so that the on-board electrical system battery can be recharged with the excess kinetic or potential energy produced. This energy is then made available to the on-board power supply during acceleration and when driving at constant speeds.

---

11. EPAS (electric power assisted steering) has already replaced the conventional hydraulic power steering system for lightweight vehicles because it is lighter and consumes less fuel as tested under NEDC conditions and on roads with few curves. In heavier vehicle applications, sufficient power assist cannot be provided using a 12 V on-board power supply; with the advent of 48 V on-board power supplies or hybrid drives, on the other hand, technology does exist to integrate an electric power steering system for these applications as well, albeit in the long term.
12. The future may see starter-alternators with considerably higher outputs and efficiency levels to reduce the parasitic effect electrical systems have on fuel consumption.

It also pays to optimize the energy consumption of consumers that require high amounts of input energy and are operated for long periods of time. Traditional halogen bulbs consume approximately 120 W, whereas gas-discharge bulbs require approximately 100 W while providing substantially more light, and the highly efficient LED daytime running lamps that have enjoyed widespread use since the introduction of daytime driving lights in February 2011 and are switched on continuously as the vehicle is operated, have an input power requirement of approximately 10 W.

3.6.4 Transmission

The entire power transmission sequence from the engine through to the drive wheels comprises the transmission, propshaft (if available), and the axle, which is made up of the differential, transfer case, and drive shafts. For longitudinally installed engines—regardless whether the vehicle has front or rear-wheel-drive—the axle must turn the driving energy 90 degrees to the wheel, which in itself creates a loss of up to 5% (cf. Table 3.6).

The efficiency with which power is transferred is of comparable significance to that of the engine, since every percentage increase in efficiency has a direct effect on overall fuel consumption. Table 3.6 lists the efficiency scores of the different types of transmission.

Table 3.6 Overall Efficiency of Different Types of Transmission			
Transmission	η	Axle	η
Manual transmission/AMT/DSG	92–97%	Differential	91–96%
Automatic transmission with torque converter lockup clutch	90–95%	Front-drive axle	93–98%

In simulation, the overall efficiency of the transmissions is depicted by way of a speed-dependent friction torque loss and a constant efficiency plateau for each gear that ranges between 96 and 99% (cf. Figure 3.23).

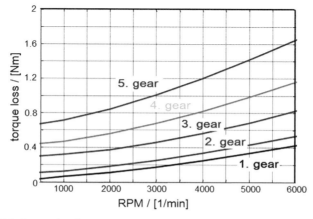

Figure 3.23 Example of transmission friction torque losses for different gears.

In addition to the efficiency of the transmission, consumption is also affected by the number of gears,[13] whereby more gears equals greater efficiency because the engine can be operated at its best operating point for each speed and load condition that can be encountered (cf. Figure 3.21). The gear ratio spread is also of importance and characterizes the gear ratios from the lowest to the highest gear. First gear is primarily used to start the vehicle from a stop (not only on a hill), whereas the highest gear provides the best fuel efficiency when driving at high speeds. The taller the highest gear, the less fuel the vehicle requires at high speeds. A taller ratio of 10% would reduce fuel consumption by 3 to 4% at constant speed, whereby manual transmissions in particular must strike a good balance between consumption and ascending ability to prevent the driver from having to downshift when the smallest of incline approaches. Automatic transmissions are more receptive to tolerating downhill shifting (cf. section 3.3.2).

The more favorable transmissions from a fuel consumption perspective are automated manual transmissions, of which there are two types: the AMT[14] and the DCT.[15] These transmissions, unlike automatic transmissions[16] and the CVT,[17] have no torque converter friction losses. Automatic transmissions can realize highly favorable gearshift programs when operated in an eco mode, for example, which very quickly shifts to the next highest gear to maximize engine load and improve efficiency (cf. Figure 3.21). For manual and automatic transmissions that encourage the driver to select a gear, gearshift recommendations are possible for NEDC fuel consumption certification that operate by providing a visual indicator in the instrument cluster to upshift or downshift accordingly. This, in turn, makes it possible to deviate from the less than optimal shifting recommendations specified by the NEDC (cf. Figure 3.13), and considerably less fuel can be consumed if drivers take note of these gearshift recommendations to facilitate fuel-efficient driving.

The efficiency drawbacks of the CVT can be somewhat compensated for by the continuously variable transfer of power as the CVT is the only transmission that forces the engine to always operate at peak efficiency.[18] The acoustic impression that the engine leaves on the driver must be taken into account, however, and continues to be a major source of complaint.

---

13. Previous manual transmissions used to have five speeds. Today, however, the standard is five or six speeds. The same holds true for automatic transmissions, which used to have three speeds and now offer five to seven speeds.
14. Automated manual transmission: The clutch is actuated and gears are shifted by electric motors—convenient, comfortable shifting but with an interruption in tractive power.
15. Dual clutch transmission: Unlike automatic transmissions that use a hydraulic torque converter, torque is transferred by one of two clutches that mechanically connect both sub-transmissions with the powertrain. Gears are shifted by having one clutch close while the other one is opened.
16. The friction losses associated with the torque converter in an automatic transmission are avoided, or prevented, in some modern vehicles from the second gear onwards via a torque converter lockup clutch.
17. Continuous variable transmission: A chain link serves as the power-transfer medium and allows for a continuously variable gear ratio between the gear wheels. Unfortunately, this type of transmission also produces higher friction losses.
18. This, when viewed on the performance map in Figure 3.16, is the "spine" or "ridge" of the efficiency curve, which is outlined with the lowest specific consumption for every operating speed relative to mean pressure.

3.6.5 Vehicle Mass

Out of all the vehicle parameters that pertain to the NEDC, vehicle mass has the greatest impact on fuel consumption due to the varying driving profile used. This can be illustrated with a view to sensitivity (cf. Figure 3.24). A mass sensitivity of 0.3 means that a weight reduction of 1% leads to a reduced fuel consumption of 0.3% under NEDC testing conditions. When the vehicle is driven at a constant speed, this value diminishes and the significance of aerodynamic drag dramatically increases (cf. section 3.6.7). These consumption sensitivities only apply to a specific vehicle with a specific engine because the relationship between aerodynamic drag and rolling resistance, for example, are inextricably linked to the type of vehicle and the engine itself has a large impact on sensitivity performance. In terms of quantifying fuel consumption in the NEDC, it must be noted that changes in mass lead to a stepped or graduated change in consumption; only when the vehicle in question reaches the next inertia weight class (cf. Table 3.5) does NEDC consumption change but by a figure of approximately 0.2 l/100 km.

Since reducing the amount of weight by one inertia weight class is typically very expensive for a vehicle concept,[19] reducing fuel consumption can usually be better targeted by investing the money elsewhere[20] (cf. section 3.7), even if the reference mass plays a key role for other vehicle functions such as driving performance, driving dynamics, crash performance, and durability.

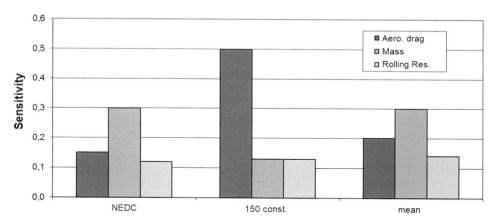

Figure 3.24 Sensitivities for consumption-relevant vehicle parameters in various testing cycles.

---

19. As a result of the platform and interchangeable parts strategies required for cost considerations, it is not possible to design each and every part specifically for one type of vehicle. The result is that many parts (bodyshell, axles, etc.) are overdimensioned because they must also be compatible with the heaviest or fastest model of a model series (typically with the most powerful engine). In order to still significantly reduce weight, the only option available is to redesign—at considerable expense—the material concept used for the bodyshell (e.g., aluminum).
20. For example, engine, ancillary components, tires, aerodynamic drag (cf. section 3.7).

3.6.6 Rolling Resistance

Although rolling resistance is a seemingly insignificant factor when compared to other driving resistance forces (cf. Figure 3.9), it continues to hold the vehicle back as soon as it starts off. Achieving lower rolling resistance thus always results in lower fuel consumption, which becomes apparent in all testing cycles in correlation with the remaining sensitivities (cf. Figure 3.24). The significance of rolling resistance plays an ever larger role when a vehicle is driven in an economical manner, whereby aerodynamic drag (by avoiding high speeds) and acceleration losses (by avoiding frequent braking) are kept to a minimum.

The difference between a standard tire and one optimized for low rolling resistance (μ_R minus 15%, cf. Figure 3.5) results in a consumption decrease of just below 2% under NEDC testing conditions. Similar reductions in the rolling resistance coefficient can also be achieved by pumping more air into the tire without exceeding the maximum operating pressure specified.[21] An increase in pressure of 1 bar corresponds to a reduction in rolling resistance of 15 to 20%. Doing this compromises ride quality, however, and it increases road and tire noise.

3.6.7 Aerodynamic Drag

Unlike rolling resistance, aerodynamic drag only plays an important role in fuel consumption when the speed of the vehicle reaches a certain level. A good rule of thumb to bear in mind in this context is that aerodynamic drag becomes increasingly prevalent as soon as the drag produced by the vehicle's speed exceeds the force attributed to rolling resistance (cf. section 3.2.6). In terms of small and lightweight vehicles, the scales can start to tip as soon as inner-city speeds are reached. The Smart City Coupé and the VW Up!, for example, reach this threshold at approximately 50 km/h, whereas larger, heavier sedans need to pass through approximately 70 km/h. This speed dependency also makes itself evident in the realm of heavily fluctuating cyclic sensitivity (cf. Figure 3.24).

Figure 3.25 clearly portrays the fuel savings realized when comparing the first B-Class launched in 2005 to the current B-Class, which debuted in 2011. Depending on the speed and model—the best predecessor vehicle had an aerodynamic drag performance figure of c_D 0.30, with the best base version currently available boasting a figure of c_D 0.26 (−13.5%) or c_D 0.24 (−20%) when equipped with the optional eco technology pack—the successor model consumes up to 1.5 l/100 km less fuel than before when traveling at a constant speed of 180 km/h solely as a result of its optimized aerodynamic drag. The two indicator marks in Figure 3.25 show the consumption savings in the NEDC (0.13

---

21. Every tire has a rated maximum operating pressure that is imprinted on the sidewall (typically specified in kPa, e.g., 350 kPa corresponds to 3.5 bar) and must not be exceeded. The reason for this specification is to limit or restrict compromises made to comfort (e.g., road and tire noise, particularly when the vehicle is traveling over pavement joints), traction (e.g., during braking), wear, and fuel consumption. Those who place high value on low fuel consumption can considerably increase their tire pressures without having to be concerned any further.

or 0.19 l/100 km), which are as low as they are due to the slow average NEDC testing speed of 33.6 km/h (cf. section 3.7). These savings, however, still lie within the range that can be achieved by lowering the inertia weight class of the vehicle by one increment (cf. section 3.7). In addition, as soon as the vehicle passes through 140 km/h, the fuel savings reach such high levels that they all but cannot be approximated by reducing weight alone.

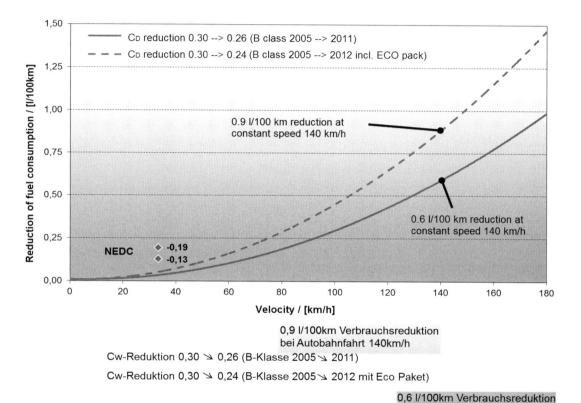

Figure 3.25 Reduction in fuel consumption via optimized aerodynamic drag as shown on the previous and current Mercedes B180 CDI.

Apart from the lower fuel consumption, reduced aerodynamic drag also offers better longitudinal dynamic performance. In reference to the preceding example ($c_D = 0.26$ versus 0.30), the top speed increases by almost 10 km/h, and acceleration and elasticity are also improved at high speeds in particular.

3.7 Reducing Aerodynamic Drag

This section provides a basic overview of measures designed to reduce aerodynamic drag (cf. chapter 4) as well as introduces several methods for comparing different consumption-reducing measures as referenced in section 3.6.

3.7.1 Possibilities for Reducing Resistance

As mentioned at the beginning, vehicles that set the aerodynamic drag benchmark in their class have implemented all cost-effective measures. These include the following design measures in addition to achieving the all-important aerodynamically favorable base form:

- Perfect rear end flow coordination with optimal trunk lid height, lateral recesses including defined airflow break-away edges, and an effective rear diffusor
- Optimized front apron for good lateral flow around the wheels, including leading and trailing airflow facilities in the rocker panel area
- Tightly sealed front-end structure with engine hood and headlamp perimeter sealing and a dense radiator surrounding to maximize the efficiency of the cooling airflow
- Smooth, flat covering of the underfloor area
- Wheel spoilers in front of the wheels with additional measures for optimizing wheel well air flow characteristics (cf. [348] und [848])
- A- and C-pillars, outside mirrors, and other add-on body parts with optimal airflow deflection.

Further improvements can also be made to today's top performing models by going beyond the aforementioned items to integrate additional measures. These measures lead to substantially increased costs, however:

- A radiator shutter that actively regulates the amount of cooling air entering the engine compartment; Δc_D potential = −0.005 up to a maximum of −0.020, depending on the cooling air requirements of the vehicle in question
- A cover for the entire underfloor area, including the axles, exhaust system, and transmission tunnel; expensive and critical with respect to exhaust heat; Δc_D potential = −0.005 up to a maximum of −0.015
- A speed-dependent ride height leveling system; very expensive if not already installed to cater to comfort or dynamic driving requirements; Δc_D potential = −0.004 for each 10 mm reduction in height
- Adaptive components such as a rear spoiler or rear wing; very elaborate in construction and very expensive

Measures can also be implemented that are intricately related to the dimensional concept but frequently detract from the design of the vehicle, at least from today's perspective:

- A rounded front end with a long overhang or reduced sweep to achieve perfect airflow around the front end
- A-pillars with a pronounced curvature, minimal pitch, and perfect transition to the side of the vehicle (objective: to further reduce air turbulence at the A-pillars)
- An overarched roof with the highest point near to the windshield

- Exaggerated recess points at the rear of the vehicle, in the Y and Z directions, in conjunction with a dramatic narrowing of the rear axle track width and elongation or stretching of the rear overhang
- Special aerodynamic wheels with minimum width and flow-through characteristics as well as optimized tires that lead to substantially better fuel consumption performance on any vehicle [892]

3.7.2 Weight Equivalency

One possibility for evaluating or assessing the impact of aerodynamic measures is to consider the weight equivalency factor. This involves comparing the consumption efficiency of a c_D-reducing measure with a reduction in weight (Table 3.7).

Table 3.7 Weight Equivalency of a Reduction in Aerodynamic Drag of $\Delta c_D = -0.010$ in Different Test Cycles

Cycle	$\Delta c_D = -0.010$ Change in consumption (l/100 km)	$\Delta m_F = -100$ kg	Weight equivalency (kg)
City	0.01	0.18	6
EUDC	0.04	0.12	33
NEDC	0.03	0.15	20
Const. 90	0.05	0.06	83
Const. 120	0.09	0.07	129
Const. 150	0.15	0.07	214

Very high values are already achieved during NEDC testing that further compound to produce extremely high values when a vehicle is driven at constant speeds. The consumption efficiency is all that is related to or taken into account, however. When more extensive changes in mass are made (which are only practical in light of the inertia mass dilemma), the effects on driving performance and any potential secondary effects such as taller gear ratios must be regarded accordingly. In extreme cases, entirely different engines can be used, at which point the above approach becomes very complex.

3.7.3 Amortization Analysis

Especially when product decisions revolve around comparably expensive aerodynamic or other consumption-reducing measures, it can prove beneficial to apply a more accurate evaluation method that compares the various measures in as transparent a fashion as possible. The amortization analysis introduced here not only facilitates a fair comparison of the measures in question but also clearly indicates how many months or years must go by before a consumption-reducing measure has paid off for the customer as verified in testing cycle performance.

All consumption-reducing measures are consolidated as paired values—for savings with respect to NEDC testing and manufacturing costs—in a double logarithmic portfolio as outlined in Figure 3.26. The gray diagonals represent different amortization periods of 15,000 (far left) to 120,000 km working under the assumption of a fuel price of € 1 and an interest or return of 5% per year. One measure that lies just above the first straight line at the left amortizes in under 15,000 km during NEDC testing. Those who frequently travel at high speeds can certainly make up for the costs of the consumption-reducing measures in even less time, provided that these measures target aerodynamic drag.

The position of the gray points with the black border indicate that currently, only those measures are implemented that amortize with the customer in less than two years. The light gray point—a weight reduction of 1 kg at a cost of € 5—indicates that this measure (if it is only designed to reduce consumption) is very expensive when compared to the aerodynamic measures and will only amortize after the vehicle has traveled more than 120,000 km. Reductions in weight also have positive "side effects," including higher agility, improved crash performance, and better durability. These positive effects can be evaluated from a cost perspective, which make the design measure more attractive when viewed in terms of consumption. The gray point can thus move to the left by the designated amount, which effectively shortens or reduces the amortization period. The same applies to many potential aerodynamic improvements. For example, cover panels that mask the underfloor not only reduce the c_D figure but also minimize the contamination of components, potential damage, and corrosion and can improve acoustic conditions. The white points with the black border indicate that many potential improvements can be made in the future that are not currently feasible with respect to the current price of energy as it relates to componentry expenses (cf. examples in section 3.7.1).

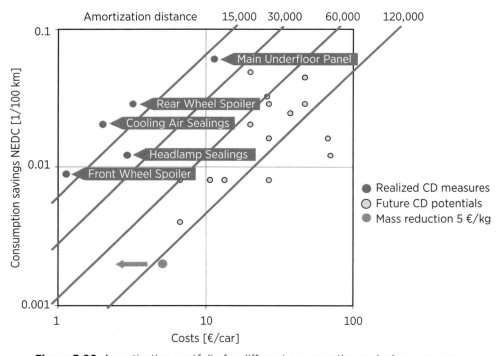

Figure 3.26 Amortization portfolio for different consumption-reducing measures.

3.8 CO$_2$ Legislation and Labels

The first emission and consumption laws originated in the United States. In response to the oil embargo of 1973, the US congress issued the Energy Policy and Conservation Act in 1975, whose primary objective was to reduce the United States' dependency on imported oil. The corporate average fuel efficiency, or CAFE (US fleet consumption), was to be almost halved by increasing fuel efficiency to 27.5 mpg, or 8.5 l/100 km, in ten years. In an effort to achieve this objective, the NHTSA put the so-called "gas guzzler tax" into effect in 1978, which targeted highly inefficient vehicles by levying a higher purchase tax (cf. section 3.8.3). The Clean Air Act issued by the California Air Resources Board (CARB) in 1990 was drafted to ensure that a minimum of 2% of newly registered vehicles were emissions-free by 1998, with 10% mandated for 2003. This law forced auto manufacturers around the world to develop emissions-free vehicles (e.g., the EV1 from GM) which, after it became clear that the law could not be implemented at the time, disappeared almost as suddenly as they entered the scene.

Fuel consumption not only affects each individual—especially in terms of expense—as the global consumption of all vehicles is playing an ever more important role. Although the share of traffic as it affects overall energy consumption and, thus, CO$_2$ emissions, barely reaches the 25% mark (Europe 20%, US 40%), this is the sector with the highest growth rates and thus deserves special attention. Following the first environmental conference in Rio de Janeiro in 1992, the first global climate summit in 1997 in Kyoto,[22] and the presentation of the fourth assessment report on global warming (4th Assessment Report, AR4) by the IPCC[23] in 2007 [395], which further underscored the role of mankind in the climate shift currently under observation, a series of new consumption laws were issued, the most important of which are described below.

3.8.1 EU Legislation

Leading up to Kyoto, the European parliament for personalized motor vehicle traffic mandated an average target value of 5 l/100 km (120 g/km CO$_2$) for 2005. Since this value could not be reached, the ACEA[24] responded with a voluntary self-commitment based

---

22. The industrial countries mutually agreed to reduce their greenhouse gas emissions by an average of 5.2% below the level measured in 1990 from 2008 to 2012. Six greenhouse gases were included: carbon dioxide (CO$_2$), methane (CH$_4$), nitrous oxide (N$_2$O), partly and fully fluorinated hydrocarbons (HFCs, CFCs), and sulfur hexafluoride (SF$_6$). Each country pursued different objectives: Japan agreed to a reduction of 6%, the United States 7%, and the EU 8%, which was distributed among the EU member countries. Germany, for example, agreed to a reduction of 21%.
23. Intergovernmental Panel on Climate Change, which was assembled in November 1988 by the United Nations Environment Programme (UNEP) and the World Meteorological Organization (WMO) to summarize and coalesce scientific research for political decision makers. The first assessment report was introduced in 1990. In 2007, the organization and former US Vice President Al Gore were presented with the Nobel Peace Prize.
24. ACEA stands for Association des Constructeurs Européens d'Automobiles or the Association of European Automobile Manufacturers. It was founded in 1991, with headquarters located in Brussels. http://www.acea.be/ The ACEA opened an office in Tokyo in 1995 and an additional office in Beijing in 2004.

on the principle of "burden sharing"[25]; the consumption of all vehicles registered in Europe was to be lowered from 7.7 l/100 km (186 g/km CO_2) in 1986 by a total of 25% to 5.8 l/100 km (140 g/km CO_2) in 2008. As this target value was met with an actual output of 154 g/km CO_2 (in Germany, 165 g/km CO_2; refer to Figure 3.1), legislators in Brussels responded with a new Regulation for Setting Emission Performance Standards for New Passenger Cars [825], which was published on April 23, 2009. Specifically, not only a more stringent target value of 130 g/km CO_2 by 2015 was required; a graduated introductory phase was also mandated. From 2012, every auto manufacturer must maintain an average limit value of 130 g/km for a specific percentage of its new car fleet: 65% from 2012, 75% from 2013, 80% from 2014, and 100% from 2015. In the event that the target values are not reached, the following penalty payments apply to each car sold between 2012 and 2018 for every gram of CO_2 above the specified average limit value:

- Euro 5 for the first gram
- Euro 15 for the second gram
- Euro 25 for the third gram

Exceeding the limit value for fleets by 3 grams leads to a combined total penalty of Euro 45 per vehicle. When the fourth gram is reached, a per gram penalty of Euro 95 is levied. From 2018 onwards, this high penalty will then apply as of the first gram that exceeds the specified limit.

A target value was also specified for 2020: 95 g/km CO_2 or minus 27% as compared to 2015. In 2013, continued negotiations for the 2020 target values are in progress with regard to the improvements made since 2009.

In addition to the laws passed at European level, several countries have already responded accordingly by implementing tax-based instruments to reduce average fuel consumption. In France, the Benelux countries, Austria, and Great Britain, for example, a very stringent CO_2-based taxation of company cars has been put in place (cf. [555]).

3.8.2 CO_2 Labels in the EU

To facilitate fleet consumption legislation, December 13, 1999 saw the implementation of Directive 1999/94/EC of the European Parliament on the Provisioning of Consumer Information Regarding the Fuel Consumption and CO_2 Emissions for Marketing Campaigns Targeting New Passenger Cars [365]. As of November 2004, salespersons are obligated to provide information about the fuel consumption and CO_2 emissions of the vehicles they have on the showroom floor as well as offer for sale or for sale in conjunction with leasing arrangements. A qualified evaluation and designation of the

---

25. Auto manufacturers collectively bear the "load" for reducing fuel consumption. To this end, manufacturers of smaller, more efficient cars must achieve a "more stringent" target value (e.g., 125 g/km CO_2) than that which applies to manufacturers of larger vehicles, while the latter (which have greater overall potential and cater to a customer base that is more willing to spend increased amounts of money for expensive fuel-reducing technologies) must counter by reducing fleet consumption by 30%, for example.

energy efficiency of motor vehicles within the respective vehicle class or segment (as for household appliances) was introduced in Germany with the passenger car energy consumption labeling ordinance (Pkw-EnVKV) on December 1, 2011 [366]. This labeling is provided in the form of an "efficiency label" as is the case in the majority of European countries and is designed to inform consumers and help them make an informed purchase decision in favor of an environmentally friendly automobile. Vehicles are broken down into eight efficiency classes: From A+ (very efficient, min. of 37% better than the CO_2 reference value) through to G (not very efficient, with an emissions output of more than 17% beyond the CO_2 reference value) (cf. Figure 3.27). As the weight of the vehicle is taken into account when calculating the CO_2 reference value such that larger vehicles with heightened fuel consumption often receive a better rating than smaller, more fuel efficient ones, this ordinance has been the target of much criticism from environmental agencies and some automotive organizations.

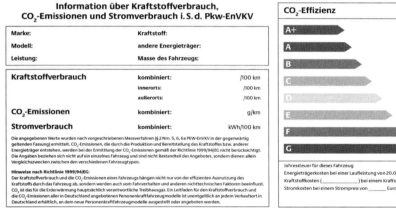
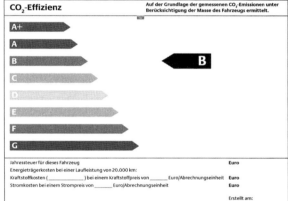

Figure 3.27 CO_2 labels in Germany.

3.8.3 Legislation in the United States

The Energy Policy and Conservation Act from 1975 and the applicable fleet consumption objective of 27.5 mpg (8.6 l/100 km) for the 1985 model year were applicable without any changes being made up to 2009 because the Bush administration refused to establish more stringent consumption limits for California and 13 other states. In March 2009, the Obama administration lowered these limit values, and in May 2009, it announced plans to implement a limit value of 34 mpg (approx. 6.7 l/100 km) for passenger cars by 2016 as proposed by California. These limit values undershoot the previously applicable values by some 30%. A further, considerable restriction took place on August 28, 2012, as the NHTSA published the CAFE and greenhouse gas regulations for model years 2017 to 2025, whereby as of 2025, new passenger cars will only be permitted to consume half as much fuel as in 2012, with an average fuel efficiency rating of a staggering 54.5 mpg (4.3 l/100 km or 109 g/km CO_2). This corresponds to an annual reduction in passenger car emissions of 5% and requires considerable engineering outlay to achieve (cf. section

3.1). With a view to "light trucks," which also include the SUVs and pickups that are very popular in the United States, less stringent regulations apply: from 2016, an annual reduction of just 3.5% is required.

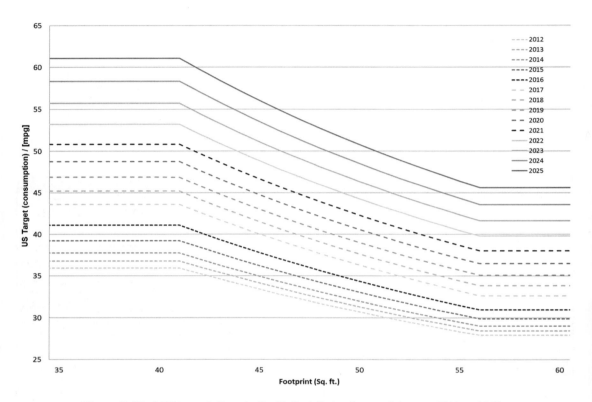

Figure 3.28 CAFE regulations in the United States for model years 2012 to 2022.

The gas guzzler tax introduced in 1991 continues to apply. When a vehicle is purchased, a penalty tax of between $1,000 (for vehicles that undershoot 22.5 mpg or consume more than 10.5 l/100 km) and $7,700 (for vehicles that undershoot 12.5 mpg or consume more than 18.8 l/100 km) must be paid—an amount that is usually absorbed by the manufacturer. The EPA website can be consulted to view gas guzzlers targeted for each year [364], and the NHTSA website is the definitive source of information regarding annual CAFE penalty payments, which have totaled almost $1 billion as of 1983.

3.8.4 CO_2 Labels in the United States

The CO_2 value forms part of the "Monroney label,"[26] which has applied to new showroom vehicles since 2008 and was redefined for the 2013 model year (cf. [815] and Figure 3.29). It not only references the estimated annual fuel costs and the projected

---

26. Named after US Senator Mike Monroney, who in 1958 became the first person to ensure that relevant information is provided as part of the Automobile Information Disclosure Act, http://www.epa.gov/carlabel/basicinformation.htm

five-year fuel cost trend as compared to an average new vehicle, but also provides information on the efficiency and environmental compatibility of the vehicle:

- Miles per gallon (mpg) during city, highway, and combined driving (in line with FTP 75)
- Fuel efficiency in relation to all new vehicles
- Fuel consumption or, inversely, fuel efficiency
- A points-based fuel efficiency and greenhouse gas assessment in comparison to all new vehicles
- CO_2 emissions in grams per mile
- An environmental assessment for air pollutants

Figure 3.29 Labeling requirements for vehicles in the United States: To the left for vehicles with a combustion engine; to right for plug-in hybrids; below for electric vehicles.

The new labeling requirements apply as of model year 2013 with the provision or option for manufacturers to voluntarily comply early on starting with model year 2012.

The fuel economy and greenhouse gas rating (1 = worst, 10 = best) is calculated using the combined mpg figure based on Table 3.8. In the example in Figure 3.29 (26 mpg or 9 l/100 km), a rating of 7 is yielded. The CO_2 value on the label (in the example, 347 g/mile or 216 g/km) is calculated in accordance with the US five-cycle system as is the mpg value (cf. Figure 3.17).

Table 3.8 Table for Calculating the Fuel Economy and Greenhouse Gas Rating

Rating	mpg	g/mi CO_2	Rating	mpg	g/mi CO_2
10	38+	0–236	5	19–21	413–479
9	31–37	237–290	4	17–18	480–538
8	27–30	291–334	3	15–16	539–612
7	23–26	335–394	2	13–14	613–710
6	22	395–412	1	0–12	711+

The EPA specifies the limit values for the next, or upcoming, model year to define the parameter range of the evaluation points. Refer to [815] for information on the methods on which this annual guideline is based.

3.8.5 Legislation in Asia

China is one of the first Asian countries to impose fuel consumption laws. Since 2012, the third phase of weight-dependent consumption limit values has been in force (cf. Table 3.9). These values mark upper restriction limits, and any vehicles that exceed these limits may no longer be sold.

Table 3.9 Consumption Limit Values in China from Model Year 2012 (Phase 3)

Vehicle mass (empty) (kg)	Consumption target manual transmission (l/100 km)	Consumption target, automatic transmission (l/100 km)
0 < m ≤ 750	5.2	5.6
750 < m ≤ 865	5.5	5.9
865 < m ≤ 980	5.8	6.2
980 < m ≤ 1,090	6.1	6.5
1,090 < m ≤ 1,205	6.5	6.8
1,205 < m ≤ 1,320	6.9	7.2
1,320 < m ≤ 1,430	7.3	7.6
1,430 < m ≤ 1,540	7.7	8.0
1,540 < m ≤ 1,660	8.1	8.4
1,660 < m ≤ 1,770	8.5	8.8
1,770 < m ≤ 1,180	8.9	9.2
1,880 < m ≤ 2,000	9.3	9.6
2,000 < m ≤ 2,110	9.7	10.1
2,110 < m ≤ 2,280	10.1	10.6
2,280 < m ≤ 2,510	10.8	11.2
2,510 < m < ∞	11.5	11.9

Chapter 4
Aerodynamic Forces and Their Influence on Passenger Vehicles

Thomas Schuetz, Lothar Krüger, Manfred Lentzen

As was shown in chapter 3, driving performance of a road vehicle is significantly determined by its aerodynamic drag. Thus, it is essential to really reach the target that has been defined in the functional requirements of a new car in terms of c_D. In the past, the drag coefficient had only to be as low as possible without unacceptably affecting the vehicle's styling, but today's aerodynamic work has to fulfill well-defined requirements. The vehicle's lift force unloads its front and rear axles and thus affects the force transmission between tire and track. Both definition and achievement of specific target lift coefficient values are absolutely necessary.

To approach the phenomenon of aerodynamic drag, some bluff bodies of the same thickness should be compared, first with the help of Figure 4.1. Such bodies have the same aspect ratio of height h or diameter d and length l. For passenger cars, h/l is about 0.3. With $0.5 > c_D > 0.15$, the passenger car is within the range of the body of revolution ($c_D \approx 0.05$) and the sharp-edged cube in a cross flow ($c_D \approx 0.9$).

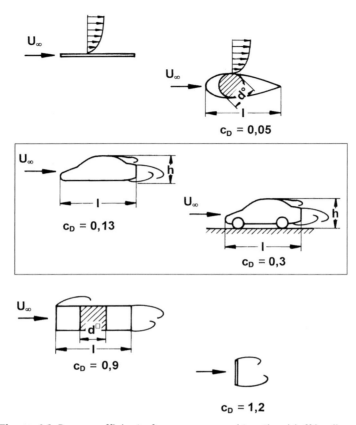

Figure 4.1 Drag coefficient of a car compared to other bluff bodies of equal solidity h/l and extreme cases.

The body of revolution even with $d/l \approx 0.3$ has almost exclusively friction drag. In its purest form, friction drag occurs in the case of a flat plate aligned with an oncoming flow. Theoretical study of friction drag has progressed considerably. In chapter 3 some basic facts are presented about the friction drag. For a slender body of given thickness d/l and volume, the shape that provides minimum drag can be computed in advance fairly well. In general terms, the flow field can be divided in a frictionless outer flow and a near-wall region, the so-called boundary layer dominated by viscous effects. The outer flow can be represented by the potential theory and determines the pressure in the boundary layer. Based on experimentally derived universal laws for the wall shear stress, the development of the turbulent boundary layer can be determined. Alternatively, the boundary layer equations can be used.

Unlike a body of revolution, a square block facing into an airflow is subject almost exclusively to pressure drag caused by flow separation and dissipation losses: $c_D \approx 0.9$. The extreme example in terms of 100% pressure drag would be the flat plate perpendicular to an oncoming flow. But even for such simple geometric bodies, the aerodynamic drag could be predicted reliably only using complex flow simulation, which is due to separation and reattachment of the flow and their strong effect on the outer flow.

Corresponding to Figure 4.1, the drag of the vehicle's exterior without wheels, with a smooth underbody, and without link to the ground is relatively close to the body of revolution because the flow remains attached on most parts of the surface and separates only in the rear part of the car. A further decrease of drag can only be achieved using a drop-shaped extension of the vehicle body in order to reduce the wake zone.

If we add the wheels, with a realistic underbody structure, and bring the body closer to the ground, the drag increases substantially, mainly due to the flow losses in the accrued separation areas. At present, despite many successes in the constant effort to reduce the aerodynamic drag of cars and to "shift" its aerodynamic properties away from a (bluff) block and toward a (slender) body of revolution, the car is still closer to the model of a block. Even without reference to the resulting c_d value, this statement nevertheless is valid at least qualitatively since the flow around a car, as in the case of all bluff bodies, is marked by separations, and the car's total drag still consists primarily of pressure drag. However, the areas of separation aren't generally determined solely by sharp edges as in the case for the cube. Additionally, longitudinal vortices occur at inclined body edges, which induce a complex pressure field on the body's surface and interfere with the wake behind the car.

While cars can be considered to be almost symmetrical with respect to their central plane (*xz* plane), top and bottom sides are of completely different shape. Hence, we will consider the *xz* plane the vehicle's plane of symmetry. A result of this is that the flow field around the vehicle generates almost no side forces but severe vertical forces. Especially coupes but also today's notchback sedans have a silhouette resembling a combination of two Clark Y airfoil sections, as shown in Figure 4.2. This airfoil generates lift at an angle of attack of $\alpha = 0°$ which increases when approaching the ground [64]. This is due to the accelerated flow on the curved upper wing surface and the resulting pressure difference to the lower surface. For the coupe, the flow structure is similar except for the wake zone behind the car, but this doesn't change anything qualitatively. Squareback vehicles (vans and compact cars) on the other hand can even produce downforce because their roofline is less curved and rather resembles an inverted wing section. Increased lift can lead to dangerous driving situations during cornering and braking and also to poorer driving dynamics. Though increasing downforce is desirable for driving dynamics, at the same time, it means higher load on the axle components and can lead to increased tire wear and to increased rolling resistance.

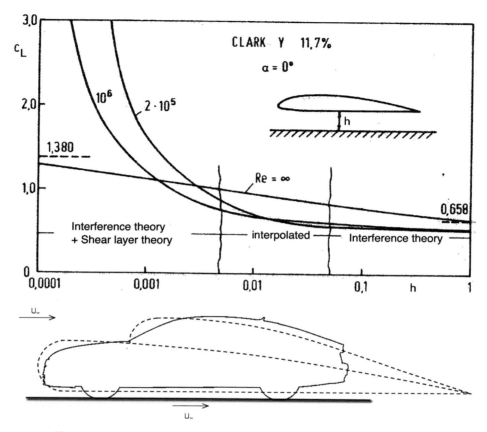

Figure 4.2 Comparison of a coupe silhouette with a Clark Y wing airfoil.

The task of predicting flow forces on road vehicles can be solved with three different approaches:

1. Experimentally using wind tunnel tests on scale or full-scale models;
2. By the use of numerical calculation based on the Navier-Stokes equations;
3. With a rating process based on empirically determined correlations between vehicle geometry parameters and aerodynamic drag.

No matter which approach is chosen, the optimization must always follow the principle of trial and error. The more experience the developer has, the faster this process converges. Therefore, the aims of this chapter about the air forces and moments acting on road vehicles are

- to examine the physical mechanisms generating aerodynamic drag and lift.
- to provide the correlations between geometric details and force action. We will try to generalize the plenty of functions that have been identified during numerous wind tunnel sessions wherever this is possible

Aerodynamic Forces and Their Influence on Passenger Vehicles

And to explain the physical processes underlying these functions, and thus prepare them for a determined application. Here we will usually have to work with qualitative explanations. Wherever individual results can be concluded from quantitative information, it is important to note that these have been developed mostly on one special car type and can therefore not be generalized.

4.1 Aerodynamic Forces and Force Coefficients

Every solid body including a car brought into an airstream experiences a force due to the air displacement and the nonreversible transformation of pressure into kinetic energy of the air. This effect may be described using either absolute force values or dimensionless parameters. The air force resulting from the pressure and shear stress distribution acts on a previously unknown point on the vehicle. Furthermore, this point is a different one for each vehicle. Therefore, in order to describe the air forces and moments, it is common practice to describe the car in a vehicle-fixed coordinate system whose origin is located in the center of gravity of the rectangle defined by the four contact points of the wheels with the roadway (Figure 4.3).

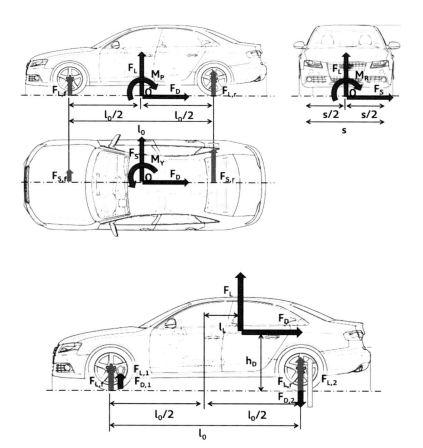

Figure 4.3 Representation of air forces and moments of the road vehicle; lift and side forces at the wheel-road contact points.

Under general flow conditions but first without crosswind, at the vehicle's reference point 0 we derive three forces and another three moment components: the drag force F_D, the lift force F_L, the side force F_S, the rolling moment M_R, the pitching moment M_P, and the yaw moment M_Y. Using l_0 as the wheelbase lift and side force together with pitch and yaw moment can be converted into front and rear axle lift and side force ($F_{L,f}$, $F_{L,r}$, $F_{S,f}$, $F_{S,r}$). For these forces acting in the wheel contact plane, the following conditions can be applied similarly to the figure.

$$F_L = F_{L,f} + F_{L,r} \tag{4.1}$$

$$F_S = F_{S,f} + F_{S,r} \tag{4.2}$$

$$M_P = \left(F_{L,f} - F_{L,r}\right)\frac{l_0}{2} \tag{4.3}$$

and

$$M_Y = \left(F_{S,f} - F_{S,r}\right)\frac{l_0}{2} \tag{4.4}$$

The pitching moment M_P, and thus the axle-wise lift forces $F_{L,f}$ and $F_{L,r}$ are obviously composed of two parts: from a portion of the pure lift component F_L (here $F_{L,1}$ and $F_{L,2}$) and the position of its point of application in x-direction and from a portion of the drag component F_D (here $F_{D,1}$ and $F_{D,2}$) and the position of its point of application in the z-direction (see Figure 4.3).

The entire pitching moment M_P is then the sum of two parts $M_{P,D}$ and $M_{P,L}$, which are calculated as follows:

$$M_{P,D} = F_D \cdot h_D \tag{4.5}$$

using

$$F_{D,1} = -F_{D,2} = \frac{M_{P,D}}{l_0} \tag{4.6}$$

and

$$M_{P,L} = -F_L \cdot l_L \tag{4.7}$$

with

$$F_{L,1} = F_L - F_{L,2} = \frac{F_L \cdot l_0 + 2M_{P,L}}{2l_0} \tag{4.8}$$

In science, the term aerodynamic drag is always understood as the force component in flow direction acting on a body which, for the vehicle aerodynamics, has been adopted from the aircraft and the definitions of Otto Liliental.[1] In crosswind under yaw angles of $\beta \neq 0°$, we want to distinguish the component of force in longitudinal direction from the

---

1. For example, in "Der Vogelflug als Grundlage der Fliegenkunst."

one in flow direction. Later, this will lead us to the tangential force F_T (see Figure 4.14). In the case of straight-on airflow ($\beta = 0°$), thus applies $F_T = F_D$.

In experiments in subsonic speed range, the resulting vector air force as well as their components can be identified to be in proportion to the density of the flow medium (here air ρ_L) for flow velocity and the expansion of the body. This expansion is represented by an arbitrarily selectable area size, usually the frontal or shadow area of the flow body A_x in longitudinal direction. With these findings, the air force can be formulated as

$$F_{\text{Air}} = c \cdot A_x \cdot \frac{\rho_L}{2} v_\infty^2 \tag{4.9}$$

Furthermore, it can be found that the proportionality constant depends exclusively on the shape of the body. Accordingly, spheres, cylinders, and all other bodies of different size each have the same constant of proportionality. Formulating the vector components of the force, the term will only differ by this coefficient. In order to calculate the air moments, an arbitrary reference length is introduced. Usually the car's wheelbase l_0 is chosen:

$$M_{\text{Air}} = c_M \cdot A_x \cdot l_0 \cdot \frac{\rho_L}{2} v_\infty^2 \tag{4.10}$$

The fraction $\frac{1}{2} \cdot \rho \cdot v_\infty^2$ in the force and moment terms is called the dynamic pressure or total pressure of the oncoming flow, thus, air forces and moments scale with the dynamic pressure. At the so-called stagnation point, the flow is divided into two parts. At the stagnation point, the speed is $v_\infty = 0$ and the pressure follows after Bernoulli:

$$p_{\text{tot},S} = p_\infty + \frac{\rho_L}{2} v_\infty^2 \tag{4.11}$$

Obviously, the biggest pressure throughout the flow at all occurs at the stagnation point. In aerodynamics, it is common to use dimensionless coefficients instead of dimensional forces, moments, and pressures, since they are independent of the speed over a wide velocity range. They are representative numbers for the aerodynamic quality of a body shape. Thus, for example, in a wind tunnel, the aerodynamic drag force is measured but then the above-mentioned proportionality constant c_D will be calculated for further use. It follows for the drag coefficient c_D and the lift coefficient c_L

$$c_D = \frac{F_D}{\frac{\rho_L}{2} v_\infty^2 A_x} \quad \text{or} \quad c_L = \frac{F_L}{\frac{\rho_L}{2} v_\infty^2 A_x} \tag{4.12}$$

and to the moment coefficients

$$c_{M,R} = \frac{M_R}{\frac{\rho_L}{2} l_0 v_\infty^2 A_x}, \quad c_{M,P} = \frac{M_P}{\frac{\rho_L}{2} l_0 v_\infty^2 A_x}, \quad \text{and} \quad c_{M,Y} = \frac{M_Y}{\frac{\rho_L}{2} l_0 v_\infty^2 A_x} \tag{4.13}$$

The pressure distribution on the vehicle surface is furthermore represented at each point x by means of the pressure coefficient c_p:

$$c_p = \frac{p_s - p_\infty}{\frac{\rho_L}{2} v_\infty^2} \qquad (4.14)$$

In the stagnation point, $c_p = 1$. Generally, the pressure of the flow around a car is within the range of about $1 > c_p > 2$. The pressure distribution in Figure 4.4 shows that the pressure on the upper side is much lower than on the under side. This is the reason for the lift that cars usually have.

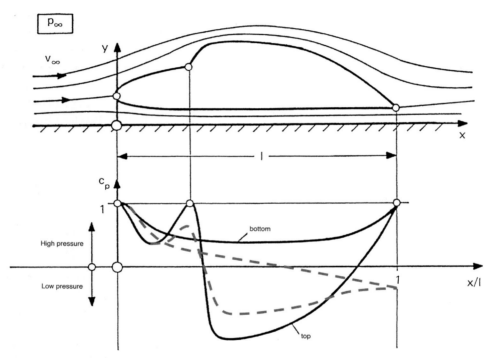

Figure 4.4 Flow around a vehicle and pressure distribution.

In frictionless flow, two stagnation points would occur, one at the front and one at the back of the vehicle. If friction is then taken into account, the second stagnation point disappears because of the flow separation at the back of the car. In frictionless flow furthermore, all pressure forces throughout the vehicle surface sum up to a drag force of $F_D = 0$. This phenomenon is called d'Alembert paradox, which doesn't allow an incompressible frictionless subsonic flow to cause any drag. In reality though, the drag is not equal to zero, which cannot explained for frictionless flows.

Using the speech of an aerodynamicist, we want to use "1 count" for a variation of the dimensionless coefficients c_D, c_L, c_p etc. each by 0.001. Hence, if we talk about 12 counts in drag, this means variation of $\Delta c_D = 0.012$.

4.1.1 Buckingham Π Theorem

We introduced the aerodynamic coefficients (c_D, c_L, c_{pr}) phenomenologically by means of a thought experiment with the result that air forces and moments behave proportional to the density ρ, the velocity v, and the size of the flow body A_x. But it is also possible to theoretically derive these number from the so-called Buckingham Π theorem in conjunction with the similitude theory.

The theorem states that a physical problem involving m physical variables expressible in terms of n physically independent quantities can be simply expressed by $m - n$ dimensionless coefficients. A very simple example is the simple gravity pendulum. Of course, its movement is fully described by complex mathematical differential equations; nevertheless, the relevant parameters can be determined by using the Buckingham Π theorem (Figure 4.5).

Buckingham's theorem:

- If a problem is described by n physical properties expressed in m basic units, it can also be described by (n − m) dimensionless Mechanics
- example: <u>Pendulum</u>
    - Physical properties m, l, t, g
    - Units length (m), mass (kg), time (s)
    - One dimensionless number P_1
- Aerodynamic example: <u>Aerodynamic drag</u>
    - Physical properties F_{air}, v_∞, l, r, n
    - Units length (m), mass (kg), time (s)
    - Two dimensionless numbers P_1 und P_2

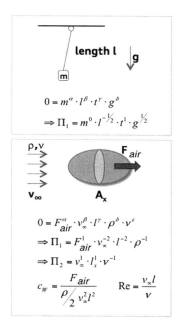

Figure 4.5 The Buckingham Π theorem applied on a pendulum.

The aerodynamic coefficients can be determined the same way. As was shown in chapter 1, aerodynamics is generally based on nonlinear partial differential equations much more complicated than for the simple gravity pendulum. The aerodynamic behavior of a flow body will certainly be determined by the flow velocity v_∞, the size of the body expressed by a characteristic length l, and a few matter constants like the density ρ_L and the viscosity v_L of air. The force F_{air} will then be resulting from the air pressure distribution, mostly written as relative pressure $p - p_\infty$.

Hence, six physical variables in terms of three physically independent quantities length (m), mass (kg), and time (s) shall be considered. After Buckingham, three dimensionless coefficients can be derived: The drag coefficient c_D, the Reynolds number Re, and the pressure coefficient c_p. Thus we also found a theoretical way leading to the presented coefficients. Further impact factors can be taken into account too, like temperature, thermal conductivity, or speed of sound resulting in further dimensionless coefficients, which, however, only have a demonstrably only a weak effect on the relevant aerodynamic coefficients.

4.2 Flow Field Around Cars

Considerable progress has been made during the last years in understanding the flow field around vehicles. The main reason for this progress is that flow observations are no longer confined to the surface of a car body but include the entire surrounding space. The result, as explained in Figure 1.4, is a readily comprehensible picture of the flow field around a car. As mentioned at the beginning, the flow around a vehicle is similar to the one around a bluff body. On the one hand, it is affected by attached flow over wide surface areas, resulting in the development of shear forces. On the other hand, separation will occur, resulting in large pressure losses, especially at the back and the underbody of the car. Previously, in Figure 1.53 some typical separation areas were identified and distinguished in terms of the rear end shape of notchback, fullback, and fastback. Two types of flow separation can be observed:

1. Flow can separate on edges running perpendicular to the local direction of flow. Vortices roll up, their axes normally being parallel to the separation line. Most of their kinetic energy is dissipated by turbulent mixing. Consequently, this first type of separation is sometimes designated as quasi–two-dimensional. Self-enclosed "separation bubbles" and zones of recirculation, frequently called dead water, figuratively go with the car.

2. The second type of separation is three-dimensional by nature. At edges around which air flows at an angle, the airstream forms cone-shaped streamwise vortices similar to those observed on aircraft wings, especially delta wings with low aspect ratio. The regions where these vortices most tend to be generated on a car are the A- and C-pillars. The axes of these vortices run essentially in the streamwise direction; unlike the quasi–two-dimensional vortices, they are very rich in energy.

In addition to separation processes, the flow field around cars is characterized by the fact that a part of the incident flow is guided through the vehicle for cooling purposes. After entering the engine compartment, this airflow passes through a complex heat exchanger package, depending on the vehicle equipment, and through the engine transmission unit, before exiting from the engine compartment, primarily into the underbody flow. These processes can be described in a first approximation by the simple case of tube flow with artificial pressure losses.

Environmental conditions are another aspect of the flow around cars. Owing to the natural wind that typically exists when driving on a road, the flow around a car with respect to its longitudinal centreline is usually asymmetrical. As will be discussed in more detail in chapter 5, the wind has boundary layer characteristics and is gusty. Both are phenomena that are increasingly receiving attention in vehicle aerodynamics but that play a minor role for the aerodynamic forces. In summary, it is possible to distinguish between four flow phenomena, which will be discussed next. Figure 4.6 provides an overview of these phenomena.

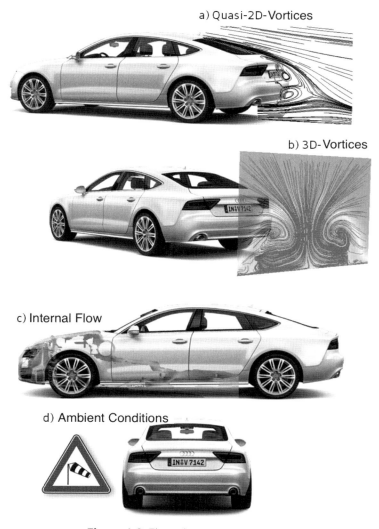

Figure 4.6 Flow phenomena on a car.

Chapter 4

4.2.1 Dead Wake

If the flow separates at lines perpendicular to the local flow direction, three forms can be distinguished, as shown schematically in Figure 4.7. The dead wake that is thereby formed is either

- non-periodic,
- periodic, or
- ring/spiral-shaped.

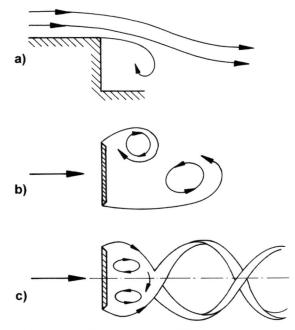

Figure 4.7 Formation of shear layers, [493]: a) on one side, nonperiodic; b) on both sides, periodic; c) rotationally symmetric.

4.2.1.1 Non-Periodic Dead Wake

As an example of nonperiodic dead wake, we shall consider the backward-facing step shown in Figure 4.8. At the point S, which does not necessarily mark a sharp edge, the boundary layer separates from the contour of the body and transitions into a shear layer. This shear layer starts to reattach again at the point R. A closed dead wake forms, within which the flow circulates.

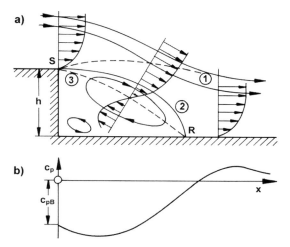

Figure 4.8 Flow at a backward-facing step, schematic: a) formation of the shear layer and the recirculating flow in the dead wake; b) the corresponding pattern of the static pressure in the dead wake.

By mixing with the nearly inviscid outer flow, the shear layer expands (line (1)). Line (2) is a streamline averaged over time; it encloses the dead wake characterized by backflow. Here, line (3) denotes the boundary between the forward flow and backflow. Downstream of R, a boundary layer forms again; the relaxation length (i.e., the distance traversed by internal flow until an equilibrium is reestablished in the boundary layer), is quite large at approximately 25 step heights ($x = 0$ at R).

According to Prandtl, separation of the flow occurs (in two-dimensional flow) when the following criteria are met:

$$\frac{\partial p}{\partial x} > 0 \tag{4.15}$$

$$\left(\frac{\partial u}{\partial y}\right)_{y=0} = 0 \tag{4.16}$$

$$\tau_w = 0 \tag{4.17}$$

At point S of the separation, the pressure gradient is positive, that is, the boundary layer flows against an increase in pressure, the velocity profile of the boundary layer has a perpendicular tangent, and, consequently, the wall shear stress at S is zero. If the point S now indicates a sharp edge, as shown, the characteristics described by the equations (4.15) to (4.17) do not apply. The pressure gradient can be any value, even negative. The velocity profile is a "normal" voluminous boundary layer profile, and the wall shear stress is different from zero. For this type of separation, Leder [493] proposes the term "sharp edge separation" (Abriss in German), a recommendation that will be followed here.

The pressure profile occurring behind the step (indicated in Figure 4.8b) is typical for a dead wake of this kind, and its universal properties, as already mentioned, are used to formulate theoretical dead wake models. Starting from the base pressure, the pressure passes through a flat minimum and then rises to equal the ambient pressure.

On cars, separation with nonperiodic dead wake can occur at all "leading" edges, as well as at the trailing edge of the roof, preferably at the notchback.

4.2.1.2 Periodic Dead Wake

Periodic dead wake is shown schematically in Figure 4.9. The two-dimensional body tends to taper toward the rear; on the top, it transitions into the base with increasing curvature, whereas the bottom side ends, after an initial tapering, in a sharp edge.

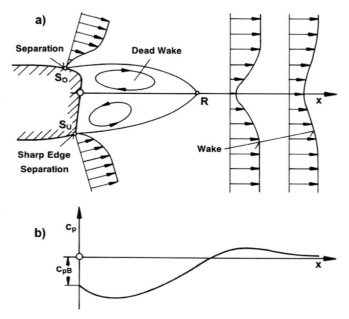

Figure 4.9 Two-sided separation or sharp edge separation at the base of a two-dimensional body, schematic: a) flow field; b) pressure distribution on the X-axis.

On the top side, the flow separates at the point SO; on the bottom side, sharp edge separation occurs at SU. Downstream of SO and SU, the boundary layers transition to free shear layers. These layers meet at point R, a free stagnation point, thereby enclosing the dead wake. In this dead wake, two counter-rotating vortices form, possibly periodically alternating, as in a Kármán vortex street. The wake flow forms to the right of R, widens downstream, and gradually fully balances out.

Periodic separation occurs at the rod aerial of a car, for example. The separation at the base of a squareback vehicle such as a bus, a container, or a passenger car is not periodic but stochastic in nature; further details are provided in section 4.2.1.4.

A periodic dead wake can be converted to a nonperiodic dead wake by introducing a flat plate into the dead wake, parallel to the X-axis. This prevents the cross exchange between the two vortices; the pressure at the base increases and the drag drops.

The flow patterns that have been considered as two-dimensional so far are in reality three-dimensional: first, because the shear layer on a perpendicular-running edge does not form simultaneously along the entire length, but only section by section, and second, because the shear layer is deformed three-dimensionally by the circulation which it encloses.

The following still remains to be added about separation with formation of cross-flow vortices: At a body of finite width, if separation occurs along a line transverse to the flow direction to form a separation bubble with cross-flow vortex, the question arises of where the circulation enclosed in the cross-flow vortex remains at the lateral edges of the bubble. According to the Helmholtz vortex theorems, vortices cannot just end anywhere. But this means that cross-flow vortices are bent in the flow direction at the lateral ends of a separation bubble to continue as longitudinal vortices to infinity—and to interact with other, both longitudinal and transverse, vortices. This must be considered, for example, for the vortex that is formed in the cowl of a car.

4.2.1.3 Ring and Spiral Vortices

Separation with cross-flow vortices also occurs in rotationally symmetric configurations, that is, if the observed flow body is conceptually extended into the third dimension. Simple examples are the wakes of a circular disk or cone. As illustrated in Figure 4.7c, averaged over time a ring vortex forms from which, perhaps in a circulating manner, free vortices separate and "float away" to the rear in a spiral shape. This type of separation can also be expected on shapes that do not deviate too much from rotational symmetry, such as exterior mirrors, but also, averaged over time, on squareback vehicles mentioned previously (see also Figure 4.23).

4.2.1.4 Unsteady Processes

By considering the flow pattern, the pressures, and the velocities up to now as averaged over time, the phenomenon of "separation" has been treated as if it were a steady process. However, that is not the reality, by any means.

The best-known unsteady phenomenon during separation is the Kármán vortex street already mentioned, which forms behind a circular cylinder in a transverse flow. The dimensionless frequency, the Strouhal number Sr, for a circular cylinder depends on the Reynolds number; how, will not be pursued here.

$$Sr = \frac{f \cdot h}{u_\infty} \qquad (4.18)$$

A Kármán vortex street is also formed behind (two-dimensional) bodies that end in a sharp edge, examples of which are a thick plate in a longitudinal flow (with a well-rounded front edge), a wedge, and its special case of a thin, sharp-edged plate in a transverse flow [493]. On these bodies, the Strouhal number, with which the vortices shed alternately, does not depend on the Reynolds number, because separation of the flow is fixed at the sharp trailing edge.

The flow in the dead wake is not only unsteady in the periodic case but also in the nonperiodic case; velocity fluctuations $\sqrt{u'}/u$ of up to 50% have been measured. This is also true for the three-dimensional case, which Duell and George [214] have studied on the body shown in Figure 4.10. This is a square prism with a rounded head in a longitudinal flow, which is located in the vicinity of a ground moving with it.

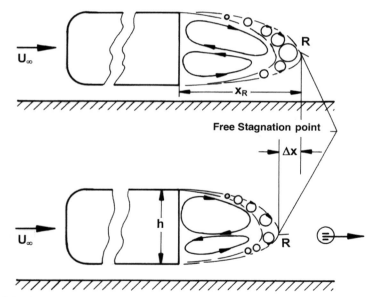

Figure 4.10 Formation of a ring vortex at the base of a prismatic body in a longitudinal flow and the occurrence of surges, schematic [214].

The two counter-rotating vortices, which can be seen in section in the dead wake, form a large ring vortex averaged over time. Small vortices form within the ring-shaped shear layer emanating from the rotating trailing edge. They shed at a dimensionless frequency $Sr = 1.157$ and grow in the flow direction by absorbing fluid from the outer flow, while their frequency decreases. At the free stagnation point R, where the shear layers from all sides come together, vortices separate periodically from the shear layer and transition into the wake. Each time a vortex is shed from the shear layer to the right, the free stagnation point jumps to the left by an amount Δx: The pressure in the dead wake then rises and forces the shear layer to the right back to its starting position—and the cycle begins again. The dead wake thus performs a surge at frequency f in the vortex shedding from the shear layer. The associated Strouhal number was measured by Duell and George from the spectrum of the base pressure as $Sr = 0.069$.

Converting this frequency for a vehicle ($h = 1.5$ m, $u_\infty = 100$ km/h) results in a frequency of $f \approx 1.3$ Hz. This frequency, which is close to the natural frequency of most vibrations of the vehicle, is the frequency at which the drag of the body also oscillates.

The unsteady nature of the flow in the dead wake behind a bluff body also becomes apparent by comparing a snapshot of the flow with its average over time, as performed by Khalighi et al. [426]. Their snapshot of a section parallel to the ground, taken using particle image velocimetry (PIV), is shown in Figure 4.11(a). The local rotation has been computed from the field of velocity vectors shown there and entered in the form of isolines on the vector field. The concentration of rotation in the shear layer is clearly visible, as well as its periodicity, indicating the formation of discrete vortices. A snapshot of the streamline pattern computed from the vector field is reproduced in Figure 4.11b. For comparison, Figure 4.11c shows the time-averaged streamline pattern, which was generated from 200 PIV snapshots. The recirculation region is clearly recognizable. Why it is not symmetrical to $y = 0$ remains unresolved.

A spectral analysis of the base pressure revealed that, in this example, the dead wake similarly performs a surge at the dimensionless frequency $Sr = 0.07$, a value that is confirmed in the literature [58]. The significance of these oscillations for a notchback is discussed in more detail in section 4.5.1.3.

If a (surrounding) fascia is attached to the base of a prismatic body, the recirculating flow moves downstream, as can be seen in Figure 4.11d. The amplitude of the surge decreases, the base pressure increases, and the drag of the body is reduced. At a depth of fascia of $t = 0.5\,h$, the drag fell by 20%.

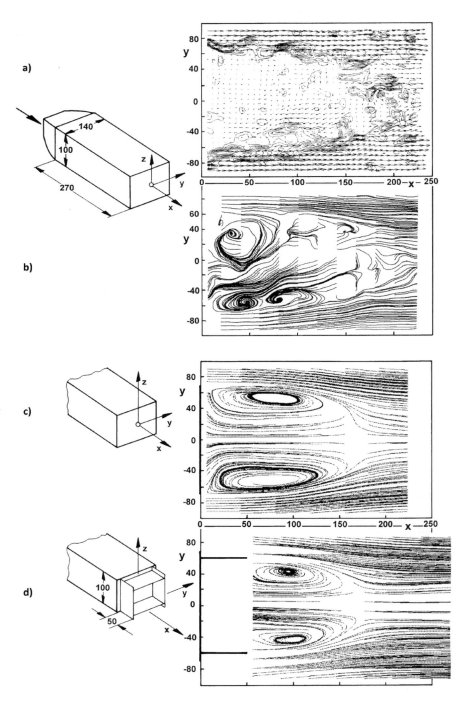

Figure 4.11 A flow field, measured using PIV, behind a prism in longitudinal flow, according to Khalighi et al. [426]: a) snapshot of the velocity vectors and lines of equal rotation; b) streamline pattern; c) time-averaged flow field behind the base; d) time-averaged flow field behind the same base, but fitted with fascia.

4.2.2 Longitudinal Vortices

Separation also occurs at edges in cross flow, as shown in Figure 1.3. The shear layers flowing toward each other from two directions separate and curl up to form vortex bags. Preferred locations for their formation are the A- and C-pillars (see Figure 4.12). The axes of these vortices are mainly oriented in the longitudinal direction. It can be assumed that the amount of circulation in them depends on the geometrical conditions, primarily on the inclination of the line (edge) at which they separate from the contour.

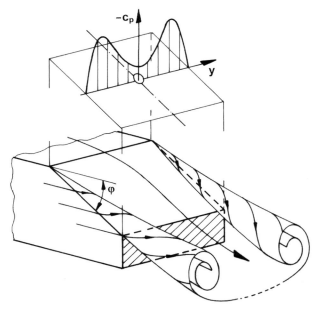

Figure 4.12 Curling up of shear layers at inclined edges and formation of a pair of longitudinal vortices; above, the pressure profile on the inclined surface, schematic.

On the surfaces they approach, these longitudinal vortices induce high negative pressures with pronounced peaks, as can be seen in Figure 4.12 from the pressure distribution over the inclined plane drawn out upwards. Their high peripheral velocity causes loud wind noise at the front side windows of a car, and the high negative pressures on the rear slope result in high levels of drag and lift at the rear axle.

Such vortex formations were first observed in slender aircraft wings. The phenomenon that occurs in such wings (i.e., these vortices burst after exceeding a certain angle φ and separation of the cross-flow vortex type occurs) is also observed in vehicles. Section 4.5.1.3.2 will discuss this in more detail.

Both types of separation, at perpendicular and inclined edges, occur at various locations on the vehicle where they interact with each other. This will be discussed in more detail when the individual vehicle shapes are described, in particular the tail shapes.

If one measures the velocity field in an x-plane closely behind a fastback model as a snapshot, for example, using PIV, the result is not an orderly field as depicted schematically in Figure 4.12. Rather, the reality is the chaotic pattern shown in Figure 4.13 (left). Counter-rotating vortices can be made out in the irregular structures. However, different patterns form continuously; the vortices constantly change their positions. Whether a harmonic part can be derived from this, as observed at the backward running head of multi-unit trains, has not been reported.

The familiar pattern of two longitudinal vortices which fits the scheme in Figure 4.12 is only revealed after taking the averate of (here) ten snapshots, to the right in Figure 4.13. That the formation is not completely symmetrical may be attributable to the fact that ten snapshots are not sufficient to form a time independent average.

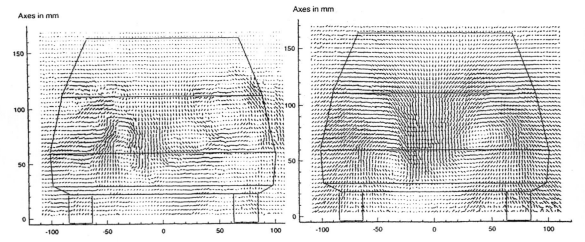

Figure 4.13 Velocity field in a plane $x =$ const. behind a fastback model, measured using PIV; according to Bearman [53]. Snapshot (left); average of ten snapshots (right).

4.2.3 Internal Flow

Power unit cooling and interior air conditioning are achieved by complex liquid cooling circuits. The coolants emit their heat to the environment via the flow through heat exchangers in the engine compartment. Hence, an airflow through the heat exchangers as large as possible is desirable for effective transfer of heat to the environment. For this purpose, a partial air mass flow is diverted into the engine compartment from the environment, which is released into the surrounding flow after passing through the heat exchangers.

The airflow through the engine compartment is characterized by friction processes at the engine compartment walls and the surfaces of the numerous components as well as in the fine air-side capillaries of the heat exchangers. These effects cause a recognizable pressure loss along a stream filament. It is therefore practical to use tube flow as a simple substitute model. The pressure loss in a tube with internal flow, of length l and diameter d, follows the relationship

$$\Delta p = \lambda \frac{l}{d} \cdot \frac{\rho}{2} v_\infty^2 \tag{4.19}$$

where the tube friction coefficient λ in fully turbulent flow can be approximated, according to Prandtl, by

$$\frac{1}{\sqrt{\lambda}} = 2.03 \cdot \lg\left(\text{Re} \cdot \sqrt{\lambda}\right) - 0.8 \tag{4.20}$$

and the tube diameter for noncircular tube cross-sections A and circumference U can be described by the hydraulic tube diameter

$$d_h = \frac{4A}{U} \tag{4.21}$$

Hence, the pressure loss depends primarily on the tube length and diameter, as well as on the flow velocity. Graphically, these relationships are depicted in the so-called tube friction chart.[2] For large Reynolds numbers, the tube friction factor λ scales approximately with $\text{Re}^{-0.25}$; the pressure loss is then proportional to $v^{1.75}$:

$$\Delta p \sim \frac{l}{d^{1.25}} \cdot \rho v^{0.25} \cdot v^{1.75} \tag{4.22}$$

Mapped to the airflow through the engine compartment, this means that a larger cooling air mass flow (greater velocity through the radiator), a more fine-meshed and larger radiator, and greater blocking of the engine compartment (both meaning a smaller hydraulic diameter) lead to greater pressure loss.

4.2.4 Environmental Influences

The driving velocity v_∞ and the velocity v_{wind} of the natural wind gives an incident flow velocity v_{res} whose vector, as shown in Figure 4.14, encloses the yaw angle β in the driving direction. In principle, this angle can take on any value. If we initially only consider the drag of cars, we only need to consider small yaw angles, because drag only plays a role and is relevant for fuel consumption if the vehicle is traveling fast—and then the angle β is small averaged over time. In addition, the increase in aerodynamic drag with the yaw angle β is usually low for small β. This is possibly the reason why the effect of side wind on the drag is often completely neglected. It should be noted that this simplification cannot be indiscriminately applied to commercial vehicles.

---

2. The tube friction chart is also referred to as the Moody chart, named after the American engineer Lewis Ferry Moody.

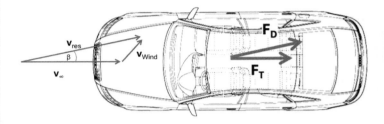

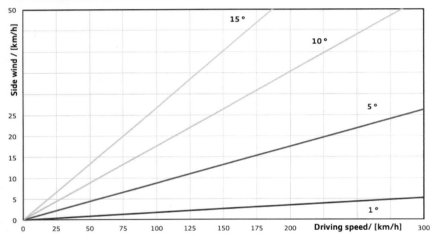

Figure 4.14 Definition of the yaw angle, side wind over driving velocity for some yaw angles, and classification of side wind strength based on the Beaufort scale.

Figure 4.15 provides some examples, which show that the tangential force coefficient increases with the yaw angle for all vehicles. The assessment made by Utz [816] for this relationship confirms that large yaw angles rarely occur at driving speeds at which the aerodynamic drag is significant. According to this assessment, it is sufficient to postpone the increase in the drag coefficient by using the yaw angle $\beta > 5°$. As can be seen in the right chart in the figure, this is usually possible.

This limitation of the yaw angles to be considered must not tempt us to ignore the other aerodynamic coefficients at large yaw angles, as, with regard to maintaining the driving direction, discussed in section 5, precisely the very brief occurrence of large yaw angles is of importance.

Especially strong gusts of wind can lead to significant changes in lift, causing dangerous driving situations. For example, on fastback vehicles that exhibit stable flow separation at the trailing edge of the roof without side wind, even small yaw angles below 5° may suddenly cause attached flow at the rear window. The associated increase in lift on the rear axle can have serious consequences when driving (cf. section 5).

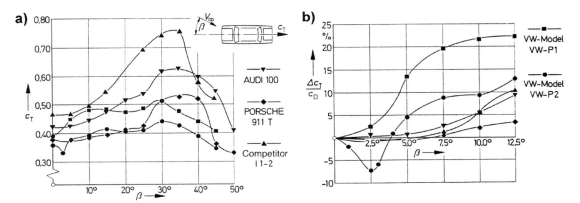

Figure 4.15 Increase in the tangential force coefficient c_T with the yaw angle β: a) absolute; b) relative, for small yaw angles [409].

The side force, which is almost completely absent when exposed to straight flow, quickly increases even at low flow velocities, resulting in correspondingly large yaw angles at strong side wind and low driving velocity. Assigning the side force to the two axles usually results in a nonuniform distribution to the front and rear axles, which leads to a yaw moment. The driver will immediately notice this from a change in direction of the vehicle.

4.2.5 Influence of the Reynolds Number

The digression in section 4.1 into Buckingham's theorem revealed that, in addition to the c-value (generally for c_D and c_L), the Reynolds number similarly describes the physical processes in the flow around the vehicle. The formulation of aerodynamic drag as a dimensionless coefficient c_D does not mean per se that it is independent of the incident flow velocity. In fact, Buckingham's theorem says that it is possible, via the dimensionless formulation of parameters, to reduce the number of dependencies. As the Reynolds number for real flows is one of the remaining parameters (which is largely determined by the incident flow velocity), according to Buckingham there may well be a functional and nonconstant relationship. Figure 4.16 shows the pattern of $c_D, c_L = f(\text{Re})$.

In fact, it is very noticeable that the c_D value can be assumed to be constant over a large range of Reynolds numbers. Incidentally, this is one reason why various industrial vehicle wind tunnels are run at different standard velocities around 150 km/h. Of special interest are the regions at low and very high Reynolds numbers, where significant deviations of the c_D value from the constant can be measured. At low velocities, the boundary layer on the surface of the vehicle is not yet fully turbulent, hence the separations are different. This affects the pressure and shear stress distribution on the vehicle surface and leads to different force effects.

Chapter 4

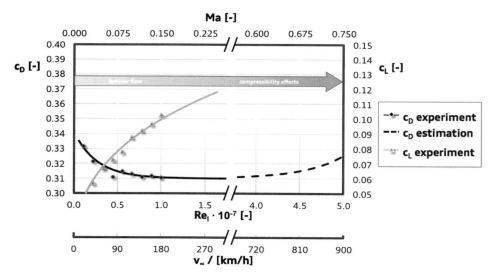

Figure 4.16 c_D value of a car as a function of the Reynolds number.

When approaching the speed of sound, the flow can no longer be regarded as incompressible. Compressibility effects such as the formation of compression waves and local speeds exceeding the speed of sound cause a rapid increase in the c_D value starting at approximately Ma = 0.7, until—at the speed of sound (Ma = 1)—a multiple of the subsonic drag is reached (cf. also Figure 9.20).

4.3 Analysis of Aerodynamic Drag Components

The following demonstrates some methods for analyzing the aerodynamic drag. First, the total drag will be divided into a pressure component and a friction component. Second, the microdrag method according to Cogotti will be outlined. Finally, the aerodynamic drag will be analyzed according to its components of (basic) form drag, induced drag, cooling aerodynamic drag, roughness drag of add-on parts, and interference drag.

4.3.1 Pressure and Friction Drag

The aerodynamic drag force can be determined by adding the surface pressure and surface shear stress values. These are also described as the pressure component (by flow separation and induced lift) and the friction component (by shear stresses). The cause of aerodynamic drag can thus be best understood by comparing the real, viscous flow with that of an ideal, inviscid fluid. The drag can be explained solely by the difference between these two flow patterns, because in inviscid flow the drag is known to be zero. Using the numerical methods described in section 14.1, it is possible to make flow calculations for basic bodies, but also for complex vehicles, to create a picture of the real flow and the ideal flow and then to compare them.

Aerodynamic Forces and Their Influence on Passenger Vehicles

On bluff bodies, such as a circular cylinder (cf. Figure 4.17) or a sphere, the pressure increase in the area of the largest body extension is so high that flow separation occurs. As a result, the pressure distribution on the body is asymmetrical, resulting in a high pressure drag.

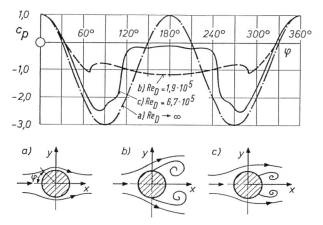

Figure 4.17 Pressure distribution and streamline pattern on a circular cylinder at different Reynolds numbers.

On the front side of the cylinder, the pressure distribution of real cases b) and c) substantially matches those of ideal inviscid flow. However, on the cylinder back there are considerable negative pressures as a result of flow separation in the real flow cases. Thus, the pressure distribution becomes asymmetrical with respect to the y-axis. The sum of the force components in the flow direction resulting from the pressure distribution is the pressure drag force.

Similar conditions are identified for the front and rear of a vehicle: While the pressures in inviscid and viscous flows are almost identical on the front—if, as assumed here, no separation occurs at this point—there are significant differences on the rear. And they are the reason why the pressure integral is different from zero. As stated above, the pressure differences between ideal and real flows are responsible for the pressure drag—not the pressure on the "stagnation surface" at the front, as postulated occasionally.

On a body in flow, the velocity gradient in the boundary layer and the molecular viscosity at each point transfer a shear stress from the fluid to the wall. The sum of the resulting force components in the direction of the flow (cf. section 2.3) is the friction force.

If there is no flow separation, the friction force is by far the predominant part of the total drag of a streamlined body. In the case of the thin plate, there is purely friction drag on both sides of the plate. Figure 4.18 shows the drag coefficient c_D over the Reynolds number Re formed using the plate length.

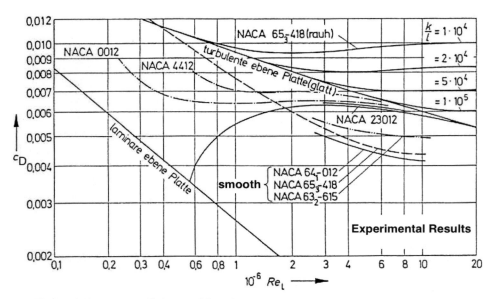

Figure 4.18 Drag coefficients of flat plates and wings as a function of the Reynolds number (according to Schlichting & Gersten [43]).

If we consider that the boundary layer in the front region of the plate is laminar and turbulent only in the rear region, we obtain the plotted transition curve. It is found that, in turbulent boundary layer flow, the friction drag is much greater than in laminar flow. This is due to the much larger velocity gradients in a turbulent velocity profile. Figure 4.18 also shows that wall roughness further increases the friction drag. Here, the drag coefficient behaves independently of the Reynolds number. Generally, it can be assumed that a vehicle has a pressure drag component between 80 and 90%, while a wing has a friction drag component of approximately 95%.

Adding the two components gives the overall drag of a body in flow. It includes all the drag components, in particular the component that is induced by any occurring edge vortices. This is emphasized here because, as explained in the next section, there are frequent misinterpretations when it comes to transferring the term "induced drag" from wing theory to the car.

You often hear that the friction component of the drag on a passenger car is small. This may have been true when $c_D = 0.5$ was regarded as a common value. Whether this also applies to vehicles with contemporary c_D values can be assessed on the basis of the computational fluid dynamics (CFD) results from Figure 4.19. Numerical flow simulation provides the ability to separate friction and pressure components, because both pressure and shear stresses are computed at each surface element.

Whether the resulting friction component is small or not depends on whether the inherent drag of the heat exchangers (radiator drag ≠ cooling aerodynamic drag!) is ascribed to the pressure drag or friction drag. If it is added to the pressure component, the friction component is just under 8%. This seems plausible because the radiator in the engine compartment—similar to a pipe baffle—is to be regarded as an additional pressure loss. Therefore, in the CFD simulation the radiator is not represented as a detailed component but as an artificial pressure drop (porous medium). However, the cause of this pressure loss is the aerodynamic drag within the radiator capillaries between the fins, so by definition it is a frictional effect. Similarly, the resulting pressure gradient in the longitudinal direction of the vehicle is probably negligible because the fins provide almost no surfaces with perpendicular components in the x-direction. So, if the radiator drag is ascribed to the friction component, its share of the total drag is as much as almost 25%, definitely not a low value.

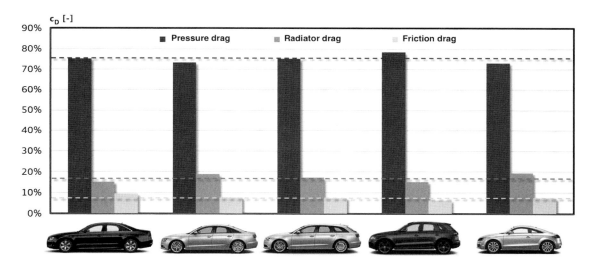

Figure 4.19 Breakdown of the total aerodynamic drag into pressure, radiator, and friction drag (percentages) for different body shapes (CFD analysis).

4.3.2 Microdrag

The perturbation effect of a moving vehicle on the environment is detectable over a large region, as symbolized in Figure 4.20. From the distribution of pressures and velocities on the boundaries of the control volume C, conclusions can be drawn about the forces and moments acting on the vehicle, in particular about drag using the momentum theorem. Cogotti ([156], [150], [160]) has developed and automated a process for measuring the variables in Gl. 4.23, and he applies this procedure during body shape development.

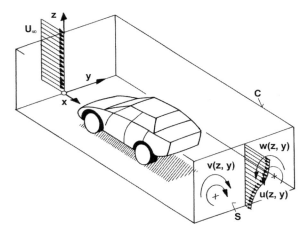

Figure 4.20 A vehicle in a control volume C in order to use Gl. (4.23).

With a 14-hole probe (see chapter 13), the components u, v, w of the velocity vector and the total pressure p_{tot} are measured over the part S of the control surface behind a vehicle. Despite the relatively high cost and complexity—a complete wake measurement requires about half an hour of wind tunnel time—this method is increasingly used for drag analysis. Using the momentum equation the total drag results after Cogotti [156] in

$$c_D \cdot A_x = \int_S \left(1 - c_{p,tot}\right) dS - \int_S \left(1 - \frac{u}{u_\infty}\right)^2 dS + \int_S \left(\left(\frac{v}{u_\infty}\right)^2 + \left(\frac{w}{u_\infty}\right)^2\right) dS \qquad (4.23)$$

Here, u, v, and w are the local components of the velocity vector, $c_{p,tot}$ is the total pressure coefficient defined similar to Gl. 4.14. Drag computed according to Gl. 4.23 turns out to be greater than the drag measured with a balance in a wind tunnel if the tunnel is equipped with a stationary ground floor; the momentum loss within the boundary layer along this floor will be accounted for, as shown in chapter 13. If drag is computed very close to the vehicle (e.g., 0.5 m behind the car), further deviation can occur compared to the value given by the wind tunnel balance due to very small velocities in the wake. The sum of the integrands in Gl. 4.23 is termed "micro drag" by Cogotti. Its distribution over the control surface S permits conclusions to be drawn about the points of generation of drag, and about the distribution of vorticity in the three planes. Here, vorticity is computed from the velocity vector as follows:

$$\Omega = \begin{pmatrix} \Omega_x \\ \Omega_y \\ \Omega_z \end{pmatrix} \text{ with } \Omega_x = \frac{1}{2}\left(\frac{\partial w}{\partial y} - \frac{\partial v}{\partial z}\right), \ \Omega_y = \frac{1}{2}\left(\frac{\partial u}{\partial z} - \frac{\partial w}{\partial x}\right) \text{ and } \Omega_z = \frac{1}{2}\left(\frac{\partial v}{\partial x} - \frac{\partial u}{\partial y}\right) \qquad (4.24)$$

Total pressure, velocities in the z, y-plane, micro drag, and vorticity are compared in Figure 4.21 for two vehicles. The CNR research car that Pininfarina presented at the 1990 auto show in Turin has low drag ($c_D = 0.19$) while the squareback vehicle has a high drag coefficient ($c_D = 0.34$). The difference between them becomes most evident when comparing their maps of total pressure and micro drag.

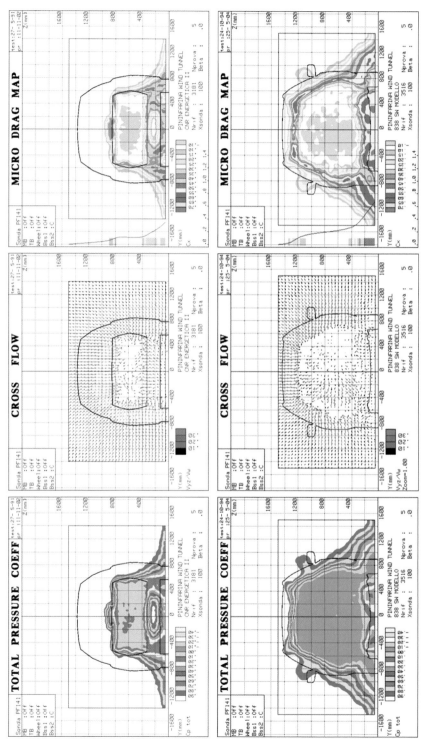

Figure 4.21 "Micro drag," velocity components, and total pressure over the control surface S (z, y-plane, $x = 0.5$ m) behind the vehicle [159]: CNR research car (upper, $c_D = 0.19$), squareback vehicle (lower $c_D = 0.34$).

4.3.3 Analysis by Individual Components

A more important—and, for the aerodynamicist, a more useful—breakdown of the aerodynamic drag will be described next. This will distinguish between form drag, induced drag, cooling air drag, roughness drag, and interference drag. Figure 4.22 provides a breakdown of the total drag into these partial drags; section 4.5 will deal with their influence.

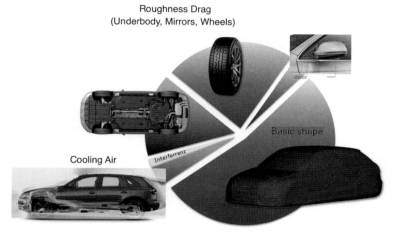

Figure 4.22 Percentage breakdown of aerodynamic drag into individual components of basic form drag, cooling air drag, roughness drag, and interference drag.

4.3.3.1 Basic Form Drag and Induced Drag

First, we will consider the basic shape of a vehicle, consisting of the vehicle's shell without wheels and cooling air, but with filled wheelhouses and smooth underbody. This therefore corresponds to a flying object near the ground, where the flow around it is dominated in wide areas by attached flow, exhibiting the separation phenomena mentioned in section 4.2 only in the rear section and in the region of the A-pillars. In Figure 4.23, the longitudinal vortices at the A-pillars and the ring vortex behind the vehicle can be identified as the main sources of loss on the basis of the isosurfaces of the total pressure. Additional pressure drops occur in the boundary layer and at the vehicle front, where the flow separates at small geometry details but reattaches rather quickly.

The comparison of such a flying object similar to a vehicle near the ground with some simple geometries of the same height and width and, if possible, the same length, quickly reveals that the shape of the vehicle basic body is already very streamlined. The overview in Figure 4.24 indicates that a squareback flying object similar to a vehicle with a c_D value of approximately 0.17 is already better than a projectile and a sphere in fully turbulent flow. Notchback-shaped flying objects even reach values of approximately $c_D = 0.13$ (not shown).

Aerodynamic Forces and Their Influence on Passenger Vehicles

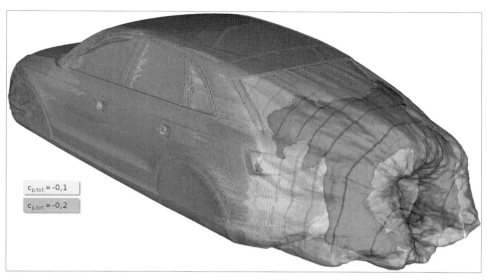

Figure 4.23 Loss regions on a squareback body, visualized using isosurfaces of the total pressure coefficient with $c_{p,tot} = -0.1$ and $c_{p,tot} = -0.2$

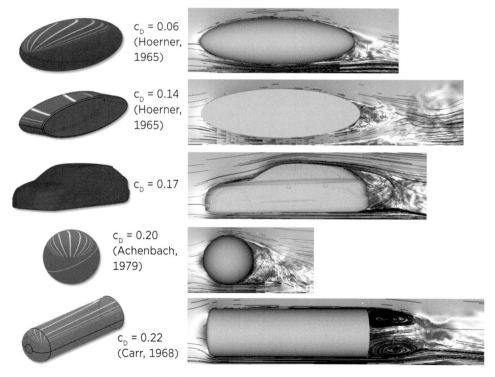

Figure 4.24 Comparison of the flow around and the drag coefficient of a squareback body with some simple geometries of the same fullness and the same Reynolds number.

227

At varying angles of attack, the squareback flying object shows constant (i.e., comparable) separation lines at the roof spoiler and the transition to the underbody rear apron. As long as these separation lines do not move, the squareback flying object near the ground behaves like a wing but with a shift toward distinct negative lift. An increase in the body's angle of attack about the transverse axis results in an increase in the drag proportional to the square of the resultant change in lift. These relationships are illustrated in Figure 4.25. In addition to the lift polar for negative and positive angles of attack, the near-wall velocities are also shown as an indicator for the separation lines, as well as an isosurface of the total pressure for $c_{p,\text{tot}} = 1$. Here you can see that in the case of a body with a positive angle of attack, longitudinal vortices are formed at the rear, which provide additional lift. One explanation for the polar shift—compared to the wing— toward downforce is the presence of the A-pillar vortices. These are always present, apparently do not change their intensity even at varying angles of attack, and seem not to mingle with the longitudinal vortices at the rear. On a cabin with a flat windshield and a well-rounded A-pillar, formation of the A-pillar vortex can be prevented. In this case, a polar similar to that of the wing should result.

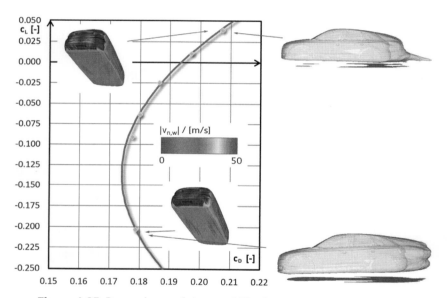

Figure 4.25 Dependence of drag and lift of a squareback flying object near the ground on its angle of attack.

Based on the theory of the wing, the total drag should therefore additively be composed of profile drag (optimum, $c_{D,0}$) and induced drag:

$$c_{D,\text{tot}} = c_{D,0} + c_{D,i} \tag{4.25}$$

For induced drag, a correlation of the form

$$c_{D,i} \sim c_L^2 \frac{A_z}{b^2} \tag{4.26}$$

Aerodynamic Forces and Their Influence on Passenger Vehicles

has been borrowed from wing theory, where b is the width and A_z is the base area of the car (= projected area in the z-direction). The quotient of A_z and b^2 denotes the aspect ratio. Visualizing this behavior graphically results in a Lilienthal polar.

The flow around a realistic vehicle is always associated with lift and, consequently, with induced drag as well. Without special measures, this is usually positive (i.e., directed upwards). This results in load reduction to the wheels, depending on the driving speed. The adverse consequences for vehicle handling are the topic of chapter 5, while chapter 9 shows what can be gained by negative lift, which is called "downforce" in racing.

By arranging the drag coefficients of cars according to their respective lift, a general trend is visible that is reminiscent of a Lilienthal polar. Figure 4.26 depicts such pairs of values for a total of 60 current production vehicles (as of 2012). Squareback cars have larger drag as a result of increased dead wake behind the car independent of lift, so that, for a comparable dataset, their drag coefficients have been corrected down by 35 counts to create the polar. The plotted best-fit polar shows optimum drag at a rear axle lift of approximately 0.06 and a significant increase in drag even at low deviations from this c_{Lr} value. The rear axle lift is chosen here because the A-pillar vortices and therefore the contribution by the front axle lift remain unchanged by most measures to influence the total lift, thereby allowing for clearer representation.

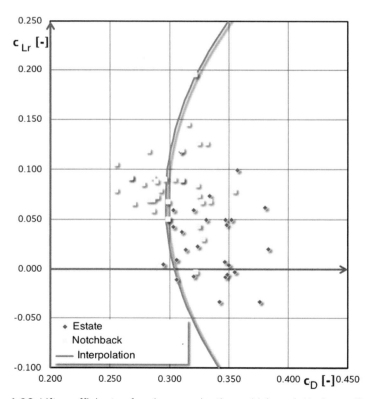

Figure 4.26 Lift coefficients of various production vehicles, plotted over their drag coefficient; measured values from the Audi aeroacoustic wind tunnel.

For clarification, Figure 4.27 shows the high-energy longitudinal vortices on a fastback body. The reason for this shift of the optimum drag compared to the wing can be explained by the presence of the wheels and the longitudinal vortices generated there. The whole vortex system at the wheel will be discussed in section 4.3.3.3 in more detail. While the A-pillar vortex at the vehicle rear apparently mixes at least to a certain extent with the C-pillar vortex, the longitudinal vortices of the wheels near the ground float away toward the rear without interacting noticeably with the longitudinal vortices of the A- and C-pillars. The resulting vorticity behind the vehicle basic body can be modified by certain measures, for example, a rear spoiler. One influence on the longitudinal vortices generated at the wheels has, however, not been reported so far, probably mainly due to the fact that such a measure would represent a fixed component located very near to the road, which would severely limit ground clearance and ramp angle. Without such measures, the lift effect of these vortices remains and an optimized body with wheels therefore always has lift at the optimum drag.

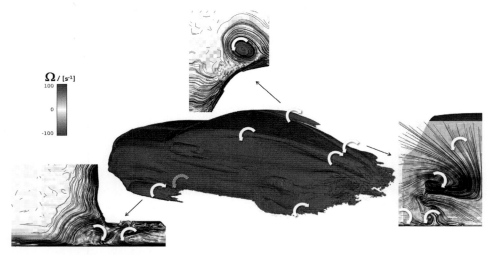

Figure 4.27 High-energy longitudinal vortices on a fastback body, visualized by an isosurface $c_{p,tot} = 0$.

With a rear wing, it is possible to influence the longitudinal vortices and thus the induced drag without adding a significant form drag. A similar study was reported by Wickern et al. [868]. On two vehicles with different rear types, a wing was mounted on the rear at different angles of attack. Figure 4.28 shows the relationship between drag and lift over the angle of attack. Increasing the lift/downforce leads to an increase in drag, where the optimum, as previously explained, is at slight lift. Some production cars, such as the Audi A7 and Porsche 911, are equipped with a retractable rear spoiler that, thanks to a significant reduction of lift, also reduces the drag.

Aerodynamic Forces and Their Influence on Passenger Vehicles

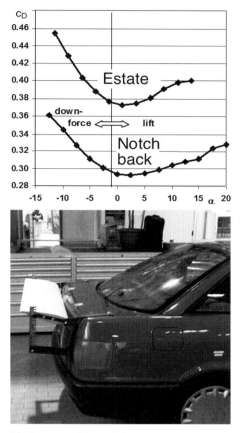

Figure 4.28 Relationship between lift and drag on two vehicles of different rear shape with variable rear wing, represented by the force coefficients over angle of attack.

The fact that lift and drag can also change in opposite directions is shown by some data compiled in Figure 4.29; they were measured on a VW 1600 (MY 1975), which was prepared as a calibration model for wind tunnel comparative measurements. Drag and lift could be varied independently of each other on this model by add-on parts. At approximately the same low drag of the configurations A and B, for B a lift was measured that is less than half as large as that of A. Similar results are shown by comparing the configurations C and D at a higher c_D value. When one finally compares D with A, the opposite trends of c_D and c_L become evident. Equation (4.25) is therefore not applicable. The opposite trends demonstrated here by one example are reflected in the large scattering, as visualized in Figure 4.26.

		c_D	c_L
A 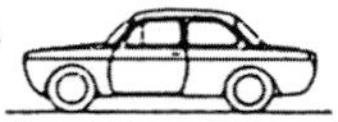	Baseline model	0,34	0,38
B 	Rear Spoiler	0,33	0,18
C 	Side Spoiler	0,38	0,48
D 	Front Spoiler	0,38	0,29

Figure 4.29 Pairs of values of drag and lift coefficients (c_L, c_D), which can be generated by different add-on parts on a car model [370]

There have been repeated attempts to break down the pressure drag from visualizations such as Figure 4.26 into lift-dependent and lift-independent components analogous to equation (4.25). However, it has not been possible to find a universally valid number for the proportionality factor in equation (4.26). As shown in the figure, the deviation of the value pairs (c_L, c_D) of the individual vehicles from the plotted parabola is so large that equation (4.25) does not even allow a rough estimate of the lift-dependent drag.

This is not surprising, as the induced component can only be calculated if the two types of drag according to equation (4.25) do not affect each other, a condition which is satisfied only for wings with a large aspect ratio, but not for cars which—in the terminology of wing theory—are bodies with an extremely small aspect ratio.

On such bodies—and also on slender wings—the interaction of the edge vortices generating drag with the flow around the middle section is very pronounced. To what extent can be derived, for example, from Figure 4.6, it is clearly visible how close the two edge vortices lie to each other. They affect the flow over the entire width (wing span) of the vehicle and thus also the profile drag. Likewise, the large dead wake—which is a typical characteristic of bluff bodies—must be expected to have a large retroactive impact on the generation and pattern of the edge vortices. While it is possible to formally break down the drag according to equation (4.25), it is not possible to calculate the drag induced by the vortices as per equation (4.26).

It should be noted at this point that measures to influence the form and cooling air drag almost always also affect the lift and thus the induced drag. For this reason, section 4.5 does not discuss the influence on the induced drag separately; rather, the resulting influence on the drag should be mentioned with regard to any influence on the basic form and cooling air drag via the influencing of lift.

4.3.3.2 Cooling Air Drag

A passenger car is crossed by several stream tubes: water-cooled radiator, condenser, oil and charge air cooler, as well as brakes must be supplied with cooling air, and the engine must be supplied with combustion air. Finally, fresh air must be passed through the passenger compartment; it passes through the heating system heat exchanger and evaporator. In part, these "tubes" are shaped like closed ducts, such as the cooling air through heat exchangers arranged in the engine compartment. But the air can also find its way unguided through the vehicle, such as at the brakes. From the aerodynamicist's perspective, these stream tubes are all "disturbing" because they are associated with losses, which in turn are reflected in increased drag.

The cooling air drag (or internal drag) is composed of the drag components from the flow through the engine compartment and the supply of cooling air to brakes, gears, and so forth. The coefficient of the cooling air drag $\Delta c_{D,C}$ is defined as

$$\Delta c_{D,C} = c_{D,\text{Grille open}} - c_{D,\text{Mock-up}} \tag{4.27}$$

For road vehicles, this is typically in the range of $0 \leq \Delta c_{D,C} \leq 0.04$, with an average of approx. 5–10% of the total aerodynamic drag. The reference here is the drag of the vehicle with closed flow openings, the so-called mock-up.

The cooling air drag is determined by four components, namely

- incident flow losses on the cooler grille and its environment,
- pressure losses in the radiator and engine compartment flow,
- impact and momentum losses at the flow outlet, and
- interactions with the flow around the vehicle, particularly with the front wheels.

The basic requirement of a cooling system is that the components of the engine can be adequately cooled at all operating conditions of the vehicle. In conventional liquid cooling, the first thing to ensure is that the coolant is not heated above its boiling point. Under these conditions, we can develop a reduced cooling air drag. The highest temperatures at the components surrounding the combustion chamber usually occur at maximum engine performance (i.e., in the region of the maximum speed of the vehicle). Other limits for the cooling system are slow ascent of a fully loaded vehicle, towing a trailer, and engine idling of a stationary car after full-load driving. Although the engine components surrounding the combustion chamber do not reach high temperatures at these operating points, the tuning of the radiator and fan must be verified to protect the coolant against overheating. These extreme points must be safeguarded at the highest ambient temperature that occurs in the specific region where the vehicle is used. The flow in the area of the vehicle radiator can vary greatly depending on the operating point.

Nonuniform flow has an unfavorable effect on the efficiency of a heat exchanger. The heat exchange is not linear with increasing flow, but degressive; on average the heat exchange becomes inefficient (cf. Figure 4.30).

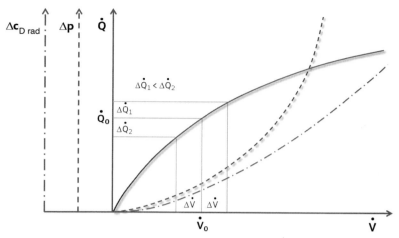

Figure 4.30 Relationships between heat dissipation Q, pressure loss, radiator flow, and radiator drag (schematic).

In particular, backflows through the radiator have a negative effect on heat exchange, which is caused by a reduction of the effective cooling surface and by heating of fresh cooling air in the intake ports. Figure 4.31 shows the flow for a stationary and a moving car. The flow velocity in a radiator is also shown for different driving states.

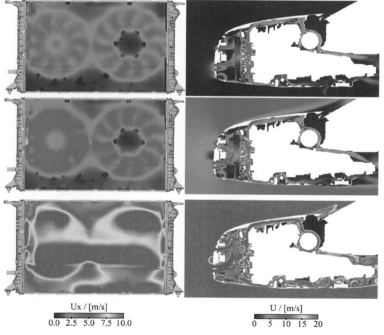

Figure 4.31 Airflow conditions in the y-center section (right) and velocity field at the coolant radiator (left) of an Audi A4 at different operating points: when stationary (top) and at 30 km/h (center) with running fan, and at vmax (bottom) without the fan.

With running fan, the goal is to achieve a through-flow condition that is independent of the driving speed. The flow velocity through the radiator at high-speed operation is typically 20–25% of the incident flow velocity.

The aforementioned first cause of cooling air drag is the inflow into the cooling air tubes (i.e., the "classic" inlet loss, as described in section 2.3.7). The inlet loss can be reduced by a factor of ten in a streamlined inlet as compared to the sharp edge. So it pays off to carefully design the inlet including the grille taking into account aerodynamic considerations.

The second cause of cooling air drag is the pressure loss caused by the radiator and other components in the engine compartment. The conditions for the radiator installation are unfavorable. The space is fixed and limited by traverses, engine, and auxiliary power units. The radiator inlet and outlet flow is hindered, which results in an additional loss of pressure. Figure 4.32 demonstrates this by the qualitative pattern of the flow and static pressure in the engine compartment.

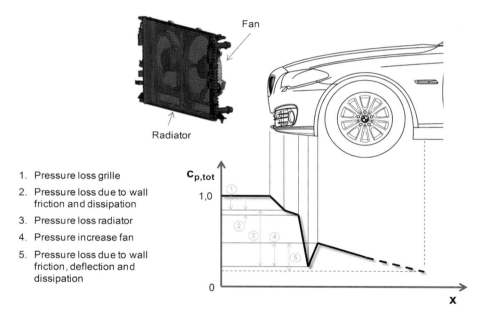

1. Pressure loss grille
2. Pressure loss due to wall friction and dissipation
3. Pressure loss radiator
4. Pressure increase fan
5. Pressure loss due to wall friction, deflection and dissipation

Figure 4.32 Flow of the cooling air and the pressure profile in the cooling system of a passenger car.

The pressure loss through the radiator can analogously be described by flow through a tube with

$$\Delta p_K = \zeta_K \frac{\rho}{2} v_K^2 \qquad (4.28)$$

The pressure loss coefficient ζ_K can be determined using the tube friction coefficient λ pipe of the fins, the radiator depth l_K, and the hydraulic tube diameter of the radiator fins, which measurements on actual radiators have shown to have the following relationship:

$$\zeta_K = \lambda_{pipe} \frac{l_K}{d_{fin}} \approx c \cdot v_K^{-0.5} \frac{l_K}{d_{fin}} \tag{4.29}$$

If it is assumed that there is no interaction with the ambient flow, the cooling air drag can be estimated on the basis of this drag coefficient—this then corresponds to the radiator drag [765]:

$$\Delta c_{D,\text{rad}} = \zeta_K \left(\frac{v_K}{v_\infty}\right)^3 \frac{A_K}{A_x} \sim v_K^{1.5} \tag{4.30}$$

The pressure loss coefficient ζ_K and the resulting through-flow velocity v_K of the radiator depend on the geometric and structural characteristics of the radiator, so, for example, its depth, pipe spacing, and fin density, as well as on the incident flow situation. A deeper and denser radiator increases the heat-exchanging surface, but also the pressure loss coefficient, reducing the throughput. As referred to a few sections above, we again refer here to Figure 4.30 and the degressive performance of heat transfer with increasing throughput. The drag of a radiator tends to become smaller with increased packing density, but it must be noted that in a denser radiator, higher fan power is required during fan operation, and thus more energy is needed.

The third cause of cooling air drag is pressure and momentum losses at the flow outlet. These are small if the pressure coefficients c_p of flow around the vehicle and through the vehicle are equal at the flow outlet and, in addition, if an outlet flow as tangential as possible with a large momentum in the vehicle's longitudinal direction is achieved. Figure 4.33 shows possible variants in this context with their influence on aerodynamic drag.

The fourth cause of cooling air drag, the interaction with the flow around the vehicle, can be explained by a change in flow around the body. This applies especially to the front end of the car. If it has an unfavorable aerodynamic design, even a negative cooling air drag may result if a separation is avoided and the drag of the ambient flow is reduced. However, if the nose is streamlined—and this applies to the majority of today's cars—this "gain" no longer occurs. Studies by Kuthada [479] show that changes to the flow topology occur even at the vehicle rear. The interaction with the ambient flow will now be illustrated by the example of the front wheels. Figure 4.34 visualizes the wake of a front wheel.

Aerodynamic Forces and Their Influence on Passenger Vehicles

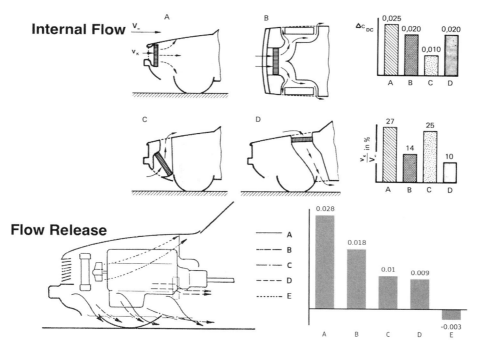

Figure 4.33 Possible solutions for cooling air routing in the engine compartment [107].

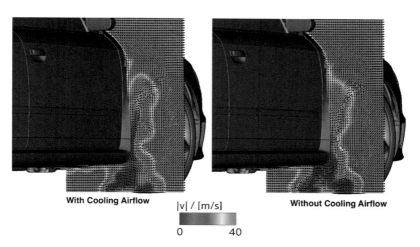

Figure 4.34 Velocity distribution in the wake of a front wheel with (left) and without (right) cooling air.

The cooling air from the engine compartment exits to a certain extent into the wheelhouses and increases the incident flow angle of the front wheels in comparison to the mock-up, so that the resulting increase in wheel wake generates more drag. Improving the flow around the vehicle is also conceivable; therefore, the entry of functional air into the ambient flow should be specifically optimized during aerodynamic design.

For a better understanding of the very complex interactions that lead to cooling air drag, let us make some basic momentum and energy statements for cooling ducts.[3] Provided that the cooling air is guided (i.e., the cooling air flows through the vehicle, entering into and exiting from defined inlet and outlet faces), the momentum theorem for the control volume can be postulated by the cooling air stream tube. This is represented in Figure 4.35 using the example of a vehicle with cooling air outlet at the underbody. By substituting the resultant force from Figure 4.35 and the definitions of c_D and c_p from section 4.1, we obtain, for the cooling air drag coefficient:

$$\Delta c_{D,C} = 2 \frac{A_K}{A_x} \frac{v_K}{v_\infty} \left(1 - \frac{v_A}{v_\infty} \cos\alpha \right) - c_{p,A} \frac{A_K}{A_x} \cos\alpha \qquad (4.31)$$

This estimation is valid where the cooling air flow does not cause a change in pressure at the boundaries of the control volume. The pressure coefficient at the outlet $c_{p,A}$ is determined by the ambient flow. The second term in equation (4.31) will hereinafter be disregarded when calculating the cooling air drag because it would predict a force even without a through-flow. However, this must be attributed to the drag of the basic form.

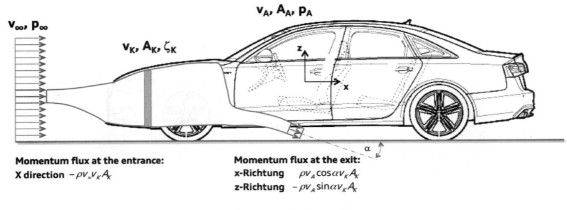

Figure 4.35 Momentum theorem for the flow through a vehicle with fully guided cooling air, according to [875].

The outlet velocity can be determined with the aid of Bernoulli's equation (cf. section 2.1.5); the following expression is obtained for the cooling air drag by substitution

---

3. The correlations were derived in detail by Wiedemann [875].

$$\Delta c_{D,C} = 2\frac{A_A}{A_x}\sqrt{\frac{1-c_{p,A}}{1+\zeta_K\left(\frac{A_A}{A_K}\right)^2}}\left(1-\sqrt{\frac{1-c_{p,A}}{1+\zeta_K\left(\frac{A_A}{A_K}\right)^2}}\cos\alpha\right) \text{ with } \frac{v_A}{v_\infty} = \sqrt{\frac{1-c_{p,A}}{1+\zeta_K\left(\frac{A_A}{A_K}\right)^2}} \quad (4.32)$$

Any deviation of the outlet angle of the cooling air from 0° results in an additional loss of momentum and thus in a higher cooling air drag. A reduction in the outlet area A_A results in a reduction in drag. Figure 4.36 (left) shows the cooling air drag related to the outlet area ratio A_A/A_x as a function of the extended pressure loss coefficient $\zeta_K(A_A/A_x)^2$ for an outlet angle $\alpha = -45°$, which is common for a car without special measures on the underbody.

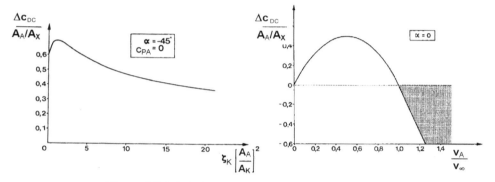

Figure 4.36 Relationships between parameters of the cooling air system and the cooling air drag [765].

The depicted curve can be divided into two parts: In the first part, the drag increases monotonically until it reaches a maximum for an extended pressure loss coefficient of one. This part of the curve corresponds to almost friction-free behavior of the flow, in which an increase in pressure loss leads to a reduced outlet momentum in the driving direction, thereby resulting in increased drag. In the second part, the behavior of the flow is dominated by friction, and a larger pressure loss results in lower flow rates and reduced drag. For an increasing pressure loss coefficient, this leads to ever smaller drag until it approaches zero. This case corresponds to the mock-up with an infinitely large extended pressure loss coefficient, which prevents any flow of cooling air.

Figure 4.36 (right) shows the curve of the drag related to the outlet area ratio over the dimensionless outlet velocity for an outlet angle of $\alpha = 0°$. In the first part of the curve, the drag increases. Initially, an increase of the cooling airflow results in increased drag. After exceeding the maximum, any further increase of the cooling airflow results in a reduction of the drag. In the case considered, it is even theoretically possible to generate negative cooling air drag. The outlet velocity can be influenced by reducing the pressure coefficient $c_{p,A}$ at the outlet or by reducing the extended pressure loss coefficient. In practice, however, a change to one of the parameters will have effects on the other parameters. They, in turn, can affect the system positively or negatively in the sum of all occurring effects. Thus, a reduction in the outlet cross section generally results in an increased pressure loss in the system, thereby reducing the volumetric flow. The thermal efficiency of the radiator must be taken into account when reducing the pressure loss by adjusting the radiator. The thermal efficiency in turn is affected by the pressure loss and the velocity distribution on the radiator.

The cooling air drag can therefore be minimized by minimizing the volumetric flow, which must be checked in terms of thermal viability, as well as by as horizontal as possible an outlet for the cooling air. The inlets should be large enough to reduce the inlet momentum, while the outlet should be small in order to increase the momentum strength in the driving direction. However, any reduction in the outlet cross section has only a minor influence on the cooling airflow. Outlets should preferably be located at the rear of the vehicle or at the vehicle sides. Finally, the radiator surface A_K should be as large as possible. In summary, it can be said that the cooling air ducts must be designed and dimensioned such that minimum losses are achieved, no leaks occur, and the trend to form backflow is prevented or reduced by using special constructions with flaps. Requirement-dependent supply of cooling air and an intelligent fan strategy are therefore of special significance.

An early implementation of intelligent cooling air ducting can be found on the UNICAR from 1981, cf. Figure 4.37. The outlet of the cooling air is particularly optimized. Outlets exist both at the vehicle sides in the fender area and at the rear, which ensures maximum momentum recovery.

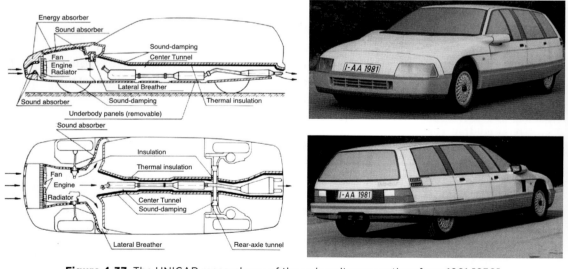

Figure 4.37 The UNICAR research car of the university consortium from 1981 [650].

Finally, Figure 4.38 compiles a few numerical values for the cooling air drag—to be determined as described above. The frontal area of each car, not the radiator frontal area, was chosen as the reference area for this comparison. As can be seen, there are significant differences from car to car—the values range from 0.006 to 0.038. This can be explained in part by the different rear shapes and the interaction with the induced drag of the basic form that is partially dependent on the rear shape. On the other hand, however, individual causes of cooling air drag, optimized to different degrees, play a role; for example, the inlet quality at the radiator grill, pressure loss at the radiator, and outlet losses, mostly on the underbody.

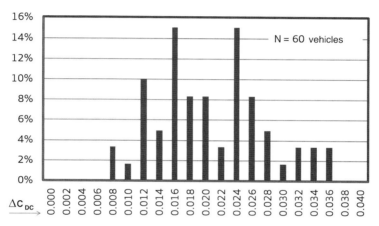

Figure 4.38 Spectrum of the cooling air drag of a total of 60 vehicles from different manufacturers, measured in the aeroacoustic wind tunnel of Audi AG.

4.3.3.3 Roughness Drag of Add-On Parts

Under roughness drag, we refer to the inherent drag of add-on parts and the surface structure under ideal incident flow conditions. This includes in particular the underbody with exhaust system, fuel tank and suspension, the wheels, and the add-on parts of the shell such as exterior mirrors, antennas, additional lights, door handles, luggage rack, windshield wipers, spoilers, wings, wheel arches, and window recesses.

The inherent drag of an exterior mirror, for example, is approximately four to ten c_D counts, a pair of external mirrors thus generates about 4% of the total vehicle drag. Most of the other add-on parts mentioned on the shell are now integrated unobtrusively into the skin, thereby contributing only small fractions to the total aerodynamic drag; only a nonpermanent component such as a roof rack with luggage box results in significant additional drag. A roof rack increases the total drag by about 10% and a rack with a ski box by about 40%.

So while the "visible" add-on parts have only limited influence, we refer again to section 4.3.3.1 and the significance of the wheels for the induced drag that is discussed there. Of course, the wheels mean generally more form and induced drag. Though they are relatively small compared to the body, their profiles and the fact that they cannot be retracted make them rather unstreamlined bodies with edges inclined to the incident flow. In addition, the nose of the vehicle must displace the air—in the vertical direction in the vehicle's center section, more in lateral direction in the side region—so the flow hits the wheels at an angle. All these aspects result in a complex vortex system behind the wheels, which Waeschle has examined in great detail (Figure 4.39).

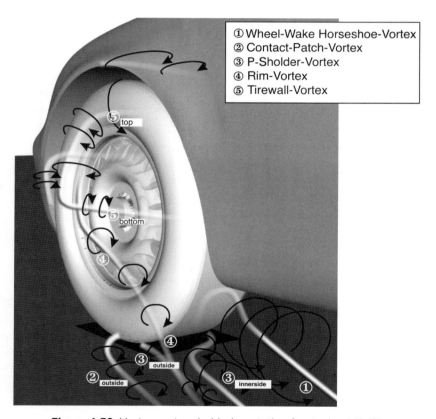

Figure 4.39 Vortex system behind a rotating front wheel [847].

The pressure and shear stress distributions at the wheel follow from the local flow topology. As with any flow body, summation on the wheel surface leads to three force and three momentum components. Assuming a precisely lateral rotational axis (i.e., located in the y-direction) of the wheels (which then have no camber and toe-in), the drag, lift, and side forces as well as the yaw and rolling moments are directly evaluated by the wind tunnel scale and added to the "total vehicle" system. However, in wind tunnels with an internal scale, the pitching moment of the wheels (i.e., the moment about the y-axis) is not detected.[4] We will return to this point in chapter 13.

Evaluation of the ventilation moment is possible in wind tunnels with an internal scale, or on a chassis dynamometer. Measurements made by Tesch [794] show a ventilation moment M_V that increases with the square of the driving speed. Since this has at any rate a braking effect on the rotary motion of the wheels during driving, it makes sense to enter an extended drag coefficient $c_D{}^*$. Here, the ventilation moment is standardized using the stagnation pressure, the face of the vehicle, and the dynamic wheel radius, and then added to the measured value without the ventilation moment:

$$c_D^* = c_D + \frac{\sum_{i=1}^{4} M_{V,i}}{\frac{\rho}{2} v_\infty^2 A_x r_{\text{dyn}}} \qquad (4.33)$$

Optimizing the wheel aerodynamics without considering the ventilation moment may therefore result in a decrease of the c_D value, while the extended drag coefficient $c_D{}^*$ increases.

To clearly distinguish between the terms ventilation effect and ventilation moment, the following should be mentioned: The ventilation moment is a ventilation effect. A second ventilation effect is the previously mentioned flow deflection in lateral direction, because the more the rim "ventilates," the greater is the deflected mass flow and thus also the loss of momentum in the vehicle's longitudinal direction. The second effect, in contrast to the first, is measured independently of a scale in a wind tunnel.

The underbody on older vehicles is very ragged, forcing the flow to stagnate and separate at countless places. On one hand, this gives rise to a very lossy flow under the vehicle. On the other hand, a rough underbody also means an upward shift of the stagnation point at the front end, causing more air to flow over the car. In particular on fastback vehicles with flow over the rear slant, this leads to increased rear lift (cf. momentum theorem section 2.1.8) and thus also to an increase in induced drag. It follows from these facts that the drag component of the underbody group is to be understood as specific to the vehicle and, especially, to the add-on parts, so that the structure of the underbody should be adapted to the vehicle shape with streamlined trims when designing the aerodynamics.

---

4. As, moreover, the rolling drag of the wheels. This is due to the surface pressure in the tire/road contact zone, which is asymmetric to the wheel center plane, cf. [272].

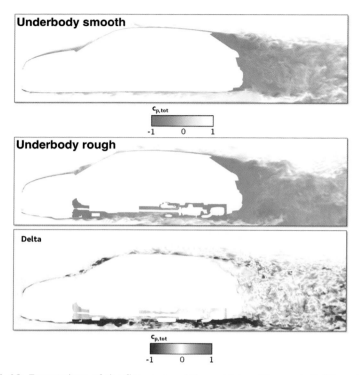

Figure 4.40 Energy loss of the flow around the vehicle with rough (left) and smooth underbody, expressed by the total pressure coefficient $c_{p,tot}$ in the y-center section.

4.3.3.4 Interference Drag

To obtain the total drag of a motor vehicle, it is not enough to add up the drags of the individual items—of basic shape, add-on parts, wheels, mirrors, and so on. In fact, it is necessary to consider their mutual influence (interference), because the flow pattern around the body is changed by add-on parts. The interference drag includes the influences that the add-on parts mentioned in section 4.3.3.3 exercise on the basic body and the basic body on the add-on parts. A typical case is the interaction between the body and the rear spoiler or between the tow vehicle and the trailer.

Both a positive and a negative interference drag are conceivable. A positive interference drag is caused by convergence of two adjacent bodies, such as vehicle body and exterior mirrors. This should first be shaped symmetrically to the incident flow direction. Without the vehicle body, the flow hits the rearview mirror evenly and with a favorable c_D value. But the influence of the body results in a higher incident flow velocity, increasing the risk of separation when the pressure rises. On the vehicle body, a shift of the stagnation point toward the mirror and an asymmetric ambient flow are identified, which increase the risk of separation (cf. Figure 4.41). One solution is to give the mirror a shell that is shaped asymmetrically to the incident flow direction.

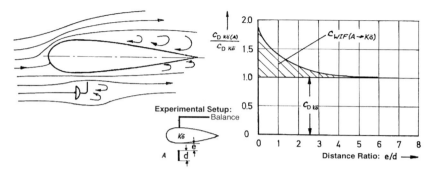

Figure 4.41 Explanation of positive interference drag on the basis of an arrangement of vehicle body (K) and exterior mirror (A) [872].

A negative interference drag is caused by convergence of two bodies arranged one behind the other. Behind each body, there is a space of reduced flow velocity (wake). A body that is in this space is subjected to lower drag corresponding to the reduced incident flow velocity. By arranging similar or different types of bodies with pressure drag in series, it is therefore possible to obtain negative interference drag. In the example described in Figure 4.42, the total drag of circular plates arranged one behind another for $\Delta x/d < 6$ is smaller than the total drag of the two plates in undisturbed incident flow. For $\Delta x/d < 2$, the total drag is even smaller than the drag of a single plate.

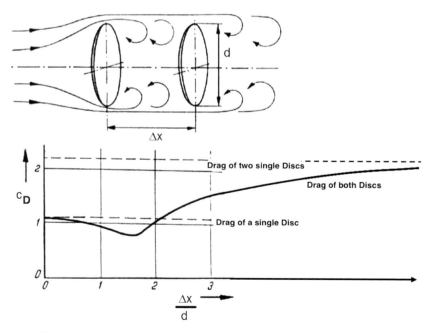

Figure 4.42 Explanation of negative interference drag on the basis of an arrangement of two circular plates [213].

Large negative interference drag may occur in particular if the induced drag of the body is reduced by an add-on part. The example of a rear spoiler will be discussed later, but let us already explain some aspects based on the extendable rear spoiler of the Audi TT that will be mentioned in section 4.5.1.3.4. The spoiler alone will of course have considerable inherent drag mainly due to flow separation at the rear side. However, the downward flow of the body, the downwash, can be reoriented in the vehicle's longitudinal direction. This reduces the vorticity of the C-pillar vortices considerably, which significantly overcompensates the inherent drag of the spoiler blade.

4.4 Other Components of Aerodynamic Force and Aerodynamic Moment

A breakdown of lift, side force, pitching, yaw, and roll moments by point of origin, as performed for the drag, is not common. Therefore, the following section is significantly shorter than the previous one. Nevertheless, we will explain at least fundamentally how these components of the total aerodynamic force come about.

4.4.1 Lift and Pitching Moment

The explanations in section 5 will show that the amount of total lift and the differences between front and rear lift resulting from the pitching moment are essential for directional stability. How they depend on the basic shape of the vehicle is shown by the diagram in Figure 4.43: A high stagnation point at the nose causes an increased front lift; a lower flow separation at the rear increases the rear lift. Noteworthy is the combination of low stagnation point at the front and high flow separation at the rear. The result is a very large positive, that is, raising, pitching moment.

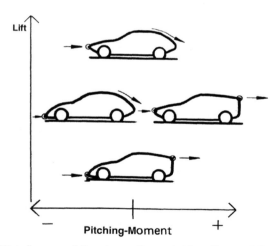

Figure 4.43 Influence of the stagnation point location and the amount of flow separation at the rear on pitching moment and lift.

Lift and pitching moment are often evaluated even at yaw angles of $\beta \neq 0$. In contrast to aerodynamic drag—which, it was deduced, is only relevant at high driving speed, but only small yaw angles occur here, and the c_D changes for them are small—the lift may vary substantially even at small yaw angles. Figure 4.44 shows a typical curve for lift and pitching moment. The lift coefficient increases disproportionately with the incident flow angle β while the pitching moment hardly changes at all. It follows that the axle lifts increase approximately in the same manner.

A round transition area from the roof to the rear window can lead to unwanted sudden flow transition for "sensitive" rear shapes and at small yaw angles. This is reflected in an altered shape of the curves, in particular by a pitching moment decreasing with β. The measures to solve this situation are discussed in section 4.5.1.3.

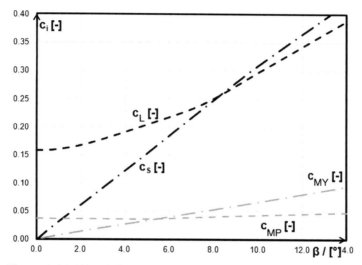

Figure 4.44 Lift, pitching moment, side force, and yaw moment as a function of the yaw angle.

Just as the cooling air affects the aerodynamic drag (cf. section 4.3.3.2), it also changes the total lift, as well as the lift components related to the axles, and therefore also the pitching moment. As the cooling air in the front end usually flows out downwards, it typically increases the front axle lift. As a result, the cooling air increases the mass flow below the vehicle, which is transported upwards by the rear diffuser; at the same time, this air mass is "missing" in the flow above the roof, thereby causing lower rear lift (cf. momentum theorem). Empirical values prove that the rear axle lift is reduced by about half of the change in the front axle lift. The front and rear axle lifts together cause a sign reversal of the moment about the vehicle's transverse axis. This situation is displayed schematically in Figure 4.45, showing the change from negative to positive pitching moment.

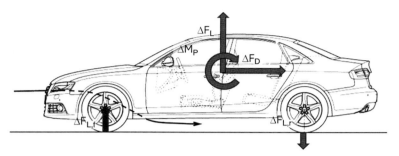

Figure 4.45 Influence of cooling air on lift and pitching moment.

4.4.2 Side Force and Yaw Moment

Similarly, side force and yaw moment affect the driving characteristics. Since a vehicle that is almost symmetrical to the y0 plane is affected ideally by no side forces and no yaw moment in straight flow, these points are also evaluated at yaw angles. For a wide range of car types, the yaw moment increases almost linearly up to a yaw angle of $\beta = 20°$ (cf. Figure 4.44 and Figure 4.94). Larger yaw angles are irrelevant to actual driving, because they occur only at lower driving speeds, where the forces are then small ($\sim v_\infty^2$). Therefore, the measured values for the yaw angle $\beta = 20°$ can very well be used as a reference for evaluating the curve of side force and yaw moment.

As shown in the diagram in Figure 4.46a, in oblique incident flow the attached flow in the nose and rear areas generates a yaw moment which rises steeply with increasing yaw angle. The quotient of the change in yaw moment to the change of the yaw angle is positive (i.e., the yaw moment falls), so in aerodynamic terms the vehicle is directionally unstable. However, if the flow separates at the rear, the instability is reduced (case b). And with a large tail fin, the rear side force can increase so much that an aerodynamically stable yaw moment occurs (case c). While this measure is only partially feasible on roadworthy vehicles, it is used for record vehicles, cars, and motorcycles.

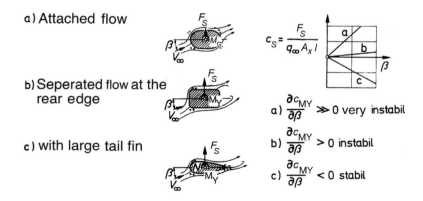

Figure 4.46 Aerodynamic directional stability in side winds [714].

Especially with box-shaped vehicles—express transporters and buses—the nose shape has a significant influence on the yaw moment, which is shown by Figure 4.47. Only a fully sharp-edged nose shape (1) is aerodynamically stable in a narrow range of yaw angles. Vehicles with rounded nose edges, as are necessary with respect to low aerodynamic drag, are aerodynamically unstable.

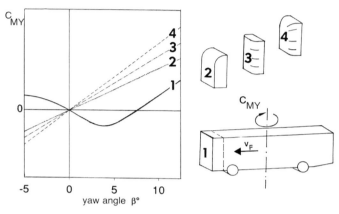

Figure 4.47 Influence of the nose shape on the yaw moment in oblique incident flow.

However, the yaw moment can be selectively influenced by the shape of the nose edges. According to Hucho [375], it is possible via controlled flow separation to terminate the linear increase in yaw moment at a certain yaw angle and to "switch" to a shallower increase in yaw moment. However, the drag then rises abruptly, as shown schematically in Figure 4.48. But if this critical angle is above the yaw angle that typically occurs during normal driving, this increase in drag has no significant impact on fuel consumption. The individual detailed changes which can be used in general to reduce the yaw moment of a car are summarized in an overview in Figure 4.49.

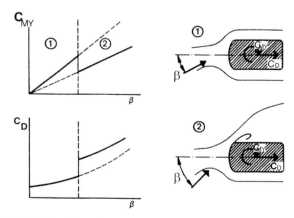

Figure 4.48 Influencing the yaw moment by controlled flow separation.

Chapter 4

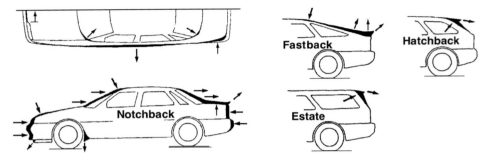

Figure 4.49 Shape changes to reduce the yaw moment [361].

4.4.3 Roll Moment

Furthermore, the roll moment does not occur in an idealized straight flow. It increases with increasing yaw angle β; typical curves for the three classical tail shapes are compiled in Figure 4.50. Owing to the approximately linear gradient of $c_{M,R}(\beta)$ in the specified angle range, specification of the coefficient at a yaw angle of $\beta = 20°$ is also sufficient here to evaluate different shapes.

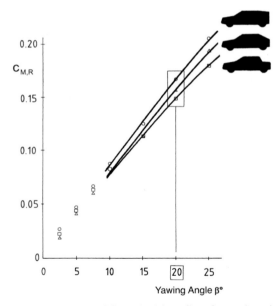

Figure 4.50 Roll moment in oblique incident flow for various basic shapes.

4.5 Influence on Aerodynamic Forces

Normally the time pressure during a development project does not allow to investigate the airflow in such detail as would be necessary. Nevertheless, the majority of the following examples are from development work on vehicles rather than research projects. The identification of the individual drag sources is based on the pragmatism of the aerodynamicist conducting his work. He detects discontinuities of the airflow (e.g., flow separations) and tries to eliminate this by modification of the body shape in order to get closer to the ideal flow behavior. The measurement of the aerodynamic drag and the remaining components is used to guarantee success, but it does not explain anything of the physical action on-site. The flow behavior can roughly be checked by the aerodynamicist by flow visualization of the streamlines only. The result of the before and after comparison will be interpreted as the individual drag of the detail which has been modified. Interference effects of the modification to the general airflow on other regions can of course not be identified.

From the pragmatic, practical approach of simply correlating the individual drag sources to the geometric changes, it will further be attempted to qualitatively explain them. Use will be made of comparison between real flow and ideal flow, and also the results of investigations of the flow field around the vehicle. Significant improvements have been possible in this area since the beginning of the 1990s, because the flow observations are now not only restricted to the surface of the body but can be extended to the whole flow field thanks to progress in specialized measuring techniques and CFD methods.

Generally, the aerodynamic development of vehicles implies the following steps: the structure of the outside surface of the body and add-on components contribute to the aerodynamic drag, as well as affecting the remaining force and moment components. In particular, the intersection between surfaces has to be smoothly executed along the vehicle. Once the styling direction is decided, the aerodynamic characteristics are roughly decided. Afterwards, the aerodynamic drag can only be improved by detailed shape optimization and development of add-on parts.

The potential areas of influence on the aerodynamic forces are described in detail in the following sections; they are ordered as shown in section 4.3 for the drag but expanded to include lift, side force, and yaw moment:

- Changes to the basic form (vehicle front, greenhouse, tail)
- Influence of modified routing of the engine cooling flow
- Changes to add-on parts (underbody, wheels, mirror, etc.)
- Influence of interference

Some of these measures to improve aerodynamic drag are roughly outlined in Figure 4.51.

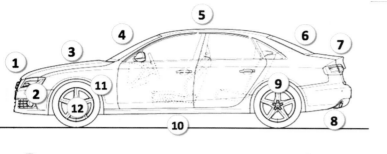
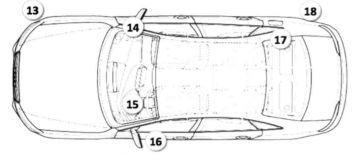

1.	Front inclined and rounded
2.	Cooling air guided and limited
3.	Engine hood inclined
4.	Windshield less steep
5.	Roof cambered
6.	Rear window less steep
7.	Trunk lid higher
8.	Diffusor applied
9.	Wheelhousing covered
10.	Underbody smoothed
11.	Wheel arch rounded
12.	Wheel disk covered
13.	Transition front – fender rounded
14.	A pillar rounded
15.	Windshield cambered
16.	External mirrors optimized
17.	Boat-tailing C-pillar
18.	Boat-tailing rear end

Figure 4.51 Shape changes that influence the aerodynamic forces.

4.5.1 The Influence of Basic Shape

In the following sections, the impact of proportion and form details of the vehicle outer skin are discussed. The sections are ordered roughly according to the progression of the flow field, from the front of the vehicle over the roof and greenhouse to the rear. Figure 4.52 supplies an overview.

Figure 4.52 Areas of the front, roof, body side, and rear of the vehicle where aerodynamic forces arise.

Part of the interference effects will therefore be hidden, since although the downstream effects of a change at the front on the flow field at the rear are captured, the reverse (i.e., the upstream effects of a change at the rear), is not. Therefore the topic of interference will be revisited in section 4.5.4.

4.5.1.1 Vehicle Front

The base form of the vehicle front—first of all without cooling airflow—can be approximately represented by a rectangular cuboid; see Figure 4.53. A stagnation point will be created at the vertical front surface. Because of the proximity of the cuboid to the ground, a positive pressure region is generated on the ground in front of the cuboid; hence the air passes preferably above the body and at the body sides. The stagnation point of a body close to the ground surface is shifted downward compared with a freely flying cuboid. The airflow is significantly deflected at the edges between the front face and the hood and fenders. Without careful design, the airflow would separate from the body surface and might reattach at a more downstream position. Separation bubbles are then generated, whereby a shear layer isolates the separated flow from the undisturbed flow. A vortex rotates inside of the separation bubble as shown in Figure 4.10. Its rotation axis is transverse to the incident flow and parallel to the edge, where the shear layer starts from (transverse eddies).

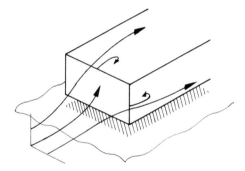

Figure 4.53 The square block as a simplified substitute model of the front end of a car.

As a consequence of this flow separation, the pressure distribution around the edges deviates from that of an inviscid flow. Inviscid flow would result in attached flow conditions with peaks of low pressure at the leading edge of the top and side surfaces, which is caused by the strong change of flow direction. In reality, this low-pressure peak is much weaker, as schematically shown in Figure 4.54 for a longitudinal middle section and a horizontal section.[5] As a consequence of this pressure distribution, a local drag source is developed due to the pressure differences that arise, while a secondary effect on the base pressure is also generated downstream.

---

5. In frictionless flow around a sharp edge $(r = 0)$: $c_p = -\infty$.

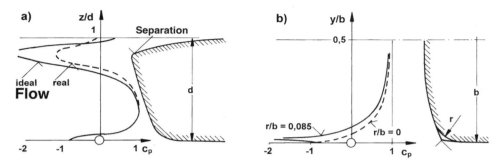

Figure 4.54 Schematic pressure distribution of a front end structure with inviscid (ideal) and viscid (real) flow: a) longitudinal cross section; b) horizontal cross section.

Translated onto a real vehicle, the implication of these results from a cuboid is that flow separations at the front need to be avoided by all means. The airflow around the front of the vehicle is significantly affected by the details, which are described in the following section, as well as by the engine cooling airflow. Figure 4.55 illustrates the pressure distribution at the front combined with some stream lines. It should be noted that such illustrations alone do not tell anything about the drag contribution of the front of the vehicle. For one thing, pressure changes caused by separations at the front also have a downstream effect on the flow over the total vehicle, and for another, the pressure distribution of the inviscid circulation must also be known in order to identify losses compared to the ideal flow, and this is normally not the case.

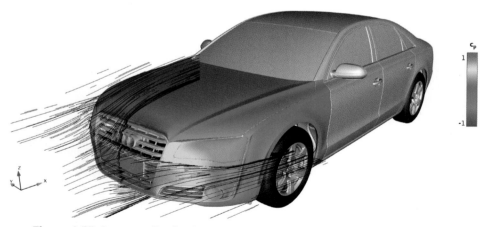

Figure 4.55 Pressure distribution and streamlines at vehicle front of an Audi A6.

For passenger cars, a boxy front end is no longer a consideration, since due to aesthetic and pedestrian protection reasons many geometrical changes are made that move away from a square-cornered cuboid shape. The main parameters are consolidated in Figure 4.56. On the centerline section these are the hood inclination, the inclination of the grille and the radii of the transitions into the hood and underbody. From the plan view the front bumper sweep, the front tapering, and another radius are added to the list. The individual parameters are to be matched together by empirical fine-tuning. In doing so, the front bumper beam and the cooling airflow, which may be achieved via several openings, need to be considered, of course.

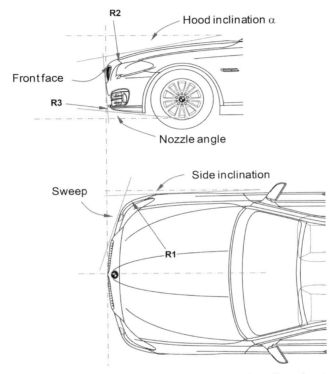

Figure 4.56 The main geometrical parameters to describe a front end.

Systematic investigations have been published for several of the parameters shown in Figure 4.56. These are the corner radii, the hood inclination, and the inclination of the grille. Of course, on real vehicles with rounded and smoothed designs these parameters are not so easily separated from each other; nevertheless the conclusions drawn here are still valid.

4.5.1.1.1 Corner Radius

The fundamental relationship between corner radius and aerodynamic drag can be found in the literature and is summarized in Figure 4.57. If the radius of a leading edge is increased step for step, the drag of the corresponding body is initially reduced rapidly. After passing a certain radius size, the drag remains constant. Flow separation no longer occurs, and the real flow behaves very similar to the ideal, inviscid flow. When applied to a car, this observation means that only minor rounding of the leading edge is required to prevent flow separation, thereby minimizing the forebody's contribution to drag. Most of today's vehicles have much bigger radii at the front of the vehicle than would be necessary from an aerodynamic point of view. How much additional potential to reduce the drag in other areas, for example at the rear, can be identified or has been achieved by doing this has not yet been published.

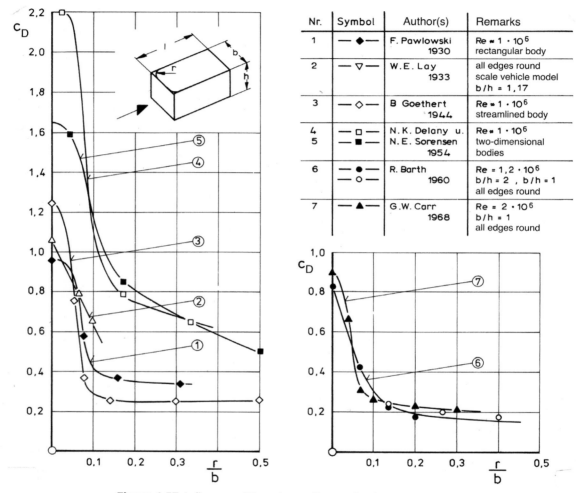

Figure 4.57 Influence of the edge radius on the drag of a squared block, compiled from [377].

It must be noted, when applying the numerical results from Figure 4.57 to real vehicles, that these figures apply to right-angled boxes, whereas the corners on a real car can just as well be blunter as well as sharper. The precise radius, where the airflow will just stay attached, has to be determined by experiment. In this case, the dependence of the radius on Reynolds number also has to be considered, as shown in chapter 13 in detail.

In a specific vehicle development project the time is hardly sufficient to vary the individual parameters mentioned again and again consecutively, but several parameters

Aerodynamic Forces and Their Influence on Passenger Vehicles

are often changed concurrently. An older example of this is shown in Figure 4.58. In this example, the specific aim was not to optimize a predetermined front end shape but to demonstrate potential variations and their impact on aerodynamic drag. The initial front end shape (1) was compared with different front end alternatives (2) to (7). Comparatively small changes in the front end geometry resulted in a major improvement of the airflow accompanied by significant drag reduction. In addition, this example shows that the c_D-target figure can usually be achieved by various different shapes.

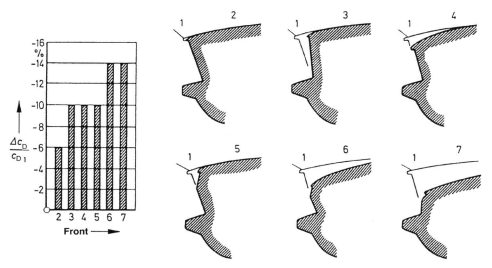

Figure 4.58 Formal variants for a front end structure and their c_D values [384].

In contrast, Figure 4.59 shows the results of a more systematic approach—it summarizes the results of the front end development of VW Golf 1. Following a proposal by Carr [126], a pre-test was conducted to determine the maximum possible front end drag reduction of the Golf. For this purpose, the front end was covered with a molding designed purely from aerodynamic principles without any consideration for stylistic or functional arguments. As shown in Figure 4.59b and d, the modification was made from two parts in order to isolate the effect of the radii to hood and fender.

This front end modification reveals that a drag improvement of $\Delta c_D = 0.05$ is attainable, as shown in the chart indicated by the bar M1 + K1. This potential was almost fully achieved by a stepwise enlargement of the radii on the real car (see bar chart with modification M3 + K3). Despite the edges still being rather sharp, the airflow around the front end of the VW Golf 1 occurs without any flow separation, as proven by the flow visualization in the photo.

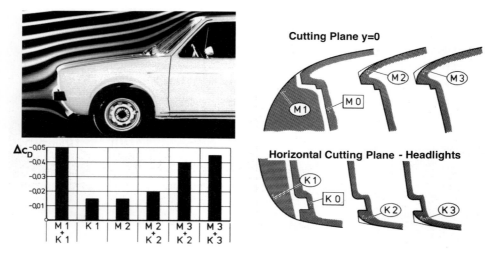

Figure 4.59 Example of optimization of a front end design, VW-Golf I [408].

The flow around an edge can also be improved by chamfering the edge instead of rounding it. An example of this is shown in Figure 4.60. By either of the two approximately equivalent measures, rounding and chamfering, the same drag improvement could be achieved as with the completely round front end. A flow separation originally occurred at the sharp edges was also entirely eliminated by a chamfer, as proved by the flow visualization in Figure 4.61.

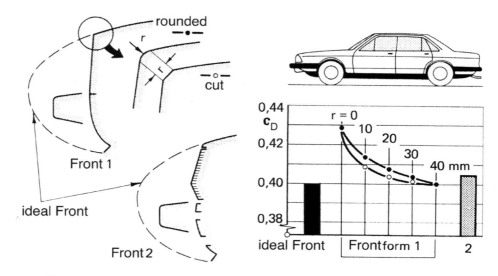

Figure 4.60 Reduction in drag by rounding or chamfering the front end edge.

Aerodynamic Forces and Their Influence on Passenger Vehicles

Figure 4.61 Flow around the front end of the VW Passat, MY 1977: Design proposal with sharp leading edge at hood (at top left), strong flow separation at this edge (at bottom left), with "ideal" front end modification (at top right), attached airflow with chamfered (optimized) leading edge (at bottom right).

The fact that it is also possible to proceed toward low drag by less spectacular means is demonstrated in Figure 4.62. The drag improvement gained by the adoption of the ideal (fully rounded) front end modification could also be achieved by fine-tuning the hood radius and the grille position.

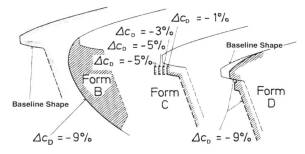

Figure 4.62 Drag reduction by fine-tuning the hood radius and grille.

In Figure 4.56 the front bumper sweep angle was also specified as an important shape parameter. Figure 4.63 shoes the result of an investigation into the effect of sweep angle on the local drag by varying the crown of the bumper corner, parameterized by the dimension Δx. Initially the airflow to the front wheels is improved with increasing Δx caused by a better guidance of the flow and smaller separation regions. In this case it results in a drag improvement of up to $c_D = 0.005$. In the case of a further enlargement of Δx toward rectangular shape, the risk of flow separation increases, and therefore after a certain point the c_D value goes through a minimum and then begins to rise again.

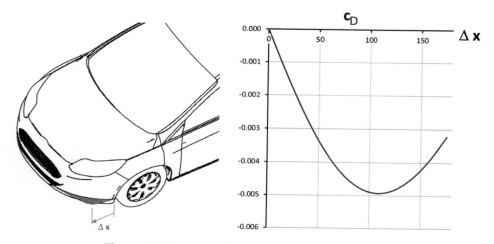

Figure 4.63 Impact of front bumper sweep on drag.

In order to show the influence of individual shape characteristics more precisely on lift force and pitching moment, the effect of rounding off the corners is next investigated (see Figure 4.64a). The individual corners of a sharp-edged generic model were rounded in sequence as indicated by the numbering.

By progressively rounding the edges, a tendency toward higher lift was observed, which can be explained as follows: rounding off the edges transverse to the flow direction reduced flow separation in the affected areas, and the flow velocity above the vehicle was increased. The static pressure at the upper surface was reduced; the lift went up. A remarkable effect was achieved by rounding the lower edge (3) of the front end. The flow entering the underbody area was increased, but decelerated further downstream due to the obstruction of the wheels. This created a pressure rise between road and underside of the vehicle and resulted in a sharp increase in front and overall lift.

Aerodynamic Forces and Their Influence on Passenger Vehicles

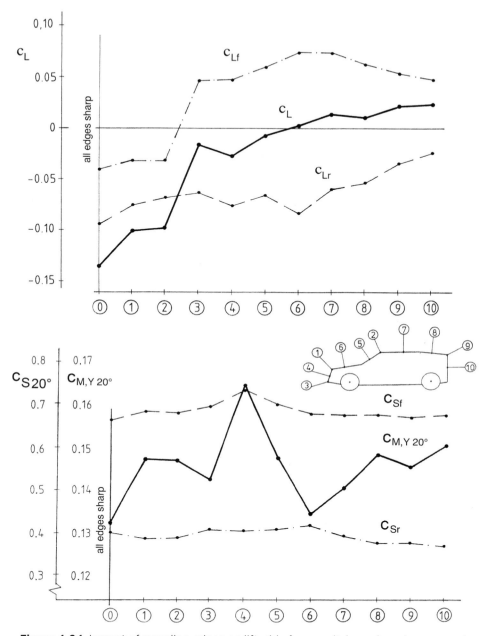

Figure 4.64 Impact of rounding edges on lift, side forces, pitch- and yawing moment.

The effect of rounding edges on the yawing moment is shown in Figure 4.64b. Rounding the leading edges of the hood and body sides leads to lower pressure at the leeward side of the front end and increases the yaw moment. On the other hand, rounding off the side edges of the hood as well as rounding the A-pillars causes a reduction of the front side force and yawing moment.

Results taken from a model with a fully smooth underbody deviate from the experience on real cars. Rounding of the lower front edge (3) on vehicles with realistic underbodies results in an increase of yaw moment, lift, and drag. A low front spoiler, on the other hand, causes a decrease of the yaw moment (see Figure 4.65).

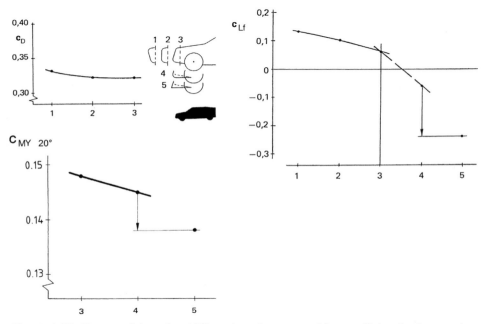

Figure 4.65 Change of drag, front lift, and yawing moment by modifying the lower edge of the front panel and by inclination of the front surface [274].

4.5.1.1.2 Position of the Stagnation Point

Depending on whether the vehicle has a smooth or a very rough underbody, more or less air is directed underneath the vehicle, which is important for aerodynamic drag. A large amount of air directed into the space between vehicle and road results, due to continuity, in high flow speed. A rough underbody will then generate high-pressure losses resulting in high drag. The position of the stagnation point determines the split between upper- and underbody flow, and this can be influenced by the hood inclination and the shape of the front spoiler.

The investigations with the fully rounded front end from Figure 4.59 were continued with a variation of its profile, shown in Figure 4.66(a). This experiment confirmed that as long as the flow past the nose takes place separation-free, almost the same drag coefficient can be achieved with completely different shapes. Setting the stagnation point as low as possible, as in the investigations of Buchheim et al. [107], whose results are reproduced in Figure 4.66, gives a slight advantage. This result is probably explained by the combination of a rough underbody and a shift in the stagnation point.

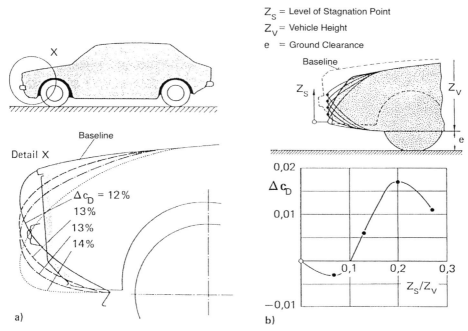

Figure 4.66 Effect of various different executions of a completely rounded front end on the c_D value; a) acc. Hucho et al. [385]; b) acc. Buchheim et al. [107].

The influence of vertical inclination of the front panel—i.e., the grille surface—on drag was reported by Gilhaus and Renn and is shown in Figure 4.65 (left). The lower the "lower lip" is pulled, the farther the stagnation point walks downward and the less air is directed into the underbody—which reduces the air resistance as already described. That the effect is relatively weak could be due to the relatively large leading edge radius fitted to the model.

The distribution of air between the over- and underbody of the car obviously has an effect on the lift, as Figure 4.65 (right) shows. Changes which reduce the amount of air flowing under the vehicle decrease the lift (due to conservation of momentum). Typical examples are a leaning back of the front face and a deep front bumper spoiler. This result has strongly affected the design of the front end of modern vehicles.

Investigations on the influence of individual shape parameters on crosswind sensitivity were published by Howell [361]. Figure 4.67 is taken from these results and shows the distribution of side force over the length of the vehicle, derived from integration of point pressure measurements. The contributions of the A-pillar and the headlight area stand out clearly on the yaw moment distribution.

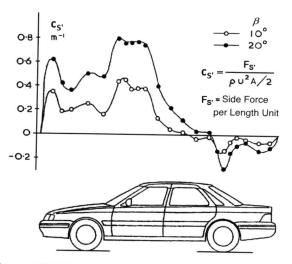

Figure 4.67 Side force distribution on a Rover 800 [361].

4.5.1.1.3 Hood Inclination

The hood inclination around an angle α has been examined in detail in the literature. Its influence on aerodynamic drag shows saturation character too; already at moderate inclination angles the aerodynamic drag does not continue to drop, as shown in Figure 4.68 (center).

This characteristic, based on a realistic, well-rounded leading edge of the hood, is a good example of an interference effect. When the front end is sufficiently rounded, a more inclined hood should mean that the flow is better directed and is able to stay attached longer downstream despite a positive pressure gradient, for example, at the base of the windshield or in the transition from windshield to roof. However, if the flow already separates the hood leading edge, changing the hood inclination will only have a minimal impact on the drag.

Aerodynamic Forces and Their Influence on Passenger Vehicles

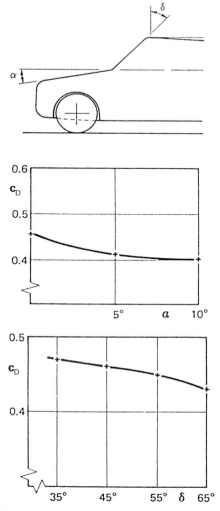

Figure 4.68 Reduction of drag with hood inclination angle α (top) and windshield inclination δ (bottom) [132].

4.5.1.2 Greenhouse and Body Side

The airflow around the windshield can be roughly approximated as a step with an inclined front surface, as shown in Figure 4.69. Flow separation is likely to occur at three different locations:

- at the base of the wind shield, in the concave region formed by its intersection with the hood (cowl)
- at the top of the screen at the transition to the roof
- at the A-pillars

265

The first two flow separations would result in transverse eddies, but at the A-pillars a longitudinal vortex (A-pillar vortex) is generated.

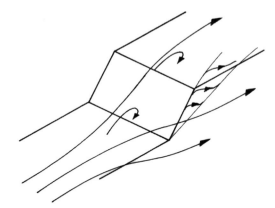

Figure 4.69 Flow along the windshield and around the A-pillar.

The flow that separates at the cowl reattaches on the windshield and forms a closed separation bubble. The transverse vortex that rotates inside the bubble exits at the sides, is diverted around the corners, and is drawn rearwards as a longitudinal vortex. It can however also occur that the circulation in the bubble is not constant, but falls toward the outside edges and disintegrates, similarly to on a wing, into longitudinal vortices distributed over the surface of the windshield. Even if the local separation in the cowl does not contribute noticeably to the overall drag, it is nevertheless undesirable, because the noise generated by the turbulence in the separation bubble can be transmitted via the lower cowl and the HVAC system into the interior of the vehicle.

At the transition between windshield and roof, if separation occurs, a closed bubble will also be formed. That a shear layer separating at the leading edge of the roof does not reattach should hardly ever be the case with a modern passenger vehicle. Instead, this edge is, on the vast majority of modern cars, so well rounded that separation is unlikely to occur at all.

At the sides the air is diverted over the A-pillars, rolling off at each side into a vortex. Due to the high velocity of the flow around the pillar and the underpressure in the vortex cores, as well as a contribution to aerodynamic drag, they cause an increased suction on the doors. This leads to an increase in wind noise on the front side windows, in other words, where the ears of the passengers are. At the top of the pillar they are deflected and flow downstream over the roof as longitudinal vortices. Their path can be easily followed on the roof of a rained-on or snow-covered car.

4.5.1.2.1 Windshield

The main shape parameters, which affect the airflow around the windshield, and its direct surroundings are emphasized in Figure 4.70. Beside the inclination of the windshield, the transverse curvature of the screen, the transition radii from the windshield

Aerodynamic Forces and Their Influence on Passenger Vehicles

to the A-pillars, and from screen to the roof are important. Two of these parameters, the windshield inclination and the radius at the A-pillars, were investigated in more detail.

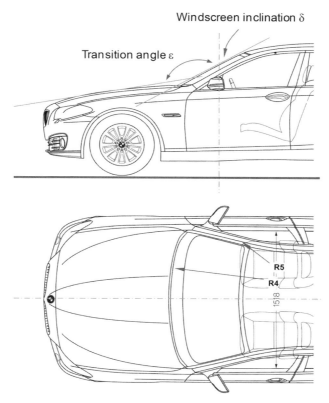

Figure 4.70 The main parameters for describing the geometry of a windshield.

The size of the separation bubble at the base of the windshield is determined by the inclination angle δ of the windshield, or more precisely by the angle ε between hood and windshield. Results shown in Figure 4.71 are for a planar windshield, a quasi–two-dimensional model investigated by Scibor-Rylski [752]. The location of separation S on the hood and the reattachment point R on the windshield are plotted versus the inclination angle δ of the windshield. With increasing windshield inclination, the separation point S on the hood moves forward and the reattachment point R on the windshield moves upward. However, these results should not be considered to indicate anything more than a trend, because the airflow in this region is of a three-dimensional character. Based on the typical windshield inclination of today's vehicles, there might not be any flow separation at the base of the windshield. In addition, this region is used to obtain fresh air for interior ventilation, which means that the air from a potentially existing separation bubble is sucked away, with the effect that the bubble becomes smaller. For drainage, this separation bubble also has a positive aspect as long as the lower reversal point of the front wipers is located within the separation bubble.

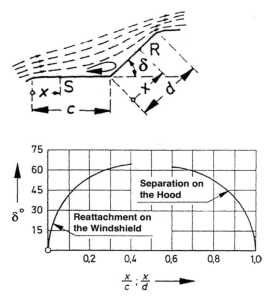

Figure 4.71 Flow separation at point S and reattachment point R as a function of the windshield inclination angle δ determined on a quasi–two–dimensional model [752].

The windshield inclination decreases the aerodynamic drag, but by far not as strongly as frequently anticipated. This is shown by Figure 4.72 and confirmed by results from a parameter study shown in Figure 4.68d and Figure 4.68c. The function $c_D(\delta)$ is of asymptotic character. For angles greater than 60° there is nearly no significant additional benefit in drag. The results of a parameter study in Figure 4.79d, which will be discussed later, also confirm this observation.

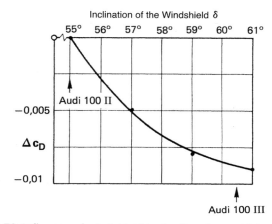

Figure 4.72 Influence of windshield inclination δ on vehicle drag [110].

The pressure distribution shown in Figure 4.73 of the longitudinal middle section of two passenger cars indicates that the c_D effect of the screen inclination is indirect (i.e., generated by an interference). A more inclined windshield leads to a less marked negative pressure peak at the transition to the roof (near to pressure tap position 20 in the figure). The subsequent positive pressure gradient is less steep; there is less loss of momentum in the boundary layer to overcome this, and thus the downstream flow has an increased pressure recovery at the rear screen.

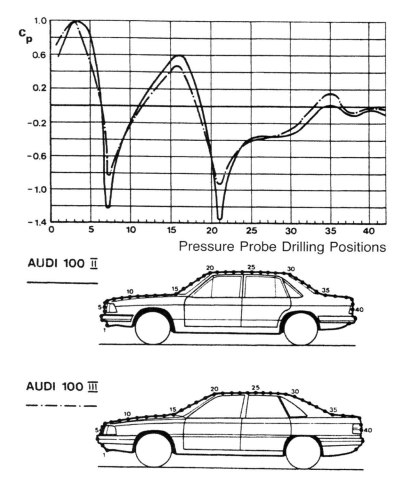

Figure 4.73 Pressure distribution in the center section of two vehicles: Audi 100 II ($c_D = 0.42$) and Audi 100 III ($c_D = 0.30$) [110].

The lift is also affected by the inclination of the windshield. A more inclined windshield results in lower lift. In the range of "faster" windshields commonly seen on today's cars there are only marginal differences. Figure 4.74 shows the effect on yaw moment, which is consistent with the experience made on real vehicles. A lowered cowl and a more inclined windshield decrease the front side force and thereby the yaw moment.

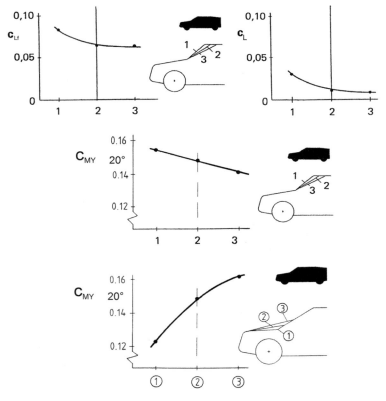

Figure 4.74 Influence of the cowl height and the windshield inclination on lift and yawing moment.

4.5.1.2.2 A-Pillar

The "faster" the windshield, the less air is pushed toward to the A-pillars, and therefore less energy is dissipated by the cone-shape vortices that form there. This aids pressure recovery further downstream. This effect is further improved by sufficient rounding of the A-pillar, as illustrated in Figure 4.75. Besides aerodynamic performance, the design of the A-pillar affects several other requirements, like low wind noise and side glass soiling, which are discussed in chapters 6 and 8.

The potential for a condition of fully attached flow around A-pillars and over the windshield is demonstrated by the research vehicle CNR/PF, as tested in the Pininfarina wind tunnel (Figure 4.76). For a long time, curving the windshield strongly in the corners has been avoided; the memory of the visual disturbances caused by panoramic windshields fashionable in the 1960s is still strong.

Aerodynamic Forces and Their Influence on Passenger Vehicles

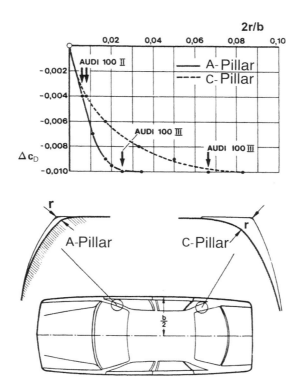

Figure 4.75 Drag reduction by rounding the A- and C-pillars [110].

Figure 4.76 Flow visualization around the research vehicle CNR/PF in the Pininfarina wind tunnel.

Reducing the strength of the A-pillar vortices has been a key aim of aerodynamic development since the early years of research. Even the Jaray-shape (see Figure 1.16) should, due to the large A-pillar radii, have had a separation-free flow field. In more recent years, the idea of strongly curved windshields is again of increased interest. The prototype of the Volvo safety car SCC, displayed at the IAA in 2001, had a windshield with large radii, with a tangential transition into the side screens. The wide, strongly rounded A-pillar was designed with a lattice construction to ensure sufficient strength and vision at the same time. The aerodynamic drag was reduced by approximately $\Delta c_D = 0.009$ compared to a conventional construction. Further examples are the SAAB 9X (2001), Skoda Roomster (2003), Renault Espace (2010), the vehicle CityEI (1987), and the Lamborghini concept S (2005). All apart from the last two are shown in Figure 4.77. Unfortunately, the latter vehicles were displayed without publishing any information on c_D effect gained by the shape of the A-pillar.

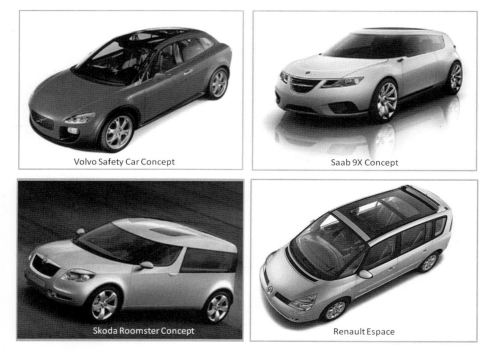

Figure 4.77 Prototypes with well-rounded A-pillars: Volvo Safety Car Concept (2001, top left), SAAB 9X (2001, top right), Skoda Roomster (2003, bottom left), Renault Espace (2010, bottom right).

The front portion of the greenhouse from the Saab is shaped similar to the cockpit of an airplane. Based on the strong inclination of the windshield and its large radius, one might expect an attached airflow around the A-pillar. The Skoda Roomster concept did not have any step between the windshield and side windows; however, this idea was not pursued in the series production of this vehicle. The CityEI was presented in 1987 as single-seat energy-saving car with a streamlined cabin. A characteristic attribute was the windshield, which was fully swept around the driver. Finally, the styling icon Lamborghini concept S was a sport convertible, whose freestanding windshield ran into the side window completely without any edges or steps.

A complete investigation of the flow around the windshield of a real car and its surroundings still does not exist; Figure 4.78 shows that the current published information is based on models that are far too simplistic. The flow around the A-pillar was investigated in detail for $\delta = 90°$ only, which corresponds to the vertical edge. An open question is at which angle δ the transition of the flow from "flow past a vertical edge" to "flow past a diagonal edge" occurs, how this depends on the radius r of the A-pillar, and what effect the individual geometrical parameters shown in Figure 4.78b might have on the c_D value, as speculated in Figure 4.78c.

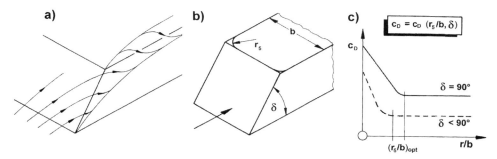

Figure 4.78 Flow around the A-pillar: a) Stream lines and A-pillar vortex, schematically; b) main geometry parameter; c) influence of rounding the A-pillar and windshield inclination, hypothetically.

4.5.1.2.3 Roof

The drag coefficient of passenger cars can be reduced by arching the roof in the longitudinal direction (see Figure 4.79a). The favorable effect of arching the roof depends on the radii at the transition from windshield to roof and roof to rear window. These radii have to be sufficiently large. Furthermore, the highest point of the roof has to be located as far forward as possible. This results in a less steep pressure raise with smaller pressure peaks and lower pressure gradients, reducing the risk of flow separation. The Audi A2, shown in Figure 4.80, is a good example. The increase in drag caused by further increasing the roof curvature can be explained by an increase in induced aerodynamic drag. The roof increasingly stimulates the air to go downward; hence rear lift is generated, which results in aerodynamic drag increase.

In order to avoid the windshield and the backlight becoming too expensive, they need to be as much as possible cylindrical or cone shaped. This implies that the curve of the roof needs to be tangential (and without sudden change of curvature, to avoid a suction peak) to the windshield and backlight where they meet. This curvature has to be implemented as much as possible in the sheet metal and not in the glass. In addition, the design of the roof arch must ensure that the frontal area of the car remains constant when curving the roof; otherwise the absolute drag ($\Delta c_D \cdot A_x$) can increase despite a reduction in a drag coefficient, as shown in Figure 4.79b and c.

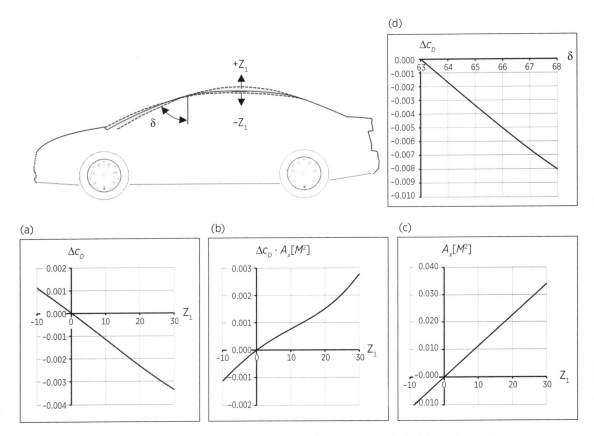

Figure 4.79 Drag reduction caused by roof curvature and windshield inclination.

Figure 4.80 Audi A2, MY 2003.

Aerodynamic Forces and Their Influence on Passenger Vehicles

In the meantime, the bow-shaped curved roof is used as a styling element by means of which the vehicle receives a coupé-like appearance; for example, see photos in Figure 4.81. The bottom of the rear screen needs to be moved backwards as far as possible, so that the headroom in the rear seats is not too compromised, although it's also important that the length of the boot/trunk doesn't become too short.

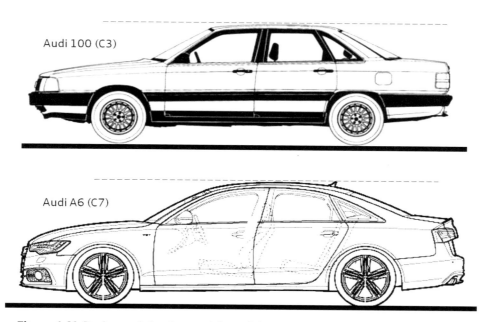

Figure 4.81 Design and development from tight to arc-shaped roof on Audi 100 / A6.

4.5.1.2.4 Body Side

The starting point for the plan view of a car is a rectangle. The airflow can be substantially improved by adding curvature to the sides. The angle from the front to the fenders is thereby made more obtuse and the transition to the boat tail is smoother with less risk of separation. This measure reduces drag, similarly to arching the roof, but only if the drag coefficient reduces faster than the frontal area increases. This measure was successfully applied on the Audi 100 III, as shown in Figure 4.82a: The c_D figure decreases faster than the frontal area increases, so that the drag area ($c_D \cdot A_x$) decreases with increasing width. However, this example is rather an exception. Usually this measure at the body sides leads to a reduction in drag only if the frontal area is kept constant, as shown in Figure 4.82b.

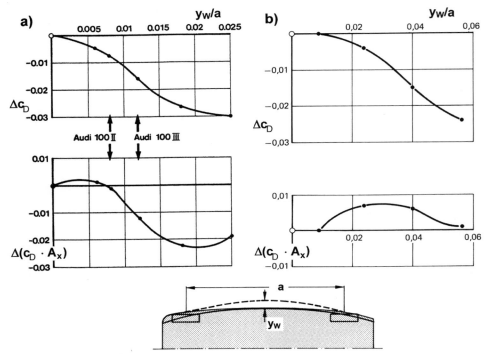

Figure 4.82 Effect of plan view curvature on aerodynamic drag: a) shown on Audi 100 II and III ([110]); b) on VW Forschungsauto 2000 ([107]) (scales are different in a) and b)).

Section 4.4.2 highlights the significance of a flow separation at the front body sides for the yawing moment. Sumitani and Yamanda [783] achieved a reduction in the yawing moment by a controlled flow separation caused by an airstream, which exits on the lee body side at the front. This airstream is generated by guiding air via a duct from the stagnation point to openings near the front of the body sides. It is led "automatically" to the correct side of the vehicle, because there is negative pressure on the lee-side opening and positive pressure on the windward side. The flow separation caused by this airstream leads to a reduction in the local suction peak and hence in a smaller yaw moment. The influence of these so-called Aero Slit® ducts on yaw moment as a function of yaw angle as well as the corresponding flow patterns are shown in Figure 4.83. The slit width (in the front bumper) was varied; the yaw moment reduces when the slit is wider.

Howell [361] investigated among other things increasing the curvature of the vehicle sides and the resulting effect on yaw moment. The effect arises from an increased speed in the area of the rear fender, which leads to a pressure drop and thus to a reduction of the yaw moment.

Aerodynamic Forces and Their Influence on Passenger Vehicles

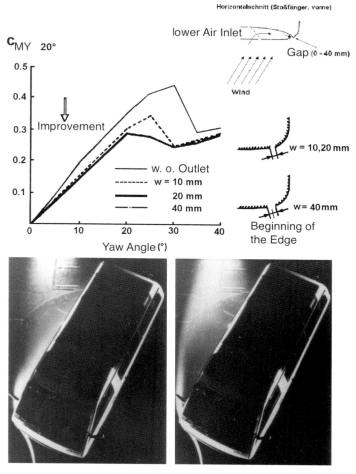

Figure 4.83 Reduction of the yawing moment by the use of a special air duct: impact of the slit width (top); comparison of the flow pattern at a yaw angle of 30 degrees yaw, at left without/at right with Aeroslit® [783].

Functional and stylistic limits as well as legal requirements limit the amount of curvature possible on the body side. The wheel base and track should be as large as possible, that is, the wheels should be positioned as close as possible to the corners of the vehicle. The wheels must also be enclosed, not least to catch the spray thrown up by the wheels. Finally, from a stylistic point of view the vehicle should not appear too narrow. The flow along the side panels of a passenger car is disturbed by three details:

- the wheels and the wheel houses;
- exterior parts, like outside mirrors and door handles;
- the side window recesses.

277

The impact of the wheels and outside mirrors is considered in section 4.5.3. Door handles do not adversely affect aerodynamics as long as they are in line with the local flow direction.

The regions in front of and behind the wheels are part of the body sides, and their design has an effect on the aerodynamic drag. Figure 4.84 shows a) the effect of the "step" between the door surface and the wheel arch and b) the shaping of the rear bumper behind the rear wheel on the c_D value, around the height of the rocker cladding in plan view. A pronounced wheel arch causes an increased c_D value, while overarching the body side (so that the body side is outboard of the wheel arch) leads to lower c_D values. Behind the wheel there is a separation region, which should be filled up by the rear bumper as much as possible. Strongly tapering the rear bumper in this area results in an increase in aerodynamic drag.

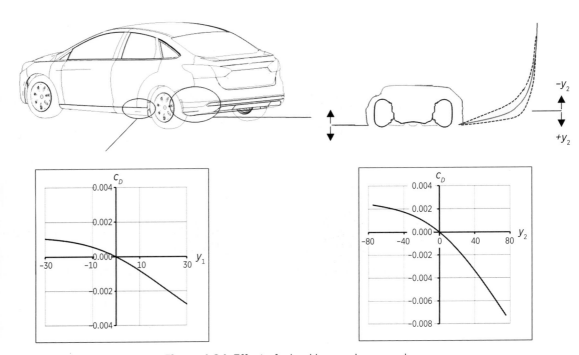

Figure 4.84 Effect of wheel house shape on drag.

The drag-increasing effect of edges perpendicular to the flow is well understood from aeronautic studies (for example [350], [885]). These findings are also applicable in the automotive industry, as in Figure 4.85, which shows the effect of the recess-depth of side windows on aerodynamic drag. The starting point is a recess-depth of $t = 15$ mm. The drag decreases with smaller step heights. Frameless doors, often seen on sport coupés and requested by stylists, have advantages here; an example is given in Figure 4.86.

Aerodynamic Forces and Their Influence on Passenger Vehicles

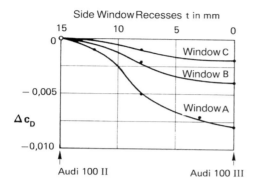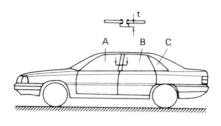

Figure 4.85 Effect of side window recess on drag [110].

Figure 4.86 Comparison of production Audi A7 with frameless doors and the original design sketch with flush glazing.

As expected the effect of the recess depth decreases to the rear. However, this fact can hardly be used; due to styling reasons the step height needs to be constant. Furthermore, it is not necessary to remove the recess depth completely on aerodynamic grounds; the last 5 mm have nearly no impact on aerodynamic drag.

Modern passenger vehicles tend to have strongly tilted side windows. In principle, this helps to reduce aerodynamic drag, as shown in Figure 4.87b. However, the flow topology at the rear end of the vehicle can be changed considerably. The well-known step effect when increasing the backlight angle beyond 30° gets smeared over a greater range of backlight angles, when the side window inclination is strongly increased, as shown in

Figure 4.102 in section 4.5.1.3.2. According to Figure 4.87c, there is no noticeable influence on the yaw moment. With increasing side window inclination the total side force is smaller, and the rolling moment is significantly reduced.

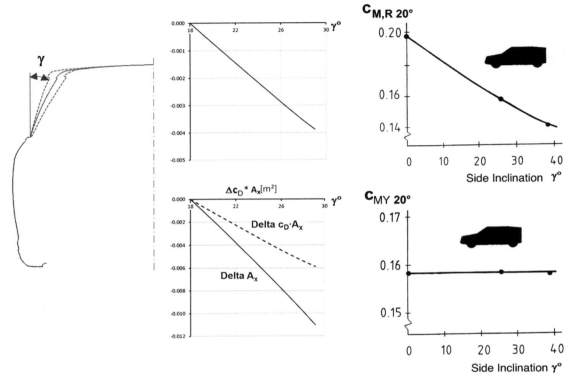

Figure 4.87 Effect of side window inclination on drag (left) and yaw and rolling moment (right).

Strictly speaking the rear pillars (C- and/or D-pillar) also belong to the greenhouse. However, since the aerodynamic forces depend on the C-pillar shape in combination with the individual rear end design, they will be discussed in the following section, which covers the rear of the vehicle.

4.5.1.3 Rear End

The aerodynamic forces acting on passenger cars are considerably affected by the flow pattern at the rear end. In the historical review of vehicle aerodynamics in section 1.2, it was seen that the rear end design of road vehicles does not allow a smooth airflow as on an airfoil, as the vehicle would be much too long. Instead various different types of flow separations occur, as discussed in section 4.2.

By means of the parameters illustrated in Figure 4.88, typical different rear ends are described and discussed in the following sections. With a given width b the fundamental dimensions are the aspect ratio $\Lambda = b/l_S$ of the rear screen and its inclination φ, as well as the length l_K and height h_K of the trunk or bootlid. Usually the cover of the trunk is also tilted slightly. Additionally, the plan view tapering δ, the roof radius r_D, and the radii r_C at the C-pillars are important. The proportion of vehicle height to vehicle width is approximately $h/b \approx 1$ for all passenger cars. Variations, although small, are seen in the inclination of the side windows γ, which was discussed in section 4.5.1.2.4.

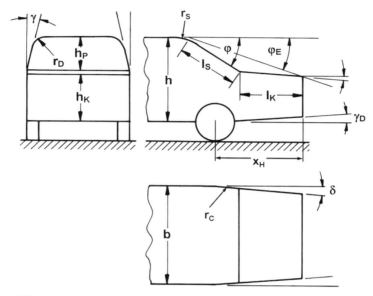

Figure 4.88 Rear end of the vehicle showing basic descriptive shape parameters.

At a wagon type rear end (also known as square back), the trunk length l_K is not applicable and the rear screen inclination is clearly larger than 30°. The back panel is not necessarily completely upright. The upper part of the back panel is often quite tilted with a large radius r_D to the roof, even if this comes at the price of reduced interior space.

On fastback rear ends, the angled backlight is inclined less than 30° and either extends down to the rear bumper or ends at the back panel (aerodynamically: base), which is in a more or less upright tilted position with a height h_K. The trunk length is not applicable here either. The height h_B on fastbacks falls into the range $0 \leq h_K/h \leq 1$. Depending on rear end styling there are two potential forms: a typical fastback with a long tilted backlight and short base, $h_K/h \approx 0$, or in reverse, a short tilted backlight with a high basis, $h_K/h \approx 1$, which in the limit is the same as the wagon type rear end.

The notchback is described by the complete parameter set; its geometry is the most complicated.

4.5.1.3.1 The Three Classical Rear Ends

The three "classical" rear ends are discussed next in more detail. The flow topologies are derived from flow observations, partly by means of wake measurements, partly by analyzing forces and pressure measurements. In all cases only time-averaged results are considered. Finally, shape modifications are discussed, by which the rear end flow can be affected to improve aerodynamic drag and reduce lift at the rear axle.

Over the course of time the three rear end forms, outlined in the upper line of Figure 4.89, have become generally accepted, each with its own characteristic flow field, which was discussed in general in section 4.2 and will be handled in this section in more detail. Only one of these rear end shapes, the so-called fastback, was driven by aerodynamics, albeit mistakenly. The other two shapes, the wagon rear end and the notchback, were adopted from carriages. The notchback is a development of a classic wagon rear end where the suit-case, which was originally strapped onto the back, is incorporated into the body.

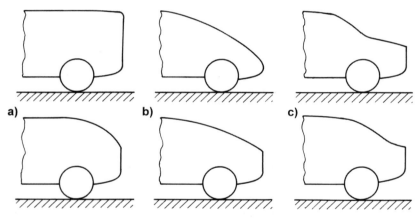

Figure 4.89 The three "classical" rear end shapes of a vehicle and forms of interpretations: a) wagon type rear end; b) fastback; c) notchback.

All three rear end forms allow the designer a relatively large range of interpretation before its characteristic flow topology will change to another one. This is illustrated in the lower part of Figure 4.89. For example, the wagon is frequently conceived with a tilted backlight in order to appear more sporty and less like a delivery van. The fastback can be almost the same as the wagon rear end, when executed with a tall rear end and a very shallow backlight. Finally, the notchback can be altered to a duck tail, if the roof is pulled backwards, as mentioned in the previous section. It can then be interpreted as a fastback with a duck tail.

The flow fields behind these rear ends were already analyzed in section 4.3.3.1 and are summarized again in the following pictures. Figure 4.90 shows the flow field on the centerline ($y = 0$), with similar wake structures to those already outlined in Figure 1.53. Compared to the flow field of the wagon type rear end (Figure 4.90a) the flow fields of fast- and notchback are completely different. The separation region is much smaller compared to the squareback, but the flow field is directed much more downwards, most obviously on the fast back. This creates lift, and in addition, induced drag is generated. The flow fields of all three rear ends show counter-rotating vortices.

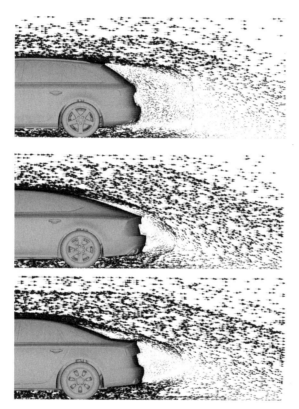

Figure 4.90 Flow field at centerline section of three "classical" rear end types.

The flow field perpendicular to the flow direction is shown in Figure 4.91; it shows just the right half in several x-sections behind the vehicle. At the back of the wagon just a small, relatively high positioned, clockwise-rotating vortex appears, which is the A-pillar vortex. Since no longitudinal vortices are generated at the rear end, the wake flow is nearly horizontal. The pictures from the fastback (b) reveal a strong, clockwise-rotating vortex, as already shown in Figure 4.27. Together with the vortex on the opposite side, this induces a strong downwash, and the flow coming across the roof is drawn downwards. The flow field of the notchback (c) is similar to the flow field of the fastback and has a clockwise rotating vortex, but it is weaker. This conclusion is indicated by the relatively fast dissipation of the vortex in the pictures taken farther away from the base of the vehicle.

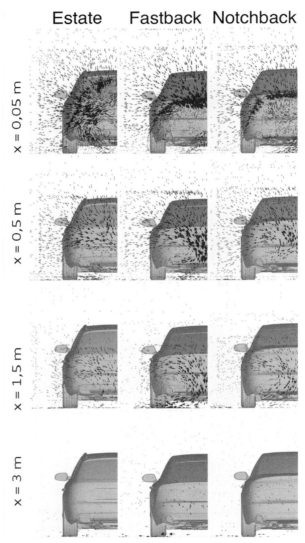

Figure 4.91 Flow field behind the wagon, fast-, and notchback in transversal sections at x = const.

Aerodynamic Forces and Their Influence on Passenger Vehicles

From these two figures, time-averaged and three-dimensional flow patterns of the vehicle wakes can be developed. The separated flow region of the wagon is schematically shown in Figure 4.92a. This region is encircled by a shear layer, whose thickness increases downstream. The shear layer develops into discrete, possibly ring-like vortices, which become more clearly structured as the flow moves downstream. The two opposite rotating vortices drawn on $Y = 0$ are traces of this large ring vortex[6] (as shown in Figure 4.23). The existence of this ring vortex (in the time-averaged flow field) has also been confirmed with CFD, as shown in Figure 4.23.

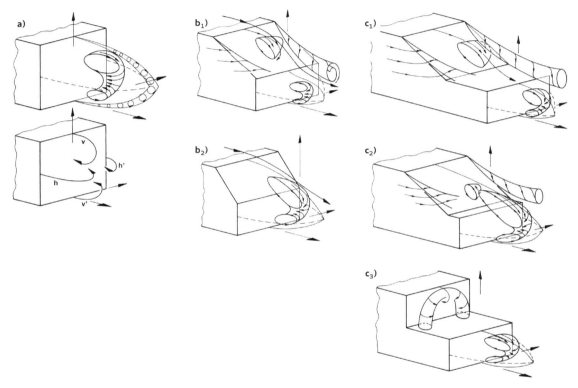

Figure 4.92 Flow around, separation, and wake flow: a) of the squareback; (b_1) of the fastback with $\varphi < \varphi_{crit}$; (b_2) of the fastback with $\varphi > \varphi_{crit}$; (c_1) of the notchback with "standard" geometry (flat rear window, long trunklid); (c_2) duck tail; (c_3) notchback with vertical rear window.

Taking the time–averaged flow field of the squareback leads to a strong simplification. A pumping oscillation (at $Sr = 0.07$) in the separated flow field was already indicated by Figure 4.10, although it has not yet been observed on real vehicles. However, when driving on a wet road behind a vehicle with a squareback rear end, it is possible to

---

6. Previously there was another perception of the flow field within the flow separation at the back, as described in earlier editions of this book. It was assumed that two counter-rotating vortices formed in the separated flow field, which flowed backwards as a horseshoe vortex.

observe that these vortices are not generated at the edges at the same time. They are more often formed stochastically as shown in Figure 4.92a, in the time order hh' and vv' as well as "crosswise" in h and v, h' and v'. Inside the separation volume the pressure is negative; the typical downstream pressure development is shown in Figure 4.9. The pressure at the base of the body is just slightly higher than the lowest pressure in the separation volume.

A schematic of the time-averaged flow field of the fastback is shown in Figure 4.92b. Two completely different flow structures are visible. The top picture represents the flow structure at relatively small slant angles of the backlight ($\varphi < \varphi_{crit}$) and the lower one shows the flow field at higher slant angles ($\varphi > \varphi_{crit}$). Only the first one represents the "real" fastback flow structure. If the transition from roof to backlight is well rounded then were will not be a flow separation, but otherwise a separation bubble is formed as sketched, which increases in size with the angle φ. The pair of vortices may not only draw the flow coming over the roof onto the backlight downwards, but may also affect the downstream separation region at the base.

For the notchback the three flow structures shown in Figure 4.92c have been derived. Picture c_1 shows the most common form of notchback. This form is characterized by a comparatively steep backlight and a long trunk. Similarly to the fastback, a pair of longitudinal vortices are generated at the C-pillars. Due to the larger aspect ratio of the backlight of the notchback ($\Lambda = b/l_S = 3$ to 4) compared with the fast back ($\Lambda \approx 1$) the pair of vortices induces a weaker downwash. Based on this a flow separation is expected at the trailing edge of the roof even at shallower backlight angles. Whether the airflow reattaches on the decklid depends on the backlight angle, the height and length of the trunk, and the shape of the trailing edge (radius, spoiler). On the other hand, flow separation at the trailing edge of the roof is less likely, if the transition from roof to backlight is smooth enough.

Experiments conducted by Gilhome et al. [281] have shown that the flow that separates on the notchback at the rear of the roof flows around the sides of the separation bubble on the rear window, forming a pair of vortices that counter-rotate with the C-pillar vortices. The time-averaged flow field is shown in Figure 4.93a, with this vortex pair labeled as "streamwise vortex." Figure 4.93b and Figure 4.94 shows the instantaneous behavior of the flow. This shows that a vortex is generated at the lower end of the separation bubble, which is periodically released from the separation region and flows farther downstream. Gilhome called this vortex a "hairpin vortex." Fluid is thereby periodically released from the bubble, similar to the separation region behind a bluff body, as reported in Figure 4.10—the bubble is "pumping."

Aerodynamic Forces and Their Influence on Passenger Vehicles

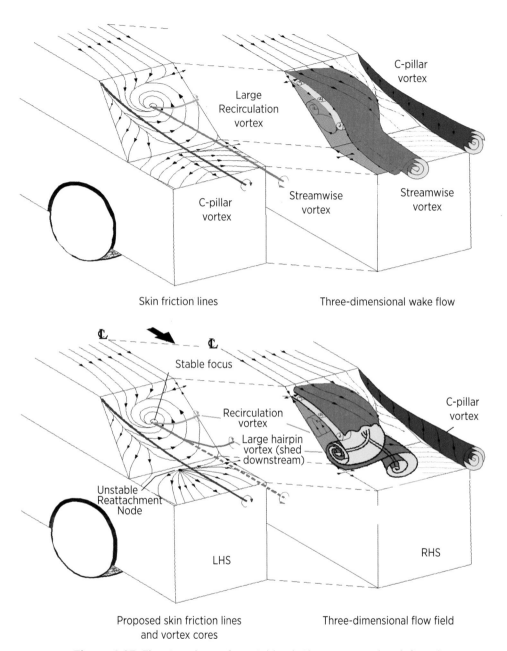

Figure 4.93 Flow topology of a notchback, time-averaged on left and time-resolved on the right [281].

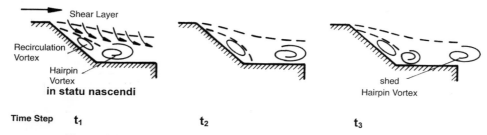

Figure 4.94 Fluid pumping periodically from the separation bubble on the decklid [281], [278].

Spectral analysis of this "pumping" reveals two dimensionless frequencies, $Sr_1 = 0.11$ and $Sr_2 = 0.42$. x_R is the distance of the reattachment point on the decklid, measured from the location of flow separation (reattachment length). In this case the start point is the trailing edge of the roof. With $u_\infty = 30$ m/s and $x_R = 1.0$ m, the "pumping" frequencies are $f_1 = 3.3$ Hz and $f_2 = 12.6$ Hz.

The former frequency is very similar to the resonant frequency of the rear axle and could be relevant for vehicle dynamics. As opposed to the pumping on the base investigated in section 4.2.1, the pressure acts on a horizontal surface, resulting in an oscillating lift on the rear axle.

The topology for the short duck tail rear end can be seen in Figure 4.92c in the middle frame. A pair of vortices also forms along the angled C-pillar on this rear end. The flow that separates at the rear edge of the roof is unable to reattach on the decklid. It is conceivable that the upper part of the ring vortex induces a separation bubble on the decklid with opposing rotation.

The bottom frame of Figure 4.92c depicts the flow on a notchback with an almost vertical rear window, commonly seen in the United States. In this case no longitudinal vortices are formed on the C-pillars. Instead, an arched vortex is formed behind the steep rear window as seen in the investigations of Sakamoto and Arie [704] on the flow around a normal plate mounted on a wall. An arched vortex was formed if the ratio of width b to height hp exceeded one, as seen in the present case ($b/h_P \approx 3$)[7]. Behind the base, there is a ring vortex as with the other notchback forms.

From the description of the flow field patterns it can be concluded that the two mechanisms involved in the development of the drag are the detachments with lateral and longitudinal vortices. The first occur on the base as well as on the rear window, if it is at a steep angle. The associated separation region is at a comparatively low pressure. The latter occur on the C-pillar, inducing a very low pressure on the rear window and a downward deflection of the flow over the roof.

---

7. For $b/h_P < 0.8$ the vortices detach only on the edges at the sides. The existence of an arched vortex can also be inferred from Jenkins's (2000) flow visualization of a notchback. The structure he marked as H5 could be the footprint of an arched vortex.

Aerodynamic Forces and Their Influence on Passenger Vehicles

The pressure on the rear end surface, the so-called base pressure, is one of the dominating sources of drag of a vehicle. If the drag is to be reduced, two possible starting points exist: either the base area is reduced in size or the pressure on this area is increased. Both measures can be addressed in various different ways, as described in section 4.5.1.3.2 and in the following sections.

An overview of the aerodynamic coefficients of a vehicle lineup with all three rear end variations can be seen in Figure 4.95. The following conclusions can be drawn, taking into account that the squareback has better starting characteristics for driving stability than the fastback and notchback shapes:

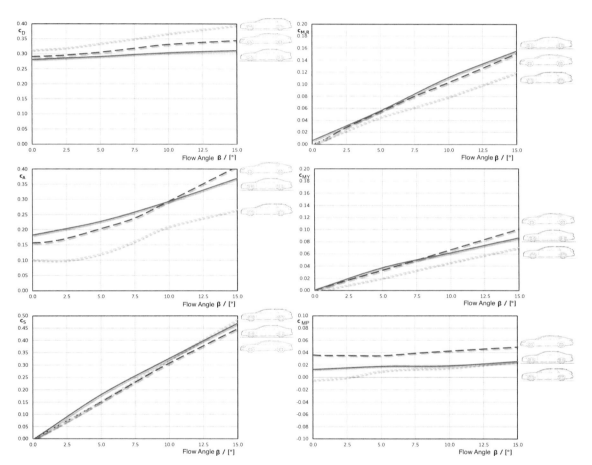

Figure 4.95 Comparison of the aerodynamic properties of the vehicles (model range with the variants notchback, fastback, and squareback).

- The differences in drag are relatively small for all three variants. The increase in drag with increasing yaw angle is steeper for the squareback than for the fastback and the notchback.

- The squareback induces significantly less lift than the notchback and the fastback. The fastback has the highest lift values. In general, the lift of all variants increases with increasing yaw angle. The gradient of the graph is the lowest for the fastback and the steepest for the squareback. At very high yaw angles ($\beta > 20°$, not shown) the differences in lift are small.

- In the whole range of yaw angles the measured side forces increase approximately proportional to the yaw angle. Notchback and fastback result in similar values. The squareback has a steeper gradient for side force with increasing yaw angle.

- The yaw moment of the fastback and the notchback also increases proportionally to the yaw angle. The squareback is characterized by a relatively small increase in yaw moment.

- The roll moment also increases almost linearly with a increasing yaw angle. The squareback has a lower roll moment than the fastback and the notchback.

- The pitch moment coefficients of all three rear end shapes stay nearly constant up to a yaw angle of approximately 15°. The squareback has the lowest pitch moment, and the absolute values are the highest for the fastback. With a increasing yaw angle the magnitude of the pitch moment declines for the fastback and the notchback. The pitch moments of all three shapes are approximately the same at a yaw angle of 20°.

4.5.1.3.2 Rear Window Inclination

For the definition of the rear window angle, we refer again to Figure 4.88. It can be seen that a squareback and a fastback only differ by the angle of the rear window; it is $\varphi = 0°$ or $\varphi = 90°$ for the squareback. All angles in between—assuming a trunk aligned with the rear window—characterize a fastback. If the flow adheres in this region, we use the term fastback; if the flow separates here, we use the term square back. The three tail flows for the squareback, fastback, and notchback were explained in Figure 4.92. For a steeply inclined slope, $\varphi > \varphi_{crit}$, there is a squareback flow. For angles of $\varphi < \varphi_{crit}$, a fastback flow occurs, the topology of which approaches the squareback flow at $\varphi \to 0$. There are therefore two questions:

- How does the flow behave at $\varphi = \varphi_{crit}$?
- What is the dependency for the aerodynamic characteristics in the angle range of the squareback flow and fastback flow?

The drag of a fastback is caused by the mechanisms described in section 4.3. First, the pair of longitudinal vortices originating from the oblique trailing edges of the C-pillars induces high negative pressures at the slope, similar to a low aspect ratio wing at an angle of attack. Second, drag is generated by separation at the trailing edge of the roof, followed by subsequent reattachment, with a separation bubble forming with enclosed cross-flow vortices. Third, comparable to the squareback, separation also occurs at the peripheral edge of the base, and dead wake also forms here, which encloses a ring vortex. Both types of vortices, longitudinally and transversely oriented, interfere with each other.

However, the interaction of these three phenomena is still unclear, especially the impact of the oblique edges on the formation of dead wake in the fastback, because, at these oblique edges, the curling up of the vortices against height is not coherent, which, according to Williamson [888], may cause the cross-flow vortices originating from there to dissolve into a number of longitudinal vortices and then disintegrate. A ring vortex may then not form at all in this case. The following discussion clarifies the situation: The pattern of the vortex structure behind a fastback is still incomplete and little is known about the time-dependent processes that were referred to in section 4.2.1.4. With regard to the studies discussed below, note that they are first performed on elementary bodies, so that no other vortices (A-pillar, cowl) disturb the pattern.

Studies on the influence of the rear window inclination on the rear flow increased, particularly during and after the development of the Golf I (cf. Figure 4.96, top). The styling model of the Golf with $\varphi = 45°$ had a very steep rear section. As can be seen from the flow patterns in Figure 1.40, the flow separated at the trailing edge of the roof, and the drag coefficient (of the 1:1 model) was $c_D = 0.40$. The gradual reduction of the inclination angle resulted in an abrupt increase in drag by 10% at $\varphi = 30°$. The separation then shifted from the roof end to the lower edge of the slant. Using a smoke probe revealed two strong, inward-rotating edge vortices at the sides of the slant. By further reducing the inclination angle, the drag dropped continuously again and the vortices became weaker. At $\varphi = 15°$, the drag was at its minimum. Finally, at even smaller angles, the original squareback flow pattern occurred again. In an angle range of $28° < \varphi < 32°$, a bi-stable state was observed: separation could occur either at the roof end or at the bottom of the slant.

The same relationships were apparent also in the development of the VW Polo I, as shown in Figure 4.96. Measuring the lift using this model confirmed the following: the change in lift is synchronous with the change in drag and is due to the formation of the vortex pair that is shed obliquely to the rear. Consequently, it is only effective on the rear axle.

So in-depth study of the flow at a fastback began after this "Golf effect" became known. It began with a simple geometric body, a cylinder with a rounded head, in free longitudinal flow, that is, without ground influence. Only the inclination angle φ of the rear slant was varied, so that its influence could be observed in isolation from other dependencies. The results obtained are summarized in Figure 4.97. Note that the steep climb of the c_D value over the inclination angle abruptly ends at $\varphi = 50°$; c_D drops abruptly and hardly depends on φ any longer at larger angles. Similar behavior is demonstrated by the lift, which also drops steeply at $\varphi = 50°$.

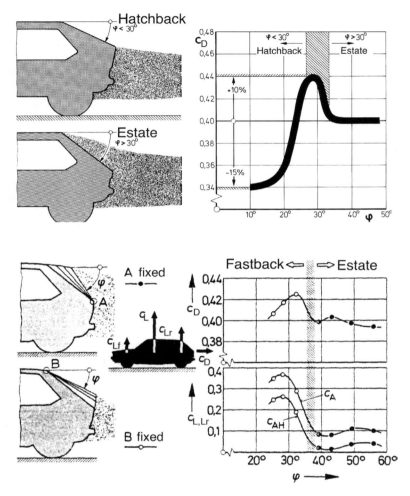

Figure 4.96 Influence of the inclination angle φ of the rear slant on drag, lift, and flow for the VW Golf I (top) and the VW Polo I (bottom) [408].

From observation of the flow using smoke and vortex probes, the pressure distribution on the slant, Figure 4.97, as well as comparisons with small aspect ratio wings, the following conclusion can be drawn for this peculiar pattern of the forces: Even at $\varphi = 20°$, the pressure coefficient c_p plotted against the width y/r in section A–A has suction peaks near the lateral edges of the section surface. They are induced apparently by the edge vortex pair that curls up at the two sharp lateral edges. With increasing inclination angle φ, these suction peaks increase corresponding to the growing intensity of the vortices. Beyond the critical inclination angle, these vortices apparently disappear again. A dead wake then forms, within which the vortices are preferably oriented transversely. This behavior is similar to the squareback flow. The static pressure on the slant is almost constant.

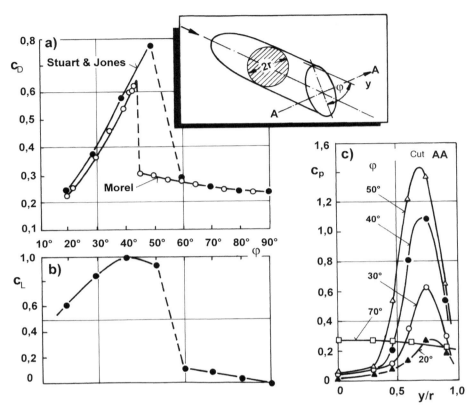

Figure 4.97 The "critical" inclination angle φ_{crit} on a cylinder (with a rounded head) in free longitudinal incident flow; according to Morel [582] and Stuart & Jones, cited by Bearman [52]: a) drag coefficient; b) lift coefficient; c) pressure distribution in section A-A.

Very similar conditions are obtained for a prism in longitudinal incident flow near the ground. Figure 4.98a shows that the critical angle is already reached at $\varphi = 30°$ in this case. This is exactly the value that was previously observed by Jansen and Hucho [408] on the Golf I. The pressure gradients in Figure 4.98b show the typical phenomena for both types of flow: When dead wake with cross-flow vortices forms, the pressure is high and nearly constant across the cross section. However, if longitudinal vortices curl up at the sides of the rear slant, typical negative pressure peaks appear near the sides and the average negative pressure is much greater.

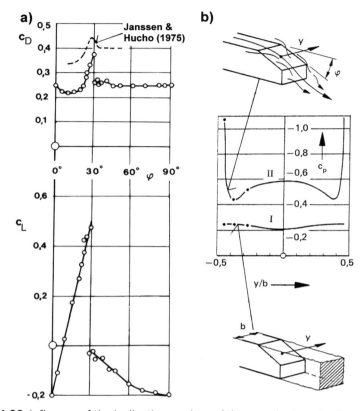

Figure 4.98 Influence of the inclination angle φ of the rear slant on the flow around a prismatic body in longitudinal incident flow near the ground; according to Morel [582]: a) coefficients of drag and lift; b) pressure distribution on the rear slant and a diagram of the flow shape. Compared to the original publication, the angle $\varphi = 90° - \alpha$ has been introduced.

How the formation of the edge vortices is influenced by the inclination angle φ of the rear slant has been documented by Ahmed and Baumert [9]; their results are shown in Figure 4.99a to d. The figures show the field of velocity vectors for different inclination angles φ in planes $x =$ const., with associated c_D values. At an inclination angle of $\varphi = 5°$, a pattern occurs that is similar to that of the squareback: A pair of vortices that are located relatively high and rotating counterclockwise can be detected, which may have its origin in the A-pillars. With increasing inclination of the slant, an increasingly strong vortex forms, which is located deeper, however, and rotates clockwise. When φ exceeds a value which is in the vicinity of $\varphi = 30°$, these vortices burst and once again the flow pattern of the squareback appears (not shown here).

The course of c_D over the inclination angle φ, which is plotted in Figure 4.99e, fits well into this picture. Up to about $\varphi = 5°$ we see a squareback flow; with increasing inclination of the rear, the drag drops to a minimum at approx. $\varphi \approx 15°$, followed by a steep increase again. At $\varphi = 30°$, exactly the angle at which the strongest tail vortices occur, the drag reaches its maximum. This also explains the hitherto unexplained high drag of many of those fastbacks—the pseudo-Jaray shapes that were mentioned in section 1.4. These cars had a rear angle near $\varphi \approx 30°$, a value that designers liked to use for a long time. For sedans, $\varphi \approx 30°$ is hardly used anymore; the vehicle would rather look like a station wagon, as shown. In this case, however, the interior is important, and $\varphi > 30°$ is more appropriate. In contrast, $\varphi \approx 15°$, the value for the drag minimum, is preferred in sports cars and coupés.

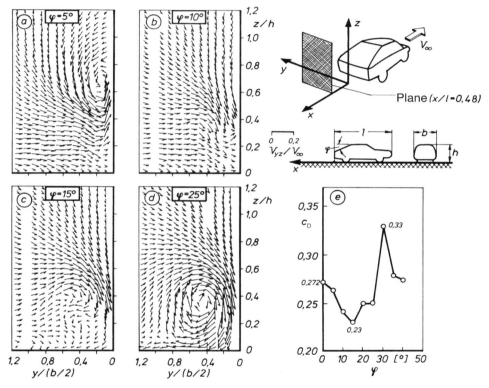

Figure 4.99 Influence of the inclination angle φ on the flow around the rear of a fastback car; according to Ahmed [6]: a) to d) field of velocity vectors in the plane $x = $ const.; (e) associated curve of the function $c_D(\varphi)$.

The drag generated by the longitudinal vortices according to Figure 4.27 is pressure drag, almost exclusively. This was proved by Bearman et al. ([55], [56]) by isolating the drag induced by the vortices from the total drag using a theory by Maskell [533]. The result is shown in Figure 4.100, where the drag components—and the circulation of the vortices—are plotted against the inclination angle φ. One can see that the increase of the total drag with the inclination angle φ is essentially due to the vortex-induced component; but the "rest" of the pressure drag[8] is only approximately constant. Likewise,

---

8. This is often referred to as profile drag with reference to wing theory.

Figure 4.100 shows that the drag measured with a scale perfectly matches the drag calculated from the wake.

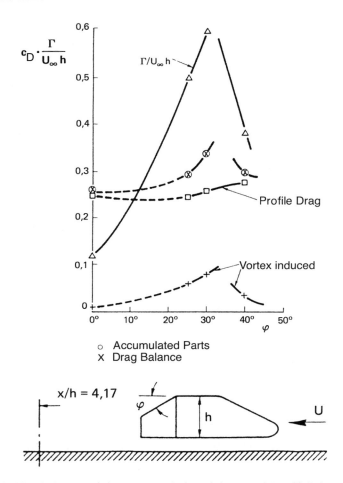

Figure 4.100 Circulation, total drag, vortex-induced drag, and "profile" drag as a function of the inclination angle φ [54].

Another way to describe the geometry of the fastback is to consider the length l_H of the slanted section. According to measurements made by Buchheim et al. [107], shown in Figure 4.101, the inclination angle φ where the drag reaches its minimum is larger the longer the tail. This is in qualitative agreement with wing theory. With increasing l_H, the aspect ratio Λ of the rear slant becomes smaller and consequently the bursting of the vortices occurs only at a larger inclination angle φ. However, according to Buchheim et al., this effect is less pronounced if the side edges of the slant, the C-pillars, are well rounded.

Aerodynamic Forces and Their Influence on Passenger Vehicles

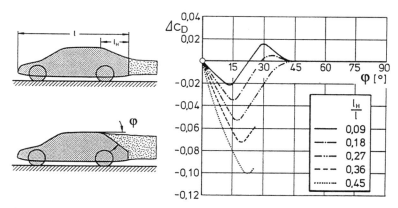

Figure 4.101 Influence of the geometrical parameters of inclination φ and aspect ratio l_H/l on the drag [107].

Unfortunately, the relationships obtained from simple geometric bodies are blurred by the effect of further shape parameters. This is underlined by the function described in Figure 4.102. The radius r_S at the transition from the roof to the slant, the radii r_C of the C-pillars, the angle γ at which the pavilion is inclined at the sides—all these shape parameters affect the flow and may completely change the "distinct" curve of the function $c_D = c_D(\varphi)$, as shown schematically in Figure 4.102b. Specific measurement results are shown in the figure on the left, following Howell [360]. By lateral inclination of the pavilion, the critical angle φ_{crit} is shifted from 30° to 37° and, after exceeding it, the decrease of the drag is less abrupt. Again, the rounding of the rear pillar (cf. section 4.5.1.3.3) is entered, so that the drag maximum is shifted even further at steeper window slants (in the figure: dotted line).

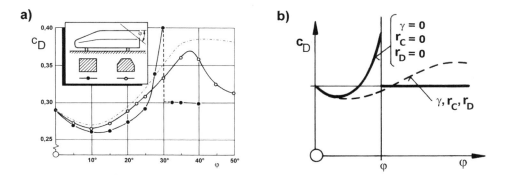

Figure 4.102 Influence of slanting the pavilion (angle γ), the C-pillar radius r_C and the roof trailing edge radius r_D on the curve of $c_D = c_D(\varphi)$; left according to Howell [360], right schematic.

For the notchback, two parameters must be considered to describe the tail inclination according to Carr [131], namely the inclination angle φ of the rear window and φ_E as the "effective" angle of the "imaginary" rear slant. By sorting the measured c_D values according to these angles, the result is the diagram shown in Figure 4.103.

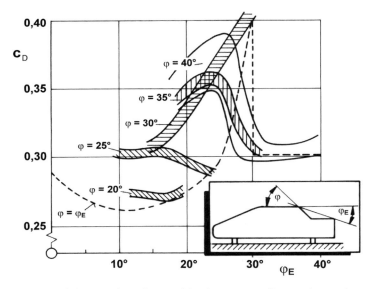

Figure 4.103 Order of the c_D value of a notchback car according to the angles φ and φ_E, as suggested by Carr [131], measurements by Howell [360].

If the two angles are equal, $\varphi = \varphi_E$, the result is a fastback, for which the measured values have been plotted as broken lines. If φ is only slightly smaller than φ_E, the drag differs little from that of the fastback; only with greater difference between the two angles do significant differences in c_D occur. Unfortunately, the c_D values can be sorted by the two angles only conditionally.[9] This must be due to the fact that, even in the notchback, further geometric parameters affect the occurrence and interactions of the various separations, such as the radii of the C-pillars and the radius at the roof edge. Their influence cannot be defined by general design rules. Rather, the geometry of a notchback must be optimized empirically by "trial and error."

As mentioned already in Figure 4.43 and Figure 4.44, differences in the rear inclination angle have a strong influence on the lift and in particular on the rear axle lift (cf. Figure 4.104). With increasing tail height and flatter rear window, the lift on the rear axle drops, and consequently therefore the total lift also falls. This effect can be explained by the fact that the streamlines are curved down less over the rear section and directed more to the horizontal. Thus, the static pressure over the vehicle is higher, while the lift is smaller.

---

9. Measured values reported by Nouzawa et al. [613] could not be fitted into this scheme at all.

Aerodynamic Forces and Their Influence on Passenger Vehicles

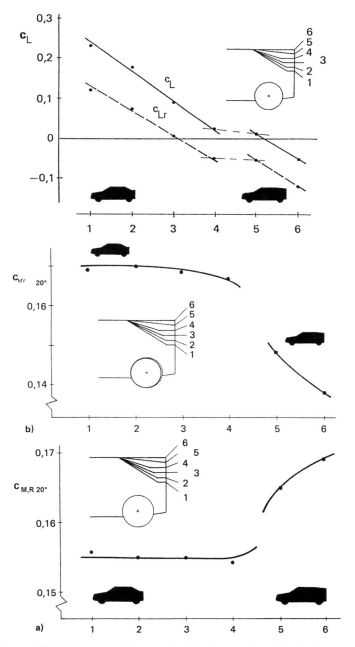

Figure 4.104 Influence of the rear inclination and the height of the rear trailing edge on lift, yaw, and roll moment.

It is also clear from the figure that the yaw moment has a slight tendency to decreasing values for shapes with higher flow separation at the rear. In this case, the notchback shape exhibits the highest and the squareback shape the lowest yaw moment. Since all three models have the same nose shape, and the front side force is therefore almost the same, the result is a reverse order of the total side force with respect to the yaw moment. It is noteworthy that relatively small changes to the side force correspond to significant differences in the yaw moment. Only in the transition to fully pronounced squareback shapes is the yaw moment significantly reduced.

The squareback has the highest roll moment, the notchback the lowest. This reverses the order compared to the yaw moment. With the increase in side forces there is an increase in the roll moment. For a wide range of rear shapes, however, the resulting roll moment displays only relatively small differences; rear extensions cause no significant changes. The same applies to the rear inclination angle and the height of the rear trailing edge. Only the transition to the squareback shape causes an increase in roll moment.

In summary, it should be noted that for rear window inclinations near φ_{crit}, the flow should be forced to squareback flow, either by increasing the inclination angle or by use of auxiliary measures. The measures that can be used for this purpose are given in section 4.5.1.3.4. For fastbacks, a rear spoiler can be used to reduce c_D and c_{Lr} without changing the rear window inclination.

4.5.1.3.3 Transition from Lateral to Rear Surfaces

From section 4.5.1.2, it remains to be added what influence the shape of the rear pillar has on the drag. For the notchback, Figure 4.75 already showed an improvement of the drag. It cannot be explained with certainty what causes this positive effect. One reason may be that the radius, similar to boat tailing, causes pressure recovery. Likewise, it could also be that the longitudinal vortices originating from the C-pillars diminish with increasing roundness of the C-pillars, thereby inducing less (pressure) drag. Structurally, the tapering of the C-pillar is limited by the demands for adequate lateral shoulder and head room in the rear and minimal blind spots when looking backward. It is believed that this behavior can be assumed in the fastback owing to flow, which similarly adheres to the rear slant. However, measured values are not available in this respect.

In the squareback, a large radius at the rear pillars could mean that even at rear window inclinations significantly greater than 30° a fastback flow is generated. This was mentioned in Figure 4.102 and might be proved by investigations by the author on the cuboid basic body in Figure 4.105 with a rear slant of 45°.

Aerodynamic Forces and Their Influence on Passenger Vehicles

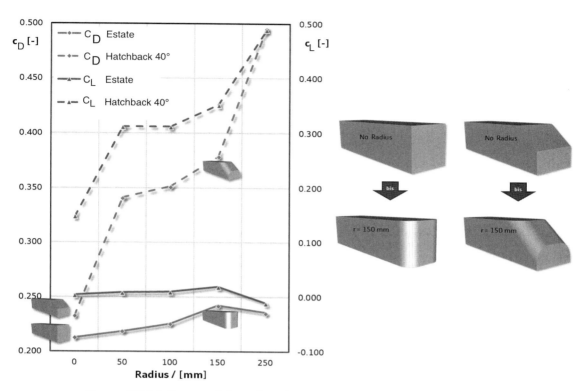

Figure 4.105 Influence of the radius of the rear pillar on drag and lift of a body.

The results show that even at small radii, drag and lift of a body increase significantly, although squareback flow would be expected at the selected rear slant and the longer adherence and resulting smaller dead wake should actually lead to reduced drag. An analysis of the balance of forces is used to explain the observations:

On an edge transition from the side to the rear slant, the flow separates at the edge. With increasing radius, however, the lateral flow is able to follow the curvature to some extent, resulting in pressure reduction (cf. section 2.3.3), which will also affect the dead wake. According to Figure 4.106, the resultant of the pressure force perpendicular to the surface and the inertia of the side flow is directed downward and inward. Apparently, according to Figure 4.105, this effect on the body increases with increasing radius, so that the roof flow is increasingly affected and directed downward by the longitudinal vortices. Despite the positive effect of reduced dead wake, this results in an increase in induced drag and increasing lift. The flow now is as for the fastback, even if the roof flow does not adhere fully or only unstably at the rear slant.

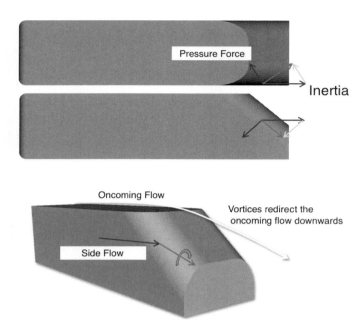

Figure 4.106 Forces acting on the rounded rear pillar and formation of longitudinal vortices

Similar conditions, albeit in a weakened form, also apply to edges perpendicular to the flow at which quasi–two-dimensional separations occur, namely in all three tail shapes. Section 4.2.1.4 already mentioned the unsteady nature of this type of separation. Considering a theoretical horizontal sectional plane at the rear of the vehicle in the area of a perpendicular vehicle edge with a large radius, we will see forward and backward shifting of the separation point due to these unsteady processes, as shown in Figure 4.107. For each position of this point, there is a slightly changed pressure distribution in the vicinity of the separation, resulting in fluctuating values for aerodynamic force components. By providing a separation edge at the point with low drag, there is no further temporal fluctuation, but the low level is maintained permanently.

Similarly, the lift is affected by the radius of the rear pillar, particularly at the rear of the vehicle. In contrast to vehicle A, vehicle B from Figure 4.44 has no defined separation edge in the transition region from the roof to the rear window. The contours in the region of the C-column were similar and had a defined separation edge. Apart from the strong asymmetry in terms of lift distribution, vehicle B shows an abrupt rise of the rear lift. This example makes it clear that in "sensitive" tail shapes and at small yaw angles, unwanted flow transition can take place, which is caused by sudden reattachment of the flow at the rear window.

Aerodynamic Forces and Their Influence on Passenger Vehicles

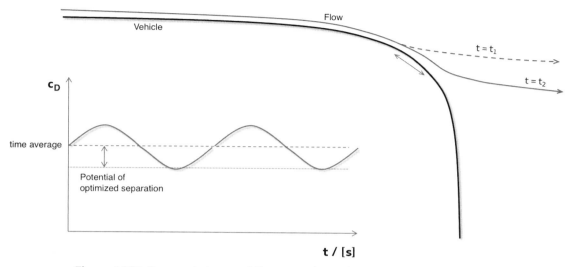

Figure 4.107 Temporal change of the separation point at the rear of a vehicle in a sectional plane through a perpendicular body edge with a large radius, and the effect on the aerodynamic drag.

A study by Howell [361] shows the yaw moment coefficient was determined at $\varphi = 15°$ as a function of the rear inclination angle. In addition, the radius of the C-pillar and the rounding of the rear head space were varied. The nonrounded model shows the lowest yaw moment coefficient across all rear inclination angles. Rounding of the rear head space only shows no deterioration of the yaw moment coefficient up to a rear inclination angle of 30°; only in conjunction with a small rounding at the C-pillar is there a significant increase in the yaw moment. This is primarily due to the reduction of the rear side force. Each additional rounding of the C-pillar then shows no further significant changes in the yaw moment coefficient.

Figure 4.95 first suggested that a reduction in the lateral projection area in the rear area reduces the rear side force, thereby increasing the yaw moment. However, this relationship does not apply to strongly rounded tail shapes. Large curves in the outline and a strongly arched rear window affect the yaw moment more than differences in the lateral projection area. Even small curves on the C-pillars generally lead to an increase in the yaw moment. This effect is most pronounced in fastback shapes. On the other hand, rounded C-pillars on the fastback cause a large reduction of aerodynamic drag. This results in a conflict when designing shapes.

For shape design, it follows from all this that the primary focus should be on reducing the vortex-induced drag, in particular because the same measures also reduce the lift on the rear axle. The fact that at a given shape a similar result can also be achieved by add-on parts is explained in section 4.5.1.3.4.

4.5.1.3.4 Separation Edges and Rear Spoilers

One conclusion from the sections on rear window inclination and transition design of the tail is, among other things, that—in particular at rear window inclinations of approximately 30°—it is important to ensure that the flow stably separates at the window top edge to form a squareback flow. Furthermore, efforts should be made to achieve controlled separation below the rear slant at the transition from the lateral surface to the base surface. This can be achieved by measures described next.

Some approaches to reducing the drag for the fastback can be found in the diagram in Figure 4.108. They are reported by Bearman [52]. The bag-shaped edge vortices that form at certain inclination angles φ—they can be recognized by the high negative pressure peaks near the edges of the base—can be disturbed in two different ways.

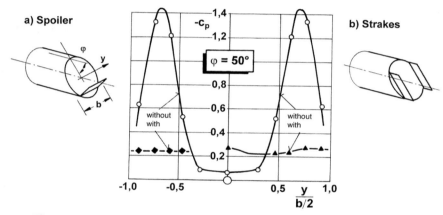

Figure 4.108 Approaches to reducing the drag of fastback vehicles [52].

The first is by attaching a flap at the slant and tilting it upward (Figure 4.108a). This imposes a pressure increase[10] on the flow over the slant and prevents the formation of longitudinal vortices.[11] As a result, the pressure increases at the slant and is approximately constant across its surface. The drag and lift on the rear axle therefore decrease. The same effect can be achieved by raising the outlined rectangular plates (Figure 4.108b). These so-called "strakes" disturb the shear layers originating from the two lateral edges such that they cannot curl up to form longitudinal vortices. The base pressure therefore remains high. The strakes can be designed to be retractable or foldable.

In practice, such ideas must be implemented in harmony with styling; radical measures as described in the previous section are often not compatible with elegant lines. Figure 4.109 shows some ways to ensure targeted flow separation at all transition surfaces. In total, c_D improved by $\Delta c_D \approx 0.1$ in the Audi Q3, in which the three measures were

---

10. The axis scaling "$-c_p$" on the diagram must be observed!
11. Occasionally, the term "burst" is used, but it should be avoided here as much as possible.

used—rear spoiler, side spoiler, and separation edges integrated into the rear lights. The figure also shows the effect of spoiler corners on the tail base pressure for the Audi A1 [737]. With the additional parts, the tail base pressure rises significantly. However, a precise numerical value for the drag reduction by the spoiler corners has not been reported.

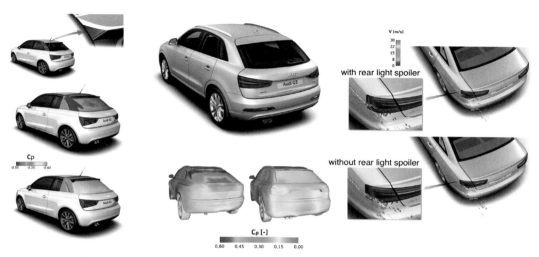

Figure 4.109 Possible ways to optimize the flow separation at the vehicle rear window in actual production vehicles.

While the spoiler corners reduce the flow situation in the corner areas of the roof spoiler at the D-pillars, the side spoiler also acts as a sidewall extension. Greater pressure recovery can therefore be generated at the rear of the vehicle. For this reason, the potential of this measure is somewhat greater than that of the spoiler corners. Schütz [747] reported another potential five to seven c_D counts for vehicles that are already in production.

As explained in the previous section, squareback flow is ensured for squarebacks with critical tail inclination angles by extending the roof and the associated increased jump back to the rear window. If this extension is shown as a separate component, we use the term "roof spoiler" to distinguish it from a rear spoiler. A rear spoiler, on the other hand, is defined as a component on fastback and notchback vehicles that is located at the end of the rear slant or at the trailing edge of the trunk lid. While the roof spoiler is used to ensure squareback flow, the rear spoiler's task is to reduce the downwash and therefore the induced drag and rear lift.

An example of the effect of the roof spoiler on a squareback with a very flat rear window of about 30° was given by Hupertz [388] for the Ford Focus. Without the roof spoiler, the vehicle exhibited attached flow on the rear window, accompanied by the deterioration of drag and lift recognized in many investigations. With the roof spoiler, however, stable squareback flow was achieved.

Chapter 4

Another function of the roof spoiler is that with increasing length it also constitutes an extension of the roof. If it extends its shear line, greater pressure recovery is achieved at the trailing edge, which also decreases drag and rear lift (Figure 4.110). In addition, an unobtrusive but efficient optimization of drag and lift can be achieved by a slight variation of the trailing edge height.[12] If the trailing edge is designed as a slight rump, the lift decreases, whereas the lift increases when there is a stronger taper than in the roof line. This allows the induced drag component to be influenced (cf. section 4.3.3.1). In the figure, it can be seen that for both shape parameters there is an optimum where the drag becomes minimal.

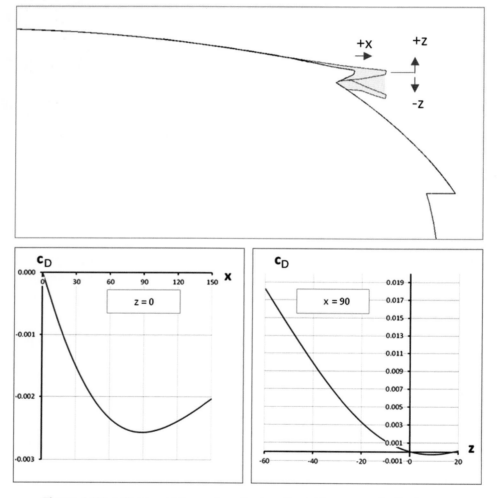

Figure 4.110 Influence of the roof spoiler length and its average inclination on drag and lift of a squareback vehicle with slanted rear window.

---

12. Such adjustment is also referred to as "aerodynamic tuning."

Roof spoilers can be deep drawn from the tailgate metal sheet or formed as a separate component, usually made of plastic. The first solution is associated with lower costs but restricts the design options with regard to the spoiler size. The fact that stylists prefer extended spoilers is another reason why the majority of squareback cars have a roof spoiler made of plastic. Other approaches to reducing the drag of squareback cars are summarized in Figure 4.111. The associated vortex system was already shown in Figure 4.92.

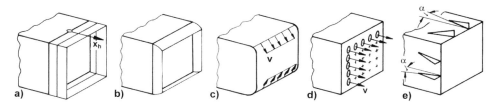

Figure 4.111 Approaches to reducing the drag of squareback vehicles.

Proposal (a) shows the extension of the tail by a box open at the rear, which was already presented in Figure 4.12. By moving the vortices in the dead wake away from the base surface, they induce less negative pressure there. This measure is particularly suitable for buses and trucks. As explained in section 10, the fascia was tested on a trial basis on cars with a box-shaped body. The result was a marked decrease in the drag. However, this has not found its way into standard production so far.

At first glance, one might think that the device by Morelli [584] as shown in Figure 4.111b is a fascia. However, this is not true; the effect of the sheets bent inwards is different according to Morelli: The ring vortex that is formed is "imprisoned" by the tapered sheets. With the sheets it produces a stable, inward flow, similar to a "cusp" as used in wing profiles with flap. The dead wake decreases, the pressure on the base increases, and thus the drag drops.

Proposal (c) comes from Geropp [270]. The planar jet that is suctioned at the top and bottom adheres to the curve of the trunk.[13] Similarly in this case, the outer flow is constricted toward the rear and the pressure on the base increases. Evidence that this measure results in an increase of the c_D value, even considering the power required for the fan, has been provided by Eberz [220] on model tests in a wind tunnel.

Proposal (d) shows planar suction, which a number of authors claim for themselves (cf. Bearman [58] and Sykes [785]). According to this proposal, air suction with a momentum that is so small that it itself induces no noticeable reduction of the drag pushes the vortices away from the base to a point where, as in case (a), they induce less negative pressure.[14]

---

13. This is the so-called Coanda effect.
14. See also Howell et al. [363], where it is found that the power to generate the suction air flow exceeds the gain in performance due to the lower drag.

Morelli ([584], [587]) combined measures (b) and (d) and demonstrated their effect on a prototype derived from the FIAT Punto 55. The metal sheet alone, surrounding and drawn in at the tail according to b), integrated flush into the shell and protruding by about 5% of the vehicle length reduced the drag by 18%. When the rear wheels were also used to promote air flow into the cavity, the drag further dropped by 2%.

Version (e) was proposed by Young [902] and verified on a two-dimensional body. The half delta wings, meshed at a small angle α, destroy—with the longitudinal vortices originating from them—the shear layer forming at the trailing edge of the bluff body, thus hindering the formation of a coherent vortex structure at the base. As a result, the pressure at the base increases.

In contrast to the roof spoiler, the rear spoiler is mounted on the end of the tailgate. It must be tailored to the shape of the rear end. Over the years, the cornered step has become a gently curved transition on notchback sedans. In particular, the flat roof has turned into a nose-shaped curved surface that merges with the rear window without any step. In some models, this shape of the roof has been combined with a very rounded trailing edge of the trunk lid.

Two designs are common for the rear spoiler: strips and wings. Strips are either attached plastic parts, commonly made from a soft foam to minimize the risk of injury on impact with a pedestrian or cyclist, or their shape is drawn out of the body sheet. Wings are usually mounted.

The development of rear spoilers focused initially on drag. Increasingly, however, the aim is to reduce the lift on the rear axle. Wings to reduce rear soiling are mainly used on squareback vehicles and in buses. They will be discussed in sections 6 and 10.

For the fastback, Ohtani et al. [620] compared the effect of a rear spoiler with the function of a wing flap, as shown in Figure 4.112 (left); there the static pressure (in dimensionless form as c_p value) on the top of the plate is plotted against its depth. As can be seen, by deflecting the flap simulating the spoiler, the pressure on the top of the plate is increased.

The isobars on the rear slant of a coupe plotted in Figure 4.112 (center) seem to confirm the above analogy. The static pressure on the rear slant is significantly higher with the mounted spoiler than without it. Plotting the pressure coefficient c_p against the vehicle height z/h—as shown in Figure 4.112 on the right for the center section—the reduction in drag becomes evident. While the pressure on the rear of the vehicle is increased by the spoiler, it remains unaffected on the front side. Integrating the changed pressure distribution in the x- and z-directions results in decreased values for drag and lift.

Aerodynamic Forces and Their Influence on Passenger Vehicles

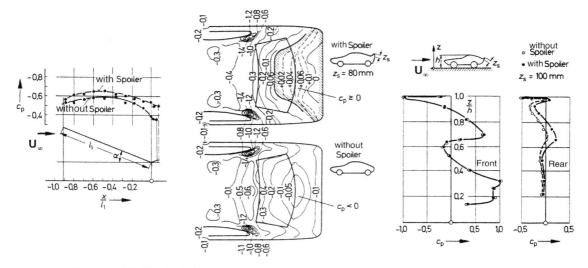

Figure 4.112 The function of a rear spoiler on a fastback, interpreted as a "wing flap" (left); increasing the static pressure on the rear slant upon disappearance of the vortices emanating from the C-pillars (center, right) [620].

However, a closer examination of the pressure distribution on the rear slant reveals a completely different working principle of this spoiler: The pattern of the isobars for the vehicle without a spoiler shows that the familiar vortex pair, originating from the C-pillars, has formed. It manifests itself by high negative pressures near the lateral edges.[15] The spoiler causes the vortices to disappear, resulting in a higher static pressure on the rear slant.

Similarly, on a notchback a rear spoiler causes an increase in the static pressure upstream. This can be seen in Figure 4.113. Since the lid of the trunk is only slightly inclined, the reduction of drag is small in this example. In contrast, the lift on the rear axle can be considerably reduced, the more so the higher the spoiler. The function of the rear spoiler on a notchback can be deduced directly from the pressure distribution plotted on the left in Figure 4.113. It is surprising that even the pressure changes so much on the bottom under the influence of the spoiler attached to the top.

In addition to this, a desired value pair $\Delta c_D / \Delta c_L$ can be adjusted with very different spoiler shapes and arrangements. It turns out that the designer has quite a range of options at his disposal in the design of such a spoiler.

---

15. At the time when Ohtani et al. conducted their studies (in 1972), the occurrence of these longitudinal vortices was not known in vehicle aerodynamics, but in wing aerodynamics.

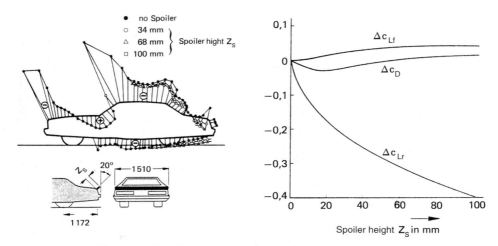

Figure 4.113 Effect of a rear spoiler on the notchback [711].

For sports cars, similar to racing cars, even a slight increase in drag is sometimes acceptable if the lift on the rear axle only is reduced significantly. A particularly striking example is presented by Figure 4.114. While the configurations (1) and (1 + 2) are favorable in terms of drag, but reduce only very little lift, the variants (3) and (4) exhibit a drastic reduction in lift, while increasing the drag.

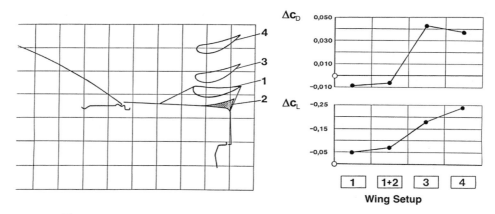

Figure 4.114 Rear wing configurations on a standard production sedan customized for competitions [193].

If a rear spoiler impairs the appearance or the rear view too much, it can also be designed to be extendable. It usually comes into play only at high driving speeds and then improves the drag and the rear axle lift that is of relevance to the dynamics of the vehicle. In this case, we are therefore looking at Bearman's idea in Figure 4.108. An example is the Audi TT shown in Figure 4.115.

The function of a rear spoiler can also be performed by a trunk lid with a suitable shape. This will be discussed in section 4.5.1.3.6.

Aerodynamic Forces and Their Influence on Passenger Vehicles

Figure 4.115 Extendable rear spoiler on the Audi TT2 and impact on the streamlines at the rear.

By providing a boardlike spoiler with a sinusoidal corrugation, as shown in Figure 4.116, its drag decreases with increasing depth s of the corrugation. This is due to the fact that the corrugation destroys the coherence of the separation across the width of the spoiler and thereby the formation of the shear layer, which causes the pressure in the dead wake behind the spoiler to increase. A similar effect is to be expected when the corrugation is applied to a flat spoiler in the perpendicular plane, either sinusoidal or jagged, as for a saw.

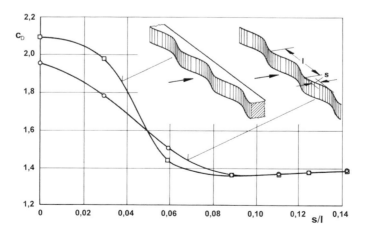

Figure 4.116 Reduction of the c_D value of a rectangular strip in transverse incident flow by corrugation [57].

4.5.1.3.5 Tail Taper

The base surface can be reduced by tapering the body on the sides, on the roof, and on the underbody to the rear, that is, by enlarging the angles φ, δ, γ_D in Figure 4.88. This method is also called boat tailing. It can also be applied to the other two tail shapes, fastback and notchback. Its limits are defined by design constraints. On the one hand, the squareback should be as wide and high as possible to provide a large through-load width and height, but on the other hand, fluid mechanics imposes some restrictions. If the taper is too great, separation occurs before the base. The effective base area will then be greater than the geometric base area and the pressure there drops. However, boat tailing is effective only if the flow separates at the transition to the rear base. This is a prerequisite that was not met, for example, by the "classic" boat tails, as is shown in Figure 1.11.

Boat tailing was first studied on simple bodies. How much the drag of a body of rotation can be reduced by tapering is shown in Figure 4.117. Note that the optimal taper angle of 22° given there is only a guideline; its precise value depends on the previous history of the flow. By extending the body into the taper that forms, the drag initially drops steeply, and further extension has hardly any impact. If the body is then cut off bluffly—a measure called "bobtailing"—the drag does not deteriorate, which is a confirmation of Kamm's idea (1934).

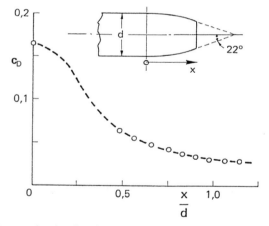

Figure 4.117 Drag reduction by "boat tailing," proved on a body of rotation [526].

These findings were transferred by Liebold et al. [498] to a record vehicle; their results are summarized in Figure 4.118 (top). The typical saturation characteristics are similarly apparent here. The length of such vehicles not intended for road traffic can be chosen with substantial freedom, so the tail can even be pointed. In contrast, on standard sedans, the tapering must be accommodated within a given total length. It is thus always at the expense of the interior, which is why it is not very popular with designers. How much the drag can be reduced by lateral tapering of the body (for a coupe) is shown by the example in the Figure 4.118. The smoke image at the bottom shows how long the flow (on a notchback) actually follows the tapered contour. Each of the three examples have in common that the selected taper angle of 10° is only about half as large as the "optimum" indicated in Figure 4.117 for the body of rotation.

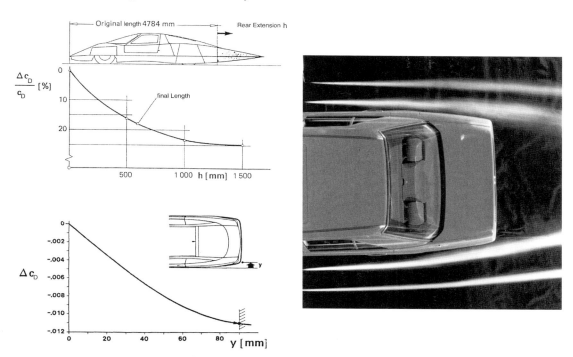

Figure 4.118 "Boat tailing": Applied to the Daimler-Benz record car C111 (top [498]), in the horizontal on the Opel Calibra Coupe (center [227]), on the Mercedes-Benz 190 (right, Daimler AG).

A very effective taper design on squareback vehicles is pulling down the roof, as shown in the example in Figure 4.119. However, this reduces the volume of the interior and the rear loading height, a restriction that is not easily accepted, especially in compact vehicles. The example also shows the limits of tapering. Enlarging the angle φ above 10° increases the drag again, because the longitudinal vortices discussed in section 4.5.1.3.2 gradually start to form.

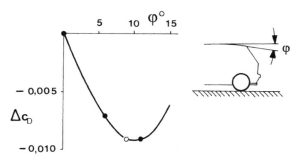

Figure 4.119 Reducing the drag by pulling down the rear roof part using the example of the Fiat Uno [525].

Similarly, at the fourth boundary surface, the bottom of the vehicle, tapering has been implemented with success and become known as a diffuser. This will be discussed in connection with the design of the vehicle's underbody, but in particular also in chapter 9.

Also the design of fastback shapes makes use of the option to taper the tail on the sides and even on the bottom. What effect this has on the drag is shown in Figure 4.120 for a prismatic body with a rear slant of $\varphi = 25°$. Pulling up the underbody (rear diffuser) and pulling in the sides (boat tailing) result in a considerable reduction in drag. The vortex-induced component of the drag decreases in particular.

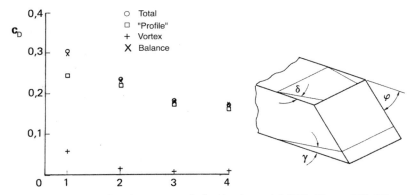

Figure 4.120 "Boat tailing" on a generic fastback model [55]: (1) $\varphi = 25°$, (2) $\varphi = 25°$, underbody pulled up by $\gamma = 10°$; (3) as (2) but with lateral tapering by $\delta = 10°$; (4) as (3) but with a beveled transition from the roof to the slant.

The effect of boat tailing can sometimes be achieved by only very small modifications; this can be deduced from Figure 4.121, which is also from the development of the VW Golf I. Both the rounding of the roof edge and the extension of the C-pillar by the amount a result in a reduction in drag of up to 9%. Why the two effects are not additive has not been clarified.

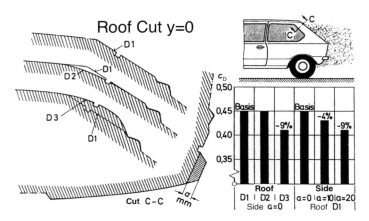

Figure 4.121 Boat tailing in detail by the modification d) to the roof and a) to the C-pillar [408].

4.5.1.3.6 Trunk Lid

If the tail continues from the roof with a large radius r_D, then, instead of the inclination angle φ, the rear height h_K would be an influencing parameter, as defined in Figure 4.122. In terms of drag, h_K passes through a minimum, which in turn is itself dependent on the shape of the C-pillars. However, the findings on the influence of the rear slant can be applied to heavily rounded tail shapes only with difficulty, because the position of the separation lines is not determined by sharp edges in these cases.

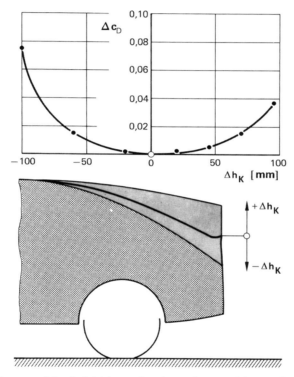

Figure 4.122 Influence of the height h_K on the drag of a car with rounded roof edge [112].

The vehicle in Figure 4.123 may serve as an example of the dependencies. The inclination angle of the rear window (a), the length of the rear lid (b), the transition from the roof to rear window (c), the height of the rear lid (d), and the z-position of the rear lid trailing edge were investigated. An optimum c_D must be conjectured for cases (a), (c), (d), and (e). The explanation for this behavior lies in the reduction of the induced drag, which is generated on short, low, and steep tails by the existing lift. The above measures reduce the lift and thereby also the drag by orienting the flow in the horizontal direction. Reducing the lift too far results in induced drag again. In case (b), the drag decreases almost linearly with the length of the rear lid, as does the rear lift. It is interesting that a higher and longer trunk reduces c_D and c_{Lr}—quite a welcome benefit that is often praised by sales people, because the trunk volume is enlarged. Another example is the influence of the trunk lid height on the drag of the Opel Calibra (cf. Figure 4.124).

Chapter 4

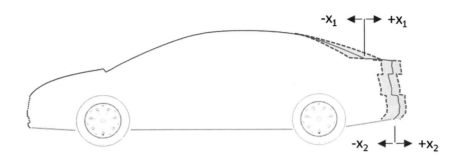

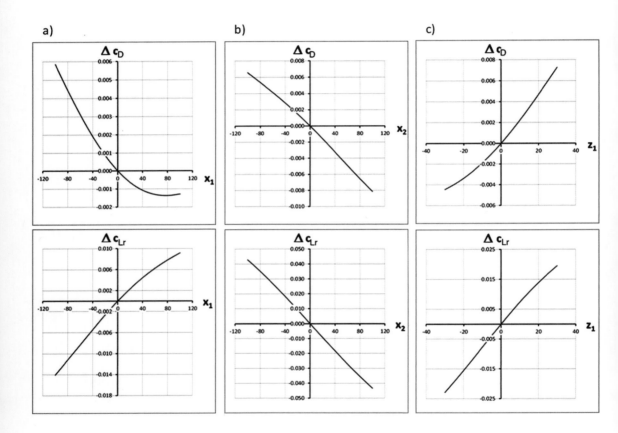

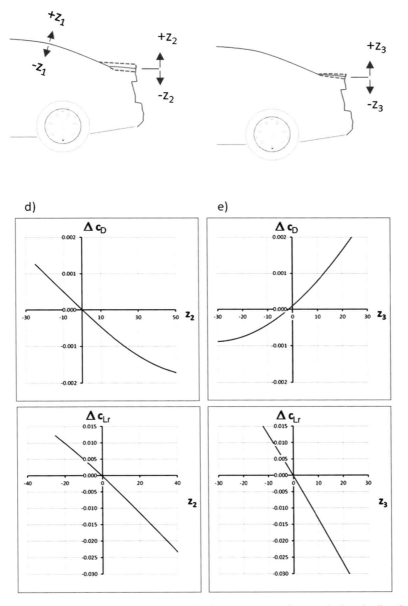

Figure 4.123 Tuning of the three geometrical parameters of rear window inclination, as well as the position, height, and length of the trunk and its trailing edge.

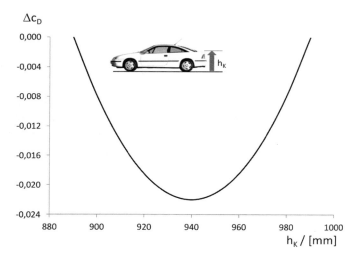

Figure 4.124 Influence of the rear lid height on the aerodynamic drag coefficient of the Opel Calibra [227].

Tail extensions, which reduce lift and drag cause an increased yaw moment, which can be deduced from Figure 4.125. This underlines that the larger lateral projection surface in the tail section does not result in an increase in the rear side force. The extension of the trunk probably causes a larger windward-side negative pressure zone.

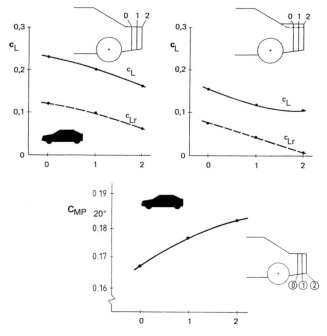

Figure 4.125 Increase of lift and yaw moment due to extension of the tail.

4.5.1.4 Systematization

The correlations presented in section 4.5.1 between geometry and the aerodynamic forces of passenger cars as well as the results of a series of parameter studies can be summarized schematically, as already provided in Figure 1.69 to Figure 1.71. In the two side-by-side, formally equal tables, the left column indicates the measure to be taken, such as "rounding," "pulling," or "curving." The center column outlines the corresponding geometric appearance, and the right column shows what effect this particular change of the geometry has on the components of the aerodynamic force. Interferences between the individual parameters are ignored. A total of twelve measures are listed here, which could be designated as primary measures. In addition, there are about 50 to 100 secondary measures, which partially "blur" the seemingly simple correlations between changes in shape and drag. Almost all functions outlined in Figure 1.69

$$\frac{c_D}{c_{D,0}} = f(\Delta \text{Geometry}) \qquad (4.34)$$

with $c_{D,0}$ as the drag before performing the geometry modification Δ, can be further attributed to four function types, which are listed schematically in Figure 4.126. These are the functions of saturation, asymptote, minimum, and jump. Here, the geometric details in question are described by the vector r_i, which is made dimensionless with a suitable quantity, in this case, the vehicle length l[16]:

$$\rho_i = \frac{\mathbf{r}_i}{l} \qquad (4.35)$$

Only the cooling air drag does not fit into this scheme.

The four function types mentioned are demonstrated by the following examples:

- The phenomenon of saturation occurs at edges in cross flow; a classic example of this is the nose of a box-type van, as shown by Figure 1.47.

- The function type asymptote can be observed in particular where inclination or tapering angles play a role; typical examples are provided by Figure 4.85 and Figure 4.123, where the focus is on the definition of angles, tapers, and lengths.

- A minimum always occurs when two opposite effects are active; a classic example of this is the tailgate height that was described in section 4.5.1.3.6.

- A jump always occurs when a modification to the geometry changes the characteristics of the flow. The jump (for the car) was discovered in the development of a squareback (VW Golf I) by studying the inclination angle φ of the rear slant over a wide range of angles. Examples have been presented in section 4.5.1.3.2.

---

16. The vehicle width b is also an appropriate reference quantity.

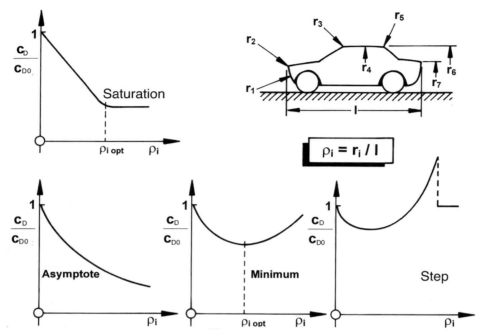

Figure 4.126 The four typical function types—saturation, asymptote, minimum, and jump—to which the functions shown in Figure 1.69 can be attributed.

Already at this point it should be noted that most of the phenomena discussed here depend on the Reynolds number. Chapter 13 will discuss this in connection with experiments on scaled-down models. There are rating systems by MIRA, but they have not been used so far by the major manufacturers, given technically advanced wind tunnels and improved CFD methods nowadays and, not least, because of the experience gained by engineers.

4.5.2 Cooling Air Effect

In section 4.3.3.2, the influence of cooling airflow on the aerodynamics of a vehicle was examined. In the following section, actions that influence the cooling airflow will be described. The sources of cooling air resistance are listed again in the following list. These will be discussed in more detail:

- Inlet losses at radiator grille and its surroundings
- Pressure losses at radiator and in the engine compartment flow
- Impact and momentum losses at flow exit
- Interaction with exterior airflow

As a result of these losses, differences in the development of the total drag as shown in Figure 4.127 will occur. Modern numerical flow simulations are able to offer this analysis as proposed by Kuthada [479]. A demonstrative assessment of the loss sources is possible with the help of the calculated flow topology images (Figure 4.132 and Figure 4.136), described in detail on the following sections.

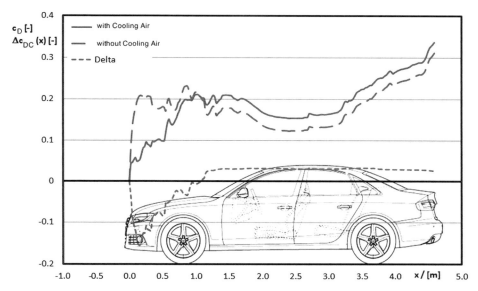

Figure 4.127 Drag development over vehicle centerline $c_D(x)$ with and without cooling air.

4.5.2.1 Inlet Losses

Inlet losses will occur when the flow separates at the entrance of the cooling airflow path, resulting in dissipation losses. The cooling airflow enters the engine compartment through the grille and flows in the direction of the heat exchangers. The grille is a major design feature which significantly defines the "face" of the vehicle, and therefore many different executions are found.

Nevertheless, there are commonalities: crash safety legal requirements restrict design freedoms, especially the height of the bumper beam. This is in most cases covered by the license plate and its bracket, which means that their position is at least implicitly defined. Furthermore, the grille must fulfill a protective function as it prevents bigger stones from hitting the heat exchangers, which in turn means that the inlet openings are usually divided into some smaller openings. For this purpose, longitudinal and lateral bars are included, which also improve the stability of the radiator grille.

Due to the displacement effect of the vehicle, the flow has to deflect in both the vertical and horizontal direction. The stagnation point is usually located in the area of the license plate, so that the flow past the bars of the radiator grille and the adjacent edges of the body is inclined. At all these locations, flow separations can occur. Figure 4.128 shows a schematic and a real representation in centerline section of a vehicle front with separation (Figure 4.128a,b) and without separation (Figure 4.128c,d). In the schematic representation it is clear that separation at the license plate bracket can be avoided by rounding the inlet. The section through the real radiator grille shows that an inclination of the longitudinal ribs in line with the flow direction would be beneficial. This modification would provide an improvement of $\Delta c_D = -0.002$ by keeping the air mass flow constant (for example, by reduction of the grille inlet area) or an increase in the cooling air mass flow of $\Delta \dot{m} = 32.9\%$. An example of grille fins positioned in the flow direction to reduce flow losses is seen on the Skoda Superb (Figure 4.128e,f).

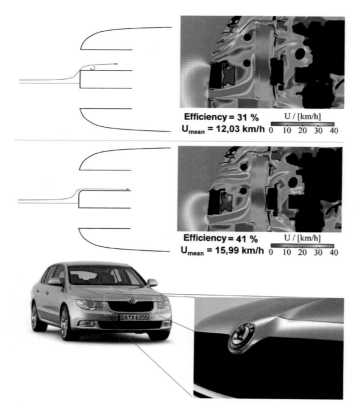

Figure 4.128 Inflow losses at radiator grille: High loss inflow (a,b), low loss inflow (c,d), (a,c) sketched, (b,d) at real vehicle; (e,f) radiator grille Skoda Superb.

4.5.2.2 Pressure Loss at the Radiator and in the Engine Compartment Flow

Significant energy losses are incurred from where the flow passes the radiator and travels through the engine compartment until it rejoins the exterior airflow. This is caused by turbulence and therefore dissipation due to separation as the air flows around obstacles. Additional pressure losses are caused by skin friction as the air flows through the small holes in the heat exchangers. As the cooling airflow required to achieve enough heat dissipation increases, these effects also increase. This means that the less efficient the radiator arrangement is, and the more leaks in the cooling airflow are present, the more losses occur.

The observations in section 4.3.3.2 showed that the cooling drag increases with increasing cooling mass airflow. Leakages in the cooling airflow ducting increase the flow into the engine compartment, which results in increased drag. This can be corrected by adding sealing in the gap between the radiator grille and the cooling pack or better installing a completely closed air duct in this area. In 2011, Audi unveiled the Q3 (Schütz and Hühnergarth [751]) for which a so-called radiator environment seal was developed (see Figure 4.129).

Figure 4.129 Radiator environment seal on Audi Q3 to reduce cooling airflow leakage losses.

The heat exchangers are particularly efficient when they are in an area of maximum possible pressure drop. This can be achieved when the cooling air openings are in the area of the stagnation point—where $c_p = 1$—and when the cooling air outlet is in a suction region. The flow is then more resistant to separation around obstacles. Therefore it makes sense to position the heat exchangers in line in the center of the vehicle. Two such arrangements are shown in Figure 4.130a.

In these examples, air in the license plate area is diverted through the radiator grille for cooling and then successively passed through the heat exchangers for engine cooling, air conditioning, and optionally charge air, which are mounted close to each other. It's less efficient to divert the air in the outboard regions, on the hood or in the under body, because the inlet pressures are significantly lower in these areas. This means that cooling air duct and inlet areas must be larger and separations occur more frequently.

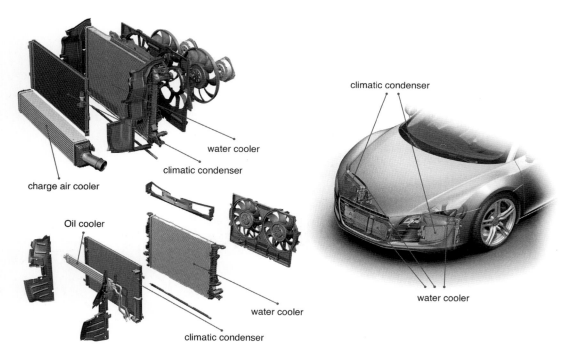

Figure 4.130 Different arrangement of heat exchangers in front end.

The heat exchangers themselves have a decisive impact on the magnitude of the pressure losses in the engine compartment. The pressure losses in the heat exchanger are scaled according to Gl. 4.30 with the through-flow velocity to the power of 1.5. In addition, the pressure loss is greater the thicker the construction of the heat exchanger. Purely from continuity considerations, it must follow that a thin heat exchanger with a larger surface area results in reduced cooling drag. However, this consideration is not sufficient for the design of a complete vehicle. Rather, it must be asked, what size heat exchanger is the most aerodynamically efficient at a constant required heat transfer. Assuming the same fin density in the heat exchanger, then from the pressure drop power:

$$P_{\Delta p} = \Delta p \cdot A_K \cdot v_K = c \cdot v_K^{-0.5} \frac{l_K}{d_{fi}} \frac{\rho}{2} v_K^2 \cdot A_K v_K = c_K l_K v_K^{2.5} \cdot A_K \text{ with } c_K = c \cdot \frac{\rho}{2 d_{fin}} \quad (4.36)$$

the continuity equation:

$$A_K \cdot v_K = A \cdot v \quad (4.37)$$

and the identity of the heat transfer of two radiator K and K_0:

$$v_K^{0.8} \cdot A_K \cdot l_K^{0.9} = v_{K,0}^{0.8} \cdot A_{K,0} \cdot l_{K,0}^{0.9} \quad (4.38)$$

it is possible to form an equation for the pressure loss power at a constant heat transfer:[17]

---

17. Assuming a heat transfer depending on the heat exchanger thickness $\sim l_K^{0.9}$ as it is shown in Figure 7.23.

$$P_{\Delta p} = c_K \cdot l_{K,0} \cdot v_{K,0}^{2.5} \cdot A_{K,0}^{(2.722)} \cdot \frac{1}{A_K^{(1.722)}} \sim A_K^{-1.722} \qquad (4.39)$$

The pressure loss power is approximately inversely proportional to the cooler surface area raised to 1.722, which confirms the previously stated assumption. With a larger radiator surface, the pressure drop decreases and thus the cooling drag—assuming the same heat transfer. The graphical representation of the pressure loss power versus the radiator surface area is shown in Figure 4.131, as well as the associated radiator mass, also assuming the boundary condition of constant heat transfer.

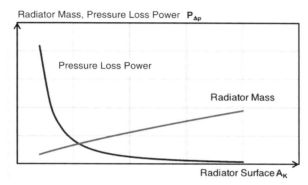

Figure 4.131 Pressure loss power in the heat exchanger and heat exchanger mass in relation to the surface area, assuming constant heat transfer.

Finally, there are many, usually not measurable flow losses behind the heat exchanger, that arise from separations around the other engine compartment components. Their contribution to the total cooling air resistance has not been studied in detail; therefore no design recommendations for aerodynamic components in the engine compartment are available. Figure 4.132 shows the many small-scale vortex structures in a snapshot of the engine compartment flow.

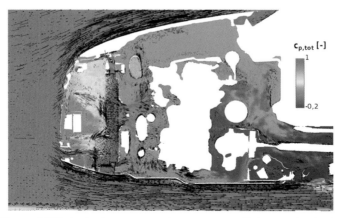

Figure 4.132 Instantaneous total pressure coefficient and velocity field in the engine compartment.

4.5.2.3 Impact and Momentum Losses at Cooling Air Outlet

In section 4.3.3, the principle of momentum was used to explain the effect of flow deflection on the rear end of a vehicle. At this point, similar observations can be done for the cooling airflow. Considering momentum and aerodynamic drag, it should be helpful when all outlet flows are directed horizontally downstream of the vehicle. This applies to the cooling air as well. The cooling air usually exits at the underbody or the wheelhouse into the flow around the vehicle body. The air flowing into the wheel arches does so in a lateral direction, and the air exiting in the underbody is typically almost vertical. Both are connected to a loss in momentum in the longitudinal direction of the vehicle, which leads to an increase in drag. Furthermore, cooling air exiting the engine compartment vertically also results in an increase in front axle lift.

An improvement in drag can be achieved, if the cooling air can be directed in the longitudinal direction of the car before it exits the cooling air duct. For this purpose, the wheelhouses should be well shielded in the area of the drivetrain and the suspension. Additionally, an extended engine undershield can deflect the cooling air from the engine compartment in a horizontal direction before it exits into the underbody airflow. The orientation of the cooling airflow is shown for an Audi Q3 in Figure 4.133. These measures together with the radiator environment seal in Figure 4.129 lead to a cooling drag of only $\Delta c_{D,C} = 0.009$ (see Schütz and Hühnergarth 2011 [751]), despite a big cooling surface area in comparison to the vehicle class. Another effect of the realignment of the flow is the reduction of the front axle lift.

Figure 4.133 Audi Q3 underbody with an extended engine undershield for a horizontal alignment of the emanating cooling air flow (Schütz and Hühnergarth 2011 [751]).

If the cooling air is directed well, it can contribute to a reduction in drag. The Ford Probe V, a publicized driveable design study, has an extremely low drag coefficient of $c_D = 0.14$! On this car, the heat exchangers were positioned on the left- and right-hand side of the rear end of the car. The cooling air entered through cooling ducts in the wheel arches, flowed through the coolers, and exited through the rear fascia. The air was guided into the wake and led to an increase of the base pressure on the rear end. The same effect was achieved with the research vehicle UNICAR (Potthoff 1982 [650]). With this car, the main part of the cooling air was guided through a tunnel to the rear end of the car and led into the wake of the car. Another smaller part was guided horizontally over the doors behind the front wheels. UNICAR and the Ford Probe are shown in Figure 4.134.

Figure 4.134 Front and rear view of the Ford Probe Concept S and the UNICAR [22].

With modern car concepts, a closed cooling airflow path as in the UNICAR concept is no longer desired, primarily because of the loss of package space. Conventional cooling ducts with exits in the area of the underbody may, as already mentioned, lead to a higher front axle lift and a lower rear axle lift with regards to the principle of momentum. These changes in the lift configuration, depending on the lift values of the car, may have an additional effect on the drag as shown on a vehicle with adjustable rear spoiler in Figure 4.135.

Chapter 4

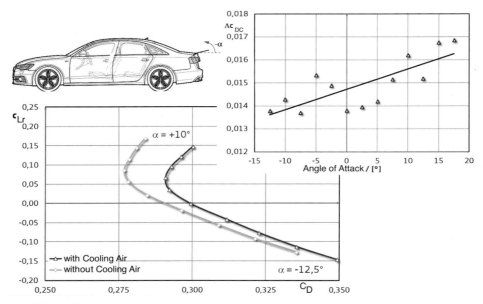

Figure 4.135 Illustration of different cooling air drag on a fastback and a hatchback.

Figure 4.26 has shown that the relationship between c_D and c_L is an almost parabolic polar. A notchback car will have its c_D and c_L values above the vertex of the parabola. A squareback is more likely to have its values below the vertex with low c_L values. If the cooling airflow is considered with both rear end shapes, compared to no cooling airflow, a translation of the parabola can be seen together with an increase of the c_D values caused by the pressure losses and turbulence in the engine compartment. This causes an increase of the cooling airflow delta with the notchback and a decrease with the squareback. This explains why the squareback usually has a smaller cooling airflow delta compared to the notchback of the same series. Based on experience, the difference between two derivatives can be quantified to approximately $\Delta c_{D,C} \approx 0.000 - 0.006$ (see Table 4.1).

Table 4.1 Cooling Airflow Resistance of Different Series and Manufacturers for the Square- and Notchback with Identical Front

Vehicle	$\Delta c_{D,C}$ Notchback	$\Delta c_{D,C}$ Squareback	Deviation $\Delta c_{D,C}$
BMW 3 series	0.024	0.018	0.006
Mercedes C-Class	0.019	0.020	-0.001
Audi A6	0.017	0.014	0.003
Ford Focus	0.010 (0.011 Hatchback)	0.007	0.003

4.5.2.4 Interaction of Cooling Airflow and External Flow

The fourth contribution to the drag is caused by the interference between the internal and external airflows. This interference occurs primarily at the cooling airflow outlet, but interferences can sometimes occur at the inlet. The main part of the cooling air exits downwards and merges with the underbody flow. The underbody flow is not flowing in a longitudinal direction but actually in an outwards direction, as discussed in more detail in section 4.5.3.3. Figure 4.136b shows that the addition of cooling air amplifies this effect. Due to this, the wheels, especially the front wheels, are subjected to an angled onset flow. According to Wiedemann [877] this causes a significant increase in drag. Porsche patented (July 26, 1986) a method by which it is possible to avoid this and even maybe convert it to a small advantage. The cooling air is ejected in front of the wheels and—according to the patent—protects the wheels from the onset flow, which would otherwise hit the wheels at high speed. The "Air Curtain," as presented in BMWs 3-Series works along a similar principle (see Figure 4.137). Another possibility in order to reduce the onset flow angle on the wheels is to seal the wheelhouses completely around the engine bay.

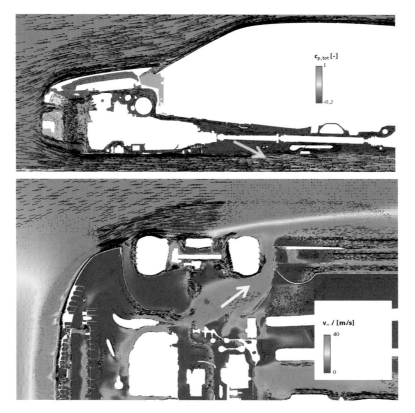

Figure 4.136 Flow of the cooling air when exiting into the wheelhouses b) and into the underbody flow a).

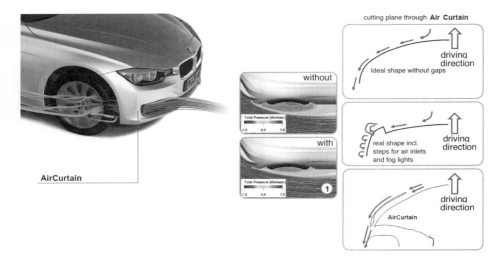

Figure 4.137 The Air Curtain on a BMW 3-Series: Ejection of air in front of the front wheel to avoid angled onset flow onto the wheels.

Even ejecting small amounts of air can induce negative effects on the aerodynamic drag, which may not necessarily occur directly at the exit point of the air. Figure 4.136a shows one example where cooling air exiting into the underbody leads to a thickening in the boundary layer and thus to an early detachment at the diffusor (not shown). A guided tangential channeling of the cooling air avoids these effects and improves the cooling air drag.

In some cases, a negative cooling air drag has been reported. This can be explained as an interaction of cooling air flow and exterior flow. Closing the cooling air path may cause a detachment on the hood or the engine undershield because of the changed pressure ratios. If the separation volume is big enough, the positive influence of the missing cooling airflow may be overcompensated and therefore a higher drag is induced compared to the flow with cooling air flow. Another example for the negative interference is the UNICAR. In this case, the cooling airflow is guided into the wake and increases the pressure on the rear end surface, resulting in a reduction of drag (Potthoff 1982 [650]).

4.5.2.5 Active Systems

The size of the cooling air opening is usually decided by the needs of the biggest engine in the series. This mostly causes a too high cooling air supply for the weaker engines with a lower need for cooling. From the aerodynamic point of view, cooling air that is not needed causes an increased drag, which results in an increased fuel consumption that might be avoidable. To enable this, earlier vehicle generations usually had smaller cooling openings for the weaker engines by means of additional blanking.

Many modern vehicles have an adaptable cooling opening that can be adjusted to the current cooling airflow requirements. Actually, cooling air flaps are not a new invention. For example, the Wartburg 311 900 had an adjustable cooling opening. Today, the pressure on continually reducing the aerodynamic drag is so high that shutter systems for the cooling opening are again being used. In some states of operation, no or just a small amount of cooling air is needed to facilitate cooling and air conditioning. For example:

- Warming phase after a cold start (engine cooling)
- Climate control in winter (air condition)
- Constant speed cruise on a flat road (engine cooling)

Alongside reduced cooling air drag, reduced fuel consumption and reduced CO_2 emission there are more advantages of the shutter system, such as faster warming of the engine, the engine oil, and the interior in cold ambient temperatures. Systems to control the cooling airflow are named differently at different manufacturers (Ford: grille shutter, VW Group: adjustable cooling air inlet, BMW: air vent control, Mercedes: cooling air shutter, etc.) and is implemented in different ways, as shown in Figure 4.138.

Figure 4.138 Different systems for cooling airflow control.

There are systems that block the cooling airflow in the grille (see Figure 4.138a), in front of the heat exchanger (see Figure 4.138b) or in the fan module (see Figure 4.138c). The ideal case for a cooling airflow control system is a 100% sealed flow path between grille and cooling air shutter. In this case, no air enters the engine compartment, and the previously named sources for drag are eliminated. As a remaining effect, there might be negative interference effects, as discussed in section 4.5.2.4.

In reality, an absolute seal is technically almost impossible, therefore leakage almost always needs to be taken into account. The leakage mass flow is the same as the additional mass flow required at the cooling inlet, which increases the further downstream the cooling airflow shutter is situated. Furthermore, a system that is positioned far downstream increases the danger of negative influences on the exterior flow (for example, in Figure 4.139). The cooling airflow enters the engine compartment below the license plate and comes back out again above the plate. Thus, a detached flow on the front edge of the hood is induced, which reduces the drag benefit gained from the shutter system.

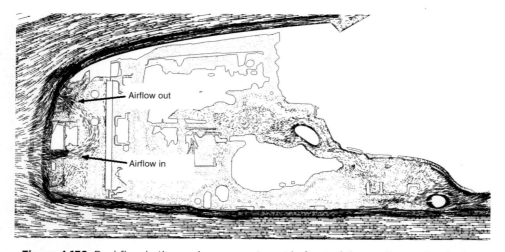

Figure 4.139 Backflow in the engine compartment in front of the cooling package when using an active system close to the fan.

The best position is therefore in the grille itself. This can lead to disadvantages in terms of the low-speed damageability rating (important for insurance costs) and with pedestrian protection. On the other hand, a system positioned far downstream has as well as a lower c_D-potential the disadvantages associated with significantly higher engine compartment pressure.

Compared to the installation position, the amount of the opening that can be closed is of greater interest since its influence on the cooling drag is higher (see Figure 4.140a). Usually the air cooling opening is subdivided in an upper and a lower cooling opening, which can be either completely or partly closed. When designing a shutter system, it is also necessary to investigate whether the shutter should open the cooling air intake completely or step by step. Alongside the opening coverage the rotational direction shall be investigated, since it influences the distribution of the airflow over the face of the heat exchangers. Figure 4.140b shows how the flow changes depending on the rotational direction of the shutter. In order to minimize fuel consumption with the use of a shutter, it is necessary to keep the shutter closed as long as possible. This depends on the engine power output and thereby the cooling air requirement. With this in mind, a multiple-stage control of the shutter should be preferred over a single-stage control. In contrary to the on/off-control, with a multiple-stage control the effective cooling opening can be adjusted more precisely to the demands of cooling air by opening the shutter step by step from fully closed to fully open. The figure shows that the drag is not linearly dependent on the opening angle. The increase in cooling drag is almost linear up to 60° opening angle. Above this, the curve is nearly flat.

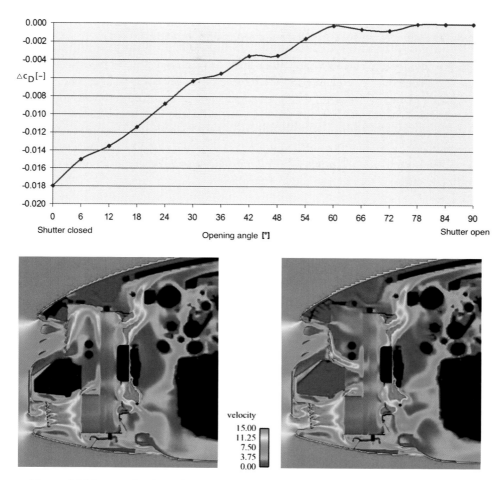

Figure 4.140 c_D-reduction when operating a grille shutter depending on the opening angle a); influence of the shutter opening direction on the cooling air b).

As seen in sections 4.4.1 and 4.5.2.3, the prevention of flow through the engine compartment has a significant influence on the lift distribution. Blocking the cooling duct decreases the lift on the front axle, if as is common a cooling air path with an outlet on the bottom of the vehicle is used, whereas the lift on the rear axle increases. Thus the aerodynamic balance of the vehicle is deteriorated. During the development of a system to control the opening of the cooling air inlet, it is also necessary to thoroughly consider the influence on the axle lift as well as on vehicle dynamics. It might be necessary to determine the opening value for the grille shutter system when operating at high vehicle speeds in order to match the required lift values for this case.

4.5.3 Add-On Parts

The roughness drag of add-on parts was mentioned in section 4.3.3.3 as the drag component, which is defined via the details on the shell of the smooth vehicle basic body in the visible and nonvisible areas under the vehicle. These details include in particular the underbody assembly, wheels, mirrors, antennas, roof rails, and handles.

The flow under the car, between the bottom sheet and the road, around the wheels, and in the wheelhouses were largely ignored by aerodynamicists for a long time. Although the bottom with its large cavities, edged bars, and protruding sheets forms a rugged surface with high drag and—not recognized until much later—a source of flow noise, functional and design arguments have long been so dominant that aerodynamics initially could do little in this area. The front spoiler could at least mitigate but not completely eliminate the problem. Similarly, wheels and wheelhouses are designed according to criteria other than aerodynamic criteria.

As the c_D value of a car approached a value of $c_D \approx 0.30$, the drag of the underside of the vehicle came to the fore. This had the attraction that there was little contact with the styling. Very soon, it became apparent that a more accurate picture of the ride on the road was required for optimization of the flow under the car than had been common so far: No matter whether in experiment or in simulation, the relative motion between the car and the road must be considered as much as the rotary motion of the wheels.

Although the importance of exterior mirrors, antenna, door handles, and roof rails overall is not nearly as large as the share of the underbody and the wheels (cf. Figure 4.22), the aerodynamic optimization of these "visible" add-on parts is a challenge because it is always in competition with styling.

4.5.3.1 Underbody Assembly

Highly simplified, the underbody can be regarded as a very rough plate. Comparison with a smooth plate (Figure 4.18) shows that the friction drag increases with the roughness. Unfortunately, this finding can only be applied to the vehicle qualitatively, because the technical details of an underbody cannot be defined by "sand roughness" or "technical roughness," as they are treated in the textbooks of fluid mechanics.

After Carr [127] had demonstrated the benefits of a completely smooth underside, increasing effort was made to achieve this ideal as closely as possible. However, the cooling airflow and thus the cooling of the various power units must not be impaired. Similar to Carr, Howell [360] also smoothed the underside in sections and additionally in a different order, with the results shown in Figure 4.141. Accordingly, a completely smooth underbody provides a drag reduction of $\Delta c_D = 0.035$. It also turned out that a smooth front end is more beneficial than simply smoothing the rear end. The reason is that a flow separated in the front end as the result of roughness is hardly likely to reattach to a trim of the rear end. And even if it does, no respectable pressure recovery would be gained up to the rear of the vehicle.

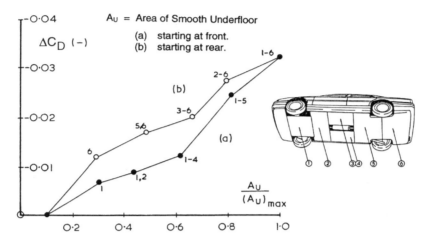

Figure 4.141 Reducing the drag by successive "smoothing" of the underbody [360].

The underside is usually smoothed by means of additional add-on parts. Their disadvantage is that they come at additional cost and weight. However, this must be viewed in the light of additional benefits: They can in fact also be used for corrosion protection or noise reduction. However, individual approaches are apparent to design components on the underside, such as muffler, fuel tank, and spare wheel recess, for optimized flow, so as to make add-on parts superfluous. How an underside can be streamlined using add-on parts will be demonstrated using the example of the Audi A8. As shown in Figure 4.142, it was possible to improve its drag by $\Delta c_D = -0.033$.

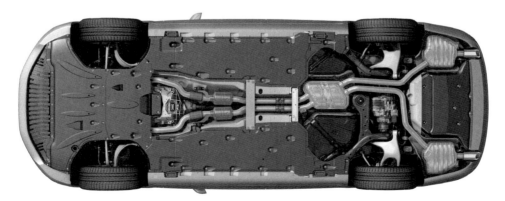

Figure 4.142 Underbody fairing of the Audi A8 [907].

The seemingly simple solution is the result of a number of steps, as demonstrated in Figure 4.143 for the Audi A2. First, the potential of perfect smoothing compared to the untreated underbody was determined: $\Delta c_D = 0.050$. By a total of 28 individual measures, just under half, namely $\Delta c_D = 0.024$, could be realized. It is nevertheless a value that should not be neglected to achieve promised fuel consumption of 3 l/100 km. In this regard, it was essential to achieve the target for the A2, $c_D = 0.25$.

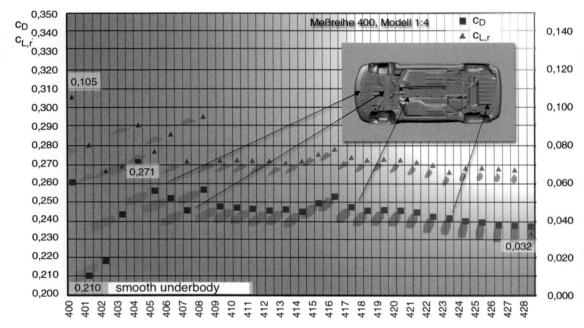

Figure 4.143 Development of the underbody of the Audi A2 [196].

An effective means of influencing the flow under the vehicle is the rear diffuser, an extension of the channel between the underbody and the road. It can be used to reduce drag and lift on the rear axle. The prerequisite for the effectiveness of a diffuser is an undisturbed inflow. This can only be achieved with a smooth underbody, not spoilers. As shown in the illustrations in Figure 4.147, at a given ground clearance there are two parameters that need to be coordinated: diffuser angle φ_D and diffuser length l_D. This has been carried out by Potthoff [650] using the example of the UNICAR (cf. Figure 4.144). It was found that long diffusers are more efficient than short ones, and the same drag reduction is achieved at a smaller angle than with a short diffuser. How such a diffuser is to be designed is described in section 9 in the context of sport and racing cars.

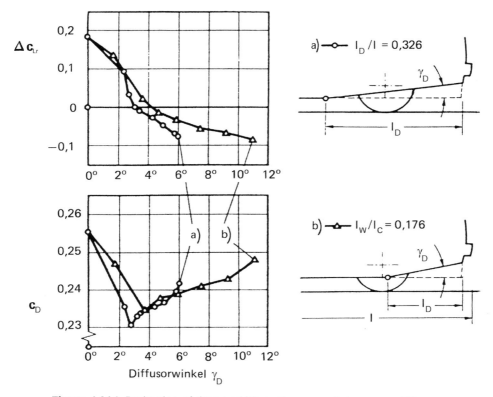

Figure 4.144 Reduction of drag and lift on the rear axle by a rear diffuser on the UNICAR research car [650].

Although the negative effect of a rough underbody on aerodynamic drag has long been known, many vehicles still have a rather ragged structure today. The reasons for this are manifold—high component costs and heat accumulation in the area of the drive and exhaust system are typically cited. Apart from the smooth surfaces described in the preceding sections, spoilers[18] in front of the most intense sources of drag may also be successful. Such sources are:

- Complete underbody if very ragged → Front spoiler
- Large gaps in the underbody surfaces → Upstream spoilers on tunnel struts and traverses
- Wheels → Wheel spoilers
- Axle parts → Upstream spoilers, attached to the fuel tank, underbody fairing, and sheet metal parts

---

18. A spoiler is literally something that "spoils" or disturbs; the original course of the flow is disturbed in order to achieve a positive effect.

- Direction of flow at the diffuser → Spoiler on rear bottom edge, attached to the bumper cover, spare wheel well, or end muffler

The front spoiler increases the surface of attack of the flow in the vertical direction, shifting the location of the stagnation point at the vehicle front downward. More air therefore flows over the vehicle and less underneath. Three effects of the front spoiler can be observed as a result:

- The drag is reduced with a rough underbody and worsened with a smooth underbody
- Reduced lift on the front axle
- Increased volumetric flow of cooling air

Depending on the task, the focus is on different aspects. While initially drag was the center of attention, it very soon shifted to lift, especially for fast cars. With regard to self-steering behavior, the distribution of lift between the front and rear axles must be considered. The increase of the cooling airflow was initially more of a side effect, at higher engine power. However, it is now used very specifically.

The drag-reducing effect of a front spoiler is based on the fact that it diminishes the air speed under a vehicle, thus attenuating the contribution of the underbody airflow to overall drag. This contribution is normally high due to the "roughness" and the non-streamlined nature of the underbody surface. However, the spoiler itself experiences drag, and so a carefully chosen design is required to achieve a positive net effect. The spoiler drag dominates in case of smooth underbody or large spoiler height. That is the reason why the car's overall drag will then increase. The drag of an underbody and spoiler combination is composed of the individual drag shares of the underbody $F_{D,B}$ and spoiler $F_{D,S}$; it can be explained with the aid of the model sketched in Figure 4.145:

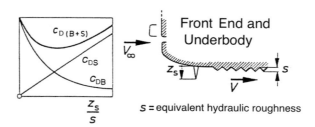

Figure 4.145 Function of a front spoiler.

In the following, underbody drag will be treated as friction drag. However, as already mentioned, underbodies are not really rough surfaces; their drag is not pure friction drag. The flow over bluff projections into the air stream (like axles, etc.) generates pressure drag. Furthermore, some elements of the underbody may be freestanding rather than projections from a surface. Consequently, flow goes all the way around them rather than just over them. Again, this will create pressure drag. To describe the drag of an underbody with a friction coefficient is a simplified, although perhaps useful, model. The underbody's friction coefficient c_f is formed by choosing as reference variables the average velocity $\bar{u}$ under a vehicle and the "wetted" surface A_B of a vehicle's underside:

$$F_{D,B} = \frac{\rho}{2} \bar{u}^2 c_f A_B \tag{4.40}$$

The area A_B of a vehicle underbody is roughly three times the frontal area A_x of the vehicle. The velocity $\bar{u}$ is proportional to the vehicle speed u_∞ and a function of the spoiler height z_S. A functional form can only expressed qualitatively: With increasing spoiler height z_S the average speed falls because the flow underneath a vehicle is "blocked" by the spoiler. Furthermore, the friction coefficient, c_f, for fully rough plate is not affected by the Reynolds number (see section 2.3) and behaves like a constant.

A spoiler can be considered as a panel held perpendicular to an airstream with one side on the ground. Its drag coefficient is $c_{D,S} \approx 1.6$ and its aerodynamic drag results in:

$$F_{D,S} \approx 1.6 \frac{\rho}{2} u_\infty^2 A_{x,S} . \tag{4.41}$$

The frontal area of the spoiler A_S is proportional to the spoiler height z_S and its drag coefficient is independent of spoiler height. Hence the drag force of the spoiler increases in direct proportion to spoiler height z_S.

The sum of the two drag components follows the thick curve in Figure 4.145. This curve has a clear minimum, which means that there is an optimum spoiler height for drag ($z_{S,opt}$). At first, increasing the spoiler height reduces the friction drag of the underbody faster than it increases the pressure drag of the spoiler. Then, however, the underbody drag is hardly reduced at all, while the spoiler drag continues to increase. With very high spoiler heights, the drag of the underbody-plus-spoiler combination can increase to even more than the drag of the underbody without spoiler. For smooth underbody structures, only slight improvement of the underbody's friction drag can be achieved. Hence, the increase of pressure drag of the spoiler dominates from the beginning.

This qualitative assessment is supported by a number of measurement results. One example is given in Figure 4.146. The underbody pressure distribution measured for different spoiler heights also provides an explanation for the reduction in lift at the front axle: The flow separates at the spoiler, and a dead-water region with low static pressure appears behind the spoiler. The absolute value of this negative pressure is increased with increasing spoiler height and spreads over a greater length of the front underbody, reducing lift. However, unlike drag, the lift does not have a minimum value. If low front axle lift is a priority, a spoiler can be made much taller than for minimum drag.

The pressure distribution in Figure 4.146 also explains why the volumetric air flow through the radiator is supported by a spoiler: The pressure difference driving the cooling airflow is increased. The simultaneous increase in rear axle lift can be traced to the fact that part of the air that can no longer flow underneath a vehicle because of the spoiler is now diverted over the vehicle, thereby further reducing the already low pressure on the upper surface of a car. This effect is more pronounced over the engine hood than over the trunk.

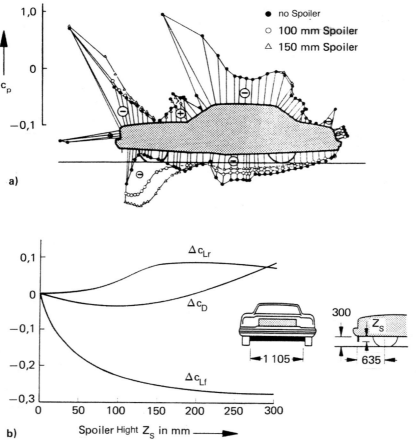

Figure 4.146 Effect of a front spoiler on a road vehicle [711]: a) pressure distribution in a cutting plane at y = 0; Druckverteilung im Längsmittelschnitt; b) drag and lift at the front and the rear axles.

Horizontally applied front spoilers called splitter plates are usually used in motorsports (see chapter 9) with focus on a severe reduction of the front lift.

The height, angular, and rearward position of a front spoiler must be experimentally adjusted to the relevant vehicle. The spoiler also has to respect the required approach angle. A parametric study of the influence of these variables on drag and lift has been made (Buchheim et al. [111]; see Figure 4.147). Instead of spoiler height z_S, Buchheim et al. have chosen a chosen the ground clearance h_S of the spoiler relative to the underbody ground clearance e as the variable.

Aerodynamic Forces and Their Influence on Passenger Vehicles

It turned out that the maximum drag reduction depends only slightly on the rear location x_S of the spoiler, but more on the reduction of lift on the front axle, Δc_{Lf}. The optimum spoiler height differs for each rear location of the spoiler. However, a generally valid trend is not evident. A spoiler at an oblique angle of attack brings only a very small improvement in drag, but it is then even more effective at reducing the lift on the front axle (not shown in the figure, cf. [409]).

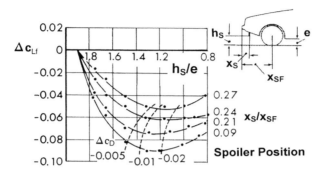

Figure 4.147 Design diagram for a front spoiler [111].

Almost all standard front spoilers produced up to the beginning of the 1990s have a roughly constant height over the full width of the vehicle. However, during development work on the Opel Calibra, Emmelmann et al. [227] were able to prove that there was room for further improvement. By cutting out the spoiler in the middle of the vehicle, as shown in Figure 4.148a, they achieved a reduction in drag. Another reduction was achieved by increasing the spoiler height in front of the wheels, as shown in Figure 4.148b. Apparently, by the former measure the obliqueness of the flow to the front wheels was attenuated while by the latter the wheels became irrelevant. Continuing this research has led to today's underbody structure, where we can find small spoilers in front of the wheels: so-called wheel spoilers. Sometimes a positive effect can also be gained when they are applied in front of the rear wheels.

Wheel spoilers in the form of a simple two-dimensional transverse flow plate improve the flow against and around the wheel, thereby improving the drag, but they cause a large increase in pressure in front of the wheel spoiler. This bow wave in front of the wheels can be avoided and the drag can be further reduced by using three-dimensionally shaped wheel spoiler trim. In this case, the curbstone clearance (i.e., the clearance in front of the front wheels) must be respected.

The fact that the effect of the spoiler is boosted by pulling down the door sill, as shown in Figure 4.148c, may at first glance be surprising. A possible explanation could be this: By pulling down, the flow under the front end is prevented from shifting sideways, which reduces the yaw angle β of the wheels. Finally, the cutout in the center section of the spoiler contributes to the fact that the flow at the end of the underbody can be delayed in a diffuser, with positive effects on drag and rear lift, (Figure 4.148d).

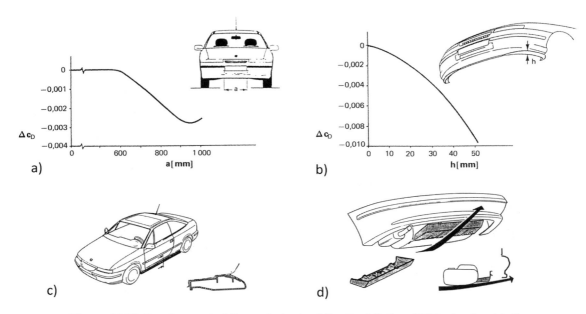

Figure 4.148 Development of the underbody of the Opel Calibra [227]: a) cut-out in the front spoiler; b) shielding of the front wheels; c) guidance of the flow through pulled-down sills; d) attaching a rear diffuser.

Spoilers at other locations on the underbody are also conceivable; an overview is given in Figure 4.149 using the example of the Audi A4. In this case, as well as the wheel spoilers at the front and rear an additional spoiler was mounted at the end of the front sealing, and two spoilers were mounted on the heel plate. Their working principle is basically similar to that of the front spoiler, but they only cause a local deceleration of the flow and hence shield larger gaps in the otherwise flat underbody. For this vehicle the effect of all underbody spoilers was $\Delta c_D = -0.017$.

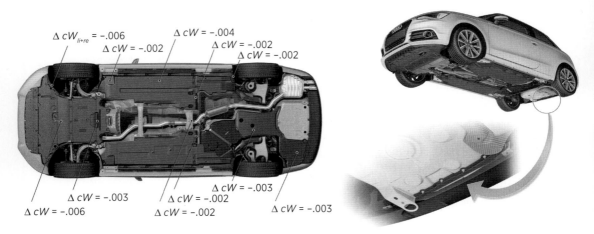

Figure 4.149 The underbody of the Audi A4 (left) and the spare wheel well spoiler of the Audi A1 (right).

For the squareback version of the car, another spoiler is added, the so-called spare wheel well spoiler. In contrast to the previously mentioned spoilers, its effect is due to reduction of the induced drag, which squareback cars usually have because of zero lift or even downforce on the rear axle. Using this spoiler, the diffuser flow is deflected downward at the end of the spare wheelwell, which generates a small lift, reaching an optimum as per section 4.3.3.1. The figure shows the Audi A1, which is equipped with such a spoiler.

In any case, the possible negative side effects of spoilers must not be overlooked. The negative effects must be compensated by special measures. For example, the shielding of the underbody impairs the cooling of the oil pan and the brakes. But this can be remedied by a controlled airflow with openings suitably placed in the spoiler.

4.5.3.2 Ground Clearance and Vehicle Position

The position of a vehicle relative to a reference system is defined in Figure 4.150. The definition of the zero value for angle of attack α has changed. Whereas in the past the design attitude was taken, reference is now made to the attitude obtained when the vehicle is loaded in the same way as for fuel consumption tests. This is analogous to EADE load (section 4.7.3), 75 kg on each front seat and another 75 kg on the back seat.

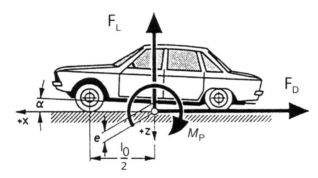

Figure 4.150 Aerodynamic forces, definition of the vehicle's position.

The effect of angle of attack on the drag and lift of cars is shown in Figure 4.151a. The ground clearance e, measured at mid-wheelbase, was maintained constant for these measurements. Lift and drag increase with angle of attack in roughly the same proportion for all vehicles. An increase of one degree in angle of attack corresponds to an increase of 2% in drag. Janssen suggested that this could be exploited in vehicle design by setting the body at a slightly negative angle to the underbody and concealing the mismatch by appropriate styling. This suggestion has been implicitly taken into account by the "wedge" shape.

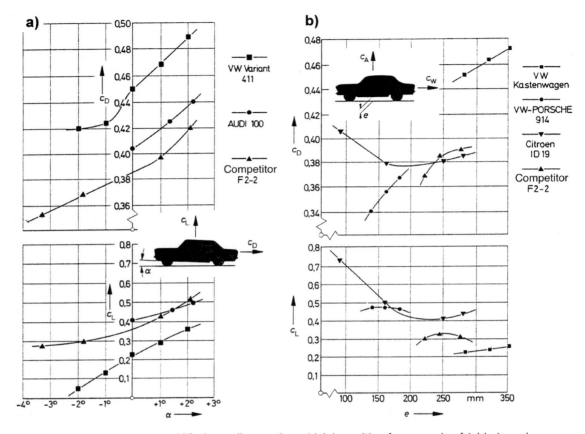

Figure 4.151 Drag and lift depending on the vehicle's position for a couple of (older) road vehicles and one transporter: a) effect of the vehicle's angle of attack α; b) effect of the vehicle's ride height [409].

The effect of ground clearance e on lift and drag is less clear-cut, as can be seen from Figure 4.151b. In "normal" vehicles (i.e., vehicles with structural roughness on the underbody), the drag decreases as a car is set closer to the ground. The Citroen DS 19 equipped with a smooth underbody has the opposite tendency. For this vehicle, the drag increases with reduced ground clearance, in the same way as for a streamlined body corresponding to Figure 1.57. This rise in drag can be traced to the increasing effective thickness of the body with reduced ground clearance, as explained in greater detail in Figure 1.58. This thickness effect is more than offset on vehicles with rough underbody, as the (high) underbody drag decreases when the flow between vehicle and road is impeded.

- Sport utility vehicles are equipped with greater ground clearance as per their intended use. The ground clearance of a conventional car is based on the so-called Italian roadway.[19] To ensure advantages regarding permitted trailer load and

---

19. In Italy, there was until a few years ago a legislative requirement according to which all registered vehicles had to comply with a ground clearance of 120 mm when fully loaded.

Aerodynamic Forces and Their Influence on Passenger Vehicles

taxation in Europe with a sport utility vehicle, they must meet a minimum ground clearance of 200 mm at a prescribed load (3 × 75 kg). Consequently, this means a higher ride height of an SUV of up to 80 mm to obtain sport utility vehicle approval and, therefore, higher values of drag and lift. Figure 4.152 shows the change of the aerodynamic drag product ($\Delta c_D \cdot A_x$) plotted against the relative drive level lowering Δe in millimeters for the Audi Q7. This confirms the findings from Figure 4.151. The curve observed is almost linear.

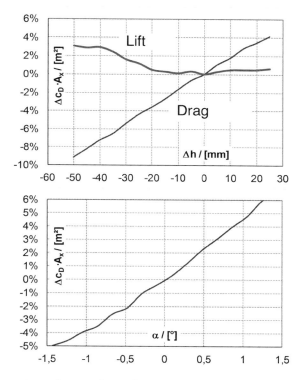

Figure 4.152 Percentage change of aerodynamic drag $\Delta c_D \cdot A_x$ and lift $\Delta c_L \cdot A_x$ of the Audi Q7 as a function of drive level Δe and angle of attack α at EADE loading.

The Audi Q7 with air spring suspension provides a lowering by $\Delta e \approx 30$ mm at speeds above 120 km/h, which means an improvement of aerodynamic drag of about 5%. With an assumed constant frontal area, this corresponds to a c_D potential of $\Delta c_D = 0.020$. It must be taken into account that lowering the vehicle results in a shift of the wheel in the wheelhouse, which has an effect on the styling. But this gets less attention from customers at higher speeds than at standstill, so this aspect can be regarded as noncritical. However, a clean design of the measure is required with regard to the impact on the driving dynamics.

In daily use of a vehicle, ground clearance and angle of attack are always varied simultaneously by the load. The trunk of a car is almost always located at the rear—the angle

345

of attack increases and the ground clearance is reduced with load. Since both factors are generally in opposition, the resulting increase in drag is small as a whole.

4.5.3.3 Wheels and Wheelhouses

The flow in the area of wheels and wheelhouses results from the superposition of flow around the vehicle, the flow through the rims, and flow in the wheelhouse. The incident flow to the wheel is under oblique incident flow, and the flow around the wheel in turn is affected by the rotation of the wheel. All together, this results in a very complex flow topology with numerous vortices, as was shown in Figure 4.39 and Figure 4.27.

Wheels and wheelhouses contribute significantly to the aerodynamic drag of passenger cars. Recent studies suggest that wheels and wheelhouses contribute up to 25% to the drag (cf. Figure 4.22); for vehicles with an extremely smooth underbody, this share of drag may be even higher. Wheels and wheelhouses thus have significant optimization potential in the aerodynamic development of future generations of vehicles.

The flow around the wheel behaves in first approximation like that of a circular cylinder whose ratio of width to diameter $w/d < 0.5$ is very small. Similar to a circular cylinder with a large span, its drag depends on the Reynolds number (see section 2.3.5.4). A guideline for cars (at a velocity $u > 30$ m/s) is $Re_{d,wheel} \geq 1 \cdot 10^6$—a Reynolds number in the supercritical, and for very fast cars, even in the transcritical range.

4.5.3.3.1 Tires

The shape of the wheel is substantially determined by the tire, whose cross section forms a bluff body. Its inherent characteristics are the more apparent, the smaller the cross-section ratio of height h to width w of the tire is; this is the exact direction in which the development of car tires has been heading over time, as can be seen from Figure 4.153.

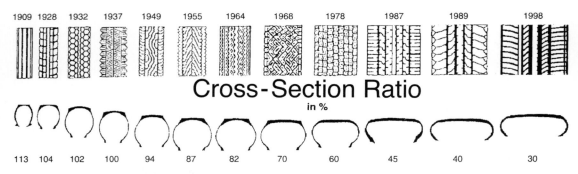

Figure 4.153 Development of the cross-section ratio of a tire [95].

The tire width is an important influencing factor for aerodynamic drag. Depending on the type of tire and the incident flow conditions on the tire, approximately three drag counts can be assumed for every 5 mm of tire width,[20] where it is not the nominal

---

20. With unchanged front face.

width but the actual width that must be taken into account. In Europe, the permitted tire dimensions are defined in the ETRTO (the European Tyre and Rim Technical Organisation) standard, which allows a wide tolerance range for each tire size. Therefore, it is possible that tires of the same type and the same size have significantly different dimensions, which results in increased aerodynamic drag in the worst case.

Other geometric parameters of the tire and their influence on the aerodynamic drag were recently investigated and presented in detail by Wittmeier et al. [892]. In addition to the tire width, the shoulder radius of the tire has been identified as an important geometrical parameter which, together with the shape of the sidewall of the tire, significantly influences the flow around the tire and, hence, the drag of the wheel.

Figure 4.154 shows a comparison of summer tires of size 205/55 R16 from various premium manufacturers and their influence on c_D, wherein the ordinate represents the deviation of the drag from the reference value. The reference value was defined to be the median of all measurement results of a vehicle so that the results of the vehicles can be compared with each other without being displaced by any outlier for a tire.

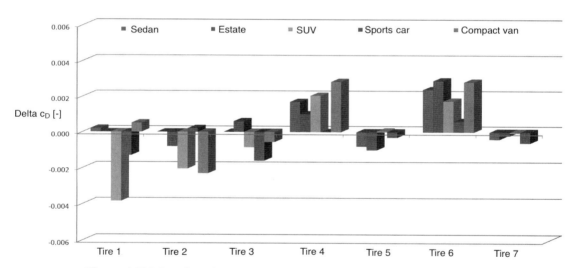

Figure 4.154 Benchmark results: aerodynamic drag coefficients of different vehicles with different tires [892].

In general, it can be stated that the differences between the examined tires are not very big and that current production tires from well-known manufacturers are at a similar level in terms of the aerodynamic characteristics of their products ($\Delta c_{D,min,max} \approx 0.006$).

Nevertheless, there are small geometric differences that are quite interesting from an aerodynamic point of view. The tires 4 and 6 show a clear and measurable increased drag compared to the other tires. On closer inspection of the tire geometry, it can be seen that tire 4 in particular, unlike any other tire, has a pronounced circumferential edge near the shoulder of the tire (cf. Figure 4.155, region 1). Tire 6, which has the largest width, is the only tire in this comparison to have a rim protector edge (region 2). Both—the peripheral edge in the shoulder region of the tire 4 and the rim protection edge of tire 6—result in flow separations and increased drag. Even the shaping of the tire sidewall lettering has an influence on the drag of the tire, as Wittmeier et al. [892] demonstrate through their work.

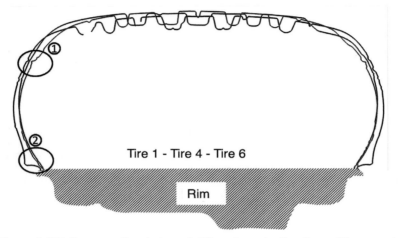

Figure 4.155 Cross-sectional view of different production tires: differences in the shoulder region (1) and the transition to the rim (2) [892].

4.5.3.3.2 Flow Topology

The following sections describe the very complex flow structure of the wheel and the associated effect mechanisms. The flow topology is explained step by step by a stationary and rotating wheel in an initially free incident flow. The flow topology of the stationary and rotating wheel is then considered, taking the wheelhouse into account. It is important to note that the flow conditions on a stationary standing wheel differ considerably from those of a rotating wheel. Despite the differences to the actual flow, the flow structure of the stationary wheel is discussed because wind tunnels without ground simulation—and thus without wheel rotation—are still used today in aerodynamics development (cf. chapter 13).

To explain the flow structures, the following sections mainly refer to the test results of Wäschle [847]. The findings gained here are based in particular on results from CFD simulations, which were supported by laser Doppler anemometry (LDA) measurements and additional oil-streak images. For the sake of clarity, the discussion is restricted to the vortex structures that are essential for the respective flow.

4.5.3.3.2.1 Stationary Single Wheel

First, the flow around an isolated, freestanding wheel on a nonmoving ground is considered. This configuration corresponds to the flow conditions around a wheel in a conventional wind tunnel test in which neither the wheel rotation nor the relative movement between the wind tunnel and the vehicle bottom are considered. The corresponding vortex structure is shown in Figure 4.156a.

① Wheel-wake horseshoe vortex
② C shoulder vortex
③ Contact patch vortex
④ Stagnation point horseshoe vortex

① Wheel wake horseshoe vortex
② Wake ring vortex
③ Contact patch vortex

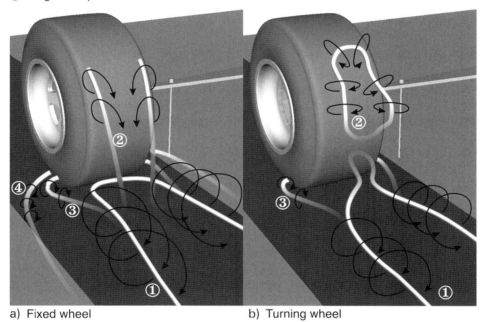

a) Fixed wheel   b) Turning wheel

Figure 4.156 Vortex systems behind a stationary a) and rotating b) single wheel [847].

The presence of the ground causes an asymmetrical flow to form between the top and bottom of the wheel. On the front of the wheel, the bottom boundary layer separates with the result that a horseshoe vortex forms around the wheel—similar to that in the flow around a flow obstacle standing on the ground. This vortex is called a stagnation point horseshoe vortex (4) according to Figure 4.156a. It is relatively weak and almost completely dissipates at the level of the rear side of the wheel.

In a stationary wheel, the wake region is essentially determined by opposing, inward-turning longitudinal (straight line) trailing vortices, which form in the lower half behind the wheel. This vortex pair whose trailing vortices are fed by a backflow close to the ground from the wheel wake (see also Figure 4.156a) is called the wheel wake horseshoe vortex (1). It is responsible for the induced drag which occurs at any finite body subject to lift.

Figure 4.157a shows the results of an LDA measurement at $x = 150$ mm in the wake of the stationary wheel. The vectors in the figure represent the course of the cross flow, from which the above-described wheel wake horseshoe vortex can be clearly seen. The given velocity component cu is defined as the ratio of the velocity component u (x-component) to the incident flow velocity u_∞. Blue-colored areas represent backflow.

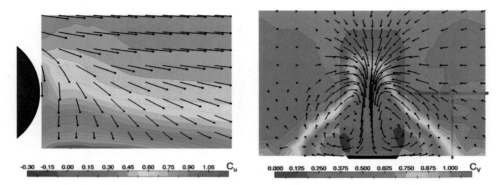

Figure 4.157 Velocity fields behind a stationary single wheel (levels at $x = 150$ mm and $y = 0$ mm) [849].

The strong, inward rotating trailing vortices of the wheel wake horseshoe vortex cause a massive downwash field that forms above the rear wheel tread of the stationary wheel (cf. Figure 4.157b). The flow above the wheel tread is accelerated, which leads to a high negative pressure on the top of the wheel and thereby to high lift forces on the wheel.

Another vortex pair, similar to the C-pillar vortex on the vehicle, forms in the upper half behind the wheel. Here, the flow separates at the rear tire shoulders and turns inward toward the center of the wheel. This vortex pair, known as the C-shoulder vortex (2) (cf. Figure 4.156a) supports the downwash field and prevents premature separation of the flow from the tread of the wheel. The downwash field is constricted by the inward-turning C-shoulder vortex, as evident in the oil-streak image (cf. Figure 4.158). According to recent findings, the two vortex centers of the C-shoulder vortex pair are influenced by the strong wheel wake horseshoe vortex to such a degree that they are deflected toward the ground.

Another vortex pair to be discussed with reference to the flow around the wheel is the so-called wheel contact patch vortex. Wheel contact patch vortices (3) (cf. Figure 4.156 a) form as the result of flow separation at the tire's front shoulders in the contact region of the tire deformed by the wheel load, the so-called wheel contact patch. Owing to high suction peaks at the tire's front shoulders in the contact region of the wheel, air is transported from the wheel wake near the ground through the localized separation regions at the sides of the wheel contact patch to the front in order to form a wheel contact patch vortex pair with the free flow.

Aerodynamic Forces and Their Influence on Passenger Vehicles

Figure 4.158 Oil-streak image of an isolated stationary wheel [847].

4.5.3.3.2.2 Rotating Single Wheel

In the following analysis, we will take the wheel rotation into account. This configuration is therefore related to the road, insofar as these conditions are comparable to the incident flow conditions of formula racing.

The wheel wake horseshoe vortex (1) behaves similarly to that of the stationary wheel and dominates the entire wheel wake (cf. Figure 4.156b). However, the vorticity of the longitudinal vortices diminishes, indicated by the reduced lift of the rotating wheel. This is also confirmed by comparing the velocity fields at $x = 150$ mm behind the stationary and the rotating wheel. The rotating wheel has a significantly less pronounced cross-flow structure (cf. Figure 4.159a).

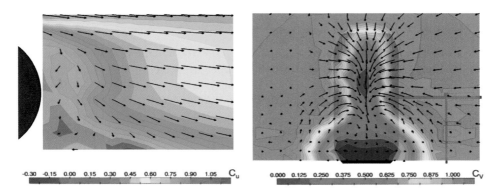

Figure 4.159 Velocity field behind a rotating single wheel
(levels at $x = 150$ mm and $y = 0$ mm

351

The flow above the tread of the rotating wheel changes fundamentally. Owing to viscosity influences, air is transported from the wheel wake against the main flow direction via the tread of the wheel from the rear to the front. Simultaneously, high-energy air is fed into the lower region of the wheel wake with the result that the base pressure rises on the back of the wheel. The higher base pressure on the back of the tire results in drag reduction of the rotating wheel (cf. Table 4.2). The two flow influences caused by the rotation of the wheel prevent the downwash field, as noted in the stationary wheel, with the flow attached to the tread (cf. Figure 4.157). Instead, the flow separates from the tread and together with the separations at the tire's rear shoulders forms a closed wake vortex (2) (cf. Figure 4.156b) in the upper half behind the wheel. This separation bubble prevents the flow being accelerated at the top side of the wheel. Compared to the stationary wheel, this leads to a significantly higher pressure, associated lower lift, and thereby the reduction of induced drag (cf. Table 4.2).

Owing to wheel rotation, the contact patch vortex (3) (cf. Figure 4.156b) has lower intensity compared to the stationary wheel. Owing to the adherence conditions, air in the region of the wheel contact patch is transported backwards, which restricts the backflow described for the stationary wheel at the sides of the wheel. On the rotating wheel, the boundary layer near the ground does not separate in front of the wheel, with the result that a stagnation point horseshoe vortex is not able to form.

As described, lift and drag decrease with the rotation of the wheel. Table 4.2 shows a summary of the results of the stationary and rotating isolated wheel according to Wäschle & Wiedemann [850]. The change in drag due to rotation is comparatively small at about 10%, whereas the lift changes drastically.

Table 4.2 Influence of Wheel Rotation on Drag and Lift of the Isolated Wheel

Wheel	Drag	Lift
Stationary	$c_D \approx 0.50$	$c_L \approx 0.30$
Rotating	$\Delta c_D \approx -0.05$	$\Delta c_L \approx -0.20$

4.5.3.3.2.3 Stationary Wheel in the Wheelhouse

Next, we will explain the flow topology of a stationary wheel in the wheelhouse, the flow around which displays the flow structure shown in Figure 4.160.

Arranging the wheel in a wheelhouse changes the flow around it fundamentally as compared with a wheel in a free incident flow. The tread of the wheel and the tire shoulders are shielded by the wheelhouse in the upper part of the wheel and are consequently no longer exposed to the direct incident flow. Separation mechanisms, as they occurred in the upper part of the stationary single wheel in free incident flow, play no role here.

Aerodynamic Forces and Their Influence on Passenger Vehicles

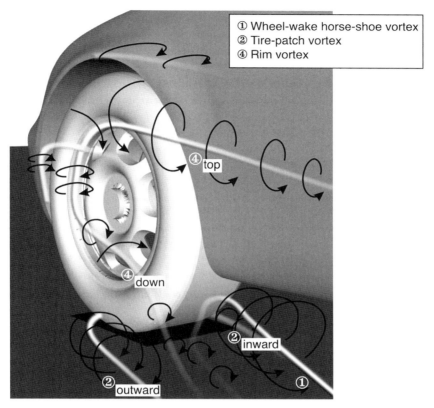

Figure 4.160 Vortex system behind a stationary front wheel [847].

The lower part of the wheel is exposed to an oblique incident flow because of the displacement effect of the vehicle front. This swept incident flow of the front wheels results in a shift of the stagnation point in the direction of the inner wheel shoulder. The magnitude of the yaw angle, and thus the relocation of the stagnation point in the direction of the inner wheel shoulder, depends on the shape of the front end, in particular on the size of the front overhang. Another factor influencing the yaw angle is the cooling air leaving the engine compartment. The stagnation point relocation to the inner wheel shoulder results in less flow deflection, lower pressure gradients occurring locally, and a smaller separation region, resulting in a weak internal wheel contact patch vortex (2). On the outside of the wheel, the flow conditions deteriorate accordingly. The flow separates earlier in the contact region of the outer wheel shoulder. Compared to the front wheel in free incident flow, the result is a pronounced outer contact patch vortex.

The wheel wake horseshoe vortex (1) also dominates the wake region of a stationary wheel in the wheelhouse, but the symmetrical horseshoe vortex in the isolated single wheel becomes unsymmetrical because of the oblique incident flow. The strength of the inner trailing vortex increases significantly, whereas the outer part of the wheel wake horseshoe vortex remains weak and dissipates shortly behind the front wheel.

Another vortex forms on the outer side of the front wheel, or more precisely, on the front rim flange. The air flowing along the wheel side separates at the apex of the wheel side, curls up toward the rim, and then forms the so-called rim vortex (4) in the shape of a horseshoe. The upper trailing vortex flows along the rim flange and is fed mainly by the air exiting between the wheel tread and the wheelhouse. In the stationary wheel, almost the entire wheelhouse is vented. The lower trailing vortex runs between the wheel hub and rim holes downstream over the tire side. The rim vortex is affected by the rim flange geometry and the flow conditions of the rim holes.

4.5.3.3.2.4 Rotating Wheel in the Wheelhouse

Finally, we will consider the flow of the rotating front wheel in the wheelhouse, which is comparable to the flow conditions of a car traveling on the road. The resulting vortex structures were presented in Figure 4.39.

The flow structures in the lower part of the rotating wheel are similar to those of the stationary wheel. Only the inner trailing vortex of the wheel wake horseshoe vortex (1) and the inner wheel contact patch vortex lose their intensity as a result of wheel rotation, and the inner contact patch vortex (2) can be hardly detected because of the momentum induced by the moving ground. As a result of the oblique incident flow, the flow conditions deteriorate on the outside of the wheel, but the separation region on the side of the contact patch near the ground decreases by the momentum of the wheel rotation in comparison to the stationary wheel.

As a result of the relatively small wheel contact patch vortex in the outer region near the ground of the rotating wheel, the free flow reaches the rear of the wheel and then separates at the rear tire shoulder. This vortex is referred to as the P-shoulder vortex (3) by Wäschle, where P stands for pes (Latin for foot) and describes the lower half of the wheel. Corresponding to the outer P-shoulder vortex rotating counterclockwise, an inner P-shoulder vortex also forms, whose direction of rotation is clockwise. As a result of the oblique incident flow, the inner rear tire shoulder is surrounded by free, high-energy air which then separates at the shoulder and, in comparison to the outer P-shoulder vortex, leads to a significantly stronger vortex on the inner side.

The rim vortex (4) also forms on the rotating front wheel shielded by the wheelhouse. However, in contrast to the stationary wheel, no upper rim vortex is formed on the rotating wheel. In the upper part of the wheel, the wall boundary layer is transported against the main flow direction, resulting in a free backflow along the tire side. The rim flange is therefore already in the backflow region so that no vortex is formed at the upper part of the rim.

Formation of the upper rim trailing vortex is prevented by the presence of the so-called flank vortex (5) on the rotating front wheel. The horseshoe-shaped flank vortex is formed as a result of wheel rotation. Part of the venting of the wheelhouse occurs via the upper trailing vortex of the flank vortex.

For completeness, it should be noted that the results of the stationary and rotating wheel in the wheelhouse are based on analyses of a simplified vehicle model in 1:4 scale. The final results in terms of the flow structure could be confirmed with good agreement on a realistic vehicle, but, owing to the complexity of the flow, the results apply so far only to sedans and cannot be transferred to other basic shapes without reservation.

4.5.3.3.3 Effects of the Rotating Wheel on the Overall Vehicle

The following section describes the effects of the rotating wheel on the total flow around the vehicle. The rotation of the wheels reduces the loss area behind the rear wheels, but also in the wake of the vehicle. The smaller wheel wakes of the rear wheels cause a better incident flow of the lateral tail tapering. The rear base pressure of the vehicle increases as a result of rotation of the wheels, which is reflected in lower aerodynamic drag of the vehicle. The underbody flow tends to be improved as a result of wheel rotation, resulting in a higher flow velocity under the vehicle and a decrease in the static pressure. This results in a decrease in the lift of the overall vehicle.

Figure 4.161 shows the influence of wheel rotation on the drag and lift coefficients of the vehicle body and the front and rear wheels. It can be clearly seen that the "global" effects on the vehicle body as a result of the interaction between the flows around the wheel and the vehicle are the largest, followed by the local effects on the front and rear wheels. This applies to both the drag and lift coefficients.

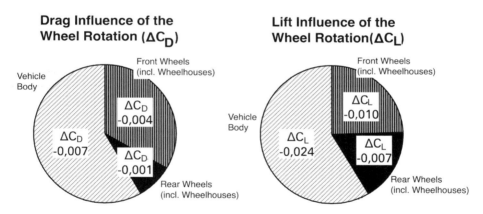

Figure 4.161 Influence of wheel rotation on the partial drag and lift coefficients of a real vehicle [847].

4.5.3.3.4 Rim

As mentioned above, the high drag of the wheels is caused by the unfavorable shape of the tires with respect to flow. In addition to the measures on the tire that were discussed at the beginning of the section, the drag can be influenced by measures on the rim.

The easiest way to reduce drag is to provide a smooth cover to the rim, so optimization of the rim would involve sealing the rim openings. However, this measure has a negative effect on the components of the braking system, since the cooling airflow to the brake disk and the brake caliper would be reduced or even prevented [748]. Furthermore, full surface coverage is frowned upon by rim designers, because it is precisely the different rim styling that is used to differentiate and visually enhance vehicles.

Another way to reduce drag is to optimize the rim geometry itself, without restricting the designer's stylistic freedom so severely as when a hub cap is used. The shape of the rim exterior as well as the ventilation characteristics of the rim are the major causes of the different drag contributed by different wheels.

Once again, we refer to section 4.3.3.3. On the one hand, the ventilation characteristics cause a deflection of the flow in the lateral direction and thus a loss of momentum in the x-direction, which leads to a measurable increase in drag. On the other hand, the rim rotation also introduces a rotation into the flow. The associated ventilation moment, braking the rotary motion of the rim, increases fuel consumption, but can often not be measured in a wind tunnel with external scale. However, it can be evaluated alternatively on a roller dynamometer and then added to the c_D value. This gives us the $c_D{}^*$ value that was introduced in section 4.3.3.3.

An approach to reducing drag on the rim was recently presented by Volvo [484]. In this study, a number of wheel rims with different geometries were tested in a wind tunnel. In parallel, the rims' shape parameters were investigated and measured and were evaluated with the wind tunnel results using statistical methods. The aim of this approach is aerodynamic evaluation of the rim using a regression model, which can then be performed solely on the basis of different shape parameters. Preliminary results show good agreement between the measured values and those predicted by the regression model. However, this method currently still has its limits on strongly three-dimensionally shaped rims because of the many possible shape parameters. The shape parameters of the rim that were classified as important during this study included, besides the rim cover, the profile and radii of the wheel spokes as well as the offset of the spokes to the rim edge.

Figure 4.162 shows the influence of different rim geometries on the rear base pressure and the aerodynamic drag, where the drag is given as a change in the c_D value for the reference measurement (cf. configuration 19). Although configurations 17 and 27 are also five-hole rims compared to the reference measurement, the different rim styling resulted in an improvement of up to ten counts, and, using a flat cover for the wheel disk, drag improvement of 18 counts was even possible.

Aerodynamic Forces and Their Influence on Passenger Vehicles

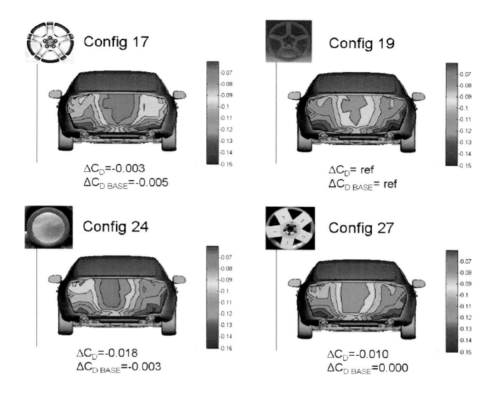

Figure 4.162 Base pressure measurements with different rim geometry [484].

These results also show an interaction of locally implemented optimization measures on the rim with the rear of the vehicle. Although the flat rim cover leads to a distinct reduction in vehicle drag, the base drag shows only a slight improvement because of the increased rear base pressure. Similarly, the measurement with the highest rear base pressure (see configuration 17) does not correlate with the lowest measured drag (see configuration 24).

A more analytical approach to optimizing the rim drag is performed by reducing the ventilation moment of the rim. Studies on roller dynamometers have shown that identical tires with different geometries require different levels of driving power, which is explained by a different ventilation moment. The profiling of the wheel spokes is an important influencing parameter, as the results of the above study show. Impact on the rim geometry is shown in Figure 4.163.

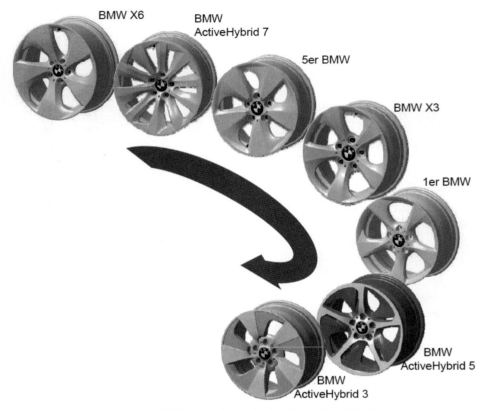

Figure 4.163 Aerodynamic rims from BMW [794].

The effectiveness of optimization measures on the rim, no matter whether simple covering the wheel disk or optimizing the wheel geometry itself, highly depends on the inflow conditions to the wheel. Drag reduction can only be achieved when the flow around the tire shoulder is streamlined.

Figure 4.164 shows the results of flow simulations with a five-hole rim compared to a fully covered rim with different shoulder shapes. On the tire with a round shoulder, the simple rim cover has a significant influence on the flow field. In contrast, the flow field on the tire with a square shoulder shows only very small differences on both rim variants. This is also evident in aerodynamic drag. In the variant with a round shoulder, the rim cover improves the drag coefficient by 0.004. In the case of the square shoulder, the aerodynamic drag is not changed by the cover.

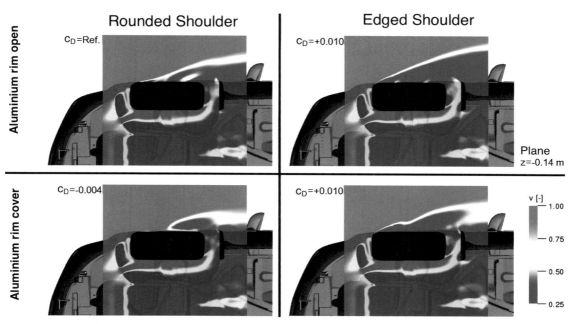

Figure 4.164 Flow around the tire with different rims and different variants of the tire shoulder [892].

The oblique incident flow of the front wheels "sharpen" the front outer tire shoulders for the flow because of the effect of displacement by the vehicle, which—in combination with square tire shoulders—can lead to flow separation on the tire shoulders. This means that the rims are already in the separated region of the flow and the rim geometry has no major aerodynamic influence anymore.

4.5.3.3.5 Wheelhouses

Finally, we will discuss the geometry of the wheelhouse and its influence on the aerodynamics of the wheel. The size or the volume of the wheelhouse and the position of the wheel in the wheelhouse are mainly determined by considering the suspension, the steering, and tire lineup to ensure the free movement of the wheel. Aerodynamic aspects play a relatively minor role, since the clearance of the wheels must be guaranteed, even when loaded or running with snow chains.

According to studies by Cogotti [151], the volume of the wheelhouse should not be greater than absolutely necessary. Drag and lift increase with increasing wheelhouse volume (VH) at constant wheel size. This statement applies to the stationary and rotating wheel in the wheelhouse, with the drag increase more pronounced for the rotating wheel.

In the front part, the air flows over the front edge of the wheelhouse, forming an upward flow (2), (cf. Figure 4.165a). The cooling air flowing through the holes from the engine compartment into the wheelhouse is another component of the inflow and it vents the engine compartment. In addition, part of the underbody flow flows via the inner tire flank from below into the rear part of the wheelhouse (see Figure 4.165b). This mass flow entering from the back of the wheelhouse is considerably smaller for the rotating wheel compared to the stationary wheel because of the absence of separation at the inner side of the wheel contact patch. In the wheelhouse, the air flows via the tread of the wheel to the front, before it is vented toward both sides along with the air flowing into the wheelhouse from the front in the upper wheelhouse gap. The air flowing into and out of the wheelhouse causes a disturbance of the flow around the vehicle, which affects the local flow as well as the total flow around the vehicle.

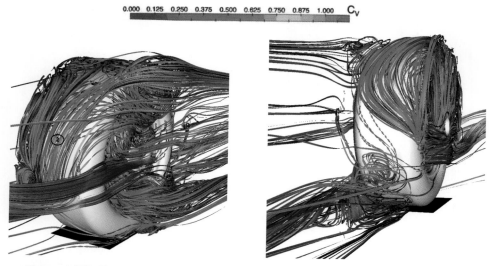

Figure 4.165 Flow structure of the rotating left front wheel on the 1:4 vehicle model: a) outer wheel view from the front; b) inner wheel view from the rear; the geometry of the wheel and vehicle body hidden each time [847].

Spoilers are used to improve the flow around the front wheels (see also the underbody assembly). They reduce the air flowing via the front edge into the wheelhouse and also influence the ventilation from the wheelhouse. Figure 4.166 provides a comparison of the total pressure distribution around the front wheel with and without wheel spoilers. The effect of the wheel spoiler on ventilation from the wheelhouse is clearly visible.

Aerodynamic Forces and Their Influence on Passenger Vehicles

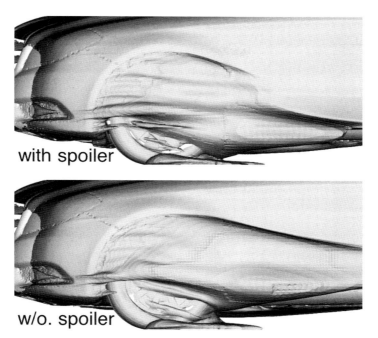

Figure 4.166 Comparison of the total pressure distribution around a front wheel with and without a wheel spoiler [847].

In some vehicles, there are vertical slots in the front wheelhouse liners on the sides in front of the front wheels from which air flows out of the front end—or the engine compartment. The exiting air causes a stabilization of the shear flow between the highly turbulent wheelhouse flow and the free flow around the front of the vehicle, which reduces the overall aerodynamic drag.

Another possibility to optimize the vehicle is to cover the rear wheel cutouts. Covering the rear wheel cutouts prevents flow separation at the rear wheelhouse and increases the effectiveness of boat tailing (cf. section 4.5.1.3.5). This measure, which was addressed in a recent case study with regard to future c_D target values, has potential for optimization of up to ten counts. Despite this relatively large potential, the rear wheel cutouts have been covered only very rarely. Justified by the need to further reduce CO_2 emissions and to improve the energy efficiency of vehicles, it would be welcomed if this measure is applied more frequently. Covering the rear wheel cutouts not only improves the aerodynamic drag of the vehicle, but also reduces spray, which would mean better visibility for a following vehicle when driving on wet roads.

4.5.3.4 Exterior Mirror

The exterior mirror is an indispensable add-on part on the vehicle and will continue to need to be installed in future years to ensure all-around visibility, despite frequent studies published in recent years that replace the mirror with sleek and streamlined booms for camera systems. Such systems, however, are currently still far from being introduced to the market in terms of production maturity, customer acceptance, and approvals.

Aspects such as mirror size, mirror shape, and body connection must be observed with regard to the aerodynamic efficiency of an exterior mirror. Sport utility vehicles need larger mirrors than other cars for adequate view, as well as because of statutory approval criteria. With regard to the attachment, a distinction is made between parapet mirrors and triangular mirrors (cf. Figure 4.167, top).

The exterior mirror in itself has a relatively large drag. The drag coefficient of the mirror on its own frontal area can be assumed to be around 0.5 for a well-shaped mirror. Mirrors resembling a sharp-edged circular disk with $c_D = 1.2$ are no longer common.

Figure 4.167 Different vehicles with parapet and triangular mirrors (top) and different mirror designs (bottom) [747].

If this is based on the vehicle's total frontal area, this results in drag coefficients in the range of $c_D = 0.004$ to 0.010. If two mirrors are mounted, their contribution to the c_D value ranges from 0.008 to 0.020, corresponding to about 5% of total drag. However, one aspect in this estimate that must not be neglected is the interference effect, because the individual drag of the exterior mirror in free flow and the drag when mounted on the vehicle are not identical. The main reason for this is the local velocity increase in the flow at the mirror caused by the displacement effect of the vehicle. An explanation for this is given by Figure 4.174 in section 4.5.4.1.

The mirror shape is largely determined by the car's styling, because the mirror needs to adapt to the overall look of the vehicle. In addition, legal requirements, as well as aeroacoustic (chapter 8) and soiling-reducing measures (chapter 6), must be considered. A variety of different mirror shapes are conceivable; Figure 4.167 (bottom) provides an overview, compiled by Schütz [747]. It includes the parapet mirrors of the Audi TT (1) and the VW Touareg II (2) and the triangular mirrors of the Audi Q7 (3) and A8 (4). While maintaining the legal requirements regarding the size of the mirror, the aerodynamic potential of the mirror shape can be estimated, for example, by morphing. The influence of the mirror connection must be taken into account. In this way, using a parapet mirror and aerodynamic shape optimization of the mirror, Schütz identified realistic potential for improvement to reduce aerodynamic drag on the exterior mirror of the Audi Q7 of $\Delta c_D = -0.004 \div -0.008$.

Suggestions for how the drag of exterior mirrors can specifically be reduced can already be found in Hoerner [350], information from which has been used to produce Figure 4.168. The base pressure and drag of the bluff body with a semi-cylindrical body at its nose can be reduced by extending the body in the direction of the flow, among other measures. However, it must be noted that the results were obtained from two-dimensional models. The reduction of drag with increased length should be much less pronounced for a three-dimensional mirror housing. Furthermore, the measures discussed in section 4.3.3.1 to reduce the base drag can also be transferred to the housings of exterior mirrors accordingly.

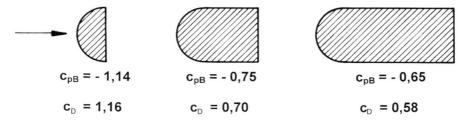

Figure 4.168 Option to reduce the c_D value of the rearview mirror [350].

Exterior mirrors have a long, wide wake that interferes with the flow on the side window and the vortex originating from the A-pillar. As described in detail in chapter 8, they contribute significantly to interior noise. Figure 4.169 shows the flow topology for a poor mirror and a mirror with favorable drag.

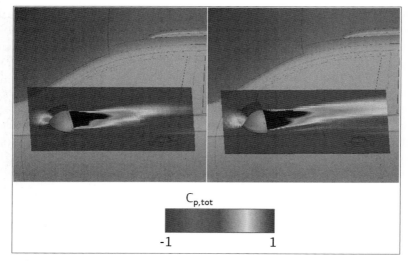

Figure 4.169 Flow around a mirror with a favorable c_D value (left) and a less favorable mirror (right) in horizontal and vertical sections.

4.5.3.5 Other Add-On Parts

Door handles, roof rail, and antenna in themselves also have high c_D values. However, their frontal areas are small compared to that of the vehicle, and consequently their share of the drag is low. More interesting than their drag is their contribution to wind noise. The contribution of the antenna can be estimated as described in section 4.5.3.4. The c_D value of a cylinder in transverse incident flow at subcritical inflow (Re < 10^5) of $c_D \approx 1.2$ and a frontal area of less than a thousandth of the vehicle face results in the following antenna drag:

$$F_{D,A} < \frac{\rho}{2} u_\infty^2 \cdot 1.2 \cdot \frac{1}{1000} A_x \quad (4.42)$$

Converted to the vehicle's frontal area, the drag coefficient of the antenna is one point, as a maximum. The drag of the antenna is thus very small indeed. Analogous estimates for railing and door handles lead to the same result.

In contrast, roof racks, ski containers, transported bicycles, surfboards, signal systems, and so forth increase the drag considerably. In the case of ski carriers, drag increases of one-third are possible; a bicycle on the roof increases the c_D value (based on the frontal area A_x of the vehicle) by about 60%. Roof boxes are increasingly designed to be streamlined in order to limit the associated increase in fuel consumption. To avoid unnecessarily wasting fuel, roof racks should be removed when they are not needed. Figure 4.170 provides an overview of all these elements.

Aerodynamic Forces and Their Influence on Passenger Vehicles

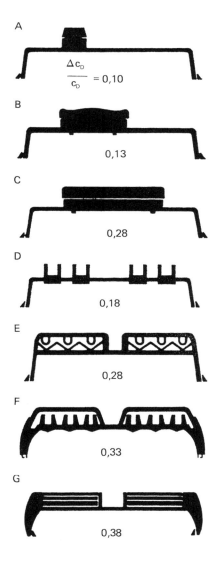

Figure 4.170 Increase of drag with ski carriers [601].

The reason for the increase in drag by roof racks and roof boxes is demonstrated impressively in Figure 4.171. Just the roof rack causes an extended wake with corresponding dissipation losses; these effects are much greater with the roof box. In the latter case, the drag increases by 33% compared to a vehicle without a roof structure, whereas the lift almost does not change at all.

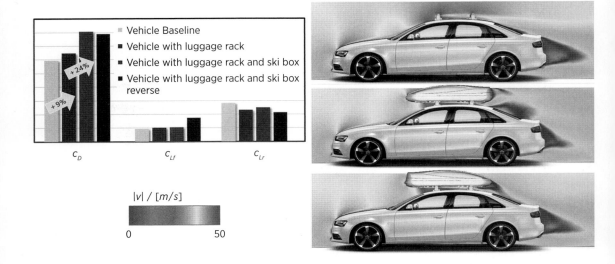

Figure 4.171 Flow velocity around a vehicle with roof structures and the impact on the force coefficients.

Roof loads also affect driving stability, for two reasons: First, the vehicle's center of gravity is relocated upward—and also usually toward the rear—and second, the aerodynamic properties are substantially modified. Some examples of the latter are compiled in Figure 4.172. Since the additional drag force acts at a relatively high position, there is a tendency toward lower load on the front axle and higher load on the rear axle. The lift is changed by the disturbed flow over the roof. It may be reduced, but also increased [678].

Under the influence of oblique incident flow, all roof loads lead to increased side force and an increase in the roll moment. Since the wind's point of attack is relocated upward, the aerodynamic force acts with greater leverage with respect to the cornering forces of the wheels. Consequently, the roll moment increases more than the lateral aerodynamic force.

Usually, roof loads cause no increase in the yaw moment. The entire roof area of the vehicle is located behind the point of attack of the vehicle's side force (center of pressure). Objects on the roof therefore move the point of attack of the aerodynamic force to the rear and, with some exceptions, increase the rear side force more than the front side force. An exception is the boat that is shown in Figure 4.172 listed as case (6); it extends far beyond the roof front edge. Owing to its fuselage shape, the additional side force acts far to the front. Aerodynamic advantages compared to roof installation may be provided by bicycle racks mounted behind the vehicle. However, appropriate measurements have not been reported so far.

Aerodynamic Forces and Their Influence on Passenger Vehicles

Roof load	c_{Lf}	c_{Lr}	c_{MY}	$c_{MP\ 20°}$	$c_{M,R,20°}$
Baseline car (1)	0.09	0.19	0.66	0.17	0.13
Luggage rack (2)	0.10	0.12	0.74	0.16	0.16
Ski (3)	0.08	0.13	0.76	0.15	0.15
Surfboard (4)	0.10	0.13	0.77	0.16	0.16
Ski box (5)	0.10	0.15	0.92	0.15	0.23
Boat (6)	0.24	-0.03	1.12	0.17	0.37
Bicycle (7)	0.19	0.03	1.00	0.12	0.32

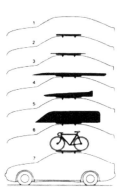

Figure 4.172 Change to aerodynamic properties by different roof loads.

4.5.4 Interference

Up to this point, we have treated the flows around each part of the car as if the other regions did not exist at all in each case. This was done despite the fact, as mentioned in section 1.2, that the interactions—in particular, those between the vehicle's front and rear ends—were discovered early on. They were then apparently ignored. The interactions between the front and the rear are taken into account only implicitly insofar as optimization of a model usually starts at the front end and ends at the rear.

However, interference occurs not only between the different surfaces of the base shape of an individual vehicle but also between the basic body and its add-on parts, the partial bodies of a vehicle combination, as well as between vehicles traveling one behind the other and when overtaking. The characteristics of the interference effects will be discussed in the next section.

4.5.4.1 Interaction of Vehicle Components

For lack of data, the analysis of the interaction phenomenon on the vehicle's basic body must be based on hypotheses. In this regard, Figure 4.173 is used to select the flow around the nose and rear of a body of rotation; three cases can be distinguished.

We assume the nose of body A has the "optimum" radius, so the flow around it would still be separation-free. At the rear part of body A, the boundary layer would only allow a moderate pressure increase, that is, the boat tailing would only allow a small angle φ_{optA}. The relationship between the referenced drag, $F_D/F_{D,0}$, and the tapering angle φ, which is outlined on the right in the figure, would be represented by the curve A. The optimum angle $\varphi_{opt\ A}$ of configuration A would be small, and the minimum drag would be comparatively high.

However, if the nose is slender, as in model C, only moderate negative pressures occur at the transition from the nose to the side. The boundary layer would have to overcome only a small pressure gradient farther downstream, allowing a wider angle $\varphi_{opt,C}$

without separating prematurely. This would lead to lower drag than in case A. Case B lies between A and C.

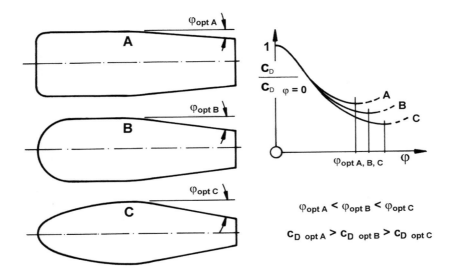

Figure 4.173 Interaction between the nose and tail in the flow around a full body, hypothetically.

Transferred to real cars, this (hypothetical) relationship would probably mean the following: Currently, the nose of cars is very much rounded, far more than required for the optimum radius. This determines the further flow around the vehicle—especially around the tail—as indicated by case C. If one day a more angular style is preferred again, this would be an approach to case A. A tail adapted to the highly rounded front end C would probably no longer "work" with a sharp-edged front end. The flow would separate earlier, with the result that the drag would increase. This increase must be defined in this case as interference drag.

To obtain the total drag of a motor vehicle, it is not enough to add up the individual drags of the individual parts. In this analysis, the interference drag covers the deviation of this sum from the total drag. This deviation is due to the fact that the basic body and the add-on parts mutually alter the flow around each other, and thus the body encounters a swept flow and/or a flow at different velocity, compared to the body in free incident flow with little loss.

In section 4.3.3.4, this was already described using the exterior mirror as an example. Figure 4.174 now shows how a mirror should be designed in order to obtain favorable drag levels. If a rotationally symmetric body with a slightly tapered tail is used as the exterior mirror housing, the oblique incident flow in the area of the A-pillar would lead to unilateral separation at the mirror. A large wake causes a large intrinsic drag, which does not equal the significantly lower drag in free and symmetrical incident flow. If the shell of the mirror is matched to the local flow conditions, the result is a body that is no longer rotationally symmetric and has a larger base area, but the flow around it is separation-free. This reduces the c_D value more than the increase in the mirror's frontal area, so the aerodynamic drag product $c_{D,S} \cdot A_{x,S}$ becomes smaller.

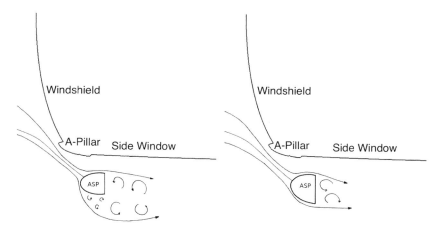

Figure 4.174 Interaction between the body and exterior mirror, schematic.

4.5.4.2 Cars with Trailers

Of the many types of trailer that can be towed by cars, only trailers (mobile homes) have been studied in depth for their aerodynamic properties. Attention was at first focused on drag, but has now increasingly shifted toward the dynamic stability of the car/trailer combination. The drag of a car-trailer combination can be about three times that of a car alone. There are two main reasons: First, a trailer has roughly twice the frontal area of a car; second, due to their box-like shape, trailers have very high drag coefficients. Design modifications can be made to reduce a trailer's drag considerably, but this reduction is usually at the expense of a less practical interior and higher manufacturing cost.

The form of a car/trailer combination generates interference effects; the underlying principles of these effects are identified in Figure 2.49. The drag of a combination is less than the sum of the drags of the two individual vehicles, as was already determined by Beauvais [58]. The drag of the car is drastically reduced by the trailer, while the trailer's drag is reduced only slightly by the car. The resulting shift of overall drag to the trailer is not very favorable. It results in a higher towing load on the drawbar, and this higher load is disadvantageous for the stability of pendulum-type swinging motions.

According to tests by Künstner [475], this drag shift increases as the towing vehicle is streamlined (see Figure 4.175). It is even possible for the drag of the towing vehicle to be negative, while that of the trailer can be greater than if it were exposed alone to the airstream. When assessing the numerical data in Figure 4.175, it should be borne in mind that the drag coefficient here is not based on the frontal area of the towing vehicle but on that of the combination. This frontal area is slightly greater than that of the trailer alone because the chassis of the car projects beyond its silhouette, as is demonstrated by Figure 4.176.

Configuration	Towing car	c_{D1} $c_{D(1+2)}$	$A_x \cdot c_{D1}$ $A_x \cdot c_{D(1+2)}$
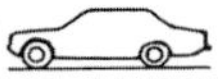	Opel Rekord C	0,452	0,87 m²
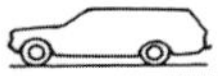	Opel Rekord C Caravan	0,435	0,84 m²
	UNICAR	0,240	0,48 m²
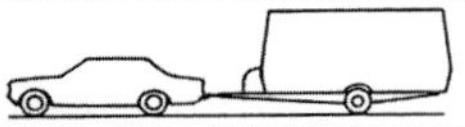	Opel Rekord C	0,764	4,10 m²
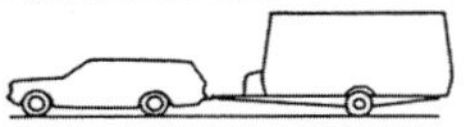	Opel Rekord C Caravan	0,864	4,63 m²
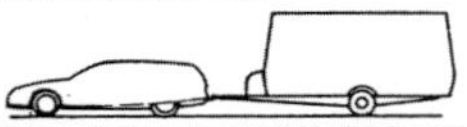	UNI-CAR	0,743	3,98 m²
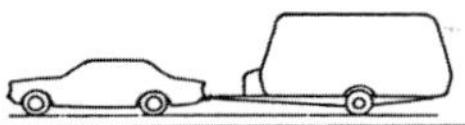	Opel Rekord C	0,605	3,24 m²
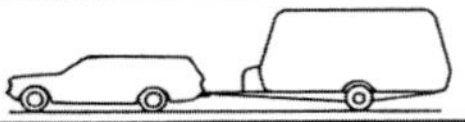	Opel Rekord C Caravan	0,562	3,01 m²
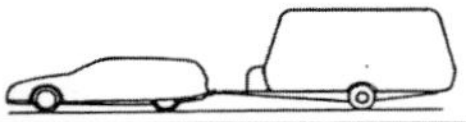	UNI-CAR	0,581	3,11 m²

Figure 4.175 Aerodynamic drag of car/trailer combination with different trailers [475].

The drag of trailers can be considerably reduced by the measures described for commercial vehicles in chapter 10. Here, it is advantageous if the required shape changes necessary to reduce trailer drag—at least in a first approximation—are "universally" applicable, that is, they are not restricted to a particular towing vehicle. One proven aid, as with buses, is the rounding of front edges. However, in order to limit manufacturing cost, this measure is almost exclusively limited to the junction between the front panel and the roof and not applied to the side-wall junctions. The associated increase in "nose-up" pitching moment relieves some of the load on the trailer coupling, which reduces the combination's stability in the case of side to side oscillations.

An example of the effectiveness of round front edges on a trailer behind a car is shown in Figure 4.100. Here, only the frontal area of the trailer is used as the reference variable. Other possibilities for reducing drag are boat-tailing at the front and rear. According to Künstner, a combination of all these measures can be expected to reduce the drag coefficient of a car/trailer combination from today's figure of $c_D = 0.6$ to $c_D = 0.4$.

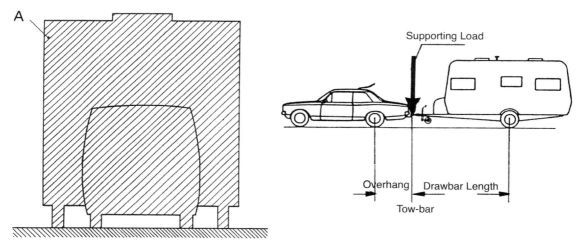

Figure 4.176 Important geometric quantities concerning drag and stability.

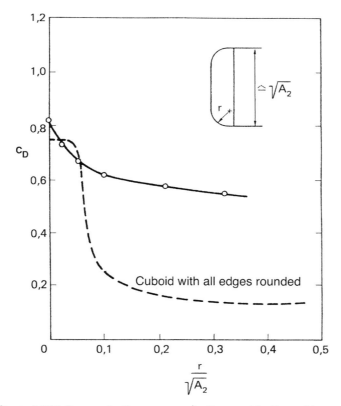

Figure 4.177 Drag reduction on a car/trailer combination with round front edges of the trailer; Waters [852].

It is also possible to improve the airflow around a car/trailer combination by modifying the flow around the towing vehicle. As shown in Figure 4.178, the use of an air deflector has a clear parallel with the semitrailer tractors considered in chapter 10. Without the air deflector, a stagnation point is formed on the face of the trailer, and the flow separates at the front edge of its roof. With a deflector, the air can be guided so that it flows tangentially to the roof, as if the roof's front edge were well-rounded. There is a minimum drag for this particular case, as shown in Figure 4.179. Here, the drag coefficient of the car/trailer combination, $c_{D(1+2)}$, is drawn versus the angle of attack of the deflector. The minimum drag can be observed at $\alpha = 28°$.

Figure 4.178 Improvement of the flow around a car/trailer combination either by rounding the roof the trailer's roof edge or by applying an airfoil to the car's roof; Künstner [475].

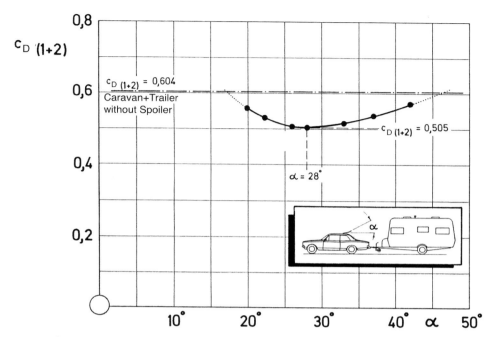

Figure 4.179 Evaluation of the best wing position and angle of attack [475].

However, the fact that even this configuration has an adverse effect on the dynamic stability of the combination was discovered by Peschke and Mankau [635] through wind tunnel measurements with full-size vehicles and road tests. As shown in Figure 4.180, the reduction of drag with the use of a deflector is associated with a "nose-up" pitching moment; with increasing vehicle speed, the drawbar load (vertical force exerted by the trailer's shaft on the car's hook) is reduced.

The negative effects of this reduced drawbar load on the dynamic stability of car/trailer combinations have been demonstrated by Zomotor et al. [906]; Figure 4.181 shows the relationships. A pendulum-type swinging motion caused by a disturbance—such as a sidewind gust or an uneven patch of road—only attenuates if the damping constant $D > 0$. However, D decreases in direct proportion to decreasing road speed, and the smaller the drawbar load, the lower the speed at which D passes through zero. In the selected example, the damping constant falls below $D = 0$ at only 120 kph (75 mph) if the drawbar load is zero: On the motorways of some European states, this speed is permissible for car/trailer combinations.

	c_D Vehicle+ Trailer	Support Load Difference in N at 80 Km/h
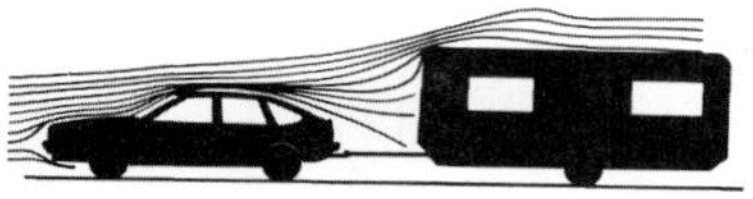	0,53	−340*
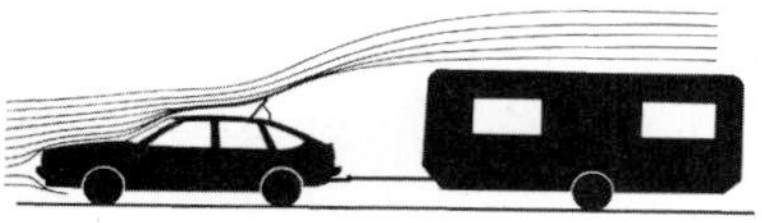	0,45	−315*
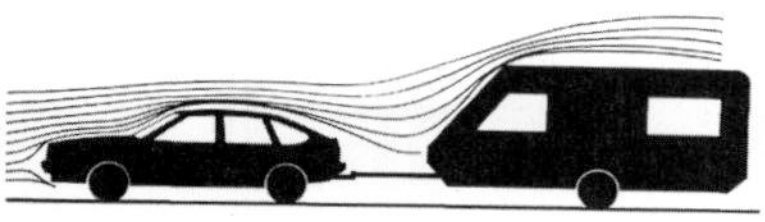	0,53	±0

Figure 4.180 Effect of the flow topology on drag and drawbar load of a car/trailer combination; after Peschke & Mankau [635].

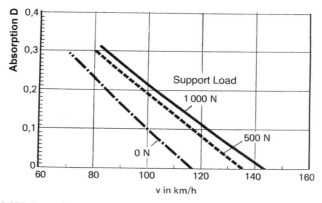

Figure 4.181 Damping constant D of the pendulum-type swinging motion of the trailer over the velocity, evaluated for different drawbar loads (measured in case of nonmoving car) [906].

Instead of a deflector on the roof of the towing vehicle, a chamfered edge on the trailer is more favorable (see Figure 4.180). Although this modification does not reduce drag compared to a vertical front, the pitching moment is reduced to zero and so the drawbar load is independent of vehicle speed. However, this shape results in an reduced interior volume.

Interference between the towing vehicle and the trailer also sometimes leads to large side forces as well as yaw and roll moments that impact the trailer in particular and are partly transferred to the towing vehicle via the drawbar. Under unfavorable wind conditions, this can lead to dangerous driving situations. The aerodynamic relationships are explained in more detail next for the two principal driving situations which occur, namely:

- Side wind in free incident flow
- Large vehicles while overtaking a vehicle combination

As shown in Figure 4.182, yaw moments turning away from the wind are caused on the towing vehicle and the trailer, with the yaw moment of the trailer opposing the yaw moment of the towing vehicle via the pivot point of the trailer hitch, thus turning the towing vehicle into the wind, causing a stabilizing effect for the combination.

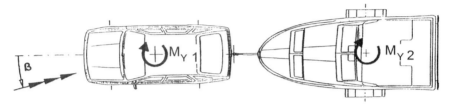

Figure 4.182 Moments acting on towing vehicle and trailer in side wind.

From the yaw velocity measured against time, it can be seen that strong dynamic effects occur. It is noticeable that the yaw velocity of the towing vehicle changes to the sign that is the opposite of the trailer. Initially, the vehicle is rotated away from the wind; shortly thereafter it is rotated into the wind. When the gust subsides, this process is repeated in reverse. This behavior is significantly influenced by the coupling overhang.

In addition to the yaw moment, the roll moment also plays a major role with regard to the driving stability of the vehicle combination. In very strong side winds, such as in the experiment shown in Figure 4.183, especially when driving on bridges, this can cause the trailer to lift the windward wheel and even overturn. For this reason, bridges are sometimes closed in strong winds for trailer combinations. Aerodynamically, this problem can hardly be solved, since it is caused essentially by the large side attack surface of the trailer.

Figure 4.183 Demonstration of the risk of overturning for an unloaded campervans when passing a side wind system.

The stability and controllability of vehicle combinations is also negatively affected when they are overtaken by large vehicles. This has been revealed by tests conducted by Kobayashi & Sasaki [446]. The yaw moment curves of both the towing vehicle and trailer show similar behavior, in which an initially positive yaw moment turns into a negative yaw moment over the entire passing maneuver.

The cause of this behavior can be illustrated by the pressure distributions. As also described in section 4.5.4.4 with regard to passing maneuvers for a single vehicle, the stagnation in front of the overtaking vehicle produces a "nose wave" that manifests itself in higher pressure in the rear of the towing vehicle, resulting in a positive yaw moment (PI). At the same level of both vehicles (sub-figure b), the positive pressure acts more in the front section, thus producing an opposing moment, which is amplified by the accelerated flow, which in turn results in lower pressures in the rear section (PII). When the truck has almost passed the towing vehicle (sub-figure c), the pressure in the front section drops and the influence of the gap between the towing vehicle and the trailer causes a weak negative yaw moment (PIII).

As shown in Figure 4.184, the distance l_1 between the two vehicles has a much greater influence on the yaw moment curve than the gap l_2 between the towing vehicle and the trailer. This can be explained by the higher velocities and thus lower pressures for smaller distances.

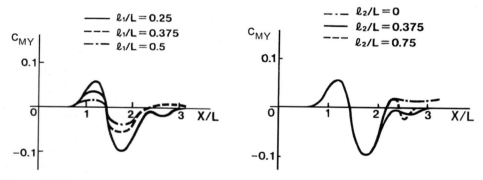

Figure 4.184 Influence of l_1 and l_2 on $c_{M,y}$, according to Kobayashi & Sasaki [446]: l_1: lateral distance between the vehicles; l_2: distance between towing vehicle and trailer.

Using simple models, the mutual influences of the two bodies can be determined by variations of the side force and the distance of the aerodynamic force point of attack to the center of gravity. According to these findings, it is recommended that aerodynamic components be developed that reduce the maximum PII on the towing vehicle and trailer by appropriate manipulation of the flow. Good results have been demonstrated in this regard by side-mounted spoilers (sub-figure d). The pressure at the tail of the towing vehicle becomes positive, as opposed to the negative pressure without add-on parts. In addition, the point of attack of the aerodynamic force moves closer to the towing vehicle's center of gravity. The function is similar when mounted on the trailer, but even slightly more efficient in terms of reducing the yaw velocity of the vehicle combination. Although a practical implementation seems difficult and an increase in the drag coefficient is also to be expected, a significant improvement in stability would be achieved.

4.5.4.3 Driving in Convoy

One possible way of making better use of limited traffic space is to group vehicles into convoys. The electronically regulated nose-to-tail distances between vehicles become so small (safety aspects are not considered here) that aerodynamic interference is active between the vehicles. The fact that slipstreaming can provide a considerable reduction in drag was first exploited in motor racing (see Romberg et al. [680]). For commercial vehicles, this effect has been analyzed in detail by Götz [291] and is described in chapter 10. Since cars have a comparatively low drag coefficient, their slipstreaming effect is less marked but still significant.

This fact has been demonstrated by Ewald [237] through measurements on reduced-scale models. An extract from this work is shown in Figure 4.185, which first examines "trains" of only two vehicles having the same drag coefficient. The drag coefficients of these vehicle pairs were varied to cover the full drag range of current and possible future cars. Ultra-low-drag cars, on the left in Figure 4.185, behave in exactly the same way as the two "convoyed" slim-profile models: The drag of the leading vehicle is significantly reduced by the approaching follower, while the drag of the follower increases. In contrast, cars with high drag, on the right in Figure 4.185, behave more like lined-up circular cylinders (see Figure 4.42). Both vehicles benefit when the distance between them is very small, although the rear vehicle benefits more than the leader. With increasing distance, only the drag of the rear vehicle is reduced, as it drives in the wake (slipstream) of the leading vehicle.

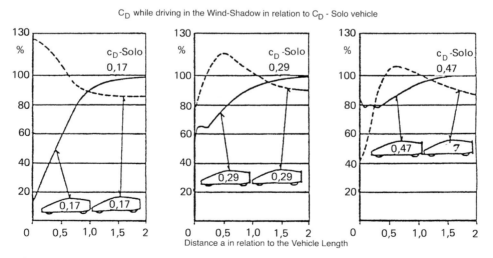

Figure 4.185 Effect of the distance between convoyed cars on aerodynamic drag [237].

The mechanism of interference has been explained using pressure distribution measurements; Figure 4.186 gives the result for the most extremely streamlined design. In each case, the differential pressure relative to the isolated vehicle is plotted; the relative change in drag is also specified. The pressure on the base of the leading car is raised by the follower; as a result, its drag is reduced. At the same time, the pressure on the nose (vertical front panel) of the follower is reduced but that on its slanted front panel (engine hood) increased, resulting in a drag increase for decreasing distance.

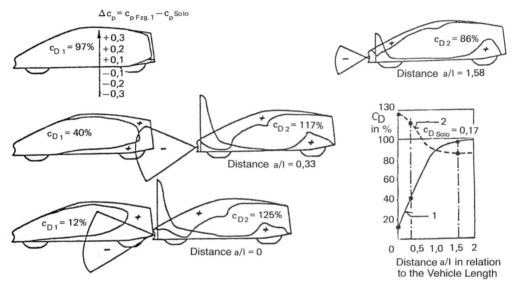

Figure 4.186 Different pressure distribution and aerodynamic drag during convoy drive [237].

Extrapolation of measurements on three-vehicle convoys to a "train" of ten vehicles is shown in Figure 4.187. With a nose-to-tail distance of one vehicle length, the improvement in drag is (on average) by 20%. Here, even for cars with an extremely low drag coefficient of $c_D = 0.17$, convoy driving would lead to significant savings in fuel consumption.

How much fuel can really be saved in a convoy has been determined by Michaelian and Browand [570] in road tests. The tests included convoys made up of two, three, and four identical cars (Buick LeSabre). Figure 4.188 provides a summary of the results. This shows the fuel consumption b/b_{solo} (referring to the fuel consumption b_{solo} of the vehicle moving alone) against the distance between the vehicles. For all configurations—number of vehicles, distance between them—the fuel economy improves the smaller the distance is. The greatest benefit from driving in a convoy is gained by the vehicles in the middle. Next are the vehicles at the rear, and the smallest gain is for the lead vehicles.

Aerodynamic Forces and Their Influence on Passenger Vehicles

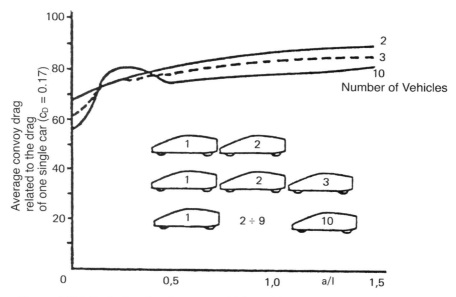

Figure 4.187 Reducing drag in convoy driving with more than two vehicles [237].

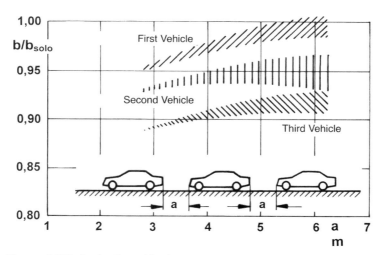

Figure 4.188 Reduction of fuel consumption when driving in a convoy, determined in road tests [570].

Yaw moment and side force are also dependent on the position within the convoy. The lead vehicle encounters the largest side force—it is of the same order of magnitude as in a single vehicle. In the trailing positions, the side forces are smaller, an effect that can be attributed to the shrinking local yaw angle.

4.5.4.4 Overtaking

The aerodynamic forces and moments exerted on a vehicle that is overtaken depend on several parameters. In addition to the lateral distance between the two vehicles from each other, the size and nose shape of the overtaking vehicle are important influencing parameters.

Drag and side force reach a maximum as soon as the nose of the bus is at the same level as the tail of the vehicle being overtaken. The side forces are defined essentially by the stagnation at the nose and the channel flow between the bus and the car. The stagnation at the nose causes a significant pressure increase in the front area of the car, whereas the channel flow between the bus and the car results in an accelerated flow and thus in a decrease in pressure. The resulting yaw moment reaches its maximum when the nose of the bus reaches the center of the vehicle being overtaken. At this point in time, the overtaken vehicle has a tendency to turn away from the bus. The strong changes in lift that also occur in this phase may cause a critical driving condition.

The results shown so far have been measured values obtained under static conditions. At low relative speeds, the dynamic effects are small so that the (transient) process of overtaking can be investigated using static positions. In contrast, at a large relative velocity there is a significant increase in yaw moment and side force compared to the value under steady conditions.

Generally, the shape of the vehicle being overtaken plays a role, as well as the size. The maximum amplitudes of yaw moment and side force are reduced by the rounding of the front of the vehicle. This finding is also used in the design of modern high-speed trains. Owing to extremely slender nose shapes, the "head wave" emanating from them is maintained as flat as possible to keep the pressure load of the windows at the nose within limits when trains pass each other and when a train enters a tunnel.

4.6 The Aerodynamic Development Process

A machine is usually developed in two stages: a computational design is produced initially. This is followed by experimental verification, which in turn is used to improve the computational model. If necessary, modifications are deduced from the test results, which are then computed and verified again by experiments. Frequently, this loop is performed several times. For complex machines, such as turbines or aircraft, this iterative procedure starts with the components and ends with the complete product. Here, the aim is to perform as many steps of the iteration as possible at the component level and to work on them in parallel, preferably numerically.

The workflow of aerodynamic product development is highly complex. All manufacturers plan their working stages as early as the beginning of the product development process, and aerodynamic support of the series only ends when production ends.

Aerodynamic development is integrated into the product development process of the entire company. Figure 4.189 shows such a product development process. Research and predevelopment (e.g., extendable rear spoiler [Audi], cooling air louver [BMW], air cap [Daimler]) are often disconnected from specific products. Consequently, the results may inform product development at the beginning of new projects but may also unavoidably become part of the process at later points in time.

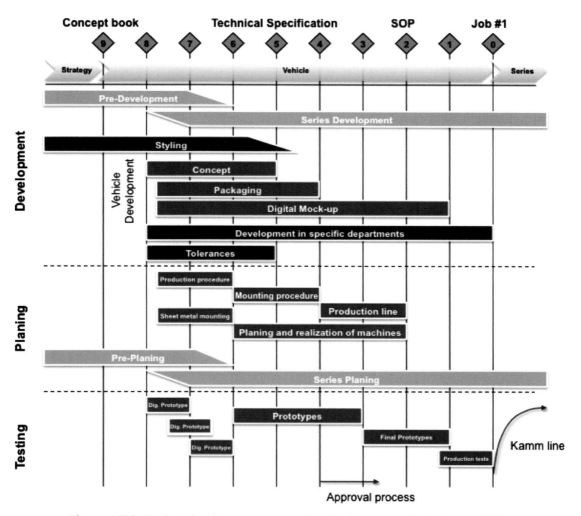

Figure 4.189 Product development process (Pep) in the automotive industry [837].

4.6.1 Goal Definition

A car is not designed according to the laws of fluid mechanics, as would be a turbine, a compressor, or an aircraft. Its shape is in fact determined by the styling. Of course, this takes into account the requirements of packaging, safety, and many other regulations. However, aesthetic arguments are dominant. Aerodynamics has the task of determining the aerodynamic drag (and also the other components of the resultant aerodynamic force and aerodynamic moment) of this design and—if it does not meet the performance specifications—to develop proposals and coordinate them with the styling.

It should be noted that the frontal area of cars in all classes grows from one model generation to the next. This is due to the fact, first, that people are getting taller by approximately 1 centimeter every ten years, and, second, because safety (such as side air bags) requires more installation space. And finally, customers want more comfort, which also includes a sense of space.

Specific targets are typically derived from a specification of the aerodynamic driving resistance. It is not always possible to compensate the increase in frontal area A_x with a lower c_D value at least to the extent that $c_D \cdot A_x$ remains constant, or even to achieve a significant reduction in the driving drag. The VW Golf may serve as an example (see Table 4.3), but it should be noted that the Golf has continued to grow a little into the next higher class with every new generation. The table also shows that the lift of the vehicles has dropped significantly, which is conducive to better handling. The same applies to reduced yaw moments.

Table 4.3 Development of Some Characteristics of the VW Golf [846]

Golf	c_D	$A_x/[m^2]$	$c_D \cdot A_x/[m^2]$	c_{Lf}	c_{Lr}	P/[kW]	Tires
I	0.42	1.83	0.77	0.08	0.11	37	155R13
II	0.35	1.89	0.67	0.04	0.06	40	175/70R13
III	0.34	1.98	0.67	0.03	0.11	44	175/70R13
IV	0.33	2.11	0.69	0.03	0.10	55	195/65R15
V	0.32	2.22	0.72	0.02	0.09	55	195/65R15
VI	0.31	2.22	0.69	0.02	0.09	59	195/65R15

The pressure of competition is also frequently a driver for improved aerodynamic characteristics. An important task at the beginning of product definition is therefore a careful analysis of the competitive situation. Owing to the difficulty of comparing different wind tunnel measurement results (chapter 13) and the partially "airbrushed" press releases regarding c_D values of new vehicles, it is necessary to analyze interesting competitors under similar conditions in the same wind tunnel. The influence of innovative components (e.g., underbody components, mirrors, cooling air guidance, etc.) is frequently investigated. Figure 4.190 illustrates the effort associated with such benchmark studies using the example of the upscale SUV class.

Figure 4.190 Competitor analysis in a wind tunnel.

4.6.2 Project Milestones and Tools

The aerodynamic development of vehicles integrates into the general product development process (Figure 4.189). An example aerodynamic development process during product development and production development is shown in Figure 4.191.

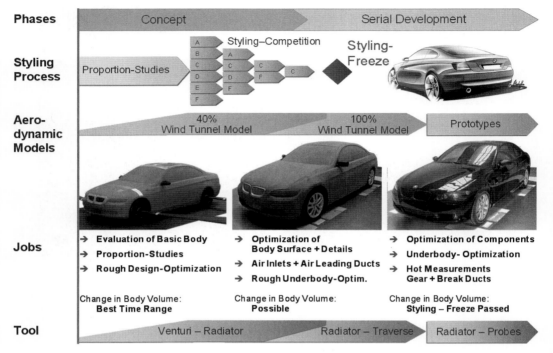

Figure 4.191 BMW AG Aerodynamic development process.

Brainstorming takes place at the beginning of a new series. Especially in revolutionary and innovative vehicle concepts, the styling is given a leading role. The response from the market and the public to new vehicles is tested very successfully with a show car at a motor show, but this is initially disconnected from other technical requirements and often far ahead of their assessment. Figure 4.192 shows how new styling ideas are sketched and finalized, based on the design study of the Pikes Peak from 2003. It was the harbinger of the Audi Q7, the first SUV of the brand which was launched in 2005.

With the beginning of the technical product definition, a rough package (main dimensions) must first be developed and a number of concept-determining modules (cooler package, axle design, and exhaust system) must be designed; the shaping of a car is also begun, based on initial styling proposals. Frequently, several designs are examined in parallel at the start of vehicle development, and sometimes external designers are invited to submit proposals. Either a design study or a predecessor vehicle is used as template, as mentioned previously. In the latter case, the design development can be understood more as an evolution of the familiar rather than a revolution.

In this way, several variants may result on paper, on screen, or as models. For these, the aerodynamicist has to

- explore the potential to achieve the drag coefficient given in the specifications; and
- ensure that none of the designs results in unacceptable aerodynamics.

Aerodynamic Forces and Their Influence on Passenger Vehicles

Figure 4.192 Product creation and first styling of a vehicle on sketches; design study as a demonstrator for the Audi Q7 (Pikes Peak) as an example.

If the aerodynamicist is fast enough, it is possible to go through this process between styling and aerodynamics iteratively even at this early stage. One example of this is shown in Figure 4.193, which has been reported by Dietz et al. [198] from the development of the Audi A4. Each bar represents the result of the aerodynamic development step, which was reached with numerous changes in details. In the subsequent stylistic revision, some of the improvements were lost again; by further optimization, the previously achieved state had to be restored or the situation had to at least be improved again if possible.

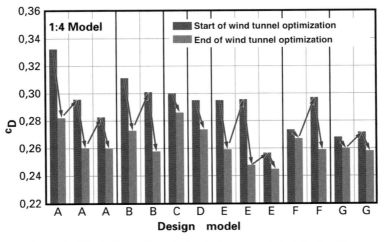

Figure 4.193 Optimization of the designer models A to G of the Audi A4 with models at 1:4 scale [198].

385

At the beginning of design discovery, there are often five to ten variants under discussion. For a long time, purely experimental methods were used on scale models made from plasticine or foam, because they can be modified and converted relatively easily and quickly. And since they require only a small wind tunnel for their study, this approach is also cost-effective. But this is fraught with risks, as described in chapter 13. The target for the c_D value of the scale model must always be calculated from the value specified in the performance specifications. An example is given in Table 4.4; it comes from the development of the Audi A2, about which Dietz [196] has reported.

Table 4.4 Determination of the Goal for the c_D Value of the 1:4 Model	
Specification for the vehicle	$c_D(1:1) = 0.250$
Cooling air	$\Delta c_{D,C} = 0.015$
Two exterior mirrors	$\Delta c_{D,S} = 0.015$
Joints, gaps	$\Delta c_{D,F} = 0.010$
Goal for the 1:4 model	$c_D(1:4) = 0.210$

The aerodynamic measurements are carried out on a full-scale model at an advanced stage of development. The prerequisite is that the design model is based on a realistic chassis. To save time in modeling, 1:1 models are often shaped with different right and left sides, which makes them almost useless for wind tunnel tests. Moreover, designers do not like to let their 1:1 models out of their hands. But copying is still a lengthy process. And since working with 1:1 models is time-consuming and expensive, it is very difficult to obtain the required data at each stage of development.

Therefore, the aerodynamics are increasingly analyzed using numerical methods (CFD) during the brainstorming phase. As shown in detail in chapter 14, some of the commercially available codes are so advanced, as comparative calculations have shown, that at least the c_D value of a car can be determined with an accuracy of ±1%. The fact that lift and pitching moment are predicted less accurately indicates that the flow details are still not reproduced accurately enough, so that the results of the calculation are subject to a degree of risk.

One advantage is that numerical computation gives very detailed information about the spatial flow field around the vehicle and, in particular, on its surface, that is, data which is determined experimentally only with great effort. The aerodynamicist is in a position to make modifications to the shape in a deterministic manner. The fact that the optimization process—the interaction between shape modification and computation of the ambient flow—can even be carried out automatically in a closed loop has been demonstrated initially by Singh [763]. This makes it possible to explore the aerodynamic potential of several design variants before the model decision.

After some time, a design is selected, after which only one or two design variants are considered further. The outer contours of these variants become more and more mature. Therefore, 1:1 models (from plasticine, foam, or fiberglass) are now used for the aerodynamics and the optimization moves from rough estimates to increasingly greater detail. Optimization at this stage focuses on air intakes, the total cooling air path, and/or the underbody design, but also on wind noise and minimizing soiling.

Prototypes are built after the "styling freeze" milestone. In addition to aerodynamic optimization of the underbody assembly and other components, the aeroacoustic design and optimization of vehicle soiling are now also becoming increasingly important. In particular, optimization of various add-on parts, especially the exterior mirror, is pursued here. Prototypes provide the opportunity to verify the result reached so far and, if necessary, to develop corrections that can still enter series production. Finally, vehicles from pre-production and the initial series production must be checked. Sometimes, small deviations from the prototype, such as additional sheet metal, may mean that the results obtained during development cannot be reproduced in the production vehicle.

Figure 4.194 shows the development of the aerodynamic drag coefficient during development of the Audi Q5. The foundation is laid for aerodynamically optimal tuning during concept and design discovery. During this time, the drag coefficient was reduced by 65 counts. Further improvement by 25 counts was then obtained on prototypes and pilot series vehicles.

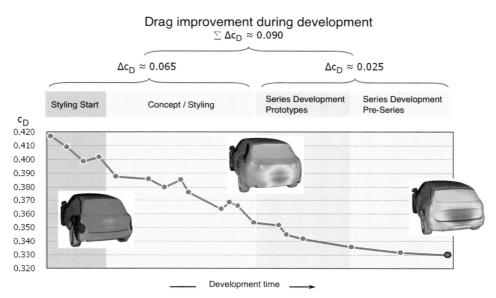

Figure 4.194 The development of the drag coefficient during product development of the Audi Q5 [835].

During the development process, a number of interfaces on a wide variety of topics arise with the other development disciplines and specialist areas. The frequency of contact with other departments is estimated in Figure 4.195. Main interfaces are coordination with styling (exterior shape) and concept development (dimensions, package). But there are also points of contact with chassis development (axle design, exhaust system, driving level, brake cooling), power unit cooling development, and body development (seals, exterior mirrors). Overall vehicle development is responsible for setting fuel consumption targets and for overall acoustic tuning. Cooperation with quality assurance is mainly based on ensuring high aeroacoustic quality during production up to end of production (EOP).

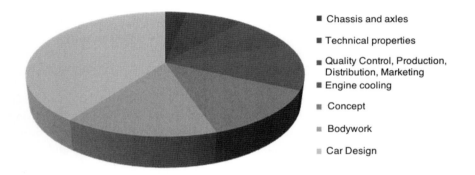

Figure 4.195 Interfaces within the company for aerodynamic development [506].

Even this brief description of the development process reveals the importance of cooperation between styling and aerodynamics. The aerodynamicist is the more likely to succeed, the more he knows the mindset of the designers—without wanting to become a designer himself.

4.6.3 Examples

Figure 4.196 from the development of the Audi A2 provides insight into one shape optimization step. In the 3-liter (consumption) version of the vehicle, the performance specification required $c_D = 0.25$, a c_D value that had to be realized by all means, otherwise the fuel consumption goal would not have been achieved. Shown here are the c_D value (measured without cooling airflow) and the lift coefficient at the rear axle, c_{Lr}, by number of the series of measurements. A total of 80 modifications were studied on the full-size model. Even the smallest improvements were considered. For example, smoothing the outside of the tires (the brand logo, information about the size, etc.); this work was "rewarded" by a drag gain of $\Delta c_D = 0.007$. The lowest c_D value, $c_D = 0.218$, was achieved with the measurement series #66, which also demonstrated the lowest lift at the rear axle, $c_{Lr} = 0.102$.

Aerodynamic Forces and Their Influence on Passenger Vehicles

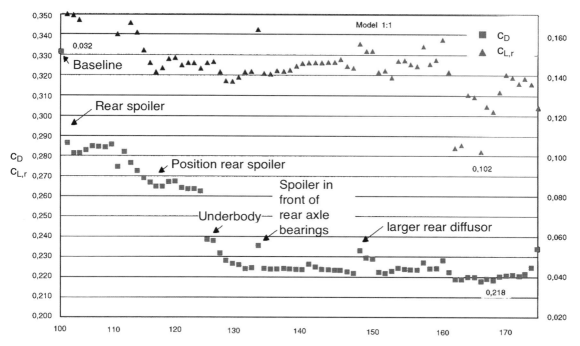

Figure 4.196 Optimization of drag and lift (at the rear axle) of the Audi A2 on a full-size model [196].

To avoid starting from scratch with every new development, aerodynamicists have come up with so-called "hard points," derived from parameter studies, whose dimensions must be strictly adhered to for a certain basic shape of the vehicle. The result of the full aerodynamic treatment of the DaimlerChrysler E-Class W 211 is summarized and compared with its predecessor, W 210, in Table 4.5. The optimization up to the current E-Class (W212) was achieved by further detail optimization while maintaining these hard points (Nebel et al. [602]).

Table 4.5 The Major Aerodynamic Data of the Mercedes-Benz E-Class, Current Vehicle W 212, Predecessors W 211 (MY 2002) and W 210 (MY 1995)

	W 212 (MY 2009)	W 211 (MY 2002)	W 210 (MY 1995)
c_D	0.25–0.27	0.26–0.28	0.27–0.29
Front face A_x/[m²]	2.32–2.33	2.21–2.23	2.16–2.17
Drag surface area $c_D \cdot A_x$ [m²]	0.58–0.63	0.57–0.62	0.58–0.63
Front axle lift c_{Lf}	0.06	0.09–0.12	0.07–0.12
Rear axle lift c_{Lr}	0.08	0.09–0.10	0.10–0.13

However, when it comes to reducing drag, it is not just about the "global" shape. Improvements also have to be achieved using unobtrusive measures. The more so as the models "delivered" by the styling department are often already of remarkable aerodynamic quality. Following are three examples.

For the Porsche Boxter, it was possible, by using the extendable rear spoiler, to reduce the lift coefficient on the rear axle, c_{Lr}, by 31%, accompanied by a c_D reduction by 4%: $c_D = 0.31$, $c_{Lf} = 0.13$; $c_{Lr} = 0.10$. Using front wheel spoilers, the c_D value of the Ford Mondeo could be reduced by 0.002. It was found that this measure was correctly detected only with moving ground and rotating wheels. With rigid ground and stationary wheels, the drag reduction was 0.005. By parallel "pulling out" of the wheel covers from the wheel disk, the c_D value of the Audi A6 could be reduced by 0.01. It was possible to ensure, with the selected design, that the wheel cover will not be scratched if the wheel comes into contact with the curb.

4.7 Drag and Lift of Passenger Cars in Production

In this section, we analyze the aerodynamic forces of real production vehicles. A comparison of the aerodynamic forces of different vehicles reveals a wide spread, which is mainly due to individual technical targets of the manufacturers, number, and quality of their tools used during development.

Conventionally, a single value of each, drag and lift, coefficient is quoted for any vehicle type. In reality, however, a type covers a wide range of c_D and c_L values because some of its features, such as vehicle concept, engine power, tires, spoilers, ground clearance, and so forth, can vary greatly. Similarly, the state of the vehicle—such as windows and sunroof open or closed—has an effect on drag. Convertibles represent a special case. With the roof up, their shape differs more or less from the basic passenger car; with the roof down, the airflow pattern is totally altered.

4.7.1 Overview of Competitors by Vehicle Class

Different objectives in the development of one vehicle to another naturally result in a wide spread of achieved aerodynamic characteristics. Figure 4.197 includes the value pairs of c_D and A_x for a total of 79 vehicles of different vehicle classes. The class names commonly used in the VW Group are used here: A0 stands for small car, A for compact car, B for mid-range, C for upper mid-range, and D for luxury car. Value pairs belonging to a market segment are each marked with a representative field.

The fields associated with the vehicles of the A0, A, B, C, and D segments lie approximately on an imaginary driving resistance hyperbola $c_D \cdot A_x = $ const., that is, the larger—especially longer, "drop-shaped"—vehicles compete with the increased frontal area by an improved drag coefficient.

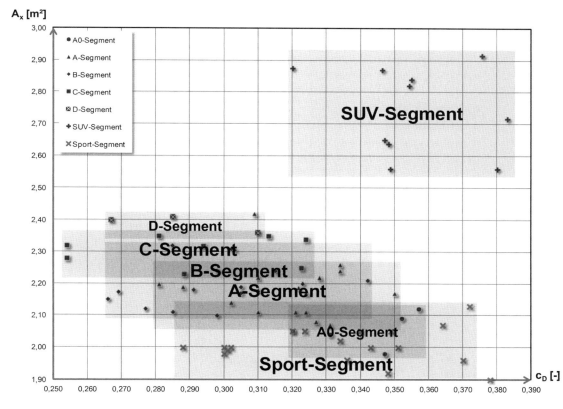

Figure 4.197 Measured aerodynamic drag values for $N = 79$ vehicles of different market segments, presented on the c_D-A_x chart.

In the sports segment, there is a larger c_D range in which the vehicles are located. The main reason is that some cars are designed specifically for downforce, which is associated with an increased induced drag. This does not apply to cars from other segments, so better c_D values are the result. Vehicles in the SUV segment generally have a much larger frontal area. In addition, their c_D value is slightly higher compared to sedans because of the larger wheels, an aerodynamically unfavorable axle, and a more ragged underbody.

Considering the lift values of the vehicles in Figure 4.198, each vehicle class is in a comparable c_L range for its frontal area. This follows from the development boundary conditions of "minimized c_D value," which, according to section 4.3.3.1, leads to slightly positive lift that is, however, limited due to driving dynamics. Sports cars are an exception again, similarly for the reason already mentioned that in this segment some vehicles are designed for noticeable downforce.

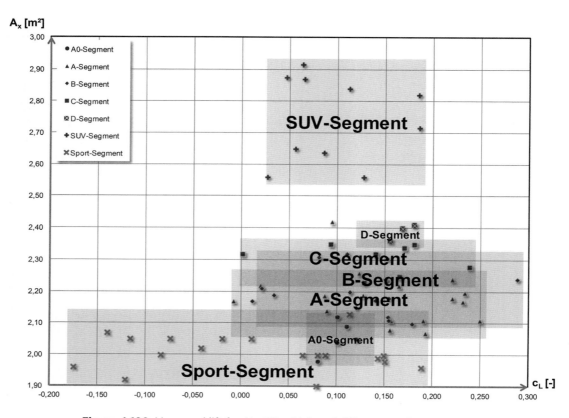

Figure 4.198 Measured lift for $N = 79$ vehicles of different market segments, presented on the c_L–A_x chart.

4.7.2 Drag Surface Area $c_D \cdot A_x$

Cars have a tendency to "grow." As already documented by Figure 3.2: Not only their weight, but also the frontal area usually increases from one generation to the next.[21] The fact that aerodynamicists struggle to counteract the growth in frontal area by decreasing c_D values is evident from the product ranges of three car manufacturers, which are summarized in Figure 4.199.

---

21. The S-Class falls out of this trend in Figure 4.199; the reason is that the size of the 1993 model was only cautiously accepted on the market, so that its successor in 1998 was again a little smaller.

Aerodynamic Forces and Their Influence on Passenger Vehicles

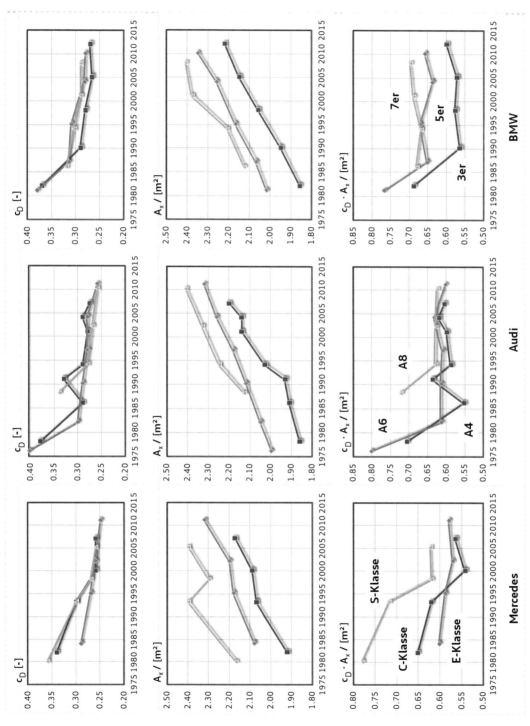

Figure 4.199 Development of the c_D value, frontal area A_x, and "drag surface area" $c_D \cdot A_x$ of sedans from Audi, BMW, and Mercedes in the mid-, upper-mid-, and luxury segments.

Audi no longer changed the drag surface area, $c_D \cdot A_x$, from one generation to the next; for Mercedes, a descending trend can be observed, which is all the more pronounced the greater the drag surface area of the preceding generation was. At BMW, the continuous improvement of the c_D value did not compensate for the steep increase in frontal area. Let us again consider the example of the VW Golf (cf. Table 4.3).

From the overview in section 4.7.1, it is possible to calculate an average drag surface area across the vehicle categories. This is shown in Figure 4.200. The impression from Figure 4.197 is confirmed that, with the exception of the SUV class, all vehicles have approximately the same $c_D \cdot A_x$ value.

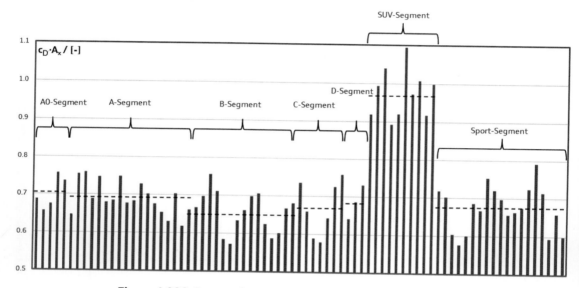

Figure 4.200 Drag surface area $c_D \cdot A_x$ of various current vehicles, ordered by vehicle category.

4.7.3 Intercomparison in Accordance with EADE

From time to time, the EADE (European Aero Data Exchange) association conducts so-called intercomparisons. This process was in fact designed to review the comparability of various European vehicle wind tunnels. The c_D and A_x values of European cars, coupes, and SUVs on the market are determined in different wind tunnels and compared. However, this data also forms a very meaningful database to track the development of aerodynamics over the years. A comparison with the data from 1991 suggests a further decrease in the c_D value. But it is not sufficient (on average) to compensate for the increase in frontal area A_x. The average aerodynamic drag of passenger cars has actually increased over the years (cf. Figure 4.201). It can be seen that the c_D value decreases slightly with increasing vehicle size in such a way that, on average, the drag surface area $c_D \cdot A_x$ is approximately the same for all cars. This is consistent with the findings in the previous sections.

Aerodynamic Forces and Their Influence on Passenger Vehicles

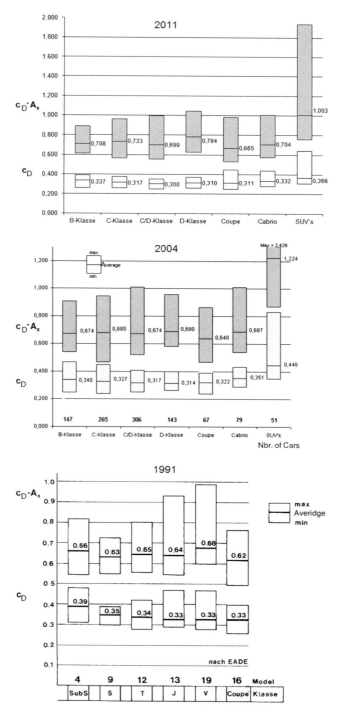

Figure 4.201 c_D values, frontal areas A_x and $c_D \cdot A_x$ of European cars, EADE data, current, from 2004 [286] and 1991 [528].

"All-terrain" vehicles, so-called SUVs, play a special role. At $c_D = 0.37$, their mean drag coefficient is at the level that was reached by cars in the mid-1970s—at a time when the optimization of details (according to the definition in section 1.9.1) had not yet been adopted by all manufacturers. Along with their large frontal areas, SUVs have a drag surface area that, at $c_D \cdot A_x \approx 1.0$, is almost 50% greater than that of traditional passenger cars. A striking characteristic is the wide distribution of their c_D values. This is due to the fact that some SUVs are styled to be square-edged, based on older models such as, for example, the Land Rover Defender (1993) or the Mercedes-Benz G400 (2001). In contrast, the models styled similar to passenger cars have coefficients of $c_D < 0.40$, that is, rather close to those of cars.

4.7.4 Influence of Vehicle Concepts

The rough concept of a new vehicle already determines its aerodynamic characteristics to a large extent. For example, section 4.5.1.3 identified the importance of the tail shape for the aerodynamic drag: The drag of squareback vehicles is greater than that of notchback and fastback cars; for lift—in particular at the rear—the situation is reversed. Owing to the carline, convertibles with a fabric soft top have a less favorable roof contour, because it is separation-prone, compared to coupes, and sports car aerodynamics often focus on generating downforce at the expense of low drag. For details, see section 6.2 and chapter 9. The drive concept and the off-road capability of SUVs also have an effect on the aerodynamic characteristics of the vehicles. These points will be discussed next.

4.7.4.1 Drive Concept

The architecture of the body is largely determined by the drivetrain selected for the vehicle. This consequentially implies the position of the engine, the torque-speed converter, the input, prop-, and drive shafts, as well as of the exhaust system and the fuel tank. If these components run from the front to the back through the vehicle, the underbody structure is substantially rougher than when this is not the case. In addition, this predetermines which of the two axles will be driven and where air has to be taken for cooling purposes from the underbody flow. An overview of the common drive trains is provided in Figure 4.202.

The front drive concepts are shown on the left. Both transversely (e.g., VW Golf) and longitudinally installed engines (Audi front-wheel drive from mid-range cars and up) deliver the same picture from an aerodynamic point of view. Since only the front axle is driven, no prop shaft to the rear end is required. Hence, the prop shaft tunnel can be omitted and only the exhaust system has to be routed to the rear. Large-area underbody fairing is possible and adequate cooling only has to be provided for the exhaust pipes and the silencers. In particular, fairing of the very rough area around the rear axle is possible, because it is not driven. This provides for a low-vortex incident flow to the rear end, which can then be shaped as a very effective diffuser. If necessary, even a fully faired underbody can be realized, if the underbody elements are provided with heat shield plates in the area of the exhaust system. All these measures contribute to low aerodynamic drag.

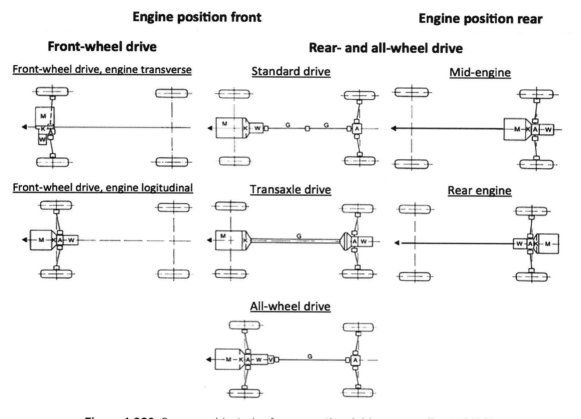

Figure 4.202 Common drivetrains for conventional drives, according to [496].

The front-wheel drive system must put the entire traction force onto the road through the front axle, so—compared with rear wheel drive—traction imposes stringent requirements in terms of front axle lift. Moreover, the center of gravity in front-wheel drive vehicles is more to the front than in rear wheel drive cars, especially those with a rear engine. The side force in oblique incident flow therefore usually causes an opposing yaw moment, so front-wheel drive vehicles are considered to be relatively insensitive to side winds.

Drive trains with rear-wheel drive and front engine are listed in the center of the figure. Manual transmission is at the front in standard drive (BMW mid-range and up) and all-wheel drive (Audi Quattro), while in the transaxle concept (various models of Aston Martin and Ferrari) it is in the rear end in front of the axle drive. Transmitting the torque from the engine to the driven axle requires a cardan shaft and, therefore, a cardan tunnel. The shaft and tunnel must be more rugged than in the transaxle design for both standard and all-wheel drives because of the higher torque to be transmitted. Together with the exhaust system running along the underbody from the front to the rear and the fuel tank, which may need to be designed as a two-chamber saddle tank to ensure sufficient fuel capacity, the standard drive in particular restricts the use of large-scale underbody covers. At the rear axle, the axle drive and (on the transaxle principle) the manual transmission must be exposed to flow for cooling reasons. A rear axle cover could be implemented only at great expense. Under these conditions, a diffuser built into the rear end does not achieve its theoretically possible performance because of the poorer inflow. This is a disadvantage, particularly for the rear axle lift, but also for the drag of the vehicle.

Since the traction force is transmitted to the rear axle (or to both axles in four-wheel drive), a high front axle lift is less critical than for the front-wheel drive. Conversely, the demands in terms of rear axle lift are more stringent, especially for the standard drive with a center of gravity far to the front.

The right of the figure shows rear-drive concepts, which differ only in the position of the engine. Both center and rear engine concepts (Lotus Elise or Porsche 911) are comparable with respect to their aerodynamic properties. Since only the rear axle is driven, no prop shaft to the rear end is required. So the tunnel can be omitted, and the exhaust system is also located exclusively in the rear of the vehicle. This makes it possible for the underbody to be a completely closed. An example is the Audi R8 (cf. Figure 4.203), where there are only a few NACA ducts to ventilate the engine compartment. Such a complete fairing also allows the best possible diffuser effect at the rear end, which overall significantly reduces drag and lift. For sports cars this is used specifically to generate downforce.

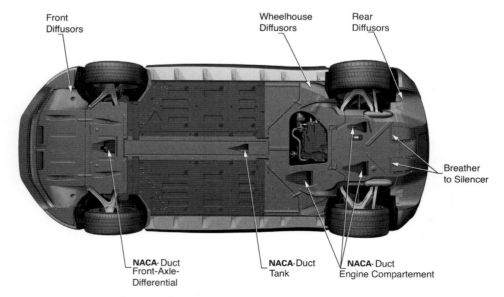

Figure 4.203 Fully faired underbody of the Audi R8.

Rear-wheel drive requires low rear axle lift for power transmission to the rear axle, but this can be achieved by appropriate shaping of the rear. In addition, the center of gravity is significantly farther to the rear than in the drive trains discussed previously. Rear-wheel drive vehicles are therefore the most sensitive to side winds, because, as a result of this position of the center of gravity, a side force will always become yawing at larger yaw angles (with amplifying effect). This must be taken into special consideration in vehicle development.

4.7.4.2 SUVs

Knowledge of the boundary conditions is important to develop a competitive SUV to production. The resulting challenges for the aerodynamic development of an SUV are varied. For example, the customer should be offered greater permissible trailer load. This is regulated in EU Directive M1 (Annex II 70/156/EEC). Among other things, this requires that an increased ground clearance and a larger approach angle be observed, implying per se a deterioration of the drag coefficient (cf. Figure 4.204).

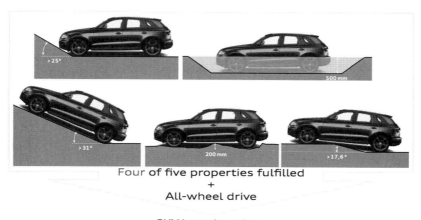

Figure 4.204 SUV requirements for approval as a sport utility vehicle with regard to approach angle, driving through water, maximum gradient, ground clearance, and ramp brakeover angle.

How the average aerodynamic drag viewed across a driving cycle can be reduced using automatic level control was described in section 4.5.3.2 using the example of the Audi Q7. In addition, a large approach angle increases the front axle lift because the stagnation point at the vehicle front is moved upward and more air passes into the space between the vehicle and the road. Large wheels are also used for robust handling and sporty design, which represent an additional challenge because of the massive flow losses in the wheel wake.

The influence of the wheels on aerodynamics was explained in section 4.5.3.3. The large frontal area resulting from the spacious interior with all its benefits for the customer also increases the aerodynamic drag product $c_D \cdot A_x$. The dynamic design of the shell suitable for a sporty vehicle also requires a greater degree of optimization, particularly at the rear. However, for pure off-road cars, such as the Mercedes G-Class or the Range Rover, this aspect is significantly less important than the all-terrain characteristics.

4.7.5 Influence of Equipment and Engine

Specifying a c_D value for all variants of the same vehicle type is not strictly correct because the customer can now configure each car almost at will and thus almost put together a custom car.

Nevertheless, manufacturers often specifically advertise a low c_D value for a series, whether for a volume model—i.e., a model with large market shares—or an equipment variant that allows such optimum values, that is, a kind of "lighthouse project," and rightly so. After all, the manufacturers want to show that they occupy a leading position in all technological fields. And this is true even though in one or the other case the equipment option associated with the low c_D value is not at all configurable.

And yet this is not primarily the true property of "streamlining." Rightly so, because customers are usually not interested in the c_D value of the desired vehicle but—among other things—its fuel consumption. Therefore, it makes little sense to specify one specific c_D value for a vehicle, because every type in fact covers a whole range of drags. This "scattering" can be attributed to the following categories of very different circumstances:

- Features such as engine performance, tires, spoilers, etc.
- Position of the vehicle and thus the angle of attack and the ground clearance (cf. section 4.5.3.2)
- Condition of the vehicle, such as windows and sunroof open or closed[22]
- Radiator shutter open or closed
- Convertibles are a special case. If their shape is derived from a sedan, it differs from this more or less significantly in closed position and, with the folding top down, the ambient flow totally changes.

All these very different factors will be separated from each other below, to the extent possible with the available data. We are compelled to have some recourse to results that were obtained on types where production has been discontinued for a long time, because the above influences on the c_D value for newer types have not been published.

The equipment of a vehicle has a marked effect on the c_D value. Evidence of this may be found in the BMW 5 Series; its figures are summarized in Table 4.6. Conclusion: The more powerful the engine, the higher the c_D value. This is because of the increased

---

22. Open windows cause 20–30 counts of increase in the c_D value according to the author's observations.

cooling air required by the engine and brakes, but also because of wider tires and "ragged" alloy wheels.

Table 4.6 c_D Values of the BMW 5 Series [468]

Model	c_D
520i with sports suspension	0.25
520i	0.26
520d	0.27
525i/525d	0.28
530i/530d	0.28
535i/545i	0.29

The BMW 5 Series therefore has a range of $0.25 < c_D < 0.29$ for the c_D value. The previous model showed a similar range, and comparable bandwidths can be found for model series of other manufacturers as well. So if one really wants to compare c_D values of models from different manufacturers, this only makes sense if the equipment—if aerodynamically relevant—is comparable. It is also apparent from these figures that a "c_D competition" run at thousands of a unit makes little sense, at least in terms of fleet fuel consumption.

The geometry of a given vehicle is—within limits, of course—variable. Opening windows and sunroof and extending retractable headlights or spoilers changes the shape and thus the ambient flow. With the exception of open windows, the impact of these variations on drag are usually small; they are in themselves rarely more than $\Delta c_D = +0.01$.

The car's load also plays a role (cf. Table 4.7). First, the suspension of a heavily loaded vehicle lowers more than an unloaded one. This results in reduced ground clearance, which is always accompanied by reduced aerodynamic drag (cf. section 4.5.3.2). The lift also changes as a result. Second, the tire is deformed more in the contact area with the road surface with increasing load—it increasingly bulges.

Table 4.7 Influence of Load on c_D and c_L (Without Changing the Tire Contact Patch)

	Sedan		SUV		Sports car	
Load	c_D	c_L	c_D	c_L	c_D	c_L
empty	0.27	0.10	0.34	0.05	0.33	0.08
full	0.26	0.13	0.32	0.02	0.31	0.09

Opening the folding top of a convertible results in a large increase in drag. Figures have been compiled in Table 4.8 for three very different roadster classes. The flow snapshots of the MB SL 500 reproduced in Figure 4.205 show that very good flow management is possible even when "open."

Chapter 4

Figure 4.205 Flow around the MB SL 500 roadster with folding steel roof: $A_x = 2.0$ m², $c_D = 0.29$ (top, roof closed); $A_x = A_x$,closed, $c_D = 0.34$ (bottom, roof open).

Table 4.8 c_D Values of Convertibles			
	Drag coefficient c_D		
Folding top	Ford Street Ka	MB SLK ($A_x = 1.93$ m²)	MB SL ($A_x = 2.0$ m²)
closed	0.435	0.32	0.29
open	0.491	0.37	0.34

The fact that opening the folding top sometimes considerably increases the drag is only of secondary importance. Drivers typically do not drive "open" and, if so, they usually drive more slowly, so the higher fuel consumption is then slight. It is important, however, to ensure that draft in the interior is not too strong when the folding top is down.

4.7.6 Driving on the Ceiling?

This question is asked every now and then, and basically it would be possible to design a production vehicle with a focus on downforce to such an extent that it is capable of driving on the ceiling—at least at vmax. In this case, the vertical aerodynamic force must balance out the weight force. In a vehicle with a mass of 1.7 tons, a top speed of 250 km/h, and a frontal area of 2.3 m², this results in a lift coefficient of

$$c_L = -\frac{2 m_F g}{\rho v^2 A_x} = -\frac{2 \cdot 1700 \cdot 9.81}{1.205 \cdot (250/3.6)^2 \cdot 2.3} = -2.5 \tag{4.43}$$

Such downforce values are not nearly achieved, even in very sporty road vehicles (cf. Figure 4.198). Although any downforce value could be obtained using wings, ground measures, and so forth, the aerodynamic drag would increase to a level that is not marketable for production vehicles (cf. induced drag, section 4.3.3.1). Such downforce values can only be achieved in racing vehicles—this will be discussed in detail in chapter 9.

According to Figure 4.198, sporty production vehicles reach downforces of approximately $c_L \approx -0.1$. With a lightweight body of 1500 kg, the vehicle would be able to overcome its own gravity only at a speed of 1174 km/h; if it traveled at maximum 300 km/h, it could have a mass of only 98 kg. These figures show that the idea of vehicles traveling or flying on the ceiling is unrealistic for mass production vehicles (these considerations apply also with inverted sign to lifting off the road).

4.8 Future Development

Figure 4.206 plots the historical development of the drag coefficient c_D of production cars over the years from 1900 on. As can be seen, c_D values were successfully reduced to 0.25 without the use of exotic measures. This is demonstrated by Mercedes E-class and Audi A2, whose developers applied only features such as those examined in section 4.5. It must be assumed that the drag coefficient will stagnate at a level of about $c_D = 0.23$ when all optimization measures listed there are utilized.

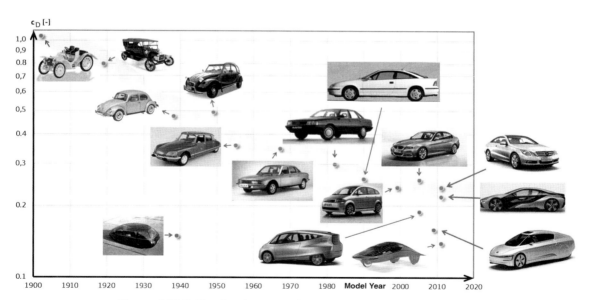

Figure 4.206 The development of c_D over time for production cars.

The following section will discuss what has to be done in order to further reduce the aerodynamic drag of road vehicles without asking whether it makes sense to reduce c_D for vehicle-technical reasons (fuel consumption, maximum speed, styling). The analysis shown in Figure 4.22 will serve as the starting point. The most promising areas for an improvement of the flow around the car are locations where separation and momentum loss still occur despite all optimization efforts:

- Basic shape including A-pillars and rear end
- Add-on parts, especially mirrors, spoilers, underbody, and wheels
- Engine compartment and cooling airflow

In order to further improve the basic vehicle shape, it is essential to avoid the longitudinal vortices generated at the slanted surfaces and to reduce the size of the quasi–two-dimensional separations. Section 4.5.1 has already discussed how to improve the flow around the A-pillar. The aim of the most promising of the approaches presented there was to achieve a radius between windshield and A-pillar that was as large as possible. Cars formed according to Jaray's lines as well as Kamm's vehicle (cf Figure 1.37) usually showed steep windshield angles but with such a large radius connecting to the side window that the A-pillar vortex was unlikely to be generated. In the 1950s, a flow around the A-pillar without separation was supposed to be achieved using a panorama windshield. But, due to the strong curvature of the windshield distorting the driver's vision, and an unusual overhang in the door opening, they soon went out of style again.

To further reduce drag, suitable measures have to be applied individually to the three tail shapes (i.e., fullback, notchback, and fastback), as was discussed in section 4.5.1. For slanted fullbacks in particular, such measures have recently been implemented in production cars (i.e., large roof spoilers, side spoilers, etc.), but measures for regular fullbacks (i.e., attica, delta wings, or air blowers) often are not applicable due to styling issues. It is in this area that the biggest potential exists for a future improvement of the aerodynamic drag.

For notchback vehicles, three different flow structures can occur: For edged contours the flow field will be a combination of those of fullback and fastback and should, consequently, be improved using the measures mentioned above, whereas it is crucial to ensure that the flow separating at the trailing edge of the roof reattaches on the trunk. Since a coupé-like styling determines the silhouette of most notchback cars today, the contours of the tail shape tend to blur into each other. This means that each new car design has to be investigated individually to determine whether the above-mentioned measures for fullbacks and fastbacks will lead to success.

The optimization of different add-on parts was discussed in section 4.5.3. It was mentioned there that optimizing the shape of the external mirrors will not lead to a significant decrease of c_D since their contribution to the aerodynamic drag is rather small. This will change if the use of external cameras instead of mirrors becomes possible through a change in the law; at present, at least one mirror is prescribed. Depending on the shape quality of the camera mount, a drag improvement of $\Delta c_D = 0.010$ is realistic. Figure 4.207 shows an example: the Audi eTron study with external cameras.

Figure 4.207 Audi eTron study with external cameras instead of mirrors.

As mentioned, underbody, wheels, and suspension parts are responsible for a considerable part of the aerodynamic drag. It should therefore be worth the effort to investigate this part of the vehicle in detail, and this may result in moving away from the idea of a flat underbody. Figure 4.208 shows some ideas on designing an aerodynamically efficient underbody. It has already been noted that the flow below the car is directed outwards. This effect is intensified by the cooling airflow exiting the engine compartment toward the wheels. Thus, the wheels and especially the front wheels are under cross flow, resulting in an increased aerodynamic drag. This was shown in Figure 4.39. Wheel cross flow should be avoided. Air guiding bodies in front of the wheels can help. Furthermore, it is common to design the rear part of the underbody as a diffuser in order to force the flow in a longitudinal direction. The volume of the wheel housings should be as small as possible, and the wheels should be designed as flat discs.

Finally, the cooling airflow: Up to now, improvements were limited to the area in front of the heat exchangers. It is also important, though, to take into account the interference between cooling and external airflow. Aerodynamic drag can be improved if the cooling airflow exit position is as close as possible to the vehicle's tail and in an almost horizontal flow angle. In any case, the amount of cooling air should never be larger than necessarily required for cooling and climatization. This requirement can be met using a cooling air shutter.

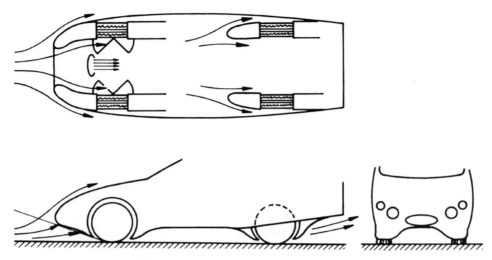

Figure 4.208 Approach to guide the airflow along the underbody.

4.9 Reference Bodies

Realistic vehicle geometries mostly are too complex for analyzing aerodynamic phenomena of interest, and varying the geometric parameters would involve lots of effort and would increase the costs of a study. Therefore, simplified and generic vehicle shapes are used for such fundamental investigations in pre-development and research. Their body shapes and basic flow field properties have been published in several papers. While different flow phenomena on realistic vehicles interact with each other (interference effects), such simplified reference bodies can be used in order to unlink and observe these phenomena. Thus, these reference bodies are often used for CFD code validation.

There are very generic reference bodies available, like the SAE body and the Ahmed body, as well as quite realistic ones such as the MIRA reference body or the DriveAer car. Further examples are the ASME body, which is mainly used by the Asian automotive industry for CFD validation and fundamental research, and the "Willy" body, which was designed to investigate side wind effects. The following survey does not claim to be complete; however, the three reference bodies that are most essential from the author's point of view are described. For all other reference bodies, the reader is referred to the advanced literature.

4.9.1 SAE Reference Body

The SAE model has been examined and documented and were archived by many aerodynamicists using different available rear end shapes such as notch back, squareback, and fast back. The flat underbody features a diffusor at its end, and cooling airflow can be investigated using a simplified cooling airflow path. The SAE model can be mounted either on turning wheels in order to investigate flow effects in the wheel arch region or on fixed struts as shown for the notchback version in Figure 4.209.

Aerodynamic Forces and Their Influence on Passenger Vehicles

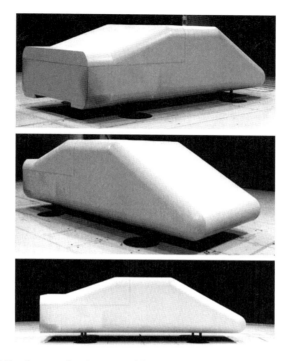

Figure 4.209 SAE reference body as notchback without cooling airflow on struts [750].

Examples of the multipurpose use of the SAE body are the investigations by Kuthada [479] and Fischer [248.] Kuthada examined the re-entrance of the cooling airflow into the external flow and its impact on the cooling air drag (Δc_{DC}, see section 4.3.3.2). A reasonable investigation of different independent parameters would be hardly feasible using a real vehicle. Figure 4.210 shows the influence of the exit angle of the cooling airflow on different force coefficients, with the cooling air drag becoming smaller when the angle decreases.

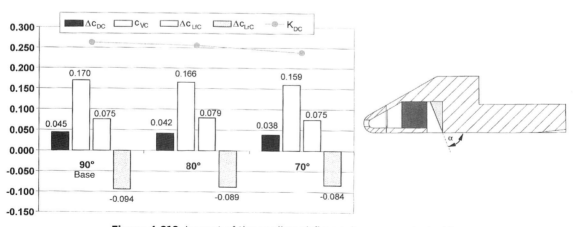

Figure 4.210 Impact of the cooling airflow exit on c_D, evaluated for the SAE reference body [479].

407

Fischer collected wind tunnel measurements on the SAE reference body to validate a CFD code. His CFD model included both wind tunnel geometry and reference body (see Figure 4.211). Furthermore, he derived much information about the flow structures leading to wind tunnel interference effects. If he had used a CAD geometry of a realistic vehicle, the runtime for each simulation would have increased enormously.

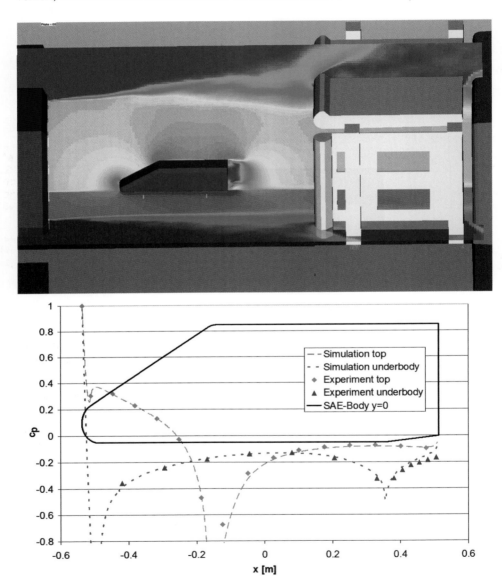

Figure 4.211 The SAE reference body in the 1:4 scale wind tunnel at the University of Stuttgart (FKFS), calculated using PowerFLOW® (upper), validation of the resulting pressure distribution (lower) [248].

4.9.2 Ahmed Body

The Ahmed reference body (Figure 4.212) was designed by Syed R. Ahmed [11], taking into account various tail shapes. It consists of a rounded front that ensures attached flow toward the body's tail, a squared main corpus, and a tail with adaptable slant angles. The Ahmed body is mounted on struts.

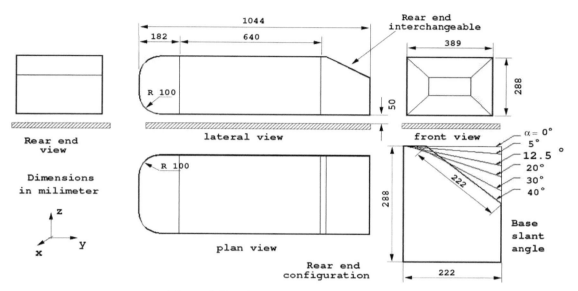

Figure 4.212 The Ahmed reference body [6].

With the help of the Ahmed reference body, some important fundamental observations have been made about the impact of the rear window slant angle on the air forces. Figure 4.213 shows the drag coefficient of the Ahmed body plotted against the rear slant angle. The contributions of rear slant, front bumper, base surface and friction are plotted separately. The investigation revealed a drag minimum at an optimal slant angle of about 10° and a discontinuous behavior at a 30° angle with a sudden decrease of c_D due to the fact that the flow changes its characteristic from a fastback into a fullback flow pattern. Many correlations described in chapter 4.5.1 are related to Ahmed's results.

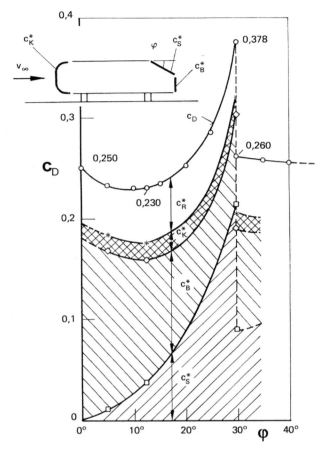

Figure 4.213 Aerodynamic drag coefficient of the Ahmed reference body plotted against the slant angle of the rear window, after Ahmed et al. [11].

4.9.3 DrivAer Body

The preceding reference bodies are highly simplified geometries whose proportions diverge widely from those of production cars. In order to acquire research results applicable to series development, the Technical University of Munich in cooperation with BMW and Audi designed a new generic vehicle body named DrivAer reference body. Its shape is based on the Audi A4 and the BMW 3 series, two typical medium-class vehicles. The proportions were derived by averaging and thus represent realistic overhangs, A- and C-pillars, roof line, bonnet and trunk lid height.

The underbody is available in either a smooth or a realistic design, the latter including exhaust system and fuel tank. The cooling airflow is accounted for by a simplified engine compartment where the cooling air can exit toward the wheels, toward the underbody, or both. The DrivAer body is available as 1:2.5 and 1:2 scale models and can be equipped with different tail shape extensions. Heft et al. [323] published complete geometrical details of the DrivAer body, merged and shown in Figure 4.214.

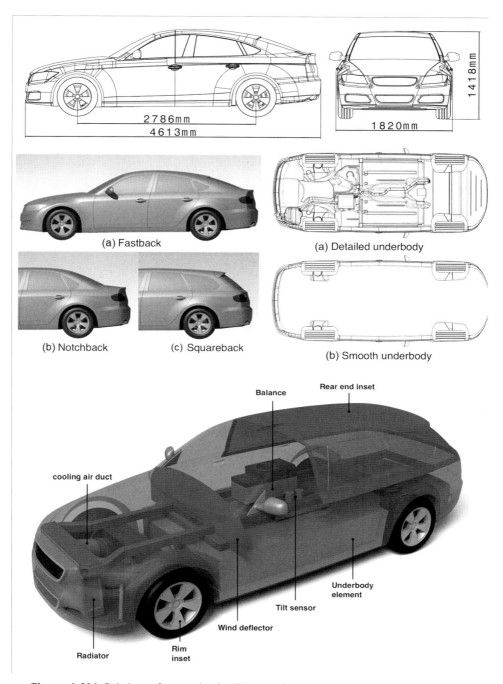

Figure 4.214 DrivAer reference body (TU Munich) including main dimensions (top), tail shapes, and underbody varieties (middle, [323]) as well as cooling airflow and balance (bottom, [518]).

The model is provided with a couple of pressure probe drillings in order to acquire flow and pressure field data for an extensive analysis of all geometric configurations. Mack et al. [518] delivered a wide range of measurement results from the scale model wind tunnel at the TU Munich. By way of example, Figure 4.215 shows the pressure distribution on the vehicle's centerline section for upper surface and underbody using different tail shape extensions. At the front and the mid-part of the car, the pressure data are similar for each tail shape; it is only in the rear part of the surface that significant changes occur.

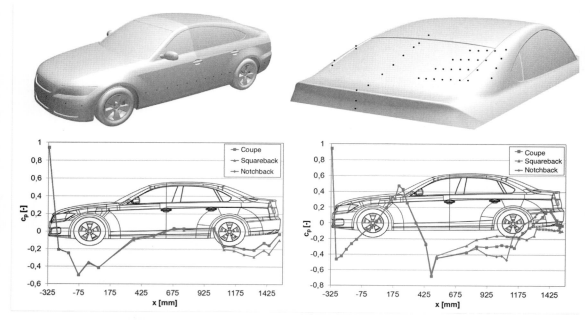

Figure 4.215 Overview of the positions of the pressure probe drillings and pressure distribution on the vehicle's centerline cut ($y = 0$) of the Driver reference body, upper side (right), and underbody (left), using various tail shapes [323].

Chapter 5
Aerodynamics and Driving Stability

David Schröck, Andreas Wagner

The impact of aerodynamic forces and moments on the vehicle has been the subject of investigation for quite some time. This is because the flow causes not only drag as discussed previously (see section 4.1) but also other forces and moments that influence the driving dynamics of the vehicle. Lift and pitching moment development was the subject of section 4.5. Their effects on the wheel loads and therewith on the vehicle performance when driving straight ahead and cornering, when load shift responses occur, and during braking maneuvers are described in section 5.2.2.

An unsymmetrical incident flow, as it occurs during crosswind conditions or when overtaking, generates a side force, a yaw moment, and a roll moment. It also influences the lift forces and pitching moment. The generation of these forces and moments and the measures to influence them were also discussed in section 4.5. The treatment of aerodynamic forces and moments under steady-state flow conditions may be justified in the case of straight, symmetric head wind. But in the case of yawed incident flow caused by side wind or during the dynamic process of overtaking, the vehicle experiences a flow situation whose variance over time no longer justifies a steady-state approach.

The fact that transient aerodynamic forces and moments show, in some respects, significant differences from those measured under steady-state conditions is shown by various studies. For vehicle manufacturers, the behavior shown by their vehicles in such a situation is of great interest, because a sudden change in the direction of flow and the associated forces may result in a path deviation of the vehicle, which must be compensated with some vigorous steering responses. In extreme cases, this may even lead to an accident. But even if the resulting vehicle performance is not necessarily assessed as a safety issue, the driver will judge an unsettled ride as a lack of comfort, because continuous steering corrections are tiring for the driver over an extended period.

The directional stability of the vehicle therefore represents an important criterion in vehicle performance evaluation, and aerodynamic coefficients that are determined under steady-state conditions are traditionally applied. Using this approach, however, different vehicles have been assessed as normal and almost identical, but have then been evaluated differently in actual road tests. This suggests that, besides aerodynamic effects, dynamic driving properties must also be considered in order to assess side wind behavior. The following sections therefore consider aerodynamic forces and moments under transient incident flow conditions and present applicable testing and evaluation methods. Since consideration of aerodynamic forces and moments alone is not sufficient to describe the vehicle behavior, the overall system consisting of aerodynamics, vehicle dynamics, and the driver is considered as a whole in section 5.2.3.

5.1 Unsteady Aerodynamic Forces and Moments

As mentioned in the introduction, today's normal vehicle aerodynamics development process is based on measurements in the wind tunnel which simulate traveling through still air or in the presence of a constant, homogeneous wind (i.e., the incident flow experienced by the vehicle is not subject to temporal changes). Incident flow velocity and angle of attack are kept constant for the measurement duration.

5.1.1 Overtaking Maneuvers

In section 4.5.4.4, overtaking maneuvers were treated using steady-state vehicle models in discrete positions. For small relative velocities, this approach is sufficiently accurate to determine the forces that occur. However, as relative velocities increase, dynamic effects occur that require time-resolved evaluation of the process. For this purpose, a rail system is installed in the test section of a wind tunnel, which allows a model (the overtaking vehicle) to be moved past a stationary model (the overtaken vehicle). The relative velocity and the lateral offset of the two models can be varied (see for example [103], [282], [612], [787]). Moreover, by rotating the entire setup within the test section of the wind tunnel, the impact of a steady-state yawed incident flow can be studied as well. Gilliéron et al. [282] have used this setup to conduct parameter studies on relative velocity and lateral offset and have measured the forces acting on the overtaken model. First, note that the time characteristic of the side force coefficient is qualitatively equivalent to that under steady-state conditions. This means that, at the beginning of an overtaking maneuver, a positive side force occurs, followed by a negative side force. A comparison of the results under steady-state and transient conditions as in Figure 5.1a shows that, as relative velocity increases, the side force coefficient increases at an even earlier point in time and takes on increasingly larger values. The maximum values exceed those determined under steady-state conditions by up to 120%. Negative side force coefficients undergo even greater change at the end of the overtaking maneuver than positive coefficients at the beginning. Figure 5.1b shows the side force coefficient at different lateral distances between the vehicle models. The relative velocity between the models is 10 m/s. The time offset of the force characteristic to the steady-state result is constant for all measured lateral distances. As expected, the mutual influence of the two models decreases with increasing lateral distance because the effect of the channel flow between the two vehicles decreases.

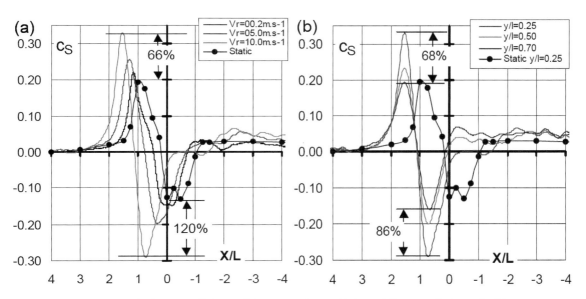

Figure 5.1 Side force coefficient of the overtaken vehicle at different relative velocities (a) and lateral offset (b) [282].

The interaction of the flow fields of the two vehicles, which is responsible for the typical characteristic of side force and yaw moment, can be seen in Figure 5.2. Here the pressure distribution on the side surface of a vehicle is shown at the beginning ($x/l = -1.2$) and end ($x/l = 0.2$) of the maneuver of overtaking a truck, whereby pressures are indicated relative to the undisturbed incident flow. The results are derived from road tests. The side facing the truck is mainly responsible for the generation of side force and yaw moment. Changes on the side facing away from the truck display smaller pressure changes. At $x/l = -1.2$, the pressure field induced by the truck acts on the front right area of the overtaking vehicle, which is indicated by lower pressures as compared to the undisturbed state. The effects on the rear area are smaller. As the position of the car advances, the pressure differences in the area of the front fender assume positive values. In addition, the interaction of the flow fields of the car and the truck also reaches the rear area of the vehicle, where negative pressure differences now occur. As the conditions on the side facing away from the truck do not change to the same extent, the yaw moment characteristic typical for overtaking maneuvers occurs.

Chapter 5

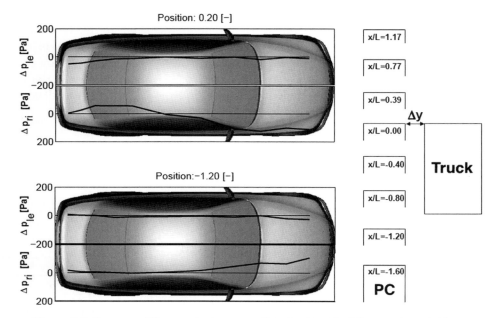

Figure 5.2 Pressure difference between undisturbed incident flow and overtaking maneuver on the side surfaces of a vehicle at $x/l = 0.2$ and $x/l = -1.2$, $v_{Car} = 160$ km/h, $v_{Truck} = 80$ km/h, $\Delta y = 1.5$ m [738].

5.1.2 Side Wind

The analysis of the directional stability under side wind is extremely complex. This is due in particular to the complicated flow processes in the atmospheric boundary layer, which also interact with the flow around the vehicle moving in this flow field. In addition, other parameters like the driving mechanics and the driver have an influence.

5.1.2.1 Natural Wind and Its Influence on the Incident Flow of a Vehicle

The flow conditions under which vehicles are moved are characterized by natural wind. In particular, strong wind affects the directional stability of the vehicle. This will therefore be the initial topic of this chapter, before its impact on the flow around a vehicle is discussed.

Extensive meteorological measurements are used as the basis to describe the weather phenomenon called wind. General statements can be made on that basis that relate mostly to an altitude of more than 10 m above the ground. This data shows that winds mainly prevail from the West in Germany. On the annual average, the wind velocity in the coastal areas is more than 5 m/s; in central Germany it is between 3 and 4 m/s; and in the South it is less than 3 m/s (see Figure 5.3).

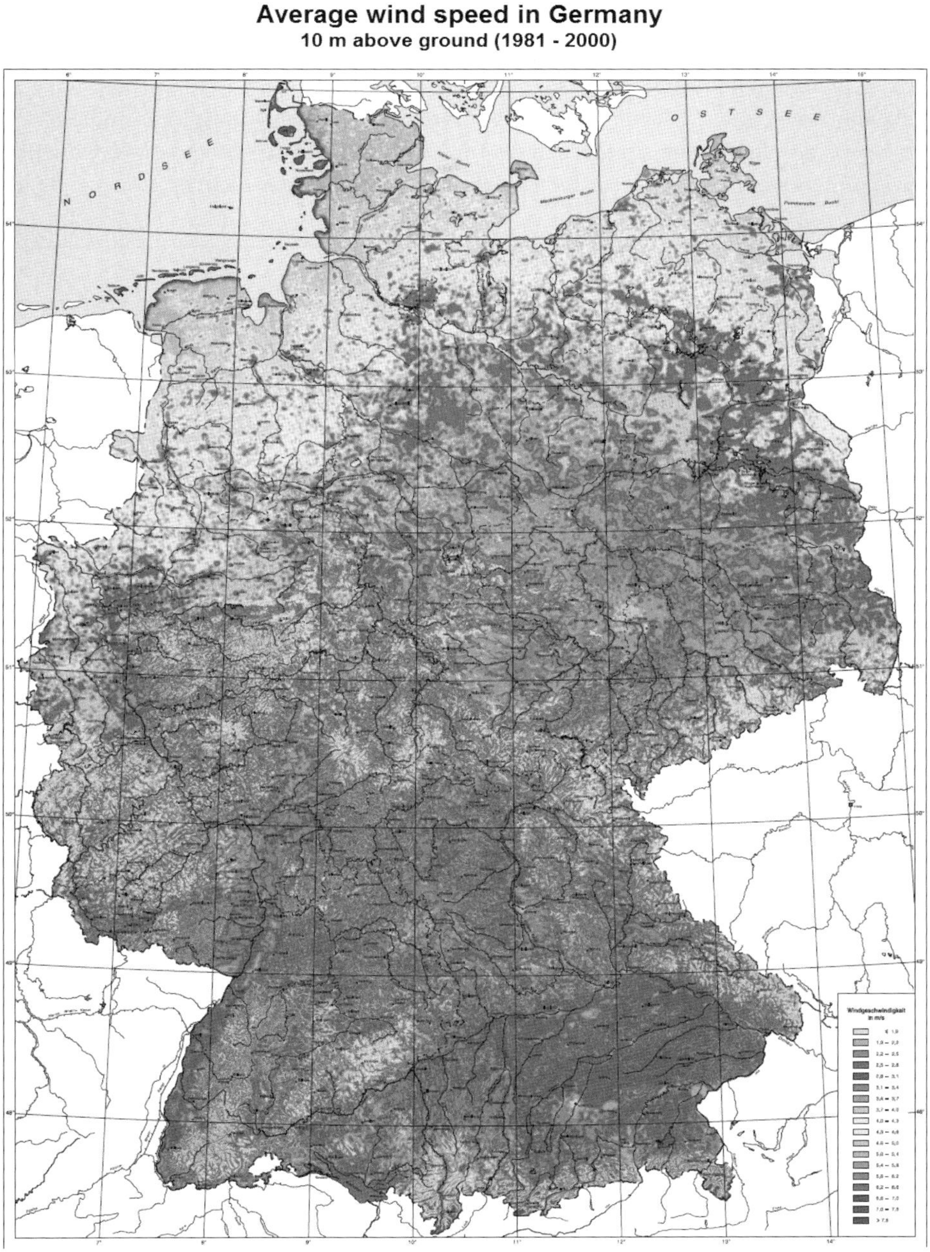

Average wind speed in Germany
10 m above ground (1981 - 2000)

Figure 5.3 Average wind velocity at 10 m altitude in Germany [367].

Two key features of wind are its strength and direction. Both parameters are subject to constant change. As can be inferred from Figure 5.4, there are both slow changes dependent on the time of the year or day and short fluctuations lasting only a few minutes or seconds. The latter are perceived as gusts of wind and are especially important from the perspective of driving safety. They are dangerous because they occur unexpectedly for the driver, and there is a strong change of wind velocity in conjunction with a change of the direction of incident flow acting on the vehicle. Particularly in the winter months, gusts may occur at remarkable velocities of more than 30 m/s. During local thunderstorms and accompanying squalls, wind peaks up to 50 m/s are even possible. Statistics show an increase in accidents caused by side wind in the winter months.

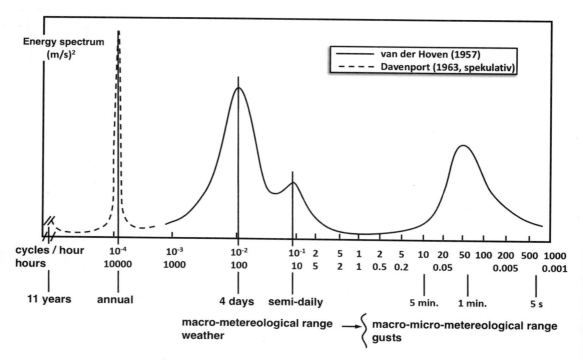

Figure 5.4 Energy spectrum of natural wind [372].

Friction between moving air and the surface causes a boundary layer to form. For atmospheric wind, this can be described by a logarithmic function, which was originally developed by Prandtl for the turbulent boundary layer over a flat plate (see also section 14.1.4). In addition to the logarithmic law of the wall, the boundary layer profile can also be described with sufficient accuracy by an exponential approach that is easier to use.

$$\frac{u(z)}{u_G} = \left(\frac{z}{z_G}\right)^\alpha \tag{5.1}$$

Here, u_G is the velocity at the upper edge z_G of the boundary layer. The influence of surface roughness on the boundary layer profile is taken into account by the roughness exponent α. It can assume values between 0.16 for open terrain and 0.4 for vegetation or buildings. Figure 5.5 describes the influence of surface roughness on the boundary layer profile.

Aerodynamics and Driving Stability

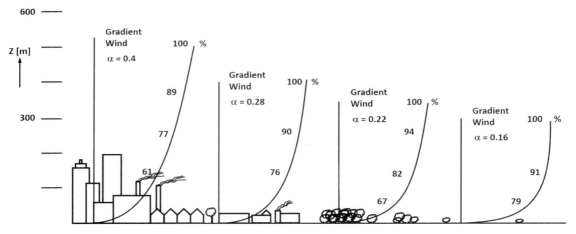

Figure 5.5 Ground boundary layer as a function of the terrain roughness [692].

A measure of wind velocity fluctuations or the gustiness of wind is the turbulence intensity. It is calculated from the square root of the variance σ^2 and the average wind velocity u_∞.

$$I_i = \frac{\sigma_i}{u_\infty} \cdot 100 \ \% \tag{5.2}$$

The index i refers the flow component in one of the three spatial directions. By definition in wind engineering, the x-axis always indicates the main wind direction. If the degree of turbulence is not dependent on the measuring position, the flow is homogeneous. A flow is isotropic if the turbulence intensity is identical for all three components. In the area near the ground that is relevant for vehicles, the flow is generally neither isotropic nor homogeneous because of interference from obstructions such as vegetation or buildings and other vehicles.

As the degree of turbulence is an integral parameter, it is possible to make a statement about the total intensity of turbulence, but mapping the fluctuation to a certain frequency requires the use of spectral analysis. For this purpose, Figure 5.4 must be supplemented by the frequency range relevant to driving dynamics with period durations of a few seconds and below. In this high-frequency range, the spectral composition of the velocity fluctuation can be described by models known from technical fluid mechanics. Three frequency or wave number ranges are distinguished in which different processes dominate that determine the characteristic shape of the spectrum. Large vortices are caused by wind shear or separation at obstacles in the flow. They obtain their energy from the average base flow. This frequency range is therefore also known as the production range. A measure of the size of the resulting vortices is the largest relevant dimension of the turbulent system. Owing to shear forces, these vortices decay to smaller vortices and enter the so-called inertia range. The input of energy from the production range and the output into the dissipation range (the third range) is in equilibrium, so that energy is neither created nor destroyed in the inertia range. Kolmogorov has demonstrated that the energy cascade (i.e., the decay of large to small vortices) is independent of the history of how the vortices occur, which leads to a universal behavior of the spectrum in this frequency range. The energy content of the spectrum is proportional to the frequency and decreases as per equation (5.3) with increasing frequency [346].

$$S_{ii}(f) \sim f^{-5/3} \tag{5.3}$$

Independently of Kolmogorov, von Kármán has developed an empirical interpolation function, which depicts the characteristic of the spectrum in the inertia range as well as in the production range. The spectra for the velocity components in the three spatial directions can be specified using the equations (5.4) and (5.5).

$$\frac{u_\infty \cdot S_{uu}(f)}{2\pi \cdot \sigma_u^2 \cdot L_u} = \frac{2}{\pi} \cdot \left(1 + 70.78 \cdot \left(\frac{f \cdot L_u}{u_\infty}\right)^2\right)^{-5/6} \tag{5.4}$$

$$\frac{u_\infty \cdot S_{vv,ww}(f)}{2\pi \cdot \sigma_{v,w}^2 \cdot L_{v,w}} = \frac{2}{\pi} \cdot \frac{1 + 187.16 \cdot \left(\frac{f \cdot L_{v,w}}{u_\infty}\right)^2}{\left(1 + 70.78 \cdot \left(\frac{f \cdot L_{v,w}}{u_\infty}\right)^2\right)^{11/6}} \tag{5.5}$$

The inertia range is followed directly by the dissipation range, in which vortices have reached a size for which the processes are no longer determined by inertia forces but by viscosity forces. This means that the kinetic energy of the vortices is dissipated in heat as a result of molecular friction.

Normalization of the measured spectra or the calculation of the spectrum according to von Kármán requires the integral length scale L_i. It provides a measure of the average size of the vortices transported in the flow. It can be calculated by the relationship

$$L_i = u_\infty \cdot \int_{\tau=0}^{\varphi_{ii}(\tau)=0} \varphi_{ii}(\tau) d\tau \tag{5.6}$$

where φ_{ii} is the correlation coefficient. The integration of the correlation function from $\tau = 0$ to τ for which $\varphi_{ii}(\tau) = 0$ applies provides the so-called integral measure, a measure of the temporal similarity of the flow. By applying Taylor's Theory of Frozen Turbulence, the integral measure of time can be converted to the integral measure of length by multiplication with the average flow velocity u_∞. This theory assumes that coherent structures floating along in the flow at the average flow velocity are retained for the amount of time they need to pass the measuring probe. Alternatively, the integral measure of length can be approximated by using the spectrum calculated from the measured data and equation (5.4) or (5.5) against variation of L_i so that the squared error is minimized. Figure 5.6 shows an example of the result of such an approximation for the spectrum of the longitudinal component of the incident flow velocity of a vehicle during a measurement in gusty wind.

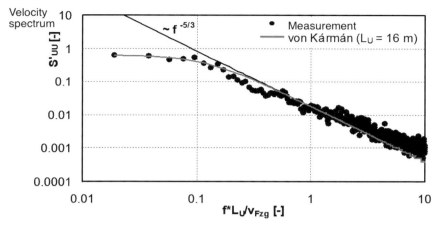

Figure 5.6 Comparison of the energy spectrum of the incident flow velocity with the von Kármán interpolation function and Kolmogorov's −5/3 law [739].

The flow velocity around a vehicle is determined by its driving velocity, because the latter is significantly higher than the wind velocity, with a few exceptions. The wind can therefore be considered as a disturbance variable, superimposed on the driving velocity. Only the lowest 2 m of the atmospheric boundary layer, which has a height of up to several hundred meters, is of significance for motor vehicles. Extrapolation of the boundary layer profile or of the velocity spectrum, based on measurements at 10 m altitude, and their vectorial superimposition on the driving velocity can provide a first theoretical assessment of the flow situation to which a vehicle is exposed. Note that the x-axis is defined as the main wind direction in the wind coordinate system. After transformation into the vehicle coordinate system, the longitudinal component of the vehicle incident flow therefore contains parts of the longitudinal and lateral components of the wind, depending on the wind direction ϕ. This makes it possible to formulate equations for the angle of attack, the degree of turbulence, and the incident flow spectrum (see [176]). The range of possible angles of attack can be calculated using equation (5.7).

$$\beta = \sin^{-1}\left(\frac{v_W \cdot \sin\phi}{\sqrt{(v_W \cdot \sin\phi)^2 + (v_W \cdot \cos\phi + v_F)^2}}\right) \quad (5.7)$$

For the situation where the wind is perpendicular to the driving direction, Equation 5.7 can then be simplified to

$$\beta = \sin^{-1}\left(\frac{v_W}{\sqrt{v_W^2 + v_F^2}}\right). \quad (5.8)$$

At a given wind velocity, the angle of attack continuously decreases as the driving velocity increases.

Equation (5.9) accounts for the dependency of the degree of turbulence intensity of the vehicle incident flow on the wind turbulence, wind velocity, and driving velocity. As the fluctuating component is no longer related to the wind velocity but to the resulting incident flow velocity, the degree of turbulence intensity is reduced from the vehicle's perspective and decreases as the driving velocity increases.

$$I_{x,\text{vehicle}} = \sqrt{v_w^2 \cdot \frac{\left(I_{x,W} \cdot \cos\phi\right)^2 + \left(I_{y,W} \cdot \sin\phi\right)^2}{\left(v_W \cdot \sin\phi\right)^2 + \left(v_W \cdot \cos\phi + v_F\right)^2}} \cdot 100\ \% \tag{5.9}$$

The wind spectrum can then be transferred into the vehicle coordinate system. Using the simplifications introduced in [176], the spectra match the von Kármán interpolation functions in equation (5.4) and (5.5), with the exception that the average wind velocity is substituted by the resulting incident flow velocity. Figure 5.7 shows the spectra for different driving velocities, a wind velocity of 10 m/s, and an integral length of 20 m. It is clear that, as driving velocity increases, the frequency at which the curves begin to fall shifts to higher values.

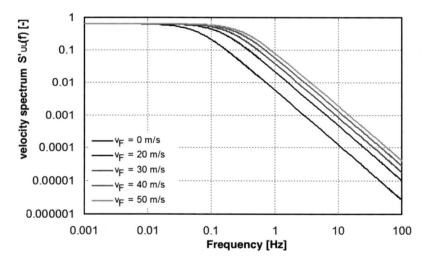

Figure 5.7 Dependency of the incident flow velocity spectrum on the driving velocity at a wind velocity of 10 m/s [739].

These theoretical assumptions may provide a way to understand the influence of the wind and the driving velocity on the resulting incident flow situation of a vehicle. However, local conditions limit the possibility to directly transfer the measurements made in greater altitude to road levels. Shielding of the road from adjacent buildings or dense forest may reduce the influence of the wind, whereas road embankments or gaps in the vegetation along the sides of a road, as shown in Figure 5.8, may cause local velocities to increase. Proximity to the ground also leads to a change in the flow, since velocity fluctuations subside asymptotically in the vertical direction to the ground, and this energy is diverted into the longitudinal and the transverse directions. It can also be assumed that the turbulence intensity calculated from equation (5.9) is too low. The gustiness of the wind is superimposed with an apparent gustiness from the vehicle's perspective, which results from the vehicle passing through areas of different wind velocities.

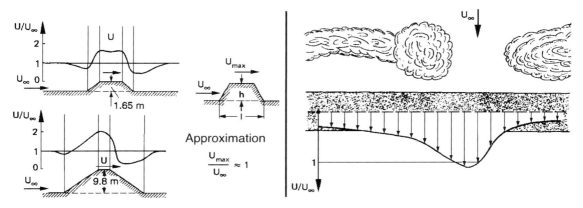

Figure 5.8 Excess velocity at road embankments (left) and through gaps in lateral vegetation (right), according to Bitzel [78].

Owing to the complexity of the flow situation near the ground, only measurements with a vehicle can provide a realistic description. Various authors have carried out measurements on public roads under different wind and traffic conditions, where the test vehicles are equipped with sensors to determine the resulting incident flow vector. Figure 5.9 shows an example of a 40-second section of a measurement in strong wind and very little traffic, so that the changes to the flow situation are primarily attributable to the wind events. The driving velocity during the measurement is 45 m/s and is constant over the measurement time. In comparison to that, fluctuations are visible in the incident flow velocity that deviate from the driving velocity by up to 10 m/s owing to the wind conditions. The time characteristic of the incident flow situation shows abrupt changes in velocity and angle, which indicates the gustiness of the wind. Owing to individual wind gusts, the angle of attack reaches maximum values of 10°. Note that in still air the incident flow velocity matches the driving velocity, and the angle of attack is constant at 0°.

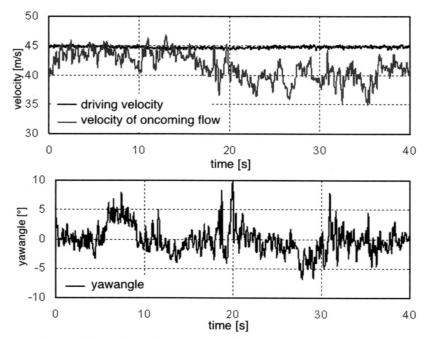

Figure 5.9 Incident flow situation of a vehicle in high wind [739].

Figure 5.10 shows the probability density functions of relative incident flow velocity and angle of attack. They reflect the bandwidth of the occurring incident flow velocities and angles of attack as well as the frequency of occurrence. The data sets have a Gaussian or normal distribution both for low and high winds. This result is consistent with experience from wind engineering (see for example [692]) and shows that the originally Gaussian-distributed velocity fluctuations of natural wind are not changed in their distribution by the movement of the vehicle (see also [739], [743], [894]). It is also evident that the distribution functions in high wind are wider than in low wind. This is due to the gustiness of the wind, which leads to a greater range of incident flow velocities and angles of attack. The charts also show that, even in high wind, the change of relative incident flow velocity hardly ever exceeds a value of 10 m/s. Similarly, the probability of exceeding an angle of attack of 10° is rather low at driving velocities greater than 160 km/h.

Figure 5.11 shows the turbulence intensity of the transverse velocity against the longitudinal velocity. As already shown by the probability density function, stronger wind causes a widening of the curve, which means an increase in the standard deviation and therewith the turbulence intensity. In strong winds, the data points are located at higher turbulence levels than in low wind. Moreover, the bandwidth of the measured turbulence intensities becomes larger, which is due to the influence of strong gusts of wind. On average, the results follow the angle bisector of the chart.

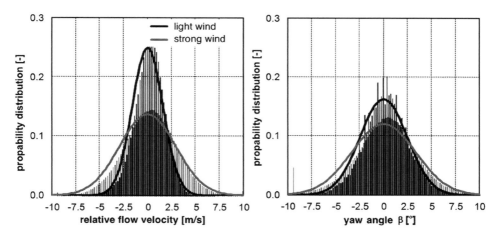

Figure 5.10 Probability density function of relative incident flow velocity (right) and angle of attack (left) under different wind conditions [739].

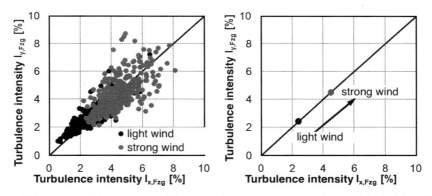

Figure 5.11 Degree of turbulence of the x- and y-components of the vehicle incident flow under different wind conditions [739].

In Figure 5.12, the results for the integral length scale in the longitudinal or transverse direction are plotted against the corresponding levels of turbulence. For both directions of incident flow, it can be seen that, as the level of turbulence increases, the integral length scale tends to increase, which is due to the presence of large-scale wind gusts in strong wind. A different situation is shown by the results in Figure 5.13, where the correlation between length and turbulence intensity is presented, dependent on the traffic situation and the lateral vegetation. Vehicles driving ahead characterize the incident flow situation of the test vehicle, which results in an increasing turbulence intensity as the vortex intensity decreases. A comparison of the factors in Figure 5.12 and Figure 5.13 illustrates the diversity of occurring incident flow situations of a vehicle and the associated relationships between the parameters describing the flow.

The integral length scale can be understood only as a factor that indicates the order of magnitude of the average vortex sizes contained in the flow. The bandwidth of occurring vortex sizes can be assessed better with the spectra of longitudinal and transverse components of the incident flow shown in Figure 5.14, because the presence of energy in a certain frequency range proves the existence of vortices of corresponding wavelength. A comparison of the spectra averaged over measurements in low and high wind shows for both components of the vehicle incident flow that, as the wind increases, the energy content of the flow also increases, approaching the proportionality described by Kolmogorov between spectral content and frequency.

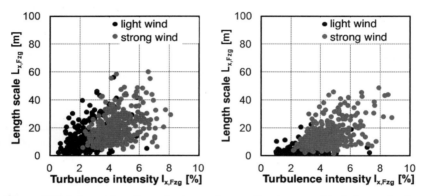

Figure 5.12 Integral length scale depending on the degree of turbulence for the longitudinal and the transverse components of the vehicle incident flow under different wind conditions [739].

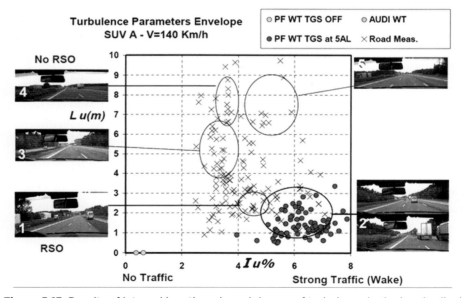

Figure 5.13 Results of integral length scale and degree of turbulence in the longitudinal direction, depending on traffic and lateral vegetation; (RSO: Road Side Obstacles) [507].

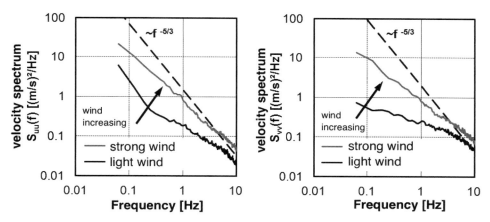

Figure 5.14 Spectrum of longitudinal and transverse components of the vehicle incident flow under different wind conditions [739].

5.1.2.2 Test and Evaluation Methods

A common approach to determine the side wind sensitivity of a vehicle, already used in early investigations by Kamm, is to drive past a side wind generator. Passing through the side wind generator results in a lane drift of the vehicle, to which the driver responds with a steering action, which is used as the criterion for the side wind sensitivity (see section 5.2.4.3). The directional stability of the vehicle is considered a problem of the entire system consisting of aerodynamics, driving dynamics, and driver. A separate analysis of the aerodynamic characteristics of the vehicle, freed from the complexity of the overall system, can only be done in the wind tunnel.

Wind tunnel measurements have the advantage of being relatively easy to perform, and it is possible to use models and prototypes already at a very early stage of the development process (see also section 4.6). However, a number of disadvantages have to be accepted, arising from simplification and idealization. For example, as shown schematically in Figure 5.15, the incident flow profile in the wind tunnel is "rectangular" if the boundary layer (measuring only a few millimeters in height) is removed. When driving on roads, however, where natural wind occurs, the incident flow profile is twisted from the superposition of driving velocity and the atmospheric boundary layer profile. In addition, it should be noted that a side wind generator generates a velocity profile that runs between the profile in the wind tunnel and the profile on the road. Further, the very low degree of turbulence of the wind tunnel flow does in no way reflect the spectrum of the velocity fluctuations within the atmospheric boundary layer.

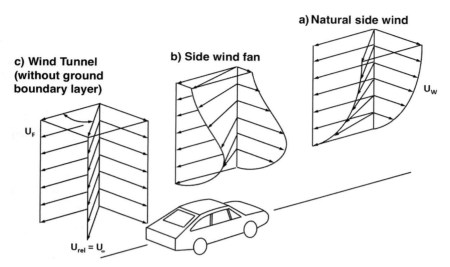

Figure 5.15 Schematic comparison of various side wind profiles according to Hucho [382].

In recent years, however, for the stated reasons, aerodynamic development has also started to focus on studying the vehicle characteristics under the unsteady flow situations described in section 5.1.2.1. A number of different experimental setups are presented in the literature indicating the complexity of the task. The following section provides an overview. An initial distinction can be made between two different methods of applying a sudden gust of wind to a model:

- Steady-state flow, moving model
- Unsteady flow, stationary model

The first approach with steady-state flow and a moving model basically simulates traveling past a side wind facility (see Figure 5.16). The vehicle model is moved on a sled, which also contains a balance to measure the forces acting on the model. A limiting factor of this method of measurement is the width of the nozzle the model crosses. The model velocity and the nozzle width together determine the time during which the model is exposed to the air stream. Owing to the typically limited rail length, the model needs be accelerated and decelerated at high rates, which puts corresponding demands on the force-measuring equipment that is used. The movement of the model and the vibrations induced reduce the signal/noise ratio, which can make an interpretation of the measured data difficult and requires special filtering of the raw signal (see [137]).

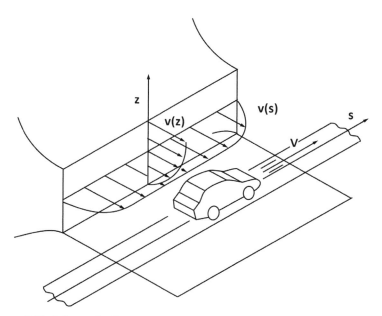

Figure 5.16 Schematic diagram of travel distance relative to the nozzle [372].

Using this approach, Chadwick et al. [138] examined the transient behavior of side force and yaw moment on simple, rectangular-shaped bodies at a scale of 1:10. They paid special attention to the influence of sharp or rounded edges on the force characteristic.

When entering and exiting the side wind stream, the yaw moments show strong overshooting compared to those determined under steady-state conditions, while the coefficients increase linearly with the yaw angle. The maximum is reached at all angles approximately one model length after entering the gust. At a yaw angle of 25° of the incident flow, the transient yaw moment coefficient increases 2.5 times compared with the steady-state value. Similarly pronounced transient behavior can be observed on the model with rounded edges, but higher steady-state yaw moments at similarly high maximum values are achieved compared to the results of the sharp-edged model, so that the ratio of maximum to steady-state value drops to 1.5.

Pressures have been measured to identify areas of the model surface contributing significantly to the development of aerodynamic forces. Figure 5.17 shows the pressure distribution at $\beta = 20°$ under steady-state conditions (a) and one model length after entering the gust (b). The comparison reveals that the leeward flow is still not fully developed and that low pressures in the front area, with a correspondingly large lever arm to the moment point of reference, are responsible for the high yaw moment at this point in time. Transient effects can also be found at the exit from the gust of wind, in particular on the lee side, when the separation bubble collapses. Figure 5.18a–c show the pressure distributions of the box model with rounded edges at the exit from the side wind. The downwind side separation region is located in the range $x/l = 0.6$. The center of pressure of the side force moves farther downstream toward the center of the vehicle. In comparison with the sharp-edged model, this leads to a 40% reduction in the yaw moment. To illustrate the differences between windward side and lee side, Figure 5.19 shows the contributions to the yaw moment of the respective sides. The transient character on the lee side is visible here, caused by separation at the front edge.

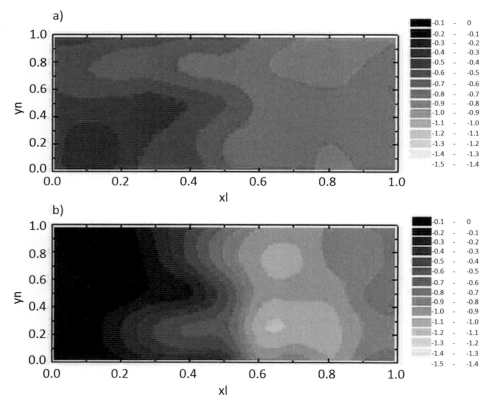

Figure 5.17 Downwind side pressure distribution of a box model with sharp edges at an effective yaw angle of 20° [138]: (a) steady-state conditions; (b) one model length after entering the wind gust.

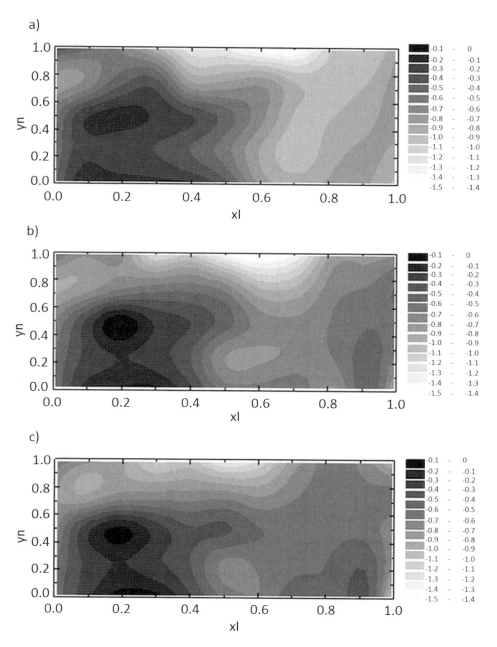

Figure 5.18 Downwind side pressure distribution: (a) when exiting the wind gust; (b) one half-model length after the exit; (c) one model length after the exit [138].

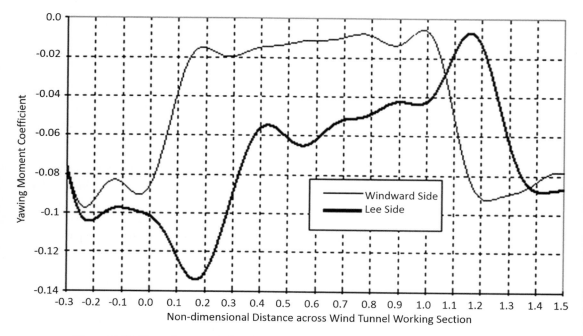

Figure 5.19 Time characteristic of upwind and downwind contributions to the yaw moment [138].

Cairns [122] performed tests on the same system using a vehicle model with different rear end shapes. When entering the side wind stream, the coefficients increase rapidly and reach the value determined under steady-state conditions. The fastback model with sharp edges shows a slower rise to the steady-state value, which according to Cairns is the result of separation that leads to a different flow around the vehicle compared to the model with rounded edges.

Kobayashi and Yamada [445] have also studied a simplified vehicle body at a scale of 1:10 in a side wind system. Their study focused on the impact of the angle of attack of the windshield, the windshield concavity in side view, the front radius, and the angle of the side taper at the vehicle rear on the transient characteristic of the yaw moment coefficient. For all measured versions, the characteristic demonstrates a strong increase approximately 1.3 model lengths after entering the side wind stream before reaching the steady-state value. When the windshield angle (a) and contour (b) are changed, the maximum and steady-state values of the yaw moment increase as the projected side surface of the model is increased. An increase in the radius at the front of the model (c) leads to a reduction of the maximum value but an increase in the steady-state value. However, a larger taper angle at the rear (d) does not affect the maximum value, but leads to a higher steady-state value. All versions have in common that the transition from maximum to steady-state value is affected. Surface pressures have been measured in addition to forces and moments. The results are consistent with those of Chadwick et al. [138]: they show that the transient characteristic of the yaw moment is caused essentially by the leeward separation at the front of the model.

Compared to the first approach, the second one involves a stationary vehicle model and an unsteady flow. The unsteadiness of the flow can be generated in several ways: for example, by a second stream of air that is blown into the main stream at an angle (see [208], [211], [694]) or by deflecting the wind tunnel flow using flaps or wings placed at the nozzle inlet or outlet (see [158], [595], [628], [739]).

Figure 5.20 schematically represents the setup of the side wind tunnel of the University of Durham. Apart from the actual wind tunnel, a second side wind tunnel is installed. Its longitudinal axis is rotated 30° from that of the first. In order to achieve a dynamic impact on the model by a gust of side wind, the nozzle of the side wind tunnel is fitted with a system of flaps that can be opened and closed by a controller. In order to keep the volume flow of the side wind fan constant, there is another flap system under the test section floor, which is always open when the corresponding upper part is closed. Figure 5.21 shows the characteristic of the angle of attack in the empty test section. A maximum angle of 30° can be achieved and the duration of the side wind gust is freely adjustable (in the example shown it is approximately 0.6 s). The results obtained with this setup essentially demonstrate the time-dependent aerodynamic effects already observed using the setup with moving models. However, it has the advantage of lower interference on the measured forces and moments and simplified measurement of surface pressures and flow fields, such as in the wake of the model.

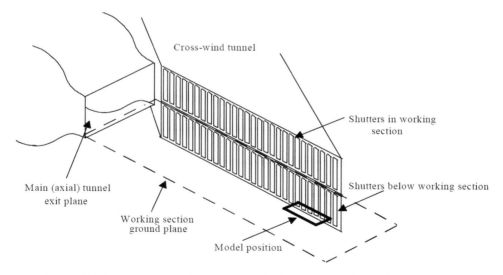

Figure 5.20 Generating a side wind gust in the Durham side wind tunnel [695].

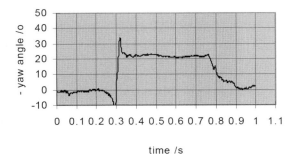

Figure 5.21 Time characteristic of the angle of attack in the empty test section [695].

Besides the experimental setups, other approaches are adopted that provide for active deflection of the main flow of the wind tunnel. This makes it possible not only to generate individual events such as generic gusts with different time characteristic and amplitudes, but also a spectrum of gusts whose transverse component covers the entire frequency range relevant from a driving dynamics point of view. In the latter method, the observation period is selected such that the flow can be viewed as statistically steady-state, that is, the measurement time is at least twice as long as the period of the lowest frequency contained in the flow. The results are no longer evaluated in the time domain but in the frequency domain.

Cogotti [158], for example, presents a so-called turbulence generation system (TGS) in which five vortex generators are arranged just behind the inlet to the nozzle but still before the start of contraction (see Figure 5.22). Each consists of a vertical bar on which a pair of rotating flaps is mounted. These five pairs of flaps can be controlled individually and perform swivel movements to reproduce both the boundary layer of the atmospheric wind above the road surface with the turbulence it contains and also dynamic changes of the angle of attack as in gusty wind. The dynamic change in the angle of attack ranges from a quasi-steady condition at a frequency of 0.01 Hz up to a frequency of 0.8 Hz. Any discrete frequency can be set by changing the flap frequency. The amplitude of the angle of attack goes up to 3° and is dependent to a small degree on the frequency and wind velocity [125].

In order to compare results for different frequencies, the occurring amplitudes of the force and moment coefficients are related to the amplitude of the corresponding angle of attack. The curves shown in Figure 5.23 are also normalized to the quasi-steady value at $f = 0.01$ Hz. They demonstrate the yaw reaction of a simplified notchback model and a squareback model to a dynamic change in the angle of attack depending on the frequency. When viewing the curves, it is apparent that they are not constant against the frequency but demonstrate a dependency, which differs between the notchback model and the squareback model. While the curve of the notchback model has an overshoot of approximately 40% compared to the steady-state value at $f = 0.1$ Hz, the curve of the squareback model does not exhibit this behavior, but continuously decreases with increasing frequency up to at least 0.5 Hz.

Aerodynamics and Driving Stability

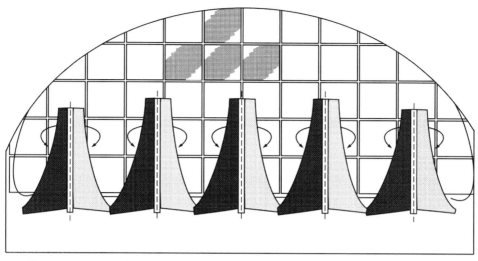

Figure 5.22 Turbulence Generation System (TGS) in the Pininfarina wind tunnel.

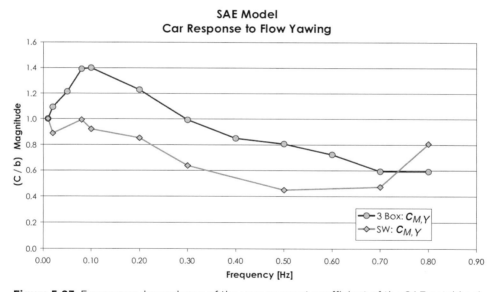

Figure 5.23 Frequency dependency of the yaw moment coefficient of the SAE notchback and squareback models, normalized to the value at quasi-steady conditions at $f = 0.01$ Hz; "3 box" = notchback, "SW" = squareback (station wagon) [125].

These results show that the quasi-steady approach commonly used is not valid to calculate the forces acting on a vehicle under unsteady incident flow conditions. This is due to the fact that changes in the incident flow occur faster than a steady-state condition of the flow field surrounding the vehicle can become established. To describe the forces under such unsteady flow conditions, Davenport [189] introduced the aerodynamic admittance. Originally developed for buildings, it is a dimensionless function that describes the frequency dependency of unsteady wind excitation and the resulting force. Applied to vehicle aerodynamics, the admittance function can be defined for side force and yaw moment according to equations (5.10) and (5.11).[1]

$$X_{a,S}(f) = \sqrt{\frac{S_{cS}(f)}{\left(\frac{dc_s}{d\beta}\right)^2 * S_\beta(f)}} \qquad (5.10)$$

$$X_{a,Y}(f) = \sqrt{\frac{S_{cM,Y}(f)}{\left(\frac{dc_{M,Y}}{d\beta}\right)^2 * S_\beta(f)}} \qquad (5.11)$$

The admittance is calculated from the spectrum of the measured force coefficients $S_{cS}(f)$ or $S_{cM,Y}(f)$, the steady-state coefficient gradient $d_{cS}/d\beta$ or $d_{cM,Y}/d\beta$ and the spectrum of the undisturbed wind excitation $S_\beta(f)$. The steady-state coefficient gradient $dc/d\beta$ describes the linear dependency of side force or yaw moment coefficient on angle of attack under steady-state conditions. For a given body, admittance can thus be understood as a measure that describes the effectiveness of the conversion of the kinetic energy of the wind excitation to side force or yaw moment, where the steady-state forces are the reference. For wind events with a very large wavelength (i.e., very small frequency), changes in the incident flow occur sufficiently slowly, so that a flow field comparable to that under steady-state conditions develops. The measured forces (numerator) therefore correspond to those calculated according to the quasi-steady-state approach (denominator), and the following applies for the admittance function:

$$\lim_{f \to 0} (X_a(f)) \to 1 \qquad (5.12)$$

If the wavelengths are of the order of magnitude of the body or smaller, the quasi–steady-state approach loses its validity because it can no longer be assumed that the flow field around the body matches that under steady-state conditions. The characteristic of the admittance function is then dependent on the shape of the body and generally falls to zero as frequency increases.

$$\lim_{f \to \infty} (X_a(f)) \to 0 \qquad (5.13)$$

The inclusion of the steady-state coefficient gradient in the calculation of the admittance function is equivalent to a vehicle-specific normalization. A vehicle configuration can therefore be evaluated relatively across the frequency range, however an absolute statement about the configuration cannot be made.

Supplementing Davenport's quasi–steady-state approach, [739] defines the transfer characteristics by an approach using linear system theory. This links the wind excitation and the reaction forces of the studied model by an input/output relation (see Figure 5.24). In contrast to the quasi–steady-state approach, this establishes a causal relationship

---

1. A detailed derivation of the admittance function can be found in [739].

between undisturbed wind excitation and response forces. The validity of this relationship can be checked using the coherence function.

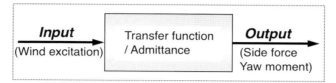

Figure 5.24 Input/output relation between wind excitation and side force or yaw moment [739].

As a complement to dimensionless aerodynamic admittance function, the aerodynamic transfer function $H_a(f)$ is defined to describe the unsteady model characteristics, which allows different models or configurations to be compared on an absolute basis. Equations (5.14) and (5.15) define the transfer functions between wind excitation and side force or yaw moment.

$$H_{a,S}(f) = \sqrt{\frac{S_{cS}(f)}{S_\beta(f)}} \tag{5.14}$$

$$H_{a,Y}(f) = \sqrt{\frac{S_{cM,Y}(f)}{S_\beta(f)}} \tag{5.15}$$

By normalization with the wind excitation spectrum, the aerodynamic transfer function at low frequencies by definition approaches the steady-state coefficient gradient of the model.

$$\lim_{f \to 0} (H_a(f)) \to \frac{dc}{d\beta} \tag{5.16}$$

According to equations (5.17) and (5.18), the coherence functions of side force and yaw moment are defined as the quotient of the cross power spectral density between wind excitation and resulting side force or yaw moment and the product of the respective power spectral densities.

$$\gamma_{\beta,S}^2(f) = \frac{|S_{\beta,S}(f)|}{S_\beta(f) \cdot S_{cS}(f)} \tag{5.17}$$

$$\gamma_{\beta,Y}^2(f) = \frac{|S_{\beta,Y}(f)|}{S_\beta(f) \cdot S_{cM,Y}(f)} \tag{5.18}$$

If the values of the coherence function in the frequency range considered are identical to 1, ideal transfer characteristics of the system can be assumed, and both the admittance and the transfer function can be used to describe the unsteady aerodynamic characteristics of the model without any restrictions. However, if the coherence function has low values, this indicates a disturbance of the input/output correlation or is due to additional components in the output signal that are uncorrelated with the input signal. The determination of admittance and transfer functions is then fraught with uncertainties.

The determination of admittance and transfer functions can be successfully used to describe the unsteady response of a vehicle to gusty wind excitation only if the measurement setup can be represented by the input/output system defined in Figure 5.24. This requires the flow to be fully coherent in a plane transverse to the main flow direction. Only then will a single input contain the entire information from the undisturbed wind excitation. For this reason, a velocity profile cannot be established in the vertical direction because the vehicle would then be subjected to varying flow conditions over its height.

Compared to the TGS presented in [158], which superimposes an atmospheric boundary layer profile with active gustiness, the systems introduced in [595], [628], and [739] establish a flow field that is completely coherent transverse to the main flow direction and thus meets the requirements described. Figure 5.25 shows a picture of the system in the model wind tunnel of the University of Stuttgart. Four blade profiles are attached to shafts at the nozzle exit plane. Each of them can be rotated by a motor to deflect the flow to the side in a specific manner. Angles of attack of ±10° and a maximum frequency of 10 Hz can be realized. The time dependency of the flow in the empty test section is depicted in Figure 5.26 for different conditions: Image (a) shows a jump-shaped deflection of the flow similar to the situation when passing a side wind generator at different flow angles and for different durations. Image (b) shows a sinusoidal curve of the angle of attack; the amplitude and frequency are freely selectable within the specified limits. Image (c) shows the flow angle over time when generating a broadband spectrum of gusts. The abrupt changes in the flow angle are clearly visible and also the maximum flow angle of ±10°. This flow condition reflects the essential characteristics of the natural gustiness of wind found in measurements on the road, which is the major cause of impaired directional stability of the vehicle (cf. section 5.1.2.1or, for example, [743]).

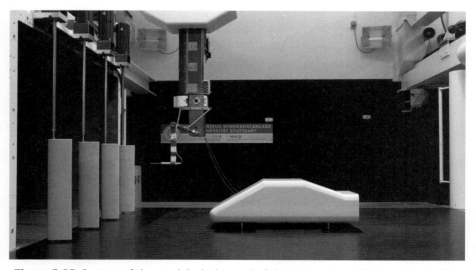

Figure 5.25 System of the model wind tunnel of the University of Stuttgart for active deflection of the main flow of the wind tunnel [739].

Aerodynamics and Driving Stability

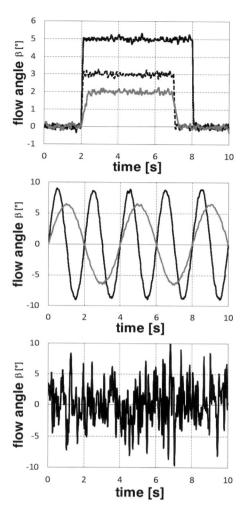

Figure 5.26 Time characteristic of the flow angle in the measuring section of the model wind tunnel of the University of Stuttgart.

Results from measurements with this system are shown in the following figures. Simplified notchback and squareback models are used as test objects. The results are plotted against the Strouhal number Sr, which is calculated from the frequency f, the wheelbase l_0, and the velocity u_∞ of the undisturbed incident flow. For illustrative purposes, it is noted that the maximum Strouhal number of $Sr = 0.15$ converts to a frequency of $f = 2.7$ Hz for a full-scale vehicle at a velocity of $u_\infty = 160$ km/h, thus the frequency range relevant to driving dynamics is covered.

$$Sr(f) = \frac{f \cdot l_0}{u_\infty} \tag{5.19}$$

To check the validity of the input/output relation, the coherence between wind excitation and side force or yaw moment is calculated. The results are shown in Figure 5.27 for the two models. It is evident that the coherence of both models assumes very high values both for the side force and for the yaw moment across the presented range of Strouhal numbers. The coherence of the side force decreases only from $Sr = 0.13$, which relates to the declining amplitude of the unsteady side force. The signal-to-noise ratio therefore decreases at the same noise level, which leads to lower coherence values. With regard to the admittance and transfer functions, it is therefore justified that both can be used without any restriction over the entire frequency range to describe the side force and yaw moment resulting from the crosswind excitation.

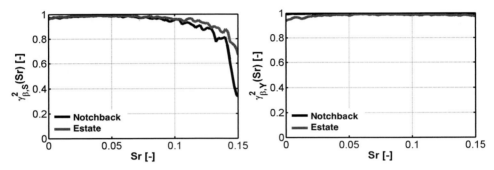

Figure 5.27 Coherence between wind excitation and resulting side force (left) and yaw moment (right).

To describe the response of the two models to a crosswind excitation, Figure 5.28 shows the admittance functions for side force and yaw moment calculated from equations (5.10) and (5.11). In accordance with the theory, the quasi–steady-state approach applies at low frequencies and the functions approach 1. For the notchback model, the admittance of the side force overshoots the steady-state value by a maximum of approximately 20% at a Strouhal number of $Sr = 0.06$, before the function falls off at higher frequencies. It is evident from the curve that, in a frequency range relevant for driving dynamics, unsteady side forces occur that are underestimated by a quasi–steady-state approach. The admittance of the yaw moment also reaches a value of 1 for low frequencies and remains constant up to a Strouhal number of $Sr = 0.05$ before the function rises up to its maximum at $Sr = 0.1$. The maximum exceeds the value determined in steady-state by approximately 30%, so that the unsteady yaw moment commonly calculated from the quasi–steady-state is also predicted too small. Compared to the notchback model, the admittance function of the side force of the squareback model hardly shows any overshooting and decreases to small values at higher frequencies with lower gradient. The admittance of the yaw moment, however, shows a more pronounced overshoot of up to 100% compared to the steady-state value.

Aerodynamics and Driving Stability

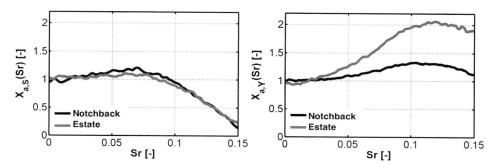

Figure 5.28 Admittance of the resulting side force (left) and of the yaw moment (right).

The quasi–steady-state forces are used to calculate the aerodynamic admittance, as explained. This is equivalent to vehicle- or configuration-specific normalization, which does not permit absolute comparison. However, as comparison is important for assessing the models, aerodynamic transfer functions are used, which are shown in Figure 5.29 for the two rear end shapes. As the transfer function for $f = 0$ Hz starts at the value of the steady-state coefficient gradient, the function of the side force of the notchback model is below that of the squareback model. As the curves of the configurations do not differ much from each other, the squareback model experiences larger unsteady side forces over the entire Strouhal number range under gusty winds than the notchback model. Regarding the yaw moment, the squareback model generally experiences smaller amplitudes, but owing to the stronger overshoot in the Strouhal number range starting from $Sr = 0.12$, they reach the same magnitude as those for the notchback model.

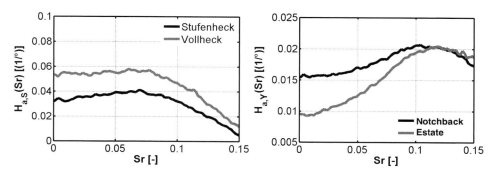

Figure 5.29 Transfer function of the side force (left) and the yaw moment (right).

The displayed results reveal that the reaction forces resulting from unsteady wind excitation differ considerably from those calculated from a quasi–steady-state approach. In particular, the overshoot of the yaw moment, which is found to be more pronounced for the squareback model than for the notchback model, may be up to 100% higher than the value determined by a steady-state approach. This must be viewed critically from a driving dynamics perspective. The cause of the frequency dependency on side force and yaw moment and the differences between the two models can be explained from the pressure distributions shown in Figure 5.30. The figure shows the values of the transfer function between the angle of attack and the difference of the surface pressure coefficient on the side surfaces of the models at $Sr = 0.015$ and $Sr = 0.1$. This illustrates the effectiveness of the conversion of the kinetic energy of the wind excitation to surface pressure and makes it possible to analyze local phenomena, depending on the model shape at different Strouhal numbers. A comparison of the front halves of the models shows that the pressure conditions are identical. The increase in amplitude at $Sr = 0.1$ in this range is also the same for both models. It is therefore not possible to identify a dependency of the result in the front half of the model on either rear end shape. However, significant differences can be seen on the rear half of the side surface of the models. At a Strouhal number of $Sr = 0.015$, higher unsteady pressures can be found in the area of the C-pillar of the notchback model than in the center of the model's side surface and in the lower half of the rear overhang, where the amplitude of the transfer function is small. Owing to these low amplitudes, the contribution of these surfaces to side force and yaw moment is also small. For the squareback model, however, the center and rear areas of the side surface exhibit an even distribution with a larger amplitude than for the notchback model. Owing to the larger surface compared to the notchback, the pressures acting on the surface make a greater contribution to the side force and yaw moment. With increasing Strouhal number, the pressure amplitude across the surface increases evenly for the squareback model, whereas in the notchback model the increase is restricted to the area in the center of the model and to the lower section of the rear overhang. The amplitudes in the area of the C-pillar remain constant, so that at $Sr = 0.1$ the distribution of the unsteady pressure amplitude is more homogeneous across the surface in the notchback model as well.

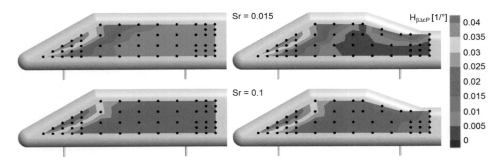

Figure 5.30 Values of the transfer function between crosswind excitation and difference of surface pressure coefficient between the side surfaces of the models at $Sr = 0.015$ and 0.1 [739].

However, the change of the pressure amplitude alone is not sufficient to fully describe the resulting forces. To do so requires the temporal relationship between the pressures occurring on the surface, which is shown in Figure 5.31 in the form of the phase relation. The pressure measuring point foremost on the side surface is used in this case as a reference, so that the phase relations are specified relative to this position. When comparing the models at $Sr = 0.015$, pressures are in phase across the surface of squareback. Only in the area of the rear overhang, a slightly larger phase angle of approximately 20° can be observed. For the notchback model, however, the front and rear halves of the model are in a phase relationship of 180° (i.e., unsteady pressures occur with opposite phases). These phase relations mean for the squareback model the yaw moment caused by the pressures occurring at the front of the model is reduced by pressures occurring in the rear section of the model, while the yaw moment is amplified for the notchback model. At $Sr = 0.1$ the squareback model shows an increase in phase relation between the front and rear parts of the side surface. This causes any pressures occurring on the surface to amplify the yaw moment, which explains the more pronounced overshoot of the yaw moment in the squareback model.

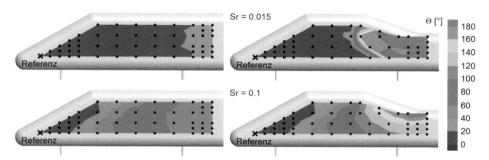

Figure 5.31 Phase relation between the first pressure measuring point (reference) and all other measuring points on the side surface of the models at $Sr = 0.015$ and $Sr = 0.1$ [739].

When the aerodynamic characteristics of a vehicle are developed, they need to be influenced specifically to minimize the side wind sensitivity of the vehicle. As examples, two measures are presented on the simplified notchback model, and their effects on the steady-state and unsteady response are examined.

The first measure consists of separation edges attached at the rear of the model, which extend beyond its side surfaces (see Figure 5.32, left). The second measure consists of a tail fin in the center line of the model, which is a two-dimensional representation of the squareback model (see Figure 5.32, right). From the literature [707], it is known that the rear side force can be increased by introducing large surfaces at the rear of the vehicle, thereby reducing the resulting yaw moment.

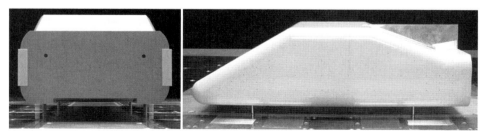

Figure 5.32 Aerodynamic modifications to the notchback model: separation edges at the rear (left); tail fin (right) [739].

Figure 5.33 shows the transfer functions of the side force and yaw moment of the notchback model, which are already known from the above observations. The impact of the modifications becomes clear via comparison. The separation edges attached at the rear exhibit no change in the steady-state gradient. There is also no influence on the unsteady response of the side force. However, the transfer function of the yaw moment can be reduced from $Sr = 0.05$ on; the maximum amplitude is approximately 8% smaller than that of the initial model. Significantly larger changes are achieved with the tail fin. It changes the steady-state coefficient gradient of side force and yaw moment and also influences the characteristic of the curves in the examined Strouhal number range. The measure increases the side force acting on the model, but the shift in the position of the center of pressure in the direction of the moment reference point reduces the resulting yaw moment. This statement is valid over the Strouhal number range under consideration. It can be seen, however, that the transfer function of the yaw moment has a larger gradient in the range of $0.05 < Sr < 0.1$ than the model without a tail fin. As a result, the relative overshoot of the yaw moment in relation to the steady-state value increases, while the maximum value is still 10% below that of the initial model.

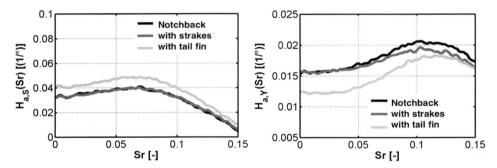

Figure 5.33 Transfer function of side force (left) and yaw moment (right) for a notchback model and its variants [739].

The results presented in this chapter for vehicles or models under unsteady incident flow conditions, as they occur during overtaking maneuvers or in gusty side wind conditions, demonstrate that the commonly used quasi–steady-state approach, which assumes frequency-independent proportionality between wind excitation and forces, is not sufficient to fully describe the unsteady behavior. The conversion of the kinetic energy of wind excitation to side force and yaw moment is frequency-dependent. It is particularly noteworthy that unsteady side forces and yaw moments occur, which exceed those calculated according to the quasi–steady-state approach. This must be viewed as critical in that these forces occur in a frequency range relevant to vehicles, thereby directly affecting the response of the vehicle and the driver.

Aerodynamics and Driving Stability

5.2 Dynamic Driving Effects

5.2.1 Single-Track Model

The following chapter discusses the influences of lift and crosswind on road dynamics. Many of the effects described there can be explained by a simplified examination of road dynamics, using a linear, flat single-track model (see [574]). The principal characteristic values and movement equations are discussed in brief next.

Note: For uniformity reasons the nomenclature customarily used in publications on road dynamics has been modified and brought into line with aerodynamic conventions in this book. When used in formulae below, the symbol β therefore describes the yaw angle rather than the vehicle's float angle. The symbol τ is used for the float angle. Figure 5.34 is a sketch of a linear, flat single-track model. In addition to steering input, the single-track model can be excited by a side force F_S, which acts on it at the vehicle's center of gravity with leverage $e_{SP,X}$.

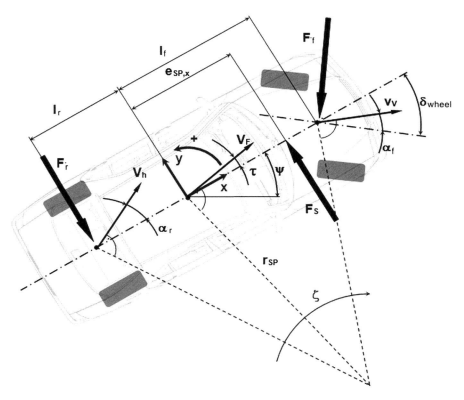

Figure 5.34 Sketch of a linear, flat single-track model [833].

The resulting lateral forces at the front and rear axles are dispersed by way of slip angles. These are combined for each axle on the linear single-track model. In reality, they largely result from wheel position changes in the following proportions:

- Angular tire skew
- Elastokinematic wheel position changes
- Roll steer at the axles
- Steering rigidity at the front axle

The following simplification has been assumed in the examination that follows:

- The relationship between lateral force and the skew angle (skew resistance) at the axle is examined linearly, that is to say, twice the lateral force results in twice the skew angle.
- The steering ratio is constant.

Rendering skew rigidity in a linear manner is permissible to within a high level of conformity for a lateral acceleration range up to approximately 0.4 g, and the vehicle's self-steering effect and reaction to crosswinds can be observed. If lateral acceleration is higher, the primary factor to be taken into account is nonlinear behavior of the tire (see section 5.2.2.1). The basic equations for the single-track model when subject to excitation by steering movement and crosswind are:

$$F_S \cdot e_{sp,x} - F_f \cdot l_f \cdot \cos\delta_{wheel} + F_r \cdot l_r \approx J \cdot \ddot{\psi} \quad \text{where } \cos\delta_{wheel} \approx 1 \tag{5.20}$$

$$\alpha_f \approx \tau - \delta_{wheel} - \frac{l_f}{r_{sp}} \quad \text{and} \quad \alpha_r \approx \tau + \frac{l_r}{r_{sp}} \tag{5.21}$$

$$F_S - F_f - F_r \approx m_F \cdot a_q = m_F \cdot \frac{v_F^2}{r_{sp}} \tag{5.22}$$

With skew angle rigidity values linearized ($F_{f/r} \sim c_{f/r} \cdot \alpha_{f/r}$), the results can be summarized as follows:

$$F_S \cdot e_{sp,x} - c_f \cdot l_f \cdot \left(\tau - \delta_{wheel} + \frac{l_f}{v_F} \cdot \dot{\psi}\right) + c_r \cdot l_r \cdot \left(\tau - \frac{l_r}{v_F} \cdot \dot{\psi}\right) \approx J_z \cdot \ddot{\psi} \tag{5.23}$$

$$F_S - c_f \cdot \left(\tau - \delta_{wheel} + \frac{l_f}{v_F} \cdot \dot{\psi}\right) + c_r \cdot \left(\tau - \frac{l_r}{v_F} \cdot \dot{\psi}\right) \approx m_F \cdot v_F \cdot (\dot{\psi} + \dot{\tau}) \tag{5.24}$$

Crosswind force F_S can be replaced by the expression (see sections 4.1 and 4.4, and equation (5.102)). By introducing the constant steering ratio

$$\delta_L = i_L \cdot \delta_{wheel} \tag{5.25}$$

the equation system for a linear, flat single-track model is obtained.

Aerodynamics and Driving Stability

$$\begin{bmatrix} \ddot{\psi} \\ \dot{\tau} \end{bmatrix} = \begin{bmatrix} -\dfrac{c_r \cdot l_r^2 - c_f \cdot l_f^2}{J_z \cdot v_F} & \dfrac{c_r \cdot l_r - c_f \cdot l_f}{J_z} \\ \dfrac{c_r \cdot l_r - c_f \cdot l_f - m_F \cdot v_F^2}{m_F \cdot v_F^2} & -\dfrac{c_r + c_f}{m_F \cdot v_F} \end{bmatrix} \cdot \begin{bmatrix} \dot{\psi} \\ \tau \end{bmatrix} \quad (5.26)$$

$$+ \begin{bmatrix} \dfrac{\rho_L \cdot A_x \cdot k_s \cdot e_{sp,x}}{2 \cdot J_z} \\ \dfrac{\rho_L \cdot A_x \cdot k_s}{2 \cdot m_F \cdot v_F} \end{bmatrix} \cdot \beta \cdot v_{res}^2 + \begin{bmatrix} \dfrac{c_f \cdot l_f}{J_z \cdot i_L} \\ \dfrac{c_f}{m_F \cdot v_F \cdot i_L} \end{bmatrix} \cdot \delta_L$$

Since this is a system capable of oscillation, it can be characterized by system-intrinsic values: eigenfrequency

$$\omega_0 = \sqrt{\dfrac{m_F \cdot v_F^2 \cdot \left(c_r \cdot l_r - c_f \cdot l_f\right) + c_r \cdot c_f \cdot \left(l_r - l_f\right)^2}{m_F \cdot v_F^2 \cdot J_z}} \quad (5.27)$$

decay constant

$$\delta = \sqrt{\dfrac{m_F \cdot \left(c_r \cdot l_r^2 - c_f \cdot l_f^2\right) + J_z \cdot \left(c_r + c_f\right)}{2 \cdot m_F \cdot v_F \cdot J_z}} \quad (5.28)$$

damped circular-track eigenfrequency

$$\omega_d = \sqrt{\omega_0^2 - \delta^2}, \quad (5.29)$$

damping value

$$\vartheta = \dfrac{\delta}{\omega_0} \quad (5.30)$$

and peak response time

$$T_{\psi,max} = \dfrac{1}{\omega_d} \cdot \arctan \dfrac{\omega_d}{\delta - \dfrac{c_r \cdot \left(l_r + l_f\right)}{m_F \cdot v_F \cdot l_f}}. \quad (5.31)$$

Increasing road speed leads to the following changes in the system's intrinsic values:

- Reduction in undamped eigenfrequency
- Reduction in decay constant
- Increase in damped eigenfrequency
- Reduced damping value
- Reduced peak response time

With regard to the behavior of the single-track model, this means that in view of the reduction in damping, all the vehicle's reactions also increase with the road speed.

This applies both to steering angle inputs and to disturbance caused by crosswinds. The principal characteristic values and interactions in the event of excitation from the steering and from crosswinds are discussed next (see also [358], [398], [574], etc.). An important value when determining the steering behavior of the single-track model is the self-steering gradient (section 5.2.2.1). This provides a clear indication of how much the driver must correct the steering wheel position when accelerating on a circular path and thus defines the vehicle's self-steering behavior. The self-steering gradient (EG) is defined as

$$EG = \frac{m_F \cdot (c_r \cdot l_r - c_f \cdot l_f)}{c_r \cdot c_f \cdot l_0} \quad (5.32)$$

with:

- $EG > 0$: understeer
- $EG = 0$: neutral
- $EG > 0$: oversteer

The float angle gradient is an indication of the margin of stability available at the rear axle (see section 5.2.2.2) and is described by

$$SG = \frac{m_F \cdot l_f}{c_r \cdot l_0} \quad (5.33)$$

Dynamic behavior of the single-track model is expressed as transfer functions in the frequency range. The amplitude ratio between yaw velocity and steering angle (section 5.2.2.2) is characteristic for behavior when steering excitation takes place and is calculated by means of equation (5.34).

$$\left|\frac{\dot{\psi}}{\delta_L}\right|(\omega) = \frac{\sqrt{D_1^2 + (D_2 \cdot \omega)^2}}{\sqrt{(C - M \cdot \omega^2)^2 + (D \cdot \omega)^2}} \quad (5.34)$$

In order to describe dynamic behavior when crosswind excitation takes place, we use the amplitude ratio of yaw velocity to crosswind excitation (section 5.2.4.3). It is calculated by means of equation (5.35).

$$\left|\frac{\dot{\psi}}{\beta \cdot v_{res}^2}\right|(\omega) = \frac{\sqrt{D_3^2 + (D_4 \cdot \omega)^2}}{\sqrt{(C - M \cdot \omega^2)^2 + (D \cdot \omega)^2}} \quad (5.35)$$

The following short forms apply to equations (5.34) and (5.35):

$$M = m_F \cdot v_F \cdot J_Z \quad (5.36)$$

$$D = m_F \cdot (c_r \cdot l_r^2 + c_f \cdot l_f^2) + J_Z \cdot (c_r + c_f) \quad (5.37)$$

Aerodynamics and Driving Stability

$$C = m_F \cdot v_F \cdot (c_r \cdot l_r - c_f \cdot l_f) + \frac{c_r \cdot c_f}{v_F} \cdot (l_r + l_f)^2 \tag{5.38}$$

$$D_1 = \frac{1}{i_L} \cdot c_r \cdot c_f \cdot (l_r + l_f) \tag{5.39}$$

$$D_2 = \frac{1}{i_L} \cdot m_F \cdot v_F \cdot c_f \cdot l_f \tag{5.40}$$

$$D_3 = \frac{\rho_L}{2} \cdot k_S \cdot A_x \cdot \left[(c_r + c_f) \cdot e_{sp,x} + (c_r \cdot l_r - c_f \cdot l_f) \right] \tag{5.41}$$

$$D_4 = \frac{\rho_L}{2} \cdot k_S \cdot A_x \cdot \left[m_F \cdot v_F \cdot e_{sp,x} \right] \tag{5.42}$$

5.2.2 Reaction to Lift Forces

The lift forces acting on a vehicle are extremely significant in road dynamic terms, since they can influence self-steering and braking behavior and thus the driver's ability to control the vehicle. This situation becomes increasingly important as road speed goes up, since lift forces—as shown in section 4.1—increase with the square of the approach velocity.

The vehicle has well-balanced road dynamic settings if its handling is neutral with a slight understeering tendency. This calls for values in the weight distribution, tire stiffness, axles, steering system, axle kinematics and axle elastokinematics areas and also aerodynamic lift forces to be carefully selected and interrelated. As road speed goes up, lift forces are superimposed increasingly on the wheel loads resulting from weight distribution. This leads to a reduction in the wheel contact forces, the results of which are discussed next (see also [574], [669], [693], etc.).

5.2.2.1 Self-Steering Behavior

Figure 5.35 illustrates the definition of a vehicle's self-steering behavior. In the driving maneuver shown in the diagram, the vehicle is driven in a circle at increasing speed so that lateral acceleration also goes up. The relationship between the steering angle needed to follow the circular path and the lateral acceleration value is an indication of self-steering behavior.

In the present example (in Figure 5.35), the steering angle increases continuously with the lateral acceleration value. In order to negotiate the circular track more rapidly and therefore with a higher lateral acceleration value, the driver must turn the steering wheel farther. When the vehicle reacts in this way, it is said to "understeer." At low lateral acceleration levels (up to approximately 0.4 g) the steering angle increases linearly, but at higher rates of lateral acceleration the increase is more marked—that is to say, the driver must turn the steering wheel more and more. When the maximum possible lateral acceleration $a_{q,max}$ is reached, turning the steering wheel farther no longer leads to any increase in lateral acceleration. The vehicle then slides sideways and can no longer follow the chosen circular path.

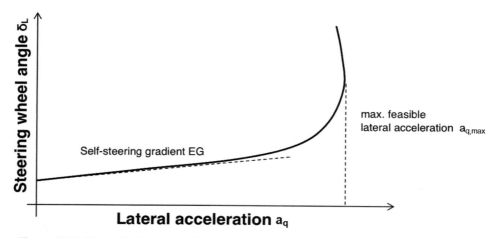

Figure 5.35 The definition of a vehicle's self-steering behavior: Steering angle vs. lateral acceleration.

A vehicle's self-steering effect is defined by the initial increase in the steering wheel angle as shown in Figure 5.35. This initial increase is referred to as the "self-steering gradient" and is described by Equation 5.32 (see section 5.2.1). Vehicles exhibiting understeer are considered to be easier to control. In critical driving situations the front axle reaches the maximum power transmission limit first, whereas the rear axle still has spare power transmission capacity. The vehicle accordingly remains directionally stable, that is to say, as the handling limit is approached it tends to run straight ahead rather than to turn into the curve. On vehicles with a tendency to oversteer, the rear axle reaches the maximum power transmission value first, and it is the front axle that still possesses spare capacity. This causes the vehicle to turn farther into the curve and to slide or spin.

If lift forces occur, they influence the self-steering gradient and the maximum lateral acceleration that can be obtained. Figure 5.36 examines the distribution of forces at the outer wheels of the car when cornering (to the right). The centrifugal force F_q caused by the vehicle's curved path calls for a lateral locating force F_S. This acts at the area of wheel contact with the road and must be withstood by the tire. Due to centrifugal force F_q there is a transfer of wheel load from the wheels on the inside to those on the outside of the curve. This proportion $\Delta F_{N,dyn.}$ is superimposed on the static wheel load constituent $F_{N,stat.}$, which is derived from gravitational force F_G. On the outside of the curve this dynamic constituent has to be added to the wheel load. The tire's lateral location potential depends on its normal force: as normal force F_N increases, the tire can transmit higher maximum lateral forces FS,max.

Aerodynamics and Driving Stability

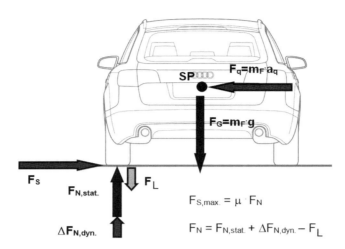

Figure 5.36 Distribution of forces at the outer wheels of the car when cornering.

If the vehicle is exposed to a lift force, the load on the tire is reduced by the amount of lift force F_L. However, centrifugal force F_q, gravitational force F_G, and the resulting reaction forces (F_S, $F_{N,stat.}$, $\Delta F_{N,dyn}$) remain unchanged, since they depend on the vehicle's mass. In order to negotiate the curve in Figure 5.36 with the same lateral acceleration, the tire should possess the same lateral locating force even if lift occurs. However, the normal force F_N is reduced by the amount of lift F_L. As a result, the tire's lateral locating potential drops, and it can only transmit lower maximum lateral forces $F_{S,max}$. In consequence, the attainable lateral acceleration value aq and the cornering speeds are also lower. Figure 5.36 uses the side of the vehicle on the outside of the curve to illustrate the effect of lift. In order to demonstrate the influence of lift on the vehicle's self-steering behavior, lift forces at the front and rear axles must be examined separately, proceeding from the following tire characteristic:

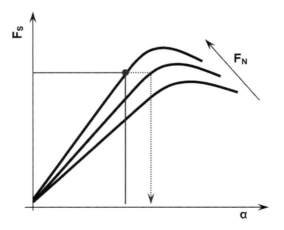

Figure 5.37 Relationship between a tire's normal force, skew angle, and lateral force transmission limit.

Figure 5.37 is a diagram showing the relationship between a tire's normal force, skew angle, and lateral force transmission limit. As the lateral force increases, the tire's skew angle becomes greater. If lateral forces are low, the tire reacts approximately linearly, so that doubling the lateral force results in the skew angle also being doubled. If lateral force continues to increase, tire behavior becomes degressive, and if the maximum lateral force that the tire can transmit is reached, it begins to slip, that is, the skew angle increases but higher lateral forces cannot be transmitted. Potential lateral location is improved when the normal force increases but deteriorates when it becomes lower.

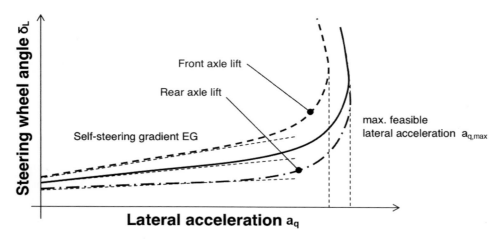

Figure 5.38 Effect on self-steering behavior under lift at the front or rear axle.

If lift forces reduce the contact pressure exerted by the tire, a higher skew angle will be needed to transmit the same lateral force (see Figure 5.37). The affected axle behaves more "softly" in the lateral direction, and the force transmission potential, meaning the maximum lateral force that can be transmitted, is also lower. Figure 5.46 illustrates the effect on self-steering behavior when there is lift at the front or rear axle. If front axle load is reduced by lift, the self-steering gradient is steeper, which means that the vehicle's handling tends toward understeer. Since the maximum lateral force that can be transmitted is reduced, maximum lateral acceleration values are also lower.

Lift at the rear axle tends to make the self-steering gradient less steep, so that the vehicle's handling tends toward oversteer, and at the same time the power transmission is reduced. Since in the present example the vehicle still understeers ($EG > 0$) although there is lift at the rear axle, the maximum attainable lateral acceleration is not reduced, but the reduction in force transmission potential has an adverse effect of rear axle stability (for this, see section 5.2.2.2).

When road-dynamic settings are chosen, it must be remembered that lift forces increase with the square of road speed. This means for example that a vehicle with high lift at the rear axle must be rated for driving safety at maximum speed and possess the necessary understeer characteristic at that speed. However, when such a vehicle is driven at lower speeds (so that the lift values are too small to take effect), it will understeer severely. This spread of values between high and low road speeds is undesirable in terms of road dynamics.

5.2.2.2 Stability and Straight-Ahead Running

Figure 5.39 illustrates the dynamic behavior of a vehicle as a result of yaw oscillation. In the driving maneuver shown here the vehicle is excited by sinusoidal steering-angle inputs at various frequencies, and the yaw velocity evaluated. The ratio of the yaw velocity and steering angle is plotted against the frequency. Three criteria are characteristic of this transmission pattern.

The amplitude ratio at frequency $f = 0$ Hz, the circular track characteristic, is equivalent to the steady yaw velocity value obtained when driving on a circular track at a predetermined steering angle. High values are characteristic of vehicles that react directly to steering inputs and need little effort at the steering wheel. The amplitude ratio at the resonance frequency provides information on what is referred to as yaw damping. The ratio is formed from the resonance and static amplitudes, that is, the increase in yaw. If a vehicle exhibits a high increase in yaw, its yaw damping is low, and when dynamic driving maneuvers are undertaken (e.g., rapid obstacle avoidance) it will exhibit severe yaw reactions and oscillations will very easily tend to increase in magnitude. This adversely influences stability and vehicle control. The resonance frequency is a parameter for vehicle agility. If a vehicle's resonance frequency is high, it will be able to respond to even rapid steering wheel inputs and will feel agile when driven.

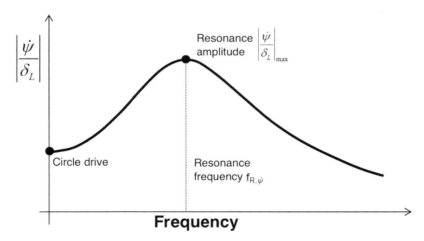

Figure 5.39 Ratio of the yaw velocity and steering angle.

Figure 5.40 shows the vehicle's stability as float angle plotted against lateral acceleration. In the maneuver undertaken here, the vehicle is driven around a circular track and its speed steadily increased. The float angle gradient shows to what extent force transmission at the rear axle, and therefore the vehicle's stability margin, is used. Gradual gradients stand for high stability, which means that rear axle force transmission potential is not fully utilized and the vehicle is less likely to depart from the chosen line in a critical situation.

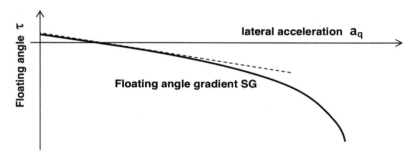

Figure 5.40 Float angle vs. lateral acceleration.

The diagrams shown here illustrate the effects of front or rear axle lift. Figure 5.41 shows the pattern of yaw velocity transmission in relation to steering inputs (by analogy to Figure 5.39) and the change in float angle in relation to lateral acceleration (by analogy to Figure 5.40) for various lift configurations. Lift at the front axle leads to lower yaw velocity amplitudes in both the static ($f = 0$ Hz) and resonance situations. This means higher steering effort compared with the reference vehicle. However, the relationship of resonance amplitude to static amplitude, that is, increased yaw, is higher than on the reference vehicle, which means that the vehicle has lower yaw damping and therefore tends more strongly to develop yaw oscillation when driven dynamically.

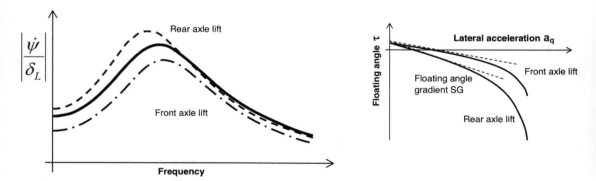

Figure 5.41 Effects of front or rear axle lift in yawing and floating angle.

In practice, the float angle gradient is not influenced by front-axle lift, which means that the stability margin is retained at the rear axle. However, as explained in section 5.2.2.1, the vehicle will exhibit stronger understeer and the maximum attainable lateral acceleration will be lower.

The effects of rear axle lift on the vehicle's stability are particularly evident when braking in a corner. Rear axle lift leads to higher yaw velocity amplitudes, that is, the vehicle needs less steering effort and responds more directly to steering angle inputs. In addition, as described in section 5.2.2.1, understeer is less pronounced. The float angle gradient is increased by lift at the rear axle. The effect of this is for the vehicle's stability margin to be reduced and for it to tend to depart from the chosen line more readily in critical driving situations.

This reaction to braking on a corner is particularly critical if there is lift at the vehicle's rear axle. The brake application, with its dynamic weight transfer from the rear to the front axle, causes a reduction in normal force at the rear axle in addition to the effect of lift. Figure 5.42 illustrates the effect on road dynamics when braking on a corner. It shows as examples the yaw velocity and float angle patterns for a stable and an unstable vehicle.

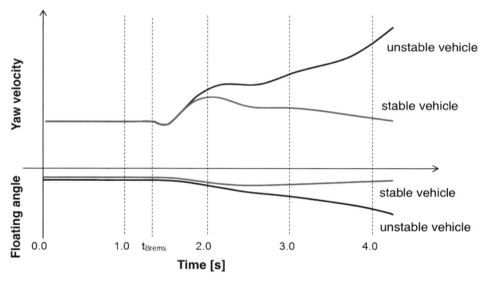

Figure 5.42 Braking on a corner for a stabile and an unstable vehicle.

For the maneuver shown here, the vehicle is driven on a circular track at a constant speed and a high rate of lateral acceleration. When time $t_{Brake} = 0$ is reached, the brakes are applied suddenly so that the vehicle is subjected to longitudinal retardation in addition to lateral acceleration. This causes dynamic shifts in wheel load from the rear to the front axle. The additional front axle load causes an increase in slip resistance (see Figure 5.37), that is, lateral forces at the front axle are built up more rapidly and yaw movement increased for a short time. In Figure 5.42, this can be seen as yaw velocity overshoot after the brake application has begun. Longitudinal retardation slows the vehicle down, so that lateral acceleration decreases and the yaw velocity subsequently drops.

The decisive conditions are those at the rear axle. On a stable vehicle the rear axle possesses sufficient force transmission potential throughout the brake application.

Dynamic load reduction due to braking leads to higher slip angles and therefore slight overshoot of the float angle in Figure 5.42, but since stability is adequate, the float angle decreases as the maneuver continues and the vehicle remains stable when braked.

If a vehicle incurs lift at the rear axle, the normal forces there are already reduced by the lift. Figure 5.42 shows this as a higher float angle for the unstable vehicle before the brakes are applied. The additional dynamic load reduction due to braking can increase total axle load reduction until the force transmission potential at the rear axle is exhausted. This means too little normal force at the tire, which is then unable to generate the necessary lateral force itself by way of an increased slip angle. Figure 5.42 shows this as an increase in float angle and yaw velocity. The rear axle slides sideways, and the vehicle begins to skid and becomes unstable. Lift forces not only influence the vehicle's stability but can also affect straight-line running. The following example illustrates how fluctuating lift forces (stimulated for instance by crosswind, overtaking other vehicles, etc.) can affect road behavior.

Figure 5.43 shows how the lateral force that the tire can transmit depends on its normal force. This is a degressive relationship: the lateral force that can be transmitted cannot be increased to the same degree as the normal force. Periodic variation of the normal force around a mean value of $F_{N,0}$ is assumed here, as caused for example by lift fluctuations. This normal force variation causes lateral locating force to vary around a mean value $F_{S,0}$. Due to the degressive relationship, the mean lateral locating force value $F_{S,0}$ is smaller than the static lateral locating force $F_{S,\text{stat}}$ that would be anticipated at the mean wheel load value. If normal forces vary, the tire can transmit only a lower mean lateral force. In order to generate the necessary lateral locating force, however, the tire adopts a larger slip angle.

If this effect occurs at the front axle, changes to the slip angle due to side wind excitation result in more severe yaw movements. Occurrence at the rear axle causes more marked lateral movements. During fast motorway journeys with many other vehicles overtaken, both situations clearly indicate that the vehicle is not pursuing a smooth course and has to be kept directionally stable by a series of corrective movements at the steering wheel. For a vehicle to be regarded as comfortable over long distances it is therefore desirable for it not to incur large amounts of lift. Furthermore, changes to the lift coefficient in response to changes in the approach angle should be low and gradual rather than increasing suddenly.

Aerodynamics and Driving Stability

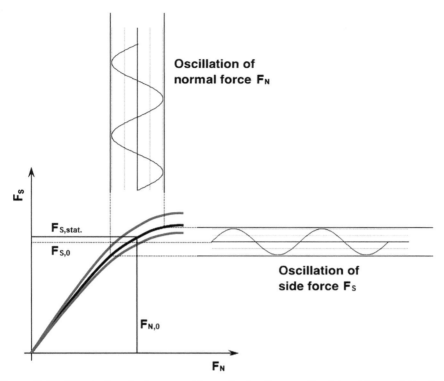

Figure 5.43 Effect of periodic normal force fluctuations on the lateral force of the tire.

5.2.2.3 Behavior when Braking

Another aspect related to lift is the influence exerted on braking behavior when driving in a straight line. Figure 5.44 is a diagram of the distribution of forces when a vehicle influenced by lift is braked. The position of the center of gravity S_P and the mass of the vehicle m_F yield the static normal forces $F_{N,f,\text{stat}}$ at the front axle and $F_{N,r,\text{stat}}$ at the rear axle. Retardation caused by braking results in a dynamic load shift from the rear to the front axle. The normal force is increased at the front axle by the dynamic constituent $\Delta F_{N,f,\text{dyn}}$ and decreased at the rear axle by $\Delta F_{N,r,\text{dyn}}$.

In order to achieve maximum braking retardation, the forces that can be transmitted at the front and rear axles must be used equally. For this to be the case, the distribution of braking force between the front and rear axles ($F_{B,f}$ and $F_{B,r}$) must have the same ratio as the normal forces at the front and rear axles $F_{N,f}$ and $F_{N,r}$. This is referred to as "ideal" brake force distribution (in this connection, see [398]) and is a permanent vehicle design feature for a predetermined braking retardation value (0.8·g).

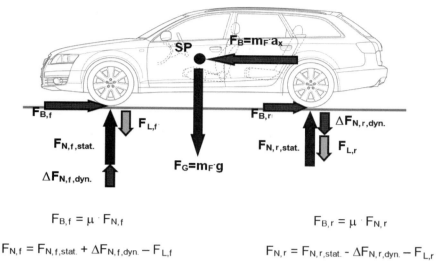

$$F_{B,f} = \mu \cdot F_{N,f} \qquad\qquad F_{B,r} = \mu \cdot F_{N,r}$$

$$F_{N,f} = F_{N,f,stat.} + \Delta F_{N,f,dyn.} - F_{L,f} \qquad\qquad F_{N,r} = F_{N,r,stat.} - \Delta F_{N,r,dyn.} - F_{L,r}$$

Figure 5.44 Distribution of forces on a vehicle influenced by lift while braking.

If the vehicle exhibits lift, the relevant forces are superimposed on the normal forces at the axles and reduce them. In this situation, the maximum braking forces that the tire can transmit are lower, and the maximum rate of retardation when braking therefore decreases. In addition, lift can influence the relationship between the normal forces at the front and rear axles to such an extent that it no longer complies with the ideal braking force distribution criteria. Lift at the rear axle reduces the maximum braking force that can be transmitted at that axle but does not influence the front axle. As a consequence of this, greater use is made of the force transmitted at the rear axle, so that its wheels tend to lock if a heavy brake application is made. Since lift forces, as explained in section 4.1, increase as the square of road speed, this tendency is more marked when braking from a high speed. Figure 5.45 shows the inter-relationships described here with the aid of a schematic braking force distribution diagram. The vehicle is assumed to incur lift at the rear axle. If the lift constituent can be disregarded, for instance at low road speeds, the ideal braking force distribution is parabolic in shape. The installed braking force distribution setting is such that at a given rate of retardation, $z = 0.8$ (equivalent to 80% of the earth's gravitational force) corresponds to the ideal braking force distribution. Above this ideal value, the vehicle is unstable when braked, since the rear wheels tend to lock first. Below it, the vehicle is stable because the front wheels lock first.

Aerodynamics and Driving Stability

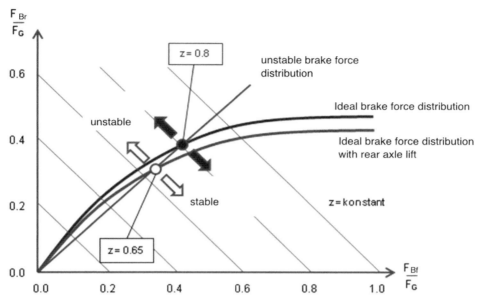

Figure 5.45 Braking force distribution diagram (exemplary for a vehicle influenced by rear axle lift).

If the vehicle described here is braked from a high speed, lift force will reduce the load on the rear axle. If this lift value is taken into account, a different ideal braking force distribution occurs and because of the above-mentioned effects, would call for lower braking force at the rear axle. The installed braking force distribution would then already be ideal at a retardation rate of $z = 0.65$, but at $z = 0.8$ the vehicle would be unstable because the rear wheels would lock first.

This "overbraking" at the rear axle has to be avoided for road safety reasons. In the example discussed here, the brake system must therefore be given settings that represent the ideal braking force distribution at a retardation rate of $z = 0.8$ when rear axle lift is taken into account. However this in turn means that when the brakes are applied at a low speed, at which lift does not exert any significant influence, the force that can be transmitted at the front axle is not fully utilized, and the vehicle's braking distance is longer.

5.2.3 Aerodynamic Axle Load Relief Settings

In order to take the interrelationships described above into account during vehicle development, the aerodynamic lift values at the front and rear must be specifically determined. For road dynamics the aerodynamic lift coefficients c_{Lf} and c_{Lr} are not themselves important but the resulting percentage axle load reductions in relation to the static axle load. Figure 5.46 illustrates this by means of a simple specimen calculation.

Chapter 5

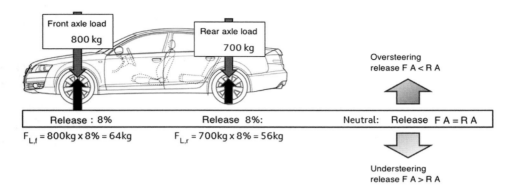

Figure 5.46 Percentage axle load reductions in relation to the aerodynamic lift coefficients.

The maximum permissible lift coefficient cLr has to be determined for various vehicle weight distributions. In the specimen calculation in Figure 5.46, a maximum permissible axle load reduction of 8% has been assumed; from this the maximum permissible load-relief forces F_{Lf} and F_{Lr}, dependent on the static front and rear axle loads can be calculated. To avoid influencing the vehicle's self-steering behavior over the entire speed range, lift forces must only lead to the same load reduction at both axles. If the load is reduced more at the front than at the rear axle, aerodynamic understeer will result; if load relief at the rear axle is greater, oversteer will result.

Load relief force F_{Lr} is used at maximum road speed vmax in the specimen calculation in order to determine maximum permissible rear axle lift c_{Lr}. For a vehicle with a rear axle load of 700 kg, the calculation yields a lift coefficient $c_{Lr} = 0.095$. If weight-saving measures reduce the static rear axle load from 700 kg to, for example, 600 kg, the calculated lift coefficient is then $c_{Lr} = 0.083$. In both cases the reduction in rear axle load is 8%, though the aerodynamic requirements differ.

This example serves to show that aerodynamic lift coefficients alone are not sufficient when assessing dynamic road behavior when exposed to lift. The resulting percentage axle load reduction has to be considered. The precise design criteria for this are specific to each manufacturer, but the following general statements can be made.

On competition cars the aerodynamic generation of high downforce forces is of great importance as a means of achieving high wheel contact forces and therefore high potential cornering speeds. On production cars it is best to generate only small amounts of lift or downforce in order to maintain the desired dynamic vehicle characteristics over the largest possible road speed range.

A marked difference in aerodynamic load reduction at the front and rear axles can lead to an undesirable oversteer or understeer tendency as a function of road speed and can also have an adverse effect on braking behavior.

Aerodynamic balance should be neutral, that is, the percentage axle load relief should ideally be the same at the front and rear axles. In practice, settings that result in mild understeer with slightly less aerodynamic load relief at the rear axle than at the front are frequently chosen for road safety reasons.

In certain circumstances, compliance with the chosen lift settings may lead to a conflict of objectives with aerodynamic drag optimization (in this connection, see section 4.5).

5.2.4 Reaction to Crosswinds

5.2.4.1 Vehicle Excitation Caused by Crosswind

Figure 5.47 outlines the approach flow conditions when a crosswind occurs [833]. The vehicle encounters a head wind v_F that has approximately the same velocity as its own road speed. A crosswind component with a velocity v_W and an approach angle ϕ is superimposed. If both component vectors are added together, the resulting combined head and crosswind has an approach velocity v_{res} and a resultant approach angle β.

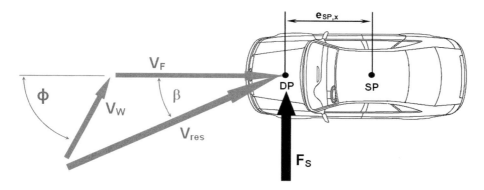

Figure 5.47 Resulting oncoming flow [833].

$$F_S = \frac{\rho_L}{2} \cdot v_{res}^2 \cdot A_x \cdot c_S \approx \frac{\rho_L}{2} \cdot A_x \cdot k_S \cdot \beta \cdot v_{res}^2 = K \cdot \beta \cdot v_{res}^2 \quad \text{for } \beta < 15° \qquad (5.43)$$

Wind flow from the side generates a lateral wind force that can be examined with the aid of equation (5.90). The aerodynamic side force coefficient can be treated as approximately linear within an approach angle range of ±15°. This is normally valid even for production vehicles with different rear end configurations. By linearizing the aerodynamic side force coefficient $c_S = k_S \cdot \beta$, the disturbance caused by a crosswind can be expressed as the product $\beta \cdot v_{res}^2$. This product depends on wind values v_W and ϕ and is therefore suitable as a means of expressing vehicle excitation due to exposure to a crosswind. It is normal to use the product $\beta \cdot v^2_{res}$ as the wind excitation parameter and as a reference value when examining vehicle transmission behavior, for example with the aid of an amplitude ratio.

Exposure to a crosswind not only generates lateral wind force F_S but also a yaw moment M_Y. This yaw moment can be taken into account with close approximation by assuming that lateral wind force F_S acts at pressure point DP with a leverage of $e_{SP,x}$. This leverage can be determined from the aerodynamic coefficients (see section 4.1) and, depending on the resultant approach angle, is obtainable from equation (5.44).

$$e_{SP,x} = \frac{c_{M,Y} \cdot l_0}{c_S} + \frac{l_0}{2} - l_r \qquad (5.44)$$

For the following discussion of behavior in crosswinds, reference is made to the linear single-track model's system of equations in section 5.2.1.

5.2.4.2 The Vehicle's Reactions to Crosswinds

Side forces occur in regular driving when a wind strikes the vehicle at an angle, causing it to deviate from the course chosen by the driver (e.g., away from a straight line). This deviation takes the form of lateral and yaw reactions. Typical situations of which the driver is aware are, for example, when a bridge is crossed or a truck overtaken at high speed, and, if stochastic gusts of wind occur, even when driving on the open road. The driver has to correct gusts of crosswind by turning the steering wheel; depending on the vehicle's sensitivity to crosswinds, this represents a reduction in comfort and convenience but can also be relevant to driving safety.

In view of its stochastic character, excitation of a vehicle by a crosswind should be examined in the frequency range. The most suitable way to "normalize" the vehicle's reaction to the crosswind is to show it as an amplitude ratio [253]. Figure 5.48 shows as an example the amplitude ratio of yaw velocity to wind excitation at various road speeds. If the crosswind is constant, the yaw velocity also has a constant value (amplitude ratio at frequency $f = 0$ Hz). Reduced damping of vehicle oscillation leads to greater yaw velocity resonance amplitudes as road speed increases.

All other vehicle reactions demonstrate a similar transmission behavior when exposed to crosswind excitation, as shown in Figure 5.48 for yaw velocity. For the driver, crosswind disturbance is sensed as a pattern made up of these vehicle reactions, the composition of which depends on vehicle design ratings or characteristics and is thus typical of the vehicle being driven [89]. This vehicle reaction pattern causes a deviation from the chosen course that has to be corrected by the driver at the steering wheel.

An important dynamic factor within this reaction pattern is yaw velocity, since it causes a visual deviation away from the intended course that is sensed by the driver. Behavior when exposed to a crosswind is frequently examined only on the basis of the yaw velocity reaction. However, with the vehicle's resonance behavior in a crosswind

Aerodynamics and Driving Stability

in mind, the vehicle reaction pattern should be considered as a whole, taking all other vehicle reactions into consideration.

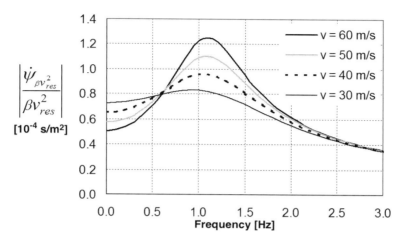

Figure 5.48 Amplitude ratio of yaw velocity to wind excitation at various road speeds [833].

Figure 5.49 shows the dependence of the damped eigenfrequency and the resonance frequencies in the relevant vehicle reaction on road speed when subject to side wind excitation. Axle assignment was deliberately chosen so that the frequency is shown on the *x*-coordinate as in Figure 5.48. Within a practically relevant speed range up to approximately 50 m/s, the resonance frequencies are scattered over a broad range. If the speed were to increase further, they would move closer to the vehicle's damped eigenfrequency value.

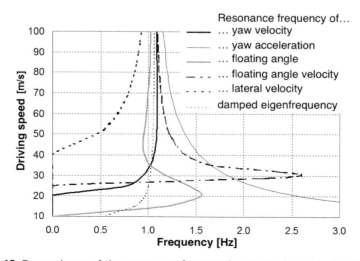

Figure 5.49 Dependence of the resonance frequencies on road speed when the vehicle subject to side wind excitation [833].

463

If gusts of wind occur at discrete intervals, the vehicle oscillates when exposed to each of them. This oscillation occurs at an intrinsic system frequency, which is the vehicle's damped eigenfrequency, and all vehicle reactions oscillate at this frequency. In this situation the vehicle's yaw velocity reaction is highly significant, since the damped vehicle eigenfrequency runs close to the yaw velocity resonance frequency. This form of excitation typically occurs when crossing bridges or overtaking trucks. In this situation, the driver is normally aware that a disturbance may occur and often makes an initial compensating steering movement before it actually takes place. Situations of this kind are to be classified as less critical.

If the vehicle is excited by a stochastic crosswind, permanent stochastic vehicle excitation is superimposed on the oscillations. In this case, unlike the sporadic vehicle excitation just described, the vehicle is excited over its entire range of resonance frequencies, so that the individual vehicle reactions resonate alternately. Here too the yaw velocity reaction is of great importance on account of the proximity of its resonance frequency to the undamped eigenfrequency. In this case, observation of the yaw velocity alone is no longer sufficient. In the event of stochastic vehicle excitation, all the vehicle's reactions should be taken into account.

5.2.4.3 Evaluation of Crosswind Behavior

The lower the lateral aerodynamic force and the yaw moment, the less marked the disturbance caused by crosswinds. An obvious aim is therefore to keep both these values as low as possible by targeted aerodynamic development. However, since disturbance from crosswinds is primarily a question of driver convenience, subjective awareness on the driver's part has to be given closer attention. A clear definition of a vehicle that is not sensitive to crosswinds, using objective criteria, is therefore much more complex to formulate. This will be discussed next.

A first step when estimating a vehicle's sensitivity to disturbance from crosswinds can be examination of the yaw angles it adopts (Figure 5.50). Compared with notchback body designs, vehicles with a two-box body have a greater surface area exposed to crosswinds at the rear. Their pressure point DP is located farther back and therefore closer to the vehicle's center of gravity SP than would be the case with a notchback body. In consequence they usually generate higher side forces but lower yaw moments; in other words, vehicles with two-box bodies incur less marked yaw reactions when crosswinds are encountered. The same applies to nose-heavy vehicles. If the vehicle's center of gravity is farther forward, it tends to suffer less severe yaw reactions in view of the reduced leverage applied by lateral wind force $e_{SP,x}$.

Aerodynamics and Driving Stability

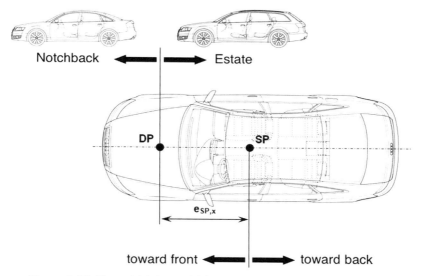

Figure 5.50 The vehicle's sensitivity to disturbance from crosswinds.

This yaw reaction estimate enables an initial statement to be made concerning a vehicle's static susceptibility to crosswinds, but in order to identify the time pattern of the vehicle's reaction to crosswinds, different approaches are necessary. One of these is to drive it past a crosswind fan ([290], [370]; see also section 5.1.2.2); this is still a frequently used method. The vehicle is exposed to a wind current generated artificially by a group of fans (Figure 5.51). Measurements can be taken by an "open loop" method, with the steering prevented from turning, or by "closed loop," in which case the driver attempts to correct the vehicle's deviation from the chosen line of travel. Evaluation is based on the time scale.

The open loop method evaluates gradient, maximum value, and overshoot of vehicle reactions when entering the crosswind airflow and the decay of vehicle oscillation when leaving the airflow. The aim is to minimize vehicle reactions and their overshoot values, in particular with regard to yaw reactions. The vehicle leaves the crosswind airflow at a permanent yaw angle and deviation from the chosen direction. These two values can be used by analogy with Figure 5.50 to determine whether a vehicle exhibits a more severe reaction to yaw or lateral movement. In the closed loop version, the action taken by the driver to neutralize the gust of wind and make a steering correction are evaluated. The evaluation criteria and target functions are specific to each manufacturer.

Evaluation of crosswind behavior by driving the vehicle past the airflow from a fan is controversial. The artificially generated side's flow profile differs from a natural gust of wind and may therefore cause a different form of vehicle excitation ([176], [694], [855]). In the closed loop, the driver's intervention cannot be assessed in realistic terms since he or she is prepared to react to the gust of wind [833]. Subjective driving impressions in an artificial and a natural crosswind often fail to coincide sufficiently [833].

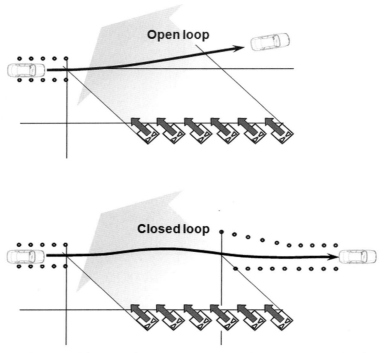

Figure 5.51 Crosswind fan experiments taken by open loop (upper) and by closed loop method (below).

Simulation by means of crosswind airflow resembles the situation when crossing a bridge or overtaking a truck. In recent years in particular, increasing value has been attached to examination of vehicle behavior in a stochastic crosswind. When assessing reactions to crosswinds, the vehicle's transmission pattern is being increasingly examined in the frequency range, with a view to achieving better correlation of objective evaluation and the driver's subjective impressions.

In [709] and [839], crosswind behavior is assessed in a natural stochastic wind while driving on public roads. The amplitude ratio of yaw velocity to wind excitation is examined (see Figure 5.48). The aim is to minimize the yaw velocity resonance amplitudes. In [839], vehicle reactions with driver influence are examined, that is, to the wind and with correction at the steering wheel. In [709], vehicle reactions without driver influence, meaning reactions to the wind only, are evaluated. In order to eliminate driver influence, single-track models are used to determine the vehicle's reactions to movements of the steering wheel and are then subtracted from the measured vehicle reactions.

These procedures permit more-detailed analysis of behavior in crosswinds than the methods previously used and result in improved agreement between objective assessment and the driver's subjective impressions. However, even in this way it is not always possible to obtain objective confirmation of subjective driving assessments. The reason is that only the vehicle's reactions to crosswinds are observed and minimized, but there are vehicles that are assessed as sensitive to crosswinds (or vice versa) although they react only slightly to stochastic gusts of wind [833]. To arrive at consistent agreement with the driver's subjective verdict, he or she must be included in the evaluation.

A procedure was developed in [833] (and subjected to further development in [462]) that takes the driver into account in the evaluation as the central link in the driver-vehicle control loop, as outlined in Figure 5.52. Wind disturbance causes vehicle reactions (index βv^2_{res}) that the driver senses and corrects by turning the steering wheel. The vehicle's reaction to the driver's steering movements (index δ_L) is superimposed on the vehicle's reaction as caused by the wind, to comprise an overall reaction that supplies the driver with information on how efficient his or her steering corrections are in compensating for wind disturbance. Equation 5.45 describes these relationships in mathematical terms.

In Figure 5.52, the driver, on account of his or her awareness of vehicle reactions (information processing), is usually regarded as a "sensor" but is thought of as an "actuator" when steering corrections are carried out. Optimized vehicle behavior immediately following the disturbance can be adversely affected by the driver's steering corrections. Alternatively, vehicle behavior that qualifies for a negative assessment objectively can be improved by the driver's influence. For this reason, in [833] and [832] the approach adopted is to optimize cooperation between driver and vehicle instead of adopting the vehicle's immediate reaction to the disturbance.

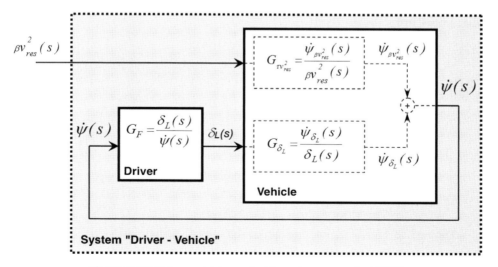

Figure 5.52 Driver-vehicle control loop in a crosswind [833].

$$\frac{\dot{\psi}}{\beta \cdot v^2_{res}}(j\omega) = \frac{\dot{\psi}_{\beta v^2_{res}}}{\beta \cdot v^2_{res}}(j\omega) + \left[\frac{\dot{\psi}_{\delta_L}}{\delta_L}(j\omega) \cdot \frac{\delta_L}{\beta \cdot v^2_{res}}(j\omega) \right] \quad (5.45)$$

The overall reaction is here to the left of the equals sign; the first summand at the right represents vehicle reaction to wind excitation, and the values in brackets are the vehicle's reactions to steering correction (left) and wind excitation (right). The driver's steering intervention is evaluated in relation to his or her processing of vehicle reaction information. All the vehicle's reactions to wind disturbance are supplied to the driver as reaction patterns within which he or she reacts to and processes the individual vehicle reactions with different levels of intensity [89]. When crosswinds are encountered, the driver is capable of steering the vehicle close to the desired course or a similar one, even if he or she is at the wheel of different vehicles. For this, the driver must match his or her steering actions to the behavior of the vehicle, in other words modify his or information processing. This modification is reflected in his or her intervention at the steering wheel [832].

Figure 5.53 shows a steering angle amplitude ratio to compensate stochastic wind disturbances, related to the wind disturbance and its internal phase shift, from a measurement obtained while exposed to stochastic crosswind (Driver reaction: Equation 5.45). The amplitude ratio enables us to analyze how intensively the driver responds to the individual reactions of the vehicle [833]. An intensive driver response to what are known as vestibular items of information, for example, yaw velocity or yaw acceleration, is reflected on the one hand in high amplitude ratios at frequencies between 0.8 and 1.5 Hz; on the other hand, the phase shift gradients are not so steep. Phase shift is an indication of how quickly the driver reacts; steep gradients are the consequence of slow driver reactions and vice versa. Intensive driver reactions to visual information, for example, deviation from the chosen line or yaw angle, lead to high amplitude ratios in a frequency range from 0 to approximately 0.6 Hz.

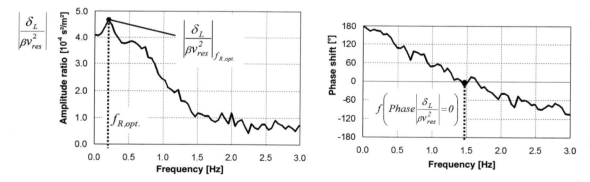

Figure 5.53 Steering-angle amplitude ratio and phase shift related to the wind disturbance [833].

The efficiency of the driver's intervention can be seen from the amplitude ratios of the vehicle's reaction to wind disturbance with and without driver influence (Figure 5.54). Vehicle reaction without driver influence is derived exclusively from excitation of the vehicle by the wind. Vehicle reaction with driver influence results from wind excitation and the driver's movement of the steering wheel with the aim of compensating for wind excitation. In a frequency range from approximately 0.5 to 2.0 Hz, the driver's interventions at the steering wheel amplify the vehicle reactions to the wind. This is typical of all vehicles and is due to the driver's reaction time and to phase shifts in the vehicle reactions.

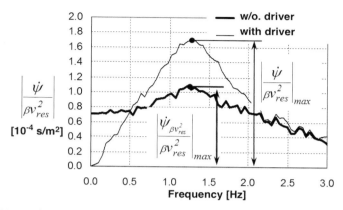

Figure 5.54 Yaw velocity amplitude ratio related to the wind disturbance with and without driver [833].

To assess crosswind behavior, four criteria are defined from Figure 5.53 and Figure 5.54 in [833], with the aid of which the driver's subjective response can be deduced: if a driver displays intensive reactions to visual information, this results in a high amplitude ratio in Figure 5.53 in the visual information processing range. In addition, it occurs at a higher frequency. Vehicles in which the driver reacts with this characteristic are sensed subjectively as "indirect" or "requiring much work at the steering wheel." Steering effort is evaluated according to equation (5.46).

$$K_L = \left|\frac{\delta_L}{\beta \cdot v_{res}^2}\right|_{f_{R,opt}} \cdot f_{R,opt} \tag{5.46}$$

If the driver reacts intensively to vestibular information, Figure 5.53 exhibits less steep phase-shift gradients between steering angle and wind disturbance, meaning more rapid driver reactions. Vehicles that encourage this characteristic form of driver intervention are sensed as "nervous" and "unpredictable." This aspect is assessed with the aid of the frequency at which the phase shift passes through zero (equation (5.47)).

$$K_P = f\left(\text{Phase}\left(\frac{\delta_L}{\beta \cdot v_{res}^2}\right) = 0\right) \tag{5.47}$$

The extent to which the driver's steering interventions strengthen the vehicle's reactions (Figure 5.54) can be expressed as the ratio of yaw velocity resonance amplitudes with and without the driver's influence. In this respect, equation (5.48) defines an amplification factor. If this is high, the driver will sense the wind disturbance as "difficult to eliminate."

$$K_V = \frac{\left|\dfrac{\dot{\psi}}{\beta \cdot v_{res}^2}\right|_{max}}{\left.\dfrac{\dot{\psi}_{\beta v_{res}^2}}{\beta \cdot v_{res}^2}\right|_{max}} \quad (5.48)$$

Equation (5.49) defines the resonance amplitude of yaw velocity with driver influence (Figure 5.54) as a further factor. If resonance amplitudes with driver influence are high, the driver senses the vehicle's reactions as "intensive."

$$K_G = \left|\frac{\dot{\psi}}{\beta \cdot v_{res}^2}\right|_{max} \quad (5.49)$$

It should be noted that the subjectively sensed intensity of wind disturbance is related to vehicle reactions with driver influence. The aim of earlier procedures was to minimize vehicle reactions without driver influence. In [833] it could be seen that the driver does not register vehicle reactions without driver influence directly. This explains why consideration of the vehicle alone when assessing crosswind behavior is often insufficient.

The individual criteria of equations (5.46) to (5.49) can be combined to obtain an overall evaluation (equation (5.50) [462], [833]), in which the various criteria are weighted to different degrees. This procedure allows for the occurrence of the driver's subjective impression: a driver assesses the vehicle's side wind behavior with the aid of several partial aspects and not on the basis of vehicle reactions. Characteristics K_P and K_V are included with a high level of weighting and exert a significant influence on the overall subjective verdict.

$$K_{Ges} = 2 \cdot [K_L] + 5 \cdot [K_P] + 3 \cdot [K_V] + 1 \cdot [K_G] \quad (5.50)$$

This procedure optimizes a vehicle's behavior when exposed to crosswinds by improving the interaction between river and vehicle. Whether or not a vehicle's behavior in crosswinds is positively assessed depends to a large extent on the characteristic with which the driver reacts to the wind disturbance in the vehicle being assessed. This evaluation method analyzes road dynamics and aerodynamics jointly, since they both affect the vehicle's reaction pattern and therefore the driver's steering intervention characteristic. To optimize crosswind behavior, a steering wheel response that is "optimal" for elimination of natural crosswinds can be derived from the individual criteria stated here. However, the running-gear measures that this calls for are strongly dependent on basic road-dynamic settings and can therefore differ from one vehicle to another or even partly act in opposite directions ([462], [832], [833]). This necessitates specific vehicle analysis and optimization. In [462] and [833], virtual methods were developed which, by suitable work on an adaptive driver model, permit optimization of driver-vehicle interaction in crosswinds even in the early concept phase of vehicle design.

5.2.4.4 Application Example

In section 5.2.1, the aerodynamic characteristics of a vehicle were described via the steady-state coefficients in a vehicle dynamics model to calculate the aerodynamic forces caused by wind excitation using a quasi–steady-state approach. The fact that this approach is applicable only with restrictions—strictly speaking, only if $f = 0$ Hz— was explained in section 5.1.2.2. The description of the aerodynamic model reaction in response to gusty wind excitation using transfer functions from system theory, as introduced by Schröck [739], allows seamless integration of the unsteady aerodynamic characteristics of a vehicle into an existing vehicle dynamics model. This means that frequency-dependent aerodynamic transfer functions are now used (see Figure 5.55) instead of the steady-state coefficients used in Figure 5.52.

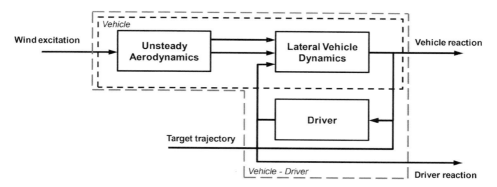

Figure 5.55 The driver-vehicle system taking into account unsteady aerodynamics [833].

In Schröck et al. [740], the effects of a vehicle modification on the characteristic of the transfer functions of side force and yaw moment were examined for the overall system. The approach developed in Wagner [832] and Krantz [462] was used in this case to evaluate the vehicle behavior from the driver's perspective by means of simulation. Figure 5.56 shows the admittance functions of side force and yaw moment for the initial vehicle and one modification. The steady-state gradients remain unchanged by the modification. The modification only affects the unsteady forces insofar as the maximum amplitude of the admittance of the side force is reduced by 6% and that of the yaw moment by 10%.

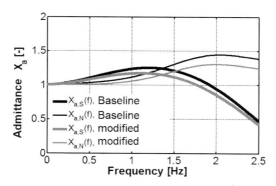

Figure 5.56 Initial state and modified aerodynamic admittance of side force and yaw moment [833].

The result of simulation of the closed control loop is illustrated in Figure 5.57. The left chart shows the amplitude of the steering wheel angle (normalized with the wind excitation), which the driver needs to apply to correct the vehicle's lane drift caused by side wind. A comparison of the results of the initial state and the modified aerodynamics shows that the steering effort in the frequency range in which the aerodynamic modification is effective can be reduced. The reason for the difference in the amplitude of the steering wheel angle is the lower reaction of the vehicle itself, as is apparent from the right chart of Figure 5.57. The chart shows the vehicle's yaw rate normalized with the wind excitation with and without the influence of the driver's steering. By separately analyzing the vehicle and the driver-vehicle system, it is possible to assess the driver's control behavior, which is of particular importance for conclusions. It is clear that the driver is able to compensate for the effect of side wind up to a frequency of approximately 0.3 Hz. Above this, the driver begins to amplify the yaw reaction of the vehicle by his steering intervention, since he is no longer able to react quickly enough as his response is delayed. The aerodynamic modification reduces the maximum yaw rate of the driver-vehicle system by approximately 8%. This means that the vehicle reactions to the side wind disturbance, as perceived by the driver, are reduced by the aerodynamic modification.

The vehicle's maximum is reduced by approximately 7%, meaning the so-called yaw amplification (ratio of the maxima with and without the driver's steering influence) is about the same. This result therefore suggests that any additional effect due to an adjustment of the driver to the vehicle with modified aerodynamic characteristics is low. This is comprehensible only if it is appreciated that, although the modified aerodynamics influence the forces and moments resulting from side wind, they have no direct influence on the interaction between driver and vehicle, as is the case when the dynamic driving properties are modified. Modification of the unsteady aerodynamic behavior can consequently contribute to improving side wind behavior.

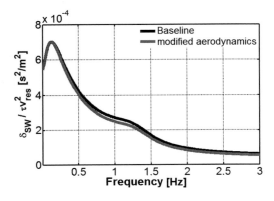 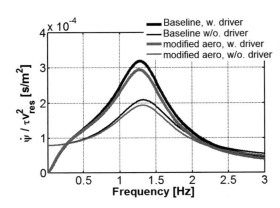

Figure 5.57 Influence of modification of the unsteady aerodynamic vehicle characteristics on the driver-vehicle system: steering angle (left); yaw rate with and without the driver's influence (right) [833].

Chapter 6
Functionality, Safety, and Comfort

Patrick Höfer, Alexander Mößner

The previous chapters discussed the global effects associated with the flow characteristics of a vehicle with a view to forces and torques and how they impact driving performance and directional stability. This section will now focus on the far-reaching influence of aerodynamics as they belong to different areas, whereby the functionality and reliability of a vehicle and its systems must be safeguarded at all times while increasing the safety and comfort of passengers.

The wind forces encountered at certain areas of the body, such as the engine hood, sunroof, and doors, are directly linked to local flow characteristics. It is not sufficient, however, to take only steady-state flow characteristics into account. Flow separations are largely transient and can also lead to dynamic deformation, or "component flutter."

Any source of airflow that enters the vehicle is frequently regarded as being negative because it invariably increases aerodynamic drag. Reducing resistance is only one of the many objectives that must be tackled by aerodynamic engineers. The most important aspect in this regard is to create a product that represents a well-balanced solution for the customer. This also includes securing the service life of all components by maintaining a sufficient temperature level, safeguarding the safety of vehicle occupants by adequately cooling the brakes, ensuring an unobstructed view of other road users (including in the rain), and providing for a comfortable cabin temperature. Passengers come into direct contact with the airflow around the vehicle when they open the convertible top or sunroof. No part or area of the vehicle may produce a draft or flutter despite pronounced transient design flow characteristics, and an open sunroof must not trigger a booming effect.

Accounting for the behavior of water on windows and contamination on vehicle surfaces goes far beyond the perception of airflow or deflection parameters. Having knowledge of particle-laden flow formations, droplet accumulation, the flow characteristics of water, and the surface properties of glass and paintwork are critical to ensuring a good result.

6.1 Component Loads

6.1.1 Component Loads and Pinpointing Them

The distribution of pressure on the surface of a vehicle can be used to determine the individual forces exerted on individual components such as the engine hood, trunk lid, doors, sunroofs, and window panes. These pressures can be sensed by taking experimental pressure measurements on component surfaces. Currently, however, greater use is being made of pressure distribution measurements as "taken" during computational fluid dynamics (CFD) simulation exercises. The benefits of the latter are that pressure distribution can be quantified within a finer scope than is possible with experimental measurements, simulation results are available in less time than can be achieved during the product engineering process because a real vehicle does not need to exist, and data can already be output in a recognized format by the CFD software so that it can be immediately processed further in other programs to determine load forces.

Different objectives are expressed that make it necessary to acquire knowledge of quasi-static and dynamic loads. A quasi-static load may be sufficient to properly design a sunroof drive motor, but the benefit of this information is limited when it comes to answering questions about component vibrations. Since CFD calculations for transient states necessitate correspondingly long computation times, experimental measurements are taken on physical vehicles. This approach allows the components to be tested in a wind tunnel using realistic materials and as they relate to other components.

Dynamic measurements in a wind tunnel are taken without making direct contact through the use of optical procedures. To this end, the surface coordinates are determined in an unloaded state, and when subjected to wind-induced forces, both states are linked via common, fixed points and the displacement vectors are ascertained. This makes it possible to visualize local deformation as Figure 6.1 illustrates, whereby the length of the arrows indicates displacement at the respective points in space. The important thing is to locate fixed points that do not become displaced under a wind-induced load. Such measurements can be taken using stereo camera systems, for example, to distribute as many measurement points as required in the room.

The temporal resolution permits sampling rates in the kHz range and not only makes movements visible but also measures the displacement vectors with an accuracy of < 0.1 mm. Complex kinematics, component deformations, oscillation curves, and relative displacements are thus easier to characterize. One of the major benefits of these systems is that they can be used on a mobile basis and, as illustrated in Figure 6.2, only a few, compact components are required. In addition, the measurement exercise can be prepared and carried out by a single person in minimal time. Evaluation results can also be obtained on site.

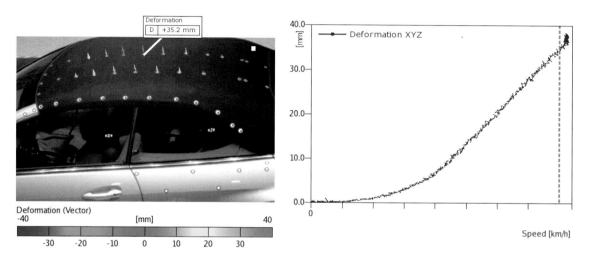

Figure 6.1 Vector depiction of the deformation analysis of a convertible soft top in relation to wind speed: to the left, a snapshot with evaluated measurement points; to the right, the progression of a measurement point over speed.

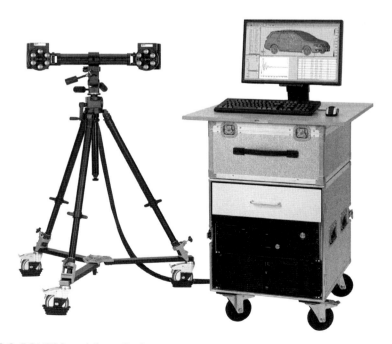

Figure 6.2 PONTOS mobile optical measurement system for analyzing 3D coordinates, 3D displacement, deformation, speed, and acceleration; frame rates of up to 5 kHz, comprising a stereo camera sensor, tripod, and computer.

6.1.2 Doors, Flaps, and Outside Mirrors

The structural rigidity of doors and flaps is important and helps to ensure proper sealing. Implementing lightweight design measures adds to such phenomena as elastic component deformation, fluttering, and body droning.

Frameless doors can be especially problematic. Although the bottom section of the doors is relatively rigid and held in position by hinges and a lock, the section above the beltline (side window) can deform much more easily. The turbulence created by the A-pillar induces a pronounced low-pressure zone at the leading edge of the side window. The relative difference in pressure between the outside and inside of the vehicle produces a force that literally pulls the window or door frame outwards. This, in turn, can cause the seals to be lifted from their resting surfaces, and occupants hear more road noise as a result. In extreme cases, very audible wind noise can be heard in localized areas. As chapter 8 explains in greater detail, this noise breaks down into two components: the first is air flowing into the gap that forms, whereby a loud tonal rustling is exhibited. The second component is the sound bridge created by the opening, through which noise generated outside the vehicle can penetrate into the passenger compartment. Fluctuations in pressure are also superimposed on the steady-state force exerted on the window pane. A window pane that is not held in place properly thus not only experiences static deflection but also vibrations that produce a low-frequency sound radiation. In addition, the throbbing of the door excites high-level, hydrodynamic pressure fluctuations in the passenger compartment. Although the gap created by the static deflection may still be able to be compensated for by the sealing strip, when superimposed vibrations or oscillations occur, it, too, separates from its resting surface. If the design construction is insufficiently rigid, the entire door can be induced into a low-frequency throbbing state.

Under side wind conditions, the downwind low-pressure zone becomes even greater and the forces acting in this area further intensify, as Figure 6.3 illustrates. According to Gilhome and Saunders [280], the rotational movement of the front wheels reduces the forces acting on the door. If wheel rotation is not tested for in the wind tunnel, the arching of the door can be easily overestimated. To ensure that the rigidity of the door and the geometry of the sealing strip are sufficient, the more critical scenario should be used.

Functionality, Safety, and Comfort

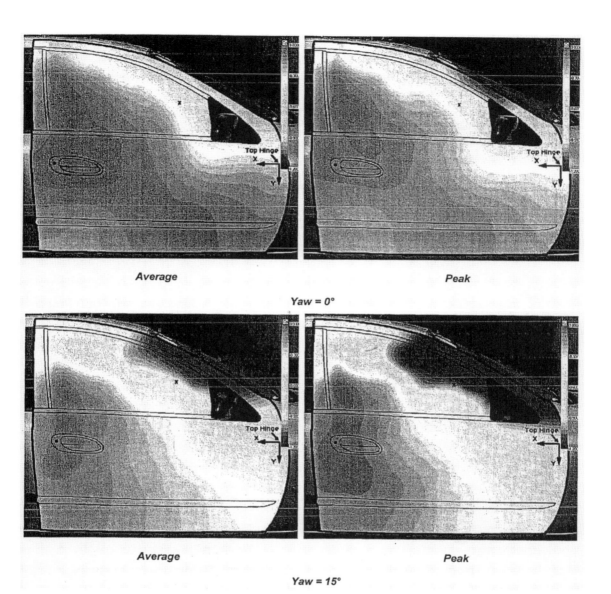

Figure 6.3 Comparison of pressure distribution on the driver door at a slip angle of 0 and 15 degrees, with an airflow velocity of 70 km/h [280]. The "X" designates the location of the pressure center point.

Some vehicles are fitted with doors that integrate into the roof structure. This design is shown in Figure 6.4b in comparison to the conventional configuration (subimage a). As shown in the cutaway section, its door frame extends across the A-pillar and roof and intersects with the windshield or roofline. Sealing occurs between the front edge of the door frame and windshield or roofline. Due to the large surface area of the door frame along the A-pillar, higher forces act in this area than with a conventional door frame. An outward-extending front edge can also reinforce turbulence at the A-pillar. In addition, the seal is placed under a load created by the air flowing into the gap between the door frame and windshield. In the most unfavorable scenario, the gap acts like a Helmholtz resonator and a tonal noise is induced.

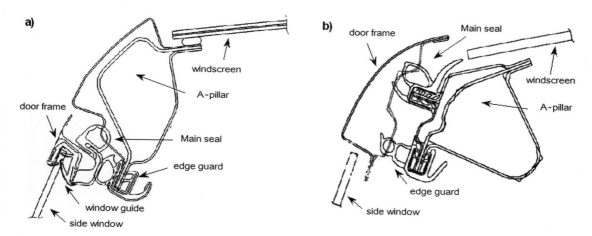

Figure 6.4 Sample depiction of a conventional (a) and an integrated (b) door concept with reference to an A-pillar cross-sectional view.

The aerodynamic load pressure on the outside mirror is exerted in the form of static wind and a transient force. Turbulence can be created at the trailing edge of the mirror housing, and a recirculating dead water flow pattern is observed. The pressure fluctuations that result cause the housing to vibrate, and visibility with respect to traffic behind the vehicle can be compromised because of this motion. This effect is especially predominant with bridge or beltline mirrors[1] that use a relatively thin connecting piece to join with the door. This construction is not very rigid by design but is frequently preferred for aerodynamic reasons. Soiling prevention and aeroacoustics profit from the pronounced constriction[2] between the mirror housing and door, whose channel flow characteristics permit the wake of the housing to be displaced away from the window. Additional benefits that aid in deflecting dirt are also provided by surfaces on which fewer water droplets accumulate and by the mirror base that is fitted below that of a mirror that attaches to the mirror triangle area of the door.

---

1. The latter are also referred to as wing mirrors.
2. Cf. section 6.3 or Banisters [37]. Mankau [530] provides an overview of the requirements for outside mirrors.

Functionality, Safety, and Comfort

A pronounced low-pressure zone forms at the leading edge of the engine hood when subjected to wind, as Figure 6.5a indicates. The resulting force can slightly raise the hood and cause it to flutter. With an angled airflow, pressure is distributed asymmetrically; the low-pressure zone moves to the windward side (Figure 6.5b). To ensure that the engine hood does not raise on this side, it can be reinforced accordingly or fitted with two locking mechanisms. Both measures increase weight and make the vehicle more expensive to manufacture, however. Should the hood ever not be properly fastened, a catch hook is installed to provide for additional safety. This hook must reliably prevent the hood from becoming airborne, even when the vehicle is traveling at its rated top speed. As the potential for a sudden airborne scenario is possible, very high peak loads are encountered in this area. Wind tunnel testing must therefore be carried out to provide verifiable evidence of the reliability of this function, whereby special safety precautions must be taken at this time.

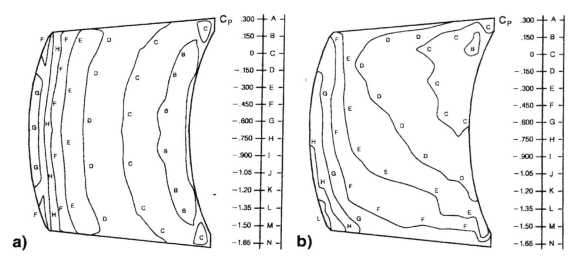

Figure 6.5 Distribution of static pressure on an engine hood: (a) at a flow angle of 0 degrees;(b) and 30 degrees. The isobars are designated: Lines of equal pressure.

If deflection at the leading edge of the hood is too intense, the airflow partially separates and regroups farther downward. The flow current in the separation pocket created is turbulent, and the hood can start to flutter. A sufficiently dimensioned edge radius can prevent this, however (section 4.5.1.1).

Sunroof designs can come in the form of conventional, relatively small openings and the frequently much larger panoramic sunroof configurations. Since local airflow speeds over the roof are very pronounced, sizable forces and torques are present in this area. Mounted roof loads further add to this as even higher speeds and, thus, lower pressures, form between the roof and the roof box. The roof system must also be capable of handling these loads when it is partially or fully opened. In the closed position, proper sealing becomes the important factor. If the sunroof separates along the entire seal or only in local areas, disturbing wind noise results. When opened, interactions occur that can create drafts in the passenger compartment or a throbbing or booming effect near the sunroof (cf. chapter 8).

6.1.3 Windshield Wipers

The windshield wipers on a vehicle must always be able to remove the water droplets and contamination that accumulate on the windshield so that the driver has a clear view forward. This should also be the case during heavy rainfall and when driving at high speeds and requires coordinated aerodynamic development of the wiper systems with respect to the vehicle. The necessary airflow phenomena that affect a wiper are explained next. Chapter 8 focuses on the wind noise generated at the wiper assembly, among other factors.

When the wipers are moved downward, to their resting position, they should not be directly hit by the wind. This is achieved by arranging the wipers below the trailing edge of the engine hood. The wake area that forms here is shown in Figure 6.6. If the wipers rest in this recirculation zone, they influence the airflow characteristics of the vehicle only marginally and thus have negligible aerodynamic and aeroacoustic impact.

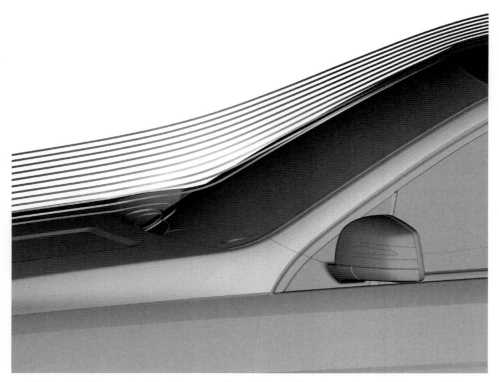

Figure 6.6 The optimal resting or park position of the wipers is at the bottom of the windshield or in the wake area of the engine hood's trailing edge.

The aerodynamic inflow at the wipers changes during a wiping cycle, whereby the vector of the airflow velocity incorporates the speed vectors of the vehicle airflow and the wiper motion. As the wipers approach their resting position, their angle of approach is almost perpendicular to the airflow. At top dead center, the airflow is all but parallel to the wiper blades and the wiper assembly linkage. When the wipers travel upwards, they move toward the direction of flow and vice versa in the downwards direction (i.e., the local speed at the wipers is greater during the downward stroke). The forces acting on the wipers and the wind noise generated by the flow separation and gap flow currents are thus at their peak in this phase of the cycle. The force of the air that occurs when it flows over the wipers consists of three components, as illustrated in Figure 6.7:

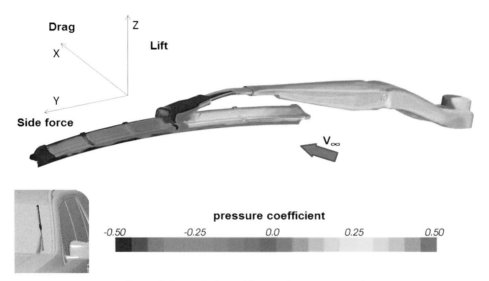

Figure 6.7 Depiction of forces that act on a wiper.

- Downforce (normal with respect to the windshield): Air deflection as influenced by the spoiler lip on the top side of the flat-bar wipers results in a force directed toward the windshield that ensures that the wiper blades are firmly pressed down even at high speeds. With conventional wipers, the airflow over the top side of the individual components produces low-pressure zones. These combine into an uplift force that can lead to the wipers rising off of the windshield, especially at high speeds.
- Resistance force (normal with respect to the wipers): The relative difference in pressure between the front and rear sides of the spoiler lip gives rise to a force that acts in parallel to the windshield. The turbulence behind the wipers (Figure 6.8) amplifies this difference in pressure. When the wipers travel upwards, resistance

decreases due to the reduced frontal area and the decreasing pressure on the windward side as a result of the angled airflow.

- Lateral force (along the wiper blades): The lateral force is comparably small in relation to the resistance and lifting forces.

Figure 6.8 Visualization of the turbulence behind a wiper.

The flat-bar wiper as shown in Figure 6.9 has represented the state of the art for years now. Unlike conventional wiper blade assemblies, these wipers do not use a bow, together with springs, to apply force to the rubber blade; instead, the blade is pressed onto the windshield by two curved spring rails integrated into the wiper blade itself. The wiper blade thus adapts to every curvature of the windshield, with equal pressure distributed across the entire length of the blade. This compact design offers several benefits:

Figure 6.9 Flat-bar wiper. Photo: Bosch.

- The wiper system has a low profile: Not only can the wipers be more easily hidden below the trailing edge of the engine hood; the lower design height also provides for a reduced cross-sectional area and thus reduced aerodynamic drag.

- The wipers are attached without the need for complex mechanicals: By eliminating various bows, the weight of the wipers is reduced by approximately 50% as compared to a conventional wiper blade assembly. A mechanical construction that can underlie to the effects of a winter ice-over is avoided.

- The spring rails generate an even pressure: The full length of the aero wiper always has contact with the windshield, as is denoted in the uniform distribution of linear contact force Figure 6.10, while the conventional wiper only establishes contact at various points and exhibits greatly fluctuating contact force. This has to do with the predominant point-type contact forces applied at the connections to the claw-hoop bows, and a smearing effect can occur between these contact points.

- The integrated spoiler ensures equal distribution of contact pressure: The downforce required for proper functioning is generated by a spoiler integrated in the wiper blade. This spoiler provides for an almost uniform downforce along the entire length of the wiper blade.

- The flat construction is quieter: The flat construction of the aero wiper not only makes it quieter than conventional wipers in the resting position; the lack of the multi-component bow connections also reduces noise during operation.

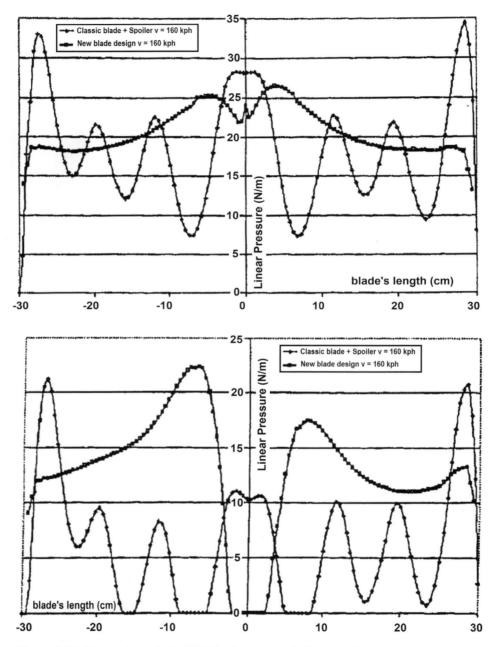

Figure 6.10 Comparison of windshield wiper concepts based on linear contact pressure: Without a wind load (top), with a wind load of 160 km/h (bottom); as per Billot et al. [77].

Functionality, Safety, and Comfort

The flat-bar wiper not only brings with it benefits, however:

- The spring rails integrated in the wiper blade must be adapted to the curvature of the windshield. The coordinated aero wiper thus only works on a certain vehicle and only for a certain curved windshield.
- Although conventional wipers include several parts, they are less costly to manufacture and can be fitted to a vehicle more easily. The high technical outlay required for the flat-bar wiper in order to manufacture the spring rails and the fact that its design must be coordinated for a specific vehicle increase the purchase price.

The connection of the wiper blade to the wiper arm influences the flow of air onto the blade and thus the contact force on the windshield. If the connection is above the blade (i.e., a top-arm connection), the effect of the spoiler geometry in this area is reduced. Figure 6.11 shows the influence of the wiper arm on the distribution of contact force. Below the wiper arm, the airflow is reduced, which increases local pressure between the arm and blade and generates uplift on the arm. The effect of the spoiler on the side of the connection (in Figure 6.11, along the positive abscissa) becomes less effective, and the total contact force of the wiper is lowered.

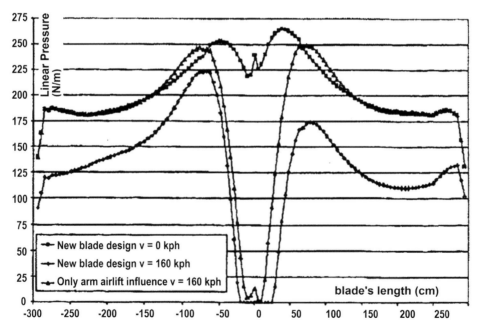

Figure 6.11 Comparison of influential variables for the flat-bar wiper: Reduction in contact force via airflow velocity of 160 km/h and asymmetrical or further reduced uplift distribution via the wiper connection; as per Billot et al. [77].

485

The following two design constructions can improve the effectiveness of the wiper: For one, the wiper arm can be arranged adjacent to the wiper blade (side-pin connection) to maintain the spoiler effect of the blade. The downside of this construction, however, is the effective widening of the wiper assembly, which restricts the outward view of the driver. The second design construction involves inserting openings in the main bow to reduce uplift on the arm and improve the effect of the spoiler geometry.

The following references the comparison with previous, conventional windshield wipers for the sake of completeness. Figure 6.12 shows a comparison of pressure distribution for the two types of wiper. The larger difference in pressure between the front and rear side of the conventional wiper indicates a higher resistance than with the flat-bar wiper. The uplift generated by the suction tip at the leading edge of the bow is even completely eliminated by the flat-bar wiper.

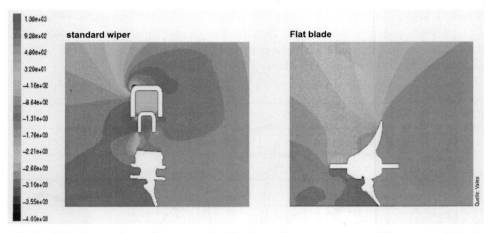

Figure 6.12 Comparison of pressure distribution with respect to two different windshield wipers; to the left, a conventional wiper; to the right, a flat-bar wiper.

The individual rubber blade and wiper arm components in a conventional wiper assembly contribute the most to the previously mentioned uplift and resistance forces simply as a result of their size (refer to Figure 6.13). According to observations made by Jallet et al. [401], the intermediate bows produce a downward airflow onto the top side of the wiper blade. Without this influence, the airflow would separate at the blade starting at the leading edge of the bow, and a local low-pressure zone would be created. The claw-hoop bows can be designed so that the airflow can follow the curvature of the bows and only separate at the end to reduce uplift.

Functionality, Safety, and Comfort

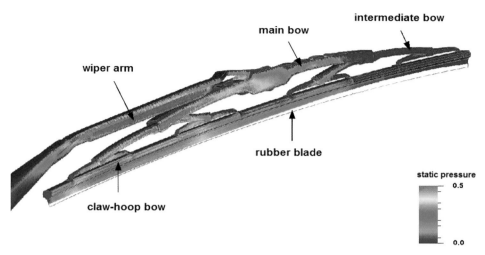

Figure 6.13 Conventional windshield wiper: The individual component designation and distribution of pressure on the wiper are illustrated.

Openings at the main bow not only reduce the effective uplift surface area parallel to the windshield but also allow pressure compensation between the upper and lower side of the bracket. The single-sided connection point of the windshield wiper to the wiper arm produces a nonuniform airflow and thus a nonuniform distribution of pressure along the wiper blade.

At high speeds (from approximately 140 km/h), the uplift force at the wiper on the driver side can become so great that the spring in the wiper arm is no longer sufficient in pressing the wiper onto the windshield. Integrating a spoiler upstream of the wiper blade can prevent this. This spoiler interrupts the airflow around the wiper and thus reduces the low-pressure zones on the bows. It also generates downforce and increases resistance by up to 25% as compared to a wiper without a spoiler.

6.2 Comfort When Driving with the Top Down

6.2.1 Objective

Open vehicles have enjoyed ever more market resonance for several years now. The spectrum ranges from highly puristic vehicles to very comfortable ones and from small, two-seat roadsters through to five-seater convertibles derived from sedans that offer a great deal of space. The expectations of passengers when it comes to comfort thus also widely differ. For some, the interior can't be breezy enough, while others want the cabin to be as quiet as possible. It is in this context that the question must be asked: What is comfort and how is it perceived? The answer goes in many directions.

The methods applied for evaluating the climate in vehicles with an enclosed passenger compartment are very far developed. For open compartments, however, such as an open convertible top or sunroof, this is not exactly the case, as these methods either do not exist or they have not been published. The evaluation methods that apply to enclosed passenger compartments are extrapolated to an extent but without having been safeguarded or verified beforehand by conducting corresponding investigation work. What this means or signifies becomes clear when the question regarding which flow conditions are required or must be tolerated in the vicinity of a person is asked. Aeroacoustics must also be taken into account, although or because noise levels are considerably higher than in enclosed vehicles when the roof is opened or closed.

The term "open vehicle" has previously been used with almost exclusive reference to convertibles and roadsters. As the sunroof has been further developed to become a complex roof opening system that offers a large upward view, however, the term must be expanded in scope. The goal of development must be to convey an unrestricted, unfiltered impression of the outside environment when the roof is opened and to feel the wind and experience an airy feel in the cabin. At the same time, this contact with the outside must be controlled in such a way that it does not become bothersome for occupants. Achieving both of these goals to the satisfaction of the target group and realizing appropriate optimization measures on the vehicles are nothing short of a set of formidable challenges.

In the process, efforts are also spent to extend the application of the convertible to include colder weather. When it gets too hot, the air conditioner is barely sufficient, and when sunlight becomes too intense, the only option is to close the roof. When it gets too cold, on the other hand, the heater usually provides for ample heating, although the vents may not be optimally positioned. Disturbing airflows must be reduced as far as possible from a development perspective. There are customers, however, who do not like "too much comfort;" these individuals can then do away with such comfort measures by removing the wind deflector or lowering the side windows, for example. This also allows the draft effect to be adapted to different driving speeds and climatic conditions.

6.2.2 Airflow with Convertible Top Open

When the convertible top is opened, a flow separation occurs at the top edge of the roof frame, above the windshield and, depending on the position of the side windows (open or closed), at the A-pillar or top and bottom edges of the side window. A turbulent shear layer forms downstream of the roof frame, and backflows enter the passenger compartment. This is also indicated by the CFD-computed stagnation streamlines in Figure 6.14, whereby pronounced spatial and temporal fluctuations in speed occur. If the roof frame is too low, the passenger is hit directly by high-speed wind at the front of his head. The higher the roof frame is, the greater the backflow effect as caused by the larger separation area behind the windshield.

When the convertible top is opened, aerodynamic drag considerably increases; $\Delta c_D \approx 0.050$ can be used as a guideline value. Due to the shorter journeys traveled and the lower speeds involved when the roof is put down, however, this fact is simply accepted.

Functionality, Safety, and Comfort

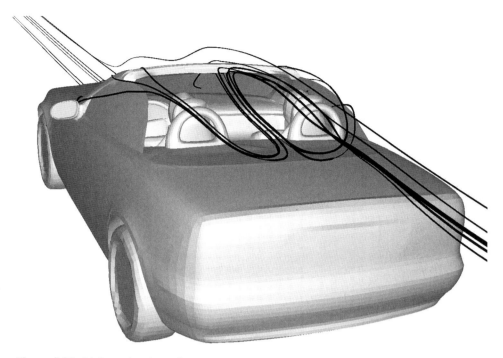

Figure 6.14 Airflow structures in an open convertible; backflow into the cabin originating at the outer roof frame and the side windows (stagnation streamline progression from steady-state CFD simulation).

6.2.3 Wind Noise

Wind noise currently must be regarded as making an individual contribution to the comfort associated with open driving, as there is still too little information available about how acoustic sensitivity interacts with the factors that characterize thermal comfort. The objective is clear: Every source of noise must be minimized as far as possible. With respect to open convertibles, flow separation noises can be heard at the roof frame or roof frame seal and side window edges, and through-flow and flow-around noises can be heard at the wind deflector and on rough or sharply angled surfaces.

6.2.4 Thermal Comfort

If passengers complain about a lack of thermal comfort when driving with the cabin open, this is generally commented on by saying, "There's a draft in here." A distinction between a mechanical flow aspect—excessively high or rapidly fluctuating flow speed here or there—and a thermal one—too cold or too warm—is not always made. Whereas the approach previously adopted for optimizing the climate in an open cabin focused on configuring the flow paths,[3] today air temperature is becoming an increasingly important influential factor. This can take the form of adapted ventilation nozzles and air

---

3. Refer to Cogotti [154] in particular.

distribution configurations for the heater/air conditioner as well as additional heating elements based on convection or radiation. The first such application made its way into the 2004 Mercedes-Benz SLK and was marketed as the AIRSCARF® system, which used a ventilation nozzle in the headrest to blow heated air onto the neck of the passenger (Figure 6.15). Other manufacturers have also utilized this concept.

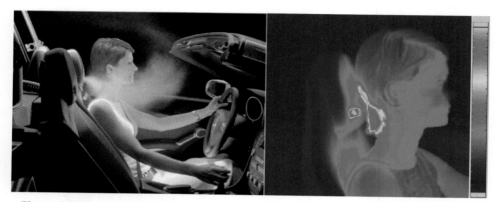

Figure 6.15 Improvement in comfort during open driving; ventilation of heated air onto the neck of passengers relative to the driving speed (air quantity) and heat level selected; to the left, airflow visualization using smoke; to the right, thermographic snapshot of skin being warmed by the heated air. Mercedes-Benz SLK, AIRSCARF®.

According to DIN (Deutsche Industrie Norm) guideline standard 33403, thermal comfort is sufficient, or given, when a person perceives the air temperature, air humidity, air movement, and heat radiation in its immediate vicinity as optimal and does not desire or want either warmer, colder, drier, or more humid air. According to Bradtke and Liese [94], thermal comfort is an "expression of harmony between individuals and their environment, or in a stricter sense, between them and their surrounding climatic environment."

To ensure that thermal comfort can be measured and reproduced in an enclosed space or room (especially with regard to a vehicle), three things are required:

- Identification of the variables that determine the climate
- Quantification of these variables
- Evaluation of these variables

Several constraints must be observed in this context: the size and shape/contours of the surrounding space or room and the structural composition of its walls. Comprehensive basic knowledge has been acquired for properly controlling the climate in buildings and enclosed motor vehicles.[4] The approaches described there for evaluating or assessing comfort have only limited applicability to open cabins, however, and to convertibles

---

4. Refer for example to Nilsson et al. [607], Grossmann [297], and DIN ISO 7730.

Functionality, Safety, and Comfort

in particular. One of the reasons why quantitative treatment or coverage of the climate phenomenon is so difficult is that assessing a particular state brings with it pronounced variation or dispersion from an:

- Interindividual perspective (one and the same thermal state is evaluated very differently by different people)
- Intraindividual perspective (one and the same person does not always assess a specific thermal state in the same manner). The reasons for this primarily have to do with fluctuations in personal preferences and different clothing.

As Figure 6.16 indicates, the percentage share of dissatisfied individuals is 5% even for a mean neutral climate as judged by the persons in question (not too warm, not too cold). When a specific mean climate is characterized by all evaluators as being neutral or comfortable, 5% are still dissatisfied.

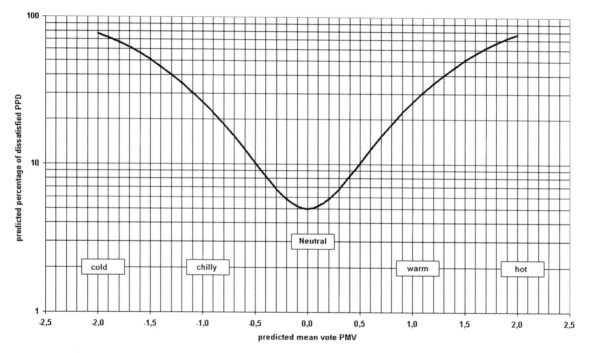

Figure 6.16 Predicted percentage of dissatisfied individuals (PPD) as a function of the predicted mean vote (PMV); DIN ISO 7730, applies for enclosed rooms.

Adding to this is the fact that local discomfort can considerably impact the thermal comfort of a person. It is exactly these local influences that, in a vehicle with its special constraints such as direct sunlight, inhomogeneous temperature, and airflow patterns, and frequent proximity to enclosing or bordering surfaces of varying temperature, play an important role. Individual body parts also react differently to how they perceive temperature. Mayer [538], for example, has collected direct statements in conjunction with the (thermal) satisfaction of individuals and the temperature of their necks (Figure 6.17). Minor deviations from the ideal temperature (with the lowest portion or percentage being dissatisfied; $T = 33.4$ °C) lead to a rapid increase in dissatisfaction (not everyone is satisfied here either).

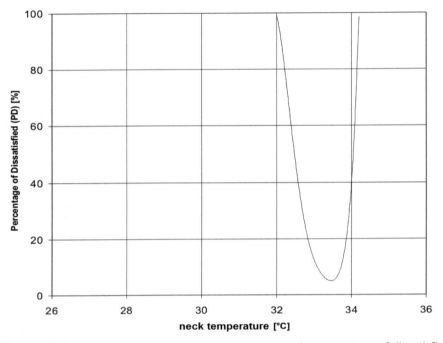

Figure 6.17 Influence of measured neck temperature on the percentage of dissatisfied individuals in relation to their thermal comfort; as per Mayer [538].

The expectations of a person sitting in a convertible influence how the vehicle climate is perceived. A comparison with the expectations for a room-based climate makes this apparent. In this context, the wind speed according to DIN ISO standard 7730 should range between 0 and 1 m/s—a value that is almost impossible to achieve in a convertible. While according to Nemecek and Grandjean [603], approximately 30% of people who work in large offices[5] complain about a draft, even when the mean air speed does not exceed 0.2 m/s, the number of those dissatisfied in comparison to everyone else is much greater than when they drive in a convertible with the convertible top down. Perception is thus based on changes according to the situation at hand.

A range of differently equipped climate-measuring manikins are used to evaluate or perceive conditions on an experimental basis. These manikins can be subdivided into

---

5. At industrial workstations, two-thirds of employees complain about drafts [294], and when low to moderately cold air temperatures are encountered, up to 100% of individuals are dissatisfied [804].

those that comprise several large, heated body segments and those that are fitted with single sensors. The benefits of the first generation lie in the pronounced, integrated areas, which not only create the spatial obstruction provided by a body but also reflect the thermal influence back into the environment as a result of the heated segments. A disadvantage is the typical lag in response times; measurement periods in excess of 30 minutes are frequently too long for practical testing.

This is where the second-generation manikins come in, which are used to eliminate this drawback by fitting smaller, individual sensors that offer a lower thermal inertia. This allows the measurement period to be reduced to less than one minute. The disadvantage here, however, is that the small individual sensors can cause the result obtained to be strongly bound or limited to their position on the manikin. The sensors also do not reflect the thermal influence of the manikin as it relates to the environment, which does not play a significant role in convertibles. The sensors themselves are sensitive and can frequently be damaged during testing. Also of mention is the fact that, depending on the sensor design, it is not always possible to determine the various measurement parameters all at the same time, at the same location; the sensor for the airflow speed, for example, is not located exactly where the sensor for the temperature is positioned.

The manikins conceived for vehicle passenger compartments are designed specifically to facilitate the recording of an equivalent temperature with heated sensors.[6] For higher flow speeds in body proximity (in convertibles, typically between 1 and 10 m/s), a different approach is followed as flow variables such as air speed and air temperature are quantified.[7] One example of just such a measuring manikin is "TANJA" (refer to Figure 6.18), who is equipped with 16 constant temperature anemometry (CTA) sensors for measuring flow speed and temperature in body proximity. With respect to comfort-based assessments in an open convertible, the assumption is made that especially near areas of exposed skin, the mechanical irritation of the highly transient flow dominates the thermal cold shock factor—at least as it pertains to the upper half of the body, where flow speeds are the highest.

Speed and temperature fields in the immediate environment of the occupants are being increasingly determined using CFD technology. Since media flows in the passenger cabin are highly transient and the perception of comfort is very dependent on these transient effects, simulation exercises must also be calculated or computed in a transient manner for the accompanying observation work. When spatial and temporal resolution become extremely fine and detailed, the calculation runs for a configuration can take so long that this method must be characterized as being in its initial stages when viewed in terms of actively applying it to general development processes. It is in exactly this context, however, that an understanding of flow paths is facilitated.

---

6. As per SAE J2234.
7. Mößner [51] provides detailed further information.

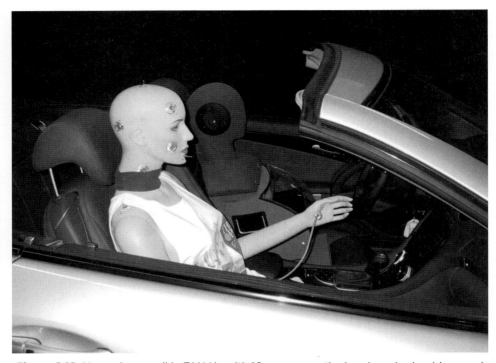

Figure 6.18 Measuring manikin TANJA, with 16 sensors on the head, neck, shoulders, and arms to take temperature and flow speed measurements for determining thermal comfort during open driving; together with an acoustic artificial head.

The goal of all of these investigations is to develop measures that minimize draft. In the process, it must be noted that there is no satisfactory solution available when a vehicle is driven with the side windows open and no wind deflector is mounted. Only after the side windows are closed is a considerably improved level attained, as shown. This figure plots measured flow speeds at 16 positions in the vicinity of a driver for four vehicle configurations. The aerodynamic effect of the wind deflector can only be observed in the lower body regions; the flow speed from the shoulders down drops. Although turning the heater on increases flow speeds somewhat, because temperatures are considerably elevated (Figure 6.20), comfort is improved at lower outside temperatures. Of considerable note is the fact that the increased temperature difference outside the vehicle corresponds to the direct application or impact of air from the outer ventilation nozzles. After this, the height difference between the head of the passenger and the upper edge of the roof is the predominant factor that determines to what extent the shear layer that forms downstream of the roof frame approximates the head and how intense the draft effect experienced at the head becomes.

Functionality, Safety, and Comfort

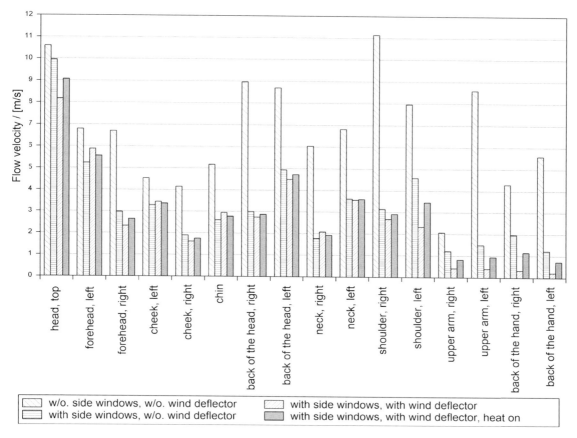

Figure 6.19 Flow speeds near the body in a convertible with the convertible top down for different vehicle configurations; TANJA measuring manikin; right front passenger seat; airflow velocity of 120 km/h.

Chapter 6

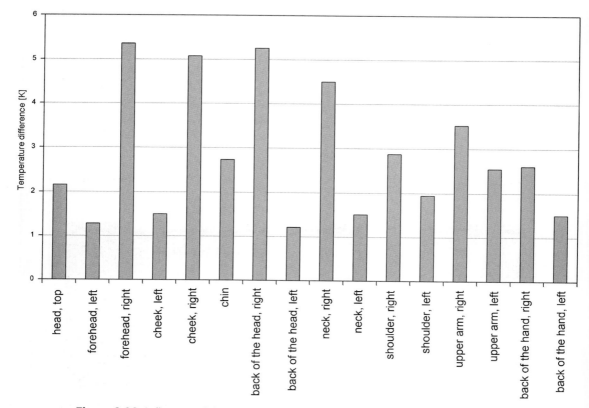

Figure 6.20 Influence of the vehicle heater on the air temperature near the body in a convertible with the convertible top down, closed side windows, and installed wind deflector; TANJA measuring manikin; right front passenger seat; $T_\infty = 28$ °C; airflow velocity of 120 km/h.

6.2.5 Design Solutions for Convertibles

Convertibles can be characterized by the type of construction used for the convertible top. The roof types available in series-production applications have been augmented by a new type since 1996, when a folding hard top was offered in addition to the conventional soft top. The soft top is easier to stow away but brings with it disadvantages such as ballooning and even fluttering as speed increases. This effect, however, can be counteracted by appropriately contouring the load-bearing structure and fitting a better textile covering or by fitting insert bars made from high-strength materials. The wind noise associated with the soft top is frequently higher than with a hard top. The performance of a soft top also hinges on the covering material selected, the design construction (number of layers, material composition and thickness), the design of the rear window (material, thickness), and the number of gap joints.

For both types, the wind noise experienced is very much dependent on the quality of the seals used. In the case of folding hard tops, this factor becomes more pronounced, as the noise that permeates the sheet metal surfaces is lower than with a textile construction such that any leaks or air gaps are easier to detect. Hard tops are offered for operation in the winter because they provide better insulation and transform the convertible into a coupé during this time.

Folding hard tops (Figure 6.21), which have become established in the marketplace and enjoy further widespread use due to ever more refined folding and stowage mechanisms, can base their success on the fact that they offer almost identical performance with that of a fixed roof. The only drawbacks are increased weight and a more expensive design construction.

Figure 6.21 Mercedes-Benz SLK with "vario roof."

To counteract air draft at the back of the open cabin, wind deflectors or draft stops (Figure 6.22) are used that typically comprise a permeable covering stretched across a frame or perforated panels. Figure 6.23 illustrates the effect. The partially permeable, elastic fabric reduces mean flow speeds and ensures a more even distribution of airflow, which is generally perceived as being more comfortable than pronounced flow gradients. Extreme gradients also do not form around the perimeter of the wind deflector. A drawback, however, is that rearward visibility is somewhat obstructed by the visually thick or dense material.

Figure 6.22 Different designs of wind deflector: lower left, BMW 6 Series Convertible; upper right, Audi TT Convertible.

Figure 6.23 Improved passenger comfort through minimization of recirculating current in convertibles with an open convertible top via a wind deflector; Mercedes-Benz SL; to the left with and to the right without a wind deflector.

Impermeable wind deflectors made from transparent material have undisputed visual benefits: Their appearance is usually deemed more aesthetically pleasing and visibility is not as compromised as with a fabric-based deflector. There is, however, a high probability that, unlike the latter, a more pronounced speed gradient across the perimeter of the deflector will be perceived and thus lead to discomfort. The reasons for this have to do with higher draft currents and louder wind noise in the direct vicinity of the passengers. To counteract this conflict with respect to rearward visibility, wind deflectors are being made to be more movable or mobile (i.e., retractable or foldable) so that when they are not needed, visibility remains entirely unobstructed. Examples of this can be found in the Porsche 911 Convertible and the Mercedes-Benz SL (Figure 6.24).

Figure 6.24 Folding, movable fabric wind deflectors; to the upper left a Porsche 911 convertible; to the lower right a Mercedes-Benz SL.

Another design approach for a wind deflector is based on diverting the rearward air away from passengers while still allowing as much air as possible to enter the cabin. The greater this quantity of air, the less pressure drop in the cabin and the more the shear layer at the roof frame is displaced upwards, away from the head area. This is a very notable aspect, especially as it pertains to small roadsters. With the AIRGUIDE® (Figure 6.25) system as found in the Mercedes-Benz SLK, the folding installation of the two side windows near the rollover bars also resolves the problem of stowing the deflector when the system is not needed.

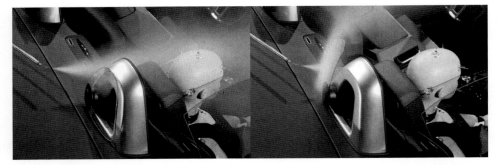

Figure 6.25 Mercedes-Benz SLK, AIRGUIDE® comfort system, left: stowage behind the rollover bar when not used, right: folded out, rearward air is diverted away from the passengers; airflow made visible through the use of a smoke tube in the wind tunnel.

Wind deflectors are used for two-seater and four-seater convertibles. Since wind deflectors can only be effective in relative proximity to passengers, installation in a four-seater car means that a deflector must protect the front passengers by partitioning the space to rear passengers or be positioned closely to the front seats in the upper area and stretch horizontally rearward over the rear seats in the lower area (Figure 6.22, lower left). Although these seats then cannot be occupied, accessible stowage space in the rear seat area is afforded. This conflict surrounding a relatively draft-free environment at the front seats but no usable rear seats and four-seat occupancy but with no draft-free comfort zone has been resolved. The load factor with regard to the rear seats is primarily based on two components: The air flowing in from the front of the vehicle that separates at the roof frame and the pressure balance or compensatory airflow at the rear. The AIRCAP® comfort system, which was launched in 2010 in the Mercedes-Benz E-Class Convertible, comprises two main components: a retractable and extendable slat at the roof frame that is used in conjunction with stretched webbing, and a wind deflector made from partially permeable fabric that extends together with the headrests behind the rear seats. The idea behind this strategy was thought of long ago; implementing it, however, was directly linked to the way in which the mechanisms could be retracted and extended with a view to aesthetic concerns. The airflow that normally separates at the roof frame is displaced further upwards, the current between the roof slat and roof frame is slowed down by the webbing, and the rearward current is kept away from passengers by the wind deflector and the large head restraints (Figure 6.26). Design efforts must take several conflicting interests into account: comfort for the front and rear-seat passengers, pronounced airflow deflection at the roof frame and the intensity of the backflow airflow, permeability of the stretched webbing at the roof slat and the wind deflector, and the relationship between wind noise and draft-free driving comfort.

Functionality, Safety, and Comfort

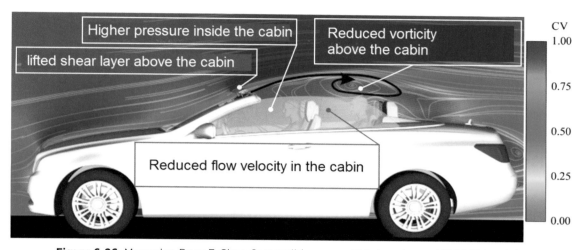

Figure 6.26 Mercedes-Benz E-Class Convertible, convertible top down, AIRCAP® (roof slat and wind deflector behind the rear-seat passengers extended), airflow speed at the center of the vehicle (blue: high, red: low speed).

6.2.6 Design Solutions for Sunroofs

Sunroofs are continuing to graduate from small openings in the roof, primarily made from sheet steel, to much larger openings that involve more complex systems made from glass, metal, fabric, and plastic. If the roof is long enough, several sunroofs can be combined. The negative affects associated with this increase, however. Booming becomes more prevalent the longer an opening is and arises when the frequency of the turbulence that separates at the leading edge of the opening coincides with the eigenfrequency of the air volume in the interior. The resonance effect can be very unpleasant at low speeds, since very high noise levels occur at lower frequencies.[8]

A rectification measure is a wind deflector (Figure 6.27) that is installed at the leading edge of the opening. Notches or grooves with irregular spacing are formed at the top edge of this deflector. These notches or grooves disrupt the shear layer and the formation of the turbulence system and noise levels are greatly reduced as a result. When designed, attention must be paid to ensure that no high-frequency noise is triggered by the notches or grooves themselves.

---

[8]. A detailed description is provided by George et al. [269].

Figure 6.27 Sunroof with wind deflector, Mercedes-Benz E-Class.

The larger a sunroof is, the greater the risk of the rear-seat passengers in particular from becoming discomforted by the incoming air. Even when the roof is closed, however, large systems made from glass (Figure 6.28, the transparent front section of the roof slides underneath the rear window) now also have a notable effect on the interior climate. This is why these systems are being increasingly outfitted with shading mechanisms such as a blind on the inside.

Figure 6.28 Porsche 911 Targa, glass sunroof, outfitted with a shading blind.

6.3 Prevention of Vehicle Soiling

6.3.1 Basics of Vehicle Soiling

In addition to the traditional topics of aerodynamics such as airflows and wind noise, vehicle soiling also represents one of the aerodynamic design tasks tackled during vehicle development. The most important aspect in this context is perceptual safety. The relevant vehicle surfaces should also remain free of contaminants so that customers do not become dirty when entering and exiting the vehicle by using the door handle, for example.

When traveling in the rain, partially contaminated water droplets collect on the surface of the vehicle and move toward the outside mirror glass, side windows, and rear window. Especially when traveling at night, these droplets can refract light in such a way that it considerably reduces visibility. When the side windows and outside mirror glasses are heavily contaminated, this effect can impede lateral and rearward visibility to the extent that other road users are entirely overlooked. If additional contamination is trapped in the water droplets, such as road salt in the winter, the contamination layer can dry on the surface and visibility is limited for an extended period of time or until the vehicle goes through the car wash at the next gas station.

The following section focuses on the causes of vehicle contamination and potential measures for reducing or preventing it. The trial testing required for this under wind tunnel conditions is described in section 13.3.6.[9]

6.3.1.1 Distinction Between External and Vehicle-Induced Contamination

The contamination of a vehicle is caused by two factors:

- External contamination is the result of (soiled) water droplets thrown up by surrounding traffic or falling rain that collect on the vehicle.
- Vehicle-induced contamination refers to the water droplets and contamination particles propagated by the vehicle itself—especially with regard to the spinning wheels—that collect on the side and rear areas.

Reducing external contamination is the main area of focus when optimizing vehicle soiling characteristics in passenger car development. However, much more effort is spent in this context than in vehicle-induced contamination. Things are different when it comes to commercial vehicle development, where increased outlay is expended to reduce vehicle-induced contamination (cf. section 10.9).

6.3.1.2 Technical Flow Phenomena with Respect to Contamination

To understand the contamination process better, it is helpful to become aware of the basic behavior of water droplets in an unrestricted, inhomogeneous flow pattern as the material collects on a wall or an existing layer of water (described as follows).

---

9. Cf. Hagemeier et al. [309].

Phenomena can then be more easily identified and corrective measures invented in less time.

In an unrestricted flow pattern, water droplets (depending on their specific size) describe a pathway that can vary greatly from the stagnation streamlines. In addition, these droplets can change in their shape, size, and weight. Generally speaking, six forces affect the stability of a water droplet in an unrestricted flow pattern: the inertia of liquid and gas, the viscosity of these two phases, gravitation, and the surface force of the liquid. When collecting on a wall, the angle of attack and the surface property also play an important role.

6.3.1.3 Droplets in Free Flow

The dispersion and coalescing of water droplets in free flow conditions influence the size of the droplets that contact the vehicle. The coalescence of the droplets takes place in relation to the contact force energy and the respective adhesion force. Not every droplet-to-droplet contact leads to a convergence; the rate at which water droplets converge is much lower than the rate at which they contact each other. The way in which droplets disperse depends on the relationship between the inertia force relative to the surface force. This can be expressed by the Weber number:

$$We = \frac{\rho_F \cdot v_\infty^2 \cdot l}{\sigma} \qquad (6.1)$$

ρ_F describes the density of the liquid, v_∞ the speed, l a characteristic length, and σ the surface tension. With respect to this Weber number, droplet dispersion can be broken down into three categories[10] (which are shown in Figure 6.29):

- Bag break-up: During bag break-up, the water droplet propagates via pressure distribution and extends outward to the low-pressure areas at the bordering contour. A flat, disk-shaped water droplet perpendicular to the flow direction results. The center of the droplet expands in the form of a bubble with a ring-shaped perimeter. This thick border moves more quickly downstream than the thinner bag structure, which in turn causes the bag to "break out." The ring then also disperses to create smaller droplets. This process largely depends on the surface tension of the droplet itself and occurs under low Weber number conditions.

- Multimode break-up: The dispersion mechanism associated with multimode break-up is similar to that of bag dispersion. The higher Weber number and thus the increased amount of surface energy present cause a liquid pillar to coalesce in the center of the bag that leads to the bag rising up and stretching, similar to the way an umbrella is opened. This beam likewise disperses after the bag "breaks up."

- Shear break-up: Although a disk-shaped water droplet forms under high Weber number conditions due to the corresponding distribution of pressure, this droplet does not emerge into a bag, but instead, smaller droplets crystallize at the bordering

---

10. Details have been provided by Reitz and Diwakar [664], Nicholls [606], Samenfink [706] and [705], and Schmehl et al. [718].

contours as a result of the surrounding airflow. The droplet then dissolves from its current, limiting structure. Surface tension is not a pronounced influential factor for this mechanism.

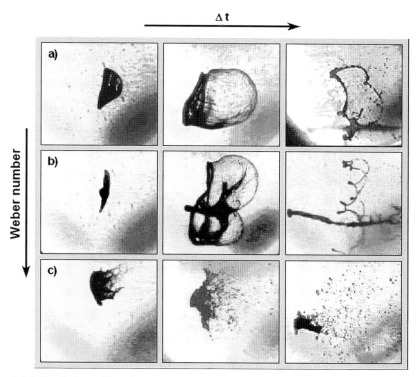

Figure 6.29 Depiction of break-up mechanisms for water droplets in an unrestricted flow pattern: (a) bag break-up; (b) multimode break-up; (c) shear break-up.

6.3.1.4 Mechanisms That Affect Water Droplets Upon Contact

The inherent behavior of a water droplet when it contacts a resisting surface is described by four contact mechanisms[11] that are shown in Figure 6.30.[12] These mechanisms relate to the following constraints: Droplet velocity and size, angle of attack, liquid properties, surface tension, surface adhesion, and, if present, thickness of the water layer. The following contact mechanisms are listed and described with respect to a rising Weber number:

- Stick: Contacting water droplets adhere to the surface in spherical form. This occurs when the contact energy involved is very low, that is, when the characteristic Weber number is also very low.

---

11. Bai and Gosman ([32], [33]) provide a descriptive summary of the contact mechanisms. The literature introduces a wide variety of models. An overview of the different approaches as based on empirical data is provided by Cossali et al. [18].
12. Refer to Stanton and Rutland [776] and Schmehl et al. [719].

- Rebound: The water droplet bounces off the wall with minimal pulse loss and all but retains its original diameter. This reflection is only observed on surfaces that are already coated with a layer of water, whereby the Weber number must be low. The layer of air trapped between the water droplet and water film leads to a sizable loss in energy during the already "gentle" contact strike and allows the droplet to rebound like an elastic collision. Secondary droplets can then form at this moment, depending on the initial energy: Small, isolated droplets separate from the larger droplet that bounces off the surface.
- Spread: During convergence, a water droplet that approaches at a moderate speed contacts the wall and disperses. A water film propagates on a dry surface. If a film of liquid is already present, the droplet combines with it. This mechanism can only occur in a specific energy field, or in a defined Weber number range. If this range is too small, the droplet rebounds; if it is too large, the droplet disperses.
- Splash: When a water droplet contacts a resisting surface at high speed and, thus, with a high amount of energy, it disperses into many smaller droplets. One part of the droplet remains on the surface, while the remaining part is reflected in the form of secondary droplets. If a layer of water is present on the contact surface, a "crater" forms and liquid lines emerge at the bordering contours, which then become unstable and disperse into finer droplets. This contact mechanism only applies to a single water droplet contact point.

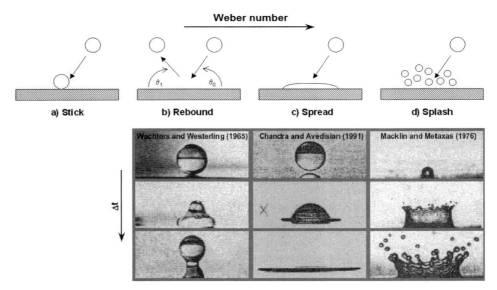

Figure 6.30 Depiction of different contact mechanisms for water droplets on a surface: (a) adhesion—stick, (b) reflection—rebound, (c) convergence—spread, (d) destruction—splash.

6.3.1.5 Influence of Surface Condition

The surface free energy influences the interaction of the water and the surface. In general, two types of surface characteristics can be distinguished, which are shown schematically in Figure 6.31.

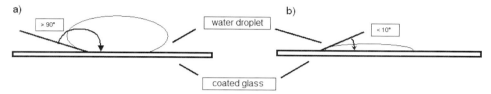

Figure 6.31 Schematic depiction of surface wetability: (a) hydrophobic coating (bordering angle greater than 90 degrees); (b) hydrophilic coating (bordering angle less than 10 degrees).

A hydrophobic surface with water-repelling characteristics points to a low surface tension. This creates a large contact angle (> 90 degrees) between the water droplet and the surface and induces a water repellent that is also referred to as the lotus effect.[13]

A hydrophilic surface with water-retention characteristics is consistent with a high surface tension that creates a small contact angle (< 10 degrees) between the water droplet and the surface.[14] A water film can thus form more easily on the surface and a running-on effect on the window pane can be prevented.

The use of hydrophobically coated windshields has been investigated many times. The water repellence is designed to have small water droplets be removed from the visible area by the wind so that the windshield wipers do not have to be used as frequently and contamination such as insects, thrown up contamination particles, and icing can be easily cleared. In an endurance test spanning several months conducted by Bauer and Schultheis [48],[15] the aforementioned positive effects were able to be observed but only within a restricted timeframe. The downside of the hydrophobic coating is its limited durability; the coating must therefore be applied regularly for it to retain its effect. During the endurance test, the coating applied to the windshield would have to be reapplied quite soon as partial wear led to smearing and droplet streaking.

Nano technology was first used successfully on side windows. Several manufacturers apply hydrophobic coatings on the front side windows in order to avoid having to implement design measures such as a water deflector rail. Such a coating also has a limited service life when applied to a side window. The friction created by the window seal during opening and closing and when the vehicle goes through an automatic car wash further restricts durability. Although a hydrophobic coating can improve visibility, the

---

13. This effect was initially observed by botanist W. Barthlott on the leaves of the lotus flower (refer to Barthlott & Neinhuis [45]).
14. An overview of these coatings is provided by Langenfeld et al. [486].
15. Refer to Chmielarz et al. [145].

benefits it provides only combat the effect and not the cause of the contamination. Water thus continues to run down and across the side window but cannot adhere to it and is transported away from the visible areas more quickly with the aid of shearing and gravitational forces.

The coating of an outside mirror glass has also been investigated many times using both possible surface properties. A hydrophobic coating causes small water droplets to form on the glass that should roll off as for the windshield and side window. The mirror glass is located in the wake area of the mirror housing and is therefore not directly wetted. The gravitational force is too insignificant to keep the glass free of droplet formation. A hydrophilic coating should cause the accumulating droplets to disperse and join up. A fully wetted mirror glass allows for visibility to be somewhat clear without leading to pronounced light refractions resulting from individual droplets. A fogging of the outside mirror glass is also avoided when the vehicle is driven through a tunnel, for example. The inherent resistance to scratching and frictional wear of the design coating speaks against an application involving a highly exposed component such as an outside mirror.

6.3.2 External Contamination

6.3.2.1 Causes of External Contamination

Contamination of the side windows can be attributed to two aerodynamic-specific causes in addition to direct contact with rain drops (refer to Figure 6.32):

- Rivulets formed by water flowing over the A-pillar
- Atomized spray or accumulated droplets caused by air flowing around the outside mirror

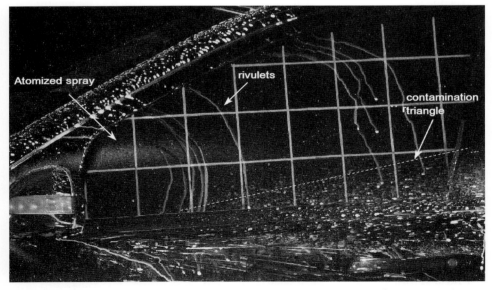

Figure 6.32 The contamination pattern of the side window at 60 km/h shows all three possible types of contamination: rivulets, contamination triangle, and atomized spray.

6.3.2.1.1 Flow over the A-Pillar

There are generally several ways in which water can flow over the A-pillar onto the side window and cause contamination in conjunction with the following individual or collective mechanisms:

- The most frequent cause is attributed to water droplets that make direct contact with the water deflector rail and A-pillar.[16] These fine, small droplets, as they fall down from the sky or are thrown up from the vehicle in front, combine with each other to form larger droplets and flow back over the A-pillar onto the side window in relation to the prevailing shearing forces. Droplets also accumulate on the windshield, which are then wiped toward the A-pillar by the wipers, and create a water film on the window. Depending on the angle of the windshield and the speed at which the vehicle is traveling, the water flows down and out the water deflector rail, channeled by gravitational forces, or moves upwards over the roof.[17] On a steeply inclined windshield impacted by a medium airflow velocity, a balance between the shearing and gravitational forces is achieved approximately at the midpoint of the A-pillar. In this range of equilibrium, the adhesion forces acting on the windshield and the water deflector rail lead to the formation of a water bubble. If accumulation upstream of the water deflector rail is sufficient enough, the shearing forces can partially overcome the surface tension and cause droplets to break away. This water then flushes over the A-pillar onto the side window and is also affected by local changes in the direction of flow as triggered by the windshield wipers when they start their downward return stroke from the vertical position. In relation to the location of the accumulated water and, thus, the overflow pattern, as well as the prevailing low flow speed,[18] these rivulets can influence visible areas that are critical to ensuring driving safety.

- The windshield wipers can also facilitate contamination of the side window, depending on the design of the water deflector rail, by flinging the droplets outwards during the return stroke as a result of inertial forces, causing the droplets to land on the side window. This effect can likewise be observed when the wipers reach their vertical position and are parallel with or angled closely to the upper leading edge of the A-pillar. A gap arises between the wiper blade and the water deflector rail that triggers a local acceleration in airflow toward the roof and amplifies the shearing strain in this direction. More water then flows up over the roof. If the water deflector rail terminates at the upper edge of the windshield, an additional quantity of water lands on the side window, since the water is no longer channeled along the roof.

---

16. Spread and splash mechanisms.
17. The more steeply inclined the windshield and the faster the airflow velocity, the more water is channeled over the A-pillar onto the roof of the vehicle.
18. At low flow velocities, it is primarily gravitational forces, not shearing forces, that act on the water droplets/rivulets that spread across the side window.

The effects of water flowing over the A-pillar are particularly intense in the presence of a low airflow velocity. The water travels across the side window, following the force of gravity. This allows the rivulets that form to flow directly past the visibly relevant areas, and water can enter the trailing wake of the mirror and further increase contamination of the mirror glass. At higher speeds in excess of approximately 100 km/h, these rivulets are transported in a downstream fashion. The upward-acting, shearing force components that result from turbulence present at the A-pillar counteract the gravitational force. Water takes the shape of a rivulet where both forces cancel each other out and thus does not enter the critical visible area, as Figure 6.33 shows. Overflow at the A-pillar frequently leads to an additional contamination mechanism whereby water is transported in the seals along the door or side window. Depending on the construction design of the sealing component, a gap can arise between the window and seal that then takes up some of the overflow water as a result of capillary attraction.[19] The distribution of pressure in the gap and the force of gravity move the water forwards and down in the direction of the mirror triangle. At the very latest when the area of minimum pressure is reached along the seal, or above the mirror triangle at the beginning or start of the A-pillar turbulence, the water drains and is either sprayed or travels in rivulet formation across the outside mirror glass.

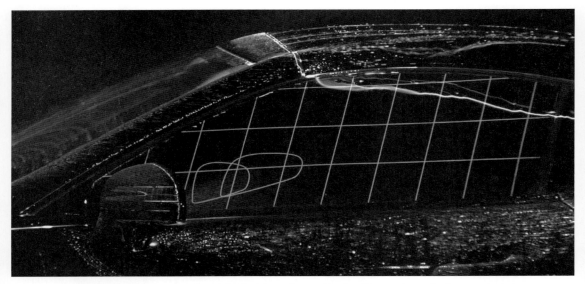

Figure 6.33 The water droplets that travel over the A-pillar coalesce to form a rivulet where shearing and gravitational forces cancel each other out as water is transported in a downstream fashion.

---

19. Adhesion forces between the liquid particles and closely arranged walls of the water runnel pull water into the gap.

6.3.2.1.2 Side Window Contamination Caused by the Outside Mirror

The outside mirror also contributes considerably to side window contamination in addition to the overflow characteristics of the A-pillar. The following types of water and particulate accumulation can compromise visibility in the rain as a result of airflow around the outside mirror:

- Water droplets on the mirror glass
- Droplets on the side window in the pattern of a contamination triangle
- Atomized spray at the front of the side window, particularly in the relevant area of visibility

As a blunt object that extends outward from the vehicle directly into the oncoming airflow, the outside mirror is hit with water droplets that angle off the vehicle's center and accumulate from overhead rainfall or from the vehicle in front. These droplets, which widely vary in terms of their size and weight, strike the surface and disperse to an extent.[20] The water is transported toward the trailing edge of the mirror housing and drips away due to the shearing forces that act on the surface. Recirculation in the dead water flow zone at the mirror carries a portion of these droplets onto the mirror glass, which greatly hinder rearward visibility. The other portion flows in a downward motion in the wake of the mirror and lands on the front and rear side windows. Collection or accumulation of the water droplets can limit the visibility offered by the mirror glass if the spread of water becomes too great.

When droplets meet the surface at a blunt angle, particularly on the inside surface area of the mirror, a misting effect occurs. The droplets burst as they hit the surface[21] and break into much smaller droplets as a result of dispersion and rebounding.[22] If these smaller droplets enter the horseshoe turbulence zone (Figure 6.34), which exists in a semicircular fashion around the base of the mirror, they are distributed in the propagating area of this turbulence system. In addition, at low speeds under 50 km/h and especially when side wind conditions are prevalent, direct rainfall can stop on the usually convex-shaped side window. This effect cannot be prevented by implementing aerodynamic measures.

---

20. Spread and splash mechanisms.
21. Splash mechanism.
22. Break-up and rebound mechanisms.

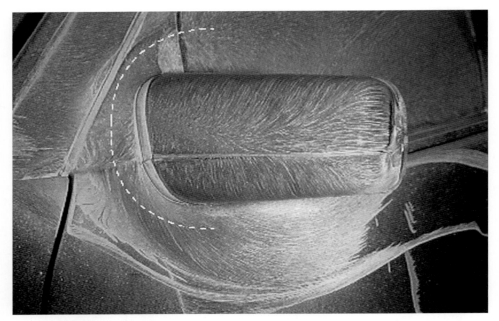

Figure 6.34 Paint test of an outside mirror from the front. Horseshoe turbulence is created on both sides of the mirror base, whose turbulence core is shown by a dotted line.

6.3.2.1.3 External Contamination of the Rear Window

Driving safety is not only promoted by a clean side window but also a clean rear window. In this context, a distinction must be made among the different shapes and contours at the rear of a vehicle. Station wagons, for example, contaminate their rear window primarily as a result of the dirt and contamination whirled up by the vehicle itself (refer to section 6.3.1.2). Hatchbacks and fastbacks, on the other hand, receive contaminated rear windows primarily as a result of the dirt and contamination whirled up by the vehicle traveling in front. The water that hits the windshield or roof flows up and back and, without corresponding design measures in place at the rear window gap would flow across the entire rear window and compromise rearward visibility, in particularly as it affects the inside rearview mirror.

6.3.2.1.4 Water Pullback on the Windshield

Another phenomenon that also relates to external contamination and frequently occurs on modern vehicles that have relatively steeply inclined windshields and tall engine hoods is the so-called water pullback effect. When this happens, the wipers pull some of the water that rebounds from the A-pillar back into the critical visibility area of the windshield and greatly compromise visibility (Figure 6.35).

Functionality, Safety, and Comfort

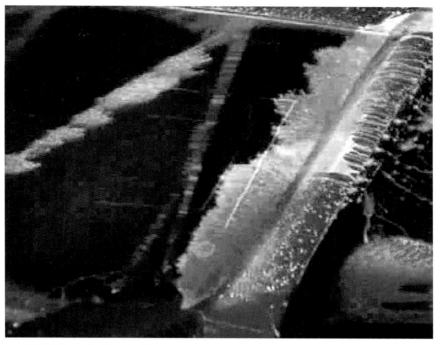

Figure 6.35 Water Pullback effect in the wind tunnel. The wipers drag water from the A-pillars back into the critical visibility area.

As already described in section 6.3.2.1, the wipers displace water on the windshield toward the A-pillar. A water film forms there, whose thickness depends on the relationship between the gravitational force and shearing force. Water then collects in relation to speed (at approximately 80 to 120 km/h) in the center area of the A-pillar. Due to the low-pressure zone that arises during the downward return stroke (refer to Figure 6.36), some of this water is sucked back into the direction of wiping or is displaced toward the wipers by the turbulence created at this time. The water follows the wipers in the local flow direction of the return stroke (refer to Figure 6.37) until the water film height reaches a critical point and the film separates. The water film then distributes itself in the outer area of the windshield until it is reintroduced with the next wiping cycle and moves back toward the A-pillar.

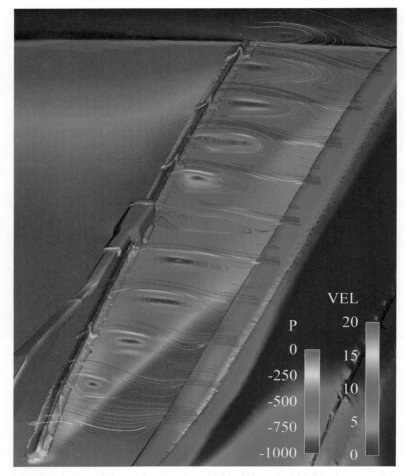

Figure 6.36 Depiction of the wake structure behind the windshield wipers as they return from their vertical position. The propagation of the wake turbulence becomes apparent in the individual sectional areas.

The water pullback can be reduced in its intensity by adapting geometric parameters such as the inclination angle and aerodynamic performance of the windshield as well as by implementing the following measures. Realizing a flat wiper geometry and coordinating the wiper arm linkage minimize the wake behind the windshield wipers. The vertical wiper position is also key, as the distance between the wiper and the A-pillar as well as the wiper speed have a direct impact on the water pullback.

Functionality, Safety, and Comfort

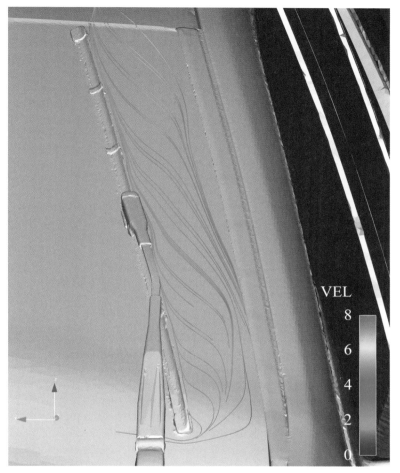

Figure 6.37 Depiction of the close-wall stagnation streamlines on the windshield during the return stroke of the windshield wipers.

6.3.2.2 Design Measures That Counteract External Contamination

When a vehicle is driven in the rain, rain drops fall directly onto the vehicle, and water droplets are also raised and land on it due to the traffic ahead. Some of this water contains dirt particles and causes the vehicle to become contaminated externally (i.e., by non-vehicle-induced factors). The side windows, outside mirror glass, and rear windows are especially targeted at this time. The following sections explain design measures that help keep relevant areas free of contamination.

6.3.2.2.1 Reduction in Overflow at the A-Pillar

It is almost impossible to keep the entire side window free of dirt accumulation. When a vehicle's soiling preventions are optimized, attention should be focused on the visible areas that are critical to ensuring driving safety. One example of such an area is shown in Figure 6.38; the border lines shown represent areas of the side window through which the driver[23] can see the outside mirror glass (visible outside mirror area) and the lateral traffic (peripheral visibility area).

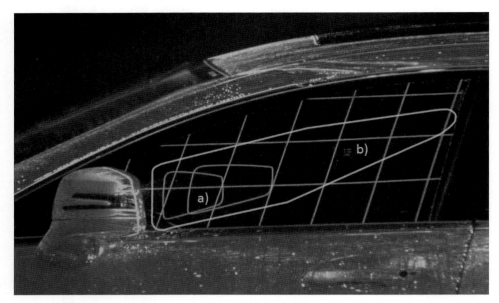

Figure 6.38 Side window contamination at 120 km/h: The contamination triangle and light mist above the mirror do not compromise the critical view areas: (a) visible outside mirror area; and (b) peripheral visibility area.

Water droplets hit the windshield when a vehicle is driven in the rain. The wipers wipe the droplets toward the A-pillar on the driver side.[24] The water traveling toward the A-pillar as set into motion by the wipers is prevented from overflowing back onto the side window by a sufficiently dimensioned water channel. In the absence of any design measure implemented at the side of the windshield, this water would run across the side window in rivulet formation, and at higher speeds, it would be caught in the turbulence present at the A-pillar and be atomized thereby greatly hindering side window visibility. To counteract this natural tendency and in accordance with the door and A-pillar design concept, vehicles are either equipped with a water deflector rail integrated in the A-pillar or with an appropriately dimensioned window encapsulation facility.[25] The water deflector rail must be designed and configured in such a way that the side window remains clean while also ensuring compliant A-pillar flow properties with respect to noise and aerodynamic resistance. The following basic design characteristics aid in engineering an effective water deflector rail, as is shown in Figure 6.39.

---

23. The lines mark the critical view areas of the side window for 5%– (women) and 95%– (men).
24. This is the case for a traditional wiper arrangement. Single-arm wipers tend to displace water toward the plenum chamber in front of the windshield.
25. A distinction will now no longer be made between the two types of water guidance or passage. The relevant design details can be applied to both design variants.

Functionality, Safety, and Comfort

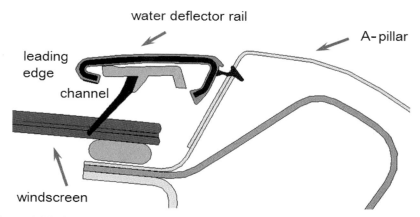

Figure 6.39 Section view of a water deflector rail. Note the leading edge radius and channel cross section.

To accommodate the quantity of water to be drawn away as dictated by the size of the windshield and the intensity of the prevailing rain, the leading edge of the runnel must be positioned at a sufficient distance to the surface of the windshield and have a correspondingly dimensioned channel cross section. In this context, the center A-pillar area is especially critical, since water can assume a bubble shape and overflow here, depending on the speed of the vehicle. On most vehicles, the water deflector rail tapers off at the upper edge of the windshield for design reasons. A sufficient cross-sectional area must still remain, however, so that water can be channeled along the A-pillar toward the roof without inducing further accumulation effects. If the water deflector rail terminates at the upper edge of the windshield or the cross-sectional area becomes smaller at this point such that the minimum dimensional requirements are undershot, water can no longer be channeled as intended and starts to overflow. This area is particularly critical on convertibles because the gap between the roof rail and convertible top is found here. Water then stops traveling completely along the roof toward the rear of the vehicle, and some of it makes its way down onto the side window.

The leading edge of the water deflector rail also determines overflow behavior onto the A-pillar in addition to the rail's cross-sectional area. Droplets in free flow formation are diverted by the windshield in such a way that they meet up with the leading edge of the rail at a rather blunt angle. It is here that the fine droplets merge to form larger droplets, which then travel down the water deflector rail farther onto the A-pillar and side window. This phenomenon can only be hindered by keeping the surface area of the water deflector rail's leading edge as small as possible (refer to Figure 6.39). Unfortunately, however, doing so frequently clashes with requirements pertaining to aerodynamic force and wind noise, which require a large leading edge radius.

The water droplets in free flow contact all areas of the vehicle that are positioned in the direct flow of the airflow; this, of course, also includes the areas of the water deflector rail and A-pillar. Here, too, droplets merge after initial contact and overflow onto the

A-pillar. Further transport onto the side window can be influenced by the design of the door seals or a trim strip in this region, for example. The design objective is to provide for additional entrapment facilities where some of the water can be retained. One example of a trim strip design is shown in Figure 6.40.

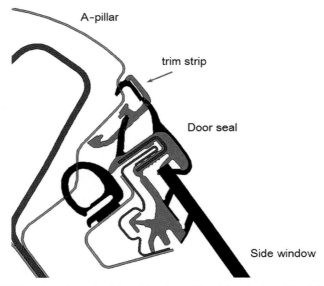

Figure 6.40 Section view of a trim strip along the side window: The trim strip forms a channel together with the seal.

6.3.2.2.2 Reduction in Contamination Caused by the Outside Mirror

As a result of the great influence of the different subareas of vehicle aerodynamics, the outside mirror represents an important component for aerodynamic engineers. As already covered in section 6.3.1.1, the outside mirror is a major factor that contributes to side window contamination. Without this mirror, the water that travels over the A-pillar would be the only cause of contamination. The contamination created by the wake of the mirror can be reduced in two ways. The transport of water on the mirror housing can be improved such that water is channeled into areas from which it can drip away without rebounding back onto the vehicle. In addition, the wake structure can be influenced to prevent droplets from accumulating in critical visibility areas. Figure 6.41 shows an outside mirror that was optimized to deflect dirt by implementing several of the following design details:

- A consistent channel that is fitted at the trailing edge of the housing in the transverse direction of the airflow can entrap water that contacts the mirror housing before it is transported to the mirror wake and guide it to a less critical area. When a correspondingly designed housing shape is created, a point can be realized at the bottom, outward edge of the mirror housing at which a local, low-pressure zone is induced. In an ideal scenario, the water that collects in the channel is transported to this area and drips away here without entering the directly adjacent wake of the mirror so that additional contact with the vehicle is avoided. Generally speaking, however, some of the droplets do make contact with the front side window. A wedge-shaped contamination pattern as shown in Figure 6.38 occurs in the lower side window area and on the upper section of the door panel. When the drip-off point is positioned correctly, this contamination pattern neither compromises visibility nor does it soil the door handle. The further inward the drip-off point is located, the sooner the droplets land on the side window and can thus enter the field of visibility or the droplets enter the mirror wake and rebound onto the mirror glass.

Figure 6.41 Sample illustration of a mirror optimized in detail to deflect dirt (water deflector rail and drip lip facility on the underside).

- Occasionally, the channel at the bottom of the mirror housing is no longer capable of trapping water. Drip lips help improve the performance of the channel. A drip lip can, depending on how it is dimensioned, alter the flow characteristics of the mirror such that the location and size of the contamination triangle are positively influenced. Airflow is diverted downward by the spoiler. Although the channel and drip lip improve soiling characteristics, they can also negatively impact aerodynamic forces and wind noise. Here, too, a compromise must be found that considers all relevant aerodynamic and design characteristics.
- The outer contour of the mirror housing can positively alter the drip-off trajectory by influencing lateral airflow. Aligning the lateral area of the mirror parallel to the window surface promotes a further downward contact point of the droplets on the side window.
- Another possibility for influencing the wake structure depends on the design of the gap between the mirror housing and the door. The mirror triangle, mirror base, and mirror housing form an upward, open channel that guides the airflow along the side window. The wake of the mirror can thus be diverted away from the window to a certain extent so that water droplets only make additional contact with

the vehicle further downstream. The channel should slightly increase in diameter as it progresses from the front to the back without causing flow separation in the channel cross section. A channel center line with a minimal outward offset also has a positive effect.

- The decisive parameter for reducing atomized spray with the described mirrors relates to the size of the inner housing radius where the channel begins, as schematically shown in Figure 6.42. Water droplets disperse at the leading edge of the mirror housing and are transported as fine droplets through the channel onto the side window. Decreasing the inner radius leads to a reduction in this critical contact surface.[26] Also, when the mirror base is designed, attention must be paid to ensure that as few droplets as possible can contact its surface at a blunt angle.

- If the mirror base is for a beltline mirror on the door, a more detailed design construction of this point will be required. The horseshoe vortex at the mirror base can induce a pronounced low-pressure zone where it transitions to the mirror housing such that the drip-off point or trajectory moves farther inward (refer to Figure 6.43). This can be prevented by coordinating the shape and positional arrangement of the mirror base or by optimizing the drip-off trailing corner point.

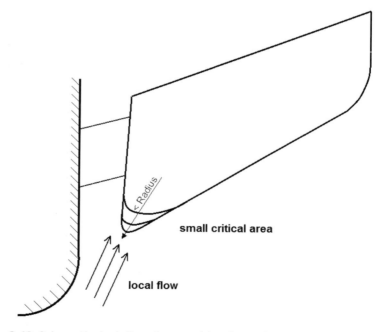

Figure 6.42 Schematic depiction of an outside mirror: The smaller the inner radius of the mirror housing, the smaller the critical surface area on which incoming droplets can accumulate and the less prevalent the side window contamination as caused by atomized spray.

---

26. Moving the outside mirror farther down the vehicle can also reduce the critical area in question or even eliminate it altogether.

Functionality, Safety, and Comfort

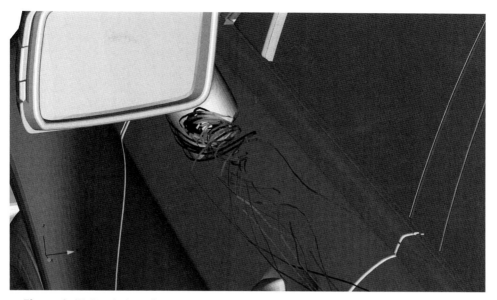

Figure 6.43 Depiction of stagnation streamlines surrounding an outside mirror that is mounted to the door panel. An intense wake turbulence forms between the mirror housing and mirror base.

The following additional design characteristics not shown in Figure 6.41 promote a further reduction in side window and mirror glass contamination:

- If the outside mirror is not mounted at the mirror triangle, but is instead mounted farther downstream, less water coming from the A-pillar contacts the housing and the critical contact surface point at the housing's leading edge.
- A relatively small frontal area of the mirror housing with respect to the local direction of flow leads to the collection of fewer water droplets.
- Elongating the mirror housing in the downstream direction beyond the perimeter of the mirror glass plane keeps the recirculating flow away from the glass to a certain extent, although the visibly relevant area of the mirror surface may be decreased in doing so.

6.3.2.2.3 Reducing Contamination on the Rear Window

The simplest and most effective measure for avoiding the accumulation of dirt and contamination on the rear window of hatchback vehicles is to integrate a channel at the upper edge of the rear window by designing the positional arrangement of the window and adhesive bead accordingly so that water can collect here and be channeled outward. On fastback vehicles that seal between the tailgate and roof, it is important that this area be planned for appropriately during the design stage. Water traveling over the roof must be able to freely flow into this water channel. An imbrication between the roof and rear window ensures that the water droplets at the trailing edge of the roof also flow into the channel and can be transported laterally away from the vehicle. This water must not be suctioned out by prevailing vacuum peaks. Covering the rear half of the water channel prevents this (refer to Figure 6.44). A laterally biased design of the rear window with similar dimensions prevents too much water from being suctioned out by the C-pillar turbulence and landing on the side of the window.

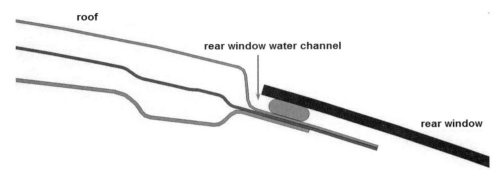

Figure 6.44 The cross-section view of a rear window as it transitions to the roof shows the half-covered water deflector rail and imbrication of the window to the roof panel.

When efforts are undertaken to optimize the soiling characteristics of the rear window, special attention must also be paid to the positional arrangement and dimensioning of the antenna. If a telephone or GPS antenna should be mounted at the rear edge of the roof or on the rear window, as is the case with the majority of newer cars, low-pressure areas can form at their perimeter contours that extract trapped water out of the rear window borderline, which then flows in rivulets over the window as Figure 6.45 shows.

Figure 6.45 Contamination pattern on the rear window at 160 km/h: Roof antenna and low-pressure peaks at the C-pillars compromise the result.

Functionality, Safety, and Comfort

Hatchback cars are also sensitive to further unhindered water drain-off (e.g., along the front trunk lid gap). If too much cross-flowing water enters the area between the rear window bottom edge and the trunk lid, it can find its way into the enclosed, turbulent separation bubble typical for hatchbacks and recontact the vehicle, on the rear window, in the form of fine droplets.

6.3.3 Vehicle-Induced Contamination

6.3.3.1 Causes of Vehicle-Induced Contamination

The main cause of vehicle-induced contamination has to do with the water droplets and contamination particles whirled up by a spinning wheel. The following highlights how this process takes place at a wheel and the effects it has. Measures that counteract this type of contamination are explained in section 6.3.3.2.

The water droplets whirled up by a spinning wheel are broken down into two categories:

- Splash water: Larger water droplets that move in parabolic curve formation close to the ground and at relatively high speeds
- Spray water: A fine mist of small water droplets that are carried downwind by the turbulent airflow

Figure 6.46 shows the spray pattern of a free-rolling, uncovered wheel as Koessler [451], Braun [98], and Clarke [148] have described. The spinning wheel with profiled tire can transport only some of the water on the road in the profile tread ribs toward the outside or in the direction of rotation. The remaining portion is displaced in front of the tire contact patch, similar to a bow wave created by a boat. Splash water then takes shape as relatively large droplets, which are dispersed forwards and to the side at an angle of up to 45 degrees. In the profile tread ribs, water rotates along with the tire due to the adhesion forces it exerts. When the circumferential speed reaches a critical threshold, however, this adhesion is no longer sufficient in resisting the centrifugal forces. Small water droplets are cast away in a tangential way in the form of spray water.

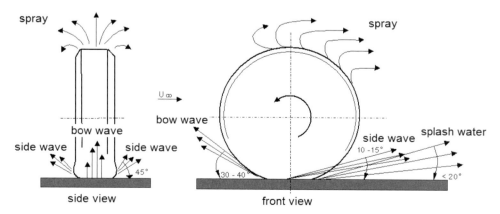

Figure 6.46 Schematic display of water separation at a spinning wheel.

523

Two independent spray zones are created: At the end of the contact area, the local circumferential wheel speed rapidly increases.[27] Some of the water below the tire contact patch sprays away in the form of fine droplets at an angle of between 20 and 30 degrees. Included in this spray mist is splash water from the profile tread ribs that is thrown outward at a shallower angle. A splash-free zone now forms in the tire circumference direction. The adhesion forces of the flat water layer remaining in the profile tread ribs exceed the centrifugal forces in the surrounding area. Only in the upper third area can the centrifugal forces overcome the adhesion, and water droplets are cast away vertically and counter to the travel direction of the vehicle.

When a wheel is covered by a fender, the droplets contact the wheel well liner and disperse. Some of the water can collect there, travel downward, and combine to form large droplets at the end of the liner. The remaining water is transported out of the wheel well by air circulating in the upper rear wheel well area and transitions into a spray mist.

In general, a portion of the spray mist at the wheel is channeled into the outside airflow by the outward flow coming out of the wheel well and in the center area of the underfloor. This mist either accumulates on the side of the vehicle and the door handle or it is transported to the rear wake and considerably contributes to vehicle-induced contamination.

Vehicle-induced contamination primarily refers to soiling of the rear of the vehicle and its side quarters. The sides and door handles are contaminated as a result of the splash water created by the spinning front wheels and the spray mist ejected from the front wheel wells, whereas the trunk lid, rear lights, and trunk lid handle are mainly contaminated by the rear wheels. Adding to these phenomena is the water that comes from the turbulent underfloor airflows. Fine water droplets, some of which have dispersed a second time here, enter the wake of the vehicle and land on the rear panels.

6.3.3.2 Design Measures That Counteract Vehicle-Induced Contamination

The most obvious measure for counteracting vehicle-induced contamination is to influence how the droplets form in the wheel well and on the underfloor. Spray mist can be reduced either by implementing aerodynamic measures or by fitting mechanical splash guard systems that collect water and guide it away. Adapting the flow topology of the front wheel wells can minimize the ejection of spray mist. Wheel spoilers can be used for this purpose.

Mechanical measures for reducing spray mist can take the shape of appropriately formed wheel well liners, whereby the wheelhouse cutout is made smaller through the use of elastic covers, for example. Lining the entire wheel well with longitudinal grooves can trap fine water droplets. The droplets collect in the grooves to become larger and flow down and off the bottom side of the wheel well. Splash guards behind the front wheels can also reduce the spray effect caused by the wheels themselves.

---

27. Caused by acceleration of the tread elements in the terminating contact area.

The following section provides detailed information about aerodynamic measures for minimizing door handle and rear window contamination.

6.3.3.2.1 Reducing Door Handle Contamination

There are two causes of door handle contamination. Water droplets that drip off the outside mirror and land on the side window and door panel: this is external contamination. If the positional arrangement of this contamination triangle is not coordinated correctly, the front door handle in particular will be heavily soiled. Another source of contamination of the side panel is the front wheel. The spray mist that ejects from the front wheel well in the wake zone of the front wheels is thrown backwards and accumulates on the side wall, as shown in section 6.3.1.2. Opening the door without getting your hands dirty then becomes impossible, depending on the size of the wake zone and the position of the door handle.

Wheel spoilers in front of the front wheels can, if positioned and dimensioned appropriately, reduce the pressure level in the wheel well such that the outward airflow is reduced and the tire wake becomes more streamlined. The accumulation pattern on the side wall moves farther downward, and the door handles lie outside the contamination zone of the front wheel. Figure 6.47 shows the influence of wheel spoilers. The designated white line marks the upper edge of the contamination.

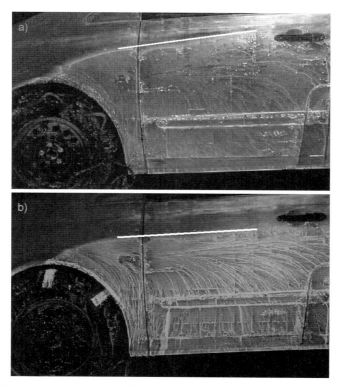

Figure 6.47 Influence of wheel spoilers on side panel contamination: (a) without wheel spoilers, the door handles become contaminated; (b) with wheel spoilers, the door handles remain free of contamination.

6.3.3.2.2 Reducing Contamination on the Rear Window

Fine water droplets that are thrown up by the wheels and the underfloor enter the wake area of the vehicle. There, they distribute themselves across all vehicle surfaces that border with the vehicle wake structure. Included among these surfaces are the taillights and the rear window. Unlike with hatchback and fastback vehicles, these areas are not largely contaminated by external contamination sources and the water that runs up over the roof as shown in section 6.3.1.1, but instead by finely distributed water droplets in the dead water flow zone.

Introducing an aerodynamic soiling prevention design is only effective if the dead water is influenced such that the contamination-laden recirculating airflow does not contact with the rear window. There have been many attempts made to affix a correspondingly shaped roof spoiler to channel the airflow along the roof in the downward direction. The close-wall flow is channeled in a targeted fashion across the rear window by the bypass opening. The fine water droplets in the turbulent dead water zone cannot strike the rear window in this area. Flow speeds drop relatively quickly. In the bottom and side areas of the rear window, the flow diversion effect has declined to such an extent that the water droplets can recontact the rear window, as Figure 6.48 shows.

Figure 6.48 Influence of a roof spoiler on rear window contamination, according to Larsson [487]: (a) with roof spoiler, (b) without roof spoiler, (c) airflow made visible in the area of the roof spoiler through the use of a smoke sensor.

Although this type of roof spoiler can keep the rear window marginally clean, the flow diversion drastically increases aerodynamic drag and rear axle uplift, which frequently means that a design implementation is not feasible. To ensure that the driver can adequately see through the rear window, the majority of vehicle manufacturers equip this window with a wiping system. Tailgate handles and reversing cameras cannot be kept clean with aerodynamic measures. The water droplets in the vehicle wake contact all exposed surfaces. If the handle is recessed in a cavity behind a cover, contamination can at least be reduced on the inner side. Reversing cameras should be handled in a similar manner; alternatively, the cameras can be actively controlled so that they extend only when the reverse gear is engaged.

Chapter 7
Cooling and Internal Flow

Ralf Neuendorf, Bernhard Zuck

7.1 Cooling Requirements

The task of vehicle cooling is to provide sufficient heat sinks for all heat sources in a vehicle. Sources of heat include the engine and gearbox, electronics, as well as the refrigeration cycle of the air conditioning system. This is to ensure the continuous operation of the vehicle without sacrificing comfort, driving dynamics, and system performance. This requires a sufficient amount of cooling air to the cooling module and the performance of its respective components to achieve this task.

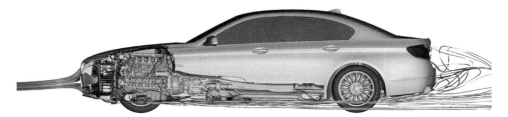

Figure 7.1 Internal flow.

This part of the vehicle aerodynamics is referred to as internal flow. Changes of the fundamental design of the overall system are highly dependent on the thermal requirements of the individual components and their position. In the following sections, these aspects and the need for optimization with respect to the overall aerodynamic vehicle context will be discussed.

Chapter 7

The cooling capacity of a vehicle is primarily determined by the driving conditions. The driving speed changes frequently between different conditions, such as urban, mountain, or highway driving. These conditions alternate between constant rate driving, acceleration, and braking phases. Therefore, all operating points of the engine map are covered, from coasting, engine idle, up to maximum power, which can be above 300 kW in modern vehicles.

To design the cooling system, a set of various operating conditions is considered, each representing different dynamic states of the vehicle. This set is to ensure the proper functionality of the vehicle in all customer relevant conditions. These operating conditions and then the cooling requirements of the major components and systems will be discussed in the following section.

7.1.1 Representative Operating Conditions

7.1.1.1 Maximum Speed

Driving at top speed over a longer period is—at least on the German Autobahn—possible. Therefore the cooling system must be designed robust enough to transfer in this steady-state situation all produced heat to the environment. At this driving condition, not only is the engine power and therefore the heat input into the cooling system high, but there is high dynamic pressure to supply cooling air.

7.1.1.2 Driving with a Trailer

When towing a trailer, the speed limit is low due to the high drag of the trailer. However, the required engine power is relatively high in relation to vehicle speed to overcome the increased drag. Therefore, this condition is also important for the analysis of the engine cooling.

7.1.1.3 Up-Hill Driving

Similar to this condition is driving in mountainous terrain with a maximum loaded vehicle. The engine power out is high while the airflow to the cooling module is low. With the addition of towing a trailer, the power output is even higher and the vehicle speed is lower still. The necessary cooling air must therefore be provided by a fan. Unlike the high-speed driving condition discussed previously, the ascent to the top of a mountain is temporary. It is useful to take this into consideration to avoid unnecessary expense and weight with an over-designed cooling module.

7.1.1.4 Idle

Although the heat output of the engine is low when the car is idling, combustion losses, internal friction, and ancillary components, such as the air compressor and the alternator, still demand a few kilowatts of cooling. Because of the absence of vehicle speed, the necessary cooling air must be supplied by a fan. Furthermore, today's engine start-stop systems shut off the coolant flow through the cooling circuit when a vehicle is stationary, which brings the aspect of thermal protection of the engine and other components into focus (see section 7.1.2.6).

7.1.2 Components and Systems
7.1.2.1 Engine

Modern vehicles are almost exclusively powered by liquid-cooled engines. These represent the main source of heat, which is why they are first considered in detail.

In a combustion engine, the chemical energy stored in fuel is turned into heat by combustion, which in turn is converted into mechanical work. While the combustion reaction usually converts 100% of the energy to heat, the efficiency of converting the thermal energy into mechanical energy is in the range between 20% and 40%. The theoretical upper limit is defined by the Carnot cycle given by the ratio of the maximum temperature to the ambient temperature. Of the remaining 60% to 80% of the energy, slightly more than half is exhausted as hot gas. The remaining heat is discharged via the engine cooling system to the air. A simple rule of thumb for a water-cooling engine in steady state is the so-called one third rule: one-third of the fuel energy is converted into mechanical energy, one-third remains in the exhaust gas, and one-third of the heat is supplied to the cooling system. This means that the cooling system converts approximately the same amount of energy that is used to propel the vehicle. While this rule is a good approximation for low engine loads, more heat is dissipated from the exhaust and less from the cooling system with increased loads.

On the other hand, for electric drive trains, including the power electronics, 80% of the energy stored is transformed into propulsion. This reduces the cooling capacity requirement accordingly. However, the maximum temperatures for electric motors and power electronics are lower than internal combustion motors. While the latter can operate with coolant temperatures up to 120 °C, electric drive trains generally only require a maximum coolant temperature of about 70 °C. At an air temperature of 30 °C for example, the driving temperature gradient reduces the radiator effectiveness by more than half, which affects the size of the required heat exchanger and cooling air mass flow.

For the design of the cooling system, an accurate knowledge of the energy flows is necessary. The starting point is to calculate the power requirements of the drive system. This is usually determined for steady driving conditions, taking into account the vehicle weight m_F, the aerodynamic drag c_D, the rolling resistance of the tires μ_R, and the gradient of the road. This is described in equation (7.1).

$$P_A = \left[c_D \cdot A_x \cdot \frac{\rho_L}{2} \cdot v_F^2 + m_F \cdot g \cdot (\mu_R + \sin(\arctan N)) \right] \cdot v_F \quad (7.1)$$

The operating point of the engine (i.e., number of revolutions n and torque M) depends on the dynamic radius of the wheel and the transmission gear ratios. Design details of the engine, especially with respect to the internal engine coolant and oil circuits, predetermine how much heat can be dissipated in both systems. The ratio is given by χ. When only using a coolant-cooled motor oil heat exchanger, the entire heat is dissipated by the main coolant radiator. Thus for the following section, only the coolant system is considered. The same principles apply for a stand-alone engine oil cooling system. The dissipated heat can be calculated according to equation (7.2).

$$\dot{Q}_M = \frac{P_A}{\eta_A \cdot \eta_M} \cdot \chi \qquad (7.2)$$

7.1.2.2 Engine Intake Air

For diesel engines, turbo charging and intercooling have been the standard for many years. This technology has become more predominant in recent years with the trend to downsize gasoline engines. The temperature of the fresh air inducted by a motor, both for naturally aspirated and turbocharged engines, has as a large influence on the combustion process and thus the engine performance. In order to have the lowest possible intake air temperature possible, it is best to draw cold air from the ambient environment. For most vehicles, this intake is located at the front of the vehicle, often in cooling module ducting before the cooling package. Considerations to take into account when locating the engine air inlet include:

- Avoiding a location where water could be sucked into the motor (destroying the motor) during a water crossing
- Minimizing the ingestion of dirt and snow directly into the intake
- Reducing the possibility of warm air from the radiator being recirculated back into the air inlet

For turbocharged engines, the temperature must be lowered significantly after the compression of air (charging) for optimum engine operation. The dissipated power from removing heat from the charged air is in the order of 20–30% of the total engine cooling requirements. The required temperature level after charging the air is considerably lower than other engine cooling requirements, and as a result requires a unique arrangement of the intercooler. For the intercooler, there are basically two different systems [908]. These are:

- Direct charge air cooling, with a direct air-air heat exchanger where the charge air is cooled directly with air from the ambient environment.
- Indirect charge air cooling, where the charge air is cooled by a coolant in a heat exchanger. The coolant then transfers heat to the air in a separate heat exchanger.

7.1.2.3 Transmission

Due to the high efficiency of modern manual transmissions, surface cooling from the transmission case is usually sufficient. With specific control of the underbody flow, the heat transfer can be increased. In automatic transmissions, the heat created from the hydraulically actuated clutches is significant and requires external cooling. In particular, the heat output from torque converters operating at low speeds is considerable.

7.1.2.4 Differential

The whole engine power is transferred to the wheels via the differential. Although the efficiency is very high, the heat has to be dissipated over a relatively small surface. The highest cooling requirements occur at high speed and at high dynamic driving conditions. While the former driving condition provides moderate oil temperatures in the differential due to a high mass flow rate of cooling air, the latter represents a special challenge to provide adequate cooling. Therfore, the underbody aerodynamics required for the cooling of the differential contribute to the operational safety of the vehicle.

7.1.2.5 Air Conditioning

In contrast to the requirements of engine cooling, climate control is independent of driving conditions and primarily dependent on the external air temperature and the desire of the occupant. This is especially true in stop-and-go traffic in major cities in Asia, the United States, Australia, and in Southern Europe, where air conditioning is an important vehicle requirement. The performance and efficiency are largely determined by the condenser. This means that a sufficient cooling airflow with a minimum temperature through the condenser is necessary. The design of the cooling module ducting is also of great importance for the air-conditioning performance, especially in slow traffic and at standstill, where hot exhaust air from the vehicle can recirculate back into the condenser, which should be prevented. A description of the climate system and an overview of the refrigerant circuit is given by Tummescheit et al. [809].

7.1.2.6 Components

The vast majority of the heat is dissipated through the heat exchangers. However, many other components, particularly in the engine compartment and underbody area, are air cooled, especially in the vicinity of the exhaust system. All these components are subject to high thermal loads. In contrast to the engine and transmission, the temperature can peak after stopping the engine due to thermal conduction, radiation, and convection from hot components if there is a lack of cooling airflow (Weidmann [858]). This leads to a correspondingly higher cooling requirement to keep the temperatures during driving conditions below permissible limits. Furthermore, electronic components increasingly require specific cooling.

7.1.3 Other Requirements

In addition to the cooling requirements, the design of the vehicle front end must consider other functional and aesthetic requirements.

7.1.3.1 Vehicle Styling

An important factor for the internal flow is the vehicle design. The design of the front of the vehicle is greatly affected by the design of the air intakes. For almost all manufactures, this is an important element of the styling. Prominent examples are the BMW double kidney grille, the horseshoe shape of Bugatti, the single-frame grille by Audi, and the Rolls Royce Temple design.

7.1.3.2 Packaging

Significant constraints for the arrangement and size of the cooling module and the airflow are given by the vehicle package. Usually, the cooling system and air ducts are not optimal because of the packaging space available. Also, the heat exchangers are usually more compact than the design requirements. This results in higher pressure loss through the cooling system and an increased cooling mass flow requirement.

7.1.3.3 Safety

Not to be neglected are safety and legal requirements of the vehicle. The front bumper must be positioned in a certain height range for safety and legal requirements. In most cases this is in front of the radiator at the position of the stagnation point and therefore can significantly hinder the airflow to the heat exchanger. However, the cooling system should be protected from minor accident damage, otherwise the vehicle will require towing, and replacing the radiator for small accidents will incur significant repair costs. There is an increasing importance on pedestrian protection in an accident, which restricts design freedom and heat exchanger packaging. Inlet opening and ducting must also be designed so that it is not possible to access the rotating fan from outside the vehicle. Typically, protective screens are used, which incurs an additional pressure loss in the cooling system.

7.1.3.4 Acoustics

Air inlets and outlets are channels for the transmission of sound to the outside of the vehicle. More noise from the engine can be transmitted, which is particularly undesirable for diesel engines. Fan noise can be predominant at low speeds, especially when there is a high fan speed. This results in trade-off for fan blade design between maximum airflow and noise emissions. A task for the aerodynamic design is to reduce the pressure losses on the air side of the cooling system in slow traffic conditions to keep the fan at a speed where the acoustic signature is acceptable.

7.2 Cooling System

7.2.1 Engine Cooling System Circuit

As already discussed, modern engines are cooled by a coolant circuit. Liquid cooling combines several advantages over air cooling:

- Higher power density
- Higher heat flux density in the motor and thus better cooling for high thermally loaded areas
- The engine has a larger operating range since the temperature can be kept constant via a thermostat
- Lower acoustic emissions

The initial disadvantages of greater weight and higher energy requirements are outweighed by the advantages. Therefore, in the following sections, only the liquid cooling system will be considered in detail.

The most important design input for the cooling system are the heat sources, the largest of which is usually represented by the engine. The motor internal cooling is assumed as a given quantity and is discussed in more detail by Pischinger et. al. [645]. The simplified diagram of the coolant cycle is shown in Figure 7.2. The coolant is conveyed by a pump in the circuit through the engine, and in parallel to the engine oil cooler (EOC), where it absorbs heat. The coolant is then conveyed to the heat exchanger, where the heat is transferred to the external air. Controlled by a thermostat, the coolant mass flow from the engine is divided into a portion that flows through the radiator, and a second portion is recycled directly to the engine inlet. This arrangement allows the mixture of warm and cool coolant to flow at constant inlet temperature into the engine, Furthermore, the transmission oil cooler (TOC) is integrated into the circuit, where the flow is also regulated by a thermostat, allowing the temperature regulation of the transmission. In addition to the drive system, there are additional cooling requirements. This includes the refrigeration cycle of the air-conditioning system, which absorbs the heat in the air-conditioning evaporator (A/C) and the compressor and emits it into the condenser. In addition, a direct charge intercooler is shown in the cooling diagram.

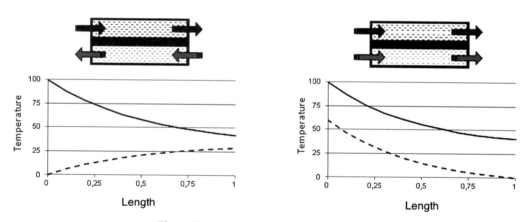

Figure 7.2 Engine cooling system circuit.

For steady-state conditions, the engine cooling can be described by the following energy balance in equation (7.3). The waste heat of the engine $\dot{Q}_M$ is absorbed by the coolant $\dot{Q}_{KMK}$ and discharged to the cooling air $\dot{Q}_{KL}$:

$$\dot{Q}_M = \dot{Q}_{KM} = \dot{Q}_{KL} \tag{7.3}$$

7.2.2 Fundamentals of Heat Transfer

This section covers the basics of heat transfer and key performance indicators. The subject is covered in more detail by Baehr and Stephan [31]. Basically, the following three transport mechanisms for heat transfer can be distinguished, where the driving force is always the temperature difference:

- Heat conduction
- Convection
- Thermal radiation

As in fluid mechanics, dimensionless numbers are also used in thermodynamics to represent dependencies among properties; therefore, the substantial numbers for heat transfer will be discussed briefly in the following section.

7.2.2.1 Heat Conduction

Heat conduction can take place in any medium, whether it be solid, liquid, or gas. According to Fourier's Law, the heat flux $\dot{q}$ through a medium is proportional to the temperature gradient. The proportionality constant λ is called the thermal conductivity:

$$\dot{q}_\lambda = -\lambda \frac{dT}{dx} \tag{7.4}$$

The thermal conductivity λ is independent of temperature in solids and liquids for a wide temperature range. In gases, however, λ increases with temperature.

7.2.2.2 Convection

In flowing fluids, heat conduction is superimposed by transport of internal energy

$$\dot{q}_h = \dot{m} \cdot c_p \cdot \Delta T . \tag{7.5}$$

Of industrial importance in particular is the heat transfer from a solid wall to a flowing fluid. The flow near the wall has a significant influence on the heat transfer. The wall heat flux can be represented by

$$\dot{q}_\alpha = \alpha \cdot (T_W - T_F) . \tag{7.6}$$

The proportionality factor α is called heat transfer coefficient. It can be represented as dimensionless value using a characteristic length l_{ref} and the Nusselt number, which represents the ratio of convective heat transfer to thermal conductivity:

$$Nu = \frac{\alpha \cdot l}{\lambda} . \tag{7.7}$$

The Nusselt number can in turn be described by the Reynolds number and the Prandtl number.

$$Nu = k \cdot Re^{k_1} \cdot Pr^{k_2} \tag{7.8}$$

These relationships allow for the calculation of heat transfer coefficients solely from the knowledge of the flow regime and material properties. For special cases, it can be determined theoretically. However, it is generally represented by semiempirical equations. For some flow regimes, the coefficients are given in the VDI Heat Atlas [824].

In heat exchangers, heat is transferred from one fluid to another via a solid wall which separates them. In this case, two instances of convective heat transfer and one instance of heat conduction occur in series. By means of the heat transfer coefficient

$$k_{1-2} = \frac{1}{\frac{1}{\alpha_1} + \frac{d_W}{\lambda} + \frac{1}{\alpha_2}} \tag{7.9}$$

The heat flux can be written similar to equation (7.6).

In addition to forced convection, there is also natural convection. This arises in fluids with inhomogeneous temperatures and therefore various densities leading to buoyancy effects. As a result, the portion of the warmer fluid rises upward while the cooler part of the fluid falls. Natural convection can be driven by the temperature difference of the fluid near a wall. This process is described by the Grashof number Gr, which represents the buoyancy force temperature difference in relationship to the viscous force of the fluid.

$$Gr = \frac{g \cdot \beta_p \cdot \Delta T \cdot l^3}{v^2} \tag{7.10}$$

Forced convection can also be determined in terms of the Reynolds and Prandtl numbers. In this case, the heat transfer can be determined as a function of the Grashof and Prandtl numbers:

$$Nu = f(Gr, Pr) \tag{7.11}$$

Also for natural convection, a number of empirical formulas for the calculation of the heat transfer are available in the literature. However, in practice for slow-moving flows, a combination of both forced and natural convection is typical.

7.2.2.3 Radiation

Between bodies, heat exchange by radiation is always present. The area-related heat flux $\dot{q}_\varepsilon$ can be calculated by the Stefan-Boltzmann radiation law, by means of the black body radiation:

$$\dot{q}_\varepsilon = \sigma_s \cdot \varepsilon_{1-2} \cdot \left(T_1^4 - T_2^4\right) \tag{7.12}$$

Since heat transfer is related to the fourth power of absolute temperature, it is significant in high-temperature areas in the vehicle. Therefore, the effects of radiation are important in the vicinity of the exhaust system, where temperatures can reach over 800K. The emission coefficient ε_{1-2} describes the surface condition, the shape, and the orientation of the radiating surfaces to each other. Gas radiation only occurs at very high temperatures or in large volumes and is therefore not considered here.

7.2.3 Design of Heat Exchanger

To transfer heat from one fluid to another, there are a variety of different heat exchangers available. Their selection is based on criteria such as temperature range and temperature difference, power density, fluid properties, volume flux and pressure losses, manufacturing processes, durability under oscillating thermal loads, and mechanical vibrations. Heat exchangers can be categorized according to the direction of flow of the two fluids into three basic forms: parallel-flow, counter-flow, and cross-flow heat exchangers.

7.2.3.1 Parallel-Flow and Counter-Flow Heat Exchanger

In Figure 7.3 the principle concept of a parallel-flow (left) and counter-flow heat exchanger (right) is shown with the corresponding temperature profiles of the fluids. As can be seen in the counter-flow heat exchanger, it is possible to have a higher temperature difference in the fluid from the inlet to the outlet, so more heat can be transferred. Moreover, due to the higher average temperature difference, it has a higher power density. Counter-flow heat exchangers are particularly effective in vehicles as oil coolers and as charge air coolers.

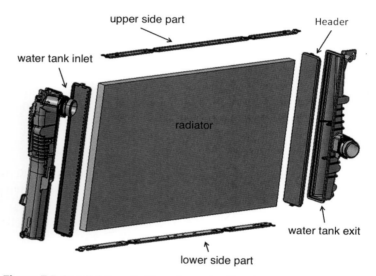

Figure 7.3 Parallel-flow (left) and counter-flow heat exchanger (right).

7.2.3.2 Cross-Flow Heat Exchanger

When the two streams of fluid cross each other, it is classed as a cross-flow heat exchanger. In this situation, the temperature change not only happens in the fluid flow direction but transversely as well. This type of heat exchanger is often used for heat transfer between a liquid and a gas, for example as a coolant radiator and condenser in a motor vehicle. The design and optimization of cross-flow heat exchangers is discussed further by Tandogan [789].

7.2.3.3 Calculation of Heat Exchanger Performance

The performance of a heat exchanger is calculated by integrating the quantities mentioned in section 7.2.2 over its surface, according to

$$\dot{Q} = \int_0^A k_{1-2} \cdot (T_1 - T_2) d\tilde{A} = k_{1-2} \cdot A \cdot \overline{\Delta T}. \qquad (7.13)$$

While for parallel-flow or counter-flow heat exchanger with a well-defined flow direction the heat transfer can be calculated by means of differential equations, in mixed forms, such as in most technical applications, this is not possible. For such designs, the so-called operating characteristic Φ is used.

$$\Phi = \mathrm{Max}\left(\frac{\Delta T_1}{\Delta T_E}, \frac{\Delta T_2}{\Delta T_E}\right) \qquad (7.14)$$

with the inlet temperature difference between the two fluids

$$\Delta T_E = T_{1E} - T_{2E}, \qquad (7.15)$$

and the temperature differences of the two fluids between the inlet and outlet given as

$$\Delta T_i = |T_{i,E} - T_{i,A}|, \qquad (7.16)$$

with

$$\Phi = \frac{\overline{\Delta T}}{\Delta T_E} \cdot \frac{kA}{\dot{w}_{min}} \qquad (7.17)$$

and

$$\dot{w}_{min} = \mathrm{Min}\left(c_{p,1} \cdot \dot{m}_1; c_{p,2} \cdot \dot{m}_2\right) \qquad (7.18)$$

equation (7.13) can be rewritten to equation (7.19), in which all variables are given:

$$\dot{Q} = \Phi \cdot \dot{w}_{min} \cdot \overline{\Delta T_E} \qquad (7.19)$$

This simple, geometric approach to calculate the heat transfer using the equations above cannot be done for a heat exchanger in a vehicle due to the complexity of the system. Therefore, Park [626] developed empirical equations for heat exchangers such as coolant radiator and air condensers based on many measurements.

7.2.4 Heat Exchangers in the Vehicle

In a vehicle, specific heat exchangers are employed for various tasks. These are described in the following section. Often, the heat exchangers are combined into one unit, called the cooling module. Typically, this includes the coolant radiator, the intercooler and/or the oil cooler, the air-conditioning condenser, and the fan.

7.2.4.1 Coolant Radiator

A critical component of the cooling system is the coolant radiator, which dissipates heat from the motor and, if necessary, the power electronics and/or the heat indirectly from the charge air to the environment. A cross-flow heat exchanger is used for this task. Figure 7.4 shows the structure and essential components:

Chapter 7

- Cooling matrix composed of tubes and fins
- Side panels and headers
- Water inlet and outlet with all the necessary connections and fastening elements

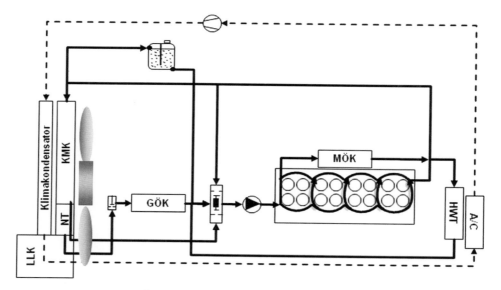

Figure 7.4 Structure of a coolant radiator.

The cooling power and the heat exchangers weight are significantly influenced by the materials used and the design of the radiator matrix. Due to its characteristics, such as low specific weight, high thermal conductivity, high strength, corrosion resistance, and excellent formability, aluminum has become the material for the matrix of modern radiator. The side boxes can also be made of aluminum and soldered directly, or plastic side pods can be mounted on the bottom plate of the matrix.

A key differentiator for the radiator core is the manufacturing process. For lower performance heat exchangers manufactured in very large quantities, a mechanically fitted cooler is used. This design consists of seamlessly drawn round or oval tubes that are mechanically joined to stamped fins. The fins are usually slotted transverse to the direction of the airflow in the form of louvers to improve heat transfer. Soldered flat tube/corrugations systems are used for higher performance requirements. This design consists of a matrix of flat tubes and rolled corrugated fins, which are also equipped with gills. Thicker matrix designs (bigger than 40 mm) are often manufactured by arranging the flat tubes in series.

On the coolant side, an increase in performance can be achieved through the use of turbulence generators. On mechanically joined radiators inserts are placed into the tubes, on brazed radiators turbulators are stamped into the pipes to increase heat transfer performance.

Typical coolant heat exchangers are approximately 15–40 mm in thickness with a fin density of 60–95 fins/dm. The coolant flow rate is typically 1.2–4.5 l/s for passenger cars, 2.5–9 l/s for commercial vehicles, whereas the maximum permissible coolant temperature is usually 115–125 °C for cars and 95–110 °C for commercial vehicles. However, this depends of the vehicle manufacturer and the operating state of the vehicle. Figure 7.5 is a radiator map showing the cooling capacity as a function of radiator coolant mass flow and cooling air mass flow.

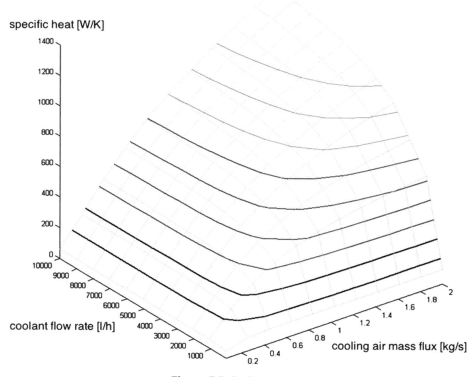

Figure 7.5 Radiator map.

Front surface area, depth, pipe spacing, and fin density are important design parameters of all coolers that are in direct interaction with the cooling airflow. These relationships and optimization are considered in more detail in section 7.4.4.

7.2.4.2 Intercooler

As well as the coolant radiator, a cross-flow heat exchanger is also used for the intercooler. This is usually placed next to or in front of the coolant heat exchanger or in front of the front wheel arches. For high cooling performance, a large intercooler is used, especially on commercial vehicles. The matrix is generally manufactured from aluminum due to the high pressures (over 3 bar) and high temperatures (200 °C) of the charge air. The matrix of the intercooler is similar to the coolant heat exchanger, except the charge air carrying tubes are considerably thicker in proportion to the coolant pipes. To increase the heat transfer surface area, the cooling tubes have fins on the inside and outside. Particular attention must be applied to the design of cooling tubes to reduce the pressure drop of the charge air, since this in turn increases the supercharging compressor requirements and thus the temperature increase due to compression. Therefore, the measure of the quality of the intercooler, density recovery η_ρ, is defined as:

$$\eta_\rho = \frac{\Delta p_{LL}}{\Delta p_{LL,max}} = \frac{\frac{T_{LL,E}}{T_{LL,A}} \cdot \left(1 - \frac{\Delta p_{LL}}{p_{LL,E}}\right) - 1}{\frac{T_{LL,E}}{T_{KL,E}} - 1} \qquad (7.20)$$

An indirect intercooler design combines several advantages, with respect to density recovery and space requirements in the front end of the vehicle. However, an additional circuit is required, which typically consists of a charge air-coolant heat exchanger, an electric pump, and a low-temperature coolant radiator. The heat of the charge air is first transferred to the coolant and then discharged in a low-temperature coolant radiator to the external air. The low-temperature coolant radiator must be placed in the front row of the cooling module to ensure the lowest external air temperature possible.

Compared to direct charge air cooling, the advantages of indirect intercooling include:

- Significantly reduced charge air pressure drop
- Improved engine dynamics with a lower volume of charged air
- More compact heat exchanger in the front end of the vehicle
- Minor restrictions on the size of the coolant radiator

7.2.4.3 Engine and Transmission Oil Cooler

At high engine speed, a substantial proportion of the heat from combustion is absorbed by the engine oil, which must be cooled. In highly loaded transmissions, especially automatic transmissions, convective cooling from the casing is insufficient to prevent excessive heating of the transmission oil. In this situation, transmission oil coolers are needed. Besides the maximum oil temperature requirement, which must not be exceeded under any circumstances, maintaining suitable temperature prevents premature aging. The maximum oil temperature, normally around 150 °C, is determined by the chemical stability of oil and other requirements.

In order to minimize the fuel consumption during the warm-up period of the engine, the cooling system must bring engine and transmission oil up to operating temperature as quickly as possible in order to minimize friction losses. Similar to charge air heat exchangers, both direct oil-air heat exchangers and oil-coolant heat exchangers can be used. The first are designed as cross-flow heat exchanger with an overall thickness between 20 mm and 40 mm. The advantage is that they can use the higher temperature difference between the oil and the external environment, which allows for a high-power density. Only oil-coolant heat exchangers that are integrated into the coolant circuit allow for the targeted heating of the oil during the warm-up period. These are usually designed as a plate heat exchanger in counter-current arrangement.

7.2.4.4 Air Conditioning

The condenser is an essential part of the refrigeration circuit for the air conditioner. The air side of the condenser is similar to the coolant heat exchanger. On the refrigerant side, they have built-in multiflow passes for the compressor gas to first cool from its superheated state to its condensation temperature, condense into a liquid, and then chill the liquid at the end. The cooling conditions are critical to the pressure that arises in the refrigerant circuit. The lowest possible exit temperature of the refrigerant forms the basis for the performance of the air-conditioning circuit as well as for determining optimal efficiency. The lower the temperature after passing through the condenser, the lower the driving power of the compressor, which minimizes the fuel consumption of the air-conditioning circuit. The preferred thickness is 12–16mm with a fin density of 55–95 fins/dm.

7.3 Internal Flow

The internal flow is a part of the overall vehicle aerodynamics and therefore may not be considered in isolation to the surrounding flow. The cooling air is generally characterized as the air that flows through the cooling module and engine compartment and then is fed back into the external flow. It produces forces on the vehicle directly through the momentum change and interference with the external flow. These interactions are dependent on the flow speed, size, and position of the intake surfaces, the cooling ducting, the cooling module, the fan, the shape of engine compartment, and the outlet opening shape and size.

7.3.1 Operating Conditions

Depending on the vehicle speed, the internal flow has very different characteristics. Figure 7.6 and Figure 7.7 show two typical flow conditions for a vehicle: one at idle and the other at high speed. These two operating points represent to some extent the extreme of all flow conditions. Therefore, they will be examined here in more detail.

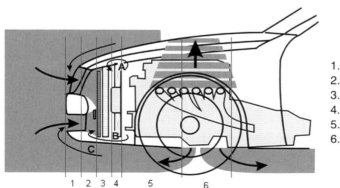

1. Grille
2. Sealings
3. Cooling module
4. Fan
5. Engine compartment
6. Outlet

Figure 7.6 Internal flow at idle.

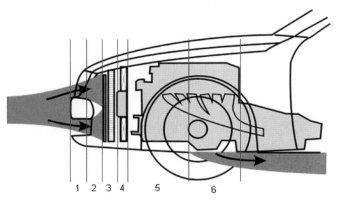

Figure 7.7 Internal flow at high speed.

1. Grille
2. Sealings
3. Cooling module
4. Fan
5. Engine compartment
6. Outlet

When the vehicle is at idle or in stop-and-go traffic, the driving wind speed is not present; hence, the fans must force the airflow through the cooling system. The fans produce a low pressure in front of the cooling module, causing air to be sucked in from the inlet surfaces (see Figure 7.6). In the engine compartment, the static pressure is slightly positive with respect to the environment. The proportion of the cooling air flowing through the upper and lower inlets is primarily due to the pressure drop and the size of the inlets. The location of the stagnation point is not crucial. However, this operating condition can be regarded as thermally challenging, since the low airspeed, static pressure upstream of the cooling module, and the various forms of backflow affect the heat dissipation significantly. Backflow circulating around the perimeter of the fan (A), reduces the mass flow through the heat exchangers. Internal recirculation caused by pressure gradient across the heat exchangers causes hot air to re-enter the front of the cooling module (B), decreasing the dissipation of heat due to a "thermal short circuit." This can largely be prevented by appropriate sealing of the cooling module and ducting. More difficult to avoid is external recirculation (C), where hot air leaves the vehicle through the appropriate outlets (6) and re-enters through the front cooling intakes, contributing to an increase in the inlet flow air temperature. A tail wind can exacerbate this phenomenon when the vehicle is stationary or slowly moving.

At high speeds, the external pressure difference between the inlet and outlet (see Figure 7.7) dominates the conditions for the internal flow. The fan, which was previously responsible for creating the mass flow through the cooling module, is now overblown and causes an additional pressure loss in the system. The external flow field not only changes the pressure on the front of the vehicle but also influences the boundary conditions in the engine compartment and the flow outlets, for example, in the wheel arch and underbody. Thus the external flow directly affects the internal flow. The change of the external flow of the two flow conditions discussed can be seen in the change of the stagnation point (SP), among other things (see Figure 7.8). In addition, the inlet ducting in condition has mostly clean flow, whereas at high speeds, there is a large separation region caused by the bumper cross member. The inlet ducting should be designed to minimize pressure loss at high speeds since the flow conditions are more demanding compared to idle conditions. The pressure loss from the inlet ducting can account for up to 30% of the total pressure loss. This is the second largest loss after the heat exchangers, which contribute nearly 60% of the total pressure loss.

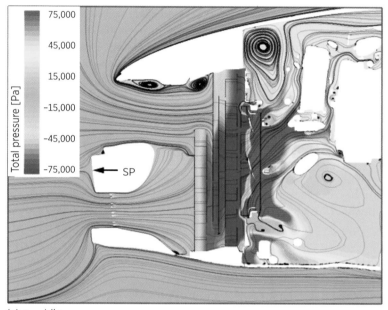

(a) $c_{p,tot}$ idle

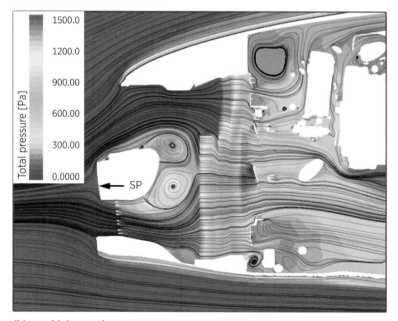

(b) $c_{p,tot}$ high speed

Figure 7.8 Effect of driving speed on internal flow.

7.3.2 Cooling Module

The main cooling module usually consists of several heat exchangers (see section 7.2.4) combined with one or more fans. If due to the high demand for cooling, one module is not enough, the cooling system is supplemented by additional heat exchangers in different positions. These are referred to as "extended coolers." All modules must be positioned (taking into account all other vehicle requirements) so that they satisfy all operating conditions to dissipate the required heat with sufficient cooling air mass flow. Not only are the size and position of the modules in the vehicle crucial, but also the arrangement of the heat exchangers within the module to ensure the most efficient heat transfer.

7.3.2.1 Positions in the Vehicle

Most vehicles have the engine located at the front and consequently the cooling system located ahead of it. Positioning the cooling module in the front center of the vehicle (see position 1 in Figure 7.9) has established itself as a good position for short tubing lengths and for a number of other reasons. First, heat exchangers with larger surface areas can be used with fans with higher efficiency. Second, the necessary air intakes are positioned close to the stagnation point for optimal utilization of the high-pressure region. Finally, the warm exhaust air (up to maximum 120 °C) from the cooling module can cool other components in the engine compartment, such as the exhaust system and the immediate surrounding environment. In addition on the sides (position 2), high-pressure differences can be achieved by the dynamic pressure of the air accelerating around the front corner of the vehicle. This installation space is often used in high-performance vehicle variants for extended coolers to supplement the central cooling module. The necessary space to accommodate a cooling module, including a fan laterally before the front wheels, is usually only for vehicles with a rear or mid-engine. Position 3 is typically used only on race cars, due to packaging space available, and minimizes the drag losses from the cooling system. In rear and mid-engine vehicles, position 4 is used to keep the tubing lengths short. The cooling air flow is entrained mostly on the side, in front or above the rear wheels. The outlet is either placed on the underbody or at the rear to utilize the low-pressure region in the wake.

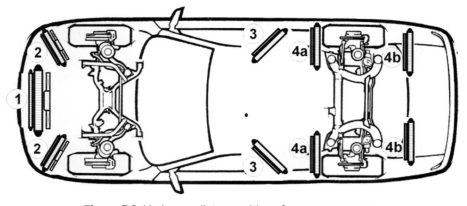

Figure 7.9 Various radiator positions for passenger cars.

7.3.2.2 Heat Exchanger Assembly

The dependencies discussed between the inlet temperature gradient of a cross-flow heat exchanger and its efficiency ϕ suggest a particular assembly design with multiple heat exchangers at different temperature levels. The most efficient order is to place the coldest heat exchanger (i.e., the air-conditioning condenser or low-temperature cooler) at the front of the cooling module, followed by the hotter heat exchangers (e.g., engine oil cooler). The more efficient the cooling module, the lower the required mass flow and corresponding cooling drag. The charge air, the air condenser, the power steering cooler, and more recently, the power electronics in hybrid electric vehicles have increased cooling air requirement at low temperatures (T_{1e} < 50 °C). Important factors to consider are whether to have a series or a parallel arrangement of the heat exchangers, the temperature level of the fluid to be cooled, and the required cooling mass flow. Heat exchangers that require a high mass flow or have a large pressure drop, such as the charge air cooler, can be arranged in parallel to the other heat exchangers. This arrangement can lead to two parallel flow streams through the cooling module with different pressure drops. To optimize the efficiency of the system, the parallel cooling paths should have a similar pressure loss.

7.3.3 Fan

As mentioned previously, at low speeds or at stop, a fan is required to provide the mass flow through the cooling module. The fan can be positioned both in front and behind the heat exchangers. A fan placed in front has a lower power demand for a constant mass flow because of the higher air density. However, it is usually possible to fit a larger diameter fan behind the cooling module. Both aspects can be easily derived from the following equation.

$$P_{Fan} = \rho_L \cdot A_{Fan} \cdot v_K^3 = \frac{\dot{m}_K^3}{A_{Fan}^2 \cdot \rho_L^2} \tag{7.21}$$

$$\rho_L = \frac{p}{R_L \cdot T_L} \tag{7.22}$$

For high cooling requirements, powerful fans can be mounted in both positions simultaneously. The risk of positioning a fan only in front of the radiator is of it freezing in winter. Snow deposited in the nearby heat exchanger will have a tendency to partially thaw and refreeze when the vehicle is stopped. This can lead to the fan being obstructed and engine overheating.

When selecting a fan, both the efficiency of the electrical motor (used in most modern cars) and of the fan blades at the desired operating point must be considered. Typical power requirements for modern fans range from 300W to 1000W. Brushed motors have been replaced with highly efficient brushless motors to decrease the power requirements. A casing fan consists of a small hub with long sickle-shaped wing blades typically made from polyamide 6.6 with different glass fiber compositions. The fan hub is directly connected to the electrical motor, which in turn is attached to the frame via struts. The struts and fan casing help to guide the flow so that air sucked by the fan can be drawn from the maximum surface area of the heat exchanger, preventing internal recirculation.

The aerodynamic properties of a fan are given by the curve in $\Delta p_{tot}/\dot{V}$ diagram, as shown in Figure 7.10. Usually this information is provided by the manufacturer and is calculated using a multichamber test rig, where the fan must draw air through a known resistance. This resistance can be adjusted by means of a continuously variable aperture. Using various operating points, the temperature, total pressure, volume flow rate, and electrical power consumption of the fan can be determined. First, the fan is tested with a completely open aperture to determine the maximum volume flow rate. The measured total pressure difference corresponds to the dynamic pressure difference across the fan. This flow rate is referred to as the "absorption capacity," but this is not reachable in a passenger car due the presence of blockages such as the motor. By continuously reducing the aperture size, the total pressure difference between the test chambers increases. More energy is converted into static pressure, increasing the efficiency. Shortly before reaching the apex of the curve, the fan reaches the maximum pressure boost in a stable condition. Increasing the system resistance further leads to a point where flow separation starts to occur on the fan blades. This is an unstable operating point with higher noise emissions and vibrations and should be avoided in the designed operating range. Increasing the system resistance further past this point will increase the pressure difference, but the flow remains detached.

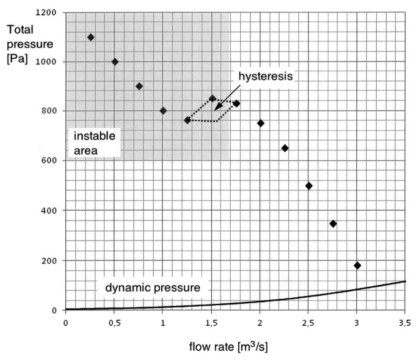

Figure 7.10 Fan characteristic curve.

7.4 Optimizing the Overall System

The task of the engineer is (see section 7.1) to optimize the entire cooling system in compliance with the overall requirements on vehicle weight, costs, and energy consumption in particular. Reducing the cooling drag also contributes to reducing the vehicle's overall energy consumption. For general application the cooling system is usually oversized and has a higher drag than necessary since the system is designed for a situation where maximum heat dissipation is required. The vast majority of journeys are covered with lower requirements, in the partial load range. For example, in environments with moderate temperatures, with flat driving conditions, the engine load is lower and therefore the cooling system dissipates considerably less heat.

In current vehicles, the cooling drag ranges from $\Delta c_{D,C} = 0.01 - 0.04$, as shown in Figure 7.11. This is determined by measuring the drag coefficient of a vehicle in a wind tunnel, with and without flow through the cooling system.

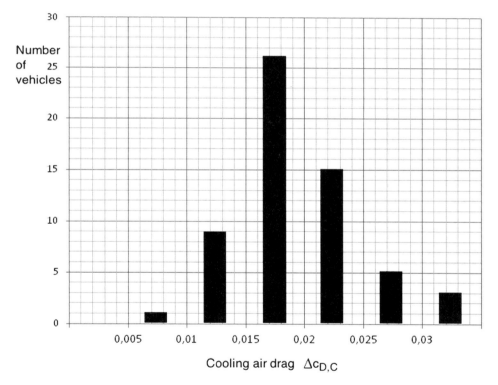

Figure 7.11 Cooling air drag of current vehicles.

However, there is no clear relationship to the vehicle drag nor the engine power or the vehicle size (see Figure 7.12). Nevertheless, just as the drag increases with cooling air mass flow, the front axle lift increases considerably in almost all cases. Figure 7.13 compares the cooling drag to the front axle lift. This graph includes information for different engine sizes, vehicles, and manufacturers and does not show a consistent trend. The inlet area, cooling ducting design, cooling modules, fans, engines, and outlet area vary greatly from vehicle to vehicle, and there is a significant interaction between these factors and the external flow field.

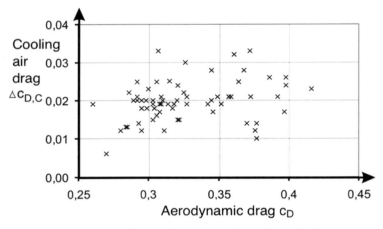

Figure 7.12 Cooling air drag vs. total aerodynamic drag.

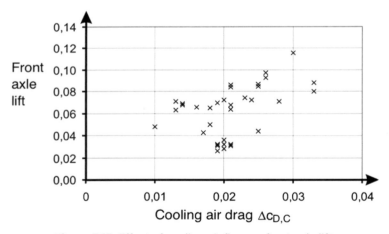

Figure 7.13 Effect of cooling air flow on front axle lift.

7.4.1 Calculation of the Cooling Air Mass Flow

The typical approach to resolve this dilemma is to separate the cooling drag into two parts. First, the "internal drag," describing the pure energy losses within the air ducts, the cooling package, and the engine compartment, and second are the losses due to the interaction with the external flow (referred to "interference drag"). Assuming a completely sealed internal cooling air flow, this seems to be an acceptable way (see also Figure 7.14).

Cooling and Internal Flow

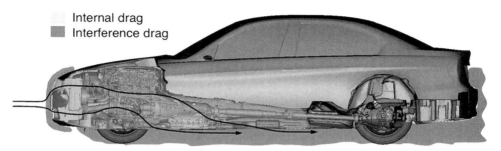

Figure 7.14 Internal drag and interference drag.

Most analytical approaches use the momentum or the energy equations as the basis for the analysis of the internal flow. In particular, for the evaluation of the inlet and outlet surfaces influence, the momentum method is recommended. Described next in its general form is the change of momentum flux by external forces.

$$\iiint_V \frac{\partial(\rho\mathbf{v})}{\partial t}dV + \oiint_S \rho\mathbf{v}(\mathbf{v}\cdot\mathbf{n})dS = \iiint_V \rho\mathbf{f}dV + \oiint_S \sigma\cdot\mathbf{n}\,dS_{\text{Fluid/Fluid-Boundaries}} + \sum F_{\text{Bodies}} \quad (7.23)$$

Neglecting the time dependent terms of the momentum flux (the first term on the left side), the flow forces (the first term on the right side) and the fluid shear forces result in the x-component of the moment equation.

$$\left(\oiint_S (\rho(\mathbf{v}\cdot\mathbf{n})dS)\cdot v_x\right)_E + \left(\oiint_S (\rho(\mathbf{v}\cdot\mathbf{n})dS)\cdot v_x\right)_E = \left(-\iint_S (p-p_\infty)n_x\,dS_E\right) + \quad (7.24)$$

$$\left(-\iint_S (p-p_\infty)n_x\,dS_A\right) + \left(-\iint_S \left(-(p-p_\infty)+\tau_{xx}+\tau_{xy}+\tau_{xz}\right)n_x\,dS_{\text{inner faces}}\right).$$

This can be rewritten to determine the internal drag W_{inner} after resolving the energy equation

$$W_{\text{inner}} = -\oiint_S \rho v_x(\mathbf{v}\cdot\mathbf{n})dS_E - \oiint_S \rho v_x(\mathbf{v}\cdot\mathbf{n})dS_A - \iint_S (p-p_\infty)n_x\,dS_E - \iint_S (p-p_\infty)n_x\,dS_A. \quad (7.25)$$

It is apparent that in order to calculate the drag, the change in momentum between the incoming and outgoing internal flow is not only the difference to take into account. There is also the difference of the pressure forces acting on the inlet and outlet cross sections and their orientation toward the direction of driving.

Ivanic et al. [400] used this method outlined to determine the influence of the position, size, and orientation of the inlet and outlet surfaces A_E and A_A on the drag of a generic blunt body. In a second approach, they improved the equations with additional terms that would account for any flow angularity at the inlet and outlet. Their study showed that rotating the entry surface away from the direction of travel always reduced the resistance for the internal flow, while changing the orientation of the outlet surface could either increase or decrease the resistance. The effect depended on the local static pressure and outlet mass flow rate. However, one part of the benefit seen had to be attributed to a reduced mass flow rate, which is rarely constructive for practical applications.

Wiedemann [875] improved Ivanic et al.'s approach by adding the relevant cooling air mass flow $v_K A_K$ but neglecting the detailed description of the inlet and outlet areas.

$$\Delta c_{D,C} = \frac{2 v_K A_K}{v_\infty A_x}\left[1 - \cos\alpha \sqrt{\frac{1 - c_{p,A}}{1 + \zeta_K (A_A/A_K)^2}}\right] \quad (7.26)$$

Most of the attempts by Barnard et al. ([40], [41]) to back up these analytical approaches with experimental results showed that without accurate experimental data of all pressures and mass flow conditions, the "internal drag" could not be separated from the interference drag.

A different approach to address the internal flow, but more complicated, is to use the energy equation for viscous fluid. Neglecting the potential energy, this can be written as:

$$\frac{\rho_L}{2} v^2 + p - \Delta p_{loss} = p_{tot} - \Delta p_{loss} = \text{const.} \quad (7.27)$$

By integrating the total pressure, p_{tot} (with data from a 3D flow simulation) along a stream tube, the energy of the flow and the pressure loss due to the installed components can be calculated. If a stream tube is tracked through the cooling system, the energy will decrease whenever there is a pressure loss. This is mainly due to the intake grilles, the cooling module, and at the fan shroud. An increase in the total pressure is only possible with the introduction of a power gain, such as mechanical energy from a fan or thermal energy from a heat exchanger. For a better overview, the stream tube through the cooling module can be divided into multiple sections and associated with various components and subsystems. Figure 7.15 shows such a division of a stream tube into six subsystems (inlet surfaces, inlet ducting, cooling module, fan, engine compartment, and outlet surfaces) and the development of the total pressure loss at the two operating points: idle and maximum speed. Both pressure loss profiles take into account not only the energy input from the fans but also the thermal energy input from the heat exchanger. In both cases, the pressure loss from the cooling module has the greatest contribution. The other losses before and after the cooling module depend on the operating point. Regions with high gradients indicate regions with higher potential for optimization. For example, there is a significant pressure loss through the inlet at maximum speed.

To summarize, from the external flow decoupled analysis of the internal flow, both approaches, the momentum and energy equation, give basic understanding of the phenomenon and provide starting points for improved internal flow concepts. However, one-dimensional analysis approaches offer little opportunity depict precise three-dimensional effects. Therefore, 3D simulation tools are indispensable in the early stage of development (see chapter 14).

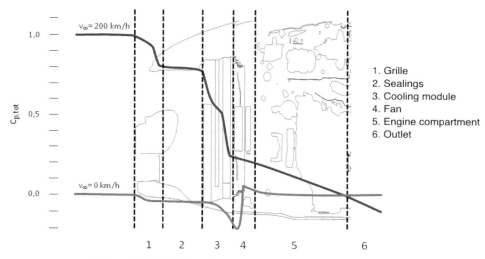

Figure 7.15 Division of a stream tube into six subsystems.

7.4.2 Influence Parameters of the Internal Flow

The previous chapters have shown that the cooling drag is primarily dependent on the mass flow $\dot{V}_K$ and the pressure loss Δp_{loss}. Therefore, to minimize the power loss, the requirement for cooling air mass flow and subsequent pressure loss Δp_{loss} in the cooling system must be kept as small as possible:

$$P_K = \dot{V}_K \cdot \Delta p_{\text{loss}} . \tag{7.28}$$

There are two options to reduce the first term on the right side. In situations where the maximum mass flow is not required, reduce the cooling mass flow by reducing the inlet surface. Otherwise, improve the heat output, $\dot{Q}$, of the cooling system and therefore reduce the cooling mass flow requirement.

$$\dot{Q} = kA_K \left(T_{L,K} - T_{W,K} \right) \tag{7.29}$$

According to equation (7.29), cooling performance can be improved by enlarging the surface area of the heat exchanger A_K, increasing the heat transfer coefficient k, or by increasing the temperature difference between the air $T_{L,K}$ and coolant temperature $T_{W,K}$ in the heat exchanger. To achieve the latter, the heat exchangers are arranged according to their temperature level with the air passing through the first being the coldest (e.g., air-conditioning condenser) and the last being the warmest (e.g., engine coolant or engine oil cooler). Another possibility to improve the thermal efficiency is to improve the distribution of the air mass flow over the heat exchanger. The distribution is described by the dimensionless parameter I, the inhomogeneity.

$$I = \frac{1}{v_K} \sqrt{\frac{1}{j} \cdot \sum_{n=0}^{i} (v_n - v_K)^2} \tag{7.30}$$

An improved airflow distribution over the heat exchanger has two positive effects. First, it reduced the pressure loss due to the quadratic relationship between pressure and velocity. Ideally when = 0 , the flow is homogeneous over the face of the heat exchanger and the minimum pressure loss is achieved. Second, the efficiency of the heat exchanger improves because of the logarithmic relationship between heat output and air mass flow (see Figure 7.16). The distribution depends strongly on the layout of the cooling system and the vehicle's operating point. However in general, high speeds and high pressure loss in the cooling module improve the homogeneity. A homogeneous velocity distribution is generally not possible in passenger vehicles with a conventional layout. However, an inhomogeneity of less than 25% is considered acceptable (see Figure 7.17).

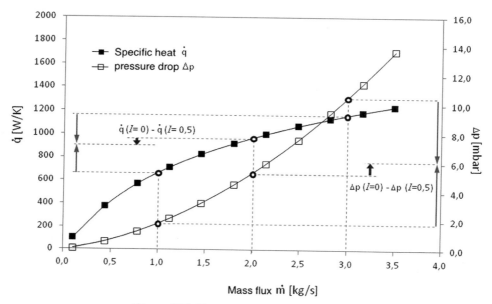

Figure 7.16 Heat output and air mass flow.

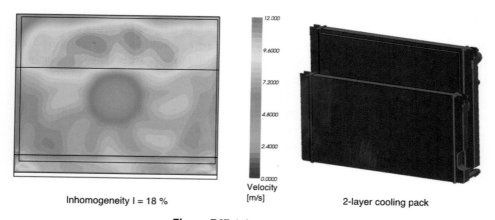

Figure 7.17 Inhomogeneity.

Cooling and Internal Flow

The previously mentioned performance-related adjustment of the cooling air mass flow by partially blocking the airflow is in general possible since there is a large difference between the design requirements for the cooling system and customer driving profile. A partial reduction can be achieved, for example, with use of shutters or flaps to close sections of the air intakes or the cooling module. This method has been used in vehicles for several years to reduce the cooling drag by up to 75%. Driver-operated flaps for the front radiators were already installed in vehicles from the 1930s. However, this was before the introduction of coolant thermostats, and the purpose was to regulate the engine temperature. In colder regions, it is still common to partially block the cooling module in winter to increase the amount of heat produced for the interior cabin warming.

The other option available to the engineer to optimize the internal flow is to minimize all energy losses of the individual components along the cooling airflow pass, they will be examined here in more detail

7.4.3 Air Intakes and Cooling Air Ducts

A typical pressure distribution and stagnation point location on the front of the vehicle is shown in Figure 7.18. The pressure distribution is an important input condition for the calculation of the cooling air mass flow and is different for each vehicle. The mass flow is not only dependent on the front shape but also on the entire vehicle proportion. The effect of the vertical stagnation point location on the total cooling mass flow is rather low, as long as the upper and lower inlet openings can communicate through the air ducts downstream. However, the cooling airflow changes the stagnation point on the vehicle front and subsequently the distribution of the external airflow over and underneath the vehicle, which accounts for a change of the aerodynamic properties and the interference drag.

Figure 7.18 Pressure distribution on the vehicle's front.

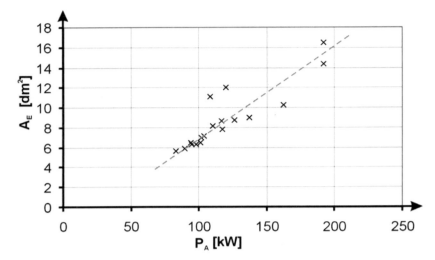
Figure 7.19 Opening size of the air inlet vs. engine power.

The second most important factor in the design of the cooling system, next to the radiator size, is the air inlet opening size and its location. This has the largest influence on the cooling airflow at high speeds as long as the flow is not throttled farther downstream, such as in the engine compartment or at the air outlets. This relationship between inlet area and engine power is shown in Figure 7.19. In general, the area of the inlet openings is approximately 30–45% of the radiator surface area, irrespective of the vehicle class. The most important requirement for the air ducts is to direct the incoming airflow to the heat exchangers with minimal pressure loss. Of particular importance is air tightness of the ducts, as was discussed in section 7.3. The requirements for tightness increase with increasing static pressure in the ducts, which are caused by denser heat exchangers or more compact engine compartments. The length of the air inlet is usually too short to have an optimum diffuser shape to achieve ideal pressure recovery, as is seen in aircraft cooling systems. Nevertheless, care must be taken to avoid flow separation with sudden steps and abrupt cross-section area changes. Vehicle components such as struts, horns, and so forth should be placed outside the air ducts if possible to avoid unnecessary pressure losses.

7.4.4 Cooling Matrix

The cooling matrix is the most important component, not only for the coolant side but also for air side. As shown in Figure 7.15, over 60% of the total pressure loss for the cooling system occurs in the cooling matrix. There are three parameters available to the engineer to improve the design: the geometric dimensions of the radiator matrix, its heat transfer properties, and the temperature difference between the cooling air and radiator coolant. As an example, these three parameters will be modified for an air/water cross-flow heat exchanger and the characteristics of the heat exchanger examined. The reference heat exchanger has a surface area A_K of 3300 cm², thickness l_K of 30mm, 65 fins/dm, and a coolant flow $\dot{V}_{KM}$ of 6000 l/h.

One of the obvious measures to increase the heat output of a heat exchanger is to increase the area of the cooling matrix A_K. As indicated in Figure 7.20, the mass flow for a constant pressure difference across the matrix is proportional to the radiator matrix area. However, the performance increase in a vehicle is less than the ideal case, since replacing the cooling module also affects the pressure boundary conditions. Furthermore, the dissipated heat depends also on the cooling air mass flow $\dot{V}_K$ and the refrigerant mass flow $\dot{V}_{KM}$. Figure 7.21 shows that the performance increase from increasing the cooling matrix area is only significant at high mass flows, in this case about 5%. At low mass flows, increasing the surface area of the reference cooler does not necessarily increase the heat output since the temperature delta between the air and the coolant diminishes. Furthermore, if the heat output at constant pressure differences is compared, which is the case if a heat exchanger is simply replaced in a fixed vehicle configuration, increasing the cooling matrix area by approximately 30% would then increase the air mass flow. and therefore the output by more than 20%, as shown in Figure 7.22.

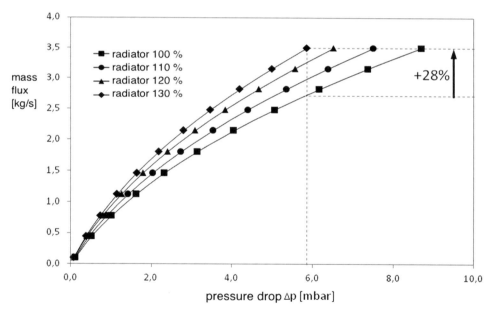

Figure 7.20 Mass flow as a function of the area of the cooling matrix and the pressure loss.

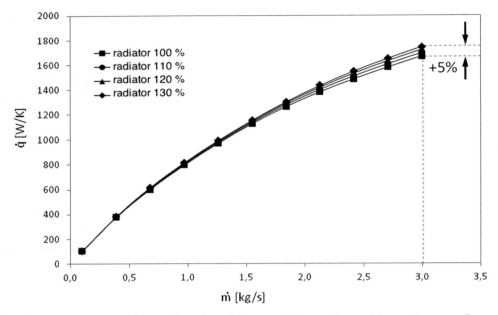

Figure 7.21 Heat transfer as a function of the area of the cooling matrix and the mass flow.

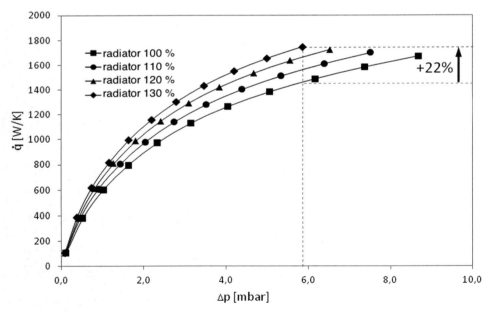

Figure 7.22 Heat transfer as a function of the area of the cooling matrix and the pressure loss.

Since most cooling modules today already use the maximum usable area between the engine side rails, the hood, and the lower ground clearance line, it is unlikely that enlarging the cooling matrix area would be a possible solution. Therefore, other measures should be explored to improve the cooling module efficiency, such as increasing the thickness of the heat exchangers l_K. As can be seen in Figure 7.23 and Figure 7.24, the change in pressure loss from the increased thickness is moderate. However, the heat output improves considerably, due to the increased surface area for heat transfer. At lower mass flows, increasing the heat exchanger thickness is less effective due to heat saturation.

If the overall thickness of cooling module is also limited, then only space neutral optimization options remain. Increasing the matrix density is an option, which can be achieved with increasing the number of cooling fins. One adverse effect, for instance, of increasing the fin density from 65 to 85 fins/dm is that the pressure loss of the heat exchanger module increases by more than 50% and reduces the mass flow accordingly. However, the heat output for an equivalent pressure difference is almost independent of fin density (see Figure 7.26). Since increased pressure loss affects air mass flow through the cooling module, particular attention should be given to the fan. As can be seen in Figure 7.25, for the same mass flow through a denser heat exchanger, the heat output increases by about 10%.

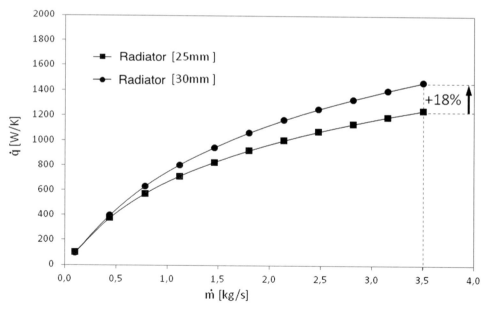

Figure 7.23 Heat transfer as a function of the thickness of the heat exchanger and the mass flow.

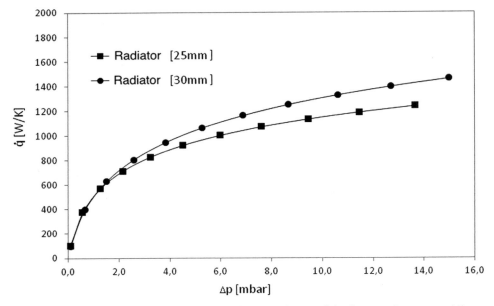

Figure 7.24 Heat transfer as a function of the thickness of the heat exchanger and the pressure loss.

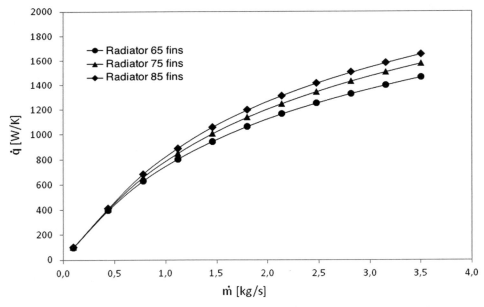

Figure 7.25 Heat transfer as a function of the fin density and the mass flow.

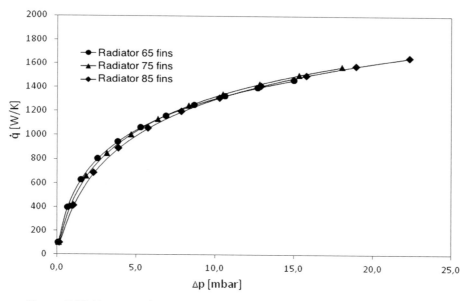

Figure 7.26 Heat transfer as a function of the fin density and the pressure loss.

7.4.5 Fan

Due to the different operating conditions of a vehicle, fan selection should not be limited to a single operating point. In modern vehicles high RPM, variable speed axial fans with good aerodynamic properties over a wide operating range are used. The fans should have an acceptable efficiency at both, low rotation speeds and high pressure, since it is inefficient to operate the fan at high RPM. The engineer has a variety of geometric properties that can be used to optimize the fan efficiency, the acoustic characteristics, the weight, and cost:

- Diameter (fan hub + fan shroud)
- Number of blades and operating speed
- Profiles for blade geometry (type, curvature, thickness, and length)
- Blade twisting: pitch of the profiles (as a function of the radius)
- Blade curvature: positioning of the profiles to each other
- Blade distribution on the circumference

A major challenge in the implementation of an efficient fan is the confined packaging space. Therefore, it is important to ensure that there is sufficient space for axial flow behind the fan. At distances less than 150 mm between the fan motor and the blockage behind it, the air undergoes a strong deflection in the radial direction. As shown in Figure 7.27, the deflection affects far upstream to the fan blade, thus changing flow around the blades and therefore its efficiency. With fan power up to 1000 W in modern cars, this presents a significant potential for energy savings. In theory, there are several options to minimize the required fan power, such as changing the pressure resistance of the complete cooling system, using a variable blade pitch angle or an adjustable inlet guide vane, or varying the fan speed. Only the later is currently practical in vehicles. Due to the fan speed variation, the angle of attack of the blades profile changes. This shifts the fan's curve diagonally on the pressure loss-mass flow diagram, shown in Figure 7.29. At high speeds, as mentioned previously, the fan actually causes a pressure loss in the cooling system. This may be because of flow separation at the fan blades itself or the supporting guide vanes, which have been optimized for low mass flow. The blockage effect is increased with the fan shroud. This can be partially overcome with bleeding flaps in the shroud, which open at high speeds to create more open area for air to pass through. This measure shifts the characteristics curve of the fan to higher mass flow rates (see Figure 7.28).

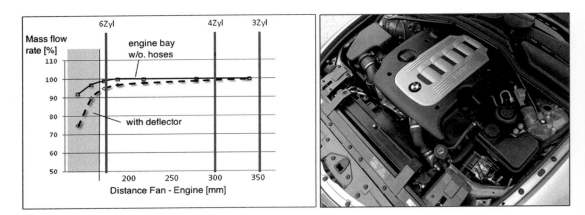

Figure 7.27 Distance between fan and engine.

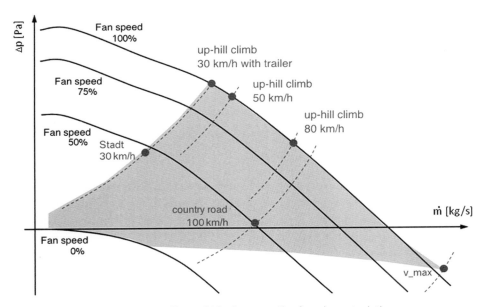

Figure 7.28 Effect of blockage on the fan characteristic.

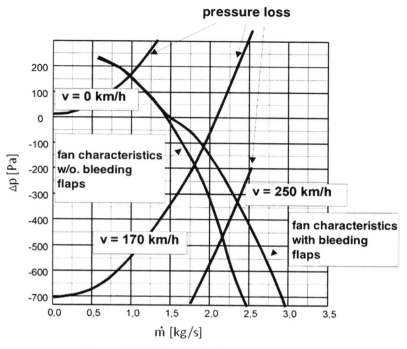

Figure 7.29 Characteristics curve of the fan.

7.4.6 Engine Compartment

As the cooling air passes through the engine compartment, this represents another pressure loss for the cooling system and thus affects the cooling air mass flow and the cooling drag. The greatest pressure drop in the engine compartment occurs immediately after the fan, which is located in close proximity to the engine. In this area, there is the narrowest cross-sectional area, and the highest air velocity exists (typically greater than 25% of the inlet velocity). As can be seen in the computational fluid dynamics (CFD) results in Figure 7.30, the air velocity and pressure losses in the remaining space are relatively low. The pressure losses due to blockage are mainly influenced by position of the engine, the front axle, and the large components of the peripherals such as the intake, the expansion tank, or the battery. This results in multiple paths to the outlet openings for the air, which have a large influence on the interference resistance. Thus, the engine compartment flow influences the cooling drag indirectly through the distribution of the cooling air to the outlets rather than from the resulting pressure loss in the engine compartment.

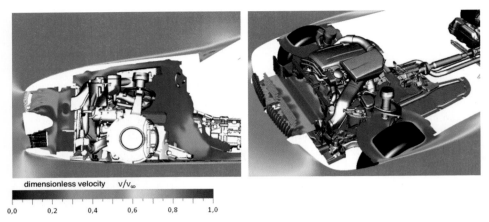

Figure 7.30 Engine compartment flow at $v = 200$ km/h.

The additional cooling effect on the engine surface is negligible at high loads due to the low heat transfer coefficients and the small temperature difference between the cooling air and engine block. Only on the oil sump can heat be dissipated to some extent. In particular, at high speeds when oil temperatures are high and plenty of cooling air mass flow is available, sufficient temperature difference allows for heat transfer. However, for this situation the air must be channeled between the engine compartment shielding, past the oil sump, and through an air outlet downstream to allow for sufficiently high air velocity over the surface of the sump. For cooling hot components, which are in the order of several hundred degrees Celsius, such as the exhaust system and their immediate environment, the warm air behind the cooling modules is sufficiently cool to be beneficial. Other components, however, can be unintentionally heated by the cooling module's exhaust air, such as air lines, and therefore should be placed away from the direct air flow or insulated.

7.4.7 Cooling Air Outlets

The cooling air outlets not only have a large influence on the cooling drag but also affect the air mass flow, in particular at high speed. Equation (7.26) shows the direct influence of the outlet opening orientation, defined by the angle α, the local pressure coefficient $c_{p,A}$, and the size A_A on the cooling drag due to the momentum loss. Furthermore, the outflow affects also the interference drag, which is in general greater than the momentum loss. In the worst case, a previously, fully attached exterior air flow is significantly disturbed by the exhausting cooling mass flow, which leads to separation and high losses. In most favorable case, the overall drag of the vehicle is reduced by selectively diverting the outflowing cooling mass flow to regions of separated flow on the underbody or the wake of the vehicle [650]. An example is described in the patent EP 0 858944 A1 from Porsche AG, where the cooling air is blown out in front of the front wheels, creating an aerodynamic lip, which significantly reduces the vehicle cooling drag. Some general cooling concepts have been investigated by Buchheim et al. [107]. The first three variants in Figure 4.33 differ mainly in the location of the outlets. While in variant A, the cooling airflow is released into the engine compartment, in variant B, it is directed to the wheel arches, and in variant C the air exits through an opening in the hood. The air outlet on the underbody behind the front axle in variant D is unfortunately connected to an air inlet at the windshield cowl, so the results are not directly comparable with the other configurations. This investigation also showed the importance of considering both the cooling mass flow and the cooling drag simultaneously. For instance, although the variants B and D show a low cooling drag, the mean velocity through the heat exchanger is significantly lower than variant A. In the vast majority of vehicles, the cooling air exits under the vehicle and out through the wheelhouses. Due to the ever improving aerodynamic design of the underbody and the resultant higher velocities beneath the vehicle, the importance of optimizing these outlets has increased. Panels under the engine compartment were first introduced for the optimization of the external flow field and for acoustics. However, the more the engine compartment is sealed, the greater the possibility to control the location and method in which the cooling air integrates back into the external flow. Unfortunately, not all locations are arbitrary. For instance, the openings around the exhaust system where the transmission tunnel connects to the underbody are mandatory for thermal reasons. Or the inside faces of the wheelhouses must be open due to suspension clearance. At those openings, the airflow velocity is low, highly turbulent, and can hardly be controlled in direction. On the other hand, for brake cooling purposes, the air mass flow and its direction are crucial and cannot be easily changed for aerodynamic benefits. These examples have shown that the cooling module and engine compartment flow is influenced by numerous factors and must be optimized individually for each vehicle.

7.5 Measurement Technology for Cooling Airflow

In todays vehicle development process, mostly global scalar parameters such as cooling drag, coolant temperatures, and component temperatures are used to find the aerodynamic and thermal optimum of a cooling system configuration . However, a better understanding of the local flow conditions in the engine compartment and near the cooler module is required for finding the optimal design of the cooling system. Unfortunately, the information about the local velocities and the pressure distribution can only be obtained by means of numerical simulation; experimental measurements are very difficult to obtain. This is mostly because of high component temperatures during driving, confined space inside the vehicle engine compartment for additional measuring equipment, and most of all due to the complex and highly three-dimensional flow regime. Nevertheless, in case of the need to measure the cooling air mass flow $\dot{m}_K$ and the inhomogeneity I, the following measurement methods are available:

- Vane anemometers
- Pressure measurements

- Laser Doppler anemometry (LDA)
- Particle image velocimetry (PIV)
- Hot-wire anemometry (HWA)

7.5.1 Vane Anemometers

Vane anemometers are a mechanical measuring method. The measured air velocity is determined from the relationship of the rotational speed of the impeller and the angle between fan direction and flow direction. Even for miniature anemometers with a diameter of about 15 mm, the minimum speed measurable even with inductive rotation speed measurement is about 0.2 m/s. Therefore, measurements below 0.4 m/s are generally not considered and tend to not make sense [16]. The main sources of error are from misalignment of the flow to the anemometer or from velocities out of the measurement range. In complex and turbulent environments such as the engine compartment, it is difficult to implement these devices for effective measurements. Nevertheless, vane anemometers are often used for qualitative results since they are fast to set up and robust in nature.

7.5.2 Pressure Measurements

A more robust method to measure the flow rate is via pressure measurements using Prandtl tubes. Due to their smaller design, they offer better spatial resolution than a vane anemometer. To reduce measurement error, it is necessary to align the probe as much as possible to the main flow direction. Particularly for static pressure measurements, the angle of attack should be kept within +/− 5° to keep the error below 1%. Because the local flow direction in front of heat exchangers is generally uneven, it is more appropriate to measure the pressure in a plane behind the heat exchanger since the flow direction is known. Note, close to the cooler tubes, the flow gradients are greater. If the gradient between the measurement points of the static and total pressure on the Prandtl tube is too large, they will no longer lie on the same streamline and the measurement will be meaningless. An estimation of the measurement error will be required in each particular case.

Another measuring principle, which is suitable for determining the cooling air mass flow, is the "cooler probe" method. It is a process developed by the FKFS Institute [479] to determine the absolute velocity by measuring the pressure difference along a stream tube within the radiator itself. This method is extremely reliable because it is very robust to oblique flows and even allows measurements with fan in operation. Moreover, the structural changes to the heat exchanger are small and the additional pressure losses due to the probes do not impact the mass flow. An enhancement of the cooler probe method is to take warm measurements as demonstrated by Thibaut [797]. With the addition of temperature probes behind the heat exchanger, the local air density can also be determined.

Cooling and Internal Flow

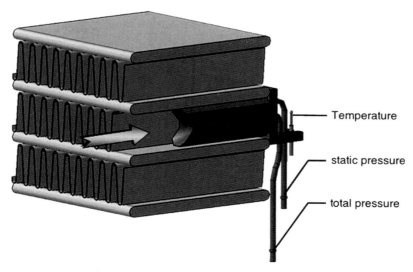

Figure 7.31 Cooler probe with thermocouple.

The measurement method shows a high repeatability and a short measurement time. However, the accuracy of the calculated total mass flow rate for the radiator is highly dependent on the number of sensors used and their spatial distribution.

7.5.3 Optical Measuring Methods

Advanced optical measuring methods, such as LDA, combine a number of advantages:

- High spatial and temporal resolution
- Interference-free speed measurements
- Measurement of up to three velocity components
- Detection of reverse flow

However, it also has major disadvantages, especially when measuring within the engine compartment. Either vehicle components must be made from glass, such as the hood when the optical equipment is located outside the vehicle, or miniature probe heads and light guides must be positioned in the engine compartment by means of a traverse. These circumstances often lead to very elaborate preparations and long measurement times. As a result, LDA measurements are more likely to be found in an academic environment, as opposed to being used in a development program of an automobile manufacturer. Optical access is also the biggest issue for PIV studies on the cooling system.

7.5.4 Hot-Wire Anemometry

In particular, single wire hot-wire anemometry operated at a constant temperature are characterized by their ease of use and good spatial and temporal resolution. However, in contrast to LDA, they offer no way to determine the direction of flow. However, the cosine-like behavior of the measuring signal at low yaw angles offers a proven method to determine the flow through the cooler in an industrial environment. At yaw angles up to 8°, the expected error is less than 1%. Investigations with direction-sensitive hot-wire probe (multiwire probes) have been used to investigate the flow field directly behind the radiator and to confirm the previously described approach with less-expensive single wires. In order to make use of the confined space behind the cooling module, a miniature traverse (see Figure 7.32 and Figure 7.33) is necessary to position the probes during the measurements. For the purpose of validation, mass flow measurements were carried out with a complete cooling package, including fan, outside a car in a dual chamber test cell with a calibrated pressure orifice. The deviation of the two measurement systems was below 2%, which gives a sufficiently high accuracy for a development tool in an industrial environment.

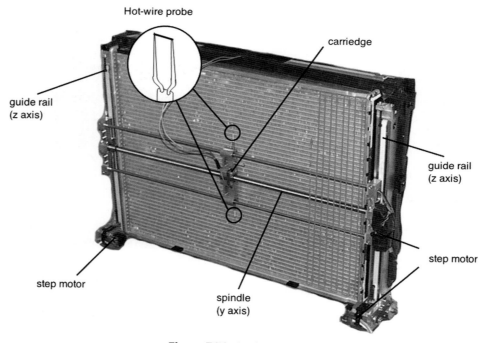

Figure 7.32 Cooler traverse.

Cooling and Internal Flow

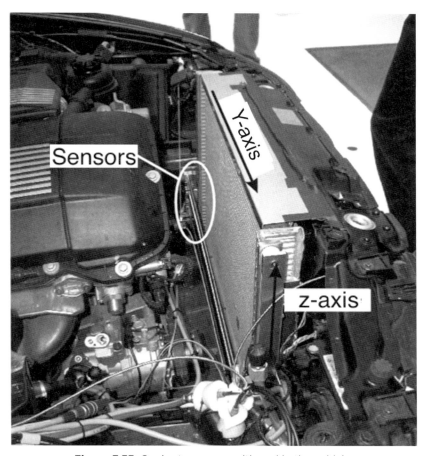

Figure 7.33 Cooler traverse positioned in the vehicle.

A typical result from a mass flow measurement of a heat exchanger in a vehicle with a fan present at 140 km/h is shown in Figure 7.34. The color map of the velocity is between 0 m/s and 15 m/s. Clearly visible are the major components such as the fan hub, the fan shroud, and the power steering cooling coil. The fine horizontal lines in the contour image show the individual cooling tubes of the heat exchanger, which are spatially well resolved by the system. Two other aspects stand out: first, large areas of low-speed flow seen in the corners. This is due to the fan shroud not having high-speed bleeding flaps. Second is the high-speed flow at the middle top and middle bottom of the radiator matrix. This leads either to unnecessary high-pressure losses or to saturated heat transfer in those regions. Both situations are to be avoided for a balanced and optimal radiator flow. The inhomogeneity for this example is very high at $I = 48\%$.

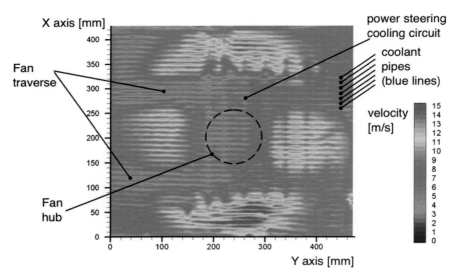

Figure 7.34 Engine compartment flow with fan at 140km/h.

η_A—efficiency factor drive train
η_M—efficiency factor engine

Chapter 8
Aeroacoustics

Martin Helfer

8.1 The Influence of Airflow on the Interior and Exterior Noise of Motor Vehicles

The operating condition is the main significant factor influencing the individual components of overall vehicle noise. At low speeds and high engine load, the noise produced by the drivetrain is the dominating contribution. If the engine load is small and vehicle speed is low, tire-road noise has the highest share of exterior noise. Particularly in trucks, this noise can play a dominating role even during full-load acceleration in the speed range around 50 km/h due to the pronounced increase in noise level with high tractive forces at the tire. In addition to the drivetrain noise and the tire-road noise coming in from outside, the rolling noise excited via structure-borne sound contributes a great share to interior noise at low speeds. At increasingly higher speeds, flow noise becomes more significant as its sound intensity increases with the fifth to sixth power of speed, whereas tire-road noise increases with only around the third power [672].

If multiple noise excitation mechanisms exist, it is highly advantageous to be able to measure individual sources separately. At the same time, the measuring setup should influence the relevant general conditions only minimally, if at all. Test stands of this kind have been in use for quite some time in the field of acoustic development for drivetrain and tire-road noise. In the case of flow noise, acoustic optimization was attempted initially in conventional wind tunnels. This turned out to be difficult, or impossible, due to the highly disturbing inherent noise of the wind tunnels, particularly during measurements of aerodynamic exterior noise. Also, the use of psycho-acoustic assessment methods met with no success, in most cases due to the high background noise.

Consequently, more and more special aeroacoustic wind tunnel facilities were taken into operation in the past years.

This has led to new findings in the field of aeroacoustics. A new clearly defined focus was the reduction of interior noise to achieve higher passenger comfort. Also the relevance of airflow noise in relation to the overall noise was examined for various vehicles. However, only in a few cases has research been done on exterior aerodynamic noise ([335], [121], [331]).

Figure 8.1 and Figure 8.2 show the contributions of the various noise sources to the interior noise of a modern upper-middle class automobile at different speeds. At 50 km/h the drivetrain noise dominates in a wide frequency range. However, in lower frequency bands, rolling noise (below 100 Hz) and tire-road noise (around 1 to 2 kHz) contribute considerably. The spectrum at 160 km/h is clearly determined by flow noise. Only in a few frequency ranges do the drivetrain noise (at typical engine orders) and the tire-road noise at around 1 to 2 kHz contribute to a certain degree.

Some results of exterior vehicle noise measurements for various vehicles on various road surfaces[1] are shown in Figure 8.3, Figure 8.4, and Figure 8.5 [331].

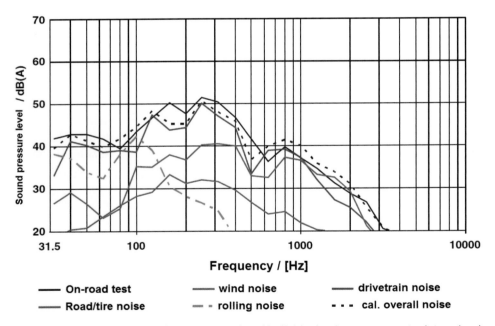

Figure 8.1 Spectra of overall noise on road and individual noise components determined on test stands in an upper-middle class passenger car at $v = 50$ km/h; dotted line: overall noise calculated summing up the individual noise components.

---

1. In this case the concrete track had a rather unusual screed-like structure, resulting in relatively low levels of tire-road noise.

Aeroacoustics

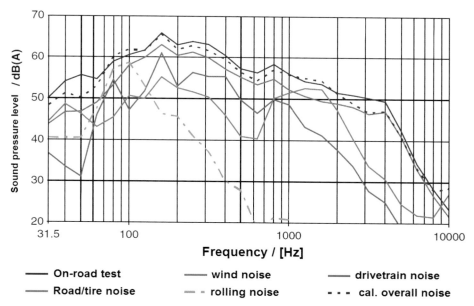

Figure 8.2 Spectra of overall noise and individual noise components determined on test stands in an upper-middle class passenger car at $v = 160$ km/h; dotted line: overall noise calculated summing up the individual noise components.

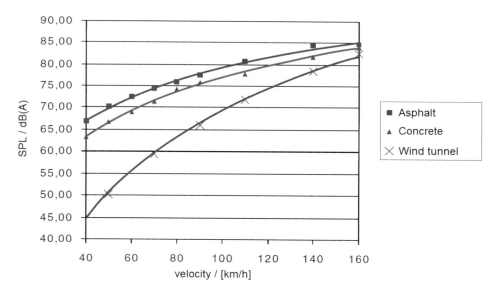

Figure 8.3 A-weighted exterior sound pressure levels from test track measurement and wind tunnel measurement for a BMW 520i.

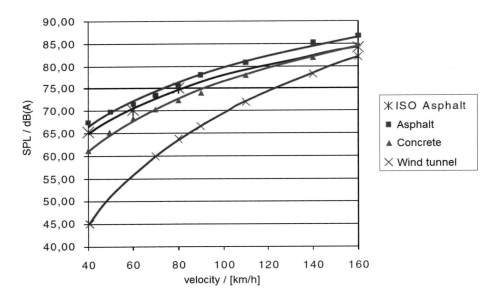

Figure 8.4 A-weighted exterior sound pressure levels from test track measurement and wind tunnel measurement for an Audi S8.

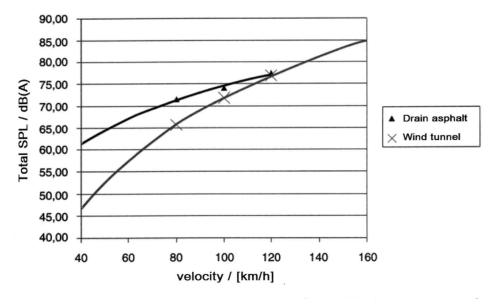

Figure 8.5 A-weighted exterior sound pressure levels from test track measurement and wind tunnel measurement for a Mercedes C-Class.

8.2 Aerodynamic Noise Generation

Aerodynamic noise is basically the result of three different noise generation mechanisms:

- mass flow sources (volume flow through small openings)
- impulse sources (fluctuating pressures impacting rigid surfaces)
- volume sources (turbulent free flow)

All these generating mechanisms are present also in vehicle aeroacoustics. Each one, however, has a different significance. Idealized approximation models can be used to characterize each of these mechanisms (so-called Ffowcs Williams-Hawkings analogy):

- Mass flows can be represented by monopole sources. Examples of this type of sound source are leaks in sealing systems or the exhaust pipe outlet of a vehicle.
- The effect of changing pressures impacting a solid surface can be represented by an acoustic dipole. This type of noise is always present when a free flow or a separated flow impacts a surface. Vehicles have a multitude of areas exposed to free or separated flows.
- Turbulent free flows generate sources with quadrupole character. Sources such as these are generated for example in turbulent shear layers or in the wake of a vehicle.

A schematic representation of the three source types is shown in Figure 8.6 [328].

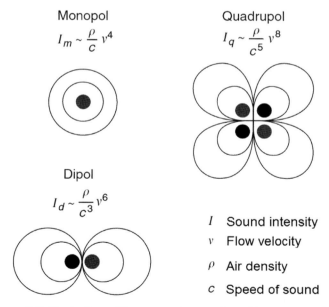

Figure 8.6 Illustration of the source types relevant in aeroacoustics.

As mentioned, the intensities of these three source types vary greatly. For a monopole source with a flow speed v, density ρ, speed of sound c, and Mach number Ma, the following applies

$$I_m \sim \frac{\rho}{c} \cdot v^4 = \rho \cdot Ma \cdot v^3, \qquad (8.1)$$

for a dipole source

$$I_d \sim \frac{\rho}{c^3} \cdot v^6 = \rho \cdot Ma^3 \cdot v^3 \qquad (8.2)$$

and for a quadrupole source

$$I_q \sim \frac{\rho}{c^5} \cdot v^8 = \rho \cdot Ma^5 \cdot v^3 \qquad (8.3)$$

Comparing the intensities, it becomes clear that at low flow speeds (Mach numbers smaller than 1) the monopole source has the most effect, followed by the dipole source. The lowest noise emission is caused by quadrupole sources, which in vehicle aeroacoustics can be neglected in most cases. We can state that in general, if there is a monopole source, it will be the loudest one. Only if all monopole sources are eliminated, one of the remaining dipole sources can dominate. As can be seen from the preceding equations, the acoustic power of a monopole source is proportional to the fourth power of the flow speed, whereas the acoustic power of a dipole source grows with the sixth power of its speed. As the aerodynamic noise generation mechanisms effective in motor vehicles can in general be represented by a combination of monopole and dipole emitters, frequently an increase of the sound power with the fourth to sixth power of the speed, depending on the generating mechanism, can be observed in experiments (as seen previously). It is therefore important that the speed be kept highly accurately in aeroacoustic measurements. Even very small variations in the settings can lead to distinct changes in noise levels. This means that aeroacoustic measurements on the road with unpredictable wind conditions are only conditionally meaningful unless the relative flow speed and direction were also recorded. As the distribution of the flow speed over the entire vehicle surface is highly irregular, the potential noise generation varies greatly, depending on the location of excitation. Assuming a dipole source, the noise generated in one place is over 10 dB louder than in an adjacent place, if the local speeds prevailing there vary by a factor of 1.5. For various areas around the vehicle, this value can be even higher. This shows that the positioning of add-on parts such as outside mirrors or antennas can be of great significance for the aeroacoustic behavior of an automobile.

8.3 Aeroacoustic Measuring Systems

8.3.1 Aeroacoustic Wind Tunnels

Unless noise-reducing measures are taken [336], the noise level in wind tunnels is quite high due to the electric noise caused by the power unit and the aerodynamic noise in the air duct and the fan [336]. First of all, the fan noise must be kept out of the test section. Built-in components (e.g., safety nets) are to be positioned in such a way that the noise impact in the test section caused by them will be as small as possible. Further measures concern the test section itself. In general, it is designed as an anechoic chamber, the walls of the plenum being covered with absorbers. It is much easier to implement this for open-jet than for closed-jet or slotted-wall test sections.

In early aeroacoustic wind tunnels, noise reduction was achieved by sound absorbers and coatings with fiber materials. In more recent wind tunnels (see for example Figure 8.7 [336]), alternative sound absorbers are mostly used. These consist of a combination of foamed material and panel or membrane absorbers. These absorbers are targeted at the lower frequency range, whereas the foam mats absorb noise components with higher frequencies. As an example, the effect of the individual measures in an aeroacoustic wind tunnel is shown in Figure 8.8 using a combination of membrane absorbers and open-cell foam panels [336].

Even though the quality of the acoustic free field decreases to a certain degree, the fiber-free implementation of noise-reducing measures in wind tunnels has definite advantages, as particles from fibrous materials can be swept along with the airflow, which affects both acoustic effectiveness and air quality.

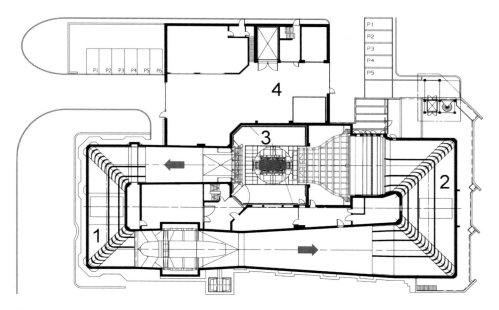

Figure 8.7 Layout of a modern aeroacoustic wind tunnel (University of Stuttgart): (1) and (2) deflection sound absorber with splitter plates designed as membrane absorbers and foam-coated turning vanes; (3) sound-absorbing lining of the plenum (see Figure 8.8); (4) acoustic labs and preparation halls.

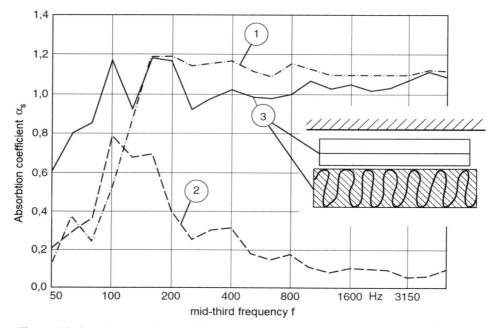

Figure 8.8 Sound absorption coefficient of various components of the plenum lining of a wind tunnel: (1) 140-mm-thick foam panel, (2) 100-mm-thick membrane absorber, (3) 100-mm-thick membrane absorber with 140-mm-thick foam panel placed 10 mm in front of the cover membrane.

8.3.2 Measuring Interior Noise

For the recording of interior aerodynamic noise, the usual measuring techniques used in vehicle acoustics are applied. As is common practice in technical acoustics, in general, artificial heads or single microphones are used here. Artificial heads are particularly suited to determine sound impact on the passengers as well as for psycho-acoustic evaluations, whereas microphones are frequently used for measurements directly at the sound admission point, where artificial heads are often awkward due to their bulky volume or when the sound field is to be scanned over numerous measuring positions.

Sound radiation areas can also be determined relatively well with sound intensity probes. However, their use basically remains restricted to the higher frequency range, where the sound field is less reverberant. Furthermore, special, usually spherically shaped microphone arrays can be used for interior noise measurements, which are suitable for locating aeroacoustic sources—likewise in the higher frequency range. Lower frequency ranges require arrays applying near-field holography techniques (see section 8.3.3.3).

8.3.3 Measuring Exterior Noise

Due to the airflow around the object to be measured, the use of conventional acoustic measuring techniques for recording exterior noise in wind tunnels is problematic.

For one, so-called pseudosound occurs at the microphones resulting from the fact that the measuring membrane is exposed to fluctuating flow pressures. In contrast to "real" sound, these changing pressures propagate approximately with the flow speed (turbulences) and not with the speed of sound. Thus, they have an interfering effect on the results. For another, flow noise and thus extraneous or background noise is also generated at microphone housings, pre-amplifiers, and fixings, which interferes during measurements. Consequently, in these cases, specifically developed measuring techniques are mostly used.

8.3.3.1 Intensity Measurements with Special Probes

Sound intensity measurements for recording aerodynamic exterior noise are generally carried out within the airflow, as intensity probes in the measuring direction feature only poor directivity, thus necessitating in most cases measurements in close proximity to the object. For the previously mentioned reasons, this cannot be accomplished with standard dual-microphone probes. Thus, special probes must be used.

Figure 8.9 shows a measuring setup of this type. The probe must be positioned in the direction of the flow. However, the application in areas with a high degree of turbulence (e.g., behind the rear view mirror) is not possible, as this would lead to pseudosound.

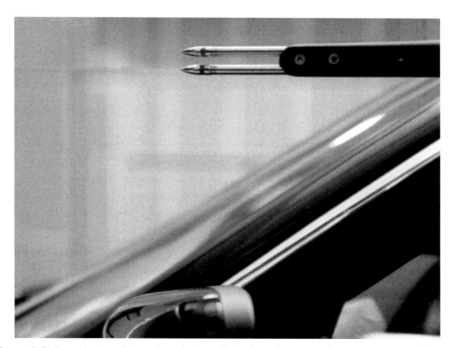

Figure 8.9 Sound intensity probe with parallel microphones and nose cones over the rear view mirror of a passenger vehicle.

Theoretically, accurate measurements of sound intensity with dual microphones can only be implemented in the absence of airflow or with planar soundwaves in a one-dimensional flow. However, aerodynamic sound sources do not emit planar waves, nor is the flow around a vehicle one-dimensional. Tests revealed, however, that errors remain very small at normal driving speeds [330].

8.3.3.2 Microphone Arrays

Microphone arrays are frequently used when it is very difficult or impossible to carry out measurements close to the object to be measured. For this reason, they have been referred to as "acoustic telescopes." They are also highly suitable for locating sound sources. For exterior noise measurements, microphone arrays consist of a number of microphones usually set up on a plane mounting surface. Virtually any kind of setup is possible, including three-dimensional, planar distributed, ring-shaped, cross-shaped, and linear arrays.

The principle of measurement is to "focus" the array on the various measuring points of the object under investigation. This is achieved by applying a time shift on each measured signal corresponding to the propagation delay from the focal point to the respective microphone. The time-corrected signals of all microphones are then added up. This results in an overall time history for the respective focal point. With this method, the sound emitted by other sources than the focal point is mostly eliminated (averaged out). However, the sound emitted at the individual focal point is enhanced [302].

The frequency range of microphone arrays is limited in the lower domain by the array size: the larger the array, the more accurate it is for localizing in the low-frequency ranges. In the upper frequency range—notably in the case of equally spaced microphones—errors increasingly occur due to pseudosound sources (aliases) that can lead to misinterpretations. These diminish when reducing the minimum microphone distances in the array.

Array techniques were initially applied mainly in rail vehicle and aeronautical engineering ([44], [571]). In the mean time, they are also common in vehicle acoustics, especially in vehicle wind tunnels ([330], [494]). They are used mainly in aeroacoustic wind tunnels with open-jet test sections. The array is then set up out of flow in the plenum chamber. Naturally, in this case the influence of the flow and the shear layer on sound propagation must be corrected mathematically and thus eliminated [674]. However, there are also arrays with microphones flush mounted into the test section walls, thus also suited for closed-jet test sections [494].

8.3.3.3 Acoustic Near-Field Holography

Acoustic near-field holography (also called STSF — spatial transformation of sound fields) is also based on measurements with several microphones. A rectangular array is usually used, including additional reference microphones positioned in close proximity to the sound sources of interest. By estimating the cross power spectra between the reference microphones themselves and between these and each microphone in the array, a so-called principal components representation of the sound field is calculated, which can be utilized for the realization of acoustic near-field holography, as well as for the Helmholtz integral equation needed to calculate the far-field parameters [313].

After measuring an array close to the sound source, acoustic holography allows the calculation of a sound field in a different plane, which can be either closer to the sound source or farther away. Noncorrelated signal components of the microphones are averaged out by calculating the cross power spectra. This usually also applies to the pseudosound with microphones in the airflow. For this reason, acoustic holography is also suited for measurements within the flow.

One possible measuring setup is shown in Figure 8.10 [330]. It consists of eight microphones with a vertical separation of 120 mm, which can be moved across the plenum chamber with the traversing system and one or several reference microphones. The holographic results then show that share of the sound which is correlated to the signals at the reference point. To be able to capture frequencies up to approximately 4 kHz, two additional measuring positions between the adjacent microphone positions must be included, so that the distance between the measuring points is reduced to 40 mm. The same measuring point density should be implemented in the vehicle's longitudinal direction. Here care should be taken that the microphones will not be positioned in zones of high turbulence (e.g., in the wake of an outside mirror), as even the averaging effect of the cross power calculation cannot eliminate the high proportion of pseudosound generated there at the microphones. Hence, near-field holography in the wind tunnel requires quite some effort. However, as a result, a complete representation of the lateral sound field of the tested vehicle is obtained.

Figure 8.10 Measuring setup for acoustic holography in the wind tunnel; microphone spacing 120 mm.

8.3.3.4 Acoustic Mirrors

The working principle of acoustic mirrors is shown in Figure 8.11 [329]. With conventional acoustic mirrors, the microphone is placed thus that it meets the point of intersection of the reflected sound beams related to one single focal point on the surface of the test object. Sound beams of sources lying outside the focal point are only included in the measurement to a very small degree. In the case of so-called array-based acoustic mirrors, the reflected sound is not only recorded by one but by several microphones positioned on a plate in front of the center of the mirror. This allows the simultaneous recording of sound sources over a larger area of the object to be measured [673]. Figure 8.12 shows such an acoustic mirror with 108 microphones embedded in the mounting plate.

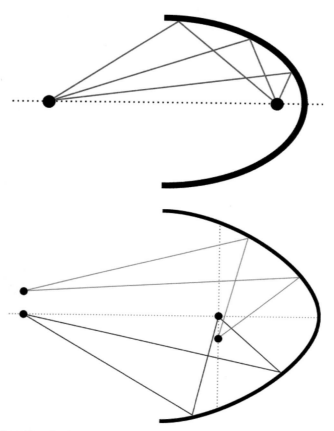

Figure 8.11 Functional principal of a conventional acoustic mirror (above) and an array-based acoustic mirror (below).

The signal amplification of an acoustic mirror depends on frequency. The smallest amplifications are obtained in the low-frequency range. In this range, greater amplifications can only be achieved with larger mirror diameters. These are also advantageous with respect to spatial resolution [329].

In wind tunnels with open-jet test sections, acoustic mirrors are used to determine exterior noise. The technology is not suited for closed-jet test sections, as in this case the system would have to be set up within the airflow.

Figure 8.12 Array-based acoustic mirror. Photo: M. Riegel.

8.3.4 Measuring Structure-Borne Sound

Structure-borne sound measurements are frequently carried out to determine transmission mechanisms in the vehicle structure and to identify body areas highly prone to vibration. In general, piezo-electric accelerometers are commonly used in this context. When dealing with lightweight sheet metal structures care must be taken that vibrational behavior is not significantly influenced by the weight of the sensor. This is why miniature sensors weighing only a few grams are used for vehicle structure measurements. If possible, however, laser Doppler vibrometry would be the preferred method, so that accelerometers can be omitted completely. Laser Doppler vibrometry uses the shift in frequency in a laser beam reflected (scattered) at the point of measurement as a measure of the vibration velocity. Hence in this case, we have an entirely nonreactive measurement. Preferably a system that permits automatic scanning of a large number of measuring points is to be used, as it significantly reduces measuring time and hence also costs, especially when considering the high hourly rates of automotive wind tunnels.

8.3.5 Sound Source Location with Special Instruments

8.3.5.1 Leakage Tests with Ultrasound

In many cases it is not so easy to identify leaks at the vehicle compartment using simple acoustic means. Thus, frequently the option of choice is to place an ultrasound transmitter inside the vehicle and to use an ultrasound receiver to look for leaks in the body or floor pan where ultrasound passes through. This can also be done outside the wind tunnel.

Commercially available transmitters for this application radiate sound semi-spherically and generate frequencies of approximately 40 kHz. In the vehicle interior they are placed such that the areas to be examined can be spread by the ultrasound in the most efficient way. Door or window seals for example can be directly scanned with the receiver housed in a handheld casing. If the sensor hits a point where ultrasound leaks to the outside, the detector emits an acoustic signal. The volume of this signal indicates the intensity of the ultrasound at this particular point. In this way, defects in seals or other air-borne sound transfer paths can be detected rapidly.

8.3.5.2 Acoustic Probes

Special acoustic probes can be used as well as stethoscopes to detect sound sources in vehicles. Such probes consist of a thin — sometimes flexible — tube that contains a microphone. This tube is connected to a housing containing a headphone amplifier with volume regulation and power supply. The signal of the microphone can be monitored with headphones, which can be connected to this housing. A sample of such a device is shown in Figure 8.13 [328].

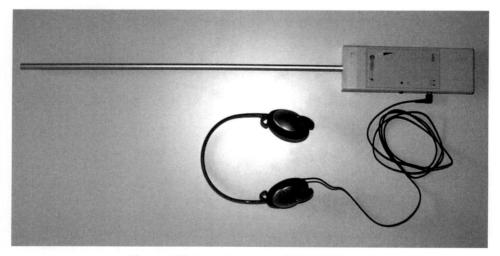

Figure 8.13 Acoustic probe with headphones.

8.4 Main Noise Sources and Options for Their Reduction

A considerable number of possible noise sources must be examined in the aeroacoustic development of motor vehicles. When optimizing details, it is advantageous when the noise source of interest can be observed in isolation. The more sound sources combine to make a noise, the more difficult it is to judge changes in one of them during measurement and subjective acoustic assessment. Consequently, in the aeroacoustic development of an exterior mirror, all window and door seals will be covered with fabric or aluminum tape to eliminate the influence of leaks (see section 8.4.1). Likewise, other add-on parts such as antenna and windshield wipers should be removed too, if necessary. If this does not significantly change the A-pillar vortex, the rain gutters can also be taped flush. Also at the mirror itself, sound sources that are not to be considered can be eliminated in this way (all gaps at the mirror base should be taped when the water drain grooves at the mirror housing are to be acoustically optimized).

When examining other sound sources, the same approach should be used in order to obtain clearer variations in the sound level. In order to assess the necessity for individual acoustic optimizations, which might possibly also be costly, ultimately tests are required to be carried out on the series production vehicle.

8.4.1 Leaks

As was mentioned in section 8.2, avoiding leaks is particularly important, due to the monopole character of the noise that they produce. In motor vehicles this mainly concerns the development of window and door seals, which must be carried out with great care. Particularly at higher speeds where the pressure difference between inside and outside increases, the risk of leaks increases, due to the fact that the doors are lifted from their seals by the high negative pressures impacting on the exterior.

The tightness of the body can already be checked outside the wind tunnel with the help of ultrasound devices, as has been explained in section 8.3.5.2. In order to determine the influence of the seals on interior noise in the wind tunnel, all gaps and grooves of the body are taped flush. By comparing the measuring results with and without tape, the overall contribution of all gaps to aeroacoustic noise can be determined. If the contribution of one individual seal section or one individual gap is to be determined, this particular section is opened exclusively, while all other gaps remain taped over. In order to test further sections, the previous gap is taped over again and the new area of interest is exposed. The respective test results are then compared with the results when all joints were completely taped flush. In this way, the contributions of the individual sections to interior noise can be determined separately and can be compared.

Figure 8.14 shows the influence of a sealing system to the sound pressure at the driver's left ear in a production vehicle in comparison to the influence of the add-on parts (exterior mirror, windshield wipers, antenna) [120]. The dominating contribution of the sealing system to interior noise can be clearly recognized. An effective but relatively expensive measure for reducing this contribution is the use of multiple sealing systems.

The seals at the mirror base as well as those between the wheel housings and in the area of the A-pillar must be designed with great care. At these points the noise is not generated directly in the passenger cell; however, there is the possibility that the noise is introduced through cavities in doors and body.

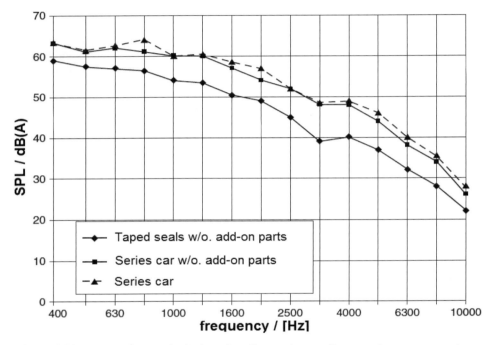

Figure 8.14 Impact of a poorly designed sealing system on the sound pressure spectrum in the vehicle interior compared to the influence of add-on parts.

8.4.2 Rear View Mirrors

A great part of aeroacoustic research deals with exterior mirrors. They are located in zones of high flow velocities and thus pose a particular acoustic problem. In vans, they can be the main sound source for exterior aerodynamic noise (Figure 8.15) [331].

The mirror shape is highly influenced by design. Functional aspects must be considered too. Measures for acoustic improvement are thus mainly focused on details as the depth and the shape of drainage grooves, folding joints, and housing drainage. Frequently, noises here have a tonal character (whistling). Vortex generators frequently help with these types of noise. They are positioned in front of the points of sound generation and interfere with its periodicity (Figure 8.16 shows an example [333]). This method is likewise applied with other sound sources, as is shown in sections 8.4.4, 8.4.6, and 8.4.7.

Aeroacoustics

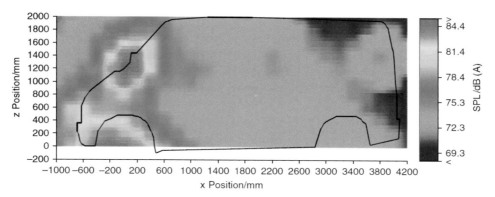

Figure 8.15 Emission patterns of a truck in the 2-kHz third octave band at 140 km/h flow speed.

Figure 8.16 Vortex generators for avoiding tonal sounds in the area of the rear view mirror.

8.4.3 Windshield Wipers

Windshield wipers, too, are highly significant in aeroacoustic optimization. In their parking position frequently they can be hidden behind the hood; in some cases, however (e.g. in minivans), they are directly exposed to the airflow, leading to a significant increase in interior noise levels [903]. Spoilers in front of the windshield guiding the airflow over the wipers can be the remedy here, as can be seen from Figure 8.17 [120]. During operation, too, the sound of windshield wipers with its distinct fluctuation can have a bothersome effect. The highest sound levels are usually generated when the wipers are in a downward movement in a position approximately perpendicular to the flow direction.

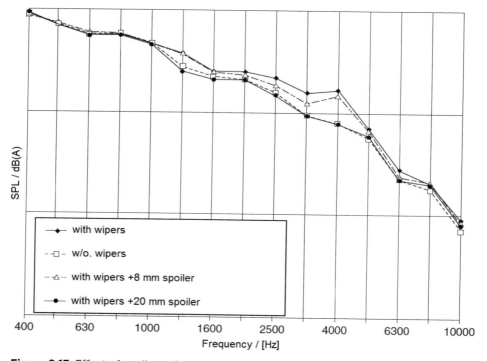

Figure 8.17 Effect of spoilers of varying height in front of the windshield of a minivan on interior noise; wipers in the parking position.

8.4.4 Antennas

Antennas can generate tonal sounds. These whistling noises, also called aeolian tones, are caused by regular flow separations taking place alternatingly to left and right of the antenna rod and forming a so-called Kármán vortex street on the lee side. Noise reduction can be implemented by putting the antenna in a highly inclined position and by a wire coil around the antenna. This measure prevents the formation of a regular vortex separation. An industrially manufactured antenna of this kind can be seen in Figure 8.18 [333].

Figure 8.18 Standard antenna with wire coil.

8.4.5 A-Pillar

The shape of the A-pillar significantly influences aerodynamic noise generation. It determines the volume and shape of the separation vortex on the side window, which in turn can affect noise generation by the exterior mirror. Moreover, the rain gutters integrated into the A-pillar are noise-generating elements. In most cases, optimization is implemented iteratively. As can be seen from Figure 8.19, the potential for improvement can be substantial [329]. The diagram shows the results of measurements with an acoustic mirror carried out from above. With this measuring method, aeroacoustic development for this body area can be started already with the clay model.

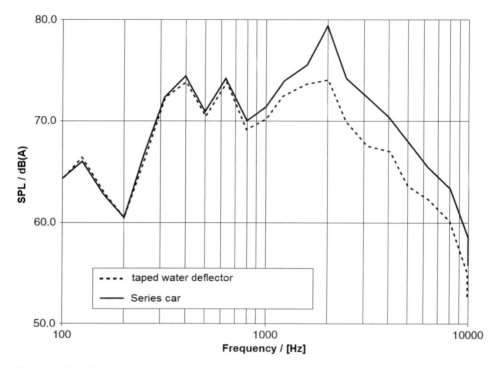

Figure 8.19 Influence of rain gutters on sound emission; measurement carried out with an acoustic mirror from above at 140 km/h, point of focus at the center of the A-pillar.

The curvature of the A-pillar is also an important parameter in aerodynamic noise generation. Figure 8.20 shows the increase of the sound pressure level at the driver's ear when the yaw angle is raised from 0° to 10° for various A-pillar curvatures [120]. The sound level differences are shown in "points," calculated from the total of all sound level differences in the individual third octave bands from 400 Hz to 10 kHz. It can be seen that under yaw condition and small curvature radii up to approximately 10 mm initially the changes in the increase of the sound levels are small. With larger radii, however, this increase drops significantly. This means that vehicles with large A-pillar radii have a less distinct acoustic reaction to yaw. Consequently, in real traffic and turbulent flows they will tend to exhibit fewer modulations in interior noise.

Aeroacoustics

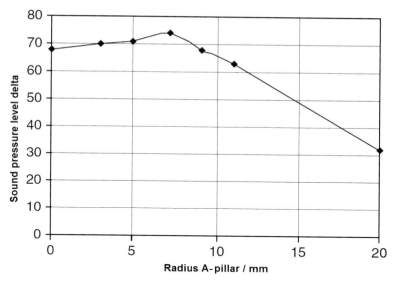

Figure 8.20 Changes in sound pressure level in the interior noise due to yaw with various A-pillar radii.

8.4.6 Cavity Resonances

In motor vehicles, two types of cavity resonances occur: First, resonances where the whole vehicle interior is excited (e.g., by an open window or an open sun roof); second, aerodynamic vibrations in small cavities such as grooves, slits, gaps, or bores. These resonances are excited in a similar way as the tones that are generated when air is blown over a bottleneck. The cavity acts like a kind of Helmholtz resonator whose natural frequency is highly dependent on the volume of the cavity. In cases of resonance, coherent vortex structures separate at the trailing edge in front of the opening, impact on the leading edge behind it, and cause pressure waves there, which excite the interior of the car and in turn lead to more vortex separations at the trailing edge.

Whether a resonance arises or not is highly dependent on the relative speed of the vortex structure, which in turn is determined by the flow speed or driving speed. Buffeting noises from an open sun roof thus only occur in a rather narrow speed range (usually somewhere between 40 and 90 km/h). If the resonance frequency of the vehicle interior is de-tuned (e.g., by changing the number of passengers), the speed range where buffeting occurs is also changed.

Sun roof buffeting generates sound pressures up to approximately130 dB at frequencies around 20 Hz and represents a considerable impairment of comfort. A lowering of the sound pressure levels can be implemented by avoiding the vortex structures impacting the leading edge of the sunroof. This can be implemented by defining a so-called comfort position of the sun roof, where the opening clearance is limited. Only when the sun roof is opened farther, buffeting occurs. Furthermore, air deflectors are installed in front of the trailing edge of the sun roof, which can shift the position of impact of the

589

vortices coming from the leading edge to the roof areas behind the sun roof opening and can also be provided with vortex generators (e.g., notches, slits, grooves, bumps, bores) that destroy the regularity of these separations (see Figure 8.21 [333]). In the case of very large sun roof dimensions (in excess of approximately 400 mm) such as panorama roofs, wind deflectors of this type are no longer sufficient. For this, net wind deflectors must be used (see Figure 8.22 [333]). Apart from these possibilities, other techniques for reducing sun roof buffeting could become significant in the future. Approaches, for instance, do exist already to disrupt the periodic excitation by a movable lip at the front edge of the sun roof, which can be moved, controlled or stochastically by actuators [326].

Figure 8.21 Wind deflector with recesses and cut-outs.

Figure 8.22 Net wind deflector in a production car.

As mentioned, smaller cavities also contribute to the generation of tonal sounds. Bores in the axle casing, for example, can cause whistling noises of several kilohertz that can be conspicuously unpleasant in the passenger compartment. To avoid this type of noise, excitation slits and bores in the outer body skin and in the underfloor should be avoided where possible, or covered.

8.4.7 Sun Roof Opening Noise

Due to the pressure differences between the vehicle interior and on the roof surface, transient, sometimes tonal noise is generated when sun roofs are opened into the lifted position. These can be prevented by a toothed strip at the back edge of the sun roof, as is shown in Figure 8.23 [333].

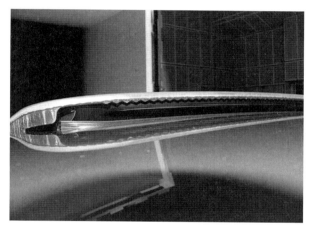

Figure 8.23 Toothed strip for preventing opening noise in a sunroof of a standard production car.

8.4.8 Wheel Housings

Notably, the front wheel housings are the primary sources for exterior aerodynamic noise over almost the entire frequency range. Owing to the—among other factors— already highly sound-proofed bulkhead, they affect interior noise only insignificantly. Figure 8.24 shows the radiation pattern of a passenger vehicle at a flow speed of 140 km/h [332]. The contribution of the front wheel housing can be seen clearly. Typically, the rear wheel housing is obviously much less implicated in noise emission. At the A-pillar, the influence of the rear view mirror can be seen.

It is not known so far how wheel rotation affects noise excitation in the wheel housings, as the rolling noise of the tires cannot be separated from aerodynamic noise when the wheels are turning. It can be assumed, however, that noise excitation more likely increases owing to the additional rotating movement of the wheel disks in the flow. There is also knowledge of cases where vortex structures generated at the front wheels cause modifications in the radiation characteristics in the front area of the vehicle and thus also have an impact on the level of the tire-road noise (see also [81], [332]).

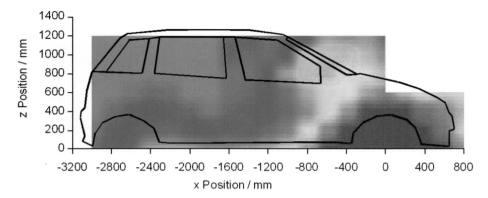

Figure 8.24 Radiation characteristics of a passenger car in the 2.5-kHz third octave at 140 km/h flow speed (measurement with an acoustic mirror).

8.4.9 Underbody

Apart from rolling noise caused by driving on road surfaces with high values of macro and mega texture, low-frequency interior noise is especially caused by aerodynamic flow over the underbody. This type of noise can be very annoying and reduce the comfort inside the vehicle considerably. The aerodynamic forces transmitted to the body at the underbody cause the whole vehicle structure to vibrate. This structure-borne sound is then radiated into the vehicle interior by various surface sections. Low front spoilers and a vehicle underside as smooth as possible can be of remedy here. Figure 8.25 shows the effect of a blockage between front bumper and wind tunnel floor on interior noise [333]. The effect of underbody flow can be perceived particularly in the frequency range below 1 kHz.

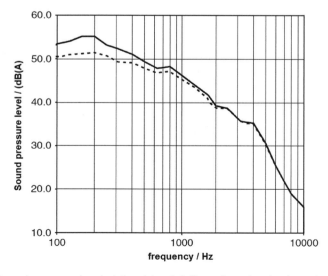

Figure 8.25 Sound pressure level at the driver's left ear in a standard production condition (———) and with blocked gap between front bumper and road (- - - -).

8.4.10 Reduction of Interior Noise by Using Special Acoustic Glass Windows

The glazing also has a considerable influence on interior noise [317]. Already by increasing the pane thickness, interior noise can be substantially reduced ([120], [903], [551]). Another option for achieving a distinct noise reduction is the use of laminated glass with plastic interlayer, as can be seen in Figure 8.26 [551]. A pane structure of this type can additionally be implemented without any weight increase.

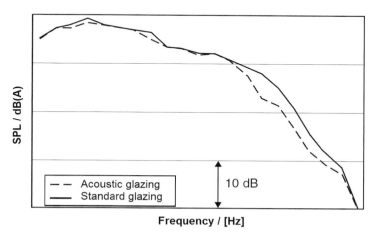

Figure 8.26 Sound pressure level in a vehicle with standard panes and acoustic panes with plastic interlayer.

8.4.11 Convertibles

In recent years definite improvements have been achieved in convertibles. For one, lined tops have increasingly come into use, which apart from having advantages in interior air conditioning also have favorable acoustic effects. For another, sewn-in cross members and higher-tensioned fabrics also prevent undesired top movements, which improves both acoustics and ballooning positively. Nevertheless, even with a sealing system of similar high quality, convertibles are still at a disadvantage compared to coupés acoustically.

In convertibles particularly, the sealing system must be designed with the greatest care. Special attention should be given to branching points, which are highly prone to leaks. With the top open, convertibles also have special acoustic idiosyncrasies. While the perceivable engine noise tends to be assessed positively, as it is usually designed appropriate to the vehicle type, extreme air draft and flow noise are usually criticized. However, precisely the systems for draft prevention (draft stops) can have an unfavorable effect on noise development [328].

8.5 Psycho-Acoustic Aspects

The annoyance of aerodynamic noise is usually assessed subjectively. There are approaches tackling the issue objectively [622]. It has been known for quite some time that the sharpness of the interior noise is almost exclusively determined by aerodynamic noise [334]. This was confirmed by later tests for other vehicles, as can be seen in Figure 8.27 [675].

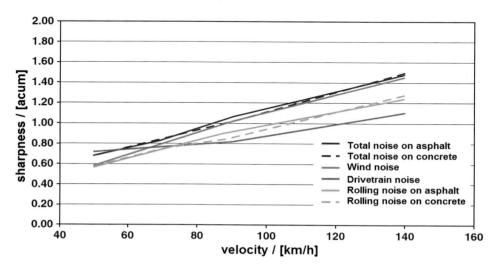

Figure 8.27 Sharpness at the driver's right ear in a vehicle of upper-middle class for various noise components over speed.

One essential aspect is that the noise impression on the road can differ substantially from that in the wind tunnel. In a vehicle wind tunnel, the flow is straightened in the settling chamber and the speed profile is adjusted over the entire cross section. Thus, the flow noise excited at the vehicle is highly homogenous. Out in the open, however, flow conditions are nonstationary due to the prevailing weather conditions (see Figure 8.28). On the road, the turbulence caused by other road users comes additionally into play. The resulting nonstationary flow profile leads to alternating flow speeds and angles over time, which are moreover dependent on location. The result is a flow noise varying in time, in its volume, and frequency composition. This kind of time-dependent structure of noise is frequently perceived as annoying.

Increasing demands for comfort therefore require that henceforth—apart from the classic stationary flow noise optimization in the wind tunnel—nonstationary wind noise should also be considered in the vehicle development process. Various methods for simulating the wind noise modulated by turbulence are employed with the help of wind tunnel measurements in order to avoid investigations on public roads. These include both stationary and active gust generators in front of the vehicle as well as methods for sound synthesis ([854], [327]).

Figure 8.28 Picture of a soap bubble to characterize the flow conditions outdoors. Photograph: Kazbeki.

The necessity for simulating realistic modulations as a part of vehicle interior noise in the aeroacoustic development of vehicles was variously realized early on (e.g., [854]). Here it is important to be able to simulate flow conditions corresponding to the conditions on the road as precisely as possible. For practical purposes, these conditions can be described best using degree and length of turbulence. As is shown in Figure 8.29, these values differ greatly regarding the situation in conventional wind tunnels and on the road [507]. Thus, possibilities for simulating road conditions in the wind tunnel were tried out in the past years.

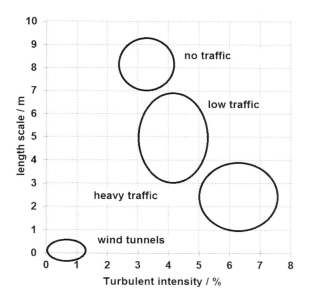

Figure 8.29 Examples of turbulence lengths and degrees on the road and in conventional wind tunnels.

8.5.1 Assessing Different Behavior Under Yaw Conditions

As turbulences at the vehicle cause alternating flow angles, a yaw angle sweep in the aeroacoustic wind tunnel can provide information about the acoustic sensitivity of a vehicle to nonstationary flow conditions. As an example, Figure 8.30 shows the interior noise perceived at the driver's left ear for two vehicles at yaw angles from $-20°$ to $+20°$ [676]. The width of the area with almost constant sound pressure level as well as the extent of level increase with growing yaw angle (downwind and upwind) can be used for assessing and comparing vehicles. An acoustically superior vehicle has a wider range with a constant level and slighter upwind and downwind level increase than a vehicle with poorer acoustic assessment.

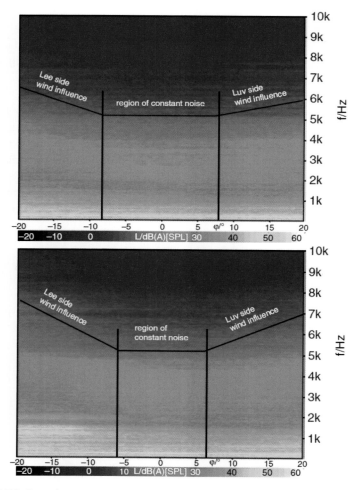

Figure 8.30 Sound pressure spectrum at the driver's left ear over the yaw angle for vehicles with aeroacoustically positive (above) and negative (below) rating in turbulence on the road.

8.5.2 Simulation with Static Vortex Generators

A number of tests were carried out with static vortex generators. In this context, either special components in the wind tunnel nozzle were used (for example, a plate measuring 4 × 6 m which was placed on the screens [854] [854] in the settling chamber or alternatively a compact car in the nozzle mouth ([854], [461]). In this way it is relatively easy to generate high degrees of turbulence, as are found on roads with high traffic; turbulence lengths, however, are rather small.

8.5.3 Simulation with Dynamic Vortex Generators

Dynamic systems generating vortices are technically the most complex approach. In order to reproduce road conditions, they must be able to generate yaw angles of up to 10° and—with smaller angles—frequencies of up to almost 10 Hz. These systems consist of wing profiles built into the nozzle mouths of wind tunnels. Systems implemented so far are able to generate high degrees of turbulence. The turbulence lengths, however, are limited here too [507].

8.5.4 Noise Synthesis

Computer-aided synthesis of stationary flow noise is based on stationary interior noise measurements in the wind tunnel specifically for each vehicle, for varying flow speeds and yaw angles. Individual noise sequences of the stationary measurements are assembled to become a nonstationary wind noise in accordance with flow characteristics measured on the road (speed and angle). Figure 8.31 illustrates this procedure schematically [461].

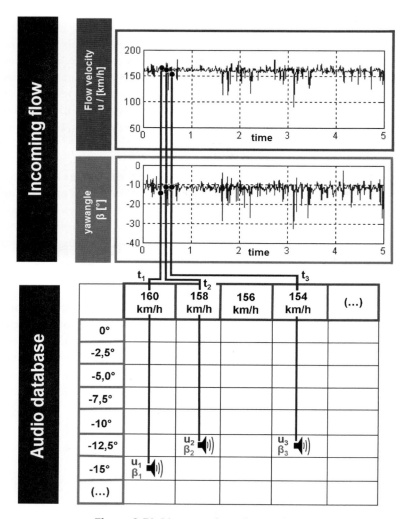

Figure 8.31 Diagram of a noise synthesis.

When the sampling rate of the flow characteristics, the length of the individual audio-clips, and suitable fading methods at the clip endings (cross-fades) are selected optimally, the subjective noise impression on the road can be approximated quite well by the synthesis. Also the modulation spectra of the synthetic noise and the road noise are very much alike as well in their frequency composition and temporal structure (see Figure 8.32) [461]. This method has yet to be optimized and validated.

A similar approach is to first record a basic noise in the wind tunnel with a stationary flow at a defined speed and flow angle. In addition, the level shift of this noise due to different yaw angles and flow speeds is measured, resulting, for example, in a dependency as shown in Figure 8.33.

Aeroacoustics

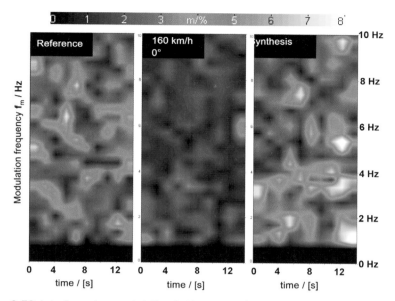

Figure 8.32 Interior noise modulation in the octave band at 5.6 kHz on the road (left), in the wind tunnel (middle), and for the synthetic noise (right), which was determined nonsynchronous to the road measurement.

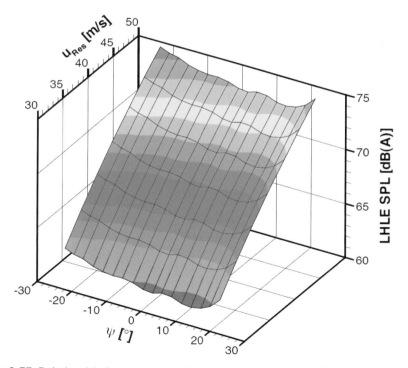

Figure 8.33 Relationship between overall sound pressure level at driver's ear, yaw angle, and flow velocity at steady flow in the wind tunnel for a specific vehicle [618].

If the nonstationary outside wind noise with certain flow characteristics is to be simulated, first the time signals of yaw angle and flow speed must be measured on the road. After that one must determine—based on the diagram in Figure 8.33—which level progression would occur according to these time signals. Then the initial stationary basic noise has to be amplitude modulated according to the determined level history in question. In this way, different vehicles can be assessed and compared in acoustic tests while using identical simulated flow characteristics [618].

Chapter 9
High-Performance Vehicles

Michael Pfadenhauer

9.1 Introduction
9.1.1 Definition
The category of high-performance vehicles includes a number of very different automobiles, namely:

- Sports cars, which are licensed for road use and that offer the driver high performance without compromising the essential requirement of everyday usability.
- Racing cars, whose sole purpose is to be used on the race track in competition. This also includes those competition cars that are derived from production cars.
- Record vehicles for different purposes such as
    - Highest speed
    - Lowest consumption
    - Longest range

The requirements of the aerodynamics are differentiated accordingly. Low drag is a general requirement common to all these vehicles. If they are to be designed for high speeds on winding tracks, this generates an additional demand for low aerodynamic lift. The measures by which both requirements are fulfilled are often in opposition to each other. How these are balanced is a focus of this discussion.

Not to be overlooked, additional tasks that must be solved with the help of aerodynamics are:

- Ensuring directional stability
- Response to driving in the slipstream
- Cooling of all systems
- Creation of acceptable comfort for the driver

9.1.2 Preview

The chapter begins with a brief summary of the history, limited to the field of aerodynamics. This is followed by an overview of the various types of high-performance vehicles: their design, and the relevant regulations therefore, are explored only as far as they are relevant to the aerodynamics, which is the main focus.

The influence of the shape and of aerodynamic devices on drag, lift, and the other forces and moments, as well as the cooling system, and naturally also in unsymmetrical flows, are described. A special section takes up the subject of ground effect.

Finally, by means of real examples, it is demonstrated how the elements of aerodynamics are used to meet the requirements of a particular case. It is made clear how different boundary conditions can lead to a different set of compromises.

9.2 Outline of the History

9.2.1 Racing Cars

The development of high-performance vehicles began with racing cars. Their aerodynamic development began with vehicles of open cockpit design, which despite open wheels attempted to penetrate the air with as little resistance as possible. A pointed-cooler, a slender body, and a pointed tail were seen as a medium for achieving this. An example of this is the Blitzen-Benz, from 1911, shown in Figure 9.1. Only the smooth surface and the pointed tail make some effort to recognize aerodynamics. However, whether the latter was effective, in view of the driver, passenger, and other more forward obstacles placed in front of it, is rather doubtful.

Figure 9.1 Blitzen-Benz racing car from 1911. Archives of R. J. F. Kieselbach.

In contrast, the Benz Tropfenwagen of 1923 (Figure 9.2), based on the patented Rumpler design, was largely aerodynamically formed throughout, except for the open wheels. The underbody also appears to have been smooth. With a ratio of length to diameter of $l/d \approx 6.8$, the car body is unusually slim.

Figure 9.2 Benz Tropfenwagen from 1923. With 66 kW (90 hp), it reached about 185 km/h. Photo Daimler-Benz AG.

The shortcoming of open wheels was overcome with the Bugatti "tank." The profile of its streamlined structure allowed the inclusion of relatively large wheels, and it was well adapted to the road. However, the three-dimensional nature of the flow around it was not accounted for and so the positioning of the driver and passenger so disturbed the flow that the sloping rear end remained ineffective.

At this time, fully enclosed vehicles were limited to individual examples. Most manufacturers of racing cars still held fast to the classic design, with a large cooler and exposed wheels. Not until later did fully enclosed, streamlined racing cars regularly start to appear on race tracks, with cars such as the Daimler-Benz (Figure 9.3) and the Auto Union (Figure 9.4), both from 1937.

Figure 9.3 Streamlined racing car of Daimler-Benz AG on the Avus circuit in Berlin 1937. Image Daimler-Benz AG.

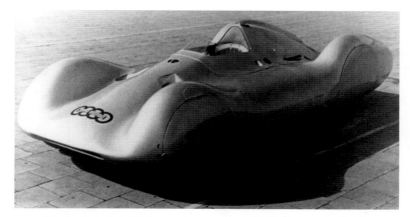

Figure 9.4 Auto Union racing car from 1937. Picture Audi AG.

The cars shown in Figure 9.2 and Figure 9.3 can be regarded as the fathers of the two "tribes" of racing cars that still exist side by side today: the open-wheeled single-seater and the car with covered wheels, from which emerged the sports cars.

The development of the enclosed-wheel racing car was revived only in the 1950s. An example shown in Figure 9.5 is the Mercedes-Benz 300 SLR, which in 1955 competed in the Le Mans 24-hour race. A distinctive feature is the retractable spoilers, which improved deceleration from high speeds by raising the drag coefficient of $c_D = 0.44$ to $c_D = 1.09$.

Figure 9.5 Mercedes-Benz 300 SLR. This 1955 Le Mans car had an air-brake spoiler on the rear, which raised the drag coefficient of $c_D = 0.44$ to $c_D = 1.09$. Image Daimler-Benz AG.

Until the late 1960s, the designers of racing cars primarily pursued the goal of low air resistance. A good example of this is the long-tail Porsche of 1966 (Figure 9.6). Further aerodynamically refined were the Le Mans cars of Panhard and Peugeot designed by Charles Deutsch. These reached top speeds of 220 and 245 km/h with, respectively, only 63 and 105 hp (Figure 9.7). The large tail fins were required to compensate for the cross-wind sensitivity of the streamlined body.

High-Performance Vehicles

Figure 9.6 Porsche Carrera 6 Langheck racing car from 1966. With 162 kW (220 hp), the vehicle reached 265 km/h. The drag coefficient was $c_D = 0.33$. Photo Dr. -Ing. H.C. F. Porsche AG.

Figure 9.7 CD Peugeot 66 racing car from 1966. It achieved a speed of 246 km/h with 78 kW (105 hp). Image Automobiles Peugeot.

That they reached high speeds is a characteristic common to all these streamlined vehicles. But in these cases the resulting lift forces[1] reduced driving stability and cornering limit speed. The Chaparral 2C of 1965 was the first racing car [422] to use a wing to generate negative lift, (i.e., downforce) with the aim of increasing the lateral roadholding so as to improve handling and stability. As a consequence, enclosed-wheel racing cars were generally equipped with wings and spoilers, as seen in Figure 9.8 and Figure 9.9.

Figure 9.8 Porsche 917/ 30 racing car from 1973. It reached a top speed of 370 km/h with 809 kW (1100 hp). The drag coefficient was $c_D = 0.57$. The front spoiler and rear wing are clearly visible. Photo Dr.-Ing. H.C. F. Porsche AG.

---

1. For racing vehicles, it has become common to describe negative lift as "downforce." The downforce is therefore treated as an independent variable; a downward-facing air force is positive.

Figure 9.9 Porsche 935/78 race car "Moby Dick," raced at Le Mans in 1978. It reached 366 km/h with 552 kW (750 hp). The drag coefficient was $c_D = 0.36$. Photo Dr.-Ing. H.C. F. Porsche AG.

Open-wheeled racing cars also followed this development. The first of these single-seaters were streamlined with bodies as narrow as possible to reduce drag. Little attention was given to the generation of downforce at first. From 1968, wings were first connected to the suspension at the front and rear axle. In order to impinge on the most undisturbed flow, they were placed high above the vehicle (Figure 9.10). When in some cases the wing supports broke off during the race, this design was banned. For the following season, it was required that the wings be fixed to the body.

Figure 9.10 Formula 1 car with a high rear wing.

The Chaparral 2J of 1969 had a totally different concept of generating downforce (Figure 9.11). Two motor-driven fans sucked air from under the vehicle, resulting in a net force pushing the vehicle onto the road. This "vacuum effect" was however quickly banned by a corresponding change in the rules. A similar experiment was done with the Brabham Formula 1 racing car in 1978. Here the well-proportioned cooling fan generated the low pressure. Supposedly, it was agreed in a meeting between the team owners and the motor sport governing body not to race the car, even though it was not clearly excluded by the regulations. Almost all current technical regulations now explicitly prohibit the use of movable aerodynamic aids. An exception to this is Formula 1, in which overtaking opportunities are currently boosted using movable wings.

Figure 9.11 Chaparral 2J, 1979.

In 1977, the Lotus team developed an under-floor that exploited the so-called "ground effect." It was first used in a Formula 1 racing car. A description of its function follows in section 9.7.4. Due to the dramatic increase in lateral accelerations that were possible with these vehicles, the physical strain on the driver escalated strongly. In 1983, the Formula 1 regulations changed to prescribe a flat underbody running along a plane between the axes, so that the downforce was limited and therefore the achievable lateral acceleration was reduced. In 1988, the Porsche 962, with ground effect, raced at Le Mans (Figure 9.12). Finally, in the 1990s, the under-floor effect for these vehicles was also restricted by the regulations.

Figure 9.12 The Porsche 962, a racing car equipped with ground effect, which raced at Le Mans in 1988. With a 529 kW (720 hp) engine, the car reached 385 km/h. Photo Dr.-Ing. H.C. F. Porsche AG.

The result of these rule changes was the restrengthening of development work into the form and location of the front and rear wings. Particularly in Formula 1, an extreme effort has been made to provide sufficient downforce. An example is shown in Figure 9.13. True works of art of wing design, with up to nine different elements, were developed. Details will follow in section 9.7.2. Finally, in 2001, the Formula 1 regulations were changed to restrict this: A maximum of only three wing elements were allowed. The result was that less downforce could be produced at the rear axle. To balance this, a certain minimum distance was set for the front wing from the ground, to reduce downforce at the front axle. "Exotic" forms of front wing were then tried, to compensate for this regulations-related disadvantage.

Figure 9.13 Formula 1 racing car BMW-Williams 2004. Photo BMW AG.

Even in other series, such as in the German Touring Car Championship (DTM), clear limits for the aerodynamics were set within the rules, mostly concerning rear downforce. The permissible wing span and platform area for the wings were limited and the airfoil profile was prescribed. This was done in order to keep the loads on man and materials within tolerable limits and to ensure safety. The trend of developing ever higher lateral accelerations should be compensated by the regulations. The success of this is evident from Figure 9.14, in which the lateral acceleration attained, is plotted versus time. The notches show clearly the effect of this regulatory measure, but it can also be seen how quickly the respective changes of the regulations have once again been offset by other technical means.

High-Performance Vehicles

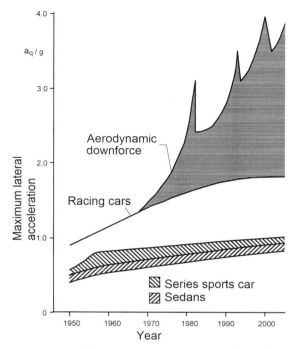

Figure 9.14 Time evolution of the maximum lateral acceleration of sedans, sports, and racing cars.

9.2.2 Record-Breaking Vehicles

9.2.2.1 Maximum Speed

Even before the turn of the century, the importance of aerodynamics was recognized for record-breaking vehicles. At first, this term was understood always to describe such vehicles that one could drive as fast as possible. The first car driven faster than 100 km/h was the La Jamais Contente by C. Jenatzy from 1899, shown in Figure 1.9 The vehicle, powered by an electric motor, had a cigar-shaped body, whose shape was obviously influenced by the airship.

Subsequent record-breaking cars were then fitted with more or less streamlined bodies. Very soon aerodynamic aids were also used for stabilization. Thus, the Opel Rak 2 rocket car from 1928 (shown in Figure 1.43) was the first vehicle equipped with two horizontal wings to generate downforce. Vertical tail fins were used to increase the longitudinal stability. The Golden Arrow, with which Henry Segrave reached a speed of 372.456 km/h in 1929, represents the first example of this (Figure 9.15).

Figure 9.15 Record Car Golden Arrow from the year 1929. Maximum speed 372.456 km/h. Image: Auto, Motor & Sport.

A proposal was made to compensate for crosswind deviation by using automatic fins, as sketched in Figure 9.16, and Kamm showed the way toward "full stabilization" of vehicles. In the Mercedes-Benz T80 of 1939, for the first time a streamlined body, vertical tail fins, and wings generating downforce (Figure 9.17) were combined. Unfortunately, the car was never used due to the start of the Second World War. With 2200 kW (3000 hp) available, the car should have achieved over 650 km/h.

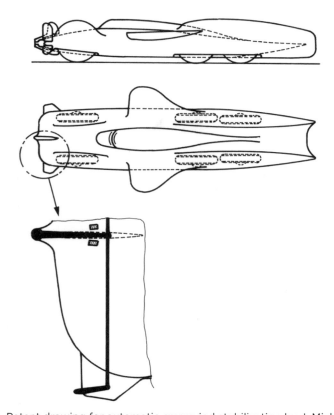

Figure 9.16 Patent drawing for automatic crosswind stabilization by J. Mickl from 1939.

High-Performance Vehicles

Figure 9.17 Mercedes-Benz T80 record car of 1939. Photo Mercedes-Benz AG.

In 1947, John Cobb attained a new record of 634.386 km/h, which was not surpassed until the 1960s, with the Railton Mobil Special. The world record for wheel-driven vehicles was set by the Golden Rod in 1965 at 658.649 km/h. Between 1963 and 1970, the absolute speed record for reaction-driven land vehicles was raised to about 1000 km/h. The LPG rocket-powered car Blue Flame (Figure 9.18) of 1970 reached a speed of 1001.671 km/h. Stan Barret broke through the sound barrier in 1979 at 1190.23 km/h with Budweiser Rocket, a rocket car similar to Blue Flame, but the record was not recognized. In 1983, a new record of 1019.7 km/h was achieved by Richard Noble in the gas-turbine-powered vehicle Thrust 2. The current speed record over 1 mile (1.609 km) was established in 1997 by Andy Green in his Thrust SSC (Super Sonic Car), propelled by two jet engines. He reached a speed of 1227.985 km/h in the Black Rock Nevada desert (Figure 9.19).

Figure 9.18 Rocket-driven record vehicle Blue Flame. In 1970 it reached a speed of 1001.671 km/h. Image Auto & Technik Museum Sinsheim.

Figure 9.19 The Thrust SSC holds the current speed record with 1227.985 km/h. Image G.T. Bowsher.

9.2.2.2 Transonic Speeds

In the design of record vehicles for transonic speeds, the compressibility of the air must be considered, and special design measures are required to accommodate the displacement of the point of action of lift or downforce, respectively, when passing through the speed of sound.[2] From preliminary investigations it was determined that between a Mach number of Ma = 0.5 and Ma = 1.2 the drag coefficient would increase by 200%.[3] This increase, as Figure 9.20 (left) shows, is fundamentally influenced by the thickness ratio d/l of the body.

For this reason, the body of Blue Flame was designed as an extremely slender ovoid body. Cross-section changes caused by protrusions or cavities produce shock waves at supersonic speeds that significantly increase drag. The so-called "area rule" therefore recommends a continuous increase in the cross-sectional area from the front of the vehicle to its tail. It is only of secondary importance, if this cross section is composed of one body (e.g., a fuselage) or more bodies (fuselage, chassis, engine nacelles, etc.). In order to decrease the frontal area, the nose wheels are integrated in the body. The relationship of the drag coefficient versus Mach number is plotted in Figure 9.20 (right).

---

2. They were discussed by Torda & Morel for the example of the Blue Flame.
3. This is the so-called "sound barrier."

High-Performance Vehicles

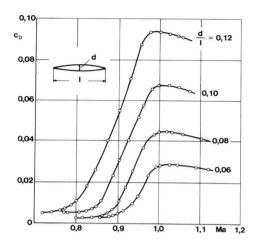 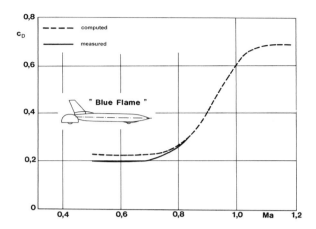

Figure 9.20 Left: Drag increase versus Mach number for a symmetrical airfoil as a function of the thickness ratio, after Malavard [529]. The slimmer the profile, the smaller is the drag increase during passage through the speed of sound. Right: Drag coefficient of the Blue Flame as a function of Mach number, after Torda [806].

To ensure the control of the vehicle at any speed, lift was generally avoided. However, be aware that when passing through the sound barrier, a new phenomenon occurs and was mentioned briefly above: The negative pressure under the car becomes positive and creates lift. The explanation for this is that in the subsonic range, as the air between the vehicle underbody and the road surface is accelerated, a vacuum is created. In the supersonic range, however, a shock wave in generated from the vehicle nose. Behind this, a higher pressure prevails and causes lift. The change in the sign of the lift force as the speed of sound is transgressed changes the angle of attack of the vehicle and can lead to instability.

Since this phenomenon occurs particularly strongly with planar underbodies, the vehicle was given a rounded triangular cross section, wherein one of the corners faces the ground. In addition, the longitudinal axis of the vehicle was set at a negative angle of 1.5° and the nose designed to be slightly sloping. The trim of the vehicle is carried out by small fins, so-called "canards" (literally, duck wings), at the front end. The steering stability was improved by a vertical fin at the stern. Through these measures, Blue Flame behaved stably up to the transonic range.

9.2.2.3 Other Record Goals

The "other" record goals, other than just high top speed, are varied: Consumption, range, alternative propulsion, and so on. The forms to which they have led are accordingly manifold.

In response to the unilateral requirement for maximum safety developments laid down under the framework of the U.S. program of experimental safety cars (Experimental Safety Vehicle, ESV) also a variety of record breaking cars—a series of research and concept vehicles was developed in the late 1970s and early 1980s. The focus was thereby turned back to cars that included a balanced, customer-relevant shape among their characteristics. Particularly outstanding was the UNICAR developed by some German universities (vgl. Figure 4.138), which had a drag coefficient of $c_D = 0.25$. In the 1990s this again led to developments toward a priority objective, namely the lowest possible drag coefficient.[4] The Ford Probe V ($c_D = 0.15$), GM Aero 2002 ($c_D = 0.14$), and Pininfarina CNR E2 ($c_D = 0.19$) were the outstanding results.

From time to time record vehicles are built to explore and expand the boundaries of certain attributes. In these cases, absolute top speed is not the goal. More often, it is, for example, to achieve a given top speed with a particular engine, to minimize the consumption, or to maximize the range achieved on a given quantity of fuel. It is generally accepted that these cars are very far removed from the current production status. It is common in the development of such vehicles that there is usually very little time available and the objective is only achievable with special organizational measures: formation of a small, multifunctional project team supported by an appropriately skilled project manager.

The Mercedes Benz C III 111 (1978) represents a "classic" example of such a record-breaker. (Figure 9.21).[5] This car was used as a test vehicle for the Wankel rotary engine. When evaluating its very low drag of $c_D = 0.18$, two details must be considered:

- The large ratio of length to height (compared to standard), of $l/h = 4.94$
- The fairing within the body of both the front and rear wheels

From a comparison with contemporary sedans (which is performed in Figure 9.21b) using the longitudinal center section, one recognizes that in form the C 111 has much in common with a sports car.

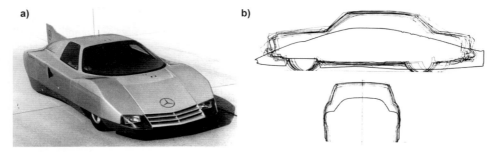

Figure 9.21 Mercedes-Benz C 111 III: a) view; b) contour comparison with contemporary production cars [498].

---

4. As reported in previous editions of this book.
5. Liebold et al. (1978, [498]).

With the ARVW[6] experimental vehicle presented in 1982, Volkswagen AG demonstrated what is achievable in a vehicle, when an extremely low air resistance is combined with a relatively low engine power. With $l/h = 5.93$, the ARVW was even slimmer than the C111. How far removed it is from what is possible in series production is shown in Figure 9.22. With $c_D = 0.15$ the drag coefficient of previous record cars was undercut; the frontal area amounted to 0.73 m2. The vehicle reached a top speed of 360 km/h with an engine power of only 129 kW, and the fuel consumption of the diesel engine amounted to only 13.6 l/100 km.[7]

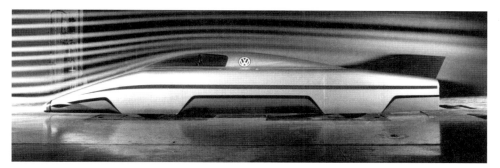

Figure 9.22 ARVW research car from Volkswagen [610].

A car with which extremely low fuel consumption was realized, the Sparmobil from VW, can be seen in Figure 9.23 With $A_x = 0.32$ m² and $c_D = 0.15$, a very low drag was achieved. The Sparmobil successfully set a world record: A distance of 1491.3 km was covered using 1 liter of diesel fuel, and the average speed amounted to 16.9 km/h.

Figure 9.23 VW Sparmobil 1982 $c_D = 0.15$, $A_x = 0.32$. Photo and data Volkswagen AG.

The VW 1L, the 1-liter car of Volkswagen AG, is a very different kind of record car (Figure 9.24). With a slenderness ratio of $l/h = 3.56/1.1 = 3.2$ it is close to a "normal"

---

6. Nitz et al. ([610], 1982); ARVW stands for Aerodynamic Research Volkswagen.
7. From the aerodynamic perspective, neither the C111 nor the ARVW represent meaningful developments. That low drag can be achieved with slender bodies needs no further proof. As is generally known, the "art" lies in achieving it with a bluff body.

car; the early expansion of the cabin form, which begins at the level of the front axle, as well as the gently rounded silhouette with sloping rear, enabled a very low drag with $c_D = 0.15$ and a frontal area of 1 m². As with the legendary Messerschmidt bubble car, the driver and passenger sit one behind the other. Occupied by two people, this car was driven a distance of 237 km with an average speed of 95 km/h, the consumption being 0.99 liters of diesel per 100 km.

Figure 9.24 One-liter car from Volkswagen in 2002, $A_x = 1\,m^2$, $c_D = 0.159$. Photo and data Volkswagen AG.

To prove the effectiveness of a new generation of diesel engines, Adam Opel AG constructed a record vehicle and the special characteristics of this engine were demonstrated under extreme conditions: Power, torque, and fuel consumption. Thus, it was proven that sportiness and low fuel consumption can be very well realized in one and the same vehicle.

The shape of the said Eco-Speedster vehicle, unveiled at the Auto Salon in Paris in 2002, was derived from the sporty Opel Speedster.[8] In Figure 9.25, both types are compared. Because of the short time available, the aerodynamic development was mainly carried out numerically. A particular focus was the optimization of the cooling airflow. The achieved values for drag and lift are shown in Table 9.1, comparing measurement and calculation together. While a good agreement was shown for the drag coefficient, it was not the case for the lift.

---

8. Kleber [18] has reported on this metamorphosis.

Table 9.1 Comparison of Measured and Computed Drag and Lift Coefficients, Frontal Area 1.36 m².

	Measurement 1:1	CFD
c_D	0.213	0.214
$c_D \times A_x$ / m²	0.290	0.291
$c_{L,f}$	-0.111	-0.225
$c_{L,r}$	0.073	0.025

With the 1.3 L 82 kW CDTI diesel engine, the target values were achieved: top speed 250 km/h, consumption (with respect to the NEDC) 2.5 l/100 km.

Figure 9.25 Opel Speedster: (left) 2003 production; (right) 2003 Eco-Speedster record vehicle. Photos Adam Opel AG.

Solar cars represent a very special type of record car,[9] for which the aerodynamic requirements are extreme. In order to achieve a minimum resistance, laminar flow is required over the longest possible portion of the upper-body surface. An example is the vehicle Spirit of Biel-Bienne III from 1993, developed by students of the Technical University of Biel-Bienne. Figure 9.26 shows version II. With a frontal area of $A_x = 1.1$ m² it had a c_D of 0.105, and the surface covered with solar cells was 7.9 m². The Flying Dutchman Numa II (Figure 9.27), developed by students and faculty of the Technical University of Delft and the Erasmus University of Rotterdam, won in 2003 in Australia. The distance of 3010 km was covered at an average speed of 96.8 km/h, the top speed was $V_{MAX} = 110$ km/h.

Figure 9.26 Solar car Spirit of Biel-Bienne. Photo University of Biele/Bienne.

---

9. Tamai [48] presented a comprehensive account of the aerodynamics of solar vehicles.

Figure 9.27 Solar car of the universities of Delft and Erasmus in the German Dutch Wind Tunnel DNW-LST. Photo Technical University of Delft and the Erasmus University of Rotterdam, 2003.

In such extreme vehicles, comfort, handling, and driving characteristics are subordinate to reaching a low drag. Likewise, other aerodynamic parameters such as lift or downforce, crosswind sensitivity, and aerodynamic balance are of only secondary importance.

9.2.3 Sports Cars

Sports cars for road use were subject to development similar to racing cars. Before the war, the car companies preferred forms with large coolers and separate fenders, which today are often labeled as classic. The contribution of aerodynamics was at first ignored. The Mercedes-Benz 720 SSK of 1928 shown in Figure 9.28 serves as a good example, with a drag coefficient of $c_D = 0.91$ and a frontal area of $A_x = 1.57 \text{m}^2$.

High-Performance Vehicles

Figure 9.28 Mercedes-Benz 720 SSK from 1928. Drag coefficient $c_D = 0.91$ at 1.57 m² frontal area. Image, Daimler-Benz AG.

Only ten years later, the first aerodynamic optimizations in the bodywork shape were already clearly visible. These noticeable aerodynamic measures would later become the defining design features of the sports car. A streamlined body was developed and based on the DKW F8 (Figure 9.29). The 1939 vehicle developed 50 hp with a 700 ccm supercharged engine. Only three of these vehicles were built before the project was canceled.

Figure 9.29 DKW Streamliner, based on the DKW F8. Only three examples of this vehicle were built, in 1939. Picture Audi AG.

In the developments made in the immediate aftermath of the Second World War, great emphasis was placed on low drag coefficient and small frontal area. The open-top Porsche 356/1 of 1948 (Figure 9.30) had a drag coefficient of only $c_D = 0.46$ with a 1.41 m² frontal area. The closed Porsche 356 A of 1950 (Figure 9.31) even reached $c_D = 0.28$ with a frontal area of $A_x = 1.68$ m². However, the lift coefficient of $c_L = 0.26$ was very high. This value could be tolerated, because the top speed was not high enough to noticeably affect the driving characteristics.

Figure 9.30 Porsche 356/1 from the year 1948. The drag coefficient was c_D = 0.46 with 1.46 m² frontal area. Photo Dr. -Ing. H.C. F. Porsche AG.

Figure 9.31 Porsche 356 A, known internally as "Ferdinand," from the year 1950. Drag coefficient c_D = 0.28 with 1.68 m² frontal area. Photo Dr. -Ing. H.C. F. Porsche AG.

In the following years, efforts to improve handling through elaborate suspension, wider track, and wider tires and at the same time to provide occupants with more comfort through increased interior space led to larger frontal areas and higher drag coefficients. The increasing cooling air requirements demanded by higher engine outputs also played a role.

In the 1970s, front and rear spoilers were also introduced to reduce drag and lift in the production sports car. The Porsche 911 Turbo (Figure 9.32) illustrates this approach. At the end of the 1980s, the application to sports cars of trends and knowledge from racing aerodynamics was accelerated. The Porsche 959 (Figure 9.33), with covered underbody and integrated rear wing, is given as an example here.

High-Performance Vehicles

Figure 9.32 Porsche 911 Turbo from 1983. With 221 kW (300 hp) of power, the car reached 260 km/h. The drag coefficient was $c_D = 0.40$ with 1.87 m² frontal area. Photo Dr. -Ing. H.C. F. Porsche AG.

Figure 9.33 Porsche 959 from 1987 with integrated rear wing and enclosed underbody. The drag coefficient was $c_D = 0.31$ with 1.92 m² frontal area. Photo Dr. -Ing. H.C. F. Porsche AG.

In the 1990s, the prevailing fashion was for more visually discreet aerodynamic accessories. To nevertheless ensure delivery of the aerodynamic characteristics necessary for performance and drivability, it became appropriate to use spoilers or wings that extended only at high speeds—and automatically. An example of this is the Lamborghini Gallardo (see Figure 9.34).

Figure 9.34 The 2003 model year Lamborghini Gallardo in the Audi aeroacoustic wind tunnel; retracted spoiler: $c_D = 0.331$; $c_{LF} = 0.036$; $c_{LR} = 0.137$; spoiler extended: $c_D = 0.353$; $c_{LF} = 0.039$; $c_{LR} = 0.042$. Images Automobili Lamborghini.

At the end of the 20th century, some manufacturers began to develop so-called "super sports cars," which were to be brought to market incorporating raw racing technology. Trademark features for such vehicles include reduced occupant comfort, a leaning toward the visual cues of Formula 1 racing cars, and sophisticated aerodynamics. So diffusers were provided front and rear, as adjustable aerodynamic elements. Typical is the Ferrari F60 Enzo (Figure 9.35). It is equipped with adjustable diffusers ahead of the front wheels and a movable rear spoiler. This allows the lift as a function of speed to be regulated in such a way that the aerodynamic balance of the car is held within a desired range.

Figure 9.35 Ferrari F60 Enzo with movable front diffuser and adjustable rear spoiler for adjusting of the lift behavior to the driving speed. Photo Audi AG.

While the adoption of active systems at the rear has been widely applied in almost all vehicle categories and price segments, active systems at the front of the vehicle are usually found only on high-priced performance sports cars. The Porsche 918 Spyder, for example, illustrates further developments of these systems. Here, two front diffusor channels can be used to adjust the vehicle character and enable the optimal aerodynamic balance to be achieved in both the low drag configuration at moderate downforce and in the performance configuration at maximum downforce. The scarp angle is tuned according to the sporty vehicle segment and is not variable.

Figure 9.36 Porsche 918 Spyder front diffusers in "standard" and "performance" positions.

9.3 Vehicle Classes

Currently, there are a multitude of vehicle classes, with which various competitive racing series are staged. The most well-known race series in Europe, Formula 1, occupies only a fraction of the total motor sport activities. Racing vehicles can be assigned to the following groups: First, those known as Formula racing cars. This includes essentially all cars with open wheels, so Formula 1, GP2, Formula 3, Formula 3000, Indy Racing League (IRL), Champ Cars (CART), and junior formula manufacturer's series, such as Formula BMW Junior or Formula Volkswagen.

The touring cars are a second group. In this class standard vehicles (e.g., sedans or coupes) are used as a starting point. Depending on the regulations, a greater or lesser degree of modification is allowed. To this group belong the German Touring Car Championship (DTM), the European Touring Car Championship (ETCC), as well as their national offshoots (e.g., BTCC in England, DTC in Germany, STCC in Sweden, etc.).

The third group is of the so-called endurance sports cars, which in turn are divided into several classes, such as prototypes, Grand Touring Cars (GT), and so on. The events range from two-hour sprint races up to the 24-hour endurance classics. Further divisions, for example Cup cars (single manufacturer racing series), rallying, historic racing events, and so forth will not be discussed here.

For the first three groups, the performance of these vehicles and their corresponding series are compared:

- From the group of Formula cars, the data of the Sauber Petronas C20 Formula 1 car from the year 2001. Figure 9.37 shows the vehicle at the Grand Prix of Malaysia. Due to the very different aerodynamic requirements required by driving on ovals, two different configurations of a racing vehicle from the Indy Racing League are selected. The two Dallara chassis from 2003 are designed for short (Figure 9.38) or long oval tracks and have a remarkably wide range of aerodynamic performance. The third representative of this group is a Formula 3 car, also from Dallara in 2001 (Figure 9.39).

- The touring cars group is represented by a DTM car. The Abt-Audi TT-R raced in and won the championship in 2002 (Figure 9.40).
- The endurance sports cars are represented by the Audi R8 sports car prototype of 2002 (Figure 9.41). In this year, the car won the classic 24-hour endurance race at Le Mans for the third time in a row, with a 1-2-3 finish.

Figure 9.37 Sauber Petronas C20 Formula 1 racing car in Malaysia 2001. Photo Sauber.

Figure 9.38 Short oval variant of the Dallara IRL race car. Image Dallara Automobili SRL.

Figure 9.39 Dallara Formula 3 racing car from 2001. Image Dallara Automobili SRL.

Figure 9.40 Abt-Audi TT-R touring car from the German Touring Car Championship (DTM), 2002. Photo Audi AG.

Figure 9.41 Audi R8 endurance sports car from 2002. Picture Audi AG.

For better performance comparability of the vehicles referred to, a generic vehicle is defined for the following considerations, which has approximately the characteristics of an aerodynamically optimized series sports car. In detail, this means a drag-area product of $c_D \cdot A_x = 0.66$ m² (i.e., about $A_x = 2.0$ m² with a drag coefficient of $c_D = 0.33$) and neither lift nor downforce (i.e., $c_L = 0$). The vehicle mass is assumed to be 1400 kg and the power at about 367 kW (500 hp).

First, the ratio of engine output is compared to vehicle mass, as described in Figure 9.42. The power-to-mass ratio of the generic standard sports car is about 0.26 kW/kg, (i.e., 0.26 kW are available to accelerate 1 kg of vehicle mass). With the assumed drag-area of 0.66 m² a top speed of around 320 km/h can be achieved. The Formula 3 racing car has a similar ratio of engine output to vehicle mass at 0.29 kW/kg. Due to the lower vehicle mass and correspondingly smaller engine, the acceleration capability of this race car is roughly comparable to that of the generic standard sports car. The top speed is 260 km/h, however, due to the much lower power.

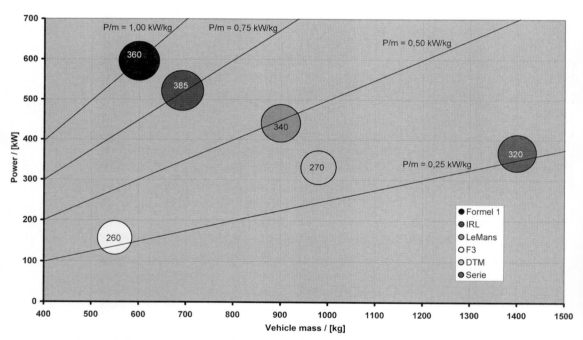

Figure 9.42 Power-to-mass diagram for selected racing vehicles, compared with a production sports car (indicating the approximate maximum speed).

The DTM race car has a power-to-mass ratio of about 0.34 kW/kg, but due to the relatively high aerodynamic drag only achieved a top speed of 270 km/h, which is comparable with that of the Formula 3 car. Much more powerful are the prototypes of the group of endurance sports cars. Here there are in the region of 0.50 kW available to accelerate each kilogram of vehicle mass. In contrast to the DTM cars, this car is designed more for high average speeds (the Le Mans race track, for example) and can reach speeds of over 340 km/hr in race trim.

The most efficient vehicles are represented by the IRL race car (P/m approximately 0.75 kW/kg) and the Formula 1 racing car (P/m approximately 1.0 kW/kg). For a low weight, a high-performance engine is installed, which can propel a Formula 1 car with up to 2.5 times the acceleration due to gravity. However, the fastest cars are the IRL race cars that run with low drag on fast oval tracks.

Figure 9.43 shows the vehicles' relative aerodynamic characteristics. The lift-area (i.e., $c_L \cdot A_x$) is plotted against the drag-area $c_D \cdot A_x$. For each vehicle, a certain adjustment range is indicated, encompassing the effects of tunable aerodynamic devices (e.g., wing elements, etc.). The average operating point is labeled with a larger symbol.

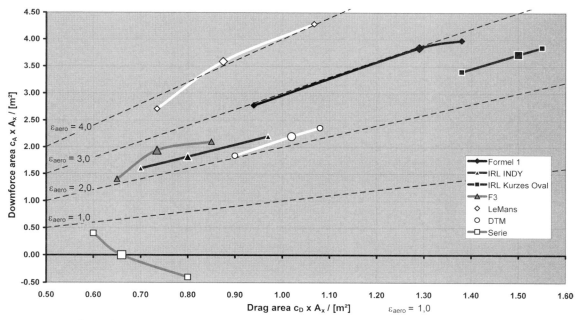

Figure 9.43 Representation of the drag- and lift-areas ($c_D \cdot A_x$ and $c_L \cdot A_x$) and their adjustment ranges for different racing vehicles, compared with a generic standard sports car.

The generic sports car with $c_L \cdot A_x = 0.00$ and $c_D \cdot A_x = 0.66$ represents an aerodynamically good state with respect to downforce and drag. The majority of existing sports cars have a similar drag value, however, generally produce lift. It is critical to consider the effect of lift on driving dynamics, particularly on the rear axle. Drag is generally increased by attempts to induce downforce on the rear axle, for example using a spoiler or rear wing. The uppermost curve in Figure 9.43 implies a direction in which series vehicles could develop. However, the trend to produce sports cars without lift is discernible for all manufacturers.

The Formula 3 race car displays a comparable drag-area value to the series sports car. However, at these resistance values significant downforce is generated. The aerodynamic efficiency (for the definition see section 9.6.5) is approximately $\varepsilon_{aero} = 2.5$. Although the power-to-mass ratio and drag are comparable with the values of the standard sports car, the driving performance with respect to braking behavior and cornering speed is considerably better than the production car as a result of aerodynamic downforce. The highest aerodynamic efficiency is generally achieved at the design point. Typical of the adjustment extremes are the interpretations for fast tracks, with low downforce and drag traded off against a compromise in aerodynamic efficiency, and so-called "high-downforce tracks" (i.e., courses with tight corners and low average speeds) at which compromises are made with respect to drag and efficiency in favor of downforce.

The drag of DTM cars, with a drag-area of up to $c_D \cdot A_x = 1.10$ m², is almost twice as high as for the production vehicles that they are based on ($c_D \cdot A_x = 0.60$ to 0.65 m²). The DTM uses the same circuits that are used for Formula 3. Since the DTM cars have a much greater power-to-mass ratio than the cars in Formula 3, their higher drag is not particularly disadvantageous. The relatively high aerodynamic drag is further accounted for by the strong restriction of aerodynamic freedom in the regulations. Rear wing position and even the use of standard wing profiles are mandatory for all vehicles. Modifications to the vehicle body, due to the desired similarity to production cars, are authorized only in specifically designated areas (e.g., wheel arches, underbody).

The two IRL vehicles, despite sharing the same chassis (as prescribed in the regulations), display a remarkably large range of aerodynamic adjustment (section 9.7). For the high-speed ovals (Indianapolis, Lausitz Ring, etc.), aerodynamic drag-areas in the range of production sports cars are realized. With high engine power outputs, peak speeds of over 380 km/h can be reached. The race series is thereby the fastest in the world. The downforce, with an efficiency of just under 2.5, assists in the achievement on the banked curves of the oval of a lateral acceleration of up to 3.5 times the acceleration due to gravity. For the short oval tracks that have extremely steep curves, the vehicle can be modified so that drag and downforce are doubled. The maximum achievable speeds are reduced by 80 km/h, to 300 km/h. The lateral accelerations can be up to 4.5 g then, however. In this series, longitudinal accelerations are of only minor importance, since only slight braking is used before the turns. The downforce is required exclusively in order attain the high lateral accelerations.

The drag of a Formula 1 car lies somewhere between the values of the two IRL cars. The downforce level of the high downforce configurations is roughly comparable to the short oval IRL car. However, the aerodynamic efficiency is somewhat higher at about $\varepsilon_{aero} = 3.0$. Again, the big adjustment between fast racetracks such as Monza, and the slow, high downforce tracks (Monaco and Hungary) is apparent. However, unlike the IRL cars, accelerating and braking maneuvers are of crucial importance to the competitiveness of the vehicle. Formula 1 cars achieve deceleration values of up to 4.0 g. The

High-Performance Vehicles

lateral acceleration can still be up to 3.5 g. Again it is apparent that for high downforce tracks compromises in drag and efficiency must be made in favor of downforce.

The prototypes of the endurance sport cars display the highest aerodynamic efficiency compared with other vehicles in Figure 9.43 The regulations allow appropriate freedom here. Their interpretation with respect to drag performance is similar to the lowest achieved by the Formula 1 cars. The main focus, when interpreting these rules, is generally the fast Le Mans racetrack. Higher downforce variants of these vehicles are driven in the FIA Championship and the American Le Mans Series. The downforce achieved is of similar magnitude to that of Formula 1. However, the transverse and longitudinal accelerations do not reach the level of Formula 1 cars due to the higher vehicle weight and reduced engine performance. Figure 9.44 summarizes the performance of the vehicles mentioned with regard to both the achievable longitudinal and lateral accelerations. The IRL vehicles for the short oval are particularly extreme, since they are designed for highest possible lateral acceleration, longitudinal acceleration not being important.

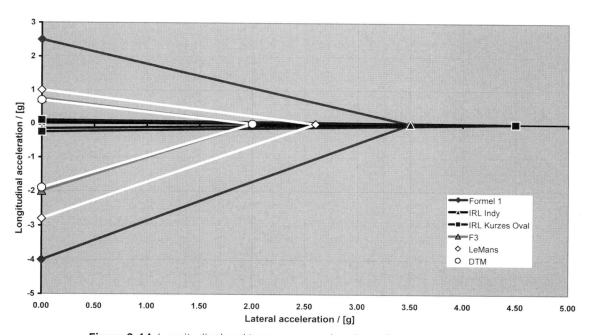

Figure 9.44 Longitudinal and transverse accelerations of selected racing cars.

9.4 Race Tracks

The variation between different race tracks will be illustrated very briefly using a selection of Formula 1 Grand Prix circuits. They are compared with each other in Figure 9.45 with reference to some typical data.

	Monza	Monte Carlo	Nürburg	Hockenheim
Circuit	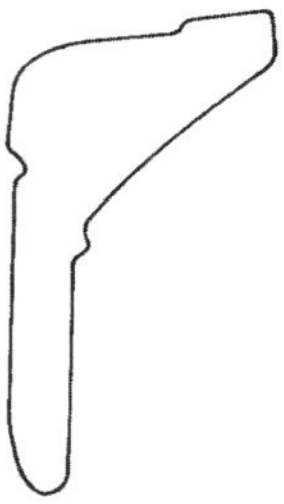	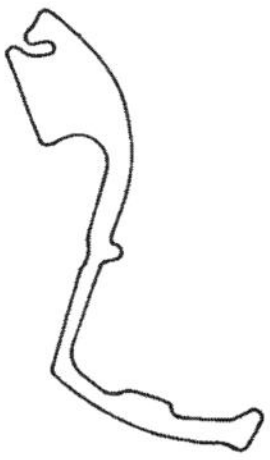		
Length km	5,973	3,340	5,148	4,574
$\bar{V}$ km/h	250	155	190	194
V_{max} km/h	359	298	310	310

Figure 9.45 Some selected Grand Prix race tracks, after Rennsport News [666] (2004). The tracks are not drawn to scale relative to each other.

At Monza, the only remaining very fast track in Formula 1, the very highest speeds are achieved, both on average and as a maximum. Three chicanes, four fast corners, and long straights characterize this track. The proportion of the track driven at full-throttle is unusually high at about 70%. Low drag, combined with high engine power and an especially pronounced stability of the vehicle when driving over the curbs in the chicanes are the key success factors here.

In contrast, the average speed in Monte Carlo, the slowest course, is low. Nevertheless, peak speeds of almost 300 km/h are achieved. The track is mainly characterized by a series of sometimes very slow corners. The Grand Hotel curve for example, is driven in first gear at only 40 km/h. It is so tight that specially adapted steering systems are used here, which are not deployed at any other circuits. Furthermore, the road course is extremely bumpy, so that increased ground clearances have to be used, which does not allow "ground-effect" to be exploited to the usual extent. As a consequence, front and rear wings are set at steep angles to compensate for the downforce loss from the increased ground clearance, at least partially. The fact that this increases the drag is accepted in exchange for a faster lap time.

The courses at Hockenheim and the Nürburgring look very different, but nearly the same values for both the maximum and average speed are achieved. Yet here also, clear differences in tuning are driven by the need to accommodate individual sectors, especially the corners, and to trim the vehicle for highest performance on these courses.

From the perspective of aerodynamics the tuning of a race car to each of these courses means that the optimal pair of values of downforce and drag has to be determined from the measured wind tunnel polars. The adjustment of the wings and flaps, and other items, to generate more favorable pairs of values for predetermined sections of track is prohibited during a lap by the regulations. That is to say that significantly better lap times would be achieved if the downforce in braking and curve phases could be set particularly high and on the straights the drag set especially low (e.g., by adjusting the wing elements depending on the speed, steering angle, longitudinal and lateral acceleration). Sensibly, this level of technical complexity has been limited by strong regulation in almost all vehicle classes. Even the allowable deformations of related components (such as front and rear wings) under wind load are prescribed to at least tightly restrict such variability. Improvements of the tune during the race are usually permitted at pit stops, at the expense however of lost time.

How such an optimization may be performed has been demonstrated by Potthoff [648]. For a generic vehicle, the lap time was calculated on a course that consisted of an equilateral triangle with rounded corners. At constant overall size, the length of the straights and the radii of the curves were varied. It was demonstrated thereby that with increasing length of the straights, the best lap times were obtained by decreasing downforce—and so also decreasing drag—whereas with increasing curve radii, increased downforce was required and thus higher drag acceptable.

In practice though, a racing car cannot be tuned with these ground rules alone, valid as they are. On the contrary, for real race tracks, which, as the sketches in Figure 9.45 show, are composed of totally different blends of straights—with chicanes—and of tight and wide curves, such calculations for the c_L, c_D pair have to be iterated numerous times to find an optimum that may then be validated and possibly further improved on the race track. Thereby, drivability criteria ("handling"), curve entry and exit speeds, deceleration ability ("out-braking"), location, and availability of convenient "overtaking zones," position in relation to competitors, and so forth, all play a role in co-determining the final tune for the race track. This may be so pronounced that theoretically slower lap times are accepted, for example, so as to be in a particularly good position to overtake, or not be overtaken, at a crucial point of the race track. The aerodynamic tuning is also influenced by a variety of other parameters of a technical or tactical kind, driver ability, and other topics which will not be further discussed here.

9.5 Regulations

Regulations form the technical foundation that is the basis of performing modifications to production cars, or which must be observed in the new construction of prototypes or vehicles that are not based on series vehicles.

The highest priority is the safety enjoyed by the driver. For this purpose, exacting rules are adopted as to how the safety cell (roll cage, monocoque) is to be constructed. A series of crash tests—front, side, and rear impact and rollover of the vehicle—is defined, with exactly prescribed test loads and their effective direction. Passing this test is a requirement to participate in competition. The sharp fall in the number of serious driver injuries in recent years is a result of the very stringent regulations in this area.

In general, the rules are based on the guidelines developed by the International Motor Sports Association Fédération Internationale de l'Automobile (FIA); they are adopted by the national associations. Specifically for the cockpit, they go so far that the outer and thus aerodynamically relevant form is significantly influenced in the rules. Aerodynamic optimization is mainly confined to the streamlined casing of the safety structures and their inclusion in the overall aerodynamic concept of the vehicle (see Figure 9.46). Front and rear crash structures essentially present a problem in terms of the available design space and the arrangement of cooling elements and air ducts, as well as for the attachment of wing elements. Depending on the design of the vehicle, these elements also determine the outer shape.

Figure 9.46 Mercedes McLaren Formula 1 racing car, 2005. Photo DaimlerChrysler AG.

Also in terms of safety, structures are restricted in terms of allowable deformations of or even movements of aerodynamic components while driving. For this purpose there are prescribed tests in which components under load may deform only within strict, narrow limits. This is especially true for wing elements and their connection to the vehicle. Active adjustment of any aerodynamic elements is prohibited in all current regulations (except specifically in the case of DRS in Formula 1, see below) for reasons of safety.[10] The same applies to elements that can actively affect stability in the transverse direction (rudder or similar) or vertical direction (e.g., amplification of downforce when driving over knolls).

The performance of the vehicle is a further safety aspect that is closely regulated. This can be done in one way by limiting engine performance, but also by means of aerodynamic regulation, such as the use of trim (or Gurney) strips of certain sizes to rear wings to regulate the top speed. Wing widths, the mounting height of wings, the number of elements used, or even the profile shape are mandatory in some regulations to specify

---

10. For example, reduced wing angle of attack while driving on straights, in order to realize greater speed, and conversely increased wing incidence while under braking or cornering in order to generate more downforce.

certain limits for downforce and drag. The purpose of limiting the vehicle performance is to maintain the strain on the driver within tolerable limits. The design of the underfloor is currently severely limited in almost all regulations (to e.g., a flat underfloor between the axles, or precise specification of the shape) to hold downforce within limits and to ensure aerodynamic stability of vehicles with proven, standardized shapes, particularly at high speeds. One example is offered in Figure 9.47.

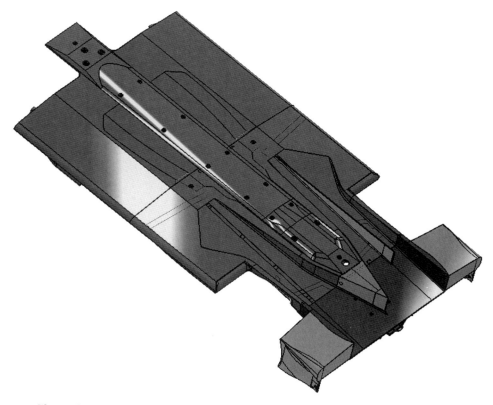

Figure 9.47 Draft regulation concerning the underfloor for sports cars for 2014.

Besides the safety aspects, equal opportunities for participating vehicles are an important aspect of the regulations. Racing will only be perceived as exciting and worth viewing by the audience if the highest possible number of participants have an equal chance to win. If a series is dominated by one vehicle type or manufacturer, measures to moderate the vehicle will be earmarked in the regulations. Aerodynamic constraints, such as the adjustment of performance-related components, for example the height and width of wings, are also a means of achieving this. An annual adjustment of regulations is necessary in order to limit the technical advances, reached through vehicle development, to a reasonable improvement in performance. Undesirable or dangerous trends can thus be stopped or controlled, to the chagrin of the inventor.

The rule changes are therefore not exclusively restrictive but can lead to quite technically complicated constructions, the implementation of which is binding. The following examples are given:

- The Formula 1 adjustable rear wing (DRS = drag reduction system) to improve overtaking
- Adjustment of the front wings during Formula 1 races, to control the aerodynamic balance for safety reasons
- Large fins at the rear of the vehicle to stabilize it when turning, driving in a slipsteam, or strong crosswinds in almost all racing series

Figure 9.48 DRS system open and closed on RB7; adjustable front wing on Michael Schumacher's Mercedes GP Formula 1 car, 2010; 2012 Audi R18 e-tron with pronounced "stabilizing fin."

Cost aspects also play a certain role in the drafting of the regulations. The racing series should remain affordable for the participants, even if there is there a considerable spectrum (e.g., Cup series through to Formula 1). The arms race in search of performance can also be controlled via regulations regarding aerodynamics. This can be done, for example, through prohibition of further aerodynamic development during the season (fixing or homologation of components at the beginning of a season that will then no longer be allowed to be changed) or be realized by use of prescribed common or "control" parts.

The design and development of a racing vehicle always starts with an intensive study of the regulations. The restrictions, also concerning aerodynamics, are so intense that the

9.6 Aerodynamics, Performance, and Handling Characteristics

9.6.1 Drag

The drag of high-performance vehicles should be particularly low. As stated in section 4.1, the aerodynamic drag is proportional to the product of the drag coefficient c_D and frontal area A_x. The simultaneous minimization of both factors often leads to target conflicts in high-performance vehicles. It is not always possible to reduce the drag coefficient to a minimum. Provisions of the regulations (such as open wheels or certain minimum size), in conjunction with other driving performance determining factors such as the tire width, cooling, and so forth, demand compromises under whose provisions the overall performance of the vehicle is optimized.

These requirements tend to increase the c_D value. The minimizing of the frontal area is limited by the large track width, which is demanded for good road holding, and the large-diameter wheels, which provide a high-energy reaction between road and tire. Furthermore, generously sized racing brakes must be accommodated within the rim diameter.

The vehicle body, the type and dimension of the wheels used, the installed cooling capacity, the drag contribution of downforce-generating elements, such as wings, spoilers, Gurneys, and the design of the floor (flat, jagged, close to production, ground effect) represent the main contributions to the drag. Depending on the class of vehicle, the individual components make a different contribution, so that no general guidelines can be given, by which the drag should be minimized. Thus, for example, a rear wing depending on the design and the vehicle, can actually lower drag, or in extreme cases generate more than 30% of the total drag.

To illustrate this, the fundamental influence of drag on a variety of optimization variables can be demonstrated. A generic racing car, with the basic data given in Table 9.2, serves this purpose:

Table 9.2 Basic Data of a Generic Racing Car	
Vehicle mass	$m = 1100$ kg
Weight split forward/rearward	47 / 53
Frontal area	$A_x = 2.0$ m²
Front lift (= Downforce)	$c_{L,F} = -0.45$
Rear lift (= Downforce)	$c_{L,R} = -0.55$

Other driving resistances, such as drivetrain losses, tire properties, wheelbase, track width, remain in this case unchanged, corresponding to the properties of a touring car. Varying the drag in a range of $c_D = 0.30$ to $c_D = 0.90$, and the power of the (generic naturally aspirated) engine from 220 kW (300 hp) to 660 kW (900 hp), we obtain the relationship between drag coefficient and maximum attainable speed shown in Figure 9.49. This is on the assumption that the highest gear is matched according to the maximum achievable speed and a long straight is available.

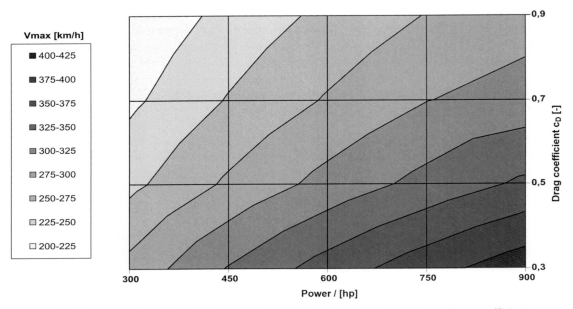

Figure 9.49 The dependency of the maximum achievable speed upon the drag coefficient and engine power of a generic racing car.

It is easy to see that low air resistance and high performance allow higher speeds. In addition, the following relationships can also be observed: With low installed power (300 hp), the top speed decreases by about 28.2% against a tripling of the drag. At higher power levels it declines by 30.2%. This relationship yields, for the optimization of drag in straight-line driving, a dependency between the achievable engine power and the achievable coefficient of drag.

From this relationship, the influence of both the engine power and drag on the overall efficiency of the vehicle can be estimated. A typical example is the cooling air requirement. That is to say that within certain limits, as the engine reaches a higher power output, the cooler it is operated. In this case, at a lower motor temperature, a power increase and a higher top speed can be reached. However, a greater cooling capacity generally demands a greater cooling air mass flow, which in turn acts to increase drag and consequently reduces the speed again. It is therefore necessary to consider whether increased cooling and the resulting higher power output represents an added gain over the necessary drag increase. For the generic racing car, this means the following: Starting with an engine power of 450 hp and a drag coefficient of $c_D = 0.50$, a cooling capacity increase which raises the drag by $\Delta c_D = 0.01$ enables the vehicle to reach a higher top speed, thanks to a power increase of at least 5.2 kW (7 hp) due to the cooler operating point of the engine.

For racing cars, long straights that allow the maximum speed to be reached are rather the exception. On circuits, typical straights are approximately between 1000 and 2000 m long, before reaching a curve that requires a reduction in speed. Generating the same diagram for a lap at the Hockenheimring with its circa 1050-meter long straights (Figure 9.49) yields the following picture.

The iso-lines of maximum speed in Figure 9.50 are substantially steeper than for the seemingly "unending" longer straight of the previous case in Figure 9.49. The engine power accordingly becomes more important. The optimization relationship for the above-mentioned cooling capacity interpretation is shifted toward higher engine performance. In the case mentioned, an increase of the cooling capacity would only need to deliver a 4.5 kW (6.2 hp) power increase for the vehicle to lap faster than with a drag reduction of $\Delta c_D = 0.01$.

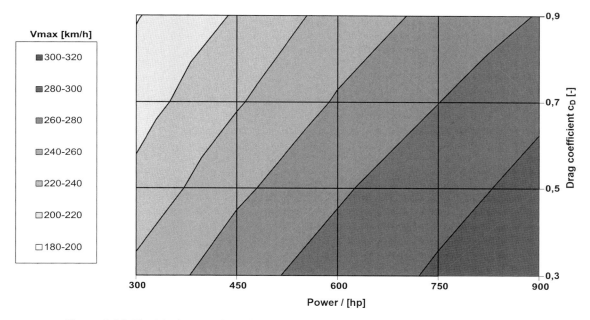

Figure 9.50 The Maximum achievable speed on the Hockenheimring as a function of the drag coefficient and the power of a generic racing car.

Another factor that is directly affected by the air resistance is the fuel consumption. Figure 9.51 shows the consumption per lap around the Hockenheimring for the example. From the respective fuel consumption, the tank volume is calculated and determines the number of fuel stops (if allowed) or the amount of fuel to be carried, which influences vehicle weight. Time penalties for fuel stops must be balanced against the lap time disadvantage from increased vehicle mass, in the context of improving lap time as a whole. This optimization task also includes the aerodynamic settings for the vehicle, and it is carried out for each individual track ([637], [638]).

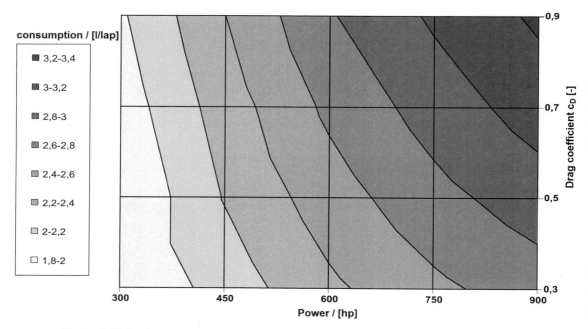

Figure 9.51 Fuel consumption as a function of engine power and the drag coefficient of a generic racing car at the Hockenheimring.

The decisive criterion is achievable lap time. The influence on it of drag coefficient and engine performance is shown in Figure 9.52 for the Hockenheimring example. It can be seen that for lap time, the importance of drag decreases with increasing engine power. Thus, for racing series with high engine power, increasing the engine performance plays a more important role than in racing series that have to make do with less engine power. For a given value of engine power, vehicles with low engine power but with a greater attention to achieving a low drag coefficient are interpreted as being equipped with high performance.

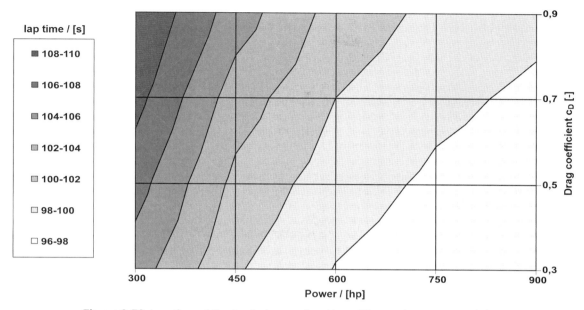

Figure 9.52 Lap time at Hockenheim as a function of the engine power and drag coefficient of a generic racing car.

9.6.2 Downforce

Another important optimization task is the downforce. The formation of its coefficient c_L is analogous to that of the drag coefficient in section 4.1. The fact that downforce is discussed here only after drag does not infer a lesser ranking in importance; conversely it is often the high priority criterion when tuning.

Through aerodynamic downforce the dynamic driving characteristics can be significantly improved. This reason is the direct influence on the wheel loads of vertical forces. Depending on the vehicle speed and downforce, wheel load changes of the magnitude of the static wheel loads may occur. Following is an example.

For the generic racing car described in the previous section, with a downforce coefficient $c_L = -1.0$ and a frontal area of 2.0 m² the wheel load doubled at about 342 km/h to 10,800 N (see Figure 9.53).

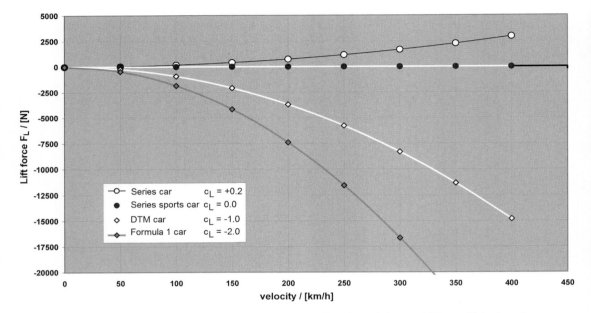

Figure 9.53 Lift forces of different vehicles as a function of the total lift coefficient and vehicle speed (frontal area of all vehicles $A_x = 2.0$ m).

In particular for sports car, the achievable cornering speed (or the maximum lateral acceleration) is an important measure of the driving characteristics. If the cornering speed (or lateral acceleration) is higher, the vehicle is considered powerful with respect to this criterion. Based on the generic vehicle with the basic data from the above section (i.e., vehicle mass $m = 1100$ kg, frontal area $A_x = 2.0$ m², weight distribution front to rear 47/53, equipped with standard racing tires), Figure 9.54 demonstrates how the transverse acceleration is dependent upon the downforce. The values are taken from a computer simulation that evaluates the lateral acceleration in a curve at the entrance to the stadium section of the Hockenheim circuit.

It is readily discernible that there is only a slight dependence on the c_D value. Besides the mechanical properties of the vehicle, the lateral acceleration depends chiefly on the downforce. Solely by better aerodynamic properties the generic racing car ($c_L = -1.0$) achieves approximately 15% higher lateral acceleration than a standard vehicle ($c_L = 0.2$, i.e., lift), which would be equipped with similar mechanics and tires. The same behavior applies to the cornering speed. Figure 9.55 shows that relationship for the same curve at the Hockenheimring.

High-Performance Vehicles

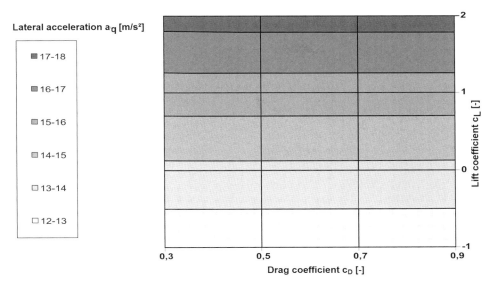

Figure 9.54 Transverse acceleration of a generic racing car as a function of the lift and the drag coefficients, on a curve entering the stadium section of the Hockenheimring.

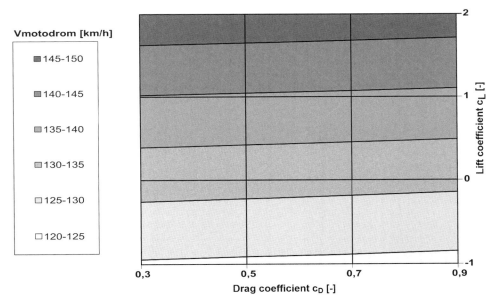

Figure 9.55 Cornering speed of a generic racing car as a function of the lift and the drag coefficients, on a curve entering the stadium section of the Hockenheimring.

The cornering speed depends primarily on the downforce. However, there is a slight influence of the drag coefficient. For lower drag coefficients, slightly higher cornering speeds can be achieved. This is due to a different driving line, arising because a vehicle with less drag is able to brake more strongly and drive more "roundly" through the curve, with a larger radius. The cornering speed increases slightly; the lateral acceleration is, as seen before, unchanged.

For racing cars, of course, the lap time is the key criterion. The achievable transverse acceleration and top speed are so matched to each other as to minimize it. Figure 9.56 shows how the lap time of the generic racing car depends on the values for c_D and c_L. Accordingly with the combination of c_D/c_L of 0.3/−0.5 the same lap time of 107–109 s can be reached as with 0.9/−2.0. Two simple dependencies can be generalized, namely, that high downforce and low drag lead to shorter lap times. The different gradients of the "iso-lap time lines" show that the optimum ratio between lift and drag coefficient varies depending on the operating point of the vehicle. If the drag coefficient is high, a correspondingly greater degree of lift change must be brought about, as for the same lap time improvement through lift change with a low drag coefficient. Depending on the structural possibilities, one or the other line of optimization is pursued.

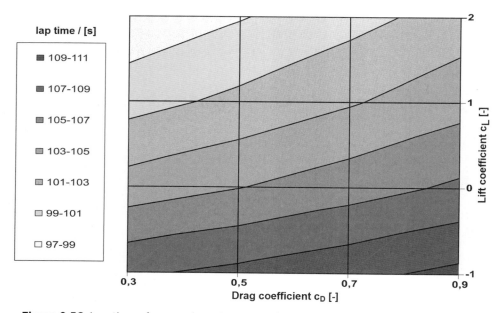

Figure 9.56 Lap time of a generic racing car at the Hockenheimring as a function of the lift and drag coefficients.

The basic elements that act to affect downforce are diffusers, wings, and spoilers on the front and rear of the vehicle but baffles and cooling elements, and so on also play a part, and their mode of action is described in section 9.7.

9.6.3 Balance

Balance is defined as the percentage distribution of the loads on the front and rear axle. We distinguish the so-called static or weight balance ω_{Stat} and aerodynamic balance ω_{Aero}, and the overall balance, which consists of aerodynamic and static balance.

The aerodynamic balance is usually stated as the percentage of the lift, or downforce, which rests on the front axle, so $\omega_{\text{Aero}} = c_{LF}/c_L$. The same applies to the static balance $\omega_{\text{Stat}} = F_{N,F}/F_{N,\text{Total}}$. Typical values for these balances lie between 35% and 55%. In special cases, for example racing cars with front wheel drive, the aerodynamic balance can amount to up to 80%.

In the case where a vehicle has different static balance and aerodynamic balance, the overall balance depends on the speed. Considering the case of a standard vehicle with a total wheel load of $F_{N,\text{Total}} = 15000$ N, and a static balance of 55% (front-heavy weight distribution), a front lift coefficient of $c_{LF} = 0.05$ and a rear lift coefficient of $c_{LR} = 0.10$ (a typical production vehicle lift distribution) with a frontal area A_x of 2.0 m², it follows that with increasing speed, due to the relatively low front lift there will be only a slightly decrease in the wheel loading at the front axle. The wheel load of the rear axle decreases more strongly with increasing speed and is about 12.5% less than the static wheel load at 300 km/h. The overall balance changes from 55.0% (static balance) at 0 km/h to 57.8% at 350 km/h. The overall balance of the vehicle becomes more front heavy with increasing speed. Figure 9.57 shows this behavior for an average production vehicle.

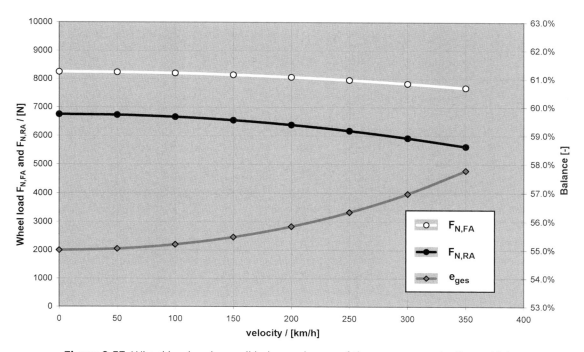

Figure 9.57 Wheel load and overall balance change of the average production vehicle ($F_{N,\text{Total}} = 15000$ N, $\omega_{\text{Stat}} = 55\%$, $c_{LF} = 0.05$, $c_{LR} = 0.10$, $A_x = 2.0$ m²).

If the vehicle shows a neutral driving behavior in its static balance, it is possible, due to the more strongly decreasing rear wheel load with increasing speed, that proportionally less transverse (and longitudinal) force is transmitted at the rear axle than at the front. In critical driving situations such as cornering, the rear axle reaches the limit of grip before the front axle and begins to slip before the front axle does. This driving behavior is known as oversteer. Since this oversteer is difficult to control, production cars are usually designed with a tendency toward understeer. This is achieved when the front axle can transmit less side force than the rear. The vehicle then slides from the front axle. Figure 9.58 shows the different driving lines of a neutral, under- or oversteering behavior.

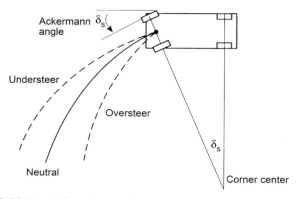

Figure 9.58 Neutral, understeering, and oversteering driving line [422].

Winkelmann [890] presented the influence of minor rear lift changes at different speeds and turning radii on driving behavior. The higher the speed, the smaller are the necessary changes in rear lift to achieve driving instability. Figure 9.59 shows that in this car at 160 km/h on a curve of radius r_K = 400 m, a rear lift coefficient of c_{LR} = 0.13 enabled stable handling. A rear lift coefficient of 0.165 prompted the vehicle to oversteer. For a speed of 200 km/h and a curve of radius r_K = 800 m, a rear lift coefficient of 0.13 was just about manageable, while a slight reduction to 0.12 already gave a significantly safer driving behavior.

The performance of racing cars is determined by, among other things, how benign is the change in driving behavior when close to the physical limits. The example described for production cars is generally true for race cars. Figure 9.60 shows a racing car with a total wheel load $F_{N,\text{Total}}$ of 10000 N, a static balance ω_{Stat} of 45% (approximately typical of mid-engined cars), a front lift coefficient c_{LF} of −0.40, a rear axle downforce of c_{LR} = −0.60, an aerodynamic balance of 40%, and a frontal area of 2.0 m² A_x.

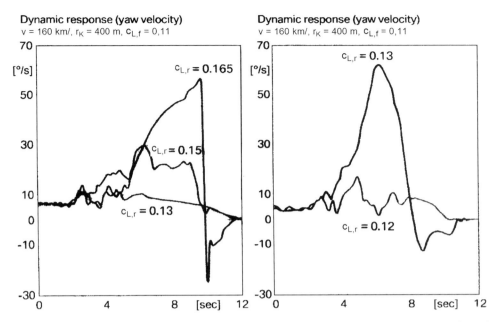

Figure 9.59 Handling (yaw rate) as a function of different speeds, different curve radii and rear lift, from [890].

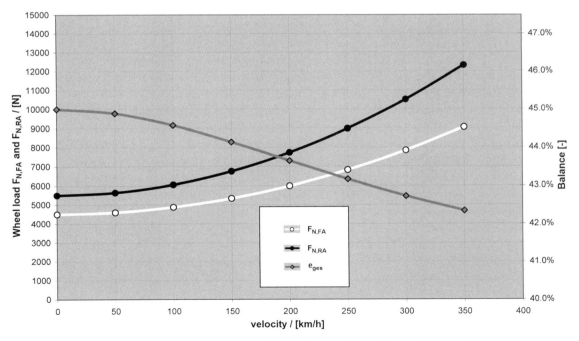

Figure 9.60 Wheel load and overall balance change of a racing car ($F_{N,\text{Total}} = 10000$ N, $\omega_{\text{Stat}} = 45\%$, $c_{LF} = -0.40$, $c_{LR} = -0.60$, $A_x = 2.0$ m^2).

The wheel loads on the front and rear axle are increased with increasing speed by the aerodynamic downforce. For the present example, the static axle load is doubled at about 349 km/h on the front axle and at 314 km/h for the rear axle. At these points, double the longitudinal or lateral forces, respectively, are then transferable in comparison to the mechanically identical vehicle without downforce. Due to the lower value of aerodynamic balance compared to static balance, the overall balance of the vehicle migrates from 45% at 0 km/h to 42.4% at 350 km/h. With increasing speed, the vehicle behavior tends more toward understeer. This interpretation, although theoretically not ideal, is preferred, since experience shows that a safer driving feel at high speeds leads to faster lap times than the achievement of a theoretically optimal limit. Particularly good racing drivers need less of a "safety margin" from the optimum and are able to drive at high speeds with nearly neutral handling on the limit.

Additional mechanical balance variations occur due to the longitudinal dynamics of the vehicle. The center of gravity of the vehicle is always at a certain height h_G above the road. Retarding or accelerating the vehicle results in a moment that puts additional load on the front axle during braking. The rear axle is relieved. In principal, such a load shift causes an oversteering driving behavior. Different configurations of the chassis can, depending upon the execution, mitigate or possibly even reverse this trend (e.g., by means of active suspension). A shift in the balance occurs, corresponding to the changes in load. The reverse occurs under acceleration. For further information, the reader is referred to the relevant literature, such as Ellis [226] and Milliken [573], which are sufficient although they deal only with basic dynamic driving problems not directly related to the aerodynamic properties.

For racing cars, where generally no active control of the chassis is allowed within the rules, these driving behaviors result in several consequences. In addition to the mechanical effects described, there are other specific aerodynamic effects. Due to the described load and balance changes, the vehicle when decelerating experiences compression at the front axle and rebound at the rear axle. The ground clearances at each axle change depending on the spring rates and the load shift. When making use of the ground effect described in the following section, the generated downforce changes with the variation in ground clearance. This dependence causes changes to the aerodynamic balance depending on the ground clearance. In general, to account for this effect the aerodynamic balance is expressed as a function of the ground clearance of the front and rear axles. Figure 9.61 shows a good example of the application of aerodynamic balance as iso-balance lines against front and rear axle ground clearance. With decreasing front axle ground clearance the balance increases and shifts toward the front axle. For the rear axle ground clearance the balance decreases with decreasing distance (i.e., shifts in the direction of the rear axle). What follows is a description of how these dynamic balance changes affect driving behavior.

High-Performance Vehicles

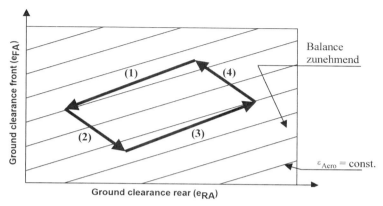

Figure 9.61 Movement pattern of a vehicle with downforce during deceleration and acceleration maneuvers [637].

Consider a typical driving situation on the racetrack. This usually consists of acceleration processes (e.g., on the straights), of a short transition phase from acceleration to braking (at the point of hitting the brake pedal), of the braking phase with strong deceleration, and with a further transition between braking and accelerating. Rolling phases, with neither strong retardation nor strong acceleration, play only a minor role in cornering, which will not be discussed here. The movement is demonstrated in a diagram in which the front axle ground clearance (e_{FA}) is plotted against the rear axle ground clearance (e_{RA}).

During the acceleration phase (1) the vehicle speed increases. The downforce increases with driving speed. The vehicle is pressed strongly to the ground, the ground clearance at the front axle and the rear axle is reduced. The vehicle moves down and to the left in the diagram. In the transition from acceleration to braking (2) an abrupt load change takes place, with the consequence that the front axle is further loaded and the rear axle unloaded (a typical front spring compression and rear spring rebounding, as in principle is also observed in series production vehicles). Accordingly, the front axle ground clearance decreases further, the rear axle ground clearance increases. The vehicle moves to the right in the diagram. During the retardation phase (3) the vehicle speed decreases again, reducing the downforce, and the front and rear axle ground clearances increase again. In the diagram, the vehicle moves up and to the right. In the short transition between deceleration and acceleration (4) the front axle rebounds and the rear axle compresses again. The vehicle moves up and to the left in the diagram.

The lines drawn in the diagram indicate a typical curve of lines of equal aerodynamic balance. To the bottom right of the diagram the balance increases. The aerodynamic downforce distribution shifts toward the front axle. As one can see, the acceleration and deceleration states run almost parallel to these lines. That is, during the deceleration and acceleration phase, the aerodynamic balance, and thus the downforce distribution between the front and rear axles, is hardly influenced. For the driver, this results in easily appraised handling from an aerodynamic point of view. The situation is different

647

within the two transition phases. Here the vehicle moves across the iso balance lines. In these phases, which are over very quickly, the aerodynamic balance varies greatly.

In the transition from acceleration to deceleration, the balance shifts toward the front axle. The weight distribution of the vehicle also shifts toward the front. The resulting higher wheel loads at the front axle allow a good deceleration but also ensure that the rear axle is unloaded accordingly. Under braking, this can easily lead to the rear of the car tending to break away (oversteer). Accordingly, the behavior is reversed in the transition from deceleration to acceleration. The aerodynamic balance and the weight balance shift toward the rear. This results in a corresponding relief of the front axle, which may then lead to an understeering driving behavior.

The size of the balance shift during these maneuvers is used as development criteria for the "pitch sensitivity." It is mainly generated by the increased downforce that arises from the narrowing of ground clearance under wing elements. The pitch sensitivity of racing cars is influenced mainly by the shaping of the downforce-generating elements at the front axle. The movement envelope of the vehicle can also be varied through the setup of the chassis. For the competitiveness of these cars, the tuning of the chassis to the aerodynamic properties and vice versa is a central development and test requirement.

Because of the high importance of a stable balance, the conceptual basis for this must already be developed before the first driving tests. Target values are already defined at the beginning of the development program and are derived from simulations and empirical values. In the context of static balance, thus targets are also defined for the aerodynamic balance. For modifications, not only the aerodynamic efficiency or lap time improvement must be considered, but also the achievability of the target value in relation to the aerodynamic balance. Based on an example this method, also known as "rebalancing," will be explained.

If we now consider a measure that generates more downforce on the front axle, for example, without increasing the drag, we would first approach view this measure as positive for the performance of the vehicle because the efficiency is increased (point 1 in Figure 9.62). The aerodynamic balance is, however, moved forward. In order to bring the vehicle into balance, this must be reacted with a downforce-increasing action at the rear. This can be achieved for example by increasing the angle of attack of the rear wing elements. In most cases, this is accompanied by an increase in drag. The corresponding point (2) can be entered in the diagram. Appropriate rules by which the balance can be corrected are obtained, for example, from the so-called rear wing polars. These are measurements from which the lift and drag behavior can be determined when adjusting the rear downforce elements. Evaluation of the overall measure, of the original front axle improvement, and the subsequent "rebalancing," can then be carried out using the iso-lap time lines. If the point lies above, the measure is an improvement; if it lies below, then the measure deteriorates the lap time.

High-Performance Vehicles

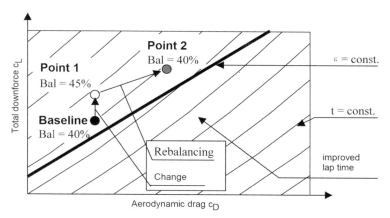

Figure 9.62 Evaluation of configuration changes with "rebalancing."

The same is true for measures at the rear end, which must be balanced with appropriate countermeasures at the front. For the adjustment of the front axle downforce the corresponding polars can be determined from the adjustment ranges of the front wing, or the addition or changing of parts that are usually mounted on the underside of the front diffuser.

9.6.4 Road Performance

Road behavior is essentially determined by the tires and the chassis; wheel loads and the longitudinal and lateral forces transferred to the road surface also have a decisive part to play. The aerodynamic effects involved (lift and downforce) produce additional wheel loads, which, particularly at high speeds, can approach the order of magnitude of the static wheel loads (see section 9.7.2)

The maximum longitudinal and lateral forces transferred depend primarily on wheel loads, friction coefficients, slip, and slip angle. Typical friction coefficients μ lie in the range of 0.5 to 0.9 for standard car tires, 1.2 to 1.5 for racing car tires, and approximately 1.4 to 1.7 for special qualification tires, which usually cannot stand up to such an amount of friction for longer than one lap. These values assume optimum slip. Slip denotes the ratio of rotation speed of the tire at the tread, that is, $\omega \cdot r$ (ω = rotation speed of wheel, r = radius of tire), to the ground speed v of the vehicle. With 0% slip, the rotation speed of the tire tread equals the ground speed of the vehicle. In a full braking maneuver with stationary wheels the slip amounts to 100%. Figure 9.63 shows friction coefficient as a function of slip when braking.

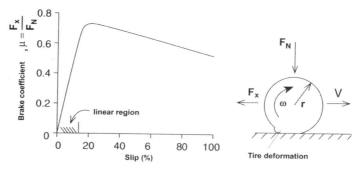

Figure 9.63 Friction coefficient μ as a function of slip when braking [422].

After an almost linear region, the maximum friction coefficient is reached at approximately 15% slip. This value is dependent on the nature of the tires. After reaching the maximum value, the friction coefficient sinks slightly with increasing slip.

For turning maneuvers with the steering wheel locked at the appropriate angle, a further parameter is involved, which primarily influences the lateral force transferred. The so-called slip angle α denotes the angular difference between the intended driving direction and the actual driving direction. This angular deviation also comes about due to slip, in this case in a direction transverse to the vehicle. Figure 9.64 shows lateral force FSF as a function of wheel load F_N and the slip angle α.

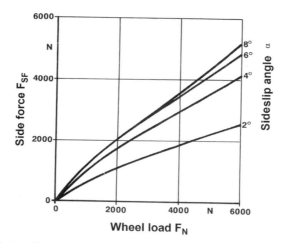

Figure 9.64 Lateral force FSF as a function of wheel load F_N and the slip angle α

Aerodynamics can be used to influence the wheel load without bringing any additional, mass-based effects into play. Among the functions shown, it can be clearly seen that aerodynamic factors have an important part to play in relation to braking and acceleration capability, as well as curve speeds achieved. Figure 9.65 shows the performance of a vehicle in terms of longitudinal and lateral acceleration with no aerodynamic influence and in a case in which the aerodynamic load is equal to the weight of the vehicle.

High-Performance Vehicles

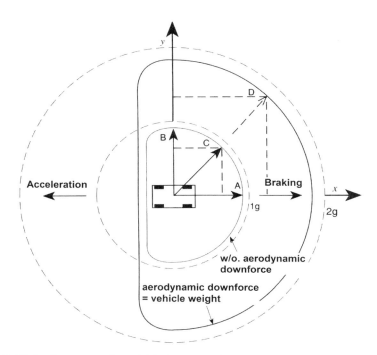

Figure 9.65 Performance of a vehicle in terms of longitudinal and lateral acceleration with no aerodynamic influence and in a case in which the aerodynamic load is equal to the weight of the vehicle [422].

The examples discussed so far assume an integral driving performance on the part of the vehicle. In practice, however, it is more often the case that, depending on driving condition, it will only be at different points in time that the front and rear axles will achieve their maximum performance in relation to transfer of longitudinal and lateral forces. The individual wheel loads—composed of static weight distribution, dynamic effects such as weight transfer due to acceleration, and the aerodynamic lift and downforce—play a crucial part in this.

9.6.5 Efficiency

When dealing with racing cars, the aerodynamic efficiency ε_{aero} is defined as the ratio of total downforce coefficient to drag coefficient: $\varepsilon_{aero} = c_L/c_D$. The aerodynamic efficiency is a measure of the performance capability of a vehicle on the racetrack. The simple determination of the efficiency out of the ratio of downforce to air drag coefficient allows the developer to rapidly assess aerodynamic changes in terms of their effectiveness on the track.

We consider now a somewhat more high-performance vehicle than that in the previous sections. The following generic racing car has a performance of approximately 400 kW (600 hp), a mass of 1000 kg, and a front surface calculated at 2.0 m². The data correspond approximately to those of the sports car prototypes deployed at the Le Mans race. The variation of downforce coefficient and air drag coefficient provides us with functions in relation to achievable lap time, which are illustrated in Figure 9.66 for the Hockenheimring.

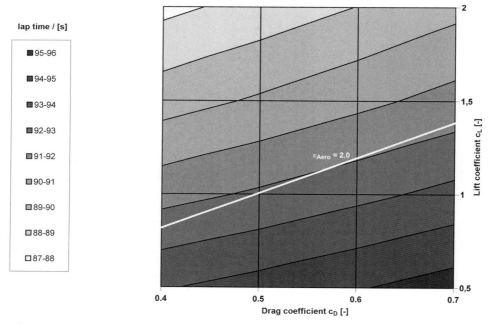

Figure 9.66 Lap time as a function of lift and drag coefficients of a generic racing car on the Hockenheimring.

In addition to the "iso lap time line" the figure also displays the "iso efficiency line" $\varepsilon_{aero} = 2.0 = \text{const}$. It will be noticed that this line runs more or less parallel to the "iso lap time line." If a vehicle has a higher aerodynamic efficiency, e.g., $\varepsilon_{aero} = 3.0$ (from $c_D = 0.5$ and $c_L = 1.5$), the achievable lap time will be better than that of a vehicle on the $\varepsilon_{aero} = 2.0$ line.

If, however, we consider the efficiency and lap time line a little more closely, it can be seen that for low drag values ($c_D = 0.4$) the "iso lap time line" for a lap time of 92 seconds lies above the efficiency line $\varepsilon_{aero} = 2.0$, but lies below it for high drag values ($c_D = 0.7$). If the downforce is increased in the same ratio as the drag increases—moving along the efficiency line from lower left to upper right—then faster lap times can be achieved. In this case we speak of a track that is neutral but mildly downforce-orientated. The specific character of a racing track is determined by the varying amount of short and long straight sections and tighter and broader turns it contains. The greater the proportion of tight turns and straight stretches, the lower the average speed and the greater the amount of downthust required.

If the same vehicle is driven on a very fast track, such as Le Mans (Figure 9.67) the situation is completely different, as shown in Figure 9.12. Due to the layout of the track (long straights, with some of the turns being very "fast") and an average speed per lap of over 220 km/h, this course is a classic example of a racing track that requires low c_D values. As can be seen in the figure, this remains the case even though the long Mulsanne Straight has been defused somewhat by the addition of three chicanes. Low c_D values lead to significantly better lap times at the same efficiency levels.

This form of representation can be used not only for development but also for fine-tuning the car to the racing track. The data gained in the wind tunnel allow the so-called vehicle polar curve to be generated. This essentially describes the curve traced by the vehicle in this diagram when downforce and drag on the vehicle are varied. This is achieved by adjusting wings and by adding or subtracting aerodynamically effective components. An adjustment of this kind requires that care be taken not to alter the ratio of front to rear axle downforce, that is, the aerodynamic balance (see section 9.6.3). Figure 9.67 contains an illustration of such a vehicle polar curve.

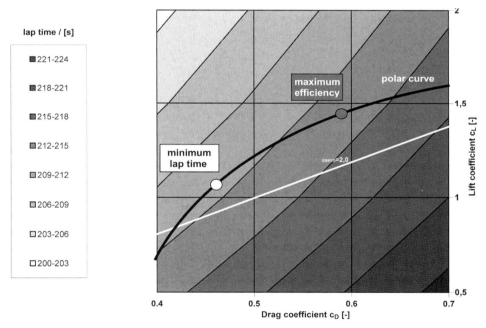

Figure 9.67 Lap time as a function of lift and drag coefficients of a generic racing car on the racing track of the 24-hour classic in Le Mans, including an illustrative vehicle polar curve.

The polar curve is characterized for low drag coefficients by the fact that small drag-increasing measures can achieve a significant lift reduction, meaning a gain in downforce. This is typical of c_D optimal vehicle forms. The aerodynamic efficiency increases in value rapidly across this range with increasing drag. In the high drag range, the vehicle polar curve flattens out noticeably. It becomes increasingly difficult to generate sufficient downforce with low increases in resistance.

For this vehicle, the maximum efficiency of ε_{aero} = 2.45 is reached at a drag coefficient of approximately c_D = 0.59 and a downforce coefficient of approximately c_L = 1.45. The theoretically achievable lap time for this vehicle in Le Mans thus amounts to approximately 3:33.5 minutes at this point. The fastest lap time point, however, is reached at a drag of c_D = 0.46 and a downforce coefficient of c_L = 1.05. The aerodynamic efficiency ε_{aero} = 2,28, the lap time 3:31.0, approximately 2.5 seconds faster when the aerodynamic efficiency is lower.

The aerodynamic efficiency is thus only a starting criterion for the evaluation of the aerodynamic performance capability of racing cars. More precise investigations require the set up of further optimization criterion, taking into account the characteristics of the particular track, the engine performance, weight of the vehicle, static wheel load distribution, and aerodynamic balance, in order to judge a vehicle's aerodynamic performance capability.

9.6.6 Cooling and Ventilation

Racing cars are equipped with a special-purpose cooling system. Inlet and outlet openings are arranged so that losses are minimal. The cooling effect is maintained by the airflow itself; fans are hardly ever used. Only rarely is there sufficient room to accommodate an "ideal" flow of the cooling air, as can be seen in Figure 9.68.

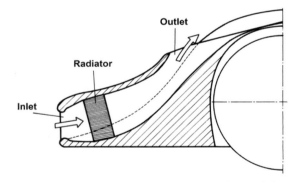

Figure 9.68 Airflow for a heat exchanger at the front of the vehicle.

Cooling air is taken from the stagnation zone in order to be able to exploit the greatest possible pressure gradient. The used air is released upwards, laterally, or downwards. When the air is released upwards, two possibilities need to be considered: the possibility of the driver's compartment being heated up and the possibility of interference with the airflow at the rear. If the air is released laterally, there may be additional drag, as the lateral airflow will be disturbed. If the cooling air is taken under the vehicle for release, the ground effect can be impeded. This last option is therefore excluded in most cases.

The fairing of the underside of the chassis makes the cooling of all other components even more critical. Normal heat insulation measures no longer suffice to protect heat-sensitive components, and targeted forced ventilation becomes necessary. In addition, there is a growing trend toward combining components with similar temperature-resistance values, thus creating zones of different temperature. Moreover, efforts are made to reuse the cooling air several times, such as when the relatively cool air exiting from an intercooler is utilized to cool brakes or transmission.

Alongside the heat exchangers, such as radiator, intercooler, and oil cooler for motor and transmission, it is becoming increasingly necessary to provide cooling air to electronic devices, sensors, and the generator. Further systems, such as fuel supply, exhaust manifold, compressor (including turbocharger), and starter, also require cooling. It goes without saying that brake cooling (steel or carbon fiber brake disks) also presents an aerodynamic optimization challenge and can significantly affect overall design.

A number of physiological studies, performed by the governing body of motor sport worldwide (FIA), have recorded a significant impact on the performance capability of the human body at high temperatures—a fact which is known from the physiology of work. Increasingly greater efforts are therefore being made to provide racing car drivers with comfortable working conditions, in particular by ensuring maximally draught-free body and helmet ventilation (cf. section 12). This is especially crucial in long-distance racing vehicles, which often demand of the driver the highest performance levels for up to four hours at a stretch in cockpit temperatures of over 60°C.

Managing the thermal balance in the vehicle for critically temperature-sensitive components has become an additional optimization goal. The large number of parts requiring cooling means that consideration must be given to these requirements right from the initial design stage of development. On the other side of the equation, the need to provide cooling provides a certain amount of freedom to undertake aerodynamically effective improvements on the vehicle in what would otherwise be a situation of particularly tight constraints.

Figure 9.69 shows an illustrative overview of the cooling elements in a racing car, using the Porsche 956 group C racing car. For reasons of space, the cooling air for the front brakes and the interior could only be taken from the stagnation zone. The cooling air for the radiator, the oil cooler, and the intercooler is inducted on the lateral surface behind the front wheel. The outlet should be directed upwards, so as not to impede utilization of the ground effect. The cooling air for the engine fan, the rear brakes, and the transmission oil cooler is inducted from the upper surface of the vehicle via NACA jets. Engine compartment ventilation is performed via air slots on the rear diffuser.

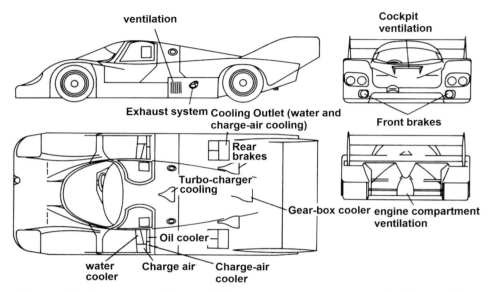

Figure 9.69 Arrangement of the ventilation and exhaust openings on the Porsche 956 racing car.

Record-breaking vehicles, which are only designed for extremely short runs and are never driven as far as the steady-state temperature, can sometimes avoid drag-increasing air recooling entirely. The necessary cooling capacity is provided in the form of water ice or dry ice, which are carried on board and melt or sublimate during use.

Due to the comparatively high engine performance levels that are designed into series sports cars, the cooling layout for these vehicles poses high demands. The test criteria are the same as for limousines (cf. chapter 7). In addition, the vehicles must be able to be deployed for races on closed-circuit tracks.

9.6.7 Oblique Incident Flow

Oblique incident flow (i.e., an incident airflow that is not longitudinal to the vehicle) occurs with crosswind and in turning. Together with the airstream caused by the vehicle's own motion, crosswind leads to an incident airflow that is not parallel to the direction of travel. When turning, with the lateral forces increasing, the tires begin to manifest a certain slip angle. This causes the vehicle body to display yaw, which is of the same degree of magnitude as the slip angle. This amounts to up to 10° for normal road tires and up to 8° for racing tires. The oblique incident airflow at yaw angle β leads to an increase in drag and a decrease in downforce. Figure 9.70 provides an example.

To ensure driving stability, and thus safety, the decrease in downforce at the rear axle should be equal to or smaller than that at the front axle. Shifts of the aerodynamic balance in the direction of the front axle affect handling, leading to oversteering, which can be difficult to bring under control. Varying the shape of the elements that cause downforce can influence this behavior significantly.

High-Performance Vehicles

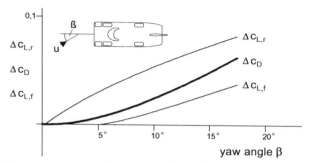

Figure 9.70 Lift and drag behavior of a vehicle subjected to oblique incident airflow.

The influence is not limited to drag and downforce but also extends to the lateral forces at the front and rear axle, thus affecting the yawing moment around the vehicle's vertical axis as well. With commonly encountered vehicle shapes, the action point (i.e., the onset point of the aerodynamic forces) is located in front of the vehicle's center of gravity and thus in front of the onset point of the lateral forces that can be absorbed by the tires; the yawing moment that is thereby produced turns the vehicle out of the wind and thus has a destabilizing effect (see also section 5). Special measures such as vertical fins or wing end plates can be used to shift the action point further toward the rear. This produces a positive yaw moment, which helps to turn the car into the wind, thus increasing stability.

The thrusting incident airflow produced by the skew of the tires during turning leads to lateral forces that are directed toward the inside of the curve and thus counteract the centrifugal forces. This stabilizing effect, however, is countered by the destabilizing effect brought about by the loss of downforce, since a lighter wheel load in the presence of lateral forces will result in a greater slip angle. Some racing series therefore permit large wing end plates, in order to sufficiently accommodate driving stability. The comparison of the Audi R sports car prototypes from 2000 and 2002 given in Figure 9.71 shows the differing size of the wing end plates.

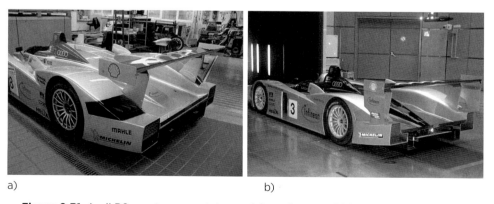

Figure 9.71 Audi R8 sports car prototype: a) from the year 2000 with small wing end plates; b) from the year 2002 with large wing end plates.

657

Further studies on racing cars show that at oblique incident airflow angles exceeding the normal levels associated with crosswind and turning (and reaching, e.g., the levels encountered in skids or spins), lift can be produced. Figure 9.72 presents a diagram showing the critical vehicle speeds (as a function of oblique incident airflow angle) at which the lift at the front or rear axle and/or the total lift is so great that it offsets the static wheel load at the front or rear axle and/or the static total wheel load due to the weight of the vehicle. If there is a measuring point on this diagram with a certain oblique incident airflow angle, it shows the speed at which aerodynamic lift exceeds static wheel load. The result is that the load on the axle concerned is nullified and the axle can thus become airborne. At an oblique incident airflow angle of 45° a speed of approximately 300 km/h is necessary in order to represent such an effect at the front axle in this example. Up to approximately ±120° the front axle must be considered to be critical. For all other angles (where the incident airflow is rather from behind) it is the rear axle that is critical. The rear wings and the rear of the vehicle are then subjected to incident airflow "the wrong way round" and there is no appreciable lift.

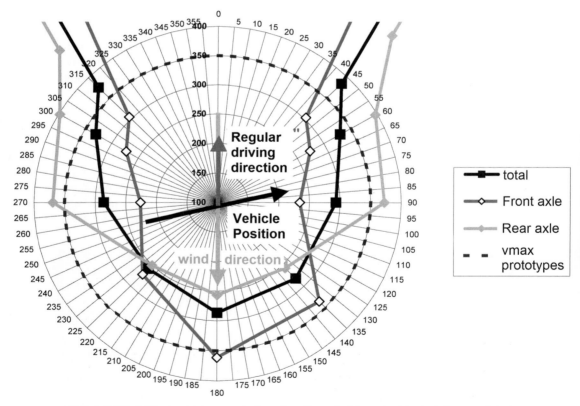

Figure 9.72 Critical vehicle speeds (as a function of oblique incident airflow angle) at which the static wheel load can be compensated by lift.

Under "normal" driving conditions (between approximately −20° and +20°) and driving situations that can be brought under control by the driver, these effects do not occur. There is no speed that generates a lift capable of completely counterbalancing the load on both axles. But beyond the normal range (e.g., in skids and spins), there are situations in which the load on an axle can be offset totally. As a result of these studies, discussions are now under way concerning changes to the regulations, designed to reduce these effects by for example redesigning the flat underside of the chassis. The goal is to ensure, via design regulations or reducing the maximum speed, that such critical ranges cease to be attainable.

In the American NASCAR racing series, the roof flaps serve to reduce the speed in the event that the driver loses control of the vehicle. The roof flaps, as the name suggests, are flaps that are installed on the roof of the vehicles. In the event that the direction of travel changes (e.g., in a spin), the wind pressure can cause them to open, thus considerably increasing the drag on the vehicle. This increased drag is used to decelerate the vehicle.[11]

9.6.8 Slipstream

Normally aerodynamic studies proceed from the assumption that the vehicle, when traveling straight ahead, is moving in an undisturbed environment. Under real driving conditions the vehicle does not move along the road on its own, but rather is influenced by the airstream in the wake of the other vehicles in front of it. Racing cars often drive "in the slipstream" in this way. In relation to the driving speed, the distances between the vehicles are significantly less than they are in normal road traffic. The phenomena that thus arise are explained in section 4.5.4 for comparatively simple bodies.[12]

To clarify the phenomena associated with slipstream driving in racing cars, Romberg et al. [680] conducted wind tunnel experiments on American stock cars.[13] In this series, slipstream driving occurs particularly frequently. The result was that depending on the distance between the vehicles, the following vehicle displayed up to 37% less drag than with undisturbed incident airstream. The effects of slipstream driving are not limited to the following car, however, but also apply to the car in front, which can experience a drag reduction of up to 30%. The base pressure is increased by the air accumulating in front of the following vehicle, which leads to a reduction in the drag on the vehicle in front. These phenomena enable groups of cars driving in close succession to travel faster than would be possible for an individual vehicle on its own. Figure 9.73 shows this effect for drag and lift forces.

---

11. For racing vehicles, it has become common to describe negative lift as "downforce." The downforce is therefore treated as an independent variable; a downward-facing air force is positive.
12. For racing vehicles, it has become common to describe negative lift as "downforce." The downforce is therefore treated as an independent variable; a downward-facing air force is positive.
13. For racing vehicles, it has become common to describe negative lift as "downforce." The downforce is therefore treated as an independent variable; a downward-facing air force is positive.

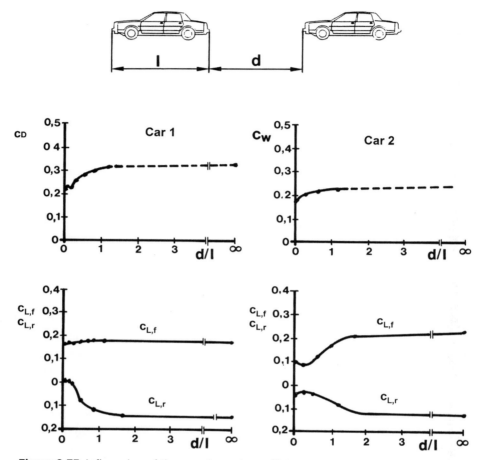

Figure 9.73 Influencing of the aerodynamic coefficients by slipstream driving [680].

In addition to influencing the drag, slipstream driving also influences the lift. In the present case the front axle lift of the following car decreases with decreasing separation distance between it and the car in front. The front axle lift of the car in front remains largely unaffected.

Studies conducted by Audi AG on sports car prototypes show that similar effects on the axles occur for vehicles with downforce as well. Here, however, the front axle downforce of the following car decreases with smaller separation distances. This is a consequence of the slipstream in the wake of the car in front, which produces only a low-energy incident airstream incapable of generating the necessary downforce. If the downforce coefficient is based on the average incident airflow speed encountered in the slipstream, there are only small variations in front axle downforce coefficient. At very small separation distances, if the main element responsible for downforce generation (e.g., front wing or front diffuser) does not experience any significant incident airflow, the front axle downforce can be reduced to values close to zero. This effect depends in part on the geometry of the front of the following vehicle and in part on the form of the slipstream in the wake of the vehicle in front. Vehicles with completely flat chassis underside generally conduct less energy into the slipstream behind them than vehicles with pronounced rear diffusers.

The rear axle lift in the studies by Romberg [680] was not much influenced by the slipstream behind. At small separation distances, the vehicle behind is additionally subject to the pitching moment due to the altered front end aerodynamics. The vehicle in front experiences a significantly greater effect on the rear downforce. At small separation distances the rear downforce is noticeably reduced. The studies performed by Audi AG essentially confirm this trend.

The airflow conditions described mean that slipstream driving is characterized by the following phenomena:

- The maximum velocity achievable rises due to the low effective incident airstream speed for the following vehicle.
- The same holds for the vehicle in front due to the drag-reducing effect (air congestion) of the following vehicle.
- Following vehicles with lift on the front axle lose lift due to the lower effective incident airstream speed and thus the handling tends toward oversteering.
- Following vehicles with downforce on the front axle lose downforce due to the lower effective incident airstream speed and thus the handling tends toward understeering.
- Vehicles experience a reduction of the rear downforce and thus the handling tends toward oversteering.

In particular, the altered drag and the altered driving performance due to the aerodynamic slipstream effects allow interesting driving and overtaking maneuvers.

9.7 Aerodynamics of Components

In contrast with the "integrated" bodies of serial production sedans, racing cars can have certain elements (such as wings or diffusers) that, to a degree, can be considered and optimized separately from each other. Their function and uses will be explained by means of examples. Here it is important not to overlook their mutual effects on one another. Finally, attention will once again be turned to the importance of regulation, the reason for this being that it is especially the aerodynamic elements that are decisively limited by regulations in terms of their form and function.

9.7.1 Basic Body

Aerodynamically relevant basic bodies can be divided into three categories:

1. Droplet-shaped basic bodies
2. Classic sports car bodies, characterized by a flat nose and a blunt, mostly cut-off tail
3. Bodies similar to those in serial production, whose aerodynamic properties are largely similar to those of serial production cars

Special rules have been arrived at for the design of front and tail.

9.7.1.1 Front

Foundational studies on the design of basic bodies have been carried out by various authors. Flegl [250] demonstrates with Figure 9.74 the effects of various front and tail shapes on drag and lift. These effects include the airstream phenomena both on the top and on the underside of the vehicle. Depending on the ground clearance, the vehicle's underside, in particular, is in a position to decisively influence the coefficients referred to. For instance, a forward-tapering nose lowers drag. The location of the tip of the nose in relation to the road surface influences the location of the stagnation point. If the tip of the nose is high, relatively high front axle lift is produced. Positive pressure is generated on the underside of the vehicle. By lowering the tip of the nose, the lift can be significantly reduced. The drag reduces up to a certain point (in this example the minimum drag is in the middle). Lowering the tip of the nose farther can lead to a considerable amount of downforce at the front axle. The drag can, via diffusion effects or flow separation along the underbody, increase once again, generating downforce. In consideration of the wing running back underneath the front, the nose of Formula 1 cars has been raised again, as can be seen in Figure 9.75.

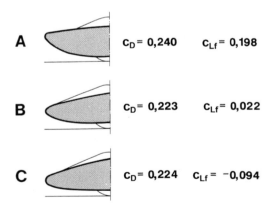

Figure 9.74 Influence of front shape on drag and thrust at the front axle.

High-Performance Vehicles

Figure 9.75 Raised nose of the Mercedes GP Formula 1 racing car at the Golden Prize of Monte Carlo 2012.

9.7.1.2 Tail

The shape of the vehicle's tail influences drag to an even greater degree. Ideal are long tail shapes tapering off like a droplet, as can be found in record-breaking cars or solar-powered cars. According to Kamm, such a droplet-shaped tail can also, as described in section 1.4, be cut off smoothly, without noticeably increasing drag. Figure 9.76 shows various design shapes of such a cut-off tail.

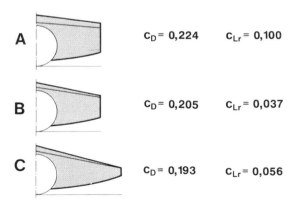

A $c_D = 0{,}224$ $c_{Lr} = 0{,}100$

B $c_D = 0{,}205$ $c_{Lr} = 0{,}037$

C $c_D = 0{,}193$ $c_{Lr} = 0{,}056$

Figure 9.76 Influence of tail shape on drag and lift at the rear axle.

Today's racing cars are subject to considerably greater constraints on the design of the basic form. Current regulations mandate criteria, above all in the area of safety (driver and co-driver protection, crash tests for frontal, rear, and lateral collisions) and in the area of aerodynamics (checking competitiveness and performance capability, vehicle safety at high speed), which basically dictate the outer shape (Figure 9.77 and Figure 9.78).

Figure 9.77 BMW racing car, winner of the 24 hours of Le Mans 1999.

Figure 9.78 Formula 1 racing car from the year 2001.   Photo: BMW AG.

Any remaining degrees of freedom in design are primarily utilized to implement technically important constraints such as downforce generation (Figure 9.75) and cooling (Figure 9.79). All these "constraints" end up assuming such prominence that basic body development in the classic sense is pursued—and can only be pursued—from very limited points of view.

Figure 9.79 Cooling air intake of the Audi RBC. The car took part in the 24 hours of Le Mans in 1999. The cooling air intake supplies the radiator, the intercooler, and the front brake. Photo: Audi AG.

When it comes to the basic shapes of serial-like cars such as the DTM, the shape of the existing serial body is taken over largely unchanged (Figure 9.80). Depending on the class and the technical regulations, there are degrees of freedom to alter the body in certain areas or to install downthrust aids (front and tail wings, spoilers, gurneys) (Figure 9.81). In the ideal case, the serial body already has advantageous aerodynamic properties (Figure 9.82), so that additional measures can largely be dispensed with.

Figure 9.80 VW Lupo: Cup vehicle with only minor aerodynamic changes relative to the serial-production car.

Figure 9.81 Audi 90 Quattro which took part in the American IMSA-GTO racing series in the year 1989. Roof, front flaps, and tail flaps correspond to the Audi 90 serial-production car. Front section, wheel arches, and tail were heavily modified in accordance with regulatory requirements.

Figure 9.82 Ferrari 360 Modena sports car.

9.7.2 Wings

9.7.2.1 Tasks

When the vehicle's weight alone is no longer sufficient to achieve the high lateral accelerations that are aimed at, aerodynamics is used to create a downforce: the downthrust. The contribution of this force, however, would remain slight, if it were to be used in association with the shape of the vehicle body alone. This is because, among other things, the effect resulting from the shape of the vehicle's underside—the key term is venturi; section 9.7.4 returns to this point—is rendered inactive by the fact that the various regulations prescribe a smooth underside at least as far as the rear axle. To gain more downthrust in spite of this, wings are employed. They have become the most conspicuous feature both of the monoposto and of the closed racing car. In sports cars, they are only used as fixed elements in extreme cases.[14] Instead, there is a preference for moveable spoilers, which are sometimes reminiscent of wings.

9.7.2.2 Function

The function of a wing can be explained with the help of airfoil theory.[15] Since airfoil theory is always concerned with lift, that is, a force directed upwards, the "overhead" position of the wings of racing cars will initially be "overlooked," meaning the wing will be considered in "normal" position and its lift will be studied.

There are three geometric features that determine the lift (and also the pitch moment) of an airfoil; they can be derived from Figure 9.83:

- The outline (the wingspan and shape of the horizontal projection)
- The profile, which can be assisted by flaps
- The lateral boundaries, such as winglets or end plates

Producing lift is always associated with drag.

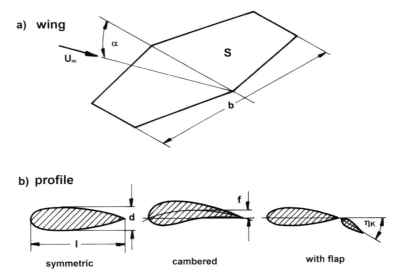

Figure 9.83 Wing and profile, definitions.

9.7.2.2.1 Outline

The outline of a wing determines how the lift is distributed over the wingspan. The essential parameter is the aspect ratio Λ:

---

14. For racing vehicles, it has become common to describe negative lift as "downforce." The downforce is therefore treated as an independent variable; a downward-facing air force is positive.
15. For racing vehicles, it has become common to describe negative lift as "downforce." The downforce is therefore treated as an independent variable; a downward-facing air force is positive.

$$\Lambda = \frac{b^2}{A} \tag{9.1}$$

As is summarized in Figure 9.83, b represents the wingspan and A the surface area of the wing in horizontal projection. In most cases, the wings of sports cars are rectangular wings; for these the aspect ratio $\Lambda = b/l$. For racing cars, $\Lambda = 5$ to 8.

9.7.2.2.2 Profile

Figure 9.84 shows schematically how lift is produce on a profile: if a symmetrical profile is placed in an airstream flowing in the direction of its chord line—the angle of attack is thus $\alpha = 0$—then no lift is produced. Lift can be produced in one of two ways, or via a combination of both ways:

- The profile is tilted by angle α
- The profile is arched by the amount f
- Both these measures are combined

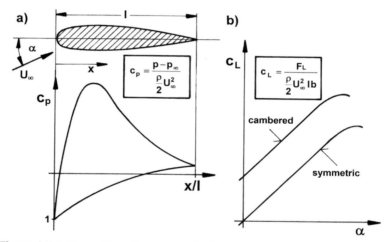

Figure 9.84 Properties of a symmetrical and an arched profile, schematic.

The streamlines around an arched profile can be seen in Figure 9.85. Airflow is faster on its upper side (distance between the streamlines is small) than on its underside (the streamlines are more widely spaced). This produces underpressure on the upper side and overpressure on the lower side.

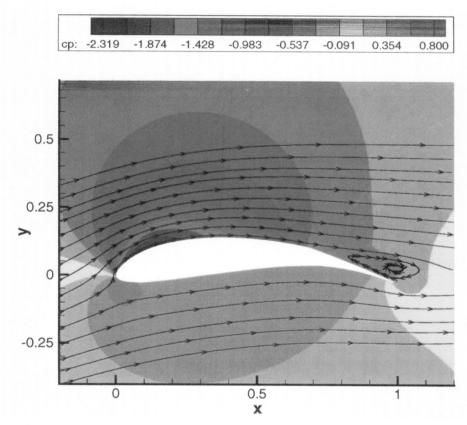

Figure 9.85 Streamlines, velocity vectors, and pressure distribution for a one-piece arched wing profile [670].

Figure 9.84a provides a schematic overview of how the static pressure p is distributed over the profile depth x/l, in the form of the dimensionless coefficient c_p, with negative c_p values at the top (as is usual).

The underpressures are, in terms of their amount, much greater than the overpressures. In total, the pressure distribution results in a lift, which, as sketched in Figure 9.86b and supported by measurements in Figure 9.86 (left), increases approximately linearly with the angle of attack. This increase comes to an end when the airstream on the upper side of the profile separates. The thicker the profile and the rounder its nose, the later the separation occurs.

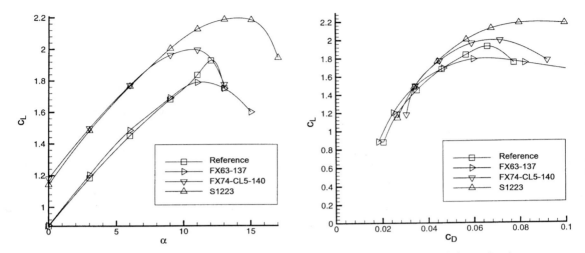

Figure 9.86 Lift over angle of attack, $c_L(\alpha)$ (left) and polar curve $c_L(c_D)$ (left) and polar curve.

9.7.2.2.3 Lateral Boundaries

The pressure difference between upper and lower surface of the wing produces a lateral airstream on the wing; this is directed on the upper side toward the center and on the underside outwards. Behind the wing, both airstreams meet and form a layer of free horizontal vortices, which in turn combine at both wing ends to an inward-turning pair of longitudinal vortices ("horseshoe vortices").

The partial equalization of pressure at the sides of the wings is associated with a loss of lift. The lift however is not constant across the wingspan; it drops off toward the sides. The vortex pair induces a downwind, which is associated with an additional drag that is called "induced drag" $F_{D,i}$ because it is "induced" by the peripheral vortices.

Small, vertically placed winglets can be used to at least partially prevent the pressure equalization already described. The same function is performed by end disks, called "end plates" in motor sport, which are an integrated part of the design of the lateral mountings of the wings, as can be seen in Figure 9.87.

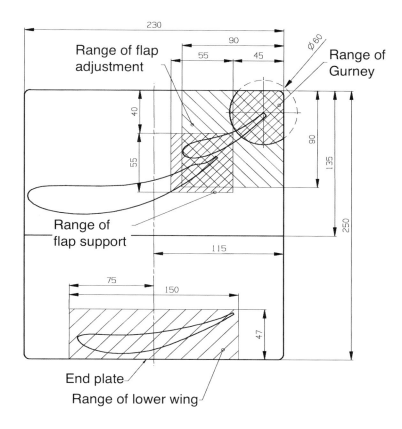
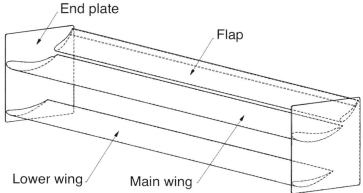

Figure 9.87 Arrangement of the prescribed rear wing elements in a special-purpose installation space from the technical regulations of the German Touring Car Masters (Deutsche Tourenwagen Masters, DTM) 2003.

9.7.2.3 Drag

An airstream flowing around a profile is associated with drag; in contrast to induced drag, which will be taken up next, this drag is known as profile drag $F_{D,0}$. It is composed additively of a pressure component and a frictional component, $F_{D,p}$ and $F_{D,f}$:

$$F_{D,0} = F_{D,p} + F_{D,f} \qquad (9.2)$$

This profile drag is recorded in Figure 9.86 (right) as a polar curve,[16] $c_L = f(c_D)$, whose typically parabolic form will be taken up next. It should already be pointed out here, however, that the form of these polar curves depends on the Reynolds Number $Re_l = U_\infty \cdot l/\nu$, a fact that should be taken into account with studies on miniaturized models.

The lift of a profile can be increased by installing one or more wingflaps. The flaps are adjusted before the race—discussion will return later to the criteria that determine how this is performed—during the race they are for the most part obligatorily nonmobile. The wing and the wingflap together form a split wing.

The effect of such a flap can be seen in the pressure distribution $c_p = f(x/l)$ in Figure 9.88: The small slit between the rear edge of the profile and the nose of the flap allows air to flow through at high speed. The boundary layer, which at the end of the profile is already thick and in danger of separating, begins at the flap anew and can thus manage another pressure increase before separating. The influence of the flap angle η_K on lift and drag can be deduced from the polar curve in Figure 9.89. If the swing of the flap is increased from 10° to 30°, the lift increases by $\Delta c_L \approx 0.9$, and der c_D value increases by $\Delta c_D \approx 0.03$. Increasing the swing of the flap even farther does not produce any further increase in drag.

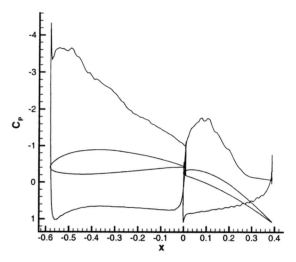

Figure 9.88 Pressure distribution of an arched profile with wing flap.

---

16. For racing vehicles, it has become common to describe negative lift as "downforce." The downforce is therefore treated as an independent variable; a downward-facing air force is positive.

High-Performance Vehicles

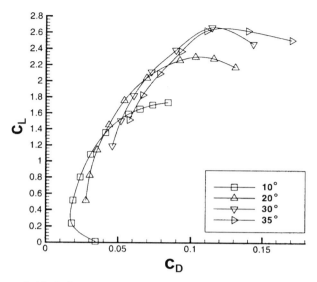

Figure 9.89 Polar curve $c_L(c_D)$ of an arched profile with wing flap.

9.7.2.4 Induced Drag

Induced drag can be explained by the curling-in of the layer of free longitudinal vortices (separation area) to the two free vortex strings:[17] "In each period of time a piece of these two free vortices must be re-formed. This requires the constant performance of work, which is contained in the vortex strings as kinetic energy." The induced drag of a wing can be calculated using Prandtl's lifting-line theory. If the lift is elliptically distributed over the wingspan, then the following holds exactly:

$$c_{D,i} = \frac{c_L^2}{\pi \Lambda} \tag{9.3}$$

This is also the minimum value that the induced drag can possibly have. The induced drag increases quadratically with the lift. This connection can be seen in Figure 9.86 and Figure 9.89; it is also an indication that the measurements reproduced there were performed on a wing of finite wingspan. This induced drag $F_{D,i}$ together with the profile drag $F_{D,0}$ accounts for the whole drag. Expressed in dimensionless coefficients, equation (4.25) already gave:

$$c_D = c_{DW,0} + c_{D,i} \tag{9.4}$$

This relationship has proven its usefulness in airfoil theory. This only applies, however, to wings with a large aspect ratio, that is, wings with a large wingspan, which is how the majority of wings employed in racing can be viewed. With small aspect ratios Λ, as occur in cars and sports cars, for example, this way of dividing up drag is, as explained in section 4.3.3, meaningless, because the profile drag is a constant.

---

17. For racing vehicles, it has become common to describe negative lift as "downforce." The downforce is therefore treated as an independent variable; a downward-facing air force is positive.

The performance capability (efficiency, $c_{L,max}$, $c_{D,min}$) of a wing is generally assessed using polar curve diagrams, as shown in Figure 9.86 and Figure 9.89. Here it is a question (depending on the particular case) of creating more lift with the same drag (polar curve is higher in the diagram than a comparable curve) or creating the same lift but with less drag (polar curve lies to the left of a comparable curve in the diagram). Normally, combinations of lift advantage and drag disadvantage have to be considered. In addition, there are secondary criteria, such as separation behavior, usable range, velocity range, as well as the degree of turbulence of the airstream, which also play a role in the selection of a wing profile. Of less importance, however, are criteria such as the resulting moment—a quantity that plays a part in aircraft construction. Wing profiles used in motor racing therefore differ in shape from those used in aviation.

9.7.2.5 Multiple Wings

Depending on the technical regulations and the vehicle class, multiple wing elements may be permitted. Combinations of several profiles often provide high performance, especially when large downthrusts are achieved. Additional optimization criteria, such as angle of attack and the distance separating one profile from another, slit streams, must then be considered. Figure 9.88 and Figure 9.89 show the pressure distribution and the Lilienthal polar curve of a two-part wing profile that was used on the Audi R8 sports car from 2000 to 2002.

There are various sources,[18] such as Althaus and Wortmann [13] and Selig [756], from which basic profiles can be taken for use as a starting point for the optimization process. For the most varied requirements, such as Reynolds number zone, laminar flow, and high thrust, there are corresponding profiles available, which can be adapted to the special application concerned. However, the functioning of such wing elements on the vehicle is limited by several factors: First, the wings used are wings of finite wingspan (the maximum width of the vehicle or narrower). The resulting loss of thrust was discussed. Technical regulations for racing cars as well as stylistic aspects for sports cars may possibly allow the use of end plates to minimize this loss. Further losses of thrust, so-called interference losses, can be caused by the immediate vicinity of components to one another: the body, wheels, exhaust pipes. These components are generally arranged so as to disturb the airstream to the wing elements as little as possible.

For racing cars, wings constitute components that strongly influence the performance capabilities of the vehicle. In the interests of safety and fairness, this area in particular is subject to very exact and heavily limiting regulations. This applies for example to the width of the wing (wingspan, in general the maximum width of the vehicle, or within the projection of the front face of the vehicle), the positioning of the rear wing with respect to the body (the closer to the vehicle, the stronger the interference and the lower the effectiveness of the wing), the chord line length (wing depth), and thus the wing area, the ground clearance for the front wings in particular, and the installation space

---

18. For racing vehicles, it has become common to describe negative lift as "downforce." The downforce is therefore treated as an independent variable; a downward-facing air force is positive.

that can be taken up by the wing including all its components. Some technical regulations even specify the profiles that are to be used. Figure 9.87 provides an example: It shows how the mounting of the rear wing elements for DTM vehicles in an exactly described installation space is prescribed (in these technical regulations, for example, the profiles for all vehicles are uniformly specified).

The description of the shape, location, and installation tolerances of wing elements typically takes up several pages in a set of technical regulations; this is done with the intention of avoiding extreme construction shapes. Figure 9.90 and Figure 9.91 show extreme shapes of this kind, which have not infrequently led to changes in the technical regulations; see also Figure 9.92 and Figure 9.93. These limitations were introduced primarily for the purpose of keeping the performance capabilities of racing cars within limits that can be managed or coped with by drivers.

The development of a wing system is only possible when the effectiveness of individual measures is transparent. Adjusting them relative to each other, and optimizing them, are traditionally done in the wind tunnel, increasingly with the aid of numerics.[19]

Figure 9.90 Pike's Peak racing car from 1987 with an arrangement of two split wings one over the other (photo: Audi AG). The current rally regulations allow one wing element.

Figure 9.91 Rally sports car Skoda Fabia WRC 2003. Photo: Skoda Auto a.s.

---

19. For racing vehicles, it has become common to describe negative lift as "downforce." The downforce is therefore treated as an independent variable; a downward-facing air force is positive.

Figure 9.92 Formula 1 racing car 1998 with nine rear wing elements (current technical regulations allow only three elements): a) view; b) section. Photos by kind permission of Giorgio Nada Editore, Milan, publisher of the Giorgio Piola yearbooks "Formula 1 Technical Analysis."

Figure 9.93 Formula 1 racing car 2001 with three elements on the rear wing. Photo: by kind permission of Giorgio Nada Editore, Milan, publisher of the Giorgio Piola yearbooks "Formula 1 Technical Analysis."

9.7.3 Spoiler and Gurneys

Spoilers are moldings on the body or on the underbody that serve to deflect the flow of air toward or away from a defined location. In contrast to (previously described) wings, spoilers do not have any underflow. The function of the spoiler has been described in section 4.5.1 with application to passenger cars. Examples in this section are provided for sports and racing cars. The reduction of drag force is the priority in the former chapter; the additional generation of downforce is the priority in the latter.

A typical competition vehicle front spoiler can be seen in Figure 9.94. Ground clearance is low; the design intent is to reduce lift force on the front axle. If the vehicle underbody is smooth, then the drag is increased.

Figure 9.94 Front spoiler with twisted cross section on the Audi A4 Super Touring in 1996. Picture by Audi AG, recorded in the wind tunnel of Pininfarina SpA.

Forwards-facing "lips" or "splitters" are also considered front spoilers when parallel to the road, as seen in Figure 9.95. The front axle downforce increases with increasing length of the splitter. This is achieved by increasing flow rates under the spoiler lip. Secondly, a zone of increased pressure on the top of the spoiler lip is generated, which further increases front axle downforce.

Figure 9.95 Vehicle with more pronounced front splitter (1999 Panoz LMP prototype).

In sports cars, the rear spoiler is intended to reduce lift on the rear axle. It can be unobtrusively molded to the body (Figure 9.96). Where its effect is not sufficient, a larger element is required, which is sometimes made movable as not to interfere with the vehicle's aesthetic design intent and is brought into an aerodynamically favorable position as a function of speed. As an example, Figure 9.97 shows the 2012 Porsche 911 Carrera S: its rear spoiler extends at 120 km/h, accordingly, the rear axle lift of $c_{L,r} = 0.14$ lowers to $c_{L,r} = 0.01$ in the extended position, with a small increase in drag force from $c_D = 0.31$ to 0.29.

Figure 9.96 Ferrari F60 Enzo production sports car with moderately shaped rear spoiler.

Figure 9.97 2012 Porsche 911 Carrera S with speed-dependent adjustable rear spoiler. Photos Dr.-Ing. H.C. F. Porsche AG.

Racing cars are prohibited from employing moving wings and spoilers in almost all racing classes. However, an altered appearance is often desirable to make the improved properties of the racing variant visible compared to the standard counterpart. Figure 9.98 shows the typical arrangement of a rear spoiler, in conjunction with a rear wing in the example of the DTM racing car Audi TT-R of 2001.

Figure 9.98 Typical arrangement of a rear spoiler, in conjunction with a rear wing in the example of the DTM racing car Audi TT-R of 2003. Photo Audi AG

The effect of this configuration can be seen from Figure 9.99. The rear spoiler without the influence of the rear wing resulted in a drag reduction of $\Delta c_D = -0.02$, while lowering the rear axle lift coefficient to $\Delta c_{L,r} = -0.13$ from point 1. For tail forms with large radii or severely sloping rear contours (large rear draft) spoilers can be particularly effectively, as they counteract lift and drag forces inherently induced owing to the shape of the tail. In the example, the effectiveness of the spoiler is particularly large owing to interactions with rear diffuser flows.

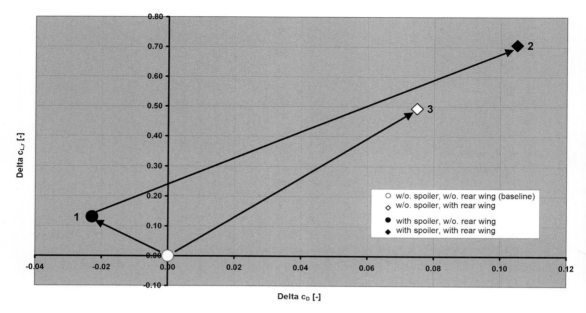

Figure 9.99 Influence of rear spoiler and rear wing on rear axle lift and drag in an Audi TT-R DTM racing car in 2001.

As can be seen from Figure 9.99, the effect of the rear wing is increased by the effects of the rear spoiler. The lift and drag changes that can be achieved by attaching a rear wing were greater in the presence of a spoiler, point 2, compared to its absence (point 3). A higher rear downforce can be realized by taking advantage of flow interactions between spoilers and wings. However, drag force is also increased in this scenario.

Another element affecting lift and drag is the "Gurney flap" (commonly abbreviated as Gurney).[20] This simple component installed perpendicular to the flow direction can be employed on many components, including wing profiles, spoilers, wing endplates, air baffle plates, general form curvature, and the like. Figure 9.100 shows the geometric design of a Gurney employed on a wing trailing edge and a sketch of the relevant potential flow field, characterized by two counter-rotating vortices downstream of the Gurney, contributing to a mean deflection of flow in the upwards direction. This contributes to an increase in (negative) lift force with an accompanying increase in drag force, as shown in Figure 9.101 in dimensionless form.

---

20. For racing vehicles, it has become common to describe negative lift as "downforce." The downforce is therefore treated as an independent variable; a downward-facing air force is positive.

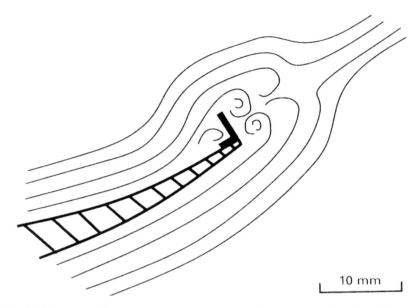

Figure 9.100 Gurney flap on a wing trailing edge with corresponding flow field [542].

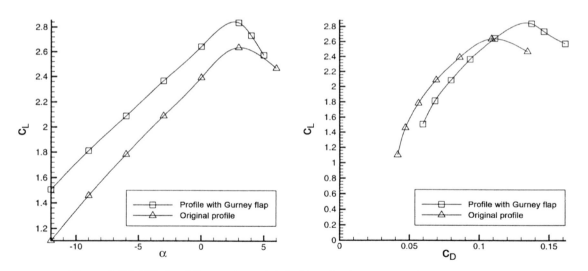

Figure 9.101 Lift about the angles: $c_L(\alpha)$ (left) and polar $c_L(c_D)$ (right) without and with Gurney flap.

A Gurney works similarly to an increased profile curvature.[21] With increased drag force, greater downforce can be realized. The device efficiency of fitting a Gurney ($\varepsilon_{Gurney} = \Delta c_L/\Delta c_D$) is, generally, worse than for the same device without fitting the

---

21. For racing vehicles, it has become common to describe negative lift as "downforce." The downforce is therefore treated as an independent variable; a downward-facing air force is positive.

Gurney. Positive effects can be achieved, especially if wing profiles, span, or element count are restricted by regulation. The key optimization parameter is considered to be the height of the Gurney, with typical heights ranging from 3 to 10 mm. Local flow conditions will determine the effectiveness of greater heights. Due to its simple construction and ease and speed of implementation, the Gurney has proven itself as an effective device, such as for balance adjustments during a racing event. Figure 9.102 shows a possible embodiment of a Gurney.

Figure 9.102 Gurney installed on the rear wing of the Audi TT-R DTM vehicle. To aid replacement, a sliding installation mechanism is employed where part of the wing endplate can be raised.

9.7.4 Ground Effect

The modification of aerodynamic properties when approaching the ground plane is called ground effect, which concerns both span-related and the length-related ground effect phenomena in racing vehicles.

The former, the span-related ground effect, plays an important role in flight vehicles; aircraft use it during takeoff and landing. When approaching the ground, a decrease in induced angle of the wings is realized by flow diversion in ground proximity.

Consequently, the induced drag force decreases and lift is increased. These influences can be noted in the design and positioning of race car front and rear wings.

The latter, the length-related ground effect, concerns the shape of the vehicle underside. The venturi effect is exploited; downforce is produced through negative pressure on the underside of the vehicle. The road surface and vehicle underbody form a combination of nozzle and diffuser, as shown schematically in Figure 9.103.

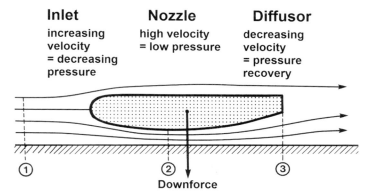

Figure 9.103 Principles of ground effect.

Incoming air is directed through the inlet under the vehicle and accelerated in the nozzle formed between the ground plane and the road surface. The subsequent diffuser slows air downstream to approximately the free stream velocity. The speed increase in area of the narrowest cross-section ensures a decrease in local pressure and an according increase in aerodynamic downforce. The pressure drop and thus the output depend essentially on the ratio of cross sections between either end of the nozzle and diffuser.[22] The optimum ratio is largely determined by the individual vehicle form and other conditions such as local suspension geometries, the location of the engine and transmission, the shape the vehicle chassis or inherent safety cell, and not least by the limitations of any class regulations.

The key parameter is the ground clearance, as seen in Figure 9.104. Negative lift increases with decreasing distance between the vehicle and the ground plane to a maximum; further decreases in ground clearance decrease negative lift, caused by reduced mass flow in the diffuser throat and by the convergence of boundary layers of road and vehicle underside. These factors lead to an increase in induced drag under the vehicle, which in turn leads to lower flow velocities, reducing the negative lift.

---

22. For racing vehicles, it has become common to describe negative lift as "downforce." The downforce is therefore treated as an independent variable; a downward-facing air force is positive.

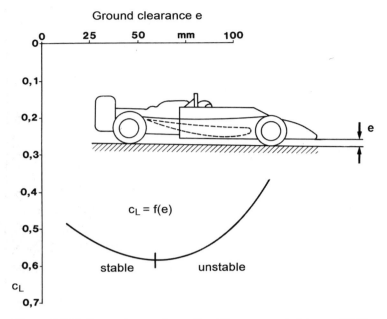

Figure 9.104 Dependence of negative lift on ground distance [895].

The great advantage of producing negative lift using the venturi effect is that induced (and thus net) drag force is hardly altered. This is clearly shown in Figure 9.105.

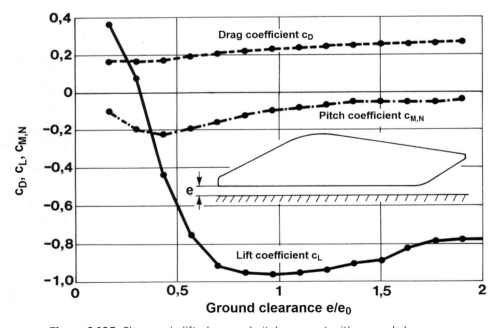

Figure 9.105 Changes in lift, drag, and pitch moment with ground clearance.

Publications from the late 1970s reported hitherto impossible vehicle aerodynamic properties. According to Wright [895], the output values of Formula 1 racing cars reached up to $c_L = 2{,}6$, up to 80% of which was achieved by ground effect. A classic example of this, and one of the first vehicles utilizing ground effect in Formula 1 racing, was the Lotus 79 (Figure 9.106).

Figure 9.106 1977 Lotus 79 Formula 1 race car. Image W. Wilhelm.

Blocking side gaps in venturi tunnels using side skirts can augment suction force acting against the ground plane. Accordingly, a gap bumper extending to the road surface was attached on both sides of the vehicle ("skirt"), as shown in Figure 9.107. The effectiveness is increased with decreasing gaps height, as per Figure 9.108.

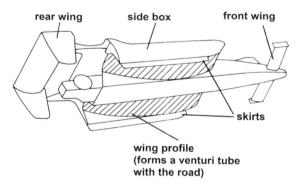

Figure 9.107 Underbody of a racing car with ground effect [488].

Due to the high effectiveness of skirting solutions and ground effect and the associated efficiency of the vehicles, critical limits were realized with respect to the longitudinal and lateral acceleration of both materials employed in vehicle construction and additionally in terms of driver resilience. For safety concerns, skirting solutions were first banned by the regulations, and limits were placed on underbody designs in Formula 1 since 1983. Most current racing regulations prescribe a level floor between the front and rear axles. Exceptions to this are some racing series in the United States (CART, IRL), as well as some formula series in which the ground effect may be used in a strictly regulated manner.

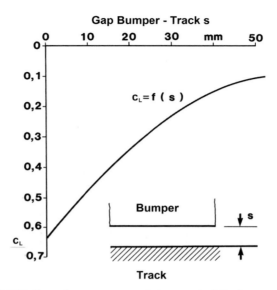

Figure 9.108 Downforce vs. ground clearance of the bumper [895].

However, recent trends revisit the intended use of ground effect, although restricted by critical geometric constraints (such as minimum ground distance). A certain basic level of output can be realized in a relatively safe manner, ensuring sufficient traction in almost all vehicle conditions, including all critical conditions. To keep the level of total negative lift within limits, other components (such as front and rear wings) are limited in their effect to a similar degree. Regulations may be prescribed that tend toward aerodynamic system solutions whereby the impact of particularly sensitive aerodynamic parts and systems may be less critically expressed in front and rear axle loads. This results in safer driving at comparable load levels. An example of this can be found in the 2014 sports car regulations (FIA and ACO); a corresponding underbody meeting these regulations is shown in Figure 9.47.

9.7.5 Diffusers

As discussed previously, diffusers find many uses in sports and racing vehicles, particularly for air intakes, and especially on the front and rear vehicle underside. Their effectiveness depends on their geometry and their proximity to the (relative to vehicle) moving ground plane.

9.7.5.1 Function

A diffusor attached to the underbody reduces the static pressure on the entire vehicle underside, thus generating downforce. Depending on the geometry of the diffuser, vehicle aerodynamic drag can also be reduced. By describing the flow conditions along the underbody, one can create some design rules for such an underbody diffuser.[23] An

---

23. For racing vehicles, it has become common to describe negative lift as "downforce." The downforce is therefore treated as an independent variable; a downward-facing air force is positive.

underbody diffuser consists of a channel formed between an upwards-drawn floor, usually toward the vehicle rear, and the nearby road. As seen in Figure 9.109, the flow field between these forms is asymmetric: one wall—that nearest to the vehicle—is pulled upwards and thus generates the downstream expanding flow cross section. The other wall—the road—moves relative to the vehicle. Such a diffuser produces a downstream increase in static pressure, as the cross-sectional area increases, and therefore, for reasons of mass flow continuity, the flow velocity in the expanding channel is reduced.

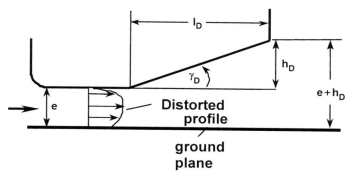

Figure 9.109 Geometry of the rear diffuser.

The ratio of outlet to inlet cross section determines the upper limit for this pressure increase, whereas the nondimensional diffuser length, l_D/e, determines the corresponding degree to which the flow cross section increases and thereby defines the opposing pressure gradient in the diffuser. For a simple, two-dimensional underbody diffuser between two parallel walls, the area ratio is defined as AR:

$$AR = \left(1 + \frac{h_D}{e}\right) = 1 + \frac{l_D}{e}\tan(\gamma_D) \tag{9.5}$$

The pressure increase generated by a diffuser in practical flows is limited by the ability of the flow to stay attached to the surrounding walls. If the opposing pressure gradient (which would be set in smooth flow) exceeds the capacity of the boundary layer to stay attached, then the flow will be removed from one of the walls—or is removed from both. In this case, the effective area ratio of the flow is less than the geometry, and at the same time, the delay of the flow is reduced as well as the corresponding pressure increased. A reduced effective area ratio can also result if the flow remains attached, when the profile of the incoming flow is not uniform. With inner currents, the nonuniformity of speed across the cross section reduces the effective cross section of the flow (flow blockage). Any nonuniformity of the inlet flow is heightened when, in the diffuser, velocity decreases and pressure increases. Flow blockage reduces the effective area ratio and thus the effectiveness of the diffuser. From considerable experimentation, it has been determined that the following parameters are relevant for the correct functioning of a diffuser:

- The cross-section ratio, AR, which sets the upper limit of the possible pressure ratio (with an absolute upper limit considered as per smooth, one-dimensional flow)
- The dimensionless length, l_D/e, the measure of what viscous effects are needed to overcome the associated adverse pressure gradient created by the expanding diffuser geometry
- Blockage and associated effects in inlet flows, which contribute to any nonuniformity of inlet velocity

9.7.5.2 Design

The following account is based on measurements on a simplified vehicle model with relevant dimensions, as shown in Figure 9.110. While no specific vehicle is evaluated, the behavior of the diffuser is considered representative. The results obtained can be applied to both road cars with a smooth underbody as well as sports and racing cars with a similarly smooth underbody. The data is less applicable to single-seater vehicles, mainly due to characteristically complex vortex formations extending from the sides of such vehicles, such as from "barge boards" or other vortex-generating systems intended to increase downforce.

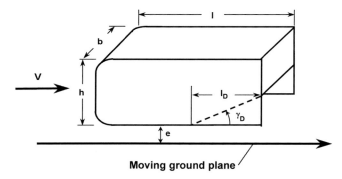

Figure 9.110 Generic vehicle model: $l/h = 2{,}40$; $b/h = 1{,}29$; $b/e \approx 20$.

9.7.5.3 Generation of Downforce

The static pressure at the base of a vehicle is nearly independent of the efficiency of the diffuser, and thus the increase in pressure which is produced by the diffuser results in a reduction of the pressure at its inlet. Consequently the average pressure in the diffuser is reduced, and the result is increased downforce. The pressure at the bottom upstream of the diffuser is also reduced, and thus the negative pressure generated is further increased. So the diffuser makes two contributions to the downforce of the vehicle. This means the ratio of the length l_D of the diffuser is a significant design parameter to the overall length l of the subfloor. Downforce is therefore generated by two mechanisms, which have no relation to the diffuser:

- Contouring upwards of the rear underbody generates a downforce at large distance from the ground (this is at about 1 AR), because it creates a negative curvature of the body. This output represents the base amount of downforce to which the output generated by the diffuser must be added.

- The interaction between body and ground at small ground clearances also generates downforce without any upwards contouring of the vehicle underbody being required.

These three components contributing to negative lift are shown in Figure 9.111 Two configurations are shown, one without and the other with raised subfloor; the independent variable is the dimensionless model ground clearance (e/h).

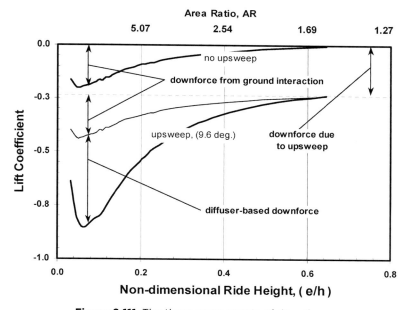

Figure 9.111 The three components of downforce.

The model in which the underfloor is not contoured or raised (no upsweep) generates little to no negative lift at a large distance from the ground because its geometry is symmetrical. But when it is brought closer to the ground plane, it produces downforce due solely to the interaction between the model and the floor. In contrast, the body with raised underbody, because of its negative curvature, generates useful downforce even at a large distance from the ground plane. Now, if this model is brought closer to the bottom, it increases the area ratio AR of the diffuser being formed, which consequently increases the pressure on the entire underbody. The resulting force is the diffuser-generated downforce. At very low ground clearance, elasticity effects reverse the output trend of both bodies.

How the total output—that is the sum of the downforce generated by the diffuser, the downforce attributable to the raising of the downstream vehicle underbody and that occurring as a result of the interaction with the ground—is dependent on the dimensionless diffuser length l_D/e and the area ratio AR is shown in Figure 9.112a and b for the model referred to in Figure 9.110. In Figure 9.112a, the relative length of the diffuser $l_{D/l} = 0.25$ is relatively small, and it is therefore likely to correspond to the ratio in an actual vehicle, where the diffuser begins at the location of the rear axle. In Figure 9.112b the relative length $l_{D/l} = 0.75$ is very large. The measurements were carried out at with a moving floor. The contour lines denote constant downforce.

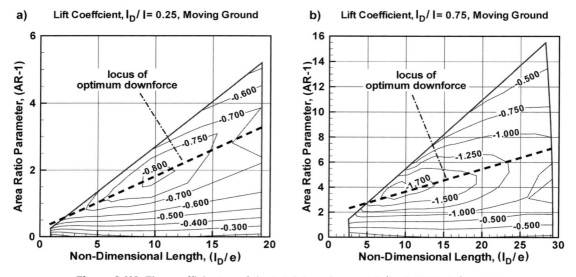

Figure 9.112 The coefficient c_L of the total downforce: a) $l_D/l = 0.25$; b) $l_D/l = 0.75$.

These diagrams can be used directly for the design of underbody diffusers. In general, the ground clearance, e, defined, and the diffuser length, l_D, is determined either by design boundary conditions, or they can be set according to the required downforce. The location of the maximum downforce is described by a straight line. When the predetermined variable is the exit altitude h_{D+e} and not diffuser length l_D, the area ratio AR of the diffuser is fixed and the downforce for this value of AR can be read at various diffuser lengths l_D/e.

If maximum downforce is the design goal, it is possible that the optimal length of the diffuser encroaches on upstream length of the underside and consequently the maximum output cannot be achieved. This influence of the length portion of the diffuser, l_D/l, is shown in Figure 9.113.

The mean effective pressure on the bottom, the dependent variable, is directly proportional to the downforce. Each of the curves applies to a given pair of values of ground clearance e/h and area ratio AR. The pair of values $e/h = 0.100$ and $AR = 2.02$ give the largest downforce, which occurs at $l_D/l = 0.50$. However, significant values for the output with other diffuser lengths l_D/l and for other combinations of e/h and AR can also be generated.

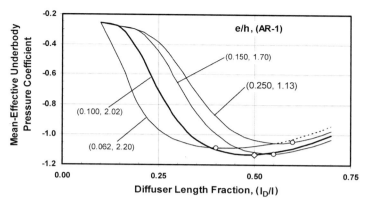

Figure 9.113 Influence of diffusion lengths on the averaged pressure coefficient cp below the vehicle.

9.7.5.4 Reduction of Drag Force

Raising the underbody will also affect induced drag. This depends mainly on the distribution of the static pressure along the projection on the rear of the vehicle caused by this modification. For the model shown in Figure 9.110, this area is composed of two parts:

- The height, h_D, of the rear projection of the raised floor in relation to the base level of the model
- The height of the base $(h-h_D)$

The drag generated by each of these two areas is the product of the mean pressure on this surface and its height. An example of how the two drag-generating components depend on each other is given in Figure 9.114.

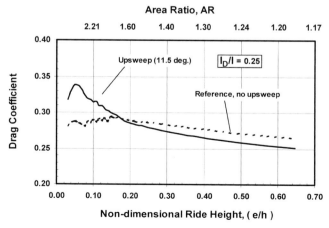

Figure 9.114 Influence of the (dimensionless) ground clearance e/h on the drag coefficient c_D.

Although this example shows the influence of an underbody diffuser on net drag, it is not typical for the design of the diffuser. In a typical scenario, the design ride height, e, and the length of the diffuser, l_D, are predetermined, and the wall angle, γ_D, which determines the area ratio AR is the free variable. How the drag coefficient of the generic model varies with AR is shown in Figure 9.115 for two different relative lengths of the diffuser, l_D/l. The dependent variable is the increase in drag $(c_D - c_{D,0})$, where $c_{D,0}$ is the drag force at the wall angle $\gamma_D = 0$ (i.e., $AR = 1$).

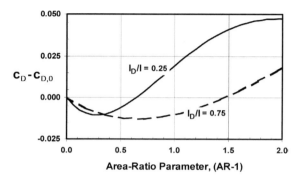

Figure 9.115 Influence of area ratio $AR - 1$ on the drag reduction.

At small AR, a small increase of pressure at the base of the vehicle is connected with a large height and thus produces a significant reduction in drag force. The concurrent reduction of the pressure on the wall of the diffuser (underfloor) is effective at only a small height h_D, thus causing only a small increase in drag force. The sum total remains a decrease in net vehicle drag.

To the extent that AR increases with at fixed ground clearance with increasing γ_D, the base pressure increases further, while the effective surface portion of the base decreases. This again reduces the base drag somewhat. In contrast, the effective pressure recovery in the diffuser reduces friction drag considerably; the reasons for this are the reduction of the mean pressure on the diffuser wall and the increase in their projected height, h_D. The net effect of the two competing trends, at a certain area ratio AR, results in a minimum of drag force, followed by an increase, which leads to an overall greater drag force than in a body without a diffuser. The optimum AR, which leads to the minimum drag, for the short diffuser is approximately 1.25, while in the extended diffuser the AR is about 1.75.

The drag-reduction charts in Figure 9.116 offer a different consideration of the potential of a diffuser to reduce drag. This diagram shows lines of constant drag changes where $(AR - 1) = f(l_D/e)$. The largest drag reduction occurs with a small AR and limited length of diffuser.

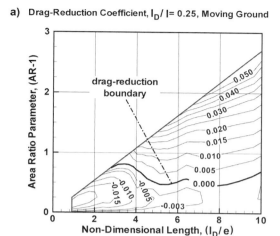 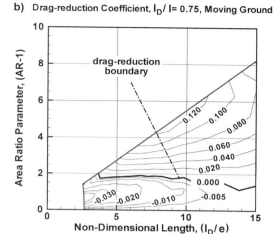

Figure 9.116 Drag reduction as a function of the dimensionless length l_D/e.

9.7.5.5 Conclusion

The reported data is based on experiments with a very basic model. The diagrams may be used as guidelines, not as precise specifications. They are equally applicable to passenger cars, sports, and racing cars having a flat underbody. The design of a system of underbody and diffuser can be guided by these principles; optimization is however dependent on CFD and wind tunnel development. In general, the intent of the diffuser is to produce downforce. However, it also allows for a reduction of drag force, but only for small area ratios. The generation of large downforce requires larger area ratios and is almost always associated with a higher vehicle drag force. Maximum downforce and minimum drag cannot be achieved simultaneously.

The function of the diffuser is very sensitive to the roughness of the upstream subfloor. The improvements communicated here have been achieved with a smooth underbody. Current racing cars have smooth subfloors. This makes it possible to develop optimum diffusers. In contrast, the underbodies of most of today's production cars are much too rough to fully exploit the advantages of a diffuser and should be "smoothed."

In racing, this means diffusers can achieve high downforce at a relatively low increase in drag force; consequently, the aerodynamic efficiency of the vehicle can be significantly increased. However, the dependence of the efficiencies of the diffusers to the ground clearance requires a high development effort and is reflected in the relatively complex forms of diffusers.

As an example, Figure 9.117 shows the front diffuser of a Lola sports car of the 24 hours of Le Mans race in 1999. Several diffuser channels of different shape and angle gradients generate high downforce on the front axle. This attempts to control the dependence of the output to the ground clearance.

Figure 9.117 Underside of the front diffuser of a Lola sports car that took part on 24 hours of Le Mans in 1999.

By contrast, Figure 9.118 shows complex diffuser channels at the rear of a Formula 1 racing car. This layout seeks to generate both downforce and control of boundary vortices at the wheels.

Figure 9.118 Complex rear diffuser channels used in a Formula 1 racing car by Brawn GP at the Australian Grand Prix in 2009. The generation of downforce and control of vortices at the wheels are key optimization criteria for the design of such rear diffusers.

High-Performance Vehicles

There are also ways to enhance the performance of a diffusor by adding high-energy flow from the exhaust gas. Two major design targets are pursued:

1. Control of the exhaust gas streams in the rear section in a manner that does not adversely affect the basic aerodynamics.
2. Active control of the rear diffuser flow through the exhaust stream to increase downforce. Here we need to consider the case of aerodynamic consequences arising when the exhaust stream is not fully available (e.g., on the overrun of the motor). This can be remedied by using a "continued burning" of the engine. The consequences of the increased fuel consumption and the resulting extended refueling times and higher vehicle weight must be considered and evaluated against the aerodynamic advantage in terms of overall performance of the vehicle.

Even with standard sports cars, diffusers are increasingly used in the underbody region. Figure 9.119 shows the design of the underbody at the rear of the Ferrari 360 Modena. The large rear diffuser channels begin immediately at the rear axle. The downforce obtained by the diffuser may be used to realize low total lift, and also offers greater flexibility in the design of rear exterior surfaces.

Figure 9.119 Using a rear diffuser on the standard sports cars Ferrari 360 Modena.

9.7.6 Inlets and Outlets

Airflows passed through the vehicle have the following three tasks:

1. Supply the engine with intake air for combustion
2. Cooling of various radiators and other heat exchangers
3. Cockpit/cabin ventilation

The inlet and outlet apertures are to be located and sized such that through flows incur minimal losses. Most regulations allow relatively large, open spaces for this purpose. In practice, they are employed to work in conjunction with other aerodynamic systems such that well-designed inlets and outlets may generate additional downforce and reduce drag force by discharging flows at the base of the vehicle—known as "base bleed."

The ideal location for an air intake is the stagnation area at the front of the vehicle (Figure 9.120). If other aspects of the vehicle design determine the shape of the inlet aperture, this does not necessarily have to lead to an aerodynamic disadvantage. If the edge of the inlet is well rounded, the influx is almost loss-free. Usually this area is used to supply air for water and oil radiators, as well as for electrical energy storage systems, the engine air intake, ventilation of the cockpit, and for cooling air to the front brakes. The opening for the cooling air inlet can be kept small, because pressure loss across the radiator results in a low-loss widening of the stream tube, which ensures a uniformly distributed inflow.

Figure 9.120 DTM racing car Audi TT-R in 2003 with provision of the cooling air for water cooler, brake cooling, and engine supply from the bow section of the vehicle.

In a number of mid-engined or rear-engined vehicles, the cooling air is drawn from both sides to the center of the vehicle. Compared to flows from the front of the vehicle, side-drawn flows are of lower energy because they are influenced by and disturbed by flows around the vehicle. Consequently, the intakes must be sized accordingly. However, by optimizing the pressure loss across the radiator, it is still possible to achieve high cooling capacities. Figure 9.121 shows such an execution. Sometimes very unconventional solutions are implemented. An example of this is the extendable cooling duct of the Lamborghini Murciélago, which provides additional cooling air when needed (Figure 9.122). Another advantage of side-mounted radiators is that less liquid coolant needs to be carried, providing a useful vehicle weight reduction.

Figure 9.121 Cooling air arrangement in the center of the vehicle in the 2005 Renault Formula 1 car. Image: Renault.

Figure 9.122 Retractable cooling duct of the Lamborghini Murciélago as an example of a solution for cooling a mid-engined vehicle. Image of Automobili Lamborghini.

Many air intakes must be arranged on the smooth surfaces of the body. For this, the "NACA" duct is a proven solution;[24] it is now also used in passenger cars, mainly on the vehicle underside. The objective in developing this form was to achieve maximum air throughput for minimum additional drag force. As shown in Figure 9.123, vortices rolling up on the sharp, outside edges induce a high-velocity inwards jet of air (a "downwind") and thus divert airflow inside.

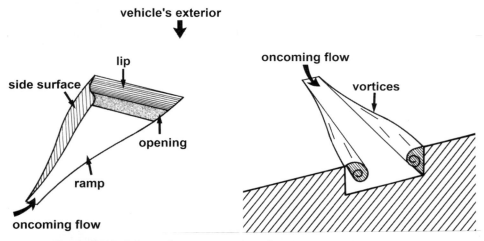

Figure 9.123 Schematic representation of a NACA inlet after Reilly [663].

If the effect of a NACA duct is not enough to produce the desired throughput, additional scoops are often attached, which either protrude from the surface of the body or which generate an inflow. In effect, these "scoop" air into the vehicle interior. These air scoops (an example is shown in Figure 9.124) can cause a significant increase in induced drag, depending on their location on the vehicle.

Figure 9.124 Scoop behind the rear wheel used for oil cooling on the Lamborghini Murciélago; Image Automobili Lamborghini.

---

24. For racing vehicles, it has become common to describe negative lift as "downforce." The downforce is therefore treated as an independent variable; a downward-facing air force is positive.

The disadvantage of this arrangement concerns the boundary layer of flows adjacent to the vehicle surface, which increases in thickness along the length of the vehicle. This limits the effect of any such scoop the farther back along the length of a vehicle, as less air is captured. To compensate for this, relevant inlet openings are elevated from the body surface in a number of racing cars; the boundary layer is then peeled off at the foot of the inlet. Depending on the inlet height above the vehicle surface, an almost uninterrupted flow into the opening can result. However, this design can result in relatively high induced drag and may generate flow phenomena that negatively influence subsequent, downstream aerodynamic systems (for example, the rear wing). In general, such inlets are used for the supply of fresh inlet air to the engine, where the resulting increased inlet pressure results in even higher cylinder pressures after compression and can be utilized for increased performance. Figure 9.125 shows a typical execution of such an intake, as used for formula car use. The height above the body is limited by the regulations.

Figure 9.125 Engine air intake on the 1978 Renault race car, which won the 24 hours of Le Mans that year.

Special openings can be found in the wheels for front brake ventilation while simultaneously reducing induced drag losses arising from cross-flow through the wheels. Figure 9.126 shows the brake air duct of a Formula 1 car. A design that has been adopted in some sports cars is shown in Figure 9.127. The key criteria here was is to limit further apertures in the outer form, as opposed to any direct aerodynamic effect.

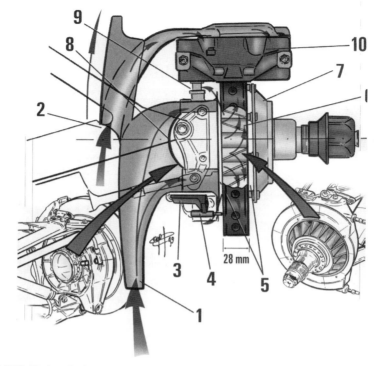

Figure 9.126 Brake air duct of a Formula 1 car. Image: Courtesy of Giorgio Nada Editore, Milan, Publisher of yearbooks Giorgio Piola "Formula 1 technical analysis."

Figure 9.127 Brake cooling system on the Ferrari F60 "Enzo."

The air outlets are located preferably in the rear of the vehicle, to inject mass flow into the separated region at the base of the vehicle and thus to increase the base pressure. Some vehicles use the entire base area to exhaust vehicle internal flows. Figure 9.128 shows a sports car prototype, in which this idea was implemented comprehensively.

Figure 9.128 Cooling air outlet at the rear of the sports car prototype R8. Almost the entire rear base surface is used as cooling air outlet for the vehicle's internal flows. Image of Audi AG.

If discharge at the rear is not possible, for example, due to front-mounted heat exchangers, then flow is vented under the vehicle in a manner analogous to production vehicles. This application is rare in race cars, as sensitive underbody flows (which can generate significant negative lift acting on the front axle) may be disturbed. Consequently, outlets may be found on the side of the vehicle (Figure 9.129) or on the front bonnet (Figure 9.130). These outlets are shaped so that outer flows are displaced to accommodate the cooling airflow. These vents can also be used also for the production of downforce, especially in the ventilation of front diffusers in touring and sports cars (Figure 9.131).

Figure 9.129 a) and b) air outlet on the sides of the vehicle. Implemented in the sports car prototype Mercedes-Benz CLK (1988).

Figure 9.130 Air outlet on the top of the vehicle as shown in the Skoda Fabia WRC rally car. Image: Skoda Auto s.a.

High-Performance Vehicles

Figure 9.131 Front diffuser ventilation to reduce lift force as shown in the sports car prototype Audi R8 in 2000. Image of Audi AG.

Further ventilation examples may be found at the top of wheel arches. These provide for an increase of negative pressure in the wheel arch and consequently suck adjacent airflows to vehicle outside surface. The decreased internal pressure increases net downforce. These vents (also louvres) can be found on many vehicles in different categories (Figure 9.132).

Figure 9.132 Short (left) and long (right) louvres on the sports car prototype Audi R8.

9.7.7 Air Guiding Elements (Vanes)

Air guiding elements or "vanes" are attached anywhere on the vehicle, where the flow must be influenced such that additional aerodynamic effects are implemented that cannot be achieved efficiently with the base form of the vehicle. This mainly involves components that generate downforce. These components are characterized by a low material thickness, often with a rounded front edge and a sharp rear edge. In addition to the this effect, they may also serve to adjust downforce and drag settings to trim aerodynamic performance to other vehicle performance parameters. Vanes on the front side of the vehicle (Figure 9.133) are a typical example.

Figure 9.133 Vanes at the front of the DTM racing car Audi TT-R in 2003 for generating downforce.

The vehicle body is equipped with side vanes (or dive planes) that can be fitted with height-adjustable Gurneys. As a consequence, the induced drag force on the side of the vehicle increases. At the same time, the vane is analogous with a wing profile; therefore there exists increased pressure on the top of the element and negative pressure on the bottom. This pressure differential across the component ensures a corresponding reduction in lift. At the outer edge of the component an intense boundary vortex is generated in the longitudinal direction of the vehicle and affects the flow conditions in the inflow to the wheelhouse and wheel. These components increase drag but reduce lift on the front axle effectively and so are popular as a simple adjustment with which to trim aerodynamic balance. Similar components can be found on the endplates of current Formula 1 front wings, as Figure 9.134 demonstrates.

High-Performance Vehicles

Figure 9.134 Use of vanes on the front wings of a Formula 1 car. Old and new "nose" of the Williams F1 BMW FW 26 Image: BMW AG.

More vanes at the front of the vehicle can often be found on the vehicle underside attached to front wings or front diffusers. These vanes can perform several tasks. They can optimize the flow volume for each of the channels formed by the vanes, so that flow separation can be avoided or systematically controlled and spatially limited, resulting in higher downforce. Cross currents can also be reduced or, depending on the arrangement of these elements at a certain angle to the flow, vortices can be generated (similar to the boundary vortices in a NACA duct), which can influence flow direction in the horizontal or vertical direction compared to the vehicle body.

Usually, additional downforce is generated by such elements, which increases with the number of elements. They act, as with all underbody aerodynamic systems, in a relatively efficient manner, meaning with only a small increase in drag. However, their relatively aggressive effect in conjunction with variable ground clearance may lead to detrimental vehicle characteristics, which are difficult to control. These elements can be used in varying configurations to trim aerodynamic balance of the vehicle. Such components can also be found in the center of the vehicle; a particular example concerns so-called "barge-boards" on formula cars, as per Figure 9.135.

Figure 9.135 "Barge-board" on a current Formula 1 car. Picture by Renault.

Vanes, in addition to lift reduction, may be used to control flows toward cooling air vents; an optimum balance between channeling and control of the approaching flow is desired. Without these vanes a larger volume would flow around and past the front of the radiator. Controlling inlet flows in the underfloor area and across the sidepod is also achieved through these components. The relatively diverse effects achievable through many tuning parameters illustrates that there exist many different combinations of adjustable aerodynamic systems able to be effected to set up a vehicle optimally.

More vanes may be found in front of the rear wheels of some formula cars (Figure 9.136) and in the rear diffusers (Figure 9.137) of almost all racing cars. The main task remains the control of flows to and from corresponding components, the spatial positioning of flow vortices to reduce lift, as well as the displacement of other sources of turbulence that could reduce the effectiveness of other aerodynamic components.

Figure 9.136 Vane used to control the inflow to the rear wheel. Image of BMW AG.

Figure 9.137 Vanes in the rear diffuser of the Panoz LMP01 sports car prototype employed to control edge vortices of the rotating wheel and for the production of rear downforce.

9.7.8 Wheels

The wheel represents a bluff body with a width of approximately 200 to 400 mm and a diameter of 600 to 750 mm. Its shape can hardly be affected aerodynamically, unless over the rim and the radius of the tire shoulder. In racing applications, a large tire is typically required to ensure the transmission of traction to the road. Additionally, the life of racing tires is influenced, in part, by their width and diameter.

Open wheels, as employed in formula racing cars, are the subject of numerous aerodynamic investigations. A discussion of typical flows can be found in section 4.5. An overview of the magnitude of lift and drag associated with open wheels is shown in

Mears et al. [549], Figure 9.138;[25] data is presented as a function of the frontal area of each wheel configuration investigated.

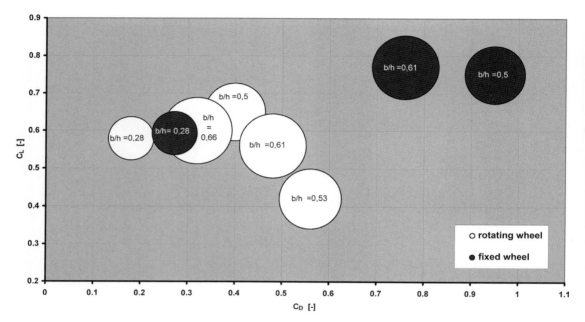

Figure 9.138 Overview of studies on open wheels and their effects on aerodynamic drag and downforce. Data from Stapleford [777], Fackrell [240], Cogotti [151], Mears [549].

For narrow wheels, as typical of production vehicles ($b/h = 0.28$), a relatively low drag force exists in an open, rotating configuration of about $c_D = 0.18$.[26] Wider wheels with a width-to-height ratio exceeding 0.5, as are common with race cars, show drag coefficients of $c_D = 0.32$ to 0.56. Wide wheels are therefore aerodynamically significantly less favorable than the narrow counterparts. Drag force is influenced by rotation, particularly for relatively wide wheels. The drag force incurred by a stationary wheel is up to twice as high as for a rotating one. Open wheels cause also lift, which varies, according to the literature, from $c_L = 0.4$ to $c_L = 0.6$. Lift is some 30% higher for a stationary wheel than for a rotating wheel.

Depending on the vehicle class and tire width used, open wheels can contribute up to 50% of the total drag of the vehicle. For wheels covered by the body on production vehicles, Wickern et al. [869] showed that the contribution of the wheels can still account for up to 25% of drag. However, race cars with well-optimized wheel arches and corresponding inflow to the wheels can sometimes achieve values well below 15%. In general,

---

25. For racing vehicles, it has become common to describe negative lift as "downforce." The downforce is therefore treated as an independent variable; a downward-facing air force is positive.
26. For racing vehicles, it has become common to describe negative lift as "downforce." The downforce is therefore treated as an independent variable; a downward-facing air force is positive.

High-Performance Vehicles

for covered wheels the drag component of the front wheels is lower than that of the rear wheels. This is mostly due to the greater freedom of the geometry before the front wheels offered by regulations. The angle of attack of the vehicles (which is usually set with a lower ride height) contributes to greater exposure of the rear wheels to underbody flows. Where applicable, greater width of the rear wheels further contributes to this phenomena.

Optimizations with covered wheels are made primarily in the areas of front and base slows, as well as the design of wheel well surroundings. Special requirements apply in the front wheel well, wherein the steering movement must also be considered in addition to the spring travel. Figure 9.139 shows basic elements that can be used to improve the wheelhouse environment, provided regulatory constraints and the ground clearance make this possible.

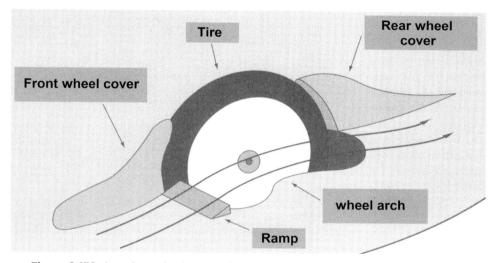

Figure 9.139 Aerodynamic elements for optimizing wheel well environment, [788].

Spoilers and vanes are not only found on vehicles with covered wheels. Even Formula cars have so-called "barge boards" (Figure 9.135) or vanes in front of the rear wheels (Figure 9.136) to effect suitable measures to have bulk flows bypass the wheels.

Other influences on wheel bypass flows and through flows can be achieved through wheel covers that can reduce drag (if allowed by the regulations) or with appropriately designed "fan" wheel rim designs that promote cross-flow through the rim and thus change the ventilation of wheel well. For better resolution of effects on the wheel, special wheel scales are used for wind tunnel measurements, allowing wheel forces to be recorded separately. Interference effects between wheel and body can then be observed. Pressure measurements offer the possibility to determine the lift force of the rotating wheel.

Because of the significant differences in the aerodynamic characteristics of stationary and rotating wheels, especially for relatively wide wheels, nearly all wind tunnel measurements with racing cars (and increasingly with production sports cars) are made with rotating wheels. A rolling road experimental arrangement covering the whole width of the test section and driving the wheels (which are mechanically separated from the chassis) has become a standard experimental setup. In such arrangements, the vehicle is usually suspended from the top as shown in Figure 9.140. More details can be found in chapter 13.

Figure 9.140 40% model ($M = 1:2.5$) with rotating wheels in the wind tunnel of Fondmetal Technologies Srl. The wide running road also serves to drive the wheels, which are mechanically decoupled from the vehicle body and are held by supports from the outside. These supports also house balances to measure the forces acting on the wheels.

Chapter 10
Commercial Vehicles

Stephan Kopp, Thorsten Frank

10.1 Target Group

The ability to operate commercial vehicles as economically and profitably as possible has always been a fundamental goal of manufacturers and suppliers. To meet this goal, they are required to make use of all available options to minimize fuel consumption. Within the context of the ongoing efforts to reduce CO_2 emissions, the aerodynamic features on commercial vehicles also play a crucial role in maximizing economy, in addition to other measures such as the further development of economical diesel engines, improvements to the tire rolling resistance, or optimization of the driveline.

This becomes evident if we consider the changes in fuel consumption over the past few decades (see Figure 10.1). From the 1970s to the 1990s, fuel consumption and thus CO_2 emissions were steadily reduced. With the introduction of the Euro 1-5 emission standards, fuel consumption stayed largely constant. It is here that aerodynamics can make a further contribution toward improving fuel economy.

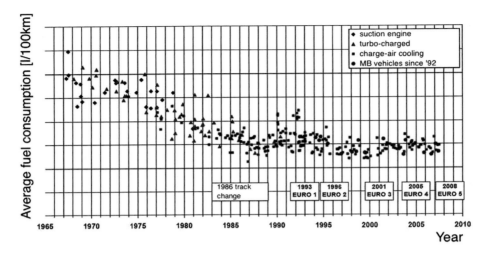

Figure 10.1 CO_2 emissions since 1965 [Lastauto Omnibus, 06/07].

In addition to the legal framework and the resultant challenges arising, it is also important to consider the overall life cycle costs (Figure 10.2). Fuel costs and wages make up the majority of these costs. On a typical long-haul truck, fuel costs account for approximately 30% of total life cycle costs, and this percentage is set to rise even further given the increasing price of crude oil. This shows that the importance of aerodynamics as a way to reduce aerodynamic drag and thus minimize fuel consumption is continuing to increase.

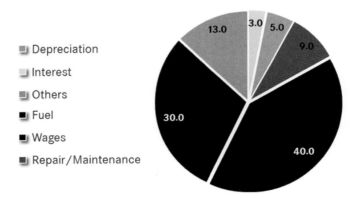

Figure 10.2 Life cycle costs.

Within this context it should be noted that, for certain commercial vehicles, aerodynamics plays only a small role due to the specific deployment profile or design of these vehicles. Examples of such vehicles include construction vehicles, municipal vehicles, or special-purpose vehicles (Figure 10.3).

Commercial Vehicles

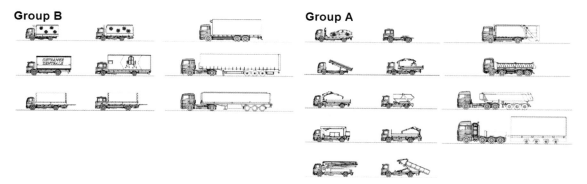

Figure 10.3 Relevant and irrelevant commercial vehicles in terms of aerodynamics.

An additional challenge facing commercial vehicle manufacturers is posed by the fact that they are generally only responsible for developing the cab, chassis, and driveline, whereas other independent companies are responsible for the industry-specific bodies, semitrailers, and trailers.

10.2 Driving Resistances and Fuel Consumption

Aerodynamics plays an important role in reducing the fuel consumption of intercity or long-haul vehicles traveling at high speeds. As such, aerodynamic measures designed to reduce fuel consumption are aimed at the most common trucks and tractor/trailer units fitted with high bodies, coaches, and express delivery vans.

When developing any measure designed to reduce aerodynamic drag and thus also fuel consumption, it is important to consider the maximum potential savings that may be achieved. It is therefore necessary to calculate the driving resistances and energy required to overcome them.

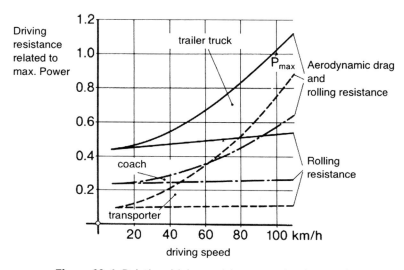

Figure 10.4 Relative driving resistances on level ground.

An analysis of a truck driven on level ground at a constant speed shows (see Figure 10.4) that the rolling resistance often exceeds the aerodynamic drag. The aerodynamic drag only becomes the primary resistance again for lightweight solo trucks or express delivery vans when driven at average speed, or for heavy-duty truck/trailer units when driven beyond the legally permitted maximum speed of 80 km/h. In spite of this, the importance of aerodynamic drag should not be disregarded, even for the latter vehicle types, if we consider that the power required by for example a 40-t semitrailer combination with a 4-m-high body to overcome aerodynamic drag is 25 kW at a speed of 60 km/h or 60 kW at a speed of 80 km/h.

If a typical European 40-t standard long-haul truck is considered, it becomes clear that the influence of aerodynamic drag on the vehicle's fuel consumption differs according to the road profile and possible driving speeds. For an Actros 1843 LS semitrailer combination, this ranges from 2.5 to approximately 40% when driving on motorways (Figure 10.5).

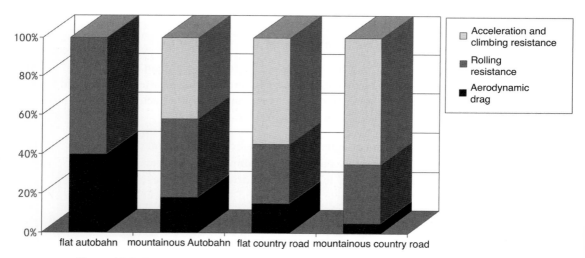

Figure 10.5 Energy required to overcome the various driving resistances for a 40-t semitrailer combination depending on different road profiles.

The impact the reduction in aerodynamic drag has on a vehicle's fuel consumption in everyday operation can be shown most clearly by using typical vehicles and driving conditions. In accordance with Mercker & Knape [563], Figure 10.6 shows the conditions for a 40-t semitrailer combination as a typical example of heavy-duty tractor/trailer units. In actual deployment under conditions such as "very difficult route" or "A road," this shows a considerably smaller fuel saving (relative to the overall consumption) when compared with the ideal of driving at a constant speed on level ground.

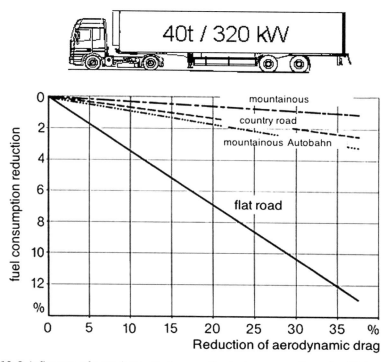

Figure 10.6 Influence of aerodynamic drag on the fuel consumption of a 40-t semitrailer combination.

The research and development work carried out by Europe's commercial vehicle manufacturers is aimed at reflecting the increased environmental awareness and complying with more stringent legal regulations. Today's commercial vehicles are quieter, cleaner, and more economical while operating at higher average speeds. Figure 10.7 shows that the fuel saving for a 40-t truck is more than 35% over 45 years ago at a considerably higher speed.

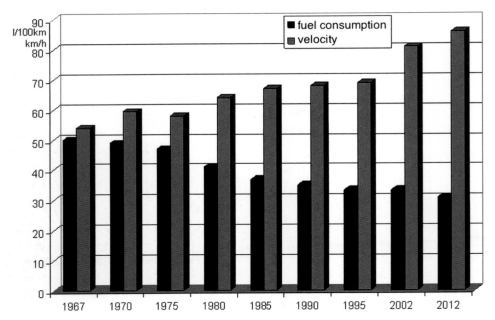

Figure 10.7 Reduced fuel consumption over time despite higher average speed for a 40-t tractor/trailer unit.

Of particular interest within this context is an analysis of the energy flow diagram for a modern long-haul truck compliant with the Euro 5 emission standard. Figure 10.8 shows how the total energy of the consumed diesel fuel is distributed among the individual consumers and losses.

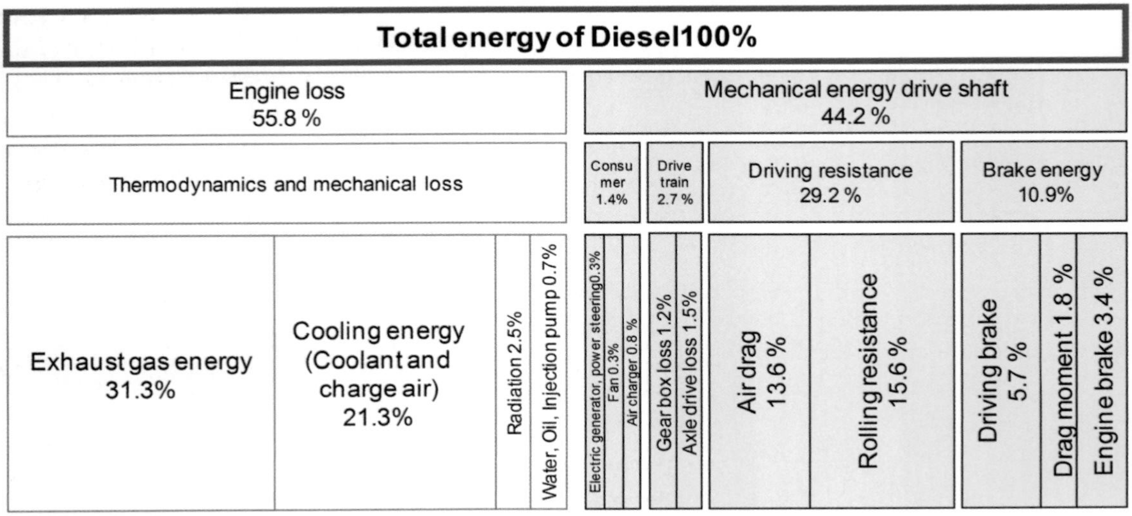

Figure 10.8 Energy flow diagram for a modern long-haul truck compliant with the Euro 5 emission standard.

First, a distinction must be made between losses in the engine and mechanical energy on the crankshaft. The losses in the engine amount to almost 56% of the total energy. The losses are distinguished between thermodynamic and mechanical losses, with the energy input in the exhaust gas and cooling system accounting for most of the losses in the engine.

Forty-four percent of the energy is available at the crankshaft for vehicle propulsion. After deducting the losses for the required auxiliary units such as the alternator, fan, and air compressor and for the braking energy, around 30% remains to overcome the driving resistances. As stated previously, aerodynamic drag accounts for approximately 40% of the total driving resistance. This shows that optimizing the aerodynamics can play a major part in reducing the vehicle's fuel consumption.

Measures designed to reduce aerodynamic drag result in excess power, which nowadays is essentially used to reduce fuel consumption instead of increasing the engine's performance. A lower fuel consumption can be achieved by making suitable modifications to the driveline configuration. The engine performance programs use empirically derived data to assign precisely defined gearshifts to specific engine speed ranges that allow lower fuel consumption. However, the most effective measure for achieving the required reduction in the engine speed is to modify the final ratio. For economic reasons, in the first instance this should be achieved by selecting a suitable ratio from those available as standard. An optimal ratio can then be implemented in subsequent development cycles and incorporated into series production. The best results are achieved if modifications to the operating conditions are developed with the aim of reaching a good compromise between improved vehicle performance and lower fuel consumption.

The example in Figure 10.9 shows an engine map. Point A shows the total driving resistance (aerodynamic drag and rolling resistance) for a 40-t semitrailer combination traveling on level ground at approximately 80 km/h. In this case, the vehicle's fuel consumption is approximately 32.8 l/100 km. The aerodynamic enhancements reduce the energy required to overcome the aerodynamic drag and thus the total driving resistance by 10 kW (point B) and reduce the vehicle's fuel consumption by approximately 2.9 l/100 km. Modifying the rear axle ratio (point C) allows the fuel consumption to be reduced by a further 0.15 l/100 km (approx.) with the same excess power as for point A.

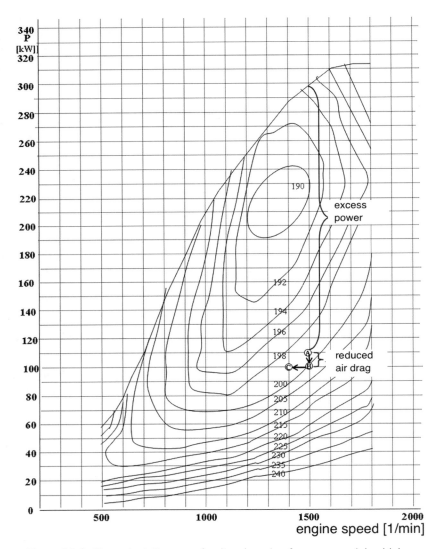

Figure 10.9 Characteristic map of a diesel engine for commercial vehicles.

10.3 History of Commercial Vehicle Aerodynamics

In the mid-1930s, construction of the German motorways meant that buses and coaches had to be as aerodynamic as possible in order to achieve the desired speeds using engines that would be considered weak by today's standards. Streamlining was the order of the day, and teardrop-shaped bodies and covered wheel housings were contemporary solutions. The focus on the aerodynamics of these buses and coaches subsequently made its way into the development of commercial vehicles (see Figure 10.10).

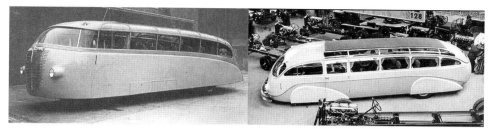

Figure 10.10 Gaubschat tram bus on a Büssing chassis and a Mercedes bus from 1935.

The first Kässbohrer streamlined buses, which were built using chassis from MAN and Büssing, already achieved a maximum speed of 130 km/h on the newly constructed motorways (see Figure 10.11).

Figure 10.11 Kässbohrer streamlined bus from 1937.

The VW Transporter, developed in 1949 and known around the world by its nickname "Bully" (Figure 10.12), was a milestone of aerodynamics in commercial vehicles. Thanks to the Transporter's rounded front, its c_D value was reduced by almost half compared with the preliminary design, which had a squared front. The VW Transporter became the prime example of commercial vehicle aerodynamics.

Figure 10.12 VW "Bully" from 1949.

When engineers first began to develop measures to improve the aerodynamics of semitrailer combinations in the 1970s, they focused in particular on roof spoilers. Roof spoilers and air deflectors rapidly became widely used, as they were very efficient, simple to retrofit, and provided a convenient way of adapting to the body height of each truck (see Figure 10.13).

Figure 10.13 MAN F90 with roof spoiler.

In-depth studies were also carried out into other measures such as tapered rear sections on semitrailers and anti-turbulence attachments in the space between the tractor vehicle and trailer. Although these measures led to major fuel savings, they were never implemented in series production.

This was due to legal restrictions and the resulting difficulties with implementation, as well as the fact that a commercial vehicle is not the responsibility of a single manufacturer but rather a number of different manufacturers. For example, the tractor vehicle (cab, chassis, and driveline) is developed by one manufacturer, but other independent body companies then add the appropriate body to complete the overall vehicle, in line with the intended deployment profile.

Morchen [579] summarized the problems associated with commercial vehicle aerodynamics with these words: "We know most of the answers, we need to apply them."

Since commercial vehicle manufacturers were unable to achieve major savings, they intensified their efforts to optimize the aerodynamics of the cab and chassis. Their work in this field and the corresponding financial outlay increased as they strived to achieve ever smaller fuel savings. It is only recently that commercial vehicle manufacturers succeeded in convincing the European Commission to rethink the legal regulations governing length dimensions in EU Directive 96/53 to enable them to develop suitable aerodynamic features on the vehicles without affecting the space available in the cargo area.

10.4 Principles of Commercial Vehicle Aerodynamics

The aerodynamic rating of a vehicle is expressed using the dimensionless aerodynamic drag coefficient c_D. However, the actual determining variable is the aerodynamic drag area $c_D \cdot A_x$. For instance, a commercial vehicle needs a much higher engine output to overcome its aerodynamic drag area at a speed of 85 km/h than a DTM or Formula 1 racing car, despite the fact that they have approximately the same c_D values (Figure 10.14).

Figure 10.14 DTM vehicles and MAN TGX from 2012.

The airflow phenomena responsible for aerodynamic drag described in section 4.1 for passenger cars can also be applied to commercial vehicles. Figure 10.15 shows the airflow separation areas on the various vehicle types, visualized by 3D airflow simulation on the basis of an isotropic area display where $p_{tot} = 0$. The airflow separates primarily at the rear but also at the front of the vehicle, at the wheels, rear view mirrors, and at the individual underbody parts (control arms, axles, and units).

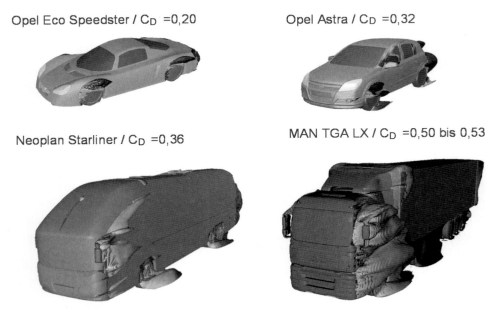

Figure 10.15 Airflow separation on various vehicle types ($p_{tot} = 0$).

Figure 10.16 shows the scatter range of c_D values for the various basic vehicle designs. The higher aerodynamic drag values for commercial vehicles compared with passenger cars are primarily due to the designers' efforts to provide the largest possible cargo area while also complying with the maximum vehicle dimensions stipulated by law. Additionally, divided bodies with poor aerodynamic properties and open load carriers such as platform bodies or tipper bodies are disadvantageous.

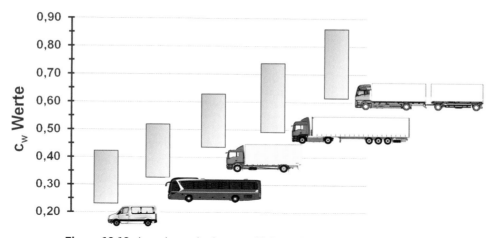

Figure 10.16 Aerodynamic drag coefficients for various vehicle types.

10.4.1 Straight/Oblique Flow

Since commercial vehicles are driven at relatively low speeds, side winds play a more important role. Figure 10.17 shows a strong side wind acting on a semitrailer combination with a 10° oblique flow.

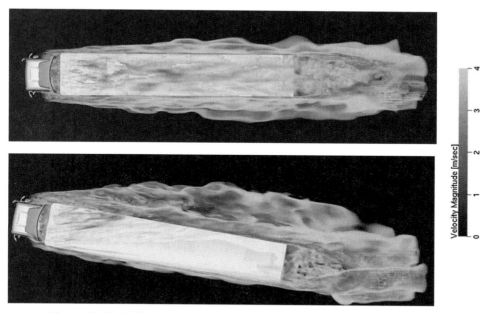

Figure 10.17 Airflow visualization with a straight flow and 10° oblique flow.

In order to take this effect into account when developing the aerodynamic features on commercial vehicles, the average aerodynamic drag coefficient c_D is increasingly being used. This more accurately indicates the aerodynamic rating of the vehicles under real conditions than a c_D value evaluated solely on the basis of a 0° airflow. The angles and speeds of a side wind are directly related with one another in geometric terms.

$$v_\infty^2 = (v_F + v_W \cos\phi)^2 + (v_W \sin\phi)^2 \tag{10.1}$$

$$\beta = \arctan\frac{v_W \sin\phi}{v_F + v_W \cos\phi} \tag{10.2}$$

The increase in aerodynamic drag compared with a symmetrical airflow depends on the flow angle β and the vehicle's basic shape. The greater the number and the dimension of the space between the body and tractor vehicle, the greater the increase in the polar with an oblique flow. This is because the flow separates over a large area and the airflow around and through the running gear is turbulent. Figure 10.18 shows how $c_D(\beta)$ changes for various vehicle types for an oblique flow. It can be seen that the increase is smaller on trucks with an enclosed body and no space behind the cab.

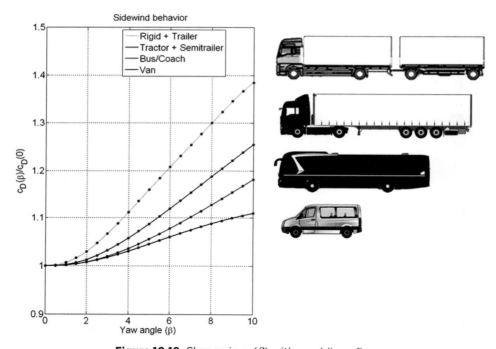

Figure 10.18 Change in $c_D(\beta)$ with an oblique flow.

In order to incorporate the effect of side winds in actual deployment, a number of different calculation formulae have been proposed that are intended to represent an aerodynamic drag coefficient $\bar{c}_D$ based on an average side wind. However, no standard definition exists for this wind-averaged aerodynamic drag coefficient. Two basic assumptions can be applied to calculate the average aerodynamic drag coefficient. First, the side wind hits the vehicle with a given probability and at a given angle and, second, a constant side wind distributed evenly over 360° can be assumed. According to Ingram

[394], the integral for calculating the average aerodynamic drag coefficient taking into consideration a probability of an assumed air flow angle is

$$\overline{c}_D = \int_0^{v_{max}} \int_0^{2\pi} c_D(\beta)\left[1+\left(\frac{v_W}{v_F}\right)^2 + 2\left(\frac{v_W}{v_F}\right)\cos\phi\right]P(v_W,\phi)d\phi dv_W \qquad (10.3)$$

where $P(v_W,\phi)$ indicates the probability that a side wind of magnitude v_W prevails at an angle ϕ to the direction of travel. $c_D(\beta)$ is the value for an oblique flow at angle β, measured in a wind tunnel or calculated using computational fluid dynamics (CFD) simulation.

According to Cooper [177], this formula is simplified if an even distribution of the side wind between a frontal head wind $\phi = 0$ and 2π and tail wind only $\phi = \pi$ is taken as the basis.

$$\overline{c}_D = \frac{1}{2\pi}\int_0^{2\pi} c_D(\beta)\left[1+\left(\frac{v_W}{v_F}\right)^2 + 2\left(\frac{v_W}{v_F}\right)\cos\phi\right]d\phi \qquad (10.4)$$

Under the assumption that the side wind v_W remains constant and occurs with equal frequency at all angles, the range between 0° and 360° can be represented by an equally distributed flow angle $\phi(j)$. A weighting factor $M(j)$ must be calculated for each side wind angle $\phi(j)$. This factor must then be multiplied by the $c_D[\beta(j)]$ values measured in the wind tunnel with an oblique flow. The weighting factor indicates whether the wind is rather a head wind or tail wind. The average value of the weighted $c_D[\beta(j)]$ values corresponds to the average aerodynamic drag coefficient c_D.

$$\overline{c}_D = \frac{1}{n}\sum_{j=1}^n M(j)c_D(j) \qquad (10.5)$$

$$M(j) = 1 + \left(\frac{v_W}{v_F}\right)^2 + 2\left(\frac{v_W}{v_F}\right)\cos\Phi(j) \qquad (10.6)$$

$$\Phi(j) = j\cdot 30° - 15° \qquad (10.7)$$

$$c_D(j) = c_D[\beta(j)] \qquad (10.8)$$

$$\beta(j) = \arctan\frac{\left(\dfrac{v_W}{v_F}\right)\sin\Phi(j)}{1+\left(\dfrac{v_W}{v_F}\right)\cos\Phi(j)} \qquad (10.9)$$

In the United States, a driving speed of $v_{F,\text{Truck}} = 55$ mph and a typical wind speed of $v_W = 7$ mph (3.1 m/s) are generally taken as the basis for the calculation [177].

If we consider the wind atlas for Europe (Figure 10.19), it is clear that the side wind values can also be applied in Europe. Average wind speeds of 5 m/s prevail at a height of 50 m. At ground level, the wind speeds are lower due to the roughness of the terrain (caused by trees and features of the landscape). Applying the Hellman exponent of $\alpha = 0.2$ results in an average wind speed of 3.0 m/s for a height of 4 m.

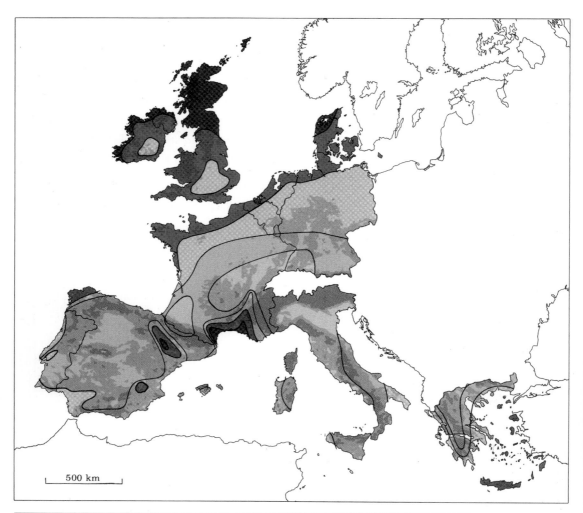

Figure 10.19 Wind atlas for Europe [www.windatlas.dk].

As such, in Europe a driving speed of $v_{F,Truck}$ = 89 km/h (24.7 m/s) and also a typical wind speed of v_W = 3.1 m/s can be taken as the basis for the calculation. These values may vary depending on the intended deployment and localization of the vehicle.

10.4.2 Legislative Framework

The development of design and aerodynamic features on commercial vehicles is subject to strict restrictions. In the case of trucks, this concerns in particular the lengths and heights prescribed by law, which generally have to be used to their full extent in order to provide the largest possible cargo area. The definitive legal framework in this regard is provided by the EU Directives 96/53/EC and 97/27/EC (Figure 10.20). Moreover, in the overwhelming majority of cases, the truck manufacturer is only responsible for the development of the vehicle's chassis, driveline, and cab, while the bodies, semitrailers, and trailers required for the various different branches are designed by independent firms. The situation that has arisen out of statutory requirements and the traditional division of labor between vehicle and body manufacturers has led to the European truck market becoming dominated by the so-called cab-over-engine type of truck, which offers only a restricted amount of space for aerodynamic design.

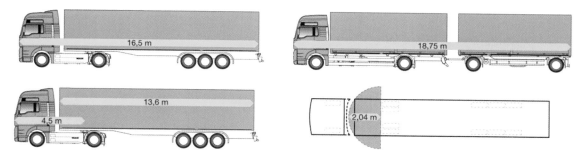

Figure 10.20 Lengths prescribed in EU Directive 96/53/EC.

In this restricted space—a length of 2.3 m—an attractively designed cab with sufficient space for two drivers to work and live in has to be harmonized with the aerodynamics. What is more, increasing demands with regards to engine cooling, interior air-conditioning, dirt on the vehicle, and aeroacoustics also have to be taken into consideration. This balancing act is achieved by optimizing many of the details of individual components, most of which remain invisible to the untrained eye.

However, the rear of semitrailer combinations offers the greatest potential for optimizing aerodynamics. Studies into front and rear extensions designed to improve aerodynamics have shown that the rear offers twice as much potential as the front.

Nevertheless, commercial vehicles offer only limited scope to improve aerodynamic drag through modifications to the exterior design of the body, as a tapered rear cuts into the cargo area and thus reduces the cost-effectiveness of the vehicle. A tapered rear would also significantly impair loading and unloading and/or require a costly flap mechanism.

Buses and coaches offer much greater scope for optimizing aerodynamics than trucks on account of their legal and commercial framework. For example, both the front and rear of buses and coaches can be used for optimizing the vehicle's aerodynamics (see section 10.7, Optimizing Aerodynamic Drag on Buses and Coaches).

10.5 Tools for Optimizing Commercial Vehicle Aerodynamics

10.5.1 Challenges Posed by Commercial Vehicles

The main challenges associated with optimizing the aerodynamics of commercial vehicles lie not only in the strict legislative and commercial framework but also in the actual tools provided to aerodynamics engineers:

- Model wind tunnel
- Wind tunnel measurements with scale 1:1
- CFD simulation
- Calculation of the c_D value during test drives

10.5.2 Model-Scale Wind Tunnel

Thanks to their ability to provide reproducible measurements that are unaffected by the weather, wind tunnels provide an ideal method of carrying out meaningful, systematic research into aerodynamic features. However, the tunnel dimensions required to accommodate commercial vehicles can pose problems. In the large wind tunnels available today, it is barely possible to test small trucks and buses/coaches in their actual size. Larger vehicles cause an impermissible blockage in the measuring section, both transversely and longitudinally. As a result, they need to be tested on a smaller scale. In order to obtain realistic values, the blockage in the wind tunnel must be less than 10%. The effect of side winds also plays a bigger role for commercial vehicles due to the lower driving speeds, and the blockage increases considerably if the vehicles are rotated 10° in the tunnel.

Given the enormous effort involved in conducting a 1:1 test in the wind tunnel as well as the associated costs, it is recommended to increasingly perform the tests using measurements based on a scale model. As for all model measurements, it is important to observe the Reynolds number dependence within this context. At a scale of 1:4, the airflow speed must be increased by a factor of four. As a result, measurements using a model scale of 1:4 are generally avoided, as airflow effects and model inaccuracies influence the measurement in a manner that cannot be anticipated. Smaller model scales are used for tests on overall vehicle combinations, where the focus is not on optimizing the details but rather on the interaction between the tractor vehicle and semitrailer type (Figure

10.21). Here the primary aim is to obtain information about the aerodynamic rating of a vehicle combination and how the aerodynamic attachments function.

Figure 10.21 1:4 wind tunnel models to test the interaction between the tractor vehicle and semitrailer type.

Cooper (1985) used a cuboid in a longitudinal airflow and with varying edge radii to determine the smallest Reynolds number required to optimize commercial vehicles. He calculated an optimal edge radius of approximately $r_{opt} = 0.05 \cdot \sqrt{A_x}$. The study showed that the Reynolds number for model measurements must be at least $Re = 2 \cdot 10^6$ if the edge radius $r_{opt} = 0.05 \cdot \sqrt{A_x}$ must also be achieved using the model scale. For passenger cars, their length is around three times that of the root of their frontal area, $l \approx 3 \cdot \sqrt{A_x}$. As such, in this case the Reynolds number calculated on the basis of the length must be at least $Re = 6 \cdot 10^6$. For semitrailer combinations with a characteristic length of around five times that of the root of their frontal area $l \approx 5 \cdot \sqrt{A_x}$, the Reynolds number must be greater than 10^7.

In general, a scale model of 1:2.5 is very well suited to use as a scale model (Figure 10.22). It permits airflow conditions that are true to the original (same Reynolds number as the full-scale version), as well as an airflow through the inside (radiator, engine compartment). The aerodynamic coefficients obtained in this way correspond well to those for the full-scale version; soiling simulations can also be carried out very effectively. The track width for this scale is approximately 1.0 m, which in turn corresponds to the usual width of the center belt in passenger car wind tunnels using a five-belt system. As such, this scale allows tests to be performed with rotating wheels and a moving roadway.

Figure 10.22 MAN TGX and NEOPLAN Cityliner as 1:2.5 scale models in the Pininfarina wind tunnel, Turin.

When testing using model measurements, it is also important to consider other factors besides the scale. The geometric similarity of the details affecting the vehicle's aerodynamic quality must be correct (Figure 10.23). This includes sunblinds, air deflectors on the A–pillars, and the details of the engine compartment and chassis area.

Figure 10.23 MAN TGX model details on the chassis and engine compartment components.

The underbody of the tractor vehicle and semitrailer must be as detailed as possible. This is because the chassis is not located underneath the vehicle but centrally within it. The ladder frame concept means the air flows through the chassis area and not underneath it as for passenger cars. The air should always flow through the engine compartment on both the wind tunnel model and the CFD model, as it has a major impact on the overall airflow around the vehicle. This particularly affects the design and efficiency of corner radii, window inclines, and, of course, underbody covers.

Another obstacle is the fact that the majority of passenger car wind tunnels are not designed to accommodate the lengths required for commercial vehicles. As a result, aerodynamics engineers have to manage with shortened semitrailers, depending on the length of the belt in the tunnels (Figure 10.24).

Figure 10.24 Use of different semitrailer lengths, depending on the wind tunnel.

Within this context, it is important to note that, when using shortened semitrailer bodies, the chassis of the missing semitrailers needs to be simulated using models. This is primarily done by using a corresponding body length and chassis paneling. Figure 10.25 shows the different air flow topology when using a truck model for a shortened semi-trailer without these modeling measures.

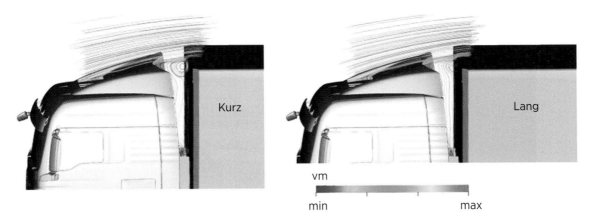

Figure 10.25 Airflow topology for a short box body in comparison with a tractor vehicle with a long semitrailer.

However, measurements with rotating wheels on the center belt in wind tunnels using a five-belt system are highly complex, as there are no weighing scales underneath the center belts in conventional wind tunnels. If the model is positioned on the center belt, the wheels are driven by the belt, and it is not possible to measure the lift or correct aerodynamic drag. However, given the high dead weight and low driving speeds of commercial vehicles, lift only plays a minor role and can therefore be disregarded for aerodynamic optimization. If the wheels are not decoupled from the model, their rolling resistance will be included in the measured resistance force. This needs to be measured using torque dynamometers inside the model and deducted from the measurement from the wind tunnel scales.

Another option would be to position the model just above the center belt and drive the wheels using motors inside the model. However, this would result in an incorrect measurement due to the gap between the belt and wheel. Nevertheless, this incorrect measurement is primarily attributable to lift, which is only of secondary importance for commercial vehicles. The error for the aerodynamic drag measurement is much smaller. Correct measurements can be ensured if the belt on which the wheels rest is integrated into the scales, meaning there is no link between the entire belt system and the ground.

Some of the wind tunnels used are aeroacoustic passenger car wind tunnels, meaning aeroacoustic tests can already be performed early on in the development stage. The wall array setup is used for this (see Figure 10.26).

Figure 10.26 Wall array measurement using scale model.

10.5.3 Full-Scale Wind Tunnel

However, the largest problem facing aerodynamics engineers for commercial vehicles is the availability of suitable wind tunnels to carry out tests using 1:1 models. There are currently only three wind tunnels worldwide that have the space to accommodate a complete semitrailer combination: the DNW in the Netherlands, the NRC wind tunnel in Canada, and the NFAC wind tunnel at the NASA Ames Research Center in California (Figure 10.27).

Figure 10.27 Wind tunnels: NFAC (California), NRC (Canada), DNW (the Netherlands).

The vehicles are linked to the scales centrally and are positioned on air cushions to ensure there is no friction with the ground. Oblique flows up to 10° can be measured, and the DNW also allows aeroacoustic tests to be carried out. Wall array measurements can also be used to locate sources of noise and thus optimize the vehicle design. However, the focus is on the measurement of sound levels using dummy heads in the vehicle interior.

Figure 10.28 shows the dimensions of the DNW and a standard semitrailer combination with a length of 16.5 m. The design of the wind tunnel means the vehicles need to be lifted into the measurement section at a height of approximately 10 m above the ground. A vehicle lift and various cranes are available for this purpose. It takes approximately four hours to change a tractor vehicle within a given measurement run. Approximately six hours are required to change a complete vehicle combination with the help of two cranes.

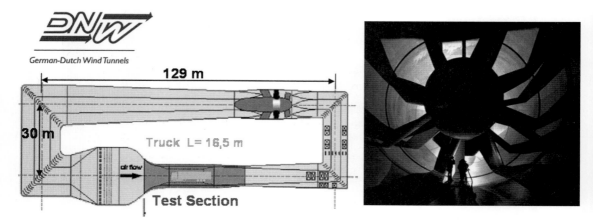

Figure 10.28 Dimensions of the DNW wind tunnel.

A further wind tunnel to carry out measurements on 1:1-scale vehicles is also available at Daimler in Untertürkheim (Figure 10.29). This tunnel has an air jet cross section of 32 m² and thus a blockage of approximately 30%. It is not possible to test a complete semitrailer combination or bus/coach, as the measurement section has a length of 12 m. As such, tractor vehicles with shortened box bodies are used instead to test aerodynamic optimizations. Regular reference measurements at the DNW provide the aerodynamics engineers with information about the corresponding aerodynamic drag of the overall vehicle combination, including the semitrailer.

Figure 10.29 Large wind tunnel at Daimler AG in Untertürkheim.

10.5.4 CFD Simulation

As described, testing commercial vehicles in wind tunnels involves considerable time and effort as well as certain compromises. As a result, numerical simulation is becoming increasingly important as a means of optimizing commercial vehicle aerodynamics, with modern data centers and CFD programs allowing rapid analysis of design changes. However, CFD tests on commercial vehicles require a large amount of computing power. Both straight flows and oblique flows need to be analyzed in order for engineers to obtain information that reflects real-world conditions as accurately as possible. As a result, each configuration usually requires two simulations to calculate the average aerodynamic drag coefficient (section 10.4.1).

In addition, the airflow around commercial vehicles is often characterized by highly variable airflow separation. This transient process poses a problem for CFD simulation if engineers wish to evaluate the aerodynamic benefit of minor design changes. The computing runs require long computing times in order to obtain meaningful average values.

Alongside these low-frequency, highly variable areas of turbulence, merely the geometric size of the vehicles and the need to include the engine compartment and chassis areas already cause long computing times (see Figure 10.30). The surface mesh of a complete semitrailer combination consists of approximately 3 million triangle surfaces. The CFD Solver PowerFlow uses this as the basis to generate around six to eight times as many surface elements (surfels) to safeguard the link to the volume mesh. The volume mesh for a semitrailer combination with air flowing through the engine compartment consists of approximately 95 million of these "voxels."

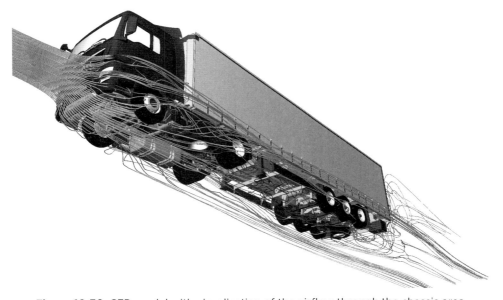

Figure 10.30 CFD model with visualization of the airflow through the chassis area.

The computing time required for one aerodynamics run including the airflow through the engine compartment amounts to approximately 4,500 CPUh. Around 5,500 CPUh are required for a thermal calculation with a rotating fan and inclusion of the heat transfer in the cooling system in the analysis. The required computing power increases by approximately 20% if the analysis is to include the rotating wheels using the sliding mesh method.

10.5.5 Test Drives with Wheel Hub Measurement Device

Two methods exist for calculating a vehicle's c_D value by measuring the driving resistances as part of an actual test drive.

- Coasting test
- Constant-speed driving

Intensive studies conducted by Graz University of Technology [224] in 2012 showed that, when calculating the c_D value, constant-speed driving provides better reproducibility of results than conventional coasting tests.

In order to calculate the c_D value using constant-speed driving, the drive torque at the driven axles and the speed of the airflow striking the vehicle need to be measured. The drive torque is measured using torque measurement hubs at the wheel rim or DMS strips on the drive shafts. The actual airflow speed and flow angle for the vehicle are measured using an anemometer fitted approximately 1.3 m above the vehicle. This height is required to ensure that the wind speed measurement is not influenced by the vehicle as far as possible. Figure 10.31 shows a representative measurement setup for DAF [829].

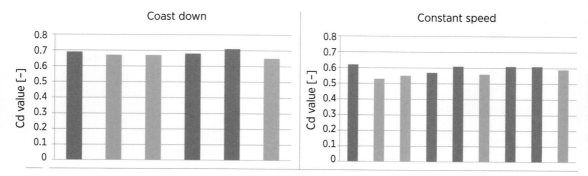

Figure 10.31 Measurement setup for constant-speed driving with anemometer and wheel hub measurement device [829].

The measurements are taken at two speeds. The first measurement is taken at a speed of 15 km/h to determine the rolling resistance. At this speed, the effect of aerodynamic drag is disregarded. The second measurement is taken at 90 km/h. Assuming that the rolling resistance remains the same when traveling at the specified speed, the rolling resistance determined at 15 km/h can now be subtracted from the drive torque of the 90 km/h measurement, resulting in the aerodynamic drag (Figure 10.32) at 90 km/h.

Figure 10.32 System for distinguishing the driving resistances.

However, it has been proven that this assumption is incorrect and causes major fluctuations in the results: the development of the rolling resistance at the specified speed varies depending on the tires used and also depends on the ambient and tire temperatures. In addition, the climatic conditions also make it very difficult to calculate the c_D value with a straight flow of 0°. According to studies by Volkers [829], side winds with an oblique flow of more than 2° occur at a frequency of approximately 40% (Figure 10.33).

In order to calculate c_D for 0°, either the test drives need to be performed when there is no wind or a generic angular polar (Figure 10.18) must be applied. This generic curve is plotted through the measured $c_D\ (\beta)$, and it is then possible to determine the theoretical $c_D\ (0°)$. To mitigate the effect of the error due to the use of the angular polar, measurements should only be performed with a side airflow of less than 3°.

Extreme care must be taken when positioning the anemometer on the vehicle and selecting the test route in order to ensure the flow angle and flow speed are correctly determined. Bushes and hedges lining the test route will result in false results regarding the actual airflow striking the test vehicle. As such, it is recommended to position an additional stationary anemometer on the test route in addition to the anemometer on the vehicle.

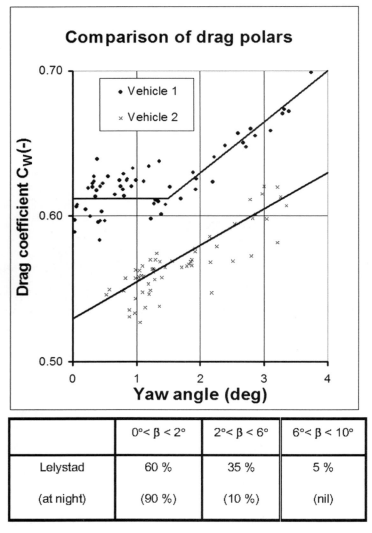

Figure 10.33 Frequency distribution of oblique flow at 90 km/h at the Lelystad test site [829].

If we compare the time and effort required for measurement, the required vehicle preparations, and the measurement accuracy with tests conducted in a wind tunnel, it is clear that test drives used to optimize vehicle aerodynamics are only useful to a limited extent.

Commercial Vehicles

10.6 Optimizing Aerodynamic Drag on Trucks

Commercial vehicle aerodynamics engineers have to work within strict limits as a result of commercial factors and legal regulations. For example, when it comes to trucks, they need to dovetail maximum loading capacity, sufficient space to work and sleep for two drivers, and aerodynamic features with one another. This balancing act is achieved by optimizing many of the details, most of which remain invisible to the untrained eye. Figure 10.34 shows the component groups on the semitrailer combination that have been optimized in terms of aerodynamics, aeroacoustics, and dirt.

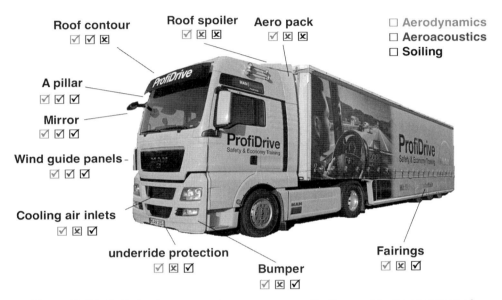

Figure 10.34 Optimized component groups on a semitrailer combination in terms of aerodynamics, aeroacoustics, and dirt.

10.6.1 Characteristic Airflow and Pressure Conditions

The air flowing around a semitrailer combination and causing aerodynamic drag is characterized by a large area of dynamic pressure at the front of the vehicle (Figure 10.35). The smaller the area of dynamic pressure, the lower the vehicle's aerodynamic drag. This becomes clear if we directly compare the various cabs within a given model series. The cabs essentially differ in their width and height. The corner radii and air deflectors are identical. The transition to the semitrailer has been located in the most favorable position for each cab.

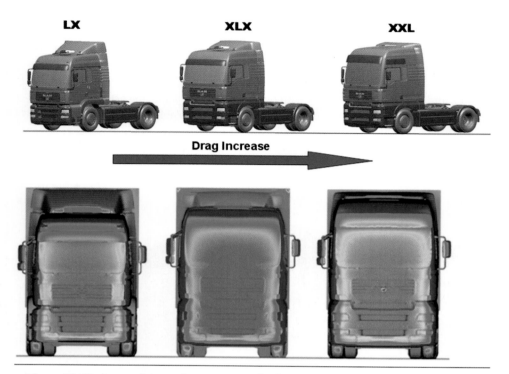

Figure 10.35 Areas of dynamic pressure at the front of the vehicle with increasing cab sizes for MAN TGA semitrailer tractors (LX and XXL with sunblind).

Hoepke [349] studied the changes in aerodynamic drag along the vehicle's longitudinal axis, and his findings clearly show where the most marked increases in aerodynamic drag occur (Figure 10.36). The dynamic pressure at the front of the vehicle causes an initial increase in aerodynamic drag, and the change in this drag up to the rear cab wall depends on the cab geometry in each case. The space between the tractor vehicle and semitrailer plays a crucial role in determining the aerodynamics of the overall vehicle combination.

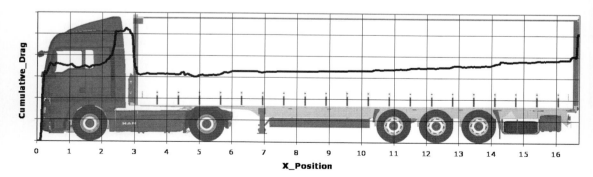

Figure 10.36 Changes in aerodynamic drag along the longitudinal axis of a semitrailer combination.

The vacuum at the rear cab wall causes a sharp increase in aerodynamic drag. However, this is largely compensated by the vacuum that simultaneously exists at the front wall of the semitrailer. The better the design and setting of the spoilers and side deflectors, the lower the increase in aerodynamic drag in this area. Moving down the truck, in particular the substructure of the semitrailer with its various axles and other attachments, results in increasing aerodynamic drag. In many cases, the rear of the semitrailer is designed as an angular box shape in order to meet the need for commercial vehicles to provide the greatest possible cargo area, and this also causes a further increase in aerodynamic drag. The airflow separation at the rear is dominated by highly pronounced turbulence structures. Figure 10.37 shows the turbulence structure at the rear of two different vehicles; this structure is even reflected in the build-up of snow at the rear of the vehicles.

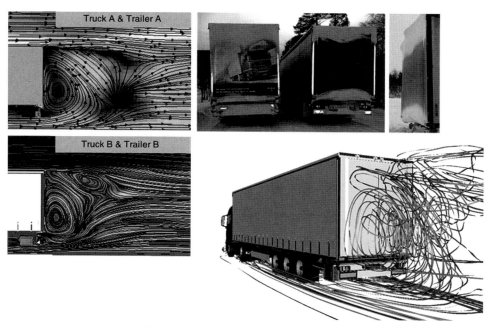

Figure 10.37 Turbulence structures at the rear of the vehicle.

10.6.2 Cab

In contrast to passenger cars, the exterior shape of commercial vehicles is determined by their intended deployment (Figure 10.38). The market is dominated by cuboid, angular bodies that meet the need to offer the greatest possible cargo area. For aerodynamics engineers, the legal restrictions on the dimensions of commercial vehicles (see previous section) mean there is only limited scope for optimization using aerodynamic shapes. However, such scope does exist when it comes to the design of the cab and the development and coordination of attachments designed to reduce aerodynamic drag.

When developing a new cab, engineers focus on the concept of "living, working and sleeping in the truck" (Figure 10.38). This means that cabs are required to meet diverse and, in some cases, conflicting requirements. First and foremost, the driver requires a functional workplace when he/she sits behind the steering wheel. The cab must also offer sufficient room for completing paperwork, space to relax during breaks or when waiting, as well as one or even two comfortable bunks for sleeping. A basic cube measuring 2.3 m by 2.5 m and with a height of approximately 2.2 m is available to meet all these requirements.

Figure 10.38 Interior of Mercedes Actros, model year 2012.

Implementing these requirements directly would result in the cab for a long-haul vehicle having a cuboid shape, which would not sufficiently meet aerodynamics requirements. However, comparing the c_D values in Figure 10.39, it is clear that a semitrailer combination with an angular cab has almost the same c_D value as one with a streamlined cab. One possible conclusion from this is that the cab shape does not affect aerodynamic drag. However, if we analyze the pressure distribution for a symmetrical flow of air to the body and cab, it can be seen that the angular cab absorbs the dynamic pressure in full and, due to the pronounced airflow separation, shields the body behind to such an extent that there is no or even a negative force effect on the body. The resistance is thus redistributed. As such, the cab shape must not be considered in isolation but always in conjunction with the relevant body.

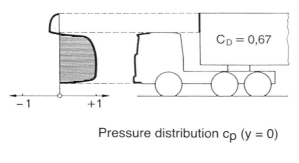

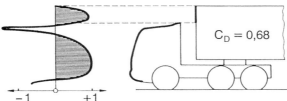

Figure 10.39 Airflow interference on cab—front of body; according to Gilhaus (1979/1980) [277].

As a rule, when developing a new long-haul vehicle today, engineers may assume a total body height of 4 m, and this requirement must be met when designing the cab and corresponding attachments. The aerodynamic quality of a cab is influenced by a number of different parameters, which result from the legal restrictions on dimensions and are crucial to defining its shape. Key parameters in this context (see Figure 10.40) include the angle of the windshield, the corner radius, and the radius of the front edge of the roof.

Previous experience in developing cabs has shown that analyzing a given parameter in isolation will not result in the required reduction in aerodynamic drag. In order to demonstrate the interaction of the individual parameters with one another, Frank [257] performed tests in a wind tunnel. The results of this parameter study clearly show that engineers must always analyze the various parameters together. The effect of a given parameter on aerodynamic drag depends on the value of the other parameters. If we examine the shape of the curve for aerodynamic drag with a windshield angle of 10°, it can be seen that increasing the radius of the front edge of the roof over approximately 200 mm does not allow any further reduction in aerodynamic drag. The positive effect of a larger radius is only seen again as the windshield angle is increased. As a result, developers must ensure that the individual parameters dovetail with one another. However, this also gives them the opportunity to compensate prescribed but suboptimal parameters for aerodynamic drag with other parameters.

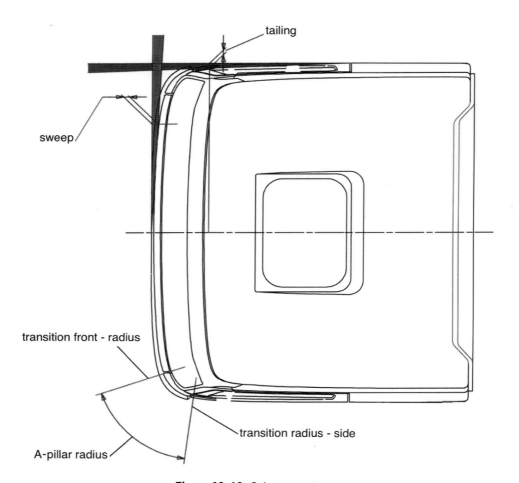

Figure 10.40 Cab parameters.

In an ideal scenario, when developing a new cab, the aerodynamics can define an envelope that must be strictly adhered to by the design. In other words, the aerodynamics engineers specify how much scope the designers have to implement their designs. However, when developing a new commercial vehicle, it is important to remember that the design department is one of the areas with the largest influence on the aerodynamics. Every curve and edge must fit into the overall design concept. One typical example for the design requirements is the radii. Preference should be given to accelerated radii as they are more dynamic. However, this places the airflow at risk of separation and thus leads to higher aerodynamic drag (Figure 10.41). Ultimately, the aerodynamics and design departments must both find a compromise to ensure the vehicle has the best possible aerodynamics but still takes into account all other relevant issues.

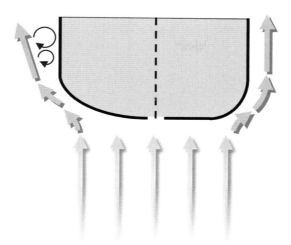

Figure 10.41 Influence of design and aerodynamics on the radii.

Given the fact that, to date, the same cabs have been used for long-haul, short-haul, as well as construction site trucks, the variety of bodies used means that a streamlined design alone is not always sufficient to achieve a favorable aerodynamic drag coefficient. For high bodies, a whole range of additional attachments have become well known as aids to reduce aerodynamic drag. Air deflectors, some of which are adjustable, have proven their worth as attachments to be mounted on the cab. These roof spoilers are easy to handle, highly effective, and favorably priced. The trail of smoke on vehicles with and without air deflectors in Figure 10.42 clearly demonstrates their impact. Depending on the body projection h, which is calculated from the difference between total cab height and total body height, the aerodynamic drag when no air deflector is fitted on the roof increases by up to 30%. Air deflectors and side deflectors constitute a formal unit together with the cab on the latest vehicle generations (see Figure 10.43).

Alongside aerodynamic attachments such as air deflectors, the aerodynamics of other attachments must also be tested and optimized when developing a vehicle or cab. Figure 10.34 lists the most important attachments affecting a vehicle's aerodynamics. These components, including sunblinds, A-pillar cladding, air vents, and mirrors, need to comply with the requirements governing aerodynamics as well as those for aeroacoustics, dirt, and cooling air.

Figure 10.42 Impact of an air deflector on the airflow at the front of the semitrailer.

Commercial Vehicles

Figure 10.43 Semitrailer tractor with integrated air deflectors.

The A-pillar is a cab area with a key role in determining the vehicle's aerodynamics. The shape of the A-pillar influences not only the aerodynamics but also the build-up of dirt and the aeroacoustics in particular. Compared with a contoured A-pillar, a smooth A-pillar (see Figure 10.44) can significantly reduce the air separation bubble, which in turn leads to an approximately 4% reduction in aerodynamic drag. However, the flow of air past the side window causes a significant increase in noise in the interior. This affects the frequency range above 500 Hz in particular (see Figure 10.45).

Figure 10.44 Airflow around cab with smooth and contoured A-pillar cladding.

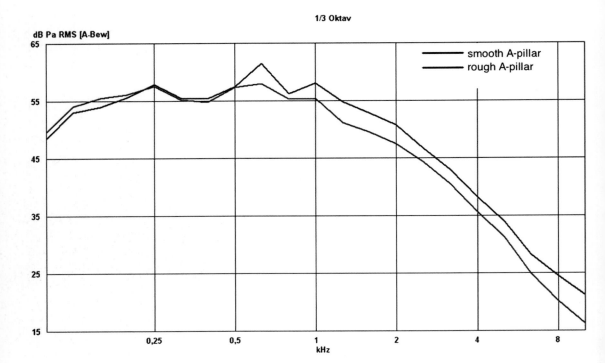

Figure 10.45 Noise level with various A-pillar claddings.

10.6.3 Mirrors and Attachments on the Cab

The vehicles may also be fitted with attachments that do not improve the vehicle's aerodynamics but are requested by the customer in the form of special equipment. Additional headlights on the roof, as shown in Figure 10.46, are located directly in the path of the airflow over the roof and increase aerodynamic drag by up to 6%. Signal horns cause a similar effect if they are located in the area of the direct airflow and increase aerodynamic drag by up to 1%. The crash guard on the bumper with a range of additional headlights directly in front of the cooling air inlets on the vehicle, as shown in Figure 10.46, reduces the cooling airflow by almost half. As a result, the fan is activated more frequently, and the power required for this causes a considerable increase in the vehicle's fuel consumption.

Figure 10.46 Vehicle with special equipment.

In contrast to mirrors on passenger cars and buses/coaches (which generally increase aerodynamic drag), mirrors on trucks can even improve aerodynamic drag as they guide the air around the vehicle. Given the angular basic shape of a truck (which negatively affects the airflow), the mirrors are used to guide the airflow along the side wall of the vehicle. However, this only works if the airflow separates as early as the A-pillar. Figure 10.47 shows this guidance of the airflow. In contrast, if the air flows past the outer shell of the vehicle behind the A-pillar, the mirrors on trucks also increase the c_D value by between 2% and 5%, depending on the design.

Figure 10.47 Airflow guidance by mirrors on truck.

Sunblinds are mainly fitted to commercial vehicles to improve their appearance. As for passenger cars, the sunblinds actually used by the driver are located inside the vehicle. Depending on the design of the roof and sunblinds, they may affect the vehicle's aerodynamic drag by up to 1%, both positively and negatively. Special attention must be paid to the aeroacoustic effect of the sunblind on noise levels inside the cab. Figure 10.48 shows the pressure fluctuations of wake turbulence caused by a sunblind.

Figure 10.48 Pressure fluctuations caused by the installation of a sunblind.

10.6.4 Airflow Through the Engine Compartment

The requirements concerning the cooling output of commercial vehicles are become ever-more stringent. As the engine output requirements increase, the available space in the engine compartment decreases as a result of the larger engines. The amount of heat to be dissipated increases due to the higher engine outputs as well as state-of-the-art exhaust gas after-treatment systems. The drag caused by cooling air for a semitrailer tractor is between 5% to 8% of the overall aerodynamic drag and can be reduced by using a smaller amount of cooling air. However, this leads to more frequent use of the fan and tends to result in higher fuel consumption, as the energy to power the fan is around 15% of the engine output in commercial vehicles.

When configuring the maximum cooling output, developers focus on the task of driving on steep gradients at a low speed and with maximum torque and maximum output. Due to the maximum fan speed required for this, the fan acts as the variable that determines the airflow through the engine compartment. As such, backflows often occur when the vehicle is traveling at speeds of approximately 20 km/h. These should be avoided, and the cooling air mass through the cooling air ducts should be guided through the radiator and engine compartment so as to minimize pressure loss as far as possible. Cooling air ducts and ambient air covers are highly effective in this regard (see Figure 10.49).

Figure 10.49 Guidance of cooling air on a modern commercial vehicle.

A crucial factor to consider when configuring the cooling system is the position of the fan relative to the radiator. This often causes problems as the high drive power of the fans used in commercial vehicles means they are not driven electrically as in passenger cars but are instead directly linked to the engine via an intermediate viscous clutch. This means the position of the fan is often determined by the position of the engine. Studies by Hallquist [314] have shown that the cooling output increases if the radiator is positioned centrally.

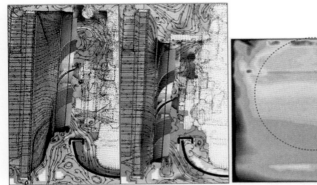
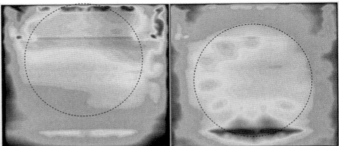

Figure 10.50 Visualization of airflow and pressure for an eccentric fan positioned centrally.

However, it is important to ensure not only that the radiator is positioned centrally, but also that the fan cowl is configured to improve the airflow. This generally requires a greater distance between the fan and radiator, which, in the confined spaces of today's designs, would result in the engine being relocated backwards. For example, moving the engine by 300 mm can increase the flow through the radiator by 25%.

More extensive studies have shown that a larger distance between the fan and engine also increases the flow through the radiator. Suitable air deflectors positioned directly between the fan and engine can improve the exit airflow and thus increase the flow through the cooling system even further. These air deflectors also reduce backflow and disruptive dust turbulence when traveling off-road. Additionally, this range of measures helps to reduce the size of the cooling system, which in turn has a positive effect on aerodynamic drag.

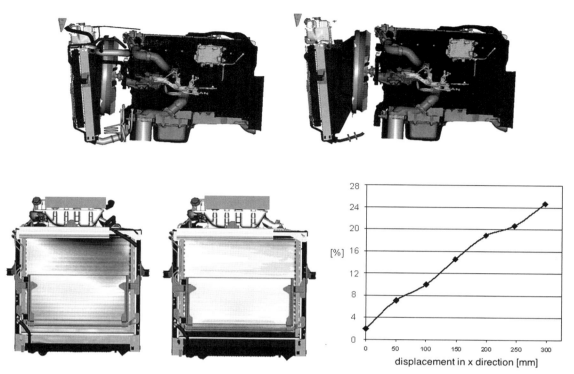

Figure 10.51 Increased flow through the radiator by increasing the distance between the fan and radiator.

Despite this, the new emission standards necessitate the use of radiator areas of up to 1.2 m² and fans with a diameter of approximately 800 mm in order to dissipate a heat quantity of approximately 200 KW at the radiator. The maximum cooling output is not required when traveling on motorways. The focus here is on optimizing fuel consumption. Modern tractor vehicles are now equipped with radiator shutters to reduce aerodynamic drag even further and to only supply cooling air to the cooling system when it is needed.

Figure 10.52 Actros air control system: left = closed and right = open.

These shutters are only fitted to long-haul vehicles. Vehicles used for construction site deployments are mostly fitted with steel bumpers. Steel bumpers are subject to certain strength requirements, meaning they have considerably smaller cooling air ducts and are therefore to be considered as more critical in terms of the cooling system. Figure 10.53 shows a vehicle fitted with a plastic bumper and with a steel bumper. The cooling air mass is 15% less with a steel bumper. As such, it is clear that vehicles deployed in mountainous regions (e.g., timber transporters) place the highest demands on the cooling system. In this case, the combination of driving on steep gradients, a steel bumper, and a high engine output in a vehicle fitted with a low cab and thus a small radiator for space reasons is applicable here.

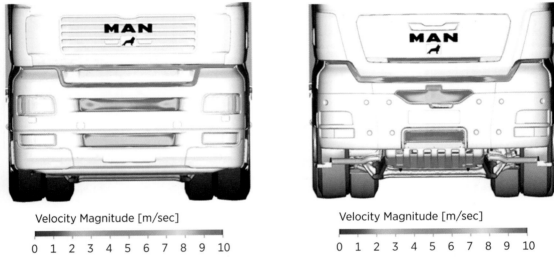

Figure 10.53 Vehicle with plastic and steel bumper.

10.6.5 Chassis

When working to optimize the aerodynamics of a truck, the focus is generally on the cab and its attachments. However, the influence of the chassis and body (section 10.6.6) on the aerodynamics must not be ignored. In essence, the optimizations on the chassis primarily concern the side skirts, underbody, and arrangement of components such as the tanks, spare wheel, or batteries on the frame. The complexity involved in developing aerodynamic measures on the chassis is due to the vast range of wheel configurations, wheelbases, and axle distances. This is particularly the case for flatbed vehicles owing to the different intended deployment areas and customer requirements (see Figure 10.54).

Commercial Vehicles

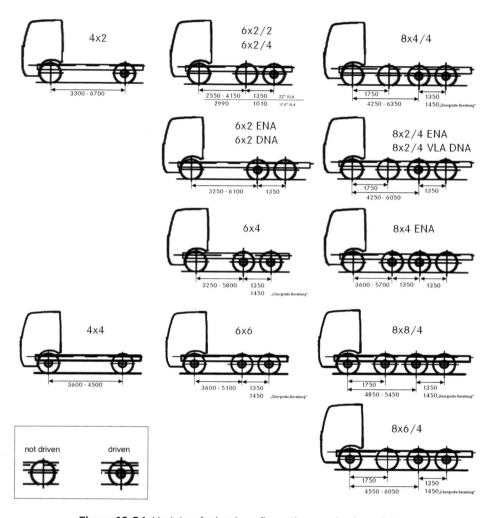

Figure 10.54 Variety of wheel configurations and axle variants.

The various contours and attachments on the frame negatively affect the aerodynamics. Small tank systems or long wheelbases result in gaps at the sides of the frame, which cause air to flow in and disrupt the flow of air around the vehicle. This is proved by studies carried out by FAT[1] on a generic drawbar combination as part of the working party on commercial vehicle aerodynamics (Figure 10.55) [252]. It is important to note here how the lateral airflow flows in, in front of the rear axle, and the turbulence caused as a result. This can be prevented through the use of side skirts or a sufficiently sized tank system, which reduces aerodynamic drag by 1–2%.

---

1. Forschungsvereinigung Automobiltechnik

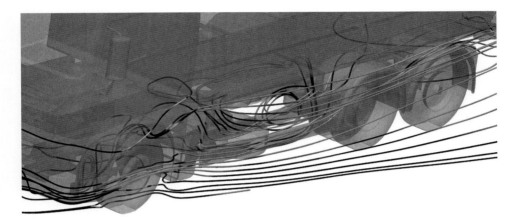

Figure 10.55 Airflow around a flatbed vehicle without side skirts.

Side skirts are nowadays a standard component on semitrailer tractors to optimize the airflow around the side of the vehicle, similar to air deflectors for the cab (Figure 10.56). The skirts between the axles effectively cover the various attachments on the frame to improve the airflow. As a result, it is possible to reduce aerodynamic drag by up to 1%. Even greater aerodynamic advantages are possible with an oblique flow in particular.

Figure 10.56 Actros semitrailer tractor with and without side skirts.

However, side skirts should not be used solely because of their positive effect on aerodynamic drag. They also considerably increase safety, as pedestrians and cyclists are "deflected" in an accident and it is more difficult for the truck to run them over. Additionally, side skirts help prevent splashing water and reduce noise.

Measures on the underbody are ideally suited for so-called wagon-carrying trailers (i.e., vehicles that do not operate off-road and have a smaller clearance between the chassis and road surface). For various reasons, smooth underbodies as are commonly found on passenger cars are not used for commercial vehicles. On account of the open ladder frame with the bolted attachments and open driveline, a smooth underbody of this nature would only be possible through the use of large covers. These would need to be set in place using sturdy auxiliary constructions. To date, partial underbody paneling has only been used on series vehicles in individual cases. Figure 10.57 shows the partial paneling on the underbody in the area in front of the front axle on an Actros semitrailer tractor.

Figure 10.57 Underbody paneling and wheel spoiler on an Actros semitrailer tractor.

Vehicles with a flatbed body developed using a holistic aerodynamic approach offer greater potential for implementing smooth underbody paneling. This has been confirmed by initial tests on prototypes (see Figure 10.58).

Figure 10.58 Mercedes-Benz aerodynamics truck study.

Another measure used to optimize underbody aerodynamics is wheel spoilers, similar to those used on many passenger cars. Designing the aerodynamics of a wheel spoiler for trucks is much more difficult, as there is much greater diversity in the available tires and chassis heights. For example, for semitrailer tractors there are at least four different frame heights and seven different tire sizes to take into account when configuring a wheel spoiler. The relevant geometric parameters in this context are the width, height, and setting angle of the wheel spoiler. If the design is tailored to the airflow underneath the vehicle, a wheel spoiler allows aerodynamic drag to be reduced by up to 1% (Figure 10.57).

Further aerodynamic losses on the chassis are caused by the airflow around the sides of the wheel arches and wheels of the vehicle. Due to the tire geometry and required spring travel and steering angles, it is difficult to implement streamlined contours in this area. The use of wheel covers can at least reduce the negative effect of the rim contour on the vehicle's aerodynamic drag (Figure 10.59). When designing such a cover, it is important to factor in the need for an adequate flow of cooling air to cool the brakes. To date, covered rear axles as can be found on some buses and coaches (section 10.7.6) have not been used on series commercial vehicles.

Figure 10.59 Wheel covers on a commercial vehicle.

10.6.6 Semitrailers and Bodies

From an aerodynamic perspective, semitrailer tractors and tractor/trailer units are units made up of different elements. When designing optimization measures, these elements must not be considered in isolation, as they affect one another. In order to better identify

Commercial Vehicles

the effect of aerodynamic measures, the air forces of the elements (with the corresponding assignment) are determined separately.[2]

Figure 10.60 shows the results of such a study method by way of example. The overall resistance of a semitrailer combination has been subdivided into the individual resistances for the body, cab, and chassis. These behave approximately as 4:3:2 with a symmetrical flow of air. As the length of the space increases, a slight increase in the individual resistances can be observed. With an oblique airflow, the tangential force coefficients for the semitrailer and chassis increase markedly; this is due to the airflow separation on the lee side and the airflow around the angular chassis. In contrast, the tangential force coefficient of the comparatively streamlined cab is barely affected by side winds.

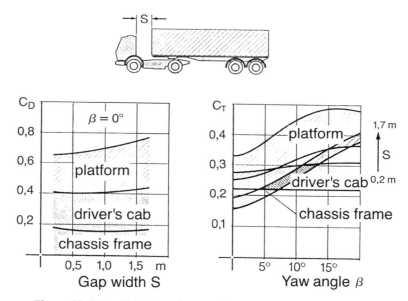

Figure 10.60 Individual resistances for a semitrailer combination.

Alternatively, as shown in Figure 10.61, it is also possible to divide the resistance for the semitrailer combination into the individual aerodynamic drag for the tractor vehicle and semitrailer. The individual tangential force coefficient for the semitrailer tractor barely changes over the airflow angle range shown. This means that, with an increasing airflow angle, the increase in the total tangential force coefficient for the semitrailer combination results solely from the increase in the tangential force for the semitrailer part. One cause for this lies in the air flowing through the gap between the cab and semitrailer at the side, with subsequent airflow separation on the lee side. Other causes are the changed exit airflow at the rear of the semitrailer with a larger wake in comparison with

---

2. Please see the studies carried out by Roshko and Koenig [684], Gilhaus et al. [277], and Morel [583].

a straight airflow and the larger volume of air flowing into the angular chassis area without side covers.

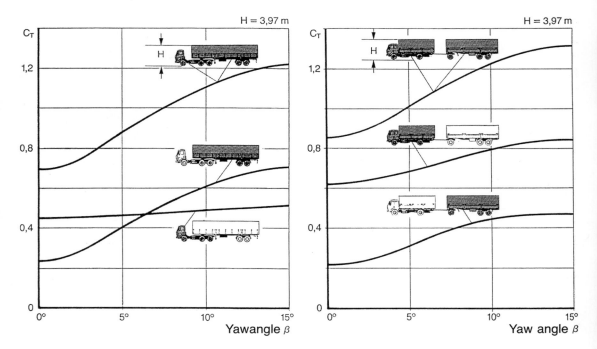

Figure 10.61 Tangential force coefficient c_T depending on the airflow angle for the semitrailer combination and tractor/trailer unit.

With symmetrical airflow, the individual resistances for the motor vehicle and trailer on the tractor/trailer unit are split roughly 70/30%. With an increasing airflow angle β, both individual tangential force coefficients increase at roughly an equal rate, meaning the motor vehicle and trailer contribute roughly in equal measure to the observed increase in the total tangential force for the vehicle combination. The reason for this is the lee-side airflow separation at the front of the vehicle. The airflow at the side through the gaps between the cab, motor vehicle body, and trailer increases with an increasing airflow angle β; this generates additional airflow separation on the lee side. Similarly to a semitrailer combination, a changed exit airflow at the rear of the trailer compared with a straight airflow leads to a larger wake and thus to an increase in the total tangential force for the tractor/trailer unit; this is also influenced by the increased volume of air flowing into the angular chassis area without side covers.

When examining these studies, it is important to note that the cab shape has a major impact on the results. A streamlined cab not only reduces resistance but also redistributes it from the tractor vehicle to the trailer, as shown in Figure 10.62.

Commercial Vehicles

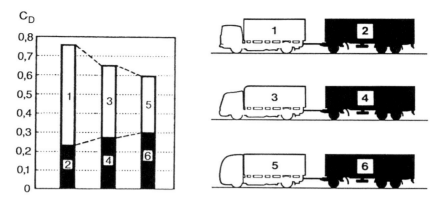

Figure 10.62 Reduction and redistribution of aerodynamic drag thanks to streamlined cab shape.

The changes in aerodynamic drag (see Figure 10.36) indicate where the airflow can be optimized on commercial vehicles. The following areas on the semitrailer offer the greatest potential for optimizing the aerodynamics: intermediate space and front wall, side area, and rear (see Figure 10.63).

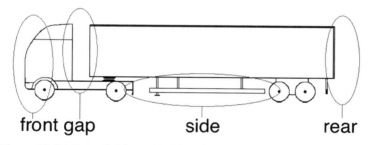

Figure 10.63 Potential for optimizing the aerodynamics of semitrailers.

The diagram also shows that, on a modern semitrailer combination, the semitrailer tractor has slightly higher aerodynamic drag than the vehicle combination overall. Figure 10.64 shows the reason for this distribution of aerodynamic drag. Adding together all vacuum and overpressure areas along the vehicle's longitudinal axis results in a lower aerodynamic drag than the total of all amounts.

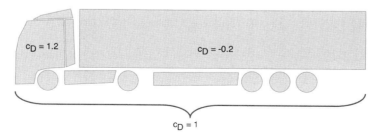

Figure 10.64 c_D value distribution on a semitrailer combination.

The process of optimizing the aerodynamics of a semitrailer begins at the front of the vehicle. Despite what is known about the advantages of the radius at the front wall edges, on conventional semitrailers and bodies they are mostly angular or designed with an angled edge and are not shaped with a corresponding radius. As shown in studies by Mercedes-Benz and Spier [772], large radii can improve aerodynamic drag by up to 7%, especially on bodies that protrude beyond the cab at the side.

Additionally, paneling or at least a smaller space between the semitrailer tractor and semitrailer are rarely implemented. Semitrailers with refrigerator units on the front wall offer relatively straightforward ways to exploit the potential to improve aerodynamics by 1–2%, through the use of additional panels.

The aerodynamics of the gap between the two vehicles was also studied for drawbar combinations. Gilhaus [276] was able to show that it is possible to use practical solutions to achieve an airflow here that is almost as good as for a smooth, enclosed transition from the motor vehicle to the trailer. However, this is not feasible from a technical standpoint. Figure 10.65 shows the steps by which the aerodynamic drag of the tractor/trailer unit was able to approach that of a configuration without any gaps.

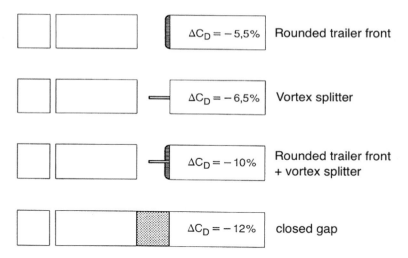

Figure 10.65 Reduction in aerodynamic drag of a tractor/trailer unit by means of attachments on the trailer.

As already described in section 10.6.5, side skirts are an important measure for optimizing the aerodynamic drag with a straight, and in particular oblique, airflow. Leuschen and Cooper [495] showed that the height of the side skirts and in particular the distance to the road surface have a significant impact on their aerodynamic effectiveness. Restrictions for day-to-day deployment lie in the ground clearance required to achieve the necessary angle of approach/departure.

Commercial Vehicles

A range of wind tunnel tests were carried out at Delft University of Technology [822] with the aim of testing various designs of side skirts on a semitrailer combination. In the process, different designs for the front and rear areas of the side skirts were tested. The c_D value was improved by up to $\Delta c_D = 0.061$ using a contour resembling a wing profile at the front approach to the side skirt (see Figure 10.66). Side skirts, as offered by a number of semitrailer manufacturers, are able to reduce aerodynamic drag by approximately 8% (Figure 10.67).

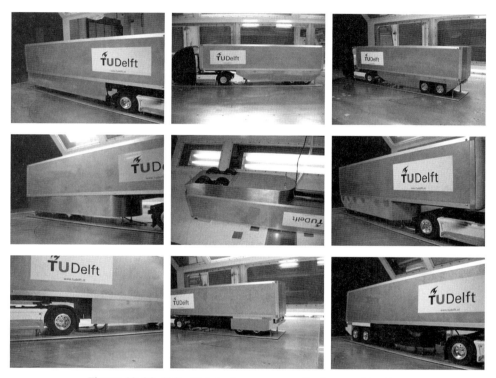

Figure 10.66 Side skirt designs on a semitrailer combination.

To date, measures for the underbody of a semitrailer have only been implemented on models or prototypes. They offer the potential to reduce aerodynamic drag by approximately 2%. When designing the chassis for a semitrailer or trailer, the individual elements such as pallet boxes or compressed-air tanks for the brake system must be arranged and configured so as to prevent unnecessary turbulence in the air flowing underneath the vehicle.

Figure 10.67 Side skirts on a semitrailer.

As is the case for passenger cars, measures at the rear of the vehicle (such as a tapered rear) offer great potential for reducing aerodynamic drag. This measure aims to increase the basic pressure on the rear surface by guiding the air flowing around the vehicle inwards and to reduce the intensity of the pronounced turbulence structures (see Figure 10.37). A number of different designs are known for creating a tapered rear (see Figure 10.68). The version with inclined extensions (b) in particular has shown itself to be a promising design compared with the other versions—straight extensions (a), curved fins (c), or full-surface covering on the rear (d).

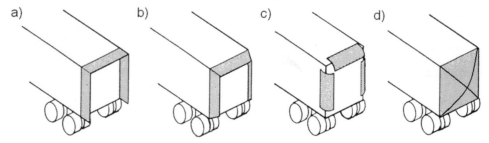

Figure 10.68 Designs for a tapered rear.

Studies carried out within the context of FAT into the length required to create such a tapered rear have shown that even a length of 400 mm in the vehicle's longitudinal direction is sufficient to exploit a potential of 7% for reducing aerodynamic drag (see Figure 10.69).

Commercial Vehicles

Figure 10.69 Studies within the context of FAT into a tapered rear design on a semitrailer combination.

Figure 10.70 Semitrailer combination with optimized aerodynamics.

Many of the optimization measures for a semitrailer combination have been well known for years. The findings from various separate studies into aerodynamics were already integrated into a semitrailer combination concept designed for long-haul transport at the start of the 1990s, as can be seen in Figure 10.70. The aim was to achieve a reduction in aerodynamic drag that is equally effective with a symmetrical airflow and with a side wind. Figure 10.71 shows the individual steps: front edge radii $r = 150$ mm on the semi-trailer, air deflectors and side deflectors on the cab roof, front apron, chassis side skirts for the semitrailer tractor and semitrailer, and a tapered rear design for the semitrailer. In total, the c_D value was reduced by 35%. The measures also proved to be effective with an oblique flow: with a $\beta = 15°$, the c_T value was improved by 40%.

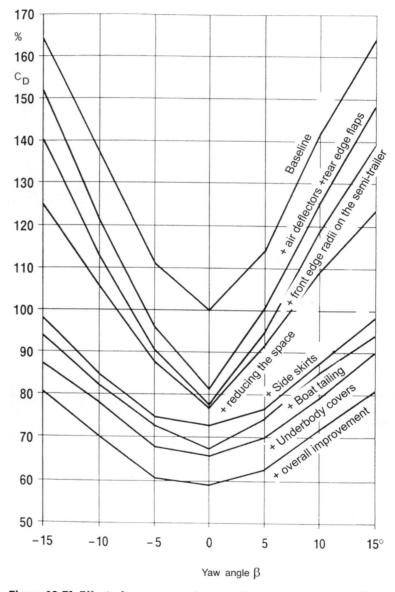

Figure 10.71 Effect of measures to improve the aerodynamics of a 15-m semitrailer combination.

Similar measures have also proven themselves for flatbed vehicles. Studies conducted by Göhring and Krämer [287] showed that the c_D value improved by 17% with air deflectors on the cab, by 4% with a front apron, and by another 6 to 8% with side skirts—in total up to 27%. Figure 10.72 shows some examples. On a light truck with a relatively large body, optimizing the aerodynamics of the cab, body, and chassis together results in an improvement in the c_D value of 36%, from 0.78 to 0.50 (see Figure 10.73 and Figure 10.58).

Figure 10.72 Truck with box body and optimized aerodynamics on transition between cab and box body at top.

Figure 10.73 Truck with box body; optimized aerodynamics on transition between cab and box body on three sides.

The Aerotrailer from Mercedes-Benz and Schmitz Cargobull, shown at IAA 2012, is a contemporary interpretation of an aerodynamic semitrailer combination (Figure 10.74). The cargo area of the trailer is completely unaffected. However, its length dimensions exceed the permissible dimensions (see section 10.4.2) by a little less than half a meter as a result of the rear extension. The Aerotrailer reduces aerodynamic drag by 18% compared with a standard semitrailer.

Figure 10.74 Aerodynamics trailer study by Mercedes-Benz.

Figure 10.75 shows further potential for optimizing aerodynamics if the length restrictions are relaxed, depending on the semitrailer design. The studies were based on the premise of retaining the cargo area volume in full. In these tests, the space lost as a result of a tapered rear was added to the semitrailer as additional length. The findings show that a vehicle with a 45% smaller c_D value at the rear needs to be 2.7 m longer than a conventional semitrailer in order to transport the same volume of cargo. However, the rear views of the vehicles give an impression of how the loading and unloading processes will be much more difficult as a result.

Commercial Vehicles

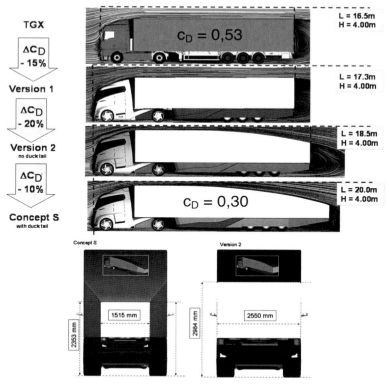

Figure 10.75 Aerodynamic influence of a tapered rear on a semitrailer.

10.6.7 Concept Vehicles

The majority of concept vehicles designed by commercial vehicle manufacturers focus on amended legislation or length restrictions. For example, the Aerotrailer from Mercedes-Benz (see Figure 10.74), the Optifuel from Renault (Figure 10.77), the IVECO Transport Concept Solution Truck (Figure 10.79), as well as the Concept S from MAN (Figure 10.76) all exceed the EU length regulations valid at the time each vehicle was unveiled. The studies are essentially always based on the following four principles:

- Reducing the dynamic pressure at the front
- Rounded corner radii
- Tapered rear designs
- Underbody paneling and side skirts on the semitrailer

The Concept S from MAN (see Figure 10.76) implemented these four principles and has a c_D value of 0.3. This roughly corresponds to a reduction of approximately 45% in the c_D value compared with a conventional semitrailer combination. As a result, fuel consumption in long-haul deployment was reduced by approximately 15%. Assuming an average fuel consumption of 30 l/100 km, this would equate to 4.5 l/100 km. The Renault Optifuel (Figure 10.77) reduces fuel consumption by 13%. Alongside the optimized driveline and rolling resistance, the aerodynamic shape of the vehicle also plays a key role in achieving this savings. A concept vehicle from Scania, shown in Figure 10.78, had a c_D value of just 0.25 and was designed as a drawbar combination or chassis vehicle.

Figure 10.76 Concept S, MAN Truck & Bus.

Figure 10.77 Optifuel, Renault.

Commercial Vehicles

Figure 10.78 Scania concept vehicle.

The IVECO Transport Solution Concept Truck reduced aerodynamic drag by 22% compared with a series vehicle [178]. Figure 10.79 clearly shows the savings that were made possible using this vehicle. A 14% reduction in aerodynamic drag was achieved on the semitrailer and an 8% reduction was possible on the tractor vehicle. The vehicle is equipped with inflatable elements between the tractor vehicle and semitrailer and at the rear of the semitrailer.

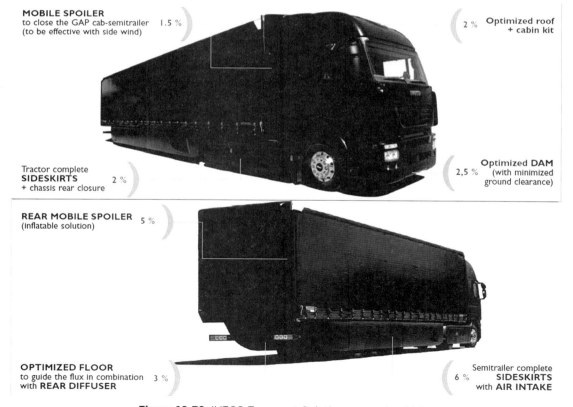

Figure 10.79 IVECO Transport Solution concept vehicle.

10.7 Optimizing Aerodynamic Drag on Buses and Coaches

For buses and coaches, the challenge lies in reconciling aerodynamics requirements with those set by the legal framework and market economy conditions. Here, designers and aerodynamics engineers are able to create the entire vehicle (i.e., the front, the sides, and the rear), as a single form-fitting unit. The efforts to achieve optimal aerodynamics are only curtailed by the customer's desire to have as many seats as possible and the corresponding luggage space. The NEOPLAN Starliner ($c_D = 0.36$) and Cityliner ($c_D = 0.35$) coaches are impressive examples of the greater design freedom available with buses and coaches. These vehicles already have aerodynamic drag coefficients similar to those of passenger cars.

10.7.1 Characteristic Airflow and Pressure Conditions

Figure 10.80 shows the pressure and airflow conditions on a conventional coach, using the example of an MAN Lion's Coach.

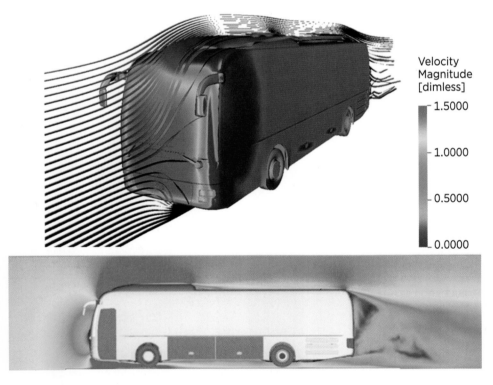

Figure 10.80 Pressure and airflow conditions on a conventional coach.

Figure 10.81 shows the breakdown of aerodynamic drag along the vehicle's longitudinal axis. This figure clearly shows the higher aerodynamic drag with the area of dynamic pressure at the front of the vehicle. It accounts for approximately 70% of the total aerodynamic drag. The air flow in the roof area and associated vacuum at the front serve to reduce aerodynamic drag, meaning section "A" at the front is responsible for around half the overall aerodynamic drag. The center section "B" is characterized by the drag from the running gear and underbody attachments, and thus accounts for an approximately 20% share of the overall aerodynamic drag. The vacuum at the rear is responsible for approximately 30%.

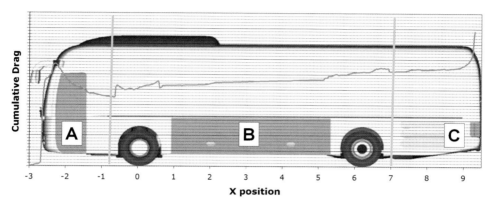

Figure 10.81 Breakdown of aerodynamic drag along the vehicle's longitudinal axis.

10.7.2 Front

The size of the area of dynamic pressure and the front edge radii have a crucial impact on the aerodynamic drag of buses and coaches. Figure 10.82 shows the results of tests where these factors were successively increased for a bus, based on an angular vehicle front design. The results show that even a radius of approximately 150 mm is sufficient to reduce the drag of a bus or coach to such an extent that no further notable improvements can be achieved, even with a pronounced streamlined frontal design. Carr [132] conducted systematic tests and drew similar conclusions (see Figure 10.83).

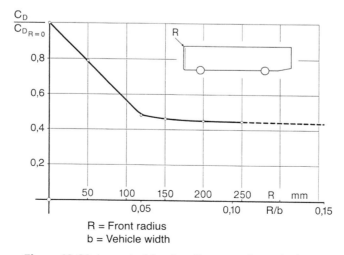

Figure 10.82 Impact of front radii on aerodynamic drag.

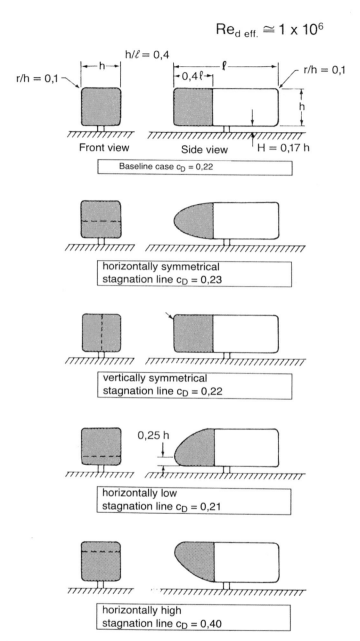

Figure 10.83 Impact of streamlined front-mounted structures on cuboids on the c_D value (at ground level), according to Carr (1968) [132].

On luxury coaches, the large radii of the corners and the pronounced angle of the roof largely prevent flow separation and reduce the area of dynamic pressure in the front. In a direct comparison with a conventional bus concept, the force acting on the front of the vehicle is reduced by approximately 30% (see Figure 10.84).

Commercial Vehicles

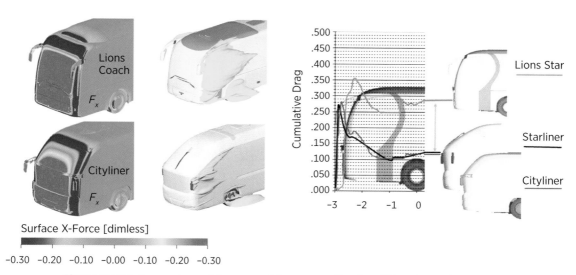

Figure 10.84 Comparison of forces and flow separation for different coach concepts [MAN Lion's Coach and NEOPLAN Cityliner/Starliner].

10.7.3 Rear View Mirrors

The majority of modern coaches and intercity buses are equipped with integrated, frontward protruding mirrors. Since most of these buses and coaches are fitted with very rounded A-pillars, these mirror types offer major advantages in terms of aerodynamics and preventing dirt on the mirror glass than conventional mirror designs attached directly to the A-pillar. As can be seen in Figure 10.85, these mirrors have a much smaller wake. In terms of fuel consumption, integrated mirrors enable a savings of approximately 2% compared to conventional mirror designs when driving at a constant speed of 100 km/h on level ground.

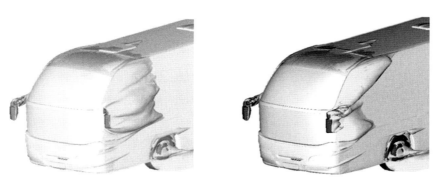

Figure 10.85 Wake for different mirror designs in combination with a large A-pillar radius.

However, they are located in an area with very high airflow speeds, meaning they have a negative effect on the vehicle's aeroacoustics (Figure 10.86). The challenge here, therefore, is to reach an optimal compromise between passenger comfort and vehicle cost-effectiveness.

Chapter 10

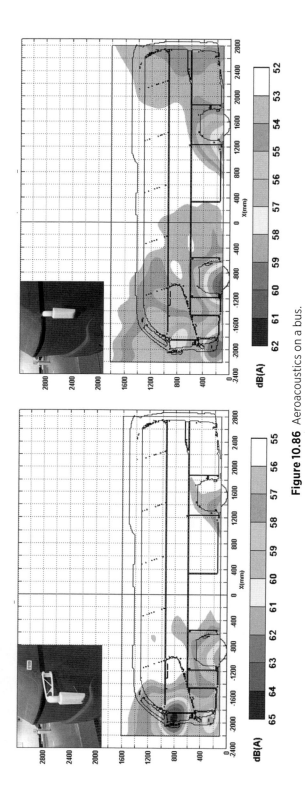

Figure 10.86 Aeroacoustics on a bus.

776

Whereas integrated, frontward protruding mirrors are beneficial with rounded A-pillars, these mirrors have a negative effect on aerodynamic drag in combination with small A-pillar angles. This is due to the airflow separation caused by the edges of the A-pillars. Side-mounted mirrors fitted close to the body are partially located in this area of airflow separation and thus have only a small influence on the airflow around the vehicle. In contrast, integrated, frontward protruding mirrors are fully located in the airflow and may substantially increase aerodynamic drag.

Therefore, for aerodynamic reasons frontward protruding mirrors should be fitted to vehicles with rounded A-pillars, and side-mounted mirrors should be fitted to vehicles with small A-pillar radii. However, the relevant legal restrictions also need to be taken into account here, some of which make it difficult to fit frontward protruding mirrors on the driver's side. The mirror must be visible through the field of the windshield wipers. If this cannot be ensured, the mirror must be attached to the side of the vehicle and must be visible through the side window.

10.7.4 Windscreen Wipers

On city buses with a virtually vertical front, windshield wipers have no influence whatsoever on the vehicle's c_D value. However, on modern coaches with their large windows and rounded A-pillars, it is important to choose the resting position of the passenger wipers with great care. Poorly positioned wipers can significantly hamper the airflow around the vehicle. The impact of the top windshield wiper on the area of dynamic pressure at the front can be seen in Figure 10.87, based on a Neoplan Skyliner.

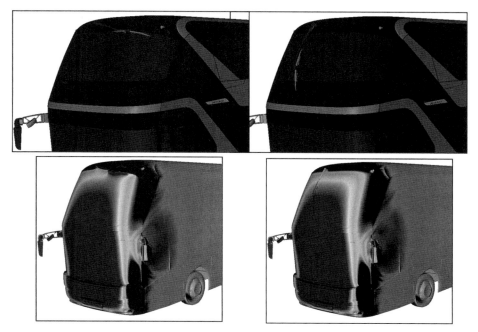

Figure 10.87 Wiper position and impact on the area of dynamic pressure at the front (NEOPLAN Skyliner).

When positioned horizontally the wiper increases the area of dynamic pressure at the front, thereby considerably increasing the vehicle's c_D value. In contrast, when positioned vertically the wiper is positioned in the direction of the airflow and has no negative impact on the airflow around the vehicle. However, this position is detrimental to passenger comfort. As a result, wiper positions that impair the airflow are often tolerated in the interests of passenger comfort.

10.7.5 Underbody

The underbody of buses and coaches is almost entirely sealed. Since the bottom of the center area simultaneously functions as a level luggage compartment, lines and frame struts are the only elements that disrupt the airflow around the vehicle in this area. The front of the vehicle, accommodating the spare wheel, offers the greatest potential for fitting underbody paneling. Figure 10.88 shows the potential available in each underbody area if they were covered with completely smooth underbody paneling. The area of vacuum at the front can clearly be seen, which is used by the vehicle's climate systems to ventilate the passenger area.

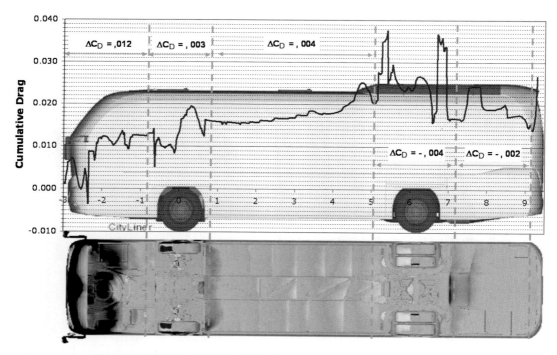

Figure 10.88 Potential of individual underbody areas on buses/coaches.

10.7.6 Wheels and Wheel Covers

On passenger cars, the rim design and wheel covers play a key role in determining the vehicle's c_D value. It is also important to observe this effect for commercial vehicles. Two types of covers are now widely accepted for buses and coaches. The most common rim cover is the one shown in Figure 10.89. The simulation clearly shows the reduction in turbulence caused by the airflow through the rim. The cover integrated into the body (Figure 10.90) is not as common but offers greater potential for savings, although this is only possible at the rear axle. However, the greatest savings are provided by wheel covers on the front axle, as there is a much larger flow of air passing through the wheel arch from the underbody to the side here than is the case for the rear axle.

Figure 10.89 Rim cover.

Figure 10.90 Integrated wheel cover in the body.

10.7.7 Airflow Through the Engine Compartment

The airflow through the engine compartment of a bus or coach is radically different from that of a truck. On buses and coaches the engine is located in the rear. No dynamic pressure is applied to the radiators; instead, they are supplied with cooling air exclusively by the fan. The engine outputs are much smaller than for trucks, as a result of which no special measures are required to safeguard the cooling output. The main problem for buses and coaches is the maximum permitted temperature for the components in the engine and adjacent body elements. In some cases, the engine compartment temperatures are so high that the paintwork on painted body components on the outer shell may occasionally be damaged. To prevent this, the flows of cooling air should be analyzed so that they can be used to selectively cool the components.

Figure 10.91 shows a characteristic airflow through the engine compartment on a city bus. There is very little scope to further develop the space in the engine compartment of a bus or coach, meaning air deflector components are very difficult to integrate. Selective openings in the underbody can help to guide the cooling air past critical components.

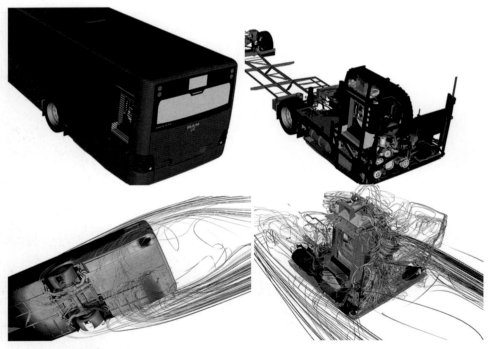

Figure 10.91 Characteristic airflow through the engine compartment on a city bus.

On hybrid buses, the electrical components and their cooling systems are located on the roof. It is possible here to selectively configure the cooling airflow, taking into account aerodynamic drag. The dynamic pressure from the head wind is used by means of targeted openings in the front. This reduces the need to activate the fan and further reduces the energy consumed by the bus or coach. Figure 10.92 shows the configuration of the cooling system (in terms of airflow) for the hybrid components on an MAN hybrid bus.

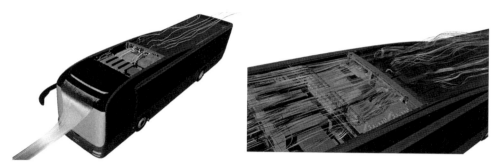

Figure 10.92 Configuration of airflow for cooling system.

10.7.8 Rear

An aerodynamically optimized, styled tapered rear is another method of reducing aerodynamic drag. Tapering considerably reduces the area of the wake and thus aerodynamic drag. Figure 10.93 shows the improvement in aerodynamic drag caused by a tapered rear design on the roof of a NEOPLAN Cityliner. The lowering of the roof by 150 mm at the rear, implemented as standard, reduces the vehicle's c_D value by 8%.

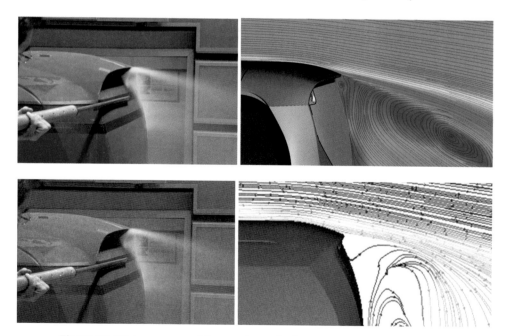

Figure 10.93 Improvement in aerodynamic drag caused by a tapered rear.

The undesired but largely unavoidable build-up of dirt at the rear of the Starliner, shown in Figure 10.94, also indirectly demonstrates the efficiency of the side tapered design. The airflow remains on the roof and at the upper side wall right up to the separation edge, which reduces the area of vacuum at the rear. A pleasant side-effect is that those parts of the body where the airflow remains show no evidence of dirt building up.

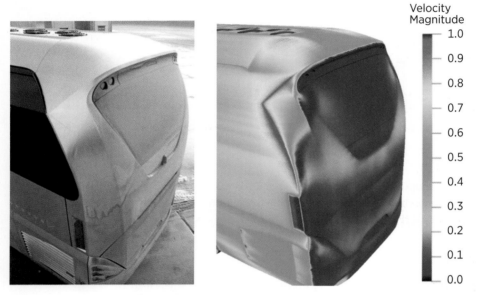

Figure 10.94 Build-up of dirt on a Starliner.

10.8 Aerodynamic Interaction

10.8.1 Nose-to-Tail Driving

Driving nose-to-tail is another way of reducing aerodynamic drag. Traffic is becoming more and more dense, and this is increasingly leading to vehicles driving nose-to-tail. As a result, several studies have already been carried out to ascertain the feasibility of using a selective convoy as a potential method of transport. The advantages of doing so clearly lie in the fuel savings achieved as a result of the aerodynamic interaction between the vehicles (cf. section 4.5.4).

Each vehicle pulls a defined wake behind it (Figure 10.95). This reduces the dynamic pressure on the following vehicle. At the same time, the area of dynamic pressure on the following vehicle has a positive effect on the wake of the vehicle in front. As such, both vehicles benefit from driving in a slipstream. This effect is particularly pronounced on commercial vehicles with their large wake areas.

The smaller the distance between the vehicles, the greater the potential savings. For example, the c_D value of the leading vehicle is reduced by 2% at a distance of 20 meters and by 8% at a distance of 10 meters (Figure 10.95). The vehicle in the middle benefits the most, with a reduction in the c_D value of 34% (at a distance of 20 m) or 40% (at a distance of 10 m). The reason for this is the aforementioned reduction in dynamic pressure at the front of the vehicle and the pushing effect from the following vehicle. However, this pushing effect is considerably smaller than the advantage of driving in a slipstream and should be disregarded when driving at a distance greater than 20 meters. Nevertheless, the last vehicle in the convoy still sees a reduction in the c_D value of approximately 32% compared with a vehicle traveling alone.

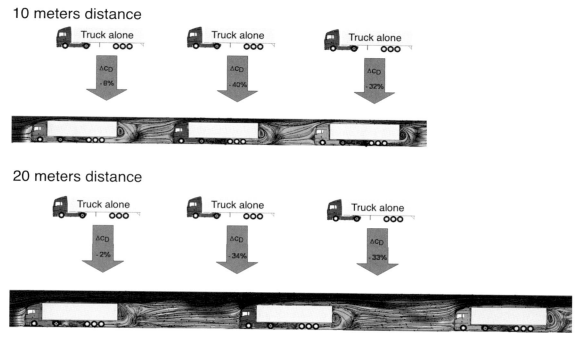

Figure 10.95 Reduction in c_D value and airflow topology when driving in a convoy at a distance of 10 m and 20 m.

In one vehicle test, a fuel savings of approximately 21% was achieved for the following truck with a gross weight of 28 t, a speed of 80 km/h, and a distance of 10 meters to the vehicle in front. At the same distance and speed, the truck in front (14.5 t) saw a fuel savings of around 7% (see Figure 10.96).

Figure 10.96 Two "Promote-Chauffeur" vehicles.

The measured fuel consumption savings and simulation results show that fuel can be saved when using an electronic drawbar at a distance of less than the length of one truck. However, no other major fuel saving was achieved when traveling at a distance of less than 10 m, meaning a distance of approximately 10 m from the truck in front is the optimal distance to achieve fuel savings.

10.8.2 Tipping and Susceptibility to Side Winds

Time and again, serious accidents are caused by buses, coaches, and trucks tipping over. This is due to the fact that, with a height of 4 m and a length of 16.5 m, they provide a contact surface of up to 66 m^2 for side winds. The vehicle's aerodynamics and driving dynamics are crucial factors in determining its tipping characteristics and susceptibility to side winds. The concept of driving dynamics includes factors such as the track width and vehicle's steering response. Driving dynamics are a key element in the phenomenon of vehicles tipping over and their susceptibility to side winds. However, this concept will not be explained further in this document (cf. chapter 5). In terms of buses and coaches, vehicles with a high design (such as double-decker buses and coaches) are particularly affected; for trucks it is the unladen, thus lightweight vehicles that are at risk.

The phenomenon occurs predominantly when traveling over bridges and on exposed motorways without shrubs or trees lining the road, where the side wind can strike the vehicle with virtually no interference. Figure 10.97 shows two vehicles shortly before they tip. It is clear that the tipping of an entire semitrailer combination is caused by the tipping of the semitrailer part.

Figure 10.97 Vehicles shortly before tipping after they have been hit by a strong side wind.

In this context, the main cause for the tipping lies in the point where the side wind's force is applied to the vehicle (located high above the vehicle's center of gravity), and thus the high roll moment M_R. Strong turbulence on the roof and at the side facing away from the wind causes a vacuum at the edges of the roof, which increases the roll moment M_R and the lift F_L of the vehicle and thus the risk of the vehicle tipping over. Cosano & Colombano [178] used simplifying assumptions to calculate a formula to determine a critical side wind vW,crit after which the vehicle will tip.

$$v_{W,\text{crit}} = \sqrt{\frac{2m_F g}{\rho_L \cdot A_x} \cdot \frac{s}{2h \cdot (c_S + c_{M,R}) + s \cdot c_L}} \quad (10.10)$$

In this formula, m_F denotes the vehicle mass; s denotes the track width; c_S, $c_{M,R}$, and c_L the aerodynamic coefficients for the side force, roll moment, and lift; A_x the characteristic vehicle surface; ρ_L denotes the air density; and h denotes the height of the vehicle's center of gravity above the road surface. Based on his calculations, Baker [35] 1987 recommended blocking exposed sections of road if wind gusts of 80 km/h or more should occur. It is clear that tipping is not dependent on the aerodynamic drag, yaw moment, or pitching moment, nor on the vehicle's wheelbase. A high vehicle mass and large track width have a positive effect on the tipping characteristics of commercial vehicles, whereas a high center of gravity and large aerodynamic moments have a negative effect.

In contrast, the vehicle's susceptibility to side winds is influenced by a force application point located next to the vehicle's center of gravity in the longitudinal direction. Vehicles featuring large contact surfaces and small A-pillar radii at the front are particularly susceptible to side winds. This susceptibility can be improved by increasing the vehicle mass. However, this in turn reduces the maximum permitted cargo volume and increases fuel consumption when accelerating, meaning this solution is not viable. A center of gravity positioned far toward the front has proven to be a very effective solution. This benefits two-axle coaches in particular, as the rear-axle load comes close to the load limit in most cases.

Changing the vehicle shape also offers great potential. The vehicle's susceptibility to side winds can be greatly improved by reducing the lateral projection area, in particular at the front. Large radii at the front not only have a positive impact on aerodynamic drag, they also reduce the vehicle's susceptibility to side winds. Both measures shift the resulting center of pressure toward the rear of the vehicle and reduce the vehicle's susceptibility to side winds. These design features were implemented as standard on the Neoplan Starliner (see Figure 10.84).

10.8.3 Aerodynamic Loads on Components

When specifying the strength of body components of corresponding size, the aerodynamic load on these components also needs to be taken into account in addition to the loads exerted in a crash. For example, when developing a roof spoiler or air deflectors, both the wind load from the front as well as from the rear must be considered (see Figure 10.98).

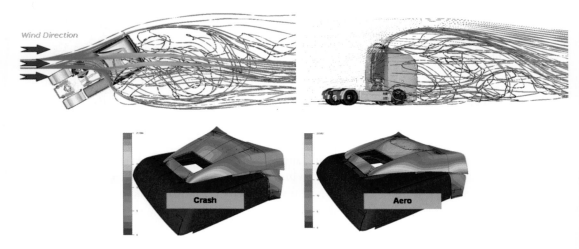

Figure 10.98 Visualization of airflow and deformation of a commercial vehicle roof spoiler: (a) in a crash and (b) with an airflow from the rear.

These wind loads are determined using CFD calculations and incorporated directly in the strength calculation as input variables. This ensures that the spoiler is not damaged either during a crash, when traveling backwards by train, or during storms.

10.8.4 Dust Turbulence

The functional aerodynamics on commercial vehicles include in particular the airflows produced by the vehicle itself (e.g., exhaust gas and engine compartment airflows). The airflow through the engine compartment should be analyzed for vehicles deployed in dusty environments in particular, in order to counteract the disruptive phenomenon of dust turbulence. The strong fans on the vehicles swirl up dust when traveling off-road, meaning the driver cannot see the rear of his semitrailer when maneuvering in reverse. As such, targeted measures are being developed to counter this disruptive dust turbulence for vehicles deployed on construction sites, and their impact on the cooling system is being evaluated. Figure 10.99 shows the function of a cover shield in terms of reducing dust turbulence.

Figure 10.99 Function of a cover shield in terms of reducing dust turbulence.

10.8.5 Intake of Warm Air

The ambient temperature directly influences the vehicle's cooling output. This not only affects the transfer of heat in the intercooler and radiator but also means that the intake air needs to be cooled down further in the intercooler. As a result, the aim is to take in air that is as cool as possible before it is heated in the compressors and then cooled again in the intercooler before finally being added to the combustion process. Since the engine's exhaust air heats up large areas of the chassis and in particular the space between the tractor vehicle and semitrailer with warm air, manufacturers position most of their air intakes in areas where the coolest possible ambient air can be taken in. These are around the roof, side wall, and at the front of the vehicle (see Figure 10.100).

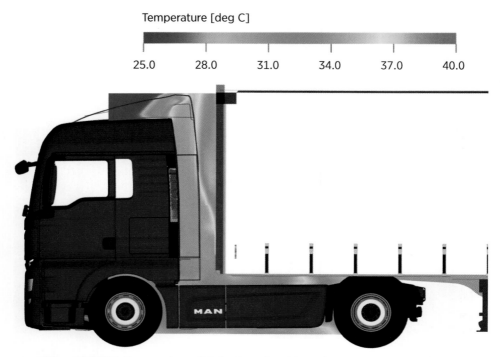

Figure 10.100 Temperature distribution of ambient air on a semitrailer tractor.

10.8.6 Management of Exhaust Gas

The new Euro 6 requirements stipulate that filter systems in the exhaust systems need to be actively regenerated at regular intervals. This process involves exhaust gas temperatures of approximately 500°C. Figure 10.101 shows such an exhaust gas stream by way of example. In order to prevent a fire risk for shrubs and trees lining the road, damage to the road surface, and a health hazard for pedestrians, particular care must be taken when channeling the exhaust gas out of the exhaust silencer. However, the frame attachments on the vehicle itself are only temperature-resistant to a certain extent, meaning the exhaust gas stream needs to be selectively diverted in a particular direction or dissipated in an area where it cannot cause any damage. Designs for exhaust systems also need to take into account the problem of dust turbulence as described in section 10.8.4 since exit flow speeds of 20 m/s with a flow of approximately 1,600 kg/h may occur.

Commercial Vehicles

Figure 10.101 Exhaust gas stream on a semitrailer tractor.

10.9 Vehicle Soiling

10.9.1 Task Description and Testing Methods

The concept of aerodynamics concerns the task of reducing not only aerodynamic drag but also the dirt that builds up on the vehicle. For safety reasons, good visibility is of crucial importance for drivers of commercial vehicles. The concept of "seeing and being seen" is affected by the dirt that builds up when traveling on wet roads. The build-up of dirt is caused by small dirt particles thrown up by the vehicle's tires. If these particles build up on the driver's own vehicle, this is referred to as self-soiling. If the dirt particles mix with the turbulent airflow in the wake and hit following or oncoming vehicles, this is generally referred to as foreign soiling (section 10.9.2).

In particular, this swirled-up dirt impairs the driver's view of the road due to a dirty windshield, side windows, and mirrors on his own vehicle, and due to spray mist in the wake trails for other road users. A layer of dirt is also deposited on various outer cab components such as the step unit or door handle.

Keeping the vehicle clear of dirt is not just important from a comfort perspective; it also makes a major contribution to active safety. The air should be channeled with the aim of preventing self- or foreign soiling such that no dirty air hits areas that must be kept clean and the proportion of dirty water in the spray produced is reduced. Compared with the task of simply optimizing the airflow around the vehicle, the solution to this task is much more complex, as it concerns a multistage airflow: air, solid particles, and a given amount of dirty water are contained in the airflow (cf. chapter 6).

When developing a new vehicle, a number of different testing methods have been established to optimize the process of keeping the vehicle clear of dirt. Combining analytical and practical procedures safeguards the results and ensures the available potential is analyzed. Tests in a wind tunnel and on the road, supported by numerical airflow simulations, have proven themselves as suitable testing methods. When developing measures to reduce the build-up of dirt, it is important to ensure they do not in turn increase aerodynamic drag.

In the wind tunnel, nozzles add water and a fluorescent substance to the air flowing around the vehicle under defined conditions such as the airflow speed and yaw angle. Figure 10.102 shows the simulation of foreign soiling from the front. In order to show self-soiling, the dirty water is supplied to the tire tread tangentially in the spray direction and thus into the airflow around the vehicle. When the wind tunnel lights are switched off, the dirty areas on the vehicle can be clearly identified under black light (see Figure 10.103).

Figure 10.102 Simulation of foreign soiling.

The dirt build-up is tested on a test course either parallel to or following on from the tests in the wind tunnel. For this purpose the vehicle being tested is driven over a course with a defined amount of dirt. To simulate foreign soiling, a second vehicle is then driven in front of the vehicle being tested to create the spray. To be sure the results can be compared with those achieved in the wind tunnel, comparable test conditions for the speed and oblique flow must be ensured.

Figure 10.103 Identification of dirt under black light.

As for all tests, the definitive results for the build-up of dirt are obtained during actual deployment. This shows whether the test conditions in the wind tunnel and on the test course were an accurate reflection of real-world deployment.

10.9.2 Foreign Soiling

External soiling is caused by spray and dirt thrown up by preceding and oncoming vehicles and water from the driver's own wiper and washer systems. Here, the build-up of dirt on the side windows and mirrors is determined by the interaction of aerodynamically profiled covers and rain channels on the A-pillar and the shape of the mirror housing. On commercial vehicles in particular, the build-up of dirt on the side window is facilitated by the exterior mirrors that protrude out as well as their holder (see Figure 10.104). Selecting a suitable and coordinated shape for the exterior mirror and A-pillar cladding makes it possible to guide the airflow around the side of the vehicle so as to ensure the mirror surfaces stay dry and to considerably reduce the spray of water onto the side window (see Figure 10.105). For foreign soiling, the dirt on the side window is assessed by splitting up the side window into a grid orientated to the driver's viewing direction. It should be noted here that the subjective perception of the build-up of dirt is quantifiable using clear methods.

Figure 10.104 Dirt on side window caused by exterior rear view mirror.

Figure 10.105 Reduced dirt on side window due to suitable shape of exterior rear view mirror and profile that channels away dirty water.

10.9.3 Self-Soiling

Self-soiling is caused by the movement of the vehicle over a wet, dirty road surface. Koessler [451] carried out studies of how the water on the road is thrown up and sprayed as the tire rolls over the road, and the results are shown in Figure 10.106. According to his studies, water is displaced to the front and sides from the contact surface of the tire (splashing water). However, it also collects in the tire tread and is ejected backwards as a result of centrifugal force, predominantly at an angle of 0° to 30° (spray water). To date, no measures have been able to effectively tackle splashing water, which occurs in the form of relatively large droplets and is dispersed at a flat angle. In contrast, a number of different solutions are known for tackling spray water on trucks and buses/coaches, and these are defined and specified by the legislator in Directive 91/226/EEC for commercial vehicles.

Commercial Vehicles

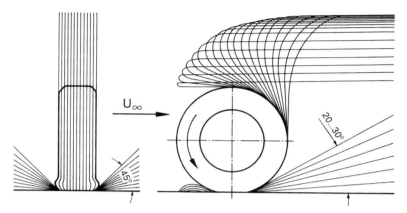

Figure 10.106 Spray of dirty water from a free-rolling wheel; in accordance with Koessler (1965) [451].

On trucks, the space between the tire and wheel cover is relatively large in order to accommodate the different tires and spring travels. As a result, standard wheel covers do not completely reduce splashing water and also do not completely stop spray water from being ejected backwards. Air vents integrated into the exterior have proven to be a highly effective method of preventing water and dirt particles from landing on the sides of the cab near the step units and door handles in day-to-day deployment (see Figure 10.107). In many cases, small guide vanes have also been integrated to increase the effectiveness of the deflector fins. The purpose of the guide vanes is to stabilize the direction of the exit airflow.

Figure 10.107 Air vents integrated into the exterior.

The numerical airflow simulation shown in Figure 10.108 demonstrates the function of the guide vanes; the air is diverted downwards. The results have been confirmed in tests in the wind tunnel, and on the road (see Figure 10.109) the cab door remains largely free of dirt.

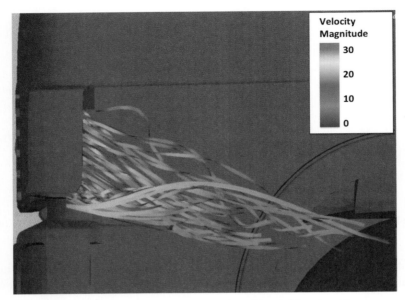

Figure 10.108 CFD tests on a deflector fin with additional guide vanes.

Figure 10.109 Wind tunnel tests for self-soiling with air vent.

Chapter 11
Motorcycle Aerodynamics

Norbert Grün, Holger Winkelmann, Frank Ullrich

11.1 Introduction

On August 29, 1885, one year before Carl Benz invented the automobile, Gottlieb Daimler received the German patent no. 36.423 for what he called the "Reitwagen," a wooden vehicle with a gas or petroleum engine (Figure 11.1). According to Schnepf [732], this vehicle had a $c_D \cdot A_x = 0.67$ m^2.

The first scientific investigations of unmotorized two-wheeler stability were published later in 1898 by Bourlet [91] and Whipple [860]. So Daimler, whose focus was on usage of an internal combustion engine, added two wooden wheels to ensure tilting stability. Nevertheless, this was the world's first motorcycle, and though it never went into serial production, it already exhibited the characteristics and proportion of today's motorcycles.

At that time, aerodynamics as a science applied to vehicles or even wind tunnels was yet unknown. It was not until the late 1920s that aerodynamic measures were used to increase the top speed for world record trials for automobiles and motorcycles.

Figure 11.1 Daimler "Reitwagen," 1885 (replica, the original burned down in 1903).

11.2 Historical Review and Current Types

11.2.1 History of Motorcycle Aerodynamics

A pioneer in vehicle aerodynamics was P. H. White. His U.S. patent 1.183.938 in 1912 was related to a single-track vehicle with a fairing (Figure 11.2) to minimize aerodynamic drag and to ensure optimal wind and weather protection (Koenig-Fachsenfeld 1946 [447]).

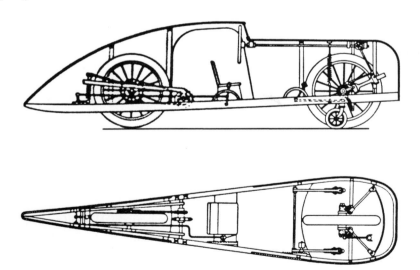

Figure 11.2 Single-track streamlined vehicle, U.S. Patent No. 1.183.938, P.H. White, 1912.

In Europe, motorcycle aerodynamics first came to attention in September 1929, when Ernst Henne set the first world speed record with 216.9 km/h near Munich on a 750cc

BMW. The motorcycle itself showed only smooth side fairings, but the pilot had to wear conical extensions on his helmet and backside to minimize flow separations (Figure 11.3, [391]). The definite effect of these measures was reported later by Sawatzki et al. [708] in 1938.

Figure 11.3 Ernst Henne in 1929 sets the world speed record at 216.9 km/h on a 750 cm^3 BMW.

Fully shrouded motorcycles with aluminum fairings, designed solely to break speed records, appeared around 1936. Again it was Ernst Henne who raised the record to 279.5 km/h on a BMW R5 (500 cm^3) in 1937, a speed that was unsurpassed for the next 14 years (Figure 11.4). In these runs, BMW faced severe stability problems, which they eventually counteracted with a kind of tail fin.

Figure 11.4 Ernst Henne in 1937 on the 500 cm^3 world record BMW (279.5 km/h).

At the same time, DKW tackled the speed records in the lower engine displacement categories and also experienced stability problems. Koenig-Fachsenfeld (1946, [447]) considered the large lateral areas (caused by the conventional seating position) as the reason when they are exposed to cross wind. He designed a "low-rider" with reduced overall height for which he received a patent in 1938. Eleven years later, NSU built such a vehicle (without engine) for which Herz et al. [341] and Trzebiatowski [807] reported a drag coefficient of $c_D = 0.14$–0.15. However, it never came into operation because at that time the driver's posture was considered as too unusual.

That would change; with the cooperation of Scholz [735], NSU developed fully shrouded record vehicles with tail fins. In 1951, Wilhelm Herz pushed the absolute world record to 290 km/h on the Delfin I, where he was highly inclined forward but still sitting at the usual position. Properties of this motorcycle were (Hütten [391], 1983) $c_D = 0.24$–0.32, $A_x = 0.54$–0.55 m². In 1956, this record was exceeded with 339 km/h by the Delfin III (Figure 11.5), whose drag coefficient was reported to be $c_D = 0.18$–0.20 with unaltered frontal area.

Figure 11.5 Wilhelm Herz in 1956 on the NSU Delfin III (World Record 339 km/h).

In parallel, even lower drag vehicles were developed by 1951 on which the pilot sat inclined backward, in an almost lying position. The originator of this idea, Baumm, wanted to set records in the small displacement categories and had recognized that not only the drag coefficient but also the frontal area are key. The first speed record runs of these Baumm'sche Liegestühle (deck chairs) were carried out in 1954 (Figure 11.6).

This time, NSU not only focused on top speed but also wanted to demonstrate fuel efficiency by optimized aerodynamics (Froede 1954 [261]; Herz & Reese 1987 [341]; Hütten 1983 [391]). The drag coefficients of various makes—due to the unusual driver's posture all being inapplicable for public use—were between 0.07 and 0.16 with frontal areas from 0.25 m² to 0.4 m². As Figure 11.6 shows, they were also fitted with tail fins to cope with stability issues.

Motorcycle Aerodynamics

Figure 11.6 Baumm'scher Liegestuhl by NSU 1954.

In 2010, the official land speed record for motorcycles of 605.697 km/h was attained by Rocky Robinson's Ack Attack (Figure 11.7), which was propelled by two turbo-charged Suzuki engines with a combined displacement of 2600 cm^3 and a power output of around 1100 hp. The highest recorded speed in a single run was 634.216 km/h. The drag coefficient and frontal area were not published.

Figure 11.7 Ack Attack driven by Rocky Robinson, world record 605.697 km/h set in 2010.

In the 1950s, aerodynamic optimization also became popular for "normal" road racing motorcycles (Trzebiatowsky 1955 [807]). First, the motorcycles were only partially fitted with fairings, as shown in Figure 11.8 and Figure 11.9. The appearance changed quickly. Soon, almost the entire motorcycle, including the front wheel, was enclosed by the fairing (Figure 11.10), made of hand-formed aluminum sheets as on the previously described speed record bikes of the same era. In particular, the integration of the front wheel guaranteed low aerodynamic drag values but for safety reasons this was later banned by the sports authorities.

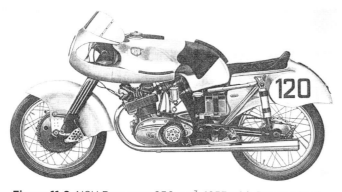

Figure 11.8 NSU Rennmax 250 cm^3, 1953 with Banana Tank.

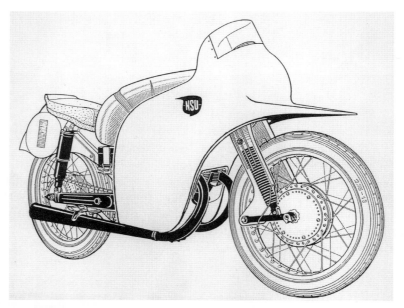

Figure 11.9 NSU Rennmax 250 cm³, 1954 with Dolphin Fairing.

Figure 11.10 NSU Rennmax 250 cm³, 1955 with Blue Whale Fairing.

Deformations, for instance after a crash, could limit or even lock the steering. Further, the aerodynamic yawing moment could become critical due to the bulky front end. Even with an intact fairing the possible steering angles were quite low. With the BMW Sports

Tourer R 100 RS in 1976 (Figure 11.11) the first motorcycle with a complete fairing, not meant for racing, was launched for series production.

The focus was not primarily to increase the top speed, but—apart from styling reasons—an increased weather protection and a relief of the driver from wind loads. Noticeable are also spoilerlike extensions on the sides, which should increase the front-wheel load for a favorable influence on stability (section 11.3.3).

Figure 11.11 BMW R 100 RS Sports Tourer 1976.

11.2.2 Current Motorcycle Categories

Contemporary motorcycles cover an enormous spectrum of weight, power, and operational purpose. Different aerodynamic properties are required according to category. Figure 11.12 places representative models of various categories in a power versus weight diagram (power units are specified here and in the following text deliberately using horsepower, hp, and not kW since this is still the most common unit used among motorcyclists).

The fact that the different categories significantly overlap indicates that often there is no sharp differentiation, and the classification of a certain motorcycle may depend on the observer. On the other hand, it is clear that sports bikes and enduro/cross bikes are distinct from the other categories. Though far from exhaustive, the following sections describe some examples of each category.

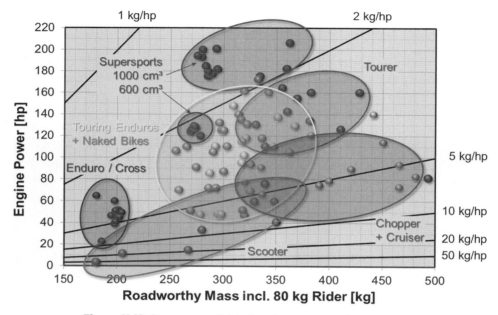

Figure 11.12 Power vs. weight of various motorcycle categories.

11.2.2.1 Off-Road Motorcycles

No demands on aerodynamics, apart from cooling, are made by off-road motorcycles, including trial, motocross, and enduro bikes (Figure 11.13 left). Even in competitions they are hardly driven at speeds above 60–80 km/h over longer periods. Trial bikes are even operated mostly at walking speed.

An exception are enduro bikes, which are used for long-distance off-road rallies like the Dakar. Top speeds up to 200 km/h were possible before the engine displacement was limited to 450 cm³ (for professionals). But even the smaller one-cylinder bikes today reach around 170 km/h. At such rallies stages of 800 km or more per day are common, and therefore these bikes are equipped with fairings (Figure 11.13 right) to improve the driving performance and to relieve the driver.

Figure 11.13 Husqvarna TE 449: (left) Enduro; (right) Rallye version.

In the power-weight diagram, all enduro and motocross bikes populate a clearly offset region. Typical key data for weight (without driver) and power are 100–120 kg and 40–60 hp, respectively. This is the only category where a noteworthy number of two-stroke engines are still represented.

11.2.2.2 Touring Enduros

The category name already indicates that these bikes are meant for touring on and off of paved roads. In 1976, this category was originated by the Yamaha XT 500, a 499 cm^3 one-cylinder four-stroke bike with at that time excellent off-road capabilities. Very soon it advanced to be the most popular bike for world tourers. In 1980, BMW followed with the R 80 G/S, the first off-road capable two-cylinder motorcycle with an 800 cm^3 engine. Bikes based on this model have won the famous Paris-Dakar rally several times. This was preceded by successes of similar prototypes at national and international competitions like the "Six-Days."

Today we find motorcycles in this category with one to four cylinders, 150 to 280 kg unloaded weight, and a power between 50 and 150 hp, where those with big displacements are often equipped with a low-maintenance cardan-shaft drive. They claim the title "Enduro," this variety requires the actual capability for off-road adventures. As Figure 11.12 denotes, the more powerful models offer a performance similar to sport bikes. The common minimum aerodynamic device in this class is an (often adjustable) windshield. As a representative example, Figure 11.14 depicts the BMW R 1200 GS, which was for years the worldwide top-selling touring enduro.

Figure 11.14 Touring Enduro BMW R 1200 GS.

11.2.2.3 Tourers

Solely designed for on-road use, the category of tourer motorcycles typically has an unloaded weight between 250 kg and 400 kg and power up to about 160 hp. At the lower end of the weight spectrum, we find the so-called sports tourer, potentially with small deficits in touring capabilities. The upper end is marked by luxury tourers with focus on maximum long-distance comfort and complete equipment (heated handle grips and seats, cruise control, navigation and audio system, electrically adjustable wind shield, etc.). As an example, Figure 11.15 shows the BMW K 1600 GTL, powered by a 1600 cm^3 six-cylinder in-line engine, delivering 160 hp.

Figure 11.15 Luxury Tourer BMW K 1600 GTL.

11.2.2.4 Scooters

Originally, scooters were pure city or short-distance vehicles typically with 50 cm^3 two-stroke engines but occasionally fitted with 125 or 250 cm^3 engines. In the meantime, particularly in southern Europe large scooters with 600–800 cm^3 four-stroke engines have become very popular. With an unloaded weight of up to 250 kg and 70 hp, they allow a performance similar to middle class motorcycles. Due to the good wind and weather protection, these large scooters are also used for long-distance traveling. Figure 11.16 shows two variants of a BMW scooter with a 600 cm^3 two-cylinder engine.

Motorcycle Aerodynamics

Figure 11.16 Scooter: BMW C650 GT (left) and C600 Sport (right).

11.2.2.5 Chopper/Cruiser

For a long time the term chopper was synonymous with Harley-Davidson (Figure 11.17). Today, many other motorcycle brands offer equivalent models. Neither sporty performance nor a notable long-distance comfort are expected in this category. Choppers populate the upper end of the weight spectrum with relatively moderate performance while cruisers are close to touring bikes and sometimes equipped with a wind shield or partial fairings.

Figure 11.17 Chopper/Cruiser: Harley-Davidson Iron 883.

11.2.2.6 Naked Bikes/Fun Bikes/Supermoto

As the name suggests, naked bikes refrain from any aerodynamic features, and at most they have a small windshield above the headlight. Their favored domain are country roads and mountain passes, but they are frequently also used on the race track. Various brand- and model-specific racing series have been established especially for these bikes. Concerning weight and performance, they cover the same spectrum as touring enduros with one- to four-cylinder engines with 600–1300 cm^3 and up to 170 hp. Supermoto bikes were originally derived from light enduros with shortened suspension travel, smaller wheels, and road tires for racing series on mixed tarmac and dirt tracks. Today they are also very popular in street-legal versions. As an example, Figure 11.18 shows the Husqvarna Nuda 900 R, driven by a 900 cm^3 two-cylinder engine with 105 hp.

Figure 11.18 Supermoto: Husqvarna Nuda 900 R.

11.2.2.7 Supersport

The prefix "super" for this category of motorcycles is necessary in order to differentiate clearly from sports tourers. They form the top of what can be used on two wheels on public roads today. Not too long ago their performance would have been sufficient to win the road-racing world championship. With an unloaded weight of less than 200 kg and a power around 200 hp, they reach top speeds of over 300 km/h. The Formula 1 on two wheels series is called the "Moto GP," where 1000 cm^3 prototypes have more than 260 hp and, depending on the track, reach top speeds around 340 km/h. At these extreme speeds of course aerodynamic performance, including drag, lift, and forces on the pilot, is key.

For a long time this class was the undoubted domain of the Japanese manufacturers Honda, Kawasaki, Suzuki, and Yamaha. In the near past, this supremacy has been challenged successfully by European producers like Aprilia, Ducati, MV Agusta, KTM, and last but not least even BMW. This is reflected in the multitude of brands on the starting grid in production-oriented racing series from the IDM (International German Championship), Superstock up to the Superbike World Championship. Similar to enduro/motocross, supersport bikes with 1000 cm^3 populate a separate region in the power-weight diagram (Figure 11.12). As an example for this class, Figure 11.19 depicts the BMW HP4 (4 cylinders, 1000 cm^3, 193 hp, unloaded 199 kg), a special version of the S 1000 RR.

Figure 11.19 Supersport: BMW HP4.

11.2.3 Special Bikes

This section describes some models that cannot be classified in the previous categories.

11.2.3.1 Motorcycle Combination

Enthusiasts of motorcycle combinations are actually driving a double-track vehicle, comparable to an automobile. The joy of riding is really the focus, and considering the small production numbers only few manufacturers can afford cost-intensive aerodynamic optimization.

Nevertheless, even small manufacturers consider drag and lift influencing measures. As shown in Figure 11.20, the sidecar mounted to a Suzuki Hayabusa exhibits spoilers on the front end, which develop a noticeable effect at the high speeds possible with this strong motorcycle. Measurements of this vehicle showed drag and lift coefficients of

$c_D = 0.637$, $c_{Lf} = 0.045$, and $c_{Lr} = -0.079$ with a frontal area of $A_x = 1.51$ m² including the driver. Despite the large frontal area (1.2–1.5 m², which is 50–70% higher than solo motorcycles) modern combinations with powerful engines may reach a top speed of more than 200 km/h.

The drag area of $c_D \cdot A_x = 0.7$–0.9 m² for series-production combinations is much higher than for modern automobiles (0.5–0.6 m²). Therefore, careful aerodynamic optimization is indispensable for combinations in road racing.

Figure 11.20 Hayabusa F1 Combination.

11.2.3.2 Three-Wheeler

Three-wheelers exist with either two wheels on the rear or on the front. The (small) ones with two wheels on the rear axle have almost disappeared from the market due to their critical tendency to tip over in turns.

Mainly the established scooter manufacturers produce vehicles with a tilting technology, which lean at cornering similar to single-track motorcycles. An example, the Piaggio Yourban, is shown in Figure 11.21, allowing lean angles up to 40°. At speeds below 10 km/h and at rest, the tilting mechanism is automatically locked. Since the track width exceeds a certain legal minimum, the Yourban may be used with an automobile driver's license.

The same holds for the powerful Can-Am Roadster from Bombardier (998 cm³, V2 engine, 106 hp), which comes without the tilting technology but with automobile-like front wheel suspension and electronic stability control.

Figure 11.21 Piaggio Yourban LT (278 cm³, 23 hp, 236 kg, Automatic Gearbox).

11.2.3.3 C1

The unique concept of the BMW C1 (Figure 11.22) combines the advantages of a single-track vehicle with the safety and comfort elements of an automobile, and it gained broad attention when it was introduced. The experience of driving a motorcycle without helmet and special protection gear yet with increased safety compared to a conventional motorcycle is a new, easy way of locomotion in urban surroundings. The drag area of $c_D \cdot A_x = 0.40$ m² despite the relatively large frontal area underlines the carefully designed aerodynamics. Adding the optional topcase even improves this value by 0.01 m².

Figure 11.22 BMW C1.

11.2.3.4 Ecomobile

Another technically sophisticated vehicle for a select circle of enthusiasts is the Ecomobile from Peraves in Switzerland. It is offered as a one- or two-seater (Figure 11.23) equipped with various engines. Its chassis is a composite monocoque enforced by roll-over bars. The installed power of 190 hp and a streamlined shape with a small frontal area allow top speeds over 300 km/h. With a drag area of $c_D \cdot A_x = 0.23$ m² (estimated from performance data) this is one of the most interesting and potentially consequential developments of the recent past regarding aerodynamics.

Figure 11.23 Peraves Ecomobile (Switzerland):left: one-seater, $l = 3135$ mm; right: two-seater, $l = 3700$ mm.

11.3 Aerodynamic Tasks

11.3.1 Aerodynamic Forces and Moments

Each body in a moving fluid experiences pressure and skin friction, that is, normal and tangential stresses on its surface. Their integration yields the resulting force and moment vectors. In ground vehicle aerodynamics, it is convenient to decompose them in the direction of a coordinate system fixed to the body (Table 11.1), because its x-axis is normally aligned with the direction of movement, regardless of the oncoming flow direction. This is in contrast to aeronautics, where the directions of movement and oncoming flow are parallel regardless of the body's orientation.

Table 11.1 Components of Forces and Moments in a Body-Fixed System

Direction	Force	Moment (Right-Handed Positive)
x (positive downstream)	F_D (Drag)	M_R (Rolling moment)
y (positive to the right)	F_S (Side force)	M_P (Pitching moment)
z (positive upward)	F_L (Lift)	M_Y (Yawing moment)

As the reference point for moments, usually the wheelbase center on the ground is selected (see also section 4.1).

Aerodynamic forces grow linearly with the fluids density and quadratically with the velocity, meaning they depend on the dynamic pressure[1] q_∞ of the undisturbed flow.

$$q_\infty = \frac{1}{2}\rho v_\infty^2 \tag{11.1}$$

This led to the introduction of dimensionless coefficients for forces and moments to describe the aerodynamic properties of a vehicle:

$$c_i = \frac{F_i}{q_\infty \cdot A_{Ref}} \text{ with } i = D, S, L \text{ and } c_{M,i} = \frac{M_i}{q_\infty \cdot A_{Ref} \cdot L_{Ref}} \text{ with } i = R, P, Y \tag{11.2}$$

The choice of the reference area A_{Ref} and the length L_{Ref} is arbitrary. For ground vehicles it is common to use the frontal area A_x and the wheelbase l_0. On motorcycles the driver (and passenger) contributes more or less to the frontal area, depending on the type of bike (Figure 11.24).

---

1. The frequently used denomination "stagnation pressure" for q_∞ is misleading since this is not the static pressure at the stagnation point but only its increase over the ambient pressure for a complete stagnation to $v = 0$.

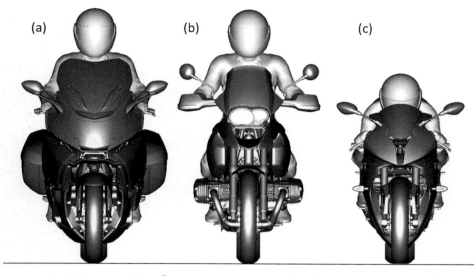

Figure 11.24 Frontal areas of luxury tourer (a), touring enduro (b), and supersport (c).

In general the force and moment coefficients depend to a certain extent on Reynolds and Mach number, so that strictly speaking not a single value can be used to calculate aerodynamic forces over the entire velocity range. However, the Mach number dependency can be neglected because even at $Ma = 0.3$, which may be reached in racing, compressibility effects are still very small.

Compared to automobiles, even faired motorcycles are jagged geometries where separation is mostly determined by edges and ridges and therefore the Reynolds dependency is quite small at velocities over 100 km/h (Figure 11.25).

The lower curves in Figure 11.25 for a motorcycle without dummy show that the drag area drops only by 2–3% from 100 km/h to 220 km/h. With floating suspension, the drag area is slightly higher than blocked and remains constant above 180 km/h (see inserted diagram in Figure 11.25). This is caused by the rebound of the front wheel suspension due the front axle lift, increasing the drag coefficient as well as the frontal area.

The measurement with dummy produces exactly the opposite trend: the drag area increases with velocity. However, this is not a Reynolds effect. Although the dummy is fixed with struts to its chest (see image in Figure 11.25), it will be deformed at high speeds, and the posture of the helmet changes. Due to this uncertainty in frontal area, motorcycle aerodynamics is typically characterized by the drag area instead of a drag coefficient.

Motorcycle Aerodynamics

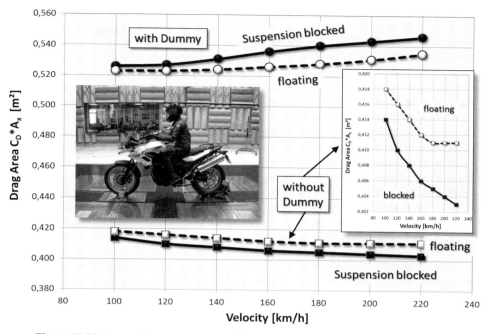

Figure 11.25 Reynolds dependency of the drag area (measurement in the BMW aeroacoustics wind tunnel).

11.3.2 Aerodynamics and Longitudinal Dynamics

11.3.2.1 Top Speed

The motion of a motorcycle in the longitudinal direction, acceleration and top speed, are governed by the sum of all longitudinal forces. If we consider a motorcycle driving at top speed on a horizontal flat road (so that no weight component has to be included) these are the following.

11.3.2.1.1 Traction Force at the Rear Wheel

The effective traction force results from the engine power P, reduced by an efficiency factor η comprising all losses from the crankshaft to the rear wheel, and the velocity.

$$F_Z = \frac{\eta \cdot P}{v_{max}} \tag{11.3}$$

11.3.2.1.2 Aerodynamic Drag

As mentioned before, the drag force depends linearly on fluid density, frontal area, drag coefficient, and quadratically on the velocity

$$F_D = c_D \cdot A_x \cdot \frac{1}{2} \rho_L v_{max}^2 \tag{11.4}$$

11.3.2.1.3 Rolling Resistance

The rolling resistance coefficient f_R normalizes the force for the tire deformation (and the resistance of wheel bearings) with the mechanical wheel load. It is constant over a wide velocity range and increases only slightly at high speeds.

$$F_R = f_R \cdot F_N = \mu_R \cdot \left(mg - c_L \cdot A_x \cdot \frac{1}{2} \rho_l v_{max}^2 \right) \tag{11.5}$$

This equation has to be used for each wheel separately because the static load as well as the aerodynamic lift or downforce and the coefficient f_R are different on front and rear.

Figure 11.26 shows aerodynamic drag and rolling resistance versus velocity for a super-sports motorcycle. Above 80 km/h, the quadratically increasing drag dominates. At 200 km/h, it is five times, at 300 km/h even 12 times the rolling resistance. Later it will be shown that nevertheless the rolling resistance has a noticeable influence on top speed.

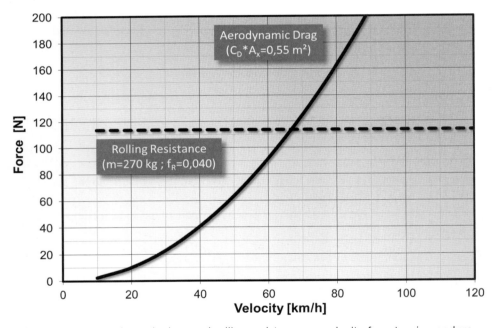

Figure 11.26 Aerodynamic drag and rolling resistance vs. velocity for a touring enduro.

The condition that at top speed the traction force is equal to the sum of all resistant forces

$$\frac{\eta \cdot P_{max}}{v_{max}} = \frac{1}{2} \rho_L v_{max}^2 \left(c_D - c_{L,f} \cdot \mu_{R,f} - c_{L,r} \cdot \mu_{R,r} \right) \cdot A_x + \left(m_f \cdot \mu_{R,f} + m_r \cdot \mu_{R,r} \right) \cdot g \tag{11.6}$$

forms a cubic equation for v_{max} with the real solution

Motorcycle Aerodynamics

$$v_{max} = \sqrt[3]{-\frac{a_0}{2} + \sqrt{\left(\frac{a_0}{2}\right)^2 + \left(\frac{a_1}{3}\right)^3}} + \sqrt[3]{-\frac{a_0}{2} - \sqrt{\left(\frac{a_0}{2}\right)^2 + \left(\frac{a_1}{3}\right)^3}} \quad (11.7)$$

where the coefficients are

$$a_0 = -\frac{\eta \cdot P_{max}}{\frac{1}{2}\rho_L A_x \left(c_D - c_{L,f} \cdot \mu_{R,f} - c_{L,r} \cdot \mu_{R,r}\right)} \quad (11.8)$$

$$a_1 = -\frac{\left(m_f \cdot f_{R,f} + m_r \cdot f_{R,r}\right) \cdot g}{\frac{1}{2}\rho_L A_x \left(c_D - c_{L,f} \cdot \mu_{R,f} - c_{L,r} \cdot \mu_{R,r}\right)} \quad (11.9)$$

Disregarding the rolling resistance, i.e., $f_{R,f} = f_{R,r} = 0$, equation (11.10) allows to express the top speed simplified as

$$v_{max} = \sqrt[3]{\frac{2\eta P_{max}}{\rho_L c_W A_x}} \quad (11.10)$$

The effect of this simplification gets clear from Figure 11.27, where the top speed of a supersports motorcycle is plotted versus engine power with and without rolling resistance. With 200 hp the difference is approximately 8 km/h (310 km/h → 318 km/h). For a touring enduro with 110 hp the difference is similar (211 km/h → 218 km/h).

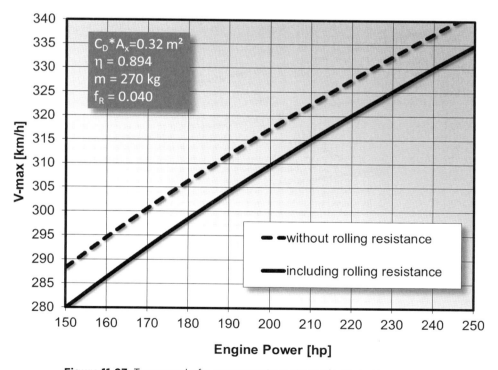

Figure 11.27 Top speed of a supersports motorcycle vs. engine power.

The impact of increasing the engine power can be estimated from equation 11.10

$$\frac{v_{max,neu}}{v_{max,alt}} = \sqrt[3]{\frac{P_{max,neu}}{P_{max,alt}}} \tag{11.11}$$

Since the ratio goes with the cubic root, the effect of engine tuning is often heavily overestimated. For instance, in this example a 10% power increase from 200 hp to 220 hp raises the top speed "only" by the factor $1.1^{1/3} \rightarrow +3.2\%$ from 310 km/h to 320 km/h.

The fact that in all expressions the drag coefficient always appears as a product with the reference area emphasizes that the drag area $c_D \cdot A_x$ [m²] is the key measure preferred for usage in motorcycle aerodynamics. This is due to the difficulty in measuring the front area including the driver, and variability of driver posture. It is customary to simply divide the measured force only by the dynamic pressure to get the drag area, without the need to know the actual frontal area.

Unfortunately, as equation (11.10) shows, the influence of the drag area on the top speed is not linear. Reducing $c_D \cdot A_x$ by half only increases v_{max} by approximately 25% (Figure 11.28).

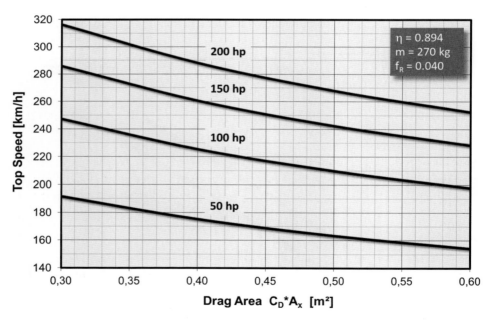

Figure 11.28 Top speed vs. drag area.

To raise the top speed significantly thus requires the combination of optimized aerodynamics and engine tuning. An example is shown in Figure 11.29, where the series version of the BMW S 1000 RR is compared to the average of purchasable competitors at the time when it was developed and its motorsport derivative which started in the superbike world championship.

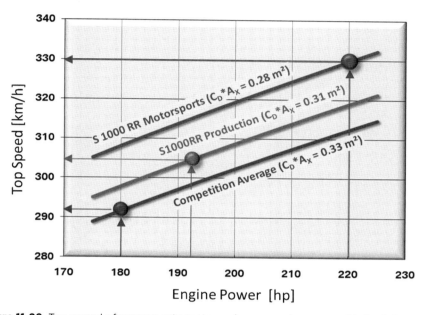

Figure 11.29 Top speed of supersports motorcycles vs. engine power (Hofer & Grün 2010 [352]).

11.3.2.1.4 Traction Force/Transmission Ratio

In order to actually reach the possible top speed from the previous considerations, it is necessary that the total transmission ratio is chosen such that in the highest gear the crankshaft rotates at the speed of maximum power output.

If the ratio is too short, the nominal rotation speed is reached too early, and if it is too long the traction force drops below the total resistance force before the theoretically possible top speed is reached. The tractive power chart in Figure 11.30 from Stoffregen (2001, [780]) shows the ideal situation, where the traction force in the highest gear matches the total drag exactly at the nominal rotation speed.

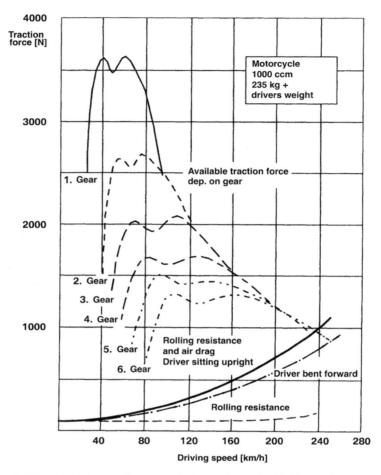

Figure 11.30 Tractive force chart of a motorcycle, Stoffregen [780].

11.3.2.2 Lift Effects

With the exception of extreme configurations, all motorcycles experience lift on the front and downforce on the rear axle. The reason becomes clear from Figure 11.31. For all types of motorcycles, the line of action of the resulting force vector is almost parallel to the ground because the total lift is small compared to the drag force. However, even for a supersports bike it is located at minimum one wheel diameter above the road. Why this is so will be explained later in the computational fluid dynamics (CFD) analysis (Figure 11.68). Therefore, the pitching moment and as a consequence the vertical forces at the axles depend predominantly on the drag force and its moment arm above the ground.

The effect of occasionally seen measures to reduce front lift (e.g., spoiler or wings on the fender, fork, or fairing) can counteract the dominating action of the drag force only to a limited extent. To work they must to be so large that they are not realized due to practical and aesthetic reasons.

Motorcycle Aerodynamics

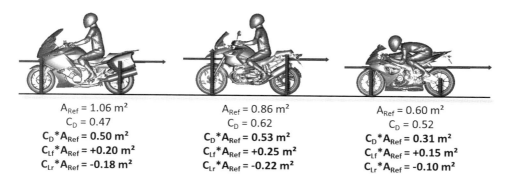

$A_{Ref} = 1.06\ m^2$	$A_{Ref} = 0.86\ m^2$	$A_{Ref} = 0.60\ m^2$
$C_D = 0.47$	$C_D = 0.62$	$C_D = 0.52$
$C_D * A_{Ref} = 0.50\ m^2$	$C_D * A_{Ref} = 0.53\ m^2$	$C_D * A_{Ref} = 0.31\ m^2$
$C_{Lf} * A_{Ref} = +0.20\ m^2$	$C_{Lf} * A_{Ref} = +0.25\ m^2$	$C_{Lf} * A_{Ref} = +0.15\ m^2$
$C_{Lr} * A_{Ref} = -0.18\ m^2$	$C_{Lr} * A_{Ref} = -0.22\ m^2$	$C_{Lr} * A_{Ref} = -0.10\ m^2$

Figure 11.31 Line of action of the total aerodynamic force and resulting lift or downforces at the axles (lift vectors drawn to scale).

Vehicle stability in maneuvers is critically affected by the front wheel load. A critical aspect of motorcycle dynamics is the weaving instability mode, an oscillation around the vertical axis, which is more predominant at higher speeds due to dynamic reduction of the effective front wheel load. In Figure 11.32, the front wheel loads are compared for luxury tourer, touring enduro, and supersports bike up to their individual top speed. The forces are based on a static front-to-rear axle load ratio of 50:50, including an 80 kg driver. Although the curve for the supersports bike has the smallest gradient due to the smallest front axle lift $c_{Lf} \cdot A_x$, only 58% of the static load remains at top speed.

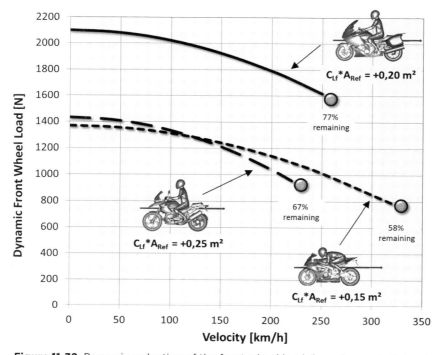

Figure 11.32 Dynamic reduction of the front wheel load (based on a static load distribution of 50:50 including 80 kg driver).

An impressive illustration of the lift-induced reduction of the contact patch at the front wheel is given in Figure 11.33, comparing the tread-shuffle at 80 km/h and 200 km/h (Weidele 1988 [857]).

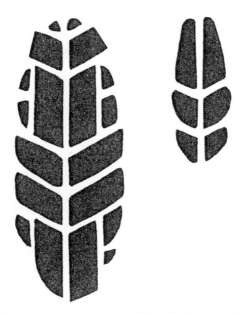

Figure 11.33 Front wheel contact patch at 80 km/h (top) and 200 km/h (bottom) (Weidele 1988 [857]).

Since the rear axle experiences a downforce which almost equals the front axle lift, the effective dynamic load balance is also affected by velocity. For the three motorcycle types in Figure 11.32, the load ratio changes from 50:50 in static conditions to 39:61 at v_{max} for the tourer, to 34:66 for the touring enduro and to 31:69 for the supersports bike. This is equivalent to an apparent shift of the center of gravity, usually with adverse consequences for driving stability. Looking at the relative change of balance over velocity in Figure 11.34, tourer and supersports bikes behave very similarly while the touring enduro has to cope with the highest dependency on speed.

Motorcycle Aerodynamics

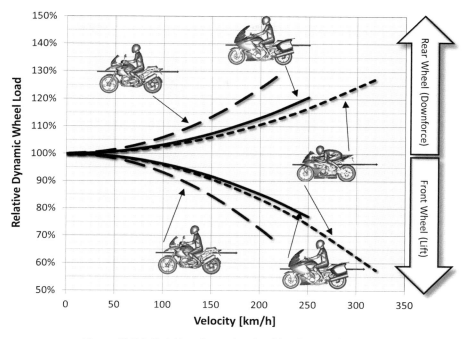

Figure 11.34 Relative dynamic wheel loads vs. velocity.

11.3.3 Aerodynamics and Lateral Dynamics

11.3.3.1 Cornering

On a straight road the motorcycle is stabilized by a combination of gyroscopic effects of the wheels, tilting motion, and steering input from the driver. Anyone who has tried to ride a bike as slowly as possible knows that the required steering amplitudes grow with decreasing velocity. Details of this mechanism and the phenomena of "shimmy" at low and "weaving" at high speeds, as well as the impact of the mass moment of inertia around the steering axis, are discussed in detail in [50], [149], [210], and [780].

A motorcycle is unlike a multitrack vehicle in that it cannot support centrifugal forces with the outside wheels when cornering. Instead it has to lean toward the inner side of the corner at an angle φ such that the resultant of weight F_G and centrifugal force F_F coincides with the line that connects the center of gravity and the contact point between tire and road.

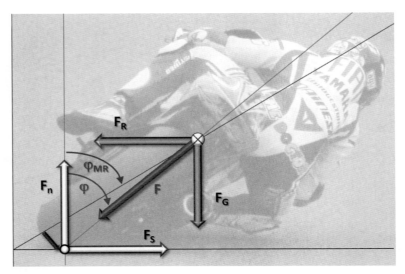

Figure 11.35 Forces and angles of a cornering motorcycle. Photo: Georg Hack,

In this figure, the center of gravity is still indicated in the symmetry plane, although the underlying photo shows a "hanging-off" pilot. The explanation follows.

Thus the required lean angle φ, which is independent of mass, follows from:[2]

$$\tan(\varphi) = \frac{F_F}{F_G} = \frac{mv^2/r}{mg} = \frac{v^2}{r \cdot g} \tag{11.12}$$

The lateral acceleration during cornering is $a_q/g = v^2/(r \cdot g) = \tan(\varphi)$. For a lean angle of 45° the lateral acceleration is 1 g; however, at 55° it increases significantly to 1.43 g.

Remarkably, the tangent of the lean angle φ depends linearly on the corner radius r, but increases quadratically with velocity v. As a consequence, doubling the speed from 40 km/h to 80 km/h on a corner with 50 m radius requires to increase the lean angle by a factor of 3.2 from 14° to 45°! See Figure 11.36.

---

2. It should be noted that the same equation holds for the bank angle of an aircraft. Since a pendulum will automatically take this orientation, each object hanging free to move from the rear view mirror in an automobile will form a simple gauge for the instantaneous lean angle of a motorcycle.

Motorcycle Aerodynamics

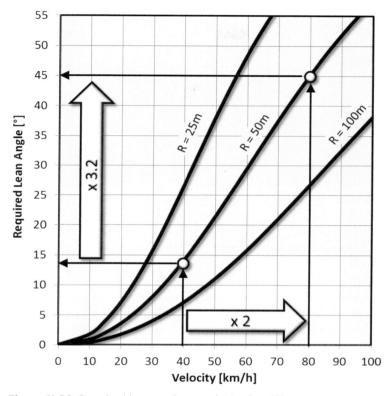

Figure 11.36 Required lean angle vs. velocity for different corner radii.

The previously discussed lean angle φ is the one of the line between the center of gravity and the center of the actual contact patch. Which lean angle the motorcycle has to take is determined by the width and the contour of the tire. As the red arrow in Figure 11.35 indicates, the contact point of the tire on the road is shifted inwards, so that the motorcycle has to be inclined farther ($\varphi_{MR} > \varphi$). The same effect is caused by lowering the center of gravity. This can lead to critically exceeding the motorcycle's clearance and foot pegs, the exhaust system, or luggage boxes may hit the road before the grip limit is even reached. Choppers and cruisers with rear tires up to 240 mm width and low center of gravity are particularly at risk. They might require up to 10° higher lean angles than dictated by physics. Therefore in motorsports sometimes the center of gravity is deliberately located higher than on the basic production bike. A spectacular means to decrease the difference between φ_{MR} and φ is "hanging-off" (Figure 11.37), which shifts the center of gravity inwards and allows the driver to fully exploit the motorcycle's clearance.

Figure 11.37 "Hanging-off" to shift the center of gravity in order to reduce the motorcycle's lean angle.

Even at moderate speeds gyroscopic effects prevent the motorcycle from being inclined just by a shifting of weight. Instead, the driver—for many, unconsciously—gives a short steering impulse in the opposite direction, producing a centrifugal force that tilts the bike in the desired direction. Once inclined, the lean angle is adjusted to varying speeds and corner radii by subtle steering inputs. Even at constant speed on a circular track (r = const.), a small steering moment is necessary to compensate for the moment generated by the lateral shift of the contact point on the front wheel. The latter is also responsible for erecting the motorcycle when braking in corners.

The centrifugal force (F_r in Figure 11.35) has to be balanced by the lateral friction force F_{SF} between tire and road,

$$F_{SF} = F_R \quad \Rightarrow \quad \mu \cdot F_N = m \frac{v^2}{r} \tag{11.13}$$

in which the friction coefficient μ together with the normal component F_N of the weight determine the maximum possible force between tire and road. As a result the maximum possible corner speed is

$$v_{K,\max} = \sqrt{\mu \cdot r \cdot g \cdot \frac{F_N}{mg}} \tag{11.14}$$

Together with equation (11.12) the corresponding lean angle at the maximum corner speed is

$$\tan(\varphi_{K,\max}) = \mu \cdot \frac{F_N}{mg}. \tag{11.15}$$

So for a given corner radius the maximum possible speed depends only on the friction coefficient and the ratio of effective normal force to weight, which is always $F_N/mg = 1$ if we disregard aerodynamics. In case of aerodynamic lift $F_N/mg < 1$ and the corner speed is reduced, whereas a downforce enables higher speeds. Of course this consideration has

Motorcycle Aerodynamics

to be applied to front and rear wheels separately because as shown, a motorcycle usually experiences lift on the front and downforce on the rear.

For a long time $\mu = 1$ and hence $\varphi = 45°$ were considered as the conceivable upper limit, but today's (sports-)tires generate higher grip ($\mu > 1$) and allow lean angles up to 50°. However, with the exception of supersports bikes, most motorcycles will scratch the road before reaching this boundary. The slick tires employed in motorsports nowadays even enable spectacular lean angles close to 60° (Figure 11.37).

So far, the considerations assumed that only lateral forces act between tire and road. If the driver accelerates or brakes while cornering, the longitudinal forces also have to be considered, and the possible side forces may be reduced. This interrelation is illustrated very clearly by the "Kamm-Circle," depicting the vector addition of lateral and longitudinal forces within the limits set by the maximum friction force (Figure 11.38). The example shows forces in a left turn, driven with a lean angle of 30° including acceleration.

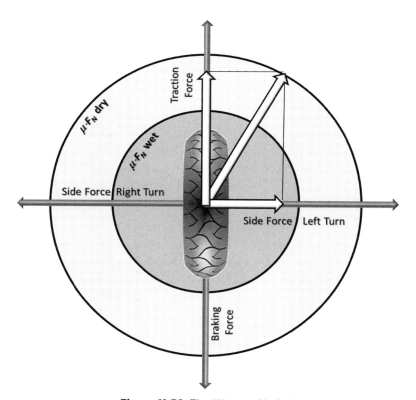

Figure 11.38 The "Kamm-Circle."

The outer circle represents the maximum possible total friction force $\mu \cdot F_N$ on a dry road; the inner circle represents a wet road with reduced friction coefficient μ. Without acceleration (or braking) the side force exploits only 50% ($\sin(\varphi)$) of the dry road potential

825

and would also be safe in wet conditions. The sketched traction force alone (driving straight) already exceeds the limit on a wet road, and in dry conditions it utilizes 87% ($\cos(\varphi)$) of the potential. The combination of cornering and acceleration in this example just reaches the limit on dry road, therefore the slightest acceleration or increase in lean angle (or a wet spot on the road) would make the tire slip.

Since the boundaries (circles) in Figure 11.38 depend on friction coefficient μ and effective normal force F_N, a positive lift will also constrain the safe range of operation.

11.3.3.2 Lift Effects

On multitrack vehicles the effective normal force can be influenced by aerodynamic means like spoiler or wings, which reduce the lift force and thus increase the possible cornering speed via the ratio F_N/mg (see equation (11.14)). As is well known, in automobile motorsports, by creating downforce the effective normal force is increased even above the static load (i.e., $F_N/mg > 1$).

On motorcycles, this is different. Since any aerodynamic device is inclined with the bike, its line of action always stays parallel to the vertical axis of the bike. Even if a downforce generating device increases the normal (perpendicular to the road) force, it will also create a horizontal force component that adds to the centrifugal force. Thus a gain in cornering speed is hardly possible. However, a higher F_N should have a favorable effect on stability.

Supposing still air, even when cornering, the instantaneous direction of the oncoming flow is parallel to the longitudinal axis with zero yaw angle. Figure 11.39 compares the total pressure distribution in x = const. planes with and without lean angle. The color image is clipped where the total pressure reaches the free stream value, meaning we see only regions of energy loss as the envelope of boundary layer and wake. Basically, this structure is tilted with the motorcycle and somewhat compressed between bike and road, however, with a more pronounced wake behind the riders back.

Also with lean angle, the line of action of the total aerodynamic force runs (with different components) above the wheels along the longitudinal axis of the motorcycle (Figure 11.40). The drag area increases by 25%, to some extent caused by the different posture of the rider (slightly hanging-off with the knee extended). This enlarges the frontal area by 9% and raises the drag coefficient by 15% from 0.452 to 0.520.

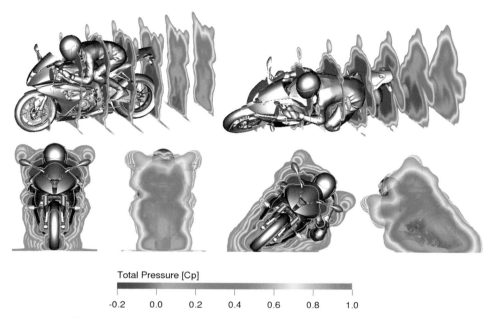

Figure 11.39 Flow field without (left) and with 45° lean angle (right).

The lift and downforce figures and vectors depicted in Figure 11.40 are the ones relevant for the effective dynamic axle load, that is, perpendicular to the road (not along the vertical axis of the tilted motorcycle). Front wheel lift increases by 12% while rear wheel downforce decreases by 13%. Both changes reduce the maximum possible cornering speed by 6–7% (equation (11.14)).

This effect is caused by the altered pressure distribution on the tilted motorcycle, which is not mirrored on either side as on a symmetric bike without lean angle. If the lift force along the vertical axis of the motorcycle would not change and just incline with the bike, its magnitude would even decrease by the factor $\cos(\varphi_{MR})$. When tilted, the bike provides a much larger working surface normal to the road (Figure 11.41), and there is a force in the vertical direction due to the pressure difference between inner and outer side—which is nearly nonexistent in straight motion. Consequently, a large, smooth fairing designed for attached flow can have an adverse impact on the maximum possible cornering speed.

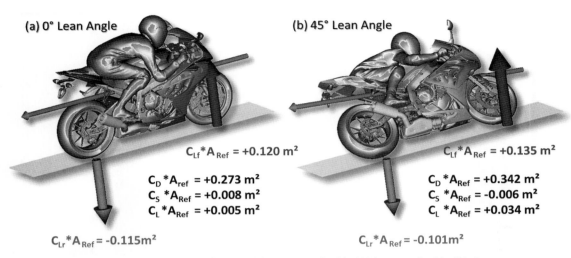

Figure 11.40 Aerodynamic forces without (a) and with 45° lean angle (b), (lift forces c_L always—also with lean angle—perpendicular to the road).

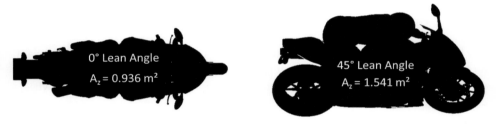

Figure 11.41 Projected areas relevant for lift (perpendicular to the road).

11.3.3.3 Yawed Flow

When assessing the characteristics of a vehicle in crosswind the range of relevant yaw angles should be considered. The relationship between vehicle speed, crosswind, and resulting yaw angle β is shown in Figure 11.42. At 100 km/h a crosswind of 27 km/h (equivalent to a wind intensity of 4–5 Beaufort) results in a yaw angle of $\beta = 15°$.
An experienced motorcyclist would have no problems adjusting to such conditions. However, to produce the same yaw angle of 15° at 200 km/h, a crosswind of 54 km/h (7 Beaufort) would be required, and in that circumstance hardly any driver would drive that fast.

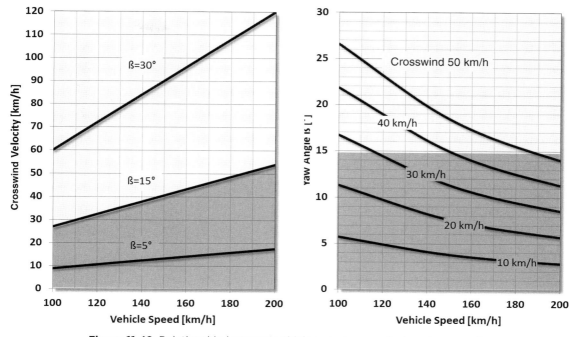

Figure 11.42 Relationship between vehicle speed, crosswind, and yaw angle.

Wojciak et al. [893] measured 163 gusts under natural conditions on public roads. The average speed was 140 km/h and the wind intensity between 4 and 7 Beaufort. Less than 1% of all gusts produced a yaw angle greater than 12°. In almost three-fourths of all gusts the yaw angle was just 6–8°. Hence it seems to be sufficient to consider $\beta = 15°$ as a worst-case design criterion.

Figure 11.43 and Figure 11.44 compare the flow fields with and without yaw angle by means of the velocity distribution in a horizontal plane and the total pressure distribution in planes normal to the motorcycle's longitudinal axis. It should be noted that the trajectories in Figure 11.43 are no streamlines in a strict sense because only in-plane velocity components have been used for their generation. The deflection of the flow behind the motorcycle and the stronger formation of a pronounced wake on the leeward side are clearly shown in the case of high yaw angle.

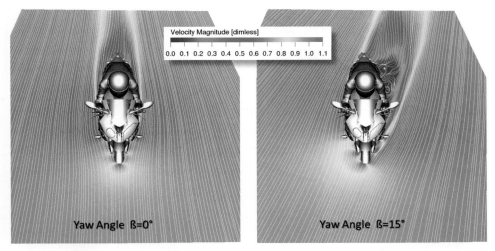

Figure 11.43 Velocity distribution without (left) and with 15° yaw angle (right).

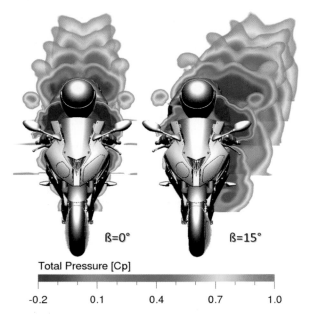

Figure 11.44 Total pressure distribution without (left) and with 15° yaw angle (right).

Forces and moments with and without yaw angle are shown in Figure 11.45. Note that these are components in the motorcycle fixed reference system (not aligned with the free stream direction). The total force increases by 73%, drag alone by 30%, while the leeward directed side force under yaw conditions reaches almost the magnitude of the drag force. Downforce at the rear wheel increases moderately by 8% while lift on the front is

34% higher as a consequence of the 22% higher pitching moment. The reference point for the moments reported in Figure 11.45 is the center of the wheelbase on the ground. A positive rolling moment tries to tilt the motorcycle toward the leeward side while at the same time a positive yawing moment generates side forces to the leeward side on the front and in the windward direction on the rear.

A rider exposed to these influences has to lean the motorcycle against the wind even on a straight road. The inevitable temporal fluctuations under natural conditions require a continual adjustment of the lean angle by appropriate steering inputs. In strong, gusty crosswinds this can lead to an unpleasant rocking motion.

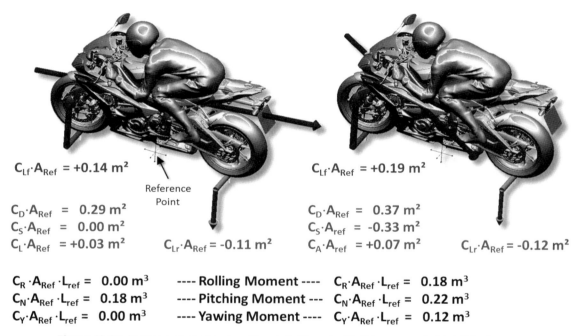

$C_{Lf} \cdot A_{Ref} = +0.14 \text{ m}^2$

$C_D \cdot A_{Ref} = 0.29 \text{ m}^2$
$C_S \cdot A_{Ref} = 0.00 \text{ m}^2$
$C_L \cdot A_{Ref} = +0.03 \text{ m}^2$

$C_{Lr} \cdot A_{Ref} = -0.11 \text{ m}^2$

$C_{Lf} \cdot A_{Ref} = +0.19 \text{ m}^2$

$C_D \cdot A_{Ref} = 0.37 \text{ m}^2$
$C_S \cdot A_{ref} = -0.33 \text{ m}^2$
$C_A \cdot A_{ref} = +0.07 \text{ m}^2$

$C_{Lr} \cdot A_{Ref} = -0.12 \text{ m}^2$

$C_R \cdot A_{Ref} \cdot L_{ref} = 0.00 \text{ m}^3$ ---- Rolling Moment ---- $C_R \cdot A_{Ref} \cdot L_{ref} = 0.18 \text{ m}^3$
$C_N \cdot A_{Ref} \cdot L_{ref} = 0.18 \text{ m}^3$ ---- Pitching Moment ---- $C_N \cdot A_{Ref} \cdot L_{ref} = 0.22 \text{ m}^3$
$C_Y \cdot A_{Ref} \cdot L_{ref} = 0.00 \text{ m}^3$ ---- Yawing Moment ---- $C_Y \cdot A_{Ref} \cdot L_{ref} = 0.12 \text{ m}^3$

Figure 11.45 Forces and moments without (left) and with 15° yaw angle (right) (reference point for moments in the center of the wheelbase on the ground).

11.3.4 Cooling and Internal Flow

As is generally known, a real Otto engine converts only approximately 30% of the fuel's chemical energy into kinetic energy of the vehicle. Most of the remaining energy is lost as thermal energy, roughly in equal parts through the exhaust gas and the engine cooling. That means the cooling system has to remove about the same amount of thermal energy from the engine as the mechanical engine power that is produced.

Up to about 100 hp this can still be achieved by pure air cooling, which requires a careful design of the airflow to cooling fins and other heat transferring surfaces. This is more challenging for instance in longitudinally installed V-engines, whose rear cylinder is prone to overheating due to its position. Less critical designs were the older BMW two-valve flat twin models, where the cylinders protrude almost unhindered into the oncoming flow. Some air-cooled engines were supported by additional oil coolers. However, the use of air cooling alone fails when the motorcycle stands still with a hot engine in high ambient temperatures, and radiation and natural convection do not suffice to cool the engine.

Nowadays high-performance motorcycles invariably rely on water cooling, often combined with oil cooling. Since the available space is limited, the aerodynamic design has to ensure the appropriate cooling airflow rate to the heat exchanger(s). Ideally, the cooling air inlet is placed in a region of high static pressure, and the discharge openings exploit low-pressure regions.

Figure 11.46 shows the arrangement of a water (top) and oil (bottom) cooler on a BMW S 1000 RR. Cooling air is discharged through openings in the side fairing. On off-road bikes space is very limited, and the middle is often blocked by the high-mounted front fender. In this case, it is common to use a split cooler and support the cooling airflow by scoops on either side of the tank (Figure 11.47). The guiding vanes on the radiator itself additionally prevent clogging by dirt and soil.

Figure 11.46 Water (upper) and oil (lower) cooler on a BMW S 1000 RR.

Motorcycle Aerodynamics

Figure 11.47 Split water cooler on an enduro (Husqvarna TE449).

An unorthodox concept was used on the Benelli Tornado 900 Tre (Figure 11.48). Air is captured on either side of the tank and guided through long ducts to a radiator which is mounted under the seat in front of the rear wheel. As in this example, almost each water-cooled motorcycle is fitted with a fan that operates on demand to minimize the risk of overheating at low speeds or at rest.

Figure 11.48 Under-seat water cooler of the Benelli Tornado 900 Tre.

A second internal flow issue is the supply of the engine with fresh air. Many motorcycles have an—often jagged—airbox with air filter under the tank or seat. With this concept the inlet openings are lying in a region of low total pressure and velocity and hence also low static pressure, which reduces the pressure difference between inlet and outlet that is used to support the flow rate into the inlet.

In motorsports and on many supersports bikes, the inlet is positioned in the stagnation region (Figure 11.49). With this so-called Ram-Air Inlet, one tries to utilize the pressure rise at stagnation for power increase. Since the effect depends on the dynamic free stream pressure, this concept is efficient only at very high speeds. An example for the ram-air effect is shown in Figure 11.50. At an ambient pressure of 100,000 Pa and a velocity of 280 km/h, the total pressure is 103,600 Pa. This represents the mechanical energy of the free stream (per volume). Starting at the inlet of the air duct, friction (and flow separation) continuously reduces the total pressure in the air duct. At the transit through the air filter, an additional strong loss occurs so that the flow has lost a potential of 1,350 Pa for supercharging when it reaches the engine. Tests on a flow bench with this engine showed that for a 600 Pa increase in the airbox, the power would increase by 1 kW. Thus, in this case approximately 4 kW were gained by the 2,250 Pa above ambient pressure remaining after frictional losses.

Figure 11.49 Ram-air inlet on the BMW S 1000 RR.

Motorcycle Aerodynamics

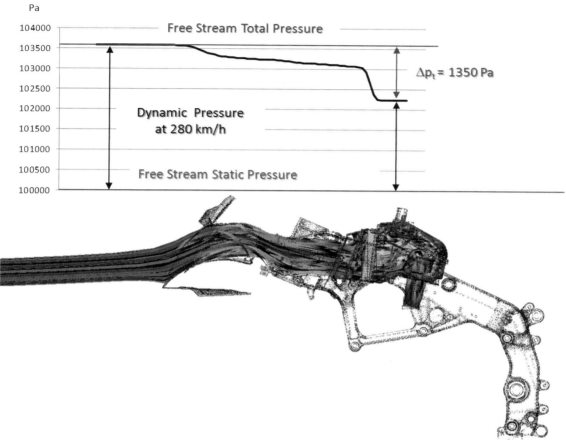

Figure 11.50 Pressure distribution in the engine air duct and airbox (Hofer & Grün 2010 [352]).

11.3.5 Wind and Weather Protection

Not only do the pilot and passenger on a motorcycle have a significant effect on the aerodynamics, they are also immediately exposed to the flow. Therefore the assessment and optimization of wind and weather protection are essential tasks of motorcycle aerodynamics. Draft phenomena are evaluated by the near surface velocity distribution. As visible in Figure 11.51, legs, arms, and hands are very well protected and therefore sheltered from cooling. In contrast there are regions of higher velocity on knees and upper arms on the touring enduro, while the hands are effectively shielded by the handlebar protectors. The supersports bike is designed rather for minimum drag and lift than for comfort. In addition, the rider will change his or her posture more often and more significantly, especially on a race track. Of higher significance on these motorcycles are the aerodynamic forces on the pilot.

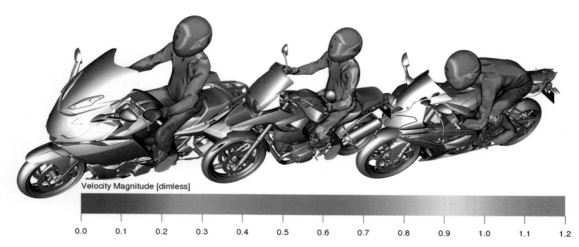

Figure 11.51 Assessment of draft phenomena by near surface velocity.

In particular on long-distance trips, the forces play an important role for the sensation of comfort. In this respect, a luxury tourer is superior to a touring enduro (factor > 3) due to its protruding fairing (Figure 11.52). The force on the supersports pilot is 50% higher than on the tourer, but it is mostly lift that is easier to endure than a high drag force. In an upright position the drag would increase by 30%, which is deliberately exploited in racing as an additional "aerodynamic brake."

Of course the forces on the rider depend very much on his or her size and posture, but they also depend on the garment (tight-fitting leather gear or flagging textile suit). Therefore during development certain standard assumptions must be employed (see section 11.4.3).

Figure 11.52 Aerodynamic forces on the rider at 200 km/h.

High helmet forces can be very bothersome, even over a relatively short time. Lift pulls on the fastening strap and depending on the helmet may even cause visibility problems. On the other hand, drag stresses the neck muscles (Figure 11.53). Apart from the helmet design itself the motorcycle's windshield is mainly responsible for this criterion. Again size and posture of the pilot are crucial. Therefore many makers of tourers and touring enduros offer variable windshields, and some can even be adjusted electrically while riding, so each pilot can find an optimum position.

Beyond the stationary forces, unsteady effects, mainly the interaction of turbulent shear layers from windshield and fairing with the pilot, are a substantial component of the rider's comfort. Even at acceptable averaged force levels, a highly transient pressure on the helmet may reduce the comfort drastically. An extreme phenomenon at high speeds is "helmet buffeting" provoked by periodic separation, which in the worst case might even affect clear visibility.

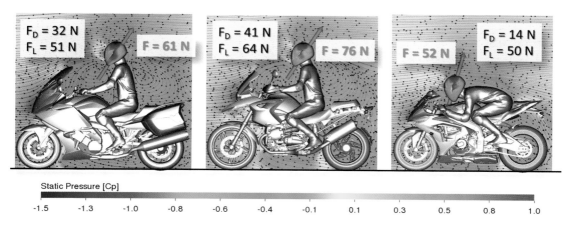

Figure 11.53 Aerodynamic forces on the helmet at 200 km/h.

11.3.6 Aeroacoustics

The transient pressure fluctuations on the helmet generate wind noise at the rider's ear. This noise is boosted significantly if the shear layer from the windshield happens to hit the helmet. Table 11.2 from Bachmann et al. (2002, [30]) lists ranges of the A-weighted sound pressure levels inside the helmet for different motorcycle categories. Almost all values are above the noise level regulated by law (90 dB), where workers would have to wear an ear protection. Apart from individually adjustable windshields, measures on the helmet itself contribute to the reduction of forces and noise load (see chapter 12).

Table 11.2 Noise Level Inside the Helmet at 100km/h (Bachmann et al., 2002 [30])	
Motorcycle Category	Sound Pressure Level
Super Sports	98–105 dB(A)
Sports Tourer	95–105 dB(A)
Tourer	85–95 dB(A)
Naked Bikes	92–95 dB(A)
Touring Enduro	95–105 dB(A)

11.4 Development Methods

11.4.1 Development Process

During the aerodynamic development of a motorcycle, roughly three phases can be distinguished where requirements are defined and then realized in more and more detail. The available tools are numerical simulation, wind tunnel, and road tests which are employed in this order in a timely overlapping manner (Figure 11.54).

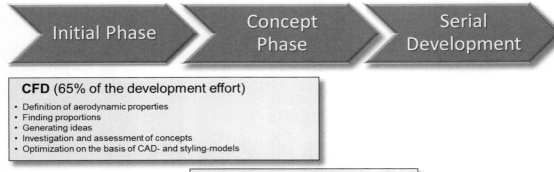

Figure 11.54 Development phases, tasks and tools (Hofer & Grün 2010 [352]; Hofer 2008 [351]).

In the initial phase, the aerodynamic properties of the motorcycle are established and the main focus is set. After proportions have been found, various concepts are investigated and their feasibility is assessed on the basis of CAD and styling data. Since at that time no hardware exists yet, numerical simulation using computational fluid dynamics (CFD) is the tool of choice. This covers approximately two-thirds of the total development effort.

When entering the concept phase, the first prototypes are available for wind tunnel tests. In this phase, detail optimizations concerning drag, lift, wind protection, and aeroacoustics can still be accomplished in a shorter time frame than per simulation. Roughly 30% of the development work is done in the wind tunnel.

At the end of the concept phase and during the entire serial development, road tests serve to confirm measures developed per simulation or in the wind tunnel under realistic conditions. According to the various operation scenarios of diverse motorcycle types, the focuses of aerodynamic development are of course also quite different. As shown in Figure 11.55, the biggest effort is invested for touring and sports motorcycles.

	Enduro Cross	Chopper Cruiser	Naked Bike	Scooter	Touring Enduro	(Luxury-) Tourer	(Super-) Sportls
Cooling	✗	✗	✗	✗	✗	✗	✗
Drag and Lift		✗	✗	✗	✗	✗	✗
Wind and Weather Protection				✗	✗	✗	✗
Soiling				✗	✗	✗	✗
Aeroacoustic					✗	✗	✗
Crosswind Behavior						✗	✗
Engine Air Supply							✗

Figure 11.55 Focuses of aerodynamic development.

11.4.2 Simulation (CFD)

Computational fluid dynamics (CFD) methods have reached an accuracy level that allows their usage as a production tool. As an example, Figure 11.56 compares the drag areas of three different windshields on a validation model. Both for absolute values and for trend prediction, CFD delivers results equivalent to physical testing. Also the velocity field between windshield and rider shows a good correlation to hot wire measurements (Figure 11.57).

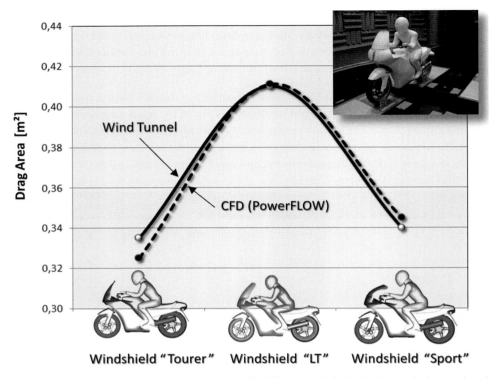

Figure 11.56 Comparison of the drag area with different windshields from wind tunnel and CFD (PowerFLOW) on a validation model.

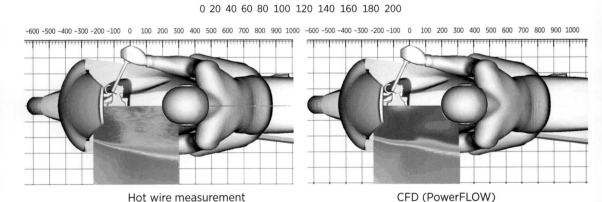

Figure 11.57 Comparison of the velocity distribution from hot wire measurement and CFD (PowerFLOW).

Together with the maturity of CFD tools, the computer capability today allows manufacturers to produce results "overnight." Simulation and wind tunnel are no longer considered as competing tools but rather used in a complementary fashion. All simulation results shown next have been created with the CFD tool PowerFLOW™ from Exa Corp, a Lattice-Boltzmann code, which is described in detail in chapter 14. Therefore in the following, only motorcycle specific processes and analysis features will be presented.

11.4.2.1 Simulation Model

Data to set up the simulation model originate from two sources. Structural components like drivetrain, frame, wheels, suspension, and so on exist as CAD data (Figure 11.58 left). They are supplemented by styling data for windshield, fairing, seat, and so on (Figure 11.58, center). Very often the styling surfaces are also available as virtual models in CAD. In some cases only clay models are available, and they must be scanned and are provided in electronic form for simulation. As input for the simulation, all parts have to be represented as individual "watertight" surface meshes. The simulation model is rounded off by a dummy, also a scanned model which is adjusted to the motorcycle by morphing.

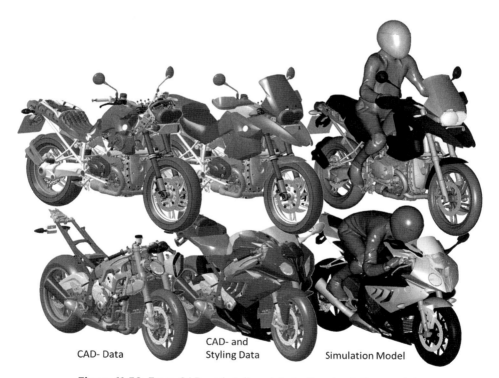

Figure 11.58 From CAD and styling data to the simulation model.

The highly detailed models nowadays require 3–5 million triangles ("facets") to describe the geometry. An impression of the fineness of this type of model is given in Figure

11.59. For the CFD tool PowerFLOW it is not necessary to create one single contiguous closed surface for the entire configuration. During case setup the individual parts are loaded separately and may touch or intersect, which facilitates and speeds up data preparation significantly.

Figure 11.59 Surface mesh (facets) as geometry input for the simulation.

The actual resolution of the simulation is not determined by the facets representing the geometry but by the size of the cubic volume cells (voxels). Immediately at the surface the typical edge length is 1–3 mm, and they grow by a factor of 2 farther away in the flow field. So-called VR (variable resolution) regions define which resolution is applied. These are either offsets on solid surfaces or simple geometric shapes. This way, a Cartesian volume mesh (lattice) is formed, indicated by planes in Figure 11.60. Typical cell counts for a motorcycle simulation are 50–100 millions.

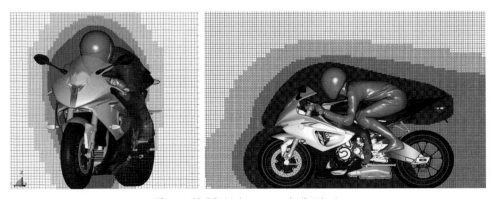

Figure 11.60 Volume mesh (lattice).

The automatic discretization intersects the lattice with the facets and thus creates surface elements ("surfels") for the simulation. Since these n-Polygons are subdivisions of the facets, their number is much higher (5–10 millions).

In contrast to the wind tunnel, after a simulation all kinds of flow quantities are available in the flow field and on the surface for further analysis. PowerFLOW always delivers unsteady, time-accurate results. For design decisions, of course time-averages of this data can be used for analysis and visualization. The following figures should demonstrate the multitude of options for analysis and visualization. All quantities, except temperature, are displayed in dimensionless notation.

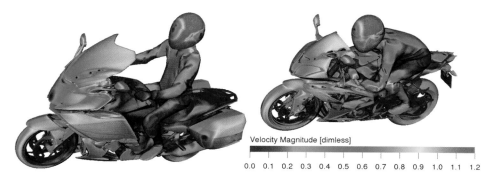

Figure 11.61 Near surface velocity distribution.

11.4.2.2 Surface Results

Due to the no-slip condition, the actual fluid velocity on the surface is 0. Figure 11.61 shows the "near-surface" velocity in the first cell above the wall, roughly comparable with the velocity at the outer edge of the boundary layer. To derive the static pressure coefficient (Figure 11.62)

$$c_p = \frac{p - p_\infty}{\frac{1}{2}\rho_l v_\infty^2} \tag{11.16}$$

from the local velocity using Bernoulli's equation requires the knowledge of the local total pressure p_{tot}. Using an analogously defined total pressure coefficient

$$c_{p,tot} = \frac{p_{tot} - p_\infty}{\frac{1}{2}\rho_l v_\infty^2} \tag{11.17}$$

for incompressible flow we get

$$c_p = 1 - \left(\frac{v}{v_\infty}\right)^2 + \left(c_{p,tot} - 1\right) \tag{11.18}$$

For a loss-free location ($p_{tot} = p_{tot\infty}$), $c_{p,tot} = 1$ and the second term in equation (11.18) vanishes. At the stagnation point with $v = 0$ the maximum (in incompressible flow) of

$C_p = 1$ is reached. Without total pressure loss in the wake with small to zero velocities, a similar high-pressure coefficient would arise and the aerodynamic forces would vanish (d'Alembert's paradox). In reality, in the wake $c_{p,\text{tot}} \approx 0$ or below so that we have slightly negative static pressure coefficients. As a consequence upstream directed forces on the rear end are missing to balance the overall pressure distribution, and a drag force remains effective. Regions of high total pressure loss are obviously visible in Figure 11.63. The fact that losses are also visible in regions of attached flow is due to the surface friction in the boundary layer.

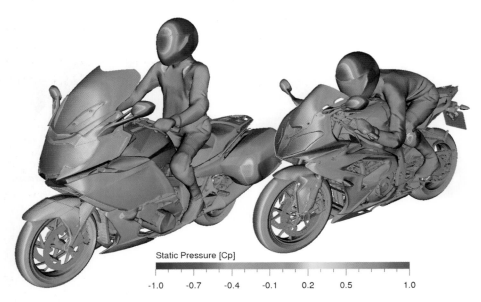

Figure 11.62 Static pressure distribution.

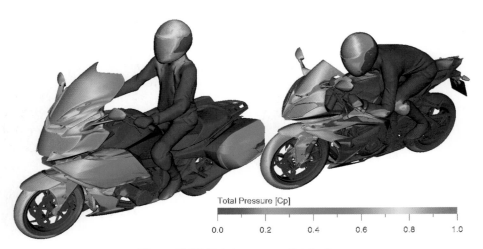

Figure 11.63 Total pressure distribution.

When integrating aerodynamic forces in addition to the static pressure the skin friction also has to be included. Its distribution in dimensionless form, calculated as

$$c_f = \frac{\tau_w}{\frac{1}{2}\rho_1 v_\infty^2} \qquad (11.19)$$

is depicted in Figure 11.64. High values only occur on the front wheel and helmet in strongly accelerated flow. Even on an aerodynamically well designed motorcycle the contribution of friction to the total drag is only 3–5 % (for comparison: the contribution is 10–15 % on an automobile, and more than 80% on a sail plane). Figure 11.64 also shows wall streamlines, comparable to oil streak measurements in a wind tunnel. They enable insight into the detailed flow topology and the occurrence of separation and reattachment. It should be noted that in 3D there is no unique separation criterion (like the disappearance of skin friction in 2D) and therefore this visualization is of valuable utility for showing flow separations on the surface.

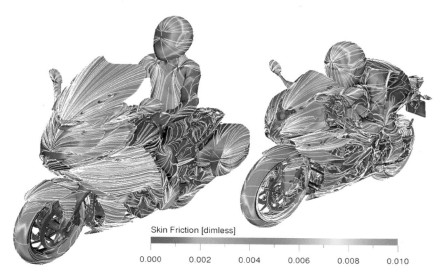

Figure 11.64 Skin friction and wall streamlines.

Looking at the static pressure distribution alone does not allow an immediate conclusion about the distribution of drag and lift generation because the surface orientation has to be included as well. For this purpose, the static pressure coefficient is multiplied by the local components of the surface normal (orientation according to Table 11.1)

$$c_x = -c_p \cdot n_x \; ; \; c_y = -c_p \cdot n_y \; ; \; c_z = -c_p \cdot n_z \qquad (11.20)$$

and quantities are obtained that enable the visualization of the three-dimensional distribution of local contributions to drag, side force, and lift (Figure 11.65 and Figure 11.66).

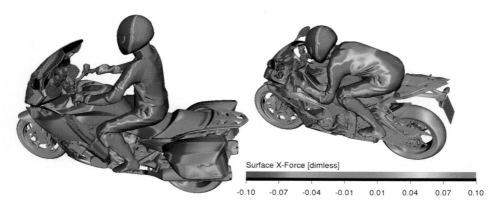

Figure 11.65 Drag distribution on the surface (red: high; blue: low).

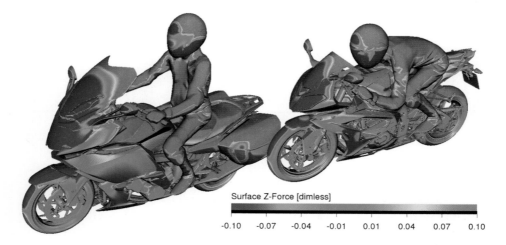

Figure 11.66 Lift distribution on the surface (red: high; blue: low).

A useful 2D analysis of drag generation is given in Figure 11.67. The motorcycle is cut into "slices" perpendicular to its longitudinal axis, and their contribution to the total drag is plotted as the bars. The curve represents the integration (accumulation) of these increments in streamwise direction. Compared to an automobile, remarkably little negative contributions ($c_p < 0$ on a surface pointing upstream, $n_x < 0$ or $c_p > 0$; on a surface pointing downstream, $n_x > 0$) show up. This method is especially insightful for identifying where a drag difference between two variants is generated by plotting the difference of the increments. It is not uncommon that one discovers that the drag difference is not generated directly at the location of the geometry difference but further downstream by interference effects.

Motorcycle Aerodynamics

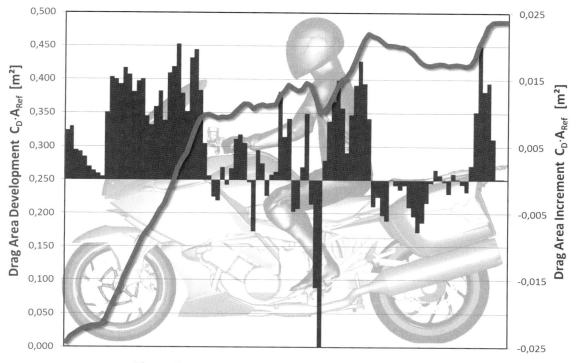

Figure 11.67 Streamwise drag distribution and generation.

When looking at the same analysis for lift on the right side of Figure 11.68, we seem to encounter a discrepancy. The largest downforce is created over the front wheel but the motorcycle experiences lift on the front? Why this is so has already been explained in section 11.3.2.2, but here it is clear from the diagram on the left in Figure 11.68 which plots the drag distribution along the vertical axis. This is the reason why the line of action of the total force, almost horizontal because the total lift nearly vanishes, is lying so high above the ground generating a nose-up pitching moment.

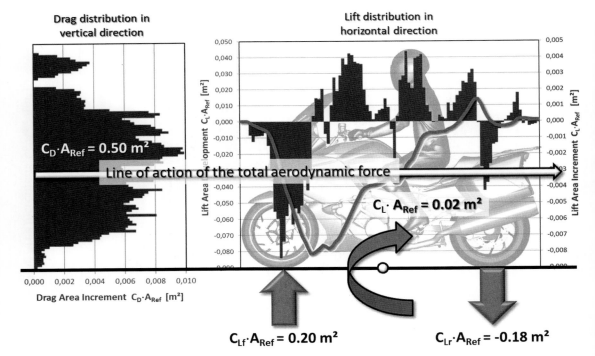

Figure 11.68 Lift distribution along the longitudinal axis, drag distribution along the vertical axis, and line of action of the total force.

Obviously measures to reduce front axle lift have to be substantial for a noticeable effect. The fact that a supersports bike has 25% less front axle lift (Figure 11.31) than a tourer is due to the lying position of the rider and the low windshield, both reducing the lever of the aerodynamic force above the road and hence the pitching moment.

11.4.2.3 External Flow Field Results

The significance of total pressure has already been emphasized with the surface results, and a visualization option was presented in Figure 11.39 and Figure 11.44. Whole regions in which a certain value is exceeded or undercut can be displayed by isosurfaces. The isosurface for $c_{p,tot} = 1$ in Figure 11.69a illustrates the envelope of attached boundary layer and wake. Outside of this isosurface the total pressure is the same as in the free stream. With $c_{p,tot} = 0$ in Figure 11.69b we see the region inside which (according to equation (11.17)) the loss is equal or higher than the dynamic pressure q_∞ of the free stream. Since vortex cores are always characterized by high losses, these isosurfaces are a vivid way to localize the generation and propagation of vortices—in addition to isosurfaces of vorticity itself. Reverse flow regions are clearly identified by isosurfaces for $v_x = 0$ (Figure 11.70).

Motorcycle Aerodynamics

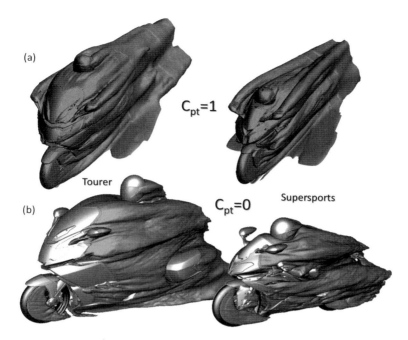

Figure 11.69 Isosurfaces of total pressure.

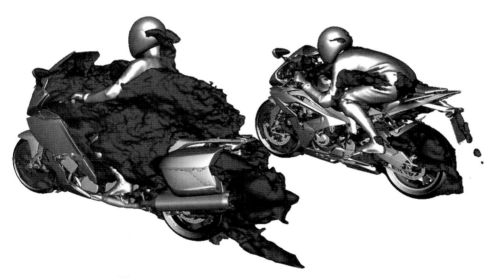

Figure 11.70 Isosurfaces of $v_x = 0$ (reverse flow).

Smoke probes are the most frequently used flow field visualization in the wind tunnel. The equivalent technique in CFD is the generation of streamlines originating at arbitrary points or rakes or grids. They can be calculated downstream, upstream, or both from the starting point. In addition, they may be colored by another quantity, for instance velocity magnitude as shown in Figure 11.71. Here these streamlines have been started upstream of the motorcycle. In interactive use, the seeding points can be moved around like the smoke probe in the tunnel to investigate interesting regions in more detail.

The advantage over smoke probes in the wind tunnel is that the origin of streamlines can be interrogated. An example is shown in Figure 11.72, where cooling air is visualized by generating streamlines up- and downstream through a rake filling exactly the heat exchangers for water and oil. This way, it is possible to identify where cooling air exits and to discover potential weak points of the flow toward the cooler.

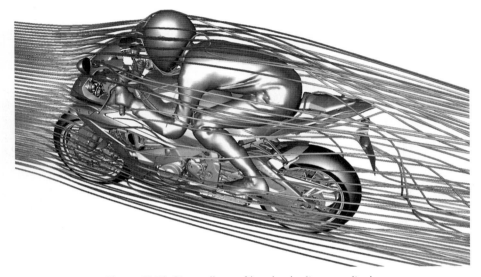

Figure 11.71 Streamlines of local velocity magnitude.

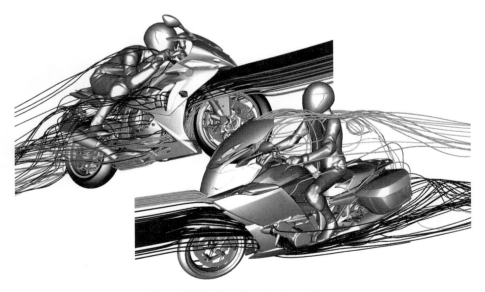

Figure 11.72 Cooling air streamlines.

Together with the temperature distribution, the cooling airflow rate determines the cooling performance. Preferably one wants a homogeneous velocity distribution, which is hardly realizable on a motorcycle. As displayed in Figure 11.73, the front wheel, fork, and fender cause strong inhomogeneity. On the tourer, the oil cooler is positioned near the stagnation point without upstream obstructions. Therefore the average velocity is nearly twice the one of the water cooler and accordingly it can be rather small. The normalized velocity does not change over a wide range of the free stream velocity so that the flow rate as the area integral increases almost linearly with speed.

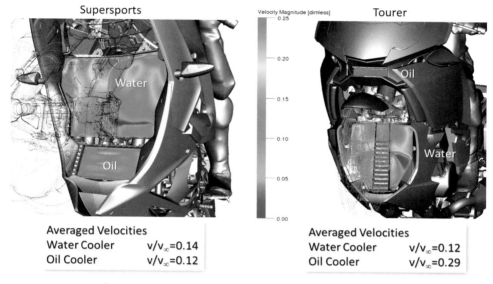

Figure 11.73 Velocity distribution on water and oil cooler.

Part of the rider's comfort is the temperature load from hot cooling air. With the oil cooler in the upper part of the fairing on the tourer, the pilot is immediately exposed to the hot air (Figure 11.74). However, due to the small amount of heat and intensive mixing with cold air, the temperature on the upper part of the body is at maximum 5°C higher than the ambient temperature. Air off the water cooler exits the lower fairing and just slightly strikes the lower leg.

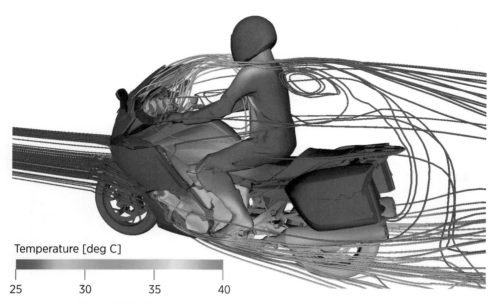

Figure 11.74 Temperature load (surface temperature) on a tourer.

On the supersports motorcycle the upper body of the pilot is not touched by hot air (Figure 11.75). Note that the temperature scales in Figure 11.74 and Figure 11.75 are quite different. It has to be ensured that no parts of the motorcycle, especially the fairing, can reach inadmissible temperatures.

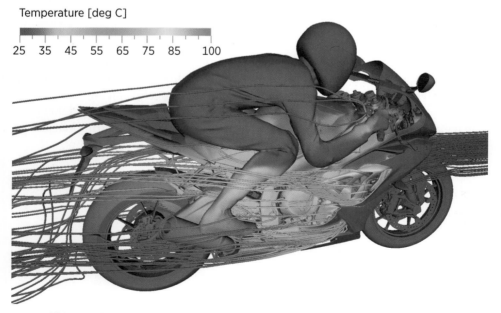

Figure 11.75 Temperature load (surface temperature) on a supersports bike.

Motorcycle Aerodynamics

Another potential heat source is the engine exhaust gas. If the outlets of the exhaust system are lying in a recirculation region, excess temperatures may arise on luggage systems or, especially with the muffler under the seat, on the back of a passenger.

11.4.3 Wind Tunnel

Structure and functionality of wind tunnels are described in detail in chapter 13. Since motorcycles are usually tested in tunnels designed for automobiles blockage (ratio of the vehicle's frontal area to the nozzle area) effects can be neglected. However, motorcycle measurements will be affected by the brackets to fix the wheels on the balance pads. Further, it has to be checked if the position off the tunnel centerline provokes unwanted asymmetries due to the proximity to the jet boundary.

Wind tunnels are also only a simulation of road conditions, and different tunnels will give different answers. Measurements of an absolutely rigid validation motorcycle in two different tunnels produced drag and lift areas that differed by 0.020 m^2 on average. Fortunately, this spread was the same for different variants so that the result of optimization work is independent of the test facility.

11.4.3.1 Stationary Wheels

Figure 11.76 shows a motorcycle measured with stationary wheels in the BMW aeroacoustics wind tunnel. On the front wheel, only lift is measured; the mount does not transfer any moments or longitudinal forces to the balance. The brackets on the front tire only prevent the wheel from sliding sideward.

Figure 11.76 Wind tunnel mounts for measurements with stationary wheels in the BMW aeroacoustics wind tunnel.

The rear mount allows measurement of both lift and drag forces. Since it is free to rotate around the lateral axis, the rear brake has to be actuated and fixed to avoid a pitching motion of the motorcycle. Depending on the type of rear wheel drive, for instance, with a double-pivot cardan shaft, minimal movements are still possible. Simulations with and without mounts showed that the drag increases by $\Delta c_D \cdot A_x \approx +0.010$ m. The difference between blocked and floating suspension has already been displayed in Figure 11.25.

11.4.3.2 Rotating Wheels

For measurements with rotating wheels, the motorcycle has to be fixed on the weighted wheel drive units (WDU). Such an arrangement in the BMW aerodynamic wind tunnel is shown in Figure 11.77. The rear wheel is not fixed at all, and its lift is measured by the WDU. On the front, the bike is fixed by a broadened axle on pillars that are mounted on the balance without force closure to the tunnel floor. Although these pillars look quite massive, simulation proved that their impact on the flow field is small due to the streamlined shape.

Figure 11.77 Wind tunnel mounts for measurements with rotating wheels in the BMW Aerodynamic Wind Tunnel.

Since this mounting also enables measurements under stationary conditions, the influence of wheel rotation could be investigated. Drag changed by $\Delta c_D \cdot A_x \approx \pm 0.004$ m and lift on the front as well as on the rear changed by $\Delta c_D \cdot A_x \approx \pm 0.005$ m, where the ± indicates that depending on the motorcycle differences in both directions have been observed. However, the effect of measures by magnitude and direction proved to be independent of wheel rotation. As Figure 11.78 shows, a difference of $\Delta c_D \cdot A_x = 0.002$ m between static and rotating wheels was conserved even over the accumulative sequence of nine different measures.

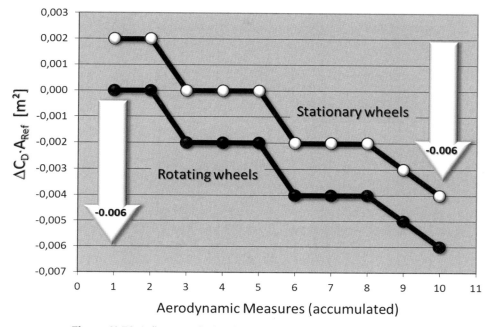

Figure 11.78 Influence of wheel rotation on the effect of measures.

No information is available to the author about the effect of the relative motion between motorcycle and road by measurements on a moving floor. In view of the very small projected area in vertical direction (compared to an automobile), no significant influences beyond the effect of wheel rotation are expected. However, this could be different at extreme lean angles.

11.4.3.3 Reproducibility, Driver Influence

The strong dependence of aerodynamic properties on size, posture, and equipment (gear, helmet) impedes reliable and reproducible measurements with human riders and passengers. Depending on the objective, 95- or 50-percentile dummies are used, whose posture is adjusted to standardized distances and angles via variable struts (Figure 11.79). To inhibit motions during the experiment, usually knees, feet, hands, and elbows are additionally strapped to the desired position.

Figure 11.79 Dummy fixing for reproducible posture

A comparison of drag and lift between upright and bent forward position is shown in Figures 11.80 and 11.81 (see also Table 11.3) for a number of motorcycles from all categories.

Motorcycle Aerodynamics

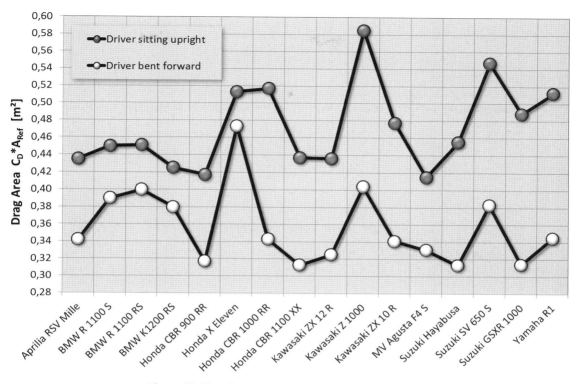

Figure 11.80 Influence of the driver's posture on drag.

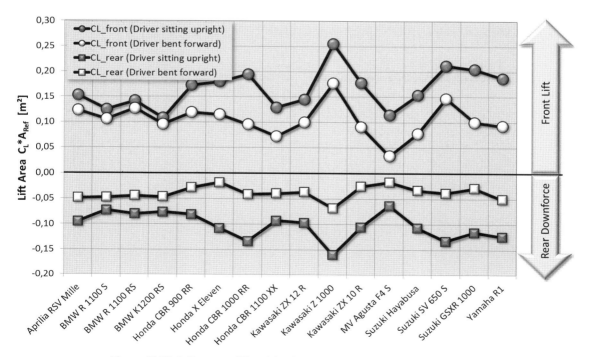

Figure 11.81 Influence of the driver's posture on front and rear axle lift.

Table 11.3 Comparison of Aerodynamic Forces for Upright and Bent Forward Driver/Dummy (All Measurements Conducted at 160 km/h in the BMW Aeroacoustics Wind Tunnel)

Motorcycle	Posture	$c_D \cdot A_{Ref}$ [m²]	$c_{Lf} \cdot A_{Ref}$ [m²]	F_{Lf} [N]	$c_{Lr} \cdot A_{Ref}$ [m²]	F_{Lr} [N]
Suzuki Hayabusa (MY 2000)	upright	0.455	0.154	173	−0.107	−120
	bent fwd.	0.313	0.078	87	−0.033	−37
Kawasaki ZX 12 R (MY 2001)	upright	0.436	0.145	163	−0.097	−109
	bent fwd.	0.325	0.100	114	−0.036	−41
Kawasaki Z 1000 (MY 2003)	upright	0.585	0.255	277	−0.160	−175
	bent fwd.	0.404	0.177	191	−0.068	−74
MV Agusta F4 S (MY 2000)	upright	0.415	0.114	128	−0.063	−71
	bent fwd.	0.331	0.034	39	−0.017	−19
Honda CBR 900 RR (MY 2000)	upright	0.417	0.173	204	−0.081	−96
	bent fwd.	0.317	0.119	136	−0.028	−34
Honda CBR 1100 XX (MY 2002)	upright	0.437	0.129	144	−0.093	−104
	bent fwd.	0.313	0.072	81	−0.039	−44
Aprilia RSV Mille (MY 1999)	upright	0.435	0.153	172	−0.095	−107
	bent fwd.	0.342	0.123	138	−0.048	−54
Honda X Eleven (MY 2001)	upright	0.513	0.180	208	−0.108	−125
	bent fwd.	0.473	0.115	130	−0.018	−21
Suzuki SV 650 S	upright	0.547	0.212	232	−0.133	−145
	bent fwd.	0.383	0.147	163	−0.038	−43
BMW K 1200 RS (MY 1999)	upright	0.425	0.108	121	−0.076	−85
	bent fwd.	0.380	0.096	110	−0.046	−53
BMW R 1100 RS (MY 1999)	upright	0.451	0.142	158	−0.080	−89
	bent fwd.	0.400	0.127	139	−0.044	−50
BMW R 1100 S (MY 1999)	upright	0.450	0.125	137	−0.073	−80
	bent fwd.	0.390	0.106	116	−0.047	−52
Ducati 916 (MY 1999)	upright	0.394	0.119	134	—	—
	bent fwd.	0.296	0.061	68	—	—
Ducati 999 (MY 2000)	upright	—	—	—	—	—
	bent fwd.	0.345	0.085	96	−0.045	−51
Aprilia RSV Mille R (MY 2002)	upright	—	—	—	—	—
	bent fwd.	0.315	0.092	101	−0.020	−22
Benelli Tornado 900 Tre (MY 2002)	upright	—	—	—	—	—
	bent fwd.	0.376	0.101	111	−0.060	−66

In addition to the pilot and his or her posture, passengers also have a significant impact on drag and lift. As is shown in Table 11.4, there is no unique trend. Both forces may increase or decrease due to the presence of a passenger.

Motorcycle Aerodynamics

Table 11.4 Comparison of Aerodynamic Forces of Various Tourer Motorcycles Without and with Passenger (All Measurements at 160 km/h in the BMW Aeroacoustic Wind Tunnel)

Motorcycle/Passenger		$c_D \cdot A_{Ref}$ [m²]	$c_{Lf} \cdot A_{Ref}$ [m²]	F_{Lf} [N]	$c_{Lr} \cdot A_{Ref}$ [m²]	F_{Lr} [N]
Harley Davidson Electra Glide Ultra Classic	w/o	0.765	0.143	163	−0.124	−141
(MY 2001)	with	0.757	0.134	153	−0.124	−141
Honda GL 1800 Goldwing	w/o	0.676	0.084	95	−0.082	−92
(MY 2002)	with	0.649	0.084	95	−0.070	−78
Yamaha Venture Royal Star	w/o	0.603	0.100	113	−0.080	−91
(MY 2002)	with	0.623	0.110	125	−0.100	−114
BMW K 1200 LT	w/o	0.519	0.149	165	−0.059	−64
(MY 2001)	with	0.527	0.139	155	−0.027	−30
BMW R 1150 RT	w/o	0.496	0.142	160	−0.087	−98
(MY 2001)	with	0.441	0.161	182	−0.099	−112

11.4.3.4 Flow Field Visualization

In a wind tunnel the options for flow field visualization are quite limited compared to CFD. Probably the oldest method is the use of yarn tufts to indicate the flow direction on the surface and to identify separation and reattachment regions (Figure 11.82). Oil streaks give a more detailed insight, but they are rarely used due to the contamination of the tunnel, especially in modern facilities with sensitive measurement equipment.

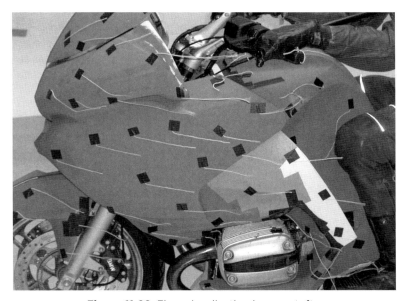

Figure 11.82 Flow visualization by yarn tufts.

Smoke probes are the most frequently employed visualization tool due to the minimal effort required. They enable the aerodynamicist to get a quick and flexible insight into which path the flow is following. Typically, they are used for investigations concerning wind and weather protection, like the example in Figure 11.83 where different wind shield positions are compared.

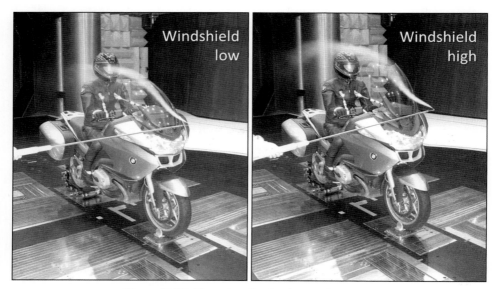

Figure 11.83 Flow visualization by smoke probe.

This method is also informative to see the extent of recirculation regions or to check where hot cooling air exits and how much the rider is exposed to it. When the probe is operated by a human for safety reasons, the tests are limited to speeds of about 60 km/h. However, in this range the flow field as well as the propagation and diffusion of smoke are sure to show a much stronger Reynolds-number dependency than what will be observed at high speeds. This reduces the accuracy of smoke probe visualizations when applied for understanding the aerodynamics at highway speed.

Particle image velocimetry (PIV) is the most elaborate method to map the velocity field in a plane. Details are described in chapter 13. As an example, Figure 11.84 shows the instantaneous wake behind a mirror. Considering the enormous effort in relation to the results (and the maturity of CFD tools today), PIV measurements are mainly conducted to collect validation data for CFD methods.

Motorcycle Aerodynamics

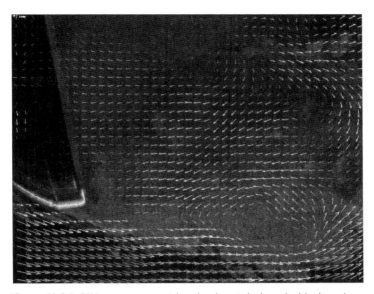

Figure 11.84 PIV measurement in a horizontal plane behind a mirror.

11.4.3.5 Aeroacoustics

The aeroacoustic perception under a helmet results from the interaction of motorcycle, rider, and helmet. To record wind noise under the helmet, dummies are used, whose structure, in particular the head, is close to the human anatomy (Figure 11.85). Ideally, these measurements are conducted in special aeroacoustics wind tunnels to exclude background noise and reflections.

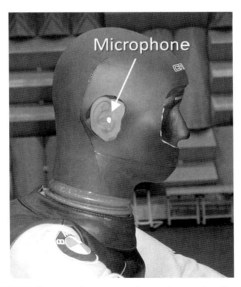

Figure 11.85 Dummy for measuring noise under the helmet.

The result of such a measurement for different helmets and variants (ventilation setting, chin spoiler, ear pads) on a naked bike is depicted in Figure 11.86. The sound pressure level shown is the average of left and right ear, which differ on average by 1–1.5 dB(A). From 100 km/h to 130 km/h, the level increases by 6–8 dB(A). The large spread of 7 dB(A) between the best and the worst value is somewhat remarkable. Measures to reduce the noise level are discussed in chapter 12.

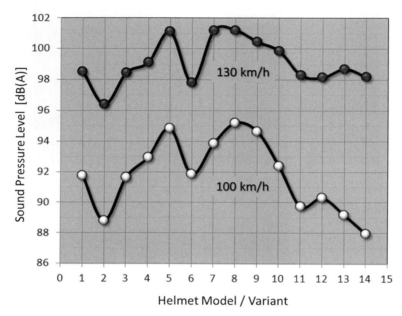

Figure 11.86 Measured noise inside the helmet (average of left and right ear) for different helmet models and variants on a naked bike.

11.4.3.6 Soiling

Due to the sensitive measurement equipment of balances, normally soiling tests showing the deposition of water and dirt from wheel sprays and other sources cannot be conducted in wind tunnels.

11.4.3.6.1 Foreign Soiling

In specially equipped tunnels, water is injected from a nozzle array in front of the motorcycle into the flow. A fluorescent agent is added to the water so that the impingement on motorcycle and rider and the propagation and accumulation on the surface become visible under UV-light (Figure 11.87).

Motorcycle Aerodynamics

Figure 11.87 Foreign-soiling test with a fluorescent agent under UV-light in the environmental test facility of BMW Group.

11.4.3.6.2 Self Soiling

To investigate the water spray from the wheels, it is of course necessary to run the test with rotating wheels. An example is shown in Figure 11.88. In the environmental test facility of the IVK at the University of Stuttgart, water is injected in front of the rotating wheels to simulate driving on a wet road. The focus is not only on self-soiling of pilot, passenger, and luggage system but also on soiling of a central shock absorber. In addition to the observation under UV-light, the images before and after soiling may also be analyzed and documented digitally using false-color displays.

Figure 11.88 Self-soiling test with a fluorescent agent under UV-light in the environmental test facility of the IVK at the University of Stuttgart.

Both facilities can only reproduce the real conditions on a road approximately and deliver preliminary guidance. The final evaluation occurs via road tests.

11.4.4 Road Test

The objective magnitude of forces and moments can be determined by simulation or in the wind tunnel, and their impact on performance and fuel consumption can be determined. However, it is difficult to assess the influence of aerodynamics, suspension, and mass distribution separately with respect to driving characteristics. Multibody simulations can be used for some assessment of driving characteristics, but in the end only a road test can show how driving characteristics will be perceived by the rider. In addition, the driver's reaction is part of the complete man-machine system (Schmieder 1991 [722]).

11.4.4.1 Driving Stability

To investigate driving stability, the motorcycle is equipped with sensors and data recording. Goniometers monitor roll- and steering-angles and an accelerometer on the rear end registers weaving motion. Speed and longitudinal acceleration are captured by high-resolution GPS receivers. During test rides with different speeds, the rider induces systematic steering inputs, and the decay in the steering response is analyzed to obtain the damping properties. However, aerodynamic influences cannot be uniquely separated. By measuring the front suspension travel together with the spring rate, the approximate front lift and hence the relieving load can be deduced. A source of irritation during such tests may be the hysteresis of the spring-damper system or bumpiness of the road.

11.4.4.2 Cross-Wind

In order to assess the sensitivity to yaw conditions, usually comparative test rides are made in a crosswind facility. The one shown in Figure 11.89 on the BMW proving ground is 30 m long and is composed of ten fans with a nozzle area of 3.0 m × 1.7 m each (fan diameter is 2.8 m). The pilot enters the measuring track with a 1-m safety distance to the yellow line and then closes the throttle. While passing the fans and over the next 15 m, no steering input is given. A light barrier triggers a camera, and the drift is read from the markers on the ground. Only very experienced test riders manage to accomplish reproducible results, which differ by less than half a meter from test to test. Each run is repeated six times, and the average is used to compare different motorcycles.

Motorcycle Aerodynamics

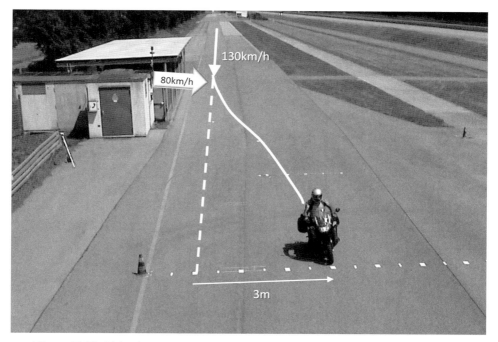

Figure 11.89 Ride along a crosswind facility (BMW proving grounds in Aschheim).

The picture in Figure 11.89 shows a test where the motorcycle rides with 130 km/h into a crosswind of 80 km/h. This is equivalent to a yaw angle of 32°, conditions which, according to Wojciak et al. (2010, [893]), hardly ever occur in nature at this speed. After 15 m, a drift of 3 m (= angular deviation 11°) is visible.

11.4.4.3 Wind and Weather Protection

The crosswind facility shown can also be used to test the water impingement on a motorcycle. A controlled water supply is fitted to the fan so that a defined amount of water can be applied to a motorcycle including rider (in rain gear) strapped in front of the fan (Figure 11.90). This surely does not replace a real road test but allows testers to quickly identify the intrusion of water and to test counter measures.

As shown, first indications on wind protection are previously provided by CFD or the wind tunnel test, but road tests with different riders and different gear and helmets complete the overall impression and enable the manufacturer to anticipate the customer's perception.

Figure 11.90 Water impingement test at 80 km/h.

11.4.4.4 Soiling

To test self-soiling a track of 80–100 m, covered by a mixture of chalk powder and water, is crossed several times at constant speed. Low velocities up to 40 km/h are suitable to check the soiling from the front wheel on fairing, cooler, and the rider's lower legs. At higher speeds beyond 80 km/h, it reveals how the dirt dispersed from the rear wheel into the wake settles on the motorcycle's rear end and the rider's back. In Figure 11.92, comparison between road test and a simulation result is shown.

Figure 11.91 Self-soiling from the front wheel.

Figure 11.92 Soiling: Road test (a); simulation (b).

Regions with higher accumulation of dirt in simulation are only approximately equivalent to the road test. The problem is not so much the calculation of particle trajectories in the flow field but the knowledge of realistic initial conditions—how much dirt is entrained by the tire for how long and at which direction the particles and droplets are ejected from the tire into the air flow.

11.4.4.5 Aeroacoustics

Measuring the noise inside the helmet while riding requires mobile equipment that can endure vibrations and humidity. Two microphones are installed in the helmet near the ears. On the track, their signal is recorded over a certain time and the average yields the frequency response (one-third octave mid-band frequency) and the overall sound pressure level. Later in the acoustic studio, the recorded impression can be reconstructed and the effectiveness of measures may be evaluated.

The noise inside the helmet is primarily influenced by the height of the windshield and the size of the rider (in particular, the distance from the windshield upper edge to the eye level of the driver); see Figure 11.93 from Bachmann et al. (2002, [30]). In addition, the diagram indicates regions of typical motorcycle categories. For instance, on a supersports bike the vertical distance is high due to the low windshield, while the horizontal distance is lower than on other categories caused by the bent forward posture of the pilot.

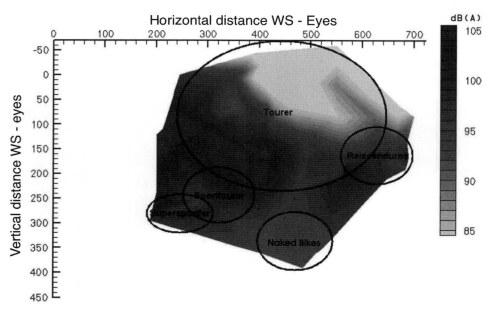

Figure 11.93 Noise inside a helmet, depending on horizontal and vertical distance between windshield and eye point (Bachmann et al. 2002 [30]).

11.4.5 Outlook—The Future of Development Methods

In the future, the distribution of development effort and tool usage shown in Figure 11.54 will surely shift toward CFD and hence to the early design phase. On the one hand, this is due to the overall progress in CAD virtualization, and on the other hand to the increasing accuracy of simulation results and the augmentation of computer performance. Further, it is possible that automatic optimization methods will produce solutions that would not be found by experience or trial-and-error. Simulation allows assessing various concepts before prototype hardware is available or when the realization in hardware would be too laborious. Simultaneously, wind tunnels will not become redundant in the foreseeable future because detail optimizations or parameter studies can be accomplished in a short time frame.

Motorcycling is primarily a leisure time activity, and the vehicle is appraised much more emotionally than an automobile. Only the final road test can capture the subjective customer's perception beyond bare figures.

11.5 Aerodynamic Design—Practical Examples

Design options of an aerodynamicist are always limited by various, often competing requirements with respect to styling, package, weight, and cost. In this chapter, some practical examples of measures to influence the aerodynamic properties of motorcycles are presented.

11.5.1 Measures to Optimize Drag and Lift

The scope of aerodynamic design varies as much as the characteristics of the different motorcycle categories. The segment of tourers has the largest spread of drag areas; nevertheless, they occupy the upper end of the total drag scale due to the larger frontal area (see Figure 11.94). Optimum values are achieved for supersports bikes, which are especially designed for performance, and they are optimized for the rider leaning forward.

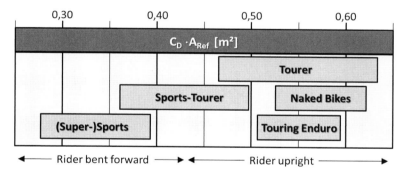

Figure 11.94 Range of drag areas for different motorcycle categories.

The main starting points for drag reduction and related potentials are shown in Figure 11.95. Among all individual measures, clearly the optimization of a partial or full fairing offers the highest potential. However, in total the sum of small measures, if feasible, may deliver a significant contribution. As an example for small measures, Figure 11.96 shows windshield variations that reduce the drag by up to 4%.

	Part / Component	$\Delta C_D \cdot A_{Ref}$ [m²]
(1)	Partial fairing	-0.020 ... -0.150
	Full fairing	-0.150 ... -0.200
(2)	Mirror	+0.010 ... +0.025
(3)	Hand protector	-0.003 ... -0.015
(4)	Front fender	-0.003 ... -0.015
(5)	Cooling air inlet	-0.010 ... -0.020
(6)	Engine spoiler	-0.010 ... -0.025
(7)	Spoiler extension	-0.005 ... -0.010
(8)	Optimized luggage boxes	-0.010 ... -0.020
(9)	Topcase & rear end fairing	-0.008 ... -0.015
(10)	Helmet	-0.010 ... -0.015

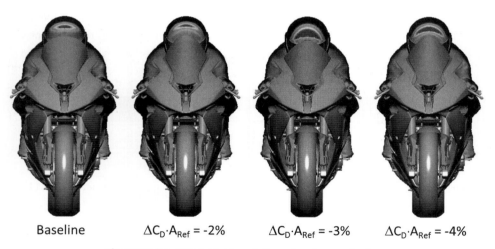

Figure 11.95 Potentials for drag reduction.

Baseline $\Delta C_D \cdot A_{Ref}$ = -2% $\Delta C_D \cdot A_{Ref}$ = -3% $\Delta C_D \cdot A_{Ref}$ = -4%

Figure 11.96 Windshield variations for drag reduction.

In motorsports, the focus is on drag reduction to increase the top speed. Figure 11.97 top depicts the closed motor spoiler that is required by the regulations to collect liquids from possible leaks. The rear end is extended downwards and to the side to reduce the incident flow on the rear wheel, similar to wheel spoilers on automobiles. In Figure 11.97 bottom left, it is visible how the passage of the front fork through the fairing has been closed. As a side effect on production motorcycles, this also helps to prevent water and dirt from being sprayed toward the rider. The changes marked in Figure 11.97 bottom right reduce the oncoming flow on the pilot's shoulders.

Motorcycle Aerodynamics

Figure 11.97 Motorsports specific variations of a production motorcycle.

Besides the rider's posture, his gear also plays a significant role. Adding a hump on the upper back is common, especially in motorsports, to fill the wake of the helmet, and this leads to an appreciable drag reduction (Figure 11.98).

Figure 11.98 Influence of posture and gear (hump).

The strong sensitivity of drag area to the pilot's posture is exploited in motor sports to support braking efficiency. Straightening up when braking can increase the drag by up to 50%, and extending the knee when entering the corner can even increase drag by more than 60% (Figure 11.99). However, as previously shown in Figure 11.40, this is simultaneously accompanied by an unwanted increase in front lift.

Figure 11.99 Changes of the drag area in braking and cornering posture.

11.5.2 Design of Internal Flows, Cooling, and Heat Protection

The ram-air concept described in section 11.3.4 works best when the engine air intake is placed closer to the stagnation point. In the concept study shown in Figure 11.100, the highest flow rate is achieved by the variant in Figure 11.100 left, although it has the smallest intake area. The other two concepts in Figure 11.100 center and right deliver considerably less air to the airbox, despite their having openings almost twice as large, because they are lying off the stagnation point in regions of lower pressure. This investigation is a good example of an efficient usage of CFD in the early phase. Conducting such a study in physical tests would require the availability of a prototype with an accordingly high effort in cost and time.

The examination of various cooling concepts including different front wheel suspensions, as shown in Figure 11.101, is not possible at all in hardware with an economically justifiable effort. In this case the curved trapezoidal heat exchanger has been selected because it delivered the requested cooling performance in the most efficient way.

Motorcycle Aerodynamics

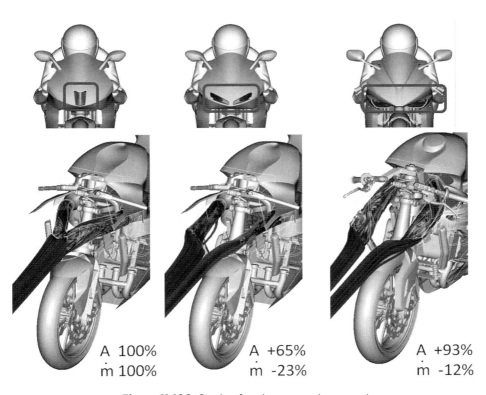

Figure 11.100 Study of various ram-air concepts.

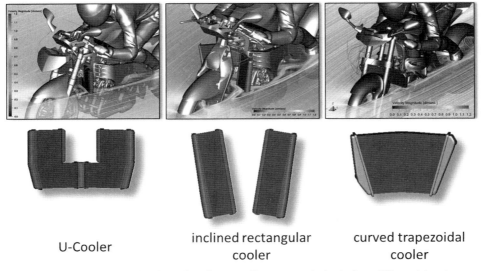

Figure 11.101 Examination of various cooling concepts including different front wheel suspensions.

Even small details on the wheel suspension upstream of the cooler can have a significant impact on the flow rate. A vertical edge along the front fork suspension in Figure 11.102 left leads to an expanded wake. Rounding this edge reduces the wake and increases the flow rate by 43%.

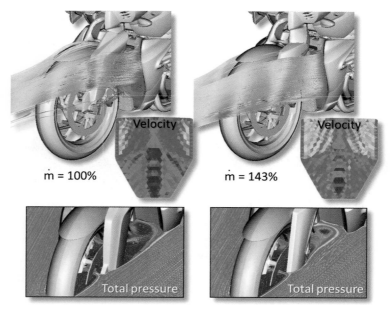

Figure 11.102 Impact of the front suspension shape on the cooling airflow rate.

Even from an isothermal simulation, heat transfer coefficients can be derived from flow field properties and allow at least qualitative conclusions. Figure 11.103 shows clearly that a caliper mounted below the axis experiences better cooling.

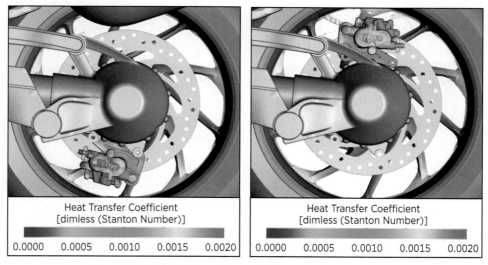

Figure 11.103 Heat transfer coefficients on the rear caliper.

11.5.3 Measures for Wind and Weather Protection

The goal of wind and weather protection is essentially to relieve the rider and passenger from exposure to high velocities. As Figure 11.104 illustrates, even small changes of the fairing can reduce the velocity at the knee and hence minimize the cooling effect. At the same time, this reduces drag. A similarly working spoiler in Figure 11.105 not only protects the foot from cooling but also from excessive rain impingement.

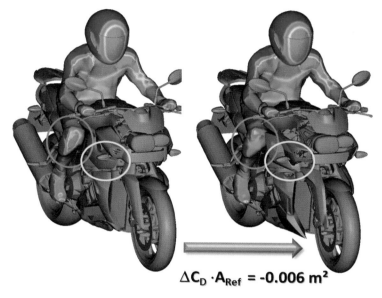

$\Delta C_D \cdot A_{Ref} = -0.006 \text{ m}^2$

Figure 11.104 Variation to reduce the oncoming flow on the rider's knee.

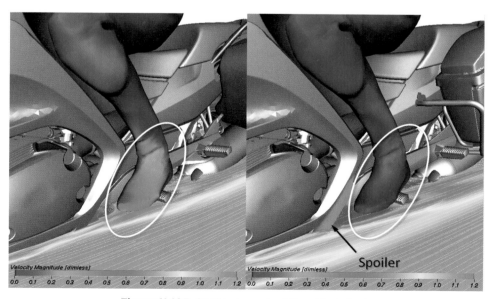

Figure 11.105 Spoiler to protect the rider's foot.

The excellent wind and weather protection on luxury tourers can turn into a disadvantage at high temperatures and moderate speeds when intentional ventilation is desired. For this reason, the BMW K 1600 GTL features foldout deflectors on the side of the fairing that direct the flow deliberately toward the rider (Figure 11.106). Developed in CFD and verified in the wind tunnel, their effect could also be confirmed clearly on the road.

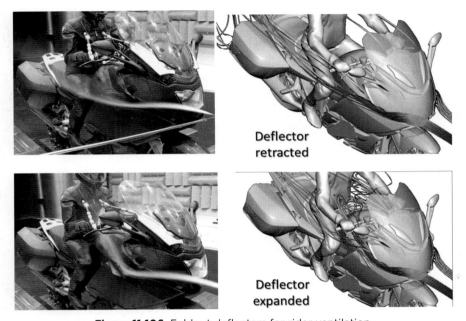

Figure 11.106 Foldout deflectors for rider ventilation.

11.6 Outlook

Today more than ever the motorcycle is a vehicle for leisure time activities. A multitude of concepts and application scenarios challenge the manufacturers. With a performance output of up to 200 hp and top speeds of around 300 km/h, aerodynamics has recently gained significance. On the other hand, tourers are becoming more and more luxurious, with increasing demands on comfort, that is, wind and weather protection, as well as aeroacoustics. Beyond the pursuit of performance nowadays, fuel consumption plays an important role for motorcycles as well.

The maturity of simulation tools allows development cycles to be shortened and allows manifold concepts to be assessed in the early phase. Aerodynamic and thermal test facilities have become very capable and help to transfer physical tests from the road into the lab. Nevertheless, road tests under customer-oriented conditions are still indispensable.

Giant leaps in aerodynamics like in the first half of the last century cannot be expected. The challenge now lies in detail optimization in order to bring the emotional experience of motorcycling to perfection.

Chapter 12
Helmets

Gerd Janke, Sebastian Reitebuch

12.1 Head Protection Technology

Motorcyclists have used head protection for over 100 years. In the beginning, linen and leather caps were worn, because the wind protection was the center of attention. Later, half bowl helmets, jet helmets, and, since the late 1960s, integral helmets were worn. In 1958, the DIN 4848 was issued, the first German standard for the homologation of motorist protective helmets. In 1975, this standard was replaced by the European standard ECE-R 22-01, which is revised at regular intervals. At present, the ECE-R 22-05 is valid.

Helmets became mandatory on motorcycles in Germany on 1.1.1976. According to Highway Code § 21a, paragraph 2, "while driving, the operators of motorcycles and their passengers must wear officially approved safety helmets." While wearing a helmet is enforced by law, the official homologation of the helmet is not. Therefore, motorcyclists may be seen on the road wearing helmets without ECE homologation. These helmets are quite small in size and in no way guarantee the necessary level of security.[1]

While motorcyclists must wear helmets all over Europe, the United States has an eventful history of introductions and suppressions of general helmet laws [814]. In 2012, the mandatory use of helmets for motorcyclists was abolished in the state of Michigan.

---

[1]. If there is evidence that her head injuries are due to an inadequate protection helmet, the blame can be partly imputed to the injured party. The insurance cover is at risk if the used helmet is not homologized according to the current ECE standard.

In addition to the ECE-R 22-05, three other standards for motorcycle helmets are widely used. For the homologation of helmets in the United States and in Australia, the U.S. DOT FMVSS 218 or the AS 1698-1988 are binding. In Japan, motorcycle helmets are homologized according to JIS T 8133:2000. For helmets used in automobile racing, the international head organization for motorsport FIA (Fédération Internationale de l'Automobile) demands homologation in accordance with the requirements of the Snell Foundation. This American foundation has established the Snell SA2000 standard, which places the highest demands on the protective effect of the helmet. For Formula 1, the FIA has specified additional requirements, which are continually being improved. Some major helmet manufacturers have also developed in-plant testing requirements, which go beyond the requirements of these standards.

Main components of the homologation according to ECE-R 22.05 are drop tests. The helmet is conditioned at +50°C or −20°C and then falls from a height of 3 m onto steel anvils of various shapes. The helmet is equipped with a standardized test head containing a tri-axial accelerometer. The maximum acceleration at impact must not exceed 275 g. Another measure of the likelihood of a serious head injury is the head injury criterion (HIC) value,[2] which is calculated from the acceleration curves. The upper limit for the HIC value is 2400.

In the ECE-R 22.05 the test points on the helmet shell are fixed, while the Snell standard checkpoints can be freely selected within a prescribed range by the examiner. In addition, the Snell standard calls for the same points being successively tested two or three times.

In addition to the drop tests, the standards place numerous supplementary demands on the helmet design. Among other things, certain areas of the head must be covered, other areas must be free, and minimum dimensions for the angle of view are required. There must not be any sharp edges, the chin strap is regulated, and also the visor is subject to extensive regulations with respect to its optical and mechanical properties.

The basic helmet structure is described in the ECE-R 22-05. "The helmet consists of a helmet shell with a hard outer surface as well as a device for damping the impact energy and a retention system." Figure 12.1 shows an example of a Formula 1 helmet where this principle has been refined to a complex composite structure, which is exceptionally light at 1000 grams and still exceeds all testing requirements over a wide temperature range. The tough, rigid outer shell consists of up to 15 fiber composite layers with optimized laying plans. The underlying soft damping shell made of expanded polystyrene (EPS) consists of 20 different bulk densities and is profiled geometrically in order to ensure an

---

2. The dimensionless quantity HIC ("head injury criterion") is a product of powers of the mean acceleration and the exposure time.

$$\text{HIC} = \left[\frac{1}{t_2 - t_1} \int_{t_1}^{t_2} a(t) dt \right]^{2,5} (t_2 - t_1) \tag{12.1}$$

The values are used in SI units in the formula; the HIC value is specified without dimension.

optimal spatial distribution of hardness. The inner comfort padding is flame resistant, as well as the retention system (chin strap and neck strap) made of woven Nomex® straps.

Figure 12.1 Design of a Schuberth Formula-1 helmet.

The art of aerodynamics, acoustics, and ventilation development is the innovative application of physical principles and new materials to improve the active and passive safety of the helmet. Active safety includes the view, the comfort, and drivability. It allows a good perception of the environment and fatigue-free driving [744]. Passive safety includes the impact absorption, penetration resistance, and the roll-off behavior of the helmet. It ensures optimum protection in an accident.

12.2 Motorcycle Helmets

12.2.1 Aerodynamics

12.2.1.1 Development Targets

A major objective in the development of motorcycle helmets is a low flow resistance, as in the aerodynamic optimization of automotive vehicles. However, motivation for this is not the optimum use of engine power, since the helmet usually contributes very little to the total resistance, as can be seen in Figure 11.68. Rather, a minimum load for the motorcyclist is sought.

Drag coefficients of motorcycle helmets range from 0.3 to 0.4, which are similar to those of cars. In the following, we use forces with dimensions, as is done in practical development work and in benchmarks. This practice ensures that the aerodynamic quality of the external geometry is not assessed separately from the total design (i.e., a good helmet is not only streamlined, but in addition is as small as compatible with good shock absorption).

According to Thöle [803], [802], not just low drag but also a low lift of the helmet is important, particularly at high speeds. Berge [71] also found that the flow resistance alone does not sufficiently describe the driving quality of a helmet, but that in addition a small lift or even a slight downforce is desirable to keep the helmet from rising.

An aerodynamically good helmet is also characterized by a low lateral force and low yaw moment at small rotations of the head. Moreover, no pitching moments and no flow-induced vibrations ("buffeting") should occur. The ideal helmet is neutral and calm under all conditions. These are determined not only by head and body posture but also by type and features of the bike. Thus, an individual aerodynamic tuning of the motorcycle/helmet/driver system can be helpful.

A special feature of helmets is the origin and orientation of the reference coordinate system. Usually, helmet wearers feel "lift" when a force aligned with the neck pulls the helmet away from the head. Due to the varying forward-leaning posture of the biker, this does not correspond to the physical definition of aerodynamic lift, as a force opposing gravity. In the following discussion and diagrams, we refer to forces acting normal to the neck as drag, forces aligned with the neck as lift, and forces acting normal to both of them as lateral forces.

The forces and moments acting on the helmet and the neck are completely described by six components. However, critical to the wearer's perception of these forces is in addition how the forces are transmitted into the head via the inner liners. A well-fitting liner distributes the force uniformly over the entire surface of the head, including the cheek area, which is very comfortable for the wearer. This way, even larger forces can be endured well. Ill-fitting liners can cause localized forces. This leads to pressure sores and permanent pain. A too-loose fitting helmet increases the perception of buffeting as well as lift, as these must then be transmitted by the chin strap and the most pressure-sensitive larynx.

12.2.1.2 Helmet Geometry

Aerodynamic considerations have been taken into account in the design of helmet geometries for over 80 years (see, e.g., Figure 11.3). Already its basic shape affects the aerodynamic characteristics of the helmet. Generally speaking, a helmet should

- Be as small as possible
- Have a good shoulder room

- Not be too long oval
- Not have too high of a spoiler

Some specific design elements are highlighted in the following. In new developments, first, limiting contours are defined, which the aesthetic design must not exceed. Various designs are then tested in the wind tunnel and, if possible, evaluated with numerical simulations (see section 12.4.5 and chapter 14).

12.2.1.2.1 Spoilers

The importance of spoilers as simple tools for targeted improvement of helmet aerodynamics was recognized at the end of the 1960s (see, e.g., US Patent US3548410). Wind-tunnel experiments[3] in the 1980s showed that spoilers have the best effect on long oval helmets on the top of the helmet (Schuberth, internal communication). This resulted in a change of the basic shapes of helmet shells. Thereafter, many variants of spoilers integrated into the helmet shell have been used in series helmets.

In addition, spoilers are often glued on helmet shells as attachments. The bond is designed so that it detaches in an accident. In this way, a good sliding behavior of the helmet on the road surface can be achieved even with sharp-edged spoilers.

Figure 12.2 shows the effect the spoiler position on drag and lift as a polar diagram. The distance between the individual points of the curve and the origin of the coordinate system gives the resulting total force. The aim of the aerodynamic optimization process is to combine a low drag, low lift, and low fluctuations, which was achieved in the present case between positions 2 and 3.

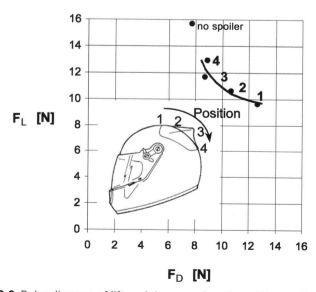

Figure 12.2 Polar diagram of lift and drag as a function of the spoiler position.

---

3. Conducted in the aeroacoustic wind tunnel of Schuberth GmbH.

Some helmet models (Arai, Schuberth SR1) feature an adjustable spoiler to enable a tuning of drag and lift. Due to technical and aesthetic reasons, the adjustment range of the spoiler is rather small, as is the aerodynamic effect.

Through a spoiler on the front lower helmet rim (Handley-Page flap), the lift and the pitching moment can be reduced (Vieri [827]). On motorcycles this spoiler causes a substantial magnification of the lateral forces when the head is turned, but in open race cars (Formula 1), lateral forces are unimportant because of the improved wind protection and restrained head turns. Therefore, in some vehicle geometries a chin spoiler can be advantageous.

12.2.1.2.2 Structured Surface

Another principle for flow control was used for the first time at the Schuberth "Speed" in 1982: A surface structured like a golf ball ("dimple shell") leads to a lower drag (see Figure 12.3).

Figure 12.3 Schuberth speed with structured surface.

The reason for this is the shift of the laminar/turbulent boundary-layer transition to smaller Reynolds numbers and the associated reduction of the separation zone, as first shown by Prandtl in his classic trip-wire experiment (see e.g., Schlichting [716]).

With the "Speed 2," the effect of the structured surface was modified with ribs transverse to the flow direction. First, with the aid of the structured surface a turbulent boundary layer is produced, which results in a more stable flow than a laminar boundary layer. This turbulent boundary layer is then brought to a controlled separation with the help of the ribs.

12.2.1.3 Direction of View

In order to get oriented in traffic, the driver must be able to turn his head safely at all speeds. Therefore, the helmet must respond "good-natured" upon changing the viewing direction: the shoulder view, the mirror view, the view of the road ahead, and the view at the display instruments must be executable without too much force or change of force. Figure 12.4 shows test results obtained with the measuring robot described in section 12.4.3. The helmet forces are plotted over the sampling time. The measuring robot sat on a sports motorcycle with medium inclination of the upper body. The head of the test dummy successively assumed ten viewing directions while, with about 100 Hz sampling frequency, the three neck-force components (drag, lift, lateral force) and the resulting total force were measured. Every position was averaged over the measurement period.

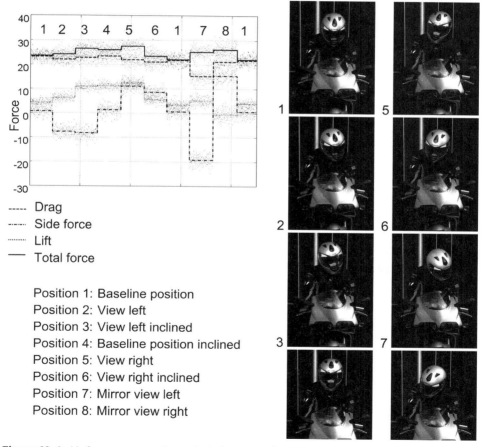

Figure 12.4 Air forces on a motorcycle helmet as a function of the viewing direction of the test dummy.

At a glance, the diagram reveals the aerodynamic characteristics of the helmet for the tested combination of motorcycle, windscreen, and riding posture. The case example shown is an aerodynamically good sports helmet with comparatively low forces acting on it.

With this tool, an overall assessment and characterization of a helmet, as a sports helmet or a touring helmet, can be drawn from a multitude of measurements with varying combinations of riding postures, motorcycle types, and accessories.

12.2.1.4 Influence of Neck Length (Shoulder Room)

Figure 12.5 shows the effect of fluid-mechanical interactions of the helmet with neck and shoulders. The closer the helmet sits on the shoulders, the higher the forces. Apparently, the obstruction of the flow underneath the helmet by the shoulders and the neck generally leads to significantly higher flow forces acting on the helmet. We can conclude that firstly, realistic measurements must take into account flow interactions. Secondly, a short lower rim and much neck and shoulder room are obviously advantageous for the aerodynamics. While much shoulder room is also beneficial for both the helmet weight and the general mobility of the head, it must of course be brought in line with the necessary coverage of the protection zones. However, it is a major disadvantage of free flow over the underside of the helmet and the neck, especially their intersection, that it is accompanied by intense turbulence, which in turn can cause considerable flow noise, as shown in section 12.2.2.2.

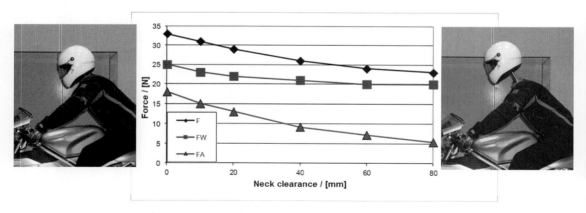

Figure 12.5 Flow forces as a function of the neck length.

12.2.1.5 Riding Position and Windshield

The flow around the helmet is strongly determined by its position relative to the windscreen, that is, by the head and body posture of the rider and by the height of the windscreen. According to Thöle [803], the lack of wind protection of many series bikes is the most common trigger for modifying a motorcycle with an accessory windscreen. This measure provides a relief to the head and upper body as well as a good rain protection. However, such modification usually has a negative effect on the ventilation. In addition, larger turbulent eddies originating from the windscreen impinge on the helmet, causing buffeting of the helmet as well as increased noise, and prevent the helmet from resting quietly and neutral in the wind.

The effect of different types of motorcycles, postures, and accessories on the helmet-flow dynamics is shown in Figure 12.6. The fairings of the touring bike provide a strong relief to the head in an upright position. On the sports bike, the windshield in connection with the stronger body inclination causes a slight relief. On the naked bike, helmet forces are maximal.

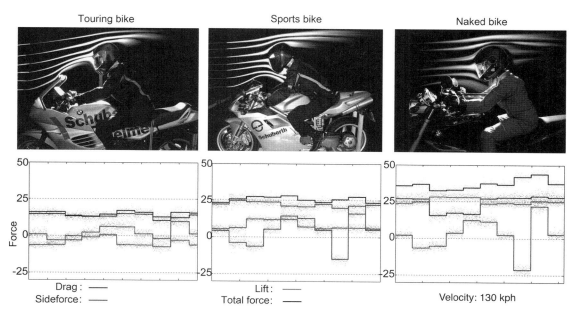

Figure 12.6 Force measurements on three different types of motorcycles; positions 1–10, Schuberth measuring robot.

The effect of body inclination is shown in Figure 12.7. Here, the helmet forces are plotted against the elevation of the rider's head above the tank. Clearly visible is the wind protection provided by a small front screen in a sporty riding position.

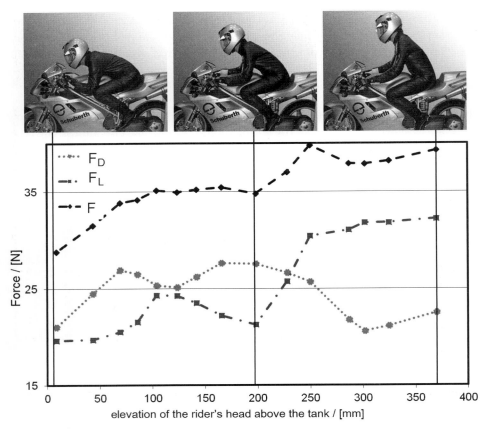

Figure 12.7 Air forces on the helmet for different rider's body inclinations.

Figure 12.8 exemplifies the influence of the windshield inclination on the helmet forces. The windscreen of the touring bike BMW K 100 RT was changed so that it could be adjusted from a completely flat to a very steep inclination. The adjustment range went far beyond the capabilities of standard adjustable windshields in order to show the possible extremes. The absolute value of the total force is plotted. If this value is taken as a measure of the aerodynamic quality of the configuration, there is the choice between two positions with low total forces, namely windshield heights 480 and 560 mm. In this case, the minimum at 560 mm was identified as optimal, since simultaneously conducted noise measurements (see section 12.2.2.2) showed lower sound levels at 560 mm than at 480 mm. With a windscreen higher than 555 mm, the total force changes its direction and there is a noticeable propulsion force at the helmet.

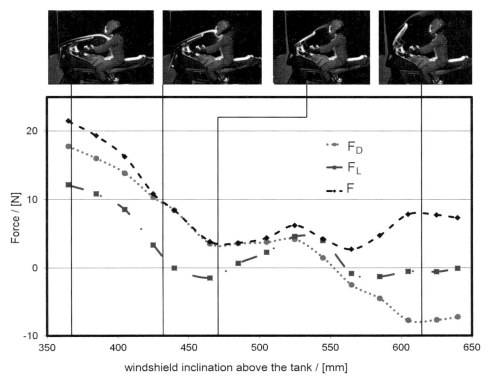

Figure 12.8 Air forces on the helmet depending on the windshield height of a modified BMW K100 RT.

12.2.2 Aeroacoustics

12.2.2.1 Development Targets and Helmet Noise

Since the early 1990s the aeroacoustics of helmets have become increasingly important, particularly with premium-class motorcycle helmets. Essentially, it's about:

- Minimizing wind noise
- Eliminating annoying noise such as whistling and rumbling
- Providing low background noise for radio communication
- Providing good perception of acoustic signals (horns, sirens, etc.) from ambient traffic

The Silencer from Baehr Tec GmbH in 1996 was the first series helmet with aeroacoustic optimization. Because at that time acoustic wind tunnels were still scarce, the Silencer was developed in extensive road tests with mobile measurement microphones.

After the Schuberth in-house wind tunnel had been upgraded to an aeroacoustic wind tunnel, a development tool specifically designed for helmet acoustics was available for the first time worldwide. In 2002 with this tool at hand, Schuberth could raise the bar for new quiet helmets with the S1 and subsequent models (C2, C3, and S1Pro). With most other helmet brands, riders are still exposed to large noise levels which, as stress factors, not only disturb the joy of riding but directly degrade the rider's performance and his ability to concentrate on road traffic (Maue [536]). Wind noise will be perceived by most riders as unpleasant, at least temporarily [102]. Furthermore, long-term hearing damage can occur. Particularly affected are professional motorcyclists, like policemen, journalists, and test drivers.[4] It is known that the harmful effect of noise is cumulative; after van Moorhem [821], a short exposure leaves mostly no damage, but many years of frequent riding often result in degraded hearing if ear protection is not worn continually (Streblow [781]).

At a speed of 100 km/h, the typical noise level of motorcycle helmets without acoustic optimization (which currently is still the majority) lies between 90 and 110 dB(A). With careful design and under favorable conditions, the noise can be lowered to below 80 dB(A) on a naked bike. It is usually dominated by wind noise, which masks the engine noise completely at speeds above 80 km/h.

Figure 12.9 shows typical A-weighted octave spectra of various helmet makes for a basic case: an upright posture on a naked bike at 100 km/h. Note the considerable range of noise levels covered by modern helmets. Nevertheless, the highest overall sound pressure levels were measured for the case without helmet. Most sound energy is contained in frequencies near 300 Hz. On faired motorcycles also significantly higher levels and lower frequencies may occur (see section 12.2.2.2).

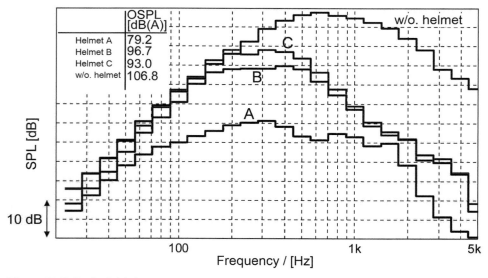

Figure 12.9 Typical third-octave spectra (A-weighted) and overall SPL of full face helmets at 100 km/h on an unfaired bike in the wind tunnel.

---

4. See, e.g., Hüttenbrink [393] and Kortesuo & Kaivola [458].

While the basic, well-defined quantities sound pressure level (SPL) in dB(A) and the loudness in sone may not capture all aspects of sound in the helmet,[5] they do give a good base for the assessment of broad-band noise and are easily communicated to the end user. Rothhämel [687] has shown that the correlation between SPL and subjective annoyance is, after all, 84%, and between loudness and annoyance it is even 88%. On the contrary, whistling and other annoying tonal peaks in wind noise as well as low-frequency "booming" are not sufficiently taken into account by these quantities. Such effects frequently occur only in specific head and body postures. They are therefore detected, tracked down, and eliminated most effectively with subjective tests in the wind tunnel.

A good two-way communication with intercom systems requires a sufficiently low noise level not only at the ear but especially at the mouth. While at the ear the signal to noise ratio can be kept favorable by controlling the speakers' SPL even in a noisy helmet, this is not possible at the mouth microphone. Usually, microphones with special directional characteristics are used. Sometimes an attempt is also made to improve the quality of the signal by digital filtering. However, both approaches cannot replace a quiet helmet (Thöle [801]).

12.2.2.2 Mechanisms and Parameters of Sound Generation

Both in its formation mechanisms and in its properties, the noise in the helmet resembles wind noise of an overflown microphone and the noise at the unprotected ear in the wind.

The numerous parameters that influence the generation and transmission of sound in the helmet are listed in Figure 12.10. The subject is complicated by the fact that many of these factors interact with each other and also are highly dependent on individual circumstances. However, some important mechanisms can be separated and considered individually.

The physical principals of the generation of wind noise are discussed in detail in chapter 8. In short, broad-band flow noise is caused by turbulent velocity fluctuations near walls. These generate local areas of alternating pressure with low spatial coherence, but—for acoustical standards—with enormous intensity. After Dobrzynski [206], the effective values of these pressure fluctuations are proportional to the square of the flow velocity, doubling the speed results in an increase of 12 dB. This slope can be confirmed experimentally. The helmet curve shown in Figure 12.27 follows a square law very closely.

---

5. After Fletcher & Munson [251] and DIN 45631 [203].

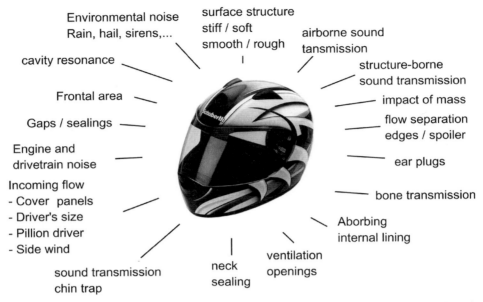

Figure 12.10 Important influence parameters on the helmet interior noise, after Heyl et al. [345].

It is well known that the turbulent velocity fluctuations are extremely high in regions of separated flow.[6] Therefore, the pressure fluctuations are also very high. The effective value of the pressure fluctuations in separated flow can be up to 10% of the dynamic pressure (Dobrzynski [206]). Particularly high rms values exist in the reattachment region (Kaltenbach & Janke [414]). Already at medium speeds of 100 km/h, this translates to a SPL close to 130 dB in these flow regions! These pressure fluctuations are sometimes called pseudosound, because the alternating pressure areas do not move with the speed of sound but approximately with the local flow velocity, and only a very small part of their power radiates as sound away from the boundary layer, as the so-called boundary layer noise. However, the helmet surface itself is hit by the full intensity of the pseudosound.[7] This sound passes as structure-borne noise through the helmet shell and in particular through openings into the inside of the helmet. The pseudosound is therefore heard only by the helmet wearer, not by his environment.

Flow noise does not only depend on the helmet geometry and structure, but especially on the individual factors of bike and rider:

- Turbulence structure of the incoming flow, in particular position and structure of the turbulent shear layer detaching from the windscreen

---

6. See, e.g., Janke [15][16].
7. Cf. Költzsch [21], according to Dobrzynski [5] a dipole source distribution is assumed in separation regions.

- Helmet fit
- Clothing (collar, neck scarf, neck brace)
- Height, seat position, and posture
- Head shape, neck thickness, and length

Even with perfectly fitting helmets, the noise exposure of different persons can vary greatly due to these factors! Wind-tunnel trials on test persons on a naked bike yielded a range of 20 dB(A) at the ear—head/helmet size left constant! The 20 dB(A) corresponds to roughly a quadrupling of the perceived noise.[8]

As an example, Figure 12.11 shows the extraordinarily large influence of the height of a windscreen on the SPL inside the helmet. The effect is well known in the literature and among motorcyclists. Heyl et al. [345] wrote: "A motorcyclist, who complains about too much helmet noise, should first check if the fairing is not a substantial cause of excessive noise." An ideal windshield position, the compromise between the ideal wind protection and low flow noise, every rider can find only for himself, as his own body proportions play a crucial role in this.

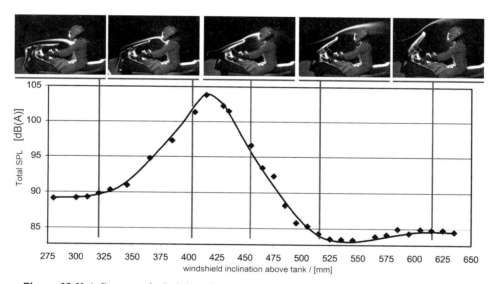

Figure 12.11 Influence of a height-adjustable windshield on the noise in the helmet at 100 km/h; BMW K 100 RT with flow visualization.

The frequency range of the flow-induced noise is associated with the size and speed of the vortex structures convected past the helmet. Heyl [344] found that low rumbling noises arise from pressure fluctuations of large-scale separation regions in the chin/

---

8. For rough estimates a level increase of 10 dB is often associated with a doubling of the perceived noise level.

neck area. Hissing noises, however, arise from small-scale turbulence inside or around the helmet.

Inside the helmet, noise is transmitted through a complex structure of sound paths. Figure 12.12 shows how wind noise and signals from the road are transmitted to the inner ear. In the acoustic optimization of a helmet, all transmission paths must be given equal consideration, since many of them work in parallel.

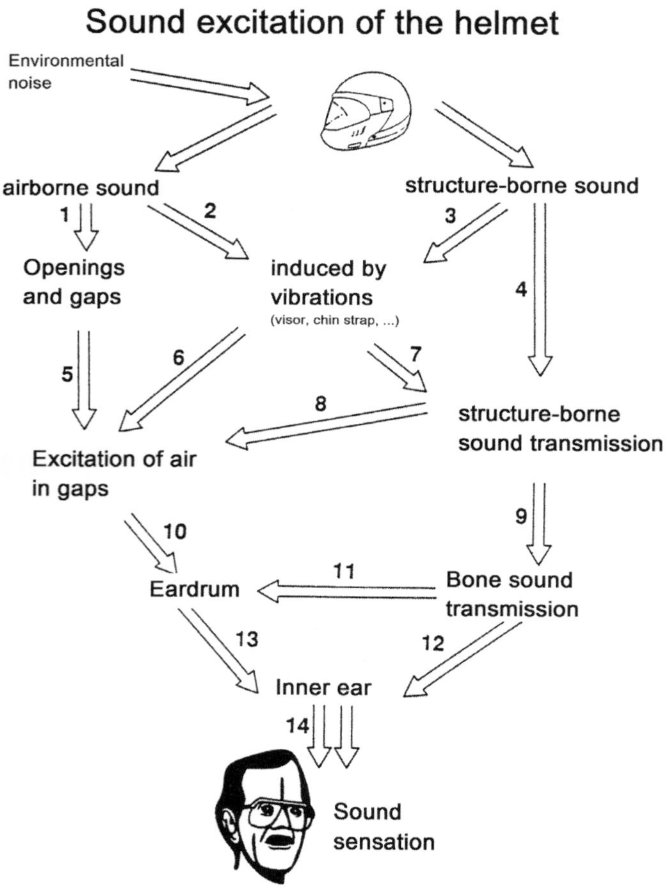

Sound excitation of the helmet

Environmental noise

airborne sound — structure-borne sound

1 — 2 — 3

Openings and gaps — induced by vibrations (visor, chin strap, ...) — 4

5 — 6 — 7 — structure-borne sound transmission

8

Excitation of air in gaps

10 — 11 — 9

Eardrum ← Bone sound transmission

13 — 12

Inner ear

14

Sound sensation

Figure 12.12 Types of noise and transmission paths, according to Heyl et al. [345].

The dominant frequencies of flow noise are well below 1000 Hz (see Figure 12.9), which relates to sound wavelengths large in comparison to the helmet diameter. Thus, inside the helmet the induced sound is not propagated like a wave.

The fit of the helmet not only determines the wearing comfort but also greatly affects the transmission paths of the sound to the ear. A helmet can only be silent if the inner liner fits tightly over the entire head area. Even small changes in the fit can result in significant changes of the overall SPL.

12.2.2.3 Perception of Environmental Signals

The helmet can have a crucial effect on the perception of the environment and change it in a way that is not intuitively comprehensible to the wearer. For example, a whistling noise caused by the flow cannot be localized by the directional hearing of the wearer. The helmet wearer is mistaken frequently if he is to specify from which point on the helmet the whistle is originating, even if he is very certain. Even extremely sensitive and experienced testers are sometimes misled by helmet phenomena. In a helmet test, an experienced Formula 1 pilot requested a tire change because he misinterpreted an aeroelastic vibration at the helm as a tire problem.

Acoustically optimized helmets immediately raise the question of whether not only the noise but also the useful signals are attenuated. Both the Highway Code as well as the standard ECE R22/05 do not provide specific guidelines on the permitted acoustic attenuation of helmets. According to §23(1) Highway Code, the driver is responsible for ensuring that his hearing is not impaired. According to the ADAC, this is often interpreted by the case law in a way that the wearing of hearing protection on the road is prohibited. The ECE-R 22.05 says in section 6.5: "The helmet must not impair the user's hearing in a dangerous manner."

In the interpretation of this requirement, however, the real purpose of §23(1) Highway Code is to be noted, namely that the driver perceives everything that is relevant for him. Therefore, one can support the view that hearing protection is prohibited in traffic only, if important acoustic information is lost. However, this is principally not the case in acoustically optimized helmets. Tests have shown that a helmet, in particular one of low noise design, reduces the noise level more than the signal level and thus improves the ambient perception. Van Moorhem [821] checked, with the help of siren noises and sounds of approaching cars, whether wearing a helmet, compared to driving without helmet, brings acoustic disadvantages. Figure 12.13 shows his most important result, namely, that on the contrary, noises are clearly more audible with helmet than without. Figure 12.13 shows that without a helmet a siren with a level of 95 dB(A) is no longer audible above 65 km/h, while when wearing a helmet, the siren can be heard up to 140 km/h.

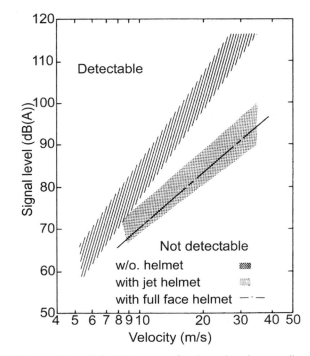

Figure 12.13 Audible and inaudible SPL range of a siren signal, according to van Moorhem [821].

Heyl et al. [345] carried out a complementary experiment. A well-sealed, quiet helmet was compared with a series helmet in relation to the perception of a police siren. Figure 12.14 shows the effect of sealing the helmet. Note that although the siren signal is attenuated, the wind noise is reduced but even more so. In the sealed helmet the siren signal clearly emerges stronger in the spectrum. This proves that the seal improves the signal/noise ratio.[9]

Krebber and Kielmann [467] note that, while stationary, the directional hearing in an acoustically optimized helmet is slightly impaired. This situation changes, however, as soon as there is movement and flow noise. Then, an acoustically optimized helmet improves the directional hearing.

---

9. These results were confirmed in Schuberth's aeroacoustic wind tunnel.

Helmets

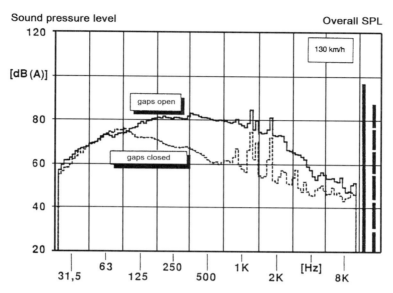

Figure 12.14 Change of siren signal by sealing the helmet, after Heyl et al. [345].

12.2.2.4 Approaches to Noise Reduction

Solid design rules or even ready blueprints for quiet helmets are not yet available. A number of approaches for noise reduction will be discussed.[10] An optimization based on aerodynamic aspects alone (smooth flow, low forces, and moments) does not necessarily lead to an audible improvement. Most important for a quiet helmet are:

- Taking acoustic optimization as a serious task during the development
- Performing intensive studies at an early stage
- Permanently monitoring the state of development in a controlled environment, like an acoustic wind tunnel
- Ongoing control of the acoustic quality in manufacturing

Though it is basically possible to control the cause of the noise, that is, to reduce the pseudosound, it is very difficult to ensure the wearing comfort in the implementation of the necessary measures.[11] Heyl et al. [345] have demonstrated the effects of individual optimization steps: sealing of gaps is very effective; flip-up helmets offer greater comfort, but due to additional openings and edges are noisier than full-face helmets in general.[12] A structure-borne noise decoupling of the outer helmet shell and the EPS-liner

---

10. Heyl [344], Heyl et al. [345], Lindemann & Hüttenbrink [503], and Kortesuo & Kaivola [459] describe such approaches.
11. See Tangorra & George [39].
12. This does not necessarily have to be the case, as demonstrated by the acoustically optimized flip-up helmets "C2" and "C3," which have a significantly lower SPL than most full-face helmets. Here, the flip-up function facilitates a good neck finish without a tight access, thereby not impairing the act of putting the helmet on.

brings only a minimal improvement in the total SPL. Windshields must be adapted to the driver. The noise load of the passenger can be even higher than that of the driver, as his helmet is exactly in the turbulent wake of the driver's helmet. Lindemann and Hüttenbrink [503] modified a helmet with a flow-separation edge raised to the ears. While this is subjectively more pleasant, it does not improve the SPL.

According to Kortesuo and Kaivola [459], an effective attenuation is achieved by the incorporation of soft earmuffs. In practice, however, there is the difficulty of positioning the capsules sufficiently well. For if the back of the head is covered with the inner liner, the touch sensitivity of the skin in the ear area is greatly reduced, so an assessment of the correct fit of the capsules is nearly impossible. Furthermore, while earmuffs can indeed reduce the noise level, they do not improve the signal-to-noise ratio, unlike a real noise optimization of a helmet.

The first patent for an inflatable helmet seal dates from 1934 (Thierry [799]). Since then, this idea is taken up again and again in patent applications. Berge [70] filed the variant of inflatable earmuffs for a patent. An adaption to a series helmet is still outstanding, among other reasons because the principle is ineffective without embedding it in an overall acoustic concept. Inflated air cushions alone are not good enough noise barriers.

Active noise reduction (ANR) using the noise cancellation principle has been a future option for a while [102]. In-ear systems with ANR exist.[13] However, a helmet-based system must reflect the fact that turbulent pressure fluctuations are of comparatively small scale. So it might take more than one microphone to capture the relevant sound event. Turbulent pressure fluctuations also have their highest level at extremely low frequencies. Thus, anti-noise of very high amplitude at very low frequencies must be generated. Current ANR headsets are not yet adequate at high speeds. Furthermore, the system must be adjusted to the position of the ears in the helmet, and for safety reasons it must not have a great overall depth.

At present, noise reduction by active control of turbulent flow, which is quite possible for simple geometries,[14] appears still far from an application at helmets. Closest to this came some promising passive-control experiments at Schuberth that employed fur patches in critical areas. They enable a "soft landing" of turbulence and thus function similarly to some microphone windshields. However, the aesthetic problem of fur at the neck role of rider's helmets prevented its use in a series helmet until now.

12.2.3 Ventilation and Rain Tests

The ventilation of motorcycle helmets has three tasks:

- Cool the head
- Supply fresh air for breathing
- Prevent visor fogging

---

13. For example, "Quietpro" Fa. Nacre.
14. See, e.g., Huppertz & Janke [13], Wengle et al. [48], and Algermissen et al. [1].

For cooling the head, a large-area flow close to the scalp is desirable, so that the heat can be carried away right from the head. However, the bulk of the heat is removed not by removal of warm air but by the much more effective process of evaporation of water. The best cooling of the head is not obtained by supplying air as cold as possible, but by supplying dry air, which can absorb the steam produced by sweating.

It is a true challenge, specific to helmets, to combine the requirements for good shock absorption with the requirements of good ventilation. A tight-fitting homogeneous shock absorption layer has to be designed, which nevertheless allows air and moisture to pass through.

To simulate the heat and moisture output of a human head, the sweating head "Alex" was developed at the Swiss Federal Laboratories for Materials Testing and Research (EMPA), Switzerland. It can be seen in Figure 12.15. The white cotton pads cover the "sweat pores" and distribute moisture released from a bore into a defined area of the head. The removal of moisture and heat of the head to the environment can be controlled and measured accurately. This is possible with and without wind.

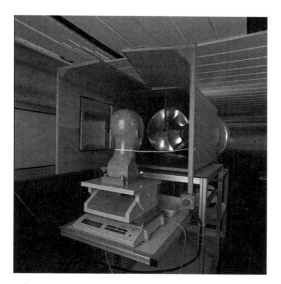
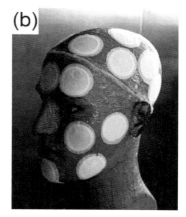

Figure 12.15 Sweating head "Alex" in the measurement position in the wind tunnel, EMPA [104].

Brühwiler [104] conducted an experimental comparison of two helmets on the sweating head. He found that with closed ventilation in both helmets, the same quantity of moisture was removed, while with open ventilation the quantities differed significantly. Thöle [802] writes about the head ventilation in a subjective helmet test: "The effect of the many knobs, sliders or scoops is limited, since the channels often end up in the inside padding. The air cannot be transported any further." A similar observation was made by Brühwiler [104] in sweating-head tests of bicycle helmets.

With the market availability of very small, inexpensive humidity sensors, a further access to the measurement of the microclimate on subjects experiments has been opened. A good combination of subjective and objective evaluation of a head ventilation's quality results from the inclusion of time histories of the temperature and the humidity inside a helmet during wear. Ideally, the measurements are taken on the entire head in order to identify ranges of varying heat and moisture removal.

Figure 12.16 shows small digitally addressable sensors that can be easily connected via a USB-like protocol to form a complete integrated monitoring network. The measuring surface for humidity and temperature is only about 2 mm². Figure 12.17 shows a view of the sensor network, as well as the representation of a set of measured values in a helmet. The lines in the upper graph show the history of the entire experiment. The two lower graphs show the interpolation of the moisture and temperature measurements at the sensors.

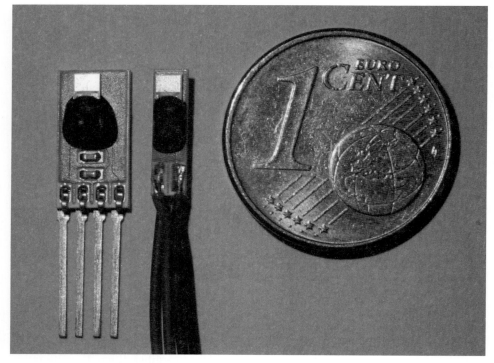

Figure 12.16 Sensors for temperature and humidity.

Helmets

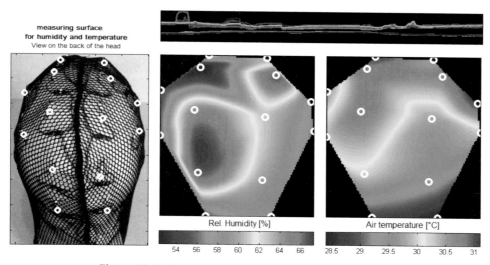

Figure 12.17 Microclimate measurements in the helmet.

Figure 12.18 shows the ventilation concept of the Schuberth S1, where great importance has been placed on damping properties, ventilation channels, and comfort padding functioning perfectly together. Through the chin vent, fresh air blows into the helmet interior and ventilates the visor. In the closed state of this flap, a minimum ventilation stays active. Through the head ventilation inlet, a ventilation of the head above the forehead can be activated. Low pressure in the region of air outlets ensures constant removal of exhaust air.

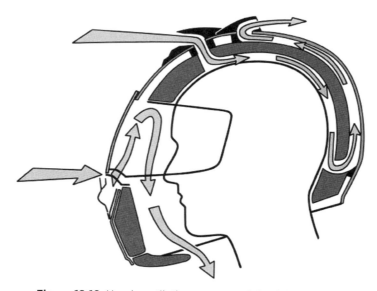

Figure 12.18 Head ventilation concept of the Schuberth S1.

The task of the visor ventilation is to avoid fogging in rain and cool weather. Here however, no unpleasant draft at the eyes should occur, not even at high speed. Wearers of contact lenses are particularly sensitive to air draft. High-quality visors are usually equipped with an anti-fogging coating that can absorb a certain amount of condensed water before it comes to the formation of drops. Often, breath deflectors are offered as accessories to keep the exhaled humid air away from the visor.

The function of ventilation when riding in the rain is investigated through driving tests and experiments in rain testing facilities. Here also leaks of ventilation ducts, seals, and so forth are detected and the vision quality through the visor can be assessed. In rain, properly functioning visor seals and close-fitting air vents are important while driving and while waiting at a light. Moreover, in the ideal case, the outer fabrics of the helmet are water-repellent and only absorb small quantities of moisture.

Benchmark tests of helmets are carried out by independent test institutes, such as the German automobile club ADAC, or motorcycle magazines. The rain test at the ADAC takes a total of 15 minutes. At city driving speeds and a water volume of 8 l/min, the helmet is sprayed at from the front in a lightly inclined position. With blotting paper, which is clamped to the test heads, a water leak can be detected inside the helmet. For the evaluation of water tightness, the following aspects are important:

- Fogging of the visor inside and tightness of the visor seal
- Wet spots on the blotting paper on the face and on the head
- Formation of drops on the helmet padding
- Tightness of the ventilation openings

Figure 12.19 shows a visor after a test in the ADAC rain facility [596]. Clearly visible is the extremely strong fogging. For good vision in rain it is also important how water drops roll off the visor. Figure 12.20 shows quality differences between wetted visors under the same condition at a free stream velocity of 100 km/h.

helmet without fogging
(Schuberth rain test)

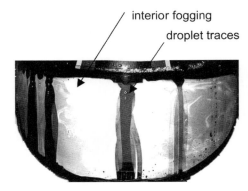

Helmet with fogging after 15 minutes of rain at weak winds (ADAC rain test < 50 km/h)

Figure 12.19 Rain tests at ADAC and Schuberth.

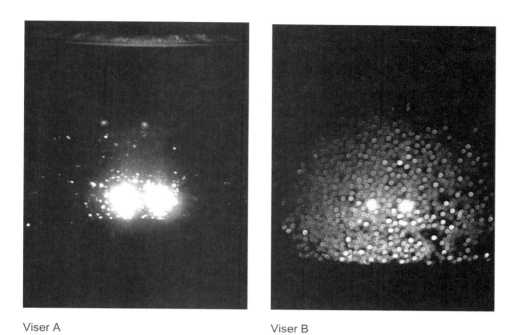

Viser A　　　　　　　　　　Viser B

Figure 12.20 View through two visors with rain wetting; wind speed was 100 km/h; simulation of an oncoming vehicle in the dark and rain.

12.3 Helmets for Open Race Cars

12.3.1 History

As early as the 1920s, solid caps were worn in automobile racing. At the famous Indianapolis 500, helmets have been mandatory since the 1940s. In 1952, the CSI, the former sports department of the FIA, passed a helmet law, but without precise design guidelines. In the 1960s, the United States saw first controls of helmet quality: an official applied a hammer blow to a helmet.[15] Today helmets are mandatory, required to have passed all standard tests prescribed in the Snell or BSI.

In addition to the advancement of crash and fire protection capabilities, significant enhancements are taking place in the areas of weight, aerodynamics, air conditioning, acoustics, and driver information systems. To augment safety, in the 2003 season the FIA introduced the "head and neck support" (HANS). It is mandatory for all teams in the Formula 1.

12.3.2 Aerodynamics and Ventilation

Even more than on motorcycles, aerodynamic interactions occur between the helmet and its environment in open race cars. These interactions have been increasingly investigated in recent years. Key to the success of an aerodynamic optimization here is a close cooperation between riders, team, and helmet manufacturers.

---

15. See Schlang [36].

Figure 12.21 gives an impression of the flow environment of a helmet in a Formula 1 race car. The helmet is half immersed in the cockpit cavity and is hit in the forehead region by a shear layer, which separates from the windshield.

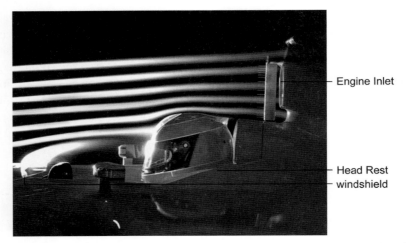

Figure 12.21 Flow visualization around helmet and cockpit and the inflow to the airbox; Formula 1 racing car with Schuberth helmet "QF1."

In some cars a strong downward wall jet over visor and chin section is formed below the stagnation point on the helmet. The wall jet can be used to control the flow with chin spoilers. In other cars, the area below the stagnation point is at rest, so that it is suitable neither for an air intake beneath the visor nor for a chin spoiler. Figure 12.22 shows a helmet with a top-down ventilation, which was developed specifically for this case. Here, both the air for cooling the head and the breathing air enter the helmet through an air inlet at the forehead.

Figure 12.22 Formula 1 helmet with top-down ventilation. Graphics: Schuberth.

The airbox and the headrest located in the direction of travel behind the helmet produce a stagnation-point flow above the helmet. Thus, the area with positive pressure gradient in the upper helmet area moves upstream in comparison to a free flow around the helmet. And accordingly, the separating tendency of the boundary layer increases in this area. The region of separated flow can be surprisingly large; it is highly undesirable, as it leads to a reduced total pressure at the intake of the engine and to reduced engine performance. By diligent tuning of the geometry of windscreen, helmet, and headrest, the total pressure can be optimized in the plane of the intake opening. Figure 12.23 shows examples of beneficial (left) and adverse total-pressure distributions in a cross section about 3 mm in front of the intake opening. The figure shows lines of constant total pressure in relation to the total pressure of undisturbed flow. The circles indicate measurement points of the pitot rakes used. Above the line $c_{p,tot} = 1$, the flow is completely lossless. In the left image, the air box receives the flow with the full total pressure.

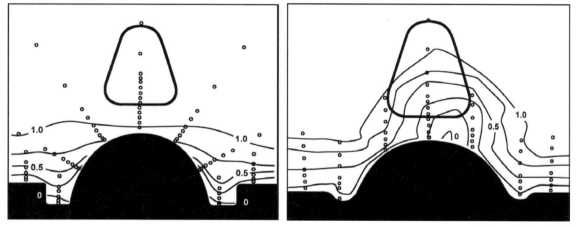

Figure 12.23 Comparison of total pressure distributions ($c_{p,tot}$) of two Formula 1 racing cars, 30 mm upstream of the engine inlets.

Due to its location in a stagnation-point flow, the helmet experiences almost no flow resistance. The only effective force is the lift, and it increases the more the helmet protrudes from the cockpit.

For many drivers, the most unpleasant effect is shaking of the helmet, also referred to as "buffeting," which originates from the low-frequency instabilities in the large-scale separated cavity flow.

12.3.3 Acoustics

The noise level in the helmet of a Formula 1 driver lies with 130 dB (A) near the pain level. Therefore, ear protection is an absolute necessity for drivers. Figure 12.24 shows on-track measurements of the noise levels at the mouth microphone and in the ear under the earplugs. At the same time, speed and engine revs were recorded. The figure shows a recording time span of approximately 2.5 rounds. In the spectrogram (Figure 12.24a), high-energy frequency components appear in red and low-energy frequency components appear in green. And obviously, in contrast to road bikes, the dominant frequency components are induced by the engine. In the spectrogram, the good correlation between the frequency peaks with the engine speed (Figure 12.24b) is visible. Distinct low-frequency components—"rumbling"—occur at speeds of 200 km/h, as can be gathered from a comparison of Figure 12.24a and c. In Figure 12.24d and e it is clear that the measured value of the total insertion loss of ear plugs is highly dependent on the weighting of the spectra.

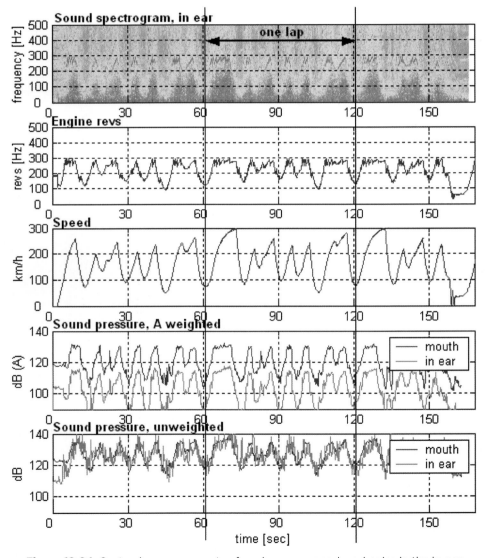

Figure 12.24 On-track measurements of engine revs, speed, and noise in the in-ear monitor and at the mouth of the driver in a Formula-1 car.

Helmets

Figure 12.25 shows the prototype of an acoustically optimized Formula-1 helmet with fan ventilation. The high sound insulation effect is realized by a multilayer sealing system and the acoustic attenuation of the incoming and the exhaust air. The large-area ring-shaped cooling of the entire back of the head ensures good air circulation and ventilation. The driver is provided with filtered and muffled breathing air by a respiratory mask.

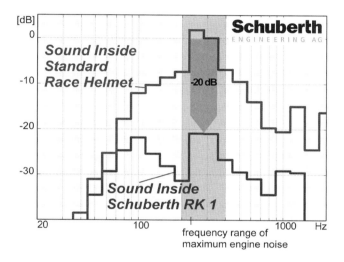

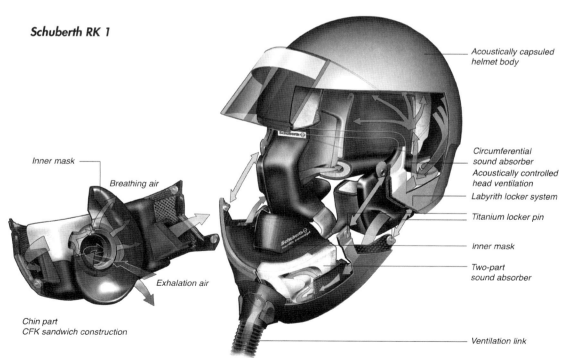

Figure 12.25 Prototype of a noise-optimized racing helmet with fan ventilation: laboratory measurement of insertion loss.

12.4 Measurement and Simulation Technology

12.4.1 Introduction

In the following section, the most important, specially designed tools for the development of helmets are presented. General methods for the detection and visualization of the pressure, velocity, and wall friction fields (see Janke [405], [404]) are described in detail in section 13. Further calculations are presented that can be useful for the development even before prototypes have been made and wind tunnel tests can be done. The methods used for this are presented in section 14.

12.4.2 Wind Tunnel

In the Schuberth wind tunnel (see Figure 12.26), helmets have been systematically investigated and optimized under reproducible conditions since 1986. The easily accessible open test section with a cross section of 1m × 1m makes it a very handy development tool to be used and maintained efficiently. Here, the flow around the helmet and its near field (i.e., upper body, arm, fittings, and front screen) can be reproduced realistically.

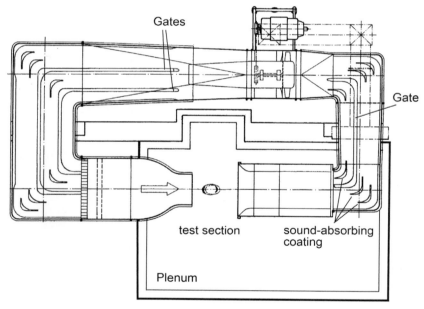

Figure 12.26 Aero-acoustic wind tunnel of Schuberth GmbH; measurement cross section 1m × 1m, maximum speed 230 km/h (short test section); drive power 120 kW.

With the conversion to an acoustic wind tunnel in 1999, a worldwide unique opportunity for the helmet development has been created. It allows us to investigate the helmet acoustics systematically and to develop and to implement new acoustic optimization methods.

Figure 12.27 (Janke et al. [406]) shows measurements of the background noise ("Out-of-Flow") before and after the conversion compared to automotive wind tunnels. A reduction of nearly 40 dB in the A-weighted sound-pressure level was achieved. As can be seen in Figure 12.27, the current background-noise level of 55 dB(A) at 100 km/h lies almost 40 dB below the interior noise of normal market helmets. For comparison: referring to Mercker and Pengel [564], an out-of-flow signal-to-noise ratio of at least 10 dB is required for aeroacoustic measurements and developments.

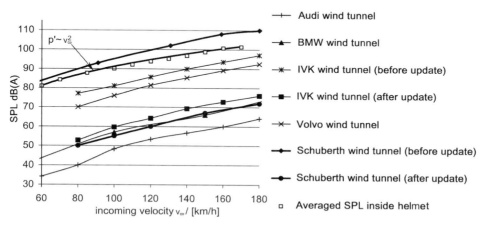

Figure 12.27 Out-of-flow sound pressure level of the Schuberth GmbH aeroacoustic wind tunnel in international comparison, and the noise level of an average volume helmet.

12.4.3 Aerodynamic Forces

For the measurement of forces affecting the helmet and the head of the driver, four methods have been used in the past 20 years at Schuberth. All methods have their place and complement each other:

- Subjective tests in the wind tunnel and on the road (see Figure 12.28): No measurement data are taken. Only perceptions are supplied that need to be compared and evaluated after considerable time delay. However, these perceptions cover the entire phenomenological experience if sufficient tests with many different test persons on different motorcycles are performed.

- Underfloor balance: The measurement data (six components) are free from external influences. Due to the comparatively little aerodynamic interference, they provide a well reproducible state. Because of the lack of interferences with the rider's upper body, the relevance of the measured values has to be regarded as rather low. This method is therefore currently not used.

- Data helmet (see Figure 12.29): Using a force transducer built into a helmet, six components (forces and moments) are recorded while the helmet is worn by a test person. So subjective perceptions can be supported with data.

- Measuring robot with computer-controlled movable head and built-in balance (see Figure 12.30): A built-in robot force transducer can record six components (forces and moments). The measuring robot has movable, lockable limbs and can be set onto different motorcycles. The movements of the head are accurately controlled by stepper motors. Different helmets or modifications of a helmet can be quickly and easily compared. Figure 12.4 shows typical results of this measurement technique. The temporal resolution of the sensor also allows the detection of unsteady forces.

Figure 12.28 Subjective tests in the aeroacoustic wind tunnels at BMW Technik GmbH and at Schuberth GmbH.

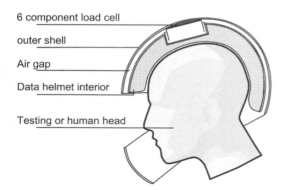

Figure 12.29 Data helmet at Schuberth GmbH; combination of force measurement and subjective perception.

Figure 12.30 Measuring robot at Schuberth GmbH.

12.4.4 Aeroacoustic and Artificial-Head-Measurement Technology

The most important tools are the following:

- Subjective tests in the wind tunnel and on the road: These are particularly suitable for the detection of whistling and acoustic phenomena in specific sitting postures.

- Artificial head measurement: Important tool for the systematic development, as it allows quick objective measurements in each step of the gradual fine-tuning of the helmet. The influence of different head shapes (with the same head circumference) on the helmet noise however is not examined.

- Ear-microphone measurements in the wind tunnel and on the road: This technique, first described by Maue [537], is well suited for proband tests in an advanced stage of the helmet development. The success of consecutive optimization steps can be validated objectively to fine-tune the interior lining of a helmet (see Figure 12.31).

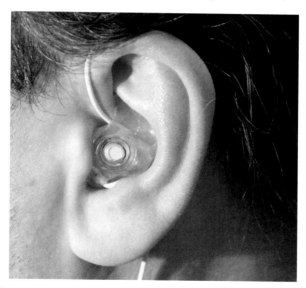

Figure 12.31 Otoplastic with miniature microphone in the ear.

12.4.5 Computational Fluid Dynamics (CFD)

Basically, it is possible to examine a helmet design toward its fluid mechanical properties before real tests with wind tunnel model are performed ([263],[442]). But concerning the significance of global variables as the drag force caution is advised: A comparison of calculation and experiment was carried out in a cooperation of Schuberth with the Flow Institute Karlsruhe. Despite the extremely careful modeling of the boundary conditions, a good match of measured and calculated wind could be achieved only for simple configurations.

Figure 12.32 shows the simulation zone: a simplified geometry of the plenum with nozzle and collector. In the first stage, a ball was placed in the measuring section. To calibrate the computational model to the reality of the wind tunnel, various boundary conditions have been adapted. The velocity distribution in the nozzle was measured and various turbulence models were tested. Finally the "best" turbulence model resulted in differences of less than 1% in the total force and less than 4% in the force angle. Also, the flow phenomenology was well recreated.

Figure 12.32 Data model of the wind tunnel test section.

In the next stage, the entire measuring robot was digitized, including helmet and motorcycle. The 3D scan was fit into the simulation model. The results are summarized in Figure 12.33. On the left side, the ball results are compared; on the right side, the helmet results are compared. The forces determined are each below the images. While the ball experiment yielded an excellent agreement of measurement and calculation, the helmet experiment resulted in deviations of the mean resistance of almost 50%.

Despite intensive research, the mismatch could not be explained and not reduced. Presumably it is due to sensitive, difficult-to-model interactions between the three-dimensional shear layers, separation areas, and stagnation point flows on the windshield, on the helmet, on the shoulders, and on the edges of the jet.

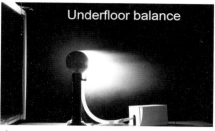
a) $F_D = 14{,}77$ N, $F_L = 10{,}35$ N, $F = 18{,}04$ N

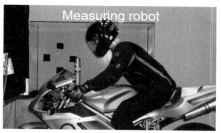
c) $F_D = 18{,}77$ N, $F_L = 18{,}93$ N, $F = 26{,}66$ N

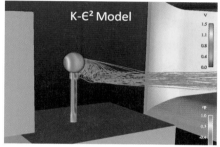
b) $F_D = 14{,}18$ N, $F_L = 11{,}38$ N, $F = 18{,}18$ N

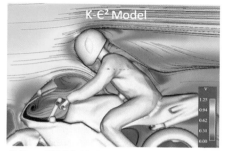
d) $F_D = 27{,}96$ N, $F_L = 24{,}08$ N, $F = 36{,}9$ N

Figure 12.33 Comparison of simulation and measurement.

Phenomena, however, could be compared very well, as shown in Figure 12.34. Here wall stream lines from experiment and calculation are superposed. Also, the modeled fluctuation terms provide an information of turbulent sound sources (Figure 12.35). This figure clearly shows the importance of the helmet bottom for the flow noise.

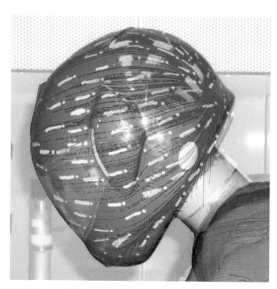

Figure 12.34 Wall stream lines from simulation (red) and measurement (white).

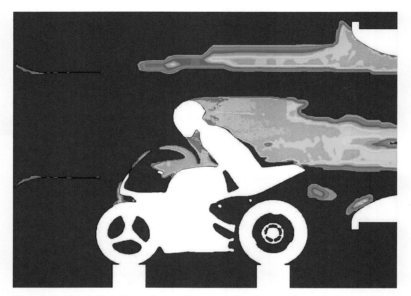

Figure 12.35 Estimation of sound source distribution from k-ε^2 modeling.

Coverage of all acoustically relevant alternating pressures requires the integration of the full, time-dependent Navier-Stokes equations down into the smallest turbulence structures, a so-called direct numerical simulation (DNS). This can be carried out for small Reynolds numbers and basic geometry.[16] In the helmet development, however, the DNS will not become a routine tool in the near future. Further details follow in chapter 14.

---

16. See Schlang [36].

Chapter 13
Wind Tunnels and Measurement Technique

Reinhard Blumrich, Edzard Mercker, Armin Michelbach, Jorg-Dieter Vagt, Nils Widdecke, Jochen Wiedemann

13.1 Scope of Wind Tunnels

From the preceding chapters of this book, it has become clear that aerodynamics is an integral part of the modern vehicle development process. Both the fulfillment of specification requirements in relation to the overall aerodynamic forces and moments, which significantly affect CO_2 emissions and driving stability for example, as well as functionality of other important components. Engine and brake cooling, air conditioning, door seals, spoilers and other attachments all require the participation of vehicle aerodynamics. The most important test facilities for this purpose are wind tunnels.

Wind tunnels simulate the on-road conditions by experimental rather than numerical means, as computational fluid dynamics (CFD) and computational aeroacoustics (CAA). The on-road conditions are always the measure of all things since the customer uses his vehicle on the road and not on test benches. For aerodynamic development work, testing on the road is not very suitable, partly due to the difficulty to create reproducible experimental conditions. As will be discussed in this chapter, wind tunnels do not have such drawbacks since they are stationary test benches. However, as with any simulation, there are limits of applicability. This requires a profound understanding of the physics of wind tunnels, which will be discussed in detail in this chapter.

Chapter 13

The availability of wind tunnels as a standard tool and their scientific understanding were not given in the early days of the automobile. The reasons for this can be found not only in the complexity of the aerodynamics and the Navier-Stokes equations but also in the fact that the experimental tools had to be developed from other areas of technology.

In fact, at an early stage wind tunnels and other fluid dynamics test benches have been used in aerospace research and introduced in other areas of mechanical engineering. However, it became evident that their applicability to automobiles was limited. The first attempts to investigate vehicle aerodynamics in wind tunnels took place in aviation wind tunnels that were retrofitted with a solid ground plane that represented the road. An example is given in Figure 13.1, which shows an Auto Union racing car. It was measured in 1934 in the wind tunnel of the German Experimental Institute for Aviation (DVL) in Adlershof, Berlin. The image reveals a bundle of metrological shortcomings that would not be acceptable today. This chapter will also deal with the prevention and correction of such shortcomings.

Figure 13.1 An Auto Union racing car in the wind tunnel of DVL in Adlershof, Berlin, 1934 [432].

On the most left side of Figure 13.1, a circular nozzle can be seen in front of the ground plane on which the vehicle is fixed. The airflow from the nozzle is divided in separate airflows above and below the ground plane. This complicates the calibration of the tunnel to determine the correct flow speed "seen" by the vehicle. Additionally, the ground plane is stationary, that is, not moving with the same velocity relative to the car as the airflow in the real world. This was mostly accepted in research and industry up to the turn of the Millennium.

Today, the so-called simulation of the road, combining a moving road with rotating wheels, is state of the art in modern automotive wind tunnels. Only in this way the aerodynamic efficiency potentials, such as fuel consumption and CO_2 emissions reductions for conventional vehicles or increase of the range of battery-electric vehicles, can be fully exploited. The same holds true for race and sports cars where the "efficiency" is usually expressed by the ratio of negative lift (i.e., downforce) to vehicle drag.

Figure 13.1 shows that the test object occupies the majority of the open test section, which is the space between the nozzle (left) and the collector (right). The rear of the vehicle is already at the collector inlet, where the ground plane also ends abruptly. Today we know (and this chapter will elaborate on this) that measurement errors can be induced in such situations and need to be corrected. Hence, it is not a surprise that such (or similar) early wind tunnel tests generated as many questions as answers. In this way uncertainty about the value of the results arose that was intensified by additional parameters that occur on the road and need to be considered in the development process. For example:

- Atmospheric wind effects (e.g., crosswinds)
- Rain, snow, and soiling
- Unsteady wind effects (atmospheric turbulence wakes of preceding vehicles) as well as stationary and nonstationary vehicle movements due to wind forces
- Ventilation effects of rotating wheels
- Temperature and humidity exposure, sunlight radiation, and comfort of the interior
- Wind noise from the vehicle exterior and its transfer into the vehicle interior
- Aero-elastic deformation of the entire vehicle or individual components
- Inclusion of longitudinal and lateral vehicle dynamics on roller or flat belt test facility

The list could be continued, which makes clear that even highly elaborated facilities can not cover all requirements in a meaningful and economical way.

Therefore today, besides smaller blowing tunnels for the investigation of vehicle components and assemblies, there are two basic types of wind tunnels in general. The first type are aeroacoustic wind tunnels with jet nozzles and a cross section of 10 m^2 to 30 m^2 to examine vehicle aerodynamics (pressure, forces, moments, deformation, flow rates etc.) and the wind noise (sound pressure and sound power levels inside and outside of the car, sound source localization, psycho-acoustical noise evaluation, acoustic transfer path analysis, body insulation, etc.).

The second type of tunnels are climatic and thermal wind tunnels. These mostly have a nozzle with a cross section between 3 m² and 12 m² and do not have the same requirements on the quality of the airflow. Instead, they are able to simulate various climatic conditions such as,

- Solar radiation
- Vehicle's foreign and self soiling
- Rain and snow
- Vehicle engine loads to measure the cooling and the heating power using rollers or flat belt test benches

Additionally, legal or customized vehicle consumption cycles can be replicated with some of these systems.

Due to significant progress made in computer technology, hardware as well as software, the CFD earned a permanent place in the product development process in the last 10 to 15 years and is therefore indispensable. Also in the computational aeroacoustics (CAA), progress has been made in the meantime. At the same time, automotive wind tunnels, more than ever, have been newly planned, commissioned, or modernized. This suggests that wind tunnels will not be dispensable in the foreseeable future. Today, both wind tunnel and CFD have their place in the development process and complement each other in a differentiation that enables the engineer to use the respective best-suited tool for a specific question. This requires an education that enables young engineers to assess the limits and possibilities of the tools available and to use them optimally.

13.2 Wind Tunnel Physics

13.2.1 Design and Function of Wind Tunnels

A wind tunnel is a technical device that generates a constant, uniform, and low-turbulence airstream that flows around a scale model or full-size vehicle in a test section. All other components of a wind tunnel only serve to provide this airflow in order to simulate the on-road conditions as best as possible. Depending on the task, it is also possible to create time-variable flow conditions. This is achieved by additional technical measures inside and outside the test section or with a variable fan control. This will be discussed next. In this chapter, wind tunnels will be discussed, which are common practice and in use for the aerodynamic development to the automotive industry. Due to their different tasks, compared to the aircraft industry, for example, they may differ significantly from a technical point of view.

Automotive wind tunnels where full-scale vehicles are measured now reach up to 300 km/h, which corresponds to a Mach number of $Ma = 0.25$. Therefore, these wind tunnels are referred to subsonic wind tunnels. However, the air density at such speeds

Wind Tunnels and Measurement Technique

is 3% different to the density at rest and thus the compressibility of the air must be taken into account when calculating aerodynamic coefficients. In particular this is important due to the fact that the local velocity around the vehicle bodywork can be significantly higher.

As illustrated in Figure 13.2, more energy is required to push a stream of air over a fixed body rather than move the same body through stationary air. However, it is always easier to carry out measurements on a body at rest. This is generally the principle on which automotive wind tunnels operate.

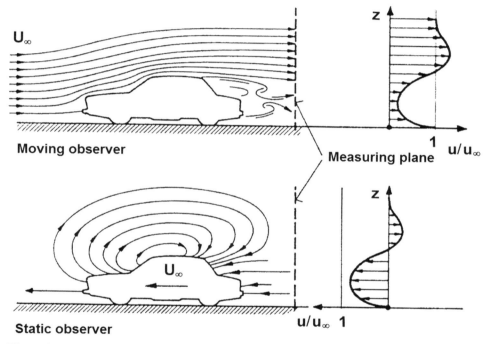

Figure 13.2 Reversing the sequence of movements with the transfer of road travel (lower figure) into wind tunnel testing (upper figure).

By the type of air return, two species of wind tunnels can be distinguished. The "Göttingen" type, first developed by Prandtl, and shown in the upper picture of Figure 13.3, circulates the air by a fan in a closed return circuit. In contrast, the Eiffel type, first built by Eiffel, and shown in the lower picture of Figure 13.3, circulates the air by a fan in a closed return circuit. After the air is passed through the test section it is discharged back out into the open atmosphere again.

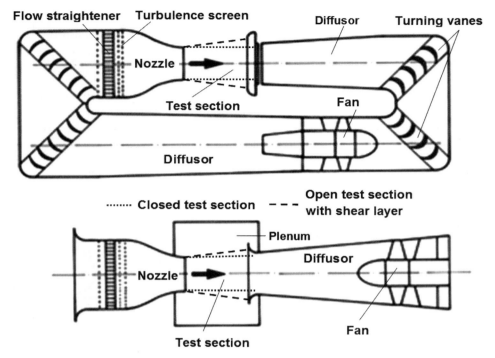

Figure 13.3 Different wind tunnel designs: upper picture, Göttinger type; lower picture, Eiffel type.

In a Göttinger type wind tunnel, the airflow will leave the test section via a tunnel section, consisting of various diffusers and four cascade turning vanes at the corners, and then be returned to the test section by a blower and a nozzle. Usually wind tunnels have an opening to the surrounding environment to make sure that the pressure remains static in the test section when the air temperature rises during a test for example. This so called "breather" is located at the end of a closed test section, or in the plenum hall for open test sections. The static pressure in the settling chamber in front of the nozzle reaches almost the same level of the overall dynamic pressure depending on the contraction ratio of the nozzle. If the airstream blew straight into the free atmosphere after it passed the test section, and no effort was made to recycle the kinetic energy (as pressure energy), the power requirement of the fan would equal slightly more than the energy stream $1/2\ \rho u^3 A$ in the test section. However, if diffusers are installed between the test section and after the fan, the speed can be reduced with minimal total pressure loss. The effect of losses within the whole wind tunnel is usually described as the ratio between the wind tunnels' input factor (fan power) and the kinetic energy flow in the test section, which results in a performance factor. Some subsonic wind tunnels reach performance factors of 0.2.

The optimal length of the diffusers, contraction ratio of nozzle, and the length of the test section lead to a minimum length of the wind tunnel facility. Depending on the space available, the wind tunnel can be placed horizontally or vertically. Vertical wind tunnels generally require more construction expenditures, and they are more expensive to build.

The biggest problem with Gottingen type wind tunnels is that circulated air stream undergoes various disturbances throughout the return circuit and adversely affects the flow quality entering the test section. To counteract these effects, various aerodynamic measures are used within the circuit. The essential element is one or several turbulence screens to reduce the flow turbulence, flow straightener to compensate the spatially different flow structures and turning vanes to reduce flow separation at the corners.

When referring to an Eiffel wind tunnel or a wind tunnel with an open return, instead of using a recirculating circuit, the entire flow is discharged at the end of the diffuser into the atmosphere and "fresh air" is drawn in via the inlet nozzle. Eiffel wind tunnels can in theory be more energy efficient than Göttingen wind tunnels. This is the case if the energy discharged from the diffuser to the environment is less than that of the friction and deflection losses that occur in a Göttingen closed return tunnel. However, this is rarely possible in practice.

Frequently, Eiffel tunnels are placed in closed buildings. This has been proven to be advantageous when the surrounding hall is designed to be as streamlined as possible. The air is then guided from the exhaust diffuser to the inlet nozzle with lower losses. However, it often proves difficult to achieve uniform inlet flow conditions over the entire nozzle entrance, which affects flow quality in the test section. While it is possible to improve the flow uniformity with the nozzle contraction to some extent, to be truly effective, the contraction ratio must be large. Furthermore, for test sections with large cross sectional area and thus with a large cross sectional area at the entrance of the first diffusor at the end of an open test section it is often not possible to accommodate the appropriate diffuser length in the building, which significantly increases the power requirements of the fan.

A pure Eiffel tunnel, those set up outdoors with no ducted recirculation loop, loses some of the advantages by the effort required to make its operation weather independent. A mesh in front of the inlet can prevent solid objects from being ingested. The influence of the wind in the atmosphere on the flow in the test section can be reduced with a large volume settling chamber between the inlet mesh and the inlet of the nozzle. Another disadvantage of the free-standing Eiffel tunnel is the noise pollution it creates. Also, the advantage of being able to exhaust gases from a vehicle engine is only possible when the tunnel is in uninhabited areas.

Section 13.5 lists examples of both types of tunnels, with Eiffel tunnels typically located inside a building. The advantages and disadvantages of the designs can be weighed against quantitative criteria only in a specific application. As a general rule, open return tunnels with optimized diffusers have low operating costs and relatively low construction costs.

The costs for the fan are greater in wind tunnels with a closed loop. Also, the costs for the tunnel construction and the additional fittings for flow conditioning as well as operating costs are higher compared to a pure Eiffel tunnel. Care must be taken to control the air temperature since it is possible for models made from plasticine to lose their strength at higher air temperatures. Therefore, a heat exchanger is often required for close return tunnels, which is expensive additional equipment. The additional pressure loss requires increased power from the fan. Alternatively, to a heat exchanger as part of the circuit, the air may be cooled during testing pauses with a bypass system. For climatic wind tunnels, only the Gottingen design is considered.

13.2.2 Wind Tunnel Nozzle

The wind tunnel nozzle has four tasks:

- Accelerate the flow
- Improve the flow uniformity at the outlet cross section
- Reduce the longitudinal turbulence in the flow
- Measure the wind velocity through a calibration routine

The demands for a uniform velocity profile and for low turbulence have not been quantified rational for automotive wind tunnels until today. However, due to the fact that it is not clear from the outset, whether the sensitivity of vehicles or components with respect to the degree of turbulence and non-uniformity of flow will need to be investigated, it should always be sought after lifelike test situations in accordance with the driving on the road in still air. The natural winds during road driving may increase the number of flow parameters rapidly, so it is not possible to simulate all these conditions in a wind tunnel at the same time. Nevertheless, to investigate the behavior of a vehicle in windy environment requires additional measures to create wind gusts for instance (see section 13.2.12).

With the careful design of a wind tunnel with a closed or open return circuit, the following values can be obtained:

- Local deviation in speed from the mean value, $\Delta u / u_\infty = (u - u_\infty)/u_\infty \leq 0.1\%$
- Angular deviation from the tunnel axis, angle of attack, and yaw angle, $\alpha, \beta \leq \pm 0.1°$
- Homogeneous, isotropic turbulence, $Tu = \sqrt{\overline{u'^2}}/u_\infty \leq 0.2\%$

Three geometric properties of the nozzle significantly affect these specifications. These are the contraction ratio, the contour of the walls, and the development of the shape of the cross section from the nozzle entrance to the test section.

The contraction ratio κ is defined as the ratio of the cross-sectional areas of nozzle outlet (A_N) to the nozzle inlet (A_S): $\kappa = A_N/A_S < 1$. Generally speaking, the greater the contraction ratio, the more evenly distributed the velocity over the exit cross section will be. Also a lower turbulence level is achieved in the wind tunnel flow field. As shown by

Bradshaw and Pankhurst [93], through the contraction, an axial velocity difference Δu_x for each streamline is reduced with the same ratio as the average velocity is increased.

In relation to the nozzle velocity u_∞ we obtain

$$\frac{\Delta u_x}{u_\infty} = \frac{1}{\kappa^2} . \tag{13.1}$$

The differences of the lateral velocity components Δu_y results in

$$\frac{\Delta u_y}{u_\infty} = \frac{1}{\sqrt{\kappa}} . \tag{13.2}$$

The axial component of the degree of turbulence results in

$$Tu_x = \frac{1}{2 \cdot \kappa^2} \cdot \sqrt{3 \cdot \left(\log\left(4\kappa^3\right) - 1\right)} \tag{13.3}$$

and accordingly, for the lateral component of the degree of turbulence we obtain

$$Tu_y = \frac{1}{2 \cdot \kappa} \cdot \sqrt{3 \cdot \kappa} . \tag{13.4}$$

The contraction ratio κ essentially determines the main dimensions of the wind tunnel and thus the level of investment that is necessary. Therefore, a greater value of κ should not be chosen than necessarily required for the quality of flow. In order to achieve a good flow quality, newer closed return wind tunnels use a contraction ratio of approximately $\kappa = 6$ in general.

For Eiffel type wind tunnels where the flow is not disturbed by the wake of the fan, diffuser, and turning vanes, a contraction ratio of $\kappa = 2$ or 3 is sufficient. However, this is only valid if the air is drawn from very large volume.

To define the contour of the nozzle, a monotonically increasing curve of the velocity distribution from the longitudinal centerline of a frictionless circular nozzle is used. With the constraint of a uniform distribution at the beginning and at the end of the nozzle, the deviation of the streamlines away from the centerline can be calculated. Each streamline is, in effect, the contour of the nozzle. All non-axial streamlines then have negative velocity gradients, with their strength increasing away from the centerline (cf. Figure 13.4).

A compromise between the length of the nozzle and velocity gradient must be found because in a real nozzle, the wall boundary layers tend to detach if the gradient is too large. This would then in turn lead to an uneven velocity distribution at the nozzle exit area. So a subsequent boundary layer calculation must be performed to iteratively come to a final nozzle contour.

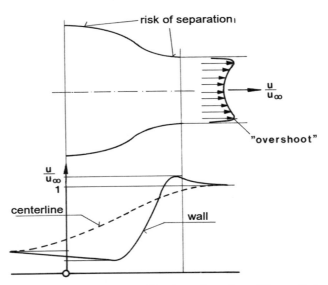

Figure 13.4 Schematic velocity profile along the axis and the wall of a nozzle.

The calculation of the nozzle contour is further complicated if the nozzle is not axially symmetric or if the ratio of the height to width of the nozzle changes along the nozzle length. For most automotive wind tunnels, this is due to structural constraints. Since there are no analytical methods to convert axially symmetric nozzle to a non-axial symmetric nozzle, they have often been the subject of pilot wind tunnel studies using experimental methods. It may then be possible with help of so-called corner fillers to prevent an increased thickness of the wall boundary layer and prevent secondary flow from wall to wall. However, this option is often dispensed with, and longitudinal corner vortices at the outlet of the nozzle are present.

The individual calculation methods described in the literature ([87], [891], [581]) differ in the means by which the contour is represented and by what method the flow is calculated. In all nozzles, the inflection point is placed at the center of the nozzle. The design of the concave section is designed to avoid detachment of the flow and the shape of the rear convex section is designed to produce the desired flow uniformity at the nozzle exit. It has been shown that the flows before and after the inflection point have a minor interaction. This is exploited when a tunnel requires nozzles of varying outlet cross sections, particularly in climatic wind tunnels. In this case, the nozzle is split at the inflection point of the nozzle contour and the downstream section can be replaced. Also, there is the possibility to use hydraulic actuators to modify the contraction with a nozzle made from steel sheet. In this way the nozzle shape can be reduced in a few minutes, such as is done at the wind tunnel at the BMW Group, where the nozzle can be reduced from 25 m² to 18 m². This increases the maximum speed from 250 km/h to 300 km/h.

The calculation for a three-dimensional nozzle was undertaken by Vooren and Sanderse [830] and Bradshaw and Pankhurst [93]. The potential equation was solved in both cases with a finite process, in which the entire flow field through the nozzle must be discretized. The solution for the potential equations was carried out according to a method of finite differences or finite elements. The disadvantage to this approach is that a lot of computational time is required. The calculation of the boundary layer was performed with a two-dimensional process.

As mentioned, the nozzle is also used to determine the flow velocity in the test section. The simplest way is to furnish a pitot-static probe (Prandtl probe) in the flow, which is connected to a pressure gauge. The dynamic pressure q_∞ is the pressure difference between total pressure p_{tot} and static pressure p_{stat}

$$p_{tot} - p_{stat} = \frac{\rho_L}{2} \cdot u_\infty^2 = q_\infty, \qquad (13.5)$$

from which the wind speed can be calculated:

$$u_\infty = \sqrt{\frac{2q_\infty}{\rho_L}}. \qquad (13.6)$$

However, when a model is located in test section, the dynamic pressure of the undisturbed flow field needs to be determined. The static pressure must also be determined in a region that is not affected by the flow around the model. This is often done by using the static pressure gradient in the nozzle as shown in Figure 13.5.

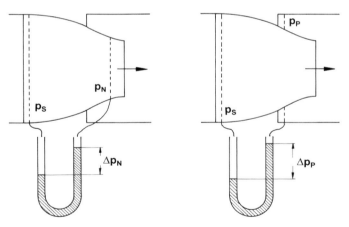

Figure 13.5 Velocity measurement in undisturbed incoming flow.

The difference in static pressure is measured between the settling chamber (p_S) and at a location in the vicinity of the nozzle end (p_N). This pressure difference Δp is then calibrated against the dynamic pressure as measured by a pitot-static pressure probe in the empty test section. The measuring probe is usually placed at the wheel base center. This results in

$$q_\infty = \frac{\rho_L}{2} \cdot u_\infty^2 = k_N \Delta p = k_N (p_s - p_N), \qquad (13.7)$$

where k_N is the so-called nozzle factor. In a first approximation, k_N is constant. However, the thickness of the wall boundary layer of the nozzle depends on the Reynolds number; it is advisable to display k_N as a function of the velocity. This type of calibration is called "calibration after the nozzle method." In closed test sections it is the only method in use.

The location of the pressure measuring tap in the nozzle must be chosen with care because the blockage of the vehicle can impact the flow in the nozzle and affect the pressure measurement tap as a result. A simple method to determine this is to run the tunnel at a constant fan speed and move the vehicle forward in the test section as described in Walter [841]. The point at which Δp changes is a good measure of the model induced influence of the pressure tap location. However, it is important that with the gradual approach of the model a so-called nozzle correction has to be applied. This type of correction is described in section 13.2.13. Due to the limit of resolution of today's pressure transducers of approximately 2 Pa, only a speed error of 0.2 km/h can be resolved.

Another method is described by Künstner et al. [476]. The pressure distribution is measured on the floor surface of the nozzle with various vehicle positions in the open test section, and then the pressure location is set where the vehicle blockage is negligible (see Figure 13.6), which is the point on the pressure curve where the deviation from the empty tunnel is tolerable.

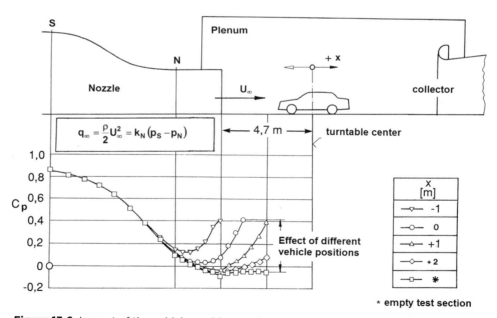

Figure 13.6 Impact of the vehicle position on the pressure distribution on the floor of the nozzle, after [476].

In open test sections, there is a second method to determine the velocity correctly as shown in Figure 13.5. Here the downstream pressure tap location p_N is placed from the nozzle into the plenum hall. This location is denoted as p_P. This results in a pressure difference

$$\Delta p = p_s - p_P \qquad (13.8)$$

for the nozzle pressure drop

$$q_\infty = k_P \cdot \Delta p . \qquad (13.9)$$

Since the plenum hall is connected via a breather channel with the external environment, atmospheric pressure is present in the plenum hall and is superimposed on the flow in the return circuit. This type of calibration is also called "calibration after the plenum method."

Both of these methods, the nozzle and the plenum method are used in open test sections. They deliver identical speeds with empty test sections, although the factors k_N and k_P are different. However, when a vehicle is installed in the test section, different speeds are calculated when the flow stagnation effect (blockage) in front of the vehicle extends into the nozzle.

During calibration the velocity profile at the exit of the nozzle is of quasi-square shape. If the vehicle blockage affects the flow in the nozzle, this profile is more or less deformed (cf. Figure 13.7).

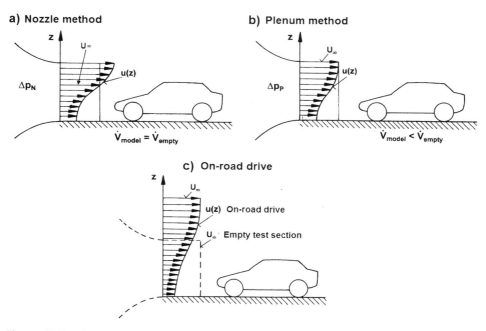

Figure 13.7 Velocity profile at the nozzle exit for a) the nozzle method; b) plenum method; c) on-road drive [370].

Using the nozzle method, an increase in velocity on the edge of the nozzle with respect to the calibrated velocity occurs if the vehicle is too close to the nozzle and the pressure tap p_N is not affected due to the nozzle blockage. This is shown in Figure 13.7a. Due to continuity, this velocity difference is greater the closer the vehicle is to the nozzle. However, regardless of the vehicle position, the volume flux in the nozzle is constant.

In the plenum method, the situation is different. Here, as described, the ambient pressure p_P of the plenum is imprinted on the test section, as shown in Figure 13.7b. With the aid of Bernoulli's equation, it can be shown that regardless of the vehicle position in the test section, the calibrated speed on the periphery of the exit nozzle is constant, even when the vehicle blockage extends into the nozzle. When a vehicle is present, the flow rate through the nozzle is less with respect to an empty test section.

If one considers the velocity profile that would occur on the road (see Figure 13.7 c), the flow velocity relative to the vehicle coordinate system by no means would match the velocity profile created by the jet nozzle.

Both methods provide an incorrect reference speed, and it must therefore be corrected. The method to perform this correction is discussed in section 13.2.13. At this point, it should be noted that both methods converge with the distance of the vehicle away from the nozzle. They converge theoretically completely at infinity. See also SAE Information Report [703] or Kücheman and Vandry [471]. To keep the corrections small, one might be inclined to plan a large distance between the vehicle and the nozzle exit area in the construction of a new wind tunnel. However, in anticipation of section 13.2.13, it can already be said that this seems unnecessary since the nozzle corrections work very reliably and thus smaller distances are very well feasible.

13.2.3 The Test Section

The test section is described by the following principal dimensions: the cross-sectional area A_N at the nozzle and the length of test section l_{TS} measured from the nozzle exit to the collector inlet area. The following qualified dimensionless quantities are relevant:

- The blockage ratio $\varphi = A_x/A_N$ with A_x as the frontal area of the vehicle
- The dimensionless length $\Lambda = l_{TS}/d_N$ of the test section with d_N as the equivalent (hydraulic) diameter of the nozzle $d_N = 4A_N/U_N$ and U_N as circumference of the nozzle for noncircular cross sections

In open sections, two more parameters are added:

- The ratio of the cross sections of the collector and the wind tunnel nozzle $\Omega = A_C/A_N$
- The ratio of the cross sections of the plenum, A_P, to the nozzle cross section, A_N: $K_P = A_P/A_N$

In previous years, it was assumed that the blockage ratio ϕ, which of course is equal to zero on the road, should be as small as possible to have the desired kinematic similarity between the situation on the road and in the wind tunnel. When building the first automotive wind tunnels, the standard in aircraft aerodynamics was to have a blockage ratio of $\phi = 0.05$. This resulted in a huge wind tunnel with a cross section of 40 m² for a standard passenger car. However, taking building and operating costs into account, certain compromises regarding size must be realized. In Buchheim et al. [109], comparative measurements between different-sized wind tunnels and road measurements were carried out. It was shown that sufficient simulation quality is not guaranteed in all wind tunnels and the effects of blockage ratios was not clearly correlated. It was found that in addition to the blockage ratio, other interference effects in the test section affected the measurement. This subject will be discussed later.

Newer designs for open-jet wind tunnels often have a nozzle cross section of about 20 m² to 25 m² and longer test sections. Using the inverse of the correction procedures, Mercker and Wiedemann [568] showed that by a suitable parameter variation of vehicle position and the test section dimensions (including nozzle and collector), drag coefficients for different vehicle types can be created, where the various interference effects cancel out. For example, with the assumptions in [568], there is an optimal nozzle cross section of about 20 m² for a typical notchback car.

The smallest wind tunnel operated for passenger cars in full size only have nozzle cross sections of 10 to 11 m², which results in a blockage ratio of about $\phi \approx 0.20$. Although absolute differences in the measured drag coefficient c_D of about ±6% may arise, the relative differences are limited when the change in the drag coefficient is observed through a configuration change on the vehicle. In this way, in many cases it is ensured that even in tunnels where the absolute drag coefficient is not correctly measured, vehicle development can still progress. Aerodynamic configuration changes are properly assessed by the wind tunnel, so that the vehicle is really evolving in the correct direction during the development process and incorrect development is not due to wind tunnel related artifacts. To do this, it is still necessary to generate a more natural ground simulation with moving floor, rotating wheels, and small floor boundary layers.

To achieve a match of the absolute drag coefficients in the different test sections, wind tunnel corrections need to be used. These are based on analytical or numerical methods, which take into account the different boundary conditions in wind tunnels and evaluate the resulting interference effects. Details about the different correction methods can be found in section 13.2.13.

With the development of correction methods, it turns out that in many cases the size of the nozzle cross section was not always the determining factor for the deviations of the measured drag coefficients from measurements. Rather, it was found that in many cases the largest deviations are generated by a pressure gradient already present in the empty test section. In contrast to the buoyancy (Archimedes principle) caused by the gravitational force, the pressure gradient forces are referred to as "horizontal buoyancy."

It does not matter if the gradients are generated by the vehicle itself or by the edge of the jet in the empty test section. Basically, generating pressure gradients create additional forces on the vehicle (interference effects) that do not occur on the road and thus need to be corrected. A special role is played by the wake region of the vehicle where flow is detached. Exposed to a pressure gradient, it can in fact change the shape of the separation bubble, and it will generate a force whose value must be determined by a correction method.

Therefore, the design of the test section is of the outmost importance to create a gradient-free flow, where corrections are not required. This is only possible with a careful design of the nozzle at the beginning of the test section and a design of a suitable flow collector at the end. Also of high importance are the length of the test section, the vehicle position, and the size of the plenum. The vehicle itself also creates pressure gradients in the test section. This "inherent" flow gradient is dependent on the type of the test section and cannot be avoided, and a correction is necessary in every case. You can find more information in section 13.2.13.

The shape of the nozzle cross sections is generally rectangular with an aspect ratio of height (h) to width (b) of about 0.6 to 0.7 for most existing wind tunnels. Section 13.2.13 showed that it is best to have a nozzle with an aspect ratio of $h/b = 0.5$ to minimize interference effects originating from the limited dimensions of the jet. Sometimes the corners of the nozzle are slightly rounded or beveled. For wind tunnels that are used exclusively for passenger cars, different cross-sectional shapes have been used in the past apart from a rectangle. Morelli [586] has realized a circle-arc shaped 11 m^2 nozzle in the Pininfarina wind tunnel in order to get a jet cross section that is nearly affine to the front surface of the car. Actually, the main reasons for this nozzle shape were to not exceed the budget for the construction and operation and to minimize the wind tunnel interference effects since reliable correction methods were only just being developed from the mid-1990s. A large contribution in this area has been made by the ECARA (European Aerodynamic Research Association) Wind Tunnel Corrections group and the SAE Subcommittee on "open throat wind tunnel adjustments."

The wall boundaries of the test section have significant affects on the flow around vehicle. They are distinguished into two types: the open and closed test sections. Special types of test section have also been developed, such as the slotted wall or adaptive wall test section. These types are outlined in Figure 13.8.

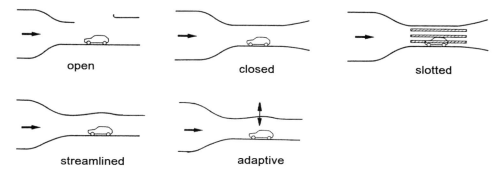

Figure 13.8 Different types of wind tunnel test sections.

The kinematics of the flow around an object differ if the flow field is restricted to a finite cross section or if there is infinitely extended space. In an open test section, the boundary condition at the edge of the jet, the static pressure p_∞ of the environment (atmospheric pressure) is present along the edge of an imaginary streamline which corresponds to the velocity u_∞ from the nozzle. Although the width of the shear layer starting from the nozzle exit increases downstream, p_∞ and u_∞ are still constant along the boundary streamline for the following considerations.

If a vehicle is introduced into the test section, the jet will move around the object and the boundary streamlines will form a curved convex. For similar on-road conditions, the pressure at the same location would be lowered and the velocity increased due to energy conservation. For the open-jet wind tunnel, the boundary condition for the ambient pressure p_∞ and a correspondingly lower velocity u_∞ must be fulfilled for the boundary streamline. That can physically happen only if the streamlines bend farther and the jet cross section increases in the region of the vehicle. In an open-jet wind tunnel, due to continuity, the vehicle is exposed to a lower velocity as obtained during the calibration of the wind tunnel. However, the latter is used for the determination of the drag coefficient. This results in an underestimate of drag coefficients.

In contrast, the streamlines are pushed together by the displacement effect of a vehicle in a closed test section. Since there is no leakage of flow out of the test section, the pressure decreases on the walls due to continuity and the flow velocity is higher in the region of the vehicle. Accordingly, the forces increase on the vehicle and the drag coefficient is overestimated.

Interference effects in closed test sections are much more significant than in open test sections. For a tunnel with aspect ratio of $h/b = 0.5$ and the same overall geometrical dimensions, the interference velocities are about four times higher in a closed section compared to an open test section. Basically, the influence of the jet edge increases with the blockage ratio ϕ.

The open test section was attributed in earlier times as having an advantage. If it were empty, the static pressure (p_{stat}) along the jet axis was supposed to be constant. Thereby, it is not sufficient that the condition p_{stat} = const. is satisfied only for the length of the vehicle but rather it must be at least until the end of the body's separation bubble (near wake region). If this boundary condition is not met, additional forces can occur in the aerodynamic measurements. In 1996, Mercker and Wiedemann [568] published a collection of data for wind tunnels. As can be seen from Figure 13.9, the condition p_{stat} = constant along a typical passenger car of 4 m to 5 m in length and along the subsequent wake region is only approximately satisfied in one of the wind tunnels.

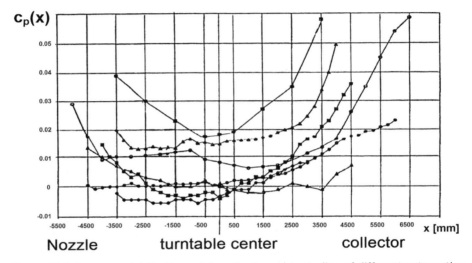

Figure 13.9 Pressure distribution $c_p(x)$ on the tunnel centerline of different automotive wind tunnels and empty test section, after [568].

Based on the work of Mercker and Wiedemann ([568], [567]), a correction method with an enhanced approach for the treatment of the static pressure distribution in the empty test section was developed by Mercker and Cooper [560] ten years later. To avoid using more complex methods, such as that described in section 13.2.13, newer wind tunnels are designed to have only a negligible static pressure gradient.

There are a number of different design parameters that have a lasting influence on the pressure distribution in the open test section:

- The test section length l related to the hydraulic diameter d_N of the nozzle, $\Lambda = l/d_N$
- The ratio of the cross section of the collector A_C to the cross section of the nozzle A_N
- The design of the nozzle
- The design of the collector
- The dimensions of the plenum

- The design of technical measures to reduce the floor boundary layer at the front of the test section

The extent to which individual parameters impact the pressure gradient still depend on the blockage ratio φ and the measuring position of the vehicle within the test section. Given the large number of parameters resulting in a significant optimization effort, often the desired result may be achieved only through a pilot wind tunnel study or by protracted CFD calculations.

The pressure gradient at the rear of the test section could be met by a correspondingly large value of Λ. However, a long test section has the ability to cause jet instabilities transverse to the flow direction, which become particularly unfavorable for unsteady measurements. Thus, this method is only recommended under certain conditions. Nevertheless, wind tunnel facilities created in the last few years, have a test section length of more than 14 m and a nozzle cross sectional area from 20 to 25 m². It can be assumed that such test sections generate jet instabilities. When shortening the test section length, the collector at the end of the test section stabilizes the flow jet in the range of the vehicle. It is therefore particularly important to ensure the optimum design of the collector and the surrounding plenum hall (see section 13.2.4 and section 13.2.5) because of the influence they have on the pressure gradient. Furthermore, for a long test section, construction and operational costs of the entire system will increase accordingly.

In 1988, Schulz-Hausmann and Vagt [746] conducted such a parameter study with regard to the aforementioned questions. They were able to show that depending on the test section length and a specified collector shape, the pressure gradient up to the nozzle could be affected. Only when the dimensionless test section length is greater than Λ = 3 were the results independent of the collector size. However, this knowledge cannot be transferred to other wind tunnels readily as a function of the geometry since there is a dependence of the degree of blockage and collector geometry, which were not varied in this study.

A further advantage of the open test section provides good accessibility features. It facilitates the introduction of measuring probes, the observation of the flow, the ability to locate of sound sources, and increased space to handle the model.

A disadvantage of the open test section is that their effective length is limited. The jet is mixed at the three free edges with the basically still air in the plenum. The cross section of the jet core in which the desired flow velocity u_∞ is present is reduced along the jet length. The mixing of the jet with the surrounding air is also associated with an increased loss factor compared to the flow bound in rigid walls. The open test section thus requires a considerably higher fan power than a closed section. Another disadvantage of the open test section is their unimpeded sound radiation. The plenum of climatic wind tunnels with open test sections must also be conditioned and larger wall surfaces must be insulated.

The advantage of the closed test section lies in their large useable length. The jet core is consumed slower since the flow boundary layer on the wall grows slower than that in free jet boundaries. However, the increasing displacement thicknesses on the walls lead to a pressure drop along the flow axis. This can be compensated by a slight expansion of the tunnel cross section in the flow direction. But this compensation is correct only for a configuration for which it was designed—usually for an empty test section and for a certain Reynolds number. With the introduction of a model, this is only approximate, since the displacement of the fluid produced by the vehicle on the walls accelerates or delays an interference flow, which affects the downstream growth of the boundary layer thickness.

Similar to an open test section, the pressure gradient in closed test sections must also be corrected mathematically, which is described in section 13.2.13. Furthermore, a large confinement of the flow may lead to separation of the flow from the test section walls as can occur when the vehicle is at large yaw angle.

However, the crucial physical aspect of a closed test section is different to open test sections. It is obvious that flow gradients are generated in the flow around a vehicle. In an infinite flow field these gradients are natural and are generated by the object to be measured. Ignoring friction drag, they determine the pressure distribution and the drag coefficient. In a flow field surrounded by solid walls, however, additional inherent flow gradients are generated in the longitudinal direction on the walls of the test section due to the flow displacement effect of the vehicle and its wake. They act on the flow and decrease toward the vehicle. The size of the gradient depends on the cross-section geometry of the test section. The effect of these gradients on the solid body of the vehicle can be calculated relatively easy and can be taken into account using corrections in the calculation of the drag coefficient. However, little attention has been paid to the region of separated flow, typically found in the wake of vehicles. According to recent work by Mercker [559] the influence can be determined with sufficient accuracy (section 13.2.13).

Classical methods for wind tunnel corrections used in aircraft aerodynamics are often used in automotive wind tunnels (Wickern [864]). However, since separated flow over aircraft wings is of lower technical interest, the gradient effects for detached flow is rarely treated and the geometric invariance of the wake due to "external constraints" is assumed.

In an experimental study on the imposed pressure gradient in flow direction in open-jet test sections, Gürtler [304] revealed that the wake is deformed behind the body. This changes the base pressure on the rearward facing separation area of the vehicle generating spurious forces. For these studies, the flow field in the open test section was suitable for measurements since p_∞, u_∞ are constant at the edge of jet, and thus no inherent flow gradient is present.

On the other hand, it was possible to determine the effect of the inherent gradient of an empty closed test section with no imposed gradients using the findings from an open test section. With this in mind, Mercker [557] in 1986 developed a correction procedure for closed test sections. The deformation of wake flow generates in return a gradient over the vehicle. In 2013, Mercker [559] showed a very good agreement for the corrected values for different levels of blockage ϕ when this pressure gradient was considered (section 13.2.13).

Since different interference effects can be negative or positive, it is possible that the overall effect of the interference (in particular in an open jet test section) could be zero. This is vehicle dependent. In detail, however, the flow around the vehicle in a wind tunnel is not represented properly, which may affect the detail optimization on a vehicle. However, the corrected integral drag coefficient may match the measurement in an interference free test section. This fact is used in some wind tunnels today and the test section is "tuned" after completion. This adjustment takes place mostly in the form of a change in the position or geometry of the collector at the end of the test section. Then, a correction of the measured values may be entirely omitted. The results from other wind tunnels or from on-road measurements are used as a guideline. This pragmatic way has been quite successful, but it is not known whether once the adjustment is fixed it is applicable to all other vehicle types.

The use of slotted walls is an attempt to combine the main advantages of the open and closed test section without having to take their respective disadvantages. The longitudinal slots in the test section walls allow pressure equalization with the environment. The aim is to achieve a constant pressure along the jet axis. The fixed part of the walls prevents the mixing of the jet with the static ambient air, which cannot be obtained in detail, since the presence of the vehicle induces a flow through the slots. As a consequence, pressure gradients are also generated in a slotted wall test section.

A reasonable area ratio of the slots to the solid walls, $\varphi = A_{open}/A_{closed}$, turned out to be about 30%. This value has good agreement compared to on-road measurements but is only derived from observations. Hoffmann et al. [353] performed an experimental investigation and showed that the slotted wall test section at high blockage ratio is unfavorable compared to the open test section, with an overestimate of the measured drag. It can be assumed that the slotted wall test section rather behaves similar to a closed wall test section.

The blockage problem of open, closed, and slotted wall test sections can be overcome with adaptive walls. However, this is complex and financially more expensive. Whitfield et al. [862] presented a design for an adaptive wall test section and tested vehicles of different scales in a model wind tunnel. They divided the walls of the test section into longitudinal bands (Figure 13.10) and its contours could be adapted to the shape of the streamline. To ensure that no leaks occur, these flexible bands were arranged perpendicular to rigid sections. The wall contours were determined iteratively using an initial calculation of the potential flow to determine the pressure distribution if the walls were placed an infinite distance away. The wall sections were then gradually deformed until the pressure distribution matched the calculated result. Even at a blockage ratio up to 20%, corrections were not required. Hoffmann et al. [353] confirmed this result experimentally.

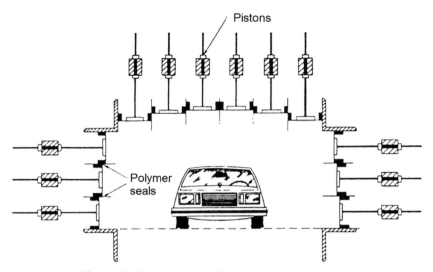

Figure 13.10 Adaptive wall test section, after [862].

In wind tunnels, in which predominantly series development is carried out on vehicles, this technology has not prevailed so far, but has for race car aerodynamics, since one single car often remains in the test section for days or weeks. Nevertheless, even here, an adaptation of the walls must be made if, for example, a Formula 1 racing car will be investigated in a high-downforce or in a low drag configuration.

13.2.4 The Collector

The fluid mechanical purpose of a collector at the end of the test section is versatile and is a particular challenge in the simulation of the free flow around a vehicle. In a closed test section the collector is often referred to as "transition" and represents the link between test section end and the first diffuser. Although in such test sections the existence of the transition can be detected in upstream direction, the influence remains moderate and can be compensated by an extension in length of the test section.

This is not the case for an open test section, and special attention has to be paid for the design of a collector. As mentioned, a shear layer starts at the edges of the nozzle and further grows in downstream direction. An entrainment flow turn up perpendicular to the main flow direction due to turbulent mixing processes with the quasi-still air in the surrounding plenum. As a consequence of the entrainment flow the volume flux of the jet in the open test section increases with the length, similar to a jet pump. The objective of the collector is to eliminate this increase of volume flow again and to generate a lateral stability of the free jet at the same time (because of the wind tunnel floor, the vertical stability of the jet plays only a minor role). This can be done in two ways:

- The inlet cross section of the collector is located within the entrainment flow
- The collector is equipped with so-called breather openings in order to discharge the additional volume flow laterally again via an annular gap or opening flaps

For acoustic reasons, an annular gap or the opening flaps are often omitted completely and the collector cross section is chosen very large so that both the core flow of the jet and the remaining entrainment flow are guided through the collector entry surface. There is however a reverse flow into the surrounding plenum along the collector walls. Furthermore, due to flow deceleration in an expanding jet, a positive pressure gradient is induced in the test section that affects the vehicle measurements. The shape and size of the collector relative to the nozzle in combination with the annular gap have a decisive influence on the pressure gradient in the test section. In existing wind tunnel facilities, convergent (cross section decreases) and divergent (increasing cross section) cross-section shapes can be found.

Looking at the side walls and the ceiling of a collector, a positive or negative flow circulation is produced around the walls, depending on the opening angle of the collector and the interaction of an annular gap (see Figure 13.11). Although the circulation around opposing walls has different signs in a mathematical sense, the volume flow is increased by a divergent collector and a flow is created in the collector gap which faces toward the plenum hall. By increasing the volume flow, an accelerated flow is generated in front of the collector in downstream direction, which is then, for example, responsible for a reduction of a positive pressure gradient in the test section.

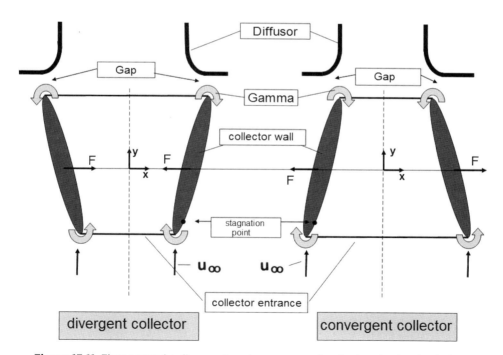

Figure 13.11 Flow around a divergent and a convergent collector due to circulation.

For the convergent collector, the situation is just the reverse. Here, the circulation changes its sign, and the volume flow through the collector decreases. The necessary flow rate for the fan is ensured by the additional inflow of air from the plenum via the annular gap. Due to the reduced volume flow through the collector a stagnation area results in front of the collector, which decays upstream. The result is a horizontal force (i.e., horizontal buoyancy), which reduces the drag force on a vehicle in the test section.

In summary, it can be stated that the circulation around the side walls and ceiling of a collector determines the flow around and through the collector. Basically, the stability of the jet is increased if the volume flow through the collector rises. Additionally, another volume flow is further added by the entrainment flow. Depending on the size of the collector in relation to the nozzle size and the test section length it is very well possible that an additional airflow enters the plenum through the annular gap in case of a convergent collector. However, this is always the case in a divergent collector.

Here, the size of an annular gap plays an important role. If the ring gap is completely closed, the circulation is zero and the additional entrainment flow rate of the jet is directed around the collector and increases the secondary flow in the plenum. The pressure gradient in the test section increases as well and the stability of the jet is decreased.

Considering the idealized case of a frictionless flow around a collector, the drag force on the collector decreases with the increase of circulation, and the result is a force F perpendicular to the direction of the flow. Ideally, the collector drag is zero (d'Almbert's Paradox). For a divergent collector this force is directed inwards, while it is directed outwards for a convergent collector.

This strategy was pursued consequently during the construction of the new wind tunnels of the BMW Group (cf. Figure 13.12). A summary of the design can be found in [216]. Walls and ceiling of the collector were built as inward-looking camberless wing profiles in order to increase the circulation. The maximum thickness of the profile amounted to approximately 25% of the chord length. The collector behaves like a convergent collector up to the closest point between two opposite sides of the collector or between floor and ceiling surface and creates a static pressure gradient in the test section. Due to the increasing cross section (divergent collector) behind this point and the increased volume flow, the overall pressure gradient in the test section is reduced again. Thus, with an adequate design of the collector the static pressure gradient remains negligible or small in the test section and in the vicinity of the collector.

Wind Tunnels and Measurement Technique

Figure 13.12 Collector with wing profiles in the AVZ of the BMW-Group, view from the first diffuser [216].

However, a second fundamental effect occurs when a vehicle is placed in the test section. Usually, vehicles produce a wake (dead water) which extends in downstream direction theoretically to infinity. In dead water regions, the flow loses momentum. If the dead water area enters the collector, this will result, for continuity reasons and due to the solid walls, in a higher velocity outside the dead water, similar to the effect in a closed test section. As a consequence of the flow acceleration, another flow gradient is induced in the open test section. This effect must be corrected and the relevant approach is described in section 13.2.13. The magnitude of the so-called collector effect depends also on whether the vehicle's near field wake enters the collector or not. This part of the wake structure is much more relevant to the displacement effect of the flow than the far field wake and is, hence, responsible for larger collector interference effects. However, the collector effect of the far field wake is present in any case.

13.2.5 Plenum

The surrounding hall of an open test section is called plenum. Depending on the nozzle size and the length of the free jet, but also on the geometry of the collector, the dimensions (width, length, height) of the plenum must be carefully chosen in order not to have interfering effects in the jet. If the plenum is too small, the jet could expand laterally (so-called Coanda effect) and a pressure gradient could be induced in the test section. This should be avoided because the measured drag on a vehicle is directly affected. The same applies to a traverse, which is installed above the test section, for the measurements of velocity profiles in the vicinity of the vehicle.

Due to the properties of an open-jet test section, secondary flows occur in the plenum in general, which should be kept as small as possible. Especially if the jet velocity is measured using the plenum method (see section 13.2.2), it is important that time independent boundary conditions are given at the location of the reference pressure tap in the plenum. However, this is not the case if the secondary flow velocities are large (see also convergent collector, section 13.2.4). It is difficult to predict how secondary flow form out in detail in the plenum without a pilot study in a model wind tunnel or calculations with aid of CFD methods.

In general, the experimental way is the usual approach in determining the dimensions of the plenum. In a pilot study the change of the drag coefficient of a vehicle is measured as a function of the dimensions of the plenum. If no change occurs, the resulting geometry is then set as plenum size. Deutenbach [194] has done this investigation the same way. But the results should not be generalized, because the geometry of the collector, as shown in section 13.2.4, is of decisive influence. Such investigations are only useful if at least the rough dimensions and position of the collector are determined previously. In this context it should also be mentioned that it is often impossible to come to a satisfactory result concerning the static pressure gradient in the empty test section by varying the geometry of collector. The reason may be a too small plenum, which induces a pressure gradient in the open test section and which may not be compensated by the collector.

The plenum surrounding the open test section is usually of rectangular shape. This means that two walls, floor, and ceiling are parallel to themselves. Such volumes have the property that acoustical standing waves may occur in them. The air particles in the plenum oscillate with different frequencies. The so-called room modes, which are in the infrasound range due to the large dimensions of the plenum, are therefore at large wavelengths. Therefore, the laws of hydromechanics apply (pressure and speed are inversely proportional) because the plenum is located in the acoustic near field.

On the walls, the pressure oscillates between minimum and maximum, whereas an oscillation node occurs in the middle of the jet and the pressure fluctuation is theoretically zero. At the edge of the jet the pressure fluctuations don't equal zero and the jet can get into oscillation, which adversely can affect the force and pressure measurements on the vehicle.

In rooms, there are six degrees of freedom for Eigen oscillations. If the integer harmonics are neglected, they can be reduced to three oscillations in good approximation. The Eigen frequencies in of the room modes can be calculated with the following equation described by Lord Rayleigh in 1896,

$$f_n = \frac{c}{2} \cdot \left[\left(\frac{m_x}{l_x}\right)^2 + \left(\frac{m_y}{l_y}\right)^2 + \left(\frac{m_z}{l_z}\right)^2 \right]^{0.5} \tag{13.10}$$

where c is the speed of sound, l_{xyz} is the distance between the walls and the ceiling, and $m_{x,y,z}$ specifies the order of the room mode. A three-dimensional field of complex sound field impedances results from the superposition of all room modes, where the room modes represent a resonance system.

According to equation (13.10), the frequency of the room modes in a plenum with large dimensions is lower compared to a smaller plenum. This would be a reason to design a rather small plenum. However, the previously mentioned Coanda effect may create a pressure gradient in the test section. Therefore, other measures must be taken to prevent low-frequency pressure fluctuations.

There would be the option to align the walls of the plenum divergent to each other. The same applies to the ceiling in relation to the test section floor. Due to the simple form of the plenum, the room modes were largely reduced, because a total reflection of the sound waves like with parallel walls is prevented. Studies on this subject were carried out by Beland at the FKFS [66]. A side wall divergence of 5° to 10° significantly reduces the intensity of the first room mode perpendicular to the jet axis.

In a closed-return wind tunnel the link of the open-jet test section to the ambient atmosphere is usually realized using breather openings. These openings are protected from rain and wind and have a major function at the beginning and the end of a wind tunnel run. Particularly fast starting and stopping results in high acceleration forces in the jet and the pressure in the plenum hall rises during these processes, and airflow enters and exits the plenum through these openings. If once the flow has reached a quasi-steady state, ambient conditions prevail in the plenum again.

It has turned out that size and location of the breather openings in the plenum is of great impact on the pressure fluctuation level. For example, if a breather opening is located directly above the nozzle, the level may drop considerably, which may not be the case when the breather is located above the collector.

The effect of an opening in the plenum can be demonstrated with another extreme example. During a pilot study for a wind tunnel (cf. Beland [66]), the ceiling of the plenum was removed. Compared to the dynamic pressure of the flow, the r-m-s value of the pressure fluctuations dropped down far below 1% under otherwise equal conditions. Obviously, the resonance phenomena of the entire system were altered by this measure. Although such a plenum is not feasible, this example shows the relevance of the porosity of the plenum.

Another important aspect about the plenum is whether the wind tunnel should also be used for acoustic tests on the vehicle. If this is the case, ceilings and walls of the plenum must be equipped with an appropriate acoustic insulation material in order to reduce the background noise. The same should also be applied to the collector and, if necessary, to the first diffuser. However, as mentioned above, these measures do not affect the room modes, of which the wave lengths are very large. Thus, the treatments at the walls can be considered to be acoustically hard.

13.2.6 Diffusers

In order to improve the energy balance of a wind tunnel linearly increasing diffusers are installed in the air return circuit for the purpose of pressure recovery. Although the ideal shape of a diffuser would be a progressive increase of its cross section, the linear shape is often chosen for constructional reasons. As a first approximation the opening angle should be approximately 5° to 7° in order to avoid flow separation on the walls. However, whether separation occurs rather depends on the turbulent boundary layer thickness on the walls and the length of the diffuser. Hence, a good surface quality of the diffuser is very important. Extensive design criteria have been provided for diffusers by Gersten and Pagendarm [273].

Due to continuity reasons the flow in a diffuser is decelerated. If additional measures are used to make the flow in the tunnel more homogeneous (screens, honeycombs, radiators, see below) the pressure losses of the installations will decrease with the square power of the velocity reduction in the diffuser.

In some wind tunnels, flow straightening measures are therefore implemented in the settling chamber because here, the cross section of the jet reaches its maximum. However, if the wind tunnel should also be used for acoustic development tasks, there are no options to reduce the sound pressure level generated by these installations with an acoustic insulation. Thus, additional installations, such as heat exchangers, are often installed between the third and fourth corner, in which the fourth corner is designed as an acoustic silencer.

So-called large angle diffusers are used to limit the length of the closed-loop air return circuit. Such large angle diffusers are used particularly in the transition of the return circuit behind the fourth corner and the settling chamber. Although the flow separates from the walls due to the large opening angle of the diffuser, the recirculation area can be limited using turbulence screens at the end of the diffuser, which also straighten the velocity distribution. However, the installation of screens requires a lower static pressure behind the screen and causes additional pressure losses in the return leg. Since the flow velocity in this part of the circuit is fairly low, though, the losses are limited. However, the aim of having a large angle diffuser located here is obtaining a maximum extension of the cross section of the settling chamber at shortest length, because this determines the contraction ratio of the nozzle. Studies on large angle diffusers were carried out (e.g., by Mehta [550]).

In many wind tunnels a further large angle diffuser can be found extending from the beginning of the return circuit to the turning vanes of the first corner. An opening angle of up to 15 degrees is not uncommon. However, without further measures the flow may separate from the diffuser walls, which should be avoided because of the high flow velocities in this part of the return circuit. For this purpose, air is blown tangentially into the ground boundary layer: the boundary layer is energized this way and flow separation can be avoided. The air that is needed is usually taken from the floor boundary layer treatment system (pre-suction or scoop) in the nozzle exit area, which is used to reduce the floor boundary layer thickness there.

13.2.7 Turning Vanes

Closed return wind tunnels use turning vanes to guide the air in the 90° corners. An appropriate design of the vanes avoids separation. The ratio of distance between the turning vanes to the chord length should be about 0.25. A 1/4 circular arc profile made of bent steel is suitable for the contour of the vanes. However, based on experimental evidence ([93]) the leading edge of the circular arc is reduced: the tangent to the leading edge of the vanes encloses an angle of 4° to 5° with the jet axis. Though, the tangent to the trailing edge is congruent with the flow axis.

Thin circular arc profiles are equivalent to thicker and curved wing profiles. However, if the wind tunnel will be used for acoustic purposes as well, thicker profiles are preferable because the profile body can be used as a sound absorber and thus provides a high contribution to minimize the noise in the test section.

In today's wind tunnels, a combination has been carried out of guiding the flow around the 90° corner and simultaneously extending the cross section behind the diversion of the flow. This is achieved by a unilateral extension of the outer wall of a corner element. But, in order to ensure non-detached flow through the turning vanes, CFD calculations should be made for confirmation. For instance, in this way cross section extensions by up to 40% can be realized in the third corner, which, in turn, avoids a large angle diffuser in the settling chamber, see e.g., [216], [841]). In up-stream direction the settling chamber extends from the beginning of the nozzle to the third corner, then. Here, non-uniformity in the flow can be effectively compensated due to friction effects in the flow as a consequence of the big yardage of the chamber.

13.2.8 Flow Conditioning Screens

A flow resistance is generated due to the installation of turbulence screens in the air return circuit. The total pressure of the flow is reduced without changing the mean velocity in the cross section in front of and behind the screen (continuity principles). As the resistance changes in areas of higher or lower speed, a cross compensation flow occurs toward the mean velocity because of local fluctuations of the static pressure. If turbulent flow is considered as a special case of the variation of the mean velocity, obviously the degree of turbulence can be influenced using screens. The biggest effect is achieved with screens with high resistance. However, for such screens the local pressure loss coefficient depends severely on small bugs in the grid. This may cause large differences in the mean velocity. For this reason, pressure loss coefficients should be lower than 3, and instead of one single screen several screens with lower pressure loss coefficients should be applied in a row at a certain distance. It can be shown that the effect of n screens in a row with a drag coefficient of ζ is greater than the effect of one single screen with a drag coefficient of $n \cdot \zeta$.

$$\frac{1}{(1+\zeta)^n} < \frac{1}{1+n\zeta} \qquad (13.11)$$

The distance between the screens is determined by the length scale of the self-made turbulence produced by the screen itself. Due to friction effects, this form of turbulence breaks up after a relatively short distance. Further investigations on the influence of wind tunnel screens can be found in Rae and Pope [655] and Bradshaw and Pankhurst [93].

13.2.9 Honeycombs

While screens and the nozzle mainly affect the longitudinal components of flow turbulence and velocity, a so-called flow straightener is installed in the tube to reduce the lateral component of the velocity and the large turbulent vortex structures. A flow straightener consists of a hexagonal honeycomb structure and effectively guides the flow in axial direction. Usually the length of a honeycomb should be about eight times the diameter of a honeycomb cell. It is obvious that a flow straightener itself generates turbulence in the wake of the honeycomb cells. Loehrke and Nagib [511] were able to show that the dimensions of the turbulence structures are the same size as the honeycomb cell diameter. In order to get a good performance of the flow straightener with low self-made turbulence, it must be avoided for the flow in the honeycomb cells to turn laminar due to a small cell size. The nonuniform velocity profile at the exit of the honeycomb would be much larger and more energetic. In contrast to a screen, the breakup of the turbulent vortex structures occurs along a longer distance. However, a large cell size would lead to rather long lasting large-scale turbulence structures due to the longitudinal scale of the turbulent eddies. Therefore, at a critical Reynolds number of the cell flow of $Re_{crit} = 2300$ a cell width has to be found that still guarantees turbulent flow conditions. This limits the flow velocity. Below this velocity limit, a poorer flow quality can be expected in axial direction.

13.2.10 Acoustic and Anti-Buffeting Measures

The study of aeroacoustics phenomena in automotive wind tunnels that lacks acoustic treatment is generally not possible because the background noise of these facilities is higher than the wind noise generated by the vehicles being tested. Even in cases where the vehicle's wind noise can be separated from the background noise with appropriate instrumentation and data processing, human judgment is still not possible. The subjective assessment of sound and its evaluation according to psycho-acoustic criteria are key components of optimizing vehicles acoustically. The goals, therefore, are that the wind tunnel's background noise should not only be low in terms of an objective measurement, but it must also be sufficiently quiet so as not to disturb the engineer's perception of the vehicle's wind noise signature. This requirement is generally met if the sound level produced by the wind tunnel is at least 10 dB below the radiated noise from the car level in a frequency band between 20 Hz and 10 kHz. To achieve this, detailed knowledge of the different sound source mechanisms in a wind tunnel must be known. See Wiedemann et al. [884].

The open-jet test section is particularly suitable for aeroacoustics work because the surrounding plenum can accommodate acoustic absorbing material that helps prevent noise emanating from the nozzle, collector, and air circuit to be amplified by reflections from hard wall surfaces. Furthermore, measurements can be performed with microphones placed outside the jet core. Thus, the acoustic signal is not distorted by the local flow conditions, and there is no self-noising generated from the microphone or its holder. Although the radiated sound of the vehicle is scattered and refracted when passing through the shear layer, there is usually only a speed-dependent offset of the sound wave that is required.

Different types of sound-absorbing materials are used to attenuate noise in wind tunnels. Membrane absorbers, also called broadband compact absorbers (BCA), are especially suitable for the plenum because they provide anechoic absorption (in the broadband sense, i.e., 1/3rd octave bands, which is sufficient for automotive wind noise) down to approximately 50 Hz. They have the form of flat boxes and usually have a thickness of about 10 cm and consist of a combination of plate oscillators and Helmholtz resonators, as shown in Figure 13.13a. These are formed by a series of hollow chambers having a volume between 0.5 and 5 liters and are covered with a perforated or slotted thin metal membrane. Over all chambers of one box, a second oscillatory metal membrane is applied. Thus, the membrane absorber forms a closed body that is made of only one material (usually steel or aluminum). The tuning of the absorber at the desired frequency range is done using the following parameters: chamber volume, thickness of the metal membrane, slit width, and spacing between the top and slit membrane.

The BCA usually consists of two absorbers in one. As shown in Figure 13.13a, there is a resonator system integrated for frequencies below approximately 125 Hz, and for higher frequencies, a conventional porous foam or fibrous layer is applied above. If mineral wool is used, a wrapping and a cover by a perforated plate are necessary to avoid the emission of fibers.

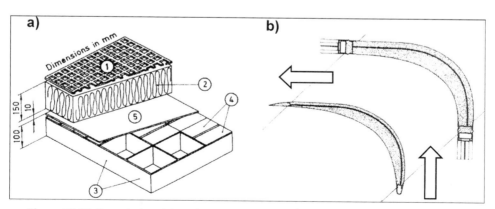

Figure 13.13 Sound-absorbing elements in the IVK/FKFS wind tunnel: a) broadband absorber (1) Perforated metal protective cover, (2) Porous polyester foam, (3) Walls of the membrane absorber cavities, (4) Resonator covers with slit openings, (5) Membrane plate; b) with polyester foam coated turning vane, after Künstner et al. [477].

Figure 13.13b shows turning vanes that have been coated with polyester foam. The use of foam in this case is advantageous because profile shapes can be generated easily. In addition, turning vanes and wall acoustic treatment made from foam are also relatively easy to install. However, foam is susceptible to impact damage unless covered by perforated plate. Foam can also age quickly in an atmospheric wind tunnel that experiences large changes in temperature and relative humidity through the seasons, or when the controlled temperature and humidity ranges are large [843]. In these cases, fiber insulation is preferable, although the fiber insulation material must be wrapped in fiberglass cloth and covered by perforated sheet metal to prevent the fibers from working loose. Once airborne, the fibers can become an irritant to the eyes and respiratory system.

One disadvantage of an open test section is that the shear layer itself generates noise due to turbulent mixing processes. This can be mitigated with acoustic lining in the plenum, as noted. Even so, the shear layer noise reflects off the hard surfaces of the vehicle, thus increasing its apparent radiated noise. This can be particularly important in small wind tunnels where the shear layers are relatively close to the vehicle.

The shear layer generated at the nozzle exit in the open-jet wind test section also creates coherent vortex structures (Figure 13.14) that affect the flow quality and the jet's stability. The free-jet instability is manifested as static pressure fluctuations in the frequency range $f < 20$ Hz and thus has negligible impact on the A-weighted sound pressure level. However, the pressure fluctuations can lock in with resonant modes of the wind tunnel test section and circuit such that the velocity-time trace in the core jet flow becomes strongly periodic (sinusoidal) instead of time-independent. This results in a modulation of the vehicle's aeroacoustics noise with the resonant frequency. Thus, although the resonant frequency is itself below the threshold of hearing, the modulation effect is audible. Resonance effects (buffeting) can be a serious problem for both psycho-acoustic evaluation and aerodynamic measurements, and will be discussed in more detail later.

As discussed in section 13.2.5, standing waves can be excited in the plenum when the wind tunnel is in operation, but they cannot be attenuated by acoustic lining on the walls because the wavelengths are too large. Even with acoustic treatment, the walls are acoustically "hard" at these wavelengths. Other measures must be taken, examples of which are shown next.

Although the excitation mechanism of such low-frequency velocity and pressure fluctuations (also known as "buffeting noise") is not yet fully understood, one cause is due to the coherent vortex structures of the shear layer, which is generated along the nozzle edge and which grows in downstream flow direction.

The vortex rings generated at the nozzle exit plane move in the direction of the collector at a convection speed R_{con} of about 65% to 70% of the jet velocity. The small-scale vortices grow in size and strength as they move downstream. Thereby they induce a flow from the surrounding plenum, which is also called entrainment flow. The required energy is drawn from the core flow of the free jet, so that the shear layers grow inward from all three sides until the potential-flow core of the jet is overwhelmed.

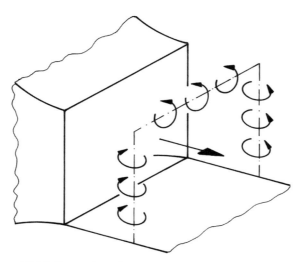

Figure 13.14 Formation of vortex rings at the nozzle exit plane.

Using a dimensional analysis approach, it can be shown that the natural frequency f_W of the passing vortex is proportional to the ratio of jet velocity u_∞ and the hydraulic diameter $d_h = 4A_N/U_N$ (A_N = nozzle area, U_N = nozzle perimeter). Using the Strouhal number Sr as proportionality constant, this results in equation (13.12)

$$f_W = \frac{Sr \cdot u_\infty}{d_h}. \tag{13.12}$$

For the open-jet test section (Figure 13.14), a typical Strouhal number is $Sr = 0.34$, which corresponds to the largest eddies and thus highest oscillation amplitudes. In reality, the Strouhal number decreases with decreasing distance to the nozzle. Taking the mirror image of the test section about its floor to create a fully open jet in a mathematical sense, $Sr = 0.48$ is also found in the literature. But this is not because of different physics but rather because the hydraulic diameter is larger by a factor of $\sqrt{2}$ in this case.

Physical changes in the Strouhal number can occur if the natural vortex formation coincides (locks in) with a spatial mode of the plenum or wind tunnel circuit. This leads to resonant phenomenon, which then determines the frequency of the vortex shedding. As discussed, this can result in high-amplitude, strong periodic pressure fluctuations that affect the force and pressure measurements on the vehicle. Four different resonance phenomena appear to occur: a plenum resonance (see section 13.2.5), a circuit duct resonance, a plenum Helmholtz resonance, and the so-called edge-tone feedback.

Considering only axial modes of the wind tunnel circuit duct, the resonant frequency f_R of its standing wave is

$$f_R = \frac{c \pm u_R}{2} \cdot \frac{m_R}{l_R} \tag{13.13}$$

Here, c again denotes the speed of sound, l_R is the length of the wind tunnel tube, and m_R is the integer order of the circuit duct mode. The speed u_R represents the average velocity in the wind tunnel duct. Because duct modes are excited by sound waves that propagate both with and against the flow direction, the appropriate sign of u_R is necessary. In the test section, pairs of excitation modes will then exist, as discussed by Duell et al. [216] and shown in Figure 13.15. Inside the circuit duct reflections of sound waves occur at the nozzle exit plane and at the collector inlet plane. Depending on the wave length (or length of the circuit duct) a standing wave pattern is established. Simultaneously, a sound wave is radiated into the plenum hall with the same frequency. Resonance phenomena between plenum and circuit duct mode will be a consequence which in turn affects the formation of vortices in the shear layer.

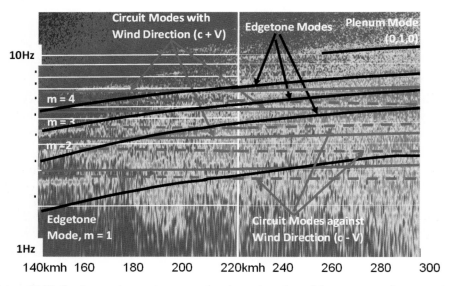

Figure 13.15 Contours of sound pressure level as a function of frequency and test section speed in the BMW Wind Tunnel, according to [216], overlaid with circuit, edge-tone, and plenum modes.

The plenum resonance mode just described is not to be confused with the so-called plenum Helmholz resonance. In this case, it is analogous to a spring-mass system is set up, wherein fluctuating pressure at the nozzle and collector causes the mass of air inside the two wind tunnel components to oscillate and be resisted by the large volume of air (spring) in the plenum hall. This process describes the essentials of how a Helmholtz resonator operates. The characteristic Helmholtz frequency can be calculated with equation (13.15). It depends only on geometrical dimensions and the speed of sound.

An additional mechanism occurring in the plenum that can affect the vortex formation frequency is the so-called edge-tone feedback. A vortex sheds from the nozzle periphery, advances downstream in the shear layer, and impinges on a downstream surface, such as the collector or rear wall of the plenum. The pressure disturbance from this impact propagates back upstream, which can then influence the frequency of the vortex formations at the nozzle The frequency of the edge-tone feedback f_E can be determined from Rossiter [686] with the following equation

$$f_E = \left(\frac{1}{m_E} \cdot \frac{l_{TS}}{R_{con} \cdot u_\infty} + \frac{l_{TS}}{c - u_\infty} \right)^{-1}, \qquad (13.14)$$

where m_E is the edge-tone mode number (1,2,3,…), l_{TS} is a representative axial length along the test section, R_{con} is the vortex convection speed (R_{con} = 0.65 to 0.7) as a fraction of the test section velocity u_∞, and c is the speed of sound.

Obviously, the plenum geometry has a big role in generating resonant phenomena. It may, therefore, be worthwhile to examine whether a nonrectangular geometry could at least reduce the plenum resonant modes.

Measurements by Duell et al. [216] in the AVZ at the BMW Group in Figure 13.15 demonstrate the relationship between sound pressure fluctuations and the various resonant modes described above in the form of a waterfall-spectrum (or frequency-interference spectrum). The figure shows contours of the sound pressure level in the empty test section as a function of test section speed and frequency. The measurements were made in the flow 6 m downstream of the turntable center.

The sound pressure levels are represented by the colors red (high) through yellow, green, and blue (low). Furthermore, using equations 13.13, 13.14, and 13.15 the different resonant modes as a function of flow velocity are shown. In the regions where the edge-tone modes and the circuit duct modes coincide, high sound pressure levels have been measured, suggesting a resonance phenomena. In most wind tunnels with open test section, these resonance phenomena occur. Typically the resonant frequencies occur from 1 Hz to 15 Hz and the pressure fluctuations can be up to 130 dB (re. 20 µPa). The mean square deviation of the pressure would then be about 60 Pa. Figure 13.15 also shows in addition the coincidence of an addition mode from the plenum with stronger sound pressures. These were reported by Duell et al. [216] to be relatively low in magnitude. This is typically the case for resonances above approximately 8 Hz–10 Hz.

To reduce the level of pressure fluctuations, various approaches have been developed. These approaches can be categorized as active or passive.

Among the first approaches taken was the so-called Seiferth-wing. Here, small vortex generators at the nozzle edge, facing both internally and externally, destroy the coherent structure of vortex rings, and the resonance of the various mode shapes of the system is prevented. An edge-tone feedback is not present and the acoustic standing wave in the circuit duct is largely suppressed. As a result, the pressure fluctuations decrease. The root-mean-square pressure fluctuation, $c_{p\,rms}$ normalized by dynamic pressure, $c_{p,rms}$, can be reduced with Seiferth wings to below 0.6%. It is worth noting that even those small fluctuations can have a negative effect on the pressure measurements at the outer shell of a vehicle.

In addition to classic Seiferth wings, there are also a number of other measures in the area of the nozzle edge (see, e.g., [482]), which all have the aim of influencing the vortex structures in order to avoid resonance phenomena. These approaches including Seiferth wings have two major drawbacks. First, they create additional noise, sometimes considerably raising the existing background noise level. Secondly, they create a pressure gradient along the test section, which is to be avoided for accurate measurements of aerodynamic forces. These approaches are therefore not recommended.

Another method to influence the vortex formation is with a system shown in Figure 13.16 (left). FKFS *besst*® (Beland Silent Stabilizer) is a system developed and patented (No. 10 2012 104 684.0) at FKFS. It is installed in the area directly upstream of the nozzle exit plane. The cambered elements serve two main purposes. First, the flow angle at the nozzle exit is changed locally, so the uniformity (two-dimensionality) of the jet shear layer is disrupted. Secondly, the pressure difference from one side of the guide element to the other generates stable axial conical vortices along the longitudinal edges of each element. This results in the coherent ring of vortex structures at the nozzle exit (see Figure 13.14) being effectively disrupted, and resonance is prevented. In contrast to Seiferth wings, any additional self-noise generated by the nozzle is negligible. With the successful reduction of the duct resonances, a reduced overall noise may even be observed in the test section. Furthermore, with an optimized design of the nozzle-guide rails, the pressure gradient in the test section can remain unaffected. Having minimal structural complexity, FKFS *besst*® system can be retrofitted cost-effectively to existing wind tunnels.

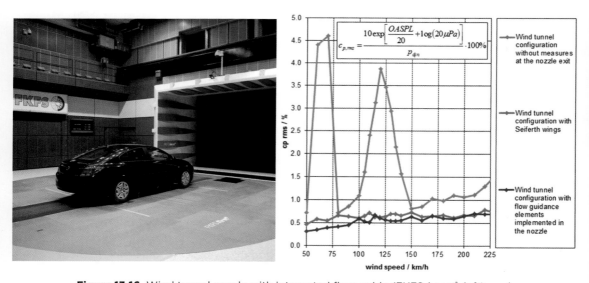

Figure 13.16 Wind tunnel nozzle with integrated flow guide (FKFS *besst*®, left), and illustration of the pressure fluctuation coefficient measured in the wind tunnel at the University of Stuttgart (right).

The right side of Figure 13.16 shows the effectiveness of the *besst* system in the wind tunnel at the University of Stuttgart. The root-mean-square pressure fluctuation normalized by dynamic pressure, $c_{p\,rms}$ is again plotted as a function of test section velocity. The configuration without any treatment at the nozzle exit has values of above 4% (dependent on speed). FKFS *besst*® system reduces the fluctuation level to approximately 0.65%. Seiferth wings at the nozzle edge achieve similarly low values, but, as mentioned, these result in unacceptable changes to the static pressure gradient, as well as higher background noise levels.

Another passive measure to reduce the fluctuating pressure levels is aimed at the standing wave in the wind tunnel duct. Similar to a wind music instrument, it is possible to put an opening in the air return duct and establish ambient pressure at this particular location in the circuit. As a result the standing wave pattern (sound column) is decomposed and the natural frequency is shifted to higher values. In principle, this measure results in a sudden change in the characteristic duct impedance and a reflection of the sound wave. The duct section is seen as acoustically shorter, so there is a higher "tone." The disadvantage of this approach is that constantly air flows out of the opening, which must be compensated from the environment via the atmospheric vent openings of the plenum hall. To avoid this, flow could be reinjected by a ducted compensation channel consisting of a large volume that leads the discharged air back into the plenum hall. However, reentry of the flow should take place via a large opening area in the plenum hall to maintain quasi-ambient pressure inside the compensation channel.

A conceivable location for such an approach would be between corners 1 and 2 or between the collector and corner 1. This was confirmed by, among others, Beland [65] using a 1:20 model of the aeroacoustics wind tunnel at the University of Stuttgart, shown in Figure 13.17.

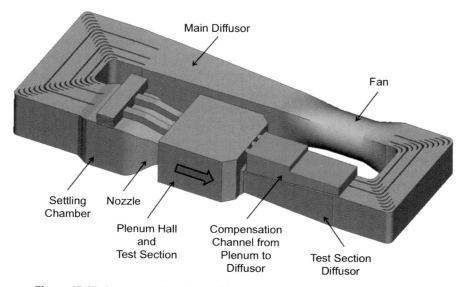

Figure 13.17 Compensation channel between the plenum and diffuser [65].

As described by Wang et al. [845], this type of compensation channel at the aeroacoustics wind tunnel of Tonji University was sealed against the plenum to create a Helmholtz resonator effect (see below).

Bergmann et al. [72] took a different approach. For an existing wind tunnel the first cross leg of the air return circuit was replaced by a section that widens continuously from corner 1 up to the fan section. Followed by a quasi-abrupt transition and area reduction of 60% into the fan housing, a mismatch of the acoustic impedance was achieved. The standing wave pattern in the duct was interrupted in this way. It seems that the natural frequencies of the duct sections were shifted to higher innocuous values. A schematic representation of the transition can be seen in Figure 13.18. To achieve the necessary cross-sectional area reduction, the portion of the airline from the collector exit all the way to the fan was replaced in connection with an upgrading of the wind tunnel. This approach yielded very low amplitudes for the pulsations over the entire speed range of the wind tunnel.

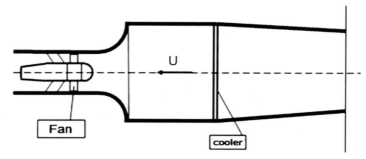

Figure 13.18 Schematic representation of the cross-sectional change in the NWB wind tunnel at DNW, after [72].

Another passive measure can also be the use of one or more Helmholtz resonators. A Helmholtz resonator comprises a cavity with an opening and a connecting neck to the main duct flow or the plenum of the wind tunnel. With proper sizing of the elements, the Helmholtz resonator can reflect acoustic waves out of phase back to the source, thus causing a cancellation, and prevent the acoustic energy from propagating downstream. Due to the importance of Helmholtz resonators, the key equations for calculating the resonant frequency of the resonator can be found in the following. More details can be taken from the work of Beland [65].

$$f_{HR} = \frac{c}{2 \cdot \pi} \cdot \left(\frac{\pi \cdot r^2}{V \cdot t_K} \right)^{0.5} \tag{13.15}$$

where c is the speed of sound, $r = \sqrt{A_H/\pi}$ the equivalent radius of the neck with the cross-sectional area A_H. V is the volume of the resonator cavity, and $t_K = (l_r + \pi r/2)$ is the equivalent neck length of the resonator, where l_r is the physical length of the neck. The bandwidth over which a Helmholtz resonator acts can be calculated with

$$\Delta f = f_o - f_u = \frac{f_{HR}}{2 \cdot \pi \cdot \sqrt{V \cdot \left(\frac{l_r}{A_H}\right)^3}} \qquad (13.16)$$

where f_o and f_u represent the upper and lower cutoff frequencies. The efficiency of a resonator can be calculated with the so-called coupling factor K, which lies between 0.02 and 0.5. A larger coupling factor results in higher efficiency

$$K = 5 \cdot 10^{-7} \cdot f_{HR}^4 \cdot V \cdot \Delta f^{-1} \cdot F_K . \qquad (13.17)$$

Here, F_K is the array factor. Since the room (plenum) to be damped has a great influence on the effectiveness of the resonator, the coupling position of the resonator to the room is important. The array factor assumes a value $F_K = 8$ when the Helmholtz resonator is connected to a corner of the plenum. Along an edge F_K is = 4, set into a wall it is 2, and under free installation (a stand-alone resonator) it is equal to 1. It is worth noting that the effect of the resonator decreases when the opening of the resonator is impacted by strong secondary flows, as is usually the case when the resonator is installed in the downstream part of the plenum hall.

In addition to passive measures to avoid resonance effects, active techniques have been developed. The Audi wind tunnel contains a large number of loudspeakers placed in a chamber connected via an opening in the first cross leg downstream of corner 1. The loudspeakers generate a pressure fluctuation of the same frequency as the duct mode to be suppressed, but out of phase with it. This is described further in Wickern et al. [867].

In this case the time signal of the fluctuating pressure in the plenum is taken as the base and provided via a signal conditioner with a phase angle of 180°. Subsequently, the signal is increased by a power amplifier to the necessary level. In this way the sound pressure level in the plenum can be lowered. For example, at a frequency of 2.5 Hz and a jet velocity of 100 km/h the rms value of the pressure fluctuation is decreased from 121 dB to 97 dB.

Again, a different approach was tested in a model-scale wind tunnel at the University of Stuttgart. Here narrow, elongated loud speaker shafts were installed flush to the lateral edges of the nozzle exit. The initial shear layer was then subjected to sound with frequencies that do not correspond to the resonant frequency. This prevented to a large extent the formation of coherent structures and thus lowered the sound pressure level in the plenum from 133 dB to 112 dB at a speed of 200 km/h. Refer to Heesen and Höpfer [321].

13.2.10.1 Acoustics of the Wind Tunnel Fan

In addition to other components of the wind tunnel, such as the settling chamber, turning vanes, and the plenum, the fan is of particular importance since it is the main sound source in aeroacoustic wind tunnels. The "traditional" measures to reduce the noise of the fan concentrate primarily on the oncoming flow, as well as the quality of the blade flow. In some projects ([745], [730]) the reduction of rotational sound noises has also been emphasized.

If a wind tunnel has to fulfill acoustic requirements, the fan must be considered as the main single sound source in the entire circuit. Instead of the efficiency, the spectral distribution of the sound power has the highest priority and has to be optimized. The objective is to achieve a sufficiently low sound pressure level so that minimum sound-absorbing measures are applied in the test section. In other words, sound that is not produced need not be reduced by expensive sound-absorbing measures. The sound pressure level that is sufficient for testing depends on the acoustic quality of the open jet and on the requirements of the vehicles to be tested. It is appropriate that the fan noise reduced by the silencer in the test section should be lower than the sum of all noise associated with the jet, such as lip-noise, turbulent mixing, and collector noise, so that any subsequent optimization on the jet noise leads to an audible and measurable improvement in the test section and the fan noise does not affect the measurement. A very detailed analysis and classification of axial fan noise is carried out in [322] and shown in Figure 13.19 and Figure 13.20.

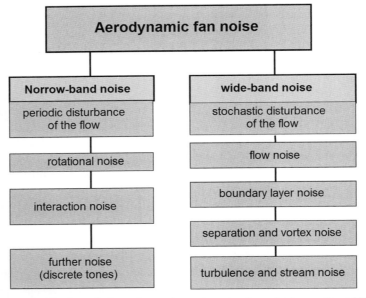

Figure 13.19 Compilation of aerodynamic noise for axial fans, after [322].

The low-frequency impact and flow noise (2) primarily result from disturbances of the incoming flow and can cause a level increase from 6 to 8 dB for circumferential rotor speeds of about 100 m/s [322]. The aerodynamic design of all supports and struts in the air circuit are especially important to help reduce the sound pressure level. The boundary layer noise (3) is usually of minor importance, and separation noise occurs preferably on axial fans without guide vanes. Thin blades with low loss factor, aerodynamically designed blade ends, and a small ring gap between the blade tip and housing help against separation noise. A relative height of 0.001·D has proven to be sufficiently low (see [322]).

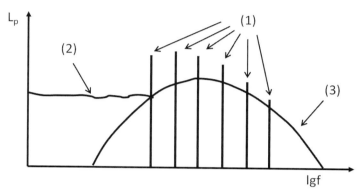

Figure 13.20 Schematic frequency spectrum of the noise of an axial fan, after [322]. (1: Rotational sound and interference noise (see section 13.3); 2: Low-frequency impact and flow noise; 3: Boundary layer and separation noise) [322].

In the following, focus especially is put on the rotational sound (1), because it is often not considered by the fan manufacturers with the necessary attention, unlike the other discussed noise contributions (cf. Figure 13.20). In order to reduce costs the reason may be that standard fans are offered to the customer with always identical number of rotor blade and guide vanes. As shown below, the acoustically ideal number of blades and their design both depend on several operating parameters which can differ quite significantly between different wind tunnels.

Rotational sound is caused by interactions between rotating blades and stationary guide vanes. It is especially well heard and thus disturbing because of its distinctive tonal character. Constructive measures can hardly prevent the generation of rotational sound noise but its propagation. The rotational sound is caused only by the fan blading. In a none-moving space the rotating blade-bounded force field produces periodic pressure fluctuations with the frequency

$$f_{BPF} = m_{LR} \cdot \Omega_{LB} \cdot n_{LS}.$$ (13.18)

Here, f_{BPF} is the blade passing frequency, $m_{LR} = 1,2,3$ is the order of harmonics, Ω_{LB} is the fan speed, and n_{LS} is the number of fan blades. The interference noise arises from the interaction between impeller and guide vanes or, in case of multistage fans, from interactions between the stages. If guide vanes are installed, the resulting pressure field doesn't necessarily rotate anymore with the frequency of the impeller Ω_{LR}, but with

$$\Omega_{Int} = \frac{n_{LS}}{m_u} \cdot \Omega_{LB}.$$ (13.19)

Chapter 13

m_u denotes the mode number and indicates the number of circumferential wavelengths of the rotating pressure field. The case $m_u = 0$ (basic mode) represents the planar wave, that is, on the circumference of the impeller, the same pressure prevails everywhere. For a fan without guide vanes $m_u = n_{LS}$ and therefore, the pressure field rotates exactly with the speed of the impeller. However, if there is a number of n_S guide vanes, the following circumferential modes will be excited:

$$m_u = |m_{LR} \cdot n_{LS} - k \cdot n_S| \text{ with } k =]-\infty;+\infty[\,. \tag{13.20}$$

If $n_{LS} = n_S$ or an integral multiple thereof, the high-energy fundamental mode ($m_u = 0$) will be excited particularly. Therefore, the rotational frequency and their higher harmonics will rather have high tonal components.

In contrast to the basic mode, not all of the calculated modes of the fan may propagate. Depending on the Mach number of the rotating pressure field, it can be determined whether a mode m_u, propagates or decays exponentially with the axial length. Therefore, the fan optimization aims to excite higher modes but only the modes that are already incapable of propagation. The corresponding calculation basics can be found in [810].

Figure 13.21 shows the influence of additional radial modes on the ability to propagate the rotational noise at a fan speed of 350 rpm for the fan/guide vane blade combination of $n_{LS}/n_S = 20/27$. All modes below the lines for $u = $ constant are incapable of propagation. This cut-off criterion is fulfilled even at higher harmonics, if there are additional radial modes characterized by m_{rad}.

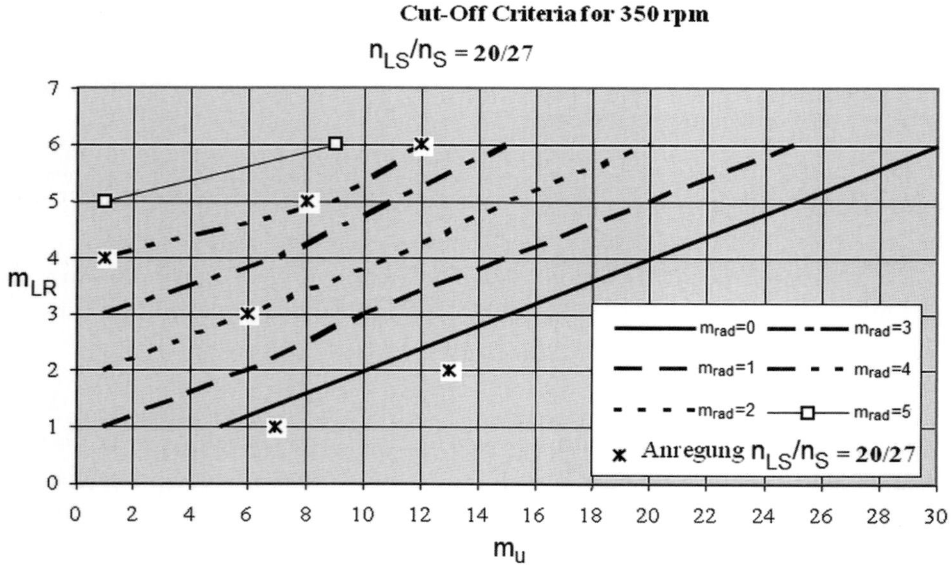

Figure 13.21 Cut-off criteria in relation to the axial and radial mode number m_{rad} for an axial fan with $d_a = 5000$ mm, $d_i = 2500$ mm, speed = 350 min⁻¹ and $n_{LS}/n_S = 20/27$, after [730].

In Figure 13.21, one can see, that the mode $m_u = 7$ (symbolized by "x") is excited (see equation (13.20)) by the basic vibration $m_{LR} = 1$, but this is still located in the cut-off range ($m_{rad} = 0$). The mode $m_u = 13$ is excited by the first rotational sound overtone ($m_{LR} = 2$) and is also in the cut-off range. The mode $m_u = 6$ is excited by $m_{LR} = 3$ and is located outside the cut-off range. The propagation of the mode $m_u = 6$ can only be prevented if the cut-off range is increased by the additional excitation of higher radial modes, approximately up to the curve of $m_{rad} = 2$. Here m_{rad} denotes the number of radial wavelengths in the blade channel. This specifies the radial mode number, similar to the circumferential order m. The more radial modes occur, the higher frequencies (characterized by n) are located in the cut-off range. In the present case, radial modes were investigated up to $m_{rad} = 5$; hence the propagation could be determined up to the harmonic $m_{LR} = 6$. Physically, radial modes can be realized for example by a phase shift between fan hub and blade tip, which requires oblique guide vanes, tapering of the rotor's trailing edge and the stator's leading edge or a radial twisting of the blades. A further summary can be found in [876].

13.2.11 Ground Simulation

The most important requirement for the aerodynamic development of a vehicle in a wind tunnel is first to provide a testing facility that matches the aerodynamic and kinematic boundary conditions on the road at its best. The extent to which this should be done has been subject of discussions for decades. Due to the large range of vehicles tested in wind tunnels and especially with regard to the ground clearance of the vehicle and the aerodynamic optimization potential in the front car area, it has become crucial that

- The oncoming floor boundary layer should be simulated as accurately as possible comparing to the one on the road
- The relative motion between the vehicle and the road should be simulated
- The wheels should rotate

Accordingly, all newer European wind tunnels are more or less equipped with technical devices that fulfill these conditions. In addition, existing older wind tunnels are retrofitted with such devices, which is associated usually with major modifications within the test section and long downtimes during the reconstruction period.

Assuming the vehicle moves on the road in still ambient air, a rectangular velocity profile from the top to the ground in the cross section of the jet would result in the wind tunnel under ideal conditions. However, this condition cannot be fulfilled a priori at the wind tunnel floor because of the no-slip condition of the flow on the ground. A boundary layer exits the nozzle with a typical thickness of about $\delta_{99} = 40$ mm to 150 mm for a full-scale wind tunnel (δ_{99} is the height above the floor where the velocity has reached 99% of the incoming flow again). Depending on the distance to the vehicle, the boundary layer grows downstream due to the turbulent mixing with the core flow. Then, the vehicle forces the boundary layer to deform due to its stagnation effect. Chassis parts that are located in the boundary layer are affected by the impulse loss; hence, the drag coefficient is altered.

Basically, the flow around the overall configuration must be considered, namely the detailed shape of the vehicle plus the floor of the test section. This aspect decides the parts of the volume flow of which the vehicle is passed underneath, above and laterally. The boundary layer might have an effect on the flow separation at the vehicle's rear end, which results in a different circulation around the vehicle, a shifted position of the stagnation point, and a different flow angle. As a consequence, drag and lift may change.

In addition, the boundary layer may also have an effect on the incoming flow angle to the front wheels. Wiedemann [877] has carried out extensive investigations on this subject. Obviously the boundary layer doesn't only change the absolute amount of aerodynamic measures at the vehicle but result in different trends for different vehicle configurations too.

The correct ground simulation in automotive wind tunnels is not only important for vehicles with low ground clearance as it was suggested in the past. It also has to be considered for all other vehicles due to the previously described issue of circulation that can affect the flow and the pressure distribution around the vehicle

Various technical methods are used to reduce the thickness of the incoming floor boundary layer. Three different systems have been established in recent years, which are often used in combination:

- Scoop
- Boundary layer suction
- Tangential blowing

A so-called scoop separates the incoming flow at the height of the boundary layer, guides the air away from the test section, and adds the mass flow again to the jet at the end of the test section. In order to compensate for the associated flow losses, a fan is arranged between inlet and outlet of the scooped flow. A large flow rate can be reached if the scoop is installed over the entire width of the nozzle. The scoop itself is an integral structure of the subsequent test section floor. Thus, the result is a step between nozzle and test section floor and a new boundary layer starts to evolve from the upper leading edge of the scoop. In poor flow conditions given for example by the stagnation effect of the vehicle, the flow could separate at the leading edge of the scoop and the boundary layer can quickly regrow again and reduce the effect of the scoop.

The flow rate that is removed by a boundary layer suction system corresponds to the displacement thickness δ_{1v} of the boundary layer (derived from measurements) in front of the suction system and the jet velocity u_∞. The suction is realized through a perforated sheet, porous plate, or an open slot, which are installed flush to the test section floor. The slot has to be designed with a streamlined inlet. The perforated plate consists of a number of rows of holes with an offset to each other in order to ensure an effective suction. The lengths of the plate in flow direction and the suction velocity determines the suction parameter

$$c_Q = \frac{v_s \cdot s}{u_\infty \cdot \delta_{1v}}.$$ (13.21)

Figure 13.22 shows the context.

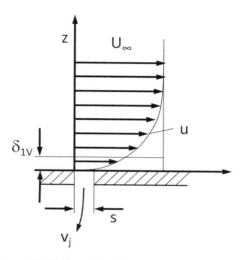

Figure 13.22 Boundary layer suction, after [370].

The suction parameter c_Q has been investigated theoretically and experimentally by several authors ([24], [866]). The ratio $\delta_{1h}/\delta_{1v} = f(cQ)$ can be well approximated with a sixth grade polynomial.

$$\frac{\delta_{1h}}{\delta_{1v}} = K_0 + K_1 \cdot c_Q^6 + \ldots + K_1 \cdot c_Q,$$ (13.22)

where δ_{1h} represents the displacement thickness behind the suction and can be calculated. For $c_Q < 15$ the coefficients of the polynomial are with $K_1 = 1.2 \cdot 10^{-7}$, $K_2 = -1.08 \cdot 10^{-5}$, $K_3 = 3.54 \cdot 10^{-4}$, $K_4 = -5.63 \cdot 10^{-3}$, $K_5 = 4.88 \cdot 10^{-2}$ and $K_6 = -2.71 \cdot 10^{-1}$. The constant is $K_0 = 0.989$. Then, the displacement thickness δ_{1h} behind the suction slot can then be calculated using equations (13.21) and (13.22).

The thickness of the perforated sheet, the number of holes, their diameter, and the porosity of the whole plate are important for the design of suction. These parameters determine the pressure loss and the associated required pump power for the suction system. They need to be interpreted with great care. In order to determine the pressure loss the calculation of the loss coefficient ζ_{LB} of the perforated plate can be found in Brauer [93].

During the design of a suction system it has to be considered that a negative flow angle (facing the road) is created by the boundary layer suction. Close to the ground the value of this flow angle is at its greatest. Mercker and Wiedemann [566] calculated these angles for different suction rates using a theoretical approach. They determined the decay behavior of the flow angles in vertical direction and behind the suction.

In addition, flow material is constantly removed from the jet by the suction along the perforated plate. Thus, the jet is decelerated up to a certain height above the sheet metal. As a consequence, a pressure increase results, which levels out again behind the suction. Thus, a pressure gradient in the flow is induced in downstream direction which, in the worst case, can be detected even at the position of the vehicle.

Both values, flow angle and pressure gradient, are affected by the suction rate and it seems only logical to keep the suction as small as possible and as big as necessary. On the other hand, it must be considered that a positive pressure gradient is also produced by the flow displacement effect of the vehicle. If the suction system is situated in the vicinity of a wheel, for instance, the pressure increase due to a stagnation effect of the wheel causes a further amount of flow ingested through the perforated sheet. The suction rate is increased unpredictably and the negative flow angle becomes larger. Therefore, it might not be desirable in certain circumstances to realize the smallest boundary layer thicknesses achievable.

Tangential blowing attempts to compensate for the momentum loss of the boundary layer. A certain amount of air is taken from the environment and is blown with some over speed into the test section tangentially to the floor via a blower. The slit width is typically only a few millimeters. Appropriate design can help to ensure that the jet stays attached on the ground and no flow separation occurs. Due to turbulent mixing the momentum loss in the boundary layer can be reduced gradually behind the gap. Due to the no-slip condition on the ground behind the suction a new boundary layer is formed. This results in a displacement which pushes the jet upwards and creates a positive flow angle. In addition, a total pressure increase of the flow on the ground will take place. Similar to the suction a horizontal buoyancy effect occurs downstream.

In order to simulate the relative movement between the vehicle and the road a moving belt is used. This steel sheet or fiber-reinforced plastic belt passes the wind tunnel floor using a roller system with an integrated electric drive.

To reduce frictional effects between belt and wind tunnel floor, air is blown below the belt and then sucked off again. The suction prevents the belt from lifting from the ground during operation. There is no general best practice concerning the choice of the free length of the belt in the test section. Basically, it can be said that a good simulation quality is achieved if the lengths of the belt in front and behind the vehicle both are approximately 0.5 to 1 m. This issue was demonstrated in a FAT project (see [236], [234] and Estrada et al. [235]).

Wind Tunnels and Measurement Technique

In the mentioned FAT project, measurements of the pressure distribution in longitudinal direction were carried out on a vehicle for various ground simulations in two wind tunnels (IVK/FKFS and an alternative wind tunnel) with open test section and on the road. The results are shown in Figure 13.23 as differences of the pressure distributions from wind tunnel and road measurements (wind tunnel minus road). Furthermore, different types of interference effects are indicated as they occur in the wind tunnel. At the front of the vehicle, the interference effects of the nozzle and the different boundary layer treatment on the ground occur in both wind tunnels. Jet expansion by the vehicle can be observed in the mid part of the test section. Here, the effect of different blockage ratios in both wind tunnels can be clearly seen. In the alternative wind tunnel this effect was higher by a factor of 2. Finally, the interference effect of the collector [234] can be seen in the rear part of the test section. Interestingly, the biggest differences in both wind tunnels occurred there if the measurements were carried out with and without center belt. It is worth noting that only negligible differences arising in the IVK/FKFS wind tunnel if the boundary layer suction system in front of the belt was switched on or off.

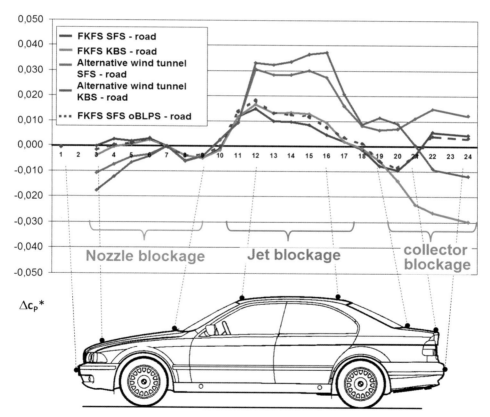

Figure 13.23 The difference of pressure distributions of wind tunnel and on-road measurements for different ground simulations and two different wind tunnels. SFS: road simulation (five-band system and suction). KBS: conventional ground simulation (fixed ground, no suction). SFS o. BLPS: road simulation without pre-suction, cf. [234].

In order to emphasize the difference in fluid mechanics for fixed and moving ground the volume flow between the vehicle and the ground shall be discussed for a steady-state laminar flow. For the fixed floor the so-called Poiseuille flow represents one of the few exact solutions of the integrated Navier-Stokes equation

$$\dot{V} = \frac{h^3}{12\mu} \cdot \left(-\frac{dp}{dx}\right) \tag{13.23}$$

The distance between the ground and the vehicle is defined as h, μ is the dynamic viscosity of the flow, and dp/dx is the pressure gradient between the inlet and the outlet. The exact solution of the so-called Couette flow represents the moving floor:

$$\dot{V} = u_\infty \cdot \frac{h}{2} + \frac{h^3}{12\mu} \cdot \left(-\frac{dp}{dx}\right). \tag{13.24}$$

In case of a negative pressure gradient in the Poiseuille flow and in the Couette flow the flow rate is increased which can be seen in equations (13.23) and (13.24). In case of a positive pressure gradient in equation (13.24), this yields two different terms of different sign. Thus, the flow rate in a Couette flow is reduced compared to the Poiseuille flow. For a given ground distance h and under otherwise equal conditions the flow rate in a Couette flow is a function of the approach flow velocity u_∞ and the pressure gradient, while for the Poiseuille flow, it is only a function of the pressure gradient. This leads to modified drag and lift coefficients.

Mercker et al. carried out measurements in the DNW LLF wind tunnel. The drag coefficient of a vehicle was determined in case of moving ground and rotating wheels as well as fixed ground and fixed wheels using an internal balance held by a strut from behind (see Figure 13.24). At a construction level of the vehicle a difference of 15 drag counts ($\Delta c_D = 0.015$) was measured between both experimental setups. It is worth mentioning that according to Figure 13.24, the wheel drive units don't produce any additional forces (are tare-measurement at target speed). However, lower drag coefficients are measured for moving ground and rotating wheels. This results from a different pressure distribution on the body generated by the rotating wheels, which adds momentum in opposite longitudinal direction and subsequently reduces the drag.

The width of the belt determines the strategy of how to mount the vehicle in the test section under wind load. Various techniques have been developed for a belt which extends the width of the vehicle.

- The vehicle is placed on the belt with its complete weight. The wheels are driven by the belt. Air bearings are installed below the belt corresponding to the positions of the tires in order to prevent from contact between belt and stationary support structure. The vehicle is fixed at the wheel hubs or at the chassis with laterally located cables or rods. Drag and lateral force can be measured with a two component balance connected to the mounting. The lift is measured through the belt with vertical single component balances at the air bearings below the tires (up to now, this technique was employed only with steel belts). The rolling resistance of the wheels must be calibrated before each measurement and subtracted from the result. As described below, this is only feasible approximately, since rolling and ventilation drag resulting from the rotation of the wheels behave non-linear for higher wind speeds. In addition, the rolling resistance depends significantly on the temperature; (cf. Mayer [539]), which can lead to a decrease of the rolling resistance by 40% between operation temperature (60°C) and ambient temperature (20°C). In addition, interference effects are generated by the model mounting.

- The vehicle is fixed in the test section using a strut. The position of the vehicle can be varied with a six axis model manipulator (hexapod) outside of the test section. The aerodynamic loads are measured with an internal six-component balance. The wheels can be driven separated from the chassis via an external wheel carrier by the belt. Drag and lift forces on the wheels are measured using two-component balances inside the wheel carriers. There is also the possibility to fix the wheels floating on the vehicle. Then, the load on the belt is only the weight of the wheels. The deformation of the tire due to the weight of the vehicle is not considered in any case and the lift of the wheels cannot be measured without additional measuring equipment.

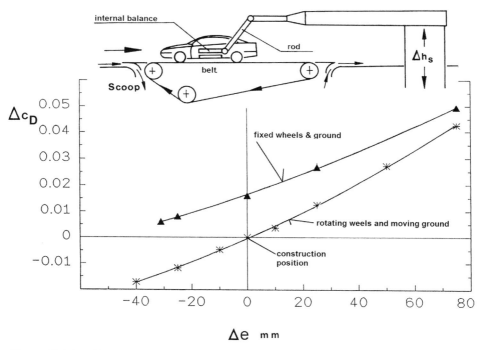

Figure 13.24 Influence of the ground clearance on the drag with stationary and moving ground (the latter with rotating wheels); Δe represents the change of ground clearance according to [556].

- The vehicle is placed on the belt with its complete weight and fixed with cables or rods. The belt and the mounting are part of the weighted part of an external balance (Talingz®). The interference forces of the belt and the mounting resulting from the aerodynamic momentum exchange between moving and static components, such as the air friction on a moving belt, must be recalibrated before every measurement session when the vehicle is changed. In this case, rolling resistance plus ventilation drag of the wheels act only as internal forces and are not measured. The lift forces are measured through the belt at the air bearings below the tires, as was described above.

The biggest challenge is to determine the variety of interference effects within the desired accuracy required. To avoid this uncertainty and to enable for a fast vehicle change, the so-called five-belt system was developed (see Potthoff et al. [652]). The large belt is located between the wheels only, which means a loss of simulation accuracy. The influence of different belt widths has been investigated by Mercker et al. [565].

The vehicle is placed on four small wheel drive units (WDU), which are short, narrow belts (so-called mini belts) and not much bigger than the footprint of the wheel. Technology-wise, they are similar to the center belt. Unlike the center belt, the wheel drive units are connected to an external underfloor balance. The vehicle is mounted on struts to hold it in position under wind load. These are referred to as rocker-panel restraint struts and connect chassis to the external balance without any contact to the wind tunnel floor. In case of non-rotating wheels, the aerodynamic loads act on the six-component balance via struts and wheel drive units. In case of rotating wheels, rolling resistance and ventilation drag of the tire and wheel rim become internal forces compensated by the torque of the WDUs and do not contribute to the overall result. As a result of the ventilation to the chassis, only the secondary forces (e.g., in the wheel arches) are measured.

However, it is often important to know the ventilation drag of the wheels, because car manufacturers have a very large number of different wheel designs and it is certainly of interest, which rim has an optimized ventilation drag with a required flow rate for brake cooling. As an example, a feasible approach should be discussed for a five-band system. The rolling resistance F_R is a function of the road load F_N and the rolling resistance coefficient μ_R:

$$F_R = \mu_R \cdot F_N \tag{13.25}$$

The rolling resistance coefficient μ_R is dependent on the material properties and the shape and the tread of the tire, the air pressure in the tires and the temperature of the tire, and the belt and its surface conditions. For a regular passenger car with summer tire, the rolling resistance coefficient μ_R today is below 0.01 under standard conditions (winter tire: $\mu_R < 0.015$). Up to a speed of about 120 km/h μ_R is approximately constant. At higher speeds, the coefficient increases due to the non-linear characteristics of tire rubber. The temperature of the tire and its tread, as noted above, plays an important role too.

As the rolling resistance and the aerodynamic ventilation drag occur together, they must be separated. Obviously, it is a complex issue when it comes to determine the absolute value of ventilation drag of a wheel. Basically, this is possible as was shown by Mayer [539] with the help of on-road measurements, but in the wind tunnel, this seems to be partly possible only. Mayer was able to show that the percentage change of ventilation drag of a tire plus its rim is independent of the driving speed at a velocity above 120 km/h (compared to a reference wheel with same tires and different rim). If all other test conditions stay the same, an investigation of different rims can be done.

The WDU itself can be mounted on a one-component balance acting in the x direction. If the vehicle is now fixed with the non-weighted part of the external balance via the restraint struts, the vehicle itself acts as a kind of wheel holder and the rolling resistance force plus the ventilation drag force can be measured under wind as a reactive force F_x on the one-component balance. To keep the error small for such measurements, it is recommended to disassemble the drive axles and keep the brakes free. Typically the drag of the wheel bearing is negligible and can be ignored in this analysis.

The topology of the flow field around and through the wheel interferes with the flow field in the wheel arch and around the vehicle. Interestingly, measurements by Mayer revealed that the change of the total aerodynamic drag (air drag plus ventilation drag) was independent on the velocity even at speeds of u_∞ > 120 km/h for different wheels.

Other interference phenomena also occur for the lift of the vehicle. Due to constructive constraints, it is not possible to build WDUs small enough to avoid free surface parts on the WDU belts besides the tire. Thus, the pressure field of the chassis and the wheels produces additional vertical forces on the free surface, which are measured with the external balance. Widdecke and Potthoff [871] developed a patented process (FKFS *pace*®) to eliminate these additional forces. Corrections methods of the so-called pad effect were presented by Wickern et al. [865] and Cogotti et al. [167].

13.2.12 Unsteady Flow and Gust Simulation

In the preceding sections, steady-state wind tunnel flow conditions have been treated. If the dynamics of a sudden excitation of a vehicle due to wind gusts is to be studied, additional equipment must be provided. Usually, the dependency of aerodynamic loads on a vehicle under crosswind is measured in a wind tunnel by rotating the vehicle around its vertical axis (yaw). For that purpose, the vehicle is installed on a turntable and the external balance must also be able to rotate in order to measure the aerodynamic forces. The crosswind sensitivity of the vehicle can then be derived from the side force and the yaw moment. However, this type of measurement provides averaged values, and all effects that occur due to an unsteady oncoming flow are neglected. But the forces in an unsteady flow (i.e., an oncoming flow), which varies temporally and spatially, could be higher than in steady flow. This has a great influence on the driving behavior of a vehicle. A methodology to simulate wind gusts has been developed by Dominy and Ryan [211]; their experimental setup is shown in Figure 13.25.

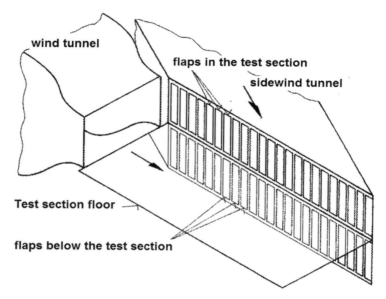

Figure 13.25 Schematic representation of a crosswind tunnel, after [211].

In addition to the wind tunnel, a crosswind tunnel is set up, of which the longitudinal axis is turned by $\beta = 30°$ compared to the axis in the first tunnel. Perpendicular to the test section floor, a vertical wall is arranged that is equipped with two rows of flaps, one above and one below the ground. These are connected to each other in such a way that whenever a flap is opened up, the underlying is closed. Therefore, the crosswind tunnel runs at nearly constant airflow. The flaps are operated in such a way that the vehicle is excited by the 30° crosswind perpendicular to its longitudinal axis for about 0.3 seconds with constant wind speed and direction. Unsteady flow effects can be studied this way.

For certain aeroacoustics problems, it is necessary to reproduce the ground boundary layer of the wind directly above the street, including its turbulence, because wind noise is modulated by the natural turbulence of the surrounding air. Cogotti [166] developed a device for this. Figure 13.26 gives an idea of the constructive solution. Five vortex generators are arranged behind the inlet to the nozzle shortly in front of the contraction.

Wind Tunnels and Measurement Technique

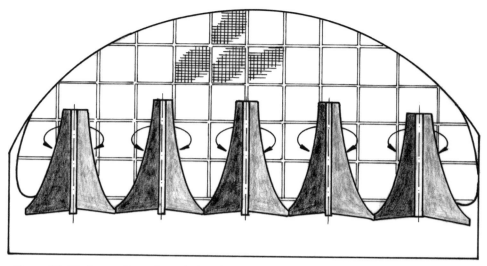

Figure 13.26 Generation of a turbulent boundary layer profile using periodically moving pairs of flaps, after [166].

Each vortex generator consists of a vertical strut on which a pair of flaps is mounted. These five pairs of flaps perform periodic oscillations. This way a velocity profile can be created, which is almost equivalent to the potential law of the urban boundary layer, of which the exponent is $0.16 < \alpha < 0.24$:

$$\frac{u}{u_\infty} = \left(\frac{z}{\delta}\right)^\alpha. \qquad (13.26)$$

The average degree of turbulence amounts to 7–9%; the average size of the turbulence structures is 0.7–1.3 m. The pairs of flaps and their drive unit are mounted on a common device. The entire assembly can be sunk in the ground when it is not needed. The limitation of the system to application on boundary layer profiles raises the question whether they are of particular interest for transient (gust) effects on vehicles or if other transient effects are rather important for the driving behavior, for example, due to vortices coming from wake areas of large vehicles moving ahead. In order to answer this question, Schröck et al. ([742], [739]) carried out extensive measurements of the flow parameters (magnitude and direction) on a vehicle in road traffic. He received a very large, statistically relevant data base, which revealed that the yaw angle is not larger than ±10° even at strong wind gusts and high speeds of up to about 160 km/h (refer to Figure 13.27).

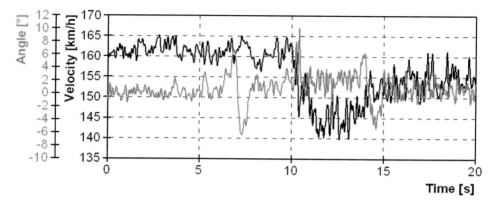

Figure 13.27 Yaw angle and relative velocity behind the vehicle during on-road measurements at a speed of 160 km/h, carried out by Schröck et al. [742].

Schröck [739] developed an alternative way to simulate side wind gusts. In these investigations, the gust was generated by vertical, symmetrical wing profiles performing periodic oscillations. In this way and with the help of a drive device the angle of attack (maximum ±10° due to the above mentioned on-road tests) of the profiles as well as the oscillation frequency (maximum 15 Hz) could be varied independently. The representation of the boundary layer of the natural wind was considered to be dispensable by Schröck as a consequence of his on-road tests. In a pilot study Schröck et al. [742] used four profiles in the nozzle exit plane. For the accomplished implementation of this technique in the 1:1-vehicle wind tunnel of the University of Stuttgart, eight profiles are provided (FKFS *swing*®); see Figure 13.28.

Figure 13.28 View into the test section of IVK/FKFS full-scale wind tunnel showing the gust generator FKFS *swing*® in the nozzle exit (only seven profiles shown). Source: FKFS.

As an evaluation criterion for the sensitivity of a vehicle on a side wind gust, Schröck et al. [742] used the so-called aerodynamic admittance function as well as the transfer function. The admittance function, introduced originally by Davenport [189] in building aerodynamics is a frequency dependent function which relates the spectrum of measured unsteady forces and moments to the spectrum of forces or moments derived from a quasi-steady theory using the unsteady wind input and the gradients of the side force or yaw moment coefficient. With this approach only a relative comparison of the unsteady forces across the frequency range is possible. A further step was taken by Schröck [739] by relating the spectrum of the measured unsteady force or moment to the spectrum of the unsteady crosswind excitation, making an absolute comparison of different vehicles or vehicle configurations possible. However, it is absolutely essential to ensure high coherence between input and output in order not to interpret flow turbulence caused by separations as a result of a dynamic approach flow. For such measurements an external balance is needed that has the ability to ensure sufficient time resolution of the measured signals. The admittance function follows as

$$X^2_{a,S/M,Y}(f) = \frac{S_{c_{S/M,Y}}(f)}{\left(\dfrac{dc_{S/M,Y}}{d\beta}\right)^2 \cdot S_\beta(f)} \tag{13.27}$$

and the transfer function as

$$H^2_{a,S/M,Y}(f) = \frac{S_{c_{S/M,Y}}(f)}{S_\beta(f)} \tag{13.28}$$

Here f denotes the frequency, $X^2_{a,S/M,Y}(f)$ the admittance function of the side force (F_S) or the yaw moment (M_Y), $H^2_{a,S/M,Y}(f)$ the transfer function of the side force or the yaw moment, $S_{c_{S/M,Y}}(f)$ the sprectrum of the side force or the yaw moment, $S_\beta(f)$ the spectrum of the yaw angle β and $dc_{S/M,Y}/d\beta$ the side force or yaw moment gradient in steady-state conditions. Further information can also be found in chapter 5.

13.2.13 Wind Tunnel Correction Methods

Because of the finite dimensions of the test section, the flow in a wind tunnel differs from an infinitely extended flow field (on the road). Depending on the boundary conditions of the jet, interference forces occur that must be corrected. The amount of interference forces depends on the geometrical properties of the vehicle in relation to the cross section of the jet (blockage ratio), the distance between nozzle and collector, the position of the vehicle in the test section and, if present, on the static pressure gradient in the empty test section.

Originating from correction procedures applied in aircraft aerodynamics, special procedures have been developed over the past 25 years for bluff bodies such as vehicles. The quality of correction procedures applied can be validated by various comparison tests using a large number of wind tunnels. Basically, two different correction procedures can be distinguished:

- "Classic" correction procedures are based on analytical flow models and require the input of geometric parameters of the vehicle and the test section.
- Wall pressure methods can only be used in closed test sections and are based on the measurement of the static pressure distribution at multiple longitudinal sections along the tunnel walls and the ceiling or at peripheral sections perpendicular to the flow direction.

Wall pressure methods are generally considered as procedures with the highest accuracy. As it will be shown, this is not always true for bluff bodies. Even with a large blockage ratio (> 15%) modern classical correction procedures yield a good match of the corrected coefficients which sometimes cannot be achieved even with wall pressure procedures.

The determination of interference effects is based on the assumption that they change the velocity field around the vehicle by an amount Δu. In addition, a force arises due to flow gradients. With regard to the velocity, we get the dimensionless interference speed ε:

$$\frac{\Delta u}{u_\infty} = \varepsilon. \tag{13.29}$$

The dynamic pressure q can then be calculated by the sum of all interference effects ε_m

$$\frac{q}{q_\infty} = \left(1 + \sum \varepsilon_m\right)^2 = n_K \tag{13.30}$$

where n_K represents the correction factor. q_∞ denotes the dynamic pressure of the undisturbed flow. The corrected aerodynamic drag coefficient c_{Dc} follows as

$$c_{Dc} = \frac{c_{Dm}}{n_K} \tag{13.31}$$

where the indices "c" and "m" stand for the corrected or measured coefficient. With the Bernoulli equation the pressure coefficient c_p results in

$$c_{pc} = \frac{c_{pm}}{n_K} + \frac{n_K - 1}{n_K} \tag{13.32}$$

The correction of the drag coefficient Δc_{DB} due to a pressure gradient dp/dx can be derived from the integral of the pressure distribution around the vehicle. Using the Gaussian theorem, the mathematical transformation of the surface integral into a volume integral results in a simple equation for the additional drag:

$$\Delta c_{DB} = G \cdot \frac{(V_F + V_T)}{A_x} \cdot \frac{dp}{dx} \tag{13.33}$$

Herein, G is the so-called Glauert factor calculated from vehicle properties. According to Glauert [283], this factor has to be included for a body exposed to an accelerated flow. In equation (13.33) the effective volume of the flow displacement of the vehicle is taken into account, which is the volume of the vehicle V_F plus the volume V_T for the separated wake region. The displacement volume of the wake region V_T can be approximated after Garry et al. [265] (see also Mercker [559] and Wickern [864]). With reference to equation (13.31) the overall corrected drag can be calculated by:

$$c_{Dc} = \frac{c_{Dm}}{n_K} + \Delta c_{DB}. \tag{13.34}$$

13.2.13.1 Closed Test Section

As already discussed in section 13.2.3, an interference velocity occurs at the solid walls of the test section due to the flow displacement effect of the vehicle and its wake. In order to calculate the interference velocity, vehicle and test section are mirror imaged mathematically at the floor. This results in a duplex nozzle exit area of $A_N = 2 \cdot h_N \cdot b_N$ (h_N = test section height; b_N = test section width) and a duplex vehicle frontal area of $A_M = 2 \cdot A_x$. Mathematically, the vehicle is then located in the center of the test section. If now the vehicle and its image both are replaced by the superposition of a potential flow source and a sink of identical strength, a three-dimensional dipole is created which has the shape of a sphere. If the dipole is mirror-imaged at a distance to one of the test section walls, a streamline is generated that corresponds to the shape of the wall. Extending the system on all four walls, the dipoles generate a curvature of the wall streamline due to the different distances to a considered point on the wall. However, continuing this procedure to infinity, we get a double-infinite series in a mathematical sense that converges to a fixed value τ_K, the so-called tunnel shape factor. The streamlines with distances of h_N and $b_N/2$ to the vehicle correspond to the shape of the straight walls again. After [864], we get to a good order of approximation for

$$\tau_K = 0.406 \cdot \left(\frac{b_N}{2h_N} + \frac{2h_N}{b_N} \right) \tag{13.35}$$

and the interference velocity of the dipole, ε_{DP} can be written as

$$\varepsilon_{DP} = \tau_K \cdot \left(\frac{A_D}{A_N} \right)^{3/2}, \tag{13.36}$$

where A_D is the cross section of the dipole. By introducing a vehicle shape factor λ_F, equation (13.36) can be specified for any frontal area, volume, and length of a solid body:

$$\varepsilon_S = \tau_K \cdot \lambda_F \cdot \left(\frac{A_M}{A_N} \right)^{3/2}. \tag{13.37}$$

Here, ε_S denotes the interference velocity, which arises as a result of the flow displacement effect of the vehicle and its wake. The effect of the inherent flow gradient generated on the walls is initially ignored because the effects cancel out behind and in front of the largest cross section of the vehicle. Because of different physical considerations, the vehicle shape factor differs in the various analytical correction methods and thus leads to different results. Refer to Wickern [864], Mercker [557], and Cooper et al. [175].

In addition to the interference velocity for the displacement ε_S of the flow, another effect occurs caused by the displacement of the far field wake. It is called ε_W and can be calculated similarly using:

$$\varepsilon_W = \frac{1}{4} \cdot \frac{A_{SEP}}{A_N} \cdot c_{Dm}. \tag{13.38}$$

Here, A_{SEP} denotes the flow separation cross section of the near field wake and c_{Dm} is the measured drag coefficient. After Mercker [559] and Hackett et al. [307], the procedure also takes into account that the near field wake of the vehicle will be deformed due to the inherent flow gradients on the walls. This is the point where the effect of flow gradients on the walls is considered. A consequence of the wake deformation is that at the point of flow separation on the vehicle the flow velocity changes and, thus, the base pressure of the vehicle is changed. Mercker [559] determines the change of the separation velocity on the vehicle using an experimentally determined constant and the blockage ratio of the wake.

$$\varepsilon_{TD} = 0.41 \cdot \frac{A_{SEP}}{A_N} \tag{13.39}$$

As a result of the altered separation velocity a pressure gradient occurs along the vehicle, and its effect can be calculated according to equation (13.33):

$$\Delta c_{DBF} = G \cdot \frac{V_F}{A_x} \cdot \frac{dp}{dx}. \tag{13.40}$$

In equation (13.40) the pressure gradient dp/dx is negative and exerts a force on the vehicle in flow direction. As explained further below, the pressure gradient can be calculated with the help of a ring vortex model. The correction equation for closed test sections follows similar to equation (13.34):

$$c_{Dc} = \frac{c_{Dm}}{n_K} - \Delta c_{DBF}. \tag{13.41}$$

The correction factor n_K follows from equation (13.30):

$$n_K = (1 + \varepsilon_S + \varepsilon_W + \varepsilon_{TD})^2. \tag{13.42}$$

If there is already a pressure gradient in the empty closed test section, we can proceed the same way, and equation (13.41) can be expanded accordingly with a further term. In contrast to the above-described procedure, Hackett and Cooper [307] proposed a global momentum analysis for the deformation of the wake. Originally derived from Maskell [534], the procedure can be found in the literature as Maskell-III.

As mentioned, besides analytic procedures, there are also wall pressure methods. In the procedure by Hackett and Wilsden [308] ("pressure signature method"), the measured longitudinal steady-state pressure distribution on the walls is approximated with the help of a potential flow model consisting of a spatially distributed source-sink system. This system is applied to the principle of the mirror-imaging technique. In this way, a more accurate representation of the model is feasible.

Another method was proposed by Ashill et al. [26] using the free-air Green's function. They were able to show that the interference velocity can be calculated directly through the pressure distribution on the walls without representing the body (vehicle) itself. The Green's function links the velocity potential around a body in an infinitely large wind tunnel with one in a finite tunnel. The integration of the equations, then, cancels out model-specific parameters. The method includes the pressure measurement on the tunnel walls in several cross sections in flow direction. The procedure is known as two-variable method in the literature and is particularly attractive when dealing with complicated body shapes.

In 1978, Mercker and Fiedler [562] investigated this idea using two-dimensional bodies based on experimental data. They could show that the velocity distribution between wall and model is always affine for various blockage ratios and model shapes when applying an appropriate normalization scale. With the help of wall pressure measurement, a blockage correction could be carried out without representing the model analytically. However, the mathematical formulation of the physical context for three-dimensional bodies is missing in [562], as it was derived by Ashill et al. [26].

Finally, it should be mentioned that a fairly good approximation of the blockage effect can often be achieved with the so-called continuity method, if the blockage ratio is < 6%. The method is based on a consideration of the continuity equation in the narrowest cross section between vehicle and tunnel wall and we get

$$c_{Dc} = c_{Dm} \cdot \left(1 + \frac{A_M}{A_N}\right)^{-2}. \tag{13.43}$$

In Figure 13.29a and b, a comparison is shown of various correction methods (see Mercker [559]). Gleason [285] computed the flow field around a passenger vehicle with different rear end shapes (squareback, hatchback) for different boundary conditions (blockage ratios) using CFD. The ordinate in Figure 13.29 arises from the difference between the drag coefficient corrected with the various procedures and the uncorrected drag coefficient extrapolated to 0% blockage (see [559]). In addition, the latter was chosen as the reference value to allow a representation in percent. The quality of each method can be evaluated if one considers that the uncorrected drag coefficient of a typical squareback vehicle with 15% blockage in a closed test section differs from the drag coefficient in free flow (0% blockage) by approximately 35%.

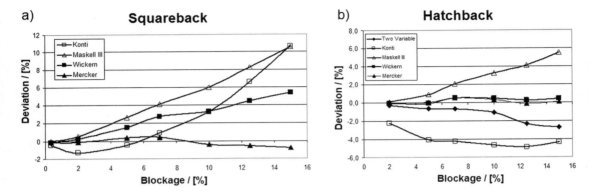

Figure 13.29 Deviation of the corrected drag coefficient to the "true" value related to the free flow as a function of the blockage. (Continuity after equation (13.43); Maskell III after [534], Wickern after [864]; Mercker after [559]; Two variable method after Ashill [26]).

Since the uncorrected values are measured too high in the closed test section compared to the road, a positive deviation means an undercorrection (corrected values are too high) while negative deviations represent an overcorrection (corrected values are too low).

In the same way, Figure 13.30 shows some results as they were achieved on commercial vehicle models in the NRC wind tunnel in Canada (see [173]). Here, the cross section of the tunnel was constant and the models were measured with different scales.

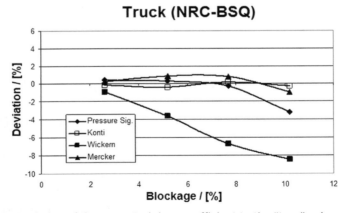

Figure 13.30 Deviation of the corrected drag coefficient to the "true" value related to the free flow as a function of the blockage ratio. (Continuity after equation (13.43); Wickern after [864]; Mercker after [559]; Pressure Sig. after [308]). Note: Due to the influence of the boundary layer on the test section floor, an adaptation of the uncorrected c_D was carried out for the model with 2.4% blockage.

13.2.13.2 The Open-Jet Test Section

Also for open test sections, the principles of classical corrections can be applied, taking into account the changing boundary conditions at the edge of the jet. As already mentioned, p_∞, u_∞ = constant are the boundary conditions along a boundary streamline of the jet (the pressure of the plenum will be imprinted on the jet and no flow gradients occur). With regard to the mirror imaging principle, the boundary conditions can be fulfilled if the signs of adjacent dipoles alternate (see [561]). For the example of two dipoles of different sign, the induced interference velocities cancel out in the mirror planes. On the other hand, a negative interference velocity remains at the position of the vehicle, when the sum of the double infinite series is computed. After [864], we get approximately

$$\tau_K = -0.03 \cdot \left(\frac{b_N}{2h_N} + \frac{2h_N}{b_N} \right)^3. \tag{13.44}$$

As mentioned in section 13.2.3, the jet expands with the presence of a vehicle more than the natural amount of displacement on the road. Since no flow gradients are generated at the edges of the jet, no deformation of the wake occurs. The blockage corrections in open test sections seem easier to be formulated. However, additional blockage effects occur in an open test section, which were treated by Mercker and Wiedemann ([568], [567]) in the mid-1990s. Ten years later the procedure by Mercker and Wiedemann was refined with a detailed examination of the pressure gradient in the empty test section (see Mercker et al. [560], [561]).

In addition to the effect of the pressure gradient in the empty test section, further effects occur, as mentioned:

- Jet expansion
- Jet deflection
- Nozzle blockage
- Collector blockage

The interference effects are represented in Figure 13.31 [370].

In addition to the jet expansion, the ram effect of the flow in front of the vehicle deflects the jet. As shown in [568], an effective nozzle cross section results that is less than the pure geometric cross section. The nozzle blockage corrections are calculated with the correspondingly smaller area.

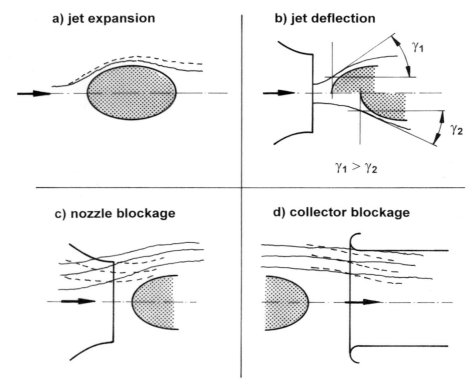

Figure 13.31 Interference effects in open-jet test sections after Mercker & Wiedemann in [370].

The nozzle blockage captures the stagnation effect of the vehicle and its effect on the determination of the reference speed u_∞. The stagnation effect itself is calculated approximately with a potential flow approach and is determined from the superposition of a potential flow source with a parallel flow. The result is a semi-infinite body of revolution, of which the source strength is chosen so that its cross-section in infinity corresponds to the frontal area of the vehicle. The integration of the induced velocity field of the body of revolution in the nozzle cross section provides the dimension free velocity deviation ε_D due to the nozzle blockage. Its value depends on whether the nozzle velocity u_∞ was evaluated with the nozzle or plenum method. In addition, the source position x_Q of the body of revolution in the test section plays a crucial role. However, by iteration, the location x_Q in the test section can be found where the nozzle blockage ε_D is the same for the plenum and nozzle method. Therefore, nozzle blockage ε_D is independent from the method of how u_∞ is determined. Interestingly a different value results for the nozzle blockage ε_D if the vehicle is measured with or without cooling airflow.

Since the vehicle is placed in the far field of the nozzle, the effect of the nozzle blockage ε_D must be calculated at the position of the vehicle. Therefore, the nozzle is replaced mathematically with a ring vortex with a circulation Γ corresponding to the nozzle blockage ε_D. With the help of the law of Biot-Savart, the far-field effect of the nozzle blockage ε_N can be calculated at the location of the vehicle. It arises (formular) with (formular) as the equivalent radius of nozzle.

$$\varepsilon_N = \varepsilon_D \cdot \left[\frac{r_D^3}{\left(x_Q^2 + r_D^2\right)^{3/2}}\right] \text{ with } r_D = \sqrt{\frac{A_N}{\pi}} \tag{13.45}$$

Similar considerations can be made for the collector as well. If the near wake reaches the collector, the situation is analogous to the closed test section. Due to the resulting flow gradients, the collector deforms the wake. In addition, the flow velocity increases due to the displacement effect of the far field wake. Using the equations (13.38) and (13.39) for the closed test section, the collector blockage and thus the dimensionless correction velocity ε_K can be calculated. The far-field effect of the collector can then be determined in analogy to equation (13.45) with the help of the potential flow model of a ring vortex. At the location of the vehicle, the interference velocity ε_C induced by the collector is

$$\varepsilon_C = \varepsilon_K \cdot \left[\frac{r_K^3}{\left(x_K^2 + r_K^2\right)^{3/2}}\right]. \tag{13.46}$$

The equivalent radius r_K is calculated from the dimensions of the collector, and x_K (in equation (13.46)) is determined by the distance to the vehicle. The ring vortex model of the nozzle and the collector is represented schematically in Figure 13.32.

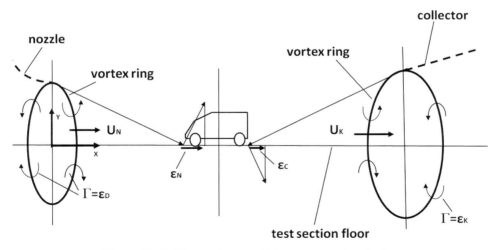

Figure 13.32 Ring vortex model for nozzle and collector.

The correction factor n_K arises for the open test section as

$$n_K = \left(1 + \varepsilon_S + \varepsilon_N + \varepsilon_C\right)^2. \tag{13.47}$$

With equation (13.47) we finally get the corrected drag coefficient

$$c_{Dc} = \frac{c_{Dm}}{n_K}. \tag{13.48}$$

As was mentioned (see section 13.2.5), open test sections may exhibit a pressure gradient even without a vehicle. This pressure gradient would interfere with the flow around the vehicle. As discussed for the closed test section, a force acts on the vehicle, which is composed of the gradient effect on the volume of the vehicle plus the deformation of the wake. As the pressure gradient does not depend on the conditions of the free jet (unlike in the closed test section), a related content between the deformation of the wake and pure vehicle and wake dimensions can only be found approximately (see [561]). Therefore, Mercker and Cooper [560] suggested an experimental procedure that proved to be sufficiently accurate (see [558], [284]).

To do this, a second different pressure gradient must be created in the empty test section. The idea behind this approach is borrowed again from the corrections in closed sections. Here, the deformation of the wake could be determined with the help of an inherent pressure gradient with the constant 0.41 (see equation (13.39)). For an open test section it is postulated in [560] that a length x_{sl} exists, which is characteristic for a vehicle and its flow about.

Considering the pressure difference using pressure coefficients for two different gradients between the front bumper at the point x_1 of the vehicle and any point x_2 downstream, an additional drag coefficient Δc_D arises for the corresponding pressure gradient. Assuming a linear pressure gradient, it follows in good approximation

$$\Delta c_D(x) = c_p(x_2) - c_p(x_1). \tag{13.49}$$

Taking into account the interference effects of the nozzle, the collector, and the jet (equations (13.47) and (13.48)), corrected drag coefficients c_{Dc} result for both pressure gradients that differ initially for values of x_2

$$c_{Dc}(x) = \frac{[c_{Dm} + \Delta c_D(x)]}{n_K}. \tag{13.50}$$

If now changing gradually the downstream horizontal position x_2, the corrected drag coefficient can be finally calculated if both pressure gradients result in the same corrected value after the iteration. Then, x_2 represents the characteristic length x_{sl} and can be used in equation (13.46). Equation (13.50) thus represents the correction equation for the drag coefficient in the open test section. This relationship is schematically represented in Figure 13.33.

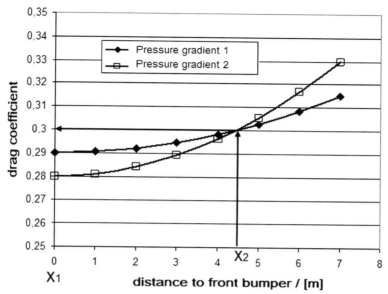

Figure 13.33 Determination of the drag coefficient due to the effect of different pressure gradients in the empty test section [560].

The generation of a second pressure gradient in the empty test section can be done for example by using additional ramps or flaps at the rear of the test section (see e.g., [841]). Geometrical changes of the collector would be another approach. The vehicle is then measured in two pressure gradients. It is important that the test section was previously calibrated for both gradients.

The effect of different pressure gradients on the wake area behind a vehicle can be seen in Figure 13.34. The diagram illustrates the total pressure difference $\Delta c_{pT} = c_{pT1} - c_{pT2}$ in an open-jet test section of a model wind tunnel. Here, the first measurement of c_{pT1} was carried out with a pressure gradient $c_p(x)$ in the empty test section, where the pressure gradient was generated by means of additional flaps in the collector. The second pressure gradient without flaps was negligible (c_{pT2}). Interestingly, the largest total pressure difference occurred around the rear base surface of the vehicle and not, as one might suspect, further downstream. This was where the differences of the pressure coefficient of the gradients in the empty test section increased. Furthermore, a distinct gradient of the total pressure difference occurs at the top of the model's near wake (Figure 13.34). Apparently, this is due to the expansion of the wake behind the vehicle, which in turn is a result of the pressure increase caused by the gradient in the empty test section. It characterizes the deformation of the wake.

Chapter 13

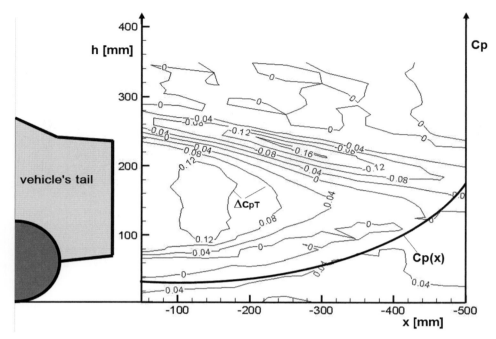

Figure 13.34 Total pressure difference ($\Delta c_{pT} = c_{pT1} - c_{pT2}$) in the wake of a vehicle in different gradients of the empty test section, after [304]. The trend of the static pressure gradient $c_p(x)$ is shown schematically for the measurement with additional flaps in the collector

The quality of the correction procedure was tested based on measurements in 13 different open-jet wind tunnels in Europe and the United States. Twelve different vehicles were measured (three vehicles in the United States). The measurements were supervised by EADE in 2010/2011 (see [558]). Regarding the drag coefficient, the result in Figure 13.35 is represented using standard deviations over all tunnels for the different vehicles. While a standard deviation of 7 drag counts ($\Delta c_D = 0.007$) can be computed for the uncorrected drag coefficients, these values decrease down to 3 drag counts after the correction.

Final note: If wind tunnel corrections should be used, it is desirable to have an instruction available in order to execute the necessary computations either via programmable worksheets of a separate computer or with a computer program in the data processing system of the wind tunnel. For the closed test section, a clear summary of the equations can be found in [559] for the procedure after Mercker (2013) and in [864] for the procedure after Wickern (2001).

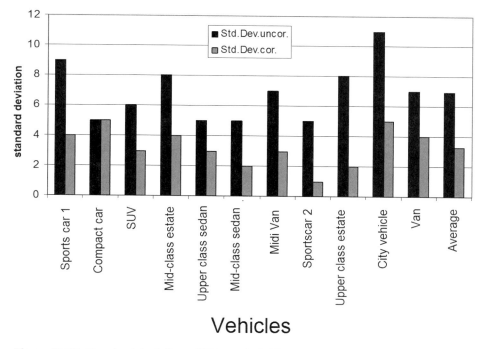

Figure 13.35 Standard deviation of 13 tunnels (without ground simulation) for the drag coefficient of various vehicles, after [558].

For the open test section procedure from Mercker and Wiedemann, a refinement of the method has taken place over the years, creating a universality of application. A clear summary of the equations can be found in a work by Walter et al. [841], (2012).

13.2.13.3 Blockage Correction Methods for Climatic and Thermal Wind Tunnels

While the previously mentioned interference corrections are used in aerodynamic wind tunnels, a correction of the approach flow velocity is omitted in climatic and thermal wind tunnels. This was assumed to be justified because in such test facilities, a proper airflow must be guaranteed in the front area of the vehicle only in order to get a correct airflow through the radiator. It has been assumed that errors in the flow downstream play only a minor role. As will be shown in the following section, this is absolutely not the case.

Since measurements in climatic and thermal wind tunnels are often carried out in test sections with smaller jet cross sections, a consideration of the interference velocity is important over the length of the vehicle and thus as a function of the axial location. For passenger cars, the bandwidth of nozzle cross sections of such tunnels is from 4 to 12 m².

If the heat rejection of certain aggregates (exhaust system, gear box and rear axle drive) in the underfloor area should be considered, a closer investigation of the interference velocities in the particular part of the vehicle is useful. As an example, the interference velocity due to the jet expansion in an open test section is shown in Figure 13.36 for a square back vehicle as a function of the axial location. Here the interference effects due to nozzle and collector blockage are not included. The effect of a horizontal pressure gradient in the empty test section was also not included.

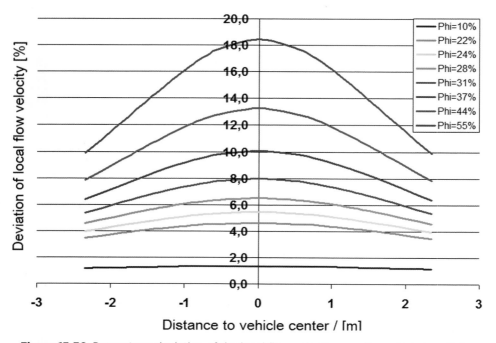

Figure 13.36 Percentage deviation of the local flow velocity over the vehicle length for various blockage conditions $\varphi = A_x/A_N$ only due to jet expansion. Position 0 means center wheelbase of the vehicle with an overall length of the vehicle of 4.8 m; vehicle volume 6.5 m³; frontal area 2.2 m².

As Figure 13.36 shows, for all blockage ratios a smaller reference velocity u_∞ is calculated compared to the empty test section. For a blockage ratio of $\varphi = 10\%$ (corresponding to a nozzle cross section of 20 m² for this vehicle) the deficit is 0.7% over the entire length of the vehicle and is more or less constant. Because of that, the assumption that a single interference velocity for the entire vehicle tested in an aerodynamic wind tunnel is valid seems belated to be justified, at least for such blockage ratios.

A blockage ratio of 22% with a nozzle cross section of almost 10 m² results in an average deviation of about 4%. The difference between the vehicle center and the front of the vehicle (e.g.) is then less than 1%. If the nozzle cross section is further reduced down to 6 m² (37% of blockage), deviations occur between vehicle center and vehicle front or vehicle end of almost 4% and the maximum deviations is almost 10% at the vehicle center. This rises to over 18%, if the nozzle cross section is only 4 m² ($\varphi = 55\%$) with the same vehicle.

In summary, it can be said that depending on the blockage ratio it seems useful to consider the interference velocities on the vehicle as a function of the axial position when it comes to the investigation of cooling issues. Here, it is not enough to change just the oncoming velocity at a particular location on the underbody using the local interference velocity, because it changes the thermal management of the engine and other components. For temperature measurement at a certain load, the real temperature must be calculated from the temperature response and the interference velocity, for example. It can be said that in an open-jet test section, the heat dissipation in the underfloor area is lower than for a road trip due to the lower speed. Here, a worst-case estimation is carried out. Reverse conditions occur in a closed test section. The heat dissipation is increased by higher flow rates. A correction can be carried out using pressure measurements and equation (13.32) for flow-dependent measurements in which the temperature plays no or only a minor role.

13.3 Wind Tunnel Measurements

After designating the previous sections to explain the physics and their resulting effects in a wind tunnel, this section illustrates specific topics regarding measurements in wind tunnels. This includes the measurement procedure, different measuring techniques, and particularities of different fields of study, such as vehicle soiling and thermal management. It should be noted that acoustic measuring technology is described in chapter 8 along with aerodynamic noise generation.

13.3.1 Test Sequence

In order to investigate the aerodynamic characteristics of a vehicle in a wind tunnel, a certain measuring procedure must be complied with. This procedure is structured in a similar fashion in most wind tunnels. Hereinafter, a procedure shall be described, which relates to simulated road travel in a wind tunnel equipped with a long center belt passing between the tires on the underside of the vehicle. A system of belts, consisting of four single-wheel rotation units, drives all four vehicle wheels. This five-belt configuration represents the current state of the art technology in all wind tunnels capable of investigating series production vehicles via modern ground simulation technology. A single-belt system, currently used to investigate racing cars, is considerably wider and longer than the vehicle itself. This system shall not be discussed in this chapter.

To ensure process reliability and to be able to move vehicles into the wind tunnel quickly, certain geometric properties such as wheelbase, track width of the front and rear axles, and tires must be known. The wheel rotation units can be positioned in the wind tunnel in advance with respect to the wheelbase and track width. The proper widths of the center belt and wheel rotation unit belts can be chosen accordingly. The value of the vehicle's frontal area is needed to calculate aerodynamic coefficients based on measurements of force and moments using the wind tunnel balance.

A vehicle being measured in a wind tunnel without being properly fixed would dislodge from the measuring position due to the wind load or the wheels driven via the wheel rotation units. It is necessary to provide proper fixing measures at least in longitudinal and lateral vehicle direction. This is typically done with rocker panel strut brackets. These fixtures are coupled directly with the balance, meaning all forces applied to the fixtures by the vehicle are fully reflected in the measurement results. The fastening of the rocker panel struts can either be achieved with the help of vehicle specific adapters, such as using a vehicle's standard jacking points, or via direct clamping on the rocker panel's fold. Figure 13.37 illustrates the aforementioned vehicle specific designs.

Figure 13.37 Mounted rocker panel struts with vehicle specific adapters (left), and rocker panel struts with clamping mechanism prior to mounting (right).

Further vehicle-specific data is required prior to the measurements in order to provide the required adapters for the rocker panel strut brackets. Information regarding the type and geometric position of a possible vehicle fixation is needed. This ensures the appropriate adapters are provided at the time of measurement and the fixtures for the rocker panel strut brackets can be placed in the wind tunnel.

Preparatory work is required before the vehicle can actually be placed in the wind tunnel. A thorough washing of the vehicle to remove any soiling on the underside of the vehicle and the wheel wells and an inspection to guarantee no loose parts or leakages is mandatory. Particular care must be taken to ensure there are no foreign objects in the tire treads that can become detached during the wind tunnel measurement and cause damage.

The vehicle can be prepared according to the desired configuration state before the actual wind tunnel measurements are conducted. The preparation work typically includes seals and masking, mounting of specific vehicle parts, or the addition of any particular measurement technology, such as pressure sensors and tubing that are mounted on the vehicle if pressure is to be measured directly at the vehicle.

Due to the fact that the vehicle's wheels are externally driven by the wheel rotation units, it is required to verify the transmission's ability to cope with this operational condition. The power flow in a manual transmission can be interrupted by shifting to neutral. In an automatic transmission, however, the drive shafts must be removed. These must be replaced by stub axles in order to ensure the preloading of the wheel bearings, which the drive shafts typically provide, is maintained.

After placing the vehicle in the wind tunnel it must be aligned on the balance. The longitudinal axis of the vehicle is aligned centrically to the balance with the help of a permanently installed laser in the wind tunnel. The longitudinal alignment is done in such a manner that the front and rear wheels can be placed centrically on the moving belts of the wheel rotation units. The rocker panel strut brackets on the vehicle and the fixtures in the wind tunnel can be simply screwed together, since the wheel rotation units and the rocker panel strut brackets are prepositioned according to the geometric data of the vehicle. The tire pressure must be inspected and, if necessary, readjusted prior to setting the vehicle height via the rocker panel struts. Subsequently, unobstructed rotation of the wheels must be guaranteed by slowly turning the wheels and, if necessary, hydraulically readjusting the vehicle height. The vehicle is now ready to be measured. Figure 13.38 shows a measurement-ready production vehicle mounted on the five-belt system of the FKFS aeroacoustic wind tunnel.

Figure 13.38 Vehicle fixed on a five-belt system of the IVK/FKFS aeroacoustic wind tunnel via rocker panel strut brackets.

The actual wind tunnel testing is fully automated. A sequence stored on a computer is executed. This includes taring of the wind tunnel balance, acceleration of wind and belt speeds, and the pre-suction and blow-out mechanisms utilized for boundary layer conditioning. These mechanisms shall not be described further in detail. The recording of force and moment measurement values on the wind tunnel balance is started as soon as the wind tunnel has achieved a stable defined set point. All measuring points can be set successively once this measurement procedure is executed. The wind tunnel will be shut down after setting all operating points. The measurement values recorded on the wind tunnel balance are recalculated in accordance with the vehicle system and saved in an extensive test protocol.

13.3.2 Measurement of Flow Velocity

The measurement of flow velocity in a wind tunnel spans various fields of responsibility. One of the main tasks involves the correct recording of the velocity inside the wind tunnel test section (cf. section 13.2.2), as well as flow velocities within and outside the test object. The measurement of the degree of turbulence can also be useful for detailed examinations.

13.3.2.1 Measurement of Flow Velocity Outside and Inside the Test Object

A combination of total pressure and static pressure probes, such as a Prandtl tube or National Physical Laboratory (NPL) probe can generally be used to determine the flow velocity. The direction of flow and probe axis are required to be arranged in parallel, and the flow is to be low in turbulence. The Prandtl tube will be described in detail in connection with the measurement of dynamic pressure in section 13.3.3.1. An analysis of possible influencing factors and the resulting errors was compiled by Chue [147].

The accuracy of these probes when determining flow velocity is no longer sufficient at low velocity (< 6 m/s). Miniature vane anemometers, as seen in Figure 13.39, can be used for these cases. The operating conditions required for accurate measurement results are comparable with those of a Prandtl tube. The tolerable angle of inflow for an anemometer comprising a cylindrical head is approximately 5° in order to ensure a measurement uncertainty of 1%. The tolerance level can be raised to an angle of 15° to 30° assuming the anemometer housing is designed accordingly (e.g., Figure 13.39, right side).

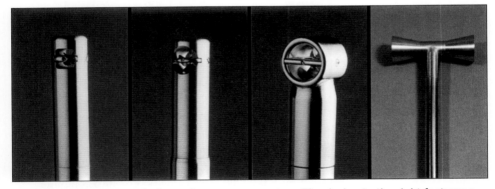

Figure 13.39 Various designs of vane anemometers. The design to the right features a head not sensitive to oblique inlet flow angles. Picture: Höntzsch GmbH.

A special design of a vane anemometer might have a diameter ranging from 40 mm to over 100 mm, albeit having a depth of only 10 mm to 12 mm. This design is well suited for measurements in gaps, mainly between condenser and the cooler matrix or between cooler matrix and fan housing, due to its thin construction. A distributed arrangement of several anemometers on the cooling area is used to achieve a distributed, albeit a roughly distributed, airflow through the cooler. The rotational direction is also registered in order to recognize back flow.[1] A more precise measurement, which nearly eliminates the influence of an angled inlet flow, is described by Kuthada [479]. The flow velocity in the cooler matrix is defined via a measurement of the pressure difference in a stream tube. The probe body features a static pressure measuring point attached to the side of a through-bore. The total pressure is recorded by a pitot tube, which is inserted through the rear of said through-bore, as seen in Figure 13.40. The right-hand side illustrates a cooler equipped with 60 probes. The hose lines of the probes are guided away to the side behind the thick tubing in order to avoid any additional blockage of the cooler.

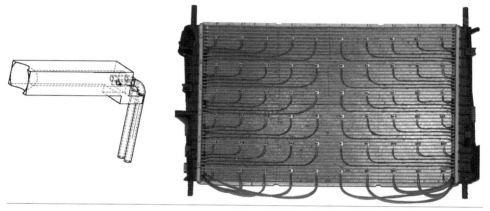

Figure 13.40 FKFS cooler probe (left side) and cooler equipped with 60 probes (right side).

Hot-wire anemometers are used for measuring strongly fluctuating velocities. The operating principle of a hot wire anemometer is based on heat loss of the electrically heated wire when exposed to air flow. The heat loss rises as wind speeds increase, leading to a decreasing wire temperature. The relation between electrical resistance and wire temperature can be utilized to determine wind speeds through the use of an integrated resistance measuring bridge, since the electrical resistance of a wire depends on its temperature. There are generally two methods.

The current in the wire of a constant current anemometer (CCA) is kept constant. This method allows accurate measurement of very low wind speeds and velocity fluctuations of up to 10 kHz. The measuring accuracy drops at higher wind speeds.

The current in a constant temperature anemometer (CTA) is regulated by the circuit bridge to allow the electrical resistance, and ergo the temperature, of the heated wire to remain constant. CTAs can cover all flow velocities that typically occur during a wind tunnel measurement and allow velocity fluctuations of up to approximately 100 kHz.

---

1. It is noted that Ng et al. [79] determined the distribution of flow velocity across the cooling area via a measurement of local pressure loss.

 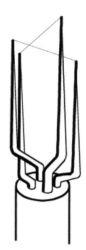 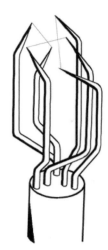

Figure 13.41 Hot-wire probes for use in velocity measurements: a) single-wire probes; b) dual-wire probe (x-wire) to determine the flow velocity vector in a 2D-plane; c) triple-wire probes to determine the flow velocity vector in 3D space.

Figure 13.41 schematically illustrates the setup of single-, dual-, and triple-wire probes. The heated wire, composed of tungsten, is welded between two electrodes, which form a sort of fork. The distance between both electrodes is approximately 2 to 5 mm, which allows the usage of small probes without disturbing local flow conditions (e.g., at curved body panels or narrow ducting and gaps).

Typical wire diameters range from 0.001 to 0.05 mm. Utilizing several wires, positioned at a 90° angle to one another, such as in a triple-wire probe, allows the measurement of the flow velocity vector with respect to the probe axis.

Additionally, hot-film probes are also widespread. Aside from the wire being replaced by a quartz rod to serve as substrate material, a hot-film probe generally features the same setup as a wire probe. The quartz rod is coated by a thin layer of platinum or gold, which serves as the actual measuring element. Generally speaking, a hot-film probe features a higher mechanical stability and robustness compared with wire probes. However, they suffer from a low frequency response of only a few kHz, which is attributed to the higher mass and, as a result, higher thermal inertia.

The calibration of hot-wire and hot-film probes is very complex and requires specific calibration wind tunnels. The flow velocity vector in the calibration tunnel can be resolved into its components u_N, u_T, and u_B relative to the probe axis for each respective wire. Each wire delivers a separate value to the probe signal.

Joergensen (1971) [412] proposed the following formula for the "effective cooling velocity":

$$u_{\textit{eff}}^2 = u_N^2 + k_1^2 u_B^2 + k_2^2 u_T^2 \tag{13.51}$$

consisting of the constants k_1 and k_2, which need to be experimentally determined, and mainly probe properties, such as wire length and wire diameter.

The calibration establishes a relation between the output voltage U and effective cooling velocity u_{eff}. In 1914, King [435] originally proposed a relation consisting of:

$$U^2 = A + B \cdot \sqrt{u_{eff}} \ . \tag{13.52}$$

However, this formula is only valid for a small range of velocities.

Simple formulae according to Siddal et al. [761] and Bruun [105] are available for the calibration of speeds up to 120 m/s. A fundamental discussion concerning the application possibilities and influencing factors in near-wall turbulent flow of hot-wire probes is found in Vagt [817].

The output voltage of all wires must be combined appropriately in a multiple-wire probe (or multiple-film probe), taking into account the angles of the wire arrangement. This leads to a rather complex calibration matrix. Different approaches for generating such a matrix that also can easily be handled are demonstrated by Karlsson (1980) and Löfdahl (1982). The measurement of flow direction will be discussed further in detail in section 13.3.2.2.

Recently, the use of laser-based flow velocity vector measuring methods have been developed. The laser Doppler anemometer (LDA) and the particle image velocimetry (PIV) are the most common laser-based methods. The flow field is not disturbed since neither of these methods requires a probe. However, both methods require particles to be inserted into the flow in order to scatter the laser light ("scattering particles").

Rudd [689] proposed an interference band model for the purpose of explaining the operating principle of an LDA. Two laser beams of identical frequency (hence the same color) are intersected at the measuring point. An interference pattern occurs in the measuring volume, which measures only a few tenths of a millimeter. The pattern depends mainly on the frequency of the laser light and the angle at which both beams are intersected. The scattering particles reflect the laser light in bright areas of the interference pattern as they pass the measuring volume. The faster the scattering particles pass through the interference pattern, the quicker are the succession of reflected light impulses, referred to as "bursts." The frequency at which the reflected light is detected by the appropriate reception optics can be regarded as a measure of flow velocity. The velocity perpendicular to the angle bisector of both laser beams is the only component measured. Flow angles and various velocity components can be measured with the help of several different pairs of laser beams of various color. Further details concerning the LDA measuring method can be found in Wiedemann [874], Ruck [688], and Durst et al. [219].

LDA enables the precise measurement of the flow velocity vector in space in a very small measuring volume without disturbing the flow field. It has been shown by Buchheim et al. [108] that LDA is capable of measuring boundary layers and recording the flow velocity field around a vehicle. The focusing of the laser beams as well as the reception optics on a tiny measuring volume and maintaining the focus during temperature changes, vibrations, and movement of the setup when shifting from one measuring point to another is very time consuming. This pertains to the adjustment as well as to the measurement process itself. Bearing this in mind, LDA systems are rarely used in ordinary development in a 1:1 wind tunnel. Several users have mounted the transmitter and reception optics, sometimes even the entire system if possible, and aligned it on a sturdy console. As a result, the focus of the laser beams and their alignment to the reception optics remain unchanged during the positioning process of the entire console from one measuring point to another, as shown by Schmitt and Wilharm [727].

The (PIV) can determine the flow velocity field not only at a single intersection point but also in a sectional plane with only one measurement, as described in Kompenhans et al. [456] and Raffel [656]. The principle can be described as follows: A laser beam is expanded in a single plane by means of a cylindrical lens and thereby creates a plane of light measuring only a few millimeters in width. The laser is not operated in a continuous but rather a pulsing manner (i.e., it emits flashes of light of a preset duration, usually in the range of nanoseconds). The time intervals between two successive flashes is controllable. Thus, the plane of light created by the laser is also done so by means of pulses. The scattering particles inserted into the flow reflect the laser light as they pass through the plane of light. The reflected light is received by a sensitive, high-resolution CCD camera. The shutter speed of the camera is synchronized with the frequency of laser pulses. In the case of the pulse frequency being set too high, a single scattering particle would be "flashed" twice at different times and positions as it passes the plane of light measuring only a few millimeters in width. The CCD camera would register two pictures of the same particle at different positions, causing the particle to be seen on two different pixels. The velocity of the particle, and ergo of the flow, can be calculated with respect to its value and direction parallel to the plane of light if the distance within the plane of light assigned to each pixel of the CCD camera and the time interval between two successive flashes of light is known, for example from the calibration process. This is applicable for nearly the entire area of the plane of light the CCD camera captures. 3D PIV measurements in a model wind tunnel are described by Ishima et al. [396].

13.3.2.2 *Measurement of Flow Direction*

As aforementioned, the prerequisite for the correct measurement of wind velocity is the knowledge of the direction of flow at the measuring point and proper alignment of the probe. Measuring errors due to angled flow at the probe can thus be prevented.

Specific technology and angled probes can be applied if the direction of flow is not known. Different types of angled probes used for measuring angles of velocity vectors of airflow in space are described thoroughly in technical literature (e.g., Gorlin & Slezinger [289], Wuest [896], Nitsche [609], and Eckelmann [222]).

These probes can be divided into two groups: in the first group, the pressure difference is measured between two symmetrically arranged pressure holes in the probe. The second group consists of hot-wire anemometers with multiple-wire and multiple-film probes, as well as LDAs and other laser-based methods.

The pressure difference is measured at pressure holes, which are affixed onto a streamlined body in an axially symmetrical manner. The pressure difference equals zero whenever the direction of flow and the symmetry plane of both measuring points coincide. Nowadays, the angle is derived from the measured pressure difference by means of a calibration curve, which is determined beforehand. Figure 13.42 shows an overview of all common designs of sensors applying this method. The 14-hole probe shown in Figure 13.45(c) as well as the 7-hole probe represent advanced developments of angled probes.

Cobra and Pyramidal probes are still applied in determining the flow velocity vector. The Cobra probe is generally designed as a four-hole probe (shown in Figure 13.43), which can measure four pressure differences. The vector of the approaching flow is determined via calibration factors. The measuring accuracy of velocities is specified as ±0.3 m/s, while the accuracy of the angle is ±1°. One distinctive feature should be emphasized: four pressure sensors are located a mere couple of centimeters from the probe body. This probe can be applied for frequencies of up to 1550 Hz due to this short distance, as shown by Schröck in [739]. This topic will be discussed again in a later chapter. Five-hole probes with a pyramidal head were applied successfully by Estrada et al. [235].

Hot-wire technology, comprising two or three sensors arranged at a 90° angle to one another, is used for measuring the direction of flow, as illustrated in Figure 13.41. A direct measurement of flow direction via LDA is not possible since only the velocity component perpendicular to the angle bisector of the intersecting laser beams can be measured. Flow direction can be calculated via simultaneous measurement of two velocity components in a 2D plane or all three components in space.

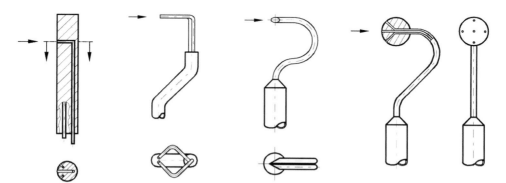

Figure 13.42 Probes used for measuring flow direction: a) Cylindrical probe with three holes; b) hooked Pitot tube; c) double Pitot tube; d) spherical probe with five holes. Pictures: Schiltknecht SIA.

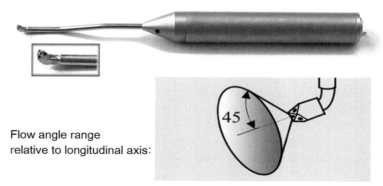

Flow angle range
relative to longitudinal axis:

Figure 13.43 Cobra probe with enlarged view of probe head. Picture: Turbulent Flow Instrumentation Pty Ltd.

13.3.3 Pressure Measurements

The measurement of static pressure distribution along the surface of a vehicle in airflow, along with determining the static and dynamic (stagnation) pressure in a flow field, are common tasks in wind tunnel operation. For this, a number of oftentimes very specific pressure probes are available.

13.3.3.1 Dynamic Pressure

Dynamic pressure is determined with the help of a measurement of total pressure and static pressure. According to Figure 13.42a, the total pressure p_{tot} is measured at the opening of the probe head, while the static pressure p_{stat} is measured at the holes or slits located farther downstream. The pressure difference $p_{ges} - p_{stat}$ is equivalent to the dynamic pressure q. The flow velocity is calculated according to equation (2.33).

The shape of the probe head has got a strong impact on the sensitivity of such systems regarding angled approaching flow; ellipsoidal and hemispherical heads are equal with respect to the sensitivity. A conical head (NPL) is much more sensitive (see Gorlin et al. [289] and Pankhurst et al. [624]). A correct alignment is particularly crucial for this type of probe.

The measuring error of a Prandtl tube with hemispherical head as a function of the approaching flow angle is illustrated in Figure 13.44 (right) (see Pope et al. [646]). The illustration concludes that the approaching flow angle of up to 12° causes a measuring error of the dynamic pressure of less than 1%, which is an acceptable measuring inaccuracy for most purposes in the measurement of vehicle aerodynamics. For the measurement of velocities less than 5 m/s using a Prandtl tube, reference is made to Eckelmann [222].

Wind Tunnels and Measurement Technique

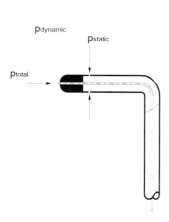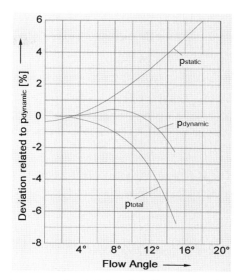

Figure 13.44 Prandtl tube; left: pressure holes for measuring total and static pressure in a flow; right: measuring error of a Prandtl tube featuring hemispherical head in angled approaching flow; according to Pope et al. [646], (Note: Static pressure is shown with a negative sign.)

If one simply wants to measure the total pressure, the use of Pitot probes is sufficient. These are reliable according to Wuest [896] for approaching flow angles of up to ±30° via appropriate tube shapes. The Kiel probe, named after Kiel [428], is regarded as a Pitot tube, which is extremely insensitive to flow direction. Its head is additionally shielded, as shown in Figure 13.45a. According to Wuest [896] it can bear an approaching flow angle of over ±60°. Kiel probes are therefore often only used in unsteady flow fields, such as wake flow.

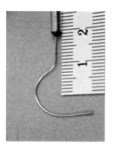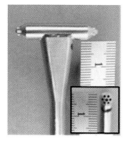

Figure 13.45 Pressure probes: a) Kiel probe, b) boundary layer probe, c) 14-hole probe (diameter of 6mm, 28mm in length), according to Cogotti [156]. The insert in illustration (c) shows the view in the direction of the probe axis. Pictures: Ford Werke GmbH.

The boundary layer probe is another specific type of Pitot tube worth mentioning (see Figure 13.45b). Its particular curved shape, as well as the leveled oval opening, enable a measurement in close proximity to surfaces without disturbing the flow.

Multiple-hole probes, such as five- and seven-hole probes, consist mainly of a set of Pitot probes. The space between each probe is filled, and the common head shared by all probes has a conical shape. The setup and calibration of a seven-hole probe is described by Cogotti [152]. It also enables the measurement of flow angles in space. Angles of up to ±70°, starting at the probe axis, can be detected.

The 14-hole probe developed by Cogotti [156] is illustrated in Figure 13.45c. It can be used for flow angles in the range of ±180°; its calibration process is very complex. They are reliable up to a minimum velocity limit of 4 m/s. Total and static pressure, as well as the velocity vector, with respect to spatial direction and value, can be measured.

13.3.3.2 Static Pressure

The distribution of static pressure on the surface of a vehicle's body in daily development is measured via thin, disc-shaped probes, also referred to as flat-probes or "bugs." Such a probe is shown as a sketch in Figure 13.46(a). The flattened connecting tube featuring an oval cross-section extends radially into the center of the disc-shaped probe and is welded or soldered to the probe. The connecting tube features a hole 0.4 mm or smaller in diameter located in the center of the probe. The static pressure at the hole is measured via the connecting tube and hose made of PVC or silicone, which lead to the pressure sensors. These "bugs" can usually be attached to the vehicle surface with ease using thin, self-adhesive foil. This avoids any damage caused to the test object. The foil is cut into circular shapes with an outer diameter larger than the probe and featuring a central hole measuring eight to ten times larger in diameter compared with the pressure hole in the probe. The local flow should not be disturbed by plastic hoses in the event that several probes are applied simultaneously.

Holes are also oftentimes drilled directly into the vehicle or model surface. The measurement point can be designed as suggested by Carr et al. [134], illustrated in Figure 13.46b. A plastic capsule featuring a large volume is adhered to a location just behind the hole. High fluctuations in pressure can be mitigated as a result of the large volume.

A falsification of results can occur when the diameter of the hole is not chosen properly. Figure 13.46c depicts the influence of the hole diameter on the magnitude of the error. A diameter of up to 1 mm is considered to be sufficient for most measurement purposes. Special emphasis is put on the edge of the pressure hole: it should be rounded or sharp-edged, in any case without burr (see Chue [147]).

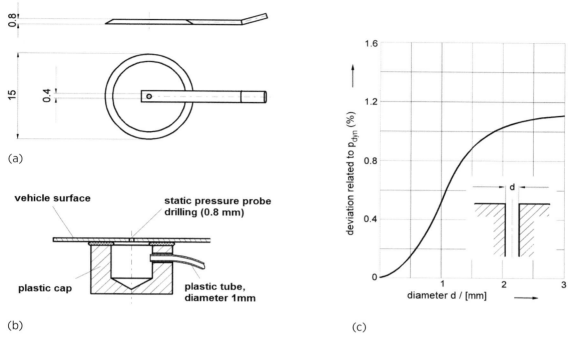

Figure 13.46 Measurement of static pressure: a) Flat-probe for measurement on the vehicle surface; b) pressure hole in body panel, according to Carr et al. [134]; c) influence of hole diameter on measurement error in a static pressure bore, according to Gorlin & Slezinger [289].

13.3.3.3 Transducers for Pressure Measurements

The measurement of pressure difference is almost exclusively the measurement of interest in wind tunnel testing, as they occur for the probes described hitherto. The pressure differences to be measured are low, usually in the range of −4000 Pa to 2000 Pa with respect to a reference pressure (in most cases the static pressure in the measuring section).

Mechanical pressure sensors are no longer used in vehicle aerodynamics; the sole type used is of electromechanical nature. The bulge of the membrane caused by the pressure difference is typically converted into an electrical signal. Capacitive, inductive, or piezoelectric elements are used as converters. The electric sensors enable a continuous recording of data and also have the advantage of using computer-friendly data collection and processing. The have almost entirely replaced "classic" measuring instruments based on a u-tube or diaphragm unit and to a certain extent enable time-resolved measurements of pressure fluctuation.

However, using thin tubes (with an inner diameter of approximately 1 to 2 mm) between probe and pressure sensor with a length of several meters in a setup as described can cause a damping of the measuring system. That is the reason time-resolved measurements of pressure fluctuations cannot be conducted with this type of measurement setup. A rather impressive example is documented by Schröck in [739]. A setup comprising a pressure measuring tube, a connecting hose, and a sensor used to determine the transfer function of the system is illustrated in Figure 13.47. The transfer function describes the change in amplitude due to resonance or damping as well as the time shift of the pressure signal (see Figure 13.48).

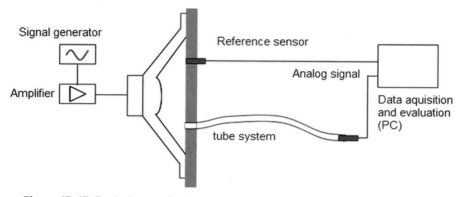

Figure 13.47 Typical setup for determining the transfer function of a system of hoses according to Schröck [739].

It is necessary for some measuring purposes, such as in aeroacoustics, to temporally resolve pressure fluctuations. In this case, the damping of the pressure signal needs to be as small as possible, which requires omitting the use of hoses. A suitable measurement setup is represented in the attachment of a small, highly sensitive pressure sensor directly at the measuring point. The membrane of the pressure sensor needs to be flush mounted to the surface of the measuring object, such as the side window or body surface. Piezoelectric pressure sensors or microphones are especially suited for this type of measurement (see Catchpole et. al [136]). The use of hoses 1 to 2 m in length is justified by frequencies up to 20 Hz, as illustrated in Figure 13.48.

Measurements consisting of many different pressure measuring points are often required, such as when measuring the pressure distribution on a vehicle's body. Scanners are used for this purpose. These machines sample (scan) the pressure existing at the pressure holes successively and transfer the respective measured signal to the data collection. Scanner developments include electromechanical as well as purely electronic scanners. Electromechanical scanners select each connected probe via an electromechanical valve and connect each pressure hole and the pressure existing there to only one pressure sensor successively. The advantage of a mechanical pressure scanner is the ability to record several pressure signals with only one measuring sensor. The disadvantage is the wear of mechanical parts in the system in the course of time, as well as time-consuming changeover and measuring processes. Changeover frequencies from one pressure probe to the next of 6 to 8 Hz is typical for a mechanical pressure scanner. It is advisable to operate several scanners in parallel for a large number of pressure measuring points.

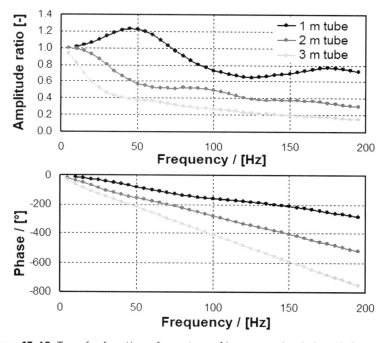

Figure 13.48 Transfer function of a system of hoses varying in length for model measurements according to Schröck [739].

Today's most common electronic scanners feature a miniature pressure sensor for each measuring position, such as a flat probe type, to which the pressure load at the measuring point is constantly applied. These miniature pressure sensors are usually consolidated into modules of 8, 16, 32, or 64 or more single sensors. The sequential processing of pressure sensor signals occurs computer-controlled with scan frequencies of up to 50 kHz. This enables a quasi-simultaneous recording of pressure curves of up to 100 measuring points.

A relatively new method of measuring the pressure distribution across the surface of a model is enabled through the use of pressure sensitive color coating; they are referred to as optical pressure sensors or "pressure sensitive paint" (PSP). This method delivers qualitative and quantitative results without flow disturbances caused by the attachment of disruptive probes.

A special bright coat of paint is applied to the model to be investigated prior to measurement, in order to achieve a uniform base color, which prevents molecule excitation on the model surface and at the same time reflects the emitted luminescent light. Appropriate luminophores are then applied via the use of a binding agent, which is spread into layers 20 to 40 µm thick. After the model is set up in the measurement section of the wind tunnel, exciting radiation, mainly blue or UV-light, is used to evenly illuminate the surface to be investigated. A CCD camera detects the local intensity of the ensuing luminescent light. The initial recording occurs without airflow and serves as a future reference. The reference picture is compared with the intensity recordings made hereafter with airflow. The pressure distribution is performed via quotient calculation of the recorded intensities with and without airflow. The results are color-coded to enable a better visualization, as shown in Figure 13.49.

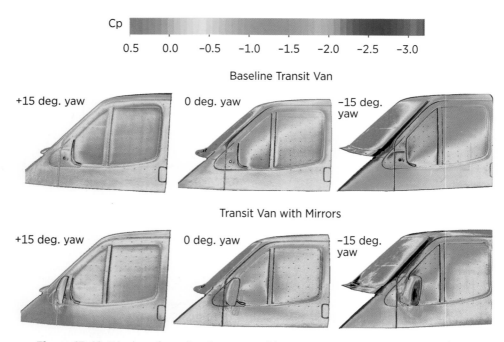

Figure 13.49 Display of results of an area-wide pressure measurement via pressure paint (PSP). Presentation of measurements of cp-values on the side window of a Ford Transit at various yaw angles, in each case with and without side mirrors; according to Duell et al. [215].

A summary of pressure sensitive coating utilized in aerodynamic studies is found in Crites [182]. A detailed description of PSP technology, including the underlying physical principle, is found in Klein [438] and Klein et al. [440]. Applications of optical pressure sensors in a wind tunnel experiment for model aircraft as well as for passenger vehicle models are described in Engler et al. [231], Klein [438], Klein et al. [439], and Duell et al. [215]. Difficulties often faced during application of PSP when experiencing low-flow velocities is examined in Duell et al. [215] as well as Coleman et al. [168].

However, due to the aforementioned extensive preparation work on the model and the potential sources of error, this method will most likely not displace the well-established pressure measuring methods in daily use in the development of vehicle aerodynamics. A measuring accuracy of ±0.1 mbar achieved with traditional pressure sensors, especially at low wind velocities, are also tenfold better. Nevertheless, measurements using pressure sensitive colors or optical pressure sensors have the advantage of offering a quantitative pressure distribution of high resolution across an area while requiring very short measuring periods and without flow disruptions caused by applying probes.

13.3.4 Measurements of Aerodynamic Loads

The measurement of forces and moments acting on the vehicle in a wind tunnel usually occurs via a multicomponent balance. Orthogonal perpendicular coordinate axes are used as a coordinate system, as defined in section 4.3.[2] Its vehicle-fixed[3] origin lies in the horizontal plane of the road in the middle of the wheelbase. This is advantageous, as the directions of the coordinate axes coincide with the axes used in driving dynamics, as agreed upon in SAE J670e [700]. This enables data collected from the wind tunnel to be used for driving dynamics calculations with identical algebraic signs.

13.3.4.1 Wind Tunnel Balances

Both external and internal balances are used. The latter shall be used discussed in a later chapter. The main purpose of a balance is to measure resulting air forces and moments acting on the vehicle and to break these down into components in all three coordinate directions (six-component balance).

Wind tunnel balances must meet several requirements if they are to carry out precise force and moment measurements:

- Parts belonging to the balance are not allowed to alter approaching flow and flow around the vehicle. The influence of any auxiliary constructions being used on results–the fastening of a model to a shaft from the top or bottom—needs to be determined beforehand in order for the results to be corrected accordingly.

- The weight force is to be nullified via taring prior to the measurement.

---

2. Coincides with definitions used in SAE J1594 [116].
3. In contrast, measurements conducted on aircraft feature a wind-fixed coordinate system.

- The vehicle's position may not be altered during the measurement. Should this happen anyway, optical measuring systems are required in order to correct the resulting effects, such as a shift in the center of gravity. Possible changes in position will be explained again later in the discussion dealing with the "Measurement with wider center belts."
- The balance needs to be constantly repositioned about the z-axis for measurements conducted with varying sideslip angle.

13.3.4.2 Decomposition of Aerodynamic Loads

The measured forces need to be split into components with the respect to the coordinate system. This occurs, depending on the type of balance, according to various methods such as described in Preusser et al. [654]. Several designs of wind tunnel balances specifically developed for the measurement of vehicles are summarized in Figure 13.50. Their essential qualities are listed to the side. The local flow field surrounding the vehicle and particularly the wheels usually features a drop in static pressure on top of the measuring platform, referred to as "boards," on which the wheels of the vehicle are resting. The resulting pressure difference between the top and the bottom of the platforms causes an additional force in the direction of the z-axis, which is falsely measured by the balance as a lifting force. The area of the measuring platforms should therefore not be significantly larger than the tire's contact patch on the road, otherwise the pressure difference acting on the platforms must be determined in order to correct the measured lift- and downforces. These effects are covered in Wickern and Beese [865] for rotating wheels and Cogotti et al. [167] for stationary wheels. The latter is a very complex and semi-empirical method, which measures the pressure distribution on an uncovered part of the board in order to determine the additional pressure force. This method is not convenient for daily wind tunnel operation due to its complexity. A further method based on force measurement is described in detail in a patent specification by Widdecke and Potthoff [871]. The vehicle, while featuring a locked wheel suspension, is fixed to the wind tunnel floor. The wheels rest on the measuring boards either in a stationary or rotating manner. The additional pressure forces acting on the board surface can be directly determined based on two measurements, with and without wind. At the same time, the possibility of adjusting the wheel base and track width must be ensured in order to enable the measurement of test vehicles featuring various dimensions. One option of implementing this adjustment mechanism is illustrated in Figure 13.51 (left). Each wheel rests on one board, which is mounted according to wheel base and track width. The residual openings on top of the measuring platforms are covered by slideable panels, which are flush mounted to the floor of the measuring section. The floor is not pictured.

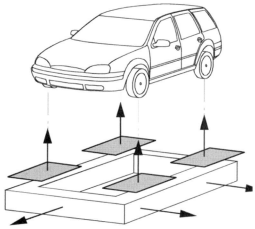

1
Each wheel is located on its own platform which measures the vertical load acting on that wheel. The platforms are connected to a floating frame which measures horizontal loads (7-component balance). Thus, horizontal and vertical aerodynamic loads are separated from each other. Possible movements of the car during the measurement do not have significant effects on the results.

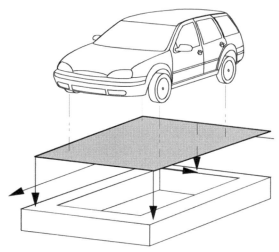

2
The vehicle is standing on a single platform. The aerodynamic loads are determined via load cells located between the platform and a ground frame (6-component balance).
Possible movements of the vehicle during the measurement can affect the results considerably. The measurement has to be corrected.

Figure 13.50 Schematic illustration of various types of vehicle balances used in wind tunnels.

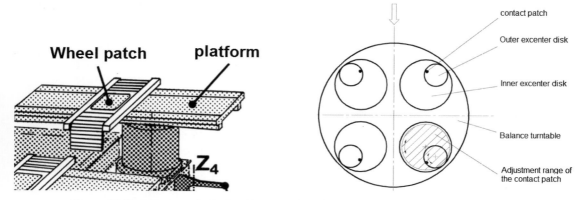

Figure 13.51 Adjustment of wheel base and track width on a balance featuring a roller shutter cover (left) via eccentric discs (right) according to Kelly et al. [424].

One alternative is illustrated on the right-hand side of Figure 13.51. Each wheel is provided with two eccentric, circular and rotatable panels. The smallest panel is the board, while the largest, outer panel is supported by the floor of the measuring section. The board can be positioned as desired by rotating the panels; every necessary combination of track width and wheel base can be achieved.

The only solution, which can be considered for a measurement section featuring rotating wheels via belt assemblies (see Figure 13.52), resembles the method first mentioned (see Figure 13.51, left). This sort of measuring section floor also describes the simulation method preferred in wind tunnel measurements of series production vehicles. The wheel rotation units are connected to the balance, as are special brackets on the underbody of the vehicle to be measured, usually via rocker panels.

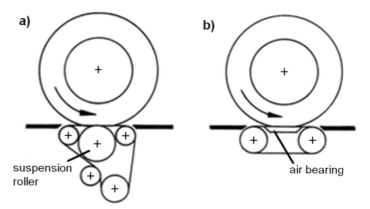

Figure 13.52 Wheel rotation unit integrated into the lift elements of a six-component balance; according to Fiedler & Potthoff [247]: a) FKFS system: Flat rubber belt featuring roller follower; b) FKFS system: MTS steel belt featuring air bearing.

Two different wheel rotating systems are illustrated in Figure 13.52. In the case of a), the wheel rotates on a flat rubber belt; the wheel load is carried by a roller follower; b) utilizes a steel belt, which is supported by a water-cooled air bearing.

Hence, the wheel rolls on a level surface, as it would on the road. This results in the same contact patch of the tire. Another advantage is the tire does not heat up more than it does on the road; the rolling resistance and noise are also identical, due to the air bearing being very quiet. However, the tire's contact patch is also smaller than the measuring platform. The air force acting on the protruding area needs to be corrected, as was mentioned. Several fundamental differences in the determination of forces with rotating wheels as opposed to an experiment with stationary wheels should be discussed. The flow through the wheel orifices as well as the flow around the vehicle is altered by the rotating wheels, leading to changes in the forces acting on the vehicle. However, the so-called ventilation force, referring to the portion of air force required to drive the wheel in the airflow relevant to fuel consumption, is not measured (see Mercker et al. [556], Wiedemann [879], Wickern et al. [869] and section 4.3.3.3). The value of the vehicle's drag is realistically seen too low when determined and needs to be supplemented by the relevant portion of ventilation force of the wheels. As discussed in section 13.2.11, this force as well as the rolling resistance appear only as an internal force within the structure of the balance in this measuring setup. Therefore, no immediate determination of these two forces and their separation from one another is possible. Various methods used to determine the ventilation resistance do however exist, of which two shall be discussed. The first method, described by Wiedemann [879], requires a force measuring device within the balance's structure. The second (see Mayer & Wiedemann [541]) necessitates elements in the vehicle that determine the torque at the wheel, which hovers above the wheel rotation unit (no rolling resistance). The wheel's circumferential velocity and belt velocity of the wheel rotation unit are identical. The wheel suspension is locked, resulting in the relative position of the wheel inside the wheel well remaining unchanged. It should be mentioned that this method results in a gap underneath the tire, which allows air to flow below the tire and can have an influence on the ventilation moment being measured. Furthermore, the systematic measuring errors will be pointed out in the following.

Figure 13.53 illustrates the forces acting on a vehicle and its wheel in a wind tunnel. The vehicle and all four wheel rotation units are connected to one another and are supported on the balance.

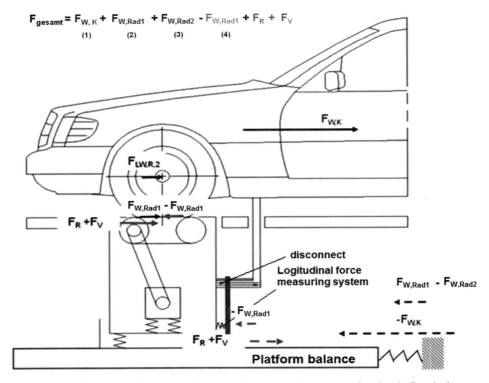

Figure 13.53 Considerations of systematic measuring errors of a classic five-belt measurement setup.

The air resistance of the vehicle body $F_{R.B}$ as well as the wheel's air resistance $F_{R,Wheel}$ is measured in this measuring arrangement (rolling resistance F_R and ventilation resistance F_{Vent} are not yet considered). This force, which acts eccentrically on the tire, can be resolved into two force components, $F_{R.Wheel1}$ and $F_{R.Wheel2}$, of which $F_{R.Wheel1}$ is applied directly to the wheel rotation unit. During real-time road travel, however, $F_{R.Wheel1}$ is applied to the road. This force is measured by the setup, since the wheel rotation units are connected to the balance. This results in a systematic error. The force comprises the equation terms (1), (2), and (3) and thereby contains one surplus term $F_{R.Wheel1}$, resulting in a force that is too large. If one were to sever the connection of the vehicle to the wheel rotation units (displayed as a vertical bar next to the wheel rotation unit) and replace it with a measuring setup for longitudinal forces, the force $F_{R.Wheel1}$ could be compensated (equation term 4). Analogously, this applies to the sum of rolling resistance F_R and ventilation force F_{Vent}, of which both are caused by wheel rotation, albeit their physical origins differing. The separation of rolling resistance and ventilation force has been previously discussed.

Aerodynamic measurements via internal balances (mounted inside the vehicle) are rare for full-size vehicles, as they require considerable modifications of the vehicle and special brackets in the wind tunnel, as illustrated in Figure 13.54. However, in conjunction with a wide belt it was possible to shed light on several fundamental issues in vehicle aerodynamics. The details pertaining to this can be found in Mercker and Knape [563], Emmelmann et al. [227], Mercker et al. [556], and Wickern et al. [869].

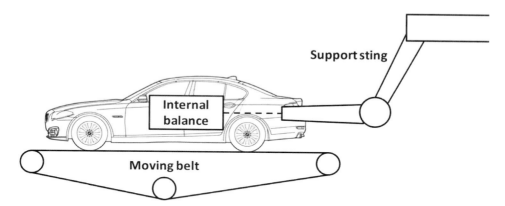

Figure 13.54 Experiment setup featuring internal balance. The wheels are driven by the belt. According to Wickern et al. [869].

Rolling resistance and ventilation resistance are reflected in the measurement results in the event that the wheels are driven by the belt and connected to the vehicle, as both forces are considered to be external forces from the balance's point of view, contrary to an external balance. The rolling resistance can be separated from the ventilation resistance via specific taring.

Most recent developments in the technology of balances enable the determination of vertical forces at the contact patch of a rotating wheel through the belt.

The belt is guided via air bearings. The measuring cells located underneath measure the forces being transmitted in the tire's contact patch. The vehicle can be fixated into a stable position via brackets located to the side of the wheels, as has been implemented in the Windshear Rolling Road Wind Tunnel in Concord, North Carolina. The brackets are connected to measuring cells located beyond the belt, which can resolve forces into x- and y-components, which correspond with resistance and lateral forces (see Figure 13.55). The details are described in Walter et al. [841].

Rolling resistance and ventilation resistance caused by wheel rotation and interference forces caused by brackets and measuring elements are also accounted for in the results of this measuring arrangement. These must be separated by means of other methods. The connection of the wheels to the vehicle's structure enables the vehicle to react to wind forces free of constraints, as it would on the road. An expansion of the tires caused by rotation would result in a change in angle of the side brackets and thus to additional vertical forces reflected in the results. Connecting the belt unit with a load cell measuring in a horizontal manner and also integrating the connection to the vehicle into this system will lead to rolling resistance and ventilation resistance being accounted for as internal forces and no longer being measured. The ventilation resistance as an air resistance component must again be defined by means of other methods and added to the total force. This measuring method was presented by Duell et al. [216] and Koremoto et al. [457] and is also known under the patented term as TALINGZ.

Figure 13.55 Vehicle connection to the measuring sensor according to Walter et al. [841].

All belt technologies feature more or less wide slits in front and behind each belt. In case the belt units belong to the measuring system, namely the balance itself, then the pressure field around the vehicle and tire can cause horizontal reaction forces, which is reflected in the results and must be corrected. These influences are type- and wind-tunnel-specific and are currently the topic of investigation (see e.g., Walker & Broniewicz [838]).

The observations made so far were based on determining forces and moments acting on the vehicle caused by stationary airflow in a wind tunnel. Time-varying forces, such as vortex separation at the rear of the vehicle, were also generally not considered due to the long averaging periods. Natural wind conditions, such as gustiness, are not represented in a wind tunnel. Therefore, a vehicle's reaction to crosswind is derived from quasi-steady examination. Recent investigations (see Schröck [739] on this topic) show that this procedure is deemed inadequate and that unsteady aerodynamic processes need to be reproduced in the wind tunnel. This implies the need to also measure unsteady forces. Conventional underbody balances in 1:1 wind tunnels are not suitable for time-resolved measurements due to their high inertia. The Eigenfrequency lies in the range of 2 Hz to 6 Hz, hence in the relevant frequency band. However, conducting unsteady measurements with the help of a special setup, a specific arrangement of load cells, seems possible.

13.3.4.3 Measurement of the Frontal Area

The measurement of frontal area A_x is necessary for wind tunnel measurements, as the absolute forces and moments, such as drag force, are measured in the wind tunnel.

$$F_D = \frac{\rho_L}{2} \cdot u_\infty^2 \cdot c_D \cdot A_x \tag{13.53}$$

The drag coefficient c_D can be determined via the measurement, provided the air density ρ_L, the velocity of the approaching flow u_∞ and also the frontal area A_x are known. The accuracy of the frontal area measurement is therefore not reflected in the determination of the drag forces; it is however reflected in the determination of the corresponding coefficients, as described in section 4.1.

The frontal area A_x is the largest area projected along the longitudinal axis of the vehicle. The oldest method of measuring frontal area is the shadow measurement. The outside contour of the vehicle is determined by use of a lamp (spotlight). The lamp is required to be located at a sufficient distance from the vehicle (200 m or more) in the vehicle's plane of symmetry at half the vehicle's height. A transparent wall is mounted behind the vehicle perpendicular to this plane of symmetry. The vehicle's shadow is displayed on this wall. Its boundary edges can subsequently be plotted and planimetered. Despite the great distance between the light source and the vehicle, the error caused by the divergence of the light rays needs to be corrected. A measuring accuracy of ±0.5% can hereby be achieved.

A measurement similar in principle is possible via a photographic image of the vehicle likewise from a great distance. Lenses with high focal lengths (1000 mm or higher) are used. The measuring accuracy is comparable to the shadow measuring method.

Methods based on laser technology have been developed that enable highly accurate frontal area measurements in a small space and without lighting divergence error. The laser (e.g., helium-neon) is shone on the vehicle parallel to the plane of symmetry. The laser is mounted on a two-dimensional traversing device featuring a movement plane arranged vertically and perpendicular to the vehicle's plane of symmetry. A laser light detector (e.g., a photodiode) is mounted behind the vehicle on a second traversing device, which can travel in a plane parallel to the first plane. The movements of the laser and the detection device are executed in a synchronous manner. The location of the detection device in the traversing plane is digitally detected in the event that the laser beam hits the detector or is blocked by the vehicle. The outside contour of the vehicle can be scanned by means of meander-shaped tracing and subsequently the frontal area can be calculated.

A retro-reflective wall can be installed behind the vehicle instead of the laser light detector in a modified method. The laser system, which features a detector integrated into its optics, is mounted on the traversing device. The outside contours of the vehicle can be detected via the laser beams hitting the wall and being reflected.

While still using a mounted CCD camera instead of a detector and attaching a diffusing screen (marata disc) between the vehicle and camera, the laser beam is expanded by the optical system (approximately 10 to 15 cm in diameter). The partial-silhouette appearing on the diffusing screen can be digitalized and stored on a computer (ISRA method). The entire frontal area can be numerically assembled and calculated from the digitalized partial-silhouettes. This setup is schematically illustrated in Figure 13.56. The achievable accuracy for an area of 2 m² is approximately 0.2%. The measurement is executed fully automated.

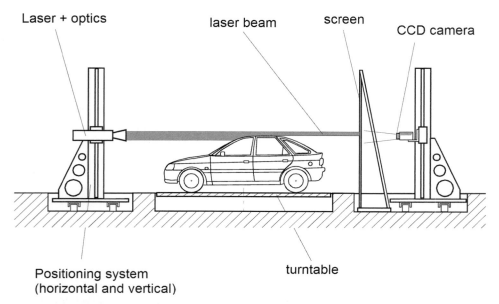

Figure 13.56 Measurement of frontal area with the help of laser technology, measuring method according to ISRA.

13.3.5 Flow Visualization

The desire to render fluid flow visible dates all the way back to the origin of fluid dynamics research itself. The insights into the details of flow movements around solid structures, which have been hereby won ever since, are documented in Van Dyke [819] in impressive fashion. Laminar and turbulent flow, the transition between the two, flow separation caused by flow around objects, and vortex formation in their wakes are made visible.

Instead of being limited to selected base bodies as shown by Van Dyke, such as spheres and cylinders, vehicle aerodynamics deals with very complex structures in a flow. The flow around a vehicle or its parts, which are made visible, are not easily interpreted under certain circumstances. The most common experimental technology provides clear images only for low velocities, whereas the actual velocities during road operation are much higher.

Nevertheless, the visualization of flow around objects plays a significant role in vehicle aerodynamics. It provides effective assistance in the mostly empirical approach during shape development. The conventional methods are summarized by Koenig-Fachsenfeld [448]. Two approaches can be distinguished: on the one hand the flow on the vehicle surface is considered, on the other hand the flow in spatial fields.

The visualization of the flow path on the vehicle body is achieved by adhering woolen threads in a grid-like structure to the body. These indicate the direction of flow when subjected to wind; separation is made evident by fluttering threads. Cone-shaped separations as seen at the A-pillars and C-pillars in the case of hatchbacks are visualized. This simple method does, however, have a disadvantage. For example, the inherent rigidity of the woolen threads themselves can falsify their orientation in areas of low-flow velocity. A further possibility exists by applying thin nylon threads impregnated with a UV-coloring agent. These threads are very visible with specific illumination. Applying the threads is very time-consuming; hence this method is rarely used anymore in development processes.

One available alternative lies in coating technology. Highly visible "tracing particles" capable of following the local flow path are added to a fluid. The result of such a simulation is illustrated in Figure 13.57.

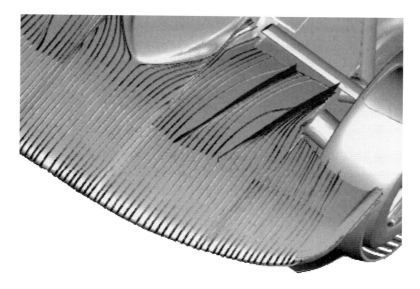

Figure 13.57 Coating method. Frontal section of a vehicle according to Hahn (2006).

Oils are usually used, which are applied by a brush or as a spray film to the surface, as seen in Cogotti [157]. Titanium dioxide (TiO$_2$) is added for improved visibility. The adhesion (viscosity) of the liquid is required to be high enough in order to prevent the paint from running back down vertical surfaces within a short period of time, caused by gravity. However, the viscosity must also be lower than air friction to allow the particles to follow the flow path.

This technology was further improved. Adding fluorescent components allows the flow patterns on surfaces under UV-light to be more visible (see Schmitt et al. [726]). However, the requirements for these substances extend beyond mere visibility and traceability. Nowadays, toxic effects and the degree of duct soiling, such as caused by the oil particles swept away by the flow, need to be considered. In this respect, considering development periods continuously becoming shorter and cost pressure, frequent cleaning processes in a wind tunnel are hardly justifiable anymore.

The second group of technology, namely the visualizations of flow fields, also uses tracing particles. However, these are inserted into the airflow and follow the fluid streamlines around the vehicle. The simplest method of creating this type of scattering particle is through the use of a so-called smoke probe. An appropriate fluid, usually a water-alcohol mixture, is pumped through a long electrically heated pipe and is evaporated in the process in such device. The steam is immediately condensated at the tip of the pipe outlet to form very fine, white droplets (aerosol), which are swept away by the airflow and create a highly visible smoke trail along the fluid streamlines, as shown in Figure 4.59. On the contrary, aerosol inserted into a region of flow separation, such as behind the side mirrors or at the rear of the vehicle, fills these areas to make the dead wake region highly visible (see Figure 1.4a). The smoke probe is relatively easy to handle, does not emit toxic products, and yields good results as a first qualitative assessment of flow conditions, albeit also only for low velocities. A typical application is illustrated in Figure 13.58. The upper picture shows the wake at the rear of a vehicle without blow-out. The lower picture illustrates the change in characteristics of the wake during blow-out as an active measure to control flow and reduce drag.

More modern methods of flow visualization utilize a sharply outlined plane of light created by a powerful halogen lamp in conjunction with an optical system or a powerful laser. The tracing particles inserted into the flow reflect the light as they cross the plane of light, which in turn is recorded by appropriate cameras. Droplets of oil produced in special generators and measuring only a few microns are usually used as tracing particles, which are inserted into the flow. The path of a single particle can be traced and recorded when larger tracing particles are used, such as helium-filled soap bubbles, as seen in Figure 13.59. The method shown in the illustration provides three-dimensional information.

Wind Tunnels and Measurement Technique

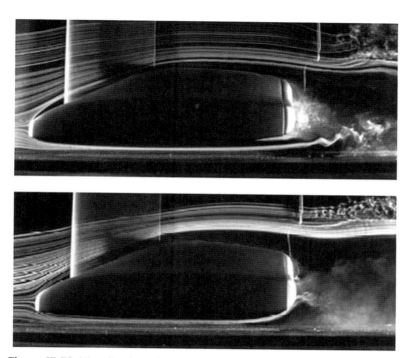

Figure 13.58 Visualization of wake with and without blow-out according to Hoffmann et al. [354].

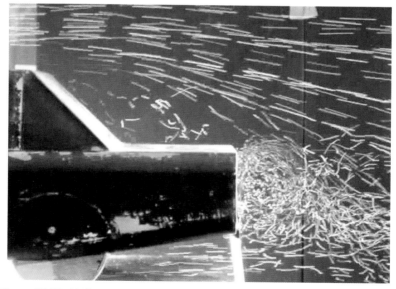

Figure 13.59 Helium bubbles in the wake of a hatchback model according to Lock et al. [510].

The particle paths are registered as lines and curves in the event that the plane of light is not created continuously but rather only for short periods of time (pulsed) and the shutter speed is synchronized accordingly. The length of each line or curve is proportional to the mean velocity of each particle. This provides a sort of quantitative recording of the flow field. Further development of this technology leads to PIV, which was described in section 13.3.2.1.

A detailed investigation of flow separation, in "dead-water" and wake flow, is only made possible once the concerned regions are selectively illuminated. A laser-light intersection technology was developed by Buchheim et al. [108] for this purpose. Various applications of this technology are illustrated in Figure 13.60. Light emitted by a high-performance laser (5 to 24 W) is first guided through a cylindrical lens and subsequently redirected by a mirror in a manner that the vehicle's entire width, length, and height can be illuminated in any desirable plane. The thickness can be adjusted between just a few millimeters and several centimeters.

The advantage of this method is the realization of every possible intersection. The method is only suitable for use at low-flow velocities and the results only provide two-dimensional information about the flow structure, which is disadvantageous.

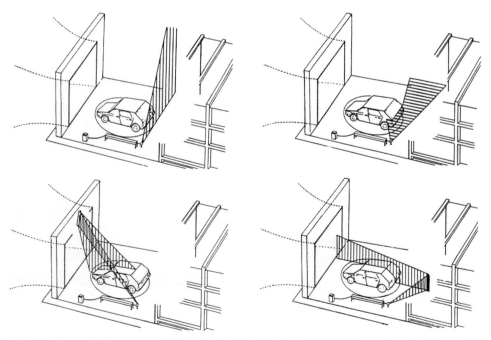

Figure 13.60 Laser-light intersection technology for the purpose of visualization of flow processes around a vehicle; according to Buchheim et al. [108].

13.3.6 Investigation of Vehicle Soiling

Vehicle soiling can be distinguished between self- and foreign soiling according to the cause of soiling (see Figure 13.61). The possibilities for examining these types of soiling in a wind tunnel will be described in the following. Both types of soiling, foreign and self, are examined using water. A coloring agent or fluorescent substance is added. The fluorescent substance absorbs light in the frequency band of UV-light and emits the absorbed energy as visible light, which involves a shift in frequency. Even the smallest water droplets become visible for photo and video analysis when illuminating the dimmed measuring section via UV-light.

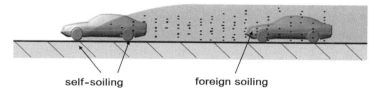

Figure 13.61 Simulation of self- and foreign soiling [773].

13.3.6.1 Simulation of Foreign Soiling

According to the definition in section 6.3, self-soiling is defined as raindrops or water droplets, which are whirled up off of the road by surrounding traffic and hit the vehicle. One main point of investigation is the dirt load carried by the droplets, which then hit the vehicle. A spraying device is installed in the wind tunnel in front of the vehicle to simulate this type of soiling. A spraying bar as well as a spraying grid can be used. The vertical position of the spraying bar can be adjusted according to the vehicle's dimensions. The soiling simulation via spraying bar is illustrated in Figure 13.62. Vehicle soiling is investigated for various approaching flow velocities and sideslip angles in order to accurately reproduce various driving situations such as city traffic, cross-country, and highway travel.

Figure 13.62 Soiling simulation in a wind tunnel at Daimler AG. The spraying bar simulates mist behind a heavy-duty truck during moderate rainfall.

Testing in a wind tunnel should not be very time consuming due to cost reasons; however, it should provide a realistic reproduction of soiling as would occur under road conditions. A thorough cleaning of the vehicle is necessary in order to ensure a good comparability. The vehicle needs to be cleaned after each measurement to guarantee reproducible initial conditions. Constant boundary conditions throughout all testing need to be ensured. Even small details of a prototype vehicle that do not match the vehicle's final design, such as differences in seals, fittings, or surface roughness, can lead to a simulation result that deviates from the soiling characteristics of the vehicle's final design.

13.3.6.2 Simulation of Self-Soiling

Self-soiling is caused by water droplets containing dirt, which are whirled up mainly by the vehicle's rotating wheels and are then deposited on the sides and rear of the vehicle (see section 6.3.3). A basic understanding of water segregation as well as droplet and spray formation during water whirl up of rotating wheels is required when analyzing self-soiling. The processes of droplet transport by the wheel to the vehicle surface and the associated deposition of dirt particles need to be understood. The FKFS test bench equipment used to analyze the spray characteristics of the wheel is illustrated in Figure 13.63. A free-standing, rotating wheel is analyzed on a specifically designed test bench in order to experimentally demonstrate the spray process and droplet and spray formation caused by a rotating wheel in detail. The analysis and visualization of droplet size distribution is performed by a Malvern Spraytec Laser Diffraction System via 15 discrete measuring positions distributed along the wheel's circumference.

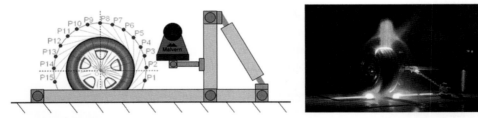

Figure 13.63 Test bench equipment to examine the spray characteristics of a free-standing wheel (left, Pi are the measuring positions) with a Malvern Spraytec Laser Diffraction System (right) [775].

The loading between the tire and the simulated road surface is an important factor for these investigations, as the load settings determine the contact patch of the tire and hence can alter the whirl up and spray process. Figure 13.64 illustrates the spray process of a rotating wheel. The left-hand side depicts the visualization provided by the experiment, and the right-hand side shows the numerically determined results.

Figure 13.64 Comparison of experiment (left) and numerically (right) determined spray characteristics of a free-standing, rotating wheel [775].

The procedure of reproducing self-soiling in a wind tunnel is similar to that of foreign soiling; however, all measurements are conducted with rotating wheels. Water endowed with a fluorescent agent is applied to the road via nozzles located in front of the wheels. The wheels are supported and driven by a belt or dynamometer. The spray results produced by an experiment using a dynamometer do, however, deviate from realistic road conditions. This is due to a difference in the tire's contact patch on the dynamometer, as this features a curved surface as opposed to road conditions. This does not occur during soiling simulation on a belt, since the contact patch represents real road conditions to a more precise extent. Comparative experimental results of self-soiling produced in a soiling simulation on a dynamometer and a belt are illustrated in Figure 13.65 and Figure 13.66. The quantitative degree of soiling, which can be defined by the c_F-value (contamination factor) via an 8-bit discretization, is depicted. The sides are soiled to a lesser extent compared with the experiment using belts, while the wake behind the rear wheels features a higher degree of soiling.

Figure 13.65 Soiling simulation on a dynamometer [774].

Figure 13.66 Soiling simulation using a belt [774].

13.3.7 Engine Cooling Tests

During the development of engine cooling systems, the requirements imposed to the simulation of the flow around a vehicle are less strict than those of measurement tests for heating and air conditioning. Since in most cases the cooling package is placed in the front end of the vehicle, the airflow in this region has to be simulated in a very precise way to obtain a good match of results between wind tunnel and road testing. In this context, slight deviations of the surface pressure distribution in the rear area of the vehicle are acceptable. Hence, engine cooling tests can be performed in smaller climatic or thermal wind tunnels, which are designed exclusively for these studies (see Figure 13.67).[4,5] However, special attention has to be turned to the calibration of the airflow velocity, since its definition becomes more and more difficult with small values of the nozzle exit area (see also section 13.2.2). Typically, the tests are carried through at high ambient air temperatures, in order to determine the thermal performance limit of the cooling system. Furthermore, solar and tail wind simulations are required frequently to account for any extreme driving conditions.

The examination of the performance of a cooling system in a wind tunnel is usually realized with three different load cases, mostly at constant speed:

- Simulation of maximum speed (v_{max})
- Simulated uphill driving with a trailer
- Following tests 1 or 2: idle operation or reheating after engine switch off

---

4. Also called thermal blowing tunnel.
5. Here too, see the experiences of Ng et al. [604].

Wind Tunnels and Measurement Technique

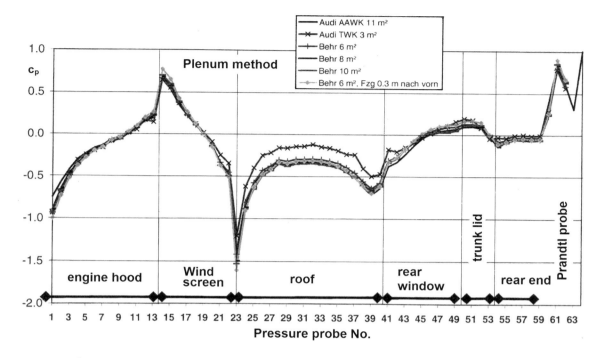

Figure 13.67 Comparison of pressure distribution in the longitudinal middle section of a passenger car, measured in different wind tunnels, determining the wind speed according to the plenum method; measurements: Audi AG,

The individual operating points of the test are run as long as the coolant and the engine oil temperature have reached steady values. In addition to critical fluid and component temperatures, the heat rejection at the radiator represents a crucial parameter. For its calculation, the coolant mass flow as well as the coolant temperature drop at the radiator have to be measured.

$$\dot{Q} = m_F \cdot c_{p,F} \cdot \left(T_{F,\text{Rad\_in}} - T_{F,\text{Rad\_out}} \right) \tag{13.54}$$

So that the heat flux rejected at the radiator is neither overestimated nor underestimated, the temperature dependency of the specific heat capacity of the coolant $c_{p,c}$ has to be taken into account. Since this dependency is fairly weak in the relevant temperature range, typically a mean value for $c_{p,c}$ is used, based on the average coolant temperature between radiator inlet and outlet.

With regard to the operational safety of motor vehicles, the main criterion for cooling systems is to observe maximum component temperature limits. Yet, there are various temperature limits for the different components and operating fluids, as for example the engine oil, transmission oil, and coolant, that all have to be monitored during measurements. At best a critical temperature is determined in advance, probably with the help of numerical simulations, which serves as a reference and, where appropriate, as a termination criterion during the measurements. In an on-road test drive, an impending exceeding of this temperature would lead to a power output limitation by the engine control unit.[6]

A methodology for assessing the performance of a cooling system is based on the contemplation of the temperature difference between a system-relevant critical temperature, such as the maximum permissible coolant temperature in the engine block $T_{F,\text{eng}}$ and the cooling air temperature $T_{L,\text{Rad\_in}}$, which is the inlet temperature of the cooling air into the radiator.

$$\Delta T = T_{F,\text{Eng}} - T_{L,\text{Rad\_in}} . \tag{13.55}$$

Depending on the number and position of other heat exchangers in the front end of the vehicle, a deviation between $T_{L,\text{Rad\_in}}$ and the ambient temperature is T_∞ possible.

As long as the opening of the thermostat and the fan speed are at the maximum, ΔT is nearly constant for different ambient temperatures.[7] This condition can be achieved, for example, by blocking the thermostat and controlling the fan at maximum speed.

The coolant in the engine starts to boil when the inlet temperature reaches the critical threshold of T_{Atb}. T_{Atb} is known as "air-to-boil-temperature" (ATB temperature). The higher this ATB temperature is, the bigger the capacity of the cooling system to dissipate heat. In this case:

$$\Delta T = T_{F,\text{Eng}} - T_{L,\text{Rad\_in}} = T_{F,b} - T_{\text{Atb}} . \tag{13.56}$$

Thus

$$T_{\text{Atb}} = T_{Fb} - \left(T_{F,\text{Eng}} - T_{L,\text{Rad\_in}}\right) = T_{Fb} - T_{F,\text{Eng}} + T_{L,\text{Rad\_in}} . \tag{13.57}$$

T_{Fb} is the boiling temperature of the coolant. It depends on the pressure in the cooling system, which itself is limited by the opening pressure of the radiator cap.

For the three previously described test scenarios, ATB temperatures can be defined that also serve as acceptance criteria. The definition of these limits is based on practical experience. For high-speed tests T_{Atb} is approximately between 48°C and 55°C. However, for the uphill trailer towing T_{Atb} has to be set at 28°C to 35°C since the coolant temperatures are significantly higher due to the higher engine load.

The ATB temperature as a criterion for assessing the efficiency of the cooling system is widely used. Alternatively, Lin et al. [502] and Ng et al. [605] proposed the specific dissipation SD, which they have defined as follows:

---

6. By supervising the temperature and its time-dependent gradient, an impending overheating can be noticed by the ECU in advance.
7. The difference between the temperature of coolant and cooling air at the inlet of the radiator, the so-called ETD value ("Entry Temperature Difference"), can be used for the assessment of the thermal efficiency of the radiator: the smaller the ETD value is, the higher is its efficiency. By defining ETD limits in dependency of the driving speed, it is possible to check the risk of boiling coolant under defined driving conditions and every gear ratio. See Eichlseder et al. [193].

$$\text{SD} = \frac{\dot{Q}}{T_{F,\text{Rad\_in}} - T_{L,\text{Rad\_in}}} = \frac{\dot{m}_F \cdot c_{p,F} \cdot (T_{F,\text{Rad\_in}} - T_{F,\text{Rad\_out}})}{T_{F,\text{Rad\_in}} - T_{L,\text{Rad\_in}}} = \frac{\dot{Q}}{\text{ETD}} \quad (13.58)$$

where $\dot{Q}$ is the radiator heat flow, $\dot{m}_F$ is the coolant mass flow through the radiator, $c_{p,F}$ is the specific heat capacity of the coolant, $T_{L,\text{Rad\_in}}$ is the air temperature at the radiator inlet, and $T_{F,\text{Rad\_in}}$, $T_{F,\text{Rad\_out}}$ are the respective coolant temperatures at the inlet and outlet of the radiator. The determination of SD therefore also requires the coolant mass flow rate (kg/s).

The main advantage of the SD method compared to the ATB temperature is that the SD method provides reliable test results, even if the air temperature in the wind tunnel cannot be maintained at a steady state during the test. Thus it is also possible to achieve reliable development results in wind tunnels that do not have air-conditioning equipment available.

An important criterion for mountain passes is the influence of the altitude. Because of the decreasing atmospheric pressure with altitude—it is only 0.74 bar at 2500 meters above sea level—and the consequent lower air density as well as the lower ambient temperature, the power output and then the thermal behavior of the engine changes. Additionally, there will be a different boiling temperature caused by the pressure conditions around the radiator cap. In a common wind tunnel, the simulation of such conditions is hardly possible, as normally just the influence of the ambient temperature can be simulated. Engine cooling testing at different ambient pressure can only be performed in especially designed test beds (altitude chamber for emission tests).

For a high-speed test and the uphill trailer towing, the driving resistances including the resistance of the slope are simulated with a dynamometer. For the simulation of the ascent with trailer, a pitch in the range from 10 to 12% is applied.

Because of finite diameter of the roller, the rolling resistance of the driven wheels on a dynamometer test bed is greater than on the road. This must be taken into account in determining the loads that have to be supplied by the test bench control. In general, the resistances to be applied are calculated via the main equation of the motor vehicle (see equation (3.6) in section 3.2.5) In order to repeat an on-road measurement on a test bench, the same heat load to the cooling system has to be ensured in both cases. Since the heat generation of an internal combustion engine depends on its operating point, both the engine speed and torque must match during the on-road test and the test bed run. Therefore, the current dynamometer load is set based on the information provided from the ECU. In this way, it is possible to reproduce the time-dependent course of the engine's operating point.

13.3.8 Heating and Climatization Tests

Prior to on-road testing, heating and air-conditioning tests are being performed in climatic wind tunnels. Due to the smaller dimensions compared to aerodynamic wind tunnels (see section 13.5), their suitability has to be ensured by the means of correlation tests. This can either occur by comparing the results to a full-balance aerodynamic wind tunnel or to aerodynamic on-road measurements where the pressure distribution on exposed regions of the bodywork is being examined. This correlation ensures that the pressure conditions around the vehicle in a chosen test environment correspond to those registered of the road. Hence, the airflow through the vehicle is portrayed realistically. If no sufficient correlation can be found, for example, if the nozzle exit area of a climatic wind tunnel is too small in relation to the test object, the flow can be "trimmed" by the installation of baffle plates in the test section. The flow velocity can therefore be adjusted and a better correlation can be achieved, as described in section 13.2.2.

A typical test chart for heat-up tests is shown in Figure 13.68. Before the beginning of the test, the vehicle is placed in a so-called "soak room" at the desired ambient temperature of the test run, until all engine compartment and passenger cabin parts, including coolants and lubricants, reach this temperature. A predefined test procedure defines the velocity-dependent braking force that is applied to the engine at the beginning of the test with the help of the dynamometer. During the test, the air temperature is recorded at several points in the passenger compartment. The positioning of the measuring points to be used are described in DIN 1946-3. Figure 13.68 shows the average temperature values for foot, chest, and head areas as well as the mean cabin temperature.

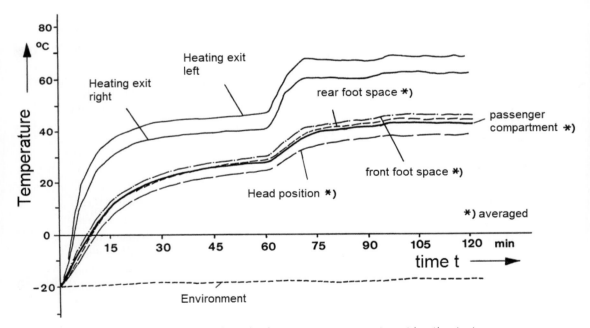

Figure 13.68 Typical result of a passenger compartment heating test.

The final assessment of the heating system performance is usually done by comparing the average values for each area or the overall average of the cabin to a predefined minimum requirement, which is mostly provided on the basis of practical experience. By the use of digital data acquisition systems and specially adapted analysis software tools, measurement results can be displayed or archived quickly and clearly.[8]

The procedure for performance tests of air-conditioning systems is very similar to that of heating tests. The vehicle is conditioned at high air temperature. The test starts with gradually increasing velocities and ends with idle operation. In addition to the high ambient temperature, special lamps simulate the thermal load resulting from solar radiation. Furthermore, the humidity is increased, respectively, in order to test the air condition under the most extreme operating conditions. Typical test conditions consist of ambient temperatures above 40°C at a relative humidity of 40% and 1 kW/m^2 simulated solar radiation intensity.

The lamps deployed to realize the solar radiation must feature a spectrum similar to the one of the sun and send out parallel directed light. Typical radiation sources are presented in Table 13.1.

Table 13.1 Comparison of the Spectral Distribution of Standard Spotlights and Sunlight; Percentage of Total Radiation

Lamp/Spotlight	UV-C	UV-B	UV-A	visible	IR-A	IR-B	IR-C
Wave length in μm	< 0.28	0.28 to 0.315	0.318 to 0.38	0.38 to 0.78	0.78 to 1,4	1,4 to 3,0	> 3,0
Terrest. Radiation after Schulze (1.12 kW/m^2)	—	0.4	3,9	51,8	31,2	12,7	—
HMI 4000 W (Osram)	0.1	0.8	6,2	53,3	32,3	7,3	-
Halogen 3400 K (Osram)	—	—	0.4	19,1	47,4	33,1	
Siccatherm 250 W (Osram)	—	—	—	4,0	42,5	50.8	2,7
MSR 1200 W HR (Philips)			3,6	53,5	30.5	12,4	

Much like the heating tests, the evaluation of cooling performance is made by comparing mean values for certain cabin temperatures with prerequisites. An example is shown in Figure 13.69, where the cooling performance of two air-conditioning configurations are compared. Both the versions were tested in a climatic wind tunnel under the portrayed test conditions.

---

8. See Hager et al. [195].

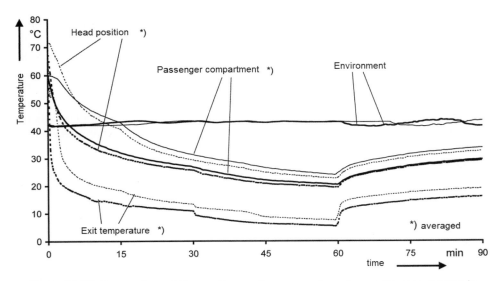

Figure 13.69 Comparison of the cooling performance of an air-conditioning system in base configuration and an optimized configuration (thick line).

13.3.8.1 Windscreen Defrosting and Dehumidifying Test

For vehicles that are driven in colder climatic regions, heating devices with good defrosting performance are a crucial requirement. In some countries a minimum power output for the defrosting of the windshield is requested. Test specifications[9] for the North American countries are specified through the FMVSS 103 (1996), and assessment demands in the European countries are described in 78/317/EEC (1978).

Defrost test: Before performing a defrost test, the vehicle is conditioned at a low temperature until all parts of the vehicle have the temperature required for the test. Test temperatures recommended in the EEC directive are between −8°C and −18°C. The test begins with the application of an ice layer with a defined thickness (e.g., 0.044 g/cm²) on the glass surfaces of the vehicle; this is mostly done by using a spray gun with water. For another 30 to 40 minutes the vehicle will be exposed to test conditions. Following, an engine heat-up is performed. In order to allow for realistic and reproducible boundary conditions for the warm-up phase, the engine speed, battery voltage, terminal voltage of the defrosting system, and so forth are specified. The defrost system is now turned on and the time course of the defrosting operation is captured by photo or video cameras. These images can be referred to for the evaluation of the defrost system. For this purpose, special zones on the windshield are defined in the above standard requirements. Those are the only zones used in the assessment of the test results.

---

9. Test preparation and test procedure refer to SAE J902 FEB99 (2003).

The minimum performance required of a defrosting system is determined as a percentage of an area of each windshield zone that has to be defrosted in a given period of time.

An infrared camera (thermography) can provide valuable information to the optimization of the defrost and dehumidification system. The temperature distributions of windscreen and other glass surfaces recorded with this measurement device under real operation conditions help the engineer to develop appropriate measures. In addition, Burch et al. [119] found out that infrared images of the windshield, recorded in a wind tunnel under room temperature conditions, are very similar to the results of a proper defrost test at low ambient temperatures. Therefore, the development tests for the defrost system can be done in advance with little effort under room temperature conditions.[10]

Dehumidification test: As described in 78/317/EEC (1978), the humidity generation of the passengers (approximately 70 g/h per person) during a dehumidification test is simulated by a steam generator specifically designed for this purpose. The test temperature is just below the freezing point (e.g., −3°C). After the conditioning phase, the steam generator is run for 5 minutes inside the vehicle in order to produce a thin layer of moisture on the glass surfaces. Then, the engine is started and the dehumidification system switched on to remove the moisture from the glass surfaces. The steam generator remains in operation until the test is finished. The outlines of the dehumidified area are recorded similar to the defrosting test or marked and documented for evaluation. The minimum capacity for European countries is defined in regulation 78/317/EEC (1978).

13.3.9 On-Road Measurements
13.3.9.1 Drag Measurement Using Coastdown Tests

The determination of the drag coefficient c_D is usually carried out in the wind tunnel. The restrictions that come with it were described in section 13.2. In order to avoid these inadequacies, procedures for the measurement of the c_D on the road were developed. Separating drag and rolling resistance as well as accounting for disturbing environmental influences such as the slope of the test track, wind, and rain are the main difficulties.

The most applicable method is the roll-out test, which is to be described on the basis of the approach proposed by Bez [75]. The test must be carried out on a long, straight, flat test track. Walston et al. [840] reported how to eliminate the influence of natural wind. It is better, however, to perform the test when there is no wind at all.

At first the test vehicle is accelerated to a high velocity, in order to roll out after stopping the engine. The time course of the vehicle velocity is recorded continuously. The

---

10. Aroussi et al. [198] applied the same method to evaluate the defrost and dehumidification behavior of the side windows.

deceleration is a result of applying driving resistances, namely the aerodynamic drag and the mechanical resistance.[11] The following relationship is applied (see equation (3.5)):

$$(1+f) \cdot m_F \cdot \frac{dv}{dt} = F_{mech} + F_W \qquad (13.59)$$

where m_F is the vehicle mass in kg, f is a coefficient that takes the rotating masses into account, $v(t)$ is the time dependent velocity in m/s. F_{mech} is the mechanical resistance in N. It consists of the rolling resistance of the wheels, the resistance of the drivetrain, and the bearings. At last, F_W is the looked for aerodynamic drag in N. The coefficient f is derived from the motion equations of the rotating masses and can be written as follows:

$$f \cdot m_F = \frac{J_d}{r_{dyn,d}^2} + \frac{J_0}{r_{dyn,0}^2} \qquad (13.60)$$

where J_d is the moment of inertia of the rotating parts of the drivetrain, including the wheels of the driven axle in Nm · s², J_o is the same item of the nondriven axle in Nm · s², r_{dyn} is the dynamic rolling radius of the tires on the driven axle (index d) or the non-driven axle (index 0) in m.

A method for the determination of the mechanical resistance is the test in a laboratory. The reliability of this approach however suffers from the difficulty to measure the rolling resistance of tires on a roller test bench accurately. Measurements on a roller—externally or internally—always cause non-negligible differences compared to an even roadway. Furthermore, the additional tire resistance due to the suspension (i.e., camber, toe-in) also needs to be considered.

A more accurate measurement of the rolling resistance in the laboratory is possible, if the wheel is tested on a belt system instead of a roller. Potthoff et al. [652] have shown that with the help of four-wheel rotation units on which the vehicle is placed, rolling resistance values of all four wheels can be measured up to a velocity of 250 km/h. The belt is made of steel with a thickness of 0.3 mm and is covered with a heat-resistant spray coating to simulate a typical road surface.

Alternatively, the mechanical resistance can also be determined in an on-road test, where the aerodynamic drag is artificially turned off. This is possible by using a hood-shaped

---

11. The total resistance can also be measured by pushing the test vehicle with another vehicle via a rod, as described by Romani [201]. The force measured at the rod equals the total resistance. However, the rod needs to be long enough in order to avoid interferences between the two vehicles. According to chapter 4, the distance between the vehicles shall be more than two vehicle lengths. Therefore, this method does not seem to be very feasible. A suggestion from Yang [202] has the test vehicle towed with a rope. A load cell at the rear end of the rope measures the total resistance. Since the distance between the cars should be at least 300 m to avoid interferences, this method is just as hardly realizable as the above.

measuring trailer, which, as sketched in Figure 13.70, is towed by a train wagon and is large enough itself to cover and enclose the test vehicle. Detailed descriptions of this technique are supplied by Carr et al. [134] and Kessler et al. [425].

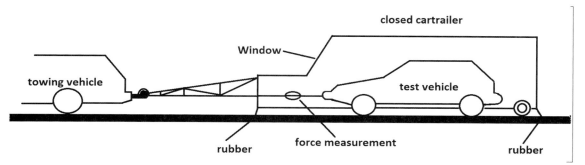

Figure 13.70 Determination of rolling resistance of a vehicle with the help of measuring trailer, schematic; after Carr et al. [134] and Kessler et al. [425].

The measuring trailer has wheels at the rear end and is connected to the tow vehicle with a trailer hitch in the front. The connecting rod between the measuring trailer and the test vehicle is equipped with a sensor for the measurement of the tractive force. The side walls of the measuring trailer end up on the ground in a rubber apron. It is constantly in contact with the road surface, so that the test vehicle is completely shielded and thus its aerodynamic drag is equal to zero.

If the measurement of the mechanical resistance is carried out immediately after the roll-out test, this method eliminates all the disadvantages connected with the previously described procedure. A disadvantages is that the driver of the test vehicle must perform subtle steering corrections in order not to depart from the course of the towing vehicle. This inevitably increases the mechanical resistance. In addition, when driving two vehicles as a team, the test speed needs to be limited for safety reasons. In order to obtain reliable results, the entire test procedure should be repeated for several times in both directions of the test section. The measurement data should then be averaged.

The rolling resistance of wheels can also be detected separately, with a measuring trailer on the road (see e.g., [311], [312], [540]). In Figure 13.71 such a trailer developed by the FKFS is shown. The airflow around the test wheel is inhibited by an apron touching the road all around the trailer. Figure 13.72 shows the internal structure of the trailer, where the front wheel is the test object. Wheel geometry and load can be varied in a wide range.

Figure 13.71 Measuring trailer of the FKFS for the detection of the rolling resistance under real conditions. Photo: FKFS.

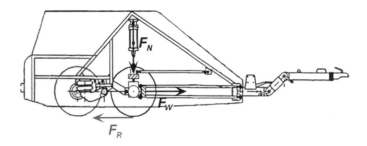

Figure 13.72 Interior of the FKFS measuring trailer. Top: variable wheel load F_N, measured longitudinal force F_L, rolling resistance $F_R = F_{L,korr}$ (see Mayer & Wiedemann [540]). Max tire size 275/35 R 20, maximum wheel load 5500 N, v_{max} = 130 km/h. Bottom: side wall opened. Wheel on the right side is the test wheel. Drawing and photo: FKFS.

Remenda et al. [665] described a method in which the aerodynamic drag can be determined as a function of the angle of attack as well as the rolling resistance during a roll-out test on a non-planar street under natural wind. Therefore, the roll-out distance as a function of time, total pressure (pressure probe), and inclination angle is measured in intervals of 0.5 seconds in order to iteratively adapt a computational model.

In a similar, more sophisticated approach, in 1995 Buckley Jr. [116] formulated the aerodynamic drag (in dependence of angle of attack and flow velocity) and rolling resistance (as a function of speed) as polynomials. The coefficients of the polynomials (maximum 15 variables) were then calculated using the measured data of roll-out tests (under the influence of natural wind) at 1-second intervals and by applying a linear regression algorithm. For a typical roll-out test, about 120 linear equations are available (ten test repetitions: 1200 equations).[12] In addition to the driving velocity, the resulting flow velocity and the angle of attack have to be recorded. Both items are measured with a vane anemometer, which is attached to a long pole at the front of the vehicle.[13]

In order to measure the fuel consumption and exhaust emissions on a chassis dynamometer, it is necessary to simulate the driving resistances realistically. One possibility is to program the test bench using vehicle data influencing the rolling resistance and aerodynamic drag, as for example vehicle mass. Another method is to first determine the total resistance of the vehicle with a roll-out test on the road (i.e., as a history of speed over time) and then to repeat the same test on the chassis dynamometer. The parameters of the test bench then need to be adjusted, so that the roll-out behavior of the vehicle on the roller matches the road measurement.

13.3.9.2 Crosswind Tests

The different possibilities to evaluate a vehicle's crosswind sensitivity were described in detail in chapter 5. In an on-road driving measurement and driving tests, such as the ones by motor journalists, passing an artificially generated crosswind gust is still preferred.[14] Figure 13.73 shows such a test facility. The lateral deflection of the vehicle's driving direction from the original course is seen as the characteristic parameter, with which the crosswind sensitivity can be quantified and assessed (see Figure 13.74).

---

12. Le Good et al. [209], [210], and Walter et al. [210] have also tried similar tests to compare c_D-values, gained from measurements in the wind tunnel and at on-road test. Passmore et al. [211] examined the bandwidth of roll-out tests results that have been carried our similarly. Furthermore, it is indicated to the SAE Standards J2263 [212] and J1263 [213], where the preparation and execution of the tests mentioned are described thoroughly.
13. In a similar method, Mayer et al. [214] used pressure probes in the front of the vehicle in order to determine the angle of attack. Furthermore, the profile of the test track was considered at the evaluation of the measurement data. It was acquired with a gyro-stabilized platform and laser distance sensors in the front and in the back of the car.
14. For the details of the method, see SAE Standard Practices J1263 [214], J2264 [216], and E/ECE/324 [216].

Figure 13.73 Crosswind test facility of Ford Werke GmbH; It consists of single fan units that can be arranged vertically or angular to the test section as needed. Photo: Ford Werke GmbH.

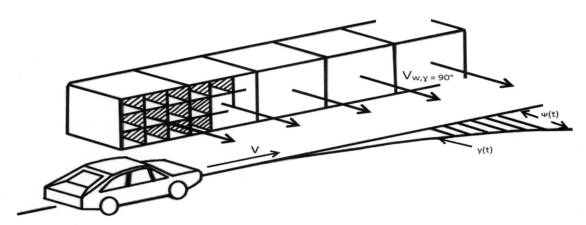

Figure 13.74 Schematic test procedure at a crosswind facility with "fixed control" steering wheel conditions.

There are two procedures available for the conduction of crosswind tests (also see section 5.1.2.2):

1. The driver does not apply any steering corrections. During the entire test the steering wheel is either held in the original position (fixed control) or it is not touched at all by the driver (free control).

2. The driver's task is to compensate for the lateral deflection of the vehicle by steering corrections and to keep the deflection as small as possible.

In the former process, only the vehicle's response to the gust is recorded. It is therefore mainly suitable for comparison tests. The latter method incorporates the reaction of the driver. The test results are therefore significantly influenced by him. The advantage of this method is that the control loop consisting of driver, vehicle, and environment is reproduced realistically. Not least, the reaction of the driver is affected by the dynamic build-up of the aerodynamic forces and torques. Unfavorable is the circumstance that the driver is anticipating the crosswind gust and consequently his response is "preprogrammed."

The lateral deflection of the vehicle can be measured in two ways:

1. The driver drives on a given guideline, parallel to the crosswind generators (see Figure 13.74) into the test section. The position of the vehicle during the test is either marked by splashes of color on the road or captured by sensors under the road. Photo sequences from the front have also been recorded for this purpose.

2. Measurement devices mounted in the vehicle acquire the driven pathway before and during the test. The equipment used to measure the vehicle's lateral movement relies on the accurate measurement of the lateral acceleration of the vehicle. The lateral deflection of the vehicle is obtained by integrating the acceleration signal twice. In order to measure the lateral acceleration correctly, the acceleration sensor is mounted on a gyro-stabilized platform. Thus, its original horizontal position is maintained in an earth-fixed coordinate system throughout the test. Finally, the vehicle speed must be measured precisely to correlate the lateral deflection of the vehicle to the position in the test section.

The process first described is quick and easy to handle; no sophisticated instruments are required. Yet, it is less accurate than the second method. Errors can arise from the fact that the driver has to eliminate the yaw angle with steering movements immediately before entering the crosswind section. This error is avoided when working with a gyro-stabilized measurement device and when the recording of the lateral deflection is started before the vehicle enters the crosswind section.

It is easy to demonstrate that the crosswind sensitivity of a vehicle depends not only on its aerodynamic properties: In order to do so, its driving behavior is influenced at two consecutive attempts by changing the tire pressures in the front and the rear. Although not changing the aerodynamic coefficients, the lateral deflection caused by the crosswind, for example, increases significantly. This is especially noticeable when the vehicle tends to oversteer due to the tire pressure changes. This happens when the tire pressures are set higher in the front and lower in the rear.

There are many different crosswind facilities in operation. Their characteristic data are collected by Klein et al. [441] and presented in an updated form in Table 13.2. Their comparison shows vast differences with regard to their length and wind speed.

Therefore, results obtained on different systems have to be compared with caution. Since the wind speed profile (i.e., wind force and wind distribution) varies with distance from the crosswind generators, a defined clearance between the vehicle and the crosswind generators has to be ensured.

The resulting wind speed vector affecting the vehicle is composed of the road speed of the vehicle and the crosswind component, as drawn in Figure 4.14. As emerging from Figure 5.15, considerable differences apply between crosswind test facility, wind tunnel, and on-road conditions. The velocity profile of the crosswind system is still heavily idealized. Figure 13.75 demonstrates how it may look at on-road driving; it is to assume that each crosswind facility has their own velocity profile. It becomes clear how difficult it is to correlate the results of such crosswind tests with the opinion of a driver under real driving conditions.

Table 13.2 Specifications of Various Crosswind Facilities, According to Klein et al. [441]

Facility	Crosswind Range (m)	Wind Speed (km/h)	Number of Fans	Drive Type
Daimler Chrysler	32	70	16	electric motors (à 45 kW)
Motor Industries Research Association (MIRA)	36	72	3	gas turbine (RR Avon, thrust 45 kN)
Toyota	44	35/70 (angular flow possible)	15	electric motors (à 95 kW) (2 fixed speeds)
Nissan Motor Co.	45	18 ... 80 (angular flow possible)	15	electric motors (à 95 kW) (variable speed)
Japan Automobile Research Institute (JARI)	15	48 / 80 / 106	5	electric motors (à 320 kW) (3 feste Drehzahlen)
Ford-Werke GmbH (TNO Design)	17	... 75 (60° angular flow instead of 90°)	6	V6 gasoline engines (à 149 kW) (variable speed)
BMW	32	60/80	10	electric motors (à 55,7 kW)
Transportation Research Centre Inc. East Liberty, Ohio	> 22 (variable distance between the fans)	40/56/80 (with the help of baffle plates improved flow quality)	6	6-cylinder gasoline engines

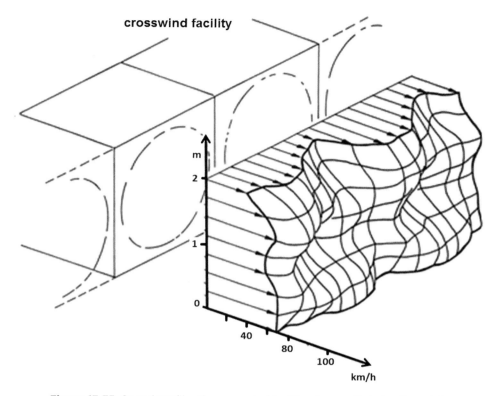

Figure 13.75 Speed profile of a crosswind facility. Image: Daimler-Benz AG.

A frequently made error during crosswind tests, especially when different cars are compared by the press, is, that a large sideslip is applied; mainly this is at 45°. At high speeds, however, where crosswinds can be dangerous, sideslip angles of this magnitude do not occur. Furthermore, at such large sideslip angles even the ranking of vehicles regarding yaw moment coefficient and lateral force coefficient can be reversed, as the depiction of yaw moment and lateral force in Figure 4.95 shows. Useful sideslip angles for testing are in the range from 20 to 30°.

13.3.9.3 On-Road Cooling Tests

A systematic study and adjustment of the cooling system in an on-road test is difficult, because of the constantly changing environment. Therefore, road tests are usually only performed in order to validate the results from the wind tunnel and to confirm their significance.

High-speed tests are easy to realize, if a long, even test track or a high-speed oval is available. An obstacle to the testing of mountain pass drives is that, at least in Europe, sections with a constant ascent that are long enough that the temperatures of cooling and lubricant reach steady-state values are hardly found. This problem can be solved by driving on flat road and towing a specially designed trailer; with the help of this, alpine mountain passes can be easily simulated (apart from the influence of the altitude). Similar to the results obtained in a wind tunnel, road testing measurements can be expressed in the form of ATB temperatures (see section 13.3.7).

Generally, the ambient air temperature changes every day; in many cases, it is lower than the temperature chosen in a thermal wind tunnel. However, the SAE standard J819 [702] recommends refraining from engine cooling testing on the road if the ambient temperature is below 24°C. This recommendation is justified by the strongly changing conditions of thermal radiation in the engine compartment, as well as changes in air density.

An important criterion for mountain passes is the influence of the altitude. Because of the decreasing atmospheric pressure with altitude—it is only 0.74 bar at 2500 meters above sea level—and the consequent lower air density as well as the lower ambient temperature, the power output and subsequently the thermal behavior of the engine changes. The simulation of such conditions is hardly possible in a common wind tunnel or in the lowlands; Engine cooling testing at different ambient pressure can only be performed in especially designed test beds (altitude chamber for emission tests). If such a test facility is not available, mountain pass test drives are indispensable.

13.3.9.4 Soiling of Windows and Chassis Parts

In everyday driving, dirt particles accumulate on the vehicle body (see section 6.3), which is either whirled up from the wheels of the own vehicle or of other vehicles. In order to evaluate the effectiveness of structural changes with regard to vehicle soiling, quantitative methods for measuring soiling are required. Such procedures have been developed for the application in the wind tunnels as well as on the road.

Hucho [381] described a method to examine the deposition of dirt on the chassis, which has been whirled up by the wheels during an on-road test. According to this, a test track is first covered with specific dirt particles. The body parts to be examined are equipped with small plates, which are weighed before the test. The test vehicle then drives along the test track for several times. After the test, the plates are removed and weighed again in order to determine the weight increase caused by the deposition of dirt.[15,16]

If the soiling effects, which are caused by a vehicle driving ahead of the test vehicle, are to be examined, water is sprayed in front of the test vehicle. The small plates used to measure the degree of contamination are coated with a thin layer of dry, hydrophile material.

Another technique to simulate the dust whirled up by the wheels is to add talcum powder to the air and to cover the relevant body parts with a thin layer of oil. The talcum is injected into the air flow either behind the wheel or in the wake behind the vehicle.

---

15. In a similar method of Bannister [221], another vehicle drove ahead of the test vehicle in order to gather contamination not only caused by the test vehicle but also of a different car.
16. Hentschel et al. [220] used clear foil instead of plates in a similar measurement; to evaluate the degree of contamination, the foil underwent a transparency measurement.

The talc concentration observed on the surface of the vehicle gives information about the soiling to be expected under natural conditions. This method is very sensitive.

In a similar process, Hentschel et al. [339] use a nozzle, placed in the area of the rear wheels, to simulate whirled-up dirt with titanium dioxide, dissolved in water. The concentration of titanium dioxide deposits on glass and body panels is in turn used to evaluate the vehicle soiling.

13.3.9.5 On-Road Measurements of the Wind Noise

During an on-road test drive, it is generally required to determine the contribution of wind noise to the overall noise inside the vehicle. In addition, the identification of the main noise sources outside the vehicle can be of interest.

Interior noise measurements: wind noise measurements inside the vehicle are not without problems on the road because the wind noise to be measured is superimposed by the noise of engine and tires. Furthermore, the reproducibility of the tests is affected by the constantly changing environmental conditions (natural wind, temperature, etc.). Watkins et al. [854], however, have shown, that at sufficiently high speeds on the road (160 km/h or higher) the interior noise measurements of on-road tests and wind tunnel tests show very similar results. Even the higher degree of turbulence on the road does not lead to any differences.

At lower speeds, the noise of the drivetrain can be eliminated by switching off the engine and rolling out. In order to minimize tire noises, special tires with appropriate rubber compound and a low-noise profile chosen according to the road surface can be used. If it is possible to perform the tests on a "quiet" road (porous asphalt and concrete), tire noises can be reduced even further (magnitude of about 15 dB).[17]

The evaluation of wind noises in on-road testing is usually confined to comparisons, which holds true for the measurements themselves, as well as for the subjective rating. The latter is performed using tape recordings in the laboratory where the signals from the different configurations of the vehicle are played back successively. Only when changing from one configuration to another within a short time interval is an efficient subjective comparison is possible.

Exterior sound measurements: The airflow detachment in different areas of the body work result in pressure fluctuations, which as noise sources ultimately contribute to the vehicle interior noise (as structure-borne noise or air-borne noise) as well as to the outside noise. Dobrzynski [207] has shown that for driving velocities of approximately 130 km/h onwards, the aerodynamic portion of the outside noise of a car is predominant. In order to locate each noise source and identify their impact, Barsikow et al. [43] suggested a method of measuring with a microphone array (equipped with 124 electret microphones). First results show that a detection of noise sources for a vehicle driving at velocities of up to 160 km/h can be performed reliably with this method.

---

17. See Beckenbauer et al. [226].

13.3.10 Additional Equipment in Climatic and Thermal Wind Tunnels

Although thermal and climatic wind tunnels are often not in the focus of the general discussion about wind tunnels, it should be highlighted that these test facilities are among the most complex forms of wind tunnels. This is due to the fact that such wind tunnels must cover a significantly wider range of test cases than mere aerodynamic wind tunnels.

It is obvious that a significant effort for cooling and heating must be made due to the wide range of desired temperature levels inside the plenum, which range from −40 to +50°C. In addition, the entire system must be insulated thermally; the relative humidity of the air shall be adjustable between 5% and 95%. Wind tunnels covering the entire temperature range are technically possible and have already been built, but they are certainly not desirable. Due to the high number of vehicles tested per day, "hot" or "cold" conditions can only be tested in phases. In this way, unproductive waiting times arise for the ever-shorter development phases of a vehicle and the flexibility of the system decreases. Considering the fact that it takes approximately 18 hours to cool down a test vehicle in a conditioning chamber to a temperature of −30°C, one can guess how long it takes until the wind tunnel itself is conditioned to the desired temperature. In order to eventually transport the cooled-down vehicle into the wind tunnel, an airlock with dried air is required to avoid condensation on the exterior of the vehicle. It is therefore worthwhile to distribute the vehicle testing spectrum to several wind tunnels, which are operated in different overlapping temperature ranges. It is a great advantage if the wind tunnels are identically constructed, hence transfer errors from one facility to another can be avoided.

For the dimensioning of climatic wind tunnels, the same criteria as for mere aerodynamic wind tunnels are applicable. Nevertheless, compromises can be made regarding various aspects. In particular, this concerns the flow quality of the flow at the nozzle outlet and the road simulation technique. In several proprietary, unpublished investigations it was observed that the heat dissipation of an artificially heated-up underbody of a vehicle was practically independent from the tests being carried out with or without (purely fluid mechanical) road simulation. Prerequisite for this result, however, was that the boundary layer at the nozzle outlet is small. This is usually realized by conventional boundary layer reducing methods, as described in section 13.2. One exception to this is the environmental wind tunnel at the EVZ of the BMW Group (see Bender et al. [68]), where it is possible to install a smaller belt for motorcycle tests in the middle floor section of the plenum.

The floor section beneath the vehicle can be realized in various forms. There are floors with shutter elements, that allow for a measuring probe to be moved under the vehicle in the direction of the flow, as well as heated floor sections that simulate the effect of a street that is heated by the sun to up to 70°C.

An important element in the test section of a climatic wind tunnel is the chassis dynamometer, which makes it possible to dissipate the power output delivered by the engine via eddy current brakes. The vehicle is fixed onto the rollers by chains or cables. The rollers can be bigger than 2 m in diameter. On the other hand, the rollers can also drive the wheels themselves, as it is necessary for brake cooling tests. Recent climatic wind tunnels provide a chassis dynamometer for each axle of the vehicle. The position of the front axle rollers is usually defined and unchangeable. The rollers for the rear axle can be moved in axial direction, in order to fit different wheelbases of vehicles. In some climatic wind tunnels only the wheels of the rear axle are placed on a roller dynamometer. Testing vehicles with front-wheel drive accordingly requires the vehicle to be moved in axial direction of the wind tunnel. With this comes the disadvantage of a bigger boundary layer due to the larger distance to the vehicle position and thus distortions of the underbody flow. The inflow angle at the wheels is also affected by this method.

The majority of measurements in a climatic wind tunnel are carried out with the engine running. The resulting exhaust gases are extracted at the tailpipe and lead outwards in a flexible pipe system. Basically, there are two approaches:

1. The exhaust gases are captured and sucked off by a collecting funnel in the immediate vicinity of the tailpipe. The suction flow rate needs to be bigger than the exhaust gas flow rate emitted by the vehicle in order to include unintentionally escaped carbon monoxide.
2. The suction pipe is connected hermetically to the tailpipe of the exhaust system. The sucked-off volume is redirected into plastic bags and subsequently undergoes a gas analysis in order to determine the concentration of each pollutant.

Both systems suffer from the disadvantage that the gas removal pipes represent a blockage in the rear end of the vehicle and disturb the flow, especially in the layer near the ground.

Since it is the aim to recreate all possible driving conditions of a vehicle in the climatic wind tunnel and to transfer the road with all its atmospheric conditions to the "lab" to a certain extent, some climatic wind tunnels are equipped with the ability to recreate the atmospheric pressure at a height of up to 4000 m above and 50 m below sea level. Appropriate pumps generate the desired atmospheric pressure in the wind tunnel. The plenum itself can only be accessed by a special pressure gate between the control room and the test section.

In order to examine the influence of solar radiation on the vehicle and the passenger compartment, large lamps simulating the spectrum of sunlight are installed in the plenum. The lights are attached to a frame structure, so the distance to the vehicle can be varied. The lamps have a defined distance to each other so that the entrainment flow (see section 13.2.4) in the direction of the shear layer will only be disturbed in an acceptable range, assuming an open test section. If the influence of solar radiation is not to be investigated, the frame structure can be elevated to the top of the plenum.

An important feature of the test spectrum of a climatic wind tunnel is the generation of rain and snow in the test section. Driving in the rain is simulated by several water spray nozzles that are installed on a grid frame. If demanded, the frame can be placed in front of the wind tunnel nozzle. It has to be accepted that the flow quality in the test section is affected by the frame. The water spray nozzles themselves are adjustable, so that the amount of water as well as the droplet size can be controlled. The water allocating in the test section is led away by a drainage system provided in the ground.

In this context, it is worth noting that soiling investigations of vehicles are also carried through in climatic wind tunnels. In the thermal wind tunnel of the University of Stuttgart, a special method has been developed. For further detail, see section 13.3.6.

The simulation of snowfall in a climatic wind tunnel is realized in a similar way. A snow lance, which can be positioned within the nozzle if necessary, generates snowflakes using a "snow machine." In effect, these machines are similar to the ones used for the generation of artificial snow for winter sports. Size and quantity of the snowflakes can be adjusted, so that the full range from wet snow to fine, fluffy snowfall can be created.

In climatic or thermal wind tunnels, it is also desirable to simulate driving cycles (switching between acceleration, constant drive, and deceleration or even mountain rides while towing a trailer). For this purpose, the wind speed and roller speed must be adapted accordingly. In this context, a special fan control is required as well as a fan with a low moment of inertia by means of a lightweight construction, in order to ensure quick changes of the fan speed. This type of fan control is called FFD (fan-follows-dyno). With the help of a programmable controller for acceleration and deceleration, which is implemented in the car, a predefined driving cycle can be run. However, experienced testing engineers are also able to rerun such driving cycles on a chassis dynamometer.

Another special driving cycle is the so-called stop-and-go cycle. In this case, the vehicle is decelerated to a standstill and subsequently accelerated again. The special aspect of this cycle is that during the standstill phase, the wind velocity may be negative, and the vehicle thus experiences reversed flow from rear to front. This can be realized by operating the fan in the opposite direction of rotation once the rotational speed reaches zero. In this operational mode, the efficiency of the fan decreases enormously, but it can generate wind speeds of up to 30 km/h. For the cases demanded, this is mostly sufficient. In this way, fast highway travel with sudden traffic jams can be simulated, where in extreme cases the direction of the natural wind is in the direction of travel. The main development aspect is to avoid heat accumulation that can occur at the vehicle in such cases.

Summarizing, it can be said that the tasks of a thermal or climatic wind tunnel can be extremely vast and complex. The planning of such test facilities already represents a major logistical challenge.

13.4 Model Testing—Dimensionless Numbers

Reduced scale models are often used for aerodynamic measurements. They present a number of advantages compared to 1:1 (i.e., full-scale models), in particular in the early phase of vehicle development. At the beginning of development there are usually not any prototypes on a 1:1 scale available. Added to which, creating 1:1 models consumes considerable time and cost. The cost of aerodynamic studies in a wind tunnel on reduced scale models is much lower compared to 1:1 models. Investigations on a reduced scale model must be portable to real vehicles, so:

- A real and modeled vehicle must correspond as closely as possible in detail (geometric similarity)
- The flow relationships on a modeled scale and 1:1 must be the same (kinematic similarity)

To meet these requirements attempts to simulate the 1:1 vehicle should be as accurately in detail as possible. Depending on the purpose of investigations and need for accuracy, outside mirrors, windshield wipers, door handles, and possibly also airflow through the engine compartment are modeled. Figure 13.76 gives an impression of this, showing a vehicle model with a detailed underfloor on a 1:5 scale.

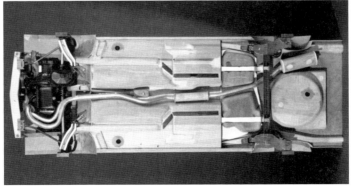

Figure 13.76 Car model scaled 1:5, Adam Opel AG.

Furthermore, to achieve the same flow relationships on a model scale and in 1:1 studies, the Mach number Ma (compressibility) and Reynolds number Re (laminar or turbulent flow character) must correspond in both cases. Taking u_∞ as flow velocity, c as speed of sound, l_{char} as characteristic length, and ν as kinematic viscosity:

$$Ma = \frac{u_\infty}{c} \tag{13.61}$$

and

$$Re = \frac{u_\infty \cdot l_{char}}{\nu}. \tag{13.62}$$

The characteristic length used in equation (13.62) for example can be the wheelbase of a vehicle. Consequently, when experimenting on a reduced scale, it is necessary to match flow velocity proportionally to the scaling to obtain the same Reynolds number as on a 1:1 scale.

The requirement for the same Reynolds number means, for example, four times the air velocity on a 1:4 model as for the 1:1 model. To simulate a measurement at 140 km/h, a 1:4 model would then have to be "attacked" at an air velocity of 560 km/h however.

Increasing flow velocity automatically means a higher Mach number according to equation (13.61). So typically it is not possible to satisfy the requirement for matching Reynolds number and Mach number on a reduced scale model and 1:1 scale at the same time using an identical flow medium.

In model experiments therefore, it is only possible to create approximately similar flow relationships compared to a 1:1 scale if only one of the two criteria is satisfied. Comparable Reynolds numbers are of primary importance, however, because laminar or turbulent flow is produced as a function of the Reynolds number. Exact adherence to Mach number is secondary in vehicle aerodynamics, at least for Mach numbers < 0.2, because the familiar compressibility effects as a function of Mach number are only noticeable at higher numbers.

For more precise adherence to the Reynolds number there is the possibility of influencing it through the kinematic viscosity of the working fluid $\nu = \mu/\rho$ by

- Variation of pressure, or
- Adapting flow temperature.

In high-pressure and cryogenic wind tunnels, these physical approaches are implemented in practice. In the cryogenic wind tunnel of the German Aerospace Center in Cologne (KKK), it is possible to conduct investigations in aircraft construction or long-distance rail traffic on a reduced scale model, whereby airflow is replaced by a pure nitrogen atmosphere that can be cooled down to a temperature of −173°C [195].

In aerodynamic investigations on reduced scale models there is the possibility that a laminar boundary layer will form instead of the expected turbulent boundary layer. The boundary layer of laminar flow is more inert and separates much earlier than that of turbulent flow; this leads to a marked increase in aerodynamic drag. To avoid this effect, the Reynolds number should not be allowed to drop below $3 \cdot 10^6$ when measuring drag and buoyancy forces on reduced scale models.

Figure 13.77 [873] shows the coefficient of drag as a function of the Reynolds number, varied here by altering the flow velocity. Aerodynamic investigation of a 1:1 model at a flow velocity of 150 km/h requires, according to the Reynolds number criterion, a flow velocity of 600 km/h on a 1:4 scale. But the diagram shows that similar conditions exist at 216 km/h already (i.e., a much lower flow velocity). The Reynolds number is much smaller for this velocity on a 1:4 scale than for a 1:1 model, but the error span width σ is only in a range of 1 to 2% because the Reynolds number for the reduced scale model is already far into the more than critical region.

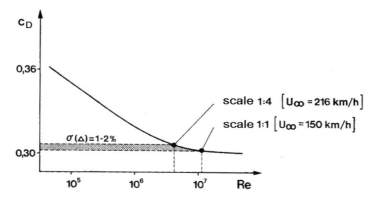

Figure 13.77 Model measurements for different Reynolds numbers [873].

This applies to many vehicles, with the exception of a few exhibiting a critical geometric feature. These include, according to Hucho [370], in particular automobiles with a hatchback or a critical radius in the front. An automobile of this kind was used by Wiedemann and Ewald [881] for investigations in which a higher Reynolds number was simulated by a greater degree of turbulence of the flow relationships.

A screen is attached in front of the nozzle to increase the degree of turbulence (cf. Figure 13.78).[18] Air flowing through this screen produces turbulence, that is, the turbulent kinetic energy of the flow increases (cf. equation (14.18)). Although this has no effect on the actual Reynolds number, the flow reacts as for a higher Reynolds number because of the increased turbulent energy [881].

---

18. This kind of turbulence screen is not to be confused with the turbulence screens inside the piping that serve to reduce the larger scaled flow turbulence (cf. section 13.2.1).

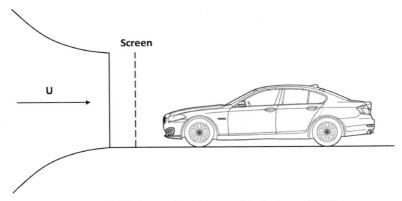

Figure 13.78 Increasing degree of turbulence [880].

The effect of greater turbulent energy on the measured pressure distribution has been investigated by comparing the pressure distribution in the center longitudinal section of a reduced scale model with varied Reynolds number with and without screen [881]. Whereas the variation without screen shows differences in particular in the region of the front of the car, along the engine bonnet and underbody (higher Reynolds number show greater pressure or higher vacuum), the variation with screen shows virtually no effect on pressure distribution. The flow without a screen is still laminar in the front region of the car, with a screen it is turbulent over the entire length of the model. Figure 13.79 shows the influence on the drag coef!cient. M stands for mesh width and d for the wire diameter of the screen.

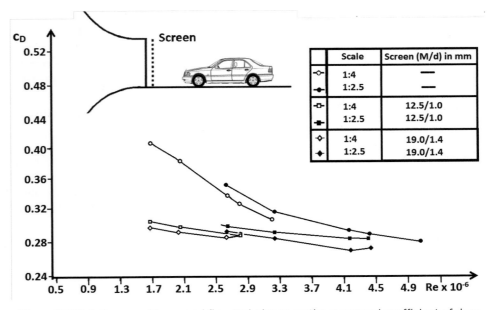

Figure 13.79 Influence of increased flow turbulence on the measured coefficient of drag on a reduced scale model (*M*: mesh width; *d*: wire diameter of screen) [881].

For measurements in model wind tunnels, this means that by increasing the degree of turbulence—such as by a screen—it is possible to create a turbulent flow for low Reynolds numbers already. It is then possible to work either with reduced scale models or with less flow velocity.

Instead of using a screen, so-called tripwires can be attached locally at critical points of the vehicle. One example is the flow circulating around a "sharp edge" of a reduced scale model. If the degree of flow turbulence is increased ahead of such a sharp edge by a tripwire, the flow is better able to follow the contour and may not separate, as can be observed on a 1:1 scale too.

When investigating models with airflow through the engine compartment, it can be observed that on heat exchangers in particular there is a much stronger dependence on Reynolds number than in air circulation of the vehicle. The circumstances of this are illustrated in Figure 13.80. It is especially clear that heat exchangers on different scales can lead to deviating assumptions of the loss coefficients and thus also to noncomparable cooling air mass flows.

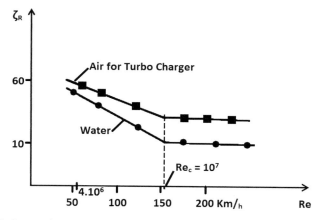

Figure 13.80 Dependence of pressure loss on Reynolds number on a heat exchanger [878].

One solution to this problem is the general-purpose cooler simulator developed by Wiedemann [878]. By this approach the heat exchanger is not adopted on an altered scale, but instead the desired volume rate of flow is set corresponding to the scaled volume rate of flow of the 1:1 model. This is done by setting the appropriate pressure loss by different pieces of perforated metal or a valve. The volume rate of flow can be calibrated once for different settings in the general-purpose heat exchanger and set accordingly once built in [873], [878].

13.5 Existing Wind Tunnels for Motor Vehicles

After the discussion of the physics of wind tunnels, the measurement techniques, and the scale-model techniques in the previous chapters, the present chapter describes and lists examples of existing wind tunnels. Distinctions are made between wind tunnels for real vehicles or full-scale (1:1) models (section 13.5.1), wind tunnels for reduced scale vehicle models (section 13.5.2), as well as climatic and thermal wind tunnels (section 13.5.3). Among the full-scale wind tunnels, both purely aerodynamic wind tunnels and wind tunnels suitable for aeroacoustic studies are listed.

The purpose of this chapter is to provide an overview of existing wind tunnels rather than a complete list. The latter would go beyond the constraints of the book. Further details of older full-scale wind tunnels can be found in the SAE Information Report J2071. An updated version of this report, dated 1994, is in preparation.

Meanwhile most automobile manufacturers operate their own facilities to conduct aerodynamic, aeroacoustic, climatic and thermal investigations. The manufacturers also use facilities of research institutes for automotive engineering and aviation. When building and using wind tunnels, it is necessary to consider many different effects and interactions (interference effects) to obtain reliable results that are comparable to the situation on the road. These effects are discussed in section 13.2.

Many tunnels built for full-scale cars and vans with an open or slotted test section use a cross-section of the nozzle outlet of about 25 m². Even large trucks and buses are tested in such wind tunnels, however, by means of scale models (usually 1:2.5 scale). The Reynolds number must be satisfied when transferring the results to actual vehicles, which requires matching the flow velocity accordingly. For a scaling factor of 1:2.5 a real driving speed of 80 km/h requires a flow velocity of $u_\infty \approx 200$ km/h in the wind tunnel. The extent to which these model measurements produce accurate and reliable results can be read in the previous section.

Climatic and thermal wind tunnels intended for the investigation of full-scale cars and vans usually have a smaller nozzle outlet. Here, the requirements on the quality of the flow around the test object are considered to be lower than in aerodynamic and aeroacoustic wind tunnels. Additionally, the high demand for cooling power may lead to smaller nozzle outlets compared to aerodynamic and aeroacoustic wind tunnels for economic reasons. Usually 10 to 14 m² were considered to be adequate for the envisaged flow quality, in facilities solely for cars 6 m² even. The fact that these relatively small nozzle outlets can be problematic is discussed in section 13.2. In the case of nozzle outlets smaller than 4 m² the wind tunnels usually are called blowing tunnels. They are not considered in this chapter.

Thermal wind tunnels are classified as those facilities which cover only the positive, that is, usually increased, temperature range, while climatic wind tunnels cover also negative temperatures and enable the simulation of winter conditions. Both kinds of wind tunnels are indispensable for examining thermal management, that is, the intelligent interaction of heat flow and other energy flow in a vehicle for the purpose of air-conditioning, heating, and cooling with respect to the powertrain and other vehicle components.

13.5.1 Full-Scale Wind Tunnels

The oldest full-scale wind tunnel designed for aerodynamic investigations is located in Stuttgart-Untertürkheim (Germany). As can be seen from Figure 13.81, it comprises a closed-loop (so-called Goettinger design) and a fan in the return path. The wind tunnel construction started in 1939 by Kamm for the Research Institute for Automotive Engineering and Vehicle Engines (FKFS). In 1970 Daimler-Benz AG acquired the wind tunnel and upgraded it comprehensively. With $v_{\infty,max}$ = 270 km/h it still provides adequate wind velocities.

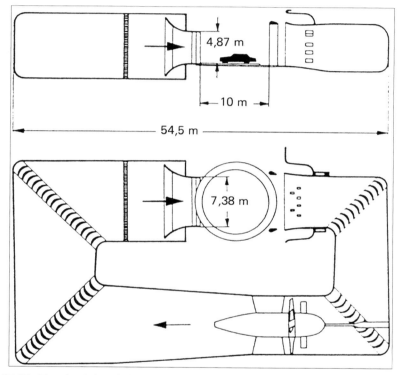

Figure 13.81 Sketch of the old full-scale wind tunnel of Daimler AG; nozzle outlet 32.6 m², maximum wind velocity 270 km/h, fan power 4 MW, Kuhn [473].

Due to the relatively high wind velocity and the large nozzle cross-section of more than 32 m², this tunnel also enables testing of truck and bus models scaled to 1:2.5 (see above). The full-scale Reynolds number can be satisfied and the product of frontal area and drag coefficient of the 1:2.5-model of a heavy-duty ¬truck of approx. 0.7 m² is only a slightly larger than that for a car. Thus, the same correction methods can be used as for cars (see section 13.2.13).

The largest wind tunnel for automobile aerodynamic testing was built by General Motors in Warren, Michigan and commissioned in 1980; also a tunnel with closed loop. The dimensions of its closed test section, A_N = 56.2 m², l_{TS} = 21.3 m, were chosen to enable also testing of large commercial vehicles in natural size. The maximum wind velocity of more than u_∞ = 250 km/h also allows studies on racing cars. The two sidewalls of the test section diverge by 0.24° to ensure uniformity of the static pressure in the empty test section. The wind tunnel is equipped with two six-component balances. The front balance serves for tests on reduced scale models, the rear balance on full-size vehicles. Wheelbase and track can be set by eccentric disks (see Figure 13.51, right). A boundary layer suction system at the nozzle outlet controls the boundary layer. A large heat exchanger, placed between the corners three and four, allows for controlling the air temperature between 16 and 29°C. In this way clay or plasticine models do not soften during the tests.

In contrast to the two wind tunnels described above, serving only for aerodynamic studies, Volkswagen AG in Wolfsburg (Germany) for the first time combined a full-scale wind tunnel with a climatic wind tunnel. The wind tunnel is fully air-conditioned from −35 to +45°C and wind velocities of up to u_∞ = approx. 200 km/h are appropriate for vehicle development. Dynometers are installed to turn the wheels with a braking or driving power of 185 kW and 150 kW, respectively. These characteristics qualify the full-scale climatic wind tunnel in Wolfsburg for many applications in vehicle aerodynamics plus thermal management.

Completed in 1965, this full-scale climatic wind tunnel remained the only one of its kind for a long time. Later wind tunnels were designed especially either for aerodynamic studies or for thermal management. Advantages and drawbacks of the different approaches are discussed for example by Hucho [381] or Bengsch [69]. However, the concept of special-purpose installations now dominates because the demand for testing time for a vehicle manufacturer with multiple car series and possibly vans is that large, that for capacity reasons only a number of facilities are needed. Hence, different special-purpose wind tunnels provide a better testing quality and are cheaper to operate than general-purpose wind tunnels.

More recently, due to the large demand for testing time, whole wind tunnel centers with facilities for the different purposes discussed above have been built or are under construction. One example is the wind tunnel center used by Ford Motor Company in Allen Park, Michigan. The center consists of two smaller climatic wind tunnels and three roller dynamometer test benchs with airflow. One of these is designed as an altitude test bench. A large aeroacoustic climatic wind tunnel was added, internally referred to as "WT8" (see Walter et al. [843]). This combination of aeroacoustics and climate (0 to +55°C) is found worldwide only rarely and aims at wind noise in dependence of temperature. Usually it is assumed that in normal conditions this dependency can be neglected. In the design of the climatic wind tunnel of Behr AG however, emphasis was placed on low sound levels in the test section in order to study the noise of air-conditioning systems and fans (Schmiederer et al. [723]).

In addition to these wind tunnels with nozzle outlets of more than 20 m², for full-scale passenger cars also wind tunnels with relatively small nozzle outlets of about

10 m² can be found. The wind tunnel of Pininfarina in Turin, Italy, built in 1972 and designed by Morelli ([586], [588]), is such an example. Originally intended as an Eiffel-type wind tunnel, it was encased by a building to reduce the demand on energy for the airflow. Other tunnels with relatively small nozzle outlets followed, for example, at BMW in Munich (Germany) (1988, A_N = 10 m²) and Audi in Ingolstadt (Germany) (1999, A_N = 11 m²). A moving belt and wheel rotation units were subsequently added to the Pininfarina wind tunnel. It was also improved by acoustic measures to enable aero-acoustic studies. Furthermore, in the meantime turbulence generators enable defined turbulent flow simulating the natural boundary layer character (Cogotti [158], [162], [163]) (see also section 13.2.12).

The BMW wind tunnel mentioned above was the first one built specifically for aero-acoustic studies. It was erected in an existing, relatively narrow building with vertical return path of the air. The wind tunnel is characterized by the patented, mushroom-shaped silencer which is located directly downstream of the collector. This silencer at the same time ensures the conversion of the flow from the rectangular cross-section of the test section to the circular ring ahead of the fan. The return contains wide-span splitter silencers know also from climatic engineering. The tunnel is illustrated in Figure 13.82.

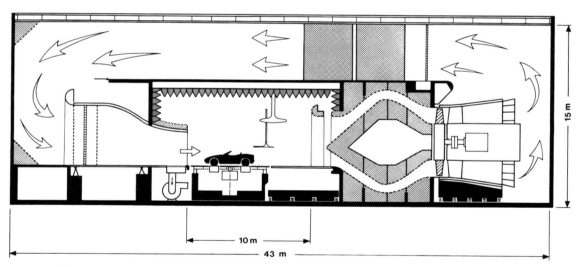

Figure 13.82 Aeroacoustic wind tunnel of BMW Group in Munich, design by L.J. Janssen, A_N = 10 m², max. wind velocity 250 km/h, L_p 66 dB(A) at 140 km/h [504].

An example of a wind tunnel designed for both aerodynamic and aeroacoustic studies can be found with Audi AG in Ingolstadt (Germany). It is outlined in Figure 13.83. The turning vanes in the second and third corners were realized as splitter silencers (in red). In this way the noise of the fan, which has been optimized regarding rotational sound (see section 13.2.1), is additionally damped. Active noise control (ANC, see section 13.2.1) was implemented to reduce low-frequency pulsations. The loudspeakers for the ANC are housed in a chamber connected to the test section diffusor behind the third corner. This system suppresses the low-frequency pulsations almost entirely. Details of this tunnel with initial experience of its operation are given by Lindener and Wickern [508].

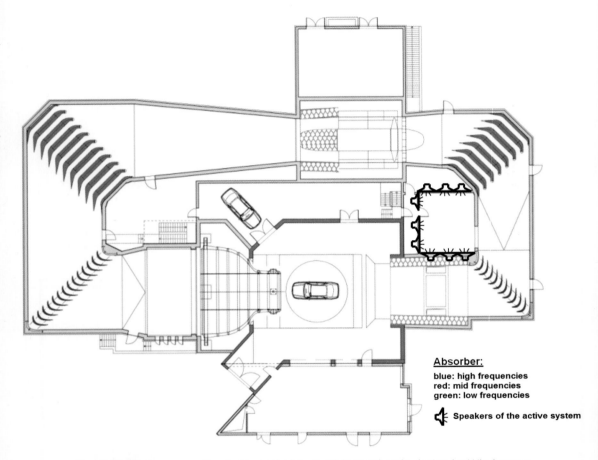

Figure 13.83 Aeroacoustic wind tunnel of Audi AG in Ingolstadt, design by Wiedemann, Wickern, and WBI GmbH (Wiedemann et al. [883]), $A_N = 11$ m², max. wind velocity 300 km/h, L_p 61 dB(A) at 160 km/h, rolling road system (center-belt and wheel rotation units) implemented.

To meet the growing demand for testing time in aeroacoustic wind tunnels, a number of conventional wind tunnels have been upgraded for aeroacoustic measurements. One example is given by the full-scale wind tunnel of the Institute for Internal Combustion Engines and Automotive Engineering (IVK) of Stuttgart University, operated by FKFS and outlined in Figure 13.84. At points 1 and 2 the air duct is split into sections with the same pressure loss and same acoustic transmission. The splitter walls have been used to integrate membrane absorbers into the tunnel. The sidewalls also have been equipped with membrane absorbers. The turning vanes have been covered with porous absorbers and the plenum has been covered with broadband absorbers (3 in Figure 13.84). In this way in 1993 the self-noise of the wind tunnel could be reduced to a level of $L_p = 69$

dB(A) out of flow at 140 km/h. Details are published in Kuenstner et al. [477]. By further measures in 2014 as for example additional absorbing material at the collector and the test section diffuror the level has been reduced further down to L_p = approx. 64 dB(A) out of flow at 140 km/h.

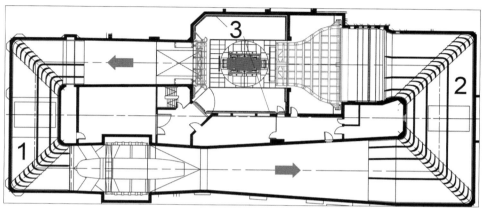

Figure 13.84 Automobile wind tunnel of Stuttgart University with measures to expand the spectrum to aeroacoustic tests at points 1, 2, and 3 (see text) (see also [477], Wiedemann et al. [883], Potthoff et al. [652], Potthoff [651]).

In 2001 the IVK/FKFS wind tunnel was upgraded by a five-belt rolling road system (MTS steel belt,) which has been replaced in 2014 by a longer and patented three- and five-belt system (FKFS *first*®, designed by MTS and FKFS). The moving belt system has a modular design that allows for a relatively fast and easy-handling change between three-belt for motorsport and five-belt for production vehicles. In 2014, also a new active side wind and turbulence simulator (FKFS *swing*®, see Figure 13.28), unique in vehicle wind tunnels to date, as well as measures to suppress low-frequency pulsations (patented flow guiding elements in the nozzle called FKFS *besst*®, see Figure 13.16, and Helmholtz resonator) have been implemented.

Another example of expanding the test spectrum to aeroacoustics is that of the wind tunnel of Ford Werke GmbH in Cologne (see Kohl [452]). In addition to the measures similar to those with the IVK/FKFS wind tunnel, the wind velocity setting has been changed from adjustment of the blade angles of the fan to control of the fan speed. The asynchronous motor could be retained. To compensate the loss of maximum wind velocity due to narrowing of the cross-sections in corners 2 and 3, the nozzle outlet was reduced in size by tapering the two top corners. Hereby, the nozzle outlet has been reduced from A_N = 24 m² to 20 m². The maximum velocity of $v_{\infty,\max}$ = 180 km/h is still feasible in continuous operation and 200 km/h for a short period.

In the recent past, despite the enormous investment costs, a number of additional wind tunnels and wind tunnel centers have been built. These are for example the Mercedes-Benz Technology Centre, including a remarkable full-scale aeroacoustic wind tunnel and climatic wind tunnel (see below), and the new full-scale aeroacoustic wind tunnel of Porsche. Others are currently under construction or in the planning phase (e.g. aeroacoustic wind tunnels at Volkswagen, Wolfsburg, and in China). This again demonstrates the importance of aerodynamics, aeroacoustics, thermal management and soiling in vehicle development. Within the present chapter, the aeroacoustic wind tunnel of Tongji University in Shanghai (China), operated within the Shanghai Automotive Wind Tunnel Center (SAWTC), and the Aerodynamic Test Center of BMW in Munich (Germany) shall serve as examples for a brief discussion of such new investments. Some € 170 million were invested in the latter.

The SAWTC comprises a full-scale aeroacoustic wind tunnel, a climatic wind tunnel plus a cold-start chamber (Tongji [805]) and was commissioned in July 2009. The aeroacoustic wind tunnel has been built in the Goettinger design with an open test section, which is common for automobile wind tunnels by now. The test section is 15 m length and the size of the nozzle outlet amounts to 27 m^2, being 6.5 m wide and 4.25 m high. These are common dimensions that have proved to be suitable. The maximum wind velocity of up to 250 km/h is also a common value for a modern automobile wind tunnel. The test section (view to nozzle) is shown in Figure 13.85.

Figure 13.85 Test section of the aeroacoustic wind tunnel of SAWTC (view to nozzle) [805].

Besides other acoustic measures, such as use of different absorbers, the fan of the tunnel was optimized with respect to rotational noise that produced one of the world's most silent full-scale wind tunnels. The level of the self-noise at 160 km/h (out of flow and taped floor) amounts to $L_p = 61$ dB(A). The fan is shown in Figure 13.86. Further characteristics are shown in Table 13.3.

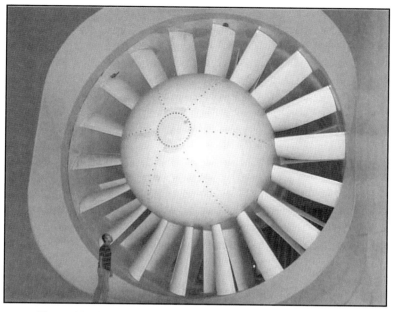

Figure 13.86 Acoustically optimized fan of the aeroacoustic wind tunnel of SAWTC [805].

At about the same time as the commissioning of the SAWTC aeroacoustic wind tunnel, in June 2009 the Aerodynamic Test Center (AVZ) of the BMW Group in Munich (Germany) came into operation. In addition to the main wind tunnel, a full-scale wind tunnel, the AVZ houses a smaller wind tunnel for specific investigations and research called Aerolab. Both wind tunnels are intended for aerodynamic tests. The plenum of the main tunnel, with a length of 22 m, width of 16 m and 13 m high, is designed on a relatively generous scale. One of the reasons for the generous scale is to allow for a possible later upgrade of the wind tunnel with acoustic linings to an aeroacoustic wind tunnel. Figure 13.12 in section 13.2 shows the plenum looking through the collector into the test section. Further features of the main wind tunnel are max. wind velocity $v_{\infty,max} = 250$ km/h for a nozzle outlet of $A_N = 25$ m² and $v_{\infty,max} = 300$ km/h for $A_N = 18$ m², a five-belt system for simulation of on-road driving (in addition to boundary layer suction and tangential blowing systems) and a six-component underfloor balance.

The smaller Aerolab is positioned vertically in the center of the air duct ring of the main tunnel. Whereas the plenum of the Aerolab is only slightly smaller than that of the main wind tunnel (20 m long, 14 m wide, 11 m high), the size of the nozzle outlet of 14 m² is significantly smaller. However, a single-belt rolling road system allows for a more comprehensive simulation of on-road driving. The Aerolab is suitable for studies on both scaled models and automobiles in their real size. Figure 13.87 shows the plenum of the

Aerolab with the single-belt system and a scaled model with fixation of the model at the wheel hubs and a Hexapod strut.

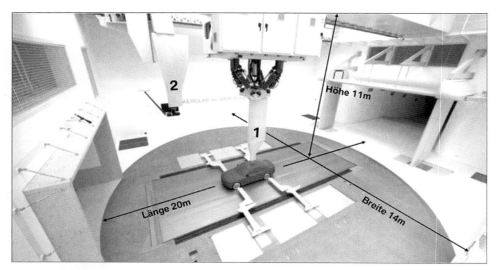

Figure 13.87 Test section of the Aerolab with vehicle model and single-belt system. Source: BMW Group.

13.5.2 Model-Scale Wind Tunnels

There are a large number of wind tunnels available for measurements on reduced scale vehicle models. Some of those wind tunnels are used for teaching and research and only occasionally for vehicle development. Others are specially dedicated to vehicle development and serve partly during the design phase of the full-scale wind tunnel as a pilot wind tunnel for the latter. The 1:4-scale wind tunnel of Porsche AG (see Vagt and Wolff [818]]) and the 3:8 scale wind tunnel of (then) DaimlerChrysler (see Romberg et al. [681]) are examples for such pilot wind tunnels. With a relatively large pilot wind tunnel larger Reynolds numbers can be achieved (because of the limited wind velocity) by which the results can be transferred more reliably to the full-scale wind tunnel. Hereby, of course it is important to keep geometric similarity between the full-scale and pilot wind tunnel. Thus, the same relations—for example, the same blockage—can be found. In vehicle development later it also simplifies the transition from the reduced-scale model to the full-scale model or prototype.

The facility of MIRA outlined in Figure 13.88 shows that a model-scale wind tunnel can be built with only relatively little effort. Details of the first configuration of this wind tunnel can be found in Carr [135]. The second stage including a moving belt is described in Carr and Eckert [133]. The relatively low-cost operation of such a wind tunnel, together with the easy handling of reduced-scale models, has resulted in more attention again being paid to scale models for vehicle development. That applies in particular to the development of racing cars.

Some larger companies which are engaged in motorsports operate their own wind tunnels in the meantime. In motorsports it is now common to state the size of a scale model by means of the percentage of the respective full scale. Fifty percent models are standard. Sauber has already moved on to 60% (see Wyss [899]).

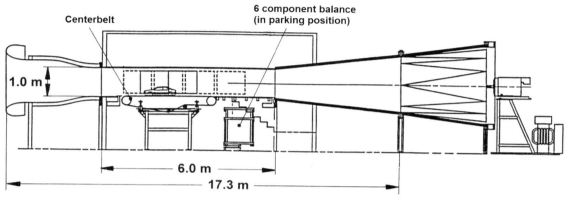

Figure 13.88 The model-scale wind tunnel of MIRA in Nuneaton, UK. Cross-section of the test section 2.12 m², maximum wind speed 90 mph (40 m/s), fan power 37 kW, road driving simulation by moving belt (Carr [128], Carr & Eckert [133]).

For flow observations on scale models of cars and on components, special low-turbulence smoke wind tunnels have been developed, for both three-dimensional (see Oda and Hoshino [615]) and only two-dimensional flow. Water tunnels are also used sometimes to visualize the flow around a vehicle model, as for example the water tunnel of the Institute of Aerodynamics and Gas Dynamics of Stuttgart University (see Speth [770]), which allows for models up to a scale of 1:4.

An example of a model-scale wind tunnel that is used both for automotive research and development as well as for research for further development of the full-scale wind tunnel is given by the model-scale wind tunnel of Stuttgart University. It is operated by FKFS (as the full-scale wind tunnel) and geometrically similar to the full-scale aero-acoustic wind tunnel including a five-belt system (MTS steel belt). More recently, it has been used for studies on unsteady aerodynamics and for the development of an appropriate active system to generate non-steady-state phenomena (Potthoff [651]). Figure 13.89 shows the current implementation of the active system FKFS *swing*® in the model-scale wind tunnel. Apart from the number of profiles (here six), the system is similar to that of the IVK/FKFS full-scale wind tunnel. Such studies on a reduced scale are much lower in costs than in a full-scale wind tunnel with original sized automobiles. However, as already mentioned the Reynolds number has to be satisfied and thus portability to full scale guaranteed.

Figure 13.89 Side wind and turbulence generator FKFS *swing*® in front of the nozzle of the IVK/FKFS model-scale wind tunnel of Stuttgart University for studies of transient aerodynamics with a 1:4-scale vehicle model (see e.g. [780a]). Source: FKFS.

13.5.3 Climatic and Thermal Wind Tunnels

Similar to the full-scale wind tunnels, climatic and thermal wind tunnels can be characterized by the size of the nozzle outlet. They can be classified into three groups:

- 10 to 12 m², primarily for vans and buses
- 6 m², primarily for cars
- 4 m², primarily for thermal management of car components

Typical of the first category are the two wind tunnels of FIAT, each of them having a nozzle outlet of 12 m² (Antonucci et al. [20]). These are part of a wind tunnel center to which also the large aerodynamic wind tunnel belongs. One of the two 12 m² wind tunnels is intended to simulate high temperatures, the other for low temperatures. The climatic wind tunnel of Ford Werke GmbH in Cologne covers a comparable temperature range alone, from −40°C through +50° C. It is also comparable in nozzle size (11 m²). Initially also aerodynamic studies were conducted in this climatic wind tunnel (Bengsch [69]). Such wind tunnels normally feature a dynamometer, adjustability of air temperature as well as air humidity and possibly solar radiation. This enables a wide range of climatic investigations necessary for the development of an automobile.

For large as well as small commercial vehicles and especially for buses since 2002 a large climatic wind tunnel is available at Rail Tec Arsenal (rta) in Vienna. It is the smaller one of two wind tunnels intended primarily for investigations on railway carriages. The nozzle outlet amounts to 16 m² the length of the test section to 30 m. Lamps to simulate sunlight are integrated in one of the sidewalls over the whole length of the test section (see Figure 13.90). It is possible to radiate between 250 and 1000 W/m². The step-shaped arrangement allows space for a fixed setting of a 30° angle of incidence (standard for rail vehicles). The dynamometer in the wind tunnel provides braking power of 280 kW. With a maximum wind speed of 120 km/h the permitted velocity for buses is exceeded, including a sufficient margin.

Figure 13.90 Simulation of solar radiation in the smaller climatic wind tunnel of rta Rail Tec Arsenal GmbH, Vienna (Austria). Flow cross-section 16 m², length of test section 30 m, maximum wind velocity 120 km/h, temperature range −50 through +60°C, solar radiation at 30° up to 1000 W/m².

For thermal management tests on passenger cars a nozzle outlet of 12 m² is larger than actually necessary. In many cases an outlet of $A_N = 6$ m² is regarded as being adequate. Even climatic wind tunnels with such a nozzle outlet size are operated, for example by Volkswagen, Denso, and Behr (for technical data and references see Table 13.3). As an example, Figure 13.91 shows Climatic Wind Tunnel 2 of Volkswagen AG, originally also designed for tests on 1:2.5-scale models.

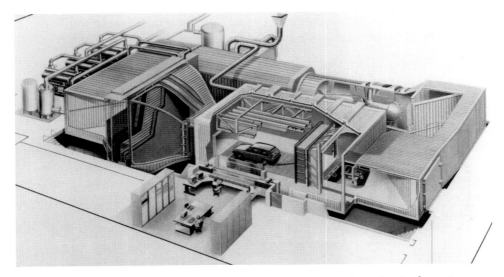

Figure 13.91 Climatic Wind Tunnel 2 of Volkswagen AG; nozzle outlet 6 m², maximum wind velocity 180 km/h; fan power 460 kW; temperature range −40 to +60°C.

An example for a thermal wind tunnel is given by the one of IVK/FKFS at Stuttgart University. It is shown in Figure 13.92. This thermal wind tunnel allows also to carry out soiling tests including rotating wheels. A further example is given by the thermal wind tunnel of Modine Europe GmbH, which is equipped with an exchangeable nozzle. The outlet cross section amounts to 5.4 m² for passenger cars and 12 m² for commercial vehicles. Maximum possible wind velocities differ correspondingly: 250 km/h for the small cross section and 130 km/h for the large cross section (see Table 13.3).

Similar to aerodynamic and aeroacoustic wind tunnels, also for new climatic and thermal wind tunnels a considerable investment has been and will be realized. In Munich for example, following the AVZ in May 2010 the BMW Group opened its Energy and Environmental Test Center (EVZ). This wind tunnel center houses an environmental, a thermal and a climatic wind tunnel. The environmental wind tunnel provides the most comprehensive simulation capabilities. It is able to simulate a combination of wind and rain, sun, or snow, including different kinds of snow. The possible temperatures range from −20 to +55°C. For further technical data see Table 13.3. These three new wind tunnels span a wide range of climatic conditions and substantially reduce the effort for testing. Tests can be carried out independently of the season of the year, the need for transportation (distance and time) is reduced, and consequently fewer prototypes are required.

Wind Tunnels and Measurement Technique

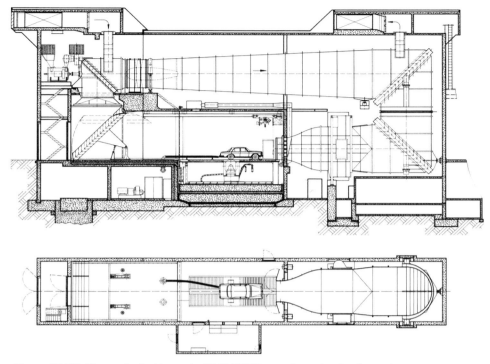

Figure 13.92 Thermal wind tunnel of IVK/FKFS; nozzle outlet 6 m^2, maximum wind speed 200 km/h, fan power 1 MW.

As a part of the Mercedes-Benz Technology Center, also Daimler AG has built two new climatic wind tunnels (see Figure 13.93). After two years of construction, they were put into operation in July 2011. One of the two climatic wind tunnels enables setting of temperature in a range from −40 to +40°C, the other from −10 to +60°C. These wind tunnels can also reproduce the effects of rain, snow and solar radiation combined with wind. The wind tunnels thus offer similar simulation capabilities than those of BMW Group. Both wind tunnels provide wind velocities up to 265 km/h and are consequently suitable for a large range of vehicle types. Air humidity can be set from 5% to 95% (see Heidrich [324]).

Figure 13.93 Test section of one of the new climatic wind tunnels of Daimler AG in Sindelfingen (see text). Source: Daimler AG.

A last example of a recent climatic wind tunnel is that of Audi AG in Ingolstadt. It started operating beginning of 2008 to complete the Audi wind tunnel center. This climatic wind tunnel can generate temperatures from −25 to +55°C and wind velocities up to 300 km/h. Simulation of sunlight and rainfall is possible, too, as well as adjustable relative humidity of the air. For further technical data see Table 13.3.

13.5.4 Overview and Correlation Measurements

After the more detailed description of the existing wind tunnels above, this chapter summarizes existing wind tunnels in the form of a table (Table 13.3). This overview includes both wind tunnels for aerodynamic and aeroacoustic studies as well as climatic and thermal wind tunnels. However, only wind tunnels for full-scale models or vehicles are listed, since the large variety of model-scale wind tunnels would exceed the bounds of this chapter. Installations with a nozzle outlet of less than 5 m² are also omitted. The listing is in descending order of nozzle outlet size.

The listed wind tunnels are solely those with return loop for the flow, either by means of a closed duct or by a surrounding building. The table does not claim to be complete. Large wind tunnels are included as much as possible; however, those which have been commissioned in 2013 or later are not yet listed. The listed climatic and thermal wind tunnels are more exemplary in nature. Further details of the listed wind tunnels including their test and measurement equipment can be found in the references stated in the table.

Table I3.3 Overview of existing Wind Tunnels (Aerodynamic, Aeroacoustic, Climatic and Thermal Wind Tunnels commissioning before 2013)

Test section: o open, g closed, s slotted walls
Ground simulation (only highest level mentioned):
LB: moving belt, bs: boundary layer removal,
ds: distributed suction, tb: tangential blowout

P_{max} maximum fan power
L_p sound pressure level at 140 km/h out of flow, in dB with A weighting

Company Location	A_N/m^2	Test section Length/m	Nozzle contraction	U_{max}/km/h	Floor simul.	Climate/C	L_p/dB(A) 140 km/h	P_{max}/ MW	Source*
DNW-LLF Marknesse (NL)	90.3/ 48.0/ 36.0	o/g 20 16	4.8 8.0 12.0	223 –547	LB	+22°	?	12.5	www.dnw.aero
General Motors Warren (USA)	57.2	g 21.7	5.0	222	Scoop bs-	—	?	3.4	Kelly et al. [424]
Volkswagen Wolfsburg (D)	37.5	o	4.0	198	—	–35°–45°	—	2.6	Mörchen [580]
Lockheed LSWT Smyrna (USA)	35.4	g 13.1	7.0	329	—	is held < 35°	—	9.0	Cummings [185]
MIRA Nuneaton (UK)	35.0	g 15	1.5	130	—	—	—	1.0	Fosbery et al. [255]
Daimler, Unter-tuerkheim (D)	32.6	o 12	3.5	250	—	—	—	5.0	Kuhn [473]
RUAG Emmen (CH)	35	g 15	4.0	245	LB	—	—	3.0	www.ruag.com
Fiat Turin (I)	31.0	o 10.5	4.0	205	LB	—	75	1.9	Antonucci et al. [20]
CSTB Aerodyn. Climatic Nantes (F)	30.0 18.0	g / 14 o / 25	2.2	290 140	— —	–25°–50°	— —	3.2	www.cstb.fr
Nissan Kanagawa (JP)	28.0 / 15.0	o 12	6.4 12	190 / 270	LB	—	66 at 100 km/h	2.2	Ogata et al. [619]
Hyundai Nam-Yang (KR)	28.0	o 18		200	bs	20°–50°	66.5	2.6	Kim et al. [433]
Chrysler Detroit (USA)	27.9	o	5.4	260	Scoop tb	—	62.3	4.7	Walter et al. [842]

(continues)

Table 13.3 Overview of existing Wind Tunnels (Aerodynamic, Aeroacoustic, Climatic and Thermal Wind Tunnels commissioning before 2013) (Cont.)

Company Location	A_N/m²	Test section Length/m	Nozzle contraction	U_{max}/km/h	Floor simul.	Climate/C	L_p/dB(A) 140 km/h	P_{max}/MW	Source*
Volvo Goeteborg (SE)	27.1	s 15.8	6.0	250	LB	–	–	5.0	Nilsson et al. [608]
SAWTC AWT (aeroacoustic) CWT (climatic) Shanghai (CN)	27 7 / 14	o / 15 o / 15		250 200/100		−20°–55°	61 (at 160 km/h)		www.sawtc.com
Mazda Miyoshi (JP)	24.0	o/g 12	6.0	200		–	–	1.6	www.mazda.com
Mitsubishi Japan	24.0	g o		216 187	bs	–	–	2.4	www.mhi.co.jp
GIE S2A Montigny (F)	24.0	o 15,1	6.0	240	LB	15°–35°	–	3.80	Waudby-Smith et al. [856]
Ford Dearborn (USA)	23.2	g	3.8	200	–	−18°–70°	–	1.87	McConnel [545]
IVK/FKFS Stuttgart (D)	22.5	o	4.4	265	LB	–	68 at 160 km/h	3.2	Potthoff et al. [651]
Porsche Weissach (D)	22.3	s/o 13.5	6.1	230	ds bs	–	–	2.60	Vagt et al. [818]
Ford Cologne (D)	20	o 10.3	4.8	200	LB	–	68	2.0	Grundmann et al. [301]
BMW wind tunnel Aerolab Munich (D)	18 / 25 14	o / 14 o / 12.6	5.8 / 8.0 5.8	300/250 300	LB LB	–	–	4.4 3.8	www.bmw.de
DTF WT8 (Ford) Detroit (USA)	18.7	o	6.0	240	tb	0°–55°C	63.7		Walter et al. [843]
Toyota	17.5	g 13.5	3.7	180		–	–	1.5	Kimura [434]
Windshear Inc. Concord (USA)	16.7	o		290	LB	25°C	–		Walter et al. [841]

Table 13.3 Overview of existing Wind Tunnels (Aerodynamic, Aeroacoustic, Climatic and Thermal Wind Tunnels commissioning before 2013) (Cont.)

Company Location	A_N/m^2	Test section Length/m	Nozzle contraction	$U_{max}/km/h$	Floor simul.	Climate/C	$L_p/dB(A)$ 140 km/h	P_{max}/MW	Source*
rta Rail Tec Arsenal SWT LWT Vienna (A)	16.1 16.1	g 33.8 100	4.0 5.7	120 300	—	−45°–60° −45°–60°	—		www.rta.co.at
Toyota Motorsport Cologne (D)	15.2	g 15		252	LB			2.3	www.toyota-motorsport.com
CNAM, S1O Saint-Cyr (F)	15.0	s 10	7.7	200	bs	—	—	1.0	www.iat.cnam.fr
Fiat Climatic warm Climatic cold Turin (I)	12.0	o 14	4.0	160	—	−10°–50° −50°–20°	—	0.56	Antonucci et al. [20]
J.A.R.I. Japan	12.0	g	4.1	205	—	—	—	1.20	Muto et al. [599]
Volvo climatic Sweden	11.2	o	2.5	95	—	−40°–50°	—	0.50	Christensen [820]
Pininfarina Italy	11.0	o 8.0	6.9	250	LB	—	78	2.0	Morelli [586]
Ford Climatic Cologne (D)	11.0	g	6.0	180	—	−40°–50°	—	1.12	Bengsch [69]
Audi Aeroacoustic Climatic Ingolstadt (D)	11.0 6.0	o / 9.5 o / 10	5.5 6.0	300 300	LB —	— −25°–55°	61 at 160 km/h	2.60 2.4	Wickern & Lindener [508] www.audi.de
BMW Aeroacoustic Munich (D)	10.0	o 10		250	LB	—	66	1.90	Lindener et al. [504]
BMW Thermo Energy engineering Environment Munich (D)	8.4 8.4 8.4	o /10 o / 10 o / 10		280 250 250	LB (motorcycle)	20°–45° −20°–55° −20°–55°		2.1 1.6 2.1	www.bmw.de

(continues)

Table 13.3 Overview of existing Wind Tunnels (Aerodynamic, Aeroacoustic, Climatic and Thermal Wind Tunnels commissioning before 2013) (Cont.)

Company Location	A_N/m²	Test section Length/m	Nozzle contraction	U_{max}/km/h	Floor simul.	Climate/°C	L_p/dB(A) 140 km/h	P_{max}/MW	Source*
Daimler Cold Hot Sindelfingen (D)	each 7/8/12	each 0/10	each 7 at 8 m2	each 265/250/1180	—	-40°–40° / -10°–60°	—	each 1.7	Heidrich [324]
IVK/FKFS Thermo Stuttgart (D)	6.0	0/15.8	4.2	210	—	20°–50°	—	1.00	Essers et al. [233]
Volkswagen 2 Wolfsburg (D)	6.0	0/7.2	6.0	180	—	-40°–60°	—	0.4	Buchheim et al. [114]
Denso Southfield (USA)	6.0	0		170	—	-30°–70°	—		Fa. (1995)
Behr Stuttgart (D)	6/8/10	0		130/100/80	—	-30°–50°	70 (at 50 km/h)	0.3	Schmiederer & Riedel [723]
Modine Filderstadt (D)	4.7/12.0	0/14.3 0/15.8/19.3		265/130	—	20°–55°	—	1.3	www.modine.com

* For the most part, original sources are stated; some tunnels were later modernized.

In a comparison of the aerodynamic and aeroacoustic wind tunnels listed in the table, it can be clearly seen that there is a relatively large difference in the size of the nozzle outlet. Disregarding the German-Dutch wind tunnel (DNW-LFT) because it was built primarily for aviation aerodynamics, there is a factor of approximately 5.6 for the difference between the largest and smallest nozzle outlet. There are two major reasons for this astounding difference in size:

1. The type of test section: closed, slotted, or open. The differing blockage effects (see section 13.2) requires a larger nozzle outlet for a closed test stretch.
2. The test spectrum: the GM wind tunnel is designed for automobiles, vans as well as heavy-duty trucks and buses. For wind tunnels that are restricted to passenger cars for example, a smaller nozzle outlet is sufficient.

To ensure the reliability of the wind tunnel test results, the operators of full-scale vehicle wind tunnels have repeatedly conducted intercomparisons by correlation measurements.

For the first intercomparison in 1980 a VW 1600 sedan reduced to a calibration model was used (smooth underbody, no bumpers). Its aerodynamic coefficients—and in particular the drag coefficient—could be altered in a reproducible way by fitting and removing spoilers. Measurements were carried out by means of the identical model in ten large wind tunnels in Europe, the United States, and Canada. These wind tunnels included such with open and closed test sections and some with slotted walls. There have been detailed reports in a number of SAE publications (see Buchheim et al. [114], Cogotti et al. [164], Costelli et al. [180], Carr [128]). The standard deviation of the drag coefficients measured in the different wind tunnels resulted in values between approx. 0.007 and 0.015, depending on the vehicle configuration, which corresponds to about 2 to 3% of the absolute value.

Somewhat greater deviations were registered for the delta measurements between two vehicle configurations. Such delta measurements can be relevant for a design decision. The attachment of a rear spoiler for example, brought about a drag reduction by 1.3% in one tunnel and 5.1% in another one. The first result would mean that the spoiler may be dispensable, while the second result could justify its application.

Calibration vehicles like the VW 1600 are important for a variety of reasons. On the one hand they serve for checking the continuity and comparability of the data measured in one wind tunnel for a long time. On the other hand they serve for comparing different wind tunnels by correlation measurements, as discussed above. These vehicles require a long-term stability, thorough maintenance and careful storage in dry environment. Such calibration vehicles are also important for comparisons between measurements in a wind tunnel and on the road. FKFS, for example, maintains three calibration vehicles:

- Between 1989 and 2014 an Opel Omega for tests with standing wheels, which ensures the long-term continuity of our data

- Since 2013 an Opel Insignia sedan as a long-term replacement for the Omega, however, also used with full road simulation
- Since 2001, a Mercedes CLK for tests with full road simulation
- A BMW E 39 sedan for wind tunnel and real road comparisons

A second and a third intercomparison test were conducted on the initiative of the European Data Exchange Committee (EADE). The aerodynamic coefficients were measured in several full-scale wind tunnels with different vehicles in running condition. The results of the second intercomparison were used to validate correction methods which had been developed by Mercker and Wiedemann [568]). Without this correction the standard deviation for the drag amounted to 0.008, similar to the standard deviation with the first intercomparison. However, using the correction method by Mercker and Wiedemann the standard deviation dropped down to 0.002. Even between wind tunnels with open and closed test section such an agreement could be found by using the correction method. Results from the third intercomparison are discussed and presented in section 13.2.13 in the context of correction methods.

Another criterion for a wind tunnel comparison can be found in the axial pressure gradient which results from the longitudinal pressure distribution in the test section (usually in the center line from nozzle to collector). This gradient can directly affect the forces measured on the vehicle. Figure 13.94 shows a comparison of such pressure gradients of a number of European wind tunnels. It can be seen that the gradients are definitely different, both at the front of the nozzle area and in the rear (the collector area). Thus, in different wind tunnels different forces act on the front and rear of the same vehicle. This produces systematically different results, for example, for the drag values, which are not a matter of so-called measurement uncertainties. The reasons for the pressure gradient, possible countermeasures, and correction methods are discussed in section 13.2.

A comparison of the A-weighted sound pressure level in the test section of the self-noise of different aeroacoustic wind tunnels (measured out of the flow) is shown in Figure 13.95. Already in the first generation of the aeroacoustic wind tunnels (e.g., aeroacoustic wind tunnel of BMW) it was possible to reduce the sound pressure level in the test section out of flow by nearly 30 dB(A) compared to typical aerodynamic wind tunnels without acoustic measures. The aeroacoustic wind tunnel of Audi AG, the first acoustic tunnel of the second generation, produced a further improvement of approx. 10 dB(A). Meanwhile, more wind tunnels provide such a low background noise (self-noise) like the one of SAWTC as well as the new ones of Daimler AG and Porsche AG. They are not shown in the figure.

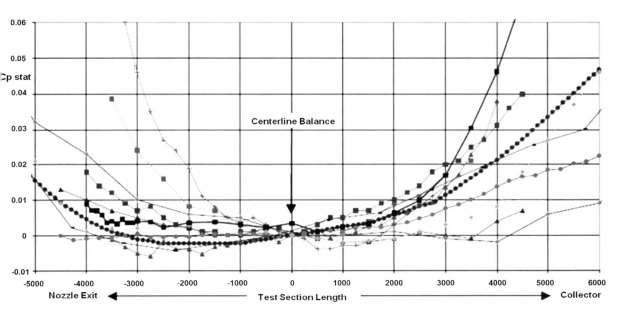

Figure 13.94 Axial pressure distribution of different European wind tunnels in comparison. Partly, there are significant differences visible (cf. discussion in section 13.2) (Mercker et al. [568]).

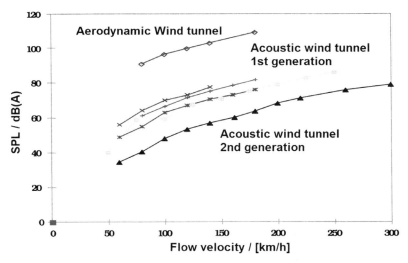

Figure 13.95 A-weighted sound pressure level of the self-noise (measured out of flow) of different aeroacoustic wind tunnels compared to one another and to a typical aerodynamic wind tunnel (Schneider et al. [730]).

Further comparisons of wind tunnels can be found in the relevant literature, such as a comparison of seven Japanese vehicle wind tunnels from 2003, published in 2005 (Maeda [523]).

13.6 Outlook

Although wind tunnels are sometimes referred to as test beds and some of the automobile manufacturer allocate them to a central test department that supports the vehicle developer in the process of creating a vehicle as a service provider, there are nevertheless marked differences from other automobile test facilities.

Vehicle, engine, acoustic or dynamometer test beds can today be purchased off the peg in many cases. Only the performance data and range of application have to be specified to identify the right test bed and ensure ready defined measuring accuracy. But wind tunnels are still the subject of research, for example by ECARA (European Car Aerodynamics Research Association), even after decades of automobile use and give rise to comparative measurements, for example by EADE (European Aerodynamic Data Exchange). This is one of the commonalities with CFD and CAA. Aeroacoustic wind tunnels are also especially high in value. They represent an investment of some € 40 million, while climatic wind tunnels can be implemented for less than € 15 million. In contrast to most other test facilities that OEMs have at their disposal, an aeroacoustic wind tunnel—or even an extended aerodynamic and thermodynamic test center—has considerable prestige assets which are often highlighted from a marketing perspective and are presented as a measure of the technical capability of the company in order to emphasize its products.

Implemented aerodynamic and aeroacoustic wind tunnels are still—with few exceptions—one of a kind. So, intensive planning goes into each project. Today the planning is often preceded by a research or pilot phase. Scale-model tests and CFD calculations are conducted to ensure that ambitious goals are achieved with the utmost probability, and that one does not fall behind of the technology status of existing installations. If a new wind tunnel exhibits operating problems (e.g., buffeting or implausible measured values), one can be sure that news of this will spread in professional circles in no time at all. Whereas during a similar problem on an engine test bed, for instance, the service team of the manufacturer would solve the problem as part of rework in a short time, such a situation is usually no easy task for the wind tunnel planner or general contractor.

One cannot too easily explore the experience of other installations, so one has to understand the often extremely complex physical problems through elaborate experimentation and calculation to find a solution or at least alleviate these problems. If, and that is usually the case, such attempts are conducted on the finalized installation and not on a model, the wind tunnel is initially blocked for later users, schedules and deadlines are delayed. Enough spare wind tunnel capacity is seldom available, or the costs are too high, so product launches can be delayed, the cost of which can easily exceed that of a wind tunnel. It is not advisable to save in the planning and pilot phase of aeroacoustic wind tunnels, especially combined with sophisticated balance and ground simulation technique, but to make use of the expertise that research produces.

In future the complexity of wind tunnel facilities—especially in conjunction with CFD and CAA—can be expected to further increase because the simulation of on-road driving should also consider non-stationary effects like turbulence and gusts. Also their influence on vehicle dynamics and acoustics, for example, should be considered in order to develop the automobile of the future in high quality. Wind tunnels, like other test facilities, will not substitute final acceptance on the road in real driving conditions. But they will lead the vehicle developer with increasing security and certainty to results that can be reproduced in driving conditions and thus avoid costly reworking and duplication of effort in the process of creating a product.

Chapter 14
Numerical Methods

Reinhard Blumrich, Norbert Grün, Thomas Schuetz

Numerical methods are nowadays employed alongside the wind tunnel in the aerodynamic and aeroacoustic design, development, and optimization of automobiles. Especially in the early phase of development, there are no real vehicles available for testing. At most, only partially complete experimental test vehicles are available. Nevertheless, assessment of aerodynamics and aeroacoustics performance of a number of competing designs is needed during this early phase. For aeroacoustics design, detailed information relating the vehicle geometry details to flow phenomena is needed so that the generation and propagation of aeroacoustic noise can be determined during each stage of vehicle development. Essentially, three reasons justify the use of numerical methods:

- Shortening the development cycle
- Reducing cost
- Increasing flexibility

The traditional approach for aerodynamic and aeroacoustic optimization of vehicles in the wind tunnel has not become obsolete; however, it is just delayed to a later phase in the development process. Numerical simulation of aerodynamics also provides some benefits due to the limitations inherent in wind tunnel testing (cf. chapter13):

- The Reynolds number is often too low when testing scale models
- Representing all geometry details on scale models is very challenging

- Finite dimensions of the wind-tunnel jet and test section impose blockage effects on the model and wake, which have to be corrected
- Simulation of the moving floor is often incomplete
- Options to represent crosswind, gusts, and turbulence are limited

Wind tunnel tests using full-scale models are expensive. The cost for measurement time as well as for model manufacturing is high. Measurements beyond weighing integral forces, for instance, recording pressure distributions on larger surfaces or velocity in space, are prohibitively time and cost intensive.

Wind tunnels and numerical methods both are simulation tools. Their characteristic differences and capabilities render them complementary rather than competitive [478]. With the increasing acceptance of numerics in the industry the focus for utilizing both tools has changed significantly. It is likely that in the future wind tunnels will be used more and more for generating the validation database for numerical methods and for performing extensive parameter variations. Numerical simulations will then be used to answer questions of why and how parameter variations influence aerodynamic and aeroacoustic properties and where on the car to look for optimization potential.

14.1 Simulation of Three-Dimensional Viscous Flows

CFD[1] methods had originally been developed to satisfy demands in the aerospace industry. The primary target for commercial airplanes is maximum lift at minimum drag, which can only be achieved simultaneously by avoiding flow separation as much as possible. In contrast, the flow around automobiles is characterized by unavoidable separation, often followed by reattachment. Initially, CFD methods in aerospace (e.g., Euler methods) were not designed for these phenomena. In the meantime, the aerospace industry must also cope with separated flow phenomena, because the maneuverability of aircraft depends highly on their behavior in the "post-stall" domain when massive separation occurs on the wings at high angles of attack.

In the past, the number of wind tunnel hours spent for automobile optimization has grown exponentially, as depicted in Figure 14.1. The expectation on CFD is to suppress this growth by limiting time and cost expenditure. This trend is supported by the continuous drop of time and cost for numerical simulations.

---

1. Computational fluid dynamics.

Numerical Methods

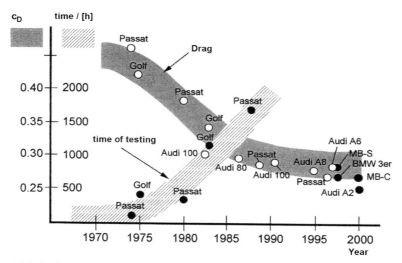

Figure 14.1 Reduction of the drag coefficient and increase of measurement time in the wind tunnel; Buchheim [113] and Hucho [371]. Unfortunately, no information is available about wind tunnel times in the past decade.

In the 1980s the CRAY-XMP computer system had a maximum performance of around 0.82 GFlops. Currently (as of November 2013[2]) the world's most powerful system, the Tianhe-2 (Milky Way 2), achieves 33,862,000 GFlops. This is a factor of more than 40 million, leading to an equivalent reduction of compute time. Today, the real bottleneck is no longer the computer performance but the preprocessing effort prior to the simulation (e.g., to prepare a high-fidelity vehicle geometry model). However, two issues have to be considered:

- The development of algorithms is not yet completed and will surely deliver further improvements in the future.
- Such powerful computer systems are not available for general use and they probably also will not become regularly available in the near future.

The strategy followed for form development both in the wind tunnel and with CFD is still mostly trial and error. CFD has the advantage that it delivers an unmatched level of flow field details compared to the wind tunnel. Driving situations can be simulated whose realization would be either too dangerous on the road or could not be reproduced in the wind tunnel, such as passing maneuvers or oncoming traffic.

On the other hand, CFD has not yet reached the same reliability as physical testing. Even trend predictions are risky. Assuming proper preparation, comprehensive parameter variations may be conducted in a wind tunnel within hours, while the time for a single full vehicle simulation is still about one day on affordable hardware.

---

2. www.top500.org

This chapter provides an overview of the capabilities and issues when applying CFD to the flow around road vehicles. The basics of various methods are described; their advantages, but also their shortcomings, are emphasized using practical examples. The description addresses automotive aerodynamics practitioners who have had no closer contact to CFD so far. For deeper details, we refer to the quoted literature. It should be noted that the development of CFD is still ongoing, supported by permanent innovations in hardware and software.

The typical characteristics of the flow around an automobile have been discussed previously in section 4.2. In order to define the demands on CFD methods, a short summary of the flow phenomena is helpful to review here. As a result of the complex geometry of a vehicle, the flow field is very complicated. It is three-dimensional, includes strong pressure gradients, also in the cross-flow direction, and produces closed and open recirculation regions. Small closed separation bubbles occur, for instance, on mirrors, ridges, door handles, and wipers. Large recirculation regions can be found on the A-pillar, on the rear end, on the jagged underbody, and on wheels and in wheel houses. On the road, a vehicle is frequently subject to natural wind, which hardly ever coincides with the driving direction. As a result, the flow field is asymmetric and may lead to wide separation regions on the leeward side. Rotating wheels and the relative motion between body and ground are specific to road vehicles and affect the flow field.

Basically, the real flow around vehicles is time-dependent. Separation and recirculation are subject to stochastic—and sometimes periodic—oscillations and gusts in the natural wind, which further contribute to the unsteadiness of the flow.

14.1.1 Requirements and Properties of CFD—Methods

Before a CFD tool can be employed in the development process, it has to meet three requirements:

- Without doubt, the same accuracy as in the wind tunnel is expected. There it is possible to identify changes of the drag coefficient by $\Delta c_D = 0{,}001$. It has to be noted that significant drag reductions are often only achieved by accumulating many of these small improvements (see examples in section 4.5).
- The expenditure of time should be comparable to the wind tunnel, of course including the time necessary for model manufacturing and all physical and numerical modifications during the development process.
- The overall cost should be comparable to the wind tunnel.

The currently available commercial tools are at best close to complying with these requirements, but it can be expected that they will outperform the wind tunnel with respect to time and cost in the next few years.

On various occasions doubts are expressed whether such tough demands on CFD methods are justified. The arguments for this are: computational methods are based on the exact equations of motion but they are only solved approximately for two reasons:

- The equations are simplified in various ways concerning physical phenomena. Essentially, that is the matter of the following sections.
- The solution is obtained numerically, that is, it is (only) an approximation.

Both result in errors, which can be difficult to quantify. The user takes a risk if he tries to keep pace with the speed of development for each new vehicle. Exact solutions of the Navier-Stokes equations (NSE) only exist for special cases, and only a few of them are relevant for automotive technology. Therefore, the aerodynamicists depend on simplified versions of the NSE for which a numerical solution is possible and feasible. Simplifications can be:

- Neglect viscous effects
- Assume irrotational flow
- Temporal averaging of all flow quantities in the entire flow field
- Spatial separation of turbulent scales in coarse and fine structures
- Model Reynolds stresses in turbulent flows

All these simplifications of the NSE idealize the physics and move away from real natural conditions. They introduce phenomenological errors in the simulation.

The second category of simplification is of mathematical nature, introduced by the way in which the selected equations are solved. In principle, the differential equations, describing a continuum, are discretized together with the specific boundary conditions (i.e., they are applied to small but finite volumes in the computational domain). Errors result from the following:

- Discretization
- Choice of the solution method for the discretized equations
- Programing and round-off errors on the computer

All CFD methods used in aerodynamics can be characterized according to their properties as listed in Table 14.1. In the past Euler methods (coupled with boundary layer methods) had been used to investigate external and internal flows on vehicles. However, today they have lost relevance for automobile aerodynamics in favor of Navier-Stokes methods and hence won't be discussed here further.

Table 14.1 General Properties of Different CFD Approaches

	Potential theory	Euler	RANS	LBM VLES	DES	LES	DNS
Simplifying Assumption	Fluid w/o. friction and rotation	Fluid w/o. friction	Fluid can be frictionless, turbulence model, wall model if necessary	Fluid only slightly compressible ($Ma \leq 0{,}3$), turbulence and wall model if necessary	Two regimes; see RANS und LES	Fluid can be frictionless, SGS turbulence model, wall model if necessary	Fluid can be frictionless
Type of equations	Laplace, linear	nonlinear differential equations, 1st order	non-linear differential equations, 2nd order	Boltzmann BGK equation: linear differential equations, 1st order	non-linear differential equations, 2nd order		

The approach to the solution is the same with all CFD methods and consists of three steps:

1. Preprocessing: Discretization of the computational domain; body surface only for Boundary Element Methods or the entire volume for Field Methods.
2. Solving: The actual numerical solution of the discretized equations.
3. Post-processing: Analysis and visualization of discrete results.

For a long time, preprocessing was the most time-consuming phase. Recently, algorithms have been developed for almost fully automatic discretization of the computational space. Nevertheless, this step still requires a significant level of experience and some basic knowledge of the expected flow field.

The raw result of a CFD simulation is a large amount of data that requires post-processing. Currently available commercial tools enable the analysis, visualization, and often animation of all quantities in the entire flow field. Despite the impressive images, one should not forget that they are based on a finite number of discrete data points. Highly resolved flow fields as shown in publications use interpolation between the discrete points. Depending on the density of the computational grid, there is always some ambiguity on the user's side when choosing how to show the flow structures. On the other hand, useful information can also get lost if the post-processing software does not allow for a clear and easy interrogation of computational results.

14.1.2 Basics of Kinetic Theory

The macroscopic appearance of a continuum flow that we perceive when looking at the flow around a vehicle is actually caused by the motion and collision of individual molecules. Gaseous flows can be characterized by the Knudsen number

$$Kn = \frac{\lambda}{L} \tag{14.1}$$

which is defined as the ratio of the mean free path λ between molecule collisions and a typical length L of the specific problem. For $Kn \ggg 1$ we have rarefied gases or free molecular flow. The opposite extreme $Kn \lll 1$ characterizes continuum flows, which is an appropriate assumption for the aerodynamics of ground vehicles.

While kinetic theory describes both cases, the Navier-Stokes equations are only valid for a continuum of Newtonian fluids. Hence, kinetic theory and discrete lattice methods based thereon constitute the more general approach to simulate flow fields. Further simplifications concerning viscous and vortical effects take us via the Euler equations to potential flow theory (Section 14.1.5). A hierarchical overview of approaches to describe gas flows is given in Figure 14.2.

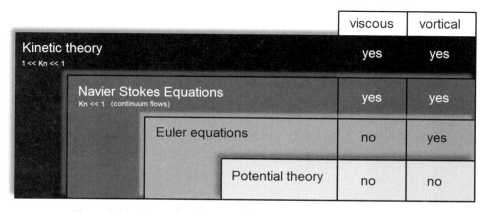

Figure 14.2 Hierarchy of approaches to describe gaseous flows.

If we confine our considerations on subsonic flows at temperatures common to automobiles, the air can be assumed to be an ideal gas (without real gas effects) and the equation of state describes the relation between pressure p, density ρ and temperature T

$$p = \rho R T \qquad (14.2)$$

with the specific gas constant $R = 287$ J/(kg · K). for dry air. Since the second half of the 19th century, the macroscopic phenomena of such flows has been explained mathematically with the motion of gas molecules by physicists like James Clerk Maxwell and Ludwig Boltzmann. They made the following assumptions:

- Molecules move in a stochastic (random) mode.
- Except elastic collisions, there is no interaction between molecules.
- The molecule diameter is small compared to the mean free path.

Since it is not feasible to consider the movement and collision of individual molecules on a microscopic scale, kinetic theory introduces a statistical description of molecule motion on a kind of mesoscopic level (still below the macroscopic Navier-Stokes approach). For this purpose, a velocity distribution function $f(x, c, t)$ is introduced that specifies how many particles (per volume) are moving with speed c at position x and time t. The generally used macroscopic quantities density ρ, momentum ρv and energy E are obtained from integrating over the entirety of all possible particle speeds c (the phase space).

$$\text{Density: } \rho(x,t) = \int f(x,c,t)\,dc \tag{14.3}$$

$$\text{Momentum: } \rho(x,t) \cdot u(x,t) = \int f(x,c,t)\,c\,dc \tag{14.4}$$

$$\text{Energy: } E(x,t) = \int f(x,c,t) \cdot (c-v)^2\,dc \tag{14.5}$$

In the flow field, the static pressure follows from the equation of state (14.2). On solid walls, the pressure results from the momentum exchange during the (elastic) collision of particles with the wall. The Boltzmann equation represents the rate of change (total differential) of the velocity distribution function f as it tends toward the thermodynamic equilibrium due to a collision operator C.

$$\frac{d}{dt}f(x,c,t) = \frac{\partial}{\partial t}f(x,c,t) + c \cdot \nabla f(x,c,t) = C(x,c,t) \tag{14.6}$$

If the collision operator itself satisfies the conservation of mass, momentum and energy, then a solution of the Boltzmann equation produces a physically correct flow field. Without a collision term on the right-hand side, the Maxwell-Boltzmann equilibrium distribution is reached immediately, which is equivalent to an inviscid and isothermal flow governed by the Euler equations.

The tendency toward equilibrium is influenced by viscosity, which has to be accounted for in the collision operator. For ideal gases viscosity increases with the mean free path and the particle speed, that is, with temperature (in contrast to liquids, where viscosity drops with increasing temperature). It is common to use the Sutherland formula

$$\mu = \mu_0 \frac{T_0 + 120}{T + 120}\left(\frac{T}{T_0}\right)^{3/2} \tag{14.7}$$

for the temperature dependency of the dynamic viscosity of air.

With the so-called Chapman-Enskog expansion of the velocity distribution function around the equilibrium state, the Navier-Stokes equation can be derived from the Boltzmann equation. It is worth noting that viscosity and thermal conductivity of a gas can be calculated in kinetic theory.

14.1.3 Lattice Methods

The term "lattice" refers to the grid used to discretize the computational domain for methods based on kinetic theory. In contrast to the meshes for Navier-Stokes methods, a lattice is always a regular, non-surface-oriented grid, in three dimensions, usually a Cartesian grid with cubic cells of aspect ratio 1.

Instead of a continuous phase space, only a finite number m of discrete particle speeds c_i with $i = 1, m$ is allowed, and the velocity distribution function is replaced by m states f_i.

Numerical Methods

$$f(x,c,t) \to f_i(x,t), i = 1, m \qquad (14.8)$$

Usually these methods are denoted as $DjQm$, where j indicates the dimensionality (2D or 3D) and m the number of discrete particle velocities. Common methods are D2Q9 and D3Q19.

The change of discrete states is now governed by the Lattice-Boltzmann equation

$$f_i(x + c_i \Delta t, t + \Delta t) = f_i(x,t) + C_i(x,t) \qquad (14.9)$$

In general, this is a simple update rule from one time step to the next, which inherently produces a time-dependent flow field. As mentioned, the key step to describe the correct fluid physics is the design of the collision operator $C_i(x, t)$

14.1.3.1 Lattice-Gas (LGA)

The first particle methods were called LGA (Lattice-Gas-Automata). In 1973, the HPP model [316] on a lattice of quadratic cells was published. Its deficits were removed in 1986 by the FHP model [260], which used hexagonal cells. Both methods were limited to two dimensions.

Common with all LGA methods is the representation of the distribution function by integer numbers and the use of logical operators in a look-up table for the redistribution of particles after collision. This guarantees the exact compliance of conservation laws without round-off errors, but on the other hand it introduces a statistical noise in the macroscopic quantities, which are obtained by a summation over the discrete states.

14.1.3.2 Lattice-Boltzmann (Example PowerFLOW™)

Lattice-Boltzmann methods (LBM) eliminate the drawbacks of LGA and lead to productively deployable tools such as PowerFLOW [140], [142], [143] from Exa Corp., a D3Q19 code which will be described in the following text.

In LBM, the particle states are represented by floating-point numbers, and instead of binary rules a collision operator is used. The so-called BGK-operator [76] is a relaxation method where the difference between the current local state f_i and the local equilibrium distribution f_i^{eq} is utilized

$$C_i(x,t) = -\omega \left(f_i(x,t) - f_i^{eq}(x,t) \right) \qquad (14.10)$$

so that the Lattice-Boltzmann equation can be written as

$$f_i(x + c_i \Delta t, t + \Delta t) = \omega \cdot f_i^{eq}(x,t) + (1 - \omega) \cdot f_i(x,t) \qquad (14.11)$$

The relaxation parameter ω has the meaning of a collision frequency and (as a result of the Chapman-Enskog expansion) is related to the kinematic viscosity $v = \mu / \rho$ by

$$\frac{v}{T} = \frac{1}{\omega} - \frac{1}{2} \qquad (14.12)$$

Chapter 14

so that the desired viscosity v is set via the time $1/\omega$ (equivalent to the mean free path) between two collisions. For a stable temporal progression of equation (14.47), a finite minimum viscosity is required (i.e. $\omega < 2$ is necessary). In general, the various LBM codes differ in the collision operator, the $DjQm$ model, and the equilibrium distribution f^{eq} they use.

14.1.3.2.1 Spatial and Temporal Discretization

For LBM methods, the computational domain is discretized using a Cartesian grid with cubic cells (the "lattice"). The size of these cells, also called voxels (as an abbreviation for volume pixels), can grow with increasing distance from the vehicle surface, but always only by a factor of 2 from one VR-region (variable resolution) to the next. Of course, the shape and location of these resolution regions as well as the voxel size will have an impact on the result quality and requires a certain experience of the user and at least a rough idea of the expected flow field. In external aerodynamics, the highest resolution regions are typically surface offsets with voxel sizes between 1 and 3 mm. An example is shown in Figure 14.3.

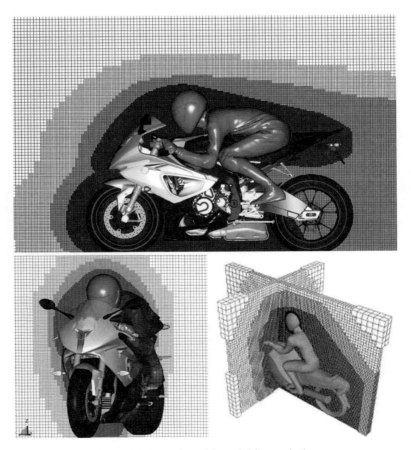

Figure 14.3 Lattice with variable resolution.

Initially, the vehicle itself is represented by a surface triangulation. It can be assembled by any number of solids, which may also intersect each other. The only requirement is that solids are described by closed facetizations so that interior and exterior can be uniquely identified. It is also possible to use "one-sided," meaning open, grids, for instance for underbody panels or guiding vanes without representing their actual volume. Figure 14.4 demonstrates the complexity of configurations that can be handled this way.

Figure 14.4 Components of a vehicle model.

Prior to the simulation, an automatic discretization is conducted. During this process the triangles representing the geometry (facets) are intersected with the lattice planes, thus creating surfels (surface elements) as fractions of the facets (Figure 14.5). In case a facet is lying entirely inside a voxel, it will completely become a surfel.

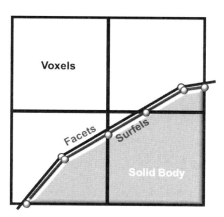

Figure 14.5 Surface discretization.

Chapter 14

The simulation resolution is not determined by the fineness of the surface triangulation (facets) but by the local voxel size. When preparing the geometry input, the user should primarily consider the facetization of the surface to ensure that the curvature is adequately resolved by the facets. For instance, the geometry of a cube would be sufficiently represented by two triangles on each face, regardless of the voxel size chosen for the simulation.

In general, surfels are n-polygons, which form the interface between the solid and those (partial) voxels that intersect the surface. This way, the solid surface is represented exactly and not as a stair-stepped approximation by just blocking complete voxels. In the course of the simulation, particles collide with surfels and are reflected again. Two extremes are conceivable for the momentum exchange between fluid and wall during this process (Figure 14.6).

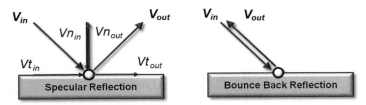

Figure 14.6 Particle reflection at the wall.

Specular reflection scatters particles like light on a mirror (i.e., the normal component of the velocity is inverted while the tangential component remains unchanged). A momentum balance results in a normal force only, equivalent to a frictionless wall (slip condition). Friction on the wall (no-slip condition) is achieved by bounce-back reflection, where both velocity components are inverted and thus also producing a tangential force on the wall. This boundary condition can be utilized without any modeling in a direct simulation. At high Reynolds numbers it has to be replaced by the wall model described later in section 14.1.3.2.2.

The iteration of equation (14.11) yields the temporal evolution of the flow field. The physical time interval per computational time step is

$$\Delta t = v_{Lattice} \frac{\Delta x}{v_{Ref}} = c_{Lattice} \cdot Ma \frac{\Delta x}{v_\infty} \left[\frac{\text{voxel}}{\text{TS}} \cdot \frac{\text{m/voxel}}{\text{m/s}} = \frac{\text{s}}{\text{TS}} \right] \qquad (14.13)$$

Since the speed of sound $c_{Lattice}$ is determined by the particle model, the time step is fixed by the voxel size Δx, the free stream velocity v_∞, and the Mach number Ma_∞.

Strictly speaking, the user has, except resolution Δx, no option—but also not the necessity—to define the time step and hence the computational effort. As an example, at $v_\infty = 50$ m/s and (at 20 °C) $Ma_\infty = 0.15$ a voxel size of $\Delta x = 2$ mm results in a time step of $\Delta t = 5 \cdot 10^{-6}$ [s/TS]. To cover a physical time of 1 second requires 200,000 computational timesteps. However, PowerFLOW has the option to deliberately simulate at another

Mach number than the one dictated by velocity and temperature. By simulating at $Ma_\infty = 0.30$ instead, only 100,000 timesteps are necessary and the computational effort is cut in half! The error one has to accept is the difference in compressibility effect between $Ma_\infty = 0.15$ and $Ma_\infty = 0.30$, which is very small in this range.

14.1.3.2.2 Turbulence Models

Most of the relevant problems in practice are lying in a Reynolds number range where turbulent phenomena play a crucial role. However, direct simulations (section 14.1.4.3) demand a resolution that is not feasible today with respect to memory requirements and acceptable compute times. Therefore, the turbulence models described in section 14.1.4.1.2 have to be employed.

This also pertains to LBM methods. To model the unresolved turbulent structures, which are assumed to be universal and isotropic, PowerFLOW uses a method called VLES (very large eddy simulation, [792]) which resembles the LES model described in section 14.1.4.2. However, not only the bulk fluid needs turbulence modeling at high Reynolds numbers but also turbulent wall boundary layers usually cannot be resolved and have to be modeled (cf. Figure 14.7). In subsonic flows, skin friction is the primary source for the generation of losses in the boundary layer and the occurrence of separation. Skin friction is proportional to the velocity gradient on the wall, and hence the correct calculation of this gradient is critical for the quality of the entire solution.

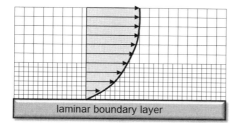
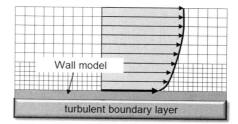

Figure 14.7 Modeling of the turbulent boundary layer profile.

For a laminar boundary layer profile (Figure 14.7 left) (i.e., at low Reynolds numbers), the resolution feasible today is sufficiently fine for using the no-slip condition ($v_W = 0$) for calculating the correct velocity gradient. However, this is not possible for a turbulent boundary with the mean velocity profile shown in Figure 14.7 (right). The grid size requirements for turbulent boundary layers are extreme due to the high-velocity gradient at the wall, which is directly related to the skin friction.

Instead of resolving these flow details, the region nearest to the wall is modeled by the "logarithmic law of the wall," as described in detail in section 14.1.4.1.3. Since this universal velocity profile has been derived under the assumption of a two-dimensional equilibrium boundary layer (no pressure gradient), its application to the highly three-dimensional flow around a vehicle which is characterized by strong pressure gradients is strictly speaking not valid. Especially to be able to predict pressure-induced separation, the wall model in PowerFLOW has been extended by an empirical term, which accounts for the increased boundary layer growth and modified separation in regions of adverse pressure gradients.

14.1.3.2.3 Simulation Process

Starting from a prescribed initial flow field, equation (14.46) is iterated for the number of time steps equivalent to the desired physical simulation time and thus produces a time accurate unsteady solution. Since the physical time per iteration depends on the voxel size (equation (14.48)), only the finest cells are updated in each (computational) time step. Voxels in the next (by a factor of 2) coarser level of resolution are updated every second step, and so on. This way, the flow field evolves synchronously in all levels and the global solution is time accurate.

In most cases, the initial condition is free stream velocity throughout the entire simulation volume, meaning also on solid walls or in cavities. This causes an initial transient oscillation to a physically correct state where solid walls are impermeable and/or the flow in cavities has come to rest. This phase can be shortened if a similar flow field is available, which can be used to seed the simulation volume. Seeding makes sense and has an effect, for instance, when exploring vehicle variants with moderate geometry changes.

A typical history of drag and lift over time (without seeding) is shown in Figure 14.8. After the initial transient, more or less pronounced—depending on the particular case—oscillations around a temporal average value are visible, like in a wind tunnel. The amplitudes that can be observed depend on the sampling rate the user has chosen (cf. the curves for 50 Hz and 1000 Hz in Figure 14.8). With time steps in the order of $\Delta t = 10^{-5}$ [s] sampling rates up to 100 kHz would be possible. However, considering typical voxel counts of 50–100 million and physical simulation times of 1–3 seconds this would generate an amount of data that cannot be feasibly stored or post-processed. In addition, even for aeroacoustic evaluation the highest frequencies of interest are much lower than the maximum possible.

Therefore, the simulation result is stored as temporal and spatial averages using sampling and averaging parameters specified by the user during simulation setup. For instance, with a sampling rate of 50 Hz, the result files contain "frames" (similar to the individual images in a movie) that represent the average over $\Delta t = 0.02$ s, in this example approximately 2000 timesteps. In addition, by averaging over 2 voxels in each direction to form a "measurement cell," the amount of data is reduced by another factor of 8. Of course, then the results can no longer be visualized with the same resolution with which they have been calculated.

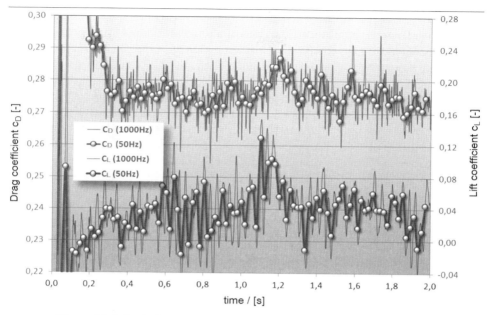

Figure 14.8 Typical time history of drag and lift from a LBM simulation.

However, it is possible to define any number of "measurements" with different spatial extension and different characteristics concerning the spatial and temporal averaging. For instance, to conduct an aeroacoustic analysis of the results on a side window one would define a measurement that stores only the static pressure on the surface of interest (if necessary, without spatial averaging). When choosing the sampling rate, it has to be considered that according to the Nyquist-Shannon theorem the maximum frequency of a spectral analysis is half the sampling rate. On the other hand, the lowest frequency depends on the physical time covered by the simulation (after the initial transient). If data for 1 second real time is available, this would contain only one period of a 1 Hz oscillation, surely not enough to draw a reliable conclusion down to 1 Hz.

In contrast to Navier-Stokes methods, a LBM solver does not integrate partial differential equations iteratively until a specified residual is reached (cp. section 14.1.9.3). Like in a wind tunnel, there is no explicit convergence criterion. Depending on the vehicle shape, a more or less unsteady flow field will arise. It is not possible to tell a priori how long a simulation must be run to achieve a nonambiguous result. To obtain the averaged drag and lift coefficients, normally a period of 1–3 seconds is sufficient. Longer runtimes are necessary for low-frequency phenomena or if unsteady boundary conditions are used, for instance simulating a passing maneuver or gusts via time-dependent free flow directions as in [796] and [795]. Cooling air mass flow rates typically require significantly less time to reach a quasi–steady-state.

The decision over which time interval to average, for instance, to obtain the drag coefficient for performance and fuel consumption calculations, is made on the basis of time histories like the one in Figure 14.8. This can be executed either pragmatically by visual inspection or through sophisticated statistical analysis. For instance, it can be requested that the fluctuation of force coefficients must be bounded to a specified interval over a certain time period. By monitoring a simulation on-line this way, it can be detected when the flow field has settled and the simulation may be terminated prematurely. Detailed examples for the visualization of PowerFLOW results are shown in section 14.1.9.4.

14.1.3.2.4–Accuracy

When validating a CFD tool by comparison with experimental data, the complete congruence of wind tunnel and simulation model geometry and boundary conditions is of paramount importance. As is well known, even small geometry changes may have a large effect on aerodynamic properties and thus corrupt the validation effort. Further, it should not be neglected that the wind tunnel is also (only) a simulation of the real-world conditions and afflicted with errors that have to be compensated for by more or less sophisticated correction methods (see section 13.2.13). Only in rare cases, validation calculations will model the entire test facility around the vehicle in detail. For these reasons it is not advised to impose excessive requirements on the correlation of numerical and experimental results.

All validation results for PowerFLOW shown next are based on the vehicle variants used by [710], where the physical and software models were manufactured from identical CAD data and hence ensure the necessary congruence. Figure 14.9 shows some variants of the modular vehicle model.

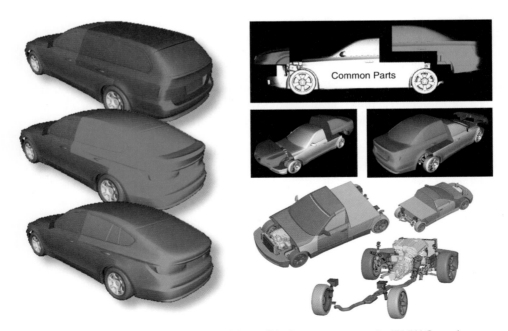

Figure 14.9 Modular vehicle model for validation measurements (BMW Group).

There were four different basic rear ends, and for each a number of subtle variations of radius and/or height of the trunk edge were available. The upper front section was also interchangeable, and the cooling air inlets were able to be covered with a panel to prevent flow through the fairly detailed engine compartment. All measurements were made in the new aerodynamic test center (AVZ) of the BMW Group, which was designed such that corrections are no longer necessary.

To assess the accuracy particularly with regard to vehicle development, a "traffic light" tri-color labeling for the difference between measurement and calculation is used (Table 14.2).

Table 14.2 Rating of the Difference Between Numerical and Experimental Data

Rating	Drag coefficient	(Axle) lift coefficient
GREEN (good)	$\Delta c_D \leq 0.005$	$\Delta c_L \leq 0.010$
YELLOW (acceptable)	$0.005 < \Delta c_D \leq 0.010$	$0.010 < \Delta c_L \leq 0.020$
RED (bad)	$\Delta c_D > 0.010$	$\Delta c_L > 0.020$

For lift, rather than the total value, the individual axle lifts are considered, because the axle lifts and the lift balance (front axle lift minus rear axle lift) are more relevant for vehicle stability characteristics. The limits for rating lift coefficients are larger than the ones for drag because the same deviation of static pressure will have a much larger effect on lift than on drag (the projected area in the vertical direction is usually much larger than the frontal area).

The following images (Figure 14.10, Figure 14.11, and Figure 14.12) compare absolute values in the diagram on the left while the plot on the right displays the difference to the wind tunnel, color coded according to the rating in Table 14.2. In the overview of all investigated variants (Figure 14.13), only 4 out of the total 42 force coefficients are rated yellow (acceptable); all others are green (good). This proves that such a code can be employed in the aerodynamic development process as a productive tool.

The aerodynamic quality of the cooling module and the cooling air routing through the engine bay is usually evaluated by the drag difference between open and closed air inlets. Figure 14.14 shows that this Δc_D from the simulation differs by only 0.001–0.003 from the data measured in the wind tunnel.

Further details of this validation study are given in Kandasamy et al. [416] and [415].

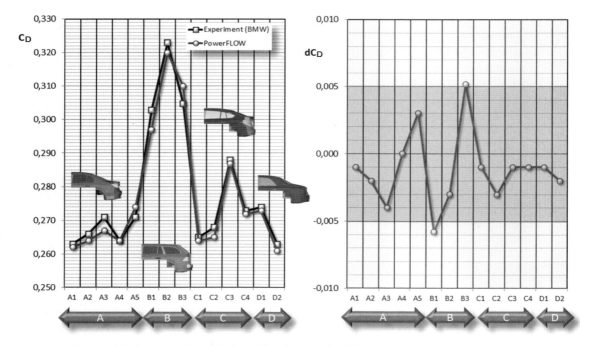

Figure 14.10 Comparison of simulation (PowerFLOW) vs. wind tunnel (BMW) for the drag coefficient.

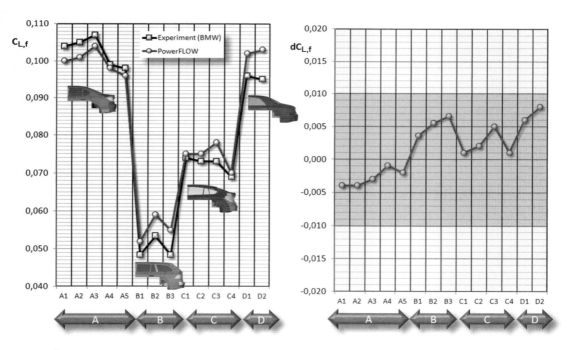

Figure 14.11 Comparison of simulation (PowerFLOW) vs. wind tunnel (BMW) for the front axle lift coefficient.

Numerical Methods

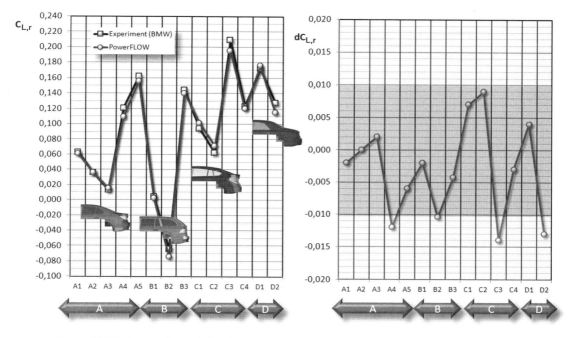

Figure 14.12 Comparison of simulation (PowerFLOW) vs. wind tunnel (BMW) for the rear axle lift coefficient.

			Cx	-0,001
A1	Base		Cz1	-0,004
			Cz2	-0,002
A2	Trunk edge raised 10 mm		Cx	-0,002
			Cz1	-0,004
			Cz2	0,000
A3	Trunk edge raised 20 mm		Cx	-0,004
			Cz1	-0,003
			Cz2	0,002
A4	Trunk edge Radius 34 mm		Cx	0,000
			Cz1	-0,001
			Cz2	-0,012
A5	Trunk edge Radius 100 mm		Cx	0,003
			Cz1	-0,002
			Cz2	-0,006

			Cx	-0,006
B1	Base		Cz1	0,004
			Cz2	-0,002
B2	Roof edge raised 30 mm		Cx	-0,003
			Cz1	0,006
			Cz2	-0,010
B3	Roof edge rounded		Cx	0,005
			Cz1	0,007
			Cz2	-0,004

			Cx	-0,001
C1	Base		Cz1	0,001
			Cz2	0,007
C2	Trunk raised 30 mm		Cx	-0,003
			Cz1	0,002
			Cz2	0,009
C3	Trunk edge rounded		Cx	-0,001
			Cz1	0,005
			Cz2	-0,014
C4	Roof edge extended 100 mm		Cx	-0,001
			Cz1	0,001
			Cz2	-0,003

			Cx	-0,001
D1	Base		Cz1	0,006
			Cz2	0,004
D2	Trunk edge sharpended		Cx	-0,002
			Cz1	0,008
			Cz2	-0,013

Figure 14.13 Accuracy overview for drag and axle lifts.

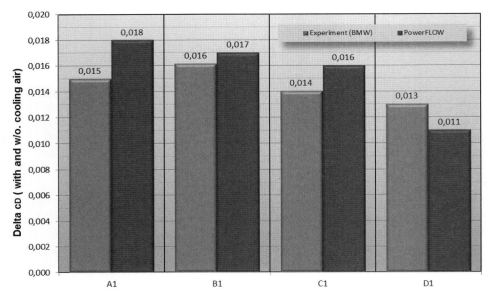

Figure 14.14 Comparison of cooling air drag increase from simulation (PowerFLOW) vs. wind tunnel (BMW).

14.1.4 Navier-Stokes Methods

The Navier-Stokes equations (NSE) have already been derived in section 2.1.7. In their most general form, they describe compressible, viscid continuum flow of Newtonian fluids without further limitations. An exact solution would produce all spatial and temporal details. However, exact solutions only exist for some special cases like Hele-Shaw creeping flow or laminar Couette and Hagen-Poiseuille flow [716]. For the majority of technical processes with turbulent flows, the solution has to be generated numerically. For this purpose, various simplifications have been made to the NSE.

In section 14.1.1, possible simplifications have already been addressed, and Table 14.1 indicates the consequences they have on the characteristics of the NSE. In the following, we will discuss some of these methods:

- RANS and URANS methods: Flow quantities are split into mean and fluctuating values and are inserted and time averaged in the NSE. This creates additional unknown turbulent shear stresses that have to be modeled (see section 14.1.4.1).
- Large-Eddy- and Detached-Eddy-Simulations (LES/DES): Flow quantities are filtered spatially according to the numerical grid size. This results in large, directly simulated structures and small, universal structures that are modeled (see section 14.1.4.2).
- Direct Numerical Simulation (DNS): The NSE are solved numerically without any modeling (see section 14.1.4.3).

- Potential Flow: The flow is assumed to be incompressible, inviscid, and irrotational. Additional models are required to include separation, wake creation, and total pressure loss (see section 14.1.5).

14.1.4.1 RANS Methods

This section gives an overview of the RANS[3] methods. First, the RANS equations are derived, followed by an explanation of the closure problem, which leads to turbulence modeling. Further, the use of wall models is discussed, which allow an acceptable computational effort. Finally, some selected results of RANS methods in vehicle aerodynamics are presented.

14.1.4.1.1 The Reynolds-Averaged Navier-Stokes Equations

For many technical problems, not the high-frequency fluctuations but the temporal mean values of flow quantities are sought, for instance, the volume flow rate in a duct or the drag of a vehicle. Even if the time-dependent NSE could be easily solved, in these cases one would average the solution over a sufficiently long time interval to obtain the mean values. So it is sensible to adopt the idea of Reynolds (1884), who simplified the NSE by applying them to the mean values of velocity and pressure. The chaotic turbulent motion of the fluid is split into a mean, time independent part and the superimposed high-frequency fluctuation (see Figure 14.15):

$$u = \bar{u} + u' \;;\; v = \bar{v} + v' \;;\; w = \bar{w} + w' \;;\; p = \bar{p} + p' \qquad (14.14)$$

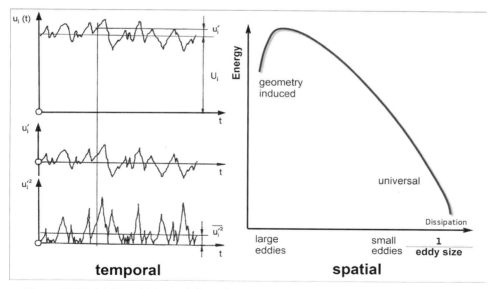

Figure 14.15 (a) Time history of the velocity in a turbulent boundary layer; (b) Energy spectrum of turbulent motion (Kolmogorov spectrum).

---

3. Reynolds Averaged Navier Stokes.

Quantities with an overbar indicate temporal mean values. The time interval for their averaging has to be chosen such that the mean value no longer depends on the averaging interval[4]:

$$\bar{u} = \frac{1}{t_1} \int_{t_0}^{t_0+t_1} u(t)dt \quad \text{with} \quad \overline{u'} = 0,\ \bar{\bar{u}} = \bar{u}\ . \tag{14.15}$$

The primed quantities in equation (14.14) are stochastic fluctuations superimposed on the mean values. By definition, their average vanishes. Inserting equation (14.14) in the complete equations of motion and averaging over time yields (cf. for instance [716]):

$$\frac{\partial \bar{u}}{\partial t} + \bar{u}\frac{\partial \bar{u}}{\partial x} + \bar{v}\frac{\partial \bar{u}}{\partial y} + \bar{w}\frac{\partial \bar{u}}{\partial z} = -\frac{1}{\rho}\frac{\partial \bar{p}}{\partial x} + \nu\left(\frac{\partial^2 \bar{u}}{\partial x^2} + \frac{\partial^2 \bar{u}}{\partial y^2} + \frac{\partial^2 \bar{u}}{\partial z^2}\right) + \left(\frac{\partial \overline{u'^2}}{\partial x} + \frac{\partial \overline{u'v'}}{\partial y} + \frac{\partial \overline{u'w'}}{\partial z}\right)$$

$$\frac{\partial \bar{v}}{\partial t} + \bar{u}\frac{\partial \bar{v}}{\partial x} + \bar{v}\frac{\partial \bar{v}}{\partial y} + \bar{w}\frac{\partial \bar{v}}{\partial z} = -\frac{1}{\rho}\frac{\partial \bar{p}}{\partial y} + \nu\left(\frac{\partial^2 \bar{v}}{\partial x^2} + \frac{\partial^2 \bar{v}}{\partial y^2} + \frac{\partial^2 \bar{v}}{\partial z^2}\right) + \left(\frac{\partial \overline{u'v'}}{\partial x} + \frac{\partial \overline{v'^2}}{\partial y} + \frac{\partial \overline{v'w'}}{\partial z}\right)$$

$$\frac{\partial \bar{w}}{\partial t} + \bar{u}\frac{\partial \bar{w}}{\partial x} + \bar{v}\frac{\partial \bar{w}}{\partial y} + \bar{w}\frac{\partial \bar{w}}{\partial z} = -\frac{1}{\rho}\frac{\partial \bar{p}}{\partial z} + \nu\left(\frac{\partial^2 \bar{w}}{\partial x^2} + \frac{\partial^2 \bar{w}}{\partial y^2} + \frac{\partial^2 \bar{w}}{\partial z^2}\right) + \left(\frac{\partial \overline{u'w'}}{\partial x} + \frac{\partial \overline{v'w'}}{\partial y} + \frac{\partial \overline{w'^2}}{\partial z}\right) \tag{14.16}$$

On the right-hand side, we now have additional terms formed by the product of two fluctuations (i.e., their mean value does not vanish). Since they act like the molecular shear stresses, they are called turbulent or apparent shear stresses and can be written in tensor form as

$$\tau'_{i,j} = \rho \cdot \begin{pmatrix} \overline{u'^2} & \overline{u'v'} & \overline{u'w'} \\ \overline{u'v'} & \overline{v'^2} & \overline{v'w'} \\ \overline{u'w'} & \overline{v'w'} & \overline{w'^2} \end{pmatrix}. \tag{14.17}$$

Equation (14.16) are denoted as Reynolds averaged Navier-Stokes equations (RANS) if the time derivative on the left-hand side is dropped, otherwise we refer to them as Unsteady RANS (URANS). Formally their left-hand sides and the first two terms on the right-hand side are identical to the original NSE (equation (2.43)); just the time-dependent quantities u, v, w, and p are replaced by their mean values. However, now we have additional terms that represent the turbulent fluctuations. They express the intensified momentum exchange caused by the turbulent mixing combined with additional stresses. The effect is the same as a—usually drastic—increase of viscosity in the fluid. Using the fluctuations, we can define the turbulent kinetic energy that will be used later as

$$k = \frac{1}{2}\cdot\left(\overline{u'^2} + \overline{v'^2} + \overline{w'^2}\right). \tag{14.18}$$

---

4. This introduces apparent stress terms in the RANS equations. When moving to LES in section 14.1.4.2, this is not the case due to the continuous Kolmogorov spectrum, but as a consequence more perturbation terms appear in the large structure equations than in RANS.

14.1.4.1.2 Turbulence Models

By averaging the time-dependent NSE, they are transformed to the (U-)RANS equations at the price of additional unknowns, namely the averaged fluctuations in the form $\overline{u'_i u'_j}$. Turbulence models[5] have been developed to match the number of available equations with the number of unknowns. These models link the turbulent fluctuations with the time averaged flow field. First of all, we distinguish between eddy viscosity models and Reynolds stress models (see Figure 14.16).

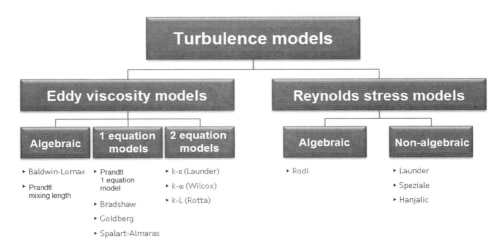

Figure 14.16 Classification of various turbulence models.

Eddy viscosity models assume isotropic turbulence, meaning all turbulent stresses $\tau'_{i,j}$ depend on the same proportionality constant, the eddy viscosity. Two examples of these models will be discussed here to demonstrate the idea behind this approach. The oldest one is the Boussinesq hypothesis, formulated in 1887 [716]. In analogy to Newton's law (equation (2.35)) he proposed the following formulation:

$$-\rho \overline{u'v'} = \tau'_{i,j} = \mu_t \frac{\partial \overline{u}}{\partial y} = \rho \nu_t \frac{\partial \overline{u}}{\partial y} \tag{14.19}$$

The constant ν_t is denoted as eddy viscosity, a quantity that is not a material property but a function of position $\nu_t(x, y, z)$. Hence it depends on the local flow field and vanishes when approaching a solid wall. The name "eddy viscosity" implies that the turbulent flow is vortical and invokes the same effect as a significant increase of viscosity.

In 1925, Prandtl introduced a "mixing length" for the turbulent fluctuations to characterize the motion of eddies across the main flow. His approach is:

$$-\rho \overline{u'v'} = \tau'_{i,j} = \rho l^2 \left|\frac{\partial \overline{u}}{\partial y}\right| \frac{\partial \overline{u}}{\partial y} \tag{14.20}$$

---

5. Not modeling turbulence as such but the Reynolds stresses.

Assuming that the mixing length l does not depend on the velocity itself, then this formulation shows that the Reynolds stresses change with the square of the mean velocity gradient. Comparing the approaches from Boussinesq (equation (14.19) and Prandtl (equation (14.20)) gives:

$$v_t = l^2 \left| \frac{\partial \overline{u}}{\partial y} \right| \tag{14.21}$$

Both models connect the turbulent shear stresses with a global quantity of the flow field, namely the gradient of the time averaged velocities. Unfortunately, the explicit calculation of eddy viscosity v_t as well as mixing length l is only possible for some special cases [716].

More commonly, these algebraic models have been replaced by multi-equation models, also known as first order closure models; the most well-known one is the k-ε model. This model also relates the eddy viscosity with gradients of the mean velocity. The eddy viscosity is obtained from two partial differential equations that describe the change of the turbulent kinetic energy k and the dissipation rate ε during their transport in the flow field. Two-equation models finally determine the eddy viscosity based on a dimensional analysis as

$$v_t = c_\mu \frac{k^2}{\varepsilon}. \tag{14.22}$$

The empirical constants in these models have been found by experiment, for instance $c_\mu = 0.09$ is a commonly used value. In the k-ε model the two PDEs for k and ε are:

$$\rho \left(\frac{\partial k}{\partial t} + \overline{u}_j \frac{\partial k}{\partial x_j} \right) = \mu_t \frac{\partial \overline{u}_i}{\partial x_j} \left(\frac{\partial \overline{u}_i}{\partial x_j} + \frac{\partial \overline{u}_j}{\partial x_i} \right) + \frac{\partial}{\partial x_j} \left(\mu \frac{\partial k}{\partial x_j} + \frac{\mu_t}{\sigma_k} \frac{\partial k}{\partial x_j} \right) - \rho \varepsilon \tag{14.23}$$

and

$$\rho \left(\frac{\partial \varepsilon}{\partial t} + \overline{u}_j \frac{\partial \varepsilon}{\partial x_j} \right) = c_{\varepsilon 1} \mu_t \frac{\varepsilon}{k} \frac{\partial \overline{u}_i}{\partial x_j} \left(\frac{\partial \overline{u}_i}{\partial x_j} + \frac{\partial \overline{u}_j}{\partial x_i} \right) + \frac{\partial}{\partial x_j} \left(\mu \frac{\partial \varepsilon}{\partial x_j} + \frac{\mu_t}{\sigma_\varepsilon} \frac{\partial \varepsilon}{\partial x_j} \right) - c_{\varepsilon 2} \rho \frac{\varepsilon^2}{k}. \tag{14.24}$$

These second order PDEs are linear in k and ε and therefore relatively easy to integrate numerically. With the last three equations, adding two more unknowns, k and ε, the closure problem is solved now. The eddy viscosity is determinable and the equation system can—at least theoretically—be solved simultaneously with the RANS equations. Further eddy viscosity models are for instance k-ω, k-L, and the Spalart-Allmaras model.

Reynolds stress models (RSM) are categorized as second order closure. For each element of the stress tensor in equation (14.17), a transport equation is derived from the Navier-Stokes equations. Since these models are very time consuming they haven't reached a broad distribution in automobile aerodynamics and won't be discussed further here.

14.1.4.1.3 Wall Modeling

When approaching a wall, the mean fluid velocity decreases as a result of friction effects. At the wall itself, the velocity is zero; this is the no-slip condition. Also the turbulent fluctuations decay to zero due to the wall proximity. The boundary layer thickness depends on the Reynolds number with the velocity gradient at the wall increasing with increasing Reynolds number and hence decreasing the boundary layer thickness. With respect to turbulence modeling, there are two ways to handle the near-wall flow (Figure 14.17).

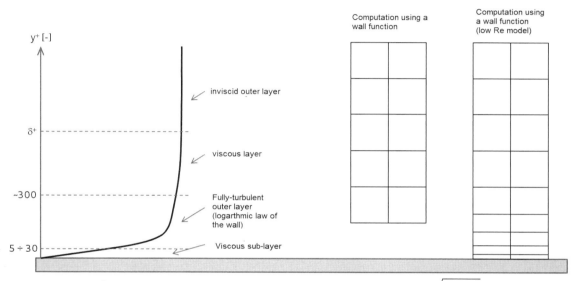

Figure 14.17 Description of the near-wall flow by means of $y^+ = y\sqrt{\tau_w/\rho v^2}$ with a wall function or a low Reynolds model [616].

Many turbulence models, for example the standard k-ε model, fail in the immediate vicinity of the wall. For this reason, but also to limit the number of grid points, very often a law-of-the-wall model is used to represent the velocity distribution in this region (center image in Figure 14.17). The outer region of the boundary layer has to be calculated numerically, which still requires a dense grid in the outer region. Since the law-of-the-wall has been derived, assuming an equilibrium boundary layer, it is strictly speaking only valid under this condition. Nevertheless, it is often also used in regions of reverse flow.

Standard wall functions represent a universal velocity profile. For example, the "log-law-of-the-wall" with the von-Karman constant $\kappa = 0.41$

$$u = \sqrt{\frac{\tau_w}{\rho}} \cdot \left[\frac{1}{\kappa} ln\left(y \cdot \sqrt{\frac{\tau_w}{\rho v^2}} \right) + C \right] \qquad (14.25)$$

will be discussed here. The empirically found value for the integration constant is $C = 5.5$. In the viscous sublayer there is a direct proportionality between velocity and wall distance

$$u = y \cdot \frac{\tau_W}{\rho \cdot v}. \qquad (14.26)$$

These wall functions have been derived assuming a flat plate flow and therefore do not hold in accelerated or decelerated flow, that is, in the presence of pressure gradients. However, pressure gradients are crucial in vehicle aerodynamics because they determine whether separation occurs or not. For these practical flow problems a correction function g is added, which accounts for pressure gradients. Further, a function h is introduced to model surface roughness ς. The extended logarithmic part of the wall function then reads

$$u = \sqrt{\frac{\tau_W}{\rho}} \cdot \left[\frac{1}{\kappa} \ln \left(\frac{y \cdot \sqrt{\frac{\tau_W}{\rho v^2}}}{1 + g(\partial p / \partial x) + h(\varsigma)} \right) + C \right]. \qquad (14.27)$$

RANS equations and bulk fluid turbulence models are valid up to the outer edge of the boundary layer, and closer to the wall these wall models are employed.

On the other hand, turbulence models can be used down to the wall (right image in Figure 14.17) if the strong decay of turbulent shear stresses and dissipative effects in the boundary layer are included. This is not described correctly by standard k-ε models. A number of alternate approaches exist. For instance, Launder and Sharma [489] use a damping function when calculating the eddy viscosity

$$v_t = f_\mu \cdot c_\mu \frac{k^2}{\varepsilon} \text{ with } f_\mu = e^{-3.4/A} \text{ and } A = \left(1 + \frac{k^2}{50 \varepsilon v} \right) \qquad (14.28)$$

and an additional term on the right-hand side of the k-equation

$$D = 2v \left(\frac{\partial}{\partial z} \left(\sqrt{k} \right) \right)^2 \qquad (14.29)$$

Since the processes in rather thick boundary layers at low Reynolds numbers can be described using such models without an unacceptable erroneous effect on the bulk flow, these approaches are also called "low-Re" turbulence models.

14.1.4.1.4 Results

The investigations of Wäschle [847] enable an assessment of the quality of RANS results. He used STAR-CD© to compute the flow around various SAE reference bodies (Figure 14.18 and Figure 14.19) and a simplified vehicle geometry (Figure 14.20). Regardless of the rear end shape of the SAE body, a very good agreement between measured and computed c_D values is achieved. The maximum difference is only 0.003 (Figure 14.18).

The analysis of lift coefficients produces a different picture (Figure 14.19). In particular for the challenging hatchback and notchback, the values deviate considerably and in addition we see contradictory trends. In experiment, the rear axle lift increases for the notchback while the simulation predicts a decrease. For the station wagon rear end the deviation is higher than $\Delta c_L = 0.050$.

Comparing the velocity field (Figure 14.20) reveals that, especially near the ground where complex turbulent structures appear, the flow field topology is not correct. Noticeable deviations to experiment also show up in the simulation at the upper edge of the wheel arch.

In summary, it can be stated that the drag is predicted very realistically while lift and flow field topology differ substantially from measurements.

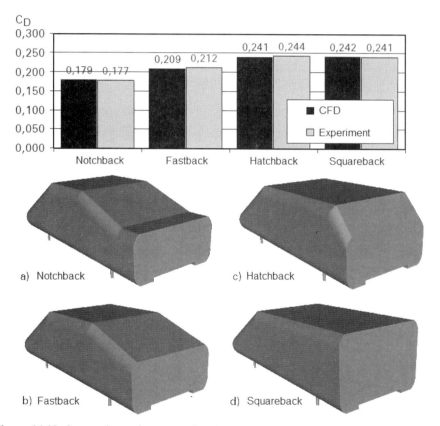

Figure 14.18 Comparison of measured and simulated drag coefficients on various SAE reference bodies, calculated with RANS [847].

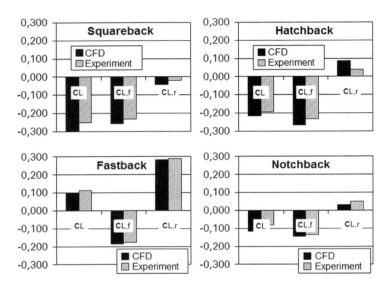

Figure 14.19 Comparison of measured and simulated lift coefficients on various SAE reference bodies, calculated with RANS [847].

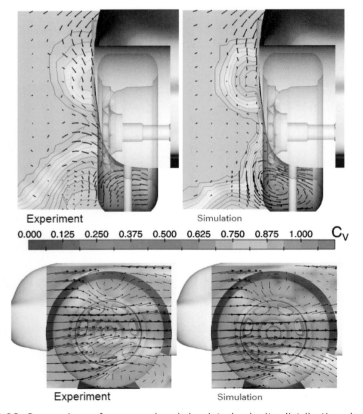

Figure 14.20 Comparison of measured and simulated velocity distributions behind and next to a wheel, calculated with RANS [847].

14.1.4.2 Large-Eddy Simulation

The time averaging of flow quantities in RANS supposes that velocity and pressure in the fluid are composed of a mean value and superimposed fluctuations. The high-frequency contributions are isolated via an adequate time averaging interval.

If we now consider the spatial instead of temporal scales, we see that the RANS approach averages over the entire spectrum of turbulent length scales, from large-scale geometry-induced structures down to the smallest dissipative scales. In contrast to the two-part spectrum resulting from assumptions made in the RANS approach, the spatial spectrum in reality is continuous (cf. Figure 14.15b). In the real spectrum, the different turbulence structures take on different roles. Large-scale formations are predominantly responsible for the transport of momentum and energy. They are anisotropic, meaning their nature is three-dimensional and their appearance is very different, depending on the shape of the body and on the Reynolds number. Small-scale turbulence dissipates kinetic to thermal energy. Its structure is similar in many flow cases; it is almost isotropic and therefore possesses a universal character. From this, it can be concluded that the RANS method is appropriate to model the small scales but not the large ones.

14.1.4.2.1 Basic Equations of Large-Eddy Simulation

A method that computes large-scale turbulence directly and models only small scales is called large-eddy simulation (LES). Since the turbulence spectrum is continuous from large to small eddies, a clear separation between these two classes is indeterminable. Capturing the large structures in LES is achieved by filtering the Navier-Stokes equations. Similar to RANS, the flow quantities are split into large-scale and small-scale contributions.

$$u = \tilde{u} + u_{SGS} \ ; \ v = \tilde{v} + v_{SGS} \ ; \ w = \tilde{w} + w_{SGS} \ \text{and} \ p = \tilde{p} + p_{SGS}. \tag{14.30}$$

Using an appropriate filter function (for instance a Gauss filter) defined by means of the cell size Δ and the local coordinate system ξ

$$F(x - \xi, \Delta) = Ae^{B} \text{ with } A = \frac{1}{\Delta\sqrt{2\pi}} \text{ and } B = -\frac{1}{2}\left(\frac{x - \xi}{\Delta}\right)^2 \tag{14.31}$$

the large-scale part is defined, for example the large-scale pressure contribution reads

$$\int_{\Delta V} F(x - \xi, \Delta) \cdot p(\xi) d\xi = \tilde{p}(x) \tag{14.32}$$

The filter function must comply with the condition

$$\int_{\Delta V} F(x - \xi, \Delta) d\xi = 1 \tag{14.33}$$

and further it has to tend to Dirac's delta-function for cell sizes approaching zero. This leads to the limiting condition

$$\lim_{\Delta \to 0} \int_{\Delta V} F(x - \xi, \Delta) \cdot p(\xi) d\xi = p(x) \tag{14.34}$$

and the subgrid-scale part (SGS) tends to zero. Figure 14.21a shows the dependency of the filter function from the cell size Δ and Figure 14.21b demonstrates the meaning of the equations shown. First the product $F \cdot P$ at position x is plotted versus ξ. The integration over ξ leads to that data point on the filtered signal curve, which results from all equivalently determined data points.

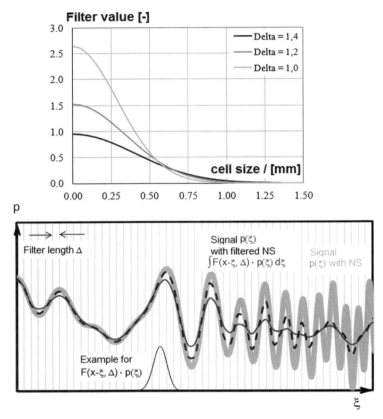

Figure 14.21 (a) Gauss filter function depending on the cells size Δ; (b) Product and integration of filter function and pressure signal as example quantity as well as resolution capability with given filter width.

A comparison with the actual signal curve shows that regions with coarse fluctuations are well represented; however, structures in the range of the filter width are poorly described. The dashed line indicates the signal with half the filter width (and same sampling rate) where also smaller structures are better resolved.

Thus, with smaller cell sizes this method converges to the direct numerical simulation (DNS). For small but finite cells, the application of the filter averages the flow quantity over the cell size Δ. Filter operators usually have the characteristic

$$\widetilde{u+v} = \tilde{u}+\tilde{v}, \quad \tilde{\tilde{u}} \neq \tilde{u} \quad \text{and} \quad \widetilde{u_{SGS}} \neq 0. \tag{14.35}$$

The last two properties are unlike the temporal RANS averaging, which is due to the continuous turbulence spectrum. By applying this averaging to the complete equations of motion (equation (2.43)), multiplying with F, and then integrating over $d\xi$, we get

$$\frac{\partial \widetilde{u}_i}{\partial t} + \widetilde{u}_j \frac{\partial \widetilde{u}_i}{\partial x_j} = -\frac{1}{\rho}\frac{\partial \widetilde{p}}{\partial x_i} + \nu \nabla^2 \widetilde{u}_i - \frac{\partial}{\partial x_j}\left(\widetilde{\widetilde{u}_i \widetilde{u}_j} - \widetilde{u}_i \widetilde{u}_j + \widetilde{\widetilde{u}_i u_{j,\text{SGS}}} + \widetilde{u_{i,\text{SGS}}\widetilde{u}_j} + \widetilde{u_{i,\text{SGS}}u_{j,\text{SGS}}}\right) \quad (14.36)$$

with

$$\widetilde{\widetilde{u}_i \widetilde{u}_j} - \widetilde{u}_i \widetilde{u}_j = L_{i,j} \; ; \; \widetilde{\widetilde{u}_i u_{j,\text{SGS}}} + \widetilde{u_{i,\text{SGS}}\widetilde{u}_j} = C_{i,j} \; ; \; \widetilde{u_{i,\text{SGS}}u_{j,\text{SGS}}} = R_{i,j}$$
$$\text{and } \tau_{i,j,\text{SGS}} = L_{i,j} + C_{i,j} + R_{i,j} \quad (14.37)$$

Formally, this is identical to the RANS equations. However, instead of the Reynolds stress tensor, the SGS stress tensor appears on the right-hand side. In the limiting case of very small cell sizes $\Delta \to 0$ the latter becomes $\tau_{i,j,\text{SGS}} = 0$. The subgrid-scale stresses are the sum of the interaction between resolved vortices $L_{i,j}$, the interaction between resolved and unresolved vortices $C_{i,j}$, and the unresolved vortices $R_{i,j}$. In case of strong streamline curvature, the latter transfer energy back to the large vortices (= backscatter).

The first and still frequently used approach to model the SGS stress tensor was proposed by Smagorinsky [764]:

$$\tau_{i,j,\text{SGS}} = -2C_S \Delta^2 \widetilde{S}_{i,j}\left(2\widetilde{S}_{i,j}\widetilde{S}_{i,j}\right)^{\frac{1}{2}} \quad (14.38)$$

with

$$\widetilde{S}_{i,j} = \frac{1}{2}\left(\frac{\partial \widetilde{u}_i}{\partial x_j} + \frac{\partial \widetilde{u}_j}{\partial x_i}\right) \quad (14.39)$$

as the filtered shear-rate tensor. In the literature values for the Smagorinsky constant C_S are between 0,005 and 0,05 depending on the particular flow problem. Since other than in the URANS method no time averaging of the turbulence is made, a considerably smaller time step has to be chosen to reproduce fluid motion in a realistic way.

14.1.4.2.2 Influence of Cell Size

In section 14.1.4.1.2, it has been pointed out that in URANS and RANS methods the most challenging regions of the flow field are treated by turbulence and wall models. Both models have in common that a refinement of the numerical grid only reduces the numerical error but does not necessarily lead to a better reproduction of the flow physics.

In contrast, a characteristic of the LES method is that with decreasing cell size, the contribution of the sub-grid-scale model to the overall method decreases and therefore not only the numerical but also the model error becomes smaller. With increasing grid-refinement, the role of the SGS turbulence model reduces but the direct simulation of turbulence is improved. Spalart (1999)[6] mentions that even a 20% change of the Smagorinsky constant is hardly noticeable in a highly resolved flow.

In fact, it is possible to combine LES methods with wall models, but this would confine one of the strengths of LES again, namely the capability to reproduce detailed phenomena in the boundary layer. On the other hand, this requires that the boundary layer has to be resolved by the numerical grid. This means the cells closest to the wall must have a dimensionless wall distance of $y^+ = 1$ or smaller. For the flow around a vehicle, cell sizes would have to be smaller by a factor of about 30 compared to a RANS simulation. Due to the enormous requirements, such computations are not feasible for vehicle development or for baseline investigations at car manufacturers. This leads to the formulation of detached-eddy simulation (DES).

14.1.4.2.3 Detached-Eddy Simulation

Detached eddy simulation (DES) uses the LES method with their SGS turbulence modeling solely in the external flow field and in wake regions. In the near-wall region, the unsteady Reynolds-averaged Navier-Stokes equations (URANS) with associated turbulence models are employed. This idea has been proposed by Spalart [767]. As long as the turbulent length scales in the boundary layer are clearly smaller than in the external flow, this approach can be considered as a good approximation. In vehicle aerodynamics, this is the case, and in addition the flow behavior near the wall is noticeably universal. Often the one-equation turbulence model of Spallart-Allmaras is employed. The transition from LES to URANS takes place at a wall distance d, which is determined with the empirical constant c_{DES} (= 0.65), the length Δ of the equivalent cell diagonal and the cell-wall distance y_W as

$$d = \min\{y_{W,SA}; \Delta \cdot c_{DES}\} \qquad (14.40)$$

If the cell wall distance is large in relation to the cell volume, then LES is applied, and otherwise URANS is applied.

14.1.4.2.4 Results of LES and DES Simulations on Automobiles

To date, the Japanese automobile industry is driving the deployment of LES in regular vehicle development. One of the examples that can be found in the literature is the study of Kitoh et al. [436], who applied LES to the external flow around the ASMO body. Figure 14.22 demonstrates that the surface pressures correlate very well with the experiment. In particular, the result quality on the vehicle's base is remarkable.

Islam et al. [397] presented a method based on detached-eddy simulation, which, in contrast to pure LES, allows acceptable turnaround times in the vehicle development process. The authors computed a number of fully detailed production cars of the Volkswagen Group and analyzed, among other quantities, the force coefficients. An overview is given in Table 14.3. The maximum differences to wind tunnel results is

---

6. Spalart, P.: Strategies for Turbulence Modelling and Simulations. In Rodi, W.; Laurence, D.: *Engineering Turbulence Modelling and Experiments* 4, Elsevier, Amsterdam 1999, S. 3–17

0.022 for the drag coefficient, 0.036 for the front axle lift, and 0.051 for the rear axle lift. In follow-up studies, these promising results were confirmed and improved by Schütz [749] on further vehicles and vehicle categories.

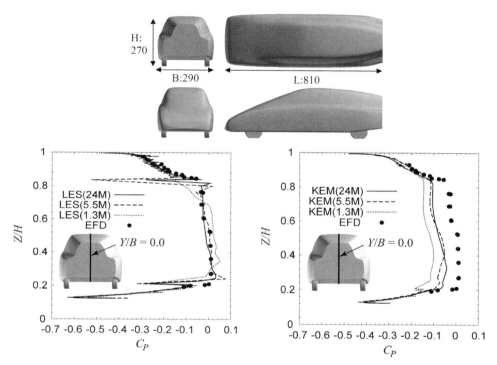

Figure 14.22 Comparison of surface pressure from experiment: LES (left) and RANS (right) simulations on a vehicle body, Kitoh et al. [436].

Table 14.3 Comparison of Force Coefficients of Various Volkswagen Group Vehicles (Figures Are Different from the Wind Tunnel Experiment)			
Vehicle	Δc_D [−]	$\Delta c_{l,f}$ [−]	$\Delta c_{l,r}$ [−]
SEAT Ibiza	0.018	−0.017	0.045
SEAT Leon	0.021	−0.005	0.030
VW Golf	0.003	0.034	0.024
VW Passat	0.011	−0.033	0.035
VW New Beetle	0.016	0.001	0.030
AUDI A3	0.007	−0.018	0.034
AUDI A5	0.011	−0.036	0.031
AUDI A6	−0.004	0.002	0.026
AUDI Q5	−0.001	−0.006	0.047
AUDI TT	−0.001	−0.006	0.051
AUDI R8	0.022	0.021	−0.012

In addition, comparisons of surface pressures on an AUDI A6 limousine published in this framework show good correlation, even in demanding regions like the rear end and underbody. This comparison is shown in Figure 14.23. Figure 14.24 displays the velocity distribution in a plane behind the AUDI A6 in comparison to a wind tunnel measurement. Small discrepancies in the reproduction of the A-pillar vortex and the wake structures near the ground indicate some potential for improvement; nevertheless, a convincing agreement can be claimed.

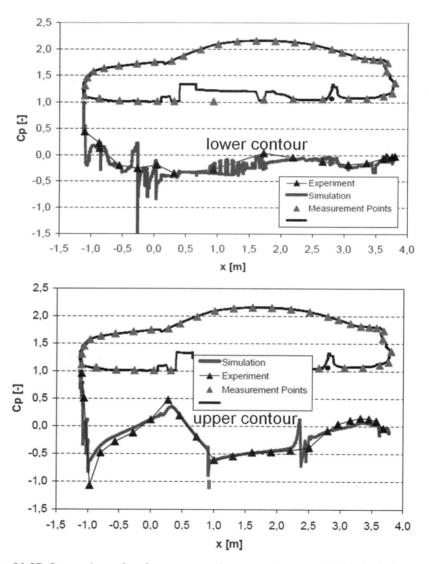

Figure 14.23 Comparison of surface pressure from experiment and DES simulation on the AUDI A6.

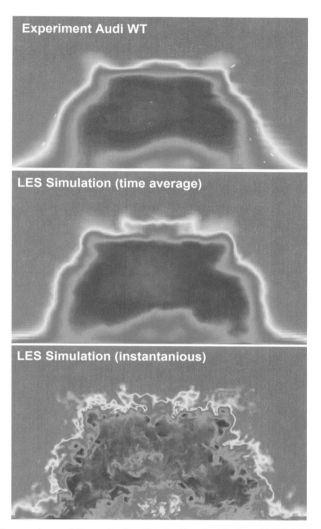

Figure 14.24 Comparison of the velocity distribution behind an AUDI A6 from experiment and DES simulation.

14.1.4.3 Direct Numerical Simulation

With direct numerical simulation (DNS), the time-dependent Navier-Stokes equations are solved without any physical simplifications. Consequently, neither a turbulence model nor a wall model is required. The goal is to resolve the entire turbulence spectrum. For this purpose, the simulation volume has to be very finely discretized, especially in the vicinity of solid walls. The number of cells increases approximately with $Re^{9/4}$ [616]. Further, DNS requires a high temporal resolution for a realistic reproduction of the unsteady flow behavior. In particular, for turbulent flows at high Reynolds numbers (i.e., for technically relevant problems), DNS is not feasible today. The DNS simulations found in literature are mostly confined to simple problems at low Reynolds numbers (Re ≤ 5000), for instance, to verify turbulence models.

Since the end of the 1980s, especially in the Japanese automobile industry, large efforts have been made to deploy DNS for vehicle aerodynamics, however, with much too low resolution in the relevant Reynolds number range. The result of Tsuboi et al. [808] depicted in Figure 14.25 was encouraging. They succeeded in reproducing the influence of the base slant angle φ on the drag of a cylindrical body (Morel [582]) quite well: with increasing slant angle φ, initially the drag increases up to a critical angle, here 44°; beyond that, the drag drops abruptly.

The flow topology is also captured correctly (Figure 14.26). For under-critical slant angles, a horseshoe-vortex forms, which expands when the critical angle is exceeded. At that time other codes were not able to produce such good results.

However, when looking at the pressure distribution in detail, we see considerable discrepancies. The suction peak near the edge of the slanted base (Figure 14.27) is not even nearly reached. Nevertheless, as the next step, the calculation of complete vehicles was tackled, using a grid with approximately 10^6 nodes in one symmetry half of the computational domain. The Reynolds number based on the length of the car was around 10^6.

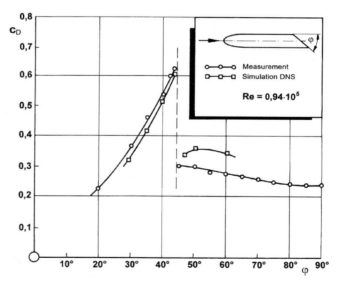

Figure 14.25 Drag coefficient c_D of a slanted body of revolution vs. slant angle. DNS result from Tsuboi et al. [808]; measurement Morel [582].

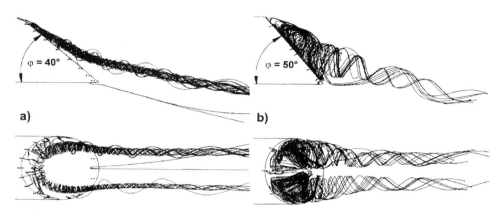

Figure 14.26 Flow field on the rear end of the body of revolution in Figure 14.25; DNS result from Tsuboi et al. [808]: a) horseshoe vortex at under-critical slant angle φ; (b) widened horseshoe vortex at overcritical slant angle φ.

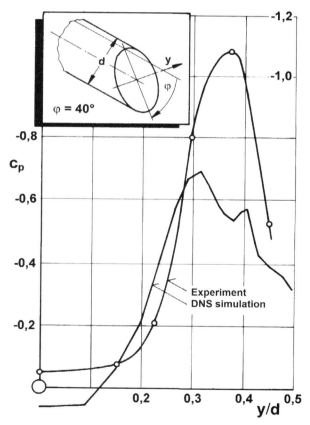

Figure 14.27 Pressure distribution on the slanted surface at $\varphi = 40°$. Comparison of the DNS result from Tsuboi et al. [808] with measurements of Bearman [52].

Out of a number of published DNS simulations, here we show one example in Figure 14.28, which was published by Kataoka et al. [421]. The vehicle was a sporty car fitted with some adjustable devices (marked in black in Figure 14.27) to increase performance and handling characteristics. Comparing simulated and measured results gave a surprising result. The difference in drag was less than 5% and the lift error was 0.02. Obviously, it was possible to predict the effect of the aerodynamic appliances quite closely.

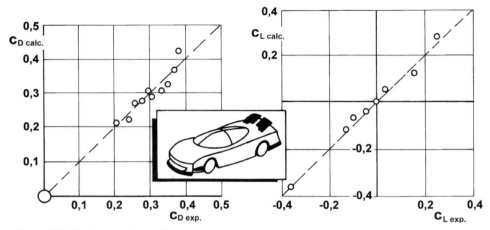

Figure 14.28 Comparison of measured drag and lift with DNS results from Kataoka et al. [421]. Drag and lift have been altered by the black components: front spoiler, wings before the front wheels, rear end flaps (adjustable), rear end lips.

Lumley [516] analyzed this result and concluded that it cannot at all be generalized. With 10^6 nodes at Re = 10^6 the grid is by two to three orders of magnitude too coarse. The boundary layer cannot be resolved and hence the friction drag cannot be predicted even roughly. According to an estimation of Jameson[7] (1996), the boundary layer of an automobile alone would require at least 10^8 cells. The question of why we see such a good correlation to measurement was answered by Lumley as follows:

- The contribution of friction drag to the total drag of an automobile is small (about 10%). Errors will not have a significant impact on the total drag.
- On cars like this, separation is often determined by the geometrical shape; it does not have to be calculated.
- The vorticity entrained into the wake is captured correctly even without a correct calculation of the boundary layer.
- Therefore, the pressure distribution and hence pressure drag and lift are predicted well.

---

7. In Ahmed [4].

Numerical Methods

According to Lumley, the calculation in this example turns out to be an unintended and uncontrolled LES; the large structures are computed and the dissipation of the small scales is a result of "numerical viscosity."

From this, we can conclude that DNS is not yet applicable to the aerodynamic development of automobiles. This also holds in particular for bodies where separation is not dictated by their shape or which have a large friction drag. For basic investigations on simple geometries and at low Reynolds numbers (Re < 10^3) DNS can do a good job.

14.1.5 Potential Flow Methods (BEM)

Potential flow theory is the strongest simplification in fluid dynamics, assuming inviscid and irrotational flow. In the 1960s, the first 3D methods (panel methods) were developed on this basis for aerospace applications [342]. Later, they were extended for the utilization on bluff bodies ([10] and [299]). Due to the highly detailed models today, they have become less important in vehicle aerodynamics. The following explanations are based on the method presented by Grün [299].

14.1.5.1 Zonal Coupling

Strictly speaking, potential flow theory can only be applied to attached flow—in other words, a potential flow method will always produce attached flow, regardless of the geometry complexity. The inevitable massive separation on the rear end of bluff bodies needs to be modeled. For this purpose different regions are coupled iteratively (Figure 14.29). This approach is also called a zonal solution.

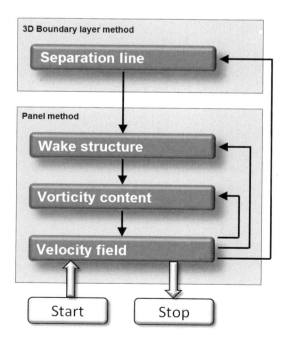

Figure 14.29 Iterative coupling of zonal solutions [299].

The starting point for potential flow theory is an ideal (i.e., inviscid and irrotational flow field). This serves as the boundary condition for the calculation of the three-dimensional boundary layer ([181], [300]) and the location of the separation line. From there, a vortex layer to model the wake is emanated, which transports the vorticity content that has been created in the attached boundary layer. In the following iterations, to calculate the velocity field the displacement effect of the boundary layer and the induction of the wake can be used as boundary conditions.

14.1.5.2 Potential Flow Theory

The condition of irrotational flow means

$$\nabla \times v = 0 \qquad (14.41)$$

Introducing a scalar velocity potential Φ

$$v = \nabla \Phi \qquad (14.42)$$

which is irrotational per definition (since $\nabla \times (\nabla \Phi) = 0$) and inserting equation (14.56) in the continuity equation

$$\nabla \cdot v = 0 \qquad (14.43)$$

we get the basic equation of linearized potential flow theory, the Laplace equation for the velocity potential

$$\nabla \cdot (\nabla \Phi) = \nabla^2 \Phi = 0 \qquad (14.44)$$

The total potential Φ can be composed of a linear combination of elementary solutions like the free stream φ_∞ and a perturbation potential φ, or expressed directly as velocity

$$v(x) = v_\infty + \nabla \varphi(x) \qquad (14.45)$$

The perturbation potential φ in turn generally consists of a linear combination of sources, sinks, and doublets that are used as necessary to model the flow (cp. section 14.1.5.3 f.).

According to Green's theorem, the volume integral of the Laplace equation can be replaced by a surface integral over the boundary of the flow field, for instance, the surface of a solid body. Since therefore only the boundaries and not the volume need to be discretized, such panel methods fall under the category boundary element methods (BEM) and require much less computational effort than field methods.

The surface integral has three different contributions (Figure 14.30) so that the resulting induced velocity at a position x is composed of

$$v(x) = v_\infty + v_\sigma(x) + v_\mu(x) + v_\partial(x) \qquad (14.46)$$

The first term after the free stream velocity is the induction of a source/sink distribution $\sigma(S)$ on the surface

$$v_\sigma(x) = \frac{1}{4\pi} \iint_S \sigma \frac{r}{r^3} dS \qquad (14.47)$$

The last two terms result from a doublet distribution $\mu(S)$.

$$v_\mu(x) = \frac{1}{4\pi} \iint_S (n \times \nabla\mu) \times \frac{r}{r^3} dS \qquad (14.48)$$

$$v_\partial(x) = \frac{1}{4\pi} \int_{\partial S} \mu \frac{dl \times r}{r^3} \qquad (14.49)$$

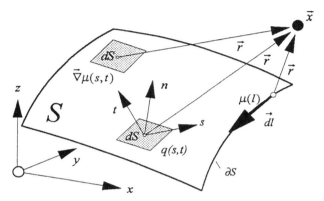

Figure 14.30 Contributions to the perturbation potential [299].

Since the integrand in equation (14.48) contains the gradient of the doublet distribution, regions with constant doublet strength μ do not induce a velocity v_μ. On the other hand, the velocity v_∂ only appears if the surface is not closed, since equation (14.49) is a line integral over the boundary ∂S of surface S. It can be shown that $v_\partial(x)$ is irrotational in any case. The doublet contributions v_∂ and v_μ are irrotational, with the exception of the singular surfaces and lines on which doublets are placed.

To demonstrate how vorticity can be modeled using doublets, we consider a volume element of a shear layer with thickness δ (Figure 14.31 left). The volumetric vorticity distribution $\mu(\xi)$ induces a velocity v_ω at position x:

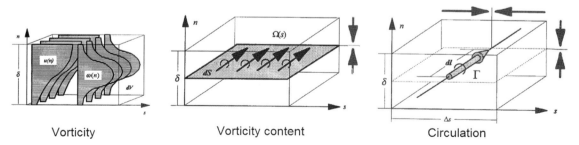

Figure 14.31 Representation of a shear layer as (a) volume, (b) surface, and (c) line element [299].

$$v_\omega(x) = \frac{1}{4\pi} \iiint_V \omega(\xi) \times \frac{r}{r^3} dV \text{ with } r = x - \xi \qquad (14.50)$$

By integrating in normal direction over the shear layer (Figure 14.31 center), we get a quantity called vorticity content

$$\Omega = \int_0^\delta \omega \, dn \tag{14.51}$$

which reduces the three-dimensional vorticity distribution to a vortex layer with vanishing thickness (equivalent to Re $\to \infty$) whose induction now only requires the evaluation of a surface integral

$$v_\Omega(x) = \frac{1}{4\pi} \iint_S \Omega \times \frac{r}{r^3} \, dS \tag{14.52}$$

A further piecewise integration of the vorticity content within the shear layer (Figure 14.31 right) creates the strongest simplification, a discrete line vortex with circulation

$$\Gamma = \int_{\Delta s} \Omega_t \, ds \tag{14.53}$$

whose induction is expressed by a line integral

$$v_\Gamma(x) = \frac{1}{4\pi} \int \Gamma(l) \frac{dl \times r}{r^3} . \tag{14.54}$$

If we now compare equations equations (14.52) and (14.48), the analogy between doublet distribution and vorticity (in local coordinates (s,t,n) according to Figure 14.31) is obvious.

$$\Omega = \begin{pmatrix} \Omega_s \\ \Omega_t \\ \Omega_n \end{pmatrix} = n \times \nabla \mu = \begin{pmatrix} 0 \\ 0 \\ 1 \end{pmatrix} \times \begin{pmatrix} \partial\mu/\partial s \\ \partial\mu/\partial t \\ 0 \end{pmatrix} = \begin{pmatrix} -\partial\mu/\partial t \\ \partial\mu/\partial s \\ 0 \end{pmatrix} \tag{14.55}$$

Equation (14.55) shows that the vorticity vector is lying in the plane of the shear layer ($\Omega_n = 0$). A panelwise linear doublet distribution is equivalent to a constant vorticity distribution (Figure 14.32 left). When choosing a constant doublet distribution, the vorticity vanishes inside the panel. Only on shared edges of neighboring panels does a line vortex remain active whose circulation is equivalent to the difference of the adjacent doublet strengths (Figure 14.32 right).

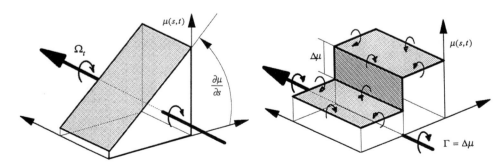

Figure 14.32 Equivalence between a) doublet and b) vorticity distribution [299].

14.1.5.3 Solid Body Simulation

Solving the Laplace equation means the determination of the distribution of sources/sinks and doublets on a panel model according to given boundary conditions. In Grün [299], a panelwise constant (unknown) σ and $\mu = 0$ is used on the vehicle surface. In turn, $\sigma = 0$ and a bilinear doublet distribution μ is set on the vortex layer modeling the wake.

Since the panel model of a vehicle (Figure 14.33) neither needs to be structured nor form a "watertight" mesh, arbitrary complex geometries can be represented with rather small effort.

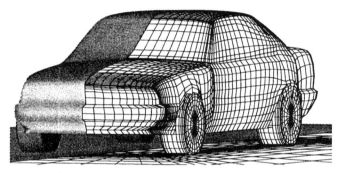

Figure 14.33 Panel model of a vehicle [299].

The boundary condition to calculate the strength of sources/sinks is the surface normal velocity component on the body's surface (the von Neumann condition)

$$n \cdot v = v_n \quad \begin{matrix} < 0 \text{ flow into the surface} \\ = 0 \text{ impermeable surface} \\ > 0 \text{ flow out of the surface} \end{matrix} \quad (14.56)$$

Inserting equation (14.46) leads to a linear system of equations to determine the unknown source strengths.

$$n \cdot v_\sigma = v_n - n \cdot \left(v_\infty + v_\mu + v_\partial \right) \quad (14.57)$$

The inductions of the wake model $v_\mu + v_\partial$ appear as known contributions on the right-hand side. After the solution, the source strengths on the panel model are known and the velocity can be evaluated at any point according to equation (14.57) to build up the wake model along streamlines. The doublet strength on the wake model cannot be solved simultaneously since its geometry is part of the solution. The interaction between wake and solid body is captured iteratively as depicted in Figure 14.29.

14.1.5.4 Wake Simulation

Modeling the wake is based on the following general idea of separation at high Reynolds numbers:

- At separation, the attached boundary layer leaves the surface and turns into a free shear layer, which transports the vorticity content that has been generated by the skin friction.

In this sense, separation always occurs if skin friction forms a boundary layer. Even the flow around an airfoil at incidence without recirculation is strictly speaking a separated flow. The flow off the trailing edge, in simulation enforced by the Kutta-condition, is a simple form of wake modeling. In the perfect inviscid case of pure potential flow, the fluid would circulate around the trailing edge to a stagnation point on the suction side and neither drag nor lift would be generated (d'Alambert's paradox).

Once separation has occurred, friction can again be neglected in the limiting case of high Reynolds numbers. The vorticity content of the free shear layer is maintained at the value that the attached boundary layer had at separation (Figure 14.34).

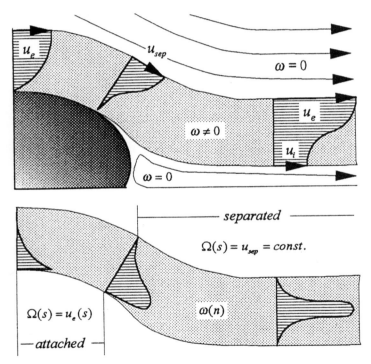

Figure 14.34 Vorticity transport at separation for high Reynolds numbers [299].

The decay of kinematic activity of the shear layer downstream due to a diffusive expansion can be modeled via a Rankine vortex with growing core radius.

If we apply the definition of vorticity content (equation (14.51)) in local coordinates (s, t, n) to the main flow profile $u(n)$ of a shear layer (Figure 14.35) we get

$$\Omega_t = \int_0^\delta \frac{\partial u}{\partial n} dn = \int_0^\delta du = u_e - u_i . \tag{14.58}$$

Numerical Methods

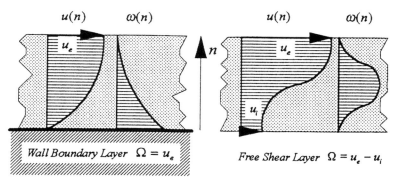

Figure 14.35 Vorticity content of a shear layer [299].

In particular, for a wall boundary layer with $u_i = 0$ the vorticity content depends only on the velocity u_e at the outer edge of the boundary layer. This implies that if separation cannot be avoided (as on bluff bodies like automobiles), it should take place at velocities as low as possible to minimize the vorticity in the wake. In addition, each variation of velocity along the separation line leads to a longitudinal vortex.

14.1.5.5 Static Pressure Calculation, Total Pressure Loss

To integrate the aerodynamic forces, the static pressure distribution on the vehicle's surface is required (apart from the skin friction from the boundary layer calculation). It can be derived from the velocity field using the Bernoulli equation (energy equation). It has to be taken into account that a total pressure loss has occurred in the wake. Therefore, the static pressure coefficient follows from

$$c_p = \frac{p - p_\infty}{q_\infty} = 1 - \left(\frac{v}{v_\infty}\right)^2 + \frac{p_{tot} - p_{t\infty}}{q_\infty} \tag{14.59}$$

Thus a total pressure loss $\Delta p_t = p_t - p_{t\infty} < 0$ reduces the static pressure at a given velocity compared to the lossless case. Typical orders of magnitude of the terms in equation (14.59) on the base and in the wake of a vehicle are

$$O(c_p) = 1 - O(0,1)^2 + O(-1) = O(0) \tag{14.60}$$

From equation (14.60) it is obvious that the total pressure loss of $O(0, 1)^2$ plays a much more important role for the correct prediction of the base pressure than the velocity magnitude of $O(-1)$. In addition, the direction of the velocity vector does not play any role at all. For an estimation of the total pressure loss we consider a shear layer of thickness δ (Figure 14.36).

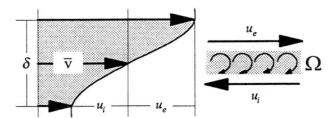

Figure 14.36 Velocity distribution in a shear layer [299].

Assuming a steady state, the static pressure on either side of the shear layer is the same while the velocity exhibits a discontinuity.

$$v_e = \bar{v} + u_e$$
$$v_i = \bar{v} - u_i \tag{14.61}$$

In the limiting case of a vanishing shear layer thickness for Re $\to \infty$ the induced velocities depend on the vorticity content of the shear layer

$$\delta \to 0 \;:\; u_e = u_i = \frac{1}{2}\Omega, \tag{14.62}$$

so that together with equation (14.61) the nondimensional total pressure loss is

$$\frac{\Delta p_{tot}}{q_\infty} = \frac{q_i - q_e}{q_\infty} = -\frac{\bar{v}}{v_\infty} \cdot \frac{\Omega}{v_\infty} \tag{14.63}$$

That means the total pressure loss vanishes if either the mean velocity $\bar{v}$ or the vorticity content Ω of the shear layer are 0. Immediately at the separation line $\Omega = v_{sep}$ and equation (14.64) simplifies to

$$\left(\frac{\Delta p_{tot}}{q_\infty}\right)_{sep} = -\left(\frac{v_{sep}}{v_\infty}\right)^2 \tag{14.64}$$

Hence, the total pressure loss at separation is equal to the dynamic pressure (at the outer edge of the boundary layer) at the separation line. This is equivalent to the pressure loss $\zeta = 1$ for flow from a container into an unconfined environment. In view of drag minimization, separation should be shifted to as low as possible velocities. For the limit of pure potential flow with a rear stagnation (or better detachment) point (i.e., $v_{sep} = 0$, the wake reduces to a streamline, and vorticity as well as total pressure loss vanish.

14.1.5.6 Results

Pressure distribution and wall streamlines from a potential flow simulation without wake modeling are depicted in Figure 14.37.

Numerical Methods

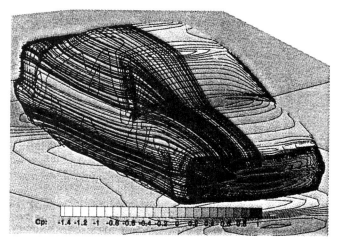

Figure 14.37 Pressure distribution and wall streamlines without wake simulation [299].

The corresponding pressure distribution along the centerline in comparison to measurements is shown in Figure 14.38. Except on the base and its immediate vicinity, there is a surprisingly good correlation to experiment considering the serious simplifications of this theory. Such results can definitely be used to calculate aerodynamic forces on parts known to be lying in attached flow.

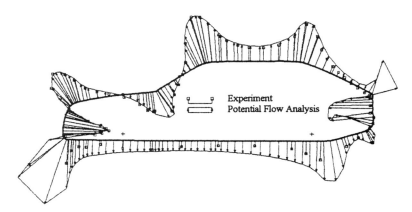

Figure 14.38 Comparison of measured and computed pressure distribution without wake modeling along the centerline [299].

The effect of wake simulation on the pressure distribution is demonstrated in Figure 14.39. While the potential flow pressure distribution shows a recompression up to positive pressure coefficients at the end of the rear window, the wake model predicts separation in the middle of the glass. This is combined with a break in the recompression and the following plateau on a slightly negative pressure coefficient level, typical for the occurrence of separation.

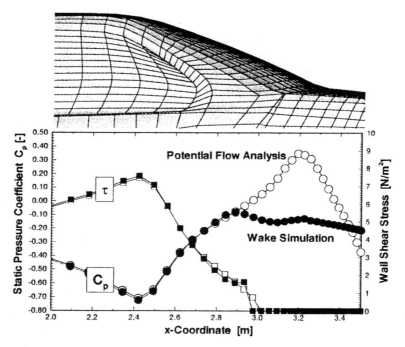

Figure 14.39 Effect of wake simulation on pressure and skin friction [299].

A comparison of measured and calculated total drag coefficients for different rear ends shows the absolute values to be lying clearly below experimental data; however, a trend prediction seems to be possible (Figure 14.40).

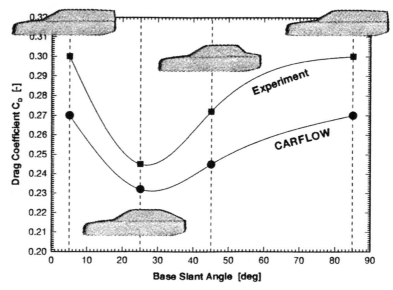

Figure 14.40 Comparison of measured and computed drag coefficient for different rear ends [299].

These results show that potential flow methods coupled with boundary and wake models can deliver quite useful output. However, on the highly detailed vehicle geometries investigated today (cp. Figure 14.4), their scope of application is by far exceeded.

14.1.6 One-Dimensional Methods for Cooling Module Design

As a result of the current tendency toward increasingly powerful vehicles, the focus of vehicle development is moving more toward thermal management in the early phases. Numerical methods are employed for simulation of thermal management attributes in the conceptual design of the engine, gear box, and air condition and heating systems. Common simplified model approaches represent the fluid and air circuits as one-dimensional networks ([74], [225]). The circuits are coupled via the heat exchanger in cooling components. Depending on the scope of applicability and boundary conditions, these models for the computational design of cooling systems can be extended as follows:

1. First, the dimensioning is completed using one-dimensional models. The dimensions of various heat exchangers like coolant, charge air, and oil cooler are fixed, and at the same time their geometrical relation to each other is defined. In this phase, the dynamic behavior of the whole system can be investigated.

2. In the next step, certain components of the cooling module are subdivided due to blockage or inhomogeneous approaching flow and integrated in the one-dimensional model. This increases the level of detail on crucial points without overly increasing the computational time.

3. The third step delivers the approaching flow field to components via the use of 3D CFD simulations. This allows the consideration of complex flow fields on the front end and the optimization of the approaching flow. The more homogeneous the velocity distribution on the inlet faces of heat exchangers, the smaller and the more cost-efficient they can be.

The calculation along stream tubes separately on the air side and coolant side is conducted using Bernoulli's equation with terms for losses and pressure sources (pumps), as described by equation (2.115). The fluid path may feature multiple branches and junctions. Air and coolant flows are coupled in the heat exchangers.

14.1.6.1 Air Side

The path of the air through a cooling system of a large car is depicted in a schematic graphic in Figure 14.41 [21]. The flow paths are sketched as lines along which the calculation takes place. The air enters the system at (1) near the stagnation point. On its way from (2) to (3) it passes the radiator grill; obstacles like a cross-traverse are included as perturbances. The next component is a condenser. Since there are two fans, the flow splits up in (3). Each of the two branches passes an air-conditioning condenser, a charge-air cooler, and finally the engine coolant radiator. After the junction (8), the air flows through the engine compartment (9) before it reaches ambient condition again in (1). Initially a homogeneous air velocity distribution is assumed on the face of heat exchangers. How this limitation can be resolved and how the transition to a three-dimensional method is carried out will be presented at the end of this chapter.

Pressure losses are basically generated on the path through the heat exchangers from (2) to (4) and (5), respectively, and during the flow around the engine to the air outlets from (8) to (1). The branching (3) and the junction (8) are assumed to happen without losses.

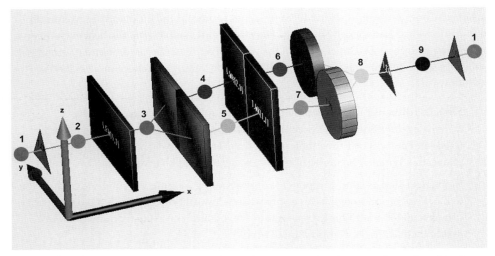

Figure 14.41 Model of an automobile cooling system for the one-dimensional tool "Kuli." Graphics: Magna Steyr.

The losses have to be known in all elements; usually they are determined by experiment. For the heat exchangers they are either described as pressure drop versus volume flow rate $\Delta p = f(\dot{V})$ or in nondimensional form $\zeta = \zeta(Re)$ as depicted in Figure 14.42. Losses occurring in the radiator grill are not insignificant; they have to be measured or taken from literature. For new vehicles, the losses generated during the flow around engine and gear box are yet unknown and have to be estimated.

The stagnation pressure of the free stream is available to overcome all combined losses, which yields the "face velocities" on the heat exchangers according to Figure 14.42. In case the resulting volume flow rate is insufficient to comply with the requirements—for instance, at slow uphill driving with trailer—a fan has to be added, whose characteristics also need to be known from measurements. An example is shown in Figure 14.43.

Numerical Methods

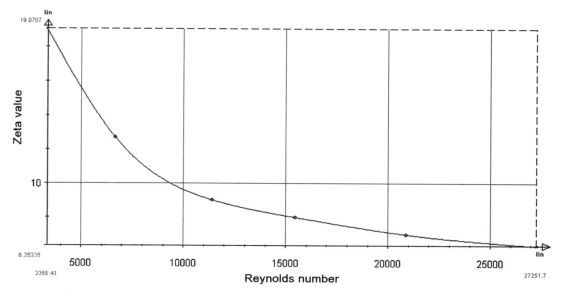

Figure 14.42 Pressure loss coefficient as a function of re-number (Magna Steyr).

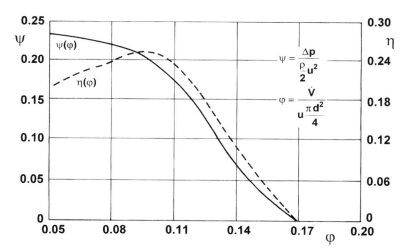

Figure 14.43 Characteristic of an axial fan, pressure coefficient $\psi\,(\varphi)$, and efficiency $\eta(\varphi)$ as a function of the discharge coefficient φ (Magna Steyr).

In order to calculate the volume flow rates, the Bernoulli equations equation (2.115) are established from node to node. For the current example, we get four equations. Two more equations result from the continuity condition:

$$\dot{V}_{1-3} = \dot{V}_{1-8} \text{ and } \dot{V}_{3-8(\text{left})} + \dot{V}_{3-8(\text{right})} = \dot{V}_{1-3} \tag{14.65}$$

Once the volume flow rates of air and coolant (the latter will be discussed in the next chapter) are known, the Nusselt number can be determined from the radiator's performance map (Figure 14.44) and the actual amount of conveyed heat can be calculated as explained in detail in chapter 7.2. The Reynolds number on the coolant side results from the revolution speed of the water pump and hence from vehicle speed and gear. Electrically driven water pumps can—within limits—be controlled independently.

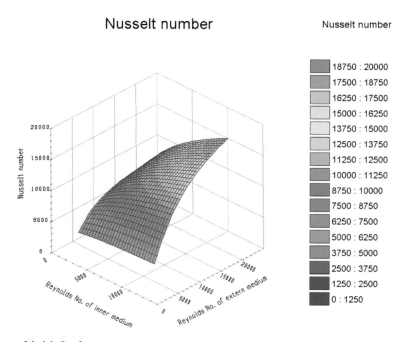

Figure 14.44 Performance map Nu(ReFlüssig, ReLuft) of a radiator (Magna Steyr).

Calculation and design are carried out iteratively until the requirements are met. All components of the cooling module are selected by experience. The calculation shows if their dimensions are sufficient to fulfill the demands. If not, they are altered until the design objective, for instance a certain water temperature, is achieved.

This iteration process has to be run for all critical load cases, typically top speed and driving up-hill. In addition, some "off-design" situations like a sudden stop after a fast drive ("thermal soak") have to be considered. Initially, these operation points are investigated statically but later also in transient mode (for instance "thermal soak"). Figure 14.45. shows an example for such a time-dependent study.

The diagram compares the coolant entry temperatures from calculation and measurement over time for an uphill drive with trailer on the Grossglockner pass in the Alps. For such a study, it is essential to account for the varying environmental conditions like slope, temperature, and ambient pressure. In this example, a temperature drop of 6.5 K per 1000 m altitude change was assumed.

Numerical Methods

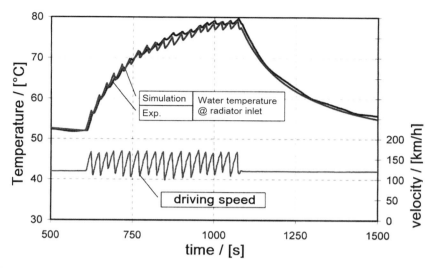

Figure 14.45 Coolant entry temperature as a function of vehicle speed. Comparison of calculation and measurement (Magna Steyr).

As mentioned, a one-dimensional method is based on a homogeneous velocity distribution on the face of heat exchangers. In reality, this is hardly ever the case. In fact, the real flow is quite inhomogeneous, as visible in Figure 4.31. Therefore, the matrix of the radiator is subdivided into elements, each representing a heat exchanger in itself. The partial heat amounts are then summed up [229].

14.1.6.2 Coolant Side

The coolant circuit of an automotive cooling system is a network of tubes. The calculation of such networks has been readily available for quite some time. Computational design of these elements has been further facilitated and accelerated through the development of user-friendly software interfaces. Loss coefficients of typical components like manifolds, T-junctions, or valves are documented in comprehensive tables ([779]), which are accessible by computer software. For their application to automotive cooling systems two extensions were necessary:

- Inclusion of vehicle specific components, for instance thermostats
- Consideration of heat transfer and transport

A typical automotive cooling system is sketched in Figure 14.46; it consists of three components:

- Engine cooling
- Oil cooling
- Heating

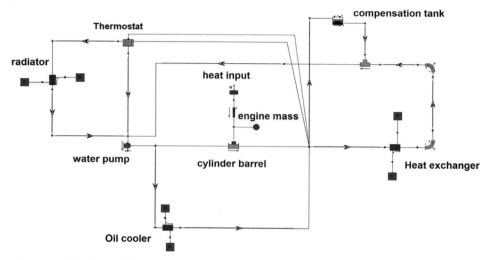

Figure 14.46 Model of the coolant circuit in a vehicle for the one-dimensional calculation with the tool Flowmaster® (Flowmaster GmbH).

For each of these circuits the Bernoulli equation (equation (2.115)) can be expressed [597]. The individual circuits are connected by the fact that at each node, branching, or junction they have the same total pressure, and the sum of all in- and outgoing volume flow rates must remain constant. It has to be considered that the properties of some components may be variable, partially including a hysteresis effect. A typical example is a thermostat, whose characteristic curve is shown in Figure 14.47.

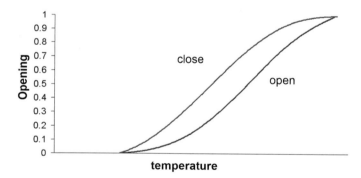

Figure 14.47 Characteristic of a thermostat as an example for a component with changing flow properties (Flowmaster GmbH).

Usually the heat rejection of the engine into this network stems from measurements or is transferred from existing engines to new ones. There are approaches to determine the heat rejection of the engine computationally from combustion characteristics and geometrical dimensions [38]. The scheme to do that is indicated in Figure 14.48.

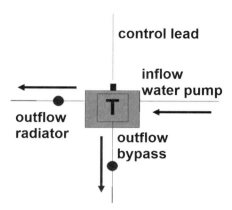

Figure 14.48 Modeling the engine as a heat source (Flowmaster GmbH).

The temperatures in the coolant circuit are calculated either in static or—in the case of heating—transient mode. An example is shown in Figure 14.49. After a cold start at −20 °C, the coolant is heated at a constant vehicle speed of 50 km/h. The correlation with experiment is very good.

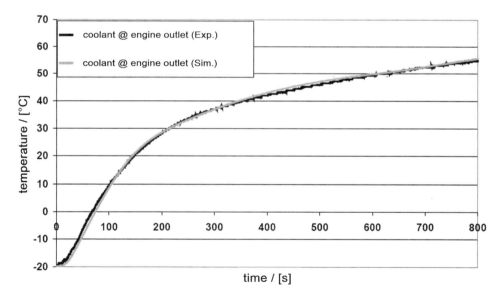

Figure 14.49 Development of the coolant entry temperature after a cold-start at −20 °C and a speed of 50 km/h. Comparison of calculation (Flowmaster®) and measurement (Flowmaster GmbH).

14.1.7 Rotating Geometries (Wheels, Fans)

Just as it is necessary to properly represent the wheel rotation in a wind tunnel to reproduce real road conditions, it is also crucial in a simulation. Another rotating part in a vehicle is the cooling fan used to ensure the required cooling airflow rate when needed.

In order to model rotating geometries, three approaches are possible, discussed in increasing order of complexity.

14.1.7.1 "Rotating Wall"

The simplest way is to prescribe the circumferential velocity resulting from the rotation as a surface boundary condition. However, this is only appropriate for perfectly axisymmetric geometries. On a wheel, this method can be accurately applied to model the (slick) tire and the rim but not to the spokes. This approach is not at all applicable for rotating fans.

14.1.7.2 Moving Reference Frame (MRF)

In general, rotation involves centrifugal and Coriolis forces. These can be accounted for if an axisymmetric fluid region, enclosing the rotating part, is calculated in a rotating frame of reference (MRF = moving reference frame) without the region actually being rotated. The coupling of stationary and rotating fluid takes places on the interface between the two regions by appropriate transformations. This method has been applied to the optimization of rim design, for instance by [57] and [58]. The drawback is that due to the stationary geometry, the result depends on the chosen position for the simulation, in particular with close proximity to non-rotating parts. For instance the rotor/stator interaction of a fan will not be captured correctly, and downstream of the blades their individual wakes show up, while in reality they are "smeared" around the circumference. A workaround in this case is a subsequent averaging in circumferential direction (Figure 14.50 from [793]).

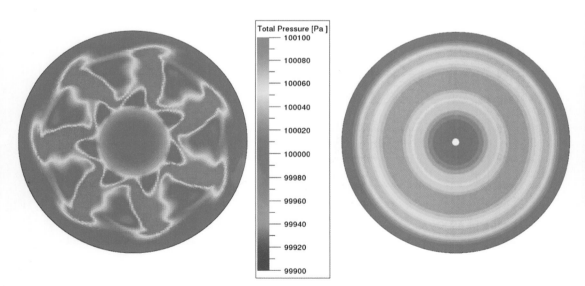

Figure 14.50 Total pressure behind a fan calculated with MRF [793]; (left) without, (right) with circumferential averaging.

14.1.7.3 Sliding Mesh

The most complete but also most elaborate option is to actually rotate the geometry. Of course this always requires an unsteady simulation with a corresponding higher computational effort. Depending on the rotational speed, it can be necessary to choose a time step, which is smaller than that without rotating geometries. In order to avoid rediscretization in each time step, most CFD codes use a sliding mesh. Similar to MRF, the rotating geometry is embedded in an axisymmetric fluid region, which this time actually rotates so that the discretization inside the sliding mesh has to be executed only once. The interface to the grid at rest is handled differently in the various codes. Actually, such a solution can only be visualized in unsteady mode as an animation. Averaging the results later does not make sense because the flow field and geometry no longer correspond to one another in a time-averaged sense. For example, on a rotating wheel streamlines would apparently run through spokes. Peric and Schreck [631] have published a method for generally overlapping grids to allow for arbitrary motions, for example, changing lanes during a passing maneuver.

14.1.8 Porous Media (Heat Exchanger)

Heat exchangers like water-, oil- and charge-air cooler or air-conditioning condenser generate a significant contribution to the total drag (cp. Figure 4.22). Further, the aerodynamics simulation approach should also be able to predict the flow rates and resulting heat rejections in these components. However, within a complete car simulation it is impossible to resolve the matrix of a heat exchanger in detail. Therefore, they are modeled as porous media (Figure 14.51).

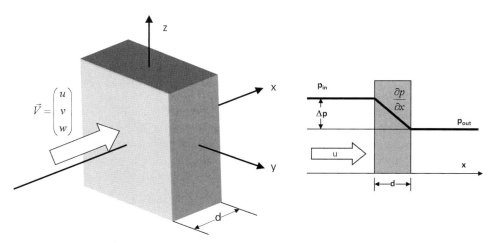

Figure 14.51 Schematic of modeling a porous medium.

The gradient of the pressure drop in the local directions ξ_i, $i = x, y, z$ is described by the Darcy-Forchheimer law.

$$-\left(\frac{dp}{d\xi_i}\right) = \rho C_V \cdot u_i + \rho C_I \cdot u_i^2 \quad . \tag{14.66}$$

The first term depending linearly on the velocity represents the viscous effect, and the second, quadratic term represents the inertial effects. Only for very small velocities, for instance a flow through sand or gravel, the viscous term alone suffices. This corresponds to the original Darcy law, which has no inertial term. On the other hand, the hydraulic loss coefficient $\zeta = \Delta p/(\rho/2 \cdot u^2)$ only considers the quadratic term.

To model the directivity of a heat exchanger, the coefficients for the local y- and z-direction are set to ∞ (or extremely high) to suppress any crossflow.

The coefficient for the local x-direction has to be determined on a test bench. In doing so, one has to be careful not to include any losses generated by the test bench itself in the measurement setup. Ideally the heat exchanger matrix fills the entire cross section, and there is no variation of the cross-sectional area up- and downstream of the heat exchanger. With such an experimental setup, the pressure drop as a function of volume flow rate (or velocity) can be measured reliably by two annular pressure tubes before and after the matrix.

Symbols in Figure 14.52 represent measured data, and the bold line represents a best-fit quadratic polynomial (equivalent to the Darcy-Forchheimer law). Additionally, three curves $\Delta p = \zeta \cdot (\rho/2 \cdot u^2)$ with different ζ-values are plotted. It is obvious that neither the pure linear (Darcy) nor the pure quadratic (hydraulic loss coefficient) approach can represent the real pressure loss correctly.

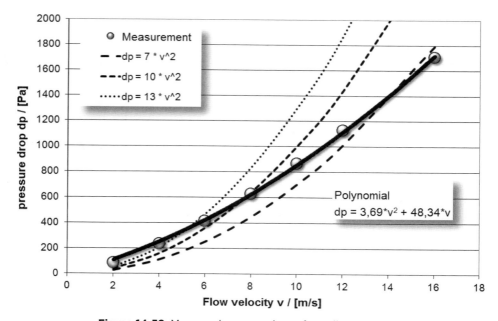

Figure 14.52 Measured pressure loss of a radiator matrix.

To calculate the pressure gradient, the pressure loss has to be divided by the thickness of the measured matrix. When using the coefficients for a simulation, it should be verified that the thickness of the porous media model in CFD is the same as the measured one. If not, the coefficients must be scaled with the thickness ratio to provoke the same pressure loss as in the experiment.

Porous media can also be used to model other geometries that cannot be resolved in detail, for example, dense grills on inlets and outlets or meshes for wind stoppers on open convertibles and deflectors on sunroofs.

14.1.9 The Solution Process

The process of a CFD-solution consists of the steps discretization/mesh-generation, solving, and post-processing, which will be described briefly in this chapter. For Lattice-Boltzmann methods, discretization and mesh generation were presented in section 14.1.3.2.

14.1.9.1 Discretization

As the first step toward the solution, the system of partial differential equations (PDEs) used by the particular CFD method is discretized. The objective is to convert the PDEs into a system of algebraic equations for discrete points in space and time. For this purpose, the fluid volume is represented by a numerical grid, and the continuous equations of motion are discretized at the nodes or cells of this grid. The most important discretization methods are these:

- finite difference methods (FDM)
- finite element methods (FEM)
- finite volume methods (FVM)

According to Ferziger and Peric [244], the concept of finite differences originated with Leonhard Euler. Differential operators in the basic equations are approximated at the grid nodes by difference quotients, for instance, the ones in Figure 14.53. A sufficient condition for arranging the nodes is an orthogonal grid with equidistant step size $\Delta \xi_i$ in each direction ξ_i. If the geometry does not allow complying with this condition, an oblique grid can be used and transformed to a Cartesian system in the computational space. However, this imposes high demands on programming and renders FDM as quite inflexible. Since orthogonal grids are always structured, each node can be identified by an index triple (i, j, k). Replacing the derivatives of first and second order in the PDEs by difference quotients leads to a large system of algebraic equations where the number of equations is equal to the number of unknowns.

Chapter 14

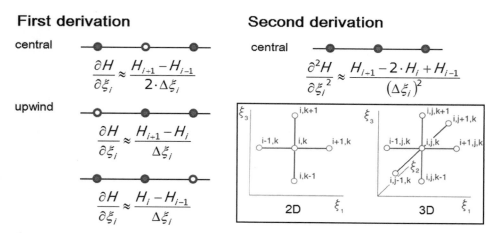

Figure 14.53 Indexing of the computational space in finite difference methods and various methods for differencing [616].

Discretization via FEM has recently gained importance. As the name indicates, the fluid volume is subdivided into elements, mostly triangles in 2D or tetrahedrons in 3D. In parallel to the global definition of node coordinates, they are defined locally by a single normalized (with values between 0 and 1) coordinate within the element. Flow quantities are represented in each element by shape functions in local coordinates. Transformations from local to global coordinates create the description of the entire flow field. When selecting shape functions, it has to be considered that they can be easily integrated when inserted in the basic equations. Like in FDM and FVM, this leads to a system of equations that can be solved stepwise in time. FEM is very flexible, since it allows for unstructured grids and easy refinement around special geometry features. However, its memory consumption per node is higher than in FDM and FVM [244].

Finite volume methods (FVM) discretize the flow field into arbitrary hexahedrons and integrate the basic equations over the entire simulation volume. With the Gauss theorem, the volume integral of each cell is transformed into six surface integrals over its faces, which linearizes the PDEs. The surface integrals use the cell center values (i.e., each quantity is assumed to be constant within the cell). Again, this leads to an algebraic system of equations that has to be solved in each time step. The advantage of FVM is the flexibility and robustness when applied to complex geometries. Further, it is not necessary to define an additional computational space that facilitates the implementation of FVM in computer programs. For these reasons, FVM is frequently used (cp. [397], [749]).

From the variety of approaches to discretize time, the explicit and implicit Euler methods will be presented briefly here. The time derivative is approximated by a difference quotient. In the explicit version, the right-hand side $f(H)$ uses only known quantities at time step t while the implicit formulation also involves data at the next time step $t + 1$. This reads:

Numerical Methods

$$\frac{\partial H}{\partial t} \approx \frac{H^{(t+1)} - H^{(t)}}{\Delta t} = f\left[H^{(t)}\right] \text{ or } \frac{\partial H}{\partial t} \approx \frac{H^{(t+1)} - H^{(t)}}{\Delta t} = f\left[H^{(t+1)}\right] \qquad (14.67)$$

With the explicit version, the new variable values at time step $t + 1$ can be computed directly, (i.e., explicitly) from values of the past. However, as with all explicit methods, the time step Δt cannot be chosen arbitrarily. For a stable integration in time the Courant number

$$Co = v_\infty \frac{\Delta t}{\Delta x} \qquad (14.68)$$

must be smaller than 1 [616]. Simply spoken, this CFL condition means that a flow quantity must not be transported farther than one cell per time step. For implicit methods, this condition does not hold. The choice of time step size has no influence on stability but of course affects accuracy. Implicit methods require more memory and computational work than explicit procedures, but normally this does not invalidate the advantage of unconditional stability.

14.1.9.2 Numerical Grids

The basis for each CFD simulation is the geometrical discretization of the computational domain. Besides the turbulence modeling and difference scheme in space and time, the grid also affects the quality of the solution. Therefore, some basic issues of grid generation will be discussed next. When initially generating the grid for the simulation, the expected flow field and the particular CFD tool have to be considered.

CFD codes can be distinguished by structured and unstructured grids [616] with consequences regarding the topological arrangement of grid cells. Structured codes require structured grids with regular arrays of grid nodes. That means there are directions in which the number of nodes are the same. In a two-dimensional case, for instance, nodes can be uniquely addressed by two indices i and j with an increment of 1 between neighboring nodes. The necessity to comply with this rule imposes limitations on the flexibility.

For unstructured codes there is no such limitation. The advantage is the freedom to use arbitrary cell types in any combination, which enables a flexible meshing of complex geometries. For very intricate geometries, the use of structured grids is mostly impossible. Also with regard to automating the very time-consuming meshing process, unstructured grids have significant advantages. The drawback of unstructured grids is higher memory consumption, because the connectivities of cell neighbors have to be stored in elaborate data structures.

Regardless of the mesh type, there are various cell types. The most important ones are triangles and quadrilaterals in two and tetrahedrons and hexahedrons in three dimensions. The choice of cell type can affect the convergence rate, the effort of mesh generation, and the capability of automation.

The quality of a computational grid can be judged by the three criteria: skewness, aspect ratio, and expansion rate. Skewness is a measure of the distortion from orthogonality of cells. The angles of a control volume should be as close as possible to a right angle because the convective and diffusive fluxes in FVM are calculated as scalar products of the flux vector with the surface normal. For this reason, it also makes sense to align the grid with the local flow direction [244].

Aspect ratio is the ratio between edge lengths of a cell. This measure influences the condition of the discretized system of equations, which in turn affects the efficiency of the solution algorithm. According to [244], the aspect ratio should not be below 0.1 and not above 10.

The expansion rate indicates the volume ratio of neighboring cells, which has an impact on the truncation error of the discretization algorithm and hence on the solution quality. Ferziger and Peric [244] recommend an expansion ratio between 0.1 and 10.

Besides the aforementioned criteria to assess computational grids, further criteria can be derived from the CFD method, in particular from the turbulence modeling. The distance of cells nearest to a solid wall, expressed as y^+, has to comply with the modeling of turbulent boundary layers. If the logarithmic wall function is employed (cp. section 14.1.4.1.3), it has to be ensured that for the first cell above the wall $y^+ > (5–30)$ (cf. Figure 14.17). However, y^+ cannot be determined a priori because its definition uses the skin friction. Thus finding the appropriate resolution is an iterative process. With the so-called low-Reynolds wall models, the boundary layer is calculated down to the wall and therefore requires a much higher resolution near the wall. In this case, y^+ should be around 1.

In the meantime, automatic meshing tools have found their way into vehicle development. For a long time it was only possible to automatically generate simply structured grids. The Lattice-Boltzmann code PowerFLOW from Exa Corp. has been the market leader in automobile aerodynamics for a long period due to its usage of block-structured Cartesian cube mesh elements (the lattice), which enabled very short turnaround times. Automatic meshing tools also exist for complex unstructured grids, for instance "Snappy Hex Mesh" from OpenCFD.

Graphical user interfaces (GUIs) support the user in designing the fluid mesh for a CFD simulation. He or she can define regions in the simulation volume and assign an index to set the level of resolution. The highest index designates the smallest cell size; toward lower indices they grow by a factor of 2. In PowerFLOW, these domains are called VR regions (VR = variable resolution). Figure 14.54 shows a virtual wind tunnel for the simulation of a complete vehicle with inlet, outlet, and the cubical outer VR regions at levels 1 to 5. Higher resolution regions are used around the chin, the rear end of the roof, and the rear fender at level 9. For a correct representation of the flow approaching the brake discs, the NACA inlet and the subsequent duct are the highest resolved regions at level 10. According to these user definitions, the lattice is then generated automatically. With a smallest cell size of 0.625 mm, this example resulted in approximately 90 million voxels.

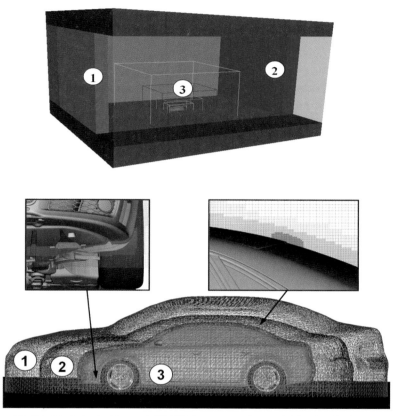

Figure 14.54 Digital wind tunnel (DWT) for a complete vehicle simulation: top: inlet (1), outlet (2), and the outer resolution regions (3, VR1 to VR5), floor (moving wall b.c.), frictionless side walls, and roof; center and bottom: inner VR-regions (1–3 = VR6–VR8) and local on the roof rear end (VR9 and front (VR10) [748].

14.1.9.3 Solution Algorithms

In Navier-Stokes-based CFD methods, discretization produces a system of n linear algebraic equations for n grid cells. There is a multitude of direct and iterative mathematical algorithms to solve such a system. However, with direct methods the effort increases rapidly with the number of cells, for instance by $O(n^3)$ for the Gaussian elimination. Direct methods are only feasible with sparse matrices as they occur, for example, in finite-difference methods. Finite-volume methods with dense matrices preferably employ iterative methods, for example, the Jacobi or Gauss-Seidel algorithms, optionally including successive over-relaxation (SOR).

The governing system of equations describing a flow problem is missing an independent equation for pressure. Its gradient appears in the momentum equation but—for incompressible flow—not in the continuity equation; this is called a coupled system of equations. In compressible flows the variable density can be determined from the continuity equation and an additional relationship, for instance the equation of state (equation (14.2)) is used to compute the pressure. This is a decoupled equations system.

For incompressible flows, a pressure correction algorithm is used to decouple the system of equations during the solution process. The basic idea is to first determine preliminary velocity components from the momentum equations. In the next step, these are corrected together with the pressure to comply with the continuity equation. This iteration is continued until both the momentum and continuity equations are fulfilled. As examples, the SIMPLE (Semi-Implicit Method for Pressure Linked Equations) and PISO (Pressure-Implicit Splitting of Operators) will be discussed. The following steps are executed:

1. Solve the momentum equations with an estimated pressure field pstart.
2. Insert the computed velocity field in a pressure correction equation and determine the pressure correction Δp_{corr} [826].
3. Compute the corrected pressure p and velocity vector u.
4. Solve all other transport equations (e.g., turbulence model).
5. Iterate this process until the pressure correction is almost 0.
6. Proceed to the next time step and set the solution as pstart.

PISO is an enhancement of the SIMPLE algorithm. A weakness of SIMPLE is that the pressure correction is only solved once, and afterward the momentum equations are not yet fulfilled. In order to increase the efficiency of the iteration, PISO executes two corrections consecutively (Figure 14.55).

The stability of the numerical tool plays an important role in CFD. So-called under-relaxation is introduced to increase the stability. Iterative algorithms can become unstable if the change of the particular quantity from one iteration to the next is not properly bounded. Limiting the change between two iterations is called under-relaxation

$$p = p_{start} + \alpha_p \cdot \Delta p_{corr} \quad \text{and} \quad u_i = u_{i,start} + (1 - \alpha_{u_i}) \cdot \Delta u_{i,corr} \tag{14.69}$$

if the relaxation parameter α is between 0 and 1. A well-considered choice of α is crucial for the efficiency of this process. Values that are too large can cause divergence, whereas values that are too small lead to unnecessarily long iterations to convergence. Therefore, methods are important, which accelerate the conventional algorithms for the solution of linear systems of equation. One option is to apply so-called multigrid methods.

Numerical Methods

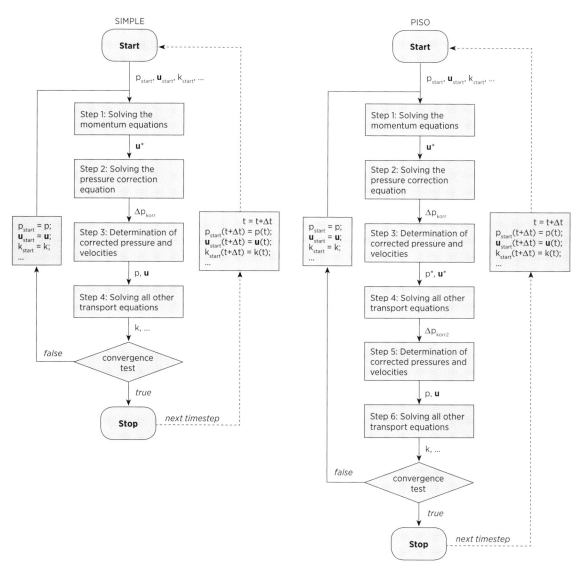

Figure 14.55 Flow chart of the solution algorithms SIMPLE and PISO.

For a multigrid method, one or more coarser versions are derived from the fine computational mesh. This is based on the idea that an iterative solution method more efficiently eliminates error components in the approximated solution whose wave length corresponds to the mesh size. For very long-wave errors relative to the mesh size, the compensation of solution errors requires many iterations, and a multigrid approach can be applied to more quickly reduce these long-wave errors. By using the various mesh sizes on different grid levels, errors are eliminated over the entire frequency spectrum faster. There exist numerous options for how many grid levels are used and how they are connected.

14.1.9.4 Post-Processing

In contrast to most wind tunnel tests, after a simulation, detailed information about velocity, pressure, temperature, and derived quantities is available in the entire flow field and on the vehicle's surface. This allows for a deep insight and understanding of flow phenomena.

The following examples for the evaluation of PowerFLOW® results with PowerVIZ® are, depending on the capabilities of their post-processing tools, of course also feasible with other CFD methods or with independent visualizing tools like EnSight® (CEI), FieldView® (Intelligent Light), or ParaView® (Kitware Inc.).

14.1.9.4.1 Analysis of Surface Results

Aerodynamic forces and moments are generated solely by normal and tangential stresses, that is, pressure and skin friction exerted from the fluid on the vehicle's surface. Their distribution is displayed as nondimensional coefficients in Figure 14.56.

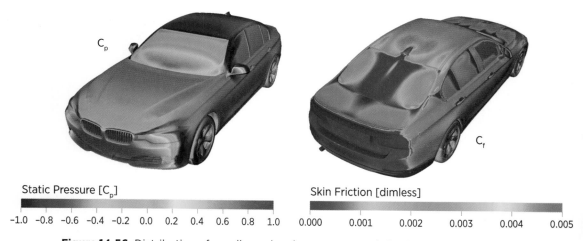

Figure 14.56 Distribution of nondimensional pressure c_p and skin friction c_f coefficients.

The pressure distribution can be used to find appropriate locations for inlets and outlets in regions of high and low pressure, respectively. To evaluate where and to which extent this contributes to the formation of forces, it is necessary to consider the local surface orientation (which is hardly possible just by visual inspection). For this purpose, the pressure coefficient is multiplied by the components of the surface normal

$$c_x = -c_{p,\text{stat}} \cdot n_x \; ; \; c_y = -c_{p,\text{stat}} \cdot n_y \; ; \; c_z = -c_{p,\text{stat}} \cdot n_z \tag{14.70}$$

so that, for instance, drag and lift distribution can be visualized as in Figure 14.57. Their integration over the surface or parts thereof finally gives the aerodynamic forces and moments.

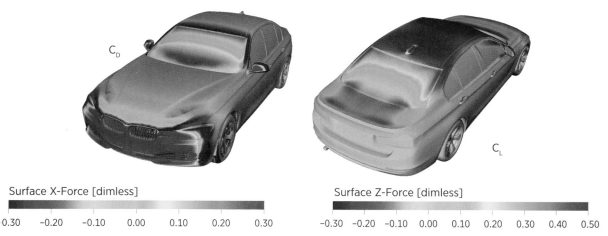

Figure 14.57 Distribution of drag c_D and lift c_L coefficients.

Further insight into the generation of forces is gained by the analysis shown in Figure 14.58 using drag, for example. The vehicle is cut into slices perpendicular to its longitudinal axis and the contribution of each slice to drag is plotted either as a distribution (bars) or as an integration (line) by accumulating in x-direction. In the same way, we can of course analyze lift or plot the distribution along the vertical axis.

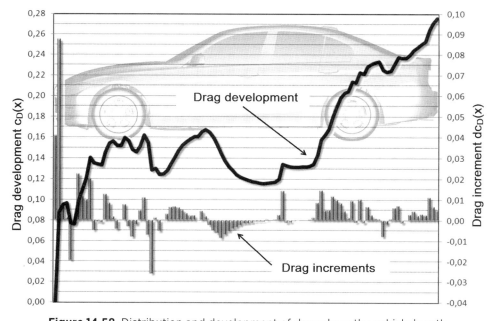

Figure 14.58 Distribution and development of drag along the vehicle length.

This analysis is useful especially to identify where, for instance, the drag difference between two variants comes from. For that purpose the difference of the drag distribution is plotted. Quite often the surprising result is that the change in drag does not occur at the location of a geometry modification itself but via interference on the rear end by an altered base pressure.

While the force components already give the direction of the aerodynamic force, the location of the line of action in space results from a minimal condition for the moment (in general there will remain a non-zero moment around the line of action itself). As Figure 14.59 (left) indicates, the question for a "force application point" makes no sense on a three-dimensional body. There may be any number of intersections of the line of action with the body. It is even possible to have no intersection at all (e.g., consider the gravity force on a horseshoe).

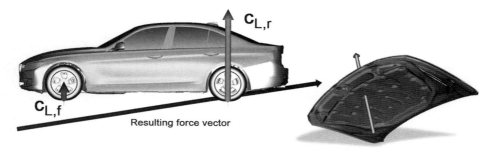

Figure 14.59 Resulting aerodynamic force vector: (left) total vehicle; (right) engine hood.

When aerodynamic loads on parts like glasses, sunroofs, or hoods are sought (Figure 14.59, right) the resulting force's vector and its intersection with the part (in this case the name "force application point" may be justified) can be used to calculate support forces, for instance on the hinges and locks of a hood.

In order to compute the three-dimensional deformation of a part under aerodynamic load, the pressure distribution can be mapped from the CFD model to the elements of a FEM model for structural analysis. Doing this once is called "loose coupling." If the flow field is recalculated for the deformed geometry (optionally iterating this interaction), we call that "tight coupling" or FSI (fluid-structure interaction).

By applying spectral analysis (e.g., FFT) on the time-dependent aerodynamic forces on these parts, it can be checked whether the aerodynamic excitation produces peaks at certain frequencies. Should this be the case and these frequencies unfortunately coincide with eigenfrequencies of the part, the probability of damage under long-time loading is high, even if the magnitude of the time-averaged force is harmless.

Wall streamlines, created on the basis of skin friction vectors, produce a vivid impression of the flow topology on the surface (Figure 14.60, left). This is equivalent to the oil flow technique in the wind tunnel. Regions of reversed flow can be identified by plotting an image of the velocity component in the longitudinal direction (Figure 14.60, right). Of course the actual "surface velocity" is 0 (due to the no-slip condition); each time a surface velocity is mentioned here, this means the value in the first cell above the wall.

Figure 14.60 Wall streamlines (left); Reverse flow regions (right).

The total pressure loss that is generated in the boundary layer (Figure 14.61) can be used as a very sensitive indicator to identify regions suitable for drag optimization.

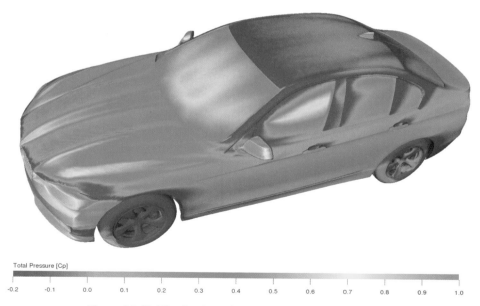

Figure 14.61 Distribution of total pressure coefficient $c_{p,tot}$.

Closely coupled with aerodynamics is the thermal operating safety of components in the engine bay or on the underbody as well as brake cooling. Even from an isothermal simulation an (extended) Reynolds analogy allows for conclusions about the heat transfer. As an example, Figure 14.62 shows the distribution of heat transfer coefficients on the underbody. Similar to structural loads, these results may be mapped onto models for further analysis of thermal behavior (e.g., the cooling of brake discs in the course of a driving profile).

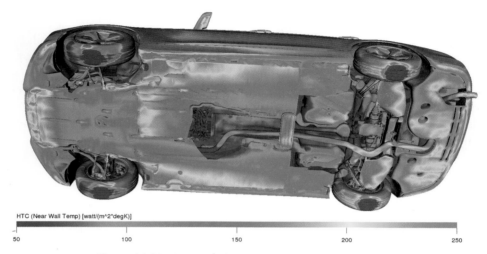

Figure 14.62 Heat transfer coefficient on the underbody.

For a simulation including thermal effects, either the wall temperature or the heat flux has to be prescribed as boundary condition and the respective other quantity will be a result of the simulation.

As already discussed in section 14.1.3.2.3, the unsteady pressures can be evaluated using spectral analysis, and the sound pressure level for selected frequencies or frequency bands may be displayed on the surface (Figure 14.63). This analysis of noise sources can be used as input for tools that compute the transmission of this excitation through the glass panels to the ears of occupants.

Figure 14.63 Sound pressure level at 500 Hz on the side windows.

14.1.9.4.2 Analysis of Fluid Results

In everyday wind tunnel testing, the flow field visualization is usually limited to the use of a smoke probe. More-sophisticated techniques like PIV (particle image velocimetry), hot-wire anemometry, or Prandtl probes are very time-consuming and provide only locally confined information. In contrast, a simulation delivers flow quantities in the entire flow field and, depending on the tool, even provides time-dependent data.

The streamlines in Figure 14.64 (top) are colored by the local velocity magnitude. They have been initiated on a line upstream of the vehicle. Those in Figure 14.64 (bottom) were started on a grid exactly covering the water cooler and have been integrated down- and upstream. Providing information not available in experiments, this enables the visualization of only those streamlines that are used for cooling, the path where this air comes from, and the downstream propagation of the heated air.

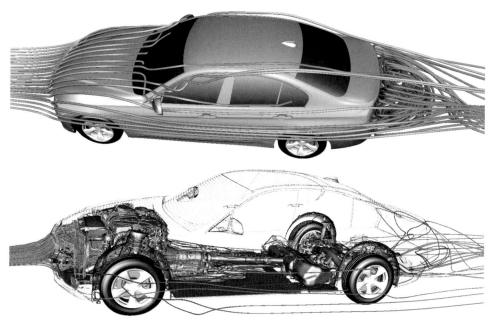

Figure 14.64 3D Streamlines: (top) external flow; (bottom) cooling air.

Plane sections can be placed in the simulation volume with arbitrary orientation and dimension. Figure 14.65 depicts the velocity distribution in the center plane. The trajectories shown in this plane are not streamlines in the strict sense because only in-plane velocity components have been used generating them. They can help to provide insight, but it has to be considered that in general there may be flow through the selected plane. As a consequence, such trajectories may seem to penetrate solid surfaces, which is of course not the case with streamlines.

Figure 14.65 Velocity distribution in the center plane.

To visualize the wake topology, Figure 14.66 displays cross flow velocity vectors, colored by the local total pressure. It is obvious that the highest losses occur in the center of the pairwise longitudinal vortices.

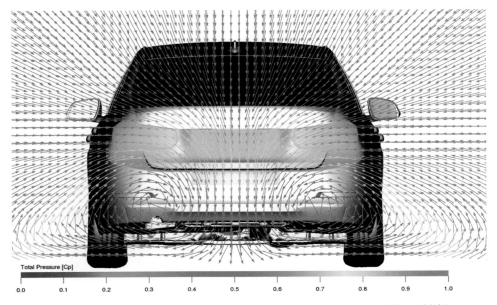

Figure 14.66 Cross flow velocity vectors in the wake (3 m behind the vehicle).

The generation and propagation of vortices along the vehicle is revealed by the total pressure maps in multiple planes perpendicular to the longitudinal axis (Figure 14.67). For improved visibility, the color images are clipped at the outer edge when the total pressure reaches the free stream value (i.e., only high-loss regions are displayed). In regions of attached flow, this technique provides visualization of the boundary layer development.

Numerical Methods

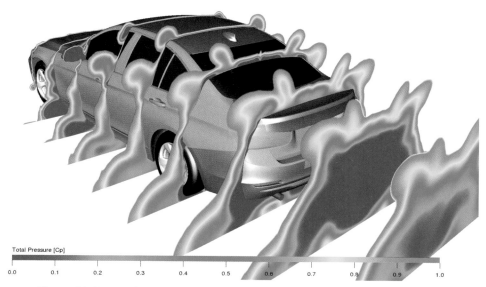

Figure 14.67 Total pressure distribution in planes perpendicular to the x-axis.

A volumetric image of high-loss regions in the fluid is obtained using isosurfaces of total pressure. In Figure 14.68 (left), the value of the isosurface is just below the free stream value so that we see the envelope of attached flow and wake. The value for the isosurface in Figure 14.68 (right) is $c_{p,tot} = 0$ (i.e., the total pressure loss on this surface is equal to the free stream dynamic pressure).

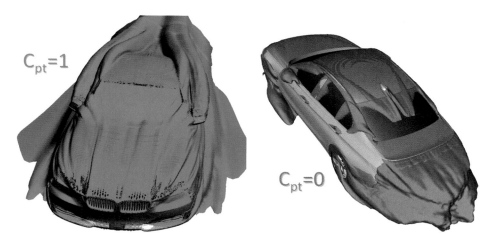

Figure 14.68 Isosurfaces of total pressure coefficient: left: $c_{p,tol} = 1$; right $c_{p,tot} = 0$

Since the highest losses occur in the core of vortices, the isosurfaces of total pressure in Figure 14.68 can easily provide a vivid way to visualize vortices, for instance, at the A-pillar. However, even more sensitive criteria for this purpose are derived quantities like vorticity (Figure 14.69 left) or the λ_2-criterion (Figure 14.69 right). Depending on the vehicle, the values for these isosurfaces have to be selected to produce meaningful illustrations.

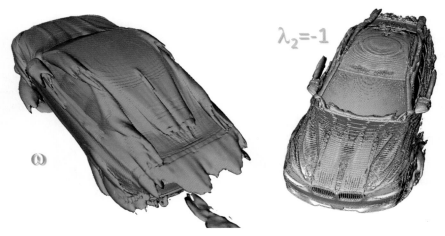

Figure 14.69 Isosurfaces of vorticity (left) and $\lambda_2 = -1$ (right).

14.1.10 Hardware and Benchmarking

The accuracy level achieved today is already sufficient to employ CFD in vehicle development. However, for an efficient integration in the development process, the turn-around time needs to be short enough to keep pace with shortened development cycles. In addition to geometry preparation and automated post-processing, computer performance plays a key role in this respect.

14.1.10.1 Computer Architectures

In high-performance computing (HPC), basically three different computer architectures (vector computers, SMP systems, and clusters/MPP) followed after each other in the course of time.

14.1.10.1.1 Vector Computers

At the end of the 1970s, vector computers formed the beginning of real HPC, headed by the manufacturers CDC and CRAY Research, later followed by NEC, Convex, Fujitsu Siemens, and others.

Instead of processing data sequentially, vector computers were able to execute an operation on all elements of an array (vector) concurrently. This made them the ideal tool for matrix operations. Provided that the programming had been adapted to this special architecture, a performance increase by orders of magnitude was possible. Although at the end of the 20th century vector computers with four or six vector processors appeared on the market, today they have been replaced by parallel systems with hundreds or thousands of processors.

14.1.10.1.2 Symmetric Multiprocessor Systems (SMP)

As on vector computers, adapted programming is necessary to utilize the advantages of multiprocessor systems. The problem is decomposed, and all processors work concurrently on these subdomains. An even load distribution between the domains to minimize latencies is key for efficient execution.

The first parallel computers were symmetric multiprocessor systems (SMP), where all processors share common memory and are controlled by a single operating system. The memory bandwidth of the processor connections via a bus could often be the bottleneck of this architecture. The leading manufacturers SGI, SUN, IBM, and HP developed their proprietary RISC-processors (reduced instruction set), which allowed for higher clock rates. However, these systems have been mostly replaced now by clusters. In the "TOP500" ranking (www.top500.org cp. section 14.1.10.1.3) of the world's largest supercomputers from November 2013, not a single SMP system is listed.

14.1.10.1.3 Clusters (MPP—Massive Parallel Systems)

Currently the predominant architecture in HPC is the cluster, composed of independent computers (nodes) that are connected either via Ethernet or special interconnects (InfiniBand, Myrinet, etc.). Each node runs its own operating system and is equipped with one or more processors. First, there were massively parallel systems (MPP), built using proprietary hardware, and then these were generally replaced by large clusters using commodity hardware. Only 15% in the TOP500 list from November 2013 are MPP systems; all others are clusters. With the introduction of multi-core processors, a bit of confusion came up between the names processor, CPU, and core. Wherever the designation, CPU appears in the following text it refers to a single core of a processor that carries out the calculation on a single domain of the decomposed problem.

Due to the commodity, hardware clusters are less expensive than SMP systems. However, since each node has its own hard disk and O/S, the risk of failure is higher.

The various CFD methods exhibit different scalability due to the different communication overhead between domains. While the performance gain of RANS methods often degrades significantly when using more than 100 CPUs, LBM-based tools feature better scalability, even when employing several hundreds of cores. Figure 14.70 plots the speed-up of PowerFLOW for cases of different size. This test was run on a cluster whose nodes were equipped with two Intel® Xeon 5600 processors (featuring six cores). When increasing the number of CPUs for a simulation from 60 to 480 (from 5 to 40 nodes) (by a factor of 8), the speed for the largest case with 82 million cells increases by 90% of the ideal linear speedup. For the smallest case with 6 million cells, only 50% of the ideal value is achieved. The reason for this is that the number of cells to be processed by one CPU is so small that the communication effort in relation to the computational work has risen adversely.

Since the memory on a cluster is distributed among the nodes, direct communication between CPUs as on a SMP system with shared memory is not possible. For this purpose, a standard called message passing interface (MPI) has to be used.

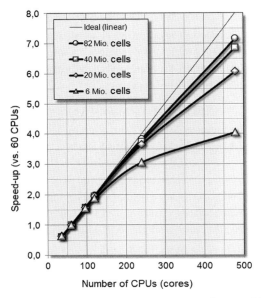

Figure 14.70 Scalability of a LBM code (PowerFLOW) on a cluster with 2x Intel Xeon 5600 six-core processor per node; InfiniBand interconnect.

The interconnect between the individual nodes is essential for the overall performance of the system. In Figure 14.71, it becomes apparent that with Gigabit-Ethernet not even twice the speed is reached when using 480 instead of 60 cores. The cluster requires InfiniBand as interconnect to achieve 90% of the ideal linear scaling.

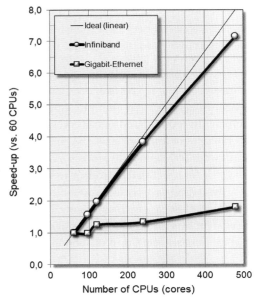

Figure 14.71 Influence of the interconnect on the scalability (PowerFLOW simulation with 82 million cells, 2x Intel Xeon 5600 6-core processor per node).

14.1.10.1.4 GPU-Computing

Graphic processors (GPUs) have been used recently to speed up the execution of simulations. They act as co-processors of the CPUs and take over the parallelizable part of a code. The pioneer of this technology is the manufacturer NVIDIA. At the beginning, conventional graphic cards of the GeForce or Quadro family were used. Later the Tesla graphic cards were developed especially for HPC. They feature several hundreds of cores per processor; for instance, model C2075 carries 448 cores. Their performance per core is inferior compared to conventional processors, but the lower cost allows operating a significantly larger cluster with a given budget. However, existing codes written in C, C++, or Fortran have to be compiled or, better, rewritten completely in a special environment (CUDA) to utilize the advantage of GPU-computing.

14.1.10.2 Computer Performance

Since computers have been employed for scientific and engineering calculations, their performance is measured by the number of floating-point operations per second (Flop/s or FLOPS). As we can see in Figure 14.72, the performance has grown exponentially.

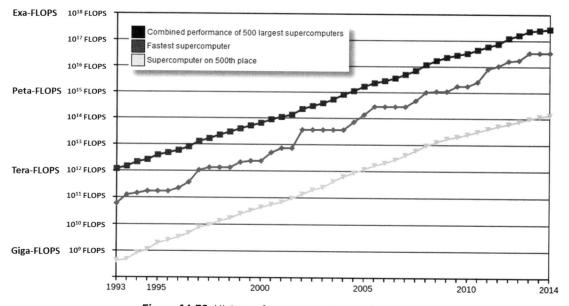

Figure 14.72 History of supercomputer performance.

Starting in 1993, a ranking of the world's largest supercomputers is published every six months (www.top500.org). It is based on a standardized Linpack benchmark for the solution of linear equations systems. Between the frontrunner in June 1993 with 60 Giga-FLOPS (on 1,024 cores) and the one in November 2013 with 33.863 Peta-FLOPS (Tianhe-2 with 3,120,000 cores; Figure 14.73), there is a performance increase by a factor of $5 \cdot 10^5$ within 20 years!

Rank	Site	System	Cores	Rmax (TFlop/s)	Rpeak (TFlop/s)	Power (kW)
1	DOE/NNSA/LLNL United States	**Sequoia** - BlueGene/Q, Power BQC 16C 1.60 GHz, Custom IBM	1572864	16324.8	20132.7	7890
2	RIKEN Advanced Institute for Computational Science (AICS) Japan	K computer, SPARC64 VIIIfx 2.0GHz, Tofu interconnect Fujitsu	705024	10510.0	11280.4	12659.9
3	DOE/SC/Argonne National Laboratory United States	**Mira** - BlueGene/Q, Power BQC 16C 1.60GHz, Custom IBM	786432	8162.4	10066.3	3945
4	Leibniz Rechenzentrum Germany	**SuperMUC** - iDataPlex DX360M4, Xeon E5-2680 8C 2.70GHz, Infiniband FDR IBM	147456	2897.0	3185.1	3422.7
5	National Supercomputing Center in Tianjin China	**Tianhe-1A** - NUDT YH MPP, Xeon X5670 6C 2.93 GHz, NVIDIA 2050 NUDT	186368	2566.0	4701.0	4040
6	DOE/SC/Oak Ridge National Laboratory United States	**Jaguar** - Cray XK6, Opteron 6274 16C 2.200GHz, Cray Gemini interconnect, NVIDIA 2090 Cray Inc.	298592	1941.0	2627.6	5142
7	CINECA Italy	**Fermi** - BlueGene/Q, Power BQC 16C 1.60GHz, Custom IBM	163840	1725.5	2097.2	821.9
8	Forschungszentrum Juelich (FZJ) Germany	**JuQUEEN** - BlueGene/Q, Power BQC 16C 1.60GHz, Custom IBM	131072	1380.4	1677.7	657.5
9	CEA/TGCC-GENCI France	**Curie thin nodes** - Bullx B510, Xeon E5-2680 8C 2.700GHz, Infiniband QDR Bull SA	77184	1359.0	1667.2	2251
10	National Supercomputing Centre in Shenzhen (NSCS) China	**Nebulae** - Dawning TC3600 Blade System, Xeon X5650 6C 2.66GHz, Infiniband QDR, NVIDIA 2050 Dawning	120640	1271.0	2984.3	2580

Figure 14.73 The top ten of the "TOP 500" supercomputers (November 2013).

However, we cannot derive the runtime of a CFD simulation from this measure. None of the existing CFD codes today is actually able to make efficient use of hundreds of thousands or even millions of cores for a single simulation.

A more practical measure for CFD is the "per-core performance"

$$p_{\text{Core}} = \frac{n_{\text{Cells}} \cdot n_{\text{Time steps}}}{n_{\text{CPU}-h}} \tag{14.71}$$

which can be determined for any code from each simulation. Once the number of cells and time steps (or iterations) is known, the runtime can be estimated depending on the number of cores n_{CPU} that will be used

$$n_h = \frac{n_{\text{Cells}} \cdot n_{\text{Time steps}}}{p_{\text{Core}} \cdot n_{\text{CPU}}}. \tag{14.72}$$

Figure 14.74 shows an example of the per-core performance for a number of PowerFLOW simulations plotted over the ratio of surface elements (surfels) to fluid elements (voxels). These simulations were run on an InfiniBand cluster, whose nodes are equipped with two Intel® Xeon® 5600 six-core processors. The data points represent real-world cases of various sizes and were executed using 108–240 cores. Since the computational effort per surfel is much higher than per voxel, the performance drops with increasing surfel to voxel ratio. After the discretization, this ratio is known, and with the per-core performance from the regression curve, the runtime can be estimated according to equation (14.72).

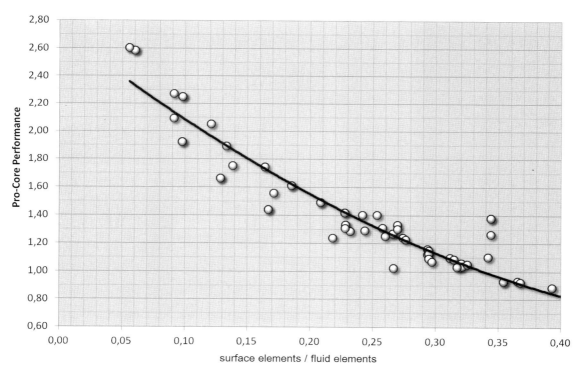

Figure 14.74 Per-core performance of PowerFLOW simulations (2x Intel Xeon 5600 six-core processor per node; InfiniBand interconnect).

In the past ten years, the per-core performance has increased by a factor of 8, as shown in Figure 14.75 for a PowerFLOW benchmark. This development is roughly exponential but still clearly below Moore's law, stating that the number of integrated circuits (not its performance—as Moore's law is often misquoted) on a processor doubles every 24 months. Intel quotes that this is quite exactly the case with their processors over the past 20 years.

The future development toward exa-scale computing, (i.e., the efficient use of supercomputers with 10^{18} FLOPS) to reduce runtimes from hours and days to minutes requires fundamental changes in the architecture of software as well as hardware and a coordinated design of both components for HPC [3]. An additional issue will be the energy consumption. Already today this is in the order of megawatts (cf. Figure 14.73) with associated consequences for cooling demand and the cost of operation.

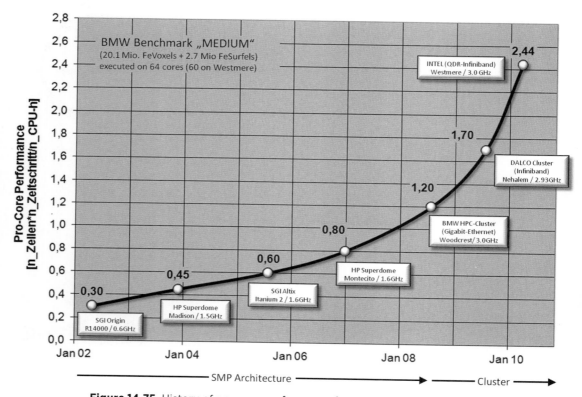

Figure 14.75 History of per-core performance for a PowerFLOW benchmark.

14.1.11 Integration of CFD in the Development Process

The processes in wind tunnel and simulation are quite different, which means that depending on the situation, either testing process may be the best choice. The requirements and processes for experiment and simulation are depicted schematically in Figure 14.76.

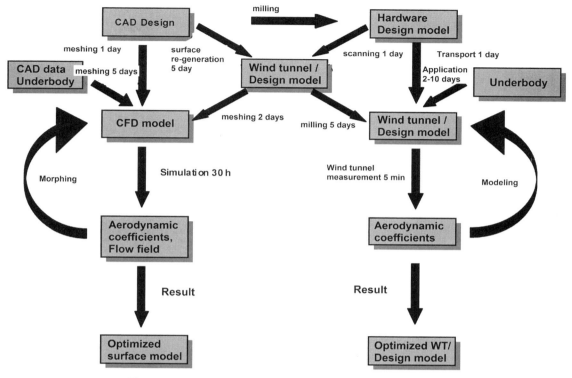

Figure 14.76 Flow chart of CFD and wind tunnel test [514].

Styling models are traditionally optimized directly in the wind tunnel, which is the most effective mode of operation in aerodynamic optimization. However, this requires a styling model that is suitable for wind tunnel testing; for instance, it has to feature a realistic underbody. Usually the basis for these models is built at the beginning of a project to interfere as little as possible with the styling process. Nevertheless, considering the shortened development cycles and accordingly limited availability of models, it is becoming increasingly difficult to retain the traditional experimental process.

An alternative is to use individual aerodynamic models. However, the fabrication and update (i.e., by re-milling to adjust to styling changes) is time- and cost-intensive. This holds especially in the early phase, where a multitude of styling variants has to be investigated. Considering the cost, a sustained aerodynamic optimization of styling variants (often more than five models) does not make sense. This is where CFD offers an efficient alternative at lower cost. In particular, for status assessments and the identification of optimization, potential CFD is advantageous.

Despite the progress in computer performance in the recent past, the actual simulation runtimes are still typically around 10–30 hours. This is clearly inferior to the wind tunnel, where about 20–50 modifications can be tested per working day. Including velocity and yaw angle sweeps, even more wind tunnel measurements are feasible. To cover a similar number of measurements in the same time, using simulations would require turnaround times for a single simulation well below one hour.

A differentiated approach is advised when comparing effort and gain of wind tunnel and simulation. If the objective is just to get integral forces and moments of an existing vehicle model, then physical testing is clearly superior. However, the picture changes if we consider the wealth of information that CFD provides without extra cost that can be exploited for a predetermined optimization.

Besides improved solution methods and accelerated computer performance, the progress in data preparation promotes the propagation of CFD. Today, tools are available to modify the geometry of a CFD model equivalent to wind tunnel testing. Local shape alterations are possible within minutes. Further, new digital tools facilitate the connection between physical and virtual shape optimization. Styling models can be captured in digital form within hours, and the post-processing of the resulting point clouds leads to a CFD model or data suitable for milling within 1–2 days. Alternatively, the return of point clouds to CAD surfaces for functional tests can be accomplished in the meantime within a few days or hours.

14.1.12 Outlook

Two conditions decide whether it makes sense to employ a numerical method (CFD) for the solution of a practical problem:

- The physics is adequately represented.
- CFD is competitive to experiment (i.e., faster or less expensive), where in many cases the criterion "faster" dominates.

Three advantages of CFD regarding its reproduction of physics—in comparison to experiment—are frequently emphasized:

- Boundary conditions are more easily complied with, and a very large simulation relative to the size of the vehicle can be used.
- The relative motion of vehicles to the road is easy to model.
- In wind tunnel tests with scaled models, the Reynolds number of the full-scale vehicle is often undercut by an order of magnitude. However, in simulation, high Reynolds numbers require more resolution, which increases the expenditure of cost and time.

A big advantage of CFD over the wind tunnel is the unrivaled wealth of information it delivers. In an experiment usually only forces and moments are measured, and flow visualization is at most confined to selected locations. Detailed pressure and field surveys require much time and are therefore rarely conducted.

When the expenditure of time for a simulation is quoted, this is often only the time for solving. However, we also have to include the time to generate the surface mesh and fluid mesh in comparison to the manufacturing of a hardware model. Solid bluff bodies are easy to produce. Further, modifications on solid models are quickly made, an advantage especially in the optimization phase. In this respect, the simulation surely cannot yet replace the experiment. In vehicle development of complex models, the CFD processes often allow for shorter turnaround time, mainly due to the laborious production of hardware models. Figure 14.77 shows a comparison of CFD to wind tunnel testing for full-scale plasticine models. Two cases are considered, a 10× complete change of the exterior styling and a 50× modification of a single part (wheel spoiler, wing mirror, etc.) in the framework of an optimization study. The plot indicates that CFD has already been competitive for years in large-scale changes. For detail optimizations CFD still falls short of the wind tunnel, in particular since more than 50 styling loops have to be completed in the course of development of a vehicle.

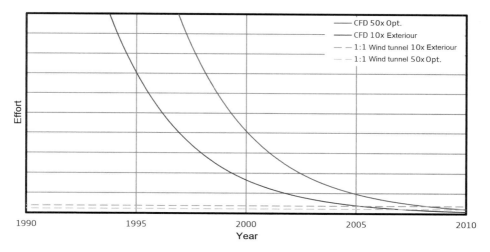

Figure 14.77 Comparison of the expenditure of time for CFD and full-scale wind tunnel testing for 10 global shape variations and 50 detail modifications of a single part.

In fact, it is essential to strive for a combination of both tools [478]:

- Numerical calculation
- Validation of the numerical model by experiment
- If necessary, modification of the numerical model
- Optimization by means of CFD,
- Including a multitude of variants (geometry, flow conditions)
- Final verification by experiment; if possible in full scale

RANS methods, which use a "universal shear stress model," don't seem to have a stake in the future applied to bodies with large, coherent wake structures (which applies to bluff bodies). It is unlikely that they can describe the entire spectrum of transient phenomena with a single model that is valid in the whole flow field. According to the current state of knowledge, there are two promising methods for the flow around bluff bodies: detached-eddy simulation (DES) and Lattice-Boltzmann methods (LBM/VLES). Advanced approaches like LES and DNS won't be applicable for technical problems in the near future, mostly due to their enormous runtimes. Hence for the time being the deployment of these methods will be limited to baseline investigations in research and academia.

14.2 Computational Aeroacoustics for Motor Vehicles

Aeroacoustic noise, caused mainly by airflow around the vehicle, plays an important role in motor vehicles (see chapter 8). Depending on the vehicle model, engine load, and tire-road combination, it dominates the overall noise from approximately 100–120 km/h and upwards, both in the interior and exterior of the vehicle. At low speeds it is either relevant or dominant in specific frequency ranges. Due to the continuing reduction of other acoustic influences, the overall noise in the passenger cell can be substantially influenced by wind noise even from approximately 75 km/h upwards [835]. Hence, wind noise will be a significant characteristic of motor vehicles in the future even if general speed restrictions will be introduced, the more so as demands for comfort will very likely increase.

14.2.1 Introduction

Aeroacoustic noise is generated primarily in the flow around obstacles or when an airflow passes over cavities, as is the case in the flow around a vehicle. In this process, turbulences are generated with fluctuations in their speed and pressure fields, which in turn generate various forms of aeroacoustic sources. Pulsating volume flows such as at the exhaust outlets or at leaks in the sealing systems are such sources. They can be represented by monopole terms. Another such source is pressure fluctuations at the vehicle's surface, which can be represented by dipole terms. These types of source can be found wherever turbulent flows impact a hard surface, such as in the wake of a side view mirror. Beyond that, turbulent shear stresses can occur, such as in the wake of a motor vehicle. These can be represented by quadrupole terms; they are however largely irrelevant for motor vehicles. A detailed description of the aeroacoustic source types and the typical aeroacoustic sound sources in motor vehicles can be found in chapter 8. Using the flow around a side view mirror as an example, Figure 14.78 illustrates the possible locations of the various source types

Research in aeroacoustics and their mathematical analysis have their roots in aerospace engineering. Here, scientists focused on the tremendous noise level of the jet engines and their environmental impact. Against this setting, the ground-breaking work of M. J. Lighthill on the theory of aeroacoustic sources was carried out between 1951 and 1954 ([499], [500]) (see section 14.2.3.1). Later, in 1967, Maestrello was the first to examine the noise generated by the flow around the outer skin of an airplane and how it is transferred to the interior cabin [524].

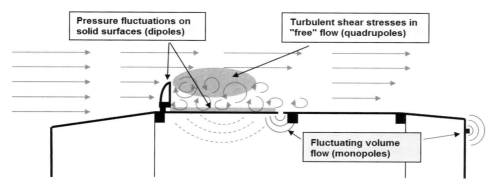

Figure 14.78 Diagram of the flow around a side view mirror as an example for areas with aeroacoustic sources at a motor vehicle (quadrupole sources can normally be neglected) [82].

Aeroacoustics in motor vehicle development received attention only much later. The research done in aerospace engineering provided the basis. Equally late on, the mathematical calculation of aeroacoustics was introduced in motor vehicle development and is thus a relatively new field. Arguments for its application are those valid for computational methods in general, and as they are described in the beginning of section 14.1. Apart from the aspired cost reduction from their use in very early stages of development, they can contribute decisively to the understanding of flow details, as well as of sound generation and sound transfer mechanisms.

Due to continually optimized algorithms and ever growing computing power, numerical methods, above all, are becoming increasingly important in motor vehicle development. Meanwhile, simulation of aerodynamics can be regarded as a standard tool, and corresponding software (computational fluid dynamics, CFD) has been commercially available for quite some time. The numerical simulation of aeroacoustics, however, is only just on its way to being deployed in motor vehicle development on a regular basis. Currently, numerical simulations in aeroacoustics typically focus on the following aspects, which are considered mainly with regard to the interior noise level:

- sunroof and side window buffeting
- under-floor noise
- side view mirror, A-pillar vortices, and side window
- windshield wipers and windshield
- HVAC systems

In the case of aeroacoustics specifically, simulations must be carried out with the greatest care with respect to the selected meshing, the initial and boundary conditions, as well as the turbulence calculation. If this is the case, the simulation of aeroacoustics can even now provide reliable data for specific sections of motor vehicles. In some years the simulation of aeroacoustics will have attained the status of a standard tool.

Beside the direct generation of aeroacoustic noise by the flow around the vehicle (or by the flow in HVAC systems), which emits noise in the fluid (basically on surfaces or at leaks, see source term description above), aeroacoustic noise can be generated indirectly. This means the excitation of vibrations in a body structure element such as window panes or doors, due to aerodynamic, fluctuating forces and the resulting radiation of noise by the vibrating structure. Here it must be differentiated whether or not a significant retroactive effect by the structural vibration on the airflow takes place. If this is not the case, one generally speaks of fluid-structure coupling; if there is a retroactive effect, one speaks of fluid-structure interaction.

Simulations should be able to take into account all these effects and thus require a combination of various calculation methods. Above all, this is true for aeroacoustic interior noise as it is dominated by body structures caused to vibrate by the flow around the motor vehicle, as long as there are no leaks in the sealing systems. Using the example shown in Figure 14.78—the airflow around a side mirror—Figure 14.79 breaks down a possible methodology for simulating the effects in question. In most cases a breaking down into flow calculation, fluid acoustics, and structural acoustics makes sense.

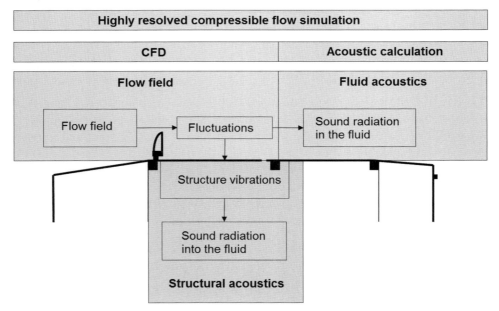

Figure 14.79 Relevant effects in vehicle aeroacoustics and possible structure of a calculation method using the flow around a side mirror as an example [84].

Based on the previous subsections, which extensively described flow simulation, the following subsections will describe several important simulation methods for the various aeroacoustic effects. They will follow the logical order from excitation to sound immission at the point of reception. The simulation methods differ in that they calculate aerodynamic noise by using various simplifying assumptions. This chapter cannot provide a complete description of all available methods. However, the references in the bibliography will be helpful here. The following section will start with the methods for calculating aerodynamic excitation, which includes the potential aeroacoustic sources in the fluid.

14.2.2 Calculation of Aerodynamic and Aeroacoustic Sources

Flow simulation with the help of CFD method is an important basis for computational aeroacoustics. The CFD provides data on pressure and velocity fluctuations generated in the flow by the vehicle, which can be a significant noise source. Here pressure fluctuations are of special interest. They are either caused by a turbulent flow over a surface or by flow separations from a surface. When leaks are taken into account in the simulation, velocity fluctuations could become important, and possibly also velocity fluctuations in wakes of, for example, the side view mirrors. However, in the usual speed range of up to 250 km/h for motor vehicles, pressure fluctuations on the surfaces—be it through turbulent flows over the surface or flow separations—are as a rule the dominating sources, as long as no monopole sources (e.g., leaks) exist.

The basic physical relations and methods upon which flow simulations are based have been explained extensively in section 14.1. On the one hand, particular mention must be made of the algorithms, based on the Navier-Stokes equations (NSE), and on the other hand, the Lattice Boltzmann method (LBM). For the treatment of the nonstationary part of the flow, as it is important for the simulation of aeroacoustics, usually various approximations are used. Either the flow fluctuations are modeled as a whole, meaning they are parameterized with a turbulence model and not calculated directly, or they will be calculated directly for larger length scales and modeled in various ways for smaller scales. Apart from the stationary part of the flow all turbulence scales can, if required, be calculated directly, however with great computational effort. The following methods are commonly used:

- Reynolds averaged Navier-Stokes (RANS): stationary method, all turbulence scales are modeled and not calculated directly; see section 14.1.4.1.
- Unsteady Reynolds averaged Navier-Stokes (URANS): nonstationary RANS method with static averaging or ensemble averaging; see also section 14.1.4.1.
- Large eddy simulation (LES): direct calculation of the larger turbulence scales and modeling of the smaller ones; see section 14.1.4.2.
- Very large eddy simulation (VLES): LES, whose dividing line between resolved (i.e., calculated) and modeled turbulence has been shifted to larger scales; see section 14.1.3.2.
- Detached-eddy simulation (DES): URANS method with areas of higher resolution, where LES is used (see section 14.1.4.2.3; as a rule, URANS at the walls and LES in the fluid volume [768]).
- Direct numerical simulation (DNS): direct calculation of all considered turbulence scales; see section 14.1.4.3.

Based on the NSE or LBM, the entire flow field including its turbulent fluctuations, as well as sound generation and sound propagation, can in theory be calculated simultaneously and directly. This would even include the propagation of sound waves of the aeroacoustic sources in the inhomogeneous flow, including refraction and scattering. This most complex calculation method is frequently called direct simulation (DS). For this, a form of equations must be chosen that includes the compressibility of air in combination with the calculation of a nonstationary solution. The required spatial resolution of the basic mesh is oriented toward relevant turbulence scales and the geometry to be described. The relevant acoustic wavelengths are as a rule greater (i.e., they do not determine spatial resolution). The resolution in time, however, must be adapted to the relatively high speed of sound to at least fulfill the Courant-Friedrichs-Lewy criterion ($c \times \Delta t / \Delta x \leq 1$, e.g., in [217]).

A DS must thus accurately represent a very large bandwidth of length, time, and energy scales. The upper section of Figure 14.80 shows an overview of the various length and energy scales, which can cover the flow with its turbulence (short lengths, high energies) and sound waves (long lengths, low energies). For this reason, the requirements on computational power are extraordinarily high, both with regard to computation speed and memory capacity. Numerical problems must be handled as well. Consequently, DS is currently only used for small, academic or low-frequency cases with relatively wide-area excitation. Usually, the above-mentioned simplifying methods are used. To what extent the various methods directly calculate the various scales of the pressure and velocity fluctuations as well as the sound waves is shown qualitatively in the lower section of Figure 14.80.

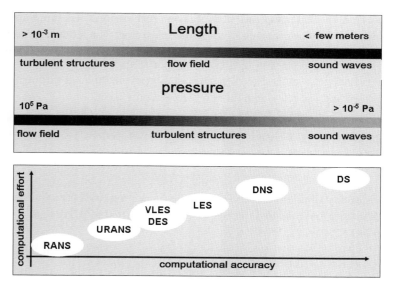

Figure 14.80 Qualitative overview of the length and energy scales of flow, turbulence, and sound waves (above) and of the degree of direct calculation of the various scales in the different methods (below) [79].

After having calculated or modeled the flow fluctuations by means of one of the mentioned methods, the acoustic sources can be determined. The time variations of the relevant fluctuations correspond to the possible acoustic frequencies to be expected. If the relevant parameters were parameterized with a turbulence model, the relevant fluctuations must be reconstructed from there. They will be approximated in time and space by pseudo-stochastic fluctuations using the statistical behavior of turbulence and the parameters from the turbulence model, for example, turbulent kinetic energy. The so-called SNGR method (stochastic noise generation and radiation; see for example [59], [34]) covers such a reconstruction.

For small Mach numbers and no (two-way) interaction between acoustics and flow field, one can assume an incompressible fluid for the calculation of the flow field. In practice, this is frequently the case for motor vehicles and simplifies simulation. However, if effects such as sunroof buffeting are to be simulated, where a coupling between vortex separation and the excited acoustic cavity resonance takes place, the compressibility of air should be included for a correct simulation, unless the interaction is covered by some specific modeling. When simulating the propagation of sound waves, compressibility must of course also be included in the calculation, as has already been mentioned.

14.2.3 Sources and Exterior Sound Field

If the flow fluctuations are calculated and the acoustic source areas have been determined, sound emission from these sources into the fluid up to the point of reception outside the cabin can be determined more or less accurately with acoustic post-processing, as long as this has not already been done in a DS. Prevalent simpler methods assume a purely geometric propagation of the sound. Hence, effects of the inhomogeneous flow on sound propagation are not taken into account. Neither are reflections off surfaces in the vicinity always taken into account. However, more-complex methods exist which consider these effects. For example, this is the case with linearized Euler equations (LEE). The following sections describe several important methods from those known for sound emission calculation.

14.2.3.1 Aeroacoustic Analogies

As has been mentioned in the introduction, aeroacoustics were first researched in the field of aerospace engineering. In this environment, Lighthill [499] [500] developed the so-called aeroacoustic analogy (AAA) in the middle of the 20th century, from which further aeroacoustic analogies have been developed.

Lighthill derived a wave equation from the compressible Navier-Stokes equations. He separated linear and nonlinear terms and moved nonlinear terms to the right side of the equation together with friction terms. This resulted in an inhomogeneous acoustic wave equation, where the terms of the right side can be interpreted as aeroacoustic sources. They are made up of pressure and velocity fluctuations as well as stress tensor and force terms. Moreover, the stationary part of the field parameters (pressure, density) was separated from the nonstationary part. The equation reads as follows:

$$\frac{\partial^2}{\partial t^2}(\rho-\rho_0)-c^2\frac{\partial^2}{\partial x_i \partial x_j}(\rho-\rho_0)=\frac{\partial^2 T_{i,j}}{\partial x_i \partial x_j} \quad (14.73)$$

$$T_{i,j} = \rho v_i v_j + \left[(p-p_0)-c^2(\rho-\rho_0)\right]\delta_{i,j} - \tau_{i,j} \quad (14.74)$$

with c being the speed of sound, ρ the density of the fluid, ρ_0 the density of the undisturbed fluid, p and p_0 the respective pressures, v_i and v_j the fluid velocity in the i- and j-directions, $\delta_{i,j}$ the Kronecker symbol, and $\tau_{i,j}$ the viscous stress tensor. The term $T_{i,j}$ is called Lighthill stress tensor. Up to this point, it is an exact theory, as no approximations were assumed.

The second spatial derivation of the stress tensor produces a quadrupole distribution. Hence, the nonstationary fluctuations of the flow are represented by a distribution of quadrupole sources in the volume under consideration. However, source and propagation terms are mixed here. In order to be able to formulate the source terms independently of the acoustic variables, approximations were introduced. For an ideal gas such as air in an isentropic flow with high Reynolds and small Mach number, the Lighthill tensor is frequently approximated as follows:

$$T_{i,j} = \rho_0 v_i v_j \quad (14.75)$$

In this form, linearized equations were derived describing the propagation of acoustic waves in a homogenous medium at rest. The waves are excited by the turbulent fluctuations, which represent the acoustic source terms. The designation "aeroacoustic analogy" originates in the fact that aeroacoustics are described by equations of classical acoustics here.

In the aeroacoustic engineering of motor vehicles, it is essential that the presence of hard surfaces is taken into account. For this, Curle [186] at first developed a formal solution of the Lighthill equation (equation (14.0)). While with Curle restrictions regarding movement of the surface (normal velocity at the surface near to zero) had to be observed, these fell away in the solution developed later by Ffowcs Williams and Hawkings [245]. As in vehicle acoustics, moving surfaces are the rule, the latter is preferably used, provided that aeroacoustic analogies are used at all. The equation of the Ffowcs-Williams-Hawkings analogy is:

$$\frac{\partial^2 \rho}{\partial t^2} - c^2 \nabla^2(\rho) = \frac{\partial^2}{\partial x_i \partial x_j}\left\{T_{i,j}H(f)\right\} + \frac{\partial F_i\delta(f)}{\partial x_i} + \frac{\partial Q\delta(f)}{\partial t} \quad (14.76)$$

with

$$F_i = -\left[\rho v_{n,i}(v_{n,j}-v_j)+p\delta_{i,j}-\tau_{i,j}\right]\frac{\partial f}{\partial x_i} \text{ and } Q_i = -\left[\rho(v_{n,j}-v_j)+\rho_0 v_j\right]\frac{\partial f}{\partial x_i} \quad (14.77)$$

The function $f(x, t)$ describes the moving surface for $f = 0$, v_n is the normal velocity of the surface, $\delta(f)$ represents the Dirac function, $H(f)$ the Heaviside function. Apart from the quadrupole terms, one can now find on the right side of this inhomogeneous wave equation—in the second and third term—the dipole and monopole terms, which are important in vehicle aeroacoustics (see Figure 14.81). Looking at the Dirac and the Heaviside functions, the different characters of the source term can be recognized. The quadrupole terms describe volume sources outside the surface (see above), and the dipole and monopole terms describe sources at a surface. Here, the monopole sources are generated by the movement of the surface, much like a loudspeaker.

So, the analogies serve to evaluate the areas where sound sources are to be expected. In case of a simulation, these areas would be determined with a flow calculation. Using the flow around a side view mirror as an example, Figure 14.81 illustrates this situation. The turbulent flow, caused by a side view mirror and calculated with CFD, creates regions of quadrupole sources in the wake and regions of dipole sources on the side window due to the pressure fluctuations. Moreover, monopole sources may appear (e.g., due to leaks in the sealing system). Usually, a linear propagation of the sound from the source in the fluid to the receiver is calculated (i.e., without special propagation effects). Basically, the analogies are only suitable for the exterior noise of cars.

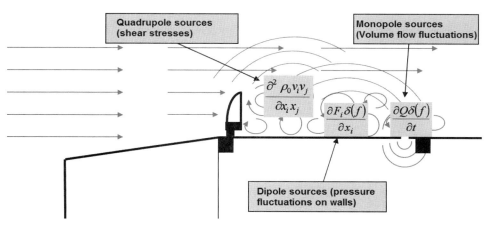

Figure 14.81 Aeroacoustic source types and relevant equation terms using the turbulent flow at a side view mirror as an example (quadrupole sources can as a rule be neglected; see Figure 14.78 [79]).

Apart from the three aeroacoustic analogies mentioned here, others do exist, some of which were or are used in vehicle acoustics. This holds true, for example, for Lilley's analogy [501] or those of Howe [359] and Möhring [577]. Due to the prevalent interest in the vehicle's interior noise, generally analogies, which primarily calculate the external emission and not the transfer to the interior, nowadays play a rather secondary role in the simulation of vehicle aeroacoustics.

14.2.3.2 Kirchhoff Integral Method

The Kirchhoff Integral method examines the source distribution on an enveloping surface that encloses the volume with potential aeroacoustic sources. For a calculation, the area to be examined will be divided into two regions:

- The nonlinear interior region, which will be calculated with a CFD simulation with low or without diffusion, dissipation and dispersion
- The linear exterior region, where an integral will be formed over the source terms on the boundary of the interior region (Kirchhoff surface), thus calculating the sound pressure in the far field

The theoretical background to this can be found in literature, for example, in Chapter 1.5 in Goldstein [288]. Compared to aeroacoustic analogies, the Kirchhoff Integral method has the advantage that only surface integrals are executed over the source enclosure and not volume integrals in the entire source area. Using the harmonic approach $(x, t) = \varphi e^{i\omega t}$, the integrals—kept in general form—are as follows:

$$\varphi(x_0) = -\frac{1}{4\pi}\int_S \frac{e^{-ikr}}{r}\frac{\partial \varphi}{\partial n}dS + \frac{1}{4\pi}\int_S \varphi \frac{\partial}{\partial n}\left(\frac{e^{-ikr}}{r}\right)dS \tag{14.78}$$

$\varphi(x_0)$ denotes the complex amplitude of the acoustic potential $\phi(x, t)$, S the enveloping surface, r the distance to the receiving point x_0, and n the surface normal. To be correct, the equation must be called Kirchhoff-Helmholtz formula. Pressure p and velocity v' of the sound field are linked to the acoustic potential via its time and space derivatives:

$$p = -\rho_0 \frac{\partial \phi}{\partial t} \tag{14.79}$$

$$v' = \text{grad } \phi \tag{14.80}$$

Originally, when using the Kirchhoff method, the surface to be integrated had to be outside the turbulent source flow and the shear flow area. More recent methods are able to deal with this; even moving surfaces can be considered [243].

14.2.3.3 Linearized Euler Equations

Linearized Euler equations (LEE) allow the calculation of the sound field from the source to the receiver, including the sound propagation in the inhomogeneous flow. For this, the volume relevant for the sound field is examined, wherein reflecting areas can also be taken into account.

The LEE are derived from the Navier-Stokes equations while neglecting nonlinear terms and viscosity. Moreover, it is assumed that the acoustic terms are very small compared to the mean flow values. Separating the scales of the mean flow values ($\bar{p}$, $\bar{u}$, $\bar{\rho}$) and the acoustic terms (p', u', ρ'), assuming adiabatic processes in sound propagation in the air, and neglecting gravitational forces, the LEE can be written as follows [85]:

$$\frac{\partial u'}{\partial t} + (\bar{u}\nabla)u' + (u'\nabla)\bar{u} + \frac{1}{\bar{\rho}}\nabla p' - \frac{1}{\kappa \bar{\rho}}\frac{p'}{\bar{p}}\nabla \bar{p} = 0 \tag{14.81}$$

$$\frac{\partial p'}{\partial t} + \bar{u}\nabla p' + u'\nabla \bar{p} + \kappa(\bar{p}\nabla u' + p'\nabla \bar{u}) = 0 \tag{14.82}$$

with $\kappa = c_p/c_v$ representing the ratio of the specific thermal capacities of air at constant pressure and constant volume, respectively. The LEE are a relatively comprehensive and generally precise simulation tool, admittedly with relative high requirements as to computational power. For this reason, they are only used for specific applications in vehicle aeroacoustics.

Figure 14.82 shows a small schematic overview of the calculation methods discussed here, with which the sound field in the fluid can be calculated. In all cases, a CFD calculation is coupled with the acoustic methods discussed here. The principle shown for the LEE also applies to an FEM calculation, if air is used as material.

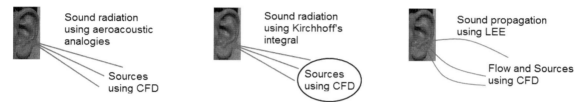

Figure 14.82 Schematic representation of the different coupled methods for calculating the sound field in the fluid. The LEE diagram basically also applies to an FEM calculation [80].

14.2.4 Transfer into the Vehicle Cabin

In the preceding discussion, the sound field on the outside of the vehicle was examined without considering the radiation of sound by vibrating structural surfaces (of the body). The latter is indeed negligible for the exterior noise. When looking at the sound field in the vehicle interior, which is the main point of interest for aeroacoustic engineering, the interaction between the flow and the structure as well as its vibration must be taken into account. Vehicle structures such as the doors, the underbody, and the roof are caused to vibrate by the aerodynamic forces and emit noise to the vehicle interior. Figure 14.83 shows the vibration pattern of a vehicle door as an example of aerodynamically caused structural vibrations (vibration mode at 122.5 Hz) at 140 km/h, measured in a wind tunnel. Here a laser scanning vibrometer was used, which determines the velocity normal to the surface.

Consequently, the entire process must be represented in a simulation:

- The generation of alternating pressures at the surfaces
- Excitation and propagation of structural vibrations
- Radiation of airborne sound

In addition, the aeroacoustic noise generated on the outside of the vehicle, for example by the side view mirrors, antennas, or roof racks, and penetrating through the structure into the vehicle interior, must be taken into account. This noise can be clearly perceptible or—at higher frequencies—even be dominant. If there are leaks in the sealing system or elsewhere, so that the noise can penetrate directly into the vehicle interior, it is normally determinant for the interior noise level. However, the leaks are usually not considered in simulations, as they are production defects and not intentional leaks.

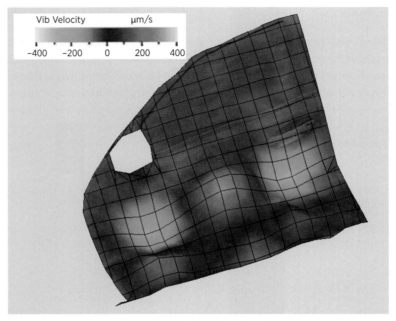

Figure 14.83 Vibration mode of a vehicle door at 122.5 Hz, excited by aerodynamic forces at 140 km/h and measured with a laser scanning vibrometer in a wind tunnel (velocity normal to the surface, red: out of the plane, green: into the plane). Source: FKFS.

When simulating the aeroacoustics of a vehicle, two different interactions between the air of the flow—the fluid—and the body structure must be considered:

- Fluid-structure coupling: Here it is assumed that structural vibrations caused by the fluid forces do not influence the flow due to relatively small amplitudes. It is also called weak coupling or one-way-coupling.

- Fluid-structure interaction: Here a strong structural vibration or deformation caused by the fluid forces has an effect on the flow, and thus a real interaction takes place. It is also called a strong coupling or two-way-coupling.

Aerodynamic noise that is transferred through the sealing systems into the vehicle interior can be regarded as a special case of fluid-structure coupling and interaction. On the one hand, the alternating pressures at the outer surface couple into the sealing structure and cause vibrations there. On the other hand, interaction takes place between the air inside the sealing system and the vibrating seal. The latter, too, influences emission into the vehicle interior. Moreover, nonlinear effects (pre-stress of the sealing) have to be taken into account. Due to the relatively soft materials and the very small areas, different boundary conditions apply in seal-specific fluid-structure coupling and interaction than in body structures. As a rule, simulation methodology is more complex here and is still in the development stage. Aerodynamic noise transferred through the seals can dominate the interior noise level, depending on the quality of the sealing system.

For the calculation of vibrations in body structures and soft structures, such as sealing systems or soft tops, methods from structural acoustics must be considered. The most frequently used are the finite-element methods (FEM, see for example [242]), the boundary element method (BEM, see for example [36]), and statistical energy analysis (SEA, see for example [517]). Figure 14.84 shows the basic cause and effect relationship of the calculation methods for the specific areas of interest, with the upper diagram showing fluid-structure coupling and the lower diagram fluid-structure interaction. These methods have been described extensively in literature (for examples see above), so they will not be discussed in detail here, but a brief overview will be provided.

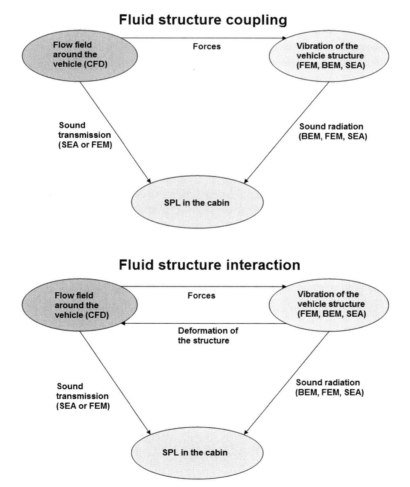

Figure 14.84 Principle of fluid-structure coupling (above) and fluid-structure interaction (below). The text in brackets indicates the calculation methods used in each case. Source: FKFS.

An FEM offers manifold possibilities. Both fluid and structure can be modeled. In this way, simulation of the structural vibrations, of sound radiation, and a direct consideration of the interaction of fluid and structure can be taken into account (not with all codes). As a consequence, FEM is also used for the development of simulation methods for sealing systems (see the preceding section). Due to detailed meshing (as each volume must first be discretized) and the necessity of a closed volume, the calculation of sound radiation and examination of the fluid with FEM are to be recommended only for relatively small air volumes. The coupling of vehicle interior and body structure in the case of sound transfer into the vehicle interior could be such an application. For sound emission into nonclosed spaces (to the outside), an infinite element method could be used (see for example [27]). However, BEM is preferably used for such cases.

In vehicle acoustics a BEM is used mainly for calculating the sound radiation of vibrating structures. It is particularly suited for this application, as only surfaces must be discretized. Hence, meshing can be kept relatively simple. Some formulations of the BEM are also suitable for the calculation of structural vibrations (see for example [4], [343]). However, FEM is better suited for this and is thus used more frequently. The strong point of BEM is the solving of acoustic problems in infinitely (nonclosed) extended areas. As has been mentioned already, this is the case with vehicles (e.g., in the emission of sound to the outside). However, the BEM is also used for calculating the sound field inside the vehicle. The relatively uncomplicated mesh is outweighed by the higher numerical effort of the BEM compared to the FEM. On the other hand, the former reduces labor costs.

Both methods, FEM as well as BEM, are as yet not practicable for frequencies above approximately 500 Hz when dealing with whole vehicles due to their respective high meshing and numerical effort. SEA, in contrast, is only suitable for higher frequencies. It deals with statistical energy flows, which require a high density of eigen-modes of the structure in question. As said, this requires high frequencies, but it is highly effective. It can cope with much simpler models and can be carried out with comparatively modest computing power. SEA models for motor vehicles consist of so-called subsystems, representing both structures as well as cavities, respectively, fluids. Structural subsystems are for example door panels, windows, and roof. Fluid subsystems can be the engine compartment, the space between road and underbody, as well the vehicle interior. Depending on the objective of the investigation, the vehicle will be modeled with 80 to 150 subsystems. SEA is used successfully for calculations of the interior noise level of motor vehicles for frequencies exceeding 500 Hz.

In order to determine the aerodynamic excitation of the structure, the various methods can be coupled with CFD software. The corresponding interfaces are available; however, not for all combinations of commercially available software. In case an interaction between fluid and structure is to be examined, the forces acting on the structure would be determined theoretically with CFD for every time step and—using the structural code—the corresponding structural response would be identified. In the next time increment, the structure's modified geometry would be entered into the CFD calculation as a new boundary condition. However, this process would be extremely complex and could cause numerical problems.

In practice, the retroaction of the structure on the flow has to be neglected, which would be a fluid-structure coupling in the sense of the previous definition, so that no iterations between CFD and structure code are required. Or alternatively, the time increment for the data exchange between CFD and structure code has to be increased appropriately, so that it is not necessary to transfer the modified boundary conditions with every time step. The time steps must be large enough to keep the computational requirements manageable but small enough to be able to approximately represent the interaction. This also applies to possible deformations caused by aerodynamic forces up to the point where they reach a stationary state under the influence of the airflow, such as the so-called ballooning of convertible soft tops (see Figure 14.85).

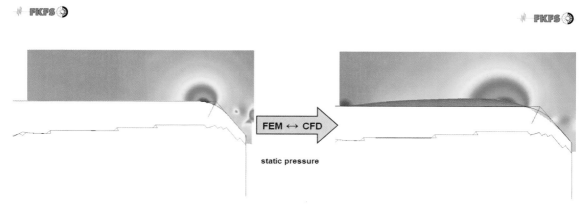

Figure 14.85 Calculation of a ballooning effect in a generic soft top of an SAE model. The calculation followed an iterative process with FEM and CFD simulations (MSC.Nastran and PowerFLOW®) [318].

Apart from the already mentioned structural acoustic methods, combinations of them can also be used for calculating structural response (e.g., FEM and BEM or FEM and SEA). Beyond that, other more recent methods exist, which have not yet been established generally. These are the wave-based method (WBM, also called wave-based technique, WBT, [191]) or the hybrid FEM-WBM method (HFE-WBM, [820]). The HFE-WBM combines the geometric flexibility of the FEM with the capability of the WBM, in order to be able to be used for higher frequencies.

14.2.5 Examples of Applications

Meanwhile, the methods of numerical simulation in vehicle aeroacoustics can be used for many of the effects occurring in the field of vehicle aeroacoustics. When modeling is done with great care, the results are to a great extent reliable. In some cases they are used for investigations of generic vehicle structures in order to continually develop the algorithms deployed. They are also used to better understand the physics of the observed effects, while complementing experimental tools. Their advantages over experimental tools are thereby put into practice. For example, a considerably larger quantity of data can be investigated in the whole space under examination. Moreover, recording of the desired data is carried out without interference. Neither flow nor acoustics are affected.

In the simulations, real, specific vehicle components and their environment at the vehicle are examined, too. Currently the main goal of such investigations is to validate the methods and to be able to understand the aeroacoustics of the considered case in detail. Complete vehicles with all their details and a comprehensive representation of the existing effects cannot yet be simulated.

In the following, only a few examples can be shown. The first example shows the investigation of coupling effects in the interior of a passenger car with sunroof buffeting. For this, a generic vehicle model in the form of an SAE model with a scale of 1:4 was used [83]. We examined the coupling of vehicle interior and trunk via air paths or via a vibrating partition wall. To permit air coupling, a rigid partition wall was installed with defined openings through which air exchange between the two volumes was possible. Figure 14.86 shows the computer model with the structure of the computational mesh in the space under investigation. The simulations were carried out with the CFD software PowerFLOW. It is based on the Lattice-Boltzmann method.

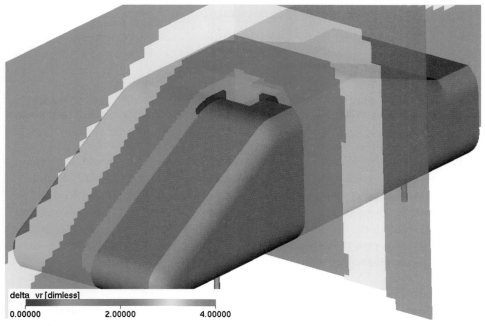

Figure 14.86 1:4 SAE model (gray) with sunroof opening and cavity (not visible) as well as sections of the volume mesh of the CFD calculation (dark blue area with finest resolution [83]).

One result of this series of tests can be seen in Figure 14.87. It compares the case without coupling between interior cabin and trunk (firmly installed partition) with the case with air coupling. The results show that the coupling effect results in a shifting of the buffeting maximum towards higher frequencies and a reduction of the maximum buffeting level. The results from experiment and simulation are largely consistent, both in the level curve versus velocity and in frequency characteristic. The latter has not been included in the diagram for reasons of clarity.

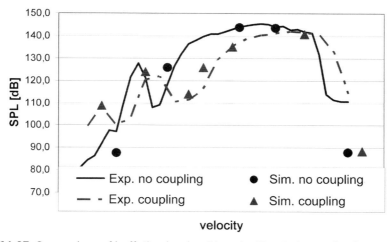

Figure 14.87 Comparison of buffeting levels with and without air coupling between cabin and trunk and comparison of experiment and simulation (microphone in the cabin of the model, see also [83]).

Due to the interaction between acoustics and flow—which is the case with buffeting phenomena—the compressibility of air was also taken into account. As far as buffeting effects are concerned, simulation techniques are relatively advanced. This is probably owing to the fact that very low frequencies are examined, and geometry can be represented relatively roughly in numerous areas. Some simulations with models of real vehicles have been carried out (see for example [184], [661], [755]). However, absorption and coupling mechanisms in the interior (see above) must be modeled with great care here.

Generic side view mirrors have been investigated many times numerically and experimentally in order to develop and verify the numerical algorithms. The area side view mirror/side window in combination with the A-pillar is aeroacoustically highly sensitive. It is also significant due to its proximity to the driver's head (see also Figure 14.81). Side view mirror and A-pillar interfere with the airflow to such an extent that distinct flow fluctuations are generated on the side window, thus contributing to the interior noise near the driver's ears.

Figure 14.88 shows the results of a simulation, calculated with the CFD software Star-CCM+®, based on the Navier-Stokes equations. Here, a detached-eddy simulation was carried out [124]. What can be seen are snapshots of shear stress on the virtual side window (horizontal area) and a cross-section of the flow speed (vertical area). In the wake of the side view mirror, distinct fluctuations can be seen both on the side window and in the volume. Such a simulation does not provide sound levels as a result but levels of the aerodynamic fluctuations of pressure and velocity. These fluctuation levels permit conclusions as to noise generation in this area, both inside and outside the vehicle.

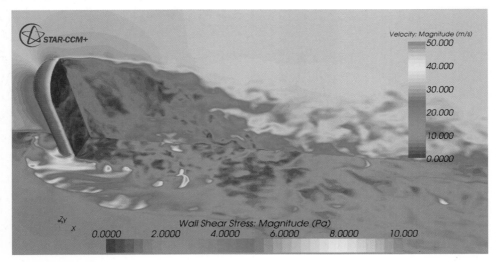

Figure 14.88 Simulated flow around a generic side view mirror. Iso-lines: static pressure. Contour areas: level of velocity. Source: CD-adapco.

The emission of sound was also calculated in this example on the basis of the simulated flow fluctuations. For this, the aeroacoustic analogy according to Ffowcs Williams-Hawkings was used and its results compared with experimental results from literature [691]. Figure 14.89 shows the frequency spectra. Experiment and simulation show the same trend with deviations up to approximately 10 dB in specific frequency ranges.

When investigating specific vehicle components, the component environment on the vehicle's body should also be modeled, as the flow must be simulated accurately. Figure 14.90 shows the flow around a windshield wiper, simulated with Fluent®, as an example. Stream lines and velocity vectors on the windshield surface visualize the flow field. Here, too, the simulation does not provide as a result sound levels in the vehicle interior or outside. Rather, the simulation is to show how strong the flow fluctuations caused by the windshield wiper are on the windshield, and thus potentially contribute to interior noise. On the windshield, vortices and fluctuations can be seen clearly in flow direction (here from right to left) behind the windshield wiper.

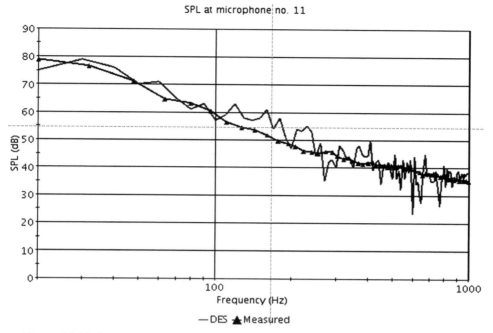

Figure 14.89 Spectral sound level, approximate distance to mirror 0.5 m (see Figure 14.10), calculated with DES and Ffowcs Williams-Hawkings analogy (red) and measured (blue, from [691]). See also [124].

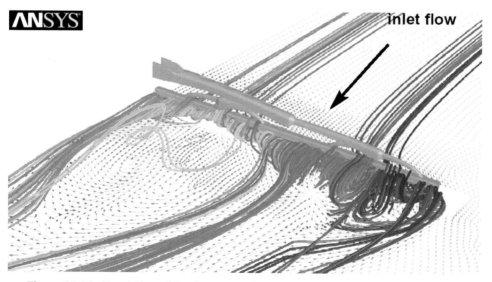

Figure 14.90 Simulation of the flow around a windshield wiper. The diagram shows streamlines and speed on the windshield surface (blue = low velocity, red = high velocity). Source: Volkswagen AG Group Research.

Due to the advantage of interference-free data collection, the simulation can be used to reveal or compensate for experimental deficits; for example, when surface microphones on a side window interfere with and modify the airflow. The surface microphones themselves generate pressure fluctuations, which means that—as a rule—they actually measure (depending on the frequency) higher values than would normally exist at the vehicle without the probes. Comparative simulations permit an assessment of the interference caused by this measuring method. In an investigation dealing with this problem carried out with the CFD software PowerFLOW, the flow was simulated and compared without surface microphones, with surface microphones, and with surface microphones with tapered housings [758]. The tapered housing simulates the attachment of the probes on the side window with tape during experiments.

Detailed views of the surface microphones in experiment and simulation can be seen in Figure 14.91. Also, the simulated pressure fluctuations on the side window—here in the octave band from 177 Hz to 354 Hz—are represented there. Comparing the simulation results reveals clearly that the surface microphones raise the fluctuating pressure in their environment, especially those without tapered housings.

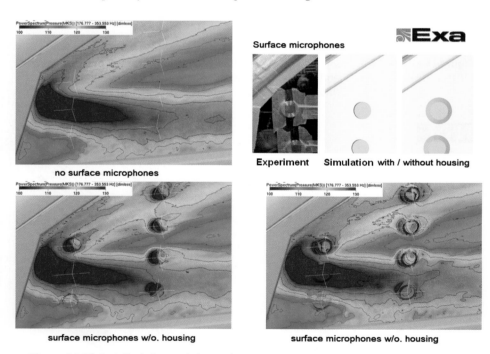

Figure 14.91 Detailed views of the surface microphones (top right) and the simulated pressure fluctuations on the side window in the octave band from 177 Hz to 354 Hz, in dB. The increase in the pressure fluctuations at the sensors can be seen clearly [758].

A frequency analysis of the data shows in which frequency ranges the pressure fluctuations are changed by the probes and to what extent. Looking at the example of two surface microphones at the A-pillar vortex, one can see that when measuring with surface microphones without housing in the range below 1 kHz, levels increased by up to 10 dB can be expected (see Figure 14.92). The housing suppresses this increase, as can be seen in the experimental results with housings (red line). Between approximately 2 kHz and 5 kHz, however, raised pressure fluctuation levels must also be expected for the surface microphones with housings.

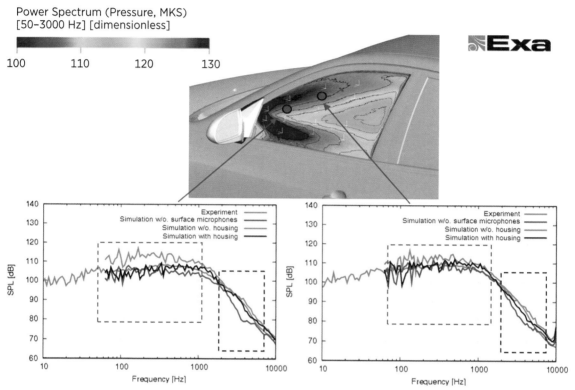

Figure 14.92 Frequency spectra of the measured and simulated pressure fluctuation levels at the location of two surface microphones (position: blue dots, upper view) for the different configurations (broken lines: areas with raised levels caused by surface microphones) [758].

The next step in the chain toward the interior noise of the vehicle is the transfer through the structure, be it windows, body, or sealing system. As an example, the simulation result of a transfer through a side window will be shown here. This transfer includes both the transmission of sound (transfer of sound from the outside through the structure) as well as the excitation of the structure by aerodynamic pressure fluctuations.

In the example shown, an SEA-based method was used, with input data from a CFD calculation with PowerFLOW. This input data is decomposed into an aerodynamic and an acoustic part in order to be able to take into account their different transfer properties. On the one hand, the method was verified with measurements (see Figure 14.93, diagram b); on the other hand it was used to assess the effect of various types of window glass on the interior noise level (see Figure 14.93, diagram c). The local maximum in the frequency range above 2.5 kHz is to be attributed to the coincidence frequency of the window, which varies with glass thickness and structure. As expected, the laminated glass delivered the lowest level (light blue). The results show that this SEA-based method permits such an assessment.

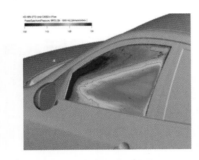
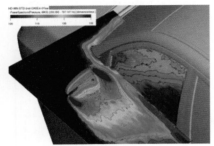
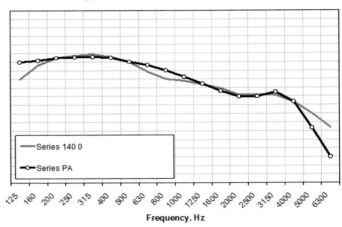
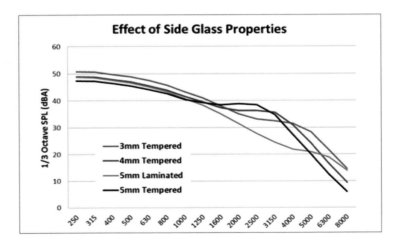

Figure 14.93 Simulation of the transfer into the vehicle interior with a coupled simulation (flow with the CFD software PowerFLOW®, transfer with the SEA method), comparison with experiment (diagram b, measurement in red) and assessment of the effect of different types of window glass on interior noise level (diagram c) [593]. Source: Exa Corp. 2012.

Another topic in vehicle aeroacoustics is air-conditioning (HVAC) systems. Here, ducts, filters, flaps, and vents are objects where the airflow can cause aeroacoustic noise. Fans even represent a kind of active sound source, as they generate their flow themselves. HVAC systems thus provide different scenarios than flow around vehicles. Their geometric dimensions are usually smaller and the considered velocities are slower, but they can be more complex. The latter is especially true when fans are to be simulated as sound sources, too. Nevertheless, numerical simulation meanwhile plays quite a significant role in the development of HVAC systems.

As a first example of an aeroacoustic simulation for HVAC systems, the simulation for a generic air vent is shown. The flow field was calculated with a CFD simulation using Star-CD®; the sound emission was subsequently determined with a BEM simulation [29]. Figure 14.94 shows the simulated velocity field and the setup of the validation experiment. In order to verify the sensitivity of the simulation with respect to the temporal resolution, various time increments were used. A comparison between the measured pressure fluctuation spectra—from a probe next to the air vent—and those from the simulations with the various temporal resolutions shows how the results depend on this simulation parameter and proves the consistency between experiment and simulation (see Figure 14.95).

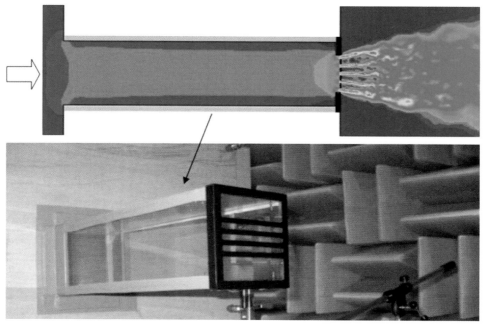

Figure 14.94 Investigation of the aeroacoustics of a generic air vent. Above: Snapshot of the magnitude of the simulated velocity field. Below: experimental setup [29].

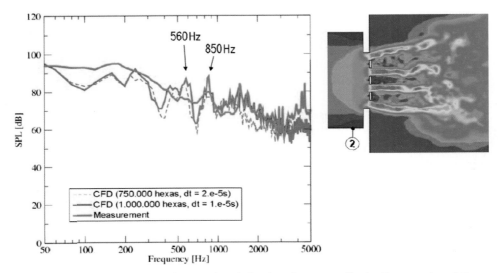

Figure 14.95 Left: measured (green) and simulated pressure fluctuation spectra at the air vent. Right: snapshot of the simulated velocity field (magnitude) with recording sensor position of the pressure fluctuations (2) [29].

As can be seen from the example shown here, this area of aeroacoustics sometimes still investigates simplified geometries in order to verify the simulations and to further develop the algorithms (see also [2]). However, geometries of real HVAC systems were indeed already investigated with numerical tools some time ago, such as the optimization of an air distributor [474]. At that time, however, stationary CFD simulations without acoustic post-processing were used in most cases. Conclusions as to the acoustic behavior were drawn based on the calculated velocity and pressure fields.

In the meantime, HVAC systems are also being investigated in more complex form and in an unsteady condition. For example, the flow through an HVAC system of a passenger vehicle has been simulated aeroacoustically from the rotating fan down to the flow entering the passenger cabin. The CFD simulations were carried out with PowerFLOW. Figure 14.96 (left side) shows the results of the simulation in the form of a snapshot of the vorticity of the outlet flow at the rotating fan; the right side shows the frequency spectra of the emitted sound. They were recorded with a microphone at 1 m distance in the axial direction and simulated for the position in question. The background noise is represented also.

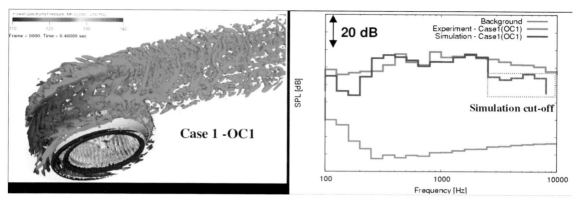

Figure 14.96 Snapshot of the simulated vorticity of the outlet flow of an HVAC system fan (left) as well as comparison of the simulated and measured sound level spectra at a distance of 1 m in direction of the fan axis (right) [633].

The flow through the outlet nozzles into the passenger cabin is illustrated in Figure 14.97. It shows the qualitative representation of the vorticity of the simulated flow into the passenger cell (left side). The right side shows frequency spectra of the noise emitted into the vehicle interior, measured (red) at a distance of 0.85 m and simulated (blue) for the same position. Between approximately 50 Hz and 3 kHz, simulation and measurement correspond in a large range within a few dB. Below approximately 50 Hz, there are deviations possibly due to the background noise during the measurements (see gray curve).

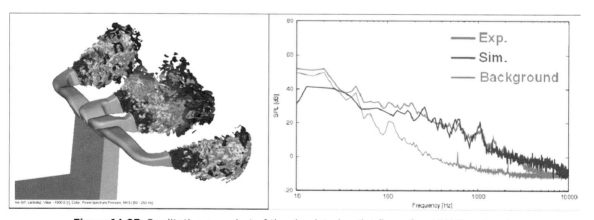

Figure 14.97 Qualitative snapshot of the simulated outlet flow of an HVAC system in the direction of the passenger cabin (left: vorticity in the range between 50 and 250 Hz, colored over pressure fluctuation levels in the range between 70 and 100 dB). Right: Comparison of measured and calculated sound level spectra (microphone in central position in front of the vent outlets at a distance of 0.85 m, 12.5 Hz bandwidth) [632].

Other examples of numerical aeroacoustic investigations of vehicles can be found for the underbody area. The flow at the underbody causes the body structure to vibrate at low frequencies, usually emitting sound into the vehicle interior (see e.g., [835], [183], [592]).

The flow around roof racks on a vehicle has also been simulated (see e.g., [759]). In the case of such add-on parts, highly annoying tonal sounds can be generated, which should be avoided. The same is true for antennas, specifically in the classic rod shape, if they are still used.

14.2.6 Conclusion and Outlook

Aerodynamically generated noise plays an important role in passenger vehicles, above all at velocities exceeding around 120 km/h. This applies both to exterior as well interior noise. In specific frequency bands, they are dominant even at lower speeds. The most significant noise sources are specific body structures such as the A-pillar or the wheel houses (the latter for exterior noise), the sealing system including door gaps, as well as add-on parts such as side mirrors. The area around the A-pillar, side view mirror, and side window is particularly prominent.

Calculating aerodynamic noise is possible today to a certain extent—usually it is based on a CFD simulation. Depending on the requested level of accuracy of the results, high to very high computational effort will be required. Also, the setup of the simulation must be carried out with great care. Acoustic results react highly sensitively to simulated flow fluctuations. They are highly dependent on the precision of turbulence calculation. Boundary conditions, too, play an important role. The vehicle's geometry as well as the air volume in the aeroacoustically sensitive areas and the flow impacting them must be spatially resolved with high precision with a correspondingly dimensioned mesh. The requirements for the spatial resolution arise from the length scales of the turbulence to be examined; the degree of temporal resolution depends on the maximum frequency to be considered (and on the speed of sound, provided that sound waves are also to be calculated).

Typically, CFD simulations are coupled with an acoustic evaluation (post-processing) of the flow fluctuations. An acoustic evaluation can be carried out using various methods. Among these are the aeroacoustic analogies, the Kirchhoff Integral method, the boundary element method, and the linearized Euler equations. As the acoustic results are highly dependent on the flow fluctuations, the modeling of the turbulence must be selected with great care. The less-refined turbulence models frequently meet only basic requirements as to accuracy in acoustics. The results must be interpreted with particular caution when absolute values of sound level and frequency are required.

Flow fluctuations and acoustic waves are directly included in the calculation when a direct simulation (DS) is carried out. They do not have to be modeled or parameterized. The acoustic results of a DS are reliable provided the setup of the simulation was selected correctly and the numerical noise of the applied algorithm is sufficiently low. Frequently, the latter was not the case in the past. For a DS of complex cases, the speed and memory requirements for the computer system are tremendous. Consequently, only smaller, less-demanding, or academic cases are calculated with the DS. More complex cases, such as the simulation of a real whole vehicle, are a task for the future.

Some commercially available CFD software has been upgraded for aeroacoustic applications. While some manufacturers concentrate on one or a few turbulence models, others offer almost the entire bandwidth of the standard turbulence models. DNS and DS can be carried out with all of these programs, given enough time and effort and taking into account the previously mentioned restrictions. Continually advancing computing power will increasingly permit the simulation of more complex cases.

In conclusion, it can be said that the simulation of flow fluctuations as a possible source of aeroacoustic noise can at present almost be considered as standard. Sound emission to the outside was also on its way there; however, it is somewhat taking a back seat due to the prevailing interest in interior noise. The reliable calculation of the transmission of sound into the vehicle interior is a topic for methodological development. Within the bounds of its possibilities, statistical energy analysis has come a long way. For simulating aeroacoustics, it is coupled with a CFD simulation.

In the simulation of aerodynamics with CFD software, inclusion of the wind tunnel environment is currently being done. This is to serve an improved comparability of experiment and simulation. In aeroacoustics, this consideration does not yet take place, as the simulation complexity and thus effort would be tremendous. Moreover, the relevance of the wind tunnel environment is distinctly lower, provided the flow speed has been determined correctly (see section 13.3) and the wind tunnel's inherent noise can be neglected.

Despite possible further driving speed restrictions, aeroacoustics will continue to play an important role in vehicle development in the future, the reason for this being a further reduction of the other noise sources and the ever-increasing demand for comfort. A special situation is given in the case of electric vehicles, where powertrain noise is usually not relevant. Here, aeroacoustics are all the more important. Simulation will continue to be adopted so that even in early stages of development, it will provide results for design decisions. Suitable tools for simulating sound transfer into the vehicle interior will be especially in demand.

Abbreviations

Abkürzung	Beschreiung
A/C	Air Condition
AAA	Aeroacoustic Analogy
ACEA	Association des Constructeurs Européens d'Automobiles
ADAC	Allgemeiner Deutscher Automobilclub (German Auto Club)
AMS	Auto Motor Sport (German Newspaper)
ANR	Active Noise Reduction
ASG	Automated Manual Transmission
ASME	American Society of Mechanical Engineers
Atb	Air to Boil
BEM	Boundary Element Method
BMW	Bayerische Motorenwerke
BSI	British Standard Institution
CAA	Computer-Aided Aeroacoustics
CAD	Computer-Aided Design
CAFE	Corporate Average Fuel Efficiency
CARB	California Air Resources Board
CCA	Constant Current Anemometer
CCD	Charge-Coupled Device
CFD	Computational Fluid Dynamics
CFL	Courant-Friedrichs-Lewy (Number)
CNR	Consiglio Nazionale delle Ricerce
CO	Carbon Monoxide
CPU	Central Processing Unit
CSI	Code Sportif International
CTA	Constant Temperature Anemometry

CVT	Continuous Variable Transmission
DB	Daimler Benz
DE	Direkteinspritzung (Direct Injection)
DES	Detached Eddy Simulation
DLR	Deutsches Zentrum für Luft und Raumfahrt
DNS	Direct Numerical Simulation
DNW	Deutsch-Niederländischer Windkanal (German Dutch Wind Tunnel)
DOT	Department of Transportation (USA)
DS	Direct Simulation
DVL	Deutsche Versuchsanstalt für Luftfahrt
EADE	European Aerodynamic Data Exchange
ECARA	European Car Aerodynamic Research Association
ECE	Economic Comission for Europe
EMPA	Eidgenössische Materialprüfungs- und Forschungsanstalt
EOP	End of Production
EPAS	Electric Power Assisted Steering
EPA	Environmental Protection Agency
EPS	Expanded Polystirene
EUDC	European Driving Cycle
EWG	Europäische Wirtschaftsgemeinschaft (European Economic Union)
FAT	Forschungsvereinigung Automobiltechnik der VDA
FDM	Finite Differencing Method
FEM	Finite Element Method
FIA	Federation Internationale de l'Automobile
FKFS	Forschungsinstitut für Kraftfahrwesen und Fahrzeugmotoren Stuttgart
FKW	Fluorkohlenwasserstoff (Perhalogenated fluorocarbon)
FLOPS	Floating-Point Operations per Second
FTP	Federal Test Procedure
FVM	Finite Volume Method
GÖK	Getriebeölkühler (Gearbox Oil Cooler)
GPU	Graphic Processing Unit
HC	Hydrocarbon
HDA	Hot Wire Anemometry
HFE-WBM	Hybrid FEM-WBM
HFKW	Hydrofluorocarbon
HIC	Head Injury Criterion
HP	Horse Power
HPC	High-Performance Computing
HVAC	Heating, Ventilation, Air Condition
IAA	Internationale Automobilausstellung (Frankfurt Motor Show)
IVK	Institut für Verbrennungsmotoren und Kraftfahrwesen
JIS	Japanese Industrial Standards
LBM	Lattice Boltzmann Methode
LDA	Laser Doppler Anemometry

Abbreviations

LEE	Linearized Euler Equations
LES	Large Eddy Simulation
LLF	Large Low-Speed Facility
MIDC	Modified Indian Driving Cycle
MÖK	Motorölkühler (Motor Oil Cooler)
MPG	Miles per Gallon
MPI	Message Passing Interface
MPP	Massive Parallel Systems
MRF	Multiple Reference Frame
NHTSA	National Highway Traffic Safety Administration
NOx	Nitrogen Oxide
NPL	National Physical Laboratory
NRC	National Research Council Canada
NSG	Navier-Stokes-Gleichungen (Navier-Stokes equations)
NWB	Low-Speed Wind Tunnel at DNW
OEM	Original Equipment Manufacturer
PD	Percentage of Dissatisfied (%)
Pep	Product Development Process
PF	Pininfarina
PISO	Pressure-Implicit with Splitting of Operators
PIV	Particle Image Velocimetry
Pkw	Personenkraftwagen (Passenger Car)
PMV	Predicted Mean Vote
PPD	Predicted Percentage of Dissatisfied
RAE	Wheel Rotation Unit
RANS	Reynolds Averaged Navier-Stokes
RSM	Reynolds Stress Model
SAE	Society of Automotive Engineering
SEA	Statistical Energy Analysis
SGS	Sub-Grid Scale
SIMPLE	Semi-Implicit Method for Pressure-Linked Equations
SKE	Steuerbarer Kühlufteinlass (Cooling Air Louvers)
SMP	Symmetric Multi-Processing Systems
SNGR	Stochastic Noise Generation and Radiation
SOP	Start of Production
SPL	Sound Pressure Level (Schalldruckpegel)
STSF	Spatial Transformation of Sound Fields
URANS	Unsteady Reynolds Averaged Navier-Stokes
VDA	Verband der Deutschen Automobilindustrie
VLES	Very Large Eddy Simulation
VR	Variable Resolution
VW	Volkswagen
WBM, WBT	Wave-Based Method/Technique
WLTP	Worldwide Harmonized Light Vehicles Test Procedure

Symbols

Symbol	Specification	Unit	In Chapter
A_A	Exit area	[m²]	
A_B	Underbody surface	[m²]	
A_C	Collector cross section	[m²]	
A_E	Entrance area	[m²]	
A_{closed}	Cross section of solid walls	[m²]	
A_H	Helmholtz Resonator Opening Section	[m²]	
A_K	Radiator cross section	[m²]	
A_{Fan}	Fan cross section	[m²]	
A_M	Double frontal area	[m²]	
A_N	Nozzle cross section	[m²]	
A_{open}	Cross section of a slotted wall test section	[m²]	
A_P	Plenum cross section	[m²]	
A_R	Area ratio	[-]	
A_{Ref}	Reference area	[m²]	
A_S	Cross section of the settling chamber	[m²]	
A_{SEP}	Cross section of the near wake	[m²]	
A_x	Frontal area	[m²]	
$A_{x,A}$	Frontal area of the antenna	[m²]	
$A_{x,K}$	Frontal area of the radiator	[m²]	
$A_{x,S}$	Frontal area of the external mirror	[m²]	
$A_{x,S}$	Frontal area of the spoiler	[m²]	
A_z	Projection area in z-direction	[m²]	
a	Acceleration	[m/s²]	3, 9
a	Distance	[m]	
a_q	Radial acceleration	[m/s²]	
b	Width	[m]	
b	Specific fuel consumption	[g/kWh]	3
b_N	Nozzle width	[m]	
b_S	Fuel consumption	[l/100km]	
C	Collision term (kinetic theory)	[s²/m⁶]	
$C_{i,j}$	Stress tensor (specifies stress by interference of resolved and unresolved vortices)	[kg/(m · s²)]	
c	Speed of sound	[m/s]	
c	Constant	[-]	2
c	Particle velocity (kinetic theory)	[m/s]	14
c_L	Lift coefficient	[-]	

---

* Only used when symbols are used twice and distinguishing is not possible

Symbols

Symbol	Specification	Unit	In Chapter*
$c_{L,r}$	Rear axle lift coefficient	[-]	
$c_{L,f}$	Front axle lift coefficient	[-]	
c_{DES}	Empirical constant for DES	[-]	
c_F	Contamination factor (soiling)	[-]	
c_f	Friction coefficient	[-]	
c_μ	Empirical constant	[-]	
$c_{M,Y}$	Yaw moment coefficient	[-]	
$c_{M,P}$	Pitch moment coefficient	[-]	
$c_{M,R}$	Roll moment coefficient	[-]	
c_p	Specific heat capacity at constant pressure	[kJ/(kg · K)]	7, 13.3
c_p	Pressure coefficient	[-]	
$c_{p,A}$	Pressure coefficient at the exit	[-]	
$c_{p,c}$	Corrected pressure coefficient	[-]	
$c_{p,K}$	Specific heat of the coolant	[kJ/(kg · K)]	
$c_{p,m}$	Measured pressure coefficient	[-]	
$c_{p,\text{stat}}$	Static pressure coefficient	[-]	
$c_{p,\text{tot}}$	Total pressure coefficient	[-]	
c_Q	Suction parameter	[-]	
c_S	Side force coefficient	[-]	
$c_{S,r}$	Rear axle side force coefficient	[-]	
$c_{S,f}$	Front axle side force coefficient	[-]	
c_T	Tangential force coefficient	[-]	
c_u	Velocity coefficient	[-]	
$c_{V,K}$	Cooling airflow rate coefficient	[-]	
c_D	Drag coefficient	[-]	
$c_{D,0}$	Drag coefficient of a profile	[-]	
$c_{D,c}$	Corrected drag coefficient	[-]	
$c_{D,i}$	Induced drag coefficient	[-]	
$c_{D,m}$	Measured drag coefficient	[-]	
c_x, c_y, c_z	Components of dimensionless air force	[-]	
D	Damping factor	[-]	
d	Distance	[m]	
d	Diameter	[m]	
d_a	Outer diameter	[m]	
$dc/d\beta$	Steady-state gradient of the angle of attack	[1/°]	

* Only used when symbols are used twice and distinguishing is not possible

Symbol	Specification	Unit	In Chapter*
d_{hyd}	Hydraulic diameter	[m]	
d_i	Inner diameter	[m]	
d_N	Diameter	[m]	
e	Ground clearance	[m]	
e_{sp}	Distance of center of gravity and side force application point	[m]	
F	Exemplary mathematical function	[-]	2
F	Force	[kg · m/s²]	
F_L	Lift force	[kg · m/s²]	
$F_{L,r}$	Rear axle lift force	[kg · m/s²]	
$F_{L,f}$	Front axle lift force	[kg · m/s²]	
F_B	Brake and acceleration force	[kg · m/s²]	
$F_{B,r}$	Brake force at the contact patches of the rear wheels	[kg · m/s²]	
$F_{B,f}$	Brake force at the contact patches of the front wheels	[kg · m/s²]	
F_D	Pressure force	[kg · m/s²]	2
F_F	Centrifugal force	[kg · m/s²]	
F_G	Vehicle's gravity	[kg · m/s²]	
F_H	Downhill-slope force	[kg · m/s²]	
F_K	Factor of alignment	[-]	
F_{mech}	Mechanic drag	[kg · m/s²]	13
F_N	Normal force	[kg · m/s²]	
$F_{N,tot}$	Total wheel load	[kg · m/s²]	
$F_{N,RA}$	Wheel load at the rear axle	[kg · m/s²]	
$F_{N,FA}$	Wheel load at the front axle	[kg · m/s²]	
F_q	Lateral force	[kg · m/s²]	
F_R	Friction force	[kg · m/s²]	2
F_R	Rolling resistance	[kg · m/s²]	
F_S	Side force	[kg · m/s²]	
$F_{S,r}$	Rear axle side force	[kg · m/s²]	
$F_{S,f}$	Front axle side force	[kg · m/s²]	
F_{SF}	Cornering force	[kg · m/s²]	
F_T	Tangential force	[kg · m/s²]	
F_{Vent}	Ventilation force	[kg · m/s²]	13
F_D	Aerodynamic drag	[kg · m/s²]	
$F_{D,D}$	Pressure component of the aerodynamic drag	[kg · m/s²]	
$F_{D,K}$	Aerodynamic drag of the body work	[kg · m/s²]	13

* Only used when symbols are used twice and distinguishing is not possible

Symbols

Symbol	Specification	Unit	In Chapter*
$F_{D,R}$	Friction component of the aerodynamic drag	[kg · m/s²]	
$F_{D,wheel}$	Aerodynamic drag at the wheel	[kg · m/s²]	13
F_Z	Traction force	[kg · m/s²]	
f	Exemplary flow field variable		2
f	Frequency	[1/s]	
f	Velocity distribution function	[s³/m⁶]	14
f	Profile camber	[m]	9
f_E	Frequency of Edgetone feedback	[1/s]	
f_{HR}	Eigenfrequency of a Helmholtz resonator	[1/s]	
f_m	Modulation frequency	[1/s]	
f_n	Eigenfrequency of room modes	[1/s]	
f_R	Eigenfrequency of the wind tunnel pipe	[1/s]	
f_o	Upper frequency limit	[1/s]	
f_u	Lower frequency limit	[1/s]	
f_W	Natural frequency of passing vortices	[1/s]	
G	Glauert factor	[-]	
Gr	Grashof number	[-]	
g	Gravity acceleration	[m/s²]	
$H(f)$	Step or Heaviside function	[-]	
$H_{a,Y}(f)$	Aerodynamic yaw moment transfer function	[1/°]	
$H_{a,S}(f)$	Aerodynamic side force transfer function	[1/°]	
h	Height	[m]	
h_D	Diffusor height	[m]	
$h_{N,g}$	Nozzle height	[m]	
h_p	Greenhouse height	[m]	
I	Intensity	[kg/s³]	
I	Moment of inertia	[kg · m²]	
I_i	Turbulence intensity	[-]	5
I_n	Sound intensity in direction n	[kg/s³]	
i	Complex number $\sqrt{-1}$	[-]	2
K	Coupling factor	[-]	
K_p	Area ratio plenum to nozzle	[-]	
Kn	Knudsen number	[-]	
K_{WK}	Cooling air efficiency number	[-]	
k	Heat transfer coefficient	[kg/K · s³]	

* Only used when symbols are used twice and distinguishing is not possible

Chapter 16

Symbol	Specification	Unit	In Chapter*
k	Axial ratio	[-]	2
k	Specific turbulent kinetic energy	[m²/s²]	14
k_N	Nozzle factor	[-]	
k_P	Plenum factor	[-]	
L	Sound pressure level	[-]	
L_i	Integral measure of length	[m]	
$L_{i,j}$	Stress tensor (specifies stress by interference of resolved vortices)	[kg/(m · s²)]	
L_P	Sound power level	[-]	
l	Length	[m]	
l_0	Wheel base	[m]	
l_1	Distance between two vehicles	[m]	
l_2	Gap between truck and trailer	[m]	
l_L	Distance between force application point and center of wheel base	[m]	
l_D	Diffusor length	[m]	
l_H	Rear end length measured from upper edge of the rear window	[m]	
l_K	Trunk length	[m]	
l_K	Depth of the radiator	[m]	4.3.3, 4.5.2, 7
l_{CO}	Clutch overhang	[m]	
l_R	Pipe length	[m]	
l_R	Throat length of the resonators	[m]	
l_{ref}	Reference length	[m]	
l_S	Length of the rear window	[m]	
l_{TS}	Test section length	[m]	
M	Moment/torque	[kg · m²/s²]	
Ma	Mach number	[-]	
M_Y	Yaw moment	[kg · m²/s²]	
M_P	Pitch moment	[kg · m²/s²]	
m	Mass	[kg]	
m	Degree of modulation	[-]	8
m_E	Order of Edgetone mode	[-]	
$\dot{m}_F$	Mass flux of the coolant	[kg/s]	
$\dot{m}_K$	Cooling air mass flux	[kg/s]	
m_F	Vehicle mass	[kg]	
m_{LR}	Harmonic order of the fan	[-]	

---

* Only used when symbols are used twice and distinguishing is not possible

Symbol	Specification	Unit	In Chapter
m_R	Order of the pipe mode	[-]	
m_{rad}	Number of radial modes	[-]	
m_U	Number of circumference modes	[-]	
m_x, m_y, m_z	Room mode order	[-]	
N	Slope	[-]	
Nu	Nusselt number	[-]	
n	Surface normal	[m]	
n_{LS}	Number of guide vanes	[-]	
n_K	Correction factor	[-]	
n_S	Number of guide vanes	[-]	
n_x, n_y, n_z	Components of surface normal	[m]	
P	Power	$[\text{kg} \cdot \text{m}^2/\text{s}^3]$	
P_A	Engine power of the vehicle	$[\text{kg} \cdot \text{m}^2/\text{s}^3]$	
Pe	Peclet number	[-]	
P_{Fan}	Fan power	$[\text{kg} \cdot \text{m}^2/\text{s}^3]$	
Pr	Prandtl number	[-]	
p	Pressure	$[\text{kg}/(\text{m} \cdot \text{s}^2)]$	
p'	Pressure fluctuation	$[\text{kg}/(\text{m} \cdot \text{s}^2)]$	
p_{LL}	Charge air pressure	$[\text{kg}/(\text{m} \cdot \text{s}^2)]$	
$p_{LL,E}$	Charge air pressure at entrance	$[\text{kg}/(\text{m} \cdot \text{s}^2)]$	
p_N	Pressure at the nozzle exit	$[\text{kg}/(\text{m} \cdot \text{s}^2)]$	
p_P	Plenum pressure	$[\text{kg}/(\text{m} \cdot \text{s}^2)]$	
p_S	Pressure in the settling chamber	$[\text{kg}/(\text{m} \cdot \text{s}^2)]$	
p_{stat}	Static pressure	$[\text{kg}/(\text{m} \cdot \text{s}^2)]$	
p_{tot}	Total pressure	$[\text{kg}/(\text{m} \cdot \text{s}^2)]$	
p_{loss}	Pressure loss	$[\text{kg}/(\text{m} \cdot \text{s}^2)]$	
p_∞	Pressure of the oncoming flow	$[\text{kg}/(\text{m} \cdot \text{s}^2)]$	
Q	Heat	$[\text{kg} \cdot \text{m}^2/\text{s}^2]$	
Q	Flow rate	$[\text{m}^3/\text{s}]$	2
$\dot{Q}$	Heat flux	$[\text{kg} \cdot \text{m}^2/\text{s}^3]$	
$\dot{Q}_{KL}$	Heat flux into the cooling air	$[\text{kg} \cdot \text{m}^2/\text{s}^3]$	
$\dot{Q}_{KM}$	Heat flux into the coolant	$[\text{kg} \cdot \text{m}^2/\text{s}^3]$	
$\dot{Q}_M$	Engine waste heat	$[\text{kg} \cdot \text{m}^2/\text{s}^3]$	
q	Dynamic pressure	$[\text{kg}/(\text{m} \cdot \text{s}^2)]$	
$\dot{q}$	Specific heat flux	$[\text{kg}/\text{s}^3]$	

* Only used when symbols are used twice and distinguishing is not possible

Symbol	Specification	Unit	In Chapter*
$\dot{q}_\alpha$	Specific heat flux due to convection	[kg/s^3]	
$\dot{q}_h$	Specific heat flux of internal energy	[kg/s^3]	
$\dot{q}_\varepsilon$	Specific heat flux due to radiation	[kg/s^3]	
$\dot{q}_\lambda$	Specific heat flux due to conduction	[kg/s^3]	
q_∞	Dynamic pressure of the oncoming flow	[kg/(m · s^2)]	
R	Specific gas constant	[m^2/(K · s^2)]	
R	Flow resistance	[1/m^2]	2
R_{con}	Convection ratio	[-]	
Re	Reynolds number	[-]	
$R_{i,j}$	Stress tensor (specifies stress by interference of unresolved vortices)	[kg/(m · s^2)]	
R_L	Specific gas constant of air	[m^2/(K · s^2)]	
r	Radius	[m]	
r_1	Radius at the transition front to fender	[m]	
r_2	Radius at the transition front to engine hood	[m]	
r_3	Radius at the transition front to underbody	[m]	
r_4	Radius of the A-pillar	[m]	
r_5	Radius of the front window curvature (xy plane)	[m]	
r_C	Radius of the C-pillar	[m]	
r_D	Radius at the transition roof to side window	[m]	
r_D	Equivalent nozzle radius	[m]	13
r_K	Corner radius	[m]	
r_K	Equivalent collector radius	[m]	13
r_S	Radius at the transition roof to rear slant	[m]	
S	Wrap surface	[m^2]	
$S_{i,j}$	Stress tensor	[kg/(m · s^2)]	
$S_{i,i}(f)$	Power spectral density		
$S_{i,j}(f)$	Cross-power spectral density		
Sr	Strouhal number	[-]	
s	Length	[m]	
T	Temperature	[K]	
Tu	Intensity of turbulence	[-]	
T'	Temperature fluctuation	[K]	
T_A	Temperature at the exit	[K]	
T_E	Temperature at the entrance	[K]	

---

* Only used when symbols are used twice and distinguishing is not possible

Symbols

Symbol	Specification	Unit	In Chapter*
T_F	Fluid temperature	[K]	
T_{Fb}	Boiling temperature of the coolant	[K]	
$T_{i,j}$	Lighthill tensors component	[K]	
T_L	Air temperature	[K]	
$T_{LL,A}$	Charge air exit temperature	[K]	
$T_{LL,E}$	Charge air entrance temperature	[K]	
$T_{x,y,x}$	Components of turbulence intensity in x,y,z direction	[K]	
T_W	Wall temperature	[K]	
T_∞	Temperature of the oncoming flow	[K]	
t	Time	[s]	
t_R	Equivalent throat length of the resonator	[m]	
U	Circumference	[m]	
U	Voltage	[V]	13
U_N	Nozzle circumference	[m]	
u, v, w	Velocity components	[m/s]	
u', v', w'	Velocity fluctuations	[m/s]	
$\bar{u}, \bar{v}, \bar{w}$	Averaged velocity components (RANS)	[m/s]	
$\tilde{u}, \tilde{v}, \tilde{w}$	Filtered velocity components (LES)	[m/s]	
u_∞, v_∞	Velocity of the oncoming flow	[m/s]	
u_G	Velocity of the upper end z_G of the atmospheric boundary layer	[m/s]	
u_m	Mean velocity	[m/s]	
u_R	Mean pipe velocity	[m/s]	
V	Volume	[m³]	
V_F	Volume of the vehicle	[m³]	
$\dot{V}_K$	Cooling airflow rate	[m³/s]	
V_T	Dead wake volume	[m³]	
v	Velocity	[m/s]	
v_A	Exit velocity	[m/s]	
v_B	Velocity of an observer	[-]	2
v_F	Driving speed	[m/s]	
v_i, v_j, v_k	Velocity components	[m/s]	
v_K	Velocity in the radiator	[m/s]	
$v_{\max}$	Maximum speed	[m/s]	
$v_{n,i}, v_{n,j}, v_{n,k}$	Components of the surface normal velocity	[m/s]	

---

* Only used when symbols are used twice and distinguishing is not possible

Chapter 16

Symbol	Specification	Unit	In Chapter*
v_S	Suction velocity	[m/s]	
v_W	Wind velocity	[m/s]	
$w_{k\max}$	Maximum overspeed	[m/s]	2
x, y, z	Cartesian coordinates	[m]	
x_H	Rear overhang	[m]	
x_i, x_j, x_k	Cartesian coordinates (index notation)	[m]	
x_R	Reattachment length	[m]	
x_S	x position of the spoiler	[m]	
z	Braking ratio	[-]	
z_G	Upper end z_G of the atmospheric boundary layer	[m]	
z_S	Spoiler height	[m]	
α	Angle of attack	[°]	
α	Exit angle of the cooling air	[°]	4.3.3, 4.5.2, 7
α	Inclination angle of the engine hood	[°]	
α	Roughness exponent	[-]	
α	Slip angle	[°]	5
α	Gradient angle	[°]	
α	Under relaxation factor	[-]	14
α	Heat transfer coefficient	[kg/(K·s³)]	
α_r	Rear axle slip angle	[°]	
α_S	Absorbance	[-]	
α_F	Front axle slip angle	[°]	
β	Angle of attack	[°]	
β_p	Thermal expansion coefficient	[1/K]	
χ	Led away heat share	[-]	
χ	Position of center of gravity related to the wheel base h/l	[-]	
$\chi_{a,Y}(f)$	Aerodynamic admittance of the yaw moment	[-]	
$\chi_{a,S}(f)$	Aerodynamic admittance of the side force	[-]	
Δ	Difference	[-]	
Δ	Laplace operator	[-]	2
Δ	Mesh size	[m]	14
$\Delta c_{L,r,K}$	Variation of the rear axle lift due to cooling	[-]	
$\Delta c_{L,f,K}$	Variation of the front axle lift due to cooling	[-]	
$\Delta c_{D,C}$	Cooling air drag coefficient	[-]	

* Only used when symbols are used twice and distinguishing is not possible

Symbols

Symbol	Specification	Unit	In Chapter*
$\Delta c_{D,\text{rad}}$	Aerodynamic drag coefficient of the radiator	[-]	
Δp_M	Pressure increase	[kg/(m·s²)]	
Δp_V	Pressure loss	[kg/(m·s²)]	2
Δp_{loss}	Pressure loss	[kg/(m·s²)]	
δ	Boundary layer thickness	[m]	
δ	Inclination angle of the front window	[°]	
$\delta(f)$	Dirac function	[-]	14
δ_1	Displacement thickness	[m]	
δ_2	Momentum thickness	[m]	
$\delta_{i,j}$	Kronecker symbol	[-]	
δ_L	Steering angle	[°]	
ε	Dissipation rate	[m]	2, 14
ε	Coefficient of emission	[-]	7
ε	Dimensionless interference velocity	[-]	13
ε	Angle between front window and engine hood	[°]	
$\varepsilon_{\text{aero}}$	Aerodynamic efficiency	[-]	
ε_C	Interference velocity of the collector at the vehicle	[-]	13
ε_D	Interference velocity at the nozzle	[-]	13
ε_{DP}	Interference velocity dipole	[-]	13
ε_K	Interference velocity at the collector	[-]	13
ε_N	Interference velocity of the nozzle at the vehicle	[-]	13
ε_S	Interference velocity displacement	[-]	13
ε_W	Interference velocity of the far wake	[-]	13
ε_{TD}	Blockage ratio of the dead wake	[-]	13
ϕ	Wind angle	[°]	
ϕ	Real part	[-]	2
ϕ_s	Skalar potential field	[-]	2
Γ	Circulation	[m²/s]	
γ	Inclination angle of the side windows	[°]	
γ	Edge angle	[°]	2
$\gamma^2(f)$	Coherence function	[-]	
γ_D	Diffusor angle	[°]	
η	Efficiency factor	[-]	
η, ξ, ζ	Local coordinates	[m]	
η_ρ	Density recovery	[-]	

* Only used when symbols are used twice and distinguishing is not possible

Chapter 16

Symbol	Specification	Unit	In Chapter*
φ	Flow coefficient	[-]	14
φ	Inclination angle of the rear window	[°]	
φ	Circumferential angle of the cylinder or ellipse	[°]	2
φ	Required lean angle of the motorcycle	[°]	11
$\varphi(x_0)$	Complex amplitude of the acoustic potential		
φ_e	Effective angle	[°]	
$\varphi_{i,i}(t)$	Correlation coefficient		
φ_{crit}	Critical inclination angle of the rear window	[°]	
φ_{MR}	Actual lean angle of the motorcycle	[°]	
κ	Isentropic exponent	[-]	
κ	von Kármám constant	[-]	2, 14
κ	Contraction ratio of the nozzle	[-]	13
Λ	Aspect ratio	[m²]	
Λ	Dimensionless length of the test section	[-]	13
λ	Free distance between molecular collisions	[m]	14
λ	Pipe friction coefficient	[-]	4
λ	Conductivity	[W · m/(s³ · K)]	7
λ	Wave length	[m]	5
λ	Exemplary scalar function	[-]	2
λ_F	Form factor	[-]	
μ	Dynamic viscosity	[kg/(m · s)]	
μ_q	Lateral friction coefficient	[-]	
μ_R	Rolling resistance coefficient	[-]	
ν	Kinematic viscosity	[m²/s]	
ρ	Density	[kg/m³]	
ρ_K	Fuel density	[kg/l]	
ρ_L	Air density	[kg/m³]	
ρ_{LL}	Charge air density	[kg/m³]	
σ	Normal stress	[kg/(m · s²)]	
σ	Standard deviation	[-]	
σ^2	Variance	[-]	5
σ_S	Stefan Boltzmann constant	[kg/(s³ · K⁴)]	
τ	Shear stress	[kg/(m · s²)]	
τ	Time shift	[s]	14

* Only used when symbols are used twice and distinguishing is not possible

Symbol	Specification	Unit	In Chapter*
$\tau_{i,j}$	Shear stress tensor entry	[m²/s²]	
τ_W	Wall shear stress	[kg/(m · s²)]	
ς	Wall roughness coefficient	[-]	
Ω	Circulation	[m/s]	
Ω	Area ratio of collector and nozzle	[-]	
Ω_{LR}	Blatt-Passier frequency	[1/s]	
$\Omega_x, \Omega_y, \Omega_z$	Components of circulation	[m/s]	
ω	Frequency	[1/s]	
ω	Relaxation parameter of the Lattice Boltzmann method	[1/s]	
ω_{aero}	Aerodynamic balance	[-]	
ω_{stat}	Static balance	[-]	
ψ	Pressure number	[-]	14
ψ	Imaginary part	[-]	2
ψ	Yaw angle	[°]	
ζ	Joukowski's mapping function	[m]	2
ζ_K	Loss coefficient of the radiator	[-]	
∇	Gradient	[-]	2

---

* Only used when symbols are used twice and distinguishing is not possible

Literature

1. Abbot, H.; Doehnhoff, E. v.: Theory of Wing Sections. New York, NY: Dover Publications, Inc., 1959.
2. Adam, J.-L.; Ricot, D.; Dubiel, F.; Guy, C.: Aeroacoustic simulation of automotive ventilation outlets. Proc. of Acoustics'08, Paris, 29.06. - 04.07.2008.
3. Adams, N.: Exascale-computing challenges and opportunities in aerodynamic simulations. 10. Tagung Fahrzeugaerodynamik, Haus der Technik, München, 2012.
4. Agnantiaris, J. P.; Polyzos, D.; Beskos, D. E.: Three-dimensional structural vibration analysis by the Dual Reciprocity BEM, Computational Mechanics, Volume 21, Issue 4/5, pp. 372–381, 1998.
5. Ahmed, S. R.: Helicopter Main- and Tail Rotor Performance in Presence of Winds. 27th European Rotorcraft Forum, Moscow, Russia, Sept. 11–14, 2001.
6. Ahmed, S. R.: Influence of Base Slant on the Wake Structure and Drag of Road Vehicles. Transactions of the ASME, Journal of Fluids Engineering, 105, 429–434, 1984.
7. Ahmed, S. R.: Numerical Flow Simulation in Road Vehicle Aerodynamics. Euromotor Short Course "Using Aerodynamics to Improve the Properties of Cars." Stuttgart: FKFS, 1998.
8. Ahmed, S. R.: Theoretical und Experimental Methods in Ground Vehicle Aerodynamics. Brüssel: Von Kármán Insitute for Fluid Dynamics. Short Course, 1986.
9. Ahmed, S. R.; Baumert, W.: The Structure of Wake Flow Behind Road Vehicles. Aerodynamics of Transportation, ASME-CSME Conference, Niagara Falls, 93–103, 1979.
10. Ahmed, S. R.; Hucho, W.-H.: The Calculation of the Flow Field Past a Van with the Aid of a Panel Method. SAE Technical Paper 770390, 1977, doi:10.4271/770390.

11. Ahmed, S. R.; Ramm, G.; Faltin, G.: Some Salient Features of the Time-Averaged Ground Vehicle Wake. SAE Technival Paper 840300, 1984, doi:10.4271/840300.
12. Algermissen, S.; Misol, M.; Unruh, O.: Reduction of Turbulent Boundary Layer Noise with Actively Controlled Carbon Fiber Reinforced Plastic Panels; Adaptive, Tolerant and Efficient Composite Structures; Springer Verlag Berlin Heidelberg.
13. Althaus, D.; Wortmann, F. X.: Stuttgarter Profilkatalog. Braunschweig, Wiesbaden: Vieweg, 1981.
14. Amato, G.: Reaction Thrust from a Vehicle Radiator, Automotive Engineering, Dezember 1980, 67–68, 1980.
15. Ambros, O.; Deussen, N.: Wärmemanagement des Kraftfahrzeugs II. Versuchsfahrzeug mit regelbarem Kühlsystem. Renningen, Expert Verlag, 2000.
16. Ambros, P.: Simulation der luftseitigen Durchströmung eines Kraftfahrzeugvorderwagens zur Auslegung des Motorkühlsystems. Dissertation Universität Stuttgart, Institut für Verbrennungsmotoren und Kraftfahrwesen, 1994.
17. Andreau, J.: Le problème de la voiture économique legere. Journal de la Société des Ingénieurs de l'Automobile, No 3 mai–juin 1946, Tome XIX, page 61, 1946.
18. Angeletti, M.; Sclafani, C.; Bella, G.; Ubertini, S.: The Role of CFD on the Aerodynamic Investigation of Motorcycles. SAE Technical Paper 2003-01-0997, 2003, doi:10.4271/2003-01-0997.
19. Anonym, Institution Valeo: Durchblick ohne Aufhebens. Automobilindustrie Dez. 2002, Der neue Audi A8, 37, 2002.
20. Antonucci, G.; Ceronetti, G.; Costelli, A.: Aerodynamic and Climatic Wind Tunnels in the FIAT Research Center. SAE Technical Paper 770392, 1977, doi:10.4271/770392.
21. Anzenberger, T.; Kronbichler, M.: Auslegung des Wärmehaushalts mit dem Programm Kuli. Interner Bericht AUDI AG, Ingolstadt, 2004.
22. Appel, H.; Breuer, B.; Essers, U.; Helling, J.; Willumeit, H.-P.: UniCar—Der Forschungspersonenwagen der Hochschularbeitsgemeinschaft. Sonderdruck 84 der Automobiltechnischen Zeitschrift, 1982.
23. Araki, J.; Gotou, K.: Development of Aerodynamic Characteristics for Motorcycles using Scale Model Wind Tunnel. SAE Technical Paper 2001-01-1851, 2001, doi:10.4271/2001-01-1851.
24. Arnold, K.: Untersuchungen über den Einfluss der Absaugung durch einen Einzelschlitz auf die turbulente Grenzschicht in anliegender und abgelöster Strömung; Institut für Strömungsmechanik, TU Braunschweig, 1965.
25. Aroussi, A.; Hassan, A.; Clayton, B. R.; AbdoulNur, B. S.: An Assessment of Vehicle Side-Window Defrosting and Demisting Process. SAE Technical Paper 2001-01-0289, 2001, doi:10.4271/2001-01-0289.
26. Ashill, P.; Hackett, J.; Mokry, M.; Steinle, F.: Boundary Measurements Methods, Wind Tunnel Wall Corrections; AGARDograph 336, 1998.
27. Astley, R. J.: Infinite Elements. Kap. 7 in: Marburg, S.; Nolte B. (eds), Computational Acoustics of Noise Propagation in Fluids—Finite and Boundary Element Methods. Springer, Berlin, 2008.
28. Aston, W. G.: Body Design and Wind Resistance. The Autocar, August 1911, 364–366, 1911.

29. Augustin, K.; Paul, M.; Späth, M.; Brotz, F.; Schrumpf, M.: Aeroakustik von Fahrzeugklimageräten. In: Tagungsband der DAGA '07, 19. bis 22. März 2007 in Stuttgart. Berlin: Deutsche Gesellschaft für Akustik e.V. ISBN 978-3-9808659-3-7, 2007.
30. Bachmann, J.; Kirst, B.; Braunsperger, M.: Segmentspezifische Einflussgrößen bei Motorrädern auf die Belastungen von Fahrer und Sozius. Tagung IFZ in Essen, 2002.
31. Baehr, H.-D.; Stephan, K.: Wärme- und Stoffübertragung. Springer Verlag, 2010.
32. Bai, C.; Gosman, A. D.: Development of Methodology for Spray Impingement Simulation, SAE Technical Paper 950283, 1995, doi:10.4271/950283.
33. Bai, C.; Gosman, A. D.: Mathematical Modelling of Wall Films Formed by Impinging Sprays. SAE Technical Paper 960626, 1996, doi:10.4271/960626.
34. Bailly, C.; Lafon, P.; Candel, S.: A Stochastic Approach to Compute Noise Generation and Radiation of Free Turbulent Flows. AIAA 95-092, 1995.
35. Baker, C. J.: Measures to Control Vehicle Movement at Exposed Sites During Windy Periods. Journal of Wind-Engineering and Industrial Aerodynamics, 25(1987) 151–161, Elsevier Science Publishers B.V.Amsterdam, 1987.
36. Banerjee, P. K.; Butterfield, R.: Boundary Element Methods in Engineering Science. McGraw-Hill Book Company, New York, 1981.
37. Bannister, M.: Drag and Dirt Deposition Mechanisms of External Rear View Mirrors and Techniques used for Optimization, SAE Technical Paper 2000-01-0486, 2000, doi:10.4271/2000-01-0486.
38. Bargende, M.: Ein Gleichungsansatz zur Berechnung der instationären Wandwärmeverluste im Hochdruckteil von Ottomotoren. Dissertation, TH Darmstadt, 1991.
29. Barnard, R. H.: Road Vehicle Aerodynamic Design. 2. Auflage. St. Albans, Hertfordshire: Mechaero Publishing, 2001.
40. Barnard, R. H.; Bullen, P. R.; Qiao, J.: Brake and Engine Cooling Flows: Influences and Interactions. MIRA International Vehicle Aerodynamics Conference, 2002.
41. Barnard, R. H.; Bullen, P. R.; Qiao, J.: Fixed and Variable Cooling Outlet Geometries for the Minimisation of Associated Drag. MIRA International Vehicle Aerodynamics Conference, 2004.
42. Barreau, M.; Boutin, L.: Réflexions sur l'énergétique des véhiclules routiers. InterAction. Cachan cedex: Institut Universitaire de Technologie, 2008.
43. Barsikow, B.; Hellmig, M.: Schallquellenlokalisierung bei Vorbeifahrten von Kraftfahrzeugen mittels eines zweidimensionalen Mikrofon-Arrays. Tagungsband der DAGA 2003. Deutsche Gesellschaft für Akustik, Oldenburg, 2002.
44. Barsikow, B.; Klemenz, M.: Diagnosis of noise sources on high-speed trains using the microphone-array technique. In: Tagungsband 16th ICA and 135th Meeting of the ASA, Seattle (USA), Vol. IV, S. 2229-2230, 1998.
45. Barthlott, W.; Neinhuis, C.: Purity of the sacred lotus, or escape from contamination in biological surfaces. Botanisches Institut der Universität Bonn, http://www.botanik.uni-bonn.de/system/planta.htm, Planta 202: 1–8, 1997.
46. Bartsch, C.: Ein Jahrhundert Motorradtechnik. 1. Aufl. Düsseldorf: VDI Verlag 1987.

47. Batista, M.; Perkovic, M.; Najdovski, D.: A simple static analysis of moving road vehicles under crosswind. University of Ljublijana, 2011.
48. Bauer, G.; Schultheis, R.: Der Einsatz von hydrophoben beschichteten Windschutzscheiben in der Automobilindustrie. In Steinmetz, E.: Glas im Automobilbau. Essen, Expert Verlag, 145–159, 2001.
49. Bayer, B.: Aerodynamik im Motorradbau—Widerstandskämpfer. PS (1987) Nr.3, S. 50/55, 1987.
50. Bayer, B.: Das Pendeln und Flattern von Krafträdern—Untersuchungen zur Fahrdynamik von Krafträdern unter besonderer Berücksichtigung konstruktiver Einflussparameter auf die Hochgeschwindigkeitsgeradeausstabilität. Dissertation Darmstadt; 2. Aufl. Bochum: Wirtschaftsverlag NW 1987 (Heft 4 der "Forschungshefte Zweiradsicherheit" des Instituts für Zweiradsicherheit e.V.), 1987.
51. Bayer, B.: Ein Modellansatz zur Beschreibung des Lenkverhaltens von Krafträdern bei stationärer Kreisfahrt. Automobil-Industrie 1985 Nr. 1, S. 33/36, 1985.
52. Bearman, P. W.: Bluff Body Flows Applicable to Vehicle Aerodynamics. In: Morel, T.; Dalton, C. (Eds.): Aerodynamics of Transportation. New York: ASME-CSME Conference, Niagara Falls, 1–11, 1979.
53. Bearman, P. W.: Near Wake Flows behind Two- an Three-dimensional Bluff Bodies. Journal of Wind Engineering and Industrial Aerodynamics, 69–71, 33–54, 1997.
54. Bearman, P. W.: Some Observations on Road Vehicle Wakes. SAE Technical Paper 840301, 1984, doi:10.4271/840301.
55. Bearman, P. W.; Davis, J. P.: Measurement of the Structure of Road Vehicle Wakes. International Journal of Vehicle Design, SP 3, 493–499, 1983.
56. Bearman, P. W.; Davis, J. P.; Harvey, J. K.: Wind Tunnel Investigation on Vehicle Wakes. International Symposium on Vehicle Aerodynamics, Wolfsburg: VW AG, 1982.
57. Bearman, P. W.; Owen, J. C.:Reduction of Bluff-body Drag and Suppression of Vortex Shedding by the Introduction of Wavy Separation Lines. Journal of Fluid Structures, 12, 123–130, 1998.
58. Beauvais, F. N.: Aerodynamic Characteristics of a Car-Trailer Combination. SAE Technical Paper 670100, 1967, doi:10.4271/670100.
59. Béchara, W.; Bailly, C.; Lafon, P.; Candel, S.: Stochastic Approach to Noise Modelling for Free Turbulent Flows. AIAA Journal, Vol. 32, No. 4, S. 455–463, August 1994.
60. Bechert, D. W.; Meyer, R.; Hage, W.: Drag Reduction of Airfoils with Miniflaps. Can We Learn from Dragonflies? AIAA-2000-2315, 2000.
61. Beckenbauer, T.; v. Blokland, G.; Huschek, S.: Einfluss der Fahrbahntextur auf das Reifen-Fahrbahn-Geräusch. Schriftenreihe Forschung Straßenbau und Straßenverkehrstechnik, Bundesministerium für Verkehr, Bau- und Wohnungswesen, Bonn, 2002.
62. Becker, E.: Technische Strömungslehre. Teubner, 1993.
63. Becker, E.; Piltz, E.: Übungen zur Technischen Strömungslehre. Teubner, 1995.
64. Beese, E.: Untersuchungen zum Einfluss der Reynoldszahl auf die aerodynamischen Beiwerte von Tragflügelprofilen in Bodennähe. Dissertation, Ruhr-Universität Bochum, 1982.

65. Beland, O.: Buffeting Suppression Technologies for Automotive Wind Tunnels Tested on a Scale Model. In Wiedemann, J. (Ed.): Progress in Vehicle Aerodynamics and Thermal Management, Expert Verlag, Renningen, ISBN 978-3-8169-2771-6, 2008.
66. Beland, O.: Untersuchungen zur Reduzierung von Raummoden in einem Pilotkanal; FKFS unveröffentlichter Bericht, 2011.
67. Bella, G.; Ubertini, S.; Desideri, U.: Experimental and Computational Analysis of the Aerodynamic Performances of a Maxi–Scooter. SAE Technical Paper 2003-01-0998, 2003, doi:10.4271/2003-01-0998.
68. Bender, T.; Hoff, P.; Kleemann, R.: The New BMW Climatic Testing Complex—The Energy and Environment Test Centre; SAE Technical Paper 2011-01-0167, 2011 doi:10.4271/2011-01-0167.
69. Bengsch, H.: Der neue Klima-Windkanal bei Ford, Köln. ATZ, 80, 17–26, 1978.
70. Berge, W.: Sturzhelm und Verfahren zum Absenken des Geräuschpegels für einen Sturzhelmträger. Offenlegungsschrift DE4207873A1, 1992.
71. Berge, W.: Untersuchungen an Motorradhelmen im Windkanal. Forschungshefte Zweiradsicherheit Nr. 5, Institut für Zweiradsicherheit, 1987.
72. Bergmann, A.; Bernardy, M.; Weber, S.: Aeroakustischer Windkanal; Offenlegungsschrift einer Patentanmeldung DE 10 2010 060 929 A1, 2012.
73. Betz, A.: Konforme Abbildung. Springer, 1948.
74. Betz, J.: Part and Potentials of the 1D Simulation in the Agggregate Cooling and Thermal-Management of a Vehicle. 98–114 in Wiedemann, J.; Hucho, W.-H.; Progress in Vehicle Aerodynamics II—Thermal-Management. Expert Verlag Renningen Malmsheim, 2002.
75. Bez, U.: Bestimmung des Luftwiderstandsbeiwertes bei Kraftfahrzeugen durch Auslaufversuch. ATZ, 76, 345–350, 1974.
76. Bhatnagar, P. L.; Gross, E. P.; Krook, M.: A Model for Collision Processes in Gases. I. Small Amplitude Processes in Charged and Neutral One-Component Systems. Physical Review, 94 (3): pp. 511–525, 1954.
77. Billot, P.; Jallet, S.; Marmonier, F.: Simulation of Aerodynamic Uplift Consequences on Pressure Repartition—Application on an Innovative Wiper Blade Design. SAE Technical Paper 2001-01-1043, 2001, doi:10.4271/2001-01-1043.
78. Bitzel, F.: Die Einwirkung von Seitenwindkräften auf den Straßenverkehr. Zeitschrift für Verkehrssicherheit, Heft 2, 1961, 1962.
79. Blumrich, R.: Berechnungsmethoden für die Aeroakustik von Fahrzeugen, ATZ/MTZ-Konferenz Akustik–Akustik zukünftiger Fahrzeuge, Stuttgart, 17.-18.05.2006.
80. Blumrich, R.: Numerische Aeroakustik, Workshop "Mess- und Analysetechnik in der Fahrzeugakustik," FKFS, Stuttgart, 09.–10. Oktober 2007.
81. Blumrich, R.: Spielen Schallausbreitungseffekte durch die inhomogene Umströmung in der Fahrzeugaeroakustik eine Rolle? In: Tagungsband der DAGA 2010, 15.-18.03.2010, Berlin. Berlin: Deutsche Gesellschaft für Akustik e.V., ISBN: 978-3-9808659-8-2, 2010.
82. Blumrich, R.: Vehicle Aeroacoustics—Today and Future Developments. 5th Int. Styrian Noise, Vibration & Harshness Congress "Optimising NVH in future vehicles" Graz, 4th–6th, June, 2008.

83. Blumrich, R.; Crouse, B.; Freed, D.; Hazir, A.; Balasubramanian, G.: Untersuchung der Kopplungseffekte Innenraum-Kofferraum beim Schiebedachwummern an Hand eines SAE-Modells. In: Tagungsband der DAGA '07, 19. bis 22. März 2007 in Stuttgart. Berlin: Deutsche Gesellschaft für Akustik e.V., ISBN 978-3-9808659-3-7, 2007.
84. Blumrich, R.; Hazir, A.: Berechnung der Fahrzeugaeroakustik–Methoden und Beispiele. 3. Seminar "Mess- und Analysetechnik in der Fahrzeugakustik," 11.-12.10.2011, Stuttgart, 2011.
85. Blumrich, R.; Heimann, D.: A Linearized Eulerian Sound Propagation Model for Studies of Complex Meteorological Effects, J. Acoust. Soc. Am. Vol. 112 (2), pp. 446–455, 2002.
86. Bönsch, H. W.: Fortschrittliche Motorrad—Technik. 1. Aufl. Stuttgart: Motorbuch Verlag, 1985.
87. Börger, G.: Optimierung von Windkanaldüsen für den Unterschallberich; ZfW 23, 1975.
88. Bosbach, J.; Lange, S.; Dehne, T.; Lauenroth, G.; Hesselbach, F.; Allzeit, M.: Alternative Ventilation Concepts for Aircraft Cabins. Bremen: Deutscher Luft- und Raumfahrt Kongress, 10–12. April, 2012.
89. Bösch, P.: Der Fahrer als Regler. Dissertation, TU Wien, 1991.
90. Bosnjakovic, F.: Technische Thermodynamik, Teil 1, Dresden: Verlag Theodor Steinkopff, 1967.
91. Bourlet, C.: Nouveau traité des bicycles et bicyclettes. Paris, 1898.
92. Bradshaw, P.: Experimental Fluid Mechanics; Pergamon Press LTD.; Library of Congress Card No. 63-18932, (1964)
93. Bradshaw, P.; Pankhurst, R. C.: The Design of Low-Speed Wind Tunnels; Prog. In Aeronautical Sci., Vol. 5, 1964.
94. Bradtke, F.; Liese, W.: Hilfsbuch für raum- und außenklimatische Messungen für hygienische, gesundheitstechnische und arbeitsmedizinische Zwecke. Berlin, Heidelberg: Springer Verlag, 1952.
95. Braess, H.-H.; Seiffert, U.: Vieweg Handbuch Kraftfahrzeugtechnik. 3. Auflage, Vieweg+Teubner Verlag, ISBN-13 978-3528231149, 2003.
96. Brandstätter, W.: Berechnung von Windkanaldüsen für den Unterschallbereich mit Hilfe der Methode der finiten Elemente: Bericht 8030, Institut f. Strömungslehre der TU Graz, 1980.
97. Brauer, H.: Grundlagen der Einphasen- und Mehrphasenströmungen; Verlag Sauerländer, Arau 1971.
98. Braun, H.: Neue Erkenntnisse über Radabdeckungen. Deutsche Kraftfahrforschung und Straßenverkehrstechnik, Heft 223, Düsseldorf: VDI Verlag, 1972.
99. Braunsperger, M.: Entwicklungstendenzen im Motorradbau aus Sicht von BMW. Internationale Fachtagung: Entwicklungstendenzen im Motorradbau, Zwickau, Haus der Technik, 2002.
100. Breuer, B.; Bayer, B.; Pyper, P.; Weidele, A.: Motorräder. Vorlesungsumdruck TH Darmstadt, 1985.

101. Bröhl, H.: Paul Jaray—Stromlinienpionier. Bern: Selbstverlag des Autors, 1978.
102. Brown, C. H.; Gordon, M. S.: Motorcycle Helmet Noise and Active Noise Reduction; The Open Acoustics Journal, 4, 14–24, 2011.
103. Brown, G. J.: Aerodynamic Disturbances Encountered on Highway Passing Situations. SAE Technical Paper 730234, 1973, doi:10.4271/730234.
104. Brühwiler, P.: Heated, Perspiring Manikin Headform for the Measurement of Headgear Ventilation Characteristics. Measurement Science and Technology, 14, 217–227, 2003
105. Bruun, H. H.: Scientific Instruments 4. Journal of Physics E, 815–820, 1971.
106. Buchheim, R. Leie, B.; Lückoff, H.-J.: Der neue Audi 100—Ein Beispiel für konsequente aerodynamische Personenwagen-Entwicklung. ATZ, 85, 419–425, 1983.
107. Buchheim, R.; Deutenbach, K.-R.; Lückoff, H.-J.: Necessity and Premises for Reducing the Aerodynamic Drag of Future Passenger Cars. SAE Technical Paper 810185, 1981, doi:10.4271/810185.
108. Buchheim, R.; Durst, F.; Beeck, M. A.; Hentschel, W. et al., Advanced Experimental Techniques and Their Application to Automotive Aerodynamics. SAE Technical Paper 870244, 1987, doi:10.4271/870244.
109. Buchheim, R.; Under, R.; Jousserandot, P.; Mercker, E. et al.: Comparison Test Between Major European and North American Automotive Wind Tunnels; SAE Technical Paper 830301, 1983, doi:10.4271/830301.
110. Buchheim, R.; Leie, B.; Lückoff, H.-J.: Der neue Audi 100–Ein Beispiel für konsequente aerodynamische Personenwagen-Entwicklung. ATZ, 85, 419–425, 1983.
111. Buchheim, R.; Maretzke, J.; Piatek, R.: The Control of Aerodynamic Parameters Influencing Vehicle Dynamics. SAE Technical Paper 850279, 1985, doi:10.4271/850279.
112. Buchheim, R.; Piatek, R.; Walzer, P.: Contribution of Aerodynamics to Fuel Economy Improvements for Future Passenger Cars. First International. Automotive Fuel Economy Conference, Washington, 1979.
113. Buchheim, R.; Röhe, H.; Wüstenberg, H.: Experiences with Computational Fluid Mechanics in Automotive Aerodynamics. 2nd Intern. ATA Symposium, Use of Supercomputers in the European Automotive Industry, Turin, Italy, 1988.
114. Buchheim, R.; Unger, R.; Carr, G. W.; Cogotti, A.; Kuhn, A.; Nilsson, L. U.: Comparison Tests between Major European Wind Tunnels. SAE Technical Paper 800140, 1980, doi:10.4271/800140.
115. Buckingham, E.: On Physically Similar Systems. In: Phys. Rev. 4, 1914.
116. Buckley, F. T. Jr.: ABCD—An Improved Coast Down Test and Analysis Method. SAE Technical Paper 950626, 1995, doi:10.4271/950626.
117. Budó, A.: Theoretische Mechanik. VEB Deutscher Verlag der Wissenschaften, 1987.
118. Burbach, U.: Fünf Tourenverkleidungen im Test—Gegen Wind und Wetter. Motorrad Nr. 9, S. 16/27, 1982.
119. Burch, S. D.; Hassani, V.; Penney, T.: Use of Infra-Red Thermography for Automotive Climate Control Analysis. SAE Technical Paper 931136, 1993, doi:10.4271/931136.

120. Burgade, L.: Mise au point aéroacoustique des véhicules chez PSA. In: Tagungsband "Interactions mécaniques entre fluides et structures." Colloques 6-2000, S. 16-21; ISSN 0018-6368, 2000.
121. Busch, J.: Verfahren zum Vergleich von Fahrzeug-Umströmungsgeräuschen auf der Straße und im Windkanal. Renningen-Malmsheim: Expert Verlag, ISBN 3-8169-1528-0, 1997.
122. Cairns, R. S.: Lateral Aerodynamic Characteristics of Motor Vehicles in Transient Crosswinds. PhD Thesis Cranfield University, 1994.
123. Calkins, D.; Chan, W.: "CDaero"—A Parametric Aerodynamic Drag Prediction Tool. SAE Technical Paper 980398, 1998, doi:10.4271/980398.
124. Caraeni, M.; Aybay, O.; Holst, S.: Tandem Cylinder and Idealized Side Mirror Far-Field Noise Predictions Using DES and An Efficient Implementation of FW-H Equation. 17. AIAA/CEAS Aeroacoustics Conference, Portland, Oregon, 5–8 June, AIAA 2011-2843, 2011.
125. Carlino, G.; Cardano, D.; Cogotti, A.: A New Technique to Measure the Aerodynamic Response of Passenger Cars by Continuous Flow Yawing. SAE Technical Paper 2007-01-0902, 2007, doi:10.4271/2007-01-0902.
126. Carr, G. W.: Aerodynamic Effects of Modifications to a Typical Car Model. MIRA-Rep. No. 1963/4, 1963.
127. Carr, G. W.: Aerodynamic Effects of Underbody Details on a Typical Car Model. MIRA Rep. No. 1965/7, 1965.
128. Carr, G. W.: Correlation of Aerodynamic Force Measurements in MIRA and Other Automotive Wind Tunnels. SAE Technical Paper 820374, 1982, doi:10.4271/820374.
1298. Carr, G. W.: Influence of Rear Body Shape on the Aerodynamic Characteristics of Saloon Cars. MIRA Rep. Nr. 1974/2, 1974.
130. Carr, G. W.: New MIRA Drag Reduction Prediction Method for Cars. Automotive Engineer, June/July 1987, 34–38, 1987.
131. Carr, G. W.: Potential for Aerodynamic Drag Reduction in Car Design. London: International Association for Vehicle Design, SP 3, 44–56, 1983.
132. Carr, G. W.: The Aerodynamics of Basic Shapes für Road Vehicles, Part I. Simple Rectangular Bodies. Motor Industry Research Association (MIRA), Report No. 1968/2, 1968.
133. Carr, G. W.; Eckert, W.: A Further Evaluation of the Ground-Plane Suction Method for Ground Simulation in Automotive Wind Tunnels. SAE Technical Paper 940418, 1994, doi:10.4271/940418.
134. Carr, G. W.; Rose, M. J.: Correlation of Full-Scale Wind Tunnel and Road Measurements of Aerodynamic Drag, MIRA Report 3, Nuneaton, 1963.
135. Carr. G. W.: The MIRA Quarter Scale Wind Tunnel. MIRA-Report 1961/11, 1961.
136. Catchpole, P.; Martin, R.; Steinmeier, E.; Dömeland, P.: Vehicle Wind Noise Optimization at BMW. Ingénieurs de l'Automobile, 120–124, 1989.
137. Chadwick, A.: Crosswind Aerodynamics of Sport Utility Vehicles. PhD Thesis, Cranfield University, 1999.

138. Chadwik, A.; Garry, K.; Howell, J.: Transient Aerodynamic Characterstics of Simple Vehicle Shapes by the Measurement of Surface Pressures. SAE Technical Paper 2000-01-0876, 2000, doi:10.4271/2000-01-0876.
139. Chandra, S.; Avedisian, C. T.: On the Collision of a Droplet with a Solid Surface. Proc. R. Soc., Vol. 432,13–41, 1991.
140. Chen, H.: Volumetric Formulation of the Lattice Boltzmann Method for Fluid Dynamics. Physical Review E, Vol. 58, 1998
141. Chen, Kandasamy, Shock, Osrzag: Extended-Boltzmann Kinetic Equation for Turbulent Flows. Science Magazine, Vol. 301, Aug 1, 2003
142. Chen, Staroselsky, Orszag, Succi: Expanded Analogy Between Boltzmann Kinetic Theory of Fluids and Turbulence. Journal of Fluid Mechanics, Vol. 519, 2004.
143. Chen, Teixeira, Molvig: Digital Physics® Approach to Computational Fluid Dynamics: Some Basic Theoretical Features. Intl. Journal of Modern Phyics C, Vol. 9, No. 4, 1997.
144. Chikada, T.; Yoshida, K.: Physical Characteristics of Accessory-Equipped Motorcycles. In: Proceedings of the "International motorcycle safety conference." Washington, 1980.
145. Chmielarz, M.; Galley, N.; Kubitzki, J.; Schneider, W.: Einfluss wasserabweisender Beschichtungen auf Windschutzscheiben im Hinblick auf Sicht und Fahrzeugsicherheit. FAT-Schriftenreihe Nr. 167, 1–149, 2001.
146. Christensen, F. M.: Die Klimaversuchsanlage von Volvo. ATZ, 75, 42–45, 1973.
147. Chue, S. H.: Pressure Probes for Fluid Measurement. Progress in Aerospace Science, Vol.16, No. 2, pp. 147–223, Pergamon Press, 1975.
148. Clarke, R. M.: Heavy Truck Splash and Spray Supression: Near and Long Term Solutions. SAE Technical Paper 831178, 1983, doi:10.4271/831178.
149. Cocco, G.: Motorradtechnik pur. Motorbuch Verlag, Stuttgart, 2001.
150. Cogotti, A.: A Strategy for Optimum Surveys of Passenger-Car Flow Fields. SAE Technical Paper 890374, 1989, doi:10.4271/890374.
151. Cogotti, A.: Aerodynamic Characteristics of Car Wheels. International Journal of Vehicle Design, SP3, Impact of Aerodynamics on Vehicle Design, 173–196, 1983.
152. Cogotti, A.: Car-Wake Imaging Using a Seven-Hole Probe. SAE Technical Paper 860214, 1986, doi:10.4271/860214.
153. Cogotti, A.: Evolution of Performance of an Automotive Wind Tunnel. Journal of Wind Engineering and Industrial Aerodynamics 96, 667–700, Elsevier Ltd., 2008.
154. Cogotti, A.: Experimental Techniques for the Aerodynamic Development of Convertible Cars. SAE Technical Paper 920347, 1992, doi:10.4271/920347.
155. Cogotti, A.: Flow Field Measurements and their Interpretation. 97–120 in Wiedemann, J.; Hucho, W.-H. (Hrsg.): Progress in Vehicle Aerodynamics—Advanced Experimental Techniques. Renningen: Expert Verlag, 2000.
156. Cogotti, A.: Flow-Field Surveys Behind Three Squareback Car Models Using a New Fourteen-Hole Probe. SAE Technical Paper 870243, 1987, doi:10.4271/870243.
157. Cogotti, A.: Flow Visualisation Techniques in the Pininfarina Full-Scale Automotive Wind Tunnel. 8th International Symposium on Flow Visualisation, 1998.

158. Cogotti, A.: Generation of a Controlled Level of Turbulence in the Pininfarina Wind Tunnel for the Measurement of Unsteady Aerodynamics and Aeroacoustics. SAE Technical Paper 2003-01-0430, 2003, doi:10.4271/2003-01-0430.
159. Cogotti, A.: Generation of a Controlled Turbulent Flow in an Automobile Wind Tunnel and its Effect on Car Aerodynamics and Acoustics. In Wiedemann, J.; Hucho, W.-H. (Eds.): Progress in Vehicle Aerodynamics III—Unsteady Flow Effects, Renningen: Expert Verlag, 150–176, 2004.
160. Cogotti, A.: Prospects for Aerodynamic Research in the Pininfarina Wind Tunnel. Paper 905149, XXIII FISITA Congress, Torino, 1990.
161. Cogotti, A.: The New Moving Ground System of the Pininfarina Wind Tunnel. SAE Technical Paper 2007-01-1044, 2007, doi:10.4271/2007-01-1044.
162. Cogotti, A.: Unsteady Aerodynamics at Pinifarina; Road Turbulence Simulation and Time Dependent Techniques. 351–382 in Bargende, M.; Wiedemann, J. (Hrsg. 2003): Karaftfahrwesen und Verbrennungsmotoren. 5. Internationales Stuttgarter Symposium. Renningen: Expert Verlag, 2003.
163. Cogotti, A.: Update on the Pininfarina "Turbulence Generation System" and its Effects on the Car Aerodynamics and Aeroacoustics. SAE Technical Paper 2004-01-0807, 2004, doi:10.4271/2004-01-0807.
164. Cogotti, A.; Buchheim, R.; Garrone, A.; Kuhn, A.: Comparison Tests between Some Full-Scale European Automotive Wind Tunnels—Pininfarina Reference Car. SAE Technical Paper 800139, 1980, doi:10.4271/800139.
165. Cogotti, A.; De Gregorio, F.: Presentation of Flow Field Investigation by PIV on a Full-Scale Car in the Pininfarina Wind Tunnel. SAE Technical Paper 2000-01-0870, 2000, doi:10.4271/2000-01-0870.
166. Cogotti, C.: Generation of a Controlled Level of Turbulence in the Pininfarina Wind Tunnel for Measurements of Unsteady Aerodynamics an Aeroacoustics; SAE Technical Paper 2003-01-0430, 2003, doi:10.4271/2003-01-0430.
167. Cogotti, F.; Widdecke, N.; Kuthada, T.; Wiedemann, J.: A New Semi-Emperical Pad Correction; Proc. 5. Internationales Stuttgarter Symposium Automobil- und Motorentechnik, Expert Verlag, Renningen, ISBN 13: 978-3-8169-2180-6, 2003.
168. Coleman, P. B.; Wallis, S. B.; Dale, G. A.; Jordan, J. D.; Watkins, N.; Goss, L.; Davis, J. C. P. N.; Walter, T. M.: A Comparisaon of Pressure.

THE ABOVE REFERENCE NEEDS PUBLISHER INFORMATION.

169. Cooper, K. R.: Influences of Ground Simulation on a Simple Body and Diffuser. EUROmotor 2000 Short Course on "Progress in Vehicle Aerodynamics—Advanced Experimental Techniques," FKFS, Stuttgart, Germany, March 29–30, 2000.
170. Cooper, K. R.: The effect of aerodynamics on the performance and stability of high speed motorcycles. In: Proceedings of the "Second AIAA Symposium on Aerodynamics of Sports and Competition Automobiles." Los Angeles, 1974.
171. Cooper, K. R.: The Effect of Handlebar Fairings on Motorcycle Aerodynamics. SAE Technical Paper 830156, 1983, doi:10.4271/830156.

172. Cooper, K. R.; Syms, J.; Sovran, G.: Selecting Automotive Diffusers to Maximise Underbody Downforce. SAE Technical Paper 2000-01-0354, 2000, doi:10.4271/2000-01-0354.

173. Cooper, K.: Closed-Test-Section Wind Tunnel Blockage Corrections for Road Vehicles. SAE Vehicle Aerodynamics Packet—2006 Edition. SAE International, VAP-2006, Warrendale, PA. 1996.

174. Cooper, K.; Bertenyi, T.; Dutil, G.; Syms, G.; et al.: The Aerodynamic Performance of Automotive Underbody Diffusers. SAE Technical Paper 980030, 1998, doi:10.4271/980030.

175. Cooper, K.; Mokry, M.; Gleason, M.: The Two-Variable Boundary-Interference Correction Applied to Automotive Aerodynamic Data; SAE Technical Paper 2008-01-1204, 2008, doi:10.4271/2008-01-1204.

176. Cooper, R. K.; Watkins, S.: The Unsteady Wind Environment of Road Vehicles, Part One: A Review of the On-road Turbulent Wind Environment. SAE Technical Paper 2007-01-1236, 2007, doi:10.4271/2007-01-1236.

177. Cooper, K. R.: SAE Wind Tunnel Test Procedure for Trucks and Buses, SAE J 1252:1981.

178. Cosano, L.; Colombano, M.: An Optimized Transport Concept for Tractor-Semitrailer Combination. VDI-Berichte Nr. 1986, 65–78, 2007.

179. Cossali, G. E.; Brunello, G.; Coghe, A.; Marengo, M.: Impact of a Single Drop on a Liquid Film: Experimental Analysis and Comparison with Empirical Models. Italian Congress of Thermofluid Dynamics UIT, 1999.

180. Costelli, A.; Garrone, A.; Visconti, A.; Buchheim, R. et al.: FIAT Research Center Reference Car: Correlation Tests between Four Full Scale European Wind Tunnels and Road. SAE Technical Paper 810187, 1981, doi:10.4271/810187.

181. Cousteix, J.: Analyse theorique et moyens de Prevision de la Couche Limite turbulente tridimensionelle. ONERA Publication No. 157, 1974.

182. Crites, R. C.: Pressure Sensitive Paint Technique. Von Karman Institute for Fluid Dynamics, Lectue Series 1993-05 Measurement Techniques, 1993.

183. Crouse, B.; Freed, D.; Senthooran, S.; Ullrich, F.; Fertl, C.: Analysis of Underbody Windnoise Sources on a Production Vehicle Using a Lattice Boltzmann Scheme, SAE Technical Paper 2007-01-2400, 2007, doi:10.4271/2007-01-2400.

184. Crouse, B.; Senthooran, S.; Balasubramanian, G.; Freed, D.: Computational Aeroacoustics Investigation of Automobile Sunroof Buffeting, SAE Technical Paper 2007-01-2403, 2007, doi:10.4271/2007-01-2403.

185. Cummings, J. B.: Lockheed-Georgia Low Speed Wind Tunnel—Some Implications to Automotive Aerodynamics. SAE Technical Paper 690188, 1969, doi:10.4271/690188.

186. Curle, N.: The Influence of Solid Boundaries upon Aerodynamic Sound. Proc. Roy. Soc. 231(A) S. 505–514, London, 1955.

187. Czichos, H.; Hennecke, M. (Hrsg.): Hütte–das Ingenieurwissen. Springer, Berlin, 2007.

188. Davenport, A. G.: Some Aspect of Wind Loading. Transaction of the Institute of Canada Paper EIC 63 CIV 3, 1963.

189. Davenport, A. G.: The Application of Statistical Concepts to the Wind Loading of Structures. Proceedings of the Institution of Civil Engineers, Vol, 19, Issue No. 4, pp. 449–472, 1961.
190. Demuren, A. O.; Rodi, W.: Calculation of Three-dimensional Turbulent Flow Around Car Bodies. Wolfsburg: International Symposium Vehicle Aerodynamics, 2–3. October, 1982.
191. Desmet, W.; Sas, P.; Vandepitte, D.: Performance of a Wave Based Prediction Technique for 3D Coupled Vibro-Acoustic Analysis. In: Proceedings of EuroNoise98, München, Vol. 3, 1998.
192. Dettki, F.: Eine Bewertungsmethode für den Geradeauslauf von Pkw unter Berücksichtigung von Seitenwind. 4. Internationales Stuttgarter Symposium "Kraftfahrwesen und Verbrennungsmotoren," Expert Verlag, Stuttgart, 2001.
193. Deutenbach, K.-R.: Persönliche Mitteilung, 1988.
194. Deutenbach, R.: Influence of Plenum Dimensions on Drag Measurements in ¾-Open-Jet Automotive Wind Tunnels; SAE Technical Paper 951000, 1985, doi:10.4271/951000.
195. Deutsches Zentrum für Luft- und Raumfahrt (DLR): "Breites Spektrum der Strömungsforschung im Kryo Kanal Köln (KKK)," http://www.dlr.de.
196. Dietz, S.: Der neue Audi A2—Ein Meilenstein in der Fahrzeugaerodynamik. ATZ-MTZ spezial, 80–91, 2000.
197. Dietz, S.: Die Auswirkung aerodynamischer Eigenschaften auf das Thermomanagement eines Kraftfahrzeuges. 647–658 in Bargende, M.; Wiedemann, J. (Hrsg. 1999): Kraftfahrwesen und Verbrennungsmotoren. 3. Stuttgarter Symposium. Renningen-Malmsheim: Expert Verlag, 1999.
198. Dietz, S.; Kolpatzik, S.; Lührmann, L.; Widmann, U.: Der neue Audi A4— Aerodynamik im Feinschliff. ATZ-MTZ Sonderausgabe, 70–77, 2000.
199. Dilgen P. G.: Berechnung der abgelösten Strömung um Kraftfahrzeuge: Simulation des Nachlaufs mit einem inversen Panelverfahren. Fortschritt-Berichte VDI, Reihe 7, 258, Düsseldorf: VDI Verlag, 1995.
200. Dilgen, P. G.; Papenfuß, H.-D.; Gersten, K.: Berechnung der abgelösten Strömung um Kraftfahrzeuge mit Hilfe der Zonenmethode im Hinblick auf die Kraftfahrzeugaerodynamik. DGLR-bericht 90-06, 236–240, 1991.
201. Dilgen, P.: Berechnung der abgelösten Strömung um Kraftfahrzeuge: Simulation des Nachlaufes mit einem inversen Panelverfahren. Reihe 7: Strömungsmechanik, No. 258. Düsseldorf: VDI Verlag, 1995.
202. DIN 33403: Klima am Arbeitsplatz und in der Arbeitsumgebung.
203. DIN 45631: Berechnung des Lautstärkepegels und der Lautheit aus dem Geräuschspektrum. Verfahren nach E. Zwicker, 1991.
204. DIN ISO 7730: Moderate thermal environments—Determination of the PMV and PPD Indices and Specification of the Conditions for Thermal Comfort. International Standards Organisation, Geneva, 1991.
205. DMSB (Deutscher Motor Sport Bund e. V.) Motorradsport Handbuch. Deutsche Motor Sport Wirtschaftsdienst GmbH, Frankfurt, 2003
206. Dobrzynski, W.: Windgeräusche am Kraftfahrzeug in: Akustik und Aerodynamik des Kraftfahrzeuges, Hrsg. Syed R. Ahmed, Renningen: expert verlag, 48–73, 1995.

207. Dobrzynski, W.: Zur Bedeutung von Strömungswechseldrücken auf der Karosserieoberfläche für den Innenlärm von Personenkraftfahrzeugen. Berlin, Technische Universität, Dissertation, 1983.
208. Docton, M. K. R.: The simulation of transient cross winds on passenger vehicles. Doctoral thesis, Durham University. Available at Durham E-Theses Online: http://etheses.dur.ac.uk/1580/, 1996.
209. Doherty, J.: Aerodynamic Design Optimization Applied to a Formula One Car. MIRA International Vehicle Aerodynamics Conference, 2002.
210. Döhring, E.: Über die Stabilität und die Lenkkräfte von Einspurfahrzeugen. Dissertation Braunschweig, 1953.
211. Dominy, R. G.; Ryan, A.: An Improved Wind Tunnel Configuration for the Investigation of Aerodynamic Cross Wind Gust Response; SAE Technical Paper 1999-01-0808, 1999, doi:10.4271/1999-01-0808.
212. Dominy, R. G.; Ryan, A.; Sims-Williams, D. B.: The Aerodynamic Stability of a Le Mans Prototype Race Car Under Off-Design Pitch Conditions. SAE Technical Paper 2000-01-0872, 2000, doi:10.4271/2000-01-0872.
213. Dubs, F.: Aerodynamik der reinen Unterschallströmung. 5. Auflage. Basel: Birkhäuser Verlag, 1987.
214. Duell, E. G.; George, A. R.: Experimental Study of a Ground Vehicle Body Unsteady Near Wake. SAE Technical Paper 1999-01-0812, 1999, doi:10.4271/1999-01-0812.
215. Duell, E.; Everstine, D.; Mehta, R.; Bell, J.; Perry, M.: Pressure-Sensitive Paint Technology Applied to Low-Speed Automotive Testing. SAE Technical Paper 2001-01-0626, 2001, doi:10.4271/2001-01-0626
216. Duell, E.; Kharazi, A.; Muller, S.; Ebeling, W.; et al.: The BMW AVZ Wind Tunnel Center; SAE Technical Paper 2010-01-0118, 2010, doi:10.4271/2010-01-0118.
217. Duncan, D. R.: Numerical Methods for Wave Equations in Geophysical Fluid Dynamics. Springer Verlag, New York, Chap. 2.2.3, 1999.
218. Duncan, L.: The Effect of Deck Spoilers and Two-Car Interference on the Body Pressures of Race Cars, SAE Technical Paper 942520, 1994, doi:10.4271/942520.
219. Durst, F.; Melling, A.; Whitelaw, J. H.: Theorie und Praxis der Laser-Doppler-Anemometrie. Karlsruhe: Verlag G. Braun, 1987.
220. Eberz, T. J.: Beiträge zur 3D-Kfz-Aerodynamik—Experimentelle und theoretische Untersuchungen der Nachlaufströmung, ihrer Modellierung und der Widerstandsreduktion. Dissertation, Universität Siegen. Aachen, Shaker Verlag, 2001.
221. Eck, B.: Technische Strömungslehre. Springer, 1961.
222. Eckelmann, H.: Einführung in die Strömungsmesstechnik. Stuttgart: Teubner Verlag, 1997.
223. Eckert, B.: Das Kühlgebläse im Kraftfahrzeug und sein betriebliches Verhalten. Deutsche Kraftfahrzeugforschung, 51, Berlin: VDI Verlag, 1940.
224. Eichlseder, H.: Evaluation of Fuel Efficiency Improvements in the Heavy Duty Vehicle Sector from Improved Trailer and Tire Designs by Application of a New Test Procedure. University of Graz, Report No. I-24/2011, 2011.

225. Eichlseder, W.; Hager, J.; Raup, M.; Dietz, S.: Auslegung von Kühlsystemen mittels Simulationsrechnung, ATZ 99, 638–647, 1999.

226. Ellis, J. R.: Vehicle Dynamics. London: Business Books, 1969.

227. Emmelmann, H.- J.; Berneburg, H.; Schulze, J.: The Aerodynamic Development of the Opel Calibra. SAE Technical Paper 900317, 1990, doi:10.4271/900317.

228. Emmelmann, H.-J.: Aerodynamic Development and Conflicting Goals of Subcompacts—Outlined on the Opel Corsa. International Symposium on Vehicle Aerodynamics. Wolfsburg: VW AG, 1982.

229. Emmenthal, K.-D.: Verfahren zur Auslegung des Wasserkühlsystems von Kraftfahrzeugen. Dissertation RWTH Aachen, 1974.

230. Emmenthal, K.-D.; Hucho, W.-H.: A Rational Approach to Automotive Radiator Systems Design. SAE Technical Paper 740088, 1974, doi:10.4271/740088.

231. Engler, R. H.; Merienne, M.-C.; Klein, C.; LeSant, Y.: Application of PSP in Low Speed Flows. Aerospace Science and Technology 6, 313–322, 2002.

232. Eppinger, C.: Tropfenwagen—Anwendung der Flugzeug-Aerodynamik. Zeitschrift für Flugtechnik und Motorluftschiffahrt, 12, 287–289, 1921.

233. Essers, U.; Thiel, E.: Institut für Verbrennungsmotoren und Kraftfahrwesen der Universität Stuttgart in neuem Gebäude. ATZ, 83, 9–14, 1981.

234. Estrada, G. E.: Das Fahrzeug als aerodynamischer Sensor. Dissertation, Renningen: Expert Verlag, ISBN 978-3-8169-3097-6, 2011.

235. Estrada, G.; Wiedemann, J.; Widdecke, N.: The Vehicle as an Aerodynamic Sensor in the Wind Tunnel and on the Road; 7. Internationales Stuttgarter Symposium "Automobil- und Motorentechnik," Band 2, Expert Verlag, Renningen, 2007.

236. Estrada, G.; Wiedemann, J.; Widdecke, N.: Vergleich verschiedener Konzepte der Bodensimulation und von drehenden Rädern zur Nachbildung der Straßenfahrt im Windkanal und deren Auswirkung auf Fahrzeuge. FAT (Forschungsvereinigung Automobiltechnik e.V.) Arbeitskreis 6 Aerodynamik, FKFS-Bericht 05/2004, Stuttgart, 2004.

237. Ewald. H.: Aerodynamische Effekte beim Kolonnenfahren (Modelluntersuchungen). Essen: Haus der Technik, 1984.

238. E/ECE/324-E/ECE/TRANS/505—Rev.1/Add.14/Rev.3/Amend.1 (June 1, 1984): Uniform Provisions Concerning the Approval of Vehicles Equipped with a Positive-Ignition Engine or with a Compression-Ignition Engine with Regard to the Emission of Gaseous Pollutants by the Engine—Method of Measuring the Power of Positive—Ignition Engines—Method of Measuring the Fuel Consumption of Vehicles (Regulation no. 15), 1984.

239. Fabijanic, J.: An Experimental Investigation of Wheel-Well Flows. SAE Technical Paper 960901, 1996, doi:10.4271/960901.

240. Fackrell, J. E.; Harvey, J. K.: The Aerodynamics of an Isolated Road Wheel. Aus Pershing, B. (Ed.): Proceedings of the Second AIAA Symposium of Aerodynamics of Sports and Competition Automobiles, 1975.

241. Fackrell, J. E.; Harvey, J. K.: The Flow Field and Pressure Distribution of an Isolated Road Wheel. In Stephens, H.S. (Ed.): Advances in Road Vehicle Aerodynamics. Cranfield: bhra fluid engineering, 155–163, 1973.

242. Fahy, F.: Sound and Structural Vibration. Academic Press, London, 1985.

243. Farassat, F.; Myers, M. K.: Extension of Kirchhoff's Formula to radiation from Moving Surfaces. Journal of Sound and Vibration 123 (3), pp. 451–460, 1988.
244. Ferziger, J. H.; Peric, M.: Computational Methods for Fluid Dynamics. 3rd edition, Springer Berlin Heidelberg, ISBN-13 978-3540420743, 2010.
245. Ffowcs Williams, J.E.; Hawkings, D.L.: Sound Generation by Turbulence and Surfaces in Arbitrary Motion. Phil. Trans. Roy. Soc. 264(A), S. 321–342, London, 1969.
246. Fiedler, F.; Kamm, W.: Steigerung der Wirtschaftlichkeit des Personenwagens. ZVDI, 84, 485–491, 1940.
247. Fiedler, R.-G.; Potthoff, J.: Vorentwicklung von Raddrehvorrichtungen für schmale Laufbandsysteme in Fahrzeugwindkanälen. 614–631 in Bargende, H.; Wiedemann, J. (Hrsg. 1999): 3. Stuttgarter Symposium Kraftfahrwesen und Verbrennungsmotoren. Renningen Malmsheim: Expert Verlag, 1999.
248. Fischer, O.; Kuthada, T.; Widdecke, N.; Wiedemann, J.: CFD Investigations of Wind Tunnel Interference Effects. SAE Technical Paper 2007-01-1045, 2007, doi:10.4271/2007-01-1045.
249. Fishleigh, W. T.: The Tear Drop Car. SAE Journal 1931-11-01. 353–362, 1931.
250. Flegl, H.: Die aerodynamische Gestaltung von Sportwagen. Christophorus, 98, 18–19, 1969.
251. Fletcher, H.; Munson, W. A.: Loudness, Its Definition, Measurement, and Calculation. Journal of the Acoustical Society of America, 5, 1933.
252. FluiDyna GmbH: Studie zur Verbrauchsreduktion an Nutzfahrzeugkombinationen durch aerodynamische Maßnahmen an einer generischen Gliederzug-Konfiguration. Finanziert durch die Forschungsvereinigung Automobiltechnik e.V. (FAT), 2011.
253. Föllinger, O.: Regelungstechnik. 7. Auflage, Hüthig Verlag Heidelberg, 1992.
254. Försching, H. W.: Grundlagen der Aeroelastik. Berlin: Springer Verlag, 1974.
255. Fosberry, R. C. A.; White, R. G. S.: The MIRA Full Scale Wind Tunnel. MIRA-Rep. Nr. 1961/8, 1961.
256. Frank, B.: Ein bildgebendes Verfahren zur Messung der Sichtweite. Dissertation, Institut für Physikalische Elektronik der Universität Stuttgart, 1997.
257. Frank, T.: Aerodynamics of commercial vehicles. Aerodynamics of Heavy Vehicles III Conference, Potsdam, September 2010.
258. Frankenberg, R. v.; Matteucci, M.: Geschichte des Automobils. Künzelsau: Sigloch Service Edition, 1973.
259. Frey, K.: Verminderung des Strömungswiderstandes von Körpern durch Leitflächen. Forschung Ingenieur Wesen, März 1933, 67–74, 1933.
260. Frisch, U.; Hasslacher, B.; Pomeau, Y.: Lattice-gas automata for the Navier-Stokes equation. Physical Review Letters, 56:pp1505–1508, 1986.
261. Froede, W.: Aus der Entwicklung des NSU-Weltrekord-Motorrads. Automobiltechnische Zeitschrift Nr. 5, S. 127/130, 1954.
262. Froede, W.: Einspurfahrzeuge mit geringstem Luftwiderstand. Automobiltechnische Zeitschrift Nr. 5/6, S. 143/147 u. 161/162, 1957.

263. Futterer, I.; Ehlen, M.: Einsatzmöglichkeiten eines Navier-Stokes-Lösers zur numerischen Strömungssimulation in der Motorradentwicklung. Entwicklungstendenzen im Motorradbau. 2. Internationale Konferenz. München, 12.-13.6.2003.
264. Garret, D.: Cooling of brakes-a conflict of interests, Braking of Road Veh. (Conf., Lougborough, 1983). Mech. Eng. Publ.; (I Mech E Conf. Publ.; 1983-2), Seite 1-5(IME Pap.; C35/83) 629.113-592/IME/83, 1983.
265. Garry, K.; Cooper, K.; Fediw, A.; Wallis, S.; et al.: The Effect on Aerodynamic Drag of the Longitudinal Position of a Road Vehicle Model in a Wind Tunnel Test Section; SAE Technical Paper 940414, 1994, doi:10.4271/940414.
266. Genger, M.; Kuthada, T.; Wiedemann, J.: Thermal Management Investigations at FKFS: Experimental Measurements and Simulation with CFD and KULI. 4th Kuli User Meeting. Steyr, Austria, 2003.
267. Genuit, K.: Kunstkopf-Meßtechnik—Ein neues Verfahren zur Geräuschdiagnose und -analyse. Zeitschrift für Lärmbekämpfung, 35, 103–105, 1988.
268. George, A. R.: Automobile Aerodynamic Noise. SAE Technical Paper 900315, 1990, doi:10.4271/900315.
269. George, A. R. (ed.): Automobile Wind Noise and Its Measurement. SAE Vehicle Aerodynamics Packet—2006 Edition. SAE International, VAP-2006, Warrendale, PA. 1996.
270. Geropp, D.: Reduktion des Strömungswiderstandes von Fahrzeugen durch aktive Strömungsbeeinflussung–Patentschrift DE 3837 729 und Leistungsbilanz. Universität Gesamthochschule Siegen, Institutsbericht, 1991.
271. Geropp, D.; Mildebrath, T.: Berechnung dreidimensionaler, abgelöster Strömungen mit Bodeneinfluss. Jahrbuch 1995 II der DGLR, 895–904, 1995.
272. Gerresheim, M.: Experimenteller und theoretischer Beitrag zu Fragen des Reifenverhaltens. Dissertation, TU München, 1974.
273. Gersten, K.; Pagendarm, H.: Diffusorströmung—Stand der Forschung; Institutsbericht, Institut für Thermo- und Fluiddynamik, TU Bochum, 1984.
274. Gilhaus, A. M.; Renn, V. E.: Drag and Driving-Stability-Related Aerodynamic Forces and their Interdependence—Results of Measurements on 3/8-Scale Basic Car Shapes. SAE Technical Paper 860211, 1986, doi:10.4271/860211.
275. Gilhaus, A.: The Main Parameters Determining the Aerodynamic Drag of Buses. Colloque construire avec le vent. Centre Scientifique et Technique du Batiment, Nantes, France, 1981.
276. Gilhaus, A.; Hau, E.: Drag Reduction on Trucks by Aerodynamic Parts and Covers. Vehicle Aerodynarnics, Int. Symposium, Volkswagen AG,1982.
277. Gilhaus, A.; Hau, E.; Künstner, R.; Potthoff, J.: Über den Luftwiderstand von Fernlastzügen, Ergebnisse aus Modellmessungen im Windkanal. Sonderdruck aus "Automobil-Industrie" Heft 3/September 1979 und Heft 3/September 1980 zum 50- jährigen Bestehen des Forschungsinstitutes für Kraftfahrwesen und Fahrzeugmotoren Stuttgart (FKFS), 1979/1980.
278. Gilhome, B. R.: Unsteady and Time-averaged Near-wake Flow over the Rear of Sedan Automobiles. PhD Thesis, Monash University, Victoria, Australien, 2002.

279. Gilhome, B. R.: Unsteady Flow Structures and Forces over/on the Rear Window and Boot Lid of Sedan Automobiles. 57–70 in Wiedemann, J.; Hucho, W.-H. (Eds.): Progress in Vehicle Aerodynamics III—Unsteady Flow Effects. Renningen: Expert Verlag, 2004.

280. Gilhome, B. R.; Saunders, J. W.: The Effect of Turbulence on Peak and Average Pressures on a Car Door. SAE Technical Paper 2002-01-0253, 2002, doi:10.4271/2002-01-0253.

281. Gilhome, B. R.; Saunders, J. W.; Sheridan, J.: Time-averaged and Unsteady Near Wake Analysis of Cars. SAE Technical Paper 2001-01-1040, 2001, doi:10.4271/2001-01-1040.

282. Gilliéron, P.; Noger, C.: Contributing to the Analysis of Transient Aerodynamic Effects Acting on Vehicles. SAE Technical Paper 2004-01-1311, 2004, doi:10.4271/2004-01-1311.

283. Glauert, H.: Wind Tunnel Interference on Wings, Bodies and Airscrews; ARC, R.&M. 1566, 1933.

284. Gleason, M.: CFD Analysis of Various Automotive Bodies in Linear Static Pressure Gradients; SAE Technical Paper 2012-01-0298, 2012, doi:10.4271/2012-01-0298.

285. Gleason, M.: Detailed Analysis of the Bluff Body Blockage Phenomenon in Closed Wall Wind Tunnels Utilizing CFD; SAE Technical Paper 2007-01-1046, 2007, doi:10.4271/2007-01-1046.

286. Glück, H.-D.: Klassierung der cD-Werte von Pkw nach EADE-Daten, 2004.

287. Göhring, E.; Krämer, W.: Seitliche Fahrgestellverkleidungen für Nutzfahrzeuge. ATZ 89, 481–488, 1987.

288. Goldstein, M. E.: Aeroacoustics. McGraw-Hill Book Company, New York, 1976.

289. Gorlin, S. M.; Slezinger, I. I.: Wind Tunnels and Their Instrumentation. Jerusalem: Israel Program for Scientific Translation, 1966.

290. Götz, H.: Crosswinds Facilities and Procedures. SAE Vehicle Aerodynamics Packet—2006 Edition. SAE International, VAP-2006, Warrendale, PA, 1996.

291. Götz, H.: Die Aerodynamik des Nutzfahrzeugs—Maßnahmen zur Kraftstoffeinsparung. Fortschritts-Berichte der VDI-Zeitschriften, Reihe 12, Nr. 31, S. 187–197, 1997.

292. Götz, H.: Nutzfahrzeuge. Kapitel 11 in Hucho, W.-H. (Hrsg. 2005, 2009): Aerodynamik des Automobils. 5. Auflage, Wiesbaden: Vieweg Verlag, 2005.

293. Graf, R.; Miehling, H.; Maier, K.; Helbrück, J.: Psychoacoustic evaluation of instationary wind noise transmitted into vehicles. Tagungsband der Internoise, Ottawa (Canada), 23.–26.08. 2009.

294. Gräff, B.; Hubert, K.; Zoller, H.-J.: Untersuchungen von Luftgeschwindigkeiten und Lufttemperaturen an industriellen Arbeitsplätzen. Schriftenreihe der Bundesanstalt für Arbeitsschutz, Fb 722, Wirtschaftsverlag NW, Bremerhaven, 1995.

295. Greaves, J. R. A.: The Development of the 3-Dimensional Motor Vehicle Aerodynamics Computer Model and Its Application to the Rover 88 Shape. 2nd International PHOENICS User Conference, 1987.

296. Grosche, G.; Zeidler, E.: Teubner-Taschenbuch der Mathematik. Teubner Verlag, 1996.

297. Großmann, H.: Pkw-Klimatisierung: Physikalische Grundlagen und technische Umsetzung. Springer Berlin Heidelberg, 1. Auflage, ISBN-13: 978-3642054945, 2010.

298. Grün, N.: Die Berechnung der Höchstgeschwindigkeit, BMW-interne Publikation, 2011.

299. Grün, N.: Simulating External Vehicle Aerodynamics with CARFLOW. SAE Technical Paper 960679, 1996, doi:10.4271/960679.

300. Grün, N.: Strömungsfeldangepasste Oberflächenkoordinaten zur Berechnung dreidimensionaler Grenzschichten. Fortschrittberichte Reihe 7, Nr. 187, VDI Verlag Düsseldorf, 1991.

301. Grundmann, R.; Kramer, C.; Konrath, B.; Kohl, W.: Der Aeroakustik-Windkanal von Ford. ATZ, 103, 932–938, 2001.

302. Guidati, S.: Mikrofonarray Technologie. In: Genuit, K. (Hrsg.): Sound-Engineering im Automobilbereich. Berlin: Springer, ISBN 978-3-642-01414-7, 2010.

303. Guidati, S.; Wagner, S.: Phased Array Measurements in a Closed Test Section Wind Tunnel. In: Wagner, S.; Ostertag, J. (Hrsg.): Third Aeroacoustics Workshop in Connection with the National Research Project SWING+, Stuttgart: Institut für Aerodynamik und Gasdynamik, 26.-27.09.2002.

304. Gürtler, T.: Some Remarks on Static Pressure Gradients in Wind Tunnel with Open Jet Test Sections; 5th ECARA Subgroup Meeting, Stuttgart, 2001.

305. Hackenberg, U.: Ein Eintrag zur Stabilitätsuntersuchung des Systems "Fahrer-Kraftrad-Straße." Dissertation Aachen: 1984/85.

306. Hackett, J. E.; Williams, J. E.; Patrick, J.: Wake Traverses Behind Production Cars and their Interpretation. SAE Technical Paper 850280, 1985, doi:10.4271/850280.

307. Hackett, J.; Cooper, K.: Extension to Maskell's Theory for Blockage Effects on Bluff Bodies in a Closed Wind Tunnel; Aeronautical Journal of the Royal Aero. Soc., Paper 14, 2001.

308. Hackett, J.; Wilsden, D.: Determination of Low Speed Wake Blockage Correction via Tunnel Wall Static Pressure Measurements; AGARD-CP-174, 1978.

309. Hagemeier, T.; Hartmann, M.; Thévenin, D.: Practice of Vehicle Soiling Investigations: A Review. International Journal of Multiphase Flow 37, 860–875, 2011.

310. Hager, J.; Heizeneder, H.; Risch, P.; Straßer, K.: Komfortable Messdatenauswertung für den Bereich Pkw-Klimatisierung. ATZ 100, 928–932, 1998.

311. Haken, K.-L.: Messung des Rollwiderstands unter realen Bedingungen. In Bargende, M.; Wiedemann, J. (Hrsg.): Kraftfahrwesen und Verbrennungsmotoren, 3. Stuttgarter Symposium. Renningen-Malmsheim: Expert Verlag, 488–504, 1999.

312. Haken, K.-L.; Stengelin, J.; Wiedemann, J.: Ermittlung des Rollwiderstands unter realen Bedingungen mit einem Messanhänger. Tagung "Fahrwerkstechnik," Haus der Technik, Essen; München, 6. und 7.6.2000.

313. Hald, J.: Use of Spatial Transformation of Sound Fields (STSF) Techniques in the Automotive Industry. Technical Review No. 1 - 1995, Nærum (DK): Brüel & Kjær Company; ISSN 007-2621, 1995.

314. Hallquist, T.: The Cooling Airflow of Heavy Trucks—a Parametric Study, *SAE Int. J. Commer. Veh.* 1(1):119–133, 2009, doi:10.4271/2008-01-1171.

315. Hansen, M.; Schlör, K.: Der AVA-Versuchswagen. Aerodynamische Versuchsanstalt Göttingen, Bericht 43 W 26, 1943.
316. Hardy, J.; Pomeau, Y.; de Pazzis, O.: Time Evolution of Two-Dimensional Model System. I. Invariant States and Time Correlation Functions. J. Math. Phys. 14 (1973), pp. 1746–1759.
317. Hartmann, M.; Ocker, J.; Lemke, T.; Mutzke, A.; Schwarz, V.; Tokuno, H.; Toppinga, R.; Unterlechner, P.; Wickern, G.: Wind Noise caused by the A pillar and the Side Mirror flow of a Generic Vehicle Model. In: Tagungsband 18th AIAA/CEAS Aeroacoustics Conference, Colorado Springs (USA), 4–6, June 2012.
318. Hazir, A.; Blumrich, R.; Crouse, B.; Freed, D.: Wind Noise Transmission into Convertibles by Fluid Structure Interaction, in: Aeroacoustics research in Europe: The CEAS-ASC Report on 2008 highlights, Journal of Sound and Vibration 328, pp. 213–242, 2009.
319. Heald, R. H.: Aerodynamic Characteristics of Automobile Models. US Dept. of Commerce, Bureau of Standards, RP 591, 285–291, 1933.
320. Heckemüller, J.: Vier 1000er im Windkanal—Sturm und Drang. Motorrad Nr. 25, S. 6/13, 1988.
321. Heesen, W. v.; Höpfer, M.: Suppression of Wind Tunnel Buffeting by Active Flow Control, SAE Technical Paper 2004-01-0805, 2004, doi:10.4271/2004-01-0805.
322. Heesen, W. v.; Reiser, P.: Lärmminderungsmaßnahmen an bergbautypischen Axialventilatoren—Teil 1. Mitteilungen der westfälischen Berggewerkschaftskasse. Heft 64, 1989.
323. Heft, A. I.; Indinger, T.; Adams, N. A.: Introduction of a New Realistic Generic Car Model for Aerodynamic Investigations. SAE Technical Paper 2012-01-0168, 2012, doi:10.4271/2012-01-0168.
324. Heidrich, M.: The Two New Climatic Wind Tunnels in the Mercedes-Benz Technology Center. Progress in Vehicle Aerodynamics and Thermal Management, in J. Wiedemann (Ed.), Proceedings of the 8th FKFS-Conference, Expert Verlag, Renningen, ISBN 978-3-8169-3116-4, 2011.
325. Heißing, B.; Ersoy, M. (Hrsg.): Fahrwerkhandbuch. ATZ-MTZ Fachbuch. ISBN: 978-3-8348-0105-0, 2007.
326. Helfer, M.: Aeroakustische Messungen an Kraftfahrzeugen in Windkanälen. In: Tagungsband "Aeroakustik" des Haus der Technik Essen, Wildau, 10.-11. Oktober, 2006.
327. Helfer, M.: Fahrzeug-Aeroakustik bei turbulenter Anströmung. Tagungsband der DAGA 2012, 19.-22. 03.2012, Darmstadt. Berlin: Deutsche Gesellschaft für Akustik e.V., ISBN: 978-3-939296-04-1, 2012.
328. Helfer, M.: General Aspects of Vehicle Aeroacoustics. Lecture Series "Road Vehicle Aerodynamics;" 30.05.-03.06.2005; Rhode-St.-Genèse, Belgien: Von Karman Institute, ISBN 2-930389-61-3, 2005.
329. Helfer, M.: Hohlspiegelmikrofone. In Genuit, K. (Hrsg.): Sound-Engineering im Automobilbereich. Berlin: Springer, ISBN 978-3-642-01414-7, 2010.
330. Helfer, M.: Localization of Sound Sources. In: Wiedemann, J.; Hucho, W.-H. (Hrsg.): Progress in Vehicle Aerodynamics—Advanced Experimental Techniques. Renningen: Expert Verlag; ISBN 3-8169-1843-3, 2000.

331. Helfer, M.: Reifen-Fahrbahn-Geräusch und Umströmungsgeräusch von Kraftfahrzeugen. In: Tagungsband der DAGA 2007, 19.-22.03.2007 in Stuttgart. Berlin: Deutsche Gesellschaft für Akustik e.V., 2007; ISBN 978-3-9808659-3-7. 2007.

332. Helfer, M.: Sound Source Localisation with Acoustic Mirrors. In: Tagungsband der NAG/DAGA 2009, 23.-26.03.2009, Rotterdam. Berlin: Deutsche Gesellschaft für Akustik e.V., ISBN: 978-3-9808659-6-8, 2009.

333. Helfer, M.: Umströmungsgeräusche. In: Genuit, K. (Hrsg.): Sound-Engineering im Automobilbereich. Berlin: Springer, ISBN 978-3-642-01414-7, 2010.

334. Helfer, M.; Busch, J.: Contribution of Aerodynamic Noise Sources to Interior and Exterior Vehicle Noise. In: Tagungsband des DGLR-Workshop "Aeroacoustics of Cars," Emmeloord (NL), 16.-17.11.1992.

335. Helfer, M.; Melchger, N.; Busch, J.: Moyens de mesure pour les bruits aérodynamiques intérieurs et extérieurs des véhicules. In: Ingénieurs de l'automobile Nr. 711, S. 43-49; ISSN 0020-1200, 1997.

336. Helfer, M.; Wiedemann, J.: Design of Wind Tunnels for Aeroacoustics. In: Riethmüller, M. L.; Lema, M. R. (Hrsg.): Experimental Aeroacoustics. Rhode-Saint-Genèse (B): von Karman Institute for Fluid Dynamics; ISBN 13 978-2-930389-70-2, 2007.

337. Heller, A.: Der neue Kraftwagen von Dr.-Ing. Rumpler. ZVDI, 39, 1011–1015, 1921.

338. Helling, J.: Krafträder. Vorlesungsumdruck 1. Aufl. Aachen: fka Verlag, 1985.

339. Hentschel, W.; Piatek, R.: Strömungssichtbarmachung in der aerodynamischen Entwicklung von Fahrzeugen. Tagung, Sichtbarmachung technischer Strömungsvorgänge, Essen: Haus der Technik, 1989.

340. Herwig, T.; v. Löhneysen, U.: Helmtest: Windkanal—Wind-Spiele. Motorrad 8/1984, S. 152–155, 1984.

341. Herz, D.; Reese, K.: Die NSU-Renngeschichte. 2. Aufl. Stuttgart: Motorbuch Verlag, 1987.

342. Hess, J. L.; Smith, A. M. O.: Calculation of Potential Flow About Arbitrary Bodies. Progress in Aeronautical Sciences, 8, New York: Pergamon Press 1–138, 1967.

343. Heuer, R.; Irschik, H.; Ziegler, F.: A BEM-Formulation of Nonlinear Plate Vibrations; in: Proc. IUTAM/IACM-Symp. on Discretization Methods in Structural Mechanics, G. Kuhn et al. (Hrsg.); Springer, Berlin, S. 341–351, 1990.

344. Heyl, G.: Fortschritte der Sicherheitstechnologie bei Motorradschutzkleidung am Beispiel des BMW-Protec-Anzugs und des BMW-Systemhelms. VDI-Berichte 779 "Motorrad." Düsseldorf: VDI Verlag, 1989.

345. Heyl, G.; Lindener, N.; Stadler, M.: Akustische/aeroakustische Eigenschaften von Motorradhelmen. VDI Fachtagung Motorrad Berlin. VDI Verlag, 1993.

346. Hinze, J. O.: Turbulence. McGraw-Hill series in Mechanical Engineering; McGraw-Hill, Inc., New York, 1975.

347. Hirt, C. W.; Ramshaw, J. D.: Prospects for Numerical Simulation of Bluff-Body Aerodynamics. In Sovran, G.; Morel, T.; Mason, W. (Eds.): Aerodynamic Drag Mechanisms of Bluff Bodies and Road Vehicles. New York, London: Plenum Press. 213–353, 1978.

348. Höfer, P.: The new B-Class—Aerodynamic Challenges of the Mercedes-Benz Front-Wheel-Drive Architecture. Progress in Vehicle Aerodynamics and Thermal Management, 2011.

349. Hoepke, E.; Breuer, S. (Hrsg.): Nutzfahrzeugtechnik. Wiesbaden: Vieweg + Teubner, 5. Auflage, ISBN: 978-3-8348-0374, 2008.

350. Hoerner, S. F.: Fluid Dynamic Drag. Midland Park, N.J.: Selbstverlag des Autors, 1965.

351. Hofer, G.: Aerodynamikentwicklung von Motorrädern. Vortrag an der FH Ingolstadt, 2008.

352. Hofer, G.; Grün, N.: Die Aerodynamikentwicklung eines Rennmotorrades der Superbike-Klasse. Haus der Technik Konferenz "Fahrzeugaerodynamik," München 2010.

353. Hoffman, J.; Martindale, B.; Arnette, S.; Williams, J.; et al.: Effect of Test Section Configuration on Aerodynamic Drag Measurements; SAE Technical Paper 2001-01-0631, 2001, doi:10.4271/2001-01-0631.

354. Hoffmann, R.; Hupertz, B.; Krueger, L.; Lentzen, M.: Active Aerodynamics on Passenger Cars. In: Wiedemann, J. (Ed.): Progress in Vehicle Aerodynamics and Thermal Management—Proceedings of the 7th FKFS-Conference. Expert Verlag, Renningen, ISBN 978-3-8169-2944-4, 2010.

355. Hoffmann, R.: Thermo Aero Systems Engineering Optimization of the Engine Compartment Airflow. 512–528 in Bargende, M.; Wiedemann, J. (Hrsg.): Kraftfahrwesen und Verbrennungsmotoren, 4. Internationales Stuttgarter Symposium. Renningen: Expert Verlag, 2001.

356. Hong, P.; Marcu, B.; Browand, F. K.; Tucker, A.: Drag Forces Experienced by Two, Full-Scale Vehicles at Close Spacing. SAE Technical Paper 980396, 1998, doi:10.4271/980396.

357. Hopf, A. Gauch, A.: Numerische Simulation der Bremsenkühlung mit CFD und FEM. VDI-Berichte 1559, Berechnung und Simulation im Fahrzeugbau, 2000.

358. Horn, A.: Fahrer-Fahrzeug-Kurvenfahrt auf trockener Straße. Dissertation, Universität Braunschweig, 1986.

359. Howe, M. S.: Contributions to the Theory of Aerodynamic Sound, with Application to Excess Jet Noise and the Theory of the Flute. J. Fluid Mech.; 71:625-673, 1975.

360. Howell, J.: Shape and Drag. In Hucho, W.-H. (Ed. 1998): Using Aerodynamics to Improve the Properties of Cars. 1st Euromotor Short Course, Stuttgart, 1998.

361. Howell, J.: Shape Features which Influence Crosswind Sensitivity. C466/036/93 IMechE, 1993.

362. Howell, J.; Le Good, G.: The Influence of Aerodynamic Lift on High Speed Stability. SAE Technical Paper 1999-01-0651, 1999, doi:10.4271/1999-01-0651.

363. Howell, J.; Sheppard, A.; Blakemore, A.: Aerodynamic Drag Reduction for a Simple Bluff Body Using Base Bleed. SAE Technical Paper 2003-01-0995, 2003, doi:10.4271/2003-01-0995.

364. Gas Guzzler Tax. United States Environmental Protection Agency. http://www.epa.gov/fueleconomy/guzzler/index.htm, 10. November 2012.

365. RICHTLINIE 1999/94/EG DES EUROPÄISCHEN PARLAMENTS UND DES RATES. http://eur-lex.europa.eu/LexUriServ/LexUriServ.do?uri=CONSLEG:1999L 0094:20081211:DE:PDF, 10. November 2012.

366. Verordnung über Verbraucherinformationen zu Kraftstoffverbrauch, CO_2-Emissionen und Stromverbrauch neuer Personenkraftwagen http://www.gesetze-im-internet.de/pkw-envkv/BJNR103700004.html, 10. November 2012.

367. Renewable Energy Concepts. http://www.renewable-energy-concepts.com/german/windenergie/standorte.html, 16. Sept. 2012.

368. DHC-13th Session. United Nations Economic Commission for Europe. http://www.unece.org/trans/main/wp29/wp29wgs/wp29grpe/wltp_dhc13.html, präsentation WLTP-DHC-13-02e.ppt, 10. November 2012.

369. Huber, S.: Windkanal-Methodik—Turbulenzen. Motorrad Nr. 9, Sonderteil "Technik aktuell," 1982.

370. Hucho, W.-H. (Ed): Aerodynamics of Road Vehicles, 4th Edition, SAE, Warrendale, PA, 1998.

371. Hucho, W.-H.: Aerodynamics of Road Vehicles—a Challenge for Computational Fluid Dynamics. Proceedings of the 1st European Automotive CFD Conference, Bingen, Germany, 25–26 June, 2003.

372. Hucho, W.-H.: Aerodynamik der stumpfen Körper, 2. Auflage. Wiesbaden: Vieweg Verlag, 2011.

373. Hucho, W.-H. (Hrsg.): Aerodynamik des Automobils. 5. Auflage, Wiesbaden, Verlag Vieweg + Teubner, ISBN 3-528-03959-0, 2005.

374. Hucho, W.-H.: Design. Verlockungen der Formen des Automobils–Teil 2: Wie wird die Technik damit fertig? Beispiel Aerodynamik. Vortrag Autosommer 2011, Stuttgart & Karlsruhe. www.aerowolf.de, 2011.

375. Hucho, W.-H.: Designing Cars for Low Drag—State of the Art and Future Potential. International Journal of Vehicle Design, SP 3, pp. 1–8, 1983.

376. Hucho, W.-H.: Die optimale Karosserieform. Volkswagen Workshop "Das Auto der 80er Jahre," 17–23, Wolfsburg: VW AG, 1978.

377. Hucho, W.-H.: Einfluss der Vorderwagenform auf Widerstand, Giermoment und Seitenkraft von Kastenwagen. ZfW,20, 341–351, 1972.

378. Hucho, W.-H.: Fahrzeugaerodynamik—Stand der Technik und Aufgaben für die Forschung. DGLR-Bericht 79-02, 401–417, 1979.

379. Hucho, W.-H.: Grenzwert-Strategie—Halbierung des cD-Wertes scheint möglich. ATZ 111, 16–23, 2009.

380. Hucho, W.-H.: The aerodynamic Drag of Cars—Current Understanding, Unresolved Problems and Future Potential. 7–44 in Sovran, G.; Morel, T.; Mason, W. T. (Hrsg.): Aerodynamic Drag Mechanisms of Bluff Bodies and Road Vehicles. New York: Plenum Press, 1978.

381. Hucho, W.-H.: Versuchstechnik in der Fahrzeugaerodynamik. Colloquium on Industrial Aerodynamics, 1–48, Aachen, 1974.

382. Hucho, W.-H.: Versuchstechnik in der Fahrzeugaerodynamik. Kolloquium Industrie-Aerodynamik, Aachen, Teil 3 Aerodynamik von Straßenfahrzeugen, 1–48, 1974.

383. Hucho, W.-H.; Emmelmann, H.-J.: Aerodynamische Formoptimierung, ein Weg zur Steigerung der Wirtschaftlichkeit von Nutzfahrzeugen. Fortschrittsberichte VDI-Z, Reihe 12, Nr. 31, 163–185, 1977.

384. Hucho, W.-H.; Janssen, L. J.: Beiträge der Aerodynamik im Rahmen einer Fahrzeugentwicklung. ATZ, 74, 1–5, 1972.

385. Hucho, W.-H.; Janssen, L. J.; Emmelmann, H.-J.: The Optimization of Body Details—A Method for Reducing the Aerodynamic Drag of Road Vehicles. SAE Technical Paper 760185, 1976, doi:10.4271/760185.

386. Hummel, D.: On the Vortex Formation Over a Slender Wing at Large Angles of Incidence. AGARD CP–247, 15–1 und 15–17, 1978.

387. Hupertz, B.: Einsatz der numerischen Simulation der Fahrzeugumströmung im industriellen Umfeld. Braunschweig: ZLR-Forschungsbericht. Dissertation TU Braunschweig, 1998.

388. Hupertz, B.: Persönliche Information, 2011.

389. Huppertz, A.; Janke, G.: Preliminary Experiments on the Control of Three-Dimensional Modes in the Flow over a Backward-Facing Step. Advances in Turbulence VI, 1996.

390. Hütten, H.: Motoren Technik—Praxis—Geschichte. Motorbuch Verlag, 1997.

391. Hütten, H.: Motorradtechnik. 1. Aufl. Stuttgart: Motorbuch Verlag, 1983.

392. Hütten, H.: Schlaglichter zum 100jährigen Motorrad. In: VDI-Bericht 577, "100 Jahre Motorrad." Düsseldorf: VDI Verlag, 1986.

393. Hüttenbrink, K.-B.: Lärmmessung unter Motorradhelmen. Zeitschrift für Lärmbekämpfung 29,182–187, 1982.

394. Ingram, K. C.: The Wind-Averaged Drag Coefficient Applied to Heavy Goods Vehicles. Transport and Road Research Laboratory Supplementary Report 392, 1978.

395. IPCC Fourth Assessment Report: Climate Change 2007 (AR4), Genf, 2007. siehe auch http://www.ipcc.ch/

396. Ishima, T.; Takahashi, Y.; Okado, H.; Baba, Y.; et al.: 3D-PIV Measurements and Visualisation of Streamlines Around a Standard SAE Vehicle Model. SAE Technical Paper 2011-01-0161, 2011, doi:10.4271/2011-01-0161.

397. Islam, M.; Decker, F.; de Villiers, E, Jackson, A.; et al.: Application of Detached-Eddy Simulation for Automotive Aerodynamics Development. SAE Technical Paper 2009-01-0333, 2009, doi:10.4271/2009-01-0333.

398. ISO 12021-1: Road vehicles—Sensitivity to lateral wind—Part 1: Open-loop test method using wind generator input. 1996.

399. ISO 20119: Road vehicles—Test method for the quantification of on center handling. 2002.

400. Ivanic, T.; Gilliéron, P.: Reduction of the Aerodynamic Drag Due to Cooling Systems: An Analytical and Experimental Approach. SAE Technical Paper 2005-01-1017, 2005, doi:10.4271/2005-01-1017.

401. Jallet, S.; Devos, S.; Maubray, D.; Sortais, J. L.; et al.: Numerical Simulation of Wiper System Aerodynamic Behavior. SAE Technical Paper 2001-01-0036, 2001, doi:10.4271/2001-01-0036.

402. Janke, G.: Experimental investigation of turbulent reattachment on a free-surface piercing body. Proc. 10th symposium on Turbulent Shear Flows, Vol. 1, 10.13–10.18, 1995.

403. Janke, G.: On the Separated Flow Behind a Swept Backward-Facing Step. Notes on Numerical Fluid Mechanics 40, 1993.

404. Janke, G.: PIV Messungen in einer abgelösten Strömung mit freier Oberfläche. Fünfte Fachtagung zu Lasermethoden in der Strömungsmechanik, TU Berlin, 1996.

405. Janke, G.: Über die Grundlagen und einige Anwendungen der Ölfilminterferometrie zur Messung von Wandreibungsfeldern in Luftströmungen. Dissertation TU Berlin, 1992.

406. Janke, G.; Grundmann, R.; Schimpf, O. (2004): Akustische Optimierung des Schuberth Windkanals.

407. Janssen, L. J.: Persönliche Information, 1998.

408. Janssen, L. J.; Hucho, W.-H.: Aerodynamische Entwicklung von VW Golf und Scirocco. ATZ, 77, 1–5, 1975.

409. Janssen, L. J.; Hucho, W.-H.: The Effect of Various Parameters on the Aerodynamic Drag of Passenger Cars. Advances in Road Vehicle Aerodynamics. Cranfield: British Hydromechanical Association, 1973.

410. Jaray, P.: Der Stromlinienwagen—Eine neue Form der Automobilkarosserie. Der Motorwagen, 17, 333–336, 1922.

411. Jerhamre, A.; Bergström, C.: Numerical Study of Brake Disc Cooling Acounting for Both Aerodynamic Drag Force and Cooling Efficiency. SAE Technical Paper 2001-01-0948, 2001, doi:10.4271/2001-01-0948.

412. Joergensen, F. E.: DISA Information No. 11, 1971.

413. Jürgensohn, T.; Müller, W.; Scheffer, T.: Verbesserte Methoden zur Objektivierung von subjektiven Bewertungen des Fahrverhaltens. Forschungsbericht Zentrum Mensch-Maschine-Systeme, Berlin, 1996.

414. Kaltenbach, H.-J.; Janke, G.: Direct Numerical Simulation of Flow Separation Behind a Swept, Rearward-Facing Step at ReH=3000. Physics of Fluids Vol. 12, No. 9, 2320–2337, 2000.

415. Kandasamy, S.; Duncan, B.; Gau, H.; Maroy, F.; et al.: Aerodynamic Performance Assessment of BMW Validation Models Using Computational Fluid Dynamics. SAE Technical Paper 2012-01-0297, 2012, doi:10.4271/2012-01-0297.

416. Kandasamy, S.; Duncan, B.; Gau, H.; Maroy, F.; Belanger, A.; Grün, N.; Schäufele, S.: Impact of Wheel Rotation on Aerodynamic Drag and Lift, FKFS Symposium, Stuttgart, 2012.

416a. Kapitza, K.: Design und Aerodynamik von Sportwagen. In Hucho, W.-H.: Design undAerodynamik im Automobilbau. Tagung Haus der Technik, Essen, 1992.

417. Karbon, K.; Longman, S.: Automobile Exterior Water Flow Analysis Using CFD and Wind Tunnel Visualization. SAE Technical Paper 980035, 1998, doi:10.4271/980035.

418. Karin, S. and Smith, N. P.: The Supercomputer Era. New York: Harcourt Brace Jovanovich, 1987.

419. Karlsson, R. I.: Studies of Skin Friction in Turbulent Boundary Layers on Smooth and Rough Walls. Ph.D. Thesis, Calmers University of Technology, 1980.
420. Kasten, H. G.: Motorrad-Aerodynamik. Motorrad Nr. 11, Sonderteil "Motorrad-Technik," 1979.
421. Kataoka, T.; China, H.; Nakagawa, K.; Yanagimoto, K.; et al.: Numerical Simulation of Road Vehicle Aerodynamics and Effect of Aerodynamic Devices. SAE Technical Paper 910597, 1991, doi:10.4271/910597.
422. Katz, J.: Race Car Aerodynamics. Cambridge, Ma.: Bentley Publishers, 1995.
423. Katz, J.; Dykstra, L.: Study of an Open-Wheel Racing-Car's Rear-Wing Aerodynamics. SAE Technical Paper 890600, 1989, doi:10.4271/890600.
424. Kelly, K. B.; Provencher, L. G.; Schenkel, F. K.: The General Motors Engineering Staff Aerodynamics Laboratory—A Full Scale Automotive Wind Tunnel. SAE Technical Paper 820371, 1982, doi:10.4271/820371.
425. Kessler, J. C.; Wallis, S. B.: Aerodynamic Test Techniques. SAE Technical Paper 660464, 1966, doi:10.4271/660464.
426. Khalighi, B.; Zhang, S.; Koromilas, C.; Balkanyi, S.R.; et al.: Experimental and Computational Study of Unsteady Wake Flow Behind a Bluff Body with a Drag Reduction Device. SAE Technical Paper 2001-01-1042, 2001, doi:10.4271/2001-01-1042.
427. Khandia, Y.; Mosquera, A. A.; Butler, M. J.: Effective Use of CFD in Vehicle Aerodynamics. 3rd MIRA International Vehicle Aerodynamics Conference, 18–19 October, 2000.
428. Kiel, G.: Total-Head Meter with small Sensitivity to Yaw. Luftfahrtforschung, Vol. XII, No.2. München, Berlin: Verlag R. Oldenbourg. NACA TM No. 775, Washington, 1935.
429. Kieselbach, R. J. F.: Stromlinienautos in Europa und USA—Aerodynamik im Pkw-Bau 1900 bis 1945. Stuttgart: Kohlhammer, 1982.
430. Kieselbach, R. J. F.: Stromlinienbusse in Deutschland—Aerodynamik im Nutzfahrzeugbau 1931 bis 1961. Stuttgart: Kohlhammer, 1983.
431. Kieselbach, R. J. F.: The Drive to Design. Stuttgart: avedition, 1998
432. Kieselbach, R. J. F.: Stromlinienautos in Deutschland—Aerodynamik im Pkw-Bau 1900 bis 1945. Kohlhammer Edition Auto Verkehr, Verlag W. Kohlhammer, Stuttgart, ISBN 3-17-007626-4, 1982.
433. Kim, M.-S.; Lee, J.-H.; Kee, J.-D.; Chang, J.-H.: Hyundai Full Scale Aero-Acoustic Wind Tunnel. SAE Technical Paper 2001-01-0629, 2001, doi:10.4271/2001-01-0629.
434. Kimura. Y. Toyota's All Weather Wind Tunnel. Company brochure.
435. King, L. V.: Royal Society, Phil.Trans. A, 214 pp. 373–432, 1914.
436. Kitoh, K; Chatani, S.; Oshima, N.; Nakashima, T.; et al.: Large Eddy Simulation on the Underbody Flow of the Vehicle with Semi-Complex Underbody Configuration. SAE Technical Paper 2007-01-0103, 2007, doi:10.4271/2007-01-0103.
437. Kleber, A.: CFD as an Integrated Part of Aerodynamic Development of the Opel Eco-Speedster. In Seibert, K. W.; Hanna, R. K. (Eds.): Proceedings of the 1st European Automotive CFD Conference, Bingen, Germany, 25–26.6 2003, 27–36, 2003.

438. Klein, C.: Einsatz einer druckempfindlichen Beschichtung (PSP) zur Bestimmung des momentanen Druckfeldes von Modellen im Windkanal. Forschungsbericht 97-55, DLR Göttingen, 1997.

439. Klein, C.: Weiterentwicklung der PSP-Technik und Erprobung an einem Windkanalmodell zur Erfassung kleiner Druckdifferenzen. Forschungsbericht 98-14, DLR Göttingen, 1998.

440. Klein, C.; Engler, R. H.; Fonov, S.; D.; Trinks, O.: Pressure Sensitive Paint (PSP) Measurements in a Low-speed Wind Tunnel. In: Wiedemann, J.; Hucho, W.-H. (Hrsg.): Progress in Vehicle Aerodynamics II- Advanced Measurement Techniques. 158–169, Renningen. Expert Verlag, 2000.

441. Klein, R. H.; Jex, H. R.: Development and Calibration of an Aerodynamic Disturbance Test Facility. SAE Technical Paper 800143, 1980, doi:10.4271/800143.

442. Kleiner, C.; Grün, N.: CFD Simulation in Motorcycle Aerodynamics at the BMW Group. Internationale Fachtagung: Entwicklungstendenzen im Motorradbau, Haus der Technik Fachbuch Nr. 28, Expert Verlag, Renningen, 2003.

443. Klemperer, W.: Luftwiderstandsuntersuchungen an Automobilmodellen. Zeitschrift für Flugtechnik und Motorluftschiffahrt, 13, 201–206, 1922.

444. Knowles, R.; Saddington, A.; Knowles, K.: Simulation and Experiments on an Isolated Racecar Wheel Rotating in Ground Contact. MIRA 4th International Vehicle Aerodynamics Conference, 2002.

445. Kobayashi, N.; Yamada, M.: Stability of a One Box Type Vehicle in a Cross-wind—An Analysis of Transient Aerodynamic Forces and Moments. SAE Technical Paper 881878, 1988, doi:10.4271/881878.

446. Kobayashi, N.; Sasaki, Y.: Aerodynamic Effects of an Overtaking Articulated Heavy Goods Vehicle on Car-Trailer—An Analysis to Improve Controllability. SAE Technical Paper 871919, 1987, doi:10.4271/871919.

447. Koenig-Fachsenfeld, R. v.: Aerodynamik des Kraftfahrzeuges. Band III + IV, Kurt Maier Verlag, Heubach,1946.

448. Koenig-Fachsenfeld, R. v.: Aerodynamik des Kraftfahrzeuges, 2 Bde. Frankfurt: Umschau Verlag, 1951.

449. Koenig-Fachsenfeld, R. v.: Windkanalmessungen an Omnibusmodellen. ATZ, 39, 143–149, 1936.

450. Koenig-Fachsenfeld, R. v.; Rühle, R.; Eckert, D.; Zeuner, A.: Windkanalmessungen an Omnibusmodellen. ATZ, 39, 143–149, 1936.

451. Koessler, P.: Kotflügeluntersuchungen. Deutsche Kraftfahrtforschung und Straßenverkehrstechnik, Heft 175, Düsseldorf: VDI Verlag, 1965.

452. Kohl, W.: Upgrade Of Aerodynamic Windtunnel, Ford-Werke AG, For Aeroacoustic Measurement. 2nd MIRA International Conference on Vehicle Aerodynamics, 1998.

453. Kokoschinski, H.: Verkleidungstest—Kanalarbeit. Motorrad, Nr. 20, S. 10/23, 1987.

454. Kollar, M.: Sichtbarmachung der Strömungen im Innenraum eines Motorradhelmes mittels eines Laserlichtschnittverfahres (Studienarbeit), Lehrstuhl für Strömungsmechanik der Friedrich-Alexander-Universität Erlangen-Nürnberg, 1984.

455. Költzsch, P.: Bemerkungen über Schall und Pseudoschall. Wissenschaftliche Zeitschrift der Technischen Hochschule Otto von Guericke Magdeburg 17 (1973), Heft 5, 567–573, 1973.

456. Kompenhans, J.; Raffel, M.; Dieterle, L.; Richard, H.; Dewhirst, T.; Vollmers, H.; Ehrenfried, K.; Willert, C.; Pengel, K.; Kähler, C.; Ronneberger, O.: Measurement of Flow Fields with Particle Image Velocimetry (PIV). 131–157 in Wiedemann, J.; Hucho, W.-H. (Hrsg.): Progress in Vehicle Aerodynamics—Advanced Experimental Techniques. Renningen: Expert Verlag, 2000.

457. Koremoto, K.; Kawamura, N.; Kuratani, N.; Nakamura, S.; Arai, T.; Galanga, F.; Walter, J.; Martindale, B.; Duell, E.; Muller, S.: The Characteristics of the Honda Full Scale Aero-acoustic Wind Tunnel Equipped with a Rolling Road System. 8th MIRA International Vehicle Aerodynamics Conference, Grove, Oxfordshire, Uk, October 2010.

458. Kortesuo, A.; Kaivola, R.: Motorcyclist´s Helmet Noise. Measurement and Attenuation. VDI Berichte Nr. 1159, 57–66,1994.

459. Kortesuo, A.; Kaivola, R.: Motorcyclist´s helmet noise, a solution for attenuation. Nordic acoustical meeting, Helsinki, June 1996.

460. Kotapati, R.; Keating, A.; Kandasamy, S.; Duncan, B.; et al.: The Lattice-Boltzmann-VLES Method for Automotive Fluid Dynamics Simulation, a Review. SAE Technical Paper 2009-26-0057, 2009, doi:10.4271/2009-26-0057.

461. Krampol, S.; Riegel, M.; Wiedemann, J.: Rechnergestützte Simulation des instationären Windgeräusches. Automobiltechnische Zeitschrift 111, Nr. 11, 2009.

462. Krantz, W.: An Advanced Approach for Predicting and Assessing the Driver's Response to Natural Crosswind. Dissertation, Stuttgart, 2012.

463. Krantz, W.; Schröck, D.; Neubeck, J.; Wiedemann, J.; Lanzilotta, E.: An enhanced single track model for evaluation of the driver-vehicle interaction under crosswind. 9th Stuttgart International Symposium, Vieweg + Teubner Verlag/GWV Fachverlage GmbH, Wiesbaden, 2009.

464. Krantz, W.; Schröck, D.; Wiedemann, J.: Fahrzeugdynamik—Das Gesamtsystem Fahrer-Fahrzeug-Regelsysteme. Themenheft Forschung der Universität Stuttgart; ISSN 1861-0269, 2010.

465. Krause, J.; Lichtenstein, C. (Hrsg.): Your Private Sky—R. Buchkminster Fuller—The Art of Design Science. Baden/Switzerland: Lars Müller Publisher, 1999.

466. Kraus-Weysser, F.: Die große Motorrad-Show. 1. Aufl. Stuttgart: Motorbuch—Verlag, 1978.

467. Krebber, W.; Kielmann, G.: Richtungshören mit Motorradhelmen. Fortschritte der Akustik DAGA, S. 740–741, Oldenburg: DEGA, 1997.

468. Krist, S.; Mayer, J.; Neuendorf, R.: Aerodynamik und Wärmehaushalt des neuen BMW 5er. ATZ extra, D 58922, 148–153, 2003.

469. Kronthaler, P.; Bayer, B.: Betrachtungen zur Kraftschlussbeanspruchung beim Motorradfahren aus der Sicht des Reifenherstellers. In: VDI—Bericht 657 "Aktive und passive Sicherheit von Krafträdern." Düsseldorf: VDI Verlag, 1987.

470. Kubisch, U.: Automobile aus Berlin—Vom Tropfenwagen zum Amphicar. Berlin: Nicolaische Verlagsbuchhandlung, 1985.

471. Küchemann, D.; Vandrey, J.: Zur Geschwindigkeitskorrektur in Windkanälen mit freier Messstrecke unter Berücksichtigung des Düseneinflusses; Jahrbuch der deutschen Luftfahrtforschung, 1941.
472. Küchemann, D.; Weber, J.: Vortex Motions. ZAMM 45, 457, 1965.
473. Kuhn, A.: Der große DB-Windkanal. ATZ, 88, 27–32, 1988.
474. Kühnel, W.; Paul, M.; Schaake, N.; Schrumpf, M.: Geräuschreduzierung in Fahrzeug-Klimaanlagen. Automobil-Technische Zeitschrift 12 / 2004.
475. Künstner, R.: Aerodynamische Untersuchungen an Personenwagen-Caravan-Zügen. ATZ, 87, 95–100, 245–255, 303–310, 1985.
476. Künstner, R.; Deutenbach, K.-R.; Vagt, J.: Measurements of the Reference Dynamic Pressure in Open-Jet Automotive Wind Tunnels; SAE Technical Paper 920344, 1992, doi:10.4271/920344.
477. Künstner, R.; Potthoff, J.; Essers, U.: The Aero-Acoustic Wind Tunnel of Stuttgart University. SAE Technical Paper 950625, 1995, doi:10.4271/950625.
478. Kuthada, T.: CFD and Wind Tunnel: Competitive Tools or Suplementary Use? 4th FKFS Conference, Stuttgart, 2007.
479. Kuthada, T.: Die Optimierung von Pkw-Kühlluftführungssystemen unter dem Einfluss moderner Bodensimulationstechniken. Dissertation IVK Universität Stuttgart, 2006.
480. Kuthada, T.; Genger, M.; Widdecke, N.; Wiedemann, J.: Joint Project for an Optimized Cooling System. In Wiedemann/Hucho (Hrsg.) Progress in Vehicle Aerodynamics III—Thermo-Management. Expert Verlag, 2002.
481. Kuttruf, T.: Was bringt Ram Air?. Motorradmagazin MO Nr. 4/2001, MO Medien Verlag GmbH, Stuttgart, 2001.
482. Lacey, J.: A Study of the Pulsation in ¾ Open-Jet Wind Tunnel; SAE Technical Paper 2002-01-0251, 2002, doi:10.4271/2002-01-0251.
483. Lamm, M.; Holls, D.: A Century of Automotive Style—100 Years of American Car Design. Stockton, Cal.: Lamm-Morada Pgl. Co., 1996.
484. Landström, C.; Josefsson, L.; Walker, T.; Löfdahl, L.: An Experimental Investigation of Wheel Design Parameters with Respect to Aerodynamic Drag. In: Progress in Vehicle Aerodynamics and Thermal Management; Proceedings of the 8th FKFS Conference, Stuttgart 2011.
485. Lange, A. A.: Vergleichende Windkanalversuche an Fahrzeugmodellen. Berichte Deutscher Kraftfahrzeugforschung im Auftrag des RVM, Nr. 31, 1937.
486. Langenfeld, S.; Friedrich, H.; Meyer, C.: Adjusted Functional Surfaces by Nanotechnology for Automotive Applications. Materials Week 2001, Internat. Congress of Adv. Materials, 2001.
487. Larsson, J.: Aerodynamic Develoment of the Volvo V 70. http://www.adapco-online.com/uconf/EU2000/volvo/index.html, 2000.
488. Larsson, L.; Hammar, L.; Nilsson, L.; Berndtsson, A.; et al.: A Study of Ground Simulation—Correlation between Wind-Tunnel and Water-Basin Tests of a Full-Scale Car, SAE Technical Paper 890368, 1989, doi:10.4271/890368.

489. Launder, B. E.; Sharma, B. I.: Application of the Energy Dissipation Model of Turbulence to the Calculation of Flow Near a Spinning Disc. Letters in Heat an Mass Transfer, vol. 1, no. 2, pp. 131–138, 1974.

490. Lay, W. E.: Is 50 Miles per Gallon Possible with Correct Streamlining? SAE Technical Paper 330041, 1933, doi:10.4271/330041.

491. Le Good, G. M.; Howell, J. P.; Passmore, M. A.; Garry, K. P.: On-Road Aerodynamic Drag Measurements Compared with Wind Tunnel Data. SAE Technical Paper 950627, 1995, doi:10.4271/950627.

492. Le Good, G. M.; Howell, J. P.; Passmore, M. A.; Cogotti, A.: A Comparison of On-Road Aerodynamic Drag Measurements with Wind Tunnel Data from Pininfarina and MIRA. SAE Technical Paper 980394, 1998, doi:10.4271/980394.

493. Leder, A.: Abgelöste Strömungen. Wiesbaden: Vieweg Verlag, 1992.

494. Leister, G.: Fahrzeugreifen und Fahrwerkentwicklung. Strategie, Methoden, Tools, Aus der Reihe: ATZ-MTZ Fachbuch, Vieweg+Teubner Verlag, ISBN: 978-3-8348-0671-0, 2009.

495. Leuschen, J.; Cooper, K. R.: Full-Scale Wind Tunnel Tests of Production and Prototype, Second-Generation Aerodynamic Drag-Reducing Devices for Tractor-Trailers. SAE Technical Paper 2006-01-3456, 2006, doi:10.4271/2006-01-3456.

496. Leyhausen, H.-J.: Die Meisterprüfung im Kfz-Handwerk. 13. Auflage, Vogel Verlag, Würzburg, ISBN 3-8023-1504-9, 1993.

497. Lichtenstein, C. Engler, F. (Hrsg.): Stromlinienform. Zürich: Verlag Lars Müller, 1992.

498. Liebold, H.; Fortnagel, M.; Götz, H.; Reinhard, T.: Aus der Entwicklung des C 111 III, Automobil-Industrie, Nr. 2, 29–36, 1979.

499. Lighthill, M. J.: On sound generated aerodynamically—I. General theory. London: Proceedings of the Royal Society, Vol. A211, 1951.

500. Lighthill, M. J.: On sound generated aerodynamically—II. Turbulence as a source of sound. London: Proceedings of the Royal Society, Vol. 222, 1951.

501. Lilley, G. M.: On the noise from air jets. In: Noise Mechanisms, AGARD CP 131, 13.1-13.12, 1973.

502. Lin, C.; Saunders, J. W.; Watkins, S.; Mole, L.: Increased Productivity—Use of Specific Dissipation to Evaluate Vehicle Engine Cooling. SAE Technical Paper 970137, 1997, doi:10.4271/970137.

503. Lindemann, J.; Hüttenbrink, K.-B: Verminderung der Windgeräusche unter Motorradhelmen. Zwanzig Jahre Geers-Stiftung, Dortmund, S. 58–59, 1996.

504. Lindener, N.; Kaltenhauser, A.: Der BMW Akustik-Windkanal. Tagungsband der Tagung Nr. T-30-341-056-2-P "Aerodynamik des Kraftfahrzeugs;" 25.-26. März; Haus der Technik; Essen, 1992.

505. Lindener, N.: persönliche Demonstration, 1996.

506. Lindener, N.: Zeitgemäße Aerodynamik Entwicklung bei Audi. Vortrag AUDI Sommerforum, 2004.

507. Lindener, N.; Miehling, H.; Cogotti, A.; Cogotti, F.; Maffei, M.: Aeroacoustic Measurements in Turbulent Flow on the Road and in the Wind Tunnel. SAE Technical Paper 2007-01-1551, 2007, doi:10.4271/2007-01-1551.

508. Lindener, N.; Wickem, G.: The Audi Aeroacoustic Wind Tunnel: Final Design and First Operational Experience, SAE Technical Paper 2000-01-0868, 2000, doi:10.4271/2000-01-0868.

509. Lock, A.: Unsteady Aerodynamics—Its Simulation, Measurement and Effect on the Driver. In: Wiedemann, J.; Hucho, W.-H. (Hrsg.) Progress in Vehicle Aerodynamics III—Unsteady Flow Effects, Renningen: expert Verlag, 2004.

510. Lock, A.; Orso Fiet, G.; Ali, M.: MIRA's New Large Volume Airflow Visualisation Technique. In: Wiedemann, J. (Ed.): Progress in Vehicle Aerodynamics and Thermal Management—Proceedings of the 7th FKFS-Conference. Expert Verlag, Renningen, ISBN 978-3-8169-2944-4, 2010.

511. Loehrke, R. I.; Nagib, H. M.: Control of Free-Stream Turbulenz by Means of Honeycombs: A Balance between Suppression and Generation. J. Fluids Engineering, 1976.

512. Löfdahl, L.: Measurements of the Reynolds Stress Tensor in the Thick Three dimensional Boundary Layer near the Stern of a Ship Model. Ph.D. Thesis, Calmers University of Technology, 1982.

513. Lower, M. C.; Hurst, D. W.; Thomas, A.: Noise levels and noise reduction under motorcycle helmets. Noise control—the next 25 years: Proceedings / Inter-Noise 96, the 1996 International Congress on Noise Control Engineering, Inst. of Acoustics, St. Albans, Liverpool: Book 2, 1996.

514. Lührmann, L.: Chancen und Grenzen der virtuellen Aerodynamik. Fahrzeugerprobung–von der Straße in den Rechner. Haus der Technik, München, 2003.

515. Lührmann, L.; Zimmermann, K.; Zörner, C.: Audi A6, Fahrzeugeigenschaften— Aerodynamik und Akustik. ATZ Sonderheft D 58922, 34–42, 2004.

516. Lumley, J.: Turbulence Modeling. 3rd International Conference, Innovation and Reliability in in Automotive Design and Testing, Florence, Italy, 8–10 April, 1381–1392, 1992.

517. Lyon, R.: Statistical Energy Analysis of Dynamical Systems, MIT Press, Cambridge, MA, 1975.

518. Mack, S.; Indinger, T.; Adams, N. A.; Blume, S; Unterlechner, P.: The Interior Design of a 40% Scaled DrivAer Body and Frst Experimental Results. In: Proceedings of the ASME Fluids Engineering Summer Meeting FEDSM 2012-72371, Puerto Rico, USA, 2012.

519. Macklin, A. R.; Garry, K. P.; Howell, J. P.: Comparing Static and Dynamic Testing Techniques for the Crosswind Sensitivity of Road Vehicles, SAE Technical Paper 960674, 1996, doi:10.4271/960674.

520. Macklin, A. R.; Garry, K. P.; Howell, J.: Assessing the Effects of Shear and Turbulence During the Dynamic Testing of the Crosswind Sensitivity of Road Vehicles. SAE Technical Paper 970135, 1997, doi:10.4271/970135.

521. Mackling, W. C.; Metaxas, G. J.: Splashing of Drops on Liquid Layers. J. of Appl. Phys., Vol 47, 3963 ff, 1976.

522. Mackrodt, P.-A.; Steinheuer, J.; Stoffers, G.: Entwicklung aerodynamisch optimaler Formen für das Rad-Schiene Versuchsfahrzeug II. AET, 35, 67–77, 1980.

523. Maeda, K.; Kitoh, K.; Nozaki, H.; Nambo, K.; et al.: Correlation Tests Between Japanese Full-Scale Automotive Wind Tunnels Using the Correction Methods for Drag Coefficient. SAE Technical Paper 2005-01-1457, 2005, doi:10.4271/2005-01-1457.

524. Maestrello, L.: Use of Turbulent Model to Calculate the Vibration and Radiation Responses of a Panel, with Practical Suggestions for Reducing Sound Level. In: Journal of Sound and Vibration, Vol. 5, No. 3, S. 13–38, 1967.

525. Maioli, M.: Function Versus Appearance: The Interaction Between Customer, Stylist and Engineer in Vehicle Design. Int. Journal of Vehicle Design, 5, 305–316, 1983.

526. Mair, W. A.: Drag-Reducing Techniques for Axi-Symmetric Bluff Bodies. in Sovran, G.; Morel, T.; Mason, W. T. (Ed. 1978): Aerodynamic Drag Mechanisms of Bluff Bodies and Road Vehicles. 161–187, New York: Plenum Press, 1978.

527. Mair, W. A.: Reduction of Base Drag by Boat-Tailed Afterbodies in Low-Speed Flow. Aeronautical Quarterly, 20, 307–320, 1969.

528. MakePeace, C.: Compilation of drag coefficients from EADE data. Rüsselsheim: Adam Opel AG, 1991.

529. Malavard, L.: Etude des écoulements transsoniques. Contrôle expérimental des règles de similitude. Jahrbuch der WGL, Nummer, 96–103, 1953.

530. Mankau, H.: Außenspiegel als Optimierungsaufgabe der Kraftfahrzeugentwicklung. 2. Stuttgarter Symposium Kraftfahrwesen und Verbrennungsmotoren, 1997.

531. Marcu, B.; Browand, F.: Aerodynamic Forces Experienced by a 3-Vehicle Platoon in a Crosswind. SAE Technical Paper 1999-01-1324, 1999, doi:10.4271/1999-01-1324.

532. Maretzke, I.; Richter, B.: Einfluss der Aerodynamik auf die Richtungsstabilität von Pkw. VDI Bericht 546, 101–116, 1984.

533. Maskell, E. C.: Progress Towards a Method for the Measurement of the Components of the Drag of a Wing of Finite Span. RAE Technical Report 72232, 1973.

534. Maskell, E.: A Theory of the Blockage Effects on Bluff Bodies and Stalled Wings in a Closed Wind Tunnel; ARC, R&M 3400, 1961.

535. Mauboussin, P.: Voitures aérodynamiques. L'Aéronautique, Nov. 1933, 239–245, 1933.

536. Maue, J. H.: Lärmbelastung für Motorradfahrer—Messergebnisse und Schutzmaßnahmen. Zeitschrift für Lärmbekämpfung 37, 15–19, 1990.

537. Maue, J. H.: Messverfahren zur Bestimmung der Lärmbelastung unter Motorradhelmen in TÜ Bd. 30 Nr. 9, 1989.

538. Mayer, E.: Ist die bisherige Zuordnung von PMV und PPD noch richtig? Klima-, Luft- und Kältetechnik 34, H.12, S. 575–577, 1998.

539. Mayer, W.: Bestimmung und Aufteilung des Fahrwiderstandes im realen Fahrbetrieb; Dissertation von der Fakultät Maschinenbau der Universität Stuttgart, 2006.

540. Mayer, W.; Wiedemann, J.: Road Load Determination Based on Driving-Torque-Measurement. SAE Technical Paper 2003-01-0933, 2003, doi:10.4271/2003-01-0933.

541. Mayer, W.; Wiedemann, J.: The Influence of Rotating Wheels on Total Road Load. SAE Technical Paper 2007-01-1047, 2007, doi:10.4271/2007-01-1047.
542. McBeath, S.: Competition Car Downforce, Haynes, 1998.
543. McBeath, S.: Formel 1 Aerodynamik. Stuttgart: Motorbuch Verlag, 2001.
544. McCallen, R.; Browand, F.; Ross, J.: The Aerodynamics of Heavy Vehicles, Springer Verlag, 2004.
545. McConnel, W. A.: Climatic Testing Indoors—Ford's Hurricane Road. SAE Technical Paper 590283, 1959, doi:10.4271/590283.
546. McCutcheon, G.; McColgan, A. H.; Grant, I.: Wake Studies of a Model Passenger Car using PIV. 3rd MIRA International Vehicle Aerodynamics Conference, Oct. 2000.
547. McKnight, A. J.; McKnight A. S.: The Effects of Motorcycle Helmets Upon Seeing and Hearing, Final Report, National Highway Traffic Safety Administration, U.S. Department of Transportation, DOT HS 808 399, 1994.
548. Mears, A. P.; Crossland, S. C.; Dominy, R. G.: An Investigation into the Flow-Field About an Exposed Racing Wheel. SAE Technical Paper 2004-01-0446, 2004, doi:10.4271/2004-01-0446.
549. Mears, A. P.; Dominy, R. G.; Sims-Williams, D. B.: The Flow About an Isolated Rotating Wheel—Effects of Yaw an Lift, Drag and Flow Structure, 4th MIRA International Vehicle Aerodynamics Conference, Warwick, 2002.
550. Mehta, R. D.: The Aerodynamic Design of Blower Tunnels with Wide-Angle Diffusors; Prog. Aerospace Sci., Vol. 18, 1977.
551. Melchger, N.: Die aeroakustische Entwicklung des Maybach. In: Bargende, M.; Wiedemann, J. (Hrsg.): 5. Internationales Stuttgarter Symposium Kraftfahrwesen und Verbrennungsmotoren 18.-20.2.2003. Renningen: Expert Verlag, ISBN 3-8169-2180-9, 2003.
552. Melchger, N.; Rosmanith, R.: Aerodynamik, Aeroakustik, Motorkühlung—Das Spiel mit dem Wind. ATZ-MTZ Extra, 114–119, 2001.
553. Mende, v. H.-U.: Styling—Automobiles Design. Stuttgart: Motorbuchverlag, 1979.
554. Mendle, T.: Hut-Probe. Tourenfahrer 10/2003, 72–78, 2003.
555. Mercedes--Benz Infobroschüre: Emissionen, Kraftstoffverbrauch—Vorschriften, Testverfahren und Grenzwerte Pkw und leichte Nutzfahrzeuge, September 2008. Siehe auch http://cis-gso.daimler.com/ (GSO—Gesetzestexte online).
556. Mercker E.; Breuer, N.; Berneburg, H.; Emmelmann, H.-J.: On the Aerodynamic Interference Due to Rolling Wheels of Passenger Cars; SAE Technical Paper 910311, 1991, doi:10.4271/910311.
557. Mercker, E.: A Blockage Correction for Automotive Testing in a Wind Tunnel with Closed Test Section. Journal of Wind Engineering & Industrial Aerodynamics, Edition 22, 1986.
558. Mercker, E.: Blockage and Interference Corrections; EADE 2010 Correlation Test, Final Report, beschränkte Veröffentlichung, 2011.
559. Mercker, E.: On Buoyancy and Wake Distortion in Closed Test Sections of Automotive Wind Tunnels. In J. Wiedemann (Ed.): Progress in Vehicle

Aerodynamics and Thermal Management—Proceedings of the 9th FKFS-Conference, Expert Verlag, Renningen (2013), to be published in 2013.

560. Mercker, E.; Cooper, K.: A Two-Measurement Correction for the Effect of a Pressure Gradient on Automotive, Open-Jet, Wind Tunnel Measurements; SAE Paper 2006-01-0568, 2006.

561. Mercker, E.; Cooper, K.; Fischer, O.; Wiedemann, J.: The Influence of a Horizontal Pressure Distribution on Aerodynamic Drag in Open and Closed Wind Tunnels; SAE Technical Paper 2005-01-0867, 2005, doi:10.4271/2005-01-0867.

562. Mercker, E.; Fiedler, H.: Eine Blockierungskorrektur für Aerodynamische Messungen in geschlossenen Unterschallwindkanälen; ZfW, Band 2, Heft 4, 1978.

563. Mercker, E.; Knape, H. W.: Ground Simulation with Moving Belt and Tangential Blowing for Full–Scale Automotive Testing in a Wind Tunnel. SAE Technical Paper 890367, 1989, doi:10.4271/890367.

564. Mercker, E.; Pengel, K.: Über den Strömungslärm in Messstrecken verschiedener Windkanäle und im Fahrgastraum eines Pkw´s. Akustik und Aerodynamik des Kraftfahrzeuges, Renningen: Expert Verlag, 1995.

565. Mercker, E.; Soja, H.; Wiedemann, J.: Experimental investigation on the Influence of Various Ground Simulation Techniques on a Passenger Car; Proceedings "Vehicle Aerodynamics Conference" der Royal Aeronautical Society, Loughborough 1994.

566. Mercker, E.; Wiedemann, J.: Comparison of Different Ground Simulation Techniques for Use in Automotive Wind Tunnels; SAE Technical Paper 900321, 1990, doi:10.4271/900321.

567. Mercker, E.; Wickern, G.; Wiedemann, J.: Contemplation of Nozzle Blockage in open Jet Wind Tunnels in View of Different 'Q' Determination Techniques; SAE Technical Paper 970136, 1997, doi:10.4271/970136, 1997.

568. Mercker, E.; Wiedemann, J.: On the correction of Interference effects in Open Jet Wind Tunnels; SAE Technical Paper 960671, 1996, doi:10.4271/960671.

569. Metternich, Graf, M. W.: Edmund Rumpler—Konstrukteur und Erfinder. München: Neuer Kunstverlag, 1985.

570. Michaelian, M.; Browand, F.: Quanitifying Platoon Fuel Savings: 1999 Field Experiments. SAE Technical Paper 2001-01-1268, 2001, doi:10.4271/2001-01-1268.

571. Michel, U.; Barsikow, B.; Hellmig, M.; Schüttpelz, M.: Schallquellenlokalisierung an landenden Flugzeugen mittels eines Mikrofon-Arrays. In: Tagungsband der DAGA 1998, Zürich, 23.-26.3.1998; ISBN 3-9804568-3-8, 1998.

572. Michelin: Der Reifen. Haftung, Komfort—mechanisch und akustisch sowie Rollwiderstand und Kraftstoffersparnis. Jeweils Michelin Reifenwerke KGaA. Karlsruhe, 2005.

573. Milliken, W.; Milliken, D.: Race Car Vehicle Dynamics. ISBN 978-1-56091-526-3, Warrendale, PA.: SAE, 1995.

574. Mitschke, M.: Dynamik der Kraftfahrzeuge, Band C, Fahrverhalten. 2. Auflage, Springer Verlag, Berlin 1990.

575. Modlinger, F.; Demuth, R.; Adams, N.: Investigations on the Realistic Modeling of the Flow around Wheels and Wheel Arches by CFD. JSAE Technical Paper 2007-08-0198, 2007.

576. Modlinger, F.; Demuth, R.; Adams, N.: New Directions in the Optimization of the Flow around Wheels and Wheel Arches. MIRA Conference 2008.
577. Möhring, W.: On Vortex Sound at Low Mach Number. Journal of Fluid Mechanics, 85:685–691,1978.
578. Möller, E.: Luftwiderstandsmessungen am VW-Lieferwagen. ATZ, 53, 153–156, 1951.
579. Mörchen, W.: Aufbau und Messsysteme des VW-Klimawindkanals. ATZ 70, 73–83, 1968.
580. Mörchen, W.: The Climatic Wind Tunnel of Volkswagenwerk AG. SAE Technical Paper 680120, 1968, doi:10.4271/680120.
581. Morel, T.: Comprehensive Design of Axissymmetric Wind Tunnel Contractions; ASME-Paper 75-Fe-17, 1975.
582. Morel, T.: The Effect of Base Slant on the Flow Pattern and Drag of Three-dimensional Bodies with Bluff Ends. In: Sovran, G.; Morel, T.; Mason, W. T. (Eds.): Aerodynamic Drag Mechanisms of Bluff Bodies and Road Vehicles. New York: Plenum Press, 191–226, 1978.
583. Morel, T.; Bohn, M.: Flow over Two Circular Disks in Tandem. Aerodynamics of Transportation. ASME-CSME-Conf. Niagara Falls, June 1979, 23–32, 1979.
584. Morelli, A.: A New Aerodynamic Approach to Advanced Automobile Basic Shapes. SAE Technical Paper 2000-01-0491, 2000, doi:10.4271/2000-01-0491.
585. Morelli, A.: Aerodynamic Basic Bodies Suitable for Automobile Applications, International Journal of Vehicle Design, SP3, 70–98, 1983.
586. Morelli, A.: General Layout Characteristics and Performance of a New Wind Tunnel for Aerodynamic and Functional Tests on Full Scale Vehicles. SAE Technical Paper 710214, 1971, doi:10.4271/710214.
587. Morelli, A.: Persönliche Information über das Gebläse-Rad, 2002.
588. Morelli, A.: The New Pininfarina Wind Tunnel for Full Scale Automobile Testing. In H. S. Stephens (Editor): Advances in Road Vehicle Aerodynamics, 335–365. Cranfield, UK: The British Hydromechanics Research Association (BHRA), 1973.
589. Morelli, A.; Di Giusto, N.: A New Step in Automobile Aerodynamics Performance Improvements and Design Implications. International Conference Vehicles and Sytems Progress. Volgograd, Russland, 7–10. September 1999.
590. Morelli, A.; Fioravanti, L.; Cogotti, A.: Sulla forma della carrozzeria di minima resistenza aerodinamica. ATA Novembre 1976, 468–476, 1976.
591. Morén, A.: Cdtool—A Parametric Aerodynamic Drag Prediction Tool. Diploma Work, Umeá University, Göteborg, 2007.
592. Moron, P.; Hazir, A.; Crouse, B.; Powell, R.; et al.: Hybrid Technique for Underbody Noise Transmission of Wind Noise. SAE Technical Paper 2011-01-1700, 2011, doi:10.4271/2011-01-1700.
593. Moron, P.; Powell, R.; Freed, D.; Perot, F.; et al.: A CFD/SEA Approach for Prediction of Vehicle Interior Noise Due to Wind Noise, SAE Technical Paper 2009-01-2203, 2009, doi:10.4271/2009-01-2203.
594. Mößner, A.: Entwicklungsmethoden zur Verbesserung des Insassenkomforts in offenen Fahrzeugen am Beispiel der SL/SLK-Klasse von Mercedes-Benz. Haus der

Technik, München, Operative und strategische Ziele der Fahrzeug-Aerodynamik, 2001.

595. Mullarkey, S. P.: Aerodynamic stability of road vehicles in side winds and gusts. Department of Aeronautics, Imperial College of Science and Technology London, 1990.

596. Müller, R.: Private Kommunikation, ADAC e.V. Bereich Fahrzeugtechnik, München, 2003.

597. Müller, U.; Klingebiel, F.: Simulation von Motorschmier- und Motorkühlkreisläufen. 634–646 in Bargende, M.; Wiedemann; J. (Hrsg. 1999): Kraftfahrwesen und Verbrennungsmotoren. 3. Stuttgarter Symposium. Expert Verlag, Renningen Malmsheim, 1999.

598. Muto, S.: The aerodynamic Drag Coefficient of a Passenger Car and Methods for Reducing It. Int. Journal of Vehicle Design, SP 3, 37–69, 1983.

599. Muto, S.; Ishihara, T.: The JARI Full Scale Wind Tunnel. SAE Technical Paper 780336, 1978, doi:10.4271/780336.

600. Mutoh, S.: Automobile Aerodynamics (Car Styling 50½ Special Edition, English & Japanese). Tokyo: Publishing Co, San'ei Shobo, 1985.

601. N. N.: Hart am Wind. Auto Motor Sport, 20, 126–130, 1982.

602. Nebel, M.; Melchger, N.; Wäschle, A.: Die Aerodynamik- und Aeroakustikentwicklung der neuen Mercedes-Benz E-Klasse. In Bargende, M.; Reuss, H.-C.; Wiedemann, J. (Hrsg.): Kraftfahrwesen und Verbrennungsmotoren. 9. Stuttgarter Symposium. Springer Vieweg Verlag, Wiesbaden, 2009.

603. Nemecek, J.; Grandjean, E.: Results of an Ergonomic Investigation of Large-Space Offices. Human Factors 15(2), 111–124, 1993.

604. Ng, E. Y.; Johnson, P. W.; Watkins, S.; Grant, L.: Wind-Tunnel Tests of Vehicle Cooling System Performance at High Blockage. SAE Technical Paper 2000-01-0351, 2000, doi:10.4271/2000-01-0351.

605. Ng. E. Y.; Watkins, S.; Johnson, P. W.; Mole, L.: Use of a Pressure-Based Technique for Evaluating the Aerodynamics of Vehicle Cooling Systems. SAE Technical Paper 2002-01-0712, 2002, doi:10.4271/2002-01-0712.

606. Nicholls, J.: Stream and Droplet Breakup by Shock Waves. NASA SP-194, 126–128, 1972.

607. Nilsson, H. (Ed.): Definition and Theoretical Background of the Equivalent Temperature. ATA 6th Int. Conf. The New Role of Experimentation in the Modern Automotive Product Development Process, 99A4082, 1999.

608. Nilsson, L.-U.; Berndtsson, A: The New Volvo Multipurpose Automotive Wind Tunnel. SAE Technical Paper 870249, 1987, doi:10.4271/870249.

609. Nitsche, W.: Strömungsmesstechnik. Berlin Heidelberg: Springer Verlag, ISBN-10: 3540544674, 1994.

610. Nitz, J.; Deutenbach, K.-R.; Poltrock, R.: ARVW—Konzept eines luftwiderstand-sarmen Rekordfahrzeugs. ATZ 84, 211–219, 1982.

611. Nizzola, C.: Modellierung und Verbrauchsoptimierung von ottomotorischen Antriebskonzepten. Dissertation ETH, Nr. 13831, Zürich, 2000.

612. Noger, C.; Gillieron, P.: Banc expérimental d'analyse des phénomènes aérodynamiques générés par le dépassement de deux véhicules automobiles. 16ième Congrès Français de Mécanique, Nice, 2003.
613. Nouzawa, T.; Haruna, S.; Hiasa, K.; Nakamura, T.; et al.: Analysis of Wake Pattern for Reducing Aerodynamic Drag of Notchback Model. SAE Technical Paper 900318, 1990, doi:10.4271/900318.
614. Nowitzki, H. J.: Rennmaschinen im Windkanal—Windspiel. Motorrad, Nr. 10, S. 6–12, 1980.
615. Oda, N.; Hoshino, T.: Three-Dimensional Air Flow Visualization by Smoke Tunnel. SAE Technical Paper 741029, 1974, doi:10.4271/741029.
616. Oertel (jr.), H.; Laurien, E.: Numerische Strömungsmechanik. Grundgleichungen—Lösungsmethoden—Softwarebeispiele. 2. Auflage, Vieweg Verlag, Wiesbaden, ISBN 3-528-03936-1, 2003.
617. Oertel (jr.), H.: Strömungsmechanik. Berlin, Heidelberg: Springer Verlag, 1995.
618. Oettle, N.; Sims-Williams, D.; Dominy, R.; Darlington, C.; et al.: The Effects of Unsteady On-Road Flow Conditions on Cabin Noise: Spectral and Geometric Dependence. *SAE Int. J. Passeng. Cars—Mech. Syst.* 4(1):120–130, 2011, doi:10.4271/2011-01-0159.
619. Ogata, N.; Iida, N.; Fuji, Y.: Nissan's Low-Noise Full Scale Wind Tunnel. SAE Technical Paper 870250, 1987, doi:10.4271/870250.
620. Ohtani, K.; Takei, M.; Sakamoto, H.: Nissan Full Scale Wind Tunnel—Its Application to Passenger Car Design. SAE Technical Paper 720100, 1972, doi:10.4271/720100.
621. Onorato, M.; Costelli, A. F.; Garrone, A.: Drag Measurements Through Wake Analysis. SAE Technical Paper 840302, 1984, doi:10.4271/840302.
622. Otto, N.; Feng, B. J.: Wind Noise Sound Quality. SAE Technical Paper 951369, 1995, doi:10.4271/951369.
623. Paefgen, F.-J.; Gush, B.: Der Bentley Speed 8 für das 24-Stunden-Rennen in Le Mans 2003. ATZ 106, 281–289, 2004.
624. Pankhurst, R. C.; Holder, D. W.: Wind-Tunnel Technique. London: Pitman, 1968.
625. Papenfuß, H.-D.; Dilgen, P.: Three-Dimensional Separated Flow Around Automobiles with Different Rear Profile: Application of the Zonal Method. In Gersten, K. (Ed.): Physics of Separated Flows—Numerical, Experimental, and Theoretical Aspects, Braunschweig: Vieweg, 241–248, 1993.
626. Park, Y.-G.: Air-Side Heat Transfer and Friction Correlations for Flat-Tube Louver-Fin Heat Exchangers. Journal of Heat Transfer, Vol. 131, 2009.
627. Passmore, M.; Le Good, G.; Howell, J.: A Practical Analysis of the Coastdown Test Technique. MIRA 2nd International Conference on Vehicle Aerodynamics, Nuneaton, 1998.
628. Passmore, M. A.; Richardson, S.; Imam, A.: An Experimental Study of Unsteady Vehicle Aerodynamics. Proc. Instn. Mech. Engrs., Vol. 215 Part D, pp. 779–788, 2001.
629. Pawlowski, F. W.: Wind Resistance of Automobiles. SAE Journal 1930-07-01, 27, 5–14, 1930.

630. Peikert, E.; Schmidt, E.-M.; Carell, G.; Koenigsbeck, A.: Leitfaden für Freunde des Gespannfahrens, 4. Auflage, Krefeld: Bundesverband der Motorradfahrer e. V., 1986

631. Peric, M.; Schreck, E.: Muzaferija, S.: Die Methode der überlappenden Rechengitter und deren Anwendung in der Fahrzeugaerodynamik. 10. Tagung Fahrzeugaerodynamik, Haus der Technik, München, 2012.

632. Pérot, F.; Kim, M.S.; Freed, D.M.; Dongkon, L.; Ih, K.D.; Lee, M.H.: Direct Aeroacoustics Prediction of Ducts and Vents Noise. AIAA paper 2010-3724, 14th AIAA/CEAS aeroacoustics conference, Stockhom, June 2010.

633. Pérot, F.; Wada, K.; Norisada, K.; Kitada, M.; Hirayama, S.; Sakai, M.; Imahigasi, S.; Sasaki, N.: HVAC Blower Aero-Acoustics Predictions Based on the Lattice Boltzmann Method, AJK 2011 Conference, Hamamatsu, Japan, July 2011.

634. Persu, A.: Luftwiderstand und Schnellwagen. Zeitschrift für Flugtechnik und Motorluftschiffahrt, 15, 25–27, 1924.

635. Peschke, W.; Mankau, H.: Auftriebskräfte am Wohnanhänger beeinflussen die Stabilität von Wohnwagengespannen. Automobil Revue, 18, 51–53, 1982.

636. Petz, R.; Charwat, M.: Das AeroLAB der BMW Group: Fahrzeugmessungen mit dem Single-Rolling Road System. 10. Tagung Haus der Technik, Fahrzeug-Aerodynamik, München, 4./5. July 2012, 2012.

637. Pfadenhauer, M.: Aerodynamikentwicklung im Rennsport am Beispiel des Audi R8, Tagung: Haus der Technik: Operative und strategische Ziele der Fahrzeug-Aerodynamik, München, 2001.

638. Pfadenhauer, M.: Aerodynamikentwicklung im Rennsport am Beispiel des Audi R8. In Bargende, M.; Wiedemann, J. (Hrsg.): Kraftfahrwesen und Verbrennungsmotoren. 4. Internationales Stuttgarter Symposium. 355–369, Renningen: Expert Verlag, 2001.

639. Pfadenhauer, M.: Konzepte zur Verringerung des Luftwiderstandsbeiwertes von Personenkraftwagen unter Berücksichtigung der Wechselwirkung zwischen Fahrzeug und Fahrbahn sowie der Raddrehung. Diplomarbeit Technische Universität München, 1995.

640. Physik-Hütte (Band 1, Mechanik), Verlag Wilhelm Ernst und Sohn, 1971.

641. Piatek, R.; Hentschel, W.: Strömungssichtbarmachung in der aerodynamischen Entwicklung von Kraftfahrzeugen. Tagung: Sichtbarmachung technischer Strömungsvorgänge, Haus der Technik, Essen, 1998.

642. Piola, G.: Formula 1 `98—Technical Analysis, Giorgio Nada Editore, 1999.

643. Piola, G.: Formula 1 `99—Technical Analysis, Giorgio Nada Editore, 2000.

644. Piola, G.: Formula 1 2001—Technical Analysis, Giorgio Nada Editore, 2002.

645. Pischinger, R.; Klell, M.; Sams, T.: Thermodynamik der Verbrennungskraftmaschine. Springer Verlag, 2002.

646. Pope, A.; Harper, J. J.: Low Speed Wind Tunnel Testing. London: J. Wiley & Sons, 1966.

647. Porsche, Dr.-Ing. e. h., AG: Europäische Patentschrift O 213 387, 1986.

648. Potthoff, J.: Aerodynamische Hilfsmittel am Rennsportwagen. Vortrag VDI-Jahrestagung Fahrzeugtechnik, Stuttgart, November 1977.

649. Potthoff, J.: Persönliche Information, 1969.
650. Potthoff, J.: The Aerodynamic Layout of UNICAR Research Vehicle. International Symposium on Vehicle Aerodynamics, Wolfsburg: VW AG, 1982.
651. Potthoff, J.; Fischer, O.; Helfer, M.; Horn, M.; Kuthada, T.; Michelbach, A.; Schröck, D.; Widdecke, N.; Wiedemann, J.: 20 Jahre Fahrzeugwindkanäle der Universität Stuttgart am Institut für Verbrennungsmotoren und Kraftfahrwesen 1989–2009. Automobiltechnische Zeitschrift 111 Nr. 12, 2009.
652. Potthoff, J.; Michelbach, A.; Wiedemann, J.: Die neue Laufbandtechnik im IVK-Aeroakustik-Fahrzeugwindkanal der Universität Stuttgart. Teil 1 ATZ, 106, 52–61; Teil 2 ATZ, 106, 150–160, 2004.
653. Potthoff, J.; Schmid, I. C.: Wunibald I. E. Kamm—Wegbereiter der modernen Kraftfahrtechnik. Berlin, Heidelberg: Springer Verlag, 2012.
654. Preusser, T.; Polansky, L.; Giesecke, P.: Advances in the Development of Wind Tunnel Balance Systems for Experimental Automotive Aerodynamics. SAE Technical Paper 890370, 1989, doi:10.4271/890370.
655. Rae, W. H.; Pope, A.: Low-Speed Wind Tunnel Testing; Verlag John Wiley & Sons, (1984).
656. Raffel, M.; Willert, C.; Kompenhans, J.; Loose, S.; Bosbach, J.: Measurement of Unsteady Flows. 177–191 in Wiedemann, J.; Hucho, W.-H. (Hrsg.): Progress in Vehicle Aerodynamics III—Unsteady Flow Effects. Renningen: Expert Verlag, 2004.
657. Ramm, G.; Hummel, D.: A Panel Method for the Computation of the Flow Around Vehicles Including Side-edge Vortices and Wake. 3rd International Conference, Innovation and Reliability in Automotive Design and Testing, Florence, 8–10. April, 1992.
658. Rauser, M.; Eberius, J.: Verbesserung der Fahrzeugaerodynamik durch Unterbodengestaltung. ATZ, 89, 535–542, 1987.
659. Rawnsley, S. M.; Glynn, D. R.: Flow around Road Vehicles. 1st International PHOENICS User Conference, 1985.
660. Rawnsley, S. M.; Tatchell, D. G.: Application of the PHOEMICS Code to the Computation of the Flow Around Automobiles. SAE Technical Paper 860217, 1986, doi:10.4271/860217.
661. Read, A.; Mendonca, F.; Scharm, Ch.; Tournour, M.: Optimal Sunroof Buffeting Predictions with Compressibility and Surface Impedance Effects—American Institute of Aeronautics and Astraunotics Paper 2005-2859, S.1-13–2005.
662. Reid, E. G.: Farewell to the Horseless Carriage. SAE Technical Paper 350095, 1935, doi:10.4271/350095.
663. Reilly, D.: NACA-Ducts—What They Are and How They Work. Road & Track, 71–74, 1979.
664. Reitz, R. D.; Diwakar, R.: Effect of Drop Break-up on Fuel Sprays. SAE Technical Paper 860469, 1986, doi:10.4271/860469.
665. Remenda, B. A. P.; Krause, A. E.; Hertz, P. B.: Vehicle Coastdown Reistance Analysis under Windy and Grade-Variable Conditions. SAE Technical Paper 890371, 1989, doi:10.4271/890371.
666. Rennsport News: Formel 1. www.rennsportnews.de, 2004.

667. Repmann, C.: Die aerodynamische Entwicklung des 1-Liter-Fahrzeugs XL1 Der Volkswagen AG. 10. Tagung Fahrzeug-Aerodynamik, 4./5. Juli 2012, München. Essen: Haus der Technik, 2012.
668. Resenhoeft, T.: Zu viel Bumm gibt Sehstörungen. Berliner Zeitung, Ressort Auto, 02.07.1999.
669. Riedel, A.; Arbinger, R.: Subjektive und objektive Beurteilung des Fahrverhaltens von Pkw. FAT-Bericht Nr. 139, Frankfurt, 1997.
670. Riederer, S.: Numerische Simulation des Strömungsfeldes zweiteiliger Hochauftriebsprofile, Diplomarbeit, Technische Universität München, 1999.
671. Riedler, A.: Wissenschaftliche Automobilbewertung. Berlin: Oldenburg Verlag, 1911.
672. Riegel, M.: Bestimmung der Anteile von Antriebs-, Umströmungs- und Rollgeräusch im Innenraum von Pkw. Renningen: Expert Verlag, ISBN 978-3-8169-3085-3, 2011.
673. Riegel, M.: Entwicklung eines akustischen Hohlspiegels mit integriertem Mikrofon-Array. In: Tagungsband der DAGA 2012, 19.-22. 03.2012, Darmstadt. Berlin: Deutsche Gesellschaft für Akustik e.V., 2012, ISBN: 978-3-939296-04-1, 2012.
674. Riegel, M.; Helfer, M.: Schallquellenortung in Windkanälen unter Berücksichtigung des Strömungsfeldes. In: Tagungsband der DAGA 2011, Düsseldorf, 21.-24.03.2011. Berlin: Deutsche Gesellschaft für Akustik e.V., 2011, ISBN: 978-3-939296-02-7, 2011.
675. Riegel, M.; Wiedemann, J.: Bestimmung des Windgeräuschanteils im Vergleich zu Antriebs- und Rollgeräusch im Innenraum von Pkw. In: Bargende, M.; Wiedemann, J. (Hrsg.): "5. Internationales Stuttgarter Symposium Kraftfahrwesen und Verbrennungsmotoren 18.-20.02.2003." Renningen: Expert Verlag, ISBN 3-8169-2180-9, 2003.
676. Riegel, M.; Wiedemann, J.; Helfer, M.: The Effect of Turbulence on In-Cabin Wind Noise—a Comparison of Road and Wind Tunnel Results. Tagungsband der "6th MIRA International Vehicle Aerodynamics Conference," Gaydon (GB), Heritage Motor Centre, 25–26.10.2006.
677. Riegels, F. W.: Aerodynamische Profile. München: R. Oldenburg, 1958.
678. Riehle, J.: Prevention Against Unsafe Transportation of Goods on Passenger Vehicle Roofs. SAE Technical Paper 2000-01-0352, 2000, doi:10.4271/2000-01-0352.
679. Romani, L.: La Mesure sur Piste de la Resistance a l'Avancement. Paper 15, Road Vehicle Aerodynamics, 1st Symposium, London, 1969.
680. Romberg, G. F.; Chianese, F.; Lajoie, R. G.: Aerodynamics of Race Cars in Drafting and Passing Situations. SAE Technical Paper 710213, 1971, doi:10.4271/710213.
681. Romberg, G. F.; Gunn, J. A.; Lutz, R. G.: The Chrysler 3/8-Scale Pilot Wind Tunnel. SAE Technical Paper 940416, 1994, doi:10.4271/940416.
682. Rompe, K.; Heißing, B.: Objektive Testverfahren für die Fahreigenschaften von Kraftfahrzeugen. Fahrzeugtechnische Schriftenreihe, 1984.
683. Roshko, A.: Perspectives on Bluff Body Aerodynamics. Journal of Wind Engineering and Industrial Aerodynamics, 49, 79–100, 1993.

684. Roshko, A.; Koenig, K.: Interaction Effects on the Drag of Bluff Bodies in Tandem. In: Sovran, G.; Morel, T.; Mason, W. T. (Ed.): Aerodynamic Drag Mechanisms of Bluff Bodies and Road Vehicles. New York: Plenum Press, 253–286, 1978.

685. Roshko, A.; Lau, J. C.: Some observations on Transition and Reattachment of a Free Shear Layer in Incompressible Flow. Proceedings of the 1965 Heat Transfer and Fluid Mechanics Institute. Stanford: University Press, 157–167, 1965.

686. Rossiter, J. E.: Wind Tunnel Experiments on the Flow over Rectangular Cavities at Subsonic and Transonic Speeds; Aeronautical Research Council Reports and Memo No. 3438, 1964.

687. Rothhämel, J.: Qualifizierung von Strömungsgeräuschen unter Motorradhelmen. Fortschritte der Akustik DAGA, Oldenburg: DEGA, 2003.

688. Ruck, B.: Laser-Doppler-Anemometrie. Stuttgart: AT-Fachverlag, 1987.

689. Rudd, M. J.: A New Theoretical Model for the Laser Doppler. J. Phys. E.: Sci. Instrum.2, 55, 1969.

690. Rumpler, E.: Das Auto im Luftstrom. Zeitschrift für Flugtechnik und Motorluftschiffahrt, 15, 22–25, 1924.

691. Rung, T.; Eschricht, D.; Yan, J.; Thiele, F.: Sound Radiation of the Vortex Flow past a Generic Side Mirror. 8. AIAA/CEAS Aeroacoustics Conference. Breckenridge, Colorado, 17–19 June, AIAA-2002-2549, 2002.

692. Ruscheweyh, H.: Dynamische Windwirkung an Bauwerken. 2 Bände. Wiesbaden, Berlin: Bauverlag; ISBN: 3-7625-2008-9, 1982.

693. Rutz, R.; Dragon, L.; Breitling, T: Fahrdynamikentwicklung in der Zukunft. Tag des Fahrwerks, Aachen, 2002.

694. Ryan, A. G.: The simulation of transient cross-wind gusts and their aerodynamic influence on passenger cars. Doctoral thesis, Durham University. Available at Durham E-Theses Online: http://etheses.dur.ac.uk/1203/, 2000.

695. Ryan, A. G.; Dominy, R. G.: Wake Surveys Behind a Passenger Car Subjected to a Transient Cross-Wind Gust. SAE Technical Paper 2000-01-0874, 2000, doi:10.4271/2000-01-0874.

696. SAE Recommended Practice SAE J1263 FEB 96: Road Load Measurement and Dynamometer Simulation Using Coastdown Techniques. SAE Standard J1263, 1996.

697. SAE Recommended Practice SAE J1594 DEC94: Vehicle Aerodynamics Terminology. SAE Standard J1594, 1994.

698. SAE Recommended Practice J2263 OCT 96: Road Load Measurements using Onboard Anemometry and Coastdown Techniques. SAE Standard J2263, 1996.

699. SAE Recommended Practice J2264 APR1995: Chassis Dynamometer Simulation of Road Load using Coastdown Techniques. SAE Standard J2264, 1995.

700. SAE Recommended Practice SAE J670e JUL76: Vehicle Dynamics Terminology. SAE Standard J670, 1976.

701. SAE Recommended Practice J902 FEB99: Passenger Car Windshield Defrosting Systems. SAE Standard J902, 1999.

702. SAE Standard J819 NOV 95: Engine Cooling System Field Test (Air-to-Boil). SAE Standard J819, 1995.

703. SAE: Surface Vehicle Information Report; SAE J 2021, 1990.
704. Sakamoto, H.; Arie, M.: Flow around a Normal Plate of Finite Width Immersed in a Turbulent Boundary Layer. Journal of Fluid Engineering, 105, 99–104, 1983.
705. Samenfink, W.: Grundlegende Untersuchung zur Tropfeninteraktion mit schubspannungsgetriebenen Wandfilmen. Dissertation, Institut für Thermische Strömungsmaschinen Universität Karlsruhe, 1997.
706. Samenfink, W.: Sekundarzerfall von Tropfen. In Atomization and Sprays, Short Course, Institut für Thermische Strömungsmaschinen Universität Karlsruhe (TH), 1995.
707. Sawatzki, E.: Die Luftkräfte und ihre Momente am Kraftwagen und die aerodynamischen Mittel zur Beeinflussung der Fahrtrichtungserhaltung. Deutsche Kraftfahrtforschung im Auftrag des Reichs-Verkehrsministeriums, Heft 50; VDI Verlag, Berlin, 1941.
708. Sawatzki, E.; Huber, L.: Luftwiderstand von Krafträdern. DKS-Heft 18. Berlin: VDI Verlag 1938.
709. Schaible, S.: Fahrzeugseitenwindempfindlichkeit unter natürlichen Bedingungen. Dissertation, TH Aachen, 1998.
710. Schäufele, S.: Validierung der neuen Windkanäle im Aerodynamischen Versuchszentrum der BMW Group und Analyse der Übertragbarkeit der Ergebnisse. Dissertation KIT Karlsruhe, 2010.
711. Schenkel, F. K.: The Origins of Drag and Lift Reductions on Automobiles with Front and Rear Spoilers. SAE Technical Paper 770389, 1977, doi:10.4271/770389.
712. Schlang, A.: Von der Kappe zum Laser-Helm in Sport1.de http://www.sport1.de/coremedia/generator/www.sport1.de/Sportarten/Formel1/Berichte/Hintergrund/Archiv1/f1geschichte_20der_20helme_20mel.html, 2002
713. Schlichting, H.; Truckenbrodt, E.: Aerodynamik des Flugzeugs. Bd. 1 und 2, Berlin, Heidelberg: Springer Verlag, 1969.
714. Schlichting, H.: Aerodynamische Untersuchungen an Kraftfahrzeugen. Hochschultag, Kassel, 1953.
715. Schlichting, H.: Grenzschicht-Theorie. Braun, 1965.
716. Schlichting, H; Gersten, K.: Grenzschichttheorie. 10. Auflage, Berlin, Heidelberg: Springer Verlag, 2006
717. Schlör, K.: Entwicklung und Bau einer luftwiderstandsarmen Karosserie auf einem 1,7-Ltr-Heckmotor-Mercedes-Benz-Fahrgestell. Deutsche Kraftfahrforschung. Zwischenbericht Nr. 48, 1938.
718. Schmehl, R.; Klose, G.; Maier, G.; Wittig, S.: Efficient Numerical Calculation of Evaporating Sprays in Combustion Chamber Flows. RTO-MP-14, Symp. On Gas Turbine Combustion, Emissions and Alternative Fuels, 1998.
719. Schmehl, R.; Rosskamp, H.; Willmann, M.; Wittig. S.: CFD Analysis of Spray Propagation and Evaporation Including Wall Film Formation and Spray/Film Interactions. International Journal of Heat and Fluid Flow, Vol. 20, 520–529, 1999.
720. Schmid, C.: Die Fahrwiderstände beim Kraftfahrzeug und die Mittel ihrer Verringerung. ATZ, 41, 465–477 und 498–510, 1938.

721. Schmidt, G.: Design. Verlockungen der Formen des Automobils—Teil 1: Einige sozial- und kulturwissenschaftliche Anmerkungen. Vortrag Autosommer 2011, Stuttgart & Karlsruhe, 2011.
722. Schmieder, M.: Die Hochgeschwindigkeitsstabilität von Motorrädern, ein Mensch / Maschine—Problem. 4. Fachtagung der VDI Gesellschaft Fahrzeugtechnik Motorrad am 5.–7. März 1991 in München, VDI Berichte 875, VDI Verlag, 1991.
723. Schmiederer, L.; Riedel, R.: Der neue Klimawindkanal von Behr. ATZ—Automobiltechnische Zeitschrift 103/11, 2001.
724. Schmitt, H.: Der Leistungsbedarf zur Kühlung des Fahrzeugmotors und seine Verminderung. Deutsche Kraftfahrforschung,45, Berlin: VDI Verlag, 1940.
725. Schmitt, J.: Einbindung der LDA-Messtechnik in den Entwicklungsprozess eines Pkw's. 579–592 in Bargende, M.; Wiedemann, J. (Hrsg.): Kraftfahrwesen und Verbrennungsmotoren. 3. Stuttgarter Symposium. Renningen-Malmsheim: Expert Verlag, 1999.
726. Schmitt, J.; Schopper, H.-D.; Breitling, T.: Flow Details, Using Aerodynamics to improve Properties of Cars. Lecture 6: International Short Course on Vehicle Aerodynamics, Stuttgart, 1998.
727. Schmitt, J.; Wilharm, K.: Measurement of Flow Fields with LDA. In Wiedemann, J.; Hucho, W.-H.: Progress in Vehicle Aerodynamics—Advanced Experimental Techniques. Renningen: Expert Verlag, 121–130, 2000.
728. Schmitz, G.; Geusen, R.: Windkanalversuche—Aerodynamische Untersuchungen an BMW—Verkleidungen. PS Nr. 1, S. 59/61, 1978.
729. Schneider, H.-J.: 125 Jahre Opel—Autos und Technik. Köln: Verlag Schneider & Repschläger, 1987.
730. Schneider, S.; Wiedemann, J.; Wickern, G.: Das Audi-Windkanalzentrum, Tagung "Aerodynamik des Kraftfahrzeuges" im Haus der Technik, Essen, 1998.
731. Schnepf, W.: Aerodynamik moderner Motorräder—In Saus und Braus. Motorrad Nr. 10, S. 28/34, 1984.
732. Schnepf, W.: Die Aerodynamik der BMW K 100 RS—Hart am Wind. Motorrad Nr. 24, S. 12/13, 1983.
733. Schnepf, W.: Die Geschichte der Motorrad—Aerodynamik—Hundert Jahre und ein bisschen weiter. Motorrad Nr. 22, S. 8/15, 1985.
734. Schnepf, W.: Motorräder im Windkanal—Dicke Luft. Motorrad Nr. 7, S. 6/21, 1987.
735. Scholz, N.: Windkanaluntersuchungen am NSU-Weltrekordmotorrad. Die Umschau, Halbmonatsschrift über die Fortschritte in Wissenschaft und Technik. Nr. 22, 691/692, 1951.
736. Scholz, N.: Windkanaluntersuchungen an Motorradmodellen. ZVDI, 95, 17, 1953.
737. Schrefl, M.; Tentrop, G.; Maier, M. J.: Der neue Audi A1—Aerodynamik, Aeroakustik, Thermomanagement und Klimatisierung. Automobiltechnische Zeitschrift, Sonderdruck, Der neue Audi A1', 2010.
738. Schrefl, M.: Instationäre Aerodynamik von Kraftfahrzeugen: Aerodynamik bei Überholvorgang und böigem Seitenwind. Dissertation TU Darmstadt; Aachen: Shaker Verlag, 2008.

739. Schröck, D.: Eine Methode zur Bestimmung der aerodynamischen Eigenschaften eines Fahrzeugs unter böigem Seitenwind. Dissertation Universität Stuttgart, Renningen: expert Verlag, ISBN 978-3-8169-3147-8, 2012.
740. Schröck, D.; Krantz, W.; Widdecke, N.; Wiedemann, J.: Instationäre aerodynamische Eigenschaften von Fahrzeugen unter böigem Seitenwind; Haus der Technik Tagung Fahrzeug-Aerodynamik, München, 2010.
741. Schröck, D.; Krantz, W.; Widdecke, N.; Wiedemann, J.: Unsteady Aerodynamic Properties of a Vehicle Model and their Effect on Driver and Vehicle under Side Wind Conditions. *SAE Int. J. Passeng. Cars—Mech. Syst.* 4(1):108–119, 2011, doi:10.4271/2011-01-0154.
742. Schröck, D.; Widdecke, N.; Wiedemann, J.: Aerodynamic Response of a Vehicle Model to Turbulent Wind. In J. Wiedemann (Ed.): Progress in Vehicle Aerodynamics and Thermal Management—Proceedings of the 7th FKFS-Conference. Expert Verlag, Renningen, ISBN 978-3-8169-2944-4, 2010.
743. Schröck, D.; Widdecke, N.; Wiedemann, J.: On Road Wind Conditions Experienced by a Moving Vehicle. 6th FKFS Conference—Progress in Vehicle Aerodynamics; Stuttgart, 2007.
744. Schüler, F.; Adolph, T.; Steinmann, K.; Ionescu, I.: Aktive Sicherheit von Helmen für Motorradfahrer; Berichte der Bundesanstalt für Straßenwesen, Unterreihe "Fahrzeugtechnik;" digitaler Bericht F64, September 2007.
745. Schulten, J. B. H. M.: Some Remarks on Pure Tone Fan Noise Suppression of the DNW Low Speed Wind Tunnel; Memorandum AV-75-010, NLR, 1975.
746. Schulz-Hausmann, F. v.; Vagt, J.-D.: Influence of Test-Section Length and Collector Area on Measurements in ¾-Open-Jet Automotive Wind Tunnels; SAE Technical Paper 880251, 1988, doi:10.4271/880251.
747. Schütz, T.: Aerodynamics of Modern Sport Utility Vehicles. MIRA International Vehicle Aerodynamics Conference, 2010.
748. Schütz, T.: Ein Beitrag zur Berechnung der Bremsenkühlung an Kraftfahrzeugen. Dissertation IVK Universität Stuttgart, 2009
749. Schütz, T.: Fortschritt der CFD-Validierung in der Entwicklung der Kraftfahrzeugaerodynamik bei Audi. 11. Internationales Stuttgarter Symposium, 2011.
750. Schütz, T.: Untersuchung der Umströmung eines SAE-Modells mit Hilfe von numerischer Simulation. Studienarbeit, IVK, Universität Stuttgart, 2005.
751. Schütz, T.; Hühnergarth, J.: Die Aerodynamik und Aeroakustik des neuen Audi Q3. Automobiltechnische Zeitschrift, Sonderdruck 'Der neue Audi Q3', 76–84, 2011.
752. Scibor-Rylski, A. J.: Road Vehicle Aerodynamics. 2nd. Edition, London: Pentech Press, 1984.
753. Sebben, S.: Challenges and Limitations of CFD in Road Vehicle Aerodynamics. Lecture Series 2005-05, Road Vehicle Aerodynamics. Brüssel: von Kármán Institute for Fluid Dynamics, 2005.
754. Seeger, H.: Transportation Design. Stuttgart: Vorlesungsskript, 2012.
755. Seibert, W.; Ehlen, M.; Sovani, S.: Simulation of Transient Aerodyanamics-Predicting Buffting, Roaring and Whistling using CFD—5th MIRA International Vehicle Aerodynamics Conference, Heritage Motor Center Warwick, October 2004.

756. Selig, M. S.: Summary of Low-Speed Airfoil Data. Vol. 1, University of Illinois, 1995.
757. Senior, A.; Zhang, X.: The Force and Pressure of a Diffusor-Equipped Bluff Body in Ground Effect. Journal of Fluids Engineering, 123, 105–111, 2001.
758. Senthooran, S.; Crouse, B.; Balasubramanian, G.; Freed, D.; Shin, S. R.; Ih, K.-D.: Effect of Surface Mounted Microphones on Automobile Side Glass Pressure Fluctuations. Proc. of 7th MIRA International Vehicle Aerodynamics Conference, Coventry 22–23 October, 2008.
759. Senthooran, S.; Duncan, B.; Freed, D.; Hendriana, D.; et al.: Design of Roof-Rack Crossbars for Production Automobiles to Reduce Howl Noise Using a Lattice Boltzmann Scheme. SAE Technical Paper 2007-01-2398, 2007, doi:10.4271/2007-01-2398.
760. Sherwood, A. W.: Wind Tunnel Test of Trailmobile Trailers. University of Maryland Wind Tunnel Report Nr. 35, 1953.
761. Siddal, R. D.; Davies, T. W.: An Improved Response Equation for Hot Wire Anemometry. Int. Journal Heat & Mass Transfer 15, 367–368, 1972.
762. Silk, G.; Anselmi, A. T. Robert, H. F.; MacMinn, S.: Automobile and Culture. New York: Harry N. Abrams Publishers, 1984.
763. Singh, R.: Automated Aerodynamic Design Optimization Process for Automotive Vehicle. SAE Technical Paper 2003-01-0993, 2003, doi:10.4271/2003-01-0993.
764. Smagorinsky, J.: General Circulation Experiments with the Primitive Equations. I: The Basic Experiment. Monthly Weather Review 91, 99–152, 1963.
765. Soja, H.; Wiedemann, J.: The interference between exterior and interior flow on road vehicles. Société des Ingénieurs de l'Automobile (S. I. A.). Journée d'etude: Dynamique du vehicule—Sécurité active, 101–105, 1987.
766. Sovran, G.; Blaser, D.: A Contribution to Understanding Automotive Fuel Economy and Its Limits. SAE Technical Paper 2003-01-2070, 2003, doi:10.4271/2003-01-2070.
767. Spalart, P. R. and Allmaras, S. R.: A One-Equation Turbulence Model for Aerodynamic Flows. AIAA Paper 92-0439, 1992.
768. Spalart, P.R.; Jou W. H.; Strelets M.; Allmaras S. R.: Comments on the feasibility of LES for wings, and on a hybrid RANS/LES approach, 1st AFOSR Int. Conf. on DNS/LES, Aug. 4–8, 1997, Ruston, LA. In: Liu, C.; Liu, Z. (Eds.): Advances in DNS/LES. Greyden Press, Columbus, OH, 1997.
769. Spalding, D. B.: An Introduction to PHOENICS. CHAM, TR 68, 1981.
770. Speth, J. F.:Der Wasserkanal—Ein Hilfsmittel bei der Fahrzeugentwicklung? Tagung 'Aerodynamik des Kraftfahrzeugs', Haus der Technik, Essen, 1984.
771. Spiegel, B.: Die obere Hälfte des Motorrads. Motorbuch Verlag, Stuttgart, 2002.
772. SPIER GmbH & Co. Fahrzeugwerk KG, Presseveröffentlichungen 2011.
773. Spruss, I.: Simulation der Fremd- und Eigenverschmutzung. Interner Bericht FKFS, Stuttgart, 2010.
774. Spruss, I.: Verschmutzungssimulation. Interner Bericht FKFS, Stuttgart, 2010.
775. Spruss, I.; Kuthada, T; Wiedemann, J.; Cyr, S.; Duncan, B.: Spray Pattern of a Free Rotating Wheel—CFD Simulation and Validation. FKFS Conference, Stuttgart 2011.

776. Stanton, D. W.; Rutland, C. J.: Multi-Dimensional Modeling of Heat and Mass Transfer of Fuel Films Resulting from Impinging Sprays. SAE Technical Paper 980132, 1998, doi:10.4271/980132.

777. Stapleford, W. R.; Carr, W. G.: Aerodynamics of Exposed Rotating Wheels, Technical Report 1970/2, MIRA, 1970.

778. Steinbach, D.: Calculation of Wall and Model-Support Interference in Subsonic Wind-Tunnel Flows. ZfW 17, 370–378, 1993.

779. Steinberg, M. O. (Hrsg.); Martynenko, O. G. (Hrsg.); Idelchik, I. G.: Handbook of Hydraulic Resistance. 3rd edition, Jaico Publishing House, 2005.

780. Stoffregen, J.: Motorradtechnik. Vieweg Verlag, Braunschweig/Wiesbaden, 2001.

780a. Stoll, D.; Kuthada, T.; Wiedemann, J.; Schütz, T.: Unsteady Aerodynamic Vehicle Properties of the DrivAer Model in the IVK Model Scale Wind Tunnel. In: Wiedemann, J. (Ed.). *Progress in Vehicle Aerodynamics and Thermal Management*. Renningen: Expert Verlag, 2015.

781. Streblow, N.: Rauschangriff. Motorrad, 14/2001.

782. Stroh, C.; Hager, J.: Optimizing Thermal Management of Vehicles Using Advanced Simulation Tools. SAE Technical Paper 2002-01-1026, 2002, doi:10.4271/2002-01-1026.

783. Sumitani, K.; Yamada, M.: Development of "Aero Slit"—Improvement of Aerodynamic Yaw Characteristics for Commercial Vehicles, SAE Technical Paper 890372, 1989, doi:10.4271/890372.

784. Summa, J. M.; Maskew, B.: Predicting Automobile Characteristics Using an Iterative Viscous/Potential Flow Technique. SAE Technical Paper 830303, 1983, doi:10.4271/830303.

785. Sykes, D. M.: The Effect of Low Flow Rate Gas Ejection and Ground Proximity on After Body Pressure Distribution. Proceedings of the 1st Symposium on Road Vehicle Aerodynamics, London: City University, 1969.

786. Széchényi, E.: Crosswind and its Simulation. In: Wiedemann, J.; Hucho, W.-H. (Hrsg.): Progress in Vehicle Aerodynamics II—Advanced Experimental Techniques. Renningen: Expert Verlag, 83–96, 2000.

787. Szechenyi, E.: The Overtaking Process of Vehicles. In: Wiedemann, J. und Hucho, W. H. (Hrsg.): Progress in Vehicle Aerodynamics III-Unsteady Flow Effects, Renningen: Expert Verlag, 2004.

788. Tamai, G.: The Leading Edge—Aerodynamic Design of Ultra-streamlined Land Vehicles. Cambridge, MA: Robert Bentley Publishers, 1999.

789. Tandogan, E.: Optimierter Entwurf von Hochleistungswärmeübertragern. Dissertation, Bochum, 2001.

790. Tangorra, J.; George, A. R.: Wind Noise of Motorcycle Helmets, Cornell University, Ithaca, New York: 1991.

791. Taylor, G. I.: Air Resistance of a Flat Plate of Very Porous Material. ARC R&M 2236, 1948.

792. Teixeira: Incorporating Turbulence Models into the Lattice-Boltzmann Method. Intl. Journal of Modern Phyics C, Vol. 9, No. 8, 1998.

793. Tesch, G.: Kühlluftführungs- und Lüfterkonzepte am Pkw bei typischen Bauraumbeschränkungen. Dissertation TU München, Verlag Dr. Hut, München, 2011.
794. Tesch, G.; Modlinger, F.: Die Aerodynamikfelge von BMW—Einfluss und Gestaltung von Rädern zur Minimierung von Fahrwiderständen. Tagung: Fahrzeug-Aerodynamik—Neue Chancen und Perspektiven für die Kraftfahrzeugaerodynamik durch CO_2-Gesetzgebung und Energiewende. Haus der Technik, München, 2012.
795. Theissen, P.: Unsteady Vehicle Aerodynamics in Gusty Crosswind, Dissertation TU München, Lehrstuhl für Aerodynamik und Strömungsmechanik, 2012.
796. Theissen, P.; Wojciak J.; Demuth R.; Adams N.A.; Indinger T.: Unsteady Aerodynamic Phenomena under Time-Dependent Flow Conditions for Different Vehicle Shapes. In Proceedings of 8th International Vehicle Aerodynamics Conference, UK, 2010.
797. Thibaut, J.: Optimierung des Fahrzeugdesigns unter Berücksichtigung der Durchströmung im aerodynamischen Entwicklungsprozesses. Dissertation, Universität Stuttgart, 2012.
798. Thibaut, J.; Wiedemann, J.: Optimization of Vehicle Design Regarding Internal Airflow in the Aerodynamic Development Process. 7. Internationales Stuttgarter Symposium—Automobil- und Motorentechnik, 2007.
799. Thierry, E. H.: Improvements in or relating to Helmets, Patent Specification 449,905, His Majesty's Stationary Office, UK, 1934.
800. Thöle, G.: Aerodynamik. Motorrad 13/2000, 154–155, 2000.
801. Thöle, G.: Jetzt aber Ruhe. Motorrad, 8/1996, 212–213, 1996.
802. Thöle, G.: Service Spezial Helmtest Praxis. Motorrad 8/2001.
803. Thöle, G.: Windjammer. Motorrad, 11/2000, 158–162, 2000.
804. Toftum, J.: A Field Study of Draught Complaints in the Industrial Work Environment. Proceedings of the Sixth International Conference of Environments, Ergonomics, pp. 252–253, Scientific Information Center, Defence & Civil Institute of Environmental Medicine, North York Ontario, Canada, 1994.
805. Tongji 2011: Broschüre der Tongji Universität: Shanghai Automotive Wind Tunnel Center; Version 20100907. www.sawtc.com; download: 20.03.2011.
806. Torda, T. P.; Morel, T. A.: Aerodynamic Design of a Land Speed Record Car. Journal of Aircraft, 8 (1971), S. 1029–1033, 1971.
807. Trzebiatowsky, H.: Motorräder, Motorroller, Mopeds und ihre Instandhaltung. 1. Aufl. Gießen: Pfanneberg Verlag, 1955.
808. Tsuboi, K.; Shirayama, S.; Oana, M.; Kuwahara, K.: Computational Study of the Effect of Base Slant. In: Marino, C. (Ed.): Supercomputer Applications in Automotive Research and Engineering Development. Minneapolis, Ma.: Cray Research Inc. Book, 257–272, 1988.
809. Tummescheit, H.; Eborn, J.; Plößl, K.; Försterling, S.; Tegethoff, W.: Pkw-Klimatisierung IV. Expert Verlag, 2007.
810. Tyler, J. M.; Sofrin, T. G.: Axial Flow Compressor Noise Studies; SAE Technical Paper 620532, 1962, doi:10.4271/620532.

811. Ubertini, S.; Desideri, U.: Aerodynamic Investigation of a Scooter in the University of Perugia Wind Tunnel Facility. SAE Technical Paper 2002-01-0254, 2002, doi:10.4271/2002-01-0254.
812. Ullrich, F.: Aeroakustik im Windkanal der BMW Group. Essen: Haus der Technik, 2011.
813. Ullrich, F.: New Possibilities for Aeroacoustic Optimization in the Underbody Region of Vehicles. In: J. Wiedemann (Ed.): Progress in Vehicle Aerodynamics and Thermal Management V. Renningen: Expert Verlag, 2008.
814. Ulmer, R. G.; Shabanowa Northrup, V.: Evaluation of the Repeal of the All-Rider Motorcycle Helmet Law in Florida, Final Report, National Highway Traffic Safety Administration, U.S. Department of Transportation, DOT HS 809 849, 2005.
815. US EPA (Environmental Protection Agency), http://www.epa.gov und DOE (US Department of Energy), http://www.fueleconomy.gov, http://www.epa.gov/carlabel/index.htm
816. Utz, H.-J.: Bestimmung der statistischen Verteilung der Anströmrichtung für Personenkraftwagen bei Autobahnfahrten in der Bundesrepublik Deutschland. Diplomarbeit, Universität Stuttgart, 1982.
817. Vagt, J. D.: Hot-Wire Probes in Low Speed Flow. Progress in Aerospace Sciences, Vol. 18, Number 4, pp. 271–323, Pergamon Press, 1979.
818. Vagt, J.-D.; Wolff, B.: Das neue Messzentrum für Aerodynamik—Zwei neue Windkanäle bei Porsche. Teil 1: ATZ, 89,121–129, Teil 2: ATZ, 89, 183–189, 1987.
819. Van Dyke, M.: An Album on Fluid Motion. Stanford, California: Parabolic Press. ISBN 978-0-915760-02-2, 2007.
820. van Hal, B.; Vanmaele, C.; Silar, P.; Priebsch, H.-H.: Hybrid Finite Element—Wave Based Method for Steady-State Acoustic Analysis. In: Proceedings of ISMA, September 2004.
821. van Moorhem, W. K. et al.: The Effects of Motorcycle helmets on Hearing and the Detection of Warning Signals. Journal of Sound and Vibration, 77 1, 39–49, 1981.
822. Van Raemdonck, G.: Aerospace Engineering, 2012.
823. VDA (Verband Deutscher Automobilhersteller): Marktgewichteter Kraftstoffverbrauch seit 1978. Tabelle vgl. www.vda.de/de/aktuell/kraftstoffverbrauch/index.html.
824. VDI Wärmeatlas, Springer-VDI Verlag, 2002.
825. Verordnung (EG) Nr. 443/2009 des Europäischen Parlaments und des Rates vom 23. April 2009, http://eur-lex.europa.eu/LexUriServ/LexUriServ.do?uri=OJ:L:2009:140:0001:01:DE:HTML.
826. Versteeg, H. K.; Malalasekera, W.: An Introduction to Computational Fluid Dynamics: The Finite Volume Method. 2nd edition, Prentice Hall, ISBN-13 978-0131274983, 2007.
827. Vieri: Considerazione di aerodinamica nella progettazione dei caschi. Giornale ed atti della Associazione Tecnica Automobile, 9, 376–379, 1977.
828. Vivarelli, C.: Linee per una storia dell' aerodinamica dell' automobile dal 1899 al 1944. Pasian di Prato: Campanotto Editore, 2009.
829. Volkers, T.: DAF Drag-meter method summary for TF1. Version 27 Jan. 2012.

830. Vooren, J. V. d.; Sanderse, A.: Finite Difference Calculation of Incompressible Flow through a Straight Channel of Varying Rectangular Cross Section, with Application to Low Speed Wind Tunnels NLR-Report TR 77109 U, 1977.

831. Wachters, L. H. J.; Westerling, N. A. J.: The Heat Transfer from a Hot Wall to Impinging Water Drops in the Spheroidal State, Chemical Engineering Science, Vol. 21, 1047–1056, 1966.

832. Wagner, A.: Die Bewertung der Fahrer-Fahrzeug Interaktion als Auslegungskriterium in der Fahrwerkentwicklung. Haus der Technik, München, 2003.

833. Wagner, A.: Ein Verfahren zur Vorhersage und Bewertung der Fahrerreaktion bei Seitenwind. Dissertation Universität Stuttgart, Band 23 der Schriftenreihe des Institutes für Verbrennungsmotoren und Kraftfahrwesen, (Hrsg.) Wiedemann; J. Stuttgart: Expert Verlag, 2003.

834. Wagner, A.: Motorrad Aerodynamik: Zusammenhänge—Messwerte—Möglichkeiten. IFZ (Institut für Zweiradsicherheit) Nr. 8, Tagung: Safety Environment Future II, Proceedings of the 1998 International Motorcycle Conference, 1998.

835. Wagner, A.; Lindener, N.: Die Aerodynamik des neuen Audi Q5. In: Schol, O. (Hrsg.): Der neue Audi Q5—Entwicklung und Technik. Springer Vieweg Verlag, ISBN 978-3-8348-0604-8, 2008.

836. Wagner, A.; Wiedemann, J.: Crosswind Behaviour in the Driver's Perspective. SAE Technical Paper 2002-01-0086, 2002, doi:10.4271/2002-01-0086.

837. Wagner, C.: Einsatz von Social Media im Product Lifecycle Management—Analyse, Konzepte und Anwendung. Master-Thesis, Universität des Saarlandes, Saarbrücken, 2012.

838. Walker, T.; Broniewicz, A.: Wind Tunnel Upgrade. Oral presentation only, abstract in Wiedemann, J. (Ed.): Progress in Vehicle Aerodynamics and Thermal Management, Expert Verlag, Renningen, ISBN 978-3-8169-2771-6, 2008.

839. Wallentowitz, H.: Fahrer—Fahrzeug—Seitenwind. Dissertation, TU Braunschweig, 1978.

840. Walston, W. H.; Buckley, F. T.; Marks, C. H.: Test Procedures for the Evaluation of Aerodynamic Drag on Full-Scale Vehicles in Windy Environments. SAE Technical Paper 760106, 1976, doi:10.4271/760106.

841. Walter, J.; Bordner, J.; Nelson, B.; Boram, A.: The Windshear Rolling Road Wind Tunnel; *SAE Int. J. Passeng. Cars—Mech. Syst.* 5(1):265–288, 2012, doi:10.4271/2012-01-0300.

842. Walter, J.; Duell, E.; Martindale, B.; Arnette, S.; et al.: The DaimlerChrysler Full-Scale Aeroacoustic Wind Tunnel. SAE Technical Paper 2003-01-0426, 2003, doi:10.4271/2003-01-0426.

843. Walter, J.; Duell, E.; Martindale, B.; Arnette, S.; et al.: The Driveability Test Facility Wind Tunnel No. 8. SAE Technical Paper 2002-01-0252, 2002, doi:10.4271/2002-01-0252.

844. Walter, J. A.; Pruess, D. J.; Romberg, G. F.: Coastdown/Wind Tunnel Drag Correlation and Uncertainty Analysis. SAE Technical Paper 2001-01-0630, 2001, doi:10.4271/2001-01-0630.

845. Wang, Y.; Yong, Z.; Li, Q.: Methods to Control Low Frequency Pulsation in Open-Jet Wind Tunnel. Applied Acoustics 73, 2012.
846. Warnecke, K.; Müller, J.: Design und Aerodynamik im Einklang. ATZ extra, October 2003, 34–35, 2003.
847. Wäschle, A.: Numerische und experimentelle Untersuchung des Einflusses von drehenden Rädern auf die Fahrzeugaerodynamik. Dissertation, Universität Stuttgart, 2006.
848. Wäschle, A.: Aerodynamik kompakt—die neue A-Klasse. In: Tagungsband 10. Tagung Fahrzeug-Aerodynamik, Haus der Technik, 2012.
849. Wäschle, A.; Cyr, S.; Kuthada, T.; Wiedemann, J.: Flow around an Isolated Wheel—Experimental and Numerical Comparison of Two CFD-Codes. SAE Technical Paper 2004-01-0445, 2004, doi:10.4271/2004-01-0445.
850. Wäschle, A.; Wiedemann, J.: Numerische Simulation der Radumströmung zur Untersuchung des Radeinflusses auf die Fahrzeugaerodynamik. VDI-Bericht 1701, Berechnung und Simulation im Fahrzeugbau, 325–352, 2002.
851. Watanabe, T.; Okubo, T.; Iwasa, M.; Aoki, H.: Establishment of an Aero-Dynamic Simulation System for Motorcycles and Its application. JSAE Technical Paper 2001-08-0357, 2001.
852. Waters, D. M.: The Aerodynamic Behavior of Car-Caravan Combinations. Paper 4, Proceedings of the 1st Symposium on Road Vehicle Aerodynamics, London: City University, 1969.
853. Watkins, S.: On the Causes of Image Blurring in External Rear View Mirrors. SAE Technical Paper 2004-01-1309, 2004, doi:10.4271/2004-01-1309.
854. Watkins, S.; Riegel, M.: The Effect of Turbulence on Wind Noise: A Road and Wind Tunnel Study. In: Bargende, M.; Wiedemann, J. (Hrsg.): 4. Internationales Stuttgarter Symposium "Kraftfahrwesen und Verbrennungsmotoren," 20.-22.02.2001. Renningen-Malmsheim: Expert Verlag, 2001, ISBN: 3-8169-1981-2, 2001.
855. Watkins, S.; Saunders, J. A: Review of the Wind Conditions Experienced by a Moving Vehicle. SAE Technical Paper 981182, 1998, doi:10.4271/981182.
856. Waudby-Smith, P.; Bender, T.; Vigneron, R.: The GIE S2A Full-Scale Aero-Acoustic Wind Tunnel. SAE Technical Paper 2004-01-0808, 2004, doi:10.4271/2004-01-0808.
857. Weidele, A.: Motorrad und Straße: Einflüsse auf die Fahrstabilität—Wetterwendisch. PS Nr. 10, S. 32/37, 1988.
858. Weidmann, E.-P.: Experimentelle und theoretische Untersuchung des Nachheizverhaltens an Kraftfahrzeugen. Dissertation, Universität Stuttgart, 2008.
859. Wengle, H.; Bärwolff, G.; Janke, G.; Huppertz, A; Fernholz, H.-H.: The Manipulated Transitional Backward-Facing Step Flow: A Comparison of the Mean Data of a Direct Numerical Simulation and an Experiment. European Journal of Mechanics B/Fluids, 20, 25–46, 2001.
860. Whipple, F. J. W.: Stability of the Motion of a Bicycle. The Quarterly Journal of Pure and Applied Mathematics. Nr. 4, S. 312/348, 1898.
861. White, R. G. S.: A Rating Method for Assessing Vehicle Aerodynamic Drag Coefficients. MIRA Report No. 1967/9, 1967.

862. Whitfield, J.; Jacocks, J.; Dietz, W.; Pate, S.: Demonstration of the Adaptive Wall Concept Applied to an Automotive Wind Tunnel; SAE Technical Paper 820373, 1982, doi:10.4271/820373.
863. Wickern, G.: Die Aerodynamik- und Aeroakustik-Entwicklung des neuen Audi A8. ATZ-MTZ Extra, D 58992, 164–170, 2002.
864. Wickern, G.: On the Application of Classical Wind Tunnel Corrections for Automotive Bodies, SAE Technical Paper 2001-01-0633, 2001, doi:10.4271/2001-01-0633.
865. Wickern, G.; Beese, E.: Computational and Experimental Evaluation of Pad Correction for a Wind Tunnel Balance Equipped for Rotating Wheels. SAE Technical Paper 2002-01-0532, 2002, doi:10.4271/2002-01-0532.
866. Wickern, G.; Dietz, S.; Lührmann, L.: Gradient Effects on Drag Due to Boundary-Layer Suction in Automotive Wind Tunnels; SAE Technical Paper 2003-01-0655, 2003, doi:10.4271/2003-01-0655.
867. Wickern, G.; von Heesen, W.; Wallmann, S.: Wind Tunnel Pulsations and Their Active Suppression; SAE Technical Paper 2000-01-0869, 2000, doi:10.4271/2000-01-0869.
868. Wickern, G.; Wagner, A.; Zöerner, C.: Induced Drag of Ground Vehicles and Its Interaction with Ground Simulation. SAE Technical Paper 2005-01-0872, 2005, doi:10.4271/2005-01-0872.
869. Wickern, G.; Zwicker, K.; Pfadenhauer, M.: Rotating Wheels—Their Impact on Wind Tunnel Test Techniques and on Vehicle Drag Results, SAE Technical Paper 970133, 1997, doi:10.4271/970133.
870. Widdecke, N.; Potthoff, J.: Method and apparatus for determining vertical forces acting on a vehicle at wind flow in a wind tunnel. FKFS, Europäisches Patent: EP1544589, Juni 2005.
871. Widdecke, N.; Potthoff, J.: Verfahren zur Bestimmung von auf ein Kraftfahrzeug in einem Windkanal unter Windströmung einwirkenden Vertikalkräften. Patent Nr. 10361314. Patentinhaber: Forschungsinstitut für Kraftfahrwesen und Fahrzeugmotoren Stuttgart (FKFS), 70569 Stuttgart, DE, 2005.
872. Wiedemann, J.: Aerodynamik I. Umdruck zur Vorlesung im Wintersemester 2008 / 2009. Institut für Verbrennungsmotoren und Kraftfahrwesen, Universität Stuttgart, 2008.
873. Wiedemann, J.: Grenzen und Möglichkeiten der Modelltechnik innerhalb der Kraftfahrzeug-Aerodynamik. Symposium Nr. T-30-905-056-7 "Aerodynamik des Kraftfahrzeugs," Haus der Technik, Essen, 1987.
874. Wiedemann, J.: Laser-Doppler-Anemometrie. Berlin: Springer Verlag, 1984.
875. Wiedemann, J.: Optimierung der Kraftfahrzeugdurchströmung zur Steigerung des aerodynamischen Abtriebes. In: Automobiltechnische Zeitschrift 88 7/8, S. 429–431, 1986.
876. Wiedemann, J.: The Design of Wind Tunnel Fans for Aero-acoustic Testing; Automobile Wind Noise and Its Measurement Part II, Callister, J. R.; George A. R. (Eds.), SAE SP-1457, 1999.

877. Wiedemann, J.: The Influence of Ground Simulation and Wheel Rotation on Aerodynamic Drag Optimization—Potential for Reducing Fuel Consumption. SAE Technical Paper 960672, 1996, doi:10.4271/960672.

878. Wiedemann, J.: Theoretical and Experimental Optimization of the Road-Vehicle Internal Flow. Von Kármán Institute for Fluid Dynamics Lecture Series 1986-05 on Vehicle Aerodynamics, Rhode-St.-Genèse (Belgium), March 17–21, 1986.

879. Wiedemann, J.: Verfahren und Windkanalwaage bei aerodynamischen Messungen an Fahrzeugen. Europäisches Patent Nr. EP 0 842 407 B1, Inhaber AUDI AG, 2000.

880. Wiedemann, J.: Windkanaltechnik. In: Vorlesung Kraftfahrzeug-Aerodynamik, Kapitel 10. Institut für Verbrennungsmotoren und Kraftfahrwesen, Universität Stuttgart, 2010.

881. Wiedemann, J.; Ewald, B.: Turbulence Manipulation to Increase Effective Reynolds Numbers in Vehicle Aerodynamics. AIAA Journal, Vol. 27, No. 6, pp. 763–769, June 1989.

882. Wiedemann, J.; Hucho, W.-H. (Hrsg.): Progress in Vehicle Aerodynamics III—Unsteady Flow Effects. Renningen: Expert Verlag, 2004.

883. Wiedemann, J.; Potthoff, J.: The New 5-Belt Road Simulation System of the IVK Wind Tunnels—Design and First Results. SAE Technical Paper 2003-01-0429, 2003, doi:10.4271/2003-01-0429.

884. Wiedemann, J.; Wickern, G.; Ewald, B.; Mattern, C.: Audi Aero-Acoustic Wind Tunnel; SAE Technical Paper 930300, 1993, doi:10.4271/930300.

885. Wieghardt, K.: Erhöhung des turbulenten Reibungswiderstandes durch Oberflächenstörungen. Techn. Berichte 10, Heft 9, 1943; siehe auch Forschungshefte Schiffstechnik, 1, 65–81, 1953.

886. Wieghardt, K.: Theoretische Strömungslehre. B. G. Teubner, 1965 und Göttinger Universitätsverlag, 2006.

887. Williams, J.: Aerodynamic Drag of Engine-Cooling Airflow with External Interference. SAE Technical Paper 2003-01-0996, 2003, doi:10.4271/2003-01-0996.

888. Williamson, C. H. K.: Three-Dimensional Wake Transition. Journal of Fluid Mechanics, 328, 345–407, 1996.

889. Willumeit, H. P.; Matheis, A.; Müller, K.: Korrelation von Untersuchungsergebnissen zur Seitenwindempfindlichkeit eines Pkw im Fahrsimulator und Prüffeld, ATZ, Vol. 93, 28–35, 1991.

890. Winkelmann, H.: Aerodynamics of the New BMW Z4, MIRA International Vehicle Aerodynamics Conference, Warwick, 2002.

891. Witoszynski, E.: Über Strahlerweiterung und Strahlablenkung. In: Karman, T. v.; Levi-Civita, H. (Hrsg.): Vorträge über Hydro und Aeromechanik, 1925.

892. Wittmeier, F.; Kuthada, T.; Widdecke, N.; Wiedemann, J.: Reifenentwicklung unter aerodynamischen Aspekten. Automobiltechnische Zeitschrift, 2012.

893. Wojciak, J.; Indinger, T.; Adams, N. A.; Theissen, P.; Demuth, R.: Experimental Study of On-Road Aerodynamics during Crosswind Gusts, MIRA Conference, 2010.

894. Wordley, S. J.: On-road Turbulence. Dissertation; Department of Mechanical Engineering Monash University, Clayton Victoria, Australia, 2009.

895. Wright, P. G.: The Influence of Aerodynamics on the Design of Formula One Racing Cars. International Journal of Vehicle Design, SP3, 158–172, 1987.
896. Wüst, W.: Strömungsmesstechnik. Braunschweig: Vieweg Verlag, 1969.
897. Wüst, W.: Verdrängungskorrekturen für rechteckige Windkanäle bei verschiedenen Strahlbegrenzungen und bei exzentrischer Lage des Modells; ZfW 9, 1961.
898. Wüsten, T.: Im Windkanal—Die Antwort kennt nur der Wind. Motorrad, Reisen und Sport Nr. 13, S. 6/17, 1985.
899. Wyss, P.: Die Wundertüte von Hinwil. Automobil Revue Nr. 25, 25, 2003.
900. Yamamoto, S.; Yanagimoto, K.; Fukudah, H.; China, H.; Nakagawa, K.: Aerodynamic influence of a Passing Vehicle on the Stability of the Other Vehicles. JSAE Review 18, 39–44, 1997.
901. Yang, W.-H.: Ein neues Verfahren zur Bestimmung der Fahrwiderstände. Improve Properties of Cars, International Short Course on Vehicle Aerodynamics, Stuttgart, 1994.
902. Young, R. A.: Bluff Bodies in a Shear Flow. Ph. D.-Thesis, University of Camebridge, 1972.
903. Zaccariotto, M.; Burgade, L.; Chanudet, P.: Aeroacoustic Studies at P.S.A. In: Essers, U. (Hrsg.): 2. Stuttgarter Symposium Kraftfahrwesen und Verbrennungsmotoren, 18–20, February 1997. Renningen: Expert Verlag, ISBN 3-8169-1522-1, 1997.
904. Zierep, J.; Bühler, K.: Grundzüge der Strömungslehre. Springer Vieweg, 2010.
905. Zomotor, A.: Fahrwerktechnik, Fahrverhalten, Würzburg: Vogel Verlag, Springer, 1987.
906. Zomotor, A.; Richter, K.-H.; Kuhn, W.: Untersuchungen über die Stabilität und das aerodynamische Störverhalten von Pkw-Wohnanhängerzügen. Automobil-Industrie, 3, 331–340, 1982.
907. Zörner, C.; Islam, M; Lindener, N.: Aerodynamik und Aeroakustik des neuen Audi A8. Automobiltechnische Zeitschrift, Sonderdruck ‚Der neue Audi A8', 2010.
908. Zuck, B.: Downsizing—Auswirkungen auf das Wärmemanagement. Renningen: Expert Verlag, 2012

The Authors
(sorted by chapter)

Dr.-Ing. **Wolf-Heinrich Hucho** studied mechanical engineering at the Technical University of Braunschweig. From 1961 to 1968 he was an assistant of Professor Schlichting, and he did his PhD on ship hydrodynamics in 1967. From 1968 he worked as a research engineer at Volkswagen AG, first as head of the large climatic wind tunnel and from 1969 to 1978 as head of the research department of drive-train technology. Then he worked as a development manager and managing director of leading suppliers. Since 1986 he has been working as a freelance consultant, journalist, and lecturer. He is the editor of the first five editions of this book and the author of *Bluff Body Aerodynamics*. He is also co-editor of the proceedings *Progress in Vehicle Aerodynamics*. He made detailed comments in *Design and Technology* and *Springer Handbook of Experimental Fluid Dynamics*. Since 2011, he has run the smoke tunnel "aero wolf."

Prof. Dr. rer. nat. Dr.-Ing. habil. **Andreas Dillmann** studied mechanical engineering at the University of Karlsruhe from 1980 to 1986. He then joined the Max-Planck Institute for Flow Research in Goettingen and received his PhD in 1989 for a thesis on the nucleation of condensing vapors. Subsequently, he joined the DLR Institute of Fluid Mechanics in Göttingen as a research associate and habilitated in 1995 at the Leibniz University Hannover with a thesis on the aerodynamics of supersonic jets. In 1996, he received a Heisenberg research scholarship of the Deutsche Forschungsgemeinschaft and in 1998, he was promoted to professor for Theoretical Fluid Mechanics at the Technical University of Berlin. Since 2003, he is director of the DLR Institute of Aerodynamics and Flow Technology and professor for Aerodynamics at the Georg-August-University in Göttingen.

The Authors

Dr.-Ing. **Teddy Woll** studied industrial engineering at the Technical University of Darmstadt from 1981 to 1987. From 1987 to 1994 he was a research assistant at the Institute of Electromechanical Design. He did his PhD and founded AKASOL, a research association that developed solar and light electric vehicles. From 1994 to 1995 he was responsible for aerodynamics and weight optimization at the Micro Compact Car GmbH and the development of alternative drive of the first Smart car. From 1996, he has been working in the advanced development at Daimler AG in Sindelfingen, and since April 1999 he is head of aerodynamics, aero-acoustics, and wind tunnels.

Dr.-Ing. **Thomas Schuetz** studied mechanical engineering at the University of Stuttgart from 2001 to 2005. He did his PhD on numerical simulation of brake cooling. From 2008 to 2013, he worked in the technical development at Audi and was responsible for the aerodynamics and aero-acoustics development of several models. Since 2014, he has been working at BMW and is currently responsible for aerodynamic concepts and motorcycle aerodynamics. Since 2012, he is a lecturer for vehicle aerodynamics at the Institute of Fluid Mechanics and Aerodynamics (SLA) at the Technical University of Darmstadt. For his academic achievements, he was awarded with the Arthur Fischer Award from the University of Stuttgart in 2006, and for his PhD thesis, with the ECARA Award from the European Car Aerodynamics Research Association (ECARA) in 2011.

Dipl.-Ing. **Lothar Krüger** studied aerospace engineering at the University of Applied Sciences of Aachen. Since 1987, he has been employed at Ford Europe/ Cologne, first from 1987 to 1990 as a test engineer in the aerodynamic wind tunnel and then from 1990 to 1999, as a development engineer in automotive aerodynamics. Since 1999, he is a technical specialist for aerodynamics, aeroacoustics, and water management.

Dipl.-Ing. **Manfred Lentzen** studied aerospace engineering at the University of Applied Sciences of Aachen. From 1989 to 1994, he worked as a designer in the structural development at Dornier. Since 1994 he has been working at Ford Europe/ Cologne as a development engineer in aerodynamics. Since 2001, he is head of the aerodynamic department.

The Authors

Dr.-Ing. **Andreas Wagner** studied vehicle technology at the University of Applied Sciences Ulm from 1994 to 1999. From 2000 to 2003 he worked as a research assistant at the University of Stuttgart. He did his PhD thesis on "A method for evaluating and predicting the driver's reaction in crosswinds." In 2003, he moved to Audi and started as a test engineer in the department of aerodynamics/ aero-acoustics. From 2008 to 2010, he was head of the development of air conditioning systems. Since 2010, he has been working in the chassis concept development as head of chassis properties (vehicle dynamics and driving comfort).

Dr.-Ing. **David Schröck** studied mechanical engineering at the University of Karlsruhe from 1999 to 2001 and mechanical engineering at the University of Stuttgart 2001 to 2005. From 2005 to 2011, he was a research associate and project manager in the wind tunnel at the Research Institute of Automotive Engineering and Vehicle Engines Stuttgart (FKFS). He did his PhD on a method for the determination of aerodynamic characteristics of a vehicle in gusty winds. Since 2011, he has been working as a development engineer in aerodynamics and aero-acoustics at the Adam Opel AG.

Dipl.-Ing. **Alexander Mößner** studied aerospace engineering at the University of Stuttgart. From 1989 to 1996, he worked at Dr.-Ing. h.c. F. Porsche AG in the aerodynamic development. From 1996 to 1998, he worked in the Micro Compact Car GmbH, also in aerodynamics. Since 1998 he has been working at Daimler AG in the department of aerodynamics and wind tunnels for Mercedes-Benz vehicles, responsible for several types in the S, SL, SLK, and E classes. Additional field of work is comfort improvement for cabriolets.

Dipl.-Ing. (FH) **Patrick Höfer**, BSc, studied automotive engineering at the University of Applied Sciences Esslingen and the University of Hertfordshire from 1996 to 2000. From 2000 to 2011, he was a development engineer in the department of aerodynamics and wind tunnels at the Daimler AG. Since 2012, he is a team leader for aerodynamics.

The Authors

Dr.-Ing. **Ralf Neuendorf** studied engineering physics at the Technical University Berlin from 1988 to 1995. From 1993, he was a research scholar at the AME department of the University of Arizona and did his PhD thesis on "Turbulent wall jet along a convex curved surface" in 1999. In 1999 he joined the BMW Group in Munich as a development engineer in the wind tunnel facility group. Since 2009, he is head of aerodynamics for mid-size and large vehicles.

Dr.-Ing. **Bernhard Zuck** studied mechanical engineering at the Technical University of Munich from 1990 to 1994. From 1995 to 1997, he was a research associate at the Institute for Thermodynamics A at the TU Munich. He did his PhD thesis on "Flow and combustion of fuel injection using micro dosing pumps." Since 1998 he has been working at the BMW Group in Munich in the aerodynamics department. From 2003 to 2008, he was a group leader for thermal management, functional concepts, and vehicle integration. Since 2008, he has been group leader for simulation and pre-development of cooling systems.

Dr.-Ing. **Martin Helfer** studied mechanical engineering at the University of Stuttgart. Since 1979, he has been a project engineer at the Research Institute of Automotive Engineering and Vehicle Engines Stuttgart (FKFS), and since 1993 he has been head of the department for vehicle acoustics and vibrations. He is a lecturer at the University of Stuttgart and organizes various events on vehicle acoustics, including the Technical Academy Esslingen (TAE). He is a member of DEGA and the normalization committee for acoustics, noise control, and vibration technology (NALS) at DIN.

Dipl.-Ing. **Michael Pfadenhauer** studied aerospace engineering at the Technical University of Munich. From 1995 to 1998, he was a research engineer for aerodynamics at Audi. From 1998 to 2005, he was responsible for the aerodynamic development of touring, sports, and racing cars as a test engineer at Audi Sport. Since 2005, he has been head of aerodynamics and thermal management in the series and motorsports at the Dr.-Ing. h.c. F. Porsche AG in Weissach.

The Authors

Dipl.-Ing. **Thorsten Frank** studied mechanical engineering at the University of Stuttgart. Since 1997, he has been responsible for aerodynamics as a test engineer for heavy trucks at Daimler AG. In 2005, he became head of the commercial vehicle aerodynamics at Mercedes-Benz Trucks and is now responsible for aerodynamics.

Dipl.-Ing. **Stephan Kopp** studied aerospace engineering at the Technical University of Munich. From 1997 to 2000, he was a research engineer in bodywork development of Adam Opel AG. From 2001 to 2004. he was a project engineer at Adam Opel AG for the aerodynamic development of the Corsa and Zafira series and the record vehicle Eco Speedster. In 2004, he moved to MAN Truck & Bus AG and was responsible for the aerodynamic development for MAN and NEOPLAN. Since 2008, he has been head of the research department bodywork, heating, and air conditioning and aerodynamics of MAN Truck & Bus and NEOPLAN.

Dr.-Ing. **Norbert Grün** studied aerospace engineering at the Technical University of Munich from 1975 to 1980. After three years as a development engineer at MBB Ottobrunn, he became a freelancer to BMW. Until 1996, he worked there on the development of a methodology for simulating the flow around the vehicle using potential theory. In 1991, he did his PhD at the Technical University of Munich on 3D boundary layer calculations. From 1996 to 2002, he worked as a technical advisor at Exa Corporation, Boston, the manufacturer of the CFD code PowerFlow. From 2002 until the beginning of his semiretirement in 2013, he was responsible for the use of simulation in the aerodynamics of cars and motorcycles at BMW.

Dipl.-Ing. **Holger Winkelmann** studied aerospace engineering at the University of Stuttgart and joined the BMW group in 1999. He held various positions including aerodynamics project engineer, team leader for aerodynamics and group leader for vehicle and model technology early in his career. From 2008 to 2009, he worked as a consultant for efficient dynamics technologies at the BMW design department. After which he was promoted to head of aerodynamics development at the BMW Group until 2014. He is now the head of rear end system development and in addition, is also responsible for glazing and wiper systems.

The Authors

Dipl.-Ing. **Frank Ullrich** studied automotive and aircraft technology at the University for Applied Sciences in Munich and at the Technical University of Munich. In 1983, he worked at MBB in Ottobrunn. In the same year, he joined BMW AG as a test engineer for vehicle aerodynamics. From 1991, he was responsible for aerodynamic and bodywork testing of HP cars at BMW Motorsport GmbH (later M GmbH). In 2000, he moved to BMW Technik GmbH as a project engineer for aerodynamics and acoustics. Since 2002, he has been a team leader of the motorcycle aerodynamics of BMW AG, and since 2004 as head of the acoustic wind tunnel and the aeroacoustics team.

Dr.-Ing. **Gerd Janke** studied mechanical engineering and physical engineering with a focus on fluid mechanics at the Technical University of Berlin and the University of Michigan. In 1992, he did his PhD on a topic in the boundary layer theory. Then, he worked as an assistant professor at the Hermann-Foettinger Institute at the Technical University of Berlin. In 1994/1995, he was a guest professor at the University of Michigan. From 1999 until 2015, he was responsible for aerodynamics, aeroacoustics, and development at Schuberth GmbH. Presently, he is responsible for acoustics at IFA Rotorion in Haldensleben.

Dipl.-Ing. **Sebastian Reitebuch** studied physical engineering at the Technical University of Berlin and worked at the Hermann-Foettinger Institute for Thermodynamics and Fluid Dynamics at the Technical University of Berlin from 1996 to 2000. Since 2000, he has been a research engineer at the acoustic wind tunnel of Schuberth GmbH.

The Authors

Prof. Dr.-Ing. **Jochen Wiedemann** received his Diploma Degree in mechanical engineering from Ruhr-Universität, Bochum, Germany in 1977. After carrying out aerodynamic research at the von Karmàn Insitute for Fluid Dynamics in Belgium and Ruhr-University Bochum he received the doctoral degree (Dr.-Ing.) in 1983 for his work on aerodynamic drag reduction. In 1984, Professor Wiedemann joined Audi AG where he held several managing positions. He was involved in many vehicle projects in aerodynamics, aeroacoustics, and driving dynamics. His final position at Audi was project manager of the newly built Audi Windtunnel-Center. In 1998, Jochen Wiedemann was appointed Chair Professor of Automotive Engineering at the Institute for Internal Combustion Engines and Automotive Engineering, (IVK) at the Stuttgart University, Germany, and he also became a member of the Board of Managing Directors of FKFS. His research work is largely associated with aerodynamics/aeroacoustics, road load, and vehicle dynamics. In 2004, Professor Wiedemann was appointed a visiting professsor at Tongji University where he gives lectures in vehicle dynamics. In appreciation for his achievements Professor Wiedemann was awarded the City of Shanghai's Magnolia Silver Award for merits about the social and economic development of Shanghai.

Dipl.-Ing. **Nils Widdecke** studied aerospace engineering at the University of Stuttgart. He worked as a research assistant at the Institute of Aerospace Thermodynamics (ITLR) and the Institute of Aerodynamics and Flow Technology (IAS) of the German Aerospace Center (DLR) in the High Enthalpy Tunnel Göttingen (HEG) in the field of shock wave research. At the Laser-Laboratorium Göttingen e.V. (LLG), he worked on the development of a laser-optical ultrafast measurement method for jet breakup and for flow velocity determination in liquid fluids and gases. Since 2000, he has been head of vehicle aerodynamics and thermal management at the Institute for Internal Combustion Engines and Automotive Engineering at the University of Stuttgart (IVK) and the Research Institute of Automotive Engineering and Vehicle Engines Stuttgart (FKFS).

Dr. rer. nat. **Reinhard Blumrich** studied physics at the Technical University of Darmstadt. After graduation, he worked as a research associate in the Faculty of Physics of Ruhr-University Bochum on acoustic and seismic surveillance of vehicle and aircraft movements. In 1998 he obtained his PhD herein. From 1998 to 2004, he worked as a researcher on traffic noise prediction at the German Aerospace Center (Oberpfaffenhofen). Since 2004, he has been working at the Research Institute of Automotive Engineering and Vehicle Engines Stuttgart (FKFS) and is responsible for the numerical simulation of vehicle acoustics. Since 2008, he has been also responsible for wind tunnel planning and consulting and was coordinating the 2014-upgrade of the aeroacoustic wind tunnel at the University of Stuttgart, operated by FKFS.

The Authors

Dipl.-Ing. **Armin Michelbach** studied mechanical engineering at the University of Stuttgart from 1982 to 1987. In 1988, he joined the newly established automotive wind tunnel at the University of Stuttgart as a chief operating engineer. Since 1998, he has been responsible for commercial operation of the scale-model wind tunnel and the aeroacoustic wind tunnel as a senior manager. He was heavily involved during the aeroacoustic retrofit of the wind tunnel and the installation of five-belt systems in both wind tunnels.

Dr.-Ing. **Edzard Mercker** studied aviation engineering at the Technical University of Berlin and did his PhD in 1981 at the department of physical engineering. He worked as a scientific assistant at the engineering department of Cambridge University, England, and at the Adelaide University, Australia. From 1982 to 2004, he was a supervisor for aerodynamic projects at the German-Dutch Wind Tunnel (DNW). From 2005 to 2009, he worked at the BMW Group, and since 2010 he has been a senior consultant at the Research Institute of Automotive Engineering and Vehicle Engines Stuttgart (FKFS). For his engineering activities, he has been awarded several times. He received the Arch. T. Colwell Merit Award from the SAE twice, a co-award from NASA, and three different co-awards from the American Helicopter Society.

Dr.-Ing. habil. **Jorg-Dieter Vagt** studied aircraft engineering at the Technical University of Berlin. From 1967 to 1983, he worked in the flow research at the Hermann-Foettinger Institute at the Technical University of Berlin and did his PhD on flow measurement technology in 1970. After his habilitation in 1979 in fluid mechanic research, he worked as a lecturer until 1983. From 1983 to 2005, he worked as a development engineer at the Dr.-Ing. h.c. F. Porsche AG in various positions (e.g., as head of the wind tunnels). Since 1998, he has been teaching at the University for Applied Sciences in Esslingen, and since 2006 he has been working at the Research Institute of Automotive Engineering and Vehicle Engines Stuttgart (FKFS).

Index

A-pillars, 228, 404
    on commercial vehicles, 747, 775, 777
    and foreign soiling, 515–518
    flow around, 266f, 509–510
    influence of, 270–273
    and noise generation, 587–588
Abbreviations, 1175–1178
Acceleration, 82, 158–159, 647–648
    and elasticity, 160–162
Ack Attack, 799f
Acoustic mirrors, 580–581
Acoustic near-field holography, 578–579
Acoustics, 532
    probes, 582
    of race car helmets, 903–905
    of wind tunnel fan, 951–955
    see also Aeroacoustics
"Actio est reaction," principle, 90, 93
Active noise reduction (ANR), 896
Actros
    1843 LS semitrailer combination, 714
    air control system, 753f
    semitrailer tractor underbody
        paneling and wheel spoiler, 757f
    semitrailer tractor with and without
        side skirts, 756f
ADAC rain test, 900
Add-on parts, *see* Components
Adler Trumpf, 36
Admittance function, 436–438, 440, 967
Aero Slit ducts, 276, 277f
Aeroacoustic wind tunnels, 915, 1042, 1043–1047, 1062
    measuring systems, 574–575

Aeroacoustics
    aerodynamic noise generation, 573–574
    audible and inaudible range of siren, 894f
    of bus, 776f
    computational, 1148–1173
        aeroacoustic analogies (AAA), 1153–1155
        application examples, 1161–1172
        calculation of sources, 1151–1153
        conclusion and outlook, 1172–1173
        introduction, 1148–1151
        Kirchhoff integral method, 1155–1156
        linearized Euler equations (LEE), 1156–1157
        transfer into vehicle cabin, 1157–1161
    of convertibles, 593
    of helmets, 909
    influence of airflow on noise, 569–572
    measuring systems
        acoustic mirrors, 580–581
        acoustic near-field holography, 578–579
        acoustic probes, 582
        for exterior noise, 576–581
        intensity measurements with special probes, 577–578
        for interior noise, 576
        leakage tests with ultrasound, 582
        microphone arrays, 578
        structure-borne sound, 581–582
        wind tunnels, 574–575

Aeroacoustics *(Cont.)*
  of motorcycles, 837, 861–862, 868
  of motorcycle helmets
    approaches to noise reduction, 895–896
    development targets and helmet noise, 887–889
    mechanisms and parameters of sound generation, 889–893
    perception of environmental signals, 893–894
  psycho-acoustic aspects, 594–600
    assessing behavior under different yaw conditions, 596
    noise synthesis, 597–600
    simulation with dynamic vortex generators, 597
    simulation with static vortex generators, 597
  sound-absorbing elements in wind tunnel, 943f
  sound pressure level (SPL), 571f, 572f, 889
    and A-pillar radius, 588, 589f
    inside helmet, 837f, 862f, 868f
    on side windows, 1134f
    of wind tunnels, 907f, 946f, 947, 1060
    and yaw angle, 596f, 599f
  source location with special instruments, 582
  speed of sound as function of temperature, 149f
  see also Noise
Aerodynamics
  basic equations in fluid dynamics
    Bernoulli equation, 83–84
    conservation laws, 75–76, 88–93
    continuity equation, 81–82
    Euler equation, 82–83
    kinematics and dynamics of flow fields, 76–80
    Navier-Stokes equation, 85–88
    potential theory, 84–85
  character of, 2–4
  compressibility effects, 148–149
  decomposition of loads, 998–1004
  density and viscosity of air, 146–147
  and design, 1–2, 60–63
  directional stability, 39–42
  dynamic driving effects
    axle load relief settings, 459–461
    braking behavior, 457–459
    reaction to crosswinds, 461–472
    reaction to lift forces, 449–459
    self-steering behavior, 449–453
    single-track model, 445–449
    straight-ahead running, 453–456
  factors influencing aerodynamic forces
    cooling air, 320–334
    exterior mirrors, 361–363
    front end shape, 253–264
    greenhouse and body side, 265–280
    ground clearance and vehicle position, 343–346
    interferences, 367–380
    rear end shape, 280–318
    systematization of, 319–320
    underbody assembly, 334–343
    wheels and wheelhouses, 346–361
  forces and force coefficients, 201–206
  frictional flow dynamics
    boundary layer separation, 115–117
    boundary layer turbulence, 118–122
    drag of simple bodies, 123–130
    multi-body systems, 131–133
    pipe systems with internal flow, 134–145
    Prandtl boundary layer concept, 112–115
    Reynolds number, 111–112
  of helmets, 907–908
  and historical development of vehicles
    "bathtub" body, 27–30
    "borrowed" shapes, 12–14
    convergence of shape by systematic development, 9f
    creating shapes, 11
    drag coefficient, 10–11
    early investigations with parameters, 25–26
    from horseless carriage to automobile, 26–34
    literature, 7

one-volume bodies, 30–34
"Panel Historique," 8f
stamping, 26–27
"streamline" era, 14–24
internal flows
	engine compartment, 47–48
	passenger compartment, 48–49
inviscid flow dynamics
	interpreting streamline patterns, 93–95
	planar model flows, 95–105
	vortex flows, 105–110
loads on components, commercial vehicle, 786
measurement of loads, 997–1006
of motorcycle helmets, 879–886
of race car helmets, 901–903
of rear-end shapes
	Kamm-back, 34–37
	fastback, 37–38
	notchback, 38
	hatchback, 39
unsteady forces and moments
	overtaking maneuvers, 414–415
	side wind, 416–444
	test and evaluation methods, 427–444
vehicle development process, 380–390
	examples, 388–390
	goal definition, 382–383
	project milestones and tools, 383–388
vehicle development strategies
	detail optimization, 49–52
	limit strategy, 57–60
	shape optimization, 53–57
vehicle development tools
	CFD methods, 70–73
	classical approach, 68–70
	rating, 65–68
	wind tunnels, 63–65
	see also Commercial vehicles, aerodynamics of; High-performance vehicles, aerodynamics of; Motorcycles, aerodynamics of
Aerodynamische Versuchsanstalt (AVA), 16, 23, 32

Aero-elastic effects, 6
Aerolab wind tunnel, 1047–1048
Aerotrailer, 768f, 769
Ahmed, Syed R., 294, 409
Ahmed body, 409
Air
	density and viscosity of, 146–147
	density as function of temperature and air pressure, 157f
	forces and moments, 201–203
Air conditioning, 182, 531, 533, 541
	aeroacoustic simulation of, 1169–1171
	performance tests, 1019
Air Curtain, 329, 330f
Air-to-boil (ATB) temperature, 1016
AIRCAP comfort system, 500
Aircraft, 2, 9, 11, 108–110, 1066, 1148
Airfoils, 15, 129–130
	Clark Y, 199, 200f
AIRGUIDE comfort system, 499
AIRSCARF, 490
"Alex" sweating head, 897f
Alpha Romeo 1900 BAT concept 1 (1953), 60f
Alternator, 182
Althaus, D., 674
Amortization analysis, 189–190
Andreau, J., 23, 24
Anemometers
	constant current (CCA), 985
	constant temperature (CTA), 985
	hot-wire, 566–567, 985, 986f
	vane, 564, 984
Angle of attack, effect on drag and lift, 228f, 231f, 343, 344f, 345f
Antennas, 364, 522, 586–587
Anti-buffeting measures, wind tunnel, 942–955
Arie, M., 288
AS 1698-1988, 878
Ascending ability, 162
Ashill, P., 971
Asia
	driving cycles, 175–176
	fuel consumption legislation, 196
	speed limits in various countries, 165t

Index

Association des Constructeurs Européens d'Automobiles (ACEA), 191–192
Asymptote, 52, 319, 320*f*
Audi
    90 Quattro, 666*f*
    100, roof design and development, 275*f*
    100 II
        effect of plan view curvature on drag, 276*f*
        influence of windshield inclination on, 268*f*
        pressure distribution in center section, 269*f*
    100 III, 56*f*, 275
        effect of plan view curvature on drag, 276*f*
        influence of windshield inclination on, 268*f*
        pressure distribution in center section, 269*f*
    A1, spare wheel well spoiler, 342*f*
    A2, 274*f*, 403
        development process, 386*t*, 389*f*
        underbody development, 336*f*
    A4
        development process, 385*f*
        front spoiler with twisted cross section, 677*f*
        underbody of, 342*f*
        velocity field at radiator, 234*f*
    A6, 56*f*, 390
        front end pressure distribution, 254*f*
        roof design and development, 275*f*
    surface pressure distribution, 1098*f*
    velocity distribution, 1099*f*
    A7, 230
        comparison with design sketch, 279*f*
    A8
        exterior mirror, 363
        underbody fairing, 335*f*
    Alpensieger, 14
    eTron, 405*f*, 634*f*
    Q3, 304–305
        radiator environment seal, 323*f*
        underbody of, 326*f*
    Q5, development process, 387*f*
    Q7, 345
    development process, 385*f*
    exterior mirror, 363
    R8 sports car, 625*f*, 657*f*, 703*f*
        fully faired underbody, 398*f*
    RBC, cooling air intake of, 665*f*
    S8, exterior sound pressure levels, 572*f*
    TT convertible, wind deflector, 498*f*
    TT2, rear spoiler on, 310, 311*f*
    TT-R, 624, 625*f*
        with cooling air from bow section, 696*f*
        gurney on rear wing, 682*f*
        rear spoiler, 679*f*, 680*f*
        vanes on, 704*f*
    Type K14, 19*f*
    wind tunnel, 951, 1043–1044
"Auto 2000," 55, 56*f*
Auto Union racing car, 604*f*, 914*f*
Automatic meshing tools, 1126
Automatic transmissions, 183
AVA, 16, 23, 32
Axles, 155, 183
    balance, 643–649
    commercial vehicle variants, 755*f*
    lift and side force, 202
    load relief settings, 459–461

Bachmann, J., 837
Baehr, H.-D., 533
Baehr Tec Silencer, 887
Bag break-up, 504
Baker, C.J., 785
Balance, aerodynamic, 622, 643–649
Balances, wind tunnel, 997–998
    internal, 1002–1003
Barnard, R.H., 550
Barreau, M., 23
Barret, Stan, 611
Barsikow, B., 1031
Base pressure, 289
"Bathtub" body, 27–30
Batista, M., 785
Bauer, G., 507
Baumert, W., 294
Baumm'sche Liegestühle, 798, 799*f*
Bearman, P.W., 295, 304, 307
Beaufort scale, 218*f*

Beauvais, F.N., 369
Beese, E., 998
Behr climatic wind tunnel, 1042
Bel Geddes, Norman, 30
Beland, O., 939, 949, 950
Benelli Tornado 900 Tre, under-seat water cooler, 833*f*
Bengsch, H., 1042
Benz Tropfenwagen (1923), 603*f*
Berge, W., 880, 896
Bergmann, A., 950
Bernoulli equation, 83–84, 85, 89
Bertone, Nucio, 60
besst (Beland Silent Stabilizer), 948–949
Betz, 48
Bez, U., 1021
Bicycles, 364, 366
Blasius, 123
Blitzen-Benz racing car (1911), 602*f*
Blockage correction methods, 979–981
Blue Flame rocket-driven vehicle (1970), 611*f*, 612
Bluff bodies, 3, 6, 115
    approaching the ground, 54*f*
    drag of, 132–133, 197
BMW
    3-Series, Air Curtain on, 329, 330*f*
    328, 1938, 19*f*
    5 Series, 400–401
    520i, exterior sound pressure levels, 571*f*
    6 Series convertible, wind deflector, 498*f*
    aerodynamic rims, 358*f*
    Aerodynamic Test Center (AVZ), 937*f*, 1047
    C1, 809
    C600 Sport, 805*f*
    C650 GT, 805*f*
    development process, 384*f*
    Energy and Environmental Test Center (EVZ), 1032, 1052
    HP4, 807*f*
    K 100 RT
        influence of windshield inclination on helmet forces, 886*f*
        influence of windshield height on noise in helmet, 891*f*
    K 1600 GTL, 804*f*
        foldout deflectors for rider ventilation, 876*f*
    modular vehicle model for validation measurements, 1080*f*
    motorcycles, 796–797
    proving grounds, 865*f*
    R 80 G/S, 803
    R 100 RS (1976), 800–801
    R 1200 GS, 803*f*
    racing car, 664*f*
    S 1000 RR
        ram-air inlet, 834*f*
        top speed vs. engine power, 817*f*
        water and oil coolers, 832*f*
    wind tunnel, 65, 937*f*, 1032, 1043, 1047, 1052
BMW-Williams Formula 1 racing car (2004), 608*f*
Boat tailing, 14, 312, 314
Boats, as roof load, 366
Body side, influence of, 275–280
Bombardier Can-Am Roadster, 809
Borgward Hansa (1949), 28*f*
Boundary element methods (BEM), 1103–1113, 1159–1160
Boundary layer
    Prandtl's concept, 112–115
    separation, 115–117
    suction, 957*f*
    thickness, 113–114, 116, 124, 1089
    turbulent, 118–122
Bourlet, C., 795
Boutin, L., 23
Brabham Formula 1 racing car (1978), 606
Bradshaw, P., 921, 942
Brake air duct of Formula 1 car, 700*f*
Brake cooling system on Ferrari F60 Enzo, 700*f*
Braking, 182, 356
    on corner, 454–456
    friction coefficient as function of slip, 650*f*
    of high performance vehicles, 628, 647, 649
    and lift, 457–459
    of motorcycles, 872

Branching losses, 138f
Brandstätter, W., 922
Braun, H., 523
Brawn GP Formula 1 racing car, rear diffuser channels in, 694f
Broadband compact absorbers (BCA), 943
Browand, F., 378
Brühwiler, P., 898
Bruun, H.H., 987
Buchheim, R., 263, 296, 340, 563, 987, 1010
Buckingham Π theorem, 205–206, 219
Buckley, J.T., Jr., 1025
Buckminster Fuller, Richard, 30
Budweiser Rocket, 611
Buildings, aerodynamics of, 6
Burch, S.D., 1021
Burden sharing, 192n
Buses, 42–43
    airflow through engine compartment, 780–781
    characteristic airflow and pressure, 772–773
    front, 773–774
    mirrors, 775–777
    optimizing drag on, 772–782
    rear, 781–782
    underbody, 778
    wheels and wheel covers, 778–779
    windscreen wipers, 777–778
Büssing chassis, 719

C-pillars, 300, 303
    influence of radius on drag, 271f, 297f
C-shoulder vortex, 350
Cab, commercial truck
    drag on, 741–747
    effect of streamlined shape, 761f
    spoiler, 45–46
Cabin, see Passenger compartment
Cairns, R.S., 432
California Air Resources Board (CARB), 191
Caliper, rear, heat transfer coefficients, 874f
Campaigns Targeting New Passenger Cars, 192
Carbon dioxide
    commercial vehicle emissions, 711
    and energy equivalents, 168–169
    legislation and labels
        in Asia, 196
        in the EU, 191–193
        in the United States, 193–196
Carr, G.W., 38, 66, 68, 257, 298, 334, 773, 992, 1023, 1048
"Cash for Clunkers" program, 151n
Catalytic converter, 151n
Cauchy-Riemann differential equations, 95
Cavity resonances, 589–591
CDTI diesel engine, 617
Ceiling, driving on the, 402–403
Center of mass theorem, 90
Chadwick, A., 429, 433
Chaparral
    2C (1965), 605
    2J (1969), 606, 607f
Characteristic curves, 136
Characteristic functions, 52
Chassis
    commercial truck, 754–758
    soiling of, 1030–1031
China, 176
    fuel consumption limit values, 196t
Choppers, 805
Chue, S.H., 984
Circular cylinder, flow around, 101–103
Citroen, 28–29
    2CV, 60
    CX (1982), 29f
    DS 19, 7, 28, 29f, 57, 344
CityEl (1987), 272–273
Clarke, R.M., 523
Classical scheme for numerical treatment, 68–70
Claveau, Emile, 30
Clean Air Act, 191
Climatic wind tunnels, 916, 1040, 1050–1054
    additional equipment in, 1032–1034
    blockage correction methods for, 979–981
Climatization tests, 1017–1019
Closed test section, 969–972

Clusters, 1139–1140
CNR research vehicles, 224, 225f, 270, 271f
Coaches, *see* Buses
CO_2, *see* Carbon dioxide
Coanda effect, 307n
Coastdown tests, 1021–1025
Coating technology, 1007–1008
Cobb, John, 611
Cobra probe, 989, 990f
Cogotti, A., 223, 224, 359, 434, 992, 1008
Cogotti, C., 964
Cogotti, F., 998
Colebrook formula, 140
Coleman, P.B., 997
Collector, wind tunnel, 931, 934–937, 975
Commercial vehicles
   aerodynamic interactions
      dust turbulence, 786–787
      loads on components, 786
      intake of warm air, 787
      management of exhaust gas, 788
      nose-to-tail driving, 782–784
      tipping and susceptibility to side winds, 784–786
   aerodynamics of, 42–46, 721–728
      CFD simulation of, 735–736
      full-scale wind tunnel for, 732–734
      history, 718–721
      legislative framework, 727–728
      model-scale wind tunnel for, 728–732
      optimizing drag on buses and coaches, 772–782
      optimizing drag on trucks, 739–771
      relevant vehicles, 711–713
      straight/oblique flow, 723–727
      test drive with wheel hub measurement, 736–738
      tools for optimizing, 728–738
   driving resistances and fuel consumption, 713–717
   life cycle costs, 712f
   soiling
      foreign, 791
      self-soiling, 792–793
      task description and testing methods, 789–791

Compensation channel between plenum and diffuser, 949f
Complex functions, use of, 95–96
Component flutter, 473
Components
   influence on aerodynamic forces
      underbody assembly, 334–343
      ground clearance and vehicle position, 343–346
      wheels and wheelhouses, 346–361
      exterior mirror, 361–362
   loads on, 786
      doors, flaps, and outside mirrors, 476–480
      pinpointing, 474–475
      windshield wipers, 480–487
   roughness drag of, 241–243
Compressibility effects, 148–149
Computational fluid dynamics (CFD), *see* Numerical simulation
Computers
   clusters and massively parallel systems, 1139–1140
   GPU computing, 1140
   symmetric multiprocessor systems (SMP), 1139
   vector, 1138–1139
   performance of, 1141–1144
Concept S, 769–770
Concept vehicles, commercial truck, 769–771
Conchoid diagram, 179, 180f
Conduction, 534
Conformal mapping, 96
Conservation laws, 75–76
   integral forms of, 88–93
Constant current anemometer (CCA), 985
Constant temperature anemometer (CTA), 985
Contact patch vortex, 350, 352, 353, 354
Continuity equation, 81–82, 84, 85
   integral form of, 89
Continuous variable transmission, 183n
Continuum mechanics, basic terms in, 76–78
Convection, 534–534
Convective derivatives, 77

Convergence, pressure losses caused by, 145
Convertibles, 400, 402t, 487–488
   airflow with top open, 488
   ballooning effect in soft top, 1161
   deformation analysis of soft top, 475f
   design solutions for, 496–500
   and noise, 593
Convoys, 377–379
   commercial vehicles, 782–784
Cooler probe, 564
Cooler traverse, 566f, 567f
Cooling system
   acoustics, 532
   components and systems, 529–531
      air conditioning, 531
      differential, 531
      engine intake air, 530
      transmission, 530
      design of air intakes, 531
   cooling matrix, 554–557
   cooling requirements, 527–532
   design methods for, 1113–1119
      air side, 1113–1117
      coolant side, 1117–1119
   drag, 233–241, 547, 549f, 550
   ducts, 553–554
   engine compartment, 561–562
      routing in, 237f
   engine cooling system circuit, 532–533
   flow measurement technology, 563–568
      hot-wire anemometry, 566–567
      optical measuring methods, 565
      pressure measurements, 564–565
      vane anemometers, 564
   fundamentals of heat transfer
      convection, 534–534
      heat conduction, 534
      radiation, 535
   heat exchanger design, 535–537
      calculation of performance, 536–537
      cross-flow, 536
      parallel-flow and counter-flow, 536
   heat exchangers in vehicle
      air conditioning, 541
      coolant radiator, 537–539
      engine and transmission oil cooler, 540–541
      intercooler, 539–540
   high-performance vehicles, 654–656
   influence on aerodynamic forces, 320–334
      active systems, 330–334
      outlet losses, 326–328
      inlet losses, 321–322
      interaction with external flow, 329–330
      pressure loss at radiator and engine compartment, 322–325
   intakes, 240, 531, 553–554, 696–703, 787, 1130
   mass flow, 551, 552f
      calculation of, 548–550
   motorcycle, 831–834
      system design, 872–874
   on-road testing, 1029–1030
   optimizing overall system, 547–563
      air intakes and cooling air ducts, 553–554
      calculation of cooling air mass flow, 548–550
      cooling air outlets, 562–563
      cooling matrix, 554–557
      engine compartment, 561–562
      fan, 559–560
      influence parameters of internal flow, 551–553
   outlets, 240, 542, 562–563, 696–703, 1130
   packaging, 532
   representative operating conditions
      driving with a trailer, 528
      idle, 528
      maximum speed, 528
      uphill driving, 528
   safety, 532
Cooper, K., 930, 969, 976
Cooper, Kevin R., 721, 725, 729, 762
Corner radius, 255–262
Cornering, 450–451
   motorcycle, 821–826
   speed of, 640–642
Corporate average fuel efficiency (CAFE), 191, 193, 194f

Count Ricotti, 13, 30
Counter-flow heat exchanger, 536
Coupe, comparison with Clark Y wing airfoil, 200*f*
Cover shield, commercial vehicle, 787*f*
Crites, R.C., 997
Critical inclination angle, 293*f*
Cross-flow heat exchanger, 536
Crosswind, 41, 446, 448, 461–472
    application example, 470–472
    evaluation of behavior, 464–470
    and motorcycles, 864–865
    tests, 1025–1029
    vehicle excitation caused by, 461–462
    vehicle reactions to, 462–464
    wind tunnel, 964
Cruisers, 805
CSI, 901
Curle, N., 1154

Daimler, Gottlieb, 795
Daimler
    SL R 231, 61*f*
    SL W 194 (1951/52), 61*f*
    wind tunnel, 734, 1053, 1054*f*, 1048
Daimler-Benz
    C111 record car, boat tailing on, 313*f*
    racing car (1937), 603*f*
    Stuttgart (1928), 20*f*
    W158, 35, 36*f*
    wind tunnel, 1041
d'Alembert's paradox, 101–103, 204
Dallara race cars, 623–624, 625*f*, 627
Darrin, Howard, 27
Data helmet, 907–908
Davenport, A.G., 435, 967
Dead wake, 116–117, 126, 132–133, 208–214
Deceleration, 647–648
Deflectors, commercial vehicle, 745, 746*f*, 747*f*, 794*f*
Deformation tensor, 79–80
Defrosting and dehumidifying test, 1020–1021
Delfin I, 798
Delfin III, 798*f*
Delft University of Technology, 763
Delta wing, 110*f*

Density, air, 146–147
Design
    and aerodynamics, 1–2, 60–63
    air intake, 531
    convertible, 496–500
    cooling module, 1113–1119
    to counteract soiling, 515–526
    diffuser, 688
    heat exchanger, 535–537
    motorcycle, 869–876
        cooling of, 872–874
        heat protection, 872–874
    sunroof, 501–502
    wind tunnel, 916–920
Detached-eddy simulation (DES), 1084, 1096–1098, 1151, 1164
Detail optimization, 49–52
Deutenbach, R., 938
Deutsch, Charles, 604
Diesel engine, commercial vehicle, characteristic map, 718*f*
Dietz, S., 385, 386
Differential, 531
Diffusers
    of high-performance vehicles, 686–695
        design, 688
        function, 686–687
        generation of downforce, 688–690
        reduction of drag force, 691–692
    wind tunnel, 940–941
Digital wind tunnel (DWT), 1127*f*
Dilgen, P.G., 69
Dimensionless numbers, 1034–1039
DIN (Deutsche Industrie Norm)
    standard 1946-3, 1018
    standard 4848, 877
    standard 7730, 492
    standard 33403, 490
    standard 70020, 171
Direct numerical simulation (DNS), 912, 1084, 1099–1103, 1173
Direct simulation (DS), 1152, 1172, 1173
Directional stability, 39–42
    in side winds, 248*f*
Displacement thickness, 113
Divergence, pressure losses caused by, 145

DKW, 798
  Streamliner (1939), 619f
DNW wind tunnel, 732, 733f, 734f, 960
Dobrzynski, W., 889, 1031
Dominy, R.G., 964
Door handles
  influence of, 364
  reducing contamination, 525
Doors, 476–480
Doublet distribution, 1106f
Downforce, of high-performance vehicles, 639–642
  generation of, 688–690
Drag, 156–157, 186–187, 197–201
  of basic bodies, 55f
  basic form drag and induced drag, 226–232
  of bluff bodies, 133f, 198f
  of buses and coaches, optimizing
    airflow through engine compartment, 780–781
    characteristic airflow and pressure conditions, 772–773
    front, 773–774
    mirrors, 775–777
    rear, 781–782
    underbody, 778
    wheels and wheel covers, 778–779
    windscreen wipers, 777–778
  of circular disk in cross-flow, 126f
  coefficient of, 10–11
  computed from drag components, 66, 223f, 226f
  computed from rating points, 65
  cooling air drag, 233–241
  development across vehicle categories, 153f
  development over time, 403f
  diffusers for reduction of, 691–692
  effect of slope angle on, 38f
  of European cars, 57, 58f
  friction drag, 220–223
  as function of Reynolds number, 220f, 222f
  future development of, 403–405
  of high-performance vehicles, 635–638
  induced drag, 226–232, 673–674
  influence of rear window inclination on, 298f
  interference drag, 131, 244–246, 548, 549f
    negative, 245
  measured values for 79 vehicles, 391f
  measurement using coastdown tests, 1021–1025
  microdrag, 223–224
  of motorcycles, 813
    measures to optimize, 869–872
  of multi-body systems
    bluff bodies in a row, 132–133
    streamlined bodies in a row, 132
    streamlined bodies side by side, 131–132
  of passenger cars in production
    drag surface area, 392–394
    driving on the ceiling, 402–403
    influence of drive concept, 396–399
    influence of equipment and engine, 400–402
    intercomparison in accordance with EADE, 394–396
    overview of competitors by vehicle class, 390–392
    SUVs, 399–400
  of plate at zero incidence, 123f
  pressure and friction drag, 220–223
  of rectangular plates, 126t
  reducing, 187–190
    amortization analysis, 189–190
    possibilities for reducing resistance, 188–189
    weight equivalency, 189
  relation to shape, 67f
  relation to lift, 231f, 232f
  roughness drag of add-on parts, 241–243
  significance of, 151–154
  of simple bodies
    flow around a sphere, 126–128
    plate at zero incidence, 123–125
    plate in cross-flow, 125–126
    streamlined body, 129–130
  of sphere, 127f

of streamlined bodies, 131f, 132f
surface area, 392–394
of trucks, optimizing
    airflow through engine compartment, 751–754
    cab, 741–747
    characteristic airflow and pressure conditions, 739–741
    chassis, 754–758
    concept vehicles, 769–771
    mirrors and attachments on cab, 748–750
    semitrailers and bodies, 758–768
of wings, 672
    induced drag, 673–674
DrivAer body, 410–412
Drive system, components of, 166f
Driver
    driver-vehicle interaction in crosswinds, 467–472
    motorcycle, influence on aerodynamics, 855–858
    noise sharpness at ear, 594f
    sound pressure at ear, 596f, 599f
Drivetrains, common, 397f
Driving cycles, 170–177
    Asian, 175–176
    history, 171
    New European Driving Cycle (NEDC), 171–173
        for hybrid drives, 173
    realistic, 177
    United States, 173–174
    WLTP—Worldwide Harmonized Light Vehicles Test Procedure, 176
Driving on the ceiling, 402–403
Driving profile, 166, 167
Driving resistance
    acceleration, 158–159
    aerodynamic drag, 156–157
    of commercial vehicles, 713–717
        measurement system for, 737f
    example of, 159
    grade resistance, 158
    overall, 159
    rolling resistance, 154–156
Driving stability

aerodynamic axle load relief settings, 459–461
behavior when braking, 457–459
motorcycle, 864
reaction to crosswinds, 461–472
reaction to lift forces, 449–459
self-steering behavior, 449–453
single-track model, 445–449
straight-ahead running, 453–456
unsteady aerodynamic forces and moments
    overtaking maneuvers, 414–415
    side wind, 416–444
    test and evaluation methods, 427–444
DTM, 608, 623, 626, 628, 721f
Dual clutch transmission, 183n
Dubonnet, Andre, 30
Duell, E., 212, 213, 946, 947, 997, 1004
Durst, F., 987
Dust turbulence, and commercial vehicles, 786–787
Dymaxion No. 1, 30f
Dynamic fuel consumption calculation, 167f
Dynamic pressure, measurement of, 990–992

EADE (European Aero Data Exchange), 394–396, 1060, 1062
Eberz, T.J., 307
ECARA (European Car Aerodynamics Research Association), 1062
    Wind Tunnel Corrections group, 928
Eck, B., 140
Eckelmann, H., 990
Eckert, W., 1048
Ecomobile, 810
Economic Commission for Europe (ECE)
    city cycle, 171, 172f
    ECE-R 22-05, 877, 878, 893
Eco-Speedster, 616, 617f
Edge radius, influence on drag, 256f
EEC directives
    78/317/EEC, 1020, 1021
    91/226/EEC, 792
    1999/94/EC, 192

Effective camber, 53
Effective cooling velocity, 987
Effective thickness, 53
Efficiency, of high-performance vehicles, 651–654
Elasticity, and acceleration, 160–162
Elliptic cylinders
　with oblique incident flow, 104
　at zero incidence, 103
Emmelmann, H.-J., 341, 1003
Emmenthal, K.-D., 48
Endurance sports cars, 623
Energy flow diagram, 177–179
Energy Policy and Conservation Act, 193
Engine, 529–530
　cooling, 540–541
　　cooling system circuit, 532–533
　　tests, 1014–1017
　efficiency and engine maps, 179–181
　influence of, 400–402
　intake air, 530
　power, 636–638
Engine compartment, 561–562
　backflow in, 332f
　internal flow, 47–48
　pressure loss in, 322–325
Engler, R.H., 997
Entry temperature difference (ETD), 1016n
ETRTO (European Tyre and Rim Technical Organisation), 347
Euler equation, 82–83, 84, 90
Euro 5 emission standard, 716
Euro 6 exhaust system requirements, 788
Euro Mix, 171
Europe
　exhaust gas limit values in, 168t
　speed limits in various countries, 165t
　wind atlas for, 726
European Aero Data Exchange (EADE), 394–396, 1060, 1062
European Union (EU)
　CO_2 legislation and labels, 191–193
　directive 96/53/EC, 721, 727f
　directive 97/27/EC, 727
　directive M1 (Annex II 70/156/EEC), 399
　fleet consumption target objective, 154
　inertia mass classes in, 169–170t

Ewald, B., 1037
Ewald, H., 377
Exhaust gas
　limit values in Europe, 168t
　management of, commercial vehicles, 788
Exterior noise
　influence of airflow on, 569–572
　measuring, 576–581
　　acoustic mirrors, 580–581
　　acoustic near-field holography, 578–579
　　intensity measurements with special probes, 577–578
　　microphone arrays, 578
External contamination, see Foreign soiling
External flow field, motorcycle, 848–853
Extra Urban Driving Cycle (EUDC), 171, 172f

Fans, 545–546, 559–560
　axial, characteristic of, 1115f
　HVAC, 1170, 1171f
　simulation of, 1119–1121
　　moving reference frame (MRF), 1120
　　"rotating wall," 1120
　　sliding mesh, 1121
　wind tunnel, acoustics of, 951–955
Fastback, 20–21, 37–38, 282–286, 289–290
　drag, 1091f
　generation of negative pressure peaks on, 110f
　high-energy longitudinal vortices on, 230f
　lift, 1092f
　rear pillar shape, 303
　rear window inclination, 291, 295–296, 298
　separation edges and rear spoilers, 304, 305, 308
　tail taper, 312, 314
　velocity distributions at wheel, 1092f
FAT (Forschungsvereinigung Automobiltechnik), 755, 764, 958–959
Ferrari
　360 Modena sports car, 666f

rear diffuser on, 695f
F60 Enzo, 622f
    rear spoiler on, 678f
    brake cooling system on, 700f
Ferziger, J.H., 1123, 1126
Ffowcs Williams, J.E., 1154, 1164
Ffowcs Williams-Hawkings analogy, 573
FIA (Federation Internationale de l'Automobile), 878, 901
FIAT, 65
    500, 26f
    508 C Mille Miglia (MM), 37f
    Barilla (MY 1932), 26f
    Uno, rear roof of, 313f
Fiedler, F., 47
Fiedler, H., 971
Finite difference methods (FDM), 1123
Finite element methods (FEM), 1124, 1159–1161
Finite volume methods (FVM), 1124
Fins, 40–41
Fischer, O., 407, 408
Fishleigh, W.T., 31–32
Fixtures, pressure losses caused by, 145
FKFS, 564
    besst, 948
    calibration vehicles, 1059–1060
    cooler and cooler probe, 985f
    measuring trailer, 1024f
    swing, 966f, 1049
    test bench equipment for wheel spray analysis, 1012f
    wind tunnel, 1041, 1044–1045, 1049, 1052
Flank vortex, 354
Flaps, 476–480
Flat-bar wipers, 482–486
Flegl, H., 662
Float angle, 448, 453–454
Flow
    at backward-facing step, 209f
    around cars, 50f, 207f, 206–220
        environmental influences, 217–219
        influence of Reynolds number, 219–220
        internal flow, 216–217
        longitudinal vortices, 215–216
        non-periodic dead wake, 208–210
        periodic dead wake, 210–211
        ring and spiral vortices, 211
        unsteady processes, 211–213
    characteristic
        buses and coaches, 772–773
        commercial trucks, 739–741
    around circular cylinder, 101–103
    and commercial vehicles, 723–727
    conditioning screens, wind tunnel, 941–942
    around elliptic cylinder, 103f, 104f
    through engine compartment, 47–48
        buses and coaches, 780–781
        commercial truck, 751–754
    free, droplets in, 504–506
    gaseous, hierarchy of approaches to describe, 1071f
    incident, 111f
    influence on noise, 569–572
    kinematics and dynamics of, 76–80
    oblique
        commercial vehicles, 723–727
        high-performance vehicles, 656–659
    planar parallel, 97f
    around planar semi-infinite body, 100f, 101f
    planar source and vortex, 99f
    planar stagnation-point, 97f
    predicting, 200
    behind prism in longitudinal flow, 214f
    around sharp edge, 98
    around sphere, 126–128
    around symmetrical wing profile, 105f
    turbulent, in a pipe, 118f
    unsteady, wind tunnel, 963–967
    and vehicle soiling, 503–504
    see also Frictional flow; Internal flow; Inviscid flow
Flow Institute Karlsruhe, 909
Flow separation, 3–4, 98, 107, 110, 115–117, 246, 265–266
    and commercial vehicles, 741, 759–760, 774f
    criteria for, 209
    at front end, 253–254
    influence on yaw moment, 249f
    types of, 206

Flow velocity
  flow direction, 988–989
  measurement of, 984–989
    outside and inside test object, 984–988
Flow visualization, 5f, 1006–1010
  motorcycle, 859–860
Flowmaster, 1118f, 1119f
Fluent, 72, 1164
Fluid-control devices, pressure losses caused by, 145
Fluid dynamics, basic equations
  Bernoulli equation, 83–84
  conservation laws, 75–76
    integral forms, 88–93
  continuity equation, 81–82
    integral form of, 89
  Euler equation, 82–83
  kinematics and dynamics of flow fields, 76–80
  Navier-Stokes equation, 85–88
  potential theory, 84–85
Fluid-structure interaction, 1158–1159
Flying Dutchman Numa II, 617
FMVSS 103, 1020
FMVSS 218, 878
Folding hard tops, 497
Ford
  Focus, 305
  Lincoln Continental (1949), 27f
  Mondeo, 390
  Probe V, 614
  Probe Concept S, 327f
  Transit, side window pressure measurements, 996f
  wind tunnel, 1042
Ford Werke wind tunnel, 1045, 1050
  crosswind test facility, 1026f
Foreign soiling, 791
  causes of, 508–514
  commercial vehicles, 791
  design measures to counteract, 515–523
  flow over A-Pillar, 509–510, 515–518
  motorcycle, 862
  of rear window, 512, 521–523
  vs. self-soiling, 503
  of side window, 511, 518–521

  simulation of, 1011–1012
  water pullback on windshield, 512–514
Formula 1, 606f, 607, 608, 634, 662, 664f, 676f
  barge-board, 706f
  brake air duct, 700f
  Brawn GP racing car, rear diffuser channels, 694f
  Grand Prix circuits, 629–630
  helmets for, 878, 879f, 901–905
  Lotus 79 race car, 685f
  Mercedes McLaren racing car, 632f
  Renault racing car, cooling air arrangement, 697f
  use of vanes on front wings, 705f
Formula racing cars, 623
Forschungsvereinigung Automobiltechnik (FAT), 755, 764, 958–959
Frank, T., 743
Free flow, droplets in, 504–506
Frey, K., 45
Friction, momentum theorem with, 92–93
Friction drag, 220–223
Frictional flow
  boundary layer separation, 115–117
  boundary layer turbulence, 118–122
  drag of multi-body systems
    bluff bodies, 132–133
    streamlined bodies, 131–132
  drag of simple bodies
    flow around a sphere, 126–128
    plate at zero incidence, 123–125
    plate in cross-flow, 125–126
    streamlined body, 129–130
  dynamics of, 111–145
  forces acting on fluid element, 87f
  multi-body systems, 131–133
  pipe systems with internal flow
    pressure loss coefficients, 138–145
    pressure loss with internal flow, 135–136
    pressure losses from change in cross section, 142–144
    pressure losses from divergence and convergence, 145
    pressure losses from fixtures and fluid-control devices, 145

pressure losses in linear flow, 139–141
redirection losses, 141–142
series and parallel connection of loss elements, 137–138
stream filament theory, 134
system characteristic and operating point, 136–137
Prandtl boundary layer concept, 112–115
Reynolds number, 111–112
Front bumper sweep, 260
Front end
buses and coaches, drag on, 773–774
effect of shape on drag coefficient, 263*f*
of high-performance vehicles, 662
influence on aerodynamic forces, 253–264
corner radius, 255–262
hood inclination, 264
position of the stagnation point, 262–264
interaction with tail in flow, 368*f*
main geometrical parameters, 255*f*
pressure distribution on, 553*f*
reduction in drag by rounding, 258*f*
Frontal area, 275, 382, 392, 635
definition of, 10
measurement of, 1005–1006
of motorcycles, 811, 812*f*
Fuel consumption, 165–170
calculating, 166–168
CO_2 legislation and labels, 191–196
of commercial vehicles, 711, 713–717
conchoid diagram of gasoline and diesel engines, 180*f*
driving cycles, 170–177
and engine output of German cars, 152*f*
measurement, 168–169
possibilities for reducing
aerodynamic drag, 186–187
ancillary components, 181–183
energy flow diagram, 177–179
engine efficiency and engine maps, 179–181
rolling resistance, 186
transmission, 183–184
vehicle mass, 185
of vehicles with plug-in hybrid drive, 173
Fuel economy and greenhouse gas rating, 196*t*
Full-scale wind tunnels, 1041–1048
for commercial vehicles, 732–734
Fun bikes, 806

Garry, K., 969
Gaubschat, 42
tram bus, 719*f*
Gauss's divergence theorem, 88–89, 90, 92
General Motors
Aero 2002, 614
wind tunnel, 1041–1042, 1059
George, A.R., 212, 213
German Aerospace Center (KKK), 1036
German Experimental Institute for Aviation (DVL), 914
German Touring Car Championship (DTM), 608
Germany, 171, 193–194, 877
average wind speed in, 416
motorways, 718–719
Geropp, D., 307
Gersten, K., 940
Gilhaus, A., 762
Gilhome, B.R., 286, 476
Gilliéron, P., 414
Glauert factor, 968
Gleason, M., 971
Glynn, D.R., 71
Göhring, E., 767
Golden Arrow record car (1929), 610*f*
Goldstein, M.E., 1156
Göttingen type wind tunnel, 917–919, 1041, 1046
Götz, H., 377, 714
Grade resistance, 158
Grandjean, E., 492
Graphic processors (GPU), 1140
Graz University of Technology, 736
Greaves, J.R.A., 71
Green, Andy, 611
Greenhouse and body side, influence of, 265–280
Grille, drag reduction by fine-tuning, 259*f*

Index

Grille shutter, effect on drag, 333f
Ground boundary layer as function of terrain roughness, 419f
Ground clearance, influence of, 343–346
Ground effect, 607
   of high-performance vehicles, 682–686
Ground simulation, wind tunnel, 955–963
Grün, N., 1103
Gurney flap, 680–682
Gurneys, of high-performance vehicles, 677–682
Gürtler, T., 932
Gust simulation, wind tunnel, 963–967

Hackett, J., 970
Hagen-Poiseuille law, 139
Hairpin vortex, 286
Halfbody, 17f, 18
Hallquist, T., 752
"Hanging-off," 822f, 823, 824f
Hanomag Kommisbrot (1924), 28f
Harley-Davidson Iron 883, 805f
Hatchback, 39, 521, 1037
   cooling air drag, 328f
   detail optimization of, 51f
   drag, 1091f
   flow around, 50f
   lift, 1092f
   velocity distributions at wheel, 1092f
   wake of, 1009f
Hawkings, D.L., 1154, 1164
Head and neck support (HANS), 901
Head injury criterion (HIC), 878n
Head protection technology, 877–879
Heald, R.H., 31–32, 40
Heat dissipation, relationship to radiator flow, 234f
Heat exchangers, 48, 535, 537–541
   air conditioning, 541
   coolant radiator, 537–539
   dependence of pressure loss on Reynolds number, 1039f
   design of, 535–537
      calculation of performance, 536–537
      cross-flow, 536
      parallel-flow and counter-flow, 536
   different arrangements of, 324f
   engine and transmission oil cooler, 540–541
   intercooler, 539–540
   simulation of, 1121–1123
Heat output, 552f
Heat protection, motorcycle, design of, 872–874
Heat transfer
   vs. area of cooling matrix and mass flow, 556f
   vs. area of cooling matrix and pressure loss, 556f
   vs. fin density and mass flow, 558f
   vs. fin density and pressure loss, 559f
   fundamentals of, 533–535
      convection, 534–534
      heat conduction, 534
      radiation, 535
   vs. thickness of heat exchanger and mass flow, 557f
   vs. thickness of heat exchanger and pressure loss, 558f
Heating and climatization tests, 1017–1019
Heft, A.I., 410
Helium bubbles in wake of model, 1009f
Helmets, 46–47, 877–912
   aeroacoustic and artificial-head measurement technology, 909
   aerodynamic forces, measurement of, 907–908
   data, 907–908
   Formula 1, 878, 879f, 901–905
   head protection technology, 877–879
   influence of windshield height on noise, 891f
   influence of windshield inclination on forces, 886f
   noise level inside, 837t, 862f, 868f
   dummy for measuring, 861f
   numerical simulation of, 909–912
   for open race cars
      acoustics, 903–905
      aerodynamics and ventilation, 901–903
      history, 901
   wind tunnel measurements, 906–907
   see also Motorcycle helmets

Helmholtz resonators, 950–951
Helmholtz vortex theorems, 106–107, 117
Henne, Ernst, 796–797
Hentschel, W., 1031
Herz, Wilhelm, 798f
Heyl, G., 891, 894, 895
High-performance computing (HPC), 1138–1144
    history of performance, 1141f
High-performance vehicles
    aerodynamics of, 661–710
        air guiding elements vanes, 703–707
        balance, 643–649
        basic body, 662–665
        cooling and ventilation, 654–656
        diffusers, 686–695
        downforce, 639–642
        drag, 635–638
        front, 662
        ground effect, 682–686
        inlets and outlets, 696–703
        spoiler and gurneys, 677–682
        tail, 663–665
        wheels, 707–710
        wings, 666–676
    definition of, 601–602
    efficiency, 651–654
    history of
        racing cars, 602–608
        record-breaking vehicles, 609–618
        sports cars, 618–622
    introduction to, 601–602
    oblique incident flow, 656–659
    performance comparison, 623–629
    race tracks, 629–631
    regulations, 631–635
    road performance, 649–651
    slipstream, 659–661
    vehicle classes, 623–629
High speed, forces acting on vehicle at, 163f
Highway Code §23(1), 893
Hirt, C.W., 71
Hockenheimring, 630
    cornering speed, 641f
    lap time as function of engine power, 639f
    lap time as function of lift and drag, 652f, 653f
    maximum achievable speed on, 637f
    transverse acceleration on curve, 641f
Hoepke, E., 740
Hoerner, S.F., 54, 130, 363
Hoffmann, J., 933
Honeycombs, wind tunnel, 942
Hood, 267, 479
    aerodynamic force vector, 1132f
    inclination of, 264
    radius of, fine-tuning, 259f
    sharp leading edge at, 259f
Horseless carriage, 26
Hot-film probes, 986
Hot-wire anemometry, 566–567, 985, 986f
Howe, M.S., 1155
Howell, J., 264, 276, 297, 303, 360
Hucho, W.-H., 48, 52, 126, 249, 293, 1030, 1037, 1042
Hühnergarth, J., 323, 326
Humidity, sensors for, 898f
Hupertz, B., 305
Husqvarna
    Nuda 900 R, 806f
    TE 449, 802f
        split water cooler, 833f
Hüttenbrink, K.-B., 896
HVAC systems, aeroacoustic simulation of, 1169–1171

ICE I, 68–69
Ideal gas equation, 146f
Idle, 528
Incident flow velocity, dependence on driving velocity, 422f
Incompressible continuity equation, 82
India, 176
Indianapolis 500, 901
Indifference point, 121–122
Induced drag, 228–232
    of wings, 673–674
Indy Racing League, 623–624
Infinite plate set in motion, velocity profile of, 112, 113f
Ingram, K.C., 724–725
Inlet losses, 235, 321–322

Index

Institute for Internal Combustion Engines and Automotive Engineering, Stuttgart University, *see* IVK
Institute of Aerodynamics and Gas Dynamics, Stuttgart University, 1049
Intake, 240, 530, 1130
  design of, 531
  of high-performance vehicles, 696–703
    on Renault race car, 699*f*
  opening size, vs. engine power, 554
  of warm air, commercial vehicles, 787
Integrated methods, 70–73
Interconnect, influence on scalability, 1140*f*
Intercooler, 539–540
Interference, 367–380
  cars with trailers, 369–377
  driving in convoy, 377–379
  interaction of vehicle components, 367–369
  overtaking, 380
Interference drag, 131, 244–246, 548, 549*f*
Intergovernmental Panel on Climate Change (IPCC), 191*n*
Interior noise
  acoustic glass windows to reduce, 593
  influence of airflow on, 569–572
  measuring, 576
Internal drag, 233–241, 547, 549*f*, 550
Internal flow, 216–217, 527, 541–546
  cooling module, 544–545
    heat exchanger assembly, 545
    positions in vehicle, 544
  engine compartment, 47–48
  fan, 545–546
  influence parameters of, 551–553
  motorcycle, 831–834
    design of, 872–874
  operating conditions, 541–543
  passenger compartment, 48–49
International Motor Sports Association Federation Internationale de l'Automobile (FIA), 631–632
Inverse panel method, 69
Inviscid flow
  dynamics of, 93–110

elementary flows, 96–99
  flow around a sharp edge, 98
  parallel flow, 96–97
  source and vortex, 98–99
  stagnation point flow, 97
flows around bodies, 99–105
  elliptic cylinders at zero incidence, 103
  elliptic cylinders with oblique incident flow, 104
  flow around circular cylinder and d'Alembert's paradox, 101–103
  semi-infinite bodies, 100–101
  wing profile, 104–105
interpreting streamline patterns, 93–95
momentum theorem for, 90–91
planar model flows, 95–105
stream filament theory, 89–90
use of complex functions, 95–96
vortex flows, 105–110
  dynamic behavior of separation planes, 108–110
  vortex filament model, 106–107
  vortex induction, 107–108
Ishima, T., 988
Islam, M., 1096
ISRA method, 1006
Italy, 344*n*
Ivanic, T., 549
IVECO Transport Concept Solution Truck, 769, 771*f*
IVK
  environmental test facility, 863
  wind tunnel, 959, 966*f*, 983*f*, 1044–1045, 1049, 1052
  sound-absorbing elements in, 943*f*

Jallet, S., 486
Janssen, L.J., 65, 293
Japan, 878
  driving cycles, 175, 176*f*
  inertia mass classes in, 169–170*t*
Jaray, Paul, 17–20, 22, 24, 53, 55, 63
Jenatzy, Camille, 12, 609
JIS T 8133:2000, 878
Joergensen, F.E., 986
Joukowski mapping function, 103

Jump, 52, 319, 320*f*

Kaiser, 27
Kaiser-Frazer (1947), 28*f*
Kaivola, R., 896
Kamm, W., 25, 26, 40, 47, 48, 64, 610
    K1 with slotted fins, 40*f*
Kamm-back, 34–37, 40
Kamm-Circle, 825*f*
Kandasamy, S., 1081
Karlsson, R.K., 987
Kässbohrer streamlined bus (1937), 719*f*
Kataoka, T., 1102
Kessler, J.C., 1023
Khalighi, B., 213
Kiel, G., 991
Kielmann, G., 894
Kinetic theory, basics of, 1070–1072
King, L.V., 987
Kirchhoff integral method, 1155–1156
Kitoh, K., 1096
Klein, C., 997, 1027
Klemperer, W., 17, 39–40, 47, 53, 55, 63
Knape, H.W., 1003
Knudsen number, 1071–1072
Kobayashi, N., 376, 432
Koenig-Fachsenfeld, R. v., 43, 798, 1007
Koessler, P., 523, 792
Kolmogorov, 420
Kompenhans, J., 988
Koremoto, J., 1004
Kortesuo, A., 896
Krämer, W., 767
Krantz, W., 471
Krebber, W., 894
Küchemann, D., 110, 926
Kuli, 1114*f*
Künstner, R., 369, 924, 1045
Kuthada, T., 236, 321, 407, 985
Kyoto summit, 191*n*

La Jamais Contente (1899), 609
Lamborghini
    concept S (2005), 272, 273
    Gallardo (2003), 621*f*
    Murcielago
        retractable cooling duct, 697*f*
        scoop behind rear wheel, 698*f*

Laminar profile NACA 634-021, 130*f*
Laminar separation point, 122
Land Rover Defender (1993), 396
Lange, A.A., 23–24
Lap time, 642
    as function of lift and drag, 652*f*
Laplace equation, 85, 94, 99
Large-eddy simulation (LES), 1084, 1093–1099
    basic equations, 1093–1095
    detached-eddy simulation, 1096
    influence of cell size, 1095–1096
    results of LES and DES simulations on automobiles, 1096–1098
Laser Doppler anemometry (LDA), 565, 987–988
Laser-light intersection technology, 1010
Lateral acceleration, 651*f*
Lateral dynamics, motorcycle
    cornering, 821–826
    lift effects, 826–827
    yawed flow, 828–831
Lateral surfaces, transition to rear surfaces, 300–303
Lattice methods, 1072–1084
    lattice-Boltzmann method (LBM), 1073–1084
        comparison with experimental data, 1080–1081
        simulation process, 1078–1080
        spatial and temporal discretization, 1074–1077
        turbulence models, 1077–1078
    lattice-gas method (LGA), 1073
Lay, W.E., 25–26, 31–32, 42
Le Corbusier Voiture Minimum (1936), 60*f*
Le Mans, 653–654
Leaks, and noise, 583–584
Lean angle, 822–824
    aerodynamic forces with and without, 828*f*
    flow field with and without, 827*f*
Leder, A., 209–210
Legislation
    CO_2 legislation and labels, 191–196
    for commercial vehicles, 727–728
Leuschen, J., 762

Liebold, H., 312
Lift, 202
    coefficient of, 203
    measured values for 79 vehicles, 392f
    motorcycle
        and lateral dynamics, 826–827
        and longitudinal dynamics, 818–820
        measures to optimize, 869–872
    negative, 101
    of passenger cars in production, 390–403
    and pitching moment, 246–247
    and rear end, 290
    vehicle's reaction to, 449–459
        behavior when braking, 457–459
        self-steering behavior, 449–453
        stability and straight-ahead running, 453–456
    at wheel-road contact points, 201f
Lighthill, M.J., 1148, 1153
Liliental, Otto, 202
Lilley, G.M., 1155
Limit strategy, 57–60
Lin, C., 1016
Lincoln Zephyr V12 (1926), 21f
Lindemann, J., 896
Lindener, N., 1044
Linear flow, pressure losses in, 139–141
Linearized Euler equations (LEE), 1156–1157
Liquid cooling, 532–533
Literature, aerodynamics, 7, 11
Load spectrum, 167n
Loehrke, R.I., 942
Löfdahl, L., 987
Lola sports car, front diffuser of, 694f
Longitudinal acceleration, 651f
Longitudinal dynamics, motorcycle, 813–820
Longitudinal vortices, 70, 215–216
Lotus 79 Formula 1 race car (1977), 685f
Lumley, J., 1102–1103

Mack, S., 412
Maestrello, L., 1148
Malvern Spraytec Laser Diffraction System, 1012f

MAN
    Concept S, 769–770
    Lion's Coach
        pressure and airflow on, 772f
        comparison of forces and flow separation, 775f
    TGA, dynamic pressure at front, 740f
    TGA LX, airflow separation on, 722f
    TGX, 721f
        scale model, 730f
    tractor-trailer truck, 46f
Manikins, climate-measuring, 492–493
Mankau, H., 373
Manual transmissions, 183
Maskell, E.C., 295
Mass, 76, 185
    conservation of, 75, 81
Mass flow, vs. area of cooling matrix and pressure loss, 555f
Massively parallel systems (MPP), 1139–1140
Mathis-Andreau 333, 23f
Mauboussin, P., 22, 24, 40, 63
Maue, J.H., 909
Maximum speed, 528, 609–611
Mayer, E., 491–492
Mears, A.P., 708
Mercedes
    Actros (2012), interior, 742f
    B SL 500, 401, 402f
    B180 CDI, reduction in fuel consumption via optimized drag, 187f
    bus (1935), 719f
    C-Class, exterior sound pressure levels, 572f
    CLA, 57
    E-class, 403
    GP Formula 1 racing car, adjustable front wing, 634f, 663f
Mercedes Benz, 28f, 762
    190, boat tailing on, 313f
    300 SLR (1955), 604f
    720 SSK (1928), 618, 619f
    aerodynamic truck prototype, 757f
    Aerotrailer, 768f, 769
    B180 CDI, driving performance and consumption values for, 164t

B-Class
    drag coefficient as function of flow
        angle, 157f
    rolling resistance, drag, accelera-
        tion, and grade resistance, 160f
    technical data, 159t
C III 111 (1978), 614f
CLK sports car prototype (1988), air
    outlet on, 702f
E-class
    convertible (2010), 500, 501f
    development process, 389t
    sunroof with wind deflector, 502f
G400 (2001), 396
sedans, increase in weight and frontal
    area of, 153f
SL convertible, wind deflector, 498f,
    wind deflector, 499f
SLK, 490
    with vario roof, 497f
    AIRGUIDE comfort system, 500f
T80 (1939), 610, 611f
Technology Centre wind tunnel, 1045–
    1046, 1053
Mercedes McLaren Formula 1 racing car,
    632f
Mercker, E., 927, 930, 932–933, 934, 960, 969,
    970, 971, 973, 976, 978, 1003, 1060
Michaelian, M., 378
Microdrag, 223–224
Microphones, 577f, 579f
    arrays, 578
    surface, 1166f, 1167f
Minimum, 52, 319, 320f
MIRA
    rating system, 65, 66, 67f
    wind tunnel, 1048
Mirrors, 241, 368–369, 508, 584
    acoustic, 580–581
    aeroacoustic source types and relevant
        terms, 1155f
    on buses and coaches, 775–777
    on commercial trucks, 748–750, 792f
    flow around, 1149f, 1150
    influence of, 361–363
    loads on, 476–480
    paint test of, 512f
    and side window contamination, 511,
        518–521
    simulated flow around, 1164f
    sound intensity probe of, 577f
Mistral, 22, 40, 63
Mixing ventilation (MV), 49
"Moby Dick" (1978), 606f
Model-scale wind tunnels, 1040,
    1048–1050
    for commercial vehicles, 728–732
Modified Indian Driving Cycle (MIDC),
    176
Modine Europe wind tunnel, 1052
Mohring, W., 1155
Möller, E., 44
Momentum conservation, 75
Momentum theorem
    application to 90° pipe bend with
        internal flow, 91
    for flow through vehicle with fully
        guided cooling air, 238f
    with friction, 92–93
    for inviscid flows, 90–91
    pressure and momentum forces, direc-
        tion of, 91f
Monroney label, 194–195
Monte Carlo, 630
Monza, 630
Morelli, A., 55, 307, 308, 928, 1043
Motorcycle helmets
    aeroacoustics of
        approaches to noise reduction,
            895–896
        development targets and noise,
            887–889
        mechanisms and parameters of
            sound generation, 889–893
        perception of environmental
            signals, 893–894
    aerodynamics of, 837f
        development targets, 879–880
        direction of view, 883–884
        geometry, 880–882
        influence of neck length (shoulder
            room), 884
        riding position and windshield,
            884–886

Motorcycle helmets *(Cont.)*
    spoilers, 881–882
    structured surface, 882
    ventilation and rain tests, 896–900
Motorcycles, 46–47
  aeroacoustics of, 837
  aerodynamics of, 819*f*
    cornering, 821–826
    design examples, 869–876
    drag, 813, 869–872
    dynamic reduction of front wheel load, 819*f*
    forces and moments, 811–812
    forces on rider, 836*f*
    history of, 796–801
    introduction, 795
    and lateral dynamics, 821–831
    lift effects, 818–820, 826–827, 869–872
    and longitudinal dynamics, 813–820
    optimizing drag and lift, 869–872
    rolling resistance, 814–817
    top speed, 813–817
    traction force at rear wheel, 813
    traction force/transmission ratio, 817
    yawed flow, 828–831
  assessment of draft phenomena, 836*f*
  BMW C1, 809
  categories of, 801–807
  CFD simulation, 839–853
    external flow field results, 848–853
    simulation model, 841–843
    surface results, 843–848
  chopper/cruiser, 805
  combination, 807–808
  cooling, 873*f*, 831–834, 872–874
  development methods, 838–869
  Ecomobile, 810
  fairings, 47, 48*f*
  front wheel suspension concepts, 873*f*
  frontal area of, 812*f*
  fun bikes, 806
  future of, 868–869
  heat protection, 872–874
  influence of driver's posture, 857*f*, 858*t*, 871*f*, 872*f*
  influence of passenger, 859*t*
  internal flows, 872–874
  naked bikes, 806
  off-road, 802–803
  outlook, 876
  potentials for drag reduction, 870*f*
  process, 838–839
  road test
    aeroacoustics, 868
    cross-wind, 864–865
    driving stability, 864
    soiling, 866–867
    wind and weather protection, 865
  scooters, 804
  special bikes, 807–810
  supermoto, 806
  supersport, 806–807
  temperature load, 852*f*
  three-wheeler, 808–809
  tourers, 804
  touring enduros, 803
  velocity distribution on water and oil coolers, 851*f*
  wheel load vs. velocity, 821*f*
  wind and weather protection, 835–837, 875–876
  wind tunnel tests
    aeroacoustics, 861–862
    flow field visualization, 859–860
    reproducibility and driver influence, 855–858
    rotating wheels, 854–855
    soiling, 862–864
    stationary wheels, 853–854
Motorsports, 870
Multigrid methods, 1128–1130
Multimode break-up, 504

NACA 634-021, 130
Nagib, H.M.S, 942
Naked bikes, 806
NASA Ames Research Center, 732
NASCAR, 659
Navier-Stokes equation, 68, 85–88
  dimensionless form of, 111
  simplifications of, 1069
Navier-Stokes methods, 1084–1103
  direct numerical simulation, 1099–1103
  large-eddy simulation, 1093–1099

basic equations, 1093–1095
detached-eddy simulation, 1096
influence of cell size, 1095–1096
results of LES and DES simulations, 1096–1098
Reynolds-averaged Navier-Stokes, 1085–1092
equations, 1085–1086
results, 1090–1091
turbulence models, 1086–1087
wall modeling, 1088–1090
Near wake, 69
Nebel, M., 389
New European Driving Cycle (NEDC), 171–173
Cycle for Hybrid Drives, 173
energy flows in, 178f
exhaust emissions following cold start, 168t
inertia mass classes, 169–170t
Negative interference drag, 245
Nemecek, J., 492
NEOPLAN
Cityliner, 772
comparison of forces and flow separation, 775f
effect of tapered rear, 781f
scale model, 730f
Skyliner, wiper position, 777f
Starliner, 772
airflow separation on, 722f
comparison of forces and flow separation, 775f
Newton's equation of motion, 75, 82, 83
Newtonian shear stress model, 86
NFAC wind tunnel, 732, 733f
Ng, E.Y., 1016
Noble, Richard, 611
Noise
inside helmet
influence parameters, 890f
types and transmission paths, 892f
interior, 1149, 1167–1168
simulation of transfer into vehicle, 1168f
sources of, 570f, 571f, 573–574
antennas, 586–587
A-pillar, 587–588
cavity resonances, 589–591
leaks, 583–584
locating, 582
mirrors, 584
sunroof, 591
underbody, 592
wheel housings, 591
windshield wipers, 586
structure-borne, measuring, 581–582
wind, on-road measurement of, 1031
see also Aeroacoustics
Notchback, 38, 282–290
drag, 1091f
lift, 1092f
rear inclination, 298f
rear window inclination, 298, 300
separation edges and rear spoilers, 305, 308, 309
shape of rear pillar, 300
tail taper, 312
velocity distributions at wheel, 1092f
Nozzle, wind tunnel, 920–926
blockage, 973–975
NRC Canada wind tunnel, 732, 733f, 972
NSU, 798
AG, 46
Rennmax 250 cm^3
with Banana tank (1953), 799f
with Dolphin fairing (1954), 800f
with Blue Whale fairing (1955), 800f
Ro 80, 29, 30f, 49, 57
Numerical simulation
of aeroacoustics
aeroacoustic analogies, 1153–1155
application examples, 1161–1172
calculation of sources, 1151–1153
introduction, 1148–1151
Kirchhoff integral method, 1155–1156
linearized Euler equations, 1156–1157
outlook, 1172–1173
sources and exterior sound field, 1153–1157
transfer into the vehicle cabin, 1157–1161
basics of kinetic theory, 1070–1072
classical approach, 68–70

of commercial vehicles, 735–736
cooling module design
  air side, 1113–1117
  coolant side, 1117–1119
grids for computation, 73f
hardware and benchmarking
  computer architectures, 1138–1140
  computer performance, 1141–1144
heat exchanger, 1121–1123
helmets, 909–912
history of, 70–73
integration into development process, 1144–1146
lattice methods, 1072–1084
  lattice-Boltzmann, 1073–1084
  lattice-gas (LGA), 1073
motorcycle, 839–853
  external flow field results, 848–853
  simulation model, 841–843
  surface results, 843–848
Navier-Stokes methods, 1084–1103
  direct numerical simulation, 1099–1103
  large-eddy simulation, 1093–1099
  Reynolds-averaged Navier-Stokes, 1085–1092
outlook, 1146–1148
potential flow methods (BEM)
  potential flow theory, 1104–1106
  results, 1110–1113
  solid body simulation, 1107
  static pressure calculation, total pressure loss, 1109–1110
  wake simulation, 1107–1109
  zonal coupling, 1103–1104
requirements and properties of methods, 1068–1070
rotating geometries (wheels, fans), 1119–1121
  moving reference frame (MRF), 1120
  "rotating wall," 1120
  sliding mesh, 1121
solution process
  discretization, 1123–1125
  fluid results, 1134–1138
  numerical grids, 1125–1127
  post-processing, 1130–1138
  solution algorithms, 1127–1130
  surface results, 1130–1134
Nürburgring, 630
NWB wind tunnel, 950f

Oblique flow
  commercial vehicles, 723–727
  high-performance vehicles, 656–659
Off-road motorcycles, 802–803
Ohtani, K., 308
Oil pump, drive torque for, 181f
On-road measurements, 1021–1031
One-volume bodies, 30–34
Opel
  Astra, airflow separation on, 722f
  Calibra
    boat tailing on, 313f
    influence of trunk lid height on, 315, 316–317f
    underbody development, 341, 342f
  Eco Speedster, airflow separation on, 722f
  GT (1968), 52, 53f
  Kapitan, (1938), 21f
  Omega MY 1986, 16
  Rak 2 rocket car (1928), 609
  Speedster, 616, 617f
Open-jet test section, 973–979
Open vehicles, 487–502
  convertibles, 487–488
    airflow with top open, 488
    design solutions for, 496–500
  sunroofs, 487–488
    design solutions for, 501–502
  thermal comfort, 489–494
  wind noise, 489
Optical measuring methods, 565
Optifuel, 769, 770f
Outlets, 240, 542, 562–563 1130
  of high-performance vehicles, 696–703
Overbraking, 459
Oversteer, 448, 450, 644
Overtaking, 380, 414–415

P-shoulder vortex, 354
Pagendarm, H., 940
Panel method, 68–69, 72f

Panel model of vehicle, 1107f
Pankhurst, R.C., 921, 942
Panoz LMP prototype, 678f, 707f
Papenfuß, H.-D., 69
Parallel connection of loss elements, 137–138
Parallel flow, 96–97
Parallel-flow heat exchanger, 536
Park, Y.G., 537
Particle image velocimetry (PIV), 860, 988
Passenger cars
    aerodynamics and design, 1–2, 60–63
    air forces and moments of, 201f
    character of vehicle aerodynamics, 2–4
    components of model, 1075f
    development strategies
        detail optimization, 49–52
        limit strategy, 57–60
        shape optimization, 53–57
    development tools
        CFD methods, 70–73
        classical approach, 68–70
        rating system, 65–68
        wind tunnels, 63–65
    directional stability, 39–42
    drag and lift of
        drag surface area, 392–394
        driving on the ceiling, 402–403
        influence of drive concept, 396–399
        influence of equipment and engine, 400–402
        intercomparison in accordance with EADE, 394–396
        overview of competitors by class, 390–392
        SUVs, 399–400
    early investigations with parameters, 25–26
    flow field around, 206–220
        environmental influences, 217–219
        influence of Reynolds number, 219–220
        internal flow, 216–217
        longitudinal vortices, 215–216
        non-periodic dead wake, 208–210
        periodic dead wake, 210–211
        ring and spiral vortices, 211
        unsteady processes, 211–213
    historical development of aerodynamics, 7–26
    from horseless carriage to automobile "bathtub" body, 27–30
        one-volume bodies, 30–34
        stamping, 26–27
    internal flows
        engine compartment, 47–48
        passenger compartment, 48–49
    panel model of, 1107f
    pressure distribution from various wind tunnels, 1015f
    rear-end shapes, 34–39
    shapes of, 11–24
    vehicle/trailer combination, 369–377
    weight of, 151–152
Passenger compartment
    air coupling with trunk, 1163f
    cabin displacement ventilation (CVD), 49
    heating test, 1018–1019
    HVAC systems, 1170–1171
    internal flow, 48–49
    noise transfer into, 1157–1161
    open vehicles, 487–502
Pavilion, influence of slanting on drag, 297f
Pawlowski, F.W., 42, 44
Peraves Ecomobile, 810f
Performance
    acceleration and elasticity, 160–162
    ascending ability, 162
    performance, 1142–1143
    top speed, 163–165
Peric, M., 1123, 1126
Perkovic, M., 785
Persu, Aurel, 31
Peschke, W., 373
Peugeot 66 racing car (1966), 605f
PHOENICS, 71, 72f
Piaggio Yourban LT, 809f
Pike's Peak racing car (1987), 675f
Pininfarina, 224
    CNR E2, 614
    wind tunnel, 1043

Pipe systems with internal flow
    pressure loss coefficients, 138–145
        losses in linear flow, 139–141
        losses due to change in cross section, 142–144
        losses caused by fixtures and fluid-control devices, 145
        losses caused by divergence and convergence, 145
        redirection losses, 141–142
    pressure loss of components, 135–136
    series and parallel connection of loss elements, 137–138
    stream filament theory including friction and energy input, 134
    system characteristic and operating point, 136–137
PISO (Pressure-Implicit Splitting of Operators), 1128, 1129f
Pitching moment, 202, 246–247
Pitot tubes, 991–992
Pkw-EnVKV, 193
Planar model flows, 95–105
Plate, drag of, 123–126
Plenum, wind tunnel, 937–939
Pontoon body, 27–28
PONTOS mobile optical measurement system, 475f
Pope, A., 942
Porsche
    356/1 (1948), 620f
    356 A (1950), 619, 620f
    911, 24f, 230
        Carrera S, adjustable rear spoiler on, 679f
        convertible, wind deflector, 499f
        Targa, glass sunroof with shading blind, 502f
        Turbo (1983), 620, 621f
    917/ 30 racing car (1973), 605f
    918 Spyder, 622, 623f
    935/78 race car"Moby Dick" (1978), 606f
    956 group C racing car, 657f
    959 (1987), 620, 621f
    962, 607f
    Boxter, 390
    Carrera 6 Langheck racing car (1966), 605f
    wind tunnel, 1046, 1048
Porsche, F.A., 61
Positive interference drag, 244
Postprocessing, 1070
Potential flow methods, 1085, 1103–1113
    potential flow theory, 1104–1106
    results, 1110–1113
    solid body simulation, 1107
    static pressure calculation, 1109–1110
    theory, 1104–1106
    total pressure loss, 1109–1110
    wake simulation, 1107–1109
    zonal coupling, 1103–1104
Potential theory, 84–85
Potthoff, J., 65, 327, 330, 336, 998, 1022
Power steering, 181–182
PowerFLOW, 73, 841, 842, 843 1126, 1130–1138, 1073–1084
Prandtl, L., 23, 124, 128, 209
Prandtl boundary layer, 112–115, 123
Prandtl tube, 990, 991f
Preprocessing, 1070
Pressure
    characteristic
        buses and coaches, 772–773
        commercial trucks, 739–741
    coefficient, 203–204
    of cooling system, 235f
    distribution
        in center section of Audi 100, 269f
        comparison of measured and computed, 1111f
        at different Reynolds numbers, 221f
        of front end, 254f
        static pressure, 992, 1109–1110
        on streamlined body, 129f
        total, 1137f
        around vehicle, 204f
        and wall streamlines, 1111f
    drag, 220–223
    gradients, 93–94
    loss, 217, 235–236
        caused by regulating and fluid-control devices, 145f
        coefficients, 138–145

of components with internal flow, 135–136
relationship to heat dissipation, radiator flow, and radiator drag, 234f
total, 1109–1110, 1133f
with unsteady cross section decrease, 144f
measurements, 564–565
dynamic pressure, 990–992
static pressure, 992
transducers for, 993–997
Pressure sensitive paint (PSP), 995–997
Preusser, T., 998
Probes, angled, 988–989
Product development process, 381f, 385f, 387f, 389f
Promote-Chauffeur, 784f
"Pseudo-Jaray," 20, 21f, 22f, 23f, 26
Pyramidal probes, 989

R8 sports car prototype, cooling air outlet on, 701f
Race tracks, 629–631
Racing cars, 602–608
helmets for, 901–905
Radiation, 535
Radiator, 537–539
commercial vehicles, 753f
measured pressure loss of, 1122f
performance map, 1116f
pressure loss at, 322–325
relationship of flow to heat dissipation, 234f
Rae, W.H., 942
Raffel, M., 988
Rail Tec Arsenal wind tunnel, 1050
Railton Mobil Special, 611
Railway aerodynamics, 5–6
Rain gutters, influence on sound emission, 588f
RAK (1928), 41f
Ram-air inlets, 834f, 873f
Ramshaw, J.D., 71
Rating systems, 65–68
Rawnsley, S.M., 71
RB7, DRS system on, 634f

Rear end
buses and coaches, 781–782
comparison of measured and computed drag, 1112f
influence on aerodynamic forces, 280–318
rear window inclination, 290–300
separation edges and rear spoilers, 304–311
tail taper, 312–314
transition from lateral to rear surfaces, 300–303
trunk lid, 315–318
Kamm-back, 34–37, 40
shapes, 34–39, 281f, 282–290
tapered
on bus, 781f
commercial vehicles, 764f, 765f
on semitrailer, 769f
see also Fastback; Hatchback; Notchback; Squareback
Rear pillar
forces acting on, 302f
influence of radius on drag and lift, 201f
Rear slant
aspect ratio, influence on drag, 297f
inclination angle, influence of, 292f, 294f, 295f, 298f
Rear window
contamination of, 512, 521–523, 526
inclination angle, 290–300, 315
tuning of, 316–317f
influence of roof spoiler on contamination, 526f
ways to optimize flow separation at, 305f
Rear wing configurations, 310f
Rebalancing, 648
Rebound, 506
Record-breaking vehicles
maximum speed, 609–611
transonic speeds, 612–613
Redirection losses, 141–142
Reference bodies
Ahmed, 409
DriveAer, 410–412
SAE, 406–408

Regulation for Setting Emission Performance Standards for New Passenger Car, 192
Regulations, for high-performance vehicles, 631–635
Reid, E.G., 32
Reitwagen, 795, 796f
Remenda, B.A.P., 1025
Renault
    Espace (2010), 272f
    Formula 1 (2005), cooling air in, 697f
    Optifuel, 769, 770f
    race car (1978), engine air intake on, 699f
Research Institute for Automotive Engineering and Vehicle Engines, *see* FKFS
Resistance, possibilities for reducing, 188–189
Resistance factor for rough tubes, 140f
Reynolds-Averaged Navier-Stokes (RANS) methods, 1084, 1085–1092
    equations, 71–72, 1085–1086
    results, 1090–1091
    turbulence models, 1086–1087
    wall modeling, 1088–1090
Reynolds number, 111–112
    critical, 119
        influence of roughness and turbulence on, 128f
    effects of, 63, 64f
    influence on flow field, 219–220
Ricotti, Count, 13, 30
Riemann mapping theorem, 96
Rim cover, commercial vehicle, 779f
Rim vortex, 354
Rims, 355–359
Ring vortices, 211
    formation at base of prismatic body, 212f
    model for nozzle and collector, 975f
Road dynamics
    aerodynamic axle load relief settings, 459–461
    of high-performance vehicles, 649–651
    reaction to crosswinds, 462–464
        application example, 470–472

        evaluation of crosswind, 464–470
        vehicle excitation, 461–462
    reaction to lift forces
        behavior when braking, 457–459
        self-steering behavior, 449–453
        stability and straight-ahead running, 453–456
    single-track model, 445–449
Road tests
    drag measurement using coastdown tests, 1021–1025
    crosswind tests, 1025–1029
    cooling tests, 1029–1030
    motorcycle
        aeroacoustics, 868
        crosswind, 864–865
        driving stability, 864
        soiling, 866–867
        wind and weather protection, 865
    soiling of windows and chassis, 1030–1031
    wind noise, 1031
Robinson, Rocky, 799
Rocker panel struts, 982f
Roll moment, 250, 300, 375
Rolling resistance, 154–156, 186
    determining with measuring trailer, 1023f, 1024f
    factors that determine, 155f
    as function of speed and type of construction, 156f
    motorcycle, 814–817
Romberg, G.F., 377, 661
Roof, 273–275, 297f, 364–367
    boxes, 364–366
    racks, 241, 364–366, 1172
    rails, 364
    spoiler, 305–307
Roshko, A., 69
Rotational sound, 953
Rothhämel, J., 889
Roughness drag of add-on parts, 241–243
Rover 800, side force distribution on, 264f
Ruck, B., 987
Rudd, M.J., 987
Rudkin Wiley, 45
Rumpler, Edmund, 15–17, 24, 62
Ryan, A., 964

Saab 9X (2001), 272f
SAE
    J670e, 997
    J2071, 1040
    reference body, 406–408
    Subcommittee on Open Throat Wind Tunnel Adjustments, 928
    Surface Vehicle Information Report, 926
Safety, 382, 418, 452, 512, 515, 532
    and commercial vehicles, 756, 789
    and high-performance vehicles, 613–614, 631–633, 664, 674
Sakamoto, H., 288
Sanderse, A., 922
Sasaki, Y., 376
Saturation, 52, 319, 320f
Sauber Petronas C20 Formula 1, 623, 624f
Saunders, J.W., 476
Sawatzki, E, 40, 797
Scania concept vehicle, 770, 771f
Schlichting, H., 44, 129, 139, 798
Schlör, K., 32–33, 63
Schlör-car (1938), 33f, 34f
Schmidt, G., 61
Schmitt, J., 988
Schmitz Cargobull, 768
Schnepf, W., 795
Schröck, D., 471, 965, 966, 967, 994
Schuberth
    head ventilation concept, 899f
    helmet, 879f, 902f
    "Speed," 882f
    wind tunnel, 807f, 906f, 909
Schultheis, R., 507
Schulz-Hausmann, F.V., 931
Schumacher, Michael, 634
Schütz, T., 305, 323, 326, 363 1097
Scibor-Rylski, A.J., 267
Scooters, 804
Sebben, S., 73
Segrave, Henry, 609
Seiferth-wing, 947–948, 949
Self-soiling
    causes of, 523–524
    commercial vehicles, 792–793
    design measures to counteract, 524–526

    door handle contamination, 525
    vs. foreign soiling, 503
    motorcycle, 863–864, 866–867
    rear window contamination, 526
    simulation of, 1012–1013
Self-steering, 448–453
Selig, M.S., 674
Semi-infinite bodies, flows around, 100–101
Semitrailers and bodies, 740f, 758–768
Sensitivities, 185
Separation bubble, 286
    formation of, 123f
    "pumping" from, 288f
Separation edges, 304–311
Separation planes
    dynamic behavior of, 108–110
    with velocity jump, models for, 108f
Series connection of loss elements, 137–138
Shanghai Automotive Wind Tunnel Center (SAWTC), 1046, 1047f
Shape
    as intersection of design and aerodynamics, 62f
    optimization of, 53–57
    relation to aerodynamic drag, 67f
Sharp edge, flow around, 98
Sharp edge separation, 210
Shear break-up, 504–505
Shear layers
    formation of, 208
    representation of, 1105f
    velocity distribution in, 1110f
    vorticity content of, 1109f
Shear stresses, 77
Sherwood, A.W., 45
Ships, aerodynamics of, 6–7, 9
Shutter systems, 331–334
Siddal, R.D., 987
Side force, 202, 219, 300
    admittance function, 436, 440, 471, 967
    distribution on Rover 800, 264f
    impact of rounding edges on, 261f
    in overtaking maneuvers, 414–415
    transfer function, 437, 441, 444, 471
    at wheel-road contact points, 201f
    and yaw moment, 248–249

Side panel, contamination of, 525f
Side skirts, 762–763, 764f
   airflow without, 756f
Side wind, 416–444
   and car/trailer combinations, 374–375
   and commercial vehicles, 784–786
   directional stability in, 248f
   over driving velocity, 218f
   on semitrailer combination, 723f
   test and evaluation methods, 427–444
Side window
   contamination of, 507–508, 511, 518–521
   effect of recess on drag, 279f
   inclination of, 280f
   simulated pressure fluctuations on, 1166f
   trim strip along, 518f
Silencer, 887
SIMPLE (Semi-Implicit Method for Pressure Linked Equations), 1128, 1129f
Simple bodies, drag of
   flow around a sphere, 126–128
   plate at zero incidence, 123–125
   plate in cross-flow, 125–126
   streamlined body, 129–130
Singh, R., 386
Single-track model, 445–449
Sink drag, 48
Ski carriers, 241, 364
Skoda
   Fabia WRC, 675f
      cooling air outlet on, 702f
   Roomster (2003), 272f
   Superb, inflow losses at grille, 322f
Slenderness, 54–55
Slip angle, lateral force as function of, 650f
Slipstream driving, 377–379
   commercial vehicles, 782–784
   high-performance vehicles, 659–661
Slope angle, 38f, 158
Smagorinsky, J., 1095
Smart City Coupé, 186
Smoke probe, flow visualization by, 860f
Snappy Hex Mesh, 1126
Snell SA2000 standard, 878

Soiling, 1011–1013
   basics of, 503–508
   commercial vehicles, 791–793
      task description and testing methods, 789–791
   droplets in free flow, 504–505
   influence of surface condition, 507–508
   mechanisms that affect water droplets, 505–506
   motorcycle, 862–864, 866–867
   on Starliner, 782f
   technical flow phenomena, 503–504
   of windows and chassis parts, 1030–1031
      rear window, 512, 521–523
      side window, 511, 518–521
   see also Foreign soiling; Self-soiling
Solar cars, 617
Solid body simulation, 1107
Sound , *see* Aeroacoustics; Noise
Source and vortex, 98–99
Spalart, P., 1096
Spalding, D.B., 71
Spatial transformation of sound fields (STSF), 578
Specific dissipation (SD), 1016–1017
Speed
   critical, as function of airflow angle, 658f
   maximum, 528, 609–611
   of sound, as function of temperature, 149f
   speed limits in various countries, 165t
   transonic, 612–613
Spier, 762
Spiral vortices, 211
Spirit of Biel-Bienne III, 617f
Splash, 506, 523
Spoilers
   front
      design diagram, 341f
      effect of, 340f
      function of, 338f
      with twisted cross section, 677f
   in front of windshield, effect on interior noise, 586f
   of high-performance vehicles, 677–682

motorcycle
   helmet, 881–882
      foot protection, 875f
   rear, 304–311
   roof
      commercial vehicle, 720f, 786f
      influence on drag and lift, 306f
      influence on rear window contamination, 526f
   and underbody, 337–343
   wheel, 360
      influence on side panel contamination, 525f
Sports cars, 618–622
Spotlights, spectral distribution of, 1019t
Spray water, 523
Spread, 506
Squareback, 282–283, 285–296, 289–290
   dependence of drag and lift on attack angle, 228f
   drag, 1091f
   flow around, 227f
   lift, 1092f
   loss regions on, 227f
   micro drag, velocity components, and total pressure, 225f
   rear pillar shape, 300–301
   rear window inclination, 290, 300
   separation edges and rear spoilers, 304, 305, 307
   tail taper, 312, 313
   velocity distributions at wheel, 1092f
Stability, vehicle, 453–456
Stagnation point, 97
   horseshoe vortex, 349
   influence of location on pitching moment, 246f
   position of, 262–264
Stamping, 26–27
Star-CD, 73, 1169, 1090
Starliner, rear soiling on, 782f
Static pressure, 1109–1110
   measurement of, 992
Stefan-Boltzmann radiation law, 535
Steinbach, D., 69
Stephan, K., 533
Stick, 506

Stoffregen, J., 817
Straight-ahead running, 453–456
Straight flow, commercial vehicles, 723–727
Strakes, 304
Stream filament theory, 90, 134
Stream surface, 80, 81f
Stream tube, 80, 81f, 89, 233
   division into six subsystems, 551f
"Streamline" era, 9f, 11, 14–24
Streamlined bodies, drag of, 129–130, 131–132
Streamlines, 80, 829, 845f, 849–850, 921, 929, 1135f
   interpreting, 93–95
Stress matrix, 77
Stresses, 76–77
Stuttgart University, 1044–1045, 1049, 1052
Sumitani, K., 276
Sunblinds, commercial vehicle, 750
Sunlight, spectral distribution of, 1019t
Sunroofs, 479–480, 487–488, 589
   computer model, 1162f
   design solutions for, 501–502
   noise from, 591
Supermoto, 806
Superposition method, 99
Supersport, 806–807
Surface forces, 76
Surface mesh, 842f
Surfels, 1075
Sutherland temperature, 147, 1072
SUVs, 396, 399–400
Suzuki Hayabusa combination, 808f
swing, 1049, 1050f
Swiss Federal Laboratories for Materials Testing and Research (EMPA), 897
Sykes, D.M., 307
Symbols, 1180–1191
Symmetric multiprocessor systems (SMP), 1139

Tail
   of high-performance vehicles, 663–665
   influence on lift and yaw moment, 318f
   interaction with nose in flow, 368f

Tail taper, 312–314
TALINGZ, 1004
Tangential force coefficient, increase with yaw angle, 219f
TANJA manikin, 493, 494f
Tatra 77, 41
Tatra 87, 11, 12f, 22, 36
Taylor, G.I., 48
"Teardrop" car, 16, 62
Temperature, sensors for, 898f
Tesch, G., 243
Test section, wind tunnel, 926–934
  closed, 929, 932–933, 969–972
  mounting vehicle in, 960–962
  open, 929–933, 973–979
  slotted wall, 933
Thermal comfort, 489–494
Thermal conductivity, 534
Thermal wind tunnels, 1050–1054
  additional equipment in, 1032–1034
  blockage correction methods for, 979–981
Thermostat, characteristic of, 1118f
Thibaut, J., 564
Thickness, influence on drag, 54
Thöle, G., 880, 885, 897–898
Three-wheeler motorcycle, 808–809
Thrust 2, 611
Thrust SSC (Super Sonic Car), 611, 612f
Tires, 346–348
  effect of periodic normal force fluctuations on lateral force, 457f
  maximum operating pressure, 186n
  relationship between normal force, skew angle, and lateral force transmission limit, 452f
  speed index, 155n
Tongji University wind tunnel, 950, 1046
Top speed, 163–165
TOP500 supercomputers, 1139, 1141, 1142f
Tourers, 804
Touring cars, 623
Touring enduros, 803
Tracing particles, 1007–1008
Tractive force, 160–161
  motorcycle, 813, 817
Tractor/trailer unit, 758–768
Trailers, 369–377, 528

Train, high-speed, ICE I, 68–69
"Tram-Bus," 42f
Transducers for pressure measurements, 993–997
Transmission, 183–184, 530
  cooling, 540–541
  friction torque losses, 183f
  overall efficiency of, 183t
Transmission ratio, motorcycle, 817
Transonic speeds, 612–613
Transport Concept Solution Truck, 769, 771f
Truckenbrodt, E., 129
Trucks
  airflow through engine compartment, 751–754
  cab, 741–747
  characteristic airflow and pressure conditions, 739–741
  chassis, 754–758
  concept vehicles, 769–771
  long-haul, energy flow diagram for, 716f
  mirrors and attachments, 748–750
  optimizing drag on, 739–771
  semitrailers and bodies, 758–768
Trunk lid, 315–318
Trzebiatowski, H., 798
Tsuboi, K., 1100
Turbulence, 594
  boundary layer, 118–122
  generation of, 965f
  increasing, 1037–1039
  modeling of, 1077, 1087–1088
  at rear of commercial vehicles, 741f
Turbulence generation system (TGS), 434
Turbulence intensity, 419, 422, 424, 425

Ultrasound leakage tests, 582
Underbody, 592
  buses and coaches, 778
  drag, 243, 244f, 778
  future development, 405
  heat transfer coefficient on, 1134f
  influence of, 334–343
  of racing car, 684f
Understeer, 448, 449, 460, 644f
UNICAR, 240f, 327f, 614

rear diffuser on, 336, 337f
United States
    CO_2 labels, 194–196
    CO_2 legislation, 193–194
    driving cycles, 173–174
    five-cycle standard, 175f
    helmet laws, 877–878
    inertia mass classes in, 169–170t
    speed limits in various states, 165t
University of Stuttgart, 863, 948f, 949, 951, 966
Unsteady flow, wind tunnel, 963–967
Unsteady RANS (URANS), 1086
Uphill driving, 528
Utz, H.-J., 218

Vacuum effect, 606
Vagt, J.D., 931, 987
Van Dyke, M. 1006
van Moorhem, W.K., 888, 893
Vandry, J., 926
Vane anemometers, 564, 984
Vanes
    of high-performance vehicles, 703–707
    wind tunnel, 941
VDI Heat Atlas, 534
Vector computers, 1138–1139
Vector surface element, 88, 89f
Vehicle classes, 623–629
    drag and lift by, 390–392
Vehicle-induced contamination, see Self-soiling
Vehicle packaging, 532
Velocity field, kinematics of, 78–80
Ventilation, of race car helmets, 901–903
Ventilation moment, 243
Viscosity, of air, 146–147
Voiture Minimum (1936), 60f
Volkers, T., 737
Volkswagen
    1600, 231
        sedan calibration model, 1059
        TL, 21f
    1L, 615–616
    ARVW experimental vehicle, 615f
    "Auto 2000," 55, 56f
    Beetle, 50
    "Bully," 719, 720f
    force coefficients of various vehicles, 1097t
    Forschungsauto 2000, effect of curvature on drag, 276f
    Golf (Rabbit), 21
        development process, 382t
    Golf I, 37, 50, 59, 291, 293, 319
        front end development, 257, 258f
        influence of inclination angle of rear slant, 292f
        boat tailing on, 314
    LT, 44–45
    Lupo, 665f
    Passat, flow around front end, 259f
    Polo I, 291, 292f
    Sirocco I (1974), 52, 53f
    Sparmobil (1982), 615f
    Touareg II, exterior mirror, 363
    Transporter, 719, 720f
    Up!, 186
    van, 44
    wind tunnel, 64–65, 1042, 1046, 1052f
    XL 1, 58, 59f
Volume mesh, 842f
Volumetric forces, 76
Volvo, 73
    PV 544 (1955), 21f
    Safety Car Concept SCC (2001), 272f
von Kármán, T., 420
Vooren, J.V.D., 922
Vortex flows, 105–110
    dynamic behavior of separation planes, 108–110
    filament model, 106–107
    induction of, 107–108
    longitudinal, 4, 70, 215–216
Vortex generators, 964–965
    noise simulation with, 597

Wagner, A., 41, 471
Wake, 3–4
    far, 68f, 70
    near, 68f, 69f
    dead, 116–117, 126, 129, 132–133, 213
        non-periodic, 208–210
        periodic, 210–211
        ring and spiral vortices, 211
    simulation of, 1107–1109

Wake (Cont.)
  unsteady processes, 211–213
  wake flow, 287f
  wheel, 349–351, 353, 354
Wallentowitz, H., 41
Walston, W.H., 1021
Walter, J., 924, 979, 1003
Wang, Y., 950
Wankel rotary engine, 614
Wäschle, A., 348, 352, 1090
Water deflector rail, 517f
Water pump, drive torque for, 181f
Watkins, S., 1031
Wave-based methods, 1161
Weber, J., 110
Weber number, 504
Weight, of Mercedes-Benz sedans, increase in, 153f
Weight equivalency, 189
Wheel covers, commercial vehicle, 758f, 778–779
Wheel drive units (WDU), 962, 963
Wheel rotation units, wind tunnel, 1000–1001
Wheel wake horseshoe vortex, 349–350, 351, 353, 354
Wheelhouse, 278f, 346, 352–355, 359–361, 591
Wheels
  commercial vehicles
    buses and coaches, drag on, 778–779
    configurations, 755f
    spray from, 793f
    test drives with hub measurement device, 736–738
  equipment for spray analysis, 1012f
  front
    velocity distribution in wake, 237f
    vortex system behind, 242f
  future development, 405
  of high-performance vehicles, 707–710
  influence on aerodynamic forces, 346–361
    effects of rotating wheel on overall vehicle, 355
    flow topology, 348–355
    rim, 355–359
    tires, 346–348

  loads on, 450, 639, 643, 649–650
  measuring rolling resistance of, 1023, 1024f
  motorcycle
    cooling concepts, 873f
    front, self-soiling, 866f
    front suspension, 873f, 874f
    rear, traction force at, 813
    wind tunnel testing, 853–855
  and self-soiling, 523–524, 1012–1013
  simulation of, 1119–1121
    moving reference frame (MRF), 1120
    "rotating wall," 1120
    sliding mesh, 1121
Whipple, F.J.W., 795
White, P.H., 796
White, R.G.S., 65
Whitfield, J., 933
Wickern, G., 230, 708, 951, 969, 978, 998, 1003, 1044
Widdecke, H., 998
Wiedemann, J., 352, 550, 927, 930, 942 956, 973, 987, 1037, 1039, 1060
Wilharm, K., 988
Williamson, C.H.K., 291
Wilsden, D., 970
Wind, 217, 1004
  influence on the incident flow of a vehicle, 416–426
  noise, 489, 496–497, 594, 597, 600
    on helmets, 889, 892f
    on-road measurement of, 1031
Wind and weather protection, motorcycle, 835–837, 865, 875–876
Wind atlas, Europe, 726
Wind deflector, 590f
Wind noise, on-road measurement of, 1031
Wind tunnels, 63–65
  acoustic and anti-buffeting measures, 942–955
  aeroacoustic, 915, 1042, 1043–1047, 1062
    measuring systems, 574–575
  aerodynamic load measurements, 997–1006
  Aerodynamic Test Center (AVZ), 937f, 1047
  Aerolab, 1047–1048

Audi, 951, 1043–1044
balances, 997–998
Behr, 1042
BMW, 65, 937f, 1032, 1043, 1047, 1052
climatic and thermal, 916, 1017–1019, 1040, 1050–1054
    additional equipment in, 1032–1034
    blockage correction methods for, 979–981
collector, 934–937
for commercial vehicles
    full-scale, 732–734
    model-scale, 728–732
compared to numerical methods, 1067, 1068, 1080–1081
correction methods, 926, 927–928, 930, 932, 967–981
    blockage, for climatic and thermal wind tunnels, 979–981
    closed test section, 969–972
    open-jet test section, 973–979
crosswind, 964, 1026f
Daimler, 734, 1048, 1053, 1054f
decomposition of aerodynamic loads, 998–1004
design and function, 916–920
diffusers, 940–941
digital wind tunnel (DWT), 1127f
DNW, 732, 733f, 734f, 960
ECARA Wind Tunnel Corrections group, 928
Eiffel type, 917, 918f, 919, 921
Energy and Environmental Test Center (EVZ), 1032, 1052
engine cooling tests, 1014–1017
existing, overview of properties, 1055–1058t
fan acoustics, 951–955
FIAT, 65
FKFS, 1041, 1044–1045, 1049, 1052
flow conditioning screens, 941–942
flow velocity measurements
    direction, 988–989
    outside and inside test object, 984–988
flow visualization, 1006–1010
Ford, 1042
Ford Werke, 1045, 1050, 1026f

foreign soiling, simulation of, 1011–1012
frontal area measurement, 1005–1006
full-scale, 64, 1039–1048
General Motors, 1041–1042, 1059
German Aerospace Center (KKK), 1036
Göttingen, 917–919, 1041, 1046
ground simulation, 955–963
heating and climatization tests, 1017–1019
helmets, 906–907
honeycombs, 942
increase of measurement time in, 1067f
intercomparisons by correlation measurements, 1059–1062
international comparison of out-of-flow SPL, 907f
IVK, 959, 966f, 983f, 1044–1045, 1049, 1052
    sound-absorbing elements in, 943f
measurements in, 981–1034
Mercedes Benz, 1045–1046, 1053
MIRA, 1048
model-scale, 438, 1034–1039, 1040, 1048–1050
    for commercial vehicles, 728–732
Modine Europe, 1052
motorcycle, 853–864
    aeroacoustics, 861–862
    flow field visualization, 859–860
    reproducibility, driver influence, 855–858
    rotating wheels, 854–855
    soiling, 862–864
    stationary wheels, 853–854
NFAC, 732, 733f
nozzle, 920–926
NRC Canada, 732, 733f, 972
NWB, 950f
on-road measurements, 1021–1031
    cooling tests, 1029–1030
    crosswind tests, 1025–1029
    drag measurement using coast-down tests, 1021–1025
    soiling of windows and chassis parts, 1030–1031

Wind tunnels *(Cont.)*
   wind noise, 1031
   outlook, 1062–1063
   overview and correlation measurements, 1054–1062
   plenum, 937–939
   Pininfarina, 1043
   Porsche, 1046, 1048
   pressure measurements
      comparison of different wind tunnels, 1015f
      dynamic, 990–992
      static, 992
      transducers for, 993–997
   Rail Tec Arsenal, 1050
   SAE Subcommittee on Open Throat Wind Tunnel Adjustments, 928
   Schuberth, 807f, 906f, 909
   scope of, 913–916
   self-soiling, simulation of, 1012–1013
   Shanghai Automotive Wind Tunnel Center (SAWTC), 1046, 1047f
   sound-absorbing elements in, 576f, 943f
   sound pressure level (SPL), 907f, 946f, 947, 1060
   test section, 926–934
      closed, 929, 932–933, 969–972
      mounting vehicle in, 960–962
      open, 929–933, 973–979
      slotted wall, 933
   test sequence, 981–984
   thermal, 1017–1019, 1040, 1050–1054
      additional equipment in, 1032–1034
      blockage correction methods for, 979–981
   Tongji University, 950, 1046
   turning vanes, 941
   University of Durham, 433
   University of Stuttgart, 438, 575f
   unsteady flow and gust simulation, 963–967
   Volkswagen, 64–65, 1042, 1046, 1052f
   vehicle soiling investigation, 1011–1013
   windscreen defrosting and dehumidifying tests, 1020–1021
   Windshear Rolling Road, 1003

Windows
   acoustic glass to reduce interior noise, 593
   rear
      contamination of, 512, 521–523, 526
      inclination angle, 290–300, 315, 316–317f
      influence of roof spoiler on contamination, 526f
      ways to optimize flow separation at, 305f
   side
      contamination of, 507–508, 511, 518–521
      effect of recess on drag, 279f
      inclination of, 280f
      simulated pressure fluctuations on, 1166f
      sound pressure level on, 996f, 1134f
      trim strip along, 518f
   soiling of, on-road measurement of, 1030–1031
Windshear Rolling Road Wind Tunnel, 1003
Windshield
   close-wall stagnation streamlines on, 515f
   defrosting and dehumidifying tests, 1020–1021
   flow along, 266f
   hydrophobically coated, 507
   influence of, 265f, 266–269, 274f
   motorcycle, variations for drag reduction, 870f
   parameters describing geometry of, 267
   water pullback on, 512–514
Windshield wipers, 480–487, 508, 513, 586
   on buses and coaches, drag on, 777–778
   flow around, 1164, 1165f
   wake structure behind, 514f
Wings, 41
   flows around, 104–105
   multiple, 674–675
   on high-performance vehicles, 605–606, 608, 630, 632–633

drag, 672
  function, 667–670
  induced drag, 673–674
  lateral boundaries, 670
  multiple wings, 674–675
  outline, 667–668
  profile, 668–669
  tasks, 666–667
 Seiferth, 947–948, 949
Winkelmann, H., 644
Wittmeier, F., 347, 348
Wojciak, J., 829, 865
Worldwide Harmonized Light Vehicles Test Procedures (WLTP), 173, 176

Wortmann, F.X., 674
WT8, 1042
Wuest, W., 991

Yamada, M., 276, 432
Yamaha XT 500, 803
Yaw angle, 217–219, 290
  and acoustic behavior, 247, 596
Yaw moment, 202, 248–249, 375–376
Yawed flow, motorcycle, 828–831
Young, R.A., 308

Zomotor, A., 373
Zonal coupling, 1103–1104